PETROLOGY

The

Study

of

IGNEOUS

SEDIMENTARY

METAMORPHIC

Rocks

PETROLOGY

The
Study
of
IGNEOUS

SEDIMENTARY

METAMORPHIC

Rocks

Loren A. Raymond
Appalachian State University

WCB
McGraw-Hill

Boston, Massachusetts   Burr Ridge, Illinios   Dubuque, Iowa
Madison, Wisconsin   New York, New York   San Francisco, California   St. Louis, Missouri

WCB/McGraw-Hill

A Division of The **McGraw·Hill** *Companies*

Book Team

Editor *Jeffrey L. Hahn*
Developmental Editor *Mary Hill*
Production Editor *Kay J. Brimeyer*
Designer *Jeff Storm*
Art Editor *Carla Goldhammer*
Photo Editor *John C. Leland*
Permissions Coordinator *Mavis M. Oeth*

Vice President and General Manager *Beverly Kolz*
Vice President, Publisher *Jeffrey L. Hahn*
Vice President, Director of Sales and Marketing *Virginia S. Moffat*
Vice President, Director of Production *Colleen A. Yonda*
National Sales Manager *Douglas J. DiNardo*
Marketing Manager *Jane Ducham*
Advertising Manager *Janelle Keeffer*
Production Editorial Manager *Renée Menne*
Publishing Services Manager *Karen J. Slaght*
Royalty/Permissions Manager *Connie Allendorf*

President and Chief Executive Officer *G. Franklin Lewis*
Senior Vice President, Operations *James H. Higby*
Corporate Senior Vice President, President of WCB Manufacturing *Roger Meyer*
Corporate Senior Vice President and Chief Financial Officer *Robert Chesterman*

Copyedited by *Martha Morss*

Original illustrations by *Tom Terranova*

Cover Photos:

Petrology: The Study of Igneous, Sedimentary, and Metamorphic Rocks:
Background—Letraset/Photone Vol. 1
Inset Top—© R. Dahlquist /Superstock/East Rift Zone Kilauea, Island of Kauai, Hawaii
Inset Middle—© Kerrick James Photography/Sandstone "Hoodoos," Bryce Canyon, Utah
Inset Bottom—© Doug Sherman/Geofile/Migmatite Exposed in the Canadian Shield near White River, Ontario
Volume I: Igneous Petrology: © R. Dahlquist/Superstock/East Rift Zone Kilauea, Island of Kauai, Hawaii
Volume II: Sedimentary Petrology: © Kerrick James Photography/Sandstone "Hoodoos," Bryce Canyon, Utah
Volume III: Metamorphic Petrology: © Doug Sherman/Geolfile/Migmatite Exposed in the Canadian Shield near White River, Ontario

The credits section for this book begins on page 727 and
is considered an extension of the copyright page.

Dedicated to
Margaret
whose patience and love during eleven years of
lost vacations,
and whose typing and encouragement
over that time,
helped make this book possible
and to
Matt
who sacrificed significant quality time
while he was growing up.

BRIEF CONTENTS

CONTENTS

Contents

Contents

xii

Part IV Metamorphic Rocks

Contents

LIST OF TABLES

PREFACE

Petrology is a required subject of study in a majority of geology programs in the United States and is an integral part of geology curricula in Canada and elsewhere in the world. The subject merits this important position because rocks make up much of the Earth. Furthermore, many subdisciplines in geology incorporate petrologic knowledge. Thus, petrology provides the foundation for the study of the Earth and its history.

Available petrology texts designed for middle-level undergraduate students (sophomores and juniors) take a variety of approaches. Some concentrate on principles, others deal only with igneous and metamorphic rocks, and still others tend to emphasize the descriptive aspects of the field (petrography). Only a few books incorporate recent advances in crystal growth studies, isotopic analyses, chemical petrology, sedimentary environmental analysis, and petrotectonics, as well as basic petrographic information.

This text, designed for the middle-level undergraduate geology major, incorporates both fundamentals and information on recent advances in our understanding of igneous, sedimentary, and metamorphic rocks. It provides an overview of the field of petrology and a solid foundation for more advanced studies. For each class of rocks—igneous, sedimentary, and metamorphic—I describe textures, structures, mineralogy, chemistry, and classification as a background to discussing representative occurrences and petrogenesis (rock origins). I have not tried to summarize all occurrences, but the explanatory notes at the end of each chapter provide additional sources the student can use to expand his or her understanding and knowledge.

Advances in petrologic knowledge come about through the efforts of individuals and groups of individuals. Individuals do not operate in a vacuum; rather, they work within a social and political environment. Whether or not an idea is accepted or has an impact on a specific scientific milieu is influenced by the quality of the idea (e.g., its ability to explain phenomena) and the quality of the data that support it; by the social, scientific, and political climate; and by the reputation of the scientist presenting the idea. Certain scientists, more than others, influence the development of a discipline. For this reason, I have chosen to cite, in the text, scientists who have been particularly influential, individuals who have made unique contributions, and a few whose work has advanced or updated earlier contributions of significance. Others who also have made significant contributions or have added to the body of knowledge in a specific area are cited in the notes. Not all of those cited have been correct in their advocacy of particular hypotheses, but each has influenced petrology by contributing major ideas or significant data or by spurring others to prove his or her ideas wrong. In citing a limited number of scientists in the text, my aim has been to make the text more readable while still including sufficient references that may also serve as a model of proper referencing and good scholarship for students.

Igneous rocks are introduced first. Here, as with the other classes of rock, the rock types that are most important volumetrically in the crust are given the greatest emphasis. Because great debates arise from time to time about the origins of these most common rock types (e.g., basalts, andesites, granites), I have chosen to discuss each rock type individually rather than as part of a tectonic association, suite, or clan, as some other texts do. Where it is essential, I do discuss associated rocks.

Conflicting points of view on rock origins are included in many chapters. The inclusion of these controversies is important for two reasons. First, they provide the student with historical background that makes contemporary theory more meaningful. One can see ideas that have come before, see where mistakes have been made (science is a self-correcting enterprise), and see why certain ideas have been abandoned. Some scientists have reinvented the wheel (independently created anew an idea that was formerly evaluated and rejected) because of a lack of knowledge of petrologic history. Second, contemporary controversies reveal that there are many unresolved questions and there is much petrologic work to be done.

Sedimentary rocks are divided into siliciclastic types (mudrocks, sandstones, conglomerates), biochemically and chemically precipitated types (e.g., limestones, cherts), and allochemical types (e.g., limestone conglomerates). Because classifications vary and devotees of one classification or another are often strongly committed to the classification and descriptive practices they follow, various classifications are presented and discussed. Mudrocks and wackes (sandstones), especially turbidites, are given more attention than is common in petrology texts, because of the abundance of mudrocks and the tectonic and economic importance of the wackes. Sedimentary facies analysis is introduced to provide the student with an understanding of the interrelationships among various sedimentary rock types. A chapter on weathering, transportation, and diagenesis link the chapters on sediments and their environments of deposition to those dealing with specific sedimentary rock types.

Metamorphic rocks are presented in the same general fashion as the other classes of rocks, with descriptive aspects preceding chapters detailing the petrogenesis of particular suites of rocks. The facies series concept of Miyashiro is used as a basis for subdividing metamorphic rock suites and for describing the distribution of these suites in metamorphic terranes. Eclogites and cataclasites are given special attention because, although the former are considered to be representative of some mantle rocks and the latter are widely distributed, both are underdescribed in most petrology texts.

The epilogue places all of the rock types into appropriate petrotectonic assemblages representing various plate tectonic sites. Plate tectonics is first discussed in the introduction and is reemphasized in many of the rock chapters where petrogenesis is cast in a plate tectonic framework. Thus, the epilogue synthesizes the broad aspects of petrogenesis.

Important terms are set in bold print in the text where they are first defined and used. Definitions are also provided in a glossary. As a teacher, I subscribe to the view that one must know the vocabulary of a subject in order to be able to effectively communicate in that subject, just as one must know the vocabulary of a foreign language if one desires to communicate in that language. Students are commonly frustrated when they make the transition from texts, which define terminology, to the professional literature, where a knowledge of terminology is assumed. The significant number of defined terms used in this text should help students with this transition.

The language (syntax and word choice) used in the text ranges from simple to moderately complex, with the intent of developing in the student an increased vocabulary and an increased facility with the English language. The presentation of ideas follows a similar (simple to complex) pattern in many of the chapters. In addition, later parts of the text assume an understanding of some earlier parts. A list of Latin words and abbreviations used in the text (e.g., *sensu lato*), as well as other abbreviations appears on page xxi. This too is presented in the hope of increasing the general literacy of students of geology. Also in To the Student on page xxi includes a list of units of measure used in the text.

REVIEWERS

Samuel E. Swanson
University of Alaska–Fairbanks

Stephen A. Nelson
Tulane University

Michael Smith
University of North Carolina–Wilmington

Daniel A. Textoris
University of North Carolina

Gail Gibson, Director
Math and Science Education Center
University of North Carolina–Charlotte

Stephan Custer
Montana State University

David Lumsden
Memphis State University

Robert Furlong
Wayne State University

Jad D'Allura
Southern Oregon State College

Gunter K. Muecke
Dalhousie University

Dexter Perkins
University of North Dakota

Calvin Miller
Vanderbilt University

Steven P. Yurkovich
Western Carolina University

Barbara Lott
University of North Carolina

Richard Heimlich
Kent State University

Edward Stoddard
North Carolina State University

PUBLISHER'S NOTE
TO THE INSTRUCTOR

Petrology: The Study of Igneous, Sedimentary and Metamorphic Rocks, has been developed to fit the special needs of your petrology course. There are several binding options that have been developed to ensure that you, the instructor, have your students purchase the material you choose to teach. Since many professors do not teach the petrology of all rock types, you can purchase the material relevant only to your particular petrology class.

Petrology: The Study of Igneous, Sedimentary and Metamorphic Rocks, by Loren Raymond is available as a complete, casebound textbook or as paperback custom separate of each individual rock type. You can purchase these textbooks individually, or have them packaged together to make a custom teaching package. To purchase the custom separates, please order the appropriate custom separate from the following table, or contact your local Wm. C. Brown Representative.

Binding Option	Description
Petrology: The Study of Igneous, Sedimentary and Metamorphic Rocks ISBN 0–697–00190–3	The complete petrology textbook covering all rock types.
Igneous Petrology ISBN 0–697–23692–7	Parts one, two, and five from *Petrology: The Study of Igneous Sedimentary, and Metamorphic Rocks*
Sedimentary Petrology ISBN 0–697–23691–9	Parts one, three, and five from *Petrology: The Study of Igneous Sedimentary, and Metamorphic Rocks*
Metamorphic Petrology ISBN 0–697–23690–0	Parts one, four, and five from *Petrology: The Study of Igneous, Sedimentary, and Metamorphic Rocks*

If you are using one of our customized textbooks from our *Petrology Series* by Loren Raymond, please note that only a portion of the table of contents applies to your textbook.

FROM THE AUTHOR

I am indebted to a large number of individuals and institutions for assistance in bringing this book to completion. I wish to thank the William C. Brown Company, including former editors Robert Stern, Cathy Di Pasquale and Ed Jaffe and present editors Jeff Hahn, Bob Fenchel, Lynne Meyers and Kay Brimeyer, Mary Hill, Carla Goldhammer, and John Leland, Mavis Oeth (Permissions Coordinator) and Jeff Storm (Designer) for making this book possible. Appalachian State University provided periodic support for my endeavors, including one semester of paid leave. I am especially indebted to all of those individuals who helped increase my understanding and knowledge of petrology, notably Professors R. N. Abbott, D. O. Emerson, C. V. Guidotti, M. E. Maddock, I. D. MacGregor, E. M. Moores, R. L. Rose, S. Skapinsky, C. H. Stevens, S. E. Swanson, and F. Webb. A number of colleagues generously shared thoughts, time for reviews, and reprints of their work.

All errors in understanding or knowledge are my own responsibility. A large number of students challenged me during my teaching, discovered sources during their own library research, or otherwise contributed to the successful completion of this project. Among those who provided particular assistance on specific projects were Paul Dahlen, Vickie Owens, Elizabeth Stevens, and Susan Wilson. Paul Dahlen assisted with photographic work. Tom Terranova, who prepared the line art for the book, provided inspiration for developing high-quality illustrations. Matt Raymond provided scale in some photographs and gave up much "quality time" over the eight-year period that I was writing the text. Margaret Raymond provided, love, support, and many hours of typing time, all of which were of immeasurable help in the completion of this project. To all of the above-named individuals and organizations, I give my heartfelt thanks.

Units of Measure Used in the Text and Conversion Factors

Pressure (P)

1 Gpa (Gigapascal) = 10^9 Pa (Pascal) = 10 kb (kilobars) = 10^4 bars

Temperature (T)

°K (degrees Kelvin) = °C (degrees Celsius) + 273°

Length (l)

1 km (kilometer) = 10^3 m (meters) = 10^5 cm (centimeters) = 10^6 mm (millimeters)

Area (A)

1 km^2 (square kilometer) = 10^6 m^2 (square meters) = 10^{10} cm^2 (square centimeters)

Volume (V)

1 km^3 (cubic kilometers) = 10^9 m^3 (cubic meters) = 10^{15} cm^3 (cubic centimeters)

Mass (m)

1 kg (kilogram) = 10^3 grams

Density (ρ)(rho)

1 kg/m^3 (kilograms per cubic meter) = 10^{-3} g/cm^3 (grams per cubic centimeter)

Acceleration of Gravity (g)

0.0098 km/sec^2 = 9.8 m/sec^2 = 980 cm/sec^2

Time (t)

1 b.y. (billion years) = 10^3 m.y.(= 10^3 ma)(million years) = 10^9 years ~ 3.16×10^{16} seconds

Phase Rule and Phase Diagrams

P = number of phases

C = least number of components necessary to define a system

F = the number of degrees of freedom = the variance

X = composition

List of Chemical Symbols Used in Text

Al—aluminum
Ar—argon
B—boron
Ba—barium
Be—beryllium
C—carbon
Ca—calcium
Ce—cerium
Cr—chromium
Cs—cesium
Eu—europium
F—fluorine
Fe—iron
Ga—gallium
Gd—gadollium
H—hydrogen

He—helium
K—potassium
La—lanthanum
Li—lithium
Lu—lutetium
Mg—magnesium
Mn—manganese
Na—sodium
Ni—nickel
Nd—neodymium
O—oxygen
Os—osmium
P—phosphorous
Pb—lead
Pr—praseodymium
Rb—rubidium

Re—rhenium
S—sulfur
Si—silicon
Sm—samarium
Sr—strontium
Ta—tantalum
Ti—titanium
Th—thorium
U—uranium
V—vanadium
W—tungsten
Y—yttrium
Yb—ytterbium
Zr—zirconium

List of Common Abbreviations and Prefixes Used in the Text and Their Meanings

blasto—to bud; to sprout; hence to form anew in a metamorphic rock

cf.—compare to

e.g.—for example

et al.—and others

i.e.—that is

in situ—in place

inter—between

intra—within

iso—the same

sensu lato—in the broad sense

sensu stricto—in the strict sense

Aleutian Arc

Muskox Intrusion•

CANADA

Duke Island

Bay-of-Islands
Ophiolite

St. Simon,
Quebec•

Lake Superior
(Sediments)

Onawa, ME•

Waterville
(SE Maine
Metamorphic Belt)

Salmo, B.C.

•Shonkin Sag

Cascade Range

Columbia River Plateau

•Stillwater
•Yellowstone
•Tin Mountain, SD

Michigan Basin
Salina Group

St. Peter
Sandstone

Rattlesnake Pluton

Franciscan Complex,
Great Valley Group

Green River
Formation

Cincinnati
Arch

Appalachian Basin

No. Appalachian Orogen

Southwestern Virginia
(Nebo; Ben Clark's Farm)

•Notch Peak, UT

UNITED STATES

Coast Ranges

Carbona Quad. and
Del Puerto Ophiolite

•Yosemite

Navajo Sandstone

Buck Creek Dunite•

So. Appalachian
Orogen

Crestmore, CA

•Grand
Canyon

Rio Grande Rift

•Laborcita
Formation,
NM

•Magnet Cove,
AR•

Brevard Zone

Atlantic Ocean

Pacific Ocean

Texas Shelf

Gulf of Mexico

Bahamas

MEXICO

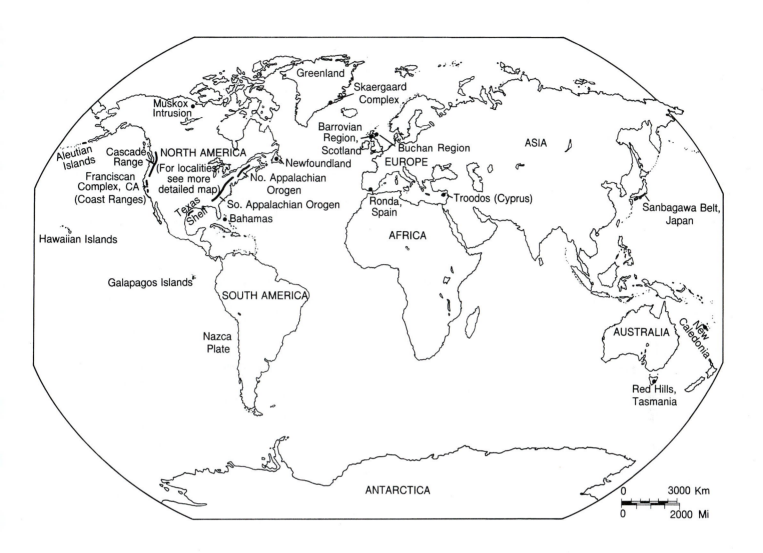

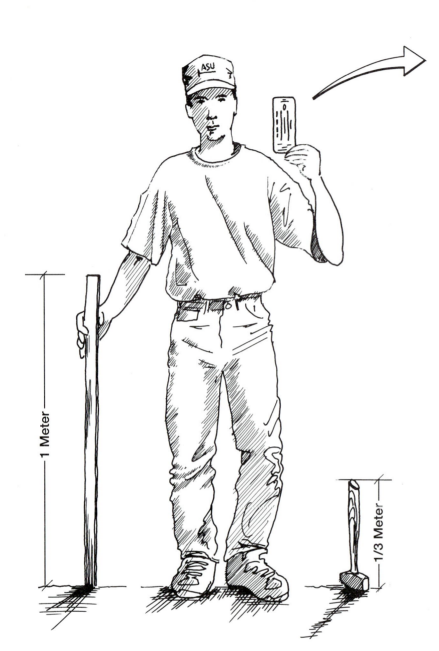

PETROLOGY

The

Study

of

IGNEOUS

SEDIMENTARY

METAMORPHIC

Rocks

Introduction

Geology is the study of the planet Earth, including its composition, origin, and history. Most of the solid part of the Earth is composed of rocks. Because the Earth is composed of rocks, an understanding of their nature, composition, origin, and histories is central to understanding Earth history. This book is an introduction to the study of rocks.

PART I

View of the Three Sisters volcanoes, Oregon, from the top of Mount Bachelor.

1

Rocks and Earth Structure

INTRODUCTION

A **rock** is a solid aggregate of mineral grains, or a solid, naturally occurring mass composed of mineral grains, glass, altered organic matter, and combinations of these components. Rocks may be studied in a variety of ways. **Petrology** is the term used to denote the overall study of rocks, including **petrography** (also called lithology), the study of the description and classification of rocks, and **petrogenesis,** the study of the histories and origins of rocks.[1] Petrography is basically an observational science. Features such as color, mineralogy, and texture are studied and used as a basis for subdividing the naturally occurring, continuous range of rock types into groups to which petrologists assign names. In some cases, petrogenetic information is also used as the basis for subdivision and classification. Petrogenetic studies combine various experimental and theoretical analyses and use inductive and deductive reasoning to arrive at conclusions about rock origins and histories.

THE THREE CLASSES OF ROCKS

All rocks may be assigned to one of three major classes—igneous, sedimentary, or metamorphic. **Igneous rocks** are rocks that form by crystallization (freezing) of melted rock materials at temperatures well above those standard at the Earth's surface. The melt, commonly referred to as **magma,** may or may not be entirely molten, as substantial amounts of solids, both crystals and rock fragments, as well as gases, occur in magmas.

Sedimentary rocks form under surface conditions and consist of accumulations of (1) chemical and biochemical precipitates; (2) fragments or grains of rocks, minerals, and fossils; or (3) combinations of these kinds of materials. Like igneous rocks, some sedimentary rocks result from crystallization, but in sedimentary environments crystallization occurs from an aqueous solution under surface or near-surface conditions of pressure and temperature. Most sedimentary rocks develop through cementation, compaction, and recrystallization of accumulated mineral, rock, and fossil fragments. In these rocks, minerals that constitute cements also crystallize from aqueous solutions.

Metamorphic rocks are rocks that formed originally as igneous or sedimentary rocks but which have been changed mineralogically, texturally, or both—without undergoing melting—in response to heat, pressure, directed stress, or chemically active fluids or gases. The pressures and temperatures that induce the changes in mineralogy and texture are generally rather different from standard surface conditions.

Nevertheless, a continuous range of conditions exists from those that prevail during sedimentation through those that produce melting at the highest grades of metamorphism.

The distinctive characteristics of each class of rocks result from the processes that form the rocks. Sedimentary rocks are typically layered, because they form by settling of materials through water or air. Igneous rocks may be structureless or they may show features that reflect movement and crystallization of hot fluids. Metamorphic rocks commonly show an alignment of grains that gives a flaky, banded, or layered appearance to the rock, and such rocks show folds or other features that reflect plastic flow. These distinctive characteristics are described in subsequent chapters dealing with the respective rock types. Mineral associations typical of each class of rocks are listed in appendix A.

The student should realize that nature provides a continuous array of rock properties. One rock type may grade into another, and the subdivisions imposed by petrologists on the continuous natural order are boundaries arbitrarily placed at seemingly significant and reasonable locations within the continuum. The boundaries between igneous and sedimentary rocks, sedimentary and metamorphic rocks, and metamorphic and igneous rocks are diffuse.

ROCK DISTRIBUTION IN THE EARTH

The Earth is a solid body, except for the outer core (figure 1.1) and some relatively small local spots within the upper mantle and crust, which are liquid. Much of the solid material is metamorphic rock. This is so because much of the rock of the inner core, mantle, and crust has been changed in texture or mineralogy in response to the high pressures, temperatures, and stresses that exist everywhere but at or near the Earth's surface. Most of these metamorphic rocks are inaccessible to study. Many of the liquids (magmas) that form in the upper crust or mantle rise to higher levels in the crust, where they crystallize as igneous rocks. Sedimentary rocks form at or near the surface. These, plus igneous and metamorphic rocks that become exposed at the surface through erosion, are accessible, have a direct impact on human endeavors, and have been studied the most.

On the land, sedimentary rocks total 66% of the exposed rocks (Blatt and Jones, 1975). Crystalline rocks, which make up the remaining 34% of the exposures, are about equally divided into igneous and metamorphic types. Below the oceans, the areal distribution of sedimentary rocks and sediments is not precisely known, but recent oceanographic work reveals that most of the ocean floor is coated with a thin veneer of sediment and sedimentary rock. Beneath the sedimentary cover, igneous and metamorphic rocks dominate the oceanic crust.

Considering the volume of the Earth as a whole ($1.083 \times 10^{21} m^3$), the core comprises about 16.2% by volume, the mantle makes up 83.2%, and the crust amounts to only 0.6% of the volume.[2] Of the crust, about 4.8% is sedimentary materials,[3] which means that those materials comprise only about 0.029% of the total volume of the Earth. These facts should be remembered in reading this book, since the text deals almost exclusively with the petrology of the rocks of the crust and upper mantle.

EARTH STRUCTURE AND PETROTECTONIC ASSEMBLAGES

Traditionally, the Earth has been recognized as having a series of layers (figure 1.1). These are defined on the basis of geophysical phenomena, with each layer distinguished from the next by a zone of marked change in seismic wave velocity called a discontinuity.[4] A solid inner core of iron-rich rocks, with a variety of probable impurities, is separated from

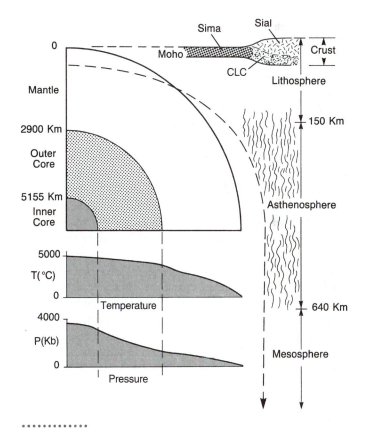

Figure 1.1 Generalized sketch of the interior of the Earth, showing major subdivisions delineated using seismic data. Pressure and temperature curves show the changing conditions towards the center of the earth. CLC = complex lower crust. *(Numerical data from Bolt, 1982)*

3

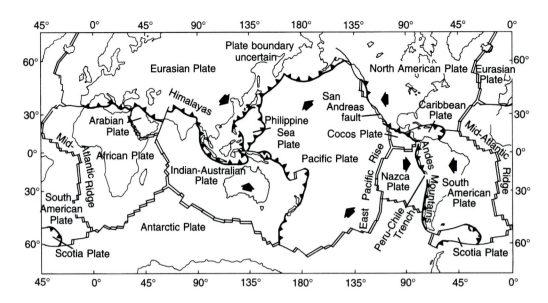

Figure 1.2 World map showing major lithospheric plates. Double lines indicate spreading centers, single lines on plate boundaries are transform faults, and heavy barbed lines are subduction zones and other zones of convergence. *(Source: After W. Hamilton, U. S. Geological Survey.)*

a liquid outer core by the Lehmann Discontinuity. Outward from the outer core, and separated from it by the Gutenberg Discontinuity, is the mantle, a layer composed of magnesium-rich rocks. The Mohorovicic Discontinuity or **Moho** separates the mantle and the crust. The crust is dominated by feldspars and other silicate minerals. Locally, the crust consists of *silicon-, iron-,* and *magnesium*-rich rocks, which are referred to as *simatic* rocks. The oceanic crust is *simatic.* In contrast, the continental crust is dominated by *silicon*- and *aluminum*-rich minerals and is said to be *sialic.* The lower continental crust is quite complex and consists of a variety of igneous and metamorphic rocks of both simatic and sialic character (Berckhemer, 1969; Finlayson, 1982; Rudnick and Taylor, 1987).[5]

Pressure and temperature increase with depth in the Earth, as shown in figure 1.1. Near the surface, the **geothermal gradient** (the increase in temperature with depth) is pronounced. In the mantle, however, the slope of the temperature curve is less steep. Although the average gradient is about 20° C/km, geographic variations in the geothermal gradient exist. Over volcanically active areas, such as in the volcanic island arcs, the temperature increases at a rate of 30–50° C/km or more. In contrast, near ocean trenches, the rate of increase may be as low as 5–10° C/km. Continental areas away from tectonically active zones have intermediate gradients.

The pressure in the Earth is given by the expression

$$P = \rho g h$$

where

P = pressure
ρ = density
g = acceleration of gravity
h = height of the column of overlying rock (depth of burial).

Average crustal values for density yield pressure gradients on the order of 0.1 Gpa/3.3 km (1 kb/3.3 km), a value useful for making quick estimates of pressure within the crust.

Increased emphasis has been placed on a threefold subdivision of the crust and mantle as a result of tectonic studies conducted during the past three decades. An upper rigid zone, the **lithosphere,** consisting of the crust and upper mantle, is separated from a lower rigid zone composed of mantle rocks, the **mesosphere,** by a plastic zone called the **asthenosphere** (figure 1.1).[6] The asthenosphere consists of partially melted mantle materials. The lithosphere is fragmented into several major and minor pieces called **plates,** the boundaries of which are marked by zones of earthquake activity (figure 1.2).

Lithospheric plates move, producing plate interactions along three types of boundaries (figure 1.3). At **spreading centers,** or divergent boundaries, plates pull apart and new crust

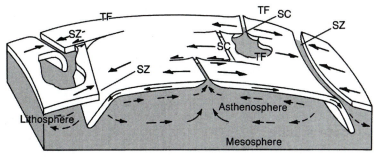

Figure 1.3 Schematic block diagram showing the three types of plate boundaries—spreading centers (SC), subduction zones (SZ), and transform faults (TF). *(Modified from Isacks et al., 1968.)*

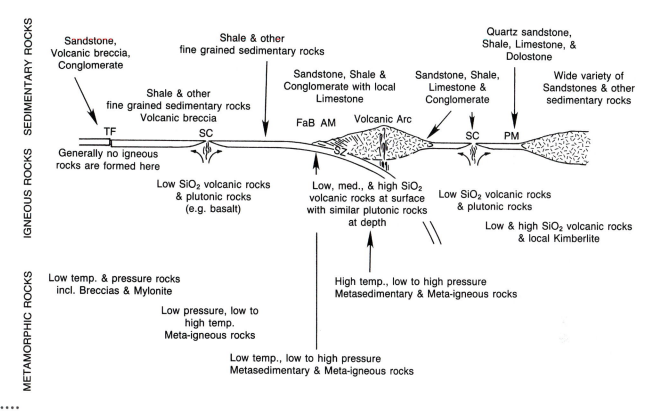

Figure 1.4 Diagram showing petrotectonic assemblages at various sites. FaB = forearc basin, SC = spreading center, SZ = subduction zone, TF = transform fault, AM = active margin, PM = passive margin.

is formed. **Subduction zones** represent convergent boundaries, where plates collide, with one plate descending beneath another. **Transform faults** mark shear boundaries, where plates slide past one another. Sites where three plate boundaries come together are called **triple junctions.** Each type of plate boundary, as well as triple junctions and sites within plates, gives rise to distinctive suites of rocks called **petrotectonic assemblages** (Dickinson, 1971) (figure 1.4). The interaction between petrologic and tectonic processes at a given plate boundary gives the rocks forming at the boundary distinctive characteristics valuable for recognizing the rocks formed at

that type boundary in the geologic past. A detailed review of petrotectonic assemblages indicative of each environment is presented in the epilogue. As a basis for discussion in the following chapters, a brief, simplified summary is presented here.

At spreading centers, one or two types of volcanic rock form and these are accompanied by various sedimentary and metamorphic rocks. Initiation of the spreading process results in the formation of graben or rift zones. On continents, these zones serve as basins for the accumulation of sand and gravel. Intrusion of magma into these sediments results in local metamorphism of low-pressure, high-temperature type. Extrusion of magma produces lava flows and other volcanic features. In ocean-floor spreading centers, volcanism is followed by circulation of seawater through hot volcanic rocks and into surrounding rocks, resulting in low-pressure chemical alteration or metamorphism. Sedimentation in the oceanic areas produces thin layers of very fine-grained sediment. Locally, however, gravels derived from the volcanic rocks are deposited.

Convergent plate boundary zones exhibit a wide range of rock types. Volcanic rocks of diverse compositions develop in volcanic mountain chains, called **arcs,** that form on the overriding plate. At depth, magma intrudes to form masses of siliceous igneous (granitoid) rock. The invading magmas cause high-temperature, low-to-high-pressure metamorphism in the surrounding rocks. Sediments derived from the erosion of arc rocks are shed both away from the plate boundary and towards it. At the plate boundary, a trench or an elongate basin forms at the collision zone and collects sediments not trapped in the arc-trench gap (also called the forearc basin) (figure 1.4). Sediments deposited on oceanic crust in the trench are then either accreted onto the overlying plate or subducted (J. C. Moore, 1989). The subducted sediments and underlying oceanic crust and mantle either become metamorphosed by the high-pressure, low-temperature conditions prevailing immediately arcward of the trench and are accreted to the overlying plate or they are carried to greater depths by the descending plate (Dewey and Bird, 1970; Ernst, 1970; 1973).

Shearing at transform fault boundaries forms metamorphic rocks. Because the rocks are deformed rapidly, they may show extensive breaking or stretching of mineral grains. Seawater percolating into such shear zones aids in the metamorphic process.

Triple junction assemblages are not easily recognized, as they consist of composite assemblages representing the various types of plate boundaries that form the triple junction. Nevertheless, some triple-junction assemblages have been preserved and recognized in the rock record.

Intraplate (within-plate) volcanism and sedimentation, like plate boundary processes, yield a variety of products. Silica-poor rocks dominate the volcanic and plutonic suites within oceanic plates. A variety of fine-grained sedimentary rocks also forms here. Within continents, at intraplate sites, unusual rocks, including diamond-bearing types, form along with a diversity of common volcanic rock types. Continental sites of sediment formation are quite varied, as are the resulting rocks. Distinct sediments form in river, lake, glacial, and other environments. Within the plates, along the tectonically quiet coastal areas, deposits of sediments yield sandstone shale, and limestone.

Because the processes and sites of rock formation vary, the petrotectonic assemblages vary. The characteristics of these assemblages and the included rocks provide the data for understanding petrogenesis, a major goal of petrology.

SUMMARY

The Earth is divided into crust, mantle, and core, with the crust and upper mantle comprising the rigid lithosphere. The lithosphere is divided into plates bounded by transform faults, subduction zones, and spreading centers. Pressure and temperature increase through the lithosphere and towards the center of the core.

The three classes of rocks—the igneous rocks, the sedimentary rocks, and the metamorphic rocks—all occur in the crust. Sedimentary rocks form a thin veneer covering 66% of the continental crust, but the igneous and metamorphic rocks are dominant volumetrically. All three classes of rock occur in petrotectonic assemblages that reflect the plate setting of their petrogenesis.

EXPLANATORY NOTES

1. Definitions of terms in this book generally follow those of the *Glossary of Geology* (Bates and Jackson, 1980, 1987). However, I have made some modifications and additions. Important (boldfaced) terms are defined in the glossary of this text.
2. Units of measure (e.g., meters, grams) used in this book are primarily SI units, although commonly used alternatives (°C for °K) are used in some cases. These units are listed in the front section of the book.
3. Ehlers and Blatt (1982, p. 6).
4. This section on the geophysical properties of the earth is based on review works, especially Wyllie (1971a), Ringwood (1975, 1979), and Bott (1982). For information on the core, for example, see Bott (1982, p. 147) and Ringwood (1979, p. 41ff.). Additional information is available in standard geophysics texts.
5. Also see Touret and Dietvorst (1983), Griffin and O'Reilly (1987), Dodge, Lockwood, and Calk (1988), and Bohlen and Mezger (1989).
6. The asthenosphere is discussed in simple terms by D. L. Anderson (1962), and its limits are shown by the work of Kanamori and Press (1970) and many others. See reviews of the seismic data in Wyllie (1971a, ch. 3), Bott (1982), and the summary discussion in Bolt (1982).

PROBLEMS

1-1. Using the generalized formula (0.1Gpa/3.3 km), calculate the pressures expected at 10 km, 20 km, and 30 km in the crust. Next, using the formula for pressure and assuming an average density of 2.70 gm/cc, calculate the pressure at 20 km. Compare your answers derived by the two calculations for 20-km depths.
2-2. Calculate the temperatures expected in the crust at a depth of 20 km, assuming (a) a geothermal gradient of 5° K/km, (b) a geothermal gradient of 20° K/km, and (c) a geothermal gradient of 50° K/km.

Igneous Rocks

Igneous rocks have played a major role in the evolution of the crust of the Earth. They are also the precursors of sedimentary and metamorphic rocks. For these reasons, igneous rocks are the subject of the first section of this book, chapters 2 through 13. The student will find that some principles learned in this section may also be applied to the study of sedimentary and metamorphic rocks in the following sections of the book.

Igneous rocks and processes are fascinating to geologists and nongeologists alike. Volcanic eruptions, especially, have provided facts and myths that have influenced human life and behavior. Both rocks and processes are described and discussed in this section. The major igneous rock types are evaluated individually and are assigned to rock associations in the epilogue of the book.

PART II

Mafic rocks in Granitoid Pluton, forming the "North American Wall" of El Capitan, (Tuolremne Intrusive Series), Sierra Nevada Batholith, Yosemite National Park, Tuolumne.

2

Igneous Rocks: Their Structures and Textures

INTRODUCTION

Magmas form in the mantle and lower crust. They rise toward the surface because they are less dense (lighter) than the surrounding rocks. These magmas may crystallize partially or entirely at various depths in the crust, or they may crystallize at or near the surface of the Earth. In either case, the products of crystallization are igneous rocks. Where magmas approach and break through to the surface, volcanic eruptions occur, producing **volcanic rocks. Plutonic rocks** crystallize at depth.[1]

The origin of rocks that crystallize during a volcanic eruption is easy to understand in general terms. Yet the geologist is faced with recognizing ancient igneous rocks and describing their origins without benefit of having observed the formative processes. In addition, the details of crystallization of volcanic rocks observed during formation are not immediately obvious. Consequently, numerous questions arise. How can igneous rocks be recognized? How are they distinguished from other kinds of rock? How does the crystallization process occur?

The answers to these and other questions are obtained through: (1) field observation of contemporary volcanic eruptions and features (the present is the key to the past), (2) field observation of features of existing igneous rocks, (3) laboratory studies of the mineralogy and textures of igneous rocks, (4) chemical analyses of igneous rocks, (5) laboratory studies of chemical processes in and crystallization behaviors of melts, and (6) the application of inductive and deductive reasoning. These topics are addressed in this and succeeding chapters.

Igneous rocks are recognized, described, named, and classified on the basis of structure, texture, and composition. The composition includes both mineral composition and chemical composition. The **texture** is the physical character of the rock, including the size, shape, orientation, and distribution of grains and the intergrain relationships. **Structures** are features of the rock, larger than grains, that result from the physical arrangement of grains, holes, fractures, or other entities in the rock mass. The textures and structures of igneous rocks are useful for distinguishing igneous rocks from rocks of other classes and they provide valuable information about the history of formation of the rocks.

RECOGNITION OF IGNEOUS ROCKS

Recognition of a rock as igneous generally begins in the field, where the rock is collected (figure 2.1). There, structures of the rock or rock body provide important clues to petrogenesis. Also in the field, a hand lens may be used to examine both the minerals and the texture of the rock. Later, laboratory

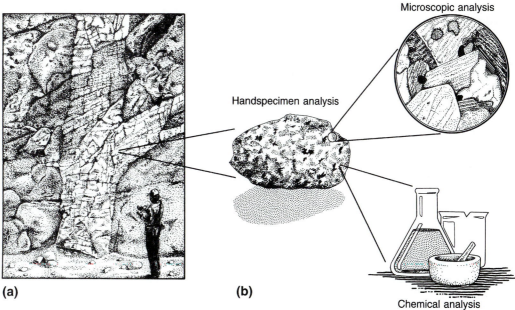

Figure 2.1 Analysis of rocks. (a) Collecting a rock in the field and observing the textures, structures, and field relationships. (b) Study of the handspecimen via laboratory analyses. *(Modified from Raymond, 1993.)*

(a)

(b)

studies, including examination of minerals and textures with a petrographic or an electron microscope and chemical analyses of the minerals and whole rock, allow greater knowledge of each rock.

Certain minerals and textures, many structures, and numerous chemical characteristics are unique to igneous rocks. Minerals typical of igneous rocks are discussed in chapter 3 and are listed in appendix A. Chemical compositions are also discussed in chapter 3. Important structures and textures are described in this chapter. Together these features allow the discrimination of igneous rocks from those of sedimentary or metamorphic character. No simple clues allow unequivocal assignment of all rocks to their respective categories.

STRUCTURES OF IGNEOUS ROCKS

The structures of igneous rocks may be divided into **extrusive** and **intrusive** types, although there is some overlap between the two groups. Extrusive structures are structures formed where magmas are forced out onto the surface of the Earth. Intrusive structures are structures formed below the surface of the earth. As will be evident, many features formed at shallow depths during volcanic eruptions are intrusive and are here described as such. Thus, there is not a one-to-one correlation between volcanic origin and extrusive features, nor between plutonic origins and intrusive features. Table 2.1 lists some criteria for distinguishing between extrusive and intrusive rocks.

Extrusive Structures

Extrusive structures are divided into three groups—major, intermediate, and minor—based on their sizes (table 2.2). The largest structures, **lava plateaus** and **basaltic plains,**[2] are tabular in overall form and are composed primarily of lava flows of silica-poor (basaltic) volcanic rock. They differ in internal structure (see below). The **lava flows**[3] are tabular to lobate masses of solidified **lava** (magma that lost gases when it erupted at the surface). At the time of formation of the plateaus and plains, lava flows formed piles of rock up to several thousand meters thick that may cover more than a million square kilometers of surface area and may include millions of cubic kilometers of rock. The relative sizes of these and other volcanic features are shown in figure 2.2. Dimensions and volumes of larger structures are listed in table 2.2.

Lava plateaus occur worldwide. In age, they range from late Precambrian to Cenozoic. Notable examples of lava plateaus include the late Jurassic to early Cretaceous Parana "basin" of Brazil, the early Cenozoic Deccan Trap Province of India, and the Miocene Columbia River Plateau of northwestern North America. Lava plateaus are composed primarily of

Table 2.1 Characteristics of Extrusive and Intrusive Rocks

Extrusive	Intrusive
1. Rock textures are glassy to fine-grained, except where coarse grains occur as phenocrysts, grains in inclusions, rock clasts in volcanic breccias, or rock fragments in pyroclastic deposits.[a]	1. Rock textures are fine-grained to coarse-grained.
2. Chilled margins (finer-grained textures) occur only at the base of the rock body.	2. Chilled margins (finer-grained textures) occur on all sides of the rock body.
3. Baked zones and minor contact metamorphism occur only below the unit.	3. Contact metamorphism may appear near all contacts. Baked zones do not exist.
4. An irregular to fragmented top of the rock body may have cracks infilled by sediment or rock of an overlying unit.	4. Smooth to irregular tops of rock bodies are characteristic, and dikes or apophyses may extend from the top (sides or bottom) of the body into surrounding rock masses.
5. Overlying rocks may contain fragments of an underlying extrusive rock body.	5. Intrusive rock bodies may contain fragments of overlying, underlying, or surrounding rock unit.
6. Vesicles or amygdules may be abundant at the top of the rock body.	6. Vesicles and amygdules are sparse or absent.
7. Little or no deformation of associated rocks is present.	7. Intrusion may cause folding or faulting in adjacent rocks.

Sources: Based on the observations of G. K. Gilbert (1880) and the author. Similar lists are provided by Grout (1932, p. 32) and Billings (1972, p. 339).
[a]Many of the terms used here have not yet been defined in this text. The student may wish to refer back to this table after learning the definitions of terms presented in this chapter.

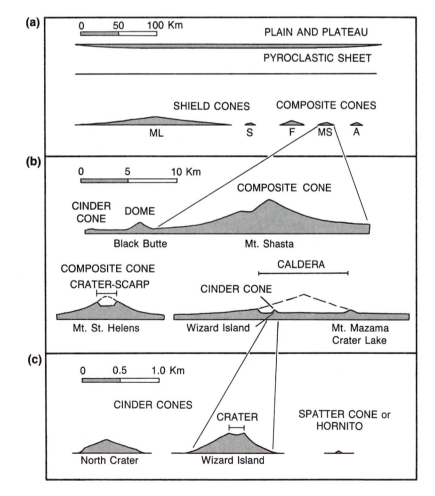

Figure 2.2 Size comparison of various volcanic features. Note the different scales in (a), (b), and (c). (a) Comparison of the relative sizes of the major volcanic structures. ML = Mauno Loa, Hawaii, S = Skjalbreidur, Iceland, F = Fujiyama, Japan, MS = Mount Shasta, California, A = Agua, Guatemala. (b) Composite cones, domes, and cinder cones. Mount Shasta is shown adjacent to a nearby dome (Black Butte) and an unnamed cinder cone. The Mount St. Helens profile shows both the pre-1980 and the post–May 18, 1980, shapes. The missing section is a combination crater and landslide scar. Mount Mazama is the name given to the prehistoric mountain that existed at the site of the present Crater Lake caldera in Oregon. (c) Cinder cones and a spatter cone. North Crater is in Craters of the Moon National Monument in Idaho. Wizard Island is a cinder cone that projects above lake level at Crater Lake, Oregon.

[(a) based on H. E. Cook (1968) and Howell Williams and McBirney (1979). (b) and (c) prepared by the author.]

Table 2.2 Structures of Extrusive Origin

Structure	Range of Dimensions[a]	Volume (km^3)
Major Structures		
Lava Plateau and	$T = <10^{-3}-12$	$6 \times 10^4 - 6.5 \times 10^5$
Basaltic Plain	$A = ?-2 \times 10^6$	
Pyroclastic Sheet	$T = <10^{-3}-2.5$	$<10-10^5$
(composite)	$A = 20-2.0^5 \times 10^5$	
Shield Cone	$T = <0.1-4.2$	$<10^3-4 \times 10^4$
(shield volcano)	$R = <1-100$	
Composite Cone	$T = <1-3.7$	$1-870$
(stratavolcano)	$R = <1-20$	
Caldera[b]	$A = 8-12,000$	——
(including cauldrons)	$R = 1.5-62$	
Intermediate Structures		
Pyroclastic Sheet	$T = <10^{-3}-1.8$	$10^{-3}-8300$
(single cooling unit)	$A = <1->10^5$	
Pyroclastic Cone	$T = .01-.46$	$1.5 \times 10^{-4}-3.25$
(cinder cone)	$R = .05-1.5$	
Lava Flow	$T = 10^{-4}-0.24$	$<0.1-1200$
	$A = <0.1-18,000$	
Dome	$T = <10^{-2}-0.8$	$3 \times 10^{-6}-4$
	$R = 10^{-2}-1.5$	
Small Structures[c]		
Crater[b]	Autolith	
Lithophysae	Pyroclastic sheet	
Lava flow	Spatter cone and hornito	
Pressure ridge	Tumulus	
Lava tube[b]	Squeeze-up	
Columnar joint	Baked zone	
Flow banding	Vesiculated flow top	
Bomb	Xenolith	

Sources: Basaltic Volcanism Study Project (1981), Billings (1972), Bultitude (1976), H. E. Cook (1968), Mackin (1961), Ollier (1969a), Settle (1979), R. L. Smith (1960a), R .L. Smith and Bailey (1968), Howell Williams and Goles (1968), and Howell Williams and McBirney (1979).

[a]A = area in km^2; T = thickness or height in kilometers; R = radius in kilometers

[b]Calderas, craters, and lava tubes differ from other features on the list, as they are voids rather than material constructions.

[c]These structures generally range from about one km (craters) down to 1.0 mm (e.g., xenoliths and lithophysae) in diameter; however, data on size ranges are generally not available.

flows, solidified masses of fluid basaltic lava that have flowed across the surface and crystallized. Typically, these lava flows are fed by magmas that come to the surface through systems of fractures, flow long distances, and accumulate to produce a layered pile (figures 2.3, 2.4).[4] Only minor amounts of **pyroclastic rock,** rock composed of fragments (clasts) of volcanic rock formed by explosive eruption, occur in the plateaus.[5]

Basaltic plains differ from plateaus in that they are made up of multiple lava flow units erupted from single, overlapping centers of eruption (Greeley, 1982). These centers have the form of **shield cones,** flat cone-shaped accumulations of lava containing minor amounts of interlayered pyroclastic materials (figure 2.3). The Snake River Plain in western North America is the type example of the basaltic plain.

Covering similar areas, but generally containing smaller volumes of material, are the **pyroclastic sheets** (table 2.2, figure 2.2). Pyroclastic sheets are accumulations of generally silica-rich, fragmental volcanic rock that form where particles settle out of the atmosphere during and after explosive eruptions. Particles may be thrown into the air and settle out to form **ash falls,** as they did in eastern Washington on May 18, 1980, after the eruption of Mount St. Helens,[6] or they may settle from a *nuée ardente*, a swiftly moving cloud of hot gas and rock particles. The rock mass formed by a *nuée ardente* is

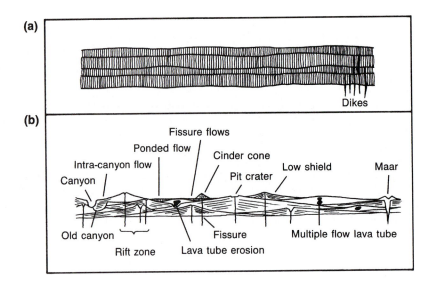

Figure 2.3 Internal structures of lava plateaus (a) and basaltic plains (b). *[(b) after Greeley, 1982]*

Figure 2.4 A well-layered sequence of lavas of the Columbia River Plateau near Sprague, Washington. The view is towards the northwest. Highway I-90, at the lower right, provides a sense of scale.

an **ignimbrite** or **ash flow**.[7] Ignimbrites are composed predominantly of fine-grained (<2 mm in diameter) pyroclastic material called **ash,** which is typically compacted by the weight of overlying material and locally annealed (welded) by the residual heat of the *nuée ardente*. Pyroclastic sheets may occur as single units cooled at one time (a single cooling unit) or as a thick set of units with a more complex cooling history (composite cooling unit). Cenozoic pyroclastic deposits may have covered more than 200,000 km^2 in the western United States.[8]

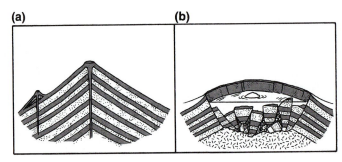

Figure 2.5 Diagrammatic sketches showing the internal structure of a composite cone (a) and a caldera (b).

Composite cones, or stratovolcanoes, derive their name from the fact that they consist of layers of both pyroclastic material and lava (figures 2.5 and 2.6a). These steep-sided cones are the classic structures the layperson associates with the word volcano (figure 2.6b). Rocks comprising composite cones range from silica-poor varieties (basalts), through intermediate types (e.g., andesite), to silica-rich types (rhyolites). The former tend to comprise the lava flows, whereas the latter tend to form pyroclastic accumulations. The Cascade Range of California, Oregon, Washington, and British Columbia; the high peaks of the Andes of South America; and the many **island arcs** (curved chains of volcanic islands) of the western Pacific Ocean each consists of a row of composite volcanoes.

A **caldera** is a large circular depression (figure 2.5b) produced by the eruption-induced collapse of a volcanic structure (Howell Williams, 1941; Bailey, Dalrymple, and Lanphere, 1976; Williams and McBirney, 1979). Where later uplift of a central domal region occurs, the caldera is called a

(a)

(b)

Figure 2.6 Composite volcanoes. (a) Internal structure of a composite volcano (stratovolcano). Note the alternating layers, with resistant lava layers forming bold outcrops and easily eroded pyroclastic layers forming more gently sloping areas. Crater Lake caldera, Oregon. The view is towards the northeast

towards the north edge of the caldera. The large peak to the right in the picture, Lauo Rock, is underlain by a thick flow. (b) Mount Shasta, California, (left) and the satellite volcano Shastina (right). Note the small pyroclastic cones near the base of the volcano. The view is towards the south.

resurgent caldera. Well-known calderas include Crater Lake, Oregon; the Valles Caldera, New Mexico; and Krakatoa, Indonesia. *Cauldron* and *resurgent cauldron* are equivalent but general terms applied to any collapse structures in volcanic rocks, regardless of the size or shape of the structure, or its association with surface volcanism (R. L. Smith and Bailey, 1968). **Craters** are also depressions, but they are produced by the direct, eruptive activity of a volcano, not by collapse. Craters are smaller than calderas, with radii of generally less than 1 km.

Pyroclastic cones, also called cinder cones, are small, steep-sided structures composed largely of pyroclasts of various sizes, with little, if any, lava. These small cones typically occur in groups and are generally associated either with major volcanoes in arcs (e.g., Wizard Island at Crater Lake, figure 2.2b) or with isolated, local volcanic terranes (e.g., the San Francisco Mountain volcanic field, Arizona)(Settle, 1979).

Volcanic domes are small, steep-sided structures shaped like inverted cups or cones. They are formed by the intrusion, extrusion, or both of thick (viscous) siliceous magmas. Typically domes are associated with major volcanoes (e.g., Mt. Shasta, figure 2.2b), but they may also occur in independent fields of intermediate to siliceous volcanic structures that lack a single major volcano, as they do in the Long Valley Caldera—Mono Lake area, California (figure 2.7).[9]

Lava flows are tabular to lobate masses of either fluid or solidified lava (figure 2.8a).[10] Solidified flows occur in many of the larger structures. Surface characteristics of lava flows vary and give rise to specific names. **Pahoehoe lavas** have relatively smooth, vesicular (full of gas holes), ropy surfaces (figure 2.8b). Rough, irregular flow tops consisting of loose fragments characterize **aa lavas** (figure 2.8c). Some petrologists[11] subdi-

Figure 2.7 Volcanic domes composed of silica-rich volcanic rock, Mono Domes, California. View is towards the southeast.

vide fragmented lavas into two groups, those with smaller vesicular fragments (aa lavas) and those with larger smooth blocks (blocky lavas). Below the flow surface, cooling creates **columnar joints,** fractures that separate polygonal, pencil-like cooling structures several centimeters across (figure 2.9).[12] **Pillow lavas** are lavas extruded under water that assume a tubular to elliptical form (figure 2.8d). Additional structures associated with extrusive rocks, such as lava tubes, spatter cones, and hornitos, are listed in table 2.2 and defined in the glossary. Weathering of a flow surface followed by extrusion of hot lava over the soil developed on the weathered surface will produce a brick-red zone of oxidized material called a **baked zone.**

(a)

(b)

(c)

(d)

Figure 2.8 Lava flows and flow features. (a) View of siliceous lava flow within the Newberry caldera, Oregon. Notice the looping ridges on the flow that reveal the flow pattern. (b) Ropy pahoehoe lava, Craters of the Moon National Monument, Idaho. (c) Aa lava flow, Homestead lava flow, Lava Beds National Monument, California. (d) Pillow lavas of the Franciscan Complex, Highway 101, Ridgewood Summit, south of Willits, California.

Figure 2.9 Columnar jointing in andesite flow, near Lost Creek Dam, Oregon. Columns are approximately 0.3 m across.

The interiors of lava flows are also characterized by several features. Fragments of the flow, previously solidified, may be broken off and included in the flow as inclusions called **autoliths.** Pieces of foreign rock occurring as inclusions are called **xenoliths** or accidental inclusions. Gases escaping from the flow will leave holes called **vesicles.** Vesicles later may become filled with minerals such as calcite, quartz, or zeolites and are then called **amygdules** (figure 2.10). Where inclusions, vesicles, or mineral grains become lined up in layers or bands due to the flow of the lava, the lava is said to have **flow banding.**

Clastic volcanic materials, fragmented as they are exploded into the air, are called **pyroclasts** and are subdivided into three groups on the basis of size (Schmid, 1981). Recall that ash is material composed of particles smaller than 2.0 mm in diameter.[13] **Lapilli** are pyroclasts 2.0–64 mm in diameter.

Figure 2.10 Amygdules in basalt, Unicoi Formation (Cambrian ?), Highway 91, on the road to Damascus, Virginia, near the Virginia-Tennessee state line.

Table 2.3 List of Structures of Intrusive Rocks

Major	
Batholith	Lopolith
Stock	Roof pendant

Intermediate	
Stock	Sill
Dike	Laccolith
Cone Sheet	Phacolith
Pipe (neck, vent)	Bysmalith
Funnel	Roof pendant
Cupola	Schlieren dome/arch

Minor	
Dike	Schlieren
Apophysis	Xenolith
Vein	Autolith
Sill	Layering
Foliation	Lineation

Larger fragments are called **bombs** or **blocks.** Bombs were at least partly molten during transport and are assigned names on the basis of surface texture and shape (e.g., breadcrust bombs).

Intrusive Structures

Like extrusive structures, intrusive structures are divided into major, intermediate, and minor types (table 2.3). These size categories do not correspond exactly in dimension to those of the extrusive structures and some structures (e.g., dikes) show a large range of sizes. Major- and intermediate-sized bodies of intrusive rock are often referred to as plutons. **Pluton** is a general term denoting a rock body composed of plutonic rock, typically with "granitic" texture (that texture characteristic of granite and related rock types).

Batholiths and lopoliths (discussed below) are the largest of the intrusive structures. **Batholiths** are defined, traditionally, as bodies of plutonic rock having map areas of 100 km^2 or more. Plutonic bodies exposed over areas of less than 100 km^2 are called **stocks.** The term batholith is usually applied to bodies of granitic rock. In the definition, however, no specifications exist for rock type, internal structure, shape, or thickness.[14]

In the older literature, batholiths were almost always depicted as steep-sided, cylinder-like bodies extending to great depths (for example, see figure 2.11c). They were seemingly bottomless. Drawing on earlier ideas and the work of several geologists who studied batholiths in western North America (notably the Boulder, Idaho, and Sierra Nevada batholiths), W. B. Hamilton and Myers (1967; 1974a; Hamilton, 1983) proposed that batholiths are lens-shaped bodies (figure 2.11d).[15] That hypothesis has been challenged on the basis of geophysical studies and relationships revealed by mapping.[16] Apparently, some

small batholiths are steep-sided (Lee, 1986). Yet the dimensions of larger batholiths support the general concept of lens-shaped batholiths. For example, the Sierra Nevada Batholith of California is about 100 km wide and 650 km long. The Sierran crust is about 50 km thick, but at least the lower 20 km or more do not appear to have the properties of the granitic rock of the batholith.[17] Thus, the batholith could not be greater than 30 km thick and must have an elongated tabular or lenticular shape (650 × 100 × 30 km). Geophysical studies confirm the interpretation that some batholiths are crudely lenticular.[18] Locally, however, within the batholiths and elsewhere, individual plutons may have steep contacts (Calkins, Huber, and Roller, 1985; Vigneresse, 1988). Typical shapes of batholiths and stocks are shown in figure 2.11.

Batholiths and stocks vary considerably in character. This fact has given rise to many debates on the origin of the rocks that compose these bodies.[19] As an outgrowth of a long debate on the origin of granite, Buddington (1959) presented a depth classification of plutons (table 2.4). In this classification, plutons are divided into **epizonal** (shallow), **mesozonal** (intermediate), and **catazonal** (deep) types. Each may be recognized on the basis of a group of criteria. For example, epizonal plutons show evidence of being forcefully injected

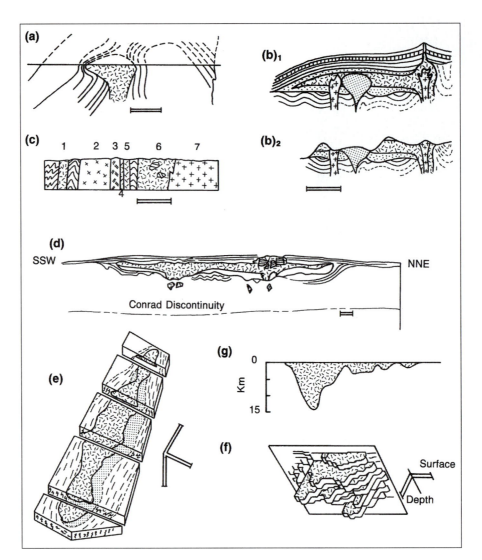

Figure 2.11 Shapes of batholiths and stocks. (a) Papoose Flat Pluton (epizonal stock), Inyo Mountains, California (modified from Sylvester, Oertel, Nelson, and Christie, 1978). (b) Coast Plutonic Complex (epizonal batholith), British Columbia, (1) before erosion and (2) in presently eroded state (from T. A. Richards and McTaggert, 1976). (c) White Mountains batholiths (epizonal and mesozonal), New Hampshire (from C. R. Williams and Billings, 1938). (d) Boulder Batholith (epizonal), Montana (from W. B. Hamilton and Myers, 1967). (e) Exeter Pluton (mesozonal batholith), New Hampshire (redrawn from Bothner, 1974). (f) Mount Waldo Batholith (mesozonal), Maine; three-dimensional view from below looking up towards the surface (modified from Sweeney, 1976). (g) Morin Anorthosite Complex (catazonal batholith), southwestern Quebec (modified from Kearey, 1978). Various hachured symbols represent different igneous bodies. Short to long lines surrounding plutons show orientation of layering in surrounding rocks. Scale bars are 5 km long.

Table 2.4 Depth Classification of Plutons

Criterion/Feature	Epizone	Mesozone	Catazone
Depth of Emplacement (km)	0–6.5(–10)[a]	(6.5–)8–14(–16)	(9–)11–19
Field Criteria[b]			
Contacts	typically discordant	variable	concordant
Homogeneity of body	homogeneous to composite	composite	homogeneous to composite
Roof Pendants	common	common	uncommon
Foliation	absent or local at contacts	common	common and parallel to regional trends
Association with volcanic rocks	common	indirect, but present locally	none
Local deformation at contact	common	present in some cases	absent
Size of pluton	small to moderate	small to large	small to large
Contact metamorphism	very common	uncommon	absent
Chilled margins	common	absent	absent
Associated dikes	aplitic, porphyritic, lamprophyric	aplitic, pegmatitic	migmatitic
Miarolitic cavities	present	absent	absent
Associated migmatites	none	minor	common
Interpretive Criteria			
Surrounding metamorphic facies	none to greenschist facies	greenschist to amphibolite facies	amphibolite to granulite facies
Temperature in country rocks	0°–450° C	250°–500° C	450°–700° C
Typical age	Cenozoic	Mesozoic–Paleozoic	Paleozoic or older

Source: Modified from Buddington (1959).
[a]Values in parentheses show ranges beyond the normal depths of these types of plutons.
[b]Most of the terms listed here are defined and described in this chapter or elsewhere in the text.

into cool rocks at shallow depths in the crust (figure 2.11a, b; figure 2.12a). In contrast, catazonal plutons appear to have developed between layers, as lenses, under high temperature conditions deep within the crust (figure 2.12c). Epizonal plutons are **discordant;** that is, they cut across the layering in surrounding rock. Mesozonal plutons may show both concordant and discordant contacts (a **contact** is the boundary between two rock masses) (figure 2.13). Catazonal plutons are typically **concordant;** i.e., the contacts parallel the layering in surrounding rocks (figure 2.12c). At the contact, epizonal plutons may have been cooled substantially, resulting in a **chilled margin,** a fine-grained sheath of rock enclosing the inner parts of the pluton. Catazonal plutons lack such margins. **Roof pendants,** masses of rock that formed part of the roof over the pluton and hung down into the magma, are isolated by later erosion of the roof (figure 2.13). Such features are common in shallower plutons.

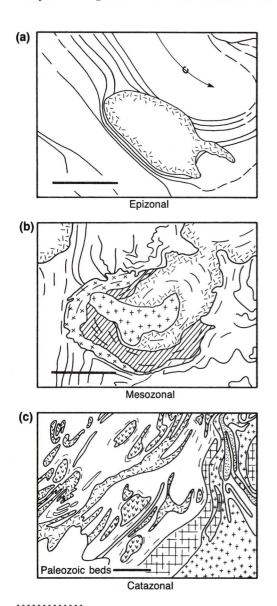

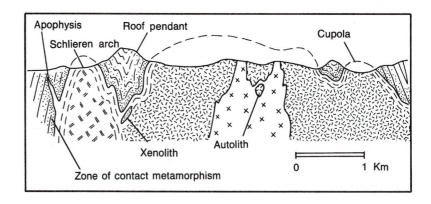

Figure 2.13 Schematic cross section through a mesozonal pluton, showing common structural features.

Figure 2.12 Map patterns of epizonal, mesozonal, and catazonal plutons. (a) Simplified sketch of the epizonal Papoose Flat Pluton, California (from Sylvester et al., 1978); cross section is shown in figure 2.11a. (b) Part of the mesozonal White Creek Batholith, British Columbia (simplified from Buddington, 1959, who modified the work of J. C. Reesor). (c) Catazonal plutons, northwest Adirondack area, New York (from Buddington, 1959). Different patterns, represent various types of igneous rock. Longer lines represent layers or foliation in the surrounding rock and plutons. Scale bars are 10 km long.

Internally, plutons also may differ in a variety of ways. Epizonal and mesozonal plutons tend to be composite, being composed of several individual intrusions of differing rock types (figures 2.11b, c; 2.13).[20] Catazonal plutons are typically homogeneous, containing but a single rock type. **Miarolitic cavities,** small holes in coarse-grained rock into

which crystals project, are present in rocks of epizonal plutons but absent in rocks of the catazone. Alignment of minerals to produce a layered appearance, a feature called **foliation,** is usually present only near the contact or is absent in epizonal plutons. In mesozonal plutons, mineral alignments and xenoliths may define archlike or dome-shaped patterns of various dimensions, as in **schlieren domes** and **arches** (Balk, 1937; Abbott, 1989)(figure 2.13). Foliations in catazonal plutons may pervade the entire body and parallel the foliation of the **country rock** (the rock surrounding the pluton). Observations of these various criteria, taken together, make it possible to assign batholiths and stocks to the appropriate depth category.

Lopoliths are large, dish-shaped (concave-up), layered intrusions (figure 2.14a). Although these structures are not particularly common, they contain a wide variety of interesting, dominantly basic to ultrabasic rocks, including some of considerable economic importance. In addition, the internal structures and compositions are such that they allow study of some important processes involved in crystallization of magmas.[21]

Laccoliths, phacoliths, and sills are all concordant structures of generally modest size (figure 2.14). **Sills,** the most common of the three structures, are tabular to elliptical bodies with width (or diameter) to thickness ratios of greater than 10.[22] **Laccoliths** are shorter and thicker than sills and tend to bow up the overlying layers. **Phacoliths** are lenticular intrusions that are emplaced into sites along fold axes. Gilbert (1877, in Gilbert, 1880) first described laccoliths in a classic report on the Henry Mountains of Utah. In that report, he showed the laccoliths connected to feeder **dikes** (discordant tabular plutons) that extended up from below (figure 2.14b). Gilbert's interpretation of the origin of laccoliths, however, has proven to be controversial in the case of the Henry Mountains plutons and other laccoliths, and has led to continuing study and debate.[23]

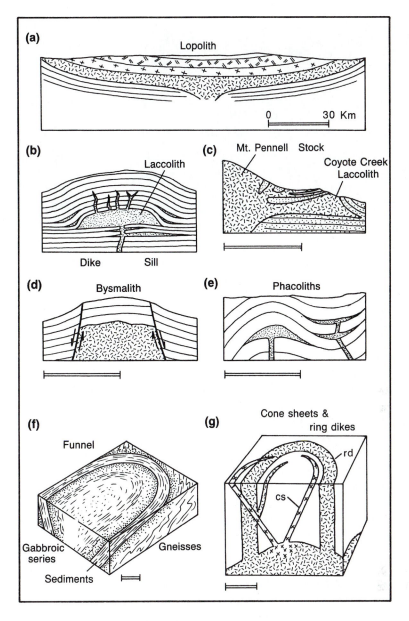

Figure 2.14 Cross sections and block diagrams of selected intrusive structures. (a) Schematic diagram of a lopolith. Note the scale. (b) Sketch of a laccolith (modified from G. K. Gilbert's classic work on the Henry Mountains of Utah, 1880). (c) Simplified section of the Coyote Creek Laccolith, Henry Mountains, Utah (Source: Charles Hunt, "Geology and Geography of the Henry Mountains Region, Utah" in *U.S. Geological Survey Professional Paper*, 228, 1953.). (d) Sketch of a bysmalith (cf. Iddings, 1898). (e) Diagrams of phacoliths (cf. G. K. Gilbert, 1880). (f) Block diagram of a funnel (modified from Balk, 1937). (g) Sketch showing cone sheet (cs) and ring dikes (rd). Scale bars in (c)–(g) are 1 km long.

Dikes are the most common type of *discordant* intrusion (figure 2.15). They occur in a variety of forms and compositions and may be simple (resulting from a single intrusion), multiple (resulting from two or more intrusions of the same magma type), or composite (resulting from intrusion of two or more magma types) (Billings, 1972). Dikes often occur together in a group of parallel structures, but they also may occur alone or in radiating groups. The tabular nature of dikes generally results from a combination of the tabular nature of the stress-induced fractures into which the dikes intrude and the relatively low viscosity of the intruding magma (Emerman and Marrett, 1990).

Ring dikes and **cone sheets** are special types of dikes. Ring dikes are often large and are vertical and cylindrical in form (figure 2.14g). These dikes, which are emplaced over

magma chambers, are commonly associated with cauldron collapse.[24] Cone sheets may occur either with ring dikes or independently. Essentially, cone sheets are dikes that occupy fractures representing segments of downward-pointing cones (figure 2.14g). Both ring dikes and cone sheets may underlie volcanic terranes and occur above batholiths or stocks.

Dikes serve as conduits for magmas rising toward the surface. Where such conduits or channels are tubular rather than tabular, eruption of lavas at the surface will occur at a point and result in the construction of a volcanic cone. Erosion of such a cone will leave a **volcanic neck** (pipe, vent). Partially eroded volcanic necks may be surrounded by a radial dike system (figure 2.16).

Intermediate-scale structures associated with batholith-to stock-sized plutons include funnels, cupolas, and schlieren

Figure 2.15 Dike of ouachitite, or biotite feldspathoidal phonolite (dark rock), cutting sandstones (light rock with slightly tilted layers), Highwood Mountains, Montana. Note DNAG scale at right side of dike.

Figure 2.16 West Spanish Peak, Colorado, an eroded volcano with radiating dikes (d), near La Veta, Colorado (view towards the south).

domes and arches. **Funnels** are solid plutonic bodies in which the layering dips inward, just as do the dikes of cone sheets (figure 2.14f). **Cupolas** are stocklike exposures of plutonic rock separated from a larger pluton by a screen of country rock but believed to be connected to the larger pluton at depth (figure 2.13). **Schlieren** (also called flow layers) are tabular, diffuse, disk-like concentrations of minerals within an igneous rock mass (Balk, 1937). Although their boundaries are diffuse, schlieren are visible because some constituent minerals are more abundant in the disks than they are in the surrounding rock. As magmas move, schlieren become oriented with their long dimensions parallel to the direction of flow, especially where they are concentrated near the margins of plutons. Where the flow of magma follows an arcuate path as it moves upward in the crust, schlieren may define arcuate patterns that take the form of arches or domes; hence the names schlieren dome and schlieren arch.[25]

Small-scale structures in intrusive rocks include a variety of features (table 2.3). Dikes, sills, foliation, and schlieren were defined above. **Apophyses** are short, irregular dikes that extend from pluton margins into the country rock (figure 2.13). **Veins,** like dikes, are crudely tabular bodies that form fracture fillings. Usually the term vein is used where the fracture filling consists of only one or two minerals (e.g., quartz or calcite) or where minerals that are not typically abundant in igneous rocks, but are of economic value, comprise the rock.

Xenoliths and autoliths, like schlieren, are small bodies of material found within plutonic rocks. Xenoliths, also called accidental inclusions, are fragments of a foreign rock body (one not derived from the same magma) that occur within a plutonic rock mass (figure 2.17). In contrast, autoliths, also called cognate inclusions, are inclusions of rock genetically related to the igneous rock in which they are included. For example, part of a magma may crystallize early and become broken by later movements of the still liquid portion within a magma chamber. The broken fragments mixed with the molten material would appear as autoliths (if they did not melt) after the entire magma solidified. Xenoliths and autoliths differ from schlieren in having sharp boundaries with the enclosing plutonic rock.

Foliation, lineation, and layering are structures that may characterize the rock of some intrusive bodies. **Foliation** is a planar or leaf-like character in a rock that results from the parallel alignment of either platy or needle-like (acicular) minerals (or fracture surfaces). **Lineation** is a feature resulting from the parallel alignment of acicular minerals or intersecting planar features.[26]

Layers also occur in igneous rocks (figure 2.18). A layer is a distinct sheetlike unit characterized by a particular mineral composition, texture, or both (Irvine, 1982).[27] Typically layers develop where silica-poor magmas cool and crystallize at rates that allow crystals to accumulate through sinking or floating in the remaining liquid. Layered features not associated with layered intrusions are called *bands* (Irvine, 1982). A variety of such features is known in igneous rocks, including flow bands in volcanic rocks and some orbicular and comb-layered structures in plutonic rocks.[28]

TEXTURES OF IGNEOUS ROCKS

Magmas are complex solutions. Because of falling temperature, changes in pressure, or changes in composition, these solutions crystallize, or if supercooled they solidify without forming crystals. The end product of crystallization or solidification is a rock composed of interlocking crystals, of crystals enveloped by a glass, or of glass. If the magma is fragmented by explosive eruption and escape of gas, the fragments of glass, crystal, and rock may accumulate and lithify to become a rock. Whatever its history, a newly formed igneous rock will

Figure 2.17 Small xenoliths within the Woodleaf Granite, Woodleaf, North Carolina.

Figure 2.18 Layers in the Stillwater Complex, along the west slope of Stillwater Canyon, Montana.

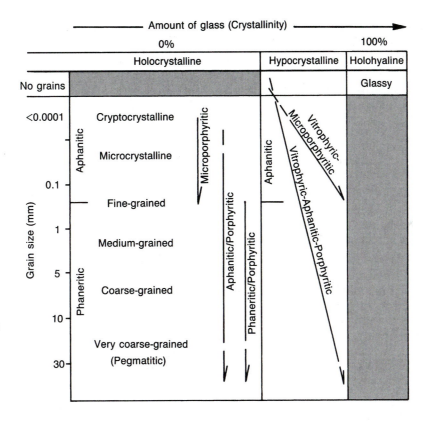

Figure 2.19 Igneous rock textures plotted in terms of crystallinity and grain size. Arrows indicate a range of grain sizes within a single rock.

consist of one or more of these elements—glass, crystals, and fragments of glass, crystal, or rock. The characteristics of and interrelationships between these materials—including grain size, grain shape, grain orientation, grain boundary relations, and the crystallinity of the rock—are together referred to as the texture of the rock.

Igneous rocks with interlocking crystals have **crystalline textures,** whereas those composed of fragments have clastic or, more specifically, pyroclastic textures. The crystallinity and dominant grain size in crystalline igneous rocks are each described by one of a series of three terms (figure 2.19). **Holocrystalline** textures are composed entirely of crystalline material. **Holohyaline** textures have components that are totally glass. Textures involving both crystals and glass are **hypocrystalline.** In terms of grain size, a rock lacking any significant crystalline material is considered to be **glassy.**

(a)

....
Figure 2.20 Porphyritic textures. (a) Phaneritic-porphyritic texture, Cathedral Peaks Granodiorite, Yosemite National Park, California. (b) Photomicrograph of aphanitic-porphyritic texture in andesite, Shasta County, California. Note euhedral hornblende (H) and rectangular plagioclase phenocrysts (P). (XN). Long dimension of photo is 6.5 mm.

(b)

Crystalline materials with grains too small to discriminate without the aid of microscope are called **aphanitic,** and those in which grains can be discerned are referred to as **phaneritic.**[29]

Some additional terms are used to describe rocks with particular grain size characteristics. For example, rocks that are very coarse-grained, dominated by grains >3 cm in length, are called **pegmatitic.** Although many petrologists reserve the term pegmatite for siliceous (granitoid) rocks of that grain size, rocks of any composition may be pegmatitic. **Porphyritic** textures are those in which there is a bimodal grain distribution; that is, there are a number of larger grains, called **phenocrysts,** surrounded by a population of grains of significantly smaller size, the **groundmass** (figure 2.20). Volcanic rocks lacking phenocrysts are said to be **aphyric** (figure 2.21b), whereas those with phenocrysts are **phyric.** In **microporphyritic** textures, both the phenocrysts and the groundmass are aphanitic. If the phenocrysts are phaneritic but the groundmass is aphanitic, the texture may be called aphanitic-porphyritic. Where both groundmass and phenocrysts are phaneritic, the texture is called phaneritic-porphyritic.

Crystal shapes in rocks are generalized when assigning descriptive terms to the textures. **Idiomorphic-granular** textures are those dominated by euhedral crystals. **Hypidiomorphic-granular** textures consist of crystals that, on the average, are subhedral in form (figure 2.22). In **allotriomorphic-granular** textures, the crystals are predominantly anhedral.

Textures characterized by certain shapes, orientations, interrelationships, or internal characteristics are given special names. Although not an exhaustive list, table 2.5 contains the names assigned to many of these special textures.

In many cases, the details of volcanic textures cannot be observed without the use of a microscope. Some of the common volcanic textures—microlitic, spherulitic, vitrophyric, intersertal, intergranular, subophitic, and ophitic—form a series. In fact, Lofgren (1983) was able to produce such a series experimentally. **Microlitic** texture is a porphyritic texture in which microlites (small fibrous crystals) are enclosed in a glassy groundmass. In **spherulitic** texture, fibrous crystals that radiate from a point are clustered in spherical groups (figure 2.21a). **Vitrophyric** texture is characterized by phenocrysts in a glassy groundmass. Porphyritic rocks in which rectangular plagioclase crystals (plagioclase laths) make up a substantial part of the rock, with the remainder being composed of glass and small crystals of other materials, are said to have **intersertal** texture. In such rocks, the plagioclase laths are not aligned. If the feldspars are aligned in subparallel arrangements, the texture is called **trachytic. Intergranular** textures (figure 2.21b) are holocrystalline textures in which grains of augite and other minerals occur between laths of plagioclase.[30] In **subophitic** texture, the augite and plagioclase assume similar sizes, with augite encompassing the ends of some plagioclase laths. In **ophitic** textures, the pyroxene exceeds the size of the plagioclase, so that several plagioclase laths are enclosed in a single pyroxene grain.

The escape of gas from a magma as it approaches the surface and erupts produces some unique textures and structures. As noted above, the violent escape of gas may fragment a rock or magma producing pyroclastic textures consisting of an aggregation of broken crystals, rocks, and glass. If the escape of gas leaves vesicles, the rock is said to be **vesicular.** Recall that filled vesicles are amygdules. Where the vesicles become elongate into fine tubular holes, as often occurs in silica-rich (rhyolitic) magmas, the rock is **pumiceous.** Vesicular, amygdaloidal, and pumiceous features are referred to as structures by some petrologists and classified as textures by

(a)

(b)

· · · · · · · · · · · · ·
Figure 2.21 Photomicrographs of two volcanic textures. (a) Spherulitic texture in quartz keratophyre, Del Puerto Formation, Coast Range Ophiolite, Carbona 15' Quadrangle, California. Radiating grains are quartz and feldspar. (XN). Long dimension of photo is 1.20 mm.

(b) Intergranular texture in basalt, collected near Barstow, California. Rectangular grains with twins (dark and light bands) are plagioclase. Granular grains are olivine, pyroxene, and magnetite. (XN). Long dimension of photo is 1.20 mm.

others. They are transitional features in that they do not involve grains (as textures do), but they are often penetrative, or pervade the rock, a characteristic of textures.

Special names for particular grain relationships are also applied to plutonic textures. A widely used term is *granitic texture*, a synonym for hypidiomorphic-granular texture (figure 2.22a, b).[31] **Poikilitic** texture is a texture in which a large crystal (an oikocryst) encloses irregularly scattered, smaller crystals of another mineral. Ophitic texture, found both in volcanic and plutonic rocks, is one particular type of poikilitic texture. **Graphic** texture (figure 2.22c) is similar to poikilitic texture in that a larger grain encloses apparently smaller grains, but this texture, which occurs in pegmatitic granitoid rock, consists of a very large crystal of alkali feldspar enclosing smaller crystals of quartz that all have the same crystallographic orientation. In some cases, it can be clearly demonstrated that the apparently separate quartz grains are all connected and are part of a large, single dendritic or

"snowflake" quartz crystal. Microscopic versions of such textures are called **micrographic.** Where wormlike, irregular grains of quartz are intergrown within sodic plagioclase, the texture is called **myrmekitic** texture. A similar texture in which the feldspar is alkali feldspar is called **granophyric** texture. Myrmekitic and granophyric textures are special types of **symplectic** texture, a general category of texture in which wormy (vermicular) or irregular intergrowths of one mineral occur in another. **Seriate** texture is a texture consisting of grains of a wide range of sizes, grading one to another.

Textures found in low-silica plutonic rocks (those with <53% SiO_2) include ophitic, subophitic, diabasic, and various cumulate textures. Ophitic and subophitic textures have been described. **Diabasic** texture, which forms the third member of the three-texture series—ophitic, subophitic, and diabasic—is a coarse-grained equivalent of the intergranular textures found in volcanic rocks. In such textures, augite and other minerals form smaller grains between laths of plagioclase. **Cumulate** textures

(a)

(c)

(b)

Figure 2.22 Two plutonic rock textures.
(a) Hypidiomorphic-granular texture in Sentinal Granodiorite, Yosemite National Park, California. (b) Photomicrograph of Sentinal Granodiorite showing hypidiomorphic-granular texture. Long dimension of photo is 6.5 mm. (c) Photomicrograph of graphic texture in pegmatitic granite, McKinney Mine, Spruce Pine District, North Carolina. Long dimension of photo is 6.5 mm.
Q = quartz, P = plagioclase, A = alkali feldspar,
H = hornblende, and B = biotite.

occur in rocks called cumulates, which are defined as igneous rocks "characterized by a cumulus framework of touching mineral crystals or grains that were evidently formed and concentrated primarily through fractional crystallization" (Irvine, 1982). After crystals cluster, some materials, called *postcumulus materials*, may crystallize between the already formed crystals. Thus, the cumulate texture is one in which accumulated crystals are welded together by postcumulus materials.

Additional textures found in plutonic rocks, listed under "Other" in table 2.5, are primarily textures centered on single grains. The most common of these is **zoned** texture. Many minerals, such as plagioclase, clinopyroxene, and garnet, exhibit zoning, a layered character resulting from changes in

Table 2.5 Names of Special Igneous Rock Textures

Volcanic[a]	Plutonic	Other
Microlitic	Poikilitic	Zoned
Spherulitic	Graphic	Corona
Vitrophyric	Ophitic	Kelphytic rim
Intersertal	Subophitic	Rapikivi
Intergranular	Diabasic	Epitaxial
Felty	Orthocumulate	Poikilitic
Pilotaxitic	Mesocumulate	
Trachytic	Adcumulate	
Subophitic	Symplectic	
Ophitic	Myrmekitic	
Dictytaxitic	Seriate	
Glomeroporphyritic	Trachytoidal	
Pyroclastic	Granophyric	
Seriate		
Spinifex		

Sources: Based on the observations of the author and, in part, on R. L. Bates and Jackson (1980), Howell Williams et al. (1982), Heinrich (1956), and Irvine (1982). For more thorough discussions and representations of textures, refer to the contemporary work of Williams et al. (1982) and the classic work of Johannsen (1939).
[a]Terms not defined in the text are defined in the glossary.

composition from the core to the rim of a mineral. A **corona** (reaction rim) is a rim of radiating minerals surrounding a core mineral. Where one mineral grows around another in such a way that their crystallographic structures are continuous from one to the other, the overgrowth is said to be **epitaxial.**

ORIGINS OF IGNEOUS TEXTURES

Because igneous rocks form from a magma, their textures are controlled by the processes that govern crystallization from a melt (Kirkpatrick, 1975, 1981; Lofgren, 1980; S. E. Swanson, 1977).[32] In chapter 5, a review of the physical-chemical processes that control crystallization is presented. In that chapter, I review phase diagrams and show the kind of mineral (phase)[33] that forms and the order of appearance of minerals that develop upon cooling of melts of particular compositions. Here I focus on the physical processes involved in crystal growth. The chemical and physical processes, however, are intimately related. Therefore, rereading this section after reading chapter 5 will be beneficial.

As melted material cools, it may pass through three general stages: (1) a stage in which all material is melted, (2) a stage in which crystals and melt coexist, and (3) a stage in which all material is solid. Figure 2.23 is a graph of temperature versus composition for the plagioclase system.[34] The three fields listed above are labelled, as are the two curves separating them. The upper curve, above which the entire mass is liquid, is called the **liquidus.** The mineral (phase) that crystallizes at the liquidus is called the liquidus phase. In figure 2.23 that phase is plagioclase. The curve below which all material is solid is the **solidus.**

As a melt cools, two processes—nucleation and crystal growth—lead to the development of a texture. **Nucleation,** the formation of nuclei, is a poorly understood process in which a few atoms assume the same relationship to one another as they would have in a solid; that is, they form a crystalline structure. A liquid may be described as a disordered solid, and nuclei are created and destroyed constantly by the random motion of atoms within the liquid. In crystallization of a melt, nucleation is the initial process that occurs, because once a structure is developed, it is possible for individual crystals to grow. Crystals will grow from the nuclei once the physical and chemical conditions favor the transformation of liquid into solid. This growth represents a transition from a higher to a lower energy state.

Wherever surfaces exist, there is an energy associated with the surface called surface energy. In order to form a crystal, energy must be used to create the surfaces that bound the crystal. Nuclei form first. Nucleation is described as either homogeneous or heterogeneous. In **homogeneous nucleation,** nuclei develop and crystals grow spontaneously within the melt. New surface must be created. **Heterogeneous nucleation,** however, occurs more readily, as it involves nuclei development on a preexisting surface, and hence requires less energy.

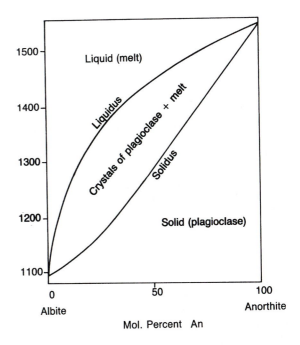

Figure 2.23 The system albite-anorthite.
(After Bowen, 1913; Kushiro, 1973b)

Nucleation is controlled by the composition of the melt, the structure of the melt, the temperature of the melt, and the cooling rate. The initial composition of the melt, of course, controls the minerals that nucleate (for example, olivine does not nucleate from a melt devoid of Fe or Mg). More subtle effects of varying composition are not well known. Naney and Swanson (1980) have shown experimentally that if iron and magnesium are introduced into a water-bearing system containing the elements Na, K, Ca, Al, and Si—a system that could be shown to crystallize plagioclase, alkali feldspar, and quartz in the absence of Fe and Mg—nucleation of quartz and the feldspars is inhibited. In some compositions, they found that plagioclase is the only tectosilicate that will nucleate.

The structure of the melt is related to the chemistry of the melt, to the maximum temperature the melt has attained during melting, and to the amount of time the melt is maintained at higher temperatures.[35] If the melt structure retains remnants of crystals, crystal growth will be much easier, occurring as a result of heterogeneous nucleation. Pure silica glass and melts form complex networks of SiO_4 tetrahedra. The addition of various ions to a melt (e.g., OH^-, Ca^{++}, Mg^{++}) breaks that structure apart (figure 2.24). Similarly, higher temperature tends to break down the structure of nuclei remaining in the liquid.[36] This reduces the possibility of, or prevents, heterogeneous nucleation.[37] Lofgren (1983) argues, however, that nuclei of crystals that melt at lower temperatures may be preserved in a melt within cavities in materials that have high melting temperatures. Consequently, he suggests that heterogeneous nucleation may be the dominant factor in the development of igneous textures. Apparently, the amount of time a melt is maintained at a

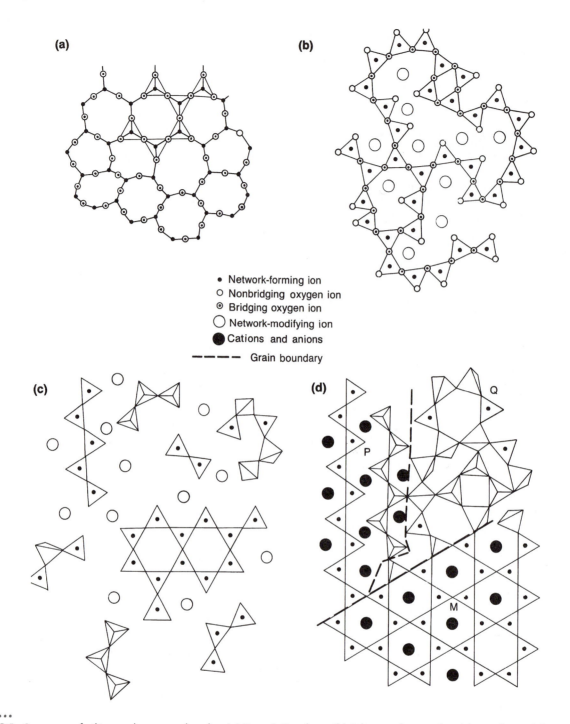

- • Network-forming ion
- ○ Nonbridging oxygen ion
- ◉ Bridging oxygen ion
- ◯ Network-modifying ion
- ● Cations and anions
- – – – – Grain boundary

Figure 2.24 Structure of silicate glasses and rocks. (a) Pure SiO_2 glass. (b) Silicate glass with additional ions (c) Partially crystallized glass. (d) Completely crystallized melt with crystals of mica (M), pyroxene (P), and quartz (Q).
((a) and (b) from Ian Carmichael, Igneous Petrology. *Copyright © 1974 McGraw-Hill, Inc., New York. Reprinted by permission.)*

given temperature also influences how many of the pre-melt nuclei are destroyed. Theoretically, if all the nuclei are destroyed, homogeneous nucleation would be important in the development of the texture. In reality, if the nuclei remain in the melt as suggested by Lofgren (1983) or if magma bodies begin to crystallize at their margins (through heterogeneous nucleation along the sides, bottom, or walls),[38] then, as Lofgren suggests, heterogeneous nucleation is the controlling process. Once some crystals have formed, heterogeneous

nucleation may also occur along the margins of early-formed crystals, particularly if local saturation in necessary chemical components develops in the melt adjacent to the crystals.[39]

Once the nuclei are formed, crystal growth can occur. Among the factors that influence growth are: (1) the composition of the melt, (2) the kind and density of nuclei present, (3) the temperature of the melt when crystallization begins (it may not be the liquidus temperature), (4) the cooling rate, (5) diffusion of chemical species through the melt, (6) reactions occurring at the melt-crystal interface, and (7) the heat flow in the region of the growing crystal.[40] Remember that the textures observed in a rock are characterized by the size, shape (morphology), orientation, and boundary relations of crystals and by the crystallinity of the rock as a whole. Which factors determine each of these characters?

Crystallinity is determined by composition and temperature factors (1, 3, and 4). Silica-rich (rhyolitic, granitic) magmas tend to be more viscous, or thicker (as honey is thicker than water). High viscosity reduces the ability of atoms to migrate through the melt, or **diffuse,** to a nucleus or a growing crystal. Low-silica magmas (basalt, gabbro) have lower viscosity, allowing greater diffusion rates. Similarly, very high cooling rates do not allow migration of materials to nuclei or growing crystal faces. In fact, melts may be cooled so quickly that they become solid, supercooled liquids (i.e., glasses).[41] High viscosity and rapid cooling combine during eruptions of high-silica magmas to produce glassy texture in the volcanic glass called **obsidian.**

Most obsidian, rather than being all glass, contains **microlites,** or tiny crystals, in the glassy matrix. Similar hypocrystalline textures occur in many other kinds of volcanic rock, as do porphyritic textures. The occurrence of textures with varying grain sizes focuses attention on those factors that influence grain size.

Hypocrystalline and other porphyritic textures have traditionally been attributed to a two-stage history of cooling. According to this two-stage hypothesis, early slow cooling, which produces phenocrysts, is followed by eruption-induced, rapid supercooling, which creates the groundmass. Multistage crystallization histories have been documented.[42] However, a two-stage history is not required to produce all porphyritic textures. Swanson (1977) discusses how such textures could be produced in siliceous plutonic rocks during single-stage cooling, and Lofgren (1980) notes that porphyritic textures commonly form in silica-poor (basaltic) melts cooled at constant rates. Both of these workers note that the kind and density of nuclei are important in producing such textures.[43]

A generalized nucleation density curve is shown in figure 2.25a. An important factor in nucleation and crystal growth is the concept of undercooling (factor 3). It is possible to cool a melt below the temperature of the liquidus. Held at that lower temperature, crystals will begin to form, after an incubation period, as equilibrium is reestablished. The difference in temperature between the liquidus temperature and the temperature of crystallization is called the *undercooling* (sometimes supercooling) and is assigned the symbol ΔT (note that $\Delta T = T_{liquidus} - T_{crystal\ growth}$). In figure 2.25a, a melt cooled to ΔT_1 will have a relatively low nucleation density (the number of nuclei/unit volume) (dashed line). The growth rate will be relatively high (solid line). As a result, the few crystals that grow will grow fast and become relatively large. A resulting texture might be pegmatitic.[44] As a second example, consider a

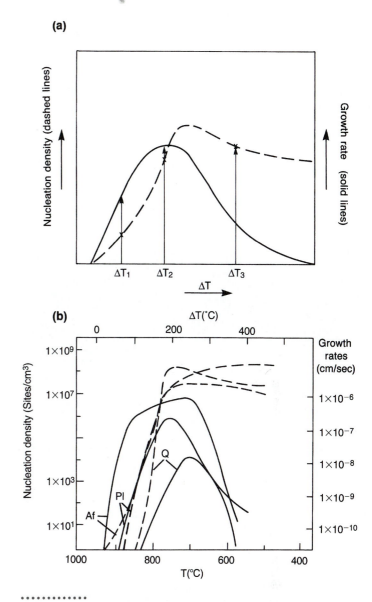

Figure 2.25
Nucleation and crystal growth rate curves. (a) Hypothetical nucleation and growth rate curves. (b) Experimentally determined nucleation and growth rate curves for a synthetic granite with 3.5 wt. % H_2O. T = temperature, ΔT = undercooling, Af = Alkali feldspar, Pl = Plagioclase feldspar, Q = quartz.
(Modified from S. E. Swanson, 1977)

melt cooled to ΔT_2. At ΔT_2, both the number of nuclei and the growth rates will be large, producing a texture with a modest density of medium-sized crystals (e.g., hypidiomorphic-granular, medium- to fine-grained texture). A rock undercooled to ΔT_3 would have a high density of nuclei but low growth rates. The resulting texture would be fine-grained to aphanitic, as in the intergranular texture common in basalts. Note that the examples cited are particular cases. The *actual grain sizes* depend on the specific growth rates and length of time over which growth occurs.

In the examples cited above, recognize that the *size* of crystals produced is not a function of *rate of cooling*, as is often assumed. Nucleation rate, and therefore, nucleation density, exert a dominant control (S. E. Swanson, 1977; Brandeis and Jaupart, 1987). Although slow cooling and crystallization at depth may result in large crystals, a combination of low nucleation densities (e.g., <1000 nuclei per cm^3) and high crystal growth rates (e.g., 3 mm to 19 m/day) may also result in the formation of large crystals.[45] Such crystals could form over very short periods of time. Some pegmatites, in fact, have crystal morphologies reflecting very rapid growth, like those formed in experimental studies (figure 2.26). Rocks composed of many small crystals develop from large undercoolings that will yield high nucleation densities and modest or low growth rates. In addition, at large ΔTs, flow or convection in the magma apparently facilitates increased nucleation (Kouchi, Tsuchiyama, and Sunagawa, 1986), contributing to the tendency to form a fine-grained rock.

Experimentally determined curves developed by Swanson (1977) for nucleation and crystal growth in a siliceous rock composition are shown in figure 2.25b. Notice that curves for several minerals are depicted. An undercooling of about 120° C would produce a porphyritic texture during a one-stage cooling process. For alkali feldspar (Af) at that ΔT,

(a)

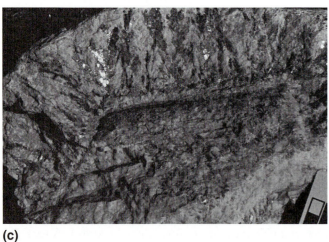

(c)

(b)

Figure 2.26 Rapid-growth crystal morphologies in pegmatite from the McKinney Mine, Spruce Pine District, North Carolina. (a) Diamond-shaped, hollow alkali feldspar in quartz. (b) "Snowflake" (dendritic) quartz in alkali feldspar. (c) Plumose (dendritic) white mica nucleated on a large alkali feldspar crystal rimmed by fine-grained garnets.

nucleation density is relatively low but growth rate is high, resulting in large crystals. For plagioclase (Pl), both nucleation density and growth rates are modest, so medium-sized crystals would develop. A few small quartz crystals (Q) would form at the same time. The resulting rock would have large phenocrysts of alkali feldspar in a matrix of finer-grained plagioclase and quartz, i.e., a porphyritic texture. Lofgren (1980) describes how crystals of two sizes but of the same mineral can form during single-stage cooling, a possibility that apparently requires that the liquidus phase crystallize alone over a range of temperatures. Similarly, Cashman and Marsh (1988) show that continuous nucleation and growth over a period of time yields a mix of crystal sizes.

In conclusion, various cooling rates, nucleation densities, growth rates, and cooling histories can result in a variety of grain sizes. In a given rock, the grain size may be uniform or grains may occur in more than one size. Large grains may result from long, slow cooling and crystallization or from rapid growth from a few nuclei at small undercoolings. Porphyritic textures may develop via single- or multistage crystallization histories.

The morphologies, and in some cases the orientation and boundary relations, of crystals depend on several factors. These include phase and melt composition, the kind and density of nuclei, the cooling rate, the undercooling, and the growth rate (factors 1, 2, 3, and 4 above). The growth rate is a function of diffusion, crystal-melt interface reactions, and heat flow near the crystal face (factors 5, 6, and 7), the details of which are beyond the scope of this text.[46] Measured growth rates vary from 2.2×10^{-2} cm/second to $<10^{-8}$ cm/second.[47] Crystals that form at high growth rates have a tendency to form in needle-like or hollow forms. Those that form at lower rates tend to fill out to form more solid shapes. With respect to undercooling, Lofgren (1974) showed, in a study of plagioclase, that tabular to equant plagioclase and pyroxene are favored at small ΔTs, acicular to skeletal crystals form at moderate ΔTs, and dendritic and sperulitic forms are favored by large ΔTs. In contrast, Kirkpatrick, Kuo, and Melchior (1981) argued, on the basis of experiments on diopside glass, that sperulitic texture is a nucleation effect.

Compositional effects on morphology are less pronounced, but Lofgren and Donaldson (1975) suggest that at a fixed cooling rate, changing the melt composition from a silica-poor (gabbroic) to a silica-rich (granitic) composition causes a change from tabular to branching crystals. Their work also demonstrated the effects of cooling rate. Low cooling rates produce tabular crystals, just as do small undercoolings. With increasing cooling rates, morphologies vary from elongate to slightly branched to complexly branched forms. Lofgren (1983) produced much of the range of basalt textures, from spherulitic to ophitic, by varying the density and kind of heterogeneous nucleations sites, and he argues (1980, 1983) that nucleation phenomena are the most critical factors in the development of textures.

The range of igneous rock textures clearly depends on variations in the complex relationships between nucleation and crystal growth. As a consequence, understanding each texture requires careful observation combined with a knowledge of crystallization processes.

SUMMARY

Igneous structures are helpful in the field identification of igneous rocks and provide valuable information about the history of formation of the rocks. Both extrusive and intrusive structures may be divided by size into major, intermediate, and minor features that range from thousands of cubic kilometers to less than a single cubic centimeter in volume. Granitoid batholiths, which are generally lenticular bodies of siliceous plutonic rock, are the largest of the intrusive structures, whereas tabular lava plateaus, composed primarily of basalts (silica-poor volcanic rocks), are the largest of the extrusive structures. Many additional structures of varied form reflect different conditions of formation.

Textures, like structures, are diverse. They are described in terms of the shape, size, orientation, and interrelationship of crystals. Plutonic rock textures are typically hypidiomorphic-granular. Volcanic rock textures are commonly porphyritic. These and other textures form when the physical-chemical conditions in the magma allow crystals to grow from nuclei, which form spontaneously through homogeneous nucleation or heterogeneously upon a previously formed crystal. Crystal shapes depend on the interplay between a variety of factors. More elongate forms (e.g., needle-like, skeletal, and branching types) are favored by rapid rates of cooling, large degrees of undercooling, high growth rates, and fewer heterogeneous nucleation sites. Tabular to equant forms are favored by silica-poorer compositions, slow cooling rates, small degrees of undercooling, low growth rates, and abundant heterogeneous nucleation sites.

EXPLANATORY NOTES

1. Rocks that crystallize at shallow depths, commonly with porphyritic textures (see definition below), have been called *hypabyssal*, especially in the older literature. In this text, hypabyssal rocks are not assigned to a separate category and the term is not used. All features formed at shallow depths are considered to be intrusive.
2. Greeley (1982) defined the basaltic plain as a new type of volcanic construction. Lava plateaus are described by T. L. Wright (1984).
3. Comprehensive discussions of these and other structures may be found in G. A. MacDonald (1972, ch. 5), Howell Williams and McBirney (1979, ch. 5), Easton and Easton (1984), and Cas and Wright (1987, ch. 4). See these works and also R. V. Fisher and Schmincke (1984) and the Basaltic Volcanism Study Project (1981) for additional discussions of various igneous structures.
4. For example, see the works of Waters (1961, 1962), D. A. Swanson and his colleagues (Swanson, 1967; Swanson, Wright, and Helz, 1975), and T. L. Wright (1984).
5. See G. A. MacDonald (1972, ch. 5), Howell Williams and McBirney (1979, ch. 5), Schmid (1981), Cas and Wright (1987, ch. 4), and especially R. V. Fisher and Schmincke (1984) for discussions of pyroclastic materials.

6. Sarna-Wojcicki et al. (1981).

7. This use of ignimbrite follows that of H. E. Cook (1968) and is equivalent to the term ash flow used in the classic work of R. L. Smith (1960a).

8. Estimate by Howell Williams, cited in H. E. Cook (1968) as personal communication.

9. Howell Williams and McBirney (1979); Bailey, Dalrymple, and Lanphere (1976); Chesterman (1968); Loney (1968); S. E. Swanson et al. (1987).

10. See MacDonald (1972, ch. 5), Howell Williams and McBirney (1979, ch. 5), and Cas and Wright (1987, ch. 4) for descriptions of lava flows. Also, Moore (1975) and Yamagishi (1985) describe pillowed flows. Kilburn and Lopes (1988) describe aa flow fields. Bonnichson and Kauffman (1987) and Manley and Fink (1987) describe the features of some silicic flows.

11. For example, Ehlers and Blatt (1980) and A. Hall (1987, p. 39ff).

12. See P. E. Long and Wood (1986) and DeGraff and Aydin (1987) for joint descriptions and a discussion of columnar joint origins. Also see A. R. Philpotts and D. H. Burkett (1987) and P. E. Long and Wood (1987).

13. R. V. Fisher (1961); Schmid (1981); R. V. Fisher and Schmincke (1984); Heiken and Wohletz (1985).

14. The term batholith (bathylith) was defined at a time when subsurface information was largely unavailable. Careful reading of the definitions in this section will make it clear that other bodies, lopoliths for example, may meet the definition of a batholith as the term is presently defined. Thus, lopoliths are more specifically defined bodies that belong to a subset of the larger set, the batholiths. By the present definition, a pluton 1 m thick exposed over an area of 100 km^2, and thus having a volume of 0.1 km^3, is a batholith. In contrast, a pluton 10 km thick exposed over an area of 99 km^2 and having a volume of 990 km^3 is not a batholith; it belongs to the category of smaller bodies called stocks. Contemporary geophysical techniques now allow three-dimensional analysis of pluton shapes. I therefore propose that batholith be redefined as a nonlayered body of intrusive rock, predominantly phaneritic in texture, with a minimum volume of 100 km^3. Similarly, a stock can be defined as a nonlayered body of intrusive rock, predominantly phaneritic in texture, with a volume of less than 100 km^3.

15. References of note on which Hamilton and Myers based their ideas and interpretations include Chamberlin and Link (1927), Lane (1931), P. C. Bateman et al. (1963), L. L. Larsen and Schmidt (1958), and Klepper (1950).

16. "The Nature of Batholiths" by W. B. Hamilton and Myers (1967) sparked an interesting series of articles revealing how both protagonists and antagonists in a geologic debate argue the evidence and modify their ideas. See W. B. Hamilton and Myers (1967, 1974a, 1974b), W. B. Hamilton (1983), Klepper, Robinson, and Smedes (1971), and Klepper et al. (1974). Also see Bothner (1974), Sweeney (1976), P. C. Bateman (1981), Lee (1986), and Vigueresse (1988) for a discussion of the shapes of some other batholiths.

17. P. C. Bateman and Wahrhaftig (1966); P. C. Bateman (1981); Ague and Brimhall (1988a).

18. Bothner (1974); Sweeney (1976); Kearey (1978); Lynn, Hale, and Thompson (1981).

19. For example, see Gilluly (1948).

20. See O. B. James (1971), Thompson, Pierson, and Lyttle (1982), and J. S. Beard and Day (1988) for examples. For excellent examples of the composite nature of major batholiths, see the works of Calkins (1930), P. C. Bateman and his coworkers (e.g., Bateman et al., 1963; Bateman, 1978, 1981, 1983), and J. G. Moore (1963), plus the numerous references cited in these works, for descriptions of the Sierra Nevada Batholith; J. S. Myers (1975), W. S. Pitcher (1978), Cobbing, Pitcher, and Taylor (1977), Cobbing et al. (1981), Cobbing and Mallick (1983), and Cobbing and Pitcher (1983) for descriptions of the Coastal Batholith of Peru; and references in W. B. Hamilton and Myers (1967; 1974a), Klepper, Robinson, and Smedes (1971), Klepper et al. (1974), and Hyndman (1983) on the Idaho and Boulder Batholiths.

21. The type lopolith is the Duluth Complex (Grout, 1918c; Lieth et al., 1935). Works on other lopoliths include the book *Layered Igneous Rocks* (Wager and Brown, 1968), the studies of H. H. Hess (1960), E. D. Jackson (1961; 1967) and N. J. Page (1977) on the Stillwater Complex, and the studies of the Muskox Intrusion by Irvine and Smith (1967) and Irvine (1980b).

22. Billings (1972) advocates this definition. However, Corry (1988) finds no mechanical or other basis for the distinction between sills and laccoliths. Nevertheless, the arbitrary distinction provides a basis for defining the two types of structures as different-shaped bodies. Corry (1988) prefers a distinction based on thickness, with sills being generally <10 m thick and laccoliths being generally >30 m thick. For additional discussions of laccoliths and their origins, see Hyndman and Alt (1987), Hunt (1988), and Jackson and Pollard (1988a, 1988b).

23. For example, see Hyndman and Alt (1983), Dixon and Simpson (1987) and Corry (1988). Jackson and Pollard (1988a) argue that all of the plutons in the southern Henry Mountains may be laccoliths.

24. J. S. Myers (1975); Sides et al. (1981); T. B. Thompson et al. (1982); W. J. Phillips (1974).

25. Balk (1937) defined these terms, the use of which is discouraged by Bates and Jackson (1980).

26. These features are more characteristic of and common in metamorphic rocks and are discussed in chapter 23.

27. Irvine (1982) discusses the terminology of layered intrusions and emphasizes that layering is a structure of a body of rock composed of individual layers, laminae, and laminations—progressively smaller primary features. Laminae are thin, sharply defined layers generally less than 3 cm in thickness, whereas lamination is a subtle feature developed on the scale of the grain size. Other discussions of layering may be found in E. D. Jackson (1967), Wager and Brown (1968), McBirney and Noyes (1979), B. D. Marsh (1988a), and M. E. Conrad and Naslund (1989).

28. For discussions of orbicular rocks and comb layering see Butler (1973), J. G. Moore and Lockwood (1973), Lofgren and Donaldson (1975), and Brigham (1983).

29. These terms, though commonly used only for igneous rocks, are nongenetic and are used in this text for all classes of rocks.

30. Johannsen (1939) used the term intergranular for basaltic rocks composed essentially of plagioclase and augite. Use of the term for rocks of similar texture, but with minerals other than augite included between the feldspar laths, is advocated by Howell Williams et al. (1982) and is recommended here.

31. Although this term is widely used, its use as a textural term is discouraged, both because it is redundant (with hypidiomorphic-granular) and because rocks other than granites have the texture.

32. This section is based on several works including Holden and Singer (1960), Brice (1965), Kirkpatrick (1975, 1981), S. E. Swanson (1977), Dowty (1980), Lofgren (1980), and Brandeis and Jaupart (1987). These works provide more information on nucleation and crystal growth.

33. Strict definitions of *phase* and *system* are given in chapter 5. Phases are homogeneous, mechanically separable parts of a system. A system can be considered to be any mass of material we wish to discuss.

34. See note 33.

35. See P. C. Hess (1980) for a review of the structure of silicate melts. Brandriss and Stebbins (1988) assess the effects of temperature on some silicate liquids.

36. The reactions involving silica in melts are complex and do not always strictly follow the simple rules stated here (see Brandriss and Stebbins, 1988).

37. Lofgren (1980, 1983).

38. For example, see Kirkpatrick (1977), T. P. Loomis and Welber (1982), and S. A. Morse (1986).

39. Lofgren (1983); Cashman and Marsh (1988); Bacon (1989). Bacon suggests that local oversaturation may cause (homogeneous) nucleation *near* the crystal-liquid interface.

40. See the works of Brice (1965), Kirkpatrick and his colleagues (Kirkpatrick, 1974, 1975, 1976, 1977, 1981; Kirkpatrick, Kuo, and Melchior, 1981; Kuo and Kirkpatrick, 1982). Lofgren and his colleagues (Lofgren, 1974, 1980, 1983; Lofgren and Donaldson, 1975), Gibb (1974), S. E. Swanson (1977), Naney and Swanson (1980), Dowty (1980), Corrigan (1982), Brandeis and Jaupart (1987), Muncill and Lasaga (1987), Cashman and Marsh (1988), and B. D. Marsh (1988b) on which this section is based.

41. See Carmichael, Turner, and Verhoogen (1974) for a discussion of the difference between supercooled liquids and glasses.

42. Detailed studies of textural histories are rare. S. E. Swanson et al. (1989) describe a multistage history for Inyo Dome lavas, but that multistage history is related to degassing, not simply to changes in cooling rate.

43. S. E. Swanson (1977); Lofgren (1983).

44. Water is thought to play a role in the development of pegmatites by reducing viscosity and by inhibiting nucleation (see Jahns and Burnham, 1969).

45. See Dowty (1980) for summary tables of rates and densities measured in the laboratory.

46. For discussions see Kirkpatrick (1975, 1981), Dowty (1980), and Lofgren (1980).

47. See Dowty (1980).

PROBLEMS

2.1 Using the schematic (dashed-line) reconstruction of Mt. Mazama in figure 2.2 and assuming a circular plan for the volcano, evaluate the volume of rock missing above the caldera. Use either a graphical-trigonometric or a calculus method of solution as requested by your instructor. *Note:* The calculated volume will represent a minimum, as the apparent horizontal floor of the caldera is the lake level. The actual floor is irregular and extends locally to nearly 620 m (2000 ft.) below the lake level. What other factors might make your analysis inaccurate?

2.2 Using available geologic maps from the region in which you live and table 2.4, determine the depth category to which the plutons on the map belong. Are all the criteria on the map consistent with your depth category assignment? Why or why not?

2.3 Using Swanson's nucleation and crystal-growth rate curves (figure 2.25b), suggest a possible origin for the texture in an alkali-feldspar, phyric, vitrophyric rhyolite.

3

Chemistry and Mineralogy
of Igneous Rocks

INTRODUCTION

Formation of minerals in an igneous rock is controlled by the chemical composition of the magma and by the physical-chemical conditions present during crystallization. Mineralogy and texture are used to describe, name, and classify rocks (chapter 4). In some cases, the chemistry alone is the criterion for classification. Thus, the study of igneous rock chemistry and mineralogy is an important descriptive aspect of the study of igneous rocks.

More important, however, is the fact that both the overall chemistry (whole-rock chemistry) and the chemistry of constituent minerals offer clues to igneous rock origins. Studies of rock chemistry reveal where magmas form and how they are modified before they solidify (chapter 6). Thus, knowledge of the subjects discussed in this chapter will help in understanding the classification and petrogenesis of igneous rocks described in subsequent chapters.

CHEMISTRY OF IGNEOUS ROCKS: AN OVERVIEW

Modern chemical analyses of igneous rocks generally include two parts, a major element analysis and a minor or trace element analysis. The earth is composed almost entirely of 15 elements, 12 of which are the dominant elements of the crust.[1] The crustal elements, considered to be the **major elements,** with oxygen, in order of decreasing abundance, are O, Si, Al, Fe, Ca, Na, Mg, K, Ti, H, P, and Mn (Mason and Berry, 1968; Ronov and Yaroshevsky, 1969).[2] Because of their local abundance, other elements such as C and S are sometimes included in the major element analysis. Major element abundances are traditionally reported in weight percent of oxides (table 3.1).

Table 3.1 lists the major element chemistry of some typical examples of several important igneous rock types. Note that the silica (SiO_2) contents lie between about 40 and 80%. Rocks with lower and higher values do exist, but they are relatively rare. Also notice the variations in other elements. Ferrous iron and magnesium are abundant in low-silica rocks, whereas potassium and sodium tend to be more abundant in high-silica rocks. Lime (CaO) is most abundant in rocks with low to intermediate silica values. LeMaitre (1976) analyzed the chemical variability of common igneous rocks and his work emphasized the wide range of chemistry in rocks assigned the same name.[3] For example, rocks called "granite" commonly show silica values between 65 and 75% by weight, with soda (Na_2O) and potash (K_2O) values ranging over several percentage points. Similarly, rocks called "basalt" typically range from 45 to 55% silica, with alumina (Al_2O_3) values between 13 and 18%.

Table 3.1 Chemical Analyses of Typical Igneous Rocks

	1	2	3	4	5	6
SiO_2	40.08[a]	49.8	58.97	59.54	69.22	77.24
TiO_2	0.01	2.6	1.04	0.14	0.48	0.20
Al_2O_3	0.29	14.0	17.17	18.60	15.50	10.81
Fe_2O_3	0.31	2.5	4.36	2.86	1.03	1.66
FeO	7.62	8.5	2.02	2.09	1.42	0.27
MnO	0.11	0.18	0.10	0.22	0.04	0.02
MgO	49.69	7.2	1.51	0.10	0.73	0.33
CaO	0.11	11.3	4.90	1.16	1.93	1.48
Na_2O	0.05	2.2	4.23	8.96	4.15	2.59
K_2O	0.01	0.62	2.90	4.24	4.42	4.12
P_2O_5	0.00	0.32	0.51	0.16	0.15	0.06
H_2O^+	—	0.25	—	1.40	—	0.37
Other	0.58	0.10	1.55	0.40	0.30	0.65
Total	98.86	99.6	99.26	99.87	99.37	99.80

1. Dunite, sample DTS-1; Flanagan (1976); analysts: R. Pouget, M. Carrier, M. Lautelin, A. Vasseur.

2. Basalt; sample BHVO-1; Flanagan et al. (1976).

3. Andesite, sample AGV-1; Flanagan (1976); analysts: R. Pouget, M. Carrier, M. Lautelin, A. Vasseur.

4. Nepheline syenite; Snavely et al. (1976); average of two samples.

5. Granite, G-2; Flanagan (1976); analysts: R. Pouget, M. Carrier, M. Lautelin, A. Vasseur.

6. Rhyolite, sample no. 6; Staatz and Carr (1964); analyst: E. D. Tomasi.

[a] All values in weight percent.

Silica and alumina contents are quite important in controlling the mineral composition of igneous rocks, as emphasized by Shand (1949). For example, rocks with abundant silica generally contain quartz, whereas those of low silica content may contain feldspathoids or magnesium-rich olivine. Shand used the term **oversaturated** for rocks that have a chemistry such that they contain quartz or another silica mineral.[4] Rocks with minerals such as nepheline that are incompatible with the silica minerals reflect an **undersaturated** chemistry. Saturated rocks are those with neither quartz nor an undersaturated mineral. Such rocks are typically rich in feldspar.

Shand (1949) also divided rocks on the basis of alumina content. A **peraluminous** chemistry is one in which the mole percent alumina is greater than the sum of lime, soda, and potash ($Al_2O_3 > CaO + Na_2O + K_2O$) (figure 3.1a). In **metaluminous** materials, the mole percent alumina is less than the

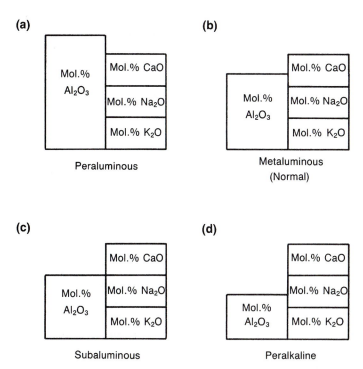

Figure 3.1 Bar graphs depicting varying degrees of alumina saturation: (a) peraluminous, (b) metaluminous, (c) subaluminous, (d) peralkaline.

sum of lime, soda, and potash, but greater than the sum of soda and potash ($CaO + Na_2O + K_2O > Al_2O_3 > Na_2O + K_2O$) (figure 3.1b). Subaluminous materials represent the special case in which alumina is about equal to soda plus potash ($Al_2O_3 \approx Na_2O + K_2O$). Alumina is less abundant than the sum of soda and potash in **peralkaline** rocks ($Al_2O_3 < Na_2O + K_2O$) (figure 3.1d). The silica and alumina saturation principles operate somewhat independently; hence granite, a silica-oversaturated rock, may be peraluminous, metaluminous, or peralkaline.

Chemical data such as those presented in table 3.1, as well as trace element data, are often presented in both tabular and graphic forms. The tables are self-explanatory. In contrast, the various graphical presentations may not be as easily interpreted. Major element data are generally plotted on either rectangular or triangular plots referred to as variation diagrams. On such diagrams, data from several analyses are plotted and examined to see whether or not systematic variations or relationships exist between the analyzed rocks. If analyses lie along a smooth curve or within a restricted area (figure 3.2), the lines or areas on the diagrams are sometimes interpreted to reflect either related origins or some distinctive characteristic of the suite of rocks under study.[5]

Figure 3.2a shows a rectangular plot, the K_2O-SiO_2 diagram, used commonly for major element data analysis. Many other parameters are also used in such plots, such as CaO and SiO_2, or MgO and $FeO + Fe_2O_3$. The K_2O and SiO_2 data from analysis 3 (table 3.1) have been used to plot a point representing that analysis on the K_2O-SiO_2 diagram. Ideally, a group of

(a)

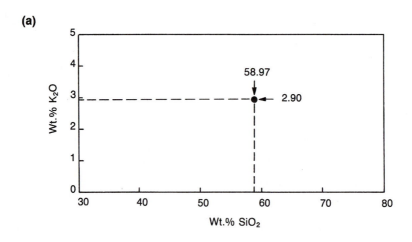

(b)

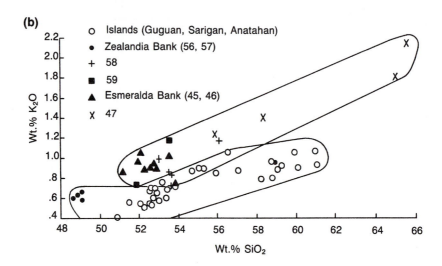

············

Figure 3.2 Rectangular variation diagrams. (a) K$_2$O-SiO$_2$ diagram showing the plotting of a point representing analysis 3 in table 3.1. (b) K$_2$O-SiO$_2$ diagram for Mariana arc seamount lavas. The data plot on two trends (outlined), a K-rich trend for southern seamounts and a K-poor trend for central islands.

(From T. H. Dixon and Stearn, 1983)

related rocks, specifically a group of rocks crystallized from the same magma or magma source, will plot as a group of points that define a curve on such a diagram (figure 3.3). A suite crystallized from a different magma or magma source might plot as a different line, perhaps with a different slope (figure 3.2b).

In general, oxides from given sources, plotted against silica, form linear arrays. A set of such plots is called a Harker diagram (figure 3.3). Note that the Harker diagram is a composite diagram in which several oxides, plotted individually on the ordinate, are compared to SiO$_2$, which is plotted on the abscissa.

Triangular variation diagrams show only the *ratios* of various oxides or elements, rather than their actual values. In order to plot points, the oxides or elements to be plotted, those represented at the corners of the triangle (figure 3.4), must add up to 100%. The plotted point from any given analysis marks a position representing each element or oxide's proportion of that total (100%). For example, analysis 3 (table 3.1) is represented on figure 3.4a as a point with the following triangular coordinates:

$$\text{wt. \% K}_2\text{O} = \frac{\text{K}_2\text{O}(\times 100)}{\text{K}_2\text{O} + \text{Na}_2\text{O} + \text{CaO}} = \frac{2.90}{2.90 + 4.23 + 4.90} \times 100 = 24\%$$

$$\text{wt. \% Na}_2\text{O} = \frac{\text{Na}_2\text{O}(\times 100)}{\text{K}_2\text{O} + \text{Na}_2\text{O} + \text{CaO}} = \frac{4.23}{2.90 + 4.23 + 4.90} \times 100 = 35\%$$

$$\text{wt. \% CaO} = \frac{\text{CaO}(\times 100)}{\text{K}_2\text{O} + \text{Na}_2\text{O} + \text{CaO}} = \frac{4.90}{2.90 + 4.23 + 4.90} \times 100 = 41\%$$

Notice that each apex of the triangle represents 100% of the element or oxide named at that point. The triangle side opposite the apex represents 0% of that element or oxide. A commonly used triangular variation diagram, the AFM diagram (alkalis-iron-magnesium), is shown in figure 3.4b. On that diagram, trend lines for different magma groups are plotted.

All elements other than the major elements are referred to as minor or **trace elements.** Some trace elements such as C, S, and Cr are quite abundant locally, where they are prin-

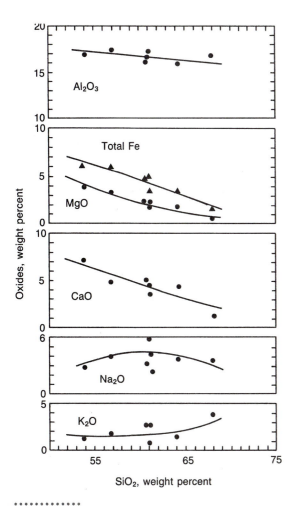

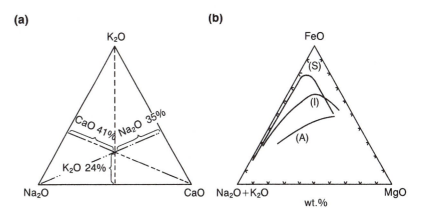

Figure 3.4 Triangular variation diagrams. (a) Sample K_2O-Na_2O-CaO diagram showing plot of analysis 3 in table 3.1. (b) AFM diagram showing tholeiitic trend line for the Skaergaard Intrusion (S), Tholeiitic trend for the rocks of the Izu-Hakone (Japan) volcano (I), and the calc-alkaline trend of rocks from the Amagi Volcano, Japan (A). Corners of the diagram are A = K_2O + Na_2O, F = total iron calculated as FeO, and M = MgO. Values are in weight percent.
(Modified from Miyashiro, 1975 d)

Figure 3.3 Harker diagram for Cenozoic volcanic rocks from Okanogan County, Washington.
(Source: C. D. Rinehart and K. F. Fox, Jr., "Bedrock Geology of the Conconully Quadrangle. Okanogan County, Washington" in USGS-B1402, 1976.)

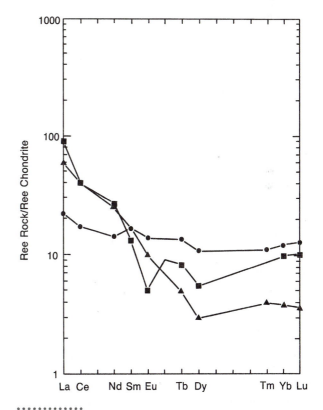

Figure 3.5 Rare earth element diagram showing curves for three common rock types. Circles = a plagioclase-rich granitoid rock. Triangles = granodiorite. Squares = a leucogranite.
(Source: Dodge, et al., in US Geological Survey, Prof. Paper 1248, 1982.)

cipal components of dominant minerals in some less common igneous rocks, but they are not as abundant as the major elements in the crust as a whole. Carbon, for example, is a major component of the igneous rock **carbonatite,** a rock composed predominantly of carbonate minerals. Other trace elements, such as Zn, Nd, and U, generally occur in minute amounts in various rock-forming minerals. Trace elements are typically reported in parts per million (ppm).[6]

Trace elements are subdivided into four groups—the rare earth elements, high field strength elements, petrologically important isotopes, and other trace elements. In studies of petrogenesis, these groups commonly are evaluated separately. Where possible, the results of the separate studies are combined with other geochemical and petrologic data in an effort to discover the petrogenesis of the rocks under investigation.

Rare earth elements (REE) are elements on the periodic chart having atomic numbers 57 to 71. REE data are plotted on rectangular plots, with the elements on the abscissa in order of increasing atomic number (figure 3.5). Light rare earth elements (LREE) plot at the left end and heavy rare earth elements (HREE) plot at the right end. On the ordinate, plotted on a log scale, is the *ratio* of measured REE abundances in the sample to REE abundances in chondrite meteorites. The latter are considered to be initial (primordial) abundances that serve as a standard. If the REE values in a rock were identical to those of chondrites, the plot would have the form of a horizontal line at a value of one.

Deviations from unity in REE curves are considered to result from various petrogenetic processes. For example, LREE enrichment in magmas is consistent with a mantle source with garnet, because garnet, which does not melt early, preferentially retains the heavier, smaller rare earths. Pyroxene likewise retains some HREE. Most REE do not substitute readily in minerals. An exception is europium. Because it will substitute for Ca, it is preferentially absorbed, relative to other rare earth elements, by crystallizing plagioclase. Thus, fractionation of plagioclase, or removal of plagioclase from a melt, produces a negative europium anomaly (a deficiency in Eu relative to adjacent REE) in the remaining melt, or the rock crystallized from the melt (see the leucogranite curve in figure 3.5). The rock containing the fractionated plagioclase, in contrast, will tend to be relatively enriched in Eu; that is, it will have a positive Eu anomaly.

High field strength elements (HFSE) are elements with a high charge/radius ratio.[7] They include Ti, Ni, Cr, Zr, Hf, Nb, Ta, and Y. Ratios of these various elements are considered to reflect genetic relationships between various rocks. For example, dikes and flows that contain the same ratios may be directly related or may have been derived from the same parent magma. In a more general use, various ratios of these elements are employed to determine particular environments of magma generation (Pearce and Cann, 1973).

In petrologic studies, **isotopes,** elements of the same atomic number but different mass number, fall into one of two groups—*stable isotopes* and *radioactive/radiogenic isotopes*. Both types may contribute to an understanding of petrogenesis. Stable isotopes used in petrology include isotopes of hydrogen, oxygen, carbon, and sulfur. Certain processes tend to concentrate either the light or the heavy isotopes: Thus, isotopic ratios provide an indication that these processes have occurred. Numerous radiogenic isotopes, plus isotope ratios involving radiogenic isotopes, their parents, or related nonradiogenic isotopes, are also indicators of certain processes or certain materials that were involved in rock- or magma-forming events. These isotopes and isotopic ratios include the several uranium-lead or lead-lead ratios (e.g., $^{206}Pb/^{204}Pb$), argon ratios, potassium-argon ratios, rubidium-strontium ratios, strontium ratios ($^{87}Sr/^{86}Sr$), neodymium ratios ($^{143}Nd/^{144}Nd$), and beryllium (^{10}Be).[8] Some of these isotopic systems also allow dating of rocks.

IGNEOUS MINERALS

Most of the minerals found in igneous rocks occur in other classes of rocks as well. Appendix A lists minerals that occur in selected rock types. **Primary minerals** are those that form when the rock first forms. **Secondary minerals** form later, via alteration or weathering. As the student of petrology is assumed to have studied mineralogy, this section of the text offers a brief review of some of the more common rock-forming minerals while focusing attention on particular occurrences in igneous rocks.[9]

The most important rock-forming minerals are the silicates. Silicates are assigned to one of six groups—nesosilicates, sorosilicates, inosilicates, cyclosilicates, phyllosilicates, and tectosilicates—on the basis of their silicate structures. If the fundamental unit of the silicates, the silica tetrahedron, occurs as isolated units connected by other ions, the mineral has a Si to O ratio of 1:4 ($-SiO_4$) and is a nesosilicate. Minerals with paired silica tetrahedra are sorosilicates. Their Si:O ranges up to 2:7 ($-Si_2O_7$). Cyclosilicates, with Si/O of 1:3 ($-Si_6O_{18}$), have rings of tetrahedra. In the inosilicates, silica tetrahedra form single or double chains having ideal Si to O ratios of 1:3 and 4:11 (e.g., $-Si_2O_6$ and $-Si_8O_{22}$), respectively. The phyllosilicates are sheet-structured minerals with Si:O of up to 2:5 ($-Si_4O_{10}$). Finally, silica tetrahedra form three-dimensional frameworks in the tectosilicates, with Si to O ratios as high as 1:2 (e.g., $-Si_3O_8$ or $-SiO_2$). The student should realize that these Si/O ratios are generalizations; Al or Ti may substitute for Si in silica tetrahedra, altering the actual Si/O values.

In addition to the silicates, several other types of minerals are important rock-forming minerals in igneous rocks. Among these are the carbonates; certain sulfides, sulfates, phosphates, and halides; and the oxides, which are present in minor to abundant quantities in some rocks. Oxides such as hematite and carbonates such as calcite are also common as secondary minerals.

Nesosilicates, Sorosilicates, and Cyclosilicates

The most important igneous rock-forming nesosilicates are the olivines. Others include garnets, zircon, topaz, and the aluminum silicates (andalusite, kyanite, and sillimanite). The olivines constitute the solid solution series Mg_2SiO_4-Fe_2SiO_4. Magnesium-rich olivines, especially forsterites, are important minerals in low-silica rocks (e.g., rock 1, table 3.1), as well as in some intermediate-silica volcanic rocks. Olivine typically forms subhedral to anhedral equant grains, but may form euhedral phenocrysts, skeletal crystals, or rare bladed crystals in volcanic rocks. Mg-rich olivines are unstable in the presence of quartz for reasons that will be made clear in chapter 5. Iron-rich olivines are somewhat less common than the magnesium-rich varieties. Fayalite [$(FeMg)_2SiO_4$], the iron-rich endmember of the series, occurs locally with quartz in some silica-oversaturated igneous rocks and is also found in iron-rich, low-silica rocks.

The remaining nesosilicates usually comprise a small percentage of the minerals in specific types of igneous rocks. Exceptionally, garnet may form local concentrations, rivaling other minerals in abundance. Aluminous garnets, topaz, and the aluminum silicates all occur in peraluminous silica-rich rocks. Zircon is ubiquitous in siliceous rocks. Sphene is a common, minor mineral, especially in plutonic rocks. All of these minerals typically form small to very small euhedral crystals.

The only common sorosilicates found in igneous rocks are minerals of the epidote group. Allanite forms microscopic grains in biotite in many intermediate- to high-silica plutonic rocks. Epidote may occur as a primary phase in the siliceous plutonic rocks (Zen and Hammarstrom, 1984; B. W. Evans and Vance, 1987), but it is usually secondary.

Cyclosilicates are not commonly abundant but are locally important igneous rock-forming minerals. Beryl occurs widely as uncommon, small euhedral crystals in silica-rich plutonic rocks and may occur as gem-quality crystals (emeralds) in siliceous pegmatitic rocks. Similarly, tourmaline, most commonly the black variety schorlite, appears as euhedral forms in these rocks. Axinite occurs locally in cavities in the siliceous plutonic rocks. Rare cordierite is a phase in both plutonic and volcanic rocks.

Inosilicates

Inosilicates are major constituents of igneous rocks. Of the inosilicates, only the pyroxenes and amphiboles are abundant. Pyroxenes of both orthorhombic character (orthopyroxenes) and monoclinic character (clinopyroxenes) occur over a range of rock chemistries. Only monoclinic amphiboles are important in igneous rocks.

The common, primary igneous pyroxenes that are calcium-rich include diopside [$(CaMg)_2Si_2O_6$], augite [$(Ca,Na,Mg,Fe^{+2},Al,Ti)_2(SiAl)_2O_6$], and hedenbergite [$(CaFe)_2Si_2O_6$].[10] These grains form subhedral to euhedral phenocrysts and anhedral grains in the groundmass of low-silica volcanic rocks and may occur in more siliceous rocks as well. In low-silica plutonic rocks, augite typically forms small to medium grains. Where the texture is ophitic, the pyroxenes may be several centimeters long.

Calcium-poor pyroxenes include the orthopyroxenes enstatite [$(MgFe)_2Si_2O_6$] and ferrosilite [$(FeMg)_2Si_2O_6$] and the clinopyroxene pigeonite [$(Mg,Fe,Ca)(Mg,Fe)Si_2O_6$]. Pi-geonite, which is stable at high temperatures, forms small grains in low-silica rocks. In plutonic rocks, pigeonite has usually "inverted," or changed to orthopyroxene containing exsolved augite. Enstatite typically occurs as subhedral grains in both volcanic and plutonic rocks of low silica content and as euhedral phenocrysts in low- to intermediate-silica volcanic rocks. Ferrosilite is reported from siliceous plutonic rocks.

Alkali pyroxenes also occur in igneous rocks. Aegirine (Acmite) ($NaFeSi_2O_6$) and aegirine-augite occur in peralkaline plutonic rocks. Euhedral to subhedral grains of these same two pyroxenes occur as phenocrysts and groundmass grains in sodic, siliceous volcanic rocks.

Hornblendes and alkali amphiboles, all of which are monoclinic, are the important amphiboles of the igneous rocks. The hornblende group, composed of calcium amphiboles, includes common hornblende [$Ca_2(Mg,Fe^{+2})_4(Al,Fe^{3+})(Si_7Al)O_{22}(OH)_2$] and amphiboles such as pargasite and ferrohastingsite. The hornblendes occur as anhedral to euhedral grains and phenocrysts in a wide range of plutonic and volcanic rocks. Magnesium-rich hornblende occurs in low-silica rocks, where it may form anhedral grains, coronas, or large poikilitic grains. Hornblende is a very common euhedral phenocryst mineral in intermediate-silica volcanic rocks. Commonly the hornblende has been oxidized to form oxyhornblende (basaltic hornblende). In the more alkali- and silica-rich rocks, more iron-rich hornblendes occur. Hornblende is a particularly important mineral in the study of plutonic rocks, because the Al content of hornblende is an indicator of pressure and therefore of depth of crystallization (Hammarstrom and Zen, 1986; Hollister et al., 1987).

Alkali-bearing amphiboles occur in soda- and potash-rich igneous rocks. Riebeckite, the only blue amphibole to occur in igneous rocks, forms euhedral phenocrysts in peralkaline volcanic rocks and euhedral to anhedral grains in plutonic rocks. Other alkali-bearing amphiboles occurring in igneous rocks include barkevikite, kaersutite, kataphorite, and arfvedsonite.[11]

Phyllosilicates

Sheet silicates, like the chain silicates, are important constituents of igneous rocks. Primary phyllosilicate minerals include muscovite, phlogopite, and biotite. Secondary phyllosilicates include minerals of the chlorite and clay groups. Biotite [$K_2(Mg,Fe)_{6-4}(Fe,Al,Ti)_{0-2}(Si_{6-5}Al_{2-3})O_{20}(OH,F)_4$], the most common of the phyllosilicates, occurs in a wide range of igneous rock types. Phlogopite is simply a low-iron, high-magnesium form of biotite that occurs in low-silica rocks. The white mica muscovite [$K_2(AlFe)_4(Si_6Al_2)O_{20}(OH,F)_4$] occurs primarily in siliceous plutonic rocks of peraluminous character. Here it forms anhedral to euhedral plates. In pegmatitic rocks, euhedral muscovite crystals may reach one-third of a meter or more in diameter. Primary muscovite does not occur in volcanic rocks. Very fine-grained white mica, seen through the petrographic microscope, has been called *sericite*, especially in the older literature. This mica typically occurs as a replacement or alteration product of feldspar.

Clay minerals and chlorites also typically occur as replacement minerals. The clay replaces feldspar as well as other minerals. Chlorite commonly replaces biotite in silica-rich plutonic rocks and it also replaces other ferromagnesium minerals.

Tectosilicates

Tectosilicates are the most important rock-forming minerals. This importance arises from the fact that the group includes the most abundant minerals of the crust (feldspars, quartz) and because minerals of this group are used as a basis for some rock classifications.

Quartz (SiO_2) is the main silica mineral in igneous rocks. Typically quartz forms anhedral grains, both in plutonic rocks and in the groundmass of many volcanic rocks. As phenocrysts, quartz may be euhedral. Other silica minerals occurring in igneous rocks include fibrous crystals of cristobalite and tridymite, which occur in siliceous volcanic rocks.

The alkali feldspars [$(K,Na)AlSi_3O_8$] are widely distributed in intermediate- to high-silica rocks, but also occur in some low-silica rocks. In porphyritic volcanic and plutonic rocks, K-rich alkali feldspar commonly forms euhedral phenocrysts. Subhedral to euhedral poikilitic grains are common in plutonic rocks, as are anhedral grains. K-rich alkali feldspar in plutonic rocks may be perthitic, containing exsolution lamellae of albite. Orthoclase and microcline are the common alkali-feldspars in plutonic rocks, whereas sanidine and anorthoclase occur typically in volcanic rocks.

Plagioclase feldspar [$(NaCa)Al_{1-2}Si_{3-2}O_8$] is the most abundant crustal mineral. It occurs as euhedral to anhedral grains in all common types of plutonic and volcanic rocks. The calcium-rich plagioclases—anorthite, labradorite, and bytownite—are present in low-silica rocks, whereas the Na-rich plagioclases—andesine, oligoclase, and albite—are typically found in intermediate- to high-silica rocks.

Feldspathoids, as a group, are far less abundant than feldspars. Nepheline [$(Na_{3-4}K_{1-0})Al_4Si_4O_{16}$], the most common of the feldspathoids, occurs typically as anhedral grains in undersaturated rocks. Leucite ($KAlSi_2O_6$) is quite rare, occurring as euhedral phenocrysts in some volcanic rocks. Similarly, sodalite and cancrinite are rare constituents of silica-undersaturated plutonic rocks.

Nonsilicates

Many of the nonsilicate rock-forming minerals that are found in igneous rocks are listed in appendix A. Among these, magnetite (Fe_3O_4) is ubiquitous and apatite [$Ca_5(PO_4)_3(OH,F,Cl)$] is very widely distributed.

Calcite ($CaCO_3$) also occurs widely, but usually as a secondary mineral. In the unusual rocks called carbonatites, however, calcite, along with dolomite and ankerite, is a principal mineral.

SUMMARY

The chemistries and minerals of the igneous rocks reflect magma compositions and rock histories. The major elements include O, Si, Al, Fe, Ca, Na, Mg, K, Ti, H, P, and Mn. Igneous rocks may be divided into silica-oversaturated, -saturated, and -undersaturated types, with peraluminous, metaluminous, and peralkaline variants. Each chemistry is reflected in the mineralogy. All elements other than the major elements are trace elements. Trace elements are subdivided into four groups—petrologically important isotopes, the rare earth elements (REE), high field strength elements (HFSE), and other trace elements. Chemical data are presented in both tabular and graphical form and are used to evaluate the materials and processes involved in magma and rock formation.

Both silicates and nonsilicates are important as rock-forming minerals, but the silicates are the most important. The feldspars and quartz are the most abundant minerals of the crust. Pyroxenes, amphiboles, and micas are also important primary constituents of igneous rocks. Clays, chlorites, and carbonate minerals are common secondary phases. Other silicates, as well as oxides, sulfides, and other types of minerals, occur as minor primary and secondary phases.

EXPLANATORY NOTES

1. Mason and Moore (1982, pp. 52–53).
2. Elemental symbols and names are listed in the To The Student Section of this book.
3. For more information on rock chemistry, see the works of LeMaitre (1976), Nockolds (1954), and Washington (1917).
4. See Shand (1949, chapter 14, p. 225 ff).
5. Words of caution about the use of these diagrams are given by Cox, Bell, and Pankhurst (1979) and Wilcox (1979).
6. Note that 10,000 ppm = 1% (by weight).
7. See Pearce and Norry (1979) and Salters and Shimizu (1988) for examples of the use of HFSE in petrogenetic analyses.
8. For example, see Cox, Bell, and Pankhurst (1979, ch. 15), and L. Brown et al. (1982) plus other articles in Wetherill (1982).

9. Readers unfamiliar with the details of mineralogy should consult standard texts such as Deer, Howie, and Zussman (1966), Mason and Berry (1968), Hurlbut and Klein (1977), and Zoltai and Stout (1984). These texts, plus the works of Heinrich (1956), Nockolds, Knox, and Chinner (1978), Hatch, Wells, and Wells (1973), and Howell Williams, Turner and Gilbert (1982), combined with the author's own experience serve as the basis for this section on mineralogy.

10. Poldervaart and Hess (1951) and Morimoto et al. (1988) discuss pyroxene nomenclature.

11. For a proposal on amphibole nomenclature, see Leake (1978).

PROBLEMS

3.1 (a) Plot the analyses from tables 3.1 and 9.1 on K_2O-SiO_2 and AFM diagrams using different symbols for the data from each table. (b) Consider the data from each table separately. Do the data form well-defined trend lines? Suggest possible explanations of why the data do or do not form such lines.

3.2 Plot the data from table 3.1 on a Harker diagram. The data in table 3.1 are from rocks that have no genetic or geographic connections. Do the data points define a trend? What general conclusion, if any, may be drawn from this exercise?

3.3 Read the caveat of LeMaitre (1976) relating to his AFM plot of data from all the igneous rock types and the discussions of T. L. Wright (1974), Robinson and Leake (1975), F. Barker (1979a), Kushiro (1979, pp. 178–87), and Cox, Bell, and Pankhurst (1979, pp. 34–36). Why are AFM diagrams still used? Do you agree that they should be used? Why or why not? (*Note*: You may wish to complete this problem after reading chapters 5 and 6.)

Classifications of Igneous Rocks

INTRODUCTION

Igneous rock classification has had a controversial and sometimes amusing history. Although many petrologists have tried to develop classifications, none has developed a universally adopted systematic classification of genetic significance. The result is that nearly every petrology book seems to feature a different classification scheme.

A number of problems have confounded attempts at igneous rock classification. A primary problem is that nature contains a continuum of rock types that we attempt to artificially subdivide into groups. Rock names provide another problem, because many commonly used names were applied to rocks before classifications were developed. Consequently, names were not created in a systematic way. Few geologists seem willing to abandon the traditional names, yet there is no consensus for redefining traditional terms.[1] Shand (1949, pp. 243–44) suggested (perhaps facetiously) that we drop names and use symbols. The alternative, he wrote,

> is to continue the witless practice of giving a new name to every rock that is slightly different from any rock seen before and thus to extend indefinitely the list of "specific" names

which already contains over six hundred items and includes such gems as katzenbuckelite and leeuwfonteinite, anabohitsite and sviatoynossite, bogusite and bugite. There is indeed another alternative which has found a few advocates: it is to return to the language of a babyhood and coin words by putting together unrelated syllables, as in neapite (nepheline-apatite rock), apaneite (apatite-nepheline rock), olpybinemelite (olivine-pyroxene-biotite-nepheline-melilite rock). There may still be more dreadful alternatives, but the writer has not been able to imagine them.

The number of rock names in the literature, as noted by Shand (1949), is also a problem. In some cases, new names were created for rocks at a specific locality. In others, new names were applied to rocks that were similar to rock types already named but which had slightly different percentages of minerals. The result of these and other practices is such a large number of rock names that, as Grout (1932, p. 49) noted long ago, remembering their meanings is nearly impossible and attempting to do so is not worthwhile. Furthermore, many geologists apparently find it troublesome to use the many rock names available. Chayes (1979) found that, although 196 different names have been proposed just for various Cenozoic volcanic rocks, only 10 are used commonly.

Many have not been used since they were suggested. For his part, Bowen (1928, pp. 321–22) recognized the problem and attempted to call attention to the fallacies of (1) considering rocks as discrete species, each of which needs a name, and (2) basing names (and classifications) primarily on mineral ratios. For the most part, his caveat has not been heeded.

A fourth problem in rock classification is the selection of a basis for classification. Proposed classifications use texture, mineralogy, chemistry, geographic location, and rock associations,[2] as well as characters derived from these. With respect to each of these categories, rocks show a continuous and wide range of characteristics.

Classifications that use geographic location as a basis for subdividing rock types, such as Harker's (1909) geographical-geochemical subdivision of volcanic rocks into Pacific and Atlantic types, are anachronistic. In contrast, classifications that employ rock associations, rock suites, rock series, or rock families are widely used.[3] A rock family is a group of rocks, each of which has certain mineralogical, chemical, and/or textural characteristics in common with other members of the family. For example, I. S. E. Carmichael, Turner, and Verhoogen (1974) indicate that the "andesite-rhyolite family includes intermediate to acid volcanic rocks. Some can be very rich in iron and peralkaline." Within the families, rock names are generally applied in a traditional (haphazard), rather than a systematic way and are based on *mineralogy, texture, chemistry,* or some combination of these features. These criteria are the most widely used in naming and classifying rocks.

TEXTURAL-MINERALOGICAL CLASSIFICATIONS

Texture is a parameter of classification used by most petrologists. The student should be aware, however, that some petrologists, especially those who advocate the use of chemical classifications (Cross et al., 1903; also see Niggli, 1931, in Rittman, 1973), tend to minimize the importance of or totally ignore texture. Those who do use texture as a classification criterion divide rocks into two, five, or more categories. For example, Travis (1955) uses five categories: phaneritic-equigranular, porphyritic with phaneritic groundmass, porphyritic with aphanitic groundmass, aphanitic-microcrystalline, and aphanitic-glassy. In recent years there has been a tendency to follow a twofold subdivision, advocated by Shand (1949), in which rocks are divided into phaneritic and aphanitic types. That practice is followed in this book, and other textures are acknowledged in rock names by using a textural term as a modifier (e.g., porphyritic granite).

Mineralogy is a second major criterion used for subdividing, naming, and classifying the range of rock types. In naming rocks, primary minerals may be divided into (1) **essential minerals**—those that must be present if a given name

is applied to the rock; (2) **characterizing accessory minerals** (sometimes called varietal minerals)—those that occur in abundances of 5% or more and are not implied by the name but are used to modify the rock name; and (3) **minor accessory minerals**—those that occur in abundances of less than 5% and are not implied by the name (Travis, 1955). Volume percentages of minerals observed in a rock are used to derive the root name. Usually, only the essential minerals are used in determining the root name. For example, gabbro is a phaneritic rock composed of calcic plagioclase with olivine, pyroxene, and other minerals as accessories. Many specialized names, however, are based on the occurrence of one or more accessory minerals. For example, hypersthene gabbro is called norite. The list of minerals observed in the rock, with their volume percentages, is called the rock **mode,** and analyses to determine the kinds and abundances of minerals are called **modal analyses.**

Chemically, rocks are diverse. This diversity is reflected by the minerals. Yet petrologists, until recently, have concentrated their efforts on subdividing only the feldspar-rich rocks, probably because those rocks are abundant on the continents. However, igneous rocks and classifications discussed in this text are divided into two main mineralogical groups: (1) the **feldspathic rocks,** in which feldspars plus quartz or feldspathoids are the essential minerals, and (2) the **ferromagnesian rocks,** in which olivines, pyroxenes, amphiboles, and some micas are the essential minerals. Classifications of phaneritic rocks are described before classifications of volcanic rocks.

The boundary line between the feldspathic rocks and the ferromagnesian rocks is based on the abundance of certain minerals. Feldspar abundance is used by some petrologists (Travis, 1955; D. W. Peterson, 1965). Peterson (1965) placed the boundary at 10% feldspar; that is, rocks with less than 10% feldspar are assigned to the ferromagnesian rock group and those with more than 10% feldspar are assigned to the feldspathic group. Other petrologists use dark-colored or ferromagnesian minerals to define the boundary. For example, a subcommittee of the International Union of Geological Sciences (IUGS) chose a value of 90% "mafic and related minerals" for the boundary (Streckeisen, 1976). Shand (1949, p. 233) and others have called this percentage of mafic or ferromagnesian minerals observed in a rock, the **color index.** The minerals used in computing color index vary,[4] but in this book only those minerals in which iron, magnesium or both are essential constituents of the chemistry—for example, forsterite [$(Mg,Fe)_2SiO_4$], but not muscovite [$K_2(Al,Fe)_4Si_6Al_2O_{20}(OH,F)_4$], in which Fe is not essential—are used to compute color index. Color index values, which range from 0 to 100 (note that color indices are written without the percent sign), have been divided by Shand (1949), the IUGS, and others into various groups (table 4.1).

Table 4.1 Some Classifications of Color Index

IUGS	Shand	Raymond
0–5 holo-leucocratic		
5–35 leucocratic	0–30 leucocratic	0–10 hyperleucocratic
35–65 mesocratic	30–60 mesotype	11–50 leucocratic
65–90 melanocratic	60–90 melanocratic	50–89 melanocratic
90–100 ultramafic	90–100 hypermelanic	90–100 hypermelanocratic

Sources: IUGS—Streckeisen (1976), Shand (1949), Raymond (1984c).

Because minerals other than plagioclase and ferromagnesian minerals may occur in a rock, a specific color index (CI), such as 90 (like that adopted by the IUGS), does not necessarily correspond exactly to a feldspar content of 10% (adopted by D. W. Peterson, 1965). Thus, a CI equal to 90 and a plagioclase content of 10% may represent different boundaries on a rock classification chart. In this book, a color index of 90 is chosen as the boundary between the two categories of rock.[5] Rocks with CI = 0–89 are assigned to the feldspathic group and rocks with a CI = 90–100 are assigned to the ferromagnesian group.

The terms felsic, mafic, and ultramafic are often applied to rocks to indicate their color or color index and their general composition. The word **felsic** refers to feldspar, feldspathoid, and silica minerals. Because these minerals are generally light-colored, the term denotes light-colored rocks. These minerals are also rich in silica and the word is used by some to imply silica-rich compositions (Hyndman, 1972). **Mafic** denotes minerals typically dark in color and low in silica that are rich in magnesium and ferric iron. Consequently, the term mafic has been used as a name for rocks composed almost entirely of ferromagnesian minerals. Such rocks have a high color index. For that reason, some petrologists attach a specific color index to the mafic-ultramafic boundary and use the term ultramafic as a name for rocks of high color index (see table 4.1). Ultramafic rocks typically have very low silica contents (40–45% SiO_2).

Classifications of Phaneritic Feldspathic Rocks

Several forms of textural-mineralogical classification of phaneritic feldspathic rocks have been used over the last several decades. Some of these classifications, modified and presented in a standardized form to allow easy comparison, are presented in figure 4.1.

The classification in figure 4.1a was constructed by the author (Raymond, 1984c), but uses boundaries like those used by Shand (1949) and several other petrologists. Hence,

it is called the RSO (Raymond, Shand, and others) classification. The main parameters of classification are the feldspar ratio—percent alkali feldspar divided by percent alkali feldspar plus percent plagioclase, or $[A/(A + P)] \times 100$ plotted on the abscissa, and the presence of minerals of silica-oversaturated character (quartz and other silica polymorphs), saturated character (feldspars), and undersaturated character (feldspathoids), plotted on the ordinate. The abundances of the latter three mineral types do not change the rock name. The distinction between diorite and gabbro, and their feldspathoid- and quartz-bearing equivalents, is based on the anorthite content of the plagioclase, with usually light-colored sodic plagioclase (An < 50) indicative of diorite, and usually dark-colored calcic plagioclase (An > 50) indicative of gabbro. The classification is generally consistent with practice, the rows have a reasonably sound physical-chemical basis (see chapter 5), and the classification is easy to remember and use, because there are only 13 rock names. Among its deficiencies are (1) the broad categories used do not allow the fine distinctions between rock types desired by some petrologists, and (2) the definitions implied by the classification are not consistent with those used by some European geologists.

Travis (1955) designed a classification published by the Colorado School of Mines (CSM). Two main differences exist between the RSO and CSM classifications. Although the parameters of classification are similar, the CSM classification expands the middle row of names (syenite, monzonite, etc.) to include rocks with up to 10% of either quartz or feldspathoids. This modification was adopted in several other classifications and is perhaps founded on the difficulty of identifying small quantities of quartz and feldspathoid in handspecimens. The second difference and a major deficiency, not at all clear in the originally published form of the CSM classification, is that certain compositions of rock are not assigned names (see Raymond, 1984c, figure 15b).

The classification shown in figure 4.1b was developed by D. W. Peterson (1965) for the American Geological Institute's (AGI) data sheets. This classification appears to be a combination of classifications like the RSO and CSM classifications. It was widely used for a time, but was

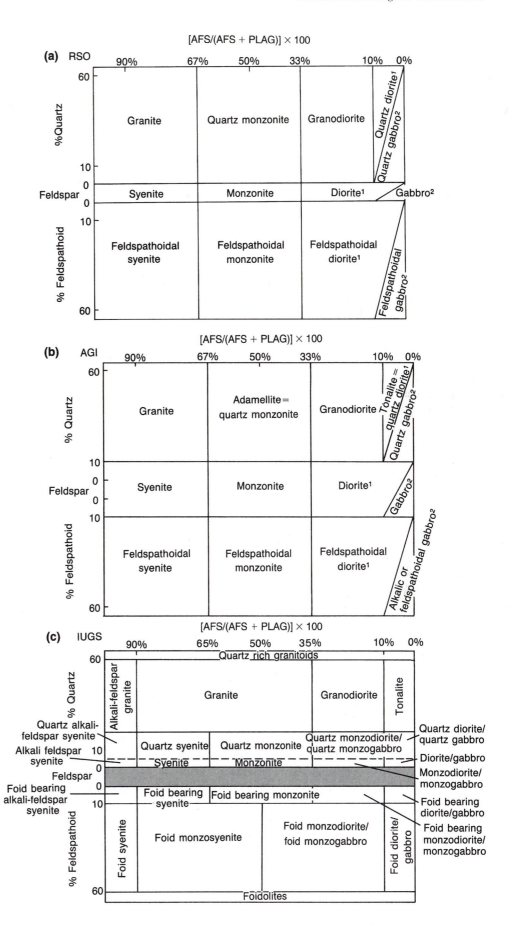

Figure 4.1 Classifications of common phaneritic feldspathic rocks (all classifications are modified to fit on grids of the same scale). (a) The RSO (Raymond, Shand, and others) classification, based on Raymond (1984c), Shand (1949), J. F. Kemp (1929), Bayly (1968), and others. (b) AGI classification of Peterson (American Geological Institute, 1965). (c) IUGS classification, rectangular version (Streckeisen, 1976). Shaded area designates area for which no names are assigned in the classification.

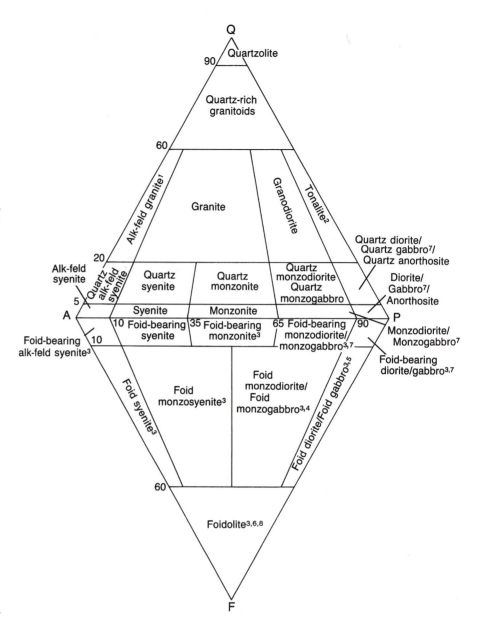

Figure 4.2 IUGS classification of phaneritic feldspathic rocks, diamond-shaped version (modified after Streckeisen, 1973, 1976). (1) Specify alkali feldspar(s) present in each case (e.g., orthoclase granite). (2) *Alaskite* may be used for light-colored alkali-feldspar granite (M = 0–10). M = mafic and related minerals. (3) *Trondhjemite* may be used for light-colored tonalites (M = 0–10) that contain oligoclase or andesine. (4) Specify feldspathoid(s) present in each case (e.g., nepheline-bearing syenite). (5) Specify feldspathoid(s) present in each case (e.g., nepheline syenite). (6) *Essexite* may be used for nepheline monzodiorite/monzogabbro. (7) *Theralite* = nepheline gabbro; *teschenite* = analcite gabbro. (8) Many special names exist; for example, nepheline-rich foidolites include *urtite, ijolite,* and *melteigite.* Diorite and gabbro (norite) are distinguished as follows: In diorite, plagioclase is oligoclase or andesine (< An$_{50}$); chief mafic minerals are hornblende and/or biotite, in some cases also augite; olivine is uncommon. In gabbro (norite), plagioclase is labradorite or bytownite (>An$_{50}$); chief mafic minerals are clinopyroxene, orthopyroxene, and olivine.

dropped by the AGI in the second edition of the data sheets (Dietrich et al., 1982) in favor of a triangular version of the IUGS classification. It shares some of the advantages and disadvantages of its predecessors.

In 1972, the International Union of Geological Sciences (IUGS) adopted a classification of plutonic rocks agreed to by a subcommittee composed of petrologists from around the world (Streckeisen et al., 1973; Streckeisen, 1976). The subcommission favored a diamond-shaped (double-triangle) figure for classification (figure 4.2), but Streckeisen (1976) also presented a rectangular version of the classification, presented here as figure 4.1c. A comparison of the rectangular IUGS and other rectangular classification schemes used in North America reveals clearly that

major changes not consistent with previous use in North America were adopted by the IUGS. One major change is that the term granite was assigned a much broader range of composition—an assignment consistent with some European usage but not with North American practice. The term tonalite was also given a compositional range unlike that of the CSM or AGI classifications.[6] Further, the IUGS classification contains several additional rows of names.

The form of classification preferred by the IUGS subcommission was a diamond (two triangles, base to base) (Streckeisen, 1976). Triangular-, diamond-, and tetrahedral-shaped classifications are not uncommon types of classification.[7] The IUGS classification, shown in figure 4.2, bases the names of rocks on the *ratios* of three minerals—alkali feldspar

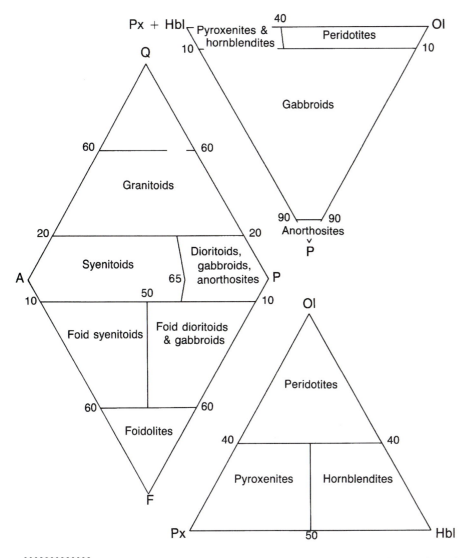

2. It provides needed discrimination between rock types.
3. It is widely used.
4. It is based on ratios of three important mineral groups—the silica minerals, feldspars, and feldspathoids.
5. It includes igneous and igneous-looking rocks.
6. It is consistent with "natural relationships."
7. It follows tradition, as much as is possible.
8. It is both simple and easy to use.[10]

Many of these advantages are open to question. While some petrologists feel that international agreement on terminology is desirable (point 1), others do not.[11] Clearly, more discrimination of rock types is provided for by the classification (point 2), but subdivision of a major group of phaneritic rocks, those composed of more than 20% quartz, abundant alkali feldspar, and modest amounts of plagioclase, is discouraged. The published literature does suggest that the classification is being used widely (point 3). Although the same important minerals are used in this classification as in others (point 4), the exclusive use of normalized mineral *ratios*, rather than observed percentages of minerals, and the positions of the arbitrary boundaries selected for the subdivisions, remove the observational and most physical-

Figure 4.3 IUGS field classification of phaneritic (plutonic) rocks. A = alkali feldspar, Hbl = hornblende, F = feldspathoids, P = plagioclase, Px = pyroxene, Ol = olivine, Q = quartz.
(After Streckeisen, 1976)

(A), plagioclase (P), and either quartz (Q) or feldspathoids (F), whichever is present. A total of 34 rock names is applied to 26 spaces on the diagram. Points are plotted in either the upper or lower triangle after normalizing the three components to 100% (i.e., Q + A + P = 100 or F + A + P = 100). The name of the rock is derived from that plot.[8] The IUGS subcommission recommends that the prefixes *leuco-* and *mela-* be attached to the root names where appropriate to designate light and dark varieties of a rock.[9]

The IUGS classification is both widely used and much debated. Among the advantages ascribed to the classification are the following.

1. It has been designed and adopted by an international organization.

chemical implications of the classification. Many petrologists consider it undesirable to include igneous-looking (metamorphic) rocks (point 5) in an igneous rock classification.[12] The IUGS subcommission apparently considered that the boundaries they chose between rock types reflected "natural relationships" (point 6), yet Le Maitre's (1976) analysis of rock chemistry shows no significant natural breaks in the continuum of rock types. The IUGS classification does follow some traditions for individual rock types (point 7)—such as the European tradition for using the term granite, the American tradition of distinguishing between feldspathic and ferromagnesian rocks, and the European tradition of a triangular classification—but it is not possible to incorporate all of the varied traditions of disparate groups into a single classification formulated by a committee. As for simplicity (point 8), that is relative. Compared to the classifications of Johannsen (1939) and Cross et al. (1903), the IUGS classification is simple; but it is not as simple as the Peterson-AGI classification. To resolve that problem the IUGS system includes a preliminary system for field use with only a few names (figure 4.3). This classification is reproduced in the third edition of the

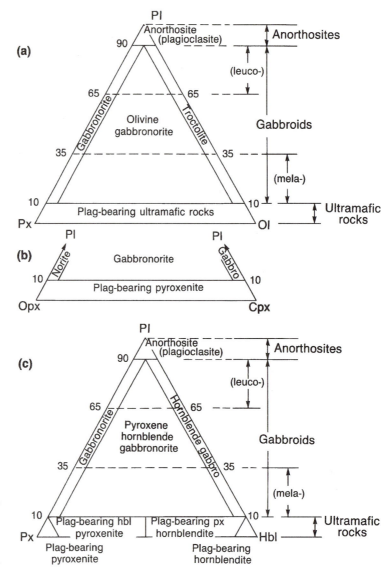

Figure 4.4 IUGS classifications of gabbroic rocks. (a) For rocks composed of plagioclase, pyroxene, and olivine. (b) For rocks composed of plagioclase, orthopyroxene, and clinopyroxene. (c) For rocks composed of plagioclase, pyroxene, and hornblende.
(From Streckeisen, 1976)

AGI data sheets (Dutro, Dietrich, and Foose, 1989). The ease-of-use advantage is similarly relative. Field use of triangular plots is not particularly convenient, because each name must be derived through a process of observing mineral percentages, adding the three essential mineral percentages, and dividing each individual percentage by the total. In short, although many of the advantages of the IUGS system are debatable, the classification is widely used.

The IUGS classification provides a separate set of plots for rocks near the feldspathic-ferromagnesian rock transition. That set of plots, shown in figure 4.4, provides names for the gabbroic rocks.

Classification of Phaneritic Ferromagnesian Rocks

Classifications for phaneritic ferromagnesian (ultramafic) rocks are not nearly as numerous as are those for feldspathic rocks. Many published discussions of terminology for ferromagnesian rocks simply list various names applied to the rocks.[13] Here, two examples of classifications are provided. Raymond's (1984c) classification, presented in figure 4.5, uses olivine versus other ferromagnesian minerals as a basis for naming rocks. Many of the names are defined in a way that is consistent with tradition, but some new names are added and

Figure 4.5 Raymond's classification of phaneritic ferromagnesian (ultramafic) rocks. *(From Raymond, 1984, 1993)*

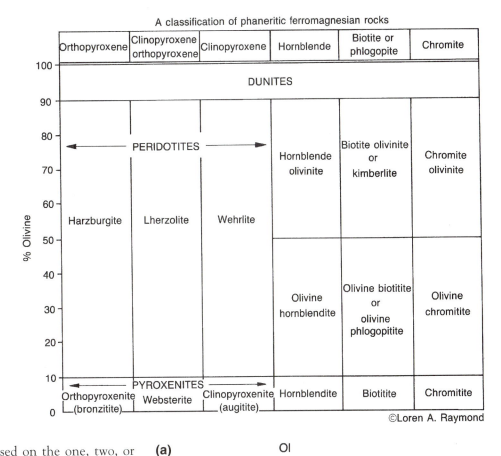

A classification of phaneritic ferromagnesian rocks

	Orthopyroxene	Clinopyroxene orthopyroxene	Clinopyroxene	Hornblende	Biotite or phlogopite	Chromite
100				DUNITES		
PERIDOTITES				Hornblende olivinite	Biotite olivinite or kimberlite	Chromite olivinite
	Harzburgite	Lherzolite	Wehrlite			
				Olivine hornblendite	Olivine biotitite or olivine phlogopitite	Olivine chromitite
PYROXENITES	Orthopyroxenite (bronzitite)	Websterite	Clinopyroxenite (augitite)	Hornblendite	Biotitite	Chromitite

% Olivine (axis: 0, 10, 20, 30, 40, 50, 60, 70, 80, 90, 100)

©Loren A. Raymond

others are modified. Names are based on the one, two, or three most abundant minerals in the rock, and the names of characterizing accessories are used as modifiers. The IUGS classification[14] is a pair of triangular plots (figure 4.6). Figure 4.6a is for rocks containing olivine, orthopyroxene, and clinopyroxene, whereas figure 4.6b is for those composed of olivine, hornblende, and pyroxene. Names are determined by plotting the mineralogy in the usual method for triangular plots (see chapter 3).

Neither the IUGS nor the Raymond classification shows all possible ferromagnesian rock compositions. The Raymond classification shows a wider range of rock types, but the IUGS names in all cases are based on *three* essential minerals rather than the one, two, or three used for names in Raymond's classification.

(a)

Ol

Dunite (90)

Peridotites

Harzburgite — Wehrlite
Lherzolite
Olivine orthopyroxenite — Olivine clinopyroxenite (40 / 40)
Olivine websterite

Pyroxenites

Websterite (10 / 10)

Opx | 10 — Orthopyroxenite ... Clinopyroxenite — 90 | Cpx

(b)

Ol

Dunite (90)

Peridotites

Pyroxene peridotite — Hornblende peridotite
Pyroxene hornblende peridotite

Olivine pyroxenite — Olivine hornblendite (40 / 40)
Olivine hornblende pyroxenite — Olivine pyroxene hornblendite (10 / 10)

Pyroxenites and hornblendites

Hornblende pyroxenite — Pyroxene hornblendite

Px | 10 — Pyroxenite ... 50 ... Hornblendite — 10 | Hbl

Figure 4.6 IUGS classification of ultramafic (phaneritic ferromagnesian) rocks. (a) For rocks with olivine and two pyroxenes. (b) For rocks with hornblende, olivine, and pyroxene. Ol = olivine, Opx = orthopyroxene, Cpx = clinopyroxene, Px = pyroxenes, Hbl = hornblende. *(From Streckeisen, 1976)*

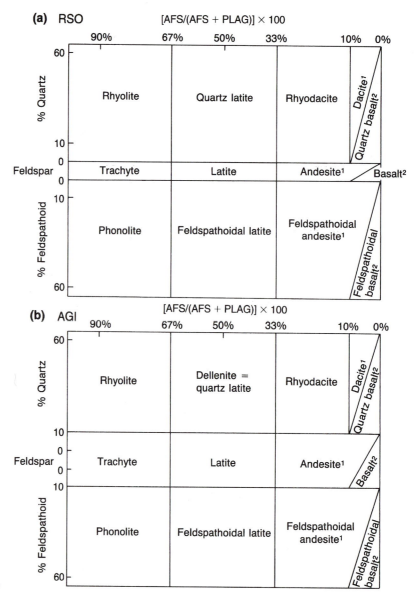

Figure 4.7 Mineralogical-textural classifications of fine-grained to glassy (volcanic) rocks. (a) RSO classification (Raymond, 1984c; based on Kemp, 1929; Shand, 1949; Bayly, 1968; and others). (b) AGI classification by Peterson (American Geological Institute, 1965). (c) Raymond's mineralogical classification of common volcanic rocks (modified from Raymond, 1984c). In (c), names based on unaided eye or hand lens identifications are preceded by an *f* (e.g., *f*-andesite); those based on thin-section petrography should be preceded by a *p* (e.g., *p*-basalt). *Obsidian is actually a textural term. Therefore, it may be used in conjunction with a rock name (e.g., *f*-rhyolite obsidian) for more specificity. Compositionally, most obsidians are siliceous and plot at the left end of the obsidian space. Notes: (1) Rocks with sodic (light-colored) plagioclase; (2) Rocks with calcic (generally dark-colored) plagioclase.

Mineralogical-Textural Classification of Fine-Grained to Glassy (Volcanic) Rocks

Volcanic rocks are typically vitrophyric, aphanitic-porphyritic, or pyroclastic. The typically aphanitic to glassy character of volcanic rocks makes them hard or impossible to identify, either in handspecimen or under the microscope, using mineralogical classifications. For this reason, textural-mineralogical classifications of volcanic rocks have been replaced almost entirely in recent years by chemical or chemically based classifications (Streckeisen and Le Maitre, 1979; Zanettin, 1984; LeBas et. al., 1986).

Three mineralogic-textural classifications are presented in figure 4.7. A word of warning about using mineralogical classifications is necessary: Because the names of rocks are based primarily either on the simple presence of a mineral, or minerals, or on the volume percent of recognizable grains, especially the phenocrysts, while ignoring glassy and aphanitic matrix materials, the names selected may not represent the bulk composition of the rock. The design of two of these charts is identical to that for the phaneritic rocks shown in figure 4.1. The classifications shown include

(c)

			Other phenocrysts or grains				
			Alkali-feldspar ±biotite	Plagioclase ± alkali-feldspar ±biotite	Hornblende ± plagioclase	Pyroxene	Olivine	Olivine+ pyroxene+ plagioclase
Porphyritic texture	Essential phenocrysts or grains	Quartz± feldspar	ƒ—Rhyolite	ƒ—Dacite			////////
		Feldspar only	ƒ—Trachyte	ƒ—Andesite (Light feldspar)		ƒ—Basalt (Dark feldspar)	
		Feldspathoids ±feldspar	ƒ—Phonolite	ƒ—Feldspathoidal andesite (Light feldspar)	ƒ— Feldspathoidal basalt (Dark feldspar)	ƒ—Basanite	
		None	ƒ—Biotite felsite	ƒ—Andesite		ƒ—Basalt	
Aphanitic texture	No visible grains Glass<50%		Felsite (Light colored)		Mafite (Dark colored)			
	Glass>50%		Obsidian*					

the RSO and AGI classifications. A handspecimen (field) classification of volcanic rocks, suggested by Raymond (1984c), is shown in figure 4.7c. Raymond recommends that names based on handspecimen analysis be preceded by a lowercase *f*, to indicate field identification. Similarly, names applied after study with a petrographic microscope should be preceded by a lowercase *p*.

The advantages of mineralogical-textural classifications are that they are useful in the field, easy to remember, and simple to use. The main disadvantage is that names that are based only on the identifiable minerals in the rock are often totally incorrect or misleading.

The observant student will have noted that ferromagnesian rocks are not included in any of the illustrated volcanic rock classifications. Ultramafic volcanic rocks, which are rare, are called *komatiites* and are discussed in chapter 7.

CHEMICAL CLASSIFICATIONS

Differences of opinion about rock classification were well developed by the end of the nineteenth century. During that century, methods of chemical analysis were applied to rocks, and chemistry was used to solve classification problems. That is what Cross, Iddings, Pirsson, and Washington (1903) tried

to do. Their system (referred to by the letters CIPW) was based on a technique of converting a chemical analysis into a standard set of minerals that was then used to classify the rock.

The CIPW approach, manipulating chemical data to form parameters used for classification, is one of two general approaches to chemical classification. The other approach involves direct use of the chemical data.

Classification of Rocks and Rock Series Using Chemical Data

One of the oldest and simplest chemical classifications, based on content of silica (generally the most abundant oxide in igneous rocks), assigns rocks to one of four categories—acid, intermediate, basic, and ultrabasic.[15] Unfortunately, the terms acid and basic were applied on the basis of chemical ideas no longer accepted and have nothing to do with pH. Nevertheless, the terms persist and are widely used in the geological literature. Table 4.2 lists some rock types characteristic of the four categories.[16]

Peacock (1931) introduced the use of the *alkali-lime index* as a way of separating the range of igneous rocks into four "rock series" (figure 4.8). Using an oxide-variation diagram, Peacock plotted two curves, one for

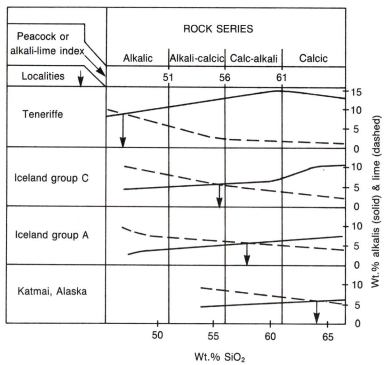

Figure 4.8 Peacock diagram showing subdivisions of rock series on the basis of alkali-lime index, with four examples of rock series curves. *(Modified from M. A. Peacock, 1931)*

Table 4.2 Classification of Igneous Rocks Based on Silica Content

SiO$_2$ (wt. %)	Group Name	Names of Typical Rocks
>66%	Acid rocks	Rhyolite, granite, quartz monzonite
66%–52%	Intermediate rocks	Andesite, quartz diorite
52%–45%	Basic rocks	Basalt, gabbro
<45%	Ultrabasic rocks	Komatiite, dunite, lherzolite

alkalis (Na$_2$O + K$_2$O) and another for lime (CaO), against the percent silica (SiO$_2$). Using several different suites of rocks, he found that the alkali and lime curves for individual suites crossed at different specific values of SiO$_2$. Peacock called the silica percentage that corresponds to the intersection of the two curves, the alkali-lime index. Rock suites with an alkali-lime index (now also called the Peacock Index) of less than 51 are called alkalic; those with indices between 51 and 56 are called alkali-calcic. Peacock Indices between 56 and 61 indicate calc-alkaline rock series, and a Peacock Index greater than 61 characterizes the calcic rock series. These terms are still commonly used, but they are often applied in a casual way, based on mineralogy or rock associations rather than on a chemical analysis.

Following Peacock's lead, numerous petrologists have used various combinations of oxides to subdivide and classify igneous rocks.[17] Some classifications apply to all rocks, others apply only to volcanic rocks or to plutonic rocks, and still others focus on a single category of volcanic or plutonic rock (e.g., basalt). Classifications of individual rock types are discussed in chapters 7–13.

Le Maitre (1976) plotted chemical data for individual rock types on several of the more commonly used oxide-oxide plots. Among the latter is the alkali-silica diagram (Na$_2$O + K$_2$O vs SiO$_2$), which Cox, Bell, and Pankhurst et al. (1979) used to show the approximate chemical ranges of nonpotassic volcanic rocks. The same parameters, total alkalis and silica, were adopted as the basis of a chemical classification by the IUGS (Zanettin, 1984; LeBas et al., 1986)(figure 4.9). In the 1986 IUGS classification, designed for use where modal analyses are not available, 15 fields are represented by 17 root names, and an additional 10 "subroot" names are proposed for K- and Na-rich rocks (LeBas et al., 1986).

Miyashiro (1975), following Kuno (1959) and Mac-Donald and Katsura (1964), also used the alkali-silica diagram to distinguish two rock series. The series basalt→hawaiite→mugearite→trachyte→phonolite,[18] which has a low Peacock Index, is designated as **alkalic.** The series basalt→andesite→dacite→rhyolite has been assigned several names, including non-alkalic, subalkalic, calcic, and calc-alkalic.[19] The two series plot separately and may therefore be distinguished on an alkali-silica diagram.

The nonalkalic rocks have been subdivided by Miyashiro (1974, 1975d) into calc-alkalic and tholeiitic types on the basis of differences in silica, total iron as FeO (FeO*), and titania versus FeO*/MgO (figure 4.10). Analysis of suites of rocks reveals that in calc-alkaline rock series, the iron percentage decreases with increase in silica (see curve A in figure 3.4b). In contrast, in tholeiitic rocks, the trend from silica-poor to silica-rich rocks shows an increase in iron followed by a decrease (see curve S in figure 3.4b). On the oxide versus FeO*/MgO diagrams, these different chemical trends plot as sloping curves, the position and slope of which reflect either a calc-alkaline or a tholeiitic character (figure 4.10).

The diagrams described here are simply examples. Others, which exist in the published literature, are used in similar ways to subdivide rocks on the basis of chemistry.

Classification Using Modified Chemical Data

Cross, Iddings, Pirsson, and Washington (1903), in addressing the problem of rock classification, decided to group all rocks of the same chemical composition together. However, instead of trying to compare chemical analyses, each composed of 10 or 12 oxides, they converted the analyses, via calculation, into a standard set of minerals.[20] These standard minerals, now called the *normative minerals*, were used to classify the rock.

The list of normative minerals with their percentages is called a rock **norm.** The process of **normative analysis,** that is, the process of calculating the norm, contrasts with modal analysis. Recall that modal analysis is essentially a process of observation and measurement used to obtain a list of the *actual minerals* present in a rock, with their volume percentages. In normative analysis, an *imaginary set of minerals* and their weight percentages are calculated using molecular proportions derived from the chemical analysis. Some or all of these minerals may not exist in the rock, but they provide standards for comparison.

The CIPW system of rock classification, based on norms, is quite detailed and consists of classes, subclasses, orders, suborders, rangs, subrangs, grads, and subgrads of rocks. Like a biological system, it tends to treat rocks like discrete entities, rather than individual members of a continuous range of types. Small differences in normative mineral composition (and chemistry) are emphasized, and separate names are given to each small subdivision. The classification contained nearly 200 rock names and sites for that many more. The system was not widely adopted. Nevertheless, the CIPW norm calculation was accepted and it or one of its successors[21] is used regularly where chemical analyses are reported.[22]

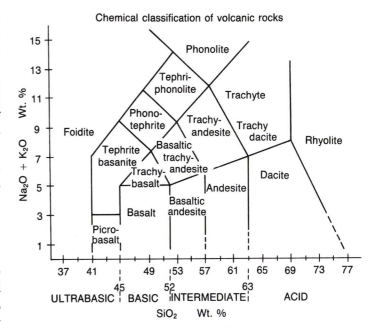

Figure 4.9 IUGS chemical classification of volcanic rocks, based on total alkalis and silica.

(From M. J. LeBas, et al., A chemical classification of volcanic rocks based on the total alkali-silica diagram" in Journal of Petrology, 27:745–50, 1986 Copyright © 1986 Oxford University Press, Oxford England. Reprinted by permission.)

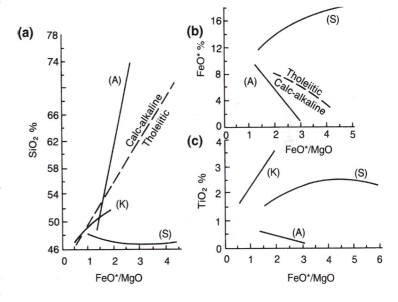

Figure 4.10 Oxide plots for distinguishing between calc-alkaline and tholeiitic rock series. (a) SiO_2 vs. FeO*/MgO, where FeO* represents total iron as FeO. (b) FeO* vs. FeO*/MgO. (c) TiO_2 vs. FeO*/MgO. Curves for Amagi Volcano, Japan (A), Kilauea Volcano, Hawaii (K), and the Skaergaard Intrusion, Greenland (S) are shown.

(Miyashiro, 1975d)

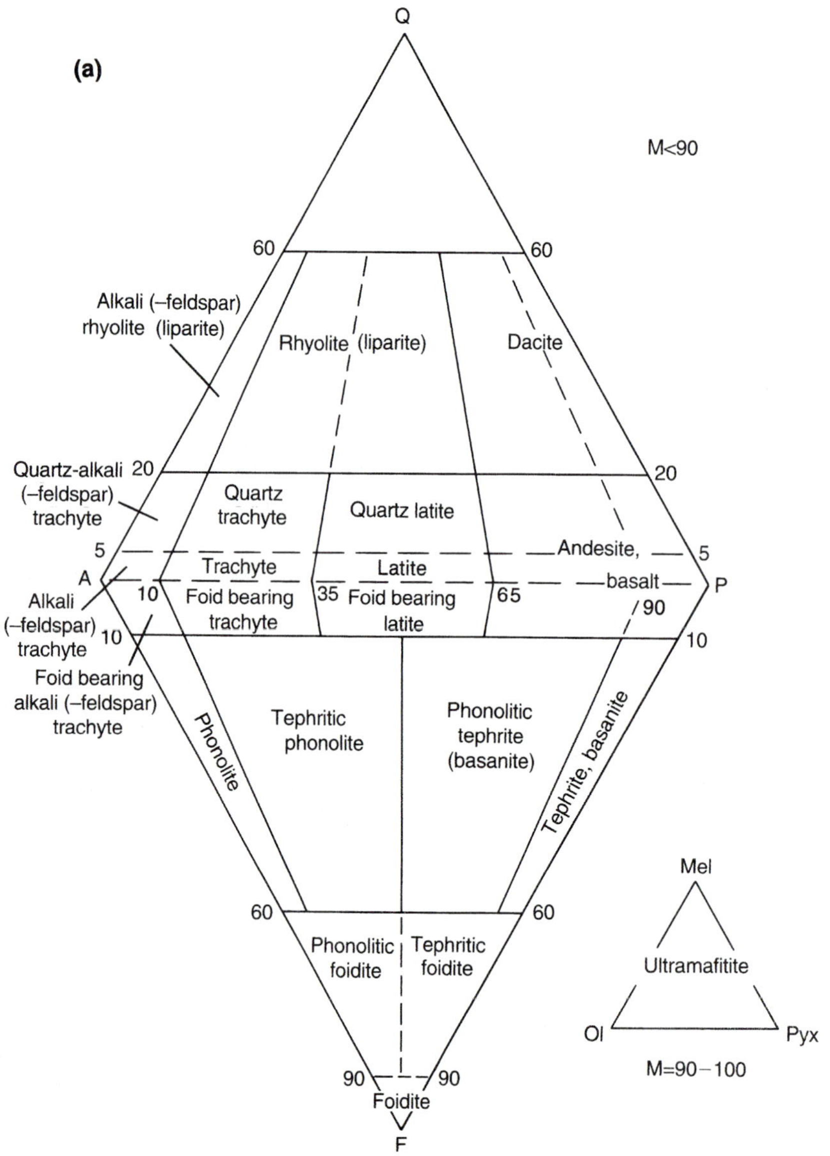

(a)

Figure 4.11 The IUGS classification of volcanic rocks (a), including a field classification (b).
(From Streckeisen, 1979)

CLASSIFICATIONS USING MULTIPLE CRITERIA

In order to classify volcanic rocks, which include holocrystalline, hypocrystalline, holohyaline, and pyroclastic types, some classifications have been developed that employ combinations of chemical data, norms, modes, textural data, and structural features. Three such classifications are the volcanic rock classifications of Irvine and Baragar (1971) and of the IUGS (Streckeisen, 1979; Streckeisen and Le Maitre, 1979) and the pyroclastic rock classification of the IUGS (Schmid, 1981).

The IUGS volcanic rock classification is shown in figure 4.11. The fields are the same as those in the IUGS plutonic rock classification, discussed above, except that some fields

show subdivision lines. Modal mineralogy is used to name and classify holocrystalline volcanic rocks following the same procedure as that outlined for plutonic rocks. For hypocrystalline and holohyaline rocks, the IUGS suggests that normative minerals be used, but the text accompanying the classification seems to indicate some ambivalence about this procedure and does not specify which type of norm should be used. A field classification is also provided (figure 4.11b). The IUGS suggests that the prefix *pheno-* be attached to rock names based on recognizable minerals (e.g., pheno-basalt). Similarly, the prefix *hyalo-* is used for vitrophyric rocks for which a chemical analysis is available (e.g., hyalo-rhyolite).

The distinction between two of the most common types of volcanic rocks, andesite and basalt, appears to present some problems in the 1979 IUGS volcanic rock classification. The

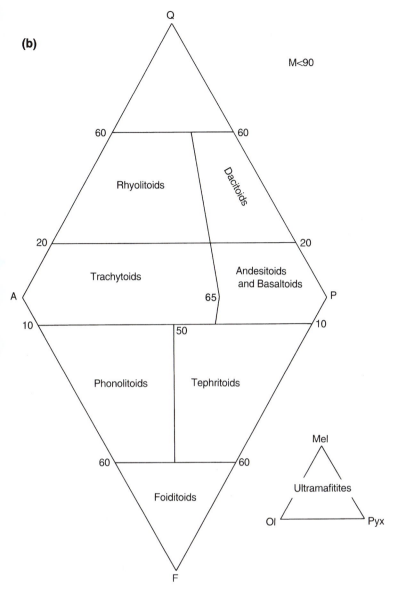

(b)

Q

M<90

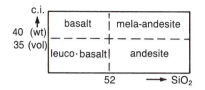

Figure 4.12 IUGS chart for distinguishing between andesite and basalt (c. i. = color index). *(After Streckeisen, 1979)*

dominated by clasts larger than 64 mm are **agglomerates** if rounded clasts predominate and **pyroclastic breccias** if angular clasts predominate. **Lapilli tuff** is a rock with pyroclasts having an average diameter between 64 mm and 2 mm. **Tuff** is pyroclastic rock dominated by grains less than 2 mm in average diameter. Tuffs are subdivided into vitric, crystal, and lithic tuffs depending on whether glass, crystals, or rock fragments, respectively, dominate the pyroclasts.

A variety of unusual igneous rocks such as carbonatites, komatiites, and lamprophyres do not fit into the classifications presented here. These rocks are discussed in later chapters.

SUMMARY

Several types of classfication, based on mineralogy, texture, and chemistry, are available for naming and organizing the wide range of igneous rocks that exists in the Earth's crust. In spite of the efforts of the IUGS subcommittees, there is not universal agreement about what classifications should be used. Each geologist is left to choose the classification most suited to his or her purpose, but it is essential that the classification chosen be specified.

In this text, the classifications used include the author's classifications of phaneritic feldspathic rocks (figure 4.1a), phaneritic ferromagnesian rocks (figure 4.5), and aphanitic feldspathic rocks (figure 4.7a). As noted, ultramafic volcanic rocks, which are important but rare, are discussed in chapter 7.

EXPLANATORY NOTES

1. Some workers have tried designating new names (see Johannsen, 1939) and others have lamented the fact that geologists seem unwilling to break with tradition (Shand, 1949). Nevertheless, many traditional names such as "granite" persist to this day. An international commission headed by A. Streckeisen has reached some agreement on the use of several terms, as discussed in the text. Reviews of older classifications may be found in Cross et al. (1903) and Johannsen (1939).
2. The word *association* has been used in different ways by different authors. Here it simply means a group of rocks that either typically occur together or have similar characteristics.

IUGS subcommission offered a chart (figure 4.12), based on normative color index and weight percent silica, for making the distinction, although it notes that plagioclase composition is used by some petrologists. Chayes (1981) analyzed application of rock names using the IUGS QAPF quadrilateral and notes that in the data base he used, andesites and basalts are "indiscriminately" mixed in their designated fields. A similar problem involving andesites in the dacite field is also noted by Chayes (1981). These and other problems, including some of the same problems that plague the IUGS plutonic rock classification, have prevented widespread acceptance of this classification.

Problems in the classification of pyroclastic rocks have also been addressed by the IUGS (Schmid, 1981). Classification of these rocks is based primarily on grain size. Rocks

3. The idea of families of rock appears to be rooted in early rock classifications of Zirkel (1866, 1877) and Rosenbusch (1900) both discussed in Johannsen (1939). The idea has been adopted by many petrologists including Johannsen (1939), Wahlstrom (1947), Turner and Verhoogen (1960), Hyndman (1972), I. S. E. Carmichael, Turner, and Verhoogen (1974), and Best (1982).

4. Shand (1949) actually used a mineral density of 2.8 rather than actual color to distinguish light from dark minerals. He argues that most minerals with densities greater than 2.8 are ferromagnesian and dark. Johannsen (1939) defined "mafite" minerals as dark minerals, regardless of whether or not they are ferromagnesian. Streckeisen (1976) lists the mafic and related minerals used in the IUGS classification as micas, amphiboles, pyroxenes, olivines, opaque minerals (e.g., oxides, sulfides), zircon, apatite, epidote, allanite, garnets, melilites, monticellite, and primary carbonates—in short, anything not a silica mineral, feldspar, or feldspathoid. Clearly, a variety of alternatives is possible.

5. A boundary of 50% ferromagnesian minerals would perhaps be more rational, but would be at odds with historical precedent and practice.

6. Bateman et al. (1963), in a triangular plot, uses a definition of tonalite like that adopted by the IUGS.

7. For example, Johannsen (1939), Bateman et al. (1963), O'Connor (1965), F. Barker (1979a). The classification of Barker is based on chemical analysis and the calculation of a norm (described in appendix B).

8. Plotting follows the same procedure as that defined for variation diagrams. See figure 3.4 and the associated text for plotting instructions.

9. Charts for determining where these prefixes should be applied are provided by Streckeisen (1976) and are reproduced in available lab manuals and texts (e.g., Raymond, 1984).

10. Based on Streckeisen (1976), the classification itself, and analyses of recently published literature.

11. Streckeisen (1976) and his subcommission clearly felt the need for such agreement, but Williams et al. (1982, p. 85) suggest that agreement on nomenclature may not be desirable.

12. For example, see Williams, Turner, and Gilbert (1982, p. 85).

13. For example, see Peterson (1965) or Wyllie (1967a).

14. Streckeisen et al. (1973), Streckeisen (1976).

15. The subdivisions were proposed in 1897 by Loewinson-Lessing (see Huang, 1962) and are still widely used (see Hyndman, 1972; I. S. E. Carmichael, Turner, and Verhoogen, 1974; and C. J. Hughes, 1982).

16. The terms ultramafic and ultrabasic are used interchangeably by many geologists. Yet as discussed by Wyllie (1967c), not all ultrabasic rocks are ultramafic. For example, a carbonatite, a rock composed essentially of calcite ($CaCO_3$) would be ultrabasic ($SiO_2 < 45\%$) but would not be ultramafic because it has a color index of < 90 or a light color. The reverse is also true; not all ultramafic rocks are ultrabasic. For example, an orthopyroxenite would have a CI > 90 and a dark color but could have a silica content in the intermediate category.

17. For example, see G. A. MacDonald and Katsura (1964), Irvine and Baragar (1971), Bernotat (1972), Miyashiro (1974, 1975d), and Brotzen (1975).

18. See the glossary for definitions of rock names not previously defined in the text.

19. See W. A. Kennedy (1933), Tilley (1950), Kuno (1960), Chayes (1966), Irvine and Baragar (1971), and Miyashiro (1974).

20. The procedure is outlined in appendix B, and every student is encouraged to calculate a few norms with pencil, paper, and calculator (rather than computer) to gain a better understanding of the calculation procedure. This exercise will also help the student understand the chemical differences between peralkaline and peraluminous rocks.

21. For alternative norm calculations, see Barth (1962, pp. 65–70), Rittman (1973), and Currie (1980). Barth (1962) describes the Niggli norm, one based on molecular rather than weight percentages.

22. In tables of chemical analyses, the analyses are typically followed by a norm for each analysis (see table 7.1).

PROBLEMS

4.1. Name a phaneritic rock composed of 16% quartz, 39% alkali feldspar, 14% plagioclase, 13% hornblende, 16% biotite, 1% magnetite, 1% sphene using all of the appropriate classifications available in this chapter.

4.2. Given the chemical analysis below, calculate a CIPW norm following the instructions in appendix B.

$SiO_2 = 47.71$	$MgO = 11.19$
$TiO_2 = 2.18$	$CaO = 9.67$
$Al_2O_3 = 14.01$	$Na_2O = 1.66$
$Fe_2O_3 = 2.28$	$K_2O = 0.07$
$FeO = 10.25$	$P_2O_5 = 0.22$
$MnO = 0.19$	$H_2O^+ = 0.61$
	Total = 100.04

4.3. (a) Name the volcanic rock represented by the analysis in problem 2 using the IUGS (1986) alkali-silica classification (figure 4.9). (b) If the chemical analysis was not available and the rock was a glassy rock with olivine phenocrysts, what names would be applied to the rock using (1) the IUGS field classification and (2) Raymond's mineralogic classification of common volcanic rocks?

5

The Phase Rule and Phase Diagrams

INTRODUCTION

When magmas crystallize, they behave according to physical-chemical laws. Chapter 2 briefly described the process of crystallization. There, the emphasis was on crystal shape and the resulting textures, as controlled by physical laws. In this chapter, emphasis is placed on the *kind of mineral* that will crystallize from melts of varying composition and on the *kind of melt* that will form from melting various rocks.

In an ideal sense, a mineral may be described as a phase. A **phase** is a homogeneous material that, because of its physical properties, can be separated by mechanical means from other phases with which it may occur. For example, ice (one phase) can be separated from water (another phase) by pouring the water-ice mixture through a sieve. Mineral grains, of course, are commonly inhomogeneous in that they are zoned or contain impurities. Consequently, the grains are not true phases, although pure mineral materials within a grain are true phases. Thinking of crystallizing minerals as phases helps us understand the crystallization of magmas. A useful method of showing the relationships between various solid (mineral) and liquid (magma) phases during crystallization or melting is to plot phase data on graphs called phase diagrams.[1] Such graphs are based on data obtained by laboratory experimentation and calculation.

SYSTEMS AND THE PHASE RULE

A basic understanding of phase diagrams requires knowledge of a few key terms and an apparently simple mathematical expression—the phase rule. When we discuss the phase rule or phase diagrams, it is important to place limits on what it is we are discussing. Rather than trying to describe the entire universe, we confine the discussion to a system. Specifically, a **system** is any part of the universe selected for study. Systems may be physically separated from their surroundings (e.g., the water-ice mixture mentioned above may be placed in a closed container) or they may be separated mentally (i.e., we can think about the water-ice system without actually doing anything with water and ice).

Systems may be open, closed, or isolated. In geology, many systems are **open systems;** that is, both material and energy may be exchanged with the surroundings (outside the system). Geologists prefer to work with **closed systems,** systems that cannot exchange material but can exchange energy with the surroundings. **Isolated systems** do not exchange energy or matter with their surroundings.

There is no limit to the size of a system. A batholith may be a system or a gram of chemical confined in a small capsule may be a system.

Systems may be either in a state of equilibrium or disequilibrium. A system at equilibrium has the minimum energy possible under the specified conditions and is said to be *stable*. There is no tendency for the system to change. The system is *unstable*, or in a state of disequilibrium, if a lower energy state for the system is possible under a given set of conditions and change is occurring in the system to bring it closer to equilibrium. *Metastable* conditions are those in which the system is not at its lowest energy state but has no tendency to change, commonly because an energy barrier blocks the change. A simple example of these three states (stable, metastable, and unstable) is depicted in figure 5.1. A skier waiting on flat ground at the top of the slope represents a metastable condition, or is in metastable equilibrium. After the skier uses energy to move from the flat ground to the slope (he or she uses energy to overcome the barrier to downslope movement), the skier's condition becomes unstable (i.e., change occurs over time) and he or she skis downhill in a state of disequilibrium. Stability is attained when the skier reaches the level ground at the foot of the slope. The skier has nowhere to go that is lower than the bottom of the slope, and so there is no tendency for the skier to move anywhere. Here the skier is at equilibrium. Addition of energy, of course, may change the state of the system. For example, energy may be used to transport the skier (via the lift) to the hilltop, where he or she again becomes metastable or unstable.

The systems of importance to petrology are usually defined in terms of three kinds of parameters—phases, components, and intensive variables. Phases were defined earlier. **Components** are chemical species, such as OH^-, H_2O, MgO, or $NaAlSi_3O_8$. When we discuss phases, the phase rule, or phase diagrams, the components of a system are assigned a number that represents *the smallest number* of chemical species needed to define the compositions of all phases in the system under consideration. In the case of the water-ice system cited previously, H_2O is the only component necessary to define (or construct) all the phases in the system. This is true, even if we also add the vapor phase (steam). H_2O is the only chemical species needed to define all of the phases—water, ice, and steam. Notice that all of these phases could be constructed from H^+ and OH^-, but the *smallest number* of chemical species needed to define or construct water, ice, and steam is one, namely H_2O. Therefore, the system is said to be a one-component system.

Systems of petrologic interest include one-component, or unary, systems; two-component systems, called binary systems; three-component (ternary) systems; and other multicomponent systems. Most laboratory studies (experimental analyses) have been done on unary, binary, and ternary systems, but some investigations of quaternary and larger systems have been undertaken. Remember that rocks typically

··············
Figure 5.1   Sketch of skiers showing relative positions of metastable equilibrium (on the flat at the top of the slope, where the skier has no tendency to move downhill but a more stable, lower potential-energy site exists—the bottom of the hill), disequilibrium (on the slope where gravity drives the skier downhill), and equilibrium (at the bottom of the slope, where the skier can go no lower and is therefore at the most stable site).

contain 10 major oxides. As a consequence, even though some important substitutions occur (e.g., MnO for FeO), experimental investigation of phase relations in systems with many components that exactly duplicate rocks are limited. Nevertheless, the study of simple systems contributes significantly to our understanding of rock origins.

The third major parameter used to define a system is the **intensive variable.** An intensive variable is a property of a system that is capable of being changed and is independent of the size of the system (i.e., independent of the amount of material or mass in the system). Pressure (P), temperature (T), and concentration of chemical species (X) are the intensive variables most commonly considered in petrologic systems. In discussing systems and intensive variables, we refer to **degrees of freedom** (F). The number of degrees of freedom (also called variance) is defined as the *smallest number* of conditions or variables that may be changed independently without changing the number or nature of phases in the system, or, in other words, the minimum number of variables needed to define a specific state of the system.

For an example of variance, consider a system consisting of a sealed can of water at standard temperature and pressure. Within limits, either pressure or temperature (or both) may be changed and the can will continue to contain only one phase—water. Thus, there are two degrees of freedom; two variables may be changed independently without changing the number or nature of phases in the system. Such a system is said to be **bivariant.** In contrast, consider a situation in which the can contains a mixture of ice and water (two phases) at equilibrium. That condition can only be maintained if, when one variable (e.g., temperature) is changed, the other variable (pressure) is also changed. Under these conditions, pressure is a *dependent variable*. In such a state, the system has one degree of freedom and is called **univariant.** Where there are no degrees of freedom (e.g., if the can contained ice, water, and vapor in equilibrium), the can system would be **invariant.** None of the intensive variables of the system could be changed without reducing the number of phases in the can.

Notice in this example that as the number of phases increased the number of degrees of freedom decreased. This apparent relationship is formalized in the **Phase Rule:**

$$F = C - P + 2 \qquad (5.1)$$

where F is the number of degrees of freedom, C is the number of components, and P is the number of phases.[2] In the water-ice example just described,

$$C = 1 \ (H_2O)$$
$$P = 2 \ (\text{water and ice})$$

Thus,

$$F = C - P + 2$$
$$F = 1 - 2 + 2$$
$$F = 1$$

The constant 2 in the phase rule refers to the two intensive variables—pressure and temperature. If a third variable controlled the number and nature of phases, the constant would become 3. In much of the older work, the vapor phase was ignored and the pressure was assumed to be fixed at one atmosphere (Wahlstrom, 1950). In such a case or in cases where the pressure is held constant, pressure is not a variable and the constant in the phase rule becomes one (1):

$$F = C - P + 1 \qquad (5.2)$$

This form of the phase rule is referred to as the *condensed phase rule*. Inasmuch as only liquid and solid phases (i.e., condensed phases) are considered in systems to which the condensed phase rule applies, such systems are called condensed systems.

UNARY SYSTEMS

The system H_2O, just discussed, is a unary system. The system is geologically significant in geomorphological studies and glaciology, where usually ephermeral rocks composed of ice are important. (Note that ice fits the definition of a rock.)

Perhaps the most important unary system in petrological studies is the system SiO_2.[3] Figure 5.2 shows the phase diagram for this system. Because $C = 1$, both pressure and temperature can be used as coordinate axes on this diagram.

The SiO_2 phase diagram consists of seven one-phase fields. Six represent ranges of pressure and temperature over which the individual silica polymorphs alpha quartz (αQ), beta quartz (βQ), tridymite (Tr), cristobalite (Cr), coesite (Co), and stishovite (St) are stable. These are bounded at high temperature by the liquid phase (or melt) field. All seven phases have the composition SiO_2. Four triple points, at which three phases coexist (in equilibrium), are present in the diagram.

Examination of possible changes in the system SiO_2 at a series of points will illustrate the phase rule while emphasizing the nature of the system. Consider point X, which lies in the coesite field at $T = 500°$ C, $P = 5.0$ Gpa (50 kb). Here, the system consists of one phase, coesite. If the pressure is changed, e.g., is increased along line $X \rightarrow 1$, neither the nature nor number of phases will change. The system still consists of coesite. Similarly, the pressure could be decreased along line $X \rightarrow 2$ (or the temperature could be increased or decreased) without creating any change in the nature or number of phases. Temperature and pressure can be varied independently without varying the phases. At point X, by the Phase Rule,

$$C = 1 \ (SiO_2)$$
$$P = 1 \ (\text{coesite})$$

and therefore, from equation 5.1,

$$F = C - P + 2$$
$$F = 1 - 1 + 2$$
$$F = 2$$

The number of degrees of freedom is 2, as we determined by mentally changing the pressure and temperature. The point X lies in a bivariant region.

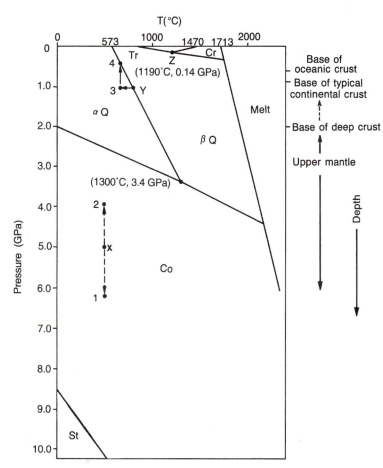

Figure 5.2 Phase diagram for the system SiO_2. αQ = alpha quartz, βQ = beta quartz, Tr = tridymite, Cr = cristobalite, Co = coesite, and St = stishovite. Points X, Y, Z, 1, 2, 3, and 4 are discussed in the text.
(Modified from Boyd and England, 1960; Ostrovsky, 1967)

Now consider point Y, which falls on the α-quartz – β-quartz curve at about 800° C, 1 Gpa (10 kb). At equilibrium, α-quartz and β-quartz both exist in the system. Suppose we now lower the temperature of the system along the line Y → 3. If the pressure remains constant, as indicated by the position of the line, the system would change. The β-quartz would disappear and only α-quartz would remain. In order to return the system to its original condition (with both α- and β-quartz), the pressure must be decreased along the line 3 → 4. At point 4, two phases coexist. An alternative path to point 4 would be along the curve Y → 4. To follow such a path, one would need to reduce temperature and pressure simultaneously. Doing so would allow both α-quartz and β-quartz to remain in the system. It should be clear now that to maintain two phases in equilibrium, only one of the intensive variables is independent; the other must be changed or the number of phases changes.

Point Y, then, by the Phase Rule (eq. 5.1) has

$$C = 1$$
$$P = 2$$

and

$$F = C - P + 2$$
$$F = 1 - 2 + 2$$
$$F = 1$$

There is one degree of freedom. The curve separating the α-quartz and β-quartz fields is an univariant line.

Point Z is one of the triple points. Here tridymite, cristobalite, and β-quartz all exist together at equilibrium. Using the Phase Rule (eq. 5.1),

$$C = 1$$
$$P = 3$$
$$F = C - P + 2$$
$$F = 1 - 3 + 2$$
$$F = 0$$

At that point, the triple point, the system is invariant. Neither pressure nor temperature can be changed without changing the number of phases.

BINARY SYSTEMS

Binary systems contain two components. To plot both components plus the pressure and the temperature, a three-dimensional diagram would be necessary, which would make plotting points more difficult. The behavior of the system may be examined in two dimensions however, by ignoring one of the variables, holding a variable constant, or projecting data from three dimensions onto a two-dimensional plane.

Not every combination of two components behaves in the same way and some combinations behave, in part, like unary systems. One example of the latter case is the system SiO_2-H_2O.

The System SiO_2-H_2O

The system SiO_2-H_2O, studied by Tuttle and England (1955), is also discussed by Tuttle and Bowen (1958) and Ernst (1976). Figure 5.3 shows two, two-dimensional phase diagrams for this system. One is a temperature-composition (T-X) plot and the other is a pressure-temperature (P-T) plot. The system is included here to emphasize three points: (1) a common result of mixing two components is the lowering of the melting temperature of one or more phases; (2) not every combination of components in a system results in typical binary-type behavior;[4] and (3) water

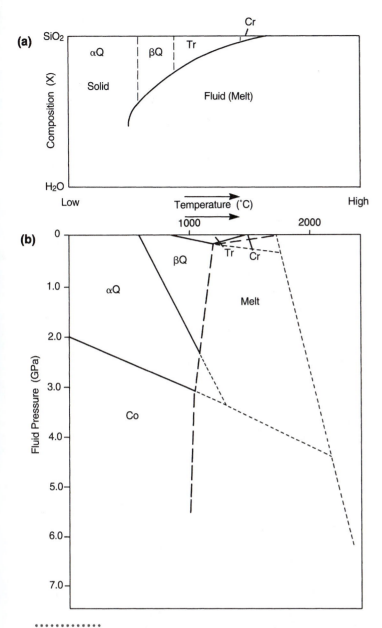

Figure 5.3 Phase diagrams for the system SiO_2-H_2O. (a) Schematic T-X diagram (modified from Tuttle and Bowen, 1958). (b) T-P_{fluid} diagram (after Tuttle and England, 1955). Dotted lines show phase fields in the dry system (from figure 5.2). Symbols are the same as in figure 5.2.

(specifically) may influence the behavior of systems, even if the water does not chemically combine with other components to create new phases.

Figure 5.3a shows part of a schematic T-X plot, from Tuttle and Bowen (1958), simplified for presentation here. The major feature is a curve separating a fluid field at high temperature from a solid field at lower temperature. On phase diagrams, curves such as this, above which only a melt phase

exists, are called **liquidus curves.** On the left side of the diagram, note that an increase in H_2O significantly reduces the temperature of the liquidus; that is, SiO_2 melts at lower and lower temperatures as the proportion of water in the mixture increases (illustrating point 1 of the preceding paragraph). A decrease in melting temperature also results from increasing pressure (under water-saturated conditions) (figure 5.3b).[5] Yet on the low-T side of the liquidus, the positions of curves bounding the stability fields of the phases are not affected by the presence of H_2O. The liquidus curve simply descends to a lower temperature, reducing the size of the stability fields of the phases (compare figures 5.2 and 5.3b). Thus, at temperatures below those of the liquidus curve, the behavior of the system is essentially unary. Nevertheless, water is an important fluxing agent and the lowering of the liquidus revealed here is a behavior that extends to other systems (Tuttle and Bowen, 1958).

Solid-Solid Eutectic Binary Systems

Many binary systems are important to petrology. Binary diagrams typically represent one of three types of systems—binary eutectic systems, binary peritectic systems, and binary solid solution systems. All three types are represented among systems of rock-forming oxides.[6]

Binary eutectic systems include SiO_2-$NaAlSi_3O_8$ and SiO_2-$CaAl_2Si_2O_8$, studied by Schairer and Bowen (1947, 1956). These systems have the basic T-X form depicted in figure 5.4, which shows an imaginary system composed of two components, A and B. Pure A and B occur as the solid phases *a* and *b*, respectively.

The imaginary system in figure 5.4 is completely labelled to clarify the nature of the diagram and the terminology used in, but commonly partly omitted from, published diagrams. The axes of the phase diagram are temperature (*T*), typically reported in degrees Celsius (C), and weight percent of the components (X). Note that the left end of the X-axis represents 100% A and 0% B, whereas the right end represents 0% A and 100% B. Points between the ends represent combinations of A and B that sum to 100%. Usually only the percentages for the right-hand component are listed, as simple subtraction will provide the corresponding value for the component on the left. Temperatures are provided for the melting of pure *a* (1250° C), the melting of pure *b* (1550° C), and the eutectic (described below) (1000° C).

Notice that the imaginary system A-B has four separate and distinct phase fields separated by three curves. At the top of the diagram (at high temperature) is a field representing a liquid phase (melt). One melt phase, composed of both A and B, exists under the T-X conditions represented by any point in this field. For example, at point 1, a single melt phase of composition A = 80% and B = 20% ($A_{80}B_{20}$) exists

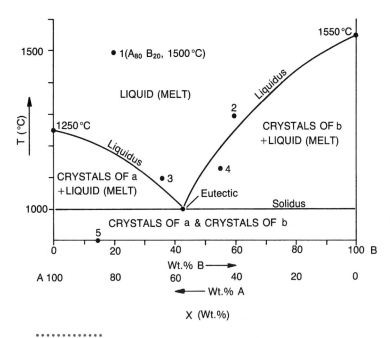

Figure 5.4 Phase diagram for the imaginary system A-B. Points and curves are discussed in the text. In this simple case, crystals of **a** have a composition identical to that of component A, and crystals of **b** have a composition identical to that of component B.

at a temperature of 1500° C. Similarly, points 2 and 3 represent positions where a single melt phase is present (e.g., point $2 = A_{40}B_{60}$, 1300° C). The curve separating the liquid field above from the field(s) below, which contain both liquid and crystals, is the *liquidus*. In general terms, the liquidus is a line (or surface, in three dimensions) defined as the locus of all points representing a correspondence between points of equilibrium between particular solid phases and liquids of specific compositions, and specific temperatures (and/or pressures).

Below the liquidus, at the right end of the diagram, is a two-phase field in which crystals of *b* coexist with a liquid composed of both A and B. Any point in the field represents both (1) a combination of crystals of *b* and a particular liquid composition, and (2) a particular *ratio* of *b* to liquid (see lever rule below). Below the liquidus, on the left, is a similar two-phase field, but one containing the phases "crystals of *a*" and "melt."

The fourth field in the A-B system is the field at the bottom of the diagram. Here, two phases—crystals of *a* and crystals of *b*—stably coexist. The curve separating this completely solid field from the fields above, in which liquids are stable, is called the *solidus*. In this diagram, it is a horizontal line at a temperature of 1000° C. In general terms, the solidus is a line (or surface, in three dimensions) defined as the locus of all points for which a correspondence exists between specific temperatures (and/or pressures) and a particular liquid in

equilibrium with one or more solid phases. Note that the liquidus in (T-X diagrams) defines the composition and temperature for liquids coexisting with a solid phase, whereas the solidus (in T-X diagrams) defines the composition and temperature of solids coexisting with a specific liquid. Any point within the solid-phase field is representative of a specific temperature and a specific proportion of crystals of *a* and crystals of *b*. These values may be read from the scale on the abscissa. For example, point 5 in figure 5.4 has coordinates of $(A_{85}B_{15},900°$ C) and corresponds to a sample containing 85% crystals of *a* and 15% crystals of *b* at a temperature of 900° C.

The **eutectic** point is the lowest point on the liquidus curve. As such, it represents the lowest possible equilibrium temperature at which liquid can exist in the system. During crystallization or melting at this temperature, the liquid has a fixed composition, the eutectic composition, and is in equilibrium with the phases *a* and *b*. The temperature of the eutectic point is called the eutectic temperature. The eutectic point exists because the addition of *b* to a liquid of composition *a* lowers the melting (or freezing or crystallization) temperature of *a*. Similar dilution of a liquid of composition *b* by *a* depresses the melting (freezing) point of *b*. The two curves intersect at a minimum—the eutectic.

Equilibrium Crystallization in the System Si_2O -$NaAlSi_3O_8$

Figure 5.5 shows the phase diagram for the system SiO_2-$NaAlSi_3O_8$. Notice that this diagram resembles that for the imaginary system in figure 5.4 except that the crystal plus liquid field (on the right) is divided into upper and lower sections. The horizontal line separating the two parts—a cristobalite-plus-liquid field and a tridymite-plus-liquid field—marks the temperature at which the polymorphic transformation of cristobalite to tridymite occurs. Other than the structural change indicated for the SiO_2 polymorphs at this temperature, the boundary has little effect on the phase relations, producing only a slight inflection in the liquidus. Note that for this diagram P is fixed; because P is a constant, the Phase Rule becomes $F = C - P + 1$.

What can be learned from this diagram? Specifically, we can learn (1) crystallization histories for various magmas of the composition represented by the diagram, and (2) melting histories for rocks composed of albite and quartz.[7] More importantly, the principles used here are applicable to all similar diagrams.

We begin by analyzing the equilibrium crystallization of a liquid (magma or melt) of composition α (figure 5.5b). From the abscissa, it is clear that α has the composition $NaAlSi_3O_8 = 40\%$, $SiO_2 = 60\%$ ($A_{40}S_{60}$). The temperature at point α is 1600° C. Point α is in a bivariant region ($F = 2 - 1 + 1 = 2$). Thus, as the temperature begins to drop and the magma cools, there are no changes in the number or types of phases. Continued cooling, however, eventually lowers the

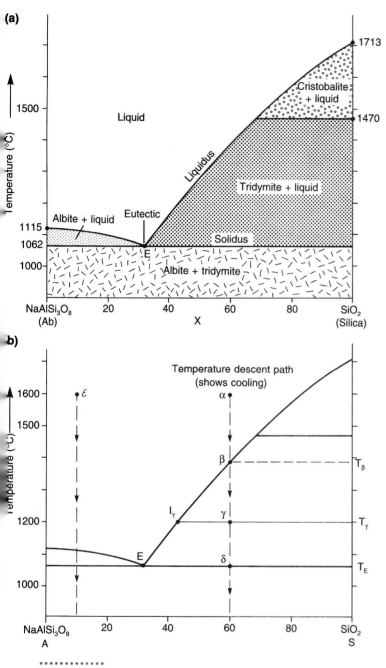

Figure 5.5 Phase diagram for the system SiO₂-NaAlSi₃O₈. (a) Labelled phase diagram. (b) Diagram showing cooling, crystallization, and melting paths discussed in the text.

(Diagram modified after Schairer and Bowen, 1947a, 1956)

Crystal formation is a consequence of supersaturation of the system with respect to the component SiO_2. Clearly, continuation of heat loss will lead along the path β towards γ, from the liquid field into the tridymite + liquid field. Thus, the crystals that begin to form must be tridymite.

Continued cooling follows the path β–γ. A **tie-line,** a line drawn *within* a phase field connecting the phases coexisting (in equilibrium) at a fixed T (or P), may be constructed at γ, as shown. The two ends of the tie-line, where they intersect boundaries, define the equilibrium compositions of the phases for a given temperature (in this case, $T_\gamma = 1200°$ C). The right-hand end of the tie-line intersects a vertical boundary, the composition of which, as read from the abscissa, is 100% SiO_2, or pure tridymite. The left-hand end of the tie line intersects the liquidus curve at l_γ. The composition of the liquid given by that point of interesection is read from the abscissa directly below and is about $A_{57}S_{43}$. In summary, at 1200° C (T_γ), pure tridymite is in equilibrium with a liquid of composition $A_{57}S_{43}$.

If we look back at point β, where the temperature was 1380° C (T_β), a tie-line constructed there indicates that the first crystals to form, also of pure tridymite, are in equilibrium with a liquid of composition $A_{40}S_{60}$. Clearly, the composition of the liquid has changed with falling temperature, from an initial composition of $A_{40}S_{60}$ at T_β to a composition of $A_{57}S_{43}$ at T_γ. Construction of tie lines at 5° C intervals between β and γ would reveal that the liquid composition has changed along a curve, the T-X curve, which defines the liquidus. Over the same temperature interval (T_β–T_γ), crystals of tridymite continue to form. As they form, that is, as SiO_2 is taken out of the liquid to become solid crystals of tridymite, the SiO_2 content of the liquid decreases. In effect, this causes the percentage of $NaAlSi_3O_8$ in the liquid to increase.

The system and its changing liquid composition is easily understood through an analogy. Imagine the system is a barrel. The initial liquid in the system consists of 100 pieces of fruit, 60 oranges and 40 apples. Upon cooling, crystals form, causing the removal of some oranges from the liquid, that is, from the total amount of fruit in the barrel. Formation of crystals equivalent to 20 oranges will leave 40 apples and 40 oranges in the barrel (liquid). Thus, although the quantity of fruit (liquid) is reduced from 100 to 80 pieces of fruit, the percentage or proportion of apples in the barrel (liquid) has increased from 40/100, or 40%, to 40/80, or 50%. In figure 5.5 $NaAlSi_3O_8$ is equivalent to apples and SiO_2 is equivalent to oranges.

It should be evident now that, if the temperature continues to fall, the liquid will continue to change composition and temperature along a path defined by the liquidus curve. Tridymite crystals will continue to form, causing the liquid to be enriched in $NaAlSi_3O_8$ (component A).

The crystallization process changes when the temperature reaches T_E (1062° C), the eutectic temperature E (corresponding to point δ). Liquid is not stable below the

temperature to point β. The cooling path followed by the material in this system (in T-X space) is shown by the dashed line connecting points α and β. At β, that path intersects the liquidus curve. The temperature is about 1380° C.

The liquidus is an univariant curve marking the boundary between the single-phase liquid field and the two-phase tridymite + liquid field. At point β, crystals begin to form.

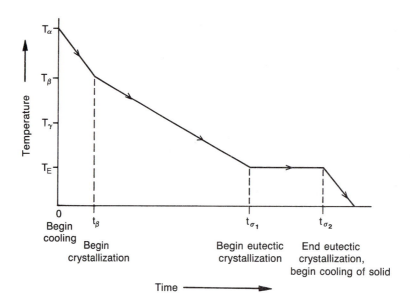

Figure 5.6   Schematic time-temperature graph showing the cooling history for the composition depicted in figure 5.5b.

solidus. As a result, the temperature will not drop below T_E until all of the liquid is gone. This is predictable via the Phase Rule. At point E, $P = 3$ (albite, tridymite, and melt) and $C = 2$. Thus, $F = 2 - 3 + 1 = 0$; that is, there are no degrees of freedom.

At the eutectic point, the liquid continues to crystallize tridymite. Were this the only phase crystallizing, the liquid would continue to be enriched in the albite component. In fact, at the eutectic, the liquid becomes saturated with respect to albite, which begins to crystallize along with tridymite. The ratio at which these two solid phases crystallize is such that the composition of the liquid remains fixed at the eutectic composition until all of it is crystallized. When the last liquid crystallizes, the solid will consist of tridymite and albite in the proportions of the original melt (Ab = 40% and Tr = 60%, for a composition $A_{40}S_{60}$). Cooling will then resume. During crystallization at T_E, heat continues to be lost from the system without concomitant lowering of the temperature. This heat represents the latent heat of crystallization.

The thermal history of the crystallization sequence described above is shown *schematically* in figure 5.6. The diagram is founded on the assumption that the *rate* of heat loss to the surroundings is constant from the beginning to the end of the crystallization sequence. As heat is lost from the melt, the temperature falls. At T_β, the liquidus is intersected and crystals begin to form. This results in a decrease in the slope of the temperature curve because, although heat continues to be absorbed by the surroundings, heat of crystallization is added to the melt by the newly forming tridymite crystals. At

temperature T_E, the temperature is stabilized, for the latent heat of crystallization of tridymite and albite is now equal to the heat lost to the surroundings. After all of the liquid is gone, the solid will then cool over time, as heat continues to be lost to the surroundings.

Crystallization histories similar to that for composition α may be generated for any composition within the system. Only the details will vary. For example, crystallization of a melt of composition ϵ, with an initial composition of 90% $NaAlSi_3O_8$ and 10% SiO_2 begins near 1100° C. The first crystals to form are albite. The liquid changes composition along the liquidus curve, as the temperature falls and the liquid becomes enriched in SiO_2. At temperature T_E (1062° C), the liquid composition reaches the eutectic, at which point crystals of both albite and tridymite form simultaneously. At this time, the temperature is stabilized until all the liquid is consumed. Then, the resulting rock, composed of albite and tridymite, continues to cool. At a lower temperature, the tridymite inverts to quartz, leaving an albite-quartz rock rich in albite, such as a quartz diorite.

Some geological implications are worth noting here. First, the *kind* of crystals that form initially from the melt (or magma), under equilibrium conditions, depends on the chemistry of the melt. For example, in the SiO_2-$NaAlSi_3O_8$ system we examined, a silica mineral, either tridymite or cristobalite, forms first *if* the melt is relatively rich in silica. Albite crystals form first if the melt is rich in $NaAlSi_3O_8$. Second, under equilibrium crystallization, the final rock has a mineralogical composition that has a bulk chemistry

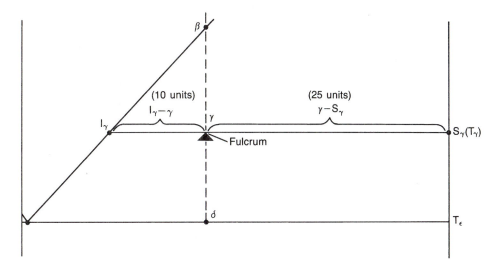

Figure 5.7 Enlargement of part of figure 5.5 showing aspects of the lever rule. See text for discussion.

identical to the composition of the original melt (in a phrase, "what you have is what you get"). Third, the later crystallizing phase will tend to fill the spaces (interstices) between the early formed crystals.

The Lever Rule

Not only is it possible to determine the sequence of crystallization and the general thermal history of a system, it is also possible to make quantitative calculations of the percentages of phases present at most points in the cooling history. This information is useful in studies of disequilibrium crystallization or calculations of changes in mass (mass balance calculations). The principle involved is borrowed from physics and is referred to as the lever principle or lever rule.

Part of figure 5.5 is enlarged and reproduced in figure 5.7. The tie line for temperature T_γ is shown intersecting the solid (tridymite) boundary (S_γ) at the right and the liquidus (l_γ) at the left. The original bulk composition of the melt is represented by the point γ.

The tie-line is loosely analogous to a lever or teeter-totter resting on a fulcrum (the point γ). The two ends of the teeter-totter are at different distances from the fulcrum. To balance a teeter-totter, the person who weighs the most sits on the shorter side. Similarly, for the tie-line to be "balanced", the end of the line (l_γ) with the shortest distance to point γ must have the most material. In other words, there is more liquid (l) than solid (S).

The segments of the tie-line are proportional to the percentages of the phases represented by its end points. Mathematically, the percentage of any phase is derived by dividing the length of the corresponding segment by the total length of the tie-line and multiplying by 100. We determined above that the amount of liquid is greater than the amount of

solid. Therefore, the length of the longer line segment (γ–S_γ) corresponds to the amount of liquid. Thus,

$$\% \text{ liquid} = \frac{\gamma - S_\gamma}{l_\gamma - S_\gamma}(100) = \frac{\text{length of larger segment}}{\text{total length}}(100)$$

$$= \frac{25}{35}(100)$$

$$\% \text{ liquid} = 71.4\%$$

Similarly, we can determine the length of the shorter segment and calculate the weight percent solid (tridymite crystals) at temperature T_γ:

$$\% \text{ Tr} = \frac{l_\gamma - \gamma}{l_\gamma - S_\gamma}(100)$$

$$= \frac{10}{35}(100)$$

$$\% \text{ Tr} = 28.6\%$$

Disequilibrium Crystallization in the System CaMgSi$_2$O$_6$-CaAl$_2$Si$_2$O$_8$

For equilibrium crystallization to occur completely, at every stage of crystallization all of the liquid must be in contact with all of the crystals *and* cooling must be slow enough for all reactions to occur. Of course, this is an ideal that cannot be met. Thus, some disequilibrium will always occur. Locally, evidence of minor disequilibrium can be observed in rocks. An example is the minor compositional variability in plagioclase feldspars.

Of concern in this section, however, is major disequilibrium in a system represented by the magma chamber and its

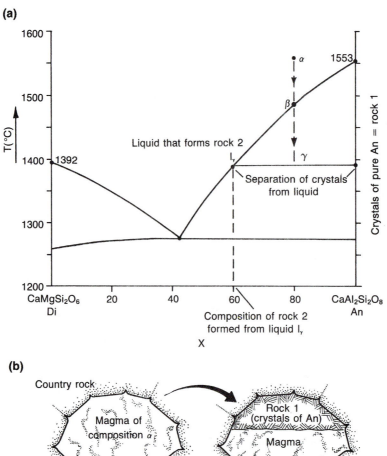

(b)

Figure 5.8 Crystallization in the system $CaMgSi_2O_6$-$CaAl_2Si_2O_8$. (a) Phase diagram for the system $CaMgSi_2O_6$-$CaAl_2Si_2O_8$. The slight slope at the left end of the solidus results from a minor deviation from strictly binary behavior (based on Bowen, 1915; Osborn, 1942). (b) Sequential disequilibrium crystallization (differentiation) of a magma of composition α, as shown in (a) above and discussed in the text.

contents. We will use the system $CaMgSi_2O_6$-$CaAl_2Si_2O_8$ (figure 5.8a) for an example, ignoring the slight deviation from strictly binary behavior that occurs because some Al combines in $CaMgSi_2O_6$ (Osborne, 1942; Schairer and Yoder, 1960; Kushiro, 1973b).

To begin, consider a liquid of composition 80% $CaAl_2Si_2O_8$ and 20% $CaMgSi_2O_6$ at α. As this liquid cools, crystals of anorthite begin to form at β, where the tempera-

ture on the liquidus is 1488° C. Continued cooling results in continued crystallization of anorthite, while the liquid changes composition along the liquidus, becoming enriched in the $CaMgSi_2O_6$ component. At point γ, the liquid l_γ has a composition of 60% $CaAl_2Si_2O_8$, 40% $CaMgSi_2O_6$. Using the lever rule, we determine that the system consists of about 50% liquid and 50% crystals of anorthite.

Suppose at this point that the liquid and crystals are entirely or almost entirely separated from one another. Geologically, this is a process of differentiation, the formation of more than one rock type from a magma. Any process in which a fraction of the magma is crystallized separately during differentiation is called fractional crystallization, and where crystals and liquid effect fractionation, the process is called crystal-liquid fractionation. Fractional crystallization may occur in several ways. One process involves the sinking or floating of all the plagioclase crystals.[8] If they accumulated at the bottom or the top of a magma chamber, they would be largely isolated from the liquid. A more effective separation would involve tectonic squeezing of the magma out of the crystal mush and into a dike, sill, or new magma chamber—a process Bowen (1928) called "differentiation by deformation" and now commonly referred to as filter pressing.[9] Alternatively, the magma could move into cracks created by deformation, in response to the lower pressure there, leaving the crystals behind. In the idealized example being described here, fractional crystallization and separation by flotation occur within the magma chamber (i.e., within the system) (figure 5.8b).

If the separated liquid subsequently crystallized completely, following a normal path of equilibrium crystallization, the final rock derived from that liquid would have the composition 60% anorthite and 40% diopside. The crystals separated earlier would be 100% An. Thus, two rocks would exist as a result of the differentiation process, one composed only of plagioclase (e.g., a hyperleucocratic gabbro or anorthosite) and another composed of plagioclase and diopside (a gabbro). Whereas equilibrium crystallization of the original magma would have produced a single rock type, a gabbro composed of 80% anorthite and 20% diopside, differentiation yields two rock types of different composition.

In nature, the processes that occur in magma chambers are far more complex than this example suggests. The final products of fractional crystallization have compositions and characters that depend on the interplay of phase equilibria, crystallization processes, and convection (Marsh, 1988b; Helz, Kirschenbaum, and Marinenko, 1989). Nevertheless, the example provides a basic view of how differentiation may affect the histories of rocks that crystallize from magmas.

Melting in Binary Eutectic Systems

Equilibrium melting follows a path exactly opposite of that followed during equilibrium crystallization. Referring back to figure 5.5, we can see that a rock of composition 40% albite plus 60% quartz, upon heating, increases in temperature until reaching point δ. During the heating, the quartz converts to tridymite at about 900° C. At the eutectic temperature (1062° C), a eutectic liquid of composition 68% $NaAlSi_3O_8$ and 32% SiO_2 forms, produced by the melting of crystals of albite and tridymite. Although heat continues to be added to the system, the temperature remains fixed because all of the added heat is the heat of fusion of the crystals. When all of the albite is melted, the temperature begins to rise. As tridymite crystals continue to melt, the liquid changes composition along the curve $E-l_\gamma-\beta$. At β, all of the crystals are gone and further additions of heat simply raise the temperature of the final liquid, which has a composition of 40% $NaAlSi_3O_8$, 60% SiO_2.

Fractional melting, sometimes also called fractional fusion, partial fusion, partial melting, or anatexis,[10] is not the reverse of fractional crystallization (Presnall, 1969). As an example, we again use the system SiO_2-$NaAlSi_3O_8$ and again we begin with a rock of composition 40% $NaAlSi_3O_8$ and 60% SiO_2 (figure 5.9a). Heat is added to the system. The temperature rises until it reaches 1062° C, the eutectic temperature. There the temperature is stabilized, as albite and tridymite begin to melt, forming a liquid (L_1) of composition 32% SiO_2, 68% $NaAlSi_3O_8$. All of the heat added contributes to fusion and is thus heat of fusion.

Several paths of fractional melting having various degrees of complexity are possible, but to keep the explanation simple we examine an admittedly special case. That case involves the separation of *all* liquid at the point in time when the last albite crystal melts. A case like this, in which blobs of magma develop and are then separated from the parent rock, is called **batch melting.** In contrast, **continuous melting** occurs where melting and separation of magma are coeval and continuous.

The liquid (L_1) that separates in this example has the eutectic composition 68% $NaAlSi_3O_8$, 32% SiO_2. After separation of the magma batch, only pure tridymite remains as a solid phase. Because pure tridymite converts to cristobalite, which melts at 1713° C, no more liquid is produced upon heating until the temperature reaches 1713° C. At that temperature, the temperature is stabilized until all cristobalite is melted to form a second liquid (L_2). Then, further heating simply raises the temperature of the liquid. A schematic temperature-time plot for this process is shown in figure 5.9b. The result of fractional melting is that two liquids of different compositions are formed, one of composition 68% $NaAlSi_3O_8$, 32% SiO_2 and one of pure SiO_2. The liquid path shows compositional breaks. This contrasts with fractional crystallization in which the crystal paths have compositional breaks but the liquid path is continuous (Presnall, 1969).

Binary Peritectic Systems

The melting of cristobalite, i.e., the transformation of solid SiO_2 to liquid SiO_2, is an example of **congruent melting.** The melted material has the same composition as the solid. In some systems, a different kind of melting, called **incongruent melting,** occurs. Substances that melt incongruently do not change from a solid of a given composition to a melt of the

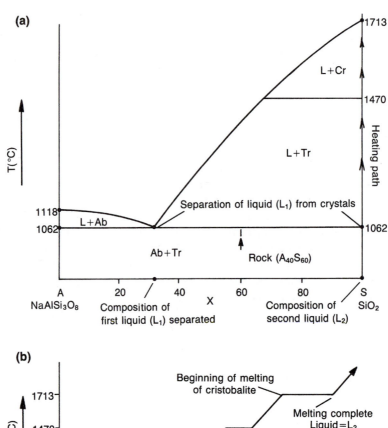

Figure 5.9   Fractional melting in the system SiO_2-$NaAlSi_3O_8$. (a) Phase diagram showing a melting path for a rock of composition $A_{40}S_{60}$ (phase diagram modified from Schairer and Bowen, 1947a, 1956). (b) Time-temperature graph for melting of the rock shown in (a).

same composition; rather, upon melting, a new solid forms as the melt develops. Both the melt and the new solid have compositions different from that of the original solid. Equilibrium crystallization in systems exhibiting incongruent melting, which are called **peritectic systems,** involves conversion of one solid to another. Two geologically important systems of this type are the systems SiO_2-$KAlSi_2O_6$ and SiO_2-Mg_2SiO_4.

The System SiO₂-KAlSi₂O₆

The peritectic system SiO_2-$KAlSi_2O_6$ is shown in figure 5.10. The Phase Rule applies here, just as it did in the systems described above. For example, at point P, the *peritectic point,*

crystals of leucite, crystals of K-feldspar, and a liquid coexist. The number of phases equals three. Pressure is constant, so the system is "condensed." The system is binary, so there are two components (two components are all that are necessary to construct all of the phases in the system). Thus,

$$F = C - P + 1$$
$$F = 2 - 3 + 1$$
$$F = 0$$

Equilibrium crystallization of compositions on the SiO_2 side of the eutectic (E) in this system, for example, a composition of $L_{30}S_{70}$ (point α), is the same as in binary eutectic systems. However, notice that the liquidus curve on the $KAlSi_2O_6$ side of the diagram has an inflection point (P), called the *peritectic point*. Equilibrium crystallization on this side of the eutectic differs in that it may follow one of four different paths.

Case 1. For compositions lying between the eutectic and peritectic compositions, between 54.4 and 42.2 weight percent SiO_2 (e.g., point β), the crystallization path is like that in a binary eutectic diagram (figure 5.10a). K-feldspar forms, beginning where the cooling path meets the liquidus at $β_1$. The liquid changes composition towards the eutectic as K-feldspar continues to crystallize. Note that the tie line only extends through a single, two-phase field, indicating equilibrium between the phases—liquid and crystals of K-feldspar. At the eutectic temperature ($β_2$), crystals of tridymite begin to form, the temperature stabilizes while K-feldspar and tridymite crystallize, and heat is lost until all the liquid is consumed. Then, the mixture of K-feldspar and tridymite continues to cool.

Case 2. Compositions between that of the peritectic (42.2% SiO_2) and the composition of K-feldspar ($KAlSi_3O_8$ = 21.6% SiO_2) crystallize in a different way. Consider equilibrium crystallization of a melt of composition SiO_2 = 30% (point γ in figure 5.10b). As the temperature drops to about 1405° C, the liquidus is encountered (point $γ_1$). Here, crystals of leucite begin to form, as indicated by the tie line. Continued cooling to 1300° C ($γ_2$) results in continued crystallization of leucite, accompanied by enrichment of the liquid in SiO_2 along the path from $γ_1$ to L_2. When the falling temperature reaches 1150° C, the peritectic temperature ($γ_3$), leucite is no longer stable. The temperature stabilizes, as leucite reacts with the liquid and K-feldspar forms. After all the leucite is gone and the new K-feldspar is formed, the temperature begins to fall again, but now, crystals of K-feldspar form directly from the liquid, which changes composition along the liquidus from P towards E (figure 5.10b). At the eutectic temperature, 990° C ($γ_5$), crystallization of tridymite joins crystallization of K-feldspar until all the liquid is consumed. During this time, the temperature is stabilized. When all the liquid is gone, the solidus is crossed and the K-feldspar–tridymite mixture cools.

Case 3. Melts with a composition of exactly 21.6% SiO_2 follow a path very similar to that in case 2. However, when the temperature reaches the peritectic temperature, there is just exactly enough liquid and $KAlSi_2O_6$ component to form K-feldspar crystals. Neither liquid nor leucite remain. The crystals of K-feldspar then continue to cool as heat is lost from the system.

Case 4. Liquids with compositions between 21.6 and 0% SiO_2 follow cooling paths like the one marked δ in figure 5.10b. At about 1630° C, crystals of leucite begin to form as the liquidus is crossed. Continued cooling renders more leucite crystals, as the liquid changes composition along the liquidus curve towards the peritectic composition. At 1150° C, the peritectic temperature, all the liquid is used up in the production of K-feldspar from leucite and the temperature remains fixed during K-feldspar crystallization. All of the leucite does not react and some leucite remains. Thus, when all of the liquid is gone, a solid composed of leucite and K-feldspar remains, the solidus has been crossed, and the temperature again begins to fall. The final product consists of crystals of leucite and crystals of K-feldspar.

Disequilibrium crystallization for compositions on the silica-rich side of the peritectic composition (42.2% SiO_2) is analogous to that in binary eutectic systems. In contrast, consider an example of disequilibrium crystallization beginning with the composition δ($L_{92}S_8$) in figure 5.10b. Just as in case 4 of equilibrium crystallization, cooling to about 1630° C will initiate the crystallization of leucite from the melt. Continued cooling yields more leucite crystals, as the liquid changes composition along the liquidus curve towards the peritectic composition. However, suppose at 1250° C ($δ_2$) the liquid and crystals are separated by one of the differentiation processes described above. The liquid has a composition of about $L_{63}S_{37}$ and the crystals are pure leucite. The liquid will now begin to crystallize as if it had an initial composition of $L_{63}S_{37}$, between the composition of K-feldspar and that of the peritectic. The crystallization sequence now follows a case 2–type path. The final product of crystallization of the liquid is a solid composed of K-feldspar and tridymite. A solid composed of the separated leucite crystals is also a product of this fractional crystallization sequence. Thus, fractional crystallization (disequilibrium) yields two rocks, one composed of leucite and one composed of K-feldspar and trydimite, whereas equilibrium crystallization would have produced a single solid composed of leucite and K-feldspar.

A similar disequilibrium crystallization sequence may be developed for case 2 compositions. Note that such compositions, under equilibrium crystallization, do not yield leucite-bearing solids (rocks). Yet early separation of leucite will yield two rocks, one composed of leucite and one composed of K-feldspar plus tridymite. Other possible products may result as well. For example, if the separated liquid has a composition on the low-silica (or leucite) side of the $KAlSi_3O_8$ composition, the final product of fractional crystallization is a pair of solids different from those in case 2.[11] In addition, separation of solid and liquid may occur continuously or more than once, a circumstance that may increase the diversity of crystallization products, i.e., resulting rock types.[12]

The geological implications of peritectic diagrams are, in part, the same as those noted for binary eutectic diagrams. Note, however, that fractional crystallization in a binary peritectic system may yield rocks that have a mineralogy different from the mineralogy that would develop from the same melt crystallized under equilibrium conditions.

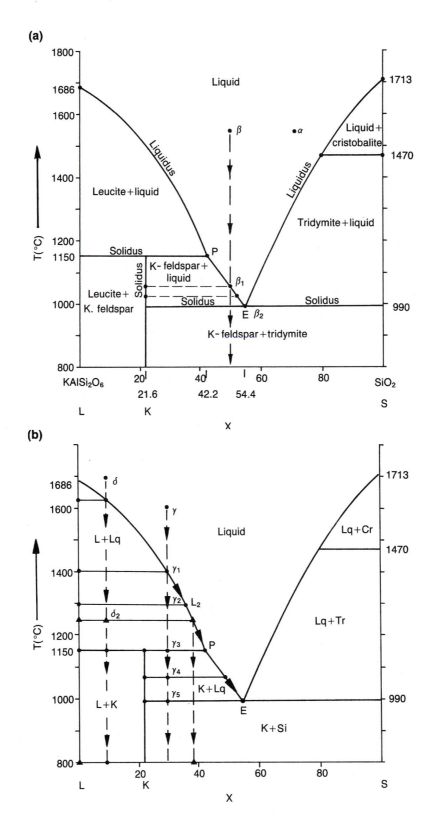

Figure 5.10   The system SiO_2-$KAlSi_2O_6$ (after Schairer and Bowen, 1947b, 1955). S = SiO_2, K = K-feldspar, L = Leucite, Lq = liquid (melt), Cr = Cristobalite, Tr = Tridymite, and Si = silica minerals (tridymite, beta quartz). α, β, γ, and δ are points discussed in the text. (a) The system SiO_2-$KAlSi_2O_6$ showing an equilibrium crystallization path for the composition β. (b) The system SiO_2-$KAlSi_2O_6$ showing equilibrium (●) and disequilibrium (▲) crystallization of melts of compositions γ and δ.

Binary peritectic systems also have important implications for the mineralogy of porphyritic volcanic rocks. Suppose that early in the crystallization of a melt of composition γ, the melt with its enclosed crystals is either extruded onto the surface or intruded at a shallow depth. The result is disequilibrium. The melt cools too quickly to react with the crystals. It may even "freeze" to form glass. Had such a melt continued to crystallize under equilibrium conditions, the final solid (rock) would have been composed of crystals of K-feldspar and tridymite. Yet, under the disequilibrium situation just described, the final rock will consist of crystals (phenocrysts) of leucite in a glassy matrix or a groundmass composed of K-feldspar ± tridymite ± glass. The phenocrysts of leucite, the most conspicuous part of the rock, may give the observer the impression that the bulk rock is quite low in silica, and has a leucite + K-feldspar bulk composition. In fact, had equilibrium been maintained, the final rock would have had no leucite, reflecting a more silicic composition. Bulk rock chemical analysis will reveal, of course, that the hand-specimen or thin-section impression is erroneous.

Disequilibrium that occurs where cooling is too rapid to allow crystals to equilibrate with the melt may also result in T-X melt paths that are erratic and cannot be predicted. Under such circumstances, phenocryst compositions may not correspond to the composition of the enclosing melt (matrix), they may show reaction zones (e.g., coronas) along their boundaries with the matrix, or they may exhibit both of these characteristics.

The point made here cannot be overemphasized for the beginning petrologist and field geologist: *Phenocryst modes do not necessarily reflect the character of the bulk chemistry of a rock, nor do they necessarily represent a coarse-grained equivalent of the groundmass mineralogy.* One final geological note is that at equilibrium, the low-silica mineral leucite does not occur with free silica (tridymite or quartz) either in this system or in rocks. This should be obvious from the phase diagram, which has no subsolidus field containing both quartz and leucite.

The System SiO₂ - Mg₂SiO₄

Melting relations in peritectic systems are analogous to those in binary eutectic systems. Equilibrium melting is the reverse of equilibrium crystallization. Fractional melting results in discontinuities in the liquid compositions, just as it does in eutectic diagrams. In this case, however, eutectic, peritectic, and endmember liquids are easily generated, as all three compositions represent temperature plateaus in the heating-history curve. In other words, assuming no solid solution, rocks will melt at invariant points—eutectics and peritectics—determined by the mineralogy of the rock. The melts thus formed will have a fixed composition. In general, only a single liquid will form in equilibrium with a given mineral assemblage.

The system SiO_2-Mg_2SiO_4 (Bowen and Anderson, 1914; Greig, 1927) also shows peritectic relations. In this system, $MgSiO_4$, $Mg_2Si_2O_6$, and SiO_2 are the three chemical species on the abscissa and the temperatures differ from those of the SiO_2-$KAlSi_2O_6$ system, but the system works in the same way. The pyroxene *clinoenstatite* melts incongruently to produce olivine (forsterite) and melt. Geologically, this system is important in considerations of both Bowen's reaction series and olivine-bearing quartz-normative basalts. At equilibrium, magnesium-rich olivine, a silica-poor mineral, does not occur with free silica (tridymite or quartz)—a situation that prevails in rocks, as well. Rather, olivine + pyroxene and pyroxene + quartz are the coexisting mineral pairs.

Binary Solid Solution Systems

Minerals that are variable in composition with respect to two or more endmember components are referred to as solid solutions. Two endmember components representing a solid solution can be used to form a **binary solid solution system.** Two important systems of this type are the systems $NaAlSi_3O_8$-$CaAl_2Si_2O_8$ (Bowen, 1913), in which the plagioclase feldspars crystallize, and Fe_2SiO_4-Mg_2SiO_4 (Bowen and Schairer, 1935), in which the olivines crystallize. Because of solid solution, mixing of the endmember components does not produce a eutectic point.

Single-Loop Systems

Equilibrium crystallization in a binary solid solution system is explained here using the system Fe_2SiO_4-Mg_2SiO_4 (figure 5.11) analyzed by Bowen and Schairer (1935). The upper curve in the diagram is the liquidus. The lower curve is the solidus. Suppose a melt of composition 20% Fe_2SiO_4 and 80% Mg_2SiO_4 at a temperature of 1900° C begins to cool (point α). At about 1820° C, the cooling path reaches the liquidus ($α_1$). A tie-line constructed between the liquidus and solidus reveals that the composition of the first olivine crystal to form (C_1) is Fa_6Fo_{94}. As the system cools, new crystals form and the old crystals react with the liquid and change composition along the solidus curve from C_1 to C_2. Iron is extracted from the melt (along with Si and O) and combined with the re-equilibrating crystals, diluting the original Mg-rich composition. While extraction of Fe would seem to enrich the melt in Mg, more Mg than Fe is extracted as new crystals form, as indicated by the curve. The result is that both curves descend towards the Fe_2SiO_4 side of the diagram, the liquid towards l_2 and the crystals toward C_2. The lever rule may be used to determine the percentages of crystals and liquid. Continued cooling brings the temperature to the solidus. Reaction with the crystals has driven the final liquid to composition l_3. The crystals end up with the composition of the initial melt ($Fa_{20}Fo_{80}$). Equilibrium melting simply follows the reverse path.

A fractional crystallization path in the binary solid solution system $NaAlSi_3O_8$-$CaAl_2Si_2O_8$ is shown in figure 5.12. Beginning with a melt of composition 20% $NaAlSi_3O_8$ and 80% $CaAl_2Si_2O_8$ at 1600° C (point β), cooling begins. At about 1518° C, the liquidus is intersected ($β_1$) and crystals

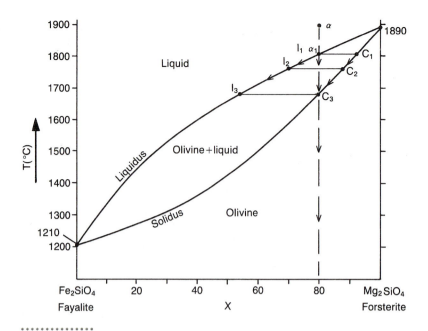

Figure 5.11 The system Fe_2SiO_4-Mg_2SiO_4 showing equilibrium crystallization of a liquid of composition α, as discussed in the text.
(Phase diagram modified from Bowen and Schairer, 1935)

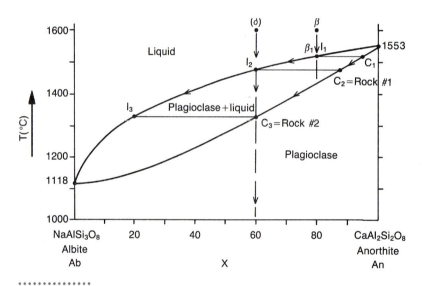

Figure 5.12 The system $NaAlSi_3O_8$-$CaAl_2Si_2O_8$ showing the fractional crystallization path for a liquid of initial composition β, as described in the text.
(Phase diagram based on the work of Bowen, 1913; Greig and Barth, 1938; and Osborn, 1942)

of plagioclase (C_1) of composition Ab_7An_{93} begin to form. As cooling continues and as long as liquid and crystals remain in contact, the liquid and crystals re-equilibrate and new crystals form, with the liquid changing composition along the liquidus toward l_2 and the crystals changing composition along the solidus toward C_2. At a temperature of 1475° C, crystals of composition C_2 ($Ab_{15}An_{85}$) are in equilibrium with a liquid of composition l_2 (40% $NaAlSi_3O_8$ and 60%

$CaAl_2Si_2O_8$). Suppose at this point that the liquid and plagioclase crystals are separated by one of the differentiation processes (e.g., filter pressing). The crystals form a solid (rock) composed of plagioclase of composition $Ab_{15}An_{85}$. The liquid begins to crystallize as if it were a liquid of original composition δ just cooling to the liquidus for the first time. New plagioclase crystals of composition C_2 ($Ab_{15}An_{85}$) then begin to form. If equilibrium now prevails until all the liquid

Figure 5.13 Photomicrograph of zoned plagioclase feldspar in andesite, Black Butte, Shasta County, California. Long dimension of photo is 3.05 mm.

is consumed, the liquid will have changed composition along the liquidus curve from l_2 to l_3. The final liquid will have a composition of 79% $NaAlSi_3O_8$ and 21% $CaAl_2Si_2O_8$. The final crystals will consist of plagioclase of composition C_3, or $Ab_{40}An_{60}$, having changed composition along the solidus from C_2 to C_3.

Had the initial liquid (β) crystallized under equilibrium conditions, the final product would consist of one solid with crystals of plagioclase of composition $Ab_{20}An_{80}$, the original composition. Instead, fractional crystallization following the path described earlier results in two solids, one composed of plagioclase crystals of composition $Ab_{15}An_{85}$ and a second composed of plagioclase crystals of composition $Ab_{60}An_{40}$. Fractional crystallization results in enrichment of one of the products in sodic plagioclase, which is the low-temperature component of the series. Multiple fractionations could greatly enrich the final product in the low-temperature component.

Fractional melting in binary solid solution systems is analogous in its products to fractional melting in binary eutectic systems. The crystals progressively change composition,

but the liquids derived from a region of melting consist of batches of magma of discrete composition.

The geological ramifications of various crystallization and melting processes in binary solid solution systems are similar, in part, to those discussed earlier. Two other points about these systems are important. First, both fractional crystallization and crystallization of melts produced by fractional fusion may result in final solid (rock) compositions that are more siliceous, more sodic, or more iron-rich (depending on the system being considered) than the original bulk composition of the parent rock or melt. Second, zoning in plagioclase feldspars may be explained.

Plagioclase feldspar crystals found in igneous rocks commonly are zoned concentrically (figure 5.13).[13] Starting at the core of the grain, successive layers surrounding the core have a more sodium-rich composition than the preceeding layer. For example, the core may have a composition of $Ab_{40}An_{60}$, each successive layer outward from this may be more sodic, and at the rim of the grain the plagioclase may have a composition of $Ab_{70}An_{30}$. Changing compositions

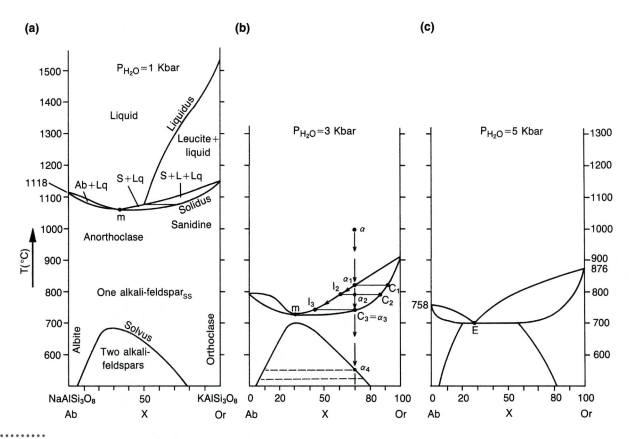

Figure 5.14 Phase diagrams for the system $NaAlSi_3O_8$-$KAlSi_3O_8$. (a) The system at P_{H_2O} = 0.1 Gpa (1 kb). (b) The system at P_{H_2O} = 0.3 Gpa (3 kb) showing equilibrium crystallization path for a liquid α, as described in text. (c) The system at P_{H_2O} = 0.5 Gpa (5 kb), ss = solid solution. *(Diagrams based on the works of Schairer, 1950; Yoder, Stewart, and Smith, 1957; Tuttle and Bowen, 1958; Orville, 1963; and S. M. Morse, 1970)*

such as this, within a single grain, may be explained by reference to the system $NaAlSi_3O_8$-$CaAl_2Si_2O_8$. During crystallization, the cores of grains (and successive rims) are unable to completely react with the liquid. The effect is one of fractional crystallization in which the cores are essentially removed from the active part of the system. Consequently, the liquid becomes more sodic and deposits a more sodic feldspar around the core. Repeated failure to attain equilibrium results in shells or zones of progressively more sodic plagioclase on the grains.

More Complex Systems: The System $NaAlSi_3O_8$-$KAlSi_3O_8$

A more complex type of binary solid solution system is represented by the system $NaAlSi_3O_8$-$KAlSi_3O_8$, which crystallizes the alkali feldspars.[14] Figure 5.14 shows the system at three different pressures—P_{H_2O} = 0.1Gpa (1 kb), P_{H_2O} = 0.3 Gpa (3 kb), and P_{H_2O} = 0.5 Gpa (5 kb). Perhaps the easiest version of the phase diagram to understand initially is that for the system at 3 kb (figure 5.14b). Obvious differences between this system and the $NaAlSi_3O_8$-$CaAl_2Si_2O_8$ system are the presence of two solid solution loops that meet at a

minimum and the presence of a curve called the solvus within the solid field. The **solvus** is a curve connecting the points representing compositions of pairs of coexisting feldspars. Above the solvus there is complete solid solution (ss) between albite and K-feldspar. Thus, for bulk compositions at temperatures that plot between the solidus and the solvus, one feldspar solid solution (feldspar$_{ss}$) is stable. Below the solvus, two feldspars are stable. During cooling of compositions that project into the subsolvus field, the one feldspar stable above the field, upon cooling, will exsolve into two feldspars.

Equilibrium crystallization of a melt of composition 30% $NaAlSi_3O_8$ and 70% $KAlSi_3O_8$ (α in figure 5.14b) follows a temperature-descent path typical for solid solution systems. When the liquidus is crossed at (α_1), crystals of K-rich alkali feldspar (Ab_7Or_{93}) begin to form. The liquid and the crystals react and re-equilibrate, and new crystals form as the temperature falls. The composition of both the crystals and the liquid change towards the minimum m. When the crystals reach a composition of $Ab_{30}Or_{70}$, the initial composition of the melt, all the liquid is consumed. The last liquid is Na-enriched, with a composition near that of the minimum. With continued heat loss, the one feldspar$_{ss}$ cools and reaches the solvus (at α_4). Here, two feldspars separate from the one. In this case, albite$_{ss}$ exsolves and crystallizes along crystallographically controlled sites in the larger (host) grain of K-feldspar$_{ss}$, thus forming *perthitic alkali feldspar* (Brown and Parsons, 1983).[15] Had the initial composition of the melt been sodic, exsolution lamellae of K-feldspar$_{ss}$ would develop in albite$_{ss}$ to form *antiperthite*. Successive tie-lines link feldspars in equilibrium, as they change composition. Exsolution will continue as long as temperatures, cooling rates, and the microstructural features of the crystals allow effective ion migration (W. L. Brown and Parsons, 1984).

Two variations of this binary diagram are shown in figures 5.14a and 5.14c. At low water-vapor pressures, the system is complicated by incongruent melting of K-feldspar to form leucite (Goranson, 1938; Tuttle and Bowen, 1958). At pressures (P_{H_2O}) slightly less than 0.5 Gpa (5 kb) and higher (figure 5.14c), the solid solution fields are depressed in temperature to the extent that they intersect the solvus, forming a eutectic (E). Crystallization in such a system results in direct crystallization of two alkali feldspars from the melt, with minor subsequent exsolution. Again, note that increases in pressure greatly depress the temperature of the liquidus and the minimum, whereas, conversely, decreases in pressure increase the temperature of the liquidus and the minimum. In nature, sudden drops in pressure (e.g., due to fracturing of the country rock around a magma body) result in sudden quenching (freezing) of the magma.

In addition to explaining perthitic and antiperthitic textures, the system $NaAlSi_3O_8$-$KAlSi_3O_8$ helps us recognize low-P_{H_2O}/high-T and high-P_{H_2O}/low-T feldspathic rocks. Tuttle and Bowen (1958) used the term *hypersolvus* to describe plutonic, feldspathic rocks characterized by one (perthitic) feldspar. The phase relations suggest that these must form at low fluid pressures (or high temperatures). *Subsolvus* feldspathic, plutonic rocks are those characterized by two alkali feldspars (plagioclase and K-feldspar), which by reference to the phase diagram form at higher fluid pressures or lower temperatures.[16] Note also that the general absence of leucite in plutonic rocks is probably explained by the marked reduction of the leucite field under elevated water-vapor pressures.

TERNARY AND OTHER MULTICOMPONENT SYSTEMS

Ternary, quaternary, and quinary systems are important to petrology because they provide more realistic approximations of rock systems. A few of these diagrams will be discussed briefly here, but the interested reader is referred elsewhere (Ricci, 1966; Ernst, 1976; Cox et al., 1979; Morse, 1980) for more detailed treatments.

Ternary (three-component) diagrams present a graphical problem. With three components plus one intensive variable to plot, it is difficult to present an easily read, two-dimensional diagram of a ternary system. If the three components are plotted as the apices of a triangle, the temperature must be plotted in a third dimension (figure 5.15a). Pressure is typically held constant. The liquidus in such a diagram consists of sloping surfaces. In order to transform this three-dimensional prism with slopes into a usable two-dimensional plot, temperature contours, equivalent to elevation contours on a topographic map, (figure 5.15b) are projected onto the composition triangle (figure 5.15c). The solidus is not projected. Each apex of the composition triangle represents 100% of the labelled component, the opposite side being equal to 0% of that component. Percentages of components at any spot on the diagram are read just as are percentages in triangular classification diagrams and variation diagrams.

The simplest type of ternary diagram is a ternary eutectic diagram in which none of the three components is involved in solid solutions. Figure 15.16 depicts an imaginary system of this type. The three components A, B, and C each form a binary eutectic system with one of the other components. Thus, each vertical side of the three-dimensional prism (figure 5.15a) is a binary eutectic system. Each binary eutectic (e) heads a thermal "valley" on the ternary diagram, marked by a **cotectic line** representing a temperature-composition curve along which *two* solid phases crystallize. The three cotectic lines (thermal valleys) meet at the ternary eutectic (E). Arrows point in down-temperature (downhill) directions. The areas separated by the cotectic lines represent liquidus conditions for which crystals of *a* plus liquid, crystals of *b* plus liquid, and crystals of *c* plus liquid, respectively, are stable (figure 5.16).

Equilibrium crystallization paths in such a system are easily traced. Consider a composition α (60% A, 30% B, 10% C) at 1400° C. As the system is cooled, the temperature drops until the cooling path (a vertical line extending down towards the page from a point in space above α) intersects the liquidus at point α (figure 5.16). Crystals now begin to form and the label in the field in which α lies indicates that those crystals are crystals of *a*. As crystals of *a* are being extracted from the melt, the composition of the melt, indicated initially by the point α, must move directly away from A (away from α along the linear extension of the straight line extending from A to α). As crystals of *a* form, the liquid composition continues to move away from A until the liquid path intersects the cotectic line at β. Here, crystals of *b* (indicated by the adjoining field) begin to crystallize along with crystals of *a*. Because both A and B are being removed from the liquid, the liquid composition follows the cotectic (down the thermal valley) away from both A and B, towards E, the ternary eutectic. When the liquid composition reaches E, all three solids—a, b, and c— crystallize together, the temperature remains at the eutectic temperature, and the composition of the liquid remains fixed at the eutectic composition until all the liquid is gone. Then the

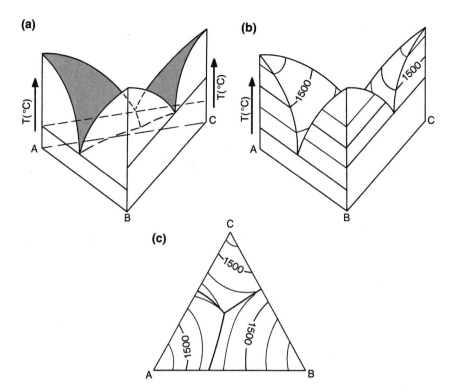

Figure 5.15 Sketches of generalized ternary systems. (a) Sketch showing the three-dimensional nature of a ternary T-X phase diagram. Note the sloping liquidus surfaces. (b) Contoured version of the ternary T-X phase diagram shown in (a). (c) Sketch showing the contoured ternary liquidus surfaces projected onto a two-dimensional triangular phase diagram (view is down the T-axis from above).

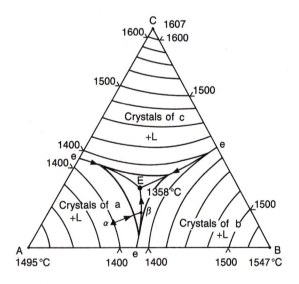

Figure 5.16 Imaginary ternary system A-B-C. Equilibrium crystallization of composition α (▲), following the path α → β → E, is discussed in the text. L = liquid (melt).

solid, composed of crystals of *a*, crystals of *b*, and crystals of *c*, simply cools down. The final bulk composition mimics the initial liquid composition; that is, its composition is 60% A, 30% B, and 10% C, and it is composed of crystals of *a*, *b*, and *c* in the ratio 60:30:10.

The foundation of understanding provided by this imaginary system is used below to explore two systems useful in understanding two major rock types—the basalts and granites. Neither of the two systems is actually ternary,[17] but a ternary-like plot may be used with some success to discuss graphical and petrologic relationships relating to these systems.

A Basalt Analogue

The system $CaMgSi_2O_6$-$CaAl_2Si_2O_8$-$NaAlSi_3O_8$ (figure 5.17), which involves plagioclase and a clinopyroxene, is a simple analogue for dry basalt (or gabbro). Note that the $CaAl_2Si_2O_8$-$CaMgSi_2O_6$ (anorthite-diopside, or An-Di) side of the diagram is represented by a binary eutectic-type diagram (figure 5.17a), as is the albite-diopside side (Ab-Di). The Ab-An side is a simple binary solid solution system (figure 5.17a). The composite diagram has a single cotectic line (thermal valley) extending from the Ab-Di binary eutectic to the Di-Ab eutectic, with the Ab end being the low-temperature end of the cotectic.

Two general types of crystallization path are possible under equilibrium conditions in such a system, one in which diopside crystallizes first and one in which plagioclase

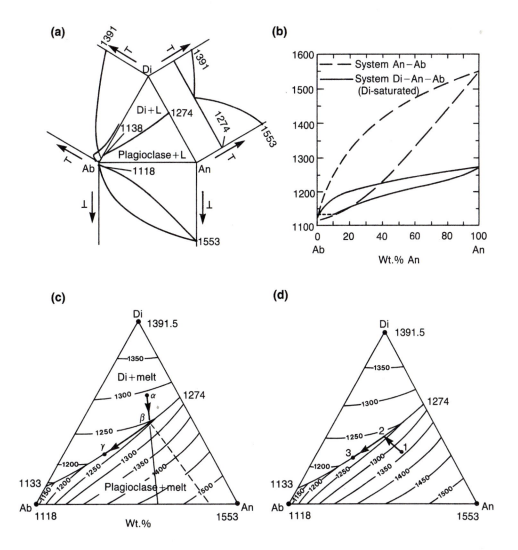

Figure 5.17 The ternary system $CaMgSi_2O_6$-$CaAl_2Si_2O_8$-$NaAlSi_3O_8$. (a) The three binary systems that form the sides of the system "unfolded" into view. (b) The system $CaAl_2Si_2O_8$-$NaAlSi_3O_8$, pure (dashed) and saturated with $CaMgSi_2O_6$ (solid), showing the depression of the solid solution loop resulting from addition of diopside to plagioclase (from S. A. Morse, 1980). (c) Crystallization path for a liquid α, rich in

the component $CaMgSi_2O_6$ (diopside). See the text for discussion. (d) Crystallization path for a liquid 1, rich in $CaAl_2Si_2O_8$. See the text for a discussion.

[Diagrams in (a), (b), and (d) based on the works of Bowen (1915), Osborn (1942), Osborn and Tait (1952), Schairer and Yoder (1960), and Kushiro (1973)]

crystallizes first. In either case, the exact compositions of the phases and in the latter case the exact liquid descent path *must be determined experimentally*.[18] Consequently, the examples given below are general approximations. One reason is that the addition of $CaMgSi_2O_6$ to the $NaAlSi_3O_8$-$CaAl_2Si_2O_8$ system drastically alters the configuration of the solid solution field (Morse, 1980) (figure 5.17b). The melting temperature of the An-Di eutectic is about 275° C lower than that of pure anorthite and the solid solution loop is compressed.

Consider equilibrium crystallization of a melt of composition 16% $NaAlSi_3O_8$, 24% $CaAl_2Si_2O_8$, and 60% $CaMgSi_2O_6$ (point α in figure 5.17c). This composition would yield a final plagioclase of composition An_{60}, as

revealed by extending a line from Di, through α, to the Ab-An join. Initial cooling will bring the melt down to the temperature of the liquidus (about 1295° C), at which point crystals begin to form. Inasmuch as the initial composition (α) projects into the diopside + melt field, the first crystals to form at the liquidus are diopside (diopside is the liquidus phase). Crystallization results in extraction of diopside component from the melt, and the melt composition moves directly away from Di down the liquidus towards β. At β, crystals of plagioclase begin to form and diopside, melt, and plagioclase coexist. Although the exact plagioclase composition must be determined experimentally, the initial plagioclase will have a composition of about An_{85}. Crystallization of An-rich plagioclase and diopside will drive the liquid composition away from both Di and An, along the cotectic towards Ab. Under the equilibrium conditions being considered, as crystallization of both diopside and plagioclase crystals continues, the plagioclase will react with the liquid to become more sodic. The newly forming crystals will also be more sodic. Thus, the plagioclase composition will change towards its final value of An_{60}. When it reaches that value, the last liquid will have been consumed, its final composition being about An_{27} (point γ in figure 5.17c). Further cooling simply cools the diopside-plagioclase rock. Usually, as in this example, the liquid composition will not reach the Di-Ab side of the triangle during equilibrium crystallization.

Equilibrium crystallization of compositions that lie on the plagioclase side of the cotectic is more complex. Consider a melt that plots at point 1 in figure 5.17d, having a composition of 28% $NaAlSi_3O_8$, 42% $CaAl_2Si_2O_8$, and 30% $CaMgSi_2O_6$. As in the previous example, the final plagioclase will have a composition of An_{60}. As cooling occurs, the temperature-descent path of the melt intersects the liquidus at point 1 at a temperature of about 1325° C. The liquidus phase, plagioclase, begins to crystallize. Again, the first crystals to form will have a composition of about An_{85}. A line connecting that composition and the initial melt will be tangent at point 1 to an arcuate melt path (determined experimentally) that will be followed by the melt as plagioclase crystallizes. The plagioclase will react with the melt to become more sodic, changing in composition towards the final composition of An_{60}. The melt will follow the arcuate path 1–2. The arcuate path develops because the melt composition always moves away from thc composition of the plagioclase, but the plagioclase composition changes continuously as it reacts with the liquid and becomes more sodic.

At point 2, the melt composition intersects the cotectic. At this point, the plagioclase has a composition of about An_{75}. Diopside begins to crystallize along with plagioclase and the melt composition moves down the cotectic. When the melt composition reaches point 3, all the liquid is consumed. The final liquid is sodium-rich and the final solid consists of 30% diopside crystals and 70% plagioclase crystals of composition An_{60}. With further cooling, there are no further changes in the amounts or types of phases. The solid simply cools. Note that, again, the final melt does not reach a composition on the $NaAlSi_3O_8$-$CaMgSi_2O_6$ join. The final product of crystallization in this system, a plagioclase-pyroxene rock, is a simple mineralogical and chemical analogue for basalt or gabbro.

Basalts and gabbros commonly contain olivine and orthopyroxene as well as intermediate plagioclase and clinopyroxene. Consequently, to describe crystallization in natural basalts, several additional and more complex systems have

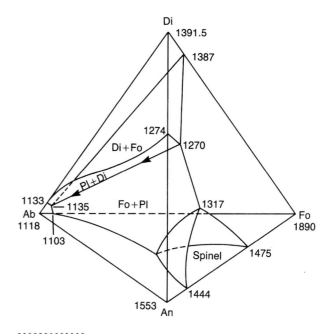

Figure 5.18   The quaternary system Mg_2SiO_4-$CaMgSi_2O_6$-$CaAl_2Si_2O_8$-$NaAlSi_3O_8$. **P** = constant.
(From H. S. Yoder, Jr., and C. E. Tilley, "Origin of Basaltic Magmas: An experimental study of natural and synthetic rock systems" in Journal of Petrology, 3:342-532, 1962. Copyright © Oxford University Press, Oxford England, Reprinted by permission.)

been studied. Of note, Yoder and Tilley (1962) discussed the origin of basalts in terms of the quaternary system Mg_2SiO_4-$CaMgSi_2O_6$-$CaAl_2Si_2O_8$-$NaAlSi_3O_8$ shown in figure 5.18. In this and similar systems, temperature axes or curves are no longer shown, pressure is held constant, and the liquidus surfaces are shown in a three-dimensional sketch with temperatures of important points labelled. Specific crystallization paths and compositions may be examined by constructing isothermal (single-temperature) planes or individual compositional planes through the tetrahedron.

In the basalt system shown here, one important feature is the four-phase curve, melt + olivine (Fo) + pyroxene (Di) + plagioclase (Pl), that extends across the diagram from a temperature of 1270° C to one of 1135° C near the albite corner. Crystallization will drive melts down that curve, enriching them in the sodic component, as in the ternary system. Fractionation, however, may produce final rock compositions of markedly different character.

Returning to the ternary systems, fractional crystallization in the system $NaAlSi_3O_8$-$CaAl_2Si_2O_8$-$CaMgSi_2O_6$, as in the binary subsystems, results in both mafic and/or calcic *and* Na-rich end products that are markedly different

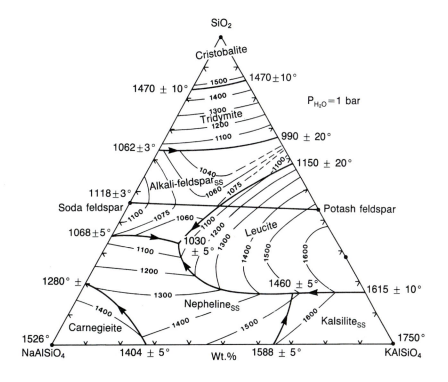

Figure 5.19 The system $NaAlSiO_4$-$KAlSiO_4$-SiO_2(-H_2O). The anhydrous system at about P= 0.1 Mpa (1 bar) modified from Schaier, 1950).

(After Zeng and MacKenzie, 1984)

in composition from those derived via equilibrium crystallization. The fractional crystallization paths are similar (but not identical) to those described for the basalt system, but the melt composition progresses down the cotectic to the minimum at the Di-Ab join. Separation of phases crystallized early results in diopside-rich and/or Ca-plagioclase-rich rocks. Such crystal-liquid fractionation may have analogues in more complex natural systems that yield pyroxene-rich and plagioclase-rich layers observed in mafic-ultramafic complexes.[19]

Petrogeny's Residua System and the Granites

The ternary system $NaAlSiO_4$-$KAlSiO_4$-SiO_2 shown in figure 5.19 encompasses the compositions of quartz, the alkali feldspars, leucite, and nepheline. These minerals include the dominant constituents of granites and rhyolites, syenites and trachytes, and nepheline syenites and phonolites. Bowen (1928, 1937, 1948) argued that the general paths of fractional crystallization lead toward melts that crystallize these phases, and thus they represent the crystallized liquid *residue* of the fractional crystallization process. Following Bowen, the system SiO_2-$NaAlSiO_4$-$KAlSiO_4$ is commonly referred to as "petrogeny's residua system."[20]

The system was studied at about 0.1 Mpa by Schairer (1950).[21] In figure 5.19, note the liquidus fields for (1) the extremely rare minerals kalsilite and carnegieite; (2) the most common feldspathoid, nepheline; and (3) the minerals leucite, alkali feldspar, and the silica polymorphs—tridymite and cristobalite. Cotectic lines lead to minima at about 990° C, involving tridymite and alkali feldspars, and at 1030° C, involving nepheline and alkali feldspars. Note also that a thermal ridge separates most of the silica-oversaturated compositions from the silica-undersaturated compositions. This ridge coincides with the alkali-feldspar join. The binary minimum in the $NaAlSi_3O_8$-$KAlSi_3O_8$ system is the lowest temperature on the ridge. Viewing figure 5.19 as a contour map with thermal topography clearly reveals that crystallization, especially fractionational crystallization, will drive compositions downhill towards one of the minima (figure 5.19). Thus, fractionation should produce feldspathoidal rocks composed principally of nepheline, leucite, and alkali feldspar, granitoid and siliceous volcanic

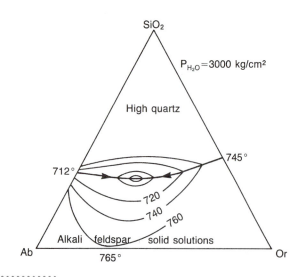

Figure 5.20   The system $NaAlSi_3O_8$-$KAlSi_3O_8$-SiO_2(-H_2O) at 0.294 Gpa (2.94 kb).
(After Tuttle and Bowen, 1958)

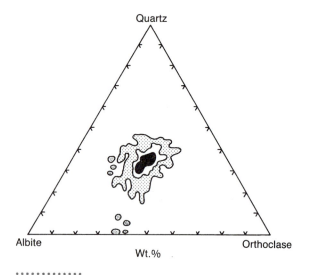

Figure 5.21   The quartz–albite–K-feldspar (orthoclase) triangle showing contours enclosing the area containing plotted points for the quartz, albite, and orthoclase ratios of all 1269 rocks, in Washington's (1917) tables, that consist of 80% or more of these minerals. Note the coincidence of the maximum concentration (dark central area) here, with the ternary minimum in the corresponding systems shown in figures 5.19a (upper triangle) and 5.20.
(From Tuttle and Bowen, 1958)

rocks composed of quartz and alkali feldspar(s), and to a lesser extent syenites and trachytes composed of alkali feldspars. The same is true for the system at 0.5 Gpa (5 kb), except that leucite is much less likely to be present because of the reduced size of the leucite field.

The effects of elevated water-vapor pressures on the SiO_2-oversaturated part of petrogeny's residua system was studied by Tuttle and Bowen (1958) as the four component system $NaAlSi_3O_8$-$KAlSi_3O_8$-SiO_2-H_2O. Their work indicates that with increasing P_{H_2O} the leucite field is reduced and then eliminated at about P_{H_2O} = 0.3 Gpa (3 kb)(figure 5.20). At higher pressures, low quartz and the alkali feldspars become liquidus phases.

Tuttle and Bowen (1958) also demonstrated that a large number of analyses of both plutonic and volcanic rocks containing 80% or more normative quartz + albite + orthoclase, if plotted on the SiO_2- $NaAlSi_3O_8$-$KAlSi_3O_8$ projection, lie near the minimum determined for low water-vapor pressures (figure 5.21). This result was later duplicated using over 7,600 analyses from around the world.[22] The implication is that most granitic and rhyolitic rocks crystallized from *magmas* at low water-vapor

pressures. Similar plots for nepheline normative rocks show that most of them, likewise, plot near the minima on and below the Ab-Or join, suggesting that they too have crystallized from magmas.[23]

BOWEN'S REACTION SERIES

Based on the phase relations he had studied, Bowen (1928) developed the "reaction principle," specifically, that during crystallization, crystals and liquid react. Systems in which crystal compositions changed continuously from one composition to another—for example, the plagioclase feldspars—were called continuous reaction series. Reaction relations in which a crystal of one composition reacts with a liquid to produce another phase of different composition (a phase that undergoes incongruent melting) belong to

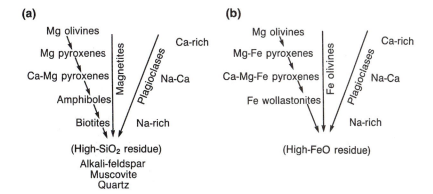

(a)

Mg olivines

Mg pyroxenes

Ca-Mg pyroxenes

Amphiboles

Biotites

Magnetites

Plagioclases

Ca-rich

Na-Ca

Na-rich

(High-SiO$_2$ residue)
Alkali-feldspar
Muscovite
Quartz

(b)

Mg olivines

Mg-Fe pyroxenes

Ca-Mg-Fe pyroxenes

Fe wollastonites

Fe olivines

Plagioclases

Ca-rich

Na-Ca

Na-rich

(High-FeO residue)

Figure 5.22 Reaction series in igneous rocks. (a) Bowen's reaction series (modified from Bowen, 1928). (b) Osborn's reaction series (modified from Osborn, 1962b; 1979).

discontinuous reaction series. The Mg–olivine→pyroxene reaction is discontinuous. Bowen suggested (rightly) that presenting reaction series in such a simplified form would facilitate understanding.

Combining known phase relations with other data, including corona textures and igneous rock-mineral sequences, Bowen (1922b, 1928) developed the reaction series now commonly called Bowen's reaction series, shown in figure 5.22a. Pyroxene coronas on olivine grains (figure 5.23) reflect known reaction relations in the system Mg$_2$SiO$_4$-SiO$_2$. Therefore, Bowen inferred that corona relations involving other pairs of minerals (e.g., amphibole coronas on pyroxene) reflect other reaction relations. In addition, he suggested that in gabbro-granite sequences, the order of appearances and disappearances of the minerals further indicates reaction series relations. It is now known that reaction series other than Bowen's reaction series exist (E. F. Osborne, 1962a, 1962b, 1979; figure 5.22b).

The reaction series reveal general mineralogical paths of fractional crystallization. They also suggest both simplified representations of early differentiates (e.g., rocks composed of Mg-olivines and Ca-plagioclase) and the compositional direction of fractional crystallization, towards enrichment in alkalis, FeO, or SiO$_2$.

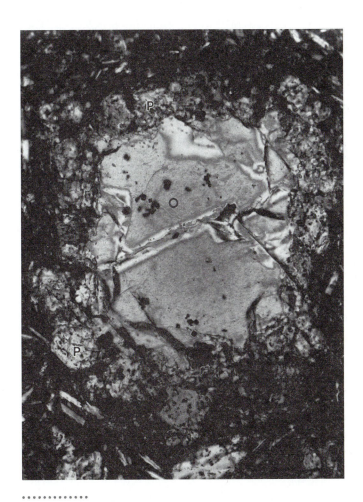

Figure 5.23 Photomicrograph of a corona texture in andesite, with pyroxene (P) rimming olivine (O). The andesite is from Makushin Volcano, Unalaska Island, Alaska. Long dimension of photo is 1.20 mm.

SUMMARY

Unary, binary, ternary, and more complex phase diagrams are useful for depicting crystallization and melting histories and the relationships between phases (magmas and minerals) and intensive variables during magma evolution. The Phase Rule, $F = C - P + 2$, simply states the relationship between phases, components, and degrees of freedom in such diagrams.

Laboratory experimentation to determine phase relations greatly aided Bowen's understanding of igneous processes and rocks. For example, the system $CaMgSi_2O_6$-$CaAl_2Si_2O_8$-$NaAlSi_3O_8$, which involves plagioclase and a clinopyroxene, is a simple analogue for dry basalt or gabbro. Similarly, the system $NaAlSiO_4$-$KAlSiO_4$-SiO_2, commonly referred to as petrogeny's residua system, is used to depict the magmatic histories of granites, rhyolites, syenites, trachytes, and feldspathoidal rocks. Those of us who follow Bowen find phase relations essential to understanding petrogenesis. They provide us with simplified models of petrologic systems, drawing our attention to the variety of possible magma compositions that can result from melting, the processes of fractional crystallization, and the paths that may be followed by natural materials as they change from rock to magma to rock.

EXPLANATORY NOTES

1. Bowen (1928) summarized much of his early work on phase diagrams in his classic work *The Evolution of the Igneous Rocks*. More detailed treatments of phase diagrams are provided by Ricci (1966), E. G. Ehlers (1972), and Ernst (1976). The last text provides a clear treatment of the subject as it relates to petrology for those students who have an introductory background in calculus, physical chemistry, or thermodynamics. Also see Muan (1979) in the 50th anniversary volume in honor of Bowen's classic work.

2. The phase rule was originally derived in the nineteenth century by the famous J. Willard Gibbs, whose works have been reprinted (see Gibbs, 1961, in Ehlers, 1972). Derivations of this rule are available in Ricci (1966), Krauskopf (1967), Kern and Weisbrod (1967), Ernst (1976), and Morse (1980).

3. Students should not get into the practice of referring to systems by mineralogical names. The system SiO_2 is *not* the "quartz system," as quartz is only one of the SiO_2 minerals that occurs in the system SiO_2.

4. Technically, some binary-type behaviors are exhibited by this system.

5. A full discussion of the thermodynamic cause for this effect is beyond the scope of this text, but the interested student with some background in chemistry and calculus will find the discussion by Ernst (1976) illuminating. Ernst shows that the reasons that water depresses the liquidus are somewhat different from the reasons involved where two solid phases are mixed, but the phenomenon is similar.

6. Convenient collections of phase diagrams are included in Morey (1964) and Levin, Robbins, and McMurdie (1964).

7. Recall that quartz inverts to tridymite at higher T, and subsequently tridymite inverts to cristobalite.

8. In older work, it was generally assumed that solid phases would sink in a magma. However, see the suggestions of Larsen et al. (1935) and Wager and Deer (1939, pp. 282, 332), the review of the floating plagioclase problem in Irvine (1980a) and the references therein, and the works of Murase (1982) and Kushiro (1980, 1982), which suggest that under certain conditions plagioclase may float rather than sink. That all the crystals would float or sink is unlikely. I. H. Campbell (1978) and McBirney and Noyes (1979) raised questions about the likelihood of *any* crystal settling in mafic magma chambers, but their conclusions have been challenged (Martin and Nokes, 1988; Martin, 1990). B. D. Marsh (1988a) discusses details of crystal fractionation and capture of floating and sinking crystals in cooling magma chambers, emphasizing the complexities that exist in nature.

9. Bates and Jackson (1980).

10. Presnall (1969) differentiates these terms. Partial fusion and partial melting denote melting of part of a preexisting rock. Similarly, in this book, anatexis refers simply to melting of preexisting rock (cf. Bates and Jackson, 1980). Fractional melting (fractional fusion) refers to a melting process in which liquids become separated from the solid rock from which they were derived. Presnall's (1969) definition of fractional fusion differs from this definition in that it specifies that during fractional fusion liquids are separated immediately, as they are formed, from the solid parent.

11. Analysis of this path of crystallization is left to the student.

12. Marsh (1988a; 1990) argues that the diversity of rock types created by fractional crystallization will be limited, a view challenged by Sparks (1990).

13. Zoning in plagioclase is often complex. Zoning and the processes that yield zoning have been described by a number of petrologists and mineralogists including Larsen and Irving (1938), Vance (1962), and Allegre, Provost, and Jaupart (1981).

14. Aspects of this system have been studied by several workers including Gorenson (1938), Schairer (1950), Bowen and Tuttle (1950), Yoder, Stewart, and Smith (1957), Tuttle and Bowen (1958), Orville (1963), Luth and Tuttle (1966), S. A. Morse (1970), I. Parsons and Brown (1983a, b), and W. L. Brown and Parsons (1983, 1984). As noted by Ernst (1976), this system is "pseudobinary" because H_2O dissolves in the melt phase and leucite is actually a part of a ternary system.

15. Other aspects of feldspar exsolution are discussed by Lofgren and Gooley (1977), I. Parsons (1978), and Brown and Parsons (1984).

16. Addition of calcic plagioclase to this and related systems influences both the crystallization process and the history (see Nekvasil, 1988; R. H. Jones and MacKenzie, 1989). Other aspects of hypersolvus versus subsolvus rocks are discussed by Tuttle and Bowen (1958, pp. 137–138).

17. Morse (1980) discusses why the system $NaAlSi_3O_8$-$CaAl_2Si_2O_8$-$CaMgSi_2O_6$ is not ternary but quinary (having five components). See Tuttle and Bowen (1958), as well as Fudali (1963), Ernst (1976), Gittins (1979), and Morse (1980), for discussion of aspects of the system SiO_2-$NaAlSiO_4$-$KAlSiO_4$.

18. See Ehlers (1972), Ernst (1976), and Morse (1980). Note that both diopsidic pyroxene and plagioclase will actually be solid solutions.

19. See discussion in chapter 10.

20. For example, see Thornton and Tuttle (1960), Krauskopf (1967), or Morse (1980).

21. Also see Schairer (1957) and Fudali (1963). Ernst (1976) and especially Morse (1980) provide good introductions to the system for students.

22. See Best (1982, p. 114, fig. 4.10), who cites the work of Le Maitre.

23. See D. L. Hamilton and MacKenzie (1965) and also Gittins (1979).

PROBLEMS

5.1. Using figure 5.5, describe (a) a crystallization history for a melt composed of 50% SiO_2 and 50% $NaAlSi_3O_8$, and (b) a crystallization history for a melt composed of 32% SiO_2 and 68% $NaAlSi_3O_8$.

5.2. Using figure 5.10, describe (a) the complete melting of a syenite, composed only of alkali feldspar (K-feldspar), and (b) the fractional melting of a syenite in which the first melt is removed as a batch at the point that melt constitutes 75% of the system. (c) What would be the composition of the melt extracted in (b)?

5.3. Through fractional melting, would it be possible to develop a $CaAl_2Si_2O_8$-rich liquid by melting albite (see figure 5.12)? Explain.

CHAPTER 6

Petrogenesis, Movement, and Modification of Magmas

INTRODUCTION

If the Earth is dominantly a solid rather than a molten mass, magmas must form somewhere by melting or partial melting of rock, that is, by anatexis. Equally obvious is the fact that molten rock moves. For example, we see movement on the Earth's surface in lava flows. Abundant evidence also indicates that magmas move through the mantle and crust. During movement, magma compositions are commonly modified. The origin, movement, and modification of magmas are the topics of this chapter.

PRIMITIVE, PRIMARY, AND PARENTAL MAGMAS

In considering the origins of magmas, it is useful to distinguish between primitive, primary, and parental magmas. **Primitive magmas** are unmodified magmas that form through anatexis of mantle rocks that have not been melted or otherwise changed in composition since they formed. **Primary magma** is any chemically unchanged, anatectic melt *derived from any kind of preexisting rock.*[1] Thus, because primary magmas

may be derived from melting mantle, as well as crustal rocks, all primitive magmas are primary, but most primary magmas that yield modern rocks are not primitive. **Parental magmas** are magmas that have given rise, in the liquid state, to other (derivative) magmas. Each primitive or primary magma may become a parental magma, if, through fractional crystallization or some other process, it yields one or more **derivative magmas** (magmas derived directly from a preexisting magma).

The distinctions made here are important in discussions of magma genesis. Some magmas that formed very early in Earth history were undoubtedly primitive. Over time, however, anatexis of mantle rock has resulted in removal of a large fraction of certain elements from large parts of the upper mantle, leaving it in a depleted state.[2] Such mantle is sometimes said to be barren, depleted, or infertile; particularly if it can no longer yield basaltic or more feldspathic magma because it lacks significant alkalis and aluminum. Any magmas derived from these depleted mantle rocks are primary but not primitive. Where primitive mantle still exists, it could yield a primitive magma. Yet, because the nature of primitive mantle rock is conjectural, demonstrating the existence of primitive magmas, or rocks derived from them, is a difficult task.

In addition to being changed through anatexis and depletion, it is likely that mantle rocks also become modified through addition of materials. This is called *enrichment* or *metasomatism*.[3] For example, lithospheric plate subduction may carry rocks rich in silicon, aluminum, and other elements down into the mantle where these elements are mixed with mantle rocks. As a consequence, discovery of primitive materials becomes even more difficult because of the difficulty of distinguishing between altered and unaltered mantle rock. Depletion and enrichment both render the mantle heterogeneous.[4]

MAGMA GENESIS

Understanding the origin of primitive and primary magmas and the development of derivative and modified magmas may be approached through a series of questions and answers. The queries are relatively easy to formulate, but the answers are more difficult. Basic questions include the following.

1. What igneous rocks are observed, the origin of which must be explained?
2. What magmas gave rise to these rocks?
3. What rocks gave rise to the magmas?
4. How were the magmas formed?

If we are to ultimately explain the origin of all the igneous rocks observed, we must also answer additional questions. These include

5. How and to where do magmas move? and
6. How are magmas modified in composition?

Note that the questions posed progress from what we can observe through a series of inferences. The evidence that provides the basis for these inferences is mineralogical, geophysical, and chemical. Isotopes, in particular, aid in discriminating between possible origins and path histories of magmas.

Igneous Rocks and Parental Magmas

What igneous rocks are observed? The answer to this question is, of course, any of the known igneous rocks (the main types were named and defined in chapter 4). Some igneous rock types are more abundant than others, however. Only 17% of the rocks exposed at the surface of the continents is igneous, and these rocks are about equally divided between intrusive and extrusive types (Blatt and Jones, 1975). Rhyolites, basalts, and andesites are particulary common eruptive rocks, and granitoid rocks are the dominant intrusive rocks. On the ocean floor the percentage of igneous rocks exposed at the surface is probably similar to that of the continents. The Deep-Sea Drilling Project (DSDP) has revealed that much of the igneous and meta-igneous rock constituting the oceanic crust is covered by layers of sediment and sedimentary rock. The igneous crustal rocks beneath that sediment are dominantly basalts and gabbros.

Estimates of rock volume are based on a combination of surface distribution data and a crustal model. Ronov and Yaroshevsky (1969) estimated that basalts and related rocks, plus granitoid rocks, dominate the crust. Other igneous rocks are not abundant. Basalts, because of their great abundance, have long been considered to represent primary magmas (Bowen, 1928).[5] Andesitic and granitoid compositions have also been considered to be primary.[6] As discussed in later chapters, other less abundant igneous rocks are generally thought to be crystallized from derivative magmas.

What magmas give rise to the known igneous rocks? Based on estimated volumes of rock, a preliminary answer is that basaltic, andesitic, and granitoid magmas yield most of the observed rocks. Under special circumstances, other primary magmas may also develop and give rise to specific rock types.[7]

Source Rocks

If magma forms by anatexis, what preexisting rock was melted to yield the magma? One approach to answering this question is to find where magmas come from and to discover what rocks exist in those regions. A second approach is to infer the source rock on the basis of the chemistry of the magma or the magmatic rocks.

Taking the first approach, we find that magmas come from the mantle and the lower crust. Support for this conclusion is derived from (1) seismic evidence, (2) studies of inclusions in basalts and other lavas, and (3) chemical data.

Seismic Indications of Magma Sources

Observations at the Hawaiian Volcano Observatory (D. A. Swanson et al., 1976) and elsewhere reveal a close association between volcanism and seismic activity. In Hawaii, pre-eruption earthquakes seem to define a conduit leading from a depth of 60 km in the upper mantle to the surface (Eaton and Murata, 1960; Koyanagi and Endo, 1971). Apparently, the magma moving up from the mantle generates earthquakes.[8]

A mantle source is also suggested by a well-known relationship that exists between volcanism and depth to the seismic zone (the Benioff-Wadati zone) in volcanic arcs (Kuno, 1959). In the arcs, more alkali-rich basaltic magmas erupt above zones of deep seismicity, whereas more siliceous basaltic magmas erupt above zones of shallower mantle-depth seismicity.

Mantle Xenoliths

Xenoliths in basalts and rocks called *kimberlites* (porphyritic phlogopite peridotites) also suggest a mantle source for some magmas. The xenoliths or **nodules**[9] (figure 6.1) include pieces of a variety of felsic to mafic and ultramafic rock types—including gabbro, eclogite, amphibolite, granulite, and various peridotites—that are enclosed in a host rock.[10]

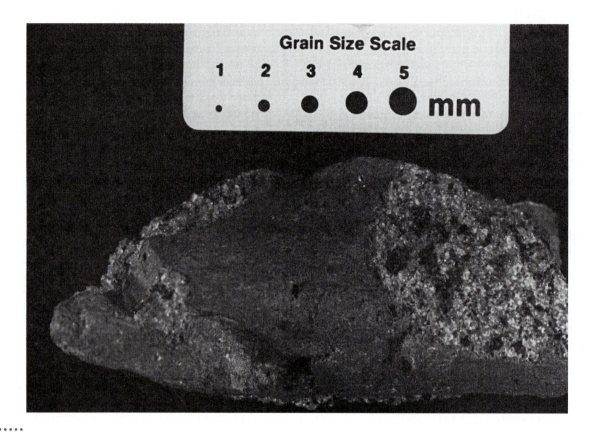

Figure 6.1 Mantle xenoliths of dunite in basalt, San Carlos, Arizona.

The rock types of the nodules may represent (1) materials crystallized earlier from the magma and then fragmented and incorporated into the magma as it moved, (2) pieces of the residual (depleted) mantle rock from which the magma was derived by anatexis, or (3) unrelated mantle and (or) crustal rocks picked up from the walls of the magma conduit by the magma as it moved toward the surface. Although it is sometimes difficult to choose between these alternatives, the phase assemblages in the rocks reveal the depths of origin of the rocks and their sources in the lower crust and mantle.

Experimental laboratory studies of appropriate systems indicate that the phase assemblages of eclogites and spinal peridotites are stable at pressures typical of upper mantle depths, depths greater than about 25 km (I. D. MacGregor, 1974).[11] Garnet peridotites are stable at mantle temperatures at pressures corresponding to depths of greater than 35 km, and may occur to depths of 120 km or more (figure 6.2). Furthermore, the occurrence of diamond in kimberlite indicates that the kimberlitic parent magmas formed at a minimum depth of 150 km (Dawson, 1981; Carswell and Gibb, 1987).

In contrast, amphibolite, granulite, and gabbro nodules have shallower, crustal sources.

The distribution of nodule occurrences is also significant, as nodules are not randomly distributed in basalts. The more siliceous, tholeiitic basalts rarely contain nodules, whereas the alkali olivine basalts commonly contain eclogite and spinel peridotite nodules.[12] Abundant nodules, including some of garnet peridotite, occur in kimberlites. These relationships suggest different depths of origin, primarily mantle depths, for different basalt and kimberlite magma types.

Chemical Constraints on Magma Sources and Magma Origins

The mantle origin of magmas, as well as a possible crustal origin for some magmas, is supported by chemical data and calculations based on those data. Major and trace element abundances, including those of radioactive isotopes, are used both to model the rocks of the source regions and to constrain possible source rock types. If the magmas were derived from the source rocks, magma chemistries and resulting

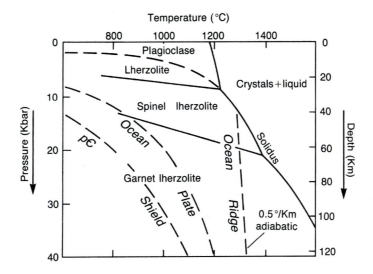

Figure 6.2 Stability fields of upper mantle rock types, with geotherms (dashed) and the peridotite solidus. (From Morse, 1980, p. 407)

rock chemistries will reflect source-rock chemistries, provided that major modifications have not occurred. For example, an obvious way to get a basalt magma would be to completely melt a basalt, an eclogite (a chemical equivalent of basalt), or a gabbro. The chemistry of the magma would duplicate the chemistry of the source rock. However, because hot magmas tend to rise as they form, and crystals tend to sink or float, it is unlikely that an entire mass of eclogite, basalt, or gabbro could be melted in place (Yoder, 1976). It is much more likely that early formed liquids would separate, resulting in fractional melting and the production of multiple magmas. Following the principles outlined in chapter 5, one would expect these early melts to be more siliceous, aluminous, and alkali-rich than either the melts formed later or the residual crystal phases. Thus, it seems reasonable that basaltic magma will form through partial melting of a parent rock type that is poorer in silica, alumina, and alkalis than is basalt.

Recall that seismic evidence suggests that melting occurs in the upper mantle. In addition, the nodules found in basalts suggest not only mantle depths but ultramafic and eclogitic mantle compositions for the parent rock. A number of additional lines of evidence are consistent with an ultramafic mantle, including geophysical characteristics, such as seismic wave velocities and rock density.[13] Also, it is commonly assumed that certain meteorites (the chondrites) reveal primordial solar system chemistry, on which models of the Earth's mantle chemistry are based.[14] *Together, these data suggest that an ultramafic mantle, perhaps containing local zones of eclogite, serves as the parent rock source for basaltic magmas.*

In an attempt to define the primitive ultramafic mantle source of basaltic magmas, Ringwood (1962a) and D. H. Green (1973) hypothesized that an artificial ultramafic mantle rock called **pyrolite**, composed of one part basalt and three or four parts peridotite or dunite, is the parental rock.[15] By definition, pyrolite melts to produce a basaltic magma, leaving a depleted, residual peridotite.

The pyrolite model has been criticized for being naively simplistic because (1) the compositional mix is artificial—the basalt and peridotite chosen for mixing are entirely unrelated to one another, and (2) post-separation modification of the basalt (e.g., by fractional crystallization) and the peridotite are not accounted for by the calculation (Harte, 1983). These criticisms are not without merit, as the current existence of primary basalt magmas at the surface and primitive upper mantle at depth is moot or doubtful, as is homogeneity in the mantle.[16] Nevertheless, the model does offer an ultramafic parent for basaltic magmas and addresses important issues relating to derivation of magmas from now depleted mantle rocks.

Ultramafic parent rocks for basaltic magmas are also proposed by most other workers. For example, Bence, Grove, and Papike (1980) argue that a careful analysis of chemistries, mineralogies, and textures of basalts allows us to recognize primary magmas derived from mantle lherzolites. Yoder (1976, 1978) suggests that the mantle is garnet peridotite and argues that 30% partial melting of this parent will yield a large volume of basaltic magma that may differentiate to produce other magmas. Experimental studies verify some aspects of these hypotheses.[17]

Among the chemical parameters useful for recognizing mantle parentage are magnesium number—$Mg^{2+}/(Mg^{2+} + Fe^{2+})$—and a number of isotope ratios, including those of lead, oxygen, helium, strontium, and neodymium. Mantle rocks have Mg numbers that typically fall in the range 0.75–0.92. Basalts with magnesium numbers of about 0.70–0.73 are likely primary melts.[18] Magmas with higher Mg numbers may result from (1) very high percentages of melting of mantle rocks, which enriches the melt in MgO; (2) melting of highly depleted, MgO-rich mantle rock (Arndt, 1977b); (3) incorporation of cumulate grains, especially forsteritic olivine, in the melt; or (4) assimilation of Mg from mantle wall rocks by magmas passing through mantle conduits (J. D. Myers, 1988). Magmas with lower Mg numbers are probably contaminated or altered.

A complete introduction to isotopes used in petrogenetic studies is beyond the scope of this text. Nevertheless, Nd and Sr isotopes are discussed here as examples of how isotopic ratios constrain magma (and rock) origins.

Rubidium and strontium are not abundant elements, but their radii and ionic charges (Rb^{+1}, Sr^{+2}) make it possible for them to easily substitute for potassium and calcium, respectively. ^{86}Sr is nonradiogenic; that is, it does not form by radioactive decay, and may therefore be used as a fixed standard of the initial abundance of Sr in the Earth. The primitive value of $^{87}Sr/^{86}Sr$, based on meteorite abundances, is 0.699.[19] Slow decay of ^{87}Rb yields ^{87}Sr, and thus the $^{87}Sr/^{86}Sr$ ratio increases slightly over time for the Earth as a whole. Because Rb substitutes for K in minerals, fractionation, which concentrates K in rocks, also concentrates Rb, especially in alkali-rich rocks (e.g., those containing K-feldspar or K-micas). In contrast, mantle rocks, which generally lack K-bearing minerals (except for minor phlogopite), are poor in Rb. The result of these geochemical conditions is that the value of $^{87}Sr/^{86}Sr$ in the mantle is very low, because little ^{87}Sr is created; whereas in sialic rocks, where Rb values are higher, more ^{87}Sr is produced and the amount of ^{87}Sr increases more per unit volume of rock, causing an associated increase in the ratio of ^{87}Sr to ^{86}Sr.

Assuming that normal chemical processes have not affected the two Sr isotopes differently, for example by concentrating or depleting one or the other, the $^{87}Sr/^{86}Sr$ ratio of a rock may be used to indicate the possible source area of its parent magma. Magmas that form in and equilibrate with the mantle will have low $^{87}Sr/^{86}Sr$ values. Those that form in, from, or equilibrate with differentiated rocks (e.g., crustal alkali-rich rocks) will have higher ratios. As shown in figure 6.3, various rock types show considerable ranges of Sr ratios. Although these values must be corrected for age (to account for the changing value of $^{87}Sr/^{86}Sr$ for the Earth as a whole over time), Sr ratio values of less than 0.704 suggest a mantle source, especially in Phanerozoic rocks. Values above 0.710 suggest significant ^{87}Sr enrichment from a crustal source. Intermediate values are subject, more than the others, to the vagaries of interpretation.

Samarium and neodymium are rare earth elements found in trace amounts in all rocks and minerals (DePaolo, 1981a, 1981b, 1981c).[20] ^{147}Sm decays to ^{143}Nd. ^{144}Nd is the standard used for comparison; thus the ratio $^{143}Nd/^{144}Nd$ may be used to measure geochemical processes. Compared to a bulk Earth value of $^{143}Nd/^{144}Nd = 0.51264$, higher Nd ratios (e.g., 0.5129–0.5132) suggest mantle sources, whereas lower values (0.5110–0.5126) reflect contamination or crustal sources (Nohda and Wasserburg, 1981; J. P. Davidson, 1987). Nd values sometimes are reported as the function ε_{Nd}, which is computed from the isotopic values and typically has values between +12 to –20.[21] High values of ε_{Nd} represent mantle sources, whereas low (negative) values reflect crustal contamination (DePaolo, 1981b; Perry, Baldridge, and DePaolo, 1987). Because Sm and Nd are fractionated during magmatic processes, the isotopic ratios reflect magmatic control. Where used in conjunction with Sr ratios, Nd ratios are especially valuable in assessing mantle sources for magmas and the amounts of crustal contamination in, or contribution to, a particular magma. Nd isotopes suggest that all oceanic volcanic rocks are derived from magmas from the upper mantle.

In summary, a number of lines or evidence suggest that basaltic magmas are derived from the mantle. This is the case for basalts from oceanic and many continental areas.

The origin of andesitic magmas, as with basaltic magmas, is attributable, at least in part, to partial melting of a parent rock. Yet, as discussed in chapter 9, a number of complicating possibilities, including contamination of magma, mixing of magmas, and fractional crystallization of basaltic magma, have been proposed to explain andesitic magma chemistry. The question here is, *where* do andesitic magmas form?

Andesites occur primarily in volcanic arcs, both on continents and on oceanic crust.[22] Because arcs can generally be shown to be associated with past or present sites of subduction; the process of subduction is clearly important to the origin of many andesitic magmas. The arcs occur above subduction zones at a location that corresponds to a depth of about 100–150 km within the zone. Thus, mantle depths are required for the initiation of the processes that eventually produce andesitic magmas. However, chemical and petrographic evidence supports the idea that crustal level processes also are involved in the formation of these magmas.[23]

Experimental work suggests the possibility that some andesitic magmas may be generated from metamorphosed crustal rocks (e.g., quartz eclogite) or from CO_2-free, hydrous

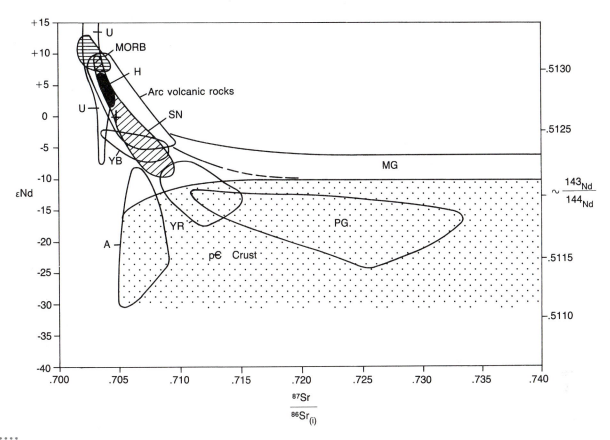

Figure 6.3 An Sr-Nd isotope diagram showing the general ranges of values for several types of rocks from selected tectonic settings. A = selected alkaline rocks, H = volcanic rocks of Hawaii (representing an oceanic plume/hot spot setting), MG = Malaysian granitoid rocks, MORB = Mid-Ocean Ridge Basalts, PG = selected peraluminous granites, SN = granitoid rocks of the Sierra Nevada Batholith, California (a typical volcanic arc batholithic core zone), U = rocks of ultramafic complexes, YB = basalts of the Yellowstone region, YR = rhyolites of the Yellowstone region (YB and YR represent rocks of a plume/hot spot setting within a continental interior).

[Based on data reported or summarized in W. E. Cameron et al. (1983); R. B. Cole and Basu (1992); DePaolo (1981a); DePaolo and Wasserburg (1977; 1979b); Hildreth and Moorbath (1988); Hildreth, Halliday, and Christiansen, et al (1991); D. E. James (1982); Kistler et al. (1986); Liew and McCulloch (1985); J. D. Morris and Hart (1983); Rautenschlein et al. (1985); Reisberg, Zindler, and Jagoutz (1989); Staudigel et al. (1984); J. D. Walker and Coleman (1991); West et al. (1987); W. M. White and Patchett (1984); and J. E. Wright and Wooden (1991)]

mantle at depths of 100 km.[24] Yet the H_2O contents required for such magma genesis may be unrealistically high (Wyllie et al., 1976). Given that crustal level processes are involved in andesitic magma formation,[25] it appears that the *magmas formed in subduction zones are altered to form most andesitic magmas.* Some high-Mg andesites may be derived directly from andesitic parental magmas of mantle origin.[26]

The formation of *granitoid* magmas may result from anatexis of both crustal and upper mantle rocks.[27] A crustal origin for many such magmas is likely. Both the chemistry and P-T conditions (2–10 kb, 650–850° C) of experimentally produced quartzo-feldspathic magmas and the phase relations in appropriate systems are consistent with crustal depths of origin (Winkler, 1976; Wyllie et al., 1976; Wyllie, 1977). Furthermore, chemical data, including high $^{87}Sr/^{86}Sr$ in some granitic rocks, and various petrologic arguments add weight to the crustal melting hypothesis.[28] As discussed in chapters 8, 11, and 12, other processes, such as fractional crystallization of more mafic magmas and assimilation, may also contribute to the origin of these magmas.

The parent magmas of some granitoid rocks (particularly quartz diorites) may form at mantle depths by anatexis of either subducted oceanic crust or the mantle overlying the subduction zone.[29] Those favoring such a hypothesis cite low

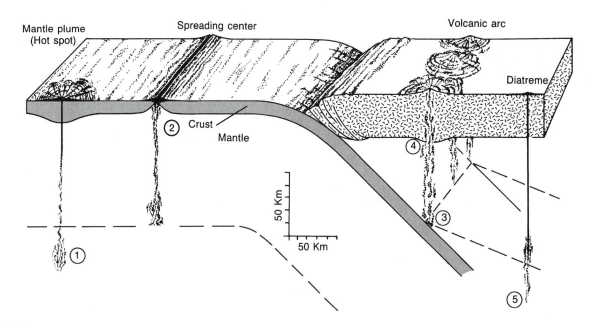

Figure 6.4   Sketch showing typical sites of magma generation. (1) A mantle hot spot (plume) source. (2) A site in mantle below a spreading center. (3) A site where the top of the subducting lithosphere intersects the base of the overlying lithosphere. (4) A site at the base of the crust. (5) A subcrustal mantle site (where kimberlitic magmas form). The dashed line is the base of the lithosphere.

Sr-isotope ratios and Nd isotope data, as well as other chemical arguments, in support of their views. However, experimental work suggests that anatectic development of typical calc-alkaline magmas of intermediate composition at deep crustal to shallow mantle depths may not be possible (Huang and Wyllie, 1986; Carroll and Wyllie, 1990).

For all three major magma types (basaltic, andesitic, and granitoid), there is strong evidence that both energy and material from the mantle are important in magma genesis (figure 6.4). Hildreth (1981, pp. 10179, 10182ff.), in fact, suggests that basaltic magmas provide the energy for *all other magmatism.*

In conclusion, then, the available evidence indicates that most igneous rocks are derived ultimately from magmas produced by melting of mantle and lower crustal rocks. The mantle rocks are generally ultramafic (low-silica) types. A range of magma compositions, generally of mafic character, forms from these mantle rocks. Intermediate to siliceous magmas are produced from the more siliceous lower crustal rocks.

MAGMA MOVEMENTS

In general, melt is less dense than rock. Consequently, magmas generated at depth, in either the deep crust or the upper mantle, tend to rise towards the suface.[30] When they reach a level at which the rocks above the magma body are less dense than the magma (the level of neutral buoyancy), the body may cease to move and will form a pluton (Walker, 1989). Alternatively, for example where magmas are under pressure at depth, they may rise beyond this level. How magmas move, how far they move from their sites of origin, and how fast they move are controlled by a number of factors, including (1) the magma composition, (2) the volatile content, (3) the phenocryst content, (4) the temperature contrast between the magma and the country rock, (5) the heat content of the magma, (6) the viscosity[31] of the magma (which depends on factors 1–3 and 5), (7) the volume increase resulting from melting, and (8) the fracture processes and stresses in the country rock. For example, magmas of high silica content are relatively viscous (sticky) compared to those of low silica content. All other factors being equal, high viscosity reduces flow velocity and inhibits lengthy distances of transport. This fact is part of the general knowledge of magma movement that has been developed, the details of which are still being investigated.[32] Magma does not move until a significant quantity of melt forms.

Melting begins at spots in the parent rock either where principal phases are in contact or around an inclusion or minor phase.[33] In a mantle garnet lherzolite, initiation of melting probably begins at a grain junction where garnet, two pyroxenes, and olivine meet. In a quartzo-feldspathic granulite deep in the crust, melting might begin where quartz and two feldspars meet. Melting occurs at these points because the solidus controls the melting, and the first melt will have the composition of the eutectic or thermal minimum (in the appropriate system). With increased volume or duration of melting, melt spreads along many grain boundaries (Maaloe, 1981).

With percentages of melting below about 10% of the rock, the rocks are neither permeable nor susceptible to filter pressing.[34] Thus, for the partial melt to accumulate and begin to move as a magma, a degree of anatexis greater than 10% is required. As melting exceeds this value and continues, the melt may accumulate in layers or pockets or, perhaps in the case of some granitic masses, may mobilize the rock mass as a crystal mush (Pitcher, 1973).[35] Zones of shear in certain tectonic regimes also may localize magma concentrations.[36] The shapes of the zones of magma accumulation in the deep lithosphere, though not well known, may be a function of tectonic position (spreading center, mantle plume, or subduction zone). Few large magma chambers exist. Proposed accumulation zones vary from networks of anastomosing layers or fractures[37] to horizontal, discoidal magma chambers[38] to long, thin, inclined linear zones and sheets[39] (figure 6.5).

Once the magma has accumulated at depth, it will move towards the surface. The causes of magma rise may vary (Rast, 1970; Philpotts, 1990, ch.3).[40] Under certain circumstances, in kimberlitic magmas, for example, gas boiling out of the magma may provide the driving force, in a process called gas fluxion or gas coring.[41] In other cases, pressures induced by tectonic forces, the weight of overlying rocks, or volume increases due to melting may cause squeezing of the magma from the rocks. The general lack of vesicles in many dikes and plutons and the widespread occurrence of magmatic activity in zones of tension argue against the general importance of these processes, though they may be significant locally.

The two most favored processes of magma movement are *diapirism* (or buoyancy) and magma movement associated with *magma-driven fracturing and crack propagation.*[42] Crack propagation allows rapid movement. Magma increases approximately 15% in volume through melting, and several petrologists have hypothesized that this increase will create pressures that fracture lithosphere rock, opening a channel along which melt can move towards the surface.[43] In addition, the buoyancy of melt trapped below heavier rigid lithosphere also provides a crack-generating force (Turcotte and Emerman, 1985). Near the surface, thermal stresses from heating and consequent expansion may cause further fracturing (Marsh, 1982). Because the lithosphere through which magma must travel towards the surface is relatively cool and rigid, fracture propagation is

considered to be a realistic process, and it has a firm foundation in theory and analogy.[44]

The movement of mafic magmas (e.g., basalt or andesite) in fractures is more likely than the movement of siliceous magmas because of the substantially lower viscosity of the former. Nevertheless, boiling (rapid escape of gas) and fragmentation of partially crystallized siliceous magmas near the surface facilitates movement of these magmas through fractures. In either case, movement must be rapid; otherwise, heat loss to the fracture walls would cause the magma to solidify (freeze).

Alternatively, magma diapirism can explain the ascent of magmas (Marsh, 1982; Mahon, Drew, and Harrison, 1985), especially granitoid types, although basaltic magmas may also rise via this mechanism.[45] The **diapir** may be envisioned as a spherical, elliptical, or tail-down drop-shaped mass that rises towards the surface as a result of its low density compared to densities of surrounding rocks. Laboratory experiments reveal that diapirs, in fact, will develop.[46] Movement of diapirs is generally slower than magma migration through cracks, yet, as with crack-controlled ascent, heat loss to the surroundings is a major controlling factor.

Marsh (1976a, 1978, 1979) has calculated cooling curves for andesitic diapirs ascending through the lithosphere. Figure 6.6 shows one set of curves plotted on a P-T grid, along with the solidus and liquidus of andesite. Magma forms at a depth of about 100 km. The ascent velocity controls the cooling path, and for rates ≥300 m per year, magmas will reach the surface. That magma diapirs remain molten during their ascent through the mantle is supported by model experiments (Ribe, 1983). Furthermore, it is much easier for subsequent diapirs of even smaller size to follow the original diapir's path to the surface, as less heat will be lost to the already warmed surroundings.

Diapirs of granitic magma have a more difficult time reaching the surface. Granitic magmas are relatively cool and wet, especially if they either are partly crystallized, which concentrates water in the residual melt, or have absorbed water from the country rock.[47] Ascent of the magma may cause the magma to lose its volatiles by "boiling" as pressure is reduced, thereby increasing the viscosity and resistance to flow (D. M. Harris, 1977). In addition, because the melting curve for wet granite has a negative slope (like that of andesite) at pressures of less than 17 kb (Stern, Huang, and Wyllie, 1975; figure 6.6), ascent with associated loss of pressure (even with constant temperature) would result in freezing (crystallization), because the magma would cross the solidus curve (Tuttle and Bowen, 1958, pp. 122–125). A granite pluton would result.

Both diapirism and flow in magma-formed cracks probably occur. The latter is favored in cases where rapid ascent is suggested by geological evidence. For example, it has been argued that alkali olivine basalts containing mantle xenoliths

(a)

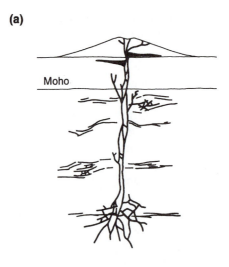

(b)

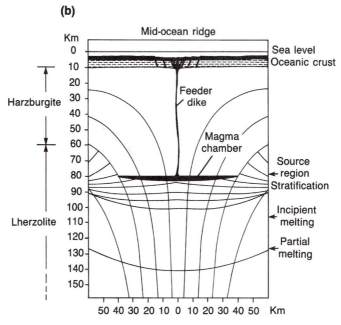

....
Figure 6.5 Proposed shapes of magma chambers in the lithosphere. (a) Fracture networks (from Wilshire and Shervais, 1975). (b) Discoidal to sheetlike magma chambers, shown in two dimensions (from S. Maaloe, "Magma accumulation in the ascending mantle" in *Journal of the*

Geological Society of London, 138:223–6, 1981. Copyright © 1981 Geological Society of London. Reprinted by permission.)

must move quickly to the surface, or the xenoliths sink or recrystallize.[48] Table 6.1 presents estimated ascent velocities of magmas based on various types of calculations. The table reveals that ascent from the base of the lithosphere via magma-driven cracking and accompanying flow would take from about a day to a few years, whereas diapiric movement requires a few to thousands of years or more. Of course, movement at very low rates means that magmas will not reach the surface, because they will crystallize. Note also that basaltic magmas rise faster than granitic magmas, because of their lower viscosities and higher temperatures.

MODIFICATION OF MAGMAS

Magmas are commonly modified. Modification occurs before magmas crystallize below the surface as plutons or erupt at the surface to produce volcanic rocks. Three major processes—differentiation, mixing of magmas, and assimilation of country rock—are responsible for these modifications. Each may leave characteristic chemical or petrographic evidence that allows the petrologist to recognize what process has taken place.

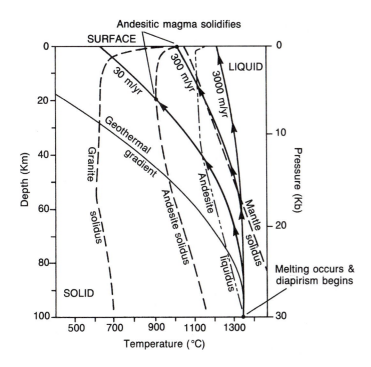

Differentiation of Magmas

Differentiation refers to a group of processes, occurring within magma bodies, that result in the creation of two or more magmas or rocks from a single homogeneous parent magma.[49] The processes include crystal-liquid fractionation, liquid immiscibility, vapor transport, and diffusion.

Crystal-liquid fractionation is the process of separating crystals from liquid in the magma chamber. Some types of crystal-liquid fractionation, including crystal settling and flotation, also called **gravitational separation,** and filter pressing, were introduced in chapter 5.[50] In addition, fractionation may occur by convective fractionation, flow differentiation, and congelation crystallization. **Convective fractionation**

....

Figure 6.6 P-T grid showing curves for ascent velocities of 0.8-km-radius andesite diapirs (arrows). Also shown are the solidus and liquidus curves for andesite, the solidus curve for the mantle, and the solidus curve for granite. (from C. R. Stern, Huang, and Wyllie, 1975. *(Modified from B. D. Marsh, 1979)*

Table 6.1 Calculated and Estimated Ascent Velocities of Magmas in the Lithosphere

Magma/Basis	Ascent Velocity (cm/sec)	Days (d) or Years (y) to Rise 100 Km	Source
xenolith settling in basalt as Bingham plastic	0.85	134 d	S. R. J. Sparks, Sigurdsson, and Wilson (1977)
xenolith settling in basalt using Stoke's law	0.83	139 d	Yoder (1976)
xenolith settling in basalt using Stoke's law	30	3.8 d	Basu (1977a)
xenolith settling in basalt	1–10	120–12 d	Maaloe (1973b)
xenolithic magma	100	1.2 d	Turcotte and Emerman (1985)
near-surface flow in 1-m dike	10–200	12–0.6 d	Delaney and Pollard (1982)
andesite in 1-km elastic crack	35	3.3 d	Marsh (1979)
Hawaiian basalt in a crack	1–10	120–12 d	Wright et al. (1979, in Wadge, 1980)
1-km andesite diapir reaching the surface	$>5.8 \times 10^{-3}$	> 54 y	Marsh and Kantha (1978)
basalt diapir	10^{-3}	316 y	Marsh (1984)
< 10-km andesite diapirs	5×10^{-7} to 1.3×10^{-5}	6.3×10^5 y to 2.4×10^4 y	Marsh (1982b)
granite diapirs	$10^{-8} - 10^{-4}$	3.2×10^7 y to 3.2×10^3 y	Marsh (1984)

involves crystallization from magmas that flow past surfaces of crystallization on the floor, walls, or roof of a magma chamber (Rice, 1981). The convective flow is caused by heat within the magma chamber. In **flow differentiation,** while magma is flowing through a crack, loose crystals are concentrated in the center of the vertical or horizontal conduit as a result of a phenomenon called grain-dispersive pressure (S. Komar, 1976; Bebien and Gagny, 1979). **Congelation crystallization** is crystallization on the walls and floor of a magma chamber that concentrates liquid enriched in certain elements toward the top or center of the chamber (C. J. Hughes, 1982).[51] The relative importance of each of these various processes is debatable and may vary for different magma types.

Liquid immiscibility refers to the formation and separation of two magmas from an initially homogeneous parent due to changes in physical properties that cause *unmixing* during cooling. As the two magmas separate, one forms drops within the other, much like mixtures of oil and vinegar in salad dressing. Although some evidence indicating liquid immiscibility in igneous rocks has been recognized in recent years, there is presently little data supporting the view that this process is a major one in magma modification.[52] Liquid immiscibility may be important locally in the development of unusual magmas, such as carbonate magmas (Treiman and Essene, 1985).[53]

Vapor transport is a process in which vapor that is enriched in certain elements (e.g., Na, K) boils out of the magma at an early stage and enriches later crystallizing fractions. Sakuyama and Kushiro (1979) demonstrated the process in the laboratory. However, vapor transport appears to be of minor importance in magma modification (Trial and Spera, 1990).

Diffusion is the migration of chemical species through a magma in response to differences (gradients) in pressure, temperature, or chemical state. Significant *local* diffusion of K and Na in response to chemical gradients has been demonstrated in the laboratory (Watson and Jurewicz, 1984). Temperature-induced migration, called the Soret effect, also occurs (Walker and DeLong, 1982). Hildreth (1979, 1981) argues that silica-rich magma chambers become zoned through a combination of diffusion and thermally driven convection. While there is theoretical, mineralogical, and chemical evidence supporting the viability of these processes,[54] the importance of diffusion in producing large-scale differentiation is debatable, because much of the data used to support various diffusion models are ambiguous or support other models equally well.[55] In addition, some theoretical arguments have been raised about the efficacy of the process (Carrigan and Cygan, 1986).

Magma Mixing

In a sense, magma mixing is the opposite of liquid immiscibility. Magma mixing is the process in which two magmas of different composition blend to form a single, more or less homogeneous, derivative magma. Depending on the proportions of the parents, the composition of the derivative magma will have some value between those of the parents. Thus, andesitic and dacitic (intermediate) magmas are commonly considered to be candidates for magmas produced by mixing of basalt and rhyolite parents, but other rock types lying between these endmembers have also been attributed to a mixing origin.

The difficulty of obtaining homogeneity in mixed magmas is an apparent obstacle to extensive mixing, especially in magmas of widely different composition (Wiebe and Wild, 1983; Frost and Mahood, 1987). Nevertheless, a growing body of chemical, petrographic, experimental, and theoretical evidence suggests that magma mixing is a widespread, if not common, process in the modification of magmas (Eichelberger, 1975b, 1981; Oldenburg et al., 1989).[56]

Assimilation

Magmas rising from their sites of origin in the mantle and lower crust may dissolve, react with, or melt and mix with the wall rocks in the conduit or magma chamber in a process called assimilation.[57] Of the three assimilation processes—melting, reaction, and dissolution—reaction is perhaps the most likely.

Evidence of melting of rock by magma is quite limited (C. J. Hughes, 1982, p. 212ff.). This is not surprising considering that melting requires considerable amounts of heat. The heat required for melting wall rock has two essential components. First, heat energy is necessary to raise the temperature of the wall rock to the melting temperature. Second, as revealed in chapter 5, the latent heat of melting (the specific heat or heat input represented by the flat part of the melting curve in figure 5.9b) must be added to the system before the rock will melt (Bowen, 1922; 1928, ch. X; 1948). In addition, heat may be required for unmixing solid solutions and other changes that must take place (J. Nicholls and Stout, 1982). If assimilation is to occur by melting, then the consequence of these heat requirements is that (1) the magma must be superheated,[58] so that it can give up substantial amounts of heat without freezing, (2) large quantities of magma must freeze in order to provide enough heat to melt smaller quantities of wall rock, or (3) only hotter (basic) magmas will assimilate cooler (siliceous) wall rocks.

Dissolution requires that the magma be undersaturated in the components to be dissolved. Undersaturation may or may not be common, but evidence of dissolution is rare.

Reaction only requires ionic exchange (diffusion) between magma and wall rock. This process is favored by chemical potential gradients in the contact zones. Reaction involves the conversion of wall-rock phases into phases compatible with those crystallizing from the magma. Major elements such as alkalis (Na and K) may be exchanged (Watson and Jurewicz, 1984), as may various isotopes and trace elements (notably Sr). Abundant evidence of reaction zones at pluton and xenolith margins occurs in the form of minerals that have compositions intermediate between those of the wall rock and the pluton. For example, hornblende-rich xenoliths in granitic magmas commonly show reaction rims rich in biotite. In cases where significant amounts of ions are absorbed by a magma, the magma composition may be substantially modified.

THE FORMATION OF ROCKS

The various factors that influence the origin, movement, and modification of magmas give magmas their character. As the magmas move toward the surface, various internal and external controls continue to influence magmatic evolution and ultimately control the final products of that evolution—the rocks.

Slow-moving, cooler, siliceous magmas will commonly boil and freeze below the surface. Later, exposure reveals that they locally have intruded passively as sills, fractured and stoped the roofs above batholithic and smaller magma chambers as they worked their way towards the surface (Bussell, Pitcher, and Wilson, 1976), or forcefully deformed the country rock via folding, stretching, and faulting before freezing to form plutons. Where siliceous magmas have broken through to the surface, explosive eruptions typically produce extensive rhyolitic tuffs that form ash sheets (see chapter 8). Where less explosive eruptions occur, steep-sided domes and flows form.

Mafic magmas are both hotter and less viscous and tend to be relatively faster-moving (via magma-generated crack propagation and flow) than siliceous magmas. After initially creating channels to the surface (Marsh, 1979, 1982b), basaltic and intermediate magmas move up through the mantle and crust. They may pause at various levels, where differentiation and/or crystallization occur, before freezing to form lopoliths and other plutons or before moving on to erupt at or near the surface to form dikes, sills, flows, minor pyroclastic deposits, domes, cones, plains, and plateaus.

Modification of magmas in transit is the norm. Where modifications are minor, the rocks that develop reflect the general chemistry of their parent magmas. Where modification is extensive, new magma compositions are formed and recognition of the parent magma may be difficult.

SUMMARY

Primitive magmas, formed by melting of undepleted mantle, and primary magmas, formed by melting of depleted mantle, lower crust, or subducted crust, move towards the surface as primary (unchanged) magmas or derivative magmas produced by modification of a parent magma. Primary magmas are probably dominantly basaltic and granitic, but andesitic and other types of primary magma are known to occur.

Mantle and lower-crustal sources for magmas are indicated by magma chemistry (including isotopes), laboratory analysis of appropriate systems, xenoliths, and seismic evidence. Once magmas form, they move towards the surface by diapirism or flow through magma-induced cracks at rates that are estimated to be between 10^{-8} cm/sec and 100 cm/sec. At these rates, the lithosphere would be traversed over time periods ranging from 32 million years to one day. Hotter, less viscous basaltic magmas move relatively rapidly, especially where they generate cracks along preweakened paths through the mantle. Cooler, viscous diapirs of granitic magma move quite slowly, commonly failing to reach the surface before they crystallize. The primary magmas are typically modified by differentiation (especially crystal-liquid fractionation), mixing of magmas of different compositions, or assimilation of country (wall) rock before they reach the surface.

Igneous rocks are formed where the various magmas crystallize. Crystallization occurs either below the surface in plutons or above the surface in volcanic constructions.

EXPLANATORY NOTES

1. This definition of primary magmas differs from some used elsewhere but is similar to that of Presnall (1979) and A. R. Philpotts (1990). Presnall defines primary magma as "magma as it exists immediately after separation from its source region" (p. 60). I. S. E. Carmichael, Turner, and Verhoogen (1974, pp. 44–46), who discuss primitive, primary, and parental magmas, specify that primary magmas are "of uniform composition rising in great volume from deep sources, with no evidence of derivation from other magma of the same igneous cycle." They reject the use of the term primitive magma because of the difficulty of determining whether or not the source rocks were previously melted or otherwise modified in composition. Some geologists use the terms primary magma and parental magma interchangeably (see Bates and Jackson, 1980). Others may use primary magma in the same sense that primitive magma is used in this text.
2. For example, see DePaolo (1983), Menzies (1983), and Michael and Bonatti (1985). Also see Dick (1977a), Boudier and Nicolas (1977), and Dick, Fisher, and Bryan (1984).

3. Metasomatism is a metamorphic process that involves any change in the chemistry of rocks, either addition or subtraction of material. The literature on mantle metasomatism commonly concentrates on additions. For discussions see Menzies and Murthy (1980), A. T. Anderson (1982), D. K. Bailey (1982), Sekine and Wyllie (1982), W. M. White and Hofmann (1982), DePaolo (1983), Menzies (1983), Dawson (1984), C. A. Evans (1985), E. M. Morris and Pasteris (1987), Nielson and Noller (1987), and Wilshire (1987). Also see papers in D. K. Bailey et al. (1980).

4. Evidence of mantle heterogeneity is discussed by Michael and Bonatti (1985), Ringwood (1985), Perry, Baldridge, and DePaolo (1987), and Lum et al. (1989), among others. Mantle mixing processes are discussed by A. W. Hofmann and White (1980), A. W. Hofmann et al. (1986) and Kellogg and Turcotte (1990).

5. Also see W. O. Kennedy (1933) and the discussions in Wyllie (1971a), Bence, Grove, and Papike (1980), and the Basaltic Volcanism Study Project (1981).

6. D. H. Green and Ringwood (1968), Wyllie (1971a), Brown and Fyfe (1972), Boettcher (1973), Kushiro (1973a, 1974a), LeBas (1978), Suzuki and Shiraki (1980), Tatsumi and Ishizaka (1982), DeVore (1983).

7. For examples, see Arndt (1977b), LeBas (1978), and Stolper (1979).

8. Available geochemical evidence suggests that Hawaiian magmas derive from mantle plumes rising from depths greater than 60 km, perhaps from as deep as 350 km (Feigenson, 1986; Wyllie, 1988).

9. *Nodule* is a nongenetic term, whereas *xenolith* and *autolith* indicate that the relationship between the inclusion and the enclosing magma is known.

10. The rock types listed here are defined in the glossary and subsequent chapters. (Eclogite is a metamorphic rock that is chemically equivalent to basalt.) Mantle xenoliths and mantle compositions are discussed in a number of sources including I. D. MacGregor (1968), J. L. Carter (1970), Wyllie (1971), Cox, Gurney, and Harte (1973), P. H. Nixon and Boyd (1973), L. H. Ahrens et al. (1975), Basu (1975), Boyd and Nixon (1975), Takahashi (1978), Boyd and Meyer (1979a, 1979b), J. Ferguson and Sheraton (1979), Dawson (1980, 1981), BVSP (1981), Berger and Vannier (1984), Gutmann (1986), Dromgoole and Pasteris (1987), S. E. Swanson et al. (1987), Goto and Yokoyama (1988), and several papers in *American Mineralogist,* v. 61, no. 7–8 (1976), *Physics and Chemistry of the Earth,* v. 9 (1975), *American Journal of Science,* Jackson Volume, pt. 2, v. 280-A (1980), Hawkesworth and Norry (1983), Kornprobst (1984b), and P. H. Nixon (1987). Crustal xenoliths are discussed by Padovani and Carter (1977), Rudnick and Taylor (1987), Dodge, Lockwood, and Calk (1988), Fodor and Vandermeyden (1988), and Roberts and Ruiz (1989).

11. I. D. MacGregor (1968), Dawson (1981), and Finnerty and Boyd (1987) summarized the implied conditions of nodule origin, and some refinements are offered by Carswell and Gibb (1987). I. D. MacGregor (1974) provides a geochemical indicator of pressure (depth), and Yoder (1976) and Morse (1980) present P-T grids.

12. Basalt types are defined and discussed in chapter 7. Nodule abundances are discussed in I. D. MacGregor (1968), BVSP (1981), and Dawson (1981). Also see Basu (1975), Ferguson and Sheraton (1979), and Dawson (1980).

13. Wyllie (1971a, 1971b), D. L. Anderson and Bass (1984), Bass and Anderson (1984), D. L. Anderson (1987), Duffy and Anderson (1989).

14. Ringwood (1975) provides a brief review of the rationale for this. A more detailed, but somewhat dated, treatment of mantle composition is provided by Wyllie (1971a). Also see Bence, Grove, and Papike (1980), BVSP (1981), and Dawson (1981).

15. Ringwood (1962a, 1962b, 1974, 1975, 1985) originally designated pyrolite as one part basalt and four parts dunite. This estimate was changed to one part basalt or eclogite to four parts dunite-peridotite (Ringwood, 1962b), one part basalt to three parts dunite (Green and Ringwood, 1963, 1967a, 1969), and one part Hawaiian tholeiite basalt to three parts alpine harzburgite (Ringwood, 1975) as refinements were made to match the chemistry of the hypothetical parent with that of the derivative basalts.

16. For example, see M. J. O'Hara (1965), Yoder (1976), Bailey, Tarney, and Dunham (1980), and Bence, Grove, and Papike (1980).

17. Mysen and Kushiro (1977), Presnall et al. (1978), BVSP (1981), Green, Falloon, and Taylor (1987).

18. Bence, Grove, and Papike (1980), Presnall and Hoover (1984).

19. See Faure (1977) and Hyndman (1972, pp. 89–94) for more information on $^{87}Sr/^{86}Sr$ in rocks. Note that rocks may initially contain some of the radiogenic ^{87}Sr, so *initial ratios* of $^{87}Sr/^{86}Sr$ must be determined for definitive analyses. See Richardson and McSween (1989, pp. 385–88) for a brief and clear explanation of determining initial Sr (and Nd) isotope ratios.

20. See DePaolo and Wasserburg (1976a, 1976b, 1979a) and DePaolo (1979, 1980a, 1980b, 1981a, 1981b) for more details on neodymium isotopes, which have become widely used in the interpretation of rock histories. The review here is based primarily on these works.

21. See DePaolo and Wasserburg (1979b) for the method of computing ε_{Nd}. Some rocks that show very low values of ε_{Nd} have been interpreted as derivatives of extremely enriched mantle (Fraser et al., 1985).

22. Non-arc oceanic andesite has been described by D. C. Stewart and Thornton (1975). Note, however, that Maaloe and Petersen (1981) call arc andesites "oceanic andesites." Robyn (1979) discusses possible non-arc andesites. Larsen and Cross (1986) describe rift zone andesites. See chapter 9 for further discussion of these and other arc and non-arc andesites.

23. For example, see Eichelberger (1975b, 1978a, 1978b), R. J. Stern and Ito (1983), Leeman (1983), Harmon et al. (1984), and Carroll and Wyllie (1990).

24. See M. J. O'Hara (1965), Green and Ringwood (1968), Boettcher (1973), Kushiro (1974a), C. R. Stern, Huang, and Wyllie (1975), and Eggler (1978); but see Stern and Wyllie (1973) and J. B. Gill (1974). Also see Gill (1981) for a thorough review.

25. See the sources in note 24.

26. Suzuki and Shiraki (1980), Tatsumi and Ishizaka (1981, 1982).

27. Wyllie (1977) reviews crustal anatexis using studies of experimental systems. H. G. F. Winkler (1965, 1976) summarizes some important experimental work on the melting of high-grade metamorphic rocks derived from wackes and shales as a means of producing granite-quartz diorite magmas. Also see the works of Brown and Fyfe (1970, 1972), G. C. Brown (1973), Kistler and Peterman (1973, 1978), Piwinskii (1973a, 1973b), Presnall and Bateman (1973), Carmichael et al. (1974, pp. 593–95), Chappel and White (1974), Leake et al. (1980), Tindle and Pearce (1983), Clemens and Wall (1984), Wyllie (1984), Grant (1985a, 1985b), Clemens, Holloway, and White (1986), Huang and Wyllie (1986), D. R. Smith and Leeman (1987), Carroll and Wyllie (1990), and the collected papers in Atherton and Tarney (1979) and Roddick (1983). Mantle origins, and possible limitations and contributions, are discussed by Chappell and White (1974), Arth and Barker (1976), Wyllie et al. (1980), DeVore (1982, 1983), Campbell and Taylor (1983), Kagami and Shuto (1983), and Hensel et al. (1985). Also see papers in Barker (1979b).

28. Kistler and Peterman (1973, 1978), R. D. Davies and Allsop (1976), A. B. Thompson and Algor (1977), Abbott and Clarke (1979), Pankhurst (1979), A. B. Thompson and Tracy (1979), Abbott (1981), E. H. Brown et al. (1981), Cullers, Koch, and Bickford (1981), Czamanske, Ishihara, and Atkins (1981), DePaolo (1981a), Ewart (1981), Hildreth (1981), Kistler, Ghent, and O'Neil (1981), Clemens and Wall (1984), Clemens (1984), Grant (1985a, 1985b), Hensel, McCulloch, and Chappell (1985), Arth, Criss, et al. (1989), Arth, Zmuda, et al. (1989). See chapters 12 and 13 for more information.

29. See papers in Barker (1979b), by McGregor, Payne and Strong, and Tarney et al. Also see P. M. Hurley et al. (1962), Moorbath (1975a), C. R. Stern, Huang, and Wyllie (1975), DePaolo (1980a,b), Leake, Brown, and Halliday (1980), DeVore (1981, 1982, 1983), and Kagami and Shuto (1983).

30. Rast (1970) and Philpotts (1990, ch. 3) review several hypotheses concerning the cause of magma ascent, including buoyancy, gas fluxion, tectonic squeezing, and volume expansion. Rast (1970) favored volume expansion. Buoyancy is now considered to be a major cause. For additional discussions of magma movements, see Ramberg (1970, 1981), Weertman (1971, 1972), Wyllie (1971c), W. J. Phillips (1974), W. S. Pitcher (1975, 1979), B. D. Marsh (1976a, 1978, 1979, 1981, 1982b), Yoder (1976), D. M. Harris (1977), Petraske, Hodge, and Shaw (1978), H. R. Shaw (1980), Spera (1980), Anderson (1981b), Turcotte (1982), Ribe (1983), Scott and Stevenson (1984), Mahon, Drew, and Harrison (1985), Spence and Turcotte (1985), Turcotte and Emerman (1985), Ida and Kumazawa (1986), Scott and Stevenson (1984), Castro (1987), Maaloe (1987), Ryan (1987), Wickham (1987), Mahon, Harrison, and Drew (1988), Glazner and Ussler (1989), J. G. Ramsay (1989), Takada (1989), and G. P. L. Walker (1989).

31. Viscosity, a property of a material, is the material's resistance to flow. Honey is more viscous than water and therefore flows with greater difficulty. Lavas may be 10^4 to 10^8 more viscous than water (see C. J. Hughes, 1982, for a review, and Kushiro, 1980).

32. See note 29.

33. Yoder (1976, p. 162), Waff and Bulau (1979), Maaloe (1981), D. S. Barker (1983, p. 124ff.), Nicolas (1986).

34. Arzi (1978), Maaloe (1981). Melt becomes interconnected where it represents 1 wt. % of the rock (Daines and Richter, 1988), but it cannot be extracted until much larger fractions of melt are formed.

35. Also see Weertman (1972), Yoder (1976, p. 168ff.), Maaloe (1981), and R. Bateman (1984).

36. Nye (1967, in Yoder, 1976), Shaw (1969, 1980), Weertman (1972), Spera (1980), Nicolas (1986).

37. Wilshire and Shervais (1975), Maaloe (1981), Nicolas (1986).

38. Maaloe (1981).

39. B. D. Marsh and Carmichael (1974), Marsh (1976a, 1979), Maaloe (1981).

40. Also see H. Ramberg (1970, 1981), Weertman (1971, 1972), Marsh (1976a, 1978, 1979, 1981, 1982b), Shaw (1980), Spera (1980), Turcotte and Emerman (1985), Ida and Kumazawa (1986), Castro (1987), Mahon, Harrison, and Drew (1988), Glazner and Ussler (1989), Takada (1989), and G. P. L. Walker (1989).

41. O. L. Anderson (1978, 1979). Also see Rast (1970) for a review.

42. Magma movements associated with crack propagation are discussed by Shaw (1980), Chouet and Julian (1985), Spence and Turcotte (1985), Emerman, Turcotte, and Spence (1986), Chouet (1986), Castro (1987), Maaloe (1987), Sammis and Julian (1987), and Takada (1989). Diapirism is discussed by Marsh (1982, 1984), Castro (1987), Rabinowicz, Ceuleneer, and Nicolas (1987), Mahon, Harrison, and Drew (1988), J. G. Ramsay (1989), and Walker (1989).

43. Yoder (1976, p. 170ff.) discusses volume increase. Rast (1970) favors this mechanism.
44. Weertman (1971), Phillips (1974), Yoder (1976), Spera (1980), Wadge (1980), Spence and Turcotte (1985), Turcotte and Emerman (1985). Also see B. D. Marsh (1976a) for a discussion of cooling models and ascent rates in fractures and pipes.
45. The diapirs discussed here are basically liquids that rise through plastic lithosphere. Students should not confuse them with solid mantle diapirs like those discussed by D. H. Green and Ringwood (1967b). See note 41 for references. For discussions and reviews of diapirism of magmas, see B. D. Marsh and Charmichael (1974), B. D. Marsh (1976a, 1978, 1979, 1984), B. D. Marsh and Kantha (1978), W. S. Pitcher (1979), Spera (1980), R. Bateman (1984), Castro (1987), Rabinowicz, Ceuleneer, and Nicolas (1987), Mahon, Harrison, and Drew (1988), J. G. Ramsay (1989), and Walker (1989).
46. Grout (1945), H. Ramberg (1967, 1970, 1981), Whitehead and Luther (1975).
47. W. S. Pitcher (1979), Eichelberger and Westrich (1981, 1983), Clemens (1984), J. A. Whitney (1988).
48. Yoder (1976, p. 189), Basu (1977b), Spera (1980), Turcotte and Emerman (1985). See Sparks, Sigurdsson, and Wilson (1977) for a contrasting view.
49. The words differentiation and fractionation are used by some authors to mean the same thing. Here, following Best (1982, p. 314ff.), C. J. Hughes (1982, p. 200), and D. S. Barker (1983, p. 129ff.), *differentiation* is used in a more general way, and *fractionation* is restricted to specific processes involving crystals. In contrast, Cox, Bell, and Pankhurst (1979, p. 2) use *fractionation* in a general sense. Also see Bowen (1928) and Yoder (1979). The works cited offer more detailed discussions of the various processes involved.
50. Also see discussions in Propach (1976), Yoder (1979), Best (1982), C. J. Hughes (1982), and D. S. Barker (1983), and papers such as Flower, Schmincke, and Ohnmacht (1977), Garcia and Jacobson (1979), Miesch (1979), Thorpe, Francis, and Morebath (1979), Irvine (1980a), C. Harris (1983), Marcelot, Maury, and Lefevre (1983), Michael (1984), C. F. Miller and Mittlefehldt (1984), Marsh and Maxey (1985), D. R. Baker and Eggler (1987), Martin and Nokes (1988), Weinstein, Yuen, and Olson (1988), Broxton, Warren, and Byers (1989), and Spell and Kyle (1989). A. Rice (1981) offers "convective fractionation" as an alternative to crystal settling. Also see B. H. Baker and McBirney (1985), McBirney et al. (1985), Nilson et al. (1985), B. D. Marsh (1988a,b), and D. Martin (1990) for related discussions. B. D. Marsh (1988) argues that crystal fractionation of basaltic magma does *not* produce a wide diversity of magma types, a thesis that is at odds with Bowen's widely accepted theory of fractional crystallization and one that will no doubt inspire vigorous reexamination of the processes of crystal-liquid fractionation.
51. See Wolff and Storey (1984) for an example.
52. J. Ferguson and Currie (1971), Roedder (1978, 1979), A. R. Philpotts (1979), A. R. Philpotts and Doyle (1983), McBirney and Naslund (1990).
53. See chapter 13.
54. Hildreth (1979, 1981), McBirney and Noyes (1979), Ludington (1981), D. Walker and DeLong (1982), Watson and Jurewicz (1984), McBirney (1985), Lesher and Walker (1988), Trial and Spera (1990).
55. For example, see Hildreth (1983a), Michael (1983a, 1983b), C. F. Miller and Mittlefehldt (1984), B. H. Baker and McBirney (1985), and Stix et al. (1988).
56. Additional discussions of magma mixing and specific examples may be found in MacDonald and Katsura (1965), T. L. Wright and Fiske (1971), A. T. Anderson (1976), Eichelberger and Gooley (1977), Oskarsson, Sigvaldson, and Steinthorsson (1979, 1982), Grove, Gerlach, and Sando (1982), W. K. Conrad, Kay, and Kay (1983), Cantagrel, Didier, and Gourgaud (1984), Vogel et al. (1984), Furman and Spera (1985), Koyaguchi (1985, 1986), Bacon (1986), S. R. Mattson, Vogel, and Wilband (1986), McMillan and Dungan (1986), R. A. Thompson, Dungan, and Lipman (1986), Visona (1986), Kenyon and Turcotte (1987), Reidel and Fecht (1987), Vogel et al. (1987), N. L. Green (1988), Juster, Grove, and Perfit (1989), and A. R. Philpotts (1990, p. 261ff.).
57. Reviews of assimilation are provided by McBirney (1979) and texts such as Best (1982, p. 328ff.), C. J. Hughes (1982, p. 211ff.), D. S. Barker (1983, p. 136ff.), A. Hall (1987, p. 260ff.) P. C. Hess (1989, pp. 34–36), and A. R. Philpotts (1990, pp. 258–61). Examples and additional discussion may be found in Bowen (1928, 1948), Whitford and Jezek (1979), H. P. Taylor (1980), James (1982), Nicholls and Stout (1982), Harmon et al. (1984), James and Murcia (1984), McDonough and Nelson (1984), J. D. Myers, Sinha, and Marsh (1984), J. P. Davidson (1985), Furlong and Myers (1985), Kelemen (1986), Campbell and Turner (1987), R. MacDonald et al. (1987), Ague and Brimhall (1988a), and J. S. Marsh (1989).
58. Bowen (1928, pp. 182–85) argues cogently that magmas are not generally superheated. Marsh (1978), in contrast, argues that andesitic magmas may be superheated over much of the length of their traverse through the lithosphere.

PROBLEMS

6.1. Calculate the magnesium numbers from the following data:

	Mantle rock	MORB*	Andesite	High-Mg andesite
MgO	40.11	6.91	6.13	10.76
FeO	6.59	7.23	4.64	6.81

*mid-ocean ridge basalt

6.2. How long would it take a granitic magma traveling at a velocity of 2.3×10^{-6} cm/sec to move from the base of the crust (at 30 km) to the surface (assuming it could do so without crystallizing)?

6.3. Using the data in table 3.1, do a simple magma-mixing experiment by mixing the basalt represented by analysis 2 and the rhyolite represented by analysis 6 (add the analyses together and divide by two). (a) What rock type is produced (use the chemical classification shown in figure 4.9 to determine the name)? (b) What differences exist between this mixed-magma composition and a similar analysis from table 3.1 or AGI Data Sheet 59 (Dutro, Dietrich, and Foose, 1989)? (c) Do the analyses match adequately? Suggest reasons for any differences that might exist between the mixed-magma composition and the like analysis from table 3.1 or AGI Data Sheet 59. Could other simple mixing ratios make the analyses more alike? Explain.

7

Basalts and Ultramafic Volcanic Rocks

INTRODUCTION

Mafic and ultramafic volcanic rocks form in diverse crustal settings. The most abundant of these rock types are the basalts. **Basalt** is fine-grained to aphanitic, generally dark-toned (mafic) igneous rock composed of plagioclase and variable amounts of other minerals (figure 7.1). Common additional minerals include Ca-rich pyroxene (e.g., augite), Ca-poor pyroxene (e.g., hypersthene), olivine, and oxides (e.g., magnetite or ilmenite). Chemically, basalts have 45–52% SiO_2. Different types of basalt are known, each having a mineralogy reflecting its chemistry. Basalts occur in various volcanic structures ranging from lapilli and bombs to shield cones, lava plateaus, and the entire upper layer of the oceanic crust.

Unlike basalts, ultramafic volcanic rocks are relatively rare. The two main types are komatiite and kimberlite. The term **komatiite** has been applied to a suite of rocks (Arndt, Naldrett, and Pyke, 1977), but here we restrict our attention to ultramafic volcanic komatiites (uv-komatiites). **Uv-komatiites** are high-Mg, low-Ti, olivine-rich volcanic rocks characterized by a distinctive herringbone-like texture called a spinifex texture. **Kimberlite** is a porphyritic ultramafic rock composed of phenocrysts of olivine, phlogopite, and minor additional minerals in a groundmass of such minerals as serpentine, calcite, and olivine. Typically kimberlite is brecciated and occurs in dike-like volcanic pipes called diatremes.

Some basaltic and ultramafic magmas are primary magmas. Basaltic magmas are also parental magmas and numerous derivative magmas and rocks develop from them.

BASALT TYPES AND DERIVATIVES

The literature on basaltic rocks is commonly confusing to geology students because different papers and books refer to basalts using different names, depending on how the author of a particular text classifies these rocks. Different modal, normative, and chemical classifications, varying sites of origin, and different interpretations of rock chemistry give rise to these different names.

Many a petrologist over the years has classified igneous rocks on the basis of their occurrence in specific tectonic or geographic regions (Harker, 1909; BVSP, 1981). The development of the plate tectonic theory allowed a refinement of the practice, so that now we often find designations such as Mid-Ocean Ridge Basalt (MORB), Ocean Island Basalt (OIB), or Back-Arc Basalt (BAB).[1] This recently refined method of separating and naming rock types has done little to overcome the

Figure 7.1 Photomicrograph of alkali olivine basalt from east of Barstow, California, showing rectangular (lath-shaped) plagioclase grains and a granular olivine phenocryst. Long dimension of photo is 3.25 mm.

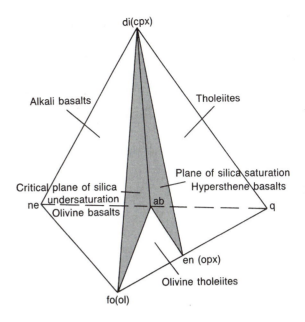

Figure 7.2 The basalt tetrahedron. The three volumes, from left to right, represent alkali basalt, olivine tholeiite, and tholeiitic basalt. The critical plane of silica undersaturation is the field of olivine basalt, whereas the plane of silica saturation is the field of hypersthene basalt. ne = nepheline, fo = forsterite (olivine), q = quartz, di = diopside (clinopyroxene), ab = albite (plagioclase), and en = enstatite (orthopyroxene).
(Modified from Yoder and Tilley, 1962, figures 1 and 2)

major flaw in naming rocks in this way, for this practice overlooks the fact that *rocks lumped together by region commonly have different chemistries and histories, whereas rocks from different regions may be similar.*

A number of petrologists choose, instead, to use modal, normative, or chemical designations. Daly (1918, 1933) divided basalts on the basis of mineralogy, which he recognized as a reflection of their chemistry. Similarly, Johannsen (1939), in his classic work on petrography, chose a mineralogical (modal) classification. These and other workers noted two main types of basalt, one containing common to abundant olivine and a Ca-pyroxene (augite) and one containing little or no olivine and a Ca-poor pyroxene (hypersthene or pigeonite). Following Kennedy (1933), the first type is now called **alkali olivine basalt** or olivine basalt and the latter is called **tholeiite** or tholeiitic basalt. Because mineralogies reflect chemistry, we generally find that tholeiites are hypersthene-normative, whereas alkali olivine basalts are olivine- and nepheline-normative.

Yoder and Tilley (1962) used the "basalt tetrahedron" to divide basalts into five types (figure 7.2). Their subdivision is based on physical-chemical principles, but is reflected by both the normative and modal mineralogies of the rocks. Two less common types of basalt, olivine basalt and hypersthene basalt, are defined by the planes of silica undersaturation and silica saturation, respectively. Tholeiite (with normative quartz), olivine tholeiite (with normative enstatite and olivine, but lacking normative quartz or nepheline), and alkali basalt (with normative nepheline) occupy the volumes in the tetrahedron.

Although the Yoder and Tilley (1962) classification has a sound chemical basis,[2] the definitions they assigned to rock names differed significantly from those already used by other

workers. This has created confusion. For example, tholeiite, as defined by Yoder and Tilley, denotes rocks with a far more limited range of chemistry than the tholeiite defined by G. A. MacDonald and Katsura (1964). Additional confusion results from the fact that the term tholeiitic is applied by some petrologists to rocks with specific alkali/silica ratios[3] or (total iron)/MgO vs. SiO_2 values (figure 7.3).[4] Similarly conflicting uses exist for the term alkali olivine basalt.

Because of the confusion, Chayes (1966) recommended abandoning the names that had conflicting uses. He recommended a two-fold division of basalts, with quartz-normative basalts being called *subalkaline basalts* and nepheline-normative basalts being called *alkaline basalts*. Rocks that lack quartz or nepheline in their norms are assigned to one of the two categories on the basis of hypersthene-diopside-olivine ratios. Irvine and Baragar (1971) adopted Chayes (1966) subdivision for their volcanic rock classification, added a *peralkaline* category to accommodate sodic volcanic rocks, and incorporated the older terms as rock types. According to these workers, Calc-alkaline Series rocks are derived from the Tholeiitic Basalt Series, and these two groups make up the Subalkaline Rocks." Various nepheline, sodic, and potassic rocks with their parent Alkali Olivine Basalt Series make up the "Alkaline Rocks."

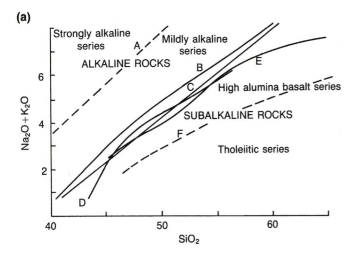

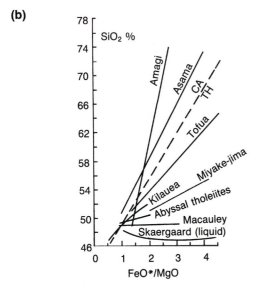

Figure 7.3 Major element discrimination diagrams for distinguishing tholeiitic series rocks from alkaline rocks. (a) Alkali/silica summary diagram (after Schwarzer and Rogers, 1974). Curve A from Saggerson and Williams (1964) separates strongly from mildly alkaline rocks. Curve B is from Irvine and Baragar (1971). Curve C is from G. A. MacDonald and Katsura (1964). Curve D is from Hyndman (1972). Curve E is from Kuno (1968). Curve F, which separates tholeiitic basalts from high-alumina basalts, is from Kuno (1968). (b) FeO*/MgO vs. SiO_2 diagram of Miyashiro (1974, 1975d). FeO* is total iron as FeO. Calc-alkaline rocks (CA) are to the left of the dashed line, whereas tholeiitic rocks (TH) plot to the right. The names on the curves apply to particular rock suites. Values plotted in weight percent.

To add to the confusion, additional special rock names based on various mineralogical or chemical parameters have been proposed. For example, in one classification high-alumina basalt is basalt with Al_2O_3 > 17% and basanite is olivine-nepheline basalt.[5] Figure 7.4 seeks to reduce the confusion by showing the approximate equivalency of terms used by various petrologists. Exact equivalency does not exist.

OCCURRENCES OF BASALTS

Basalts are known from a number of different tectonic settings. Each setting has a combination of physical characteristics, such as stress regime, P-T conditions, and source rock, that give various basalts formed in that environment a distinctive chemistry.

Settings in which basalts form are here divided into three major categories—rift settings, subduction-compressional settings, and intraplate settings. Each category may be further subdivided.

Rift Volcanism

Rift settings are environments where tensional forces cause normal faulting, minor reverse faulting, and associated volcanism. The origin of the basaltic magma that forms in these settings is generally attributed to partial melting in a mass of hot, rising mantle (a diapir) that moves up from the subsurface towards the low-pressure rift zone created by the tension. The magmas separate from the mantle and continue their movement towards the surface, where they intrude or extrude, and crystallize to form rocks. The mid-ocean ridges, spreading centers or rifts that occur on the ocean floors (see figure 6.4), are the sources of large quantities of Mid-Ocean Ridge Basalt, or **MORB** (A. E. J. Engel, Engel, and Havens, 1965).[6] Similar volcanism occurs behind volcanic arcs at sites of back-arc spreading, where Back-Arc Basalt (BAB) is generated.[7] Tensional forces may also split continental crust, forming continental rifts and causing associated continental rift volcanism.[8] Some so-called flood basalts may form in association with such a rifting event.[9] In addition, infrequently, basaltic volcanism occurs along leaky transform faults, which have tensional forces acting on them (Pilger and Henyey, 1979; C. M. Johnson and O'Neil, 1984).

Basalts formed at rifts tend to be tholeiitic but are not exclusively so, as local alkali olivine basalts are also known in these settings. Continental rifts, in particular, are characterized by a bimodal suite (two rock types) composed of tholeiitic basalt and rhyolite. At back-arc sites associated with young arcs and in other rifts, crustal contamination and fractional crystallization may yield chemically altered magmas that form andesite, dacite, and rhyodacite.[10]

Volcanism in Subduction Zone Compressional Settings

Subduction-zone compressional settings occur where tectonic plates collide to form subduction zones (figure 6.4) and include those cases where continents or arcs, transported on the plates, collide as a result of plate collisions. Continental (or arc) collisions form mountain belts. Plate collisions create volcanic arcs, either on continental margins (Andean-type) or on oceanic crust (Pacific-type). At depth in the collision zones, partial melting yields basaltic

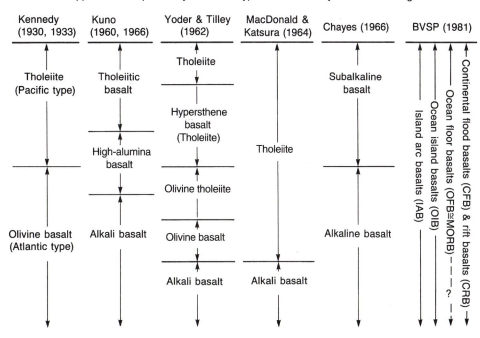

Figure 7.4
Diagram showing
approximate equivalency of
basalt types as defined by
various petrologists. BVSP = Basaltic Volcanism Study Project.
MORB = Mid-Ocean Ridge Basalt.

magmas that rise towards the surface where they, along with associated andesites, dacites, and rhyolites, erupt to form composite volcanoes. Similar volcanism occurs where crustal compression causes melting, as is proposed for some Archean Precambrian terranes.

Basalts are less abundant in arcs and compressional zones than in rifts (BVSP, 1981). Because of the more complex tectonic environment, however, arc basalts are chemically more diverse. Typically, arc basalts have calc-alkaline chemistries, but tholeiitic chemistries also are common.[11] Modally, arc basalts range from alkali olivine to tholeiitic types.

Intraplate Volcanism

Intraplate settings, environments that occur within the plates away from plate margins, include "hot spot" environments (both continental and oceanic) and flood-basalt provinces.[12] Hot spots are localized sites of volcanism above mantle diapirs or plumes that yield basaltic partial melts.[13] Where plates move over a hot spot, a linear volcanic province or a seamount chain may develop. Flood-basalt environments occur where large volumes of basaltic magma erupt over relatively short periods of time to produce lava plateaus. Eruptions occur along fractures in continental or oceanic crust that serve as conduits for basaltic magmas generated in the mantle.

Tholeiitic basalt is the dominant rock type in intraplate settings. However, like the arc settings, hot spot settings locally contain chemically and mineralogically diverse rock types.

BASALT CHEMISTRY, MINERALOGY, TEXTURES, AND STRUCTURES

The two major types of basalt differ in mineralogy and chemistry. Likewise, the textures and structures found in basalts are diverse.

Mineralogically, basalts consist of essential plagioclase and characterizing accessory minerals such as augite, magnesium-rich orthopyroxene, olivine, and magnetite. In addition, basalts may contain pigeonite or hornblende and minor accessory minerals such as apatite, sphene, ilmenite, pyrite, pyrrhotite, and pentlandite. In feldspathoidal varieties, nepheline and analcime are present.[14]

Differences in the modal abundances of minerals in different types of basalts reflect different basalt chemistries. Both major and trace element chemistry have been used to characterize, distinguish, and classify basalts.[15] Table 7.1 presents some selected major element analyses and their norms. Recall that by chemical definition, basalts contain between 45 and 52% silica. Note the range of alumina contents, the alkali values, and the general decrease in MgO values with increase in SiO_2. Pearce (1976) used data like these, in greater numbers, to statistically distinguish six basalt magma types—ocean floor basalt, low-potassium tholeiite, calc-alkaline basalt, shoshonite, ocean island basalt, and continental basalt—each representative of a particular tectonic environment.[16] Because weathering and metamorphism can cause ion migration and

Table 7.1 Major Element Analyses of Basalts with CIPW Norms

	1	2	3	4	5	6
SiO$_2$	43.28[a]	47.38	48.35	49.20	50.45	51.95
TiO$_2$	4.12	2.81	1.57	2.03	2.18	1.15
Al$_2$O$_3$	14.43	13.50	15.49	16.09	14.06	18.03
Fe$_2$O$_3$	0.70	3.18	3.26	2.72	4.75	4.97
FeO	10.92	9.74	8.05	7.77	6.91	5.20
MnO	0.13	0.19	0.17	0.18	0.16	0.12
MgO	11.68	9.24	7.03	6.44	7.83	5.52
CaO	11.22	9.92	9.92	10.46	10.50	8.70
Na$_2$O	2.49	2.58	2.76	3.01	2.41	3.35
K$_2$O	0.83	0.54	0.51	0.14	0.30	0.98
P$_2$O$_5$	0.31	0.40	0.24	0.23	0.33	0.24
H$_2$O	0.08	0.21	1.52	1.65	0.99	—
Other	0.35	0.09	0.05	—	—	0.71
Total	100.54	99.78	98.92	99.92	100.87	100.62
CIPW Norms[b]						
q	—	—	—	0.08	4.50	3.05
or	4.90	3.20	3.01	0.84	1.67	5.97
ab	7.58	21.88	23.34	25.49	20.44	28.32
an	25.74	23.71	28.38	29.96	26.69	31.27
lc	—	—	—	—	—	—
ne	7.32	—	—	—	—	—
di	22.52	18.49	15.64	16.61	18.81	8.21
hy	—	12.00	16.04	16.97	16.06	13.60
ol	22.49	9.50	2.68	—	—	7.20
mt	1.02	4.62	4.72	3.94	6.96	7.20
il	7.83	5.35	2.97	3.85	4.10	2.19
hm	—	—	—	—	—	—
cm	—	0.06	—	—	—	—
ap	0.74	0.95	0.57	0.54	0.67	0.57
Other	0.43	0.05	1.57	1.65	0.99	0.71
Total	100.57	99.81	98.92	99.93	100.89	100.62

Sources:

1. Picritic alkali olivine basalt from 1750 (?) lava flow, Haleakala Volcano, Maui, Hawaii. Analyst: M. G. Keyes (Eaton and Murata, 1960). Norm by L. A. Raymond.
2. Alkali olivine basalt, sample KLPA-1, East Molokai Volcano, Hawaii. Analysts: V. C. Smith and R. L. Rahill (Beeson, 1976).
3. Tholeiite (high-Mg Picture Gorge–type), Columbia River Basalt, Sample CP-1 (BVSP, 1981, p. 82). Norm by L. A. Raymond.
4. "Oceanic Tholeiite" (MORB), sample AD-2, Atlantic Ocean (A. E. J. Engel, Engel, and Havens, 1965). Norm by L. A. Raymond.
5. Tholeiite from Kulani Cone, Mauna Loa, Hawaii. Analyst: H. Asari (G. A. MacDonald, 1968, p. 498).
6. Tholeiite from continental arc, Laguna del Maule Volcanics, Chile. Sample LM-7 (F. A. Frey et al., 1984). Norm by David West.

[a] Values in weight percent.
[b] Norms calculated by cited author except as noted.

accompanying chemical change, classification of ancient basalts using major element analyses is restricted to rocks that are fresh and unmetamorphosed.

In contrast, some trace elements are immobile, even during alteration and metamorphism; hence, trace element chemistry has been used with some success in determining the paleotectonic settings of ancient rocks. For example, Pearce and Cann (1973) proposed the use of diagrams based on various combinations of Ti, Zr, Y, Nb, and Sr for distinguishing basalt types (figure 7.5). They separated ocean floor basalt (OFB), low-potassium tholeiite (LKT), calc-alkaline basalt (CAB), and within-plate basalts (WPB) using discriminant diagrams based on selected elemental ratios. Meschede (1986) proposed an additional Nb-Zr-Y diagram, and Ikeda (1990) introduced a Ce-Sr-Sm diagram, particularly useful for distinguishing Back-Arc Basin basalts. Pearce and Cann (1973) and others[17] recommend caution in the use of this technique, notably because of uncertainties and the fact that alteration and metamorphism can cause migration of some trace elements, altering the elemental ratios.

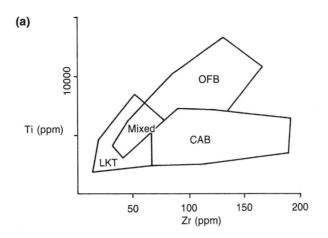

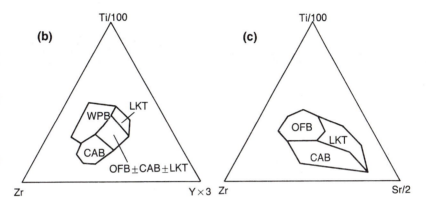

Figure 7.5 Discrimination diagrams of Pearce and Cann (1973) for use in distinguishing various basalt types. (a) Ti-Zr plot showing four fields representing basalt types. Low-potassium tholeiites (LKT) plot in fields labelled LKT and "Mixed." Ocean floor basalts (OFB) plot in "Mixed" and OFB fields. Calc-alkaline basalts (CAB) plot in "Mixed" and CAB fields. (b) Plot of Zr-Ti/100-3Y showing fields for low-potassium tholeiites (fields LKT and OFB ± CAB ± LKT), calc-alkaline basalts (fields OFB ± CAB ±LKT and CAB), ocean floor basalts (field OFB ± CAB ± LKT), and within-plate basalts (field WPB). (c) Zr-Sr/2-Ti/100 diagram showing fields for low-potassium tholeiites (LKT), calc-alkaline basalts (CAB), and ocean floor basalts (OFB). *Note:* Because Sr may be mobilized by alterations of the rock, this plot is most useful for unmetamorphosed and unweathered rocks.

In addition to allowing tectonic assignments, major and trace elements, including isotopes, provide important constraints on magma sources and subsequent alterations (e.g., through differentiation or assimilation).[18] As noted in chapter 6, Sr-isotope ratios, Nd/Sm, and other isotopic ratios, as well as rare earth elemental abundances (see chapter 3), reveal mantle versus crustal sources, the role of differentiation processes in magma evolution, and the presence or absence of contamination.

Basalts show a wide range of textures. Commonly they have porphyritic or intergranular textures, but trachytoidal, spherulitic, intersertal, dichtytaxitic, seriate, subophitic, and ophitic textures are present in some basalts.[19]

Like textures, structures are diverse in basaltic rocks. Flows with blocky to ropy surfaces characterize recent basalts, and vesicles and amygdules, especially near flow tops, occur in both Recent and ancient rocks. Columnar joints are relatively common features in basaltic flows. Lava tubes and pillow structures are more common in basaltic rocks than in any other rock type. Additional structures found in basaltic terranes include lapilli, bombs, squeeze-ups, hornitos, pressure ridges, tumuli, spatter cones, cinder cones, and craters. Major structures include shield cones, lavas plains, and lava plateaus.

SPECIFIC OCCURRENCES AND ORIGINS OF BASALTS

Recall that basalts are found in rift zones, subduction-zone compressional settings, and intraplate sites. Also recall that basaltic magmas originate in the mantle and that tholeiitic magmas form at shallower depths than do alkali olivine basalt magmas. This section describes the occurrences and the processes involved in the development of basalts, including general basalt characteristics and petrogenesis for each type of tectonic setting.

Intraplate Volcanism: Hot Spot—The Hawaiian Islands

The Hawaiian Islands[20] are a group of islands located at the southeast end of a 5600-km-long chain of volcanic structures, called the Hawaiian Island–Emperor Seamount Chain, that extends from the Aleutian Arc in the north-central Pacific Ocean to the central equatorial region of that ocean (figure 7.6a). The island of Hawaii itself is a volcanic landform that above sea level consists of five volcanic peaks (figure 7.6b). Three of these volcanoes—Hualalai,

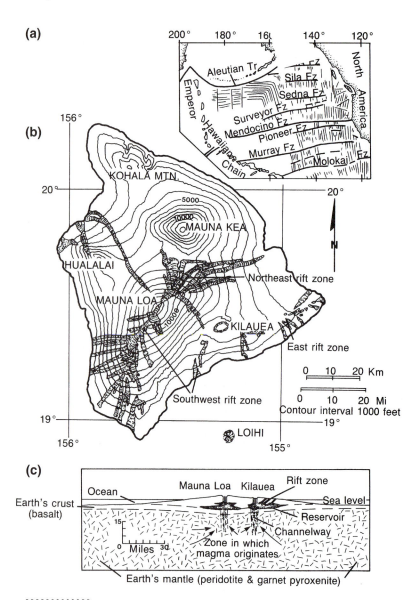

(a)

(b)

(c)

Figure 7.6 Geographic setting of Hawaiian volcanism. (a) Location of the Hawaiian Island-Emperor Seamount Chain in the Pacific Ocean (based on D. E. Hayes and Pittman, 1970; E. D. Jackson, Silver, and Dalrymple, 1972). (b) Map of Hawaii showing the five major volcanoes and historic lava flows (dark areas)(modified from G. A. MacDonald and Abbott, 1970), with the location of the Loihi Seamount (from J. G. Moore, Clague, and Normark, 1982). (c) Simplified, idealized cross section through Hawaii (from G. A. MacDonald and Abbott, 1970).

Mauna Loa, and Kilauea—have erupted in historic times. Mauna Loa and Mauna Kea, the largest of the Hawaiian volcanoes, are the tallest mountains on earth, rising more than 9 km (30,000 ft.) above the surrounding sea floor. Surprisingly, the entire volume of these volcanoes, more than 50,000 km[3], apparently was amassed in less than one million years.[21] Southeast of Hawaii, the newest volcano and future island is beginning to form, as the volcanically active Loihi Seamount (figure 7.6b).[22]

Observations

Sampling and analysis of Hawaiian volcanic rocks reveals both tholeiitic and alkalic rocks (G. A. MacDonald and Katsura, 1964; J. W. Hawkins and Melchior, 1983)[23] (figure 7.7). Tholeiitic rocks (tholeiitic basalts and their derivatives) are generally considered to make up about 99% of the volcanic rocks, with alkalic rocks (alkali olivine basalts and derivatives) comprising the remainder.[24] Locally associated with the Hawaiian tholeiites are **picritic** (olivine-rich) **basalts,** olivine tholeiites, and minor rhyodacite and granophyre (micrographic granite).[25] Associated with alkali olivine basalt are rocks such as ankaramite (pyroxene-olivine phyric basalt), hawaiite (andesine andesite with high color index, high Na/K, and commonly modal or normative olivine), mugearite (oligoclase-, alkali-feldspar-, olivine-, and pyroxene-bearing porphyritic rock), and trachyte.[26]

The eruptive history of seamounts and islands in the Hawaiian-Emperor Chain has two major aspects. First, the age of seamounts and volcanoes decreases from 80–75 m.y. at the north end to Recent age in the south (H. R. Shaw, Jackson, and Bargar, 1980; D. A. Clague and Dalrymple, 1987). Second, each individual volcano or volcano group has passed through a series of stages (Stearns, 1940; G. A. MacDonald and Abbott, 1970; D. A. Clague and Dalrymple, 1987).[27] These stages are as follows.

1. *Preshield Stage.* This stage is characterized by eruption of alkali olivine basalt, with less abundant tholeiite and basanite (nepheline-olivine basalt), to form a small shield cone. (Loihi, the newly discovered, active seamount southwest of Hawaii, represents this stage.)
2. *Shield-building Stage.* Voluminous eruptions of tholeiite in frequent eruptions construct the bulk of the shield volcano during this stage. Flank eruptions occur, producing flows on the sides of the volcano. Rare siliceous rocks such as rhyodacite are produced. During the later parts of the stage, collapse produces a caldera. The caldera is then filled with tholeiitic basalts. As the end of this phase approaches, eruptions become less frequent.
3. *Postshield Stage.* Relatively infrequent, more explosive, much less voluminous eruptions of alkali olivine basalts, with phenocrysts of pyroxene, olivine, and plagioclase, build pyroclastic cones and flows that cap the shield cone and cover the caldera. Erupted with the alkali basalts are alkaline differentiates such as hawaiite and trachyte.
4. *Rejuvenated Stage.* Following a major period of quiescence and erosion, silica-poor lavas, including alkali olivine basalt, basanite, ankaramite, nephelinite, and trachyte, are erupted. These lavas again cap the volcano.

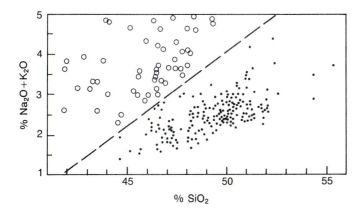

Figure 7.7 Alkali/silica diagram of MacDonald and Katsura (1964), showing data from Hawaii. Open circles represent alkali basalts; solid dots represent tholeiites. The line corresponds approximately with the plane of silica saturation of Yoder and Tilley (1962).

(From G. A. MacDonald and T. Katsura, "Chemical Composition of Hawaiian Lavas" in Journal of Petrology, 5:82–133, 1964. Copyright © 1964 Oxford University Press, Oxford England. Reprinted by permission.)

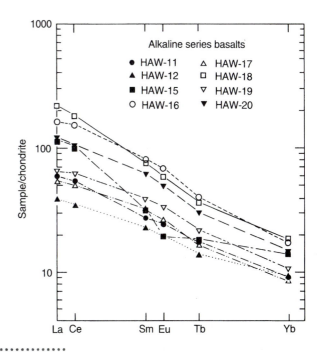

Figure 7.8
Rare earth element data for Hawaiian alkalic rocks.
(From BVSP, 1981, p. 172)

Chemically, the Hawaiian basalts generally have moderate to high magnesium numbers (0.50–0.78), although in some members of the alkali suite, the Mg number falls to less than 0.20.[28] With the exception of some alkalic rocks, TiO_2 is also generally high (> 1.5%). Lime (CaO), potash (K_2O), soda (Na_2O), ferrous oxide (FeO), and alumina (Al_2O_3) show a considerable range of values (BVSP, 1981). For example, alumina ranges from about 12 to 18 wt.% in alkali basalts and from 9 to 14 wt.% in tholeiitic rocks. Soda ranges between 2 and 5 wt.% in alkalic rocks, but is restricted to the 2–3% range in tholeiites. Tholeiitic glasses erupted early are enriched in Fe, Ca, Ti, Na, K, H_2O, CO_2, S, F, and Cl relative to glasses erupted later in the eruptive sequence (M. O. Garcia et al., 1989).

The trace element chemistry is also distinctive. Rare earth elements, especially the light rare earth elements (e.g., La and Ce), are substantially enriched in Hawaiian volcanic rocks (figure 7.8)(Lanphere and Frey, 1987; Tilling, Wright, and Millard, 1987). In addition, Nd ratios exceed 0.51262, the $^{87}Sr/^{86}Sr$ is low, falling in the range 0.70305 to 0.70420, and $^3He/^4He$ is as much as four times the values for MORBs.[29] Notably, changes in chemistry are associated with changes in eruptive stages (C. Y. Chen et al., 1991).

Interpretations and Petrogenesis

Seismic evidence (Eaton and Murata, 1960; Aki and Koyanagi, 1981) provides an initial indication that Hawaiian lavas are generated at depths of 40 km or more in the mantle (figure 7.6c). Model studies and calculated phase equilibria for presumed mantle compositions suggest further that olivine-basalt magmas must rise from depths of 80 km or more (Saxena and Eriksson, 1985; Wyllie, 1988). The laboratory experiments discussed in chapter 6 also generally suggest mantle depths of origin for basalt magmas, as do some included xenoliths (R. W. White, 1966; D. A. Clague, 1988). Thus, it is reasonable to conclude that the Hawaiian magmas were formed at mantle depths.

How might these magmas form? Two hypotheses have been proposed. The favored thesis, based on an idea presented by J. T. Wilson (1963) and developed by W. J. Morgan (1971, 1972), is that the Hawaiian Island-Emperor Seamount Chain (and other similar chains) represents the trace of a mantle hot spot on a moving lithospheric plate.[30] The volcanoes (representing the hot spot) mark the position on the surface beneath which there is a hot mantle plume, a cylindrical column of hot mantle material rising through cooler mantle materials (Burke and Wilson, 1976). The plume partially melts on the way up, providing magmas that penetrate the lithosphere to produce the observed volcanism. The best evidence for the mantle-plume hypothesis is (1) the age progression of volcanism along the chain, (2) the similar trends of the Hawaiian and other volcanic chains in the Pacific Ocean, (3) the compatibility of the plume hypothesis with observations and theories supporting the plate tectonics model, and (4) geochemical evidence, such as high He-isotope ratios and low Nd-isotope ratios, indicating that components of the magmas are derived from primitive, deeper mantle source rocks (C. Y. Chen and Frey, 1985; Feigenson, 1986).

A second model of magma formation at hot spots was developed by E. D. Jackson and H. R. Shaw and their colleagues.[31] In this model, a moving plate of lithosphere generates shear melting at its base. When the rocks fail (break) as a result of movement, heat is generated, melting occurs, and there is a "thermal feedback" that produces more melting.

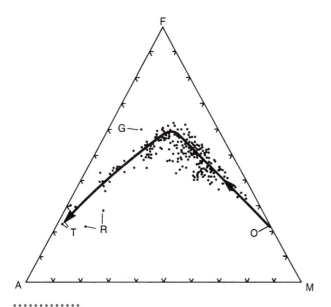

Figure 7.9 AFM plot of analyses of Hawaiian volcanic rocks. O = olivine, the composition of phenocrysts from tholeiite. The arrow shows the fractionation trend, first away from olivine, somewhat towards G, the composition of a granophyre (granitoid rock) and then towards T, the composition of some trachytes. R = composition of rhyodacites.

(Modified from G. A. MacDonald, 1968)

During melting, the residue of unmelted material sinks, creating a "gravitational anchor" and causing further flow and melting (H. R. Shaw and Jackson, 1973). Eruption of the melts at the surface produces volcanoes. This thermal-feedback, shear-melting model is favored by its ability to explain (1) an *en echelon* distribution of island groups within the volcanic chain, and (2) the time-distance-volume relations. Some geochemical and other data are also consistent with the thermal-feedback, shear-melting model (Feigenson and Spera, 1981). Nevertheless, the model fails to explain how lighter, hot residue can sink into heavier, colder pristine mantle or why some data suggest depths of origin for the magmas that are greater than the lithosphere-asthenosphere boundary.[32]

Both models require partial melting of mantle rocks to produce the magmas for Hawaiian volcanoes. Because of some similarities in major element chemical trends for tholeiitic and alkalic rock series, because of the volumes and close association of the two rock types, and because of the apparent gradation of tholeiite to alkalic rocks, Eaton and Murata (1960), G. A. MacDonald and Katsura (1962), and G. A. MacDonald (1968) concluded that an olivine tholeiite magma was the parent for all other Hawaiian magmas. Olivine-rich types would be derived by crystal settling fractionation, whereas more alkalic and siliceous magmas would represent the liquid residual melts developed through fractional crystallization. The fractionation trend is suggested by the AFM plot of Hawaiian lavas, which shows a trend away

from olivine and ultimately towards rhyodacites and trachytes (figure 7.9). Alternatively, Wyllie (1988) suggested that both tholeiitic and alkalic magma types develop from picritic (olivine-rich) parent magmas.

In conflict with the hypothesis of Eaton and Murata (1960) are data obtained by Yoder and Tilley (1962) through experimental work. That work suggested that at low pressures neither tholeiitic nor alkalic magmas could be derived from the other by fractional crystallization, because such a chemical change would require the magma composition to cross the critical silica-saturation plane of the basalt tetrahedron (see figure 7.2). They argue that the two Hawaiian magma types were independently formed. Further, they suggested that tholeiitic magmas form at shallower depths and alkalic magmas form at greater depths. Additional experiments at higher pressures are consistent with these views, but conflicting data exist.[33]

Do the chemical data resolve these problems? They do not entirely. First, factors such as the high magnesium number, Nd-isotope ratios, low values of $^{87}Sr/^{86}Sr$, and high TiO_2 suggest mantle sources for all of the magmas. Similarly, the light rare earth element (LREE) enrichment is consistent with a mantle source containing garnet.[34] This is so because garnet preferentially takes up heavier, smaller rare earths in its structure and because partial melting of garnetiferous (possibly deep-seated) mantle rock, and specifically preferential melting of a mix of the *other* minerals in a mantle rock, yields magma relatively enriched in LREE. Furthermore, the high $^3He/^4He$ values indicate a primitive (deep?) mantle source. However, none of these data clarify which of the magma types is parental.

The trace elements and isotopic ratios do suggest that various batches of tholeiite, as well as tholeiites and alkali basalts in general, are derived from a *heterogeneous mantle*. Analyses of hydrogen and oxygen isotopes suggest a mantle plume source for the H_2O present in volcanic glasses (M. O. Garcia et al., 1989). Yet significant differences in Sr ratios in different rocks from single volcanoes (e.g., Loihi, Haleakala) demand a heterogeneous source rather than a homogeneous one (Lanphere, 1983; West and Leeman, 1987). Similarly, details of the trace element distributions suggest either mixing of magmas[35] or a complex source for magmas, with at least three geochemically distinct components, notably, a deep, primitive mantle-plume component, an enriched-mantle/oceanic-lithosphere component, and a depleted-mantle/crustal-lithosphere (MORB?) component.[36] Magma mixing may also explain some of the variations in major element chemistry (Wright and Fiske, 1971). Alternatively, enrichment of early glasses in elements such as Fe, Na, and K, relative to later glasses, suggests progressive melting and depletion of a source rock (M. O. Garcia et al., 1989). In short, the available data clearly suggest considerable complexity in source rocks, magma composition, and history for Hawaiian basalts.

That history is further complicated by late-stage fractionation in the lithosphere, especially in the crust.[37] Ryan, Koyanagi, and Fiske (1981) have modelled the three-dimensional

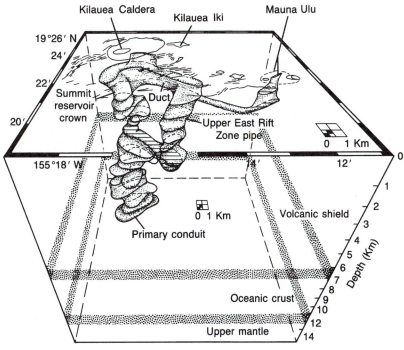

Figure 7.10 The magma transport system of Kilauea Volcano, Hawaii. Note the large areas, especially the summit reservoir crown, where magmas may undergo differentiation
(Modified from Ryan, Koyanagi, and Fiske, 1981).

internal "plumbing" of Kilauea (figure 7.10) and show that numerous shallow crustal sites of magma storage (and fractionation) are possible.[38] Geophysical data indicate that storage occurs. Observation of many Hawaiian lavas reveals that olivine is an early formed and common phenocryst phase. Phenocrysts suggest fractionation, and early crystallization of olivine controls the lava chemistry during fractionation. Both olivine and pyroxene, which crystallize early as Mg-rich forms, remove magnesium and, to a lesser extent, iron from the magma, driving the magma composition first towards the F-A side of the AFM diagram and then, as pyroxene and plagioclase form, towards A, the alkali-rich corner (figure 7.9).[39] Fractionation may occur at a variety of locations, from the asthenosphere-lithosphere boundary to very shallow sites within the volcano.[40]

In summary, based on mineralogical, petrographic, geochemical, and geophysical data, we conclude that Hawaiian magmas form by partial melting of a heterogeneous mantle plume source, perhaps containing primitive garnet lherzolite, additional enriched or depleted mantle rocks, and a crustal (MORB?) component. As the lithospheric plate moves over the plume, primitive alkalic magmas that are anatectically produced and equilibrated in the deeper mantle (> 80 km) will move towards the surface and erupt (the pre–shield stage). Some magmas erupt in unmodified form. As most magmas rise towards the surface, they pond, equilibrate with mantle rocks, assimilate mantle or crustal rocks, and fractionate before erupting. These processes continue to varying degrees during all phases of the eruptive history. Fractionation occurs via flow fractionation and by crystal-liquid fractionation. Ponding and fractionation may occur at several sites, including the asthenosphere-lithosphere boundary, shallow-level magma chambers and conduits, and planar magma-storage zones in the shallow crust.[41] Fractionation results in the development of various magmas that yield the variety of rocks present in the volcanoes.

As the plate moves across the plume, melting of the plume center at depths of 60–90 km yields great volumes of magma that swamp the early formed alkali basalt magmas. These new magmas erupt at the surface as tholeiites that build the volcano during the shield-building stage. As the eruptive site centers over the plume, eruption of tholeiites continues and local ponding of magmas within the volcano and shallow crust allows fractionation, producing more siliceous differentiates. Later during this stage, as the plate moves on, the quantities of magma diminish and magma chambers begin to cool and crystallize, leading ultimately to local, shallow-level collapse and the formation of a caldera (F. A. Frey et al., 1990). Remaining tholeiites in magma chambers in the crust and upper mantle continue to erupt and are replenished with lesser quantities of new magma from below. The frequency of eruptions diminishes. At this point, magmas may pond at the base of the crust (Ten Brink and Brocher, 1987; Watts and Ten Brink, 1989; F. A. Frey et al., 1990).

With the decrease in magma production as the eruptive center moves away from the plume center, differentiated alkalic magmas and their parents, generated at deeper levels or on plume margins, are again able to penetrate through to the surface.[42] These post–shield-stage alkalic magmas, which may also pond and fractionate, erupt less frequently than the tholeiites did, and they form a thin cap of flows and pyroclastic cones that conceals the caldera.[43] When adequate heat for melting is no longer available, volcanism ceases.

Erosion of the volcano ensues. The cause of the rejuvenated stage volcanism is unknown but it may be related to stress release (Ten Brink and Brocher, 1987). A drop in pressure due to isostatic uplift of rocks below the volcano may cause minor melting of mantle rocks, producing alkalic magma that invades the crust. Alternatively, a late pulse of deep mantle-plume material may cause the volcanism. If these late magmas reside in the crust and differentiate, ankaramite, hawaiite, mugearite, and trachyte may arise through fractionation. These materials finally erupt, erosion ensues, subsidence occurs, and the volcano eventually becomes a seamount.

The Hawaiian volcanoes are some of the most studied volcanoes in the world and are considered to be representative of hot spot volcanoes within an oceanic plate.[44] This example of volcanism demonstrates some of the ways that the variety of available information may be used to reconstruct the history of a volcano. The example also shows how complex the explanation of a volcano can be, even one in a somewhat simple environment away from the complexities of plate margin tectonics, continental erosion, and related processes.

Intraplate Volcanism: Flood Basalts—The Columbia River Plateau

The Columbia River Plateau in northwestern North America[45] is one of several provinces in the world—including the Deccan Traps province of western India, the Karro province of South Africa, and the Parana Basin of Brazil—underlain by huge volumes of tholeiitic basalt erupted over short periods of geological time.[46] The Columbia River Plateau is underlain by Miocene basalts covering an area of about 163,700 km^2 (figure 7.11) and occupying an estimated volume of 174,300 km^3 (Tolan et al., 1989). This large volume of basalt was erupted primarily during about a two-million-year period between 17 and 15 million years ago (Baksi, 1990), but diminished volcanism continued until about 6 million years ago (figure 7.12).[47]

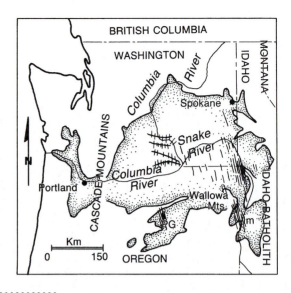

Figure 7.11 Map showing the area underlain by Columbia River basalts. North-northwest-trending lines are feeder dikes; west-northwest hachured lines are anticlinal ridges. PG = Picture Gorge region and IM = Imnaha region, where basalts of those names are localized.
(Modified from Hooper, 1982)

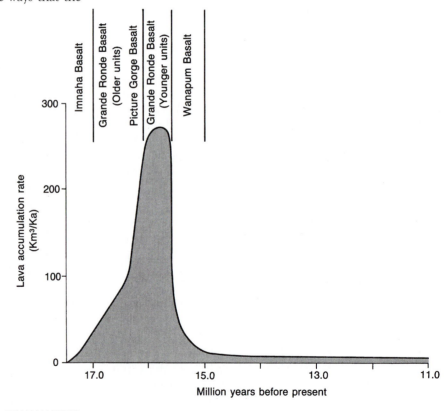

Figure 7.12
Diagram showing the accumulation rate of Columbia River basalts plotted against age. Grande Ronde Basalt here includes the Picture Gorge Basalt.
(From Baksi, 1990)

(a)

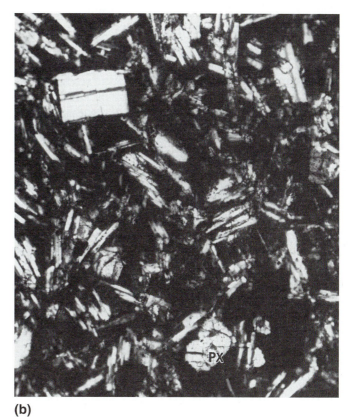

(b)

Figure 7.13 Columbia River basalts. (a) View east along Picture Gorge, Oregon. (b) Photomicrograph of Grande Ronde Basalt with plagioclase (rectangular) and pyroxene (blocky) grains (PX). Long dimension of photo is 1.00 mm.

The Columbia River basalts erupted through a series of long north-northwest-trending fractures, located primarily in the southern and eastern parts of the province. Individual flows developed quickly, as floods of fluid lava spread out as sheets, travelling long distances and covering large areas. One, two, or more flows cooled together (because they accumulated rapidly) forming *cooling units*. The cooling units together make up a thick volcanic pile composed of members, formations, and a subgroup of the Columbia River Basalt Group.[48] From oldest to youngest, the major units are the Imnaha Basalt, the Picture Gorge and Grande Ronde basalts, the Wanapum Basalt, and the Saddle Mountains Basalt (D. A. Swanson et al., 1979). Each of the units, including subunits, is distinguished on the basis of petrography, chemical composition, and magnetic polarity (Hooper, 1988a). In addition, each major unit is divisible into members, cooling units, or flows; or, on the basis of chemistry, into chemical types.

Observations

Unlike Hawaiian volcanism, the Columbia River Basalt eruptions underwent no obvious, regular petrographic and chemical variations with time. The petrologic variability is limited, and most rocks are varieties of tholeiite. Quartz tholeiite is the dominant rock type, but olivine tholeiite and tholeiitic andesite comprise minor amounts of the section. Imnaha basalts are typically porphyritic rocks with coarse plagioclase phenocrysts.[49] They commonly contain zeolite-filled amygdules. The most voluminous unit of the Columbia River Basalt Group, the Grande Ronde Basalt, is dominantly aphyric, fine-grained tholeiite. Local plagioclase phyric types do occur in this unit and some rocks contain tiny clots of plagioclase and clinopyroxene. The Picture Gorge Basalt (figure 7.13)—which is partly time-correlative with but geographically and petrographically distinct from the Grande Ronde Basalt—consists of aphyric basalt overlain and underlain by porphyritic plagioclase phyric rocks. Wanapum basalts are generally olivine-bearing, plagioclase phyric rocks. Similarly, Saddle Mountains basalts may be olivine or plagioclase phyric, but the unit is petrographically diverse, containing andesite and breccia in addition to a variety of tholeiites.

Major and trace element chemical analyses reveal that the chemistry of the basalts is variable, however the range of SiO_2 contents is limited. Magnesium numbers are moderate to low (0.30–0.65) and a correlation exists between CaO and MgO (BVSP, 1981, p. 84). Trace element–REE diagrams show relative enrichment in Ba, Th, and Nb in some flows (Brandon and Goles, 1988). The most strikingly regular chemical variation is revealed by isotopic ratios of Sr and Nd (figure 7.14).[50] The $^{87}Sr/^{86}Sr$ ratios fall in the range .703 to .7045 for the older Imnaha and Picture Gorge units, whereas progressively higher values (up to 0.715) characterize younger formations (McDougall, 1976). The ε_{Nd} isotopic ratios vary from 8 to –12.

Interpretations and Petrogenesis

These and other trace element, major element, and petrographic data suggest a three-stage magmatic history for the Columbia River Basalt Group. That history involved (1) mantle melting and ascent of magma, (2) crystal-liquid fractionation of plagioclase, olivine, and pyroxene, and (3) minor magma mixing and assimilation of deep crustal rocks. Nd and Sr isotopic data are consistent with such a complex history. Apparently, the mantle source rock was heterogeneous and may have included an enriched mantle-plume component as indicated by Ba, Th, and Nb enrichment of some lavas (Piccirillo et al., 1987; Brandon and Goles, 1988). The low to moderate Mg numbers indicate that the *erupted* magmas were not primary. The CaO vs. MgO data suggest crystal-liquid fractionation involving clinopyroxene (BVSP, 1981, p. 84), and plagioclase fractionation is indicated by an Eu anomaly.[51]

The importance of magma mixing and assimilation of crustal rocks is variable. In some local units, the chemistry suggests that mixing had a recognizable affect on the compositions of the rocks (Hooper, 1985; Reidel and Fecht, 1987), but in others a single source seems to have been involved (Hooper, 1988b). Where sources are complex, Pb isotopic data, other isotopic data, and trace element data suggest that assimilated granitic crust and other sources may have contributed material to the magmas.[52]

Major problems in understanding flood-basalt genesis in general and the origin of the Columbia River Basalt Group in particular remain unresolved. The nature of the mantle source is a primary problem for flood-basalt studies. A second major problem is that the cause of flood-basalt volcanism remains unknown. Overall causes of volcanism, such as hot spot generation, rifting origins, or bolide impact, have yet to be adequately investigated.[53] Among the proposed explanations for the Columbia River Basalt are mantle diapirism caused by plate motion changes, with resultant partial melting of (1) the subducting Juan de Fuca plate, (2) part of the asthenosphere, (3) subcontinental lithosphere, or (4) a horizontal shear zone beneath the North American plate (Snavely, MacLeod, and Wagner, 1973; McDougall, 1976; S. E. Church, 1985; Prestvik and Goles, 1985). Passage of the Yellowstone hot spot beneath the region or back-arc rifting are possible "triggers" for some of

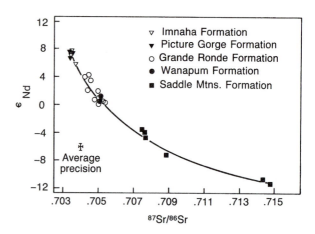

Figure 7.14 Graph of Sr-isotope ratios vs. ε_{Nd} values for the Columbia River Basalt Group (from BVSP, 1981, p. 87). See DePaolo and Wasserburg (1979a) for the method of calculating ε_{Nd}. High values plot in the mantle array. Lower, negative values reflect crustal contamination.

these melting processes, as evidence of pre-eruptive rifting is present in the subsurface (Hart and Carlson, 1987; Catchings and Mooney, 1988; Hooper, 1988a). A third major problem is, how much did magma mixing affect the compositions of Columbia River and similar flood-basalt flows?[54] Additional detailed chemical studies are needed to resolve these questions.

Rift Basalts

Rift basalts form along the mid-ocean ridge spreading centers, along continental rifts, and in leaky transform faults. The most abundant are the Mid-Ocean Ridge basalts (MORBs).[55]

Observations

Mid-ocean ridges extend for more than 65,000 km around the Earth as a chain of tectonically active, faulted, volcanic mountains (figure 1.2).[56] Igneous activity, including voluminous basaltic volcanism in the central rift zone of these ridges (figure 7.15), creates new crust that covers about 60% of the Earth's surface[57] and comprises the upper part of newly forming lithospheric plates. The tremendous area covered by the Mid-Ocean Ridge basalts and the continuing nature of MORB volcanism means that MORB is the most abundant single rock type in the crust. Yet because MORBs lie mostly below the oceans, our knowledge of them is based on (1) isolated samples drilled and dredged from the sea floor by various oceanographic expeditions and drilling projects (e.g., DSDP and IPOD),[58] and (2) detailed studies at a few sites on the Mid-Atlantic Ridge (e.g., FAMOUS) and the ridges of the eastern Pacific Ocean, in the Galápagos and the Gulf of California–Tamayo region.[59]

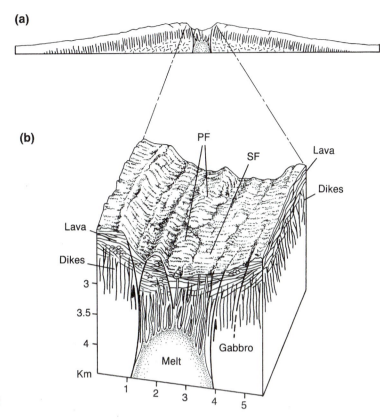

Figure 7.15 Idealized cross section of a mid-ocean ridge (a), with enlargement showing details of the rift valley structure (b). PF = pillow flows, SF = sheet flows.

(Modified from Hekinian, Moore, and Bryan, 1976; Ballard and van Andel, 1977; and Ballard, van Andel, and Holcomb, 1982)

Petrographically MORBs show some variability. They are texturally diverse, ranging from locally holohyaline to aphanitic-porphyritic, with spherulitic, intergranular, intersertal, supophitic, and ophitic textures represented. Similarly, there is some mineralogic variability. Olivine and plagioclase are the most common phenocrysts, but augite also occurs. The same three minerals with glass form the groundmass. In addition, other phases such as chrome spinel, pigeonite, magnetite, titanomagnetite, ilmenite, a few amphiboles (e.g., hornblende), and some sulfide minerals (e.g., pentlandite) also occur.[60]

Chemically, MORBs show a remarkable similarity considering the large volumes of rock involved (A. E. J. Engel, Engel, and Havens, 1965). They typically contain about 49 to 50% SiO_2, are very low in K_2O (< 0.15–0.25), have low Sr-isotope ratios (0.702–0.705), and also have low amounts of elements such as Ba, Pb, and Zr.[61] Nevertheless, some chemical measures vary. Magnesium numbers range between 0.30 and 0.76, and average values for various regions typically fall between 0.50 and 0.65.[62] Nd-isotope ratios also vary but are high, falling in the range of 0.5128 and 0.5134 at the upper end of the mantle array (figure 6.3).[63] Some chemical variations reflect important but subtle differences between MORBs from different geographic regions (Shido and Miyashiro, 1973), different tectonic settings (Bender et al., 1984), or different ridge depths (E. M. Klein and Langmuir, 1987; Brodholt and Batiza, 1989).[64] These differences reflect variable conditions of magma genesis and history.

Magma Genesis

The origin of MORB is controversial. Four contrasting models have been the focus of the debate on MORB magma genesis. The first two are as follows.

1. MORB magmas are primary magmas and their derivatives initially formed by partial melting of mantle rocks at relatively *shallow* depths at pressures < 1.0 Gpa.[65]
2. MORB magmas are fractionated derivatives of primary picritic magmas formed at *intermediate* depths under pressures of 1.5–3.0 Gpa.[66]

Both of these models accept the notion that most MORBs are derivatives developed by low-pressure fractionation of one or the other of the proposed parents.[67] This is favored by a number of lines of evidence including the Mg numbers,

111

which are predominantly well below 0.70. The third and fourth models involve magma mixing.

3. MORB magmas are the mixing products of materials derived from two *deep* sources—the asthenosphere and the bottom of the upper mantle.[68]

4. MORBs are differentiates of mixed magmas, one of which formed at a *shallow* depth (P < 1.1 Gpa) and the other of which formed at an *intermediate* depth (P = 1–2 Gpa).[69]

The chemical arguments relating to source chemistry and depth of formation are not yet adequately developed for a thorough assessment of the merits of these mixing models. Nevertheless, arguments for both models rely on chemical data, and model 4, relies specifically on analyses of phase equilibria.

Because most MORBs are formed from evolved (i.e., differentiated) magmas, arguments favoring one of the four models listed above center on two kinds of evidence. First, theoretical and/or experimental analysis (e.g., of partial melt-ing of appropriate mantle rocks) is used to determine the likely composition of melts derived by anatexis. Both major elements and trace elements, including isotopes, should occur in the right proportions and abundances if the analysis is to represent an analogue of magma genesis. Unfortunately, evidence of this first type is ambiguous at our present level of understanding. Individual studies have been used to support both shallow generation (model 1) (Presnall et al., 1979; Fujii and Scarfe, 1985) and intermediate depth of generation (model 2) (M. J. O'Hara, 1968, D. H. Green, Hibberson, and Jaques, 1979, Elthon and Scarfe, 1984).

The second kind of evidence is obtained by analyzing materials *thought* to represent the best candidates for undifferentiated melts from the mantle source. Glasses and basalts with Mg numbers of about 0.70 are the most likely materials derived from primary melts. Melting experiments with various samples (Maaloe and Jakobsson, 1980; Fujii and Bougault, 1983; Presnall and Hoover, 1987; Falloon and Green, 1987, 1988), like studies of the first type, yield conflicting interpretations that have been used to support both models 1 and 2.

One fact that favors shallow-level magma generation is that picritic parental materials (picritic basalts) and cumulates derived from them (dunites) are simply rare in the ocean crust.[70] If they are parental to MORB, it would seem likely that they would appear as major components of ocean crust. This is true, even though picrites are dense and would not commonly invade the crust and would locally experience mixing that would alter their chemistry. Yet picrite is not present, even where faulting exposes more or less complete sections through apparent oceanic crust.[71]

The magma genesis question is not resolved by data on mineral, rock, and glass chemistry. For example, M. J. Johnson, Stallard, and Lundberg (1990) argue that trace element data, from diopsides in peridotites of the oceanic lithosphere

such as depletion of Sr and distinct negative anomalies of Zr and Ti relative to REE, indicate that MORB magmas form through fractional melting over a range of depths. This view is consistent with parts of models 2 and 4.

In spite of the disagreements, most analyses do agree on two aspects of MORB genesis. First, magmas are derived by varying degrees of partial melting. Second, the mantle sources of the magma are heterogenous.[72]

Formation of the Oceanic Crust

Magmas derived from a source or sources in the mantle move towards the surface along the faults that mark the central rift region of the Mid-Ocean Ridge. Here, as evidenced by mineralogy, texture, and chemistry, fractional crystallization occurs in shallow magma chambers below the thin crust (figure 7.16).[73]

Fractional crystallization in shallow MORB magma chambers is a crystal-liquid process that produces olivine or plagioclase phenocrysts, or both.[74] Although clinopyroxene typically occurs only as a groundmass (late) phase, the chemistry of basalt suites suggests that it was also a major fractionating phase.[75] Extensive fractionation, especially in isolated magma chambers, may locally give rise to andesites (Fornari et al., 1983; C. D. Byers, Muenow, and Garcia, 1983) or more siliceous rocks (C. J. Engel and Fisher, 1975). Generally, however, magma compositions may become chemically complex but generally uniform in composition as a result of continuing replenishment by new magmas rising from below (M. J. O'Hara and Matthews, 1981).

The overall result of fractional crystallization is thought to be a layered ocean crust.[76] Surficial layers are composed of tholeiitic sheetlike flows and localized accumulations of pillow lava (Ballard and van Andel, 1977; Ballard, van Andel, and Holcomb, 1982). Beneath this is a layer composed of sheeted vertical dikes (dikes intruding dikes intruding dikes, etc.). Complexly mixed diabase-gabbro plutons, with basal, layered cumulates of gabbro and ultramafic rocks below, are believed to underlie the sheeted dike complex. Depleted mantle presumably occurs below the crustal rocks.

Continental Rift Volcanism: The Rio Grande Rift

Tensional forces acting on continents produce rifting, as is the case of tension acting at mid-ocean ridges and back-arc basins. Partial melting of the upwelling underlying asthenospheric mantle rocks, due to heating or pressure release, yields basaltic melts (Sleep, 1984; Hart et al., 1989; R. White and McKenzie, 1989). Alternatively, at depth, mantle plumes or diapirs may be associated locally with rifts, and basaltic volcanism may result from melting of the diapirs.

Whatever the cause of rift volcanism, both tholeiitic and alkali olivine basalts occur along continental rift zones. Typically, these basalts are accompanied by more siliceous volcanic rocks ranging from andesite to rhyolite in composition,[77] but the volcanic rock suites are commonly

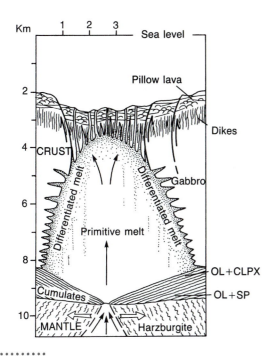

Figure 7.16 Model of a fractionating magma chamber beneath the axis of a mid-ocean ridge. OL = olivine, SP = spinel, CPX = clinopyroxene.
(After W. B. Bryan and Moore, 1977)

bimodal, consisting mainly of two types—basalt and rhyolite (R. L. Christiansen and Lipman, 1972).[78] The basalts are not usually MORB-like. Instead, many are similar to ocean island basalts (Leeman and Fitton, 1989; Moyer and Esperanca, 1989).

The Rio Grande Rift is a continental rift that extends from central Colorado south through central New Mexico and on into Mexico.[79] Although unique in some details of its history, the Rio Grande rift exhibits a number of features that typify continental rifts. The rift itself is a semicontinuous, north-south-trending basin bounded by normal faults on the east and west. Local centers of volcanism are scattered along the rift (rather than being uniformly distributed).

Volcanism during rifting was predominantly basaltic to bimodal basaltic-rhyolitic (Elston and Bornhorst, 1979). Nevertheless, andesites and other rock types occur (Zimmerman and Kudo, 1979). Especially in the north, tholeiitic basalts tend to occur in the center of the rift, whereas more alkalic rocks, such as nepheline-normative alkali olivine basalts, occur on the flanks of the rift. Locally, however, both types of basalt occur together.

Chemically, the basalts do not have all of the characteristics of primary mantle melts. They have moderate Mg numbers (0.66–0.47), but are enriched in LREE (Baldridge, 1979; Lipman and Mehnert, 1979).[80] Sr-isotope ratios fall in the range 0.70369–0.70503 and Nd-isotope ratios cluster at 0.51243—values indicative of a mantle source.[81]

Available evidence suggests that the basalts of the Rio Grande Rift were formed at a shallow to moderate level by small to large degrees of partial melting of upwelling mantle material. Since the moderate Mg numbers suggest that these rocks are not primary mantle melts, it is likely that local melting of basal crust, magma mixing, crustal contamination, and/or crystal-liquid fractionation may have played a role in the generation of various rift rocks (Baldridge, 1979; Lipman and Mehnert, 1979; Dungan, 1987).

Arc Basalts

Volcanic arcs occur in oceanic areas (e.g., the Marianas Island Arc in the western Pacific Ocean) and on continental margins (e.g., the Andes of western South America). Within these arcs, the volcanoes contain a diverse suite of volcanic rocks. Arcs basalts are a major component of the volcanic suite and they provide important clues to the origins of the arcs and the rocks that compose them.

Tectonically, arcs are interpreted as a sign of plate subduction (see figure 6.4). Subduction is marked by a zone of earthquakes, the Benioff-Wadati Zone, that dips beneath the arc. Volcanism commonly occurs in an area overlying the part of the subduction zone that is at depths between 100 and 200 km, but volcanism above shallower and deeper parts of the zone occurs in some arcs.[82]

Observations

Basalts make up about 20% of the rocks of the volcanoes in volcanic arcs.[83] Here they occur with andesites, dacites, and rhyolites (discussed in subsequent chapters of this book). Both tholeiitic and alkali olivine basalts occur in island arcs, with tholeiitic basalts occurring nearest the trench above the shallowest parts of the subduction zone (Kuno, 1959; Kushiro, 1983). Basalts of intermediate compositions (e.g., high-Al_2O_3 basalts) overlie moderate depths of the subduction zone, whereas alkali olivine basalts overlie the deepest zones (figure 7.17).

The mineralogies and chemistries of arc basalts differ in some ways from those of basalts from other tectonic environments. Plagioclase is the most abundant phenocryst mineral, but clinopyroxene and olivine also occur (Ewart, 1976). Arc-basalt chemistries are distinctive as a group, both because of their variable SiO_2 content (48 to 52%) and Mg numbers (0.44–0.70) and because of their distinctive isotopic signatures.[84] Alumina is commonly high. Sr-isotope ratios are typically low, usually in the range 0.703–0.704, whereas large ion lithophile elements (e.g., K, Rb), are enriched relative to REE.[85]

Petrogenesis

Substantial disagreement exists about the origin of arc basalts. Their origin is intimately linked to the origins of other arc rocks, particularly the andesites, which may represent about three-quarters of all arc rocks.[86] Although several aspects of arc-basalt and andesite origins are discussed in chapter 9, important models of arc-basalt petrogenesis are presented here.

113

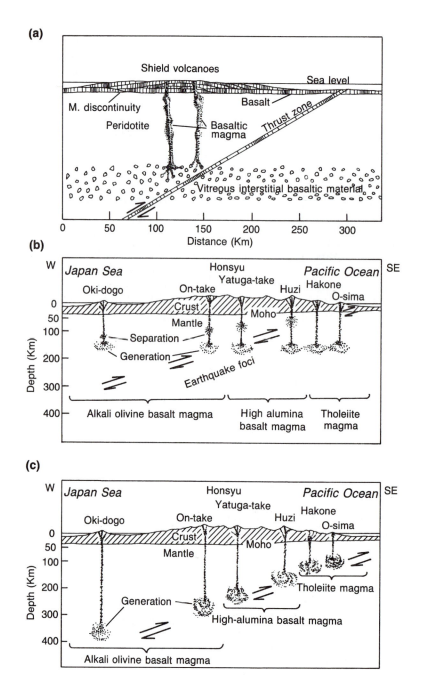

●●●●●●●●●●●●●●

Figure 7.17 Models of arc-basalt genesis. (a) Uniform-depth model of Coats (1962). (b) Depth-of-separation model of Yoder and Tilley (From H. S. Yoder and C. E. Tilley, "Origin of Basalt Magmas: An experimental study of natural and synthetic rock systems" in *Journal of Petrology*, 3:342-532, 1962. Copyright © 1962 Oxford University Press Ltd., Oxford, England. Reprinted by permission.) (c) Depth-of-generation model of Kuno (1966a, 1966b; modified after Kuno, 1966a).

Believable models should explain relevant data such as the chemical variation of basalts across the arcs, Sr and other isotopic ratios indicative of exchange with a mantle source, and variations in Mg number among arc basalts. Several models were proposed before all of these and other chemical constraints were known.[87] For example, Coats (1962) proposed a model, here called the *uniform-depth model*, in which melting near the Benioff-Wadati Zone produces a layer of basaltic magma at a uniform depth in the mantle (figure 7.17a). These magmas coalesce and rise to the surface to erupt and form the arc. Chemical variations across the arc are not explained by the uniform-depth model.

In contrast, Yoder and Tilley (1962) related major chemical variations to depth of fractionation or separation of the magma,[88] in a *depth-of-separation model* (figure 7.17b). Specifically, while they thought, as Coats did, that the

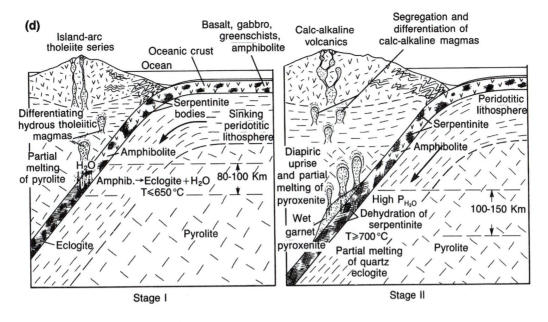

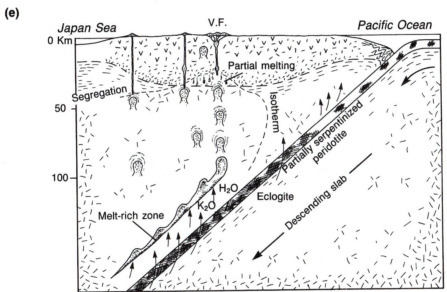

•••••••••••••••
Figure 7.17 Models of arc-basalt genesis. (continued)
(d) Two-stage pyrolite-pyroxenite model of Ringwood (From
A. E. Ringwood, "The petrological evolution of Island Arc
systems" in *Journal of the Geological Society of London,*

Part 3, 130:183–204, 1974. Copyright © 1974 Geological
Society of London. Reprinted by permission.) (e) Depth-of-
separation model of Kushiro (1983). (redrawn and modified
from original works and Yoder, 1976)

basaltic magmas all originated at about the same depth, they suggested that chemical variations were due to depth of fractionation or separation of magmas from source materials. Tholeiites, they argued, fractionate in shallow chambers, intermediate (high-alumina) basalts fractionate at intermediate depths, and alkali olivine basalts separate from parental materials at substantial depths. No particular connection to the Benioff-Wadati Zone was suggested.

Kuno (1959, 1966a, 1966b) recognized the importance of the relationship between Benioff-Wadati Zone and basalt chemistry. He related basalt chemistry to the *depth of generation* in the seismic zone (figure 7.17c). On the basis of major, trace,

rare earth, and isotopic element data, plus phase equilibria, Brophy and Marsh (1986) and J. D. Myers (1988) supported Kuno's interpretation that high-alumina basalt liquids are primary magmas. In contrast, some theoretical, chemical, and experimental work on magma generation seems inconsistent with the depth-of-generation model (Gust and Perfit, 1987; Brophy, 1989b), making the model debatable.[89]

Models that attempt to account for major element and trace element differences within arc rocks and across volcanic arcs are generally complex (Ringwood, 1974; Kushiro, 1983; Myers, 1988). For example, Ringwood (1974), following Nicholls and Ringwood (1973), presented a model, here called the *pyrolite-pyroxenite model,* in which volcanism occurs in two phases (figure 7.17d). In this model, the physical-chemical history of the mantle rock, rather than simply the depth, controls rock chemistry. In the pyrolite-pyroxenite model, early, shallow, tholeiitic volcanism results from the following series of processes:

1. Dehydration of the subducting plate;
2. Partial melting of mantle pyrolite above the subduction zone;
3. Diapiric rise of pyrolite followed by separation and differentiation of tholeiitic magma in the upper mantle;
4. Eruption of tholeiites at the surface.

Later, deeper, calc-alkaline volcanism occurs after continued subduction converts oceanic crust to quartz eclogite, in the following sequence of events.

1. Dehydration of serpentine occurs in the subducting plate (to provide water).
2. Melting of quartz eclogite produces siliceous (e.g., rhyolitic) magmas.
3. Reaction of siliceous magmas with overlying mantle pyrolite produces garnet pyroxenite.
4. Diapiric rise of wet pyroxenite masses accompanies partial melting in the mantle to produce calc-alkaline magmas.
5. Eruption of calc-alkaline magmas occurs at the surface.

This complex model requires dehydration of the subducting slab and mixing of some subducted components (fluid and melt) with mantle rock that is later melted. The likelihood of dehydration is now widely accepted (Fyfe and McBirney, 1975; Delaney and Helgeson, 1978; Tatsumi, 1989),[90] and Pb and Be isotopes in arc rocks support the involvement of subducted materials in models of magma generation (J. D. Myers and Marsh, 1987; Morris and Tera, 1989). On the negative side, melting and assimilation experiments do not support the model (Stern and Wyllie, 1981; W. L. Huang and Wyllie, 1981; Sekine and Wyllie, 1983; Johnston and Wyllie, 1989). Still, the model is evolving. In an effort to explain the origin of source rocks for intraplate magmas, Ringwood (1982, 1985, 1990; Kesson and Ringwood, 1989) extended his model. The expanded version focuses on shearing at depths below 100 km along the subduction zone and hybridization of sheared, depleted mantle rocks by magmas derived from the subducting slab.

In another complex model, Kushiro (1983) adopted a depth-of-separation approach to explain variation in basalt chemistry and volume across the Japanese arc (figure 7.17e). Some recent experimental work supports this model (Tatsumi et al., 1983; Kushiro, 1990). Basalts closest to the trench, that is, those overlying the shallower part of the subduction zone, are tholeiitic. They have relatively high values of total $Fe/(Mg + total Fe)$, or X_{Fe}, and are the most voluminous. Away from the trench (and the volcanic front[91]), X_{Fe}

decreases, as does volume. Given the apparent differentiation of tholeiites, Kushiro (1983) argues the following points.

1. Melting is concentrated in a melt rich zone above the subduction zone.
2. Magmas arise from diapirs or melts from this zone.
3. Magmas migrate up the melt-rich zone, and thus diapirs are concentrated at the upper end of this zone, creating more voluminous magmatism at the volcanic front.
4. Primary tholeiitic magmas, which are more dense, cannot penetrate the upper crust, so they pond and fractionate within the crust or at its base to produce less dense, iron-richer tholeiitic differentiates.
5. Primary alkali olivine basalt magmas, which are less dense and in equilibrium with the mantle, segregate from diapirs at greater depths and rise quickly to the surface without significant fractionation.

Based on experimental analyses and estimates of crustal compositions in the Japanese arc, Kushiro (1990) later concluded that the mantle wedge from which the magmas are derived must be convecting in order to provide the necessary volumes of fertile mantle rock that would produce the large volumes of magma required to build the arc.

Like the pyrolite-pyroxenite model, Kushiro's depth-of-separation model requires the contribution of fluid from the subducting plate, providing a clear link to the Benioff-Wadati (subduction) Zone and the crust. The fluid phase, derived from dehydration of the subducting crust, carries large-ion lithophile elements into the mantle, enriching it. Both models involve magma generation within the enriched mantle wedge overlying the subduction zone, and both emphasize diapiric rise of magma. In spite of these major similarities, the mantle sources and processes of magma generation differ.

Not all contemporary models of arc magma genesis involve diapirs. For example, Nye and Reid (1986) argue that diapirs are precluded as a source by data obtained from rocks crystallized from presumed primary magmas in the Aleutian Arc. They argue instead that the high-Mg chemistry and inferred high temperatures require that primary arc magmas form as a result of decompression of a rising mantle convection current. Mantle mixing and metasomatism produce compositional variability.[92]

Clearly, some aspects of arc magma genesis are resolved, whereas others are not. The fluids from the subducting slab obviously are involved in magma generation, yet the contribution that the fluids and the rocks make continues to be debated. Are any magmatic components derived from the subducting plate, or are chemically active fluids the only contributions derived from the plate? Also unresolved is the question of whether or not high-alumina basalts are primary or derivative magmas. In fact, the range of magma compositions that might possibly result from melting the mantle wedge above the subduction zone is still unresolved. These and other questions provide impetus for further study.

ULTRAMAFIC VOLCANIC ROCKS

Ultramafic volcanic rocks are rare. Until the last half of this century, ultramafic lavas were virtually unrecognized, although kimberlites were known. Now, komatiites and some kimberlites are recognized as volcanic equivalents of peridotites and related rocks, with the komatiites representing lava flows. The kimberlites occur only on the continents, as do most komatiites. Both rock types are associated almost exclusively with pre-Cenozoic terranes.

Komatiites

Komatiitic rocks form a series of high-Mg rocks that range from ultramafic to andesitic compositions (Viljoen and Viljoen, 1969, in Arndt and Nisbet, 1982a; Arndt and Nisbet, 1982b).[93] Structurally and chemically, the uv-komatiites, or komatiites *sensu stricto*, are flow and volcaniclastic rocks with > 18% MgO (calculated on an anhydrous basis). They are typified by spinifex-textured olivine-pyroxene phenocrysts that occur with chromite in a glassy groundmass (figure 7.18). Komatiites are 20–35% modal olivine and 10–30% normative olivine, yet many are normative basalts because they typically contain > 15% normative plagioclase. Komatiites are primarily Precambrian in age, although komatiitic (high-Mg) basalts and **boninites** (high-Mg andesites) have formed during Phanerozoic time.

Komatiite chemisty indicates a close relationship with mantle rocks. Mg numbers are commonly high (>0.70).[94] Typically, light rare earth elements are depleted,[95] suggesting either high degrees of melting of undepleted mantle or small degrees of melting of a depleted mantle source.[96] Oxygen isotopes reflect a mantle source.[97]

Experimental work indicates that komatiite liquids would have temperatures in the range of 1500–1850° C, well above those for basalts (Arndt, 1976; D. H. Green et al., 1975; Wei, Trunes, and Scarfe, 1990). Such temperatures have long been considered too high to be attained presently in the upper mantle, where most contemporary magmas are generated. Three solutions are offered. First, it is widely believed that Archean (Precambrian) geothermal gradients, and therefore temperatures, were higher. Consequently, high temperatures become plausible for Archean rocks but *not* for komatiitic Phanerozoic rocks (Upadhyay, 1982). Thus, this solution cannot account for all komatiites.

The second solution offered is that water lowered the melting point of the mantle; that is, komatiites are derived from wet mantle (Allegre, 1982). The problem here is that the mantle is generally dry. Consequently, methods of creating wet mantle at local sites or at periodic times are required, or these rocks must be derived only from wet, *primitive* mantle rocks.

A third solution involves the generation of komatiite lavas in hot mantle diapirs rising from a depth of 400 km. Experimental work by Wei, Trunes, and Scarfe (1990) on

Figure 7.18
Photomicrograph of spinifex-textured komatiite, Monroe Township, Ontario, Canada. Width of photo is 6.5 mm.

komatiite of both Al-depleted and Al-undepleted character suggests that primary komatiitic melts separated from such a diapir at depths of about 300–350 km and 150 km, respectively. If all chemical data are compatible with such a model, it may provide the solution to one komatiite genesis problem.

A second major problem relating to komatiite genesis is that high degrees of melting are seemingly required to produce such high-Mg rocks (D. H. Green et al., 1975; Arndt, 1977b). Is it likely that large masses of magma would remain at depth while this high degree of melting occurred? Some workers think not,[98] although Nesbitt (1982) proposes the existence of a highly molten zone in the Archean mantle. He suggests that komatiitic liquids are compressible and become dense enough to remain for some time at depth. Alternatively, magmas within a rising diapir could evolve through a high degree of melting over a period of time.

The details of komatiite origin remain debatable. Nevertheless, it is clear that highly magnesian, modally ultramafic magmas have existed.

Kimberlites

Kimberlites are ultramafic volcanic rocks that occur most commonly in dikelike diatremes.[99] The kimberlite diatremes are essentially confined to stable continental areas.

Mineralogically, kimberlites contain large grains (phenocrysts, megacrysts, and xenocrysts) of olivine plus additional minerals such as phlogopite, garnet, pyroxenes, or ilmenite.[100] The groundmass is composed of such minerals as calcite, serpentine, and chlorite plus some of the same minerals that comprise the large crystals. Because diamonds, stable only at high pressures, occur in kimberlites, these rocks are the object of considerable economic as well as petrologic interest. Texturally, kimberlites are brecciated.

Chemically, like the komatiites, kimberlites reflect their mantle source.[101] Mg numbers are high and Sr-isotope ratios are low. LREEs are enriched.

The occurrence of diamond, as well as the chemistries of the rocks and the nature of the mineral assemblages, requires that kimberlites form in the mantle and rise rapidly (35–100 km/hr) toward the surface.[102] Furthermore, cooling must be rapid. Calcite- and serpentine-rich matrix materials argue for involvement of a CO_2- and H_2O-rich fluid phase, probably important as a driving mechanism for cracking of the crust and emplacement of the kimberlite.[103] Rapid rise facilitates the characteristic brecciation of these rocks.

What remains unknown, or at least in dispute, is the nature of the mantle source region, the controls on the compositions of possible primary kimberlitic melts, the significance of CO_2 and H_2O, and the causes for the distribution of kimberlites.[104] Kimberlites, which may derive their water and CO_2 either from mantle sources or from recycled lithosphere, perhaps are produced by limited partial melting of the mantle.[105] Explanations of where and why that melting occurs have pointed to hot spot upwelling, gravitational instability, subduction-related processes, continental rifting, mantle diapirism, and the escape of remnant fluids from either undepleted, volatile-rich crust beneath Precambrian crust or a crystallizing Precambrian magma ocean.[106] To resolve these questions, considerable experimental work needs to be undertaken.

SUMMARY

Basalts are dark igneous rocks composed of plagioclase and additional minerals such as olivine, the pyroxenes, and magnetite. Chemically, silica contents lie between 45 and 52%. Two main basalt types are recognized—the tholeiitic basalts and the alkali olivine basalts. Tholeiitic basalts are hypersthene-normative and may contain normative quartz. Alkali olivine basalts are olivine- and nepheline-normative and typically contain abundant model olivine and augite.

Basalts occur along rifts in both oceanic and continental areas, above subduction zones in volcanic arcs, and at intraplate sites where fracturing occurs or where hot spots overlie mantle plumes. Both tholeiites and alkali olivine basalts occur in most settings, but tholeiites are generally more abundant.

The origins of basaltic magmas are widely debated. Clearly, alkali olivine basalt magmas form at greater depths, whereas tholeiitic magmas represent shallow levels of formation. Partial melting of heterogeneous, diapirically rising mantle rocks (lherzolite, eclogite, pyroxenite, or pyrolite) yields magmas that rise towards the surface. Some alkali olivine basalt magmas reach the surface relatively unaltered,

uncontaminated, and undifferentiated. In contrast, tholeiitic magmas are commonly modified before they reach the surface. Their chemistries reveal that history.

Magnesium-rich, mantle-derived magmas yield the ultramafic to mafic komatiites and the ultramafic kimberlites. The brecciated textures and high fluid contents of kimberlites, combined with the presence of diamond and mantle mineral assemblages, suggest a history involving rapid rise, emplacement, and cooling of mantle-derived, fluid-rich partial melts. A mantle origin, but one involving high degrees of high-temperature partial melting of less hydrous mantle rocks, explains the komatiitic magmas, which form pyroclastic rocks and flows with spinifex textures. Although the mantle origins for these magmas are likely, the details of magma genesis are still debated.

EXPLANATORY NOTES

1. For example, Pearce and Cann (1973) and BSVP (1981).
2. I. S. E. Carmichael, Nicholls, and Smith (1970), I. S. E. Carmichael, Turner, and Verhoogen (1974), and C. J. Hughes (1982) further show that there is a sound chemical foundation in the relationship between the chemical activity of silica and various magma types.
3. For example, G. A. MacDonald and Katsura (1964) and Kuno (1966a,b). For a review, see Schwarzer and Rogers (1974). DeLong and Hoffman (1975) discuss a problem with the use of this diagram (figure 7.3).
4. Miyashiro (1974, 1975d).
5. Kuno (1960, 1966), Carmichael, Turner, and Verhoogen (1974, pp. 33–34). See BVSP (1981) and J. D. Myers (1988) for examples of other types.
6. BVSP (1981).
7. Back-arc settings are discussed by Sleep and Toksoz (1971), Chase (1978), Tamaki (1985), and in Kokelaar and Howells (1984). Although basalts formed in back-arc settings are generally similar to MORB (J. W. Hawkins, 1976; Pineau et al., 1976; J. W. Hawkins and Melchior, 1985; but see W. K. Hart et al., 1972, and Saunders and Tarney, 1984, for additional data and a contrasting viewpoint), most differ in isotopic and trace element chemistry (Hawkesworth et al., 1977; Poreda, 1985; Sinton and Fryer, 1987; Volpe, Macdougall, and Hawkins, 1988; Hochstaedter et al., 1990; Ikeda, 1990; Ishizuka, Kawanobe, and Sakai, 1990) and some are alkaline (Nakamura et al., 1989).
8. For example, see R. L. Christiansen and Lipman (1972), Baldridge (1979), Mahood and Baker (1986), R. J. Coleman and McGuire (1988), JGR, v. 94, no. B6 (1989), and Leat et al. (1990).
9. Bellieni et al. (1986a), R. White and McKenzie (1989).
10. S. D. Weaver et al. (1979), Brandon (1989).
11. Kuno (1959, 1966a), Yoder and Tilley (1962), S. M. Kay, Kay, and Citron (1982).
12. J. T. Wilson (1963), Jackson et al. (1972), BVSP (1981), Geist, McBirney, and Duncan (1986), Hooper (1988a), DePaolo, Stolper, and Thomas (1991).

13. For discussion of plumes, see W. J. Morgan (1971, 1972) and Bercovici et al. (1989).

14. In some classifications, pyroxene is an essential mineral in basalt. See appendix A and Nockolds, Knox, and Chinner (1978) for additional minerals that occur in basalts.

15. For example, see Pearce and Cann (1973), Miyashiro (1974, 1975d), Floyd and Winchester (1975), Pearce (1975, 1976, 1987), R. E. Smith and Smith (1976), and J. C. Butler and Woronow (1986). For important cautions in using such data, see Wood et al. (1979), Holm (1982), and Prestvik (1982).

16. See Bates and Jackson (1982, 1976) for definitions of these basalt types.

17. For example, see Holm (1982) and Prestvik (1982). Also see Butler and Woronow (1986), Arculus (1987), A. R. Duncan (1987), and Pearce (1987).

18. For reviews of and comments on the uses of trace elements in petrologic analyses, see Cox, Bell, and Pankhurst (1979), Rudnick et al. (1986), P. C. Hess (1989, ch. 4), Ragland (1989, ch. 5), M. Wilson (1989, esp. ch. 2), and A. R. Philpotts (1990, ch. 13).

19. Augustithis (1978), Nockolds, Knox, and Chinner (1978), MacKenzie et al. (1982), and Bard (1986) include illustrations of these various textures.

20. The broad outline of the geology of the Hawaiian Islands presented here is based on G. A. MacDonald and Abbott (1970), MacDonald, Abbott, and Peterson (1983), Decker et al. (1987), and D. A. Clague and Dalrymple (1987), with additional information from J. T. Wilson (1963), Jackson, Silver, and Dalrymple (1972), Bargar and Jackson (1974), and Shaw, Jackson, and Bargar (1980).

21. See Dalrymple (1971) and McDougall and Swanson (1972). Also see H. R. Shaw, Jackson, and Bargar (1980), G. A. MacDonald, Abbott, and Peterson (1983), and D. A. Clague and Dalrymple (1987) for reviews.

22. J. G. Moore et al. (1982), F. A. Frey and Clague (1983), J. W. Hawkins and Melchior (1983), Staudigel et al. (1984).

23. Also see Wentworth and Winchell (1947), Eaton and Murata (1960), G. A. MacDonald (1968), T. L. Wright (1971, 1984), T. L. Wright, Swanson, and Duffield (1975), Beeson (1976), D. A. Swanson et al. (1976), Leeman et al. (1980), BVSP (1981), C. Y. Chen and Frey (1983, 1985), Feigenson (1984), A. W. Hofmann, Feigenson, and Raczek (1984), D. A. Clague and Dalrymple (1987), R. B. Moore et al. (1987), and Chen et al. (1990).

24. G. A. MacDonald (1968), MacDonald, Abbott, and Peterson (1983), Staudigel et al. (1984).

25. Eaton and Murata (1960), G. A. MacDonald and Katsura (1964), G. A. MacDonald (1968), D. A. Clague, Jackson, and Wright (1980), R. B. Moore et al. (1987).

26. See note 25.

27. Also see G. A. MacDonald (1968), J. G. Moore et al. (1982), Staudigel et al. (1984), D. A. Clague and Dalrymple (1987), T. L. Wright and Helz (1987). Different authors have used different names for the stages and some workers divide the fourth stage of this summary into two stages, making a total of five stages.

28. This discussion of chemistry is based primarily on the data recorded in BVSP (1981, p. 163ff.), with additions from the works cited in note 23, Budahn and Schmitt (1985), Lanphere and Frey (1987), Tilling et al. (1987), and Garcia et al. (1989).

29. Lanphere and Dalrymple (1980), Kurz et al. (1983), Lanphere (1983), Hofmann et al. (1984), Staudigel et al. (1984), Hofmann et al. (1987), Lanphere and Frey (1987), Tatsumoto, Hegner, and Unruh (1987), West and Leeman (1987), West et al. (1987).

30. Hot spots are discussed by W. J. Morgan (1981) and DePaolo, Stolper, and Thomas (1991). For a more thorough description of the history of the Hawaiian-Emperor Chain, see the summary provided by D. A. Clague and Dalrymple (1987).

31. E. D. Jackson, Silver, and Dalrymple (1972), H. R. Shaw (1973), H. R. Shaw and Jackson (1973), Bargar and Jackson (1974), E. D. Jackson and Shaw (1975), E. D. Jackson, Shaw, and Bargar (1975), H. R. Shaw, Jackson, and Bargar (1980), H. R. Shaw (1980).

32. BVSP (1981, p. 833), Saxena and Eriksson (1985), D. A. Clague and Dalrymple (1987).

33. Compare Kushiro (1972, 1973a) and Presnall et al. (1978) to T. H. Green, Green, and Ringwood (1967) and D. H. Green and Ringwood (1967b), who argue that high-pressure fractionation of olivine tholeiite can lead towards alkali olivine basalt compositions. Also see Wyllie (1988) for a hypothesis on the origin of very early and late-stage nepheline-normative lavas at depths of >150 km.

34. This point is debated. See, for example, Feigenson (1986) and Hofman et al. (1987).

35. Compare C. Y. Chen and Frey (1983, 1985), with Budahn and Schmitt (1985) and Ho and Garcia (1988).

36. Staudigel et al. (1984), Feigenson (1986), West et al. (1987), West and Leeman (1987), Tatsumoto, Hegner, and Unruh (1987), M. O. Garcia et al. (1989).

37. See T. L. Wright and Helz (1987), Ho and Garcia (1988), C. Y. Chen et al. (1990), J. K. Russell and Stanley (1990). M. J. O'Hara and Mathews (1981) warn about the complexities introduced by replenishment and mixing of magmas during shallow-level fractionation. Sen and Jones (1990), on the basis of cumulate xenolith compositions, suggest that ponding and fractionation of some magmas occurs at or near the base of the lithosphere. More extensive fractionation probably occurs in the crust.

38. Also see Ryan (1988) and L. Wilson and Head (1988).

39. See G. A. MacDonald (1968), Beeson (1976), Irvine (1980a), Ho and Garcia (1988), and Russell and Stanley (1990) for additional discussion of fractional crystallization.

40. Wyllie (1988), Ho and Garcia (1988), Russell and Stanley (1990), Sen and Jones (1990).

41. See G. A. MacDonald (1968), Wright and Fiske (1971), Ryan, Koyanagi, and Fiske (1981), L. Wilson and Head (1988), Ho and Garcia (1988), Mangan (1990), Russell and Stanley (1990), and Sen and Jones (1990).

42. Wyllie (1988), F. A. Frey et al. (1990).

43. D. A. Clague and Dalrymple (1987), R. B. Moore et al. (1987), C. Y. Chen et al. (1990).

44. See Geist, McBirney, and Duncan (1986), Devey et al. (1990), and Ringwood (1990) for additional discussions of hot spot volcanism.

119

45. This summary is based on several reviews and papers, including Mackin (1961), Waters (1961, 1962), Snavely, MacLeod, and Wagner (1973), G. W. Walker (1973), T. L. Wright, Grolier, and Swanson (1973), McDougall (1976), D. A. Swanson et al. (1979), Cox (1980), BVSP (1981), D. A. Swanson and Wright (1981), Hooper (1982, 1984, 1985, 1988a, 1988b), Barrash, Bond, and Venkatakrishnan (1983), M. E. Ross (1983), Hooper et al. (1984), S. E. Church (1985), Prestvik and Goles (1985), Mangan et al. (1986), Reidel and Hooper (1989), Tolan et al. (1989), and Baksi (1990), plus personal communications with C. W. Myers (1975). Also see the articles collected in Macdougall (1988a).

46. BVSP (1981), Macdougall (1988a). Prestvik and Goles (1985) suggest, however, that the Columbia River Plateau may be atypical.

47. See Watkins and Baksi (1974), McKee, Swanson, and Wright (1977), McKee et al. (1981) in Barrash, Bond, and Venkatakrishnan (1983), Hooper (1988a), and Baksi (1989, 1990).

48. D. A. Swanson et al. (1979), Hooper et al. (1984), Mangun et al. (1986), Hooper (1988a), M. M. Bailey (1989b), Reidel et al. (1989).

49. D. A. Swanson et al. (1979) describe the petrography of all the rocks summarized here.

50. BVSP (1981, pp. 85–88), R. W. Carlson, Lugmair, and MacDougall (1981).

51. See chapter 3 for an explanation of Eu anomalies.

52. I. MacDougall (1976), Hooper et al. (1984), S. E. Church (1985), R. W. Carlson and Hart (1988), Hooper (1988b). Fodor, Corwin, and Roisenberg (1985) and Fodor, Corwin, and Sial (1985) offer a similar explanation for the chemistry of Parana (Brazil) basalts.

53. Bellieni et al. (1986a,b), Hooper (1988a), White and McKenzie (1989), Baksi (1989), I. H. Campbell and Griffiths (1990).

54. Hooper (1985), Fodor, Corwin, and Roisenberg (1985), Sen (1986), Reidel and Fecht (1987).

55. There is a tremendous volume of literature on ocean-floor volcanic rocks, most of which are mid-ocean ridge basalts (MORBs). Among the particularly useful works are W. B. Bryan et al. (1976), reports of the FAMOUS project in Geological Society of America Bulletin, v. 88, nos. 4, 5 (1977), BVSP (1981), J. F. G. Wilkinson (1982), Ballard, van Andel, and Holcomb (1982), and LeRoex et al. (1983). Other good references include D. M. Christie and Sinton (1986), W. B. Bryan and Moore (1977), A. S. Davis and Clague (1987, 1990), Hekinian and Walker (1987), and E. M. Klein and Langmuir (1987).

56. Morgan (1968), BVSP (1981, p. 132).

57. J. F. G. Wilkinson (1982).

58. DSDP refers to the Deep Sea Drilling Project sponsored by the Joint Oceanographic Institutions for Deep Earth Sampling (JOIDES). IPOD refers to the International Project for Ocean Drilling.

59. FAMOUS stands for French-American Mid-Ocean Undersea Study. In this study, teams of French and American scientists studied the area in the central Atlantic Ocean southwest of the Azores and along the mid-Atlantic Ridge in considerable detail (see Heirtzler and van Andel, 1977, and the Geological Society of America Bulletin, v. 77, nos. 4, 5). For reports on the Galapagos region, see Ballard, van Andel, and Holcomb (1982), Fornari et al. (1983), Perfit and Fornari (1983), and Perfit et al. (1983). For a discussion of the MORBs of the Tamayo region of the East Pacific Rise, see Bender, Langmuir, and Hanson (1984).

60. See BVSP (1981) for a review, and also Bence, Papike, and Ayuso (1975), W. B. Bryan and Moore (1977), Czamanske and Moore (1977), Perfit and Fornari (1983), and Humphris et al. (1985).

61. For example, see A. J. Engel, Engel, and Havens (1965), BVSP (1981), W. M. White (1985), and A. S. Davis and Clague (1987).

62. BVSP (1981, p. 139), J. F. G. Wilkinson (1982), A. S. Davis and Clague (1987). Also see R. W. Kay, Hubbard, and Gast (1970) for a discussion of other MORB chemical variations.

63. O'Nions et al. (1977), LeRoex et al. (1983), W. M. White (1985), DePaolo and Wasserburg (1979a), White and Hofmann (1982), W. M. White, Hofmann, and Puchelt (1987).

64. Also see J. G. Schilling (1975), Bryan et al. (1976), Sun, Nesbitt, and Sharaskin (1979), and LeRoex et al. (1983).

65. A. E. J. Engel, Engel, and Havens (1965), D. H. Green and Ringwood (1967b), Kushiro (1973a), T. Shibata (1976), Langmuir et al. (1977), Presnall et al. (1979), Wilkinson (1982), Takahashi and Kushiro (1983), Fujii and Bougault (1983), Presnall and Hoover (1984, 1986, 1987), Fujii and Scarfe (1985), and Rhodes, Morgan, and Lilas (1990). Note that Green et al. (1979) retract the position of Green and Ringwood (1967) and adopt model 2.

66. M. J. O'Hara (1968), Green et al. (1979), Stolper (1980), Elthon and Scarfe (1984), Elthon and Casey (1985). D. L. Anderson (1981a) suggests a picritic eclogite source.

67. Shibata and Fox (1975), Hekinian, Moore, and Bryant (1976), Flower et al. (1977), Puchelt and Emmermann (1977), W. M. White and Schilling (1978), Sun, Nesbitt, and Sharaskin (1979), D. A. Wood et al. (1979), BVSP (1981), Grove and Bryan (1983), LeRoex et al. (1983), Perfit and Fornari (1983).

68. Allegre, Hamelin, and Dupre (1984), for example, following J. G. Schilling (1973).

69. D. M. Christie and Sinton (1986). E. M. Klein and Langmuir (1987) also argue for variable depths of melting (P = 0.5–1.6 Gpa).

70. BVSP (1981, p. 151), J. F. G. Wilkinson (1982).

71. Engel and Fisher (1975), Bonatti and Honnorez (1976), Wilkinson (1982).

72. For additional discussion, see O'Nions et al. (1977), Sun, Nesbitt, and Sharaskin (1979), D. A. Wood et al. (1979), Dupre and Allegre (1980), Oskarsson, Sigraldson, and Steinthorsson (1982), Grove and Bryan (1983), Bender, Langmuir, and Hanson (1984), Hamelin, Dupre, and Allegre (1984), Langmuir and Bender (1984), Humphris et al. (1985), W. M. White (1985), A. S. Davis and Clague (1987), and E. M. Klein and Langmuir (1987).

73. W. B. Bryan and Moore (1977), Michael and Chase (1987), Hekinian and Walker (1987).

74. BVSP (1981) and Wilkinson (1982) review the evidence. Typical mineralogies and textures are described in L. J. Mazzulo and Bence (1976), Hekinian, Moore, and Bryan (1976), Bryan et al. (1976), LeRoex et al. (1983), Perfit and Fornari (1983), and Humphris et al. (1985). Also see Hekinian and Walker (1987).

75. For example, Bryan et al. (1976), Bryan (1979), Walker et al. (1979), Grove and Bryan (1983), and Perfit and Fornari (1983).

76. See Dewey and Byrd (1971), the papers in the *Journal of Geophysical Research*, v. 81, no. 23 (1976), Becker et al. (1989), and the more complete discussion in chapter 11.

77. R. L. Christiansen and Lipman (1972), BVSP (1981, p. 108 ff.), Baldridge (1979), Lipman and Mehnert (1979), R. P. Smith (1979), C. Zimmerman and Kudo (1979), R. G. Coleman and McGuire (1988), Spell and Kyle (1989), Moyer and Nealy (1989).

78. This subject is discussed in chapter 8. See R. G. Coleman and McGuire (1988), Moyer and Esperanca (1989), and Pallister (1989) for examples and additional discussion.

79. See papers in Riecker (1979), *Journal of Geophysical Research*, v. 91, no. B6 (1986) and v. 94, no. B6 (1989), plus BVSP (1981), on which much of this review is based. Also see Dungan et al. (1986), Henry and Price (1986), F. V. Perry, Baldridge, and DePaolo (1987), and Leat et al. (1990). For seismically based cross sections, see L. D. Brown et al. (1980) and K. H. Olsen, Baldridge, and Callender (1987). Although the rift valley ends in central Colorado, Tweto (1979) argues that rift-related structures extend to the Wyoming border.

80. Also see BVSP (1981, pp. 115–22).

81. From Williams and Murthy (1978, 1979), in BVSP (1981). Somewhat higher values are reported by Leat et al. (1990) for the unusual Yarmony Mountain lavas.

82. See the review in J. B. Gill (1981, p. 53ff.).

83. Ewart (1976) provides this estimate.

84. BVSP (1981, p. 207), Allegre and Condomines (1982), W. M. White (1985), J. D. Myers (1988).

85. W. M. White and Patchett (1984), Arculus and Powell (1986), J. P. Davidson (1987).

86. Ewart (1976).

87. See Yoder (1976, ch. 1) and T. H. Green (1980) for reviews of some of the models summarized in this section.

88. See Kuno's (1966a) interpretation of Yoder's and Tilley's (1962) work and Yoder's (1976) discussion of both. Also see D. H. Green and Ringwood (1967b).

89. For additional discussion, see Kushiro (1983), Uto (1986), and Brophy (1989a).

90. Also see Wyllie (1979, 1983a), Tatsumi et al. (1986), Kesson and Ringwood (1989), and Ringwood (1990).

91. The volcanic front is a line delimiting the abrupt beginning of volcanism landward from the trench.

92. Also see the models described in chapter 9.

93. This section is based primarily on D. H. Green (1975) and D. H. Green et al. (1975), on Arndt (1976, 1977a, 1977b), on section 3.35 of BVSP (1981), and on the various papers in Arndt and Nisbet (1982a). The definition and description of komatiite is from Arndt and Nisbet (1982b). Also see Malyuk (1985), Cattell and Arndt (1987), R. J. Walker, Shirey, and Stecher (1988), and Wei, Trunes, and Scarfe (1990).

94. For example, BVSP (1981), Auvray et al. (1982), M. J. Viljoen, Viljoen, and Pearton (1982).

95. Sun and Nesbitt (1978), Cattell and Arndt (1987). Also see the references in note 94.

96. Green (1975), D. H. Green et al. (1975), Arndt (1977b), Sun and Nesbitt (1978).

97. Beaty and Taylor (1982), R. J. Walker, Shirey, and Stecher (1988).

98. See Green (1975), Arndt (1977b), Arndt and Nisbet (1982a, p. 477).

99. The data and discussion here are based on a number of works, notably the various papers in P. H. Nixon (1973b), Boyd and Meyer (1979), and Kornprobst (1984a), plus the works of P. H. Nixon, von Knorring, and Rooke (1963), Dawson (1980, 1981), Eggler and Wendlandt (1978), Edgar et al. (1988), and Foley (1990).

100. See Bates and Jackson (1980) and the various works cited in note 97.

101. For example, P. H. Nixon (1973a), Dawson (1980), Apter et al. (1984), Fieremans et al. (1984).

102. Mercier (1979), Artyushkov and Sobolev (1984).

103. O. L. Anderson (1979), Artyushkov and Sobolev (1984).

104. Eggler and Wendtlandt (1978), Meyer (1979), Helmstaedt and Gurney (1984), Edgar et al. (1988), Foley (1990).

105. Dawson (1971, 1980), Wyllie (1979, 1980).

106. See Helmstaedt and Gurney (1984) for a summary, and also Taylor (1984), Freund (1984), and T. Anderson (1984).

PROBLEMS

7.1. Confirm the tholeiitic and alkaline character of analyses 2 and 5 of table 7.1 by plotting them on the alkali/silica discrimination diagram (figure 7.3a). Are the analyses classified as the same type of rock by all curves on the diagram? Explain.

7.2. If analysis of a back-arc basin basalt revealed that the basalt contained 1.0% TiO_2, 65 ppm Zr, 26 ppm Y, and 250 ppm Sr, what tectonic environment would be suggested for the basalt by the Pearce and Cann diagrams (figure 7.5)? Explain your analysis and conclusion.

Rhyolites and Pyroclastic Rocks

INTRODUCTION

Rhyolites are silica-rich volcanic rocks. They are far less voluminous than basalts, yet they cover vast areas of the earth. In western North, Central, and South America sheets of white to dark gray rhyolitic tuffs cover thousands of square kilometers.[1] Similarly, in parts of Australia, in New Zealand, and in other circum-Pacific volcanic zones, rhyolitic volcanic rocks are an important constituent of the volcanic section.[2] Notable occurrences are also present in Europe, Asia, Iceland, and East Africa.[3]

Rhyolites, like basalts, occur in a variety of tectonic environments, including sites above mantle plumes, along rift zones, and in compressional zones of plate or continental collision. They are rare in oceanic settings, but common on the continents. Also, like the basaltic rocks, rhyolites are the object of some petrogenetic controversy.

MINERALOGY, TEXTURES, AND STRUCTURES OF RHYOLITES

Rhyolites commonly occur as quartz-feldspar porphyries, as glasses, and as various types of pyroclastic rocks (figure 8.1).

In addition to quartz and sanidine, other common phyric phases include biotite, plagioclase, anorthoclase, and magnetite. Among the minor to rare minerals that occur as groundmass and phenocryst phases are fayalite, hypersthene, hornblende, riebeckite, acmite, ilmenite, zircon, sphene, apatite, garnet, cordierite, andalusite, and sillimanite.[4]

The rhyolitic glasses and lavas form domes and relatively short flows in composite volcanoes, occur in separate fields of domes, are found in siliceous volcanoes associated with major shield volcanoes, and occur in and around calderas. The calderas contain abundant rhyolitic pyroclastic rocks and they served as vents for the huge pyroclastic deposits that cover thousands of square kilometers in Nevada, New Mexico, and other areas of western North America (R. L. Smith, 1960a, 1960b, 1979).[5]

Texturally, rhyolites range from holohyaline to holocrystalline. Porphyritic textures are common, especially hypocrystalline types. Hypocrystalline to holohyaline obsidian and pumice are typical locally. The most common textures, however, are pyroclastic textures, fragmental textures that characterize the tuffs (solidified deposits of ash). Pyroclastic textures consist of fragments of crystal, rock (i.e., lithic fragments), glass, or combinations of these materials. Based on size, the pyroclastic fragments are classified into three categories—ash, lapilli,

(a)

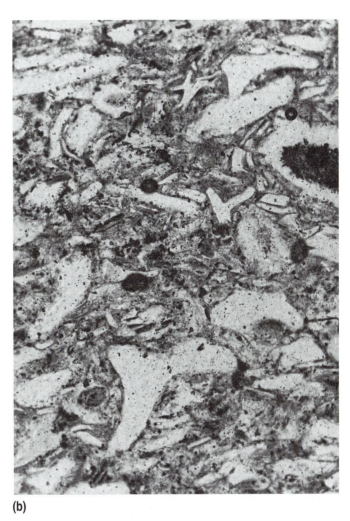

(b)

Figure 8.1 Photomicrographs of rhyolitic rocks.
(a) Porphyritic rhyolite, Yellowstone National Park. Note the phenocrysts of quartz (Q) and sanidine (S). (XN) Long dimension of photo is 3.25 mm. (b) Rhyolitic tuff, Valley Springs Formation, near Columbia, California. Curved fragments are glass shards. (PL) Long dimension of photo is 1.27 mm.

and bombs or blocks (table 8.1).[6] Corresponding rock texture names are associated with each size class. Tuffs and lapilli tuffs are most common among the rhyolitic rocks, but breccias also occur. Locally the tuffs are annealed by their own heat, after they are deposited, and become **welded tuffs.**

The physical properties of rhyolitic magmas, notably their high viscosities, result in distinctive structures formed during extrusion and flow. The high viscosity inhibits flow, resulting in the formation of short flows and domes rather than long flows and sheets of lava. Flow-banding is common in rhyolites, and associated lithophysal structures occur locally. In the pyroclastic rocks, layering is typical.

An important observation in tuff sequences is that the sequences are chemically layered. Sequences of tuff may begin with a rhyolite or high-silica rhyolite at the base, change up the section to low-silica rhyolite, and end with a less siliceous rock such as quartz latite, dacite, or andesite.[7] Rock and mineral chemistries show corresponding changes.

Table 8.1 Size Classification of Pyroclastic Materials

Size	Fragment Name	Rock Textural Name
> 64 mm	bomb, block	volcanic breccia, agglomerate
2–64 mm	lapillus	lapilli tuff
< 2 mm	ash	tuff

Sources: W. H. Parsons (1969), Fisher and Schmincke (1984), Raymond (1993)

CHEMISTRY OF RHYOLITIC ROCKS

Many rocks that appear to be rhyolites modally, such as rocks with quartz and sanidine phenocrysts, are not normative rhyolites. Analyses of both normative rhyolites and modal rhyolites are included in table 8.2, where the major element chemistry of these rocks is tabulated. Both types are reported here, because rocks typically reported in the literature as rhyolites are designated as such because of their phyric phases (phenocrysts of quartz and alkali feldspar), their silica contents (> 69% SiO_2), or the position of their rhyolitic norms in an IUGS-type classification. Normative rhyolites, determined using the AGI classification, are relatively rare. In contrast, all but one of the rocks in table 8.2 is a normative rhyolite in the IUGS classification.

Examination of table 8.2 will reveal the significant variations that occur in rhyolite chemistries. Note the range of silica contents and the corresponding variations in normative quartz. Also note the variations in alumina content. For example, analyses 3, 4, and 5 all have about 72% silica, but have Al_2O_3 values in the range of 10 to 17%. Rhyolites with low Al_2O_3 values (e.g., analysis 3) are peralkaline and contain acmite (sodic pyroxene) in their norms, whereas rhyolites with high Al_2O_3 values (e.g., analysis 5) are peraluminous and contain significant amounts of normative corundum. Magnesium numbers are typically quite low in rhyolites (< 0.30).

The trace element chemistry of rhyolites is also varied. Rare earth element patterns may show LREE enrichment relative to HREE and may have slight to pronounced negative europium anomalies (figure 8.2).[8] Although commonly ascribed to plagioclase fractionation, Eu anomalies may also result from liquid-state thermogravitational diffusion (see discussion below).[9] Sr isotope ratios vary widely.

OCCURRENCES

Rhyolites occur in the same tectonic environments as do basalts, but they are abundant only in continental areas. The chemistry of rhyolites—for example, low Mg numbers, siliceous character, and trace element characteristics—suggests different origins for rhyolites than for the basalts. The detailed chemistry also reveals that rhyolites from different environments (and even those representing different parts of an eruptive cycle within the same environment) vary.

Mantle-Plume (Hot Spot) Rhyolites

Rhyolite occurs at intraplate sites in both oceanic and continental areas. Rhyolite is quite rare in oceanic hot spot volcanoes, but locally, for example, in Iceland, rhyolite is a bit more abundant.[10] This may be due to the fact that Iceland is unique in being a large area of thick, continent-like crust lying astride a mid-ocean ridge. In Hawaii (a more typical oceanic hot spot), the Alae lava lake of Kilauea Volcano developed a rhyolitic liquid[11] and rhyodacite occurs in

Oahu.[12] The Alea Lake rhyolite is not compositionally unusual in that it contains about 1% iron, about 3% Na_2O, and approximately 5.5% K_2O (Wright and Fiske, 1971). Although equivalent rocks occur elsewhere in Hawaii, they are quite rare.

Mantle plumes that underlie continental masses may produce large amounts of rhyolite. The Yellowstone National Park area (described in greater detail below) is an example. There, three phases of rhyolitic volcanism produced thousands of cubic kilometers of rhyolitic tuff and lava (R. L. Christiansen and Blank, 1972; Doe et al., 1982).

Rift, Transform, and Triple-Junction Rhyolites

Rhyolites may be found in rifts of both oceanic and continental character. Rhyolites are quite rare, however, in most ocean rifts (i.e. along mid-ocean ridges). Siliceous phaneritic dikes reported by Engel and Fisher (1975) from the Indian Ocean indicate that under exceptional conditions fractionation will produce siliceous magmas. Such is the case along the Galapagos Rift, where plagioclase and augite phyric "rhyodacites" with SiO_2 values in the rhyolite range (> 69%) occur and where local soda rhyolite is present (Byerly, Melson, and Vogt, 1976; McBirney et al., 1985).

Rhyolites are common along some continental rift zones. The Rio Grande Rift, discussed briefly in chapter 7, is an example of a continental rift where rhyolites comprise a significant fraction of the section.[13] Similarly, in the Salton Sea Geothermal Field, along the landward extension of the East Pacific Rise, rhyolite occurs in a series of domes (P. T. Robinson, Elders, and Muffler, 1976). Rhyolites are also present as part of the bimodal volcanic suite exposed along the margins of the Red Sea Rift (Pallister, 1987; R. G. Coleman and McGuire, 1988).

The interaction of a plate triple junction with a continental mass may produce rifted (leaky) transform faults and other tectonic features that yield rhyolite-bearing volcanic sequences.[14] Perhaps the best described example of such volcanism is the Eastern Coast Range Volcanic Suite of California (C. M. Johnson and O'Neil, 1984).

Arc Rhyolites

Rhyolites are quite common in arcs. Like rift rhyolites, they are far more abundant in continental arc volcanoes than in volcanoes founded on oceanic crust. Examples of continental arc rhyolites occur in North America, both in Alaska and in the Cascade Range, which extends from British Columbia to northern California. In the Cascade Range, chemically defined rhyolites ($SiO_2 > 69\%$) were erupted from both Mount Lassen, California (Howell Williams, 1932), and Mount Mazama, Oregon, the latter being the precursor of

Table 8.2 Chemical Analyses and CIPW Norms of Rhyolites and Related Rocks

	1	2	3	4	5	6	7
SiO_2	69.03[a]	70.10	72.43	72.06	72.8	76.57	77.26
TiO_2	0.33	nr	0.25	0.28	0.02	0.26	0.18
Al_2O_3	14.64	14.51	10.50	12.38	16.3	12.23	11.54
Fe_2O_3	2.42	nr	2.00	1.00	0.29	0.68	0.85
FeO	0.97	0.77	2.58	1.44	0.30	0.94	0.13
MnO	0.10	nr	0.11	0.04	0.06	0.02	0.03
MgO	0.32	0.82	0.04	0.35	0.00	0.18	0.20
CaO	0.41	1.18	0.17	0.96	0.16	0.60	0.58
Na_2O	6.46	0.53	5.91	2.88	4.1	3.27	2.96
K_2O	4.84	11.15	4.35	5.74	3.7	5.20	4.65
P_2O_5	0.04	nr	0.01	0.04	0.55	0.05	trace
H_2O	0.14	1.38	0.40	2.68	0.70	@	1.99
Other	0.15	nr	0.07	0.14	0.10	—	0.00
Total	99.85	100.44	98.82	99.99	99.08	100.00	100.37

CIPW Norms

	1	2	3	4	5	6	7
q	13.26	20.27	28.54	30.37	34.58	35.86	40.67
or	28.60	65.91	25.71	33.90	21.88	30.73	27.50
ab	48.36	4.51	29.79	24.39	34.72	27.69	25.07
an	—	4.26	—	3.90	—	2.62	2.86
ac	5.55	—	5.79	—	—	—	—
ns	—	—	3.18	—	—	—	—
di	0.69	1.30	0.30	0.44	—	—	—
hy	1.46	2.81	4.47	2.08	0.40	1.24	—
ol	—	—	—	—	—	—	—
mt	0.72	—	—	1.46	0.42	0.97	—
il	0.63	—	0.47	0.53	0.03	0.49	0.33
hm	—	—	—	—	—	—	0.85
c	—	—	—	—	5.55	0.25	0.58
ap	0.09	—	0.02	0.10	0.30	0.13	—
Other	0.48	1.38	0.56	2.83	1.22	—	—
Total	99.84	100.44	98.83	100.00	99.10	99.98	100.35

Sources:

1. Crystalline comendite (peralkaline rhyolite), sample 33714, from southeast of Mount Lamonia, D'Entrecasteaux Islands, Papua, New Guinea (analysis and norm from I. E. M. Smith, 1976).
2. High-potash rhyolite, Homestake Mine, Lead, South Dakota. Analyst: E. W. Adams (J. A. Noble, 1948).
3. Glassy comendite (peralkaline rhyolite), sample 33741, from north of Numaruma Bay, D'Entrecasteaux Islands, Papua, New Guinea (analysis and norm from I. E. M. Smith, 1976).
4. Circle Creek rhyolite, sample 60NC145, Elko County, Nevada (Coats, 1968).
5. "Macusanite" (peraluminous rhyolite) from Macusani area, Peru, Analysts: P. Elmore, S. Botts, L. Artis (V. E. Barnes et al., 1970).
6. Hypersthene-bearing rhyolite from eastern Australia. Average of three analyses (Ewart, 1981).
7. "Lithoidal welded tuff" from Beatty, Nevada area. Analyst: G. Steiger (Cornwall, 1962).

[a]Valves in weight percent.
@ Anhydrous basis (the H_2O is deleted and the analysis is summed to 100% as if the H_2O were not there).
[nr] Not reported.

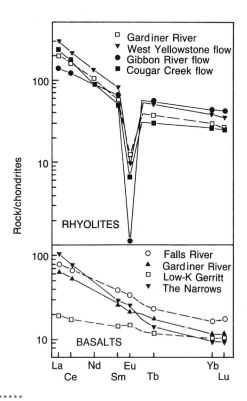

Figure 8.2 Rare earth element patterns of Yellowstone Plateau rhyolitic and basaltic volcanic rocks (chondrite normalized). Notice the slight positive europium anomaly in some basalts, the pronounced negative europium anomaly in the rhyolites, and the LREE enrichment relative to HREE.
(Unpublished data of Leeman, 1979, from B. R. Doe et al., 1982)

the Crater Lake Caldera (McBirney, 1968a). Rhyolitic tuffs, obsidians, and flows also occur at the Newberry Caldera, Oregon (Higgins and Waters, 1968), and rhyolites, including obsidian, are present in the Three Sisters Volcanic Complex (McBirney, 1968b). In Alaska, perhaps the most notable rhyolite occurrence is in Katmai National Park, where eruptions in 1912 from the Novarupta Caldera produced the famous deposits of the Valley of Ten Thousand Smokes (Curtis, 1968; Hildreth, 1983b; Eichelberger et al., 1990).

Oceanic-arc rhyolites occur locally in Indian, Atlantic, and Pacific arcs. For example, rhyolite is an important constituent of the New Britain Island Arc of Papua, New Guinea (I. E. M. Smith and Johnson, 1981). Also, rhyolitic rocks have been recognized in the Indonesian Sunda and Sangihe arcs (Van Bemmelen, 1970; Morrice et al., 1983) and in the Antilles Arc of the Caribbean Sea (H. H. Hess et al., 1966, Donnelly et al., 1971, S. N. Carey and Sigurdsson, 1980).

PETROGENESIS OF RHYOLITES

The origin and evolution of rhyolitic magmas is controversial and has been attributed to six major processes:

1. Anatexis of rocks in the mantle, either mantle rocks or rocks transported into the mantle by tectonic processes;
2. Fractional crystallization of basaltic or other magmas;
3. Anatexis of crustal rocks;
4. Assimilation accompanied by fractionational crystallization (AFC);
5. Magma mixing with fractional crystallization; and
6. Fractionation or melting followed by liquid-state thermogravitational diffusion.

Chemical data and experimental studies constrain these possibilities.

Generation of rhyolitic magmas by melting of rocks in the mantle (process 1) is not favored by the available experimental work. Ringwood (1974) had suggested that rhyolitic magmas might form by partial melting of quartz eclogite, as a part of his model for arc-basalt petrogenesis (see chapter 7). Eclogite is produced at mantle depths, during subduction, by metamorphism of basaltic oceanic crust. This process was investigated via experimental melting studies conducted by C. R. Stern and Wyllie (1981) and Huang and Wyllie (1975; 1981), but these studies indicate that primary rhyolitic melts *cannot* be generated at mantle depths by anatexis of either mantle rocks or subducted oceanic crust. Thus, the process seemingly cannot explain the origin of rhyolitic magmas.

Fractional crystallization of basaltic magma via crystal-liquid fractionation, as a method of generating silicic melts, was championed by Bowen (1928). Favoring this hypothesis, Bowen cited such evidence as the liquid "lines of descent" of magma chemistry, i.e., the path of changing liquid composition, indicated for rock suites plotted on SiO_2 vs. oxide variation diagrams. Bowen (1928) also cited the siliceous (evolved) compositions of volcanic glasses.

The process of crystal-liquid fractionation is considered by most petrologists to be a viable process of magma modification. Arguments supporting crystal-liquid fractionation include (1) arguments based on partitioning of elements between crystals and glass or between glasses;[15] (2) enrichment/depletion trends, that is, the tendency of elements to become enriched or depleted in successively more felsic rocks of a rock sequence;[16] (3) the occurrence of a large number of phases (≥ 6) in equilibrium with the liquid, which suggests that crystals and liquid were at equilibrium at a cotectic;[17] and (4) the occurrence of highly fractionated rocks chemically like rhyolite (e.g., aplites) as dikes, pods, border zones, and small stocks that intrude larger plutons.[18] The evidence confirms that crystal-liquid fractionation is a viable process for the formation of many rhyolitic rocks, even though all of the parent magmas may not have been basaltic.[19]

Some of the evidence used to support the crystal-liquid fractionation model of rhyolite formation has been questioned, and there are arguments against major quantities of rhyolite forming via the process (e.g., Hildreth, 1981; Marsh, 1990). For example, silica variation diagrams are now recognized to show little more than the general tendency of rocks to become more Ca-, Fe-, and Mg-poor with increasing silica contents,[20] a tendency dominantly controlled by the increase in silica itself.[21] Also, Hildreth (1979, 1981) has emphasized the fact that major enrichment of numerous elements occurs in the early erupted, presumably most differentiated magmas of silica-rich, rhyolite-bearing rock suites—a fact that he argues cannot be explained by crystal-liquid fractionation.[22] A third argument offered in opposition to the crystal-liquid fractionation model is that, at least for pyroclastic deposits, the huge volumes of siliceous rock erupted from single sources preclude a fractionation origin from parental basic magmas because *enormous quantities* of parental basalt would be required to yield the tuffs.[23]

A third process of formation for rhyolitic magmas, anatexis of crustal rocks, is now recognized as significant.[24] In this process, siliceous magmas are produced by melting siliceous rocks or partially melting more basic rocks. Such melting occurs at the base of or within the lower crust, as a result of the emplacement of basaltic magmas (figure 8.3). Evidence supporting the likelihood of an anatexic origin for rhyolitic magma includes (1) high Sr isotopic ratios in some rhyolites;[25] (2) correlations between siliceous basement-rock chemistry and elevated Sr isotopic ratios (i.e., where siliceous basement rock occurs, the Sr ratios are high);[26] (3) the spatial association of bimodal basalt-rhyolite volcanic suites;[27] (4) distinctive major and rare earth element compositions;[28] (5) textural relations (e.g. corroded or partly melted plagioclase and quartz crystals);[29] (6) successful modelling of combined chemical and mineralogical data to reproduce the effects of anatectic magma evolution,[30] and (7) melting experiments that show that melting of likely source rocks under appropriate conditions can yield rhyolitic liquids.[31] Alternative explanations for some of these lines of evidence, such as elevated Sr-isotope ratios, have been used to support other models of magma genesis (Hildreth, 1979, 1981). Nevertheless, in many cases, the evidence seems compelling in favor of anatexis.

Assimilation of siliceous country rocks or mixing of magmas accompanied or followed by fractionation may also yield rhyolitic magmas (Sigurdsson and Sparks, 1981). Mix-

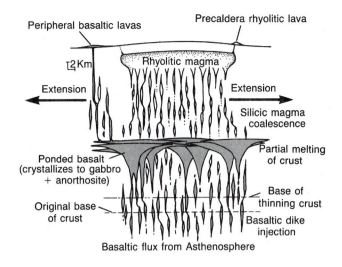

Figure 8.3 The anatectic origin of rhyolitic magmas. Heat is provided by mantle-(asthenospere-) derived basaltic magmas. The size of the magma chambers is probably exaggerated.
(From Hildreth, 1981)

ing of magmas may produce a variety of rocks, but anatectic rhyolites are generally considered to be *endmembers* (not products of mixing) that are mixed with basalt to form intermediate varieties of rock (Eichelberger, 1978a, 1978b; Eichelberger and Gooley, 1977).[32] Where mixing of rhyolite and basalt produces magmas less siliceous than rhyolite, later fractionation could yield rhyolitic magma. Similarly, assimilation of siliceous country rocks could increase the silica content and change the other chemistry of more basic magmas, but fractionation would likely be required to generate rhyolites.[33] Where assimilation and fractional crystallization (AFC) occur *simultaneously* (DePaolo, 1981d), rhyolite might develop, but experimental, isotopic, and phenocryst abundance data suggest that AFC is restricted to the lower, more mafic parts of magma chambers. In the rhyolitic roof zones, the processes of assimilation and crystallization are separated in time and space (I. H. Campbell and Turner, 1987; C. M. Johnson, 1989).

Liquid-state thermogravitational diffusion has been advocated as a method of generating a zoned magma chamber that will yield high-silica rhyolites ($SiO_2 > 75$ wt. %) (Hildreth, 1979, 1981).[34] The process involves thermal and convection-driven chemical processes that enrich the roof zone of the magma chamber in volatiles, promoting chemical redistribution of elements and the formation of high-silica rhyolites (figure 8.4). Minor exchange of elements with the wall-rocks occurs around the chamber margins. A temperature gradient, from cooler near the top to hotter at the base, exists in the magma chamber and is the primary cause of the convection.

127

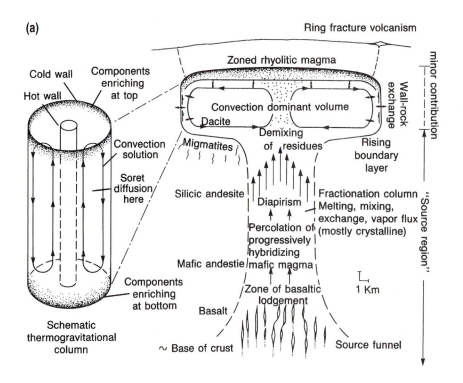

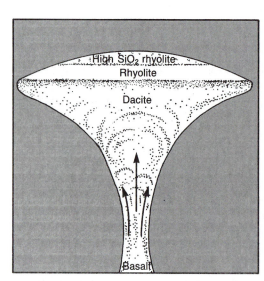

············
Figure 8.4 Zoned magma chambers and magma modification by thermogravitational diffusion.
(a) Diagrammatic representation of a shallow-level magma chamber undergoing liquid-state thermogravitational diffusion to produce a zoned (layered) chamber as shown in (b). Note that heat is provided by basaltic magmas from the mantle, initial fractionation occurs in the column feeding the magma chamber, and final magma modification occurs as a result of convection, Soret diffusion, and wall-rock exchange.
[(a) modified from Hildreth, 1981; (b) based on R. L. Smith, 1979, and Hildreth, 1981]

Recall that examination of the eruptive sequences present in ash sheets reveals that the chemistry of the tuffs typically becomes less siliceous upward. These observations suggest that (1) the magma chambers are compositionally zoned (figure 8.4), and (2) eruptive emptying of the chambers proceeds from the top down.[35] This is easily envisioned. If a magma chamber is compositionally zoned and eruption and emptying of the chamber proceed from the top down, early eruption phases will yield high-silica rhyolite from the top of the chamber and this rock will blanket the ground first. Later eruptive phases may successively yield rhyodacite, dacite, and even basalt, forming successive layers. Thus, a pile of rocks forms that varies in composition, becoming less siliceous, or less evolved, towards the top. The process requires relatively rapid emptying of the magma chamber, because periods of time between eruptions will allow continued modification of the magma, which may result in eruption of progressively *more evolved* magmas over time (Stix and Gorton, 1990a).

Evidence offered by Hildreth (1979, 1981) in support of the liquid-state thermogravitational diffusion model includes the following.

1. Gradients exist in $^{87}Sr/^{86}Sr$ and other isotopes within magma chambers, suggesting a layered or zoned magma chamber.

2. Major and trace element zoning is present in silicic magma chambers.
3. Roof zones typically have magmas with about 77% SiO_2, as suggested by the composition of first-erupted lavas.
4. Thermogravitational separations have been demonstrated in the laboratory.

In addition,

5. Diffusion can be shown to occur in experiments (D. R. Baker, 1990, 1991).

Some of these points have been challenged or the observations attributed to other mechanisms of formation.[36] For example, zoned magma chambers are considered by many petrologists to be consistent with crystal-liquid fractionation models (T P. Flood, Vogel, and Schuraytz, 1989). The most critical problem, however, is that experimental simulation of an important component of the model, the Soret effect (thermally induced chemical diffusion), has not created the expected chemical trends.[37] Thus, major bodies of magma are not likely produced by thermogravitational diffusion. Yet, inasmuch as magma chambers are dynamic bodies in which chemical and thermal gradients exist, the process may contribute to magma modification after an initial melt has formed.

Clearly, the origin of rhyolitic magmas is controversial. Which process produces rhyolites? Probably several processes are important, each within particular settings. In oceanic settings, crystal-liquid fractionation, locally combined with assimilation or anatexis, is probably of major importance. In the continental arcs, at transform fault and continent interaction sites, and in continental hot spots, anatexis, together with crystal-liquid fractionation, magma mixing, and assimilation, are important. Liquid-state thermogravitational diffusion may affect the origin of arc-magmas, evolved rift-magmas, and continental rhyolite magmas, but the degree to which this process influences magma evolution is unresolved.

RHYOLITE VOLCANISM: THE YELLOWSTONE EXAMPLE

Rhyolites are varied and abundant in the Yellowstone National Park area.[38] Their abundance, combined with the youth of the rocks, makes the province a good one for exemplifying rhyolitic volcanism. Yellowstone National Park lies within a major volcanic field at the northeast end of the Snake River Plain (figure 8.5a).

Observations

Lithologically, the Yellowstone volcanic rocks are somewhat bimodal, with Pliocene- to Pleistocene-aged rhyolite dominating the basalt-rhyolite sequence (figure 8.6). The bimodal volcanism is the recent continuation of similar volcanism that progressed from west to east across western North America, forming the Snake River Plain[39] (The plain may have resulted when plate movements caused the North American continent to override a mantle plume that provided crust-penetrating magmas (Morgan, 1972b; Iyer, 1979).

The general geology of the Yellowstone region has been mapped by Boyd (1961) and R. L. Christiansen and his colleagues (figure 8.5b).[40] The late Cenozoic Yellowstone Group volcanic rocks were erupted onto a terrain underlain by Paleozoic to Cenozoic rocks, including older, unrelated Eocene volcanic rocks of the Absaroka Volcanic Supergroup.[41] Three major calderas, the youngest centered in Yellowstone National Park and two centered to the west, served as vent areas for Yellowstone Group rhyolite flows, rhyolite tuffs, and minor basalts. Each caldera resulted from an explosive eruption that produced a major tuff unit. Caldera-forming eruptions were preceded by milder precaldera volcanism and followed by collapse and postcaldera volcanism. Each sequence of precaldera, caldera-forming, and postcaldera eruptions is called a cycle. The voluminous, explosive volcanism and caldera collapse events of Yellowstone occurred in three cycles at about 2.0, 1.3, and 0.6 million years ago.[42]

As is typical of volcanic provinces, the stratigraphy of the Yellowstone area is complex. The localized volcanic centers and preexisting topography of the region, coupled with the irregular nature of flows and periodic erosion events, renders individual units discontinuous and localized in their occurrences (figure 8.7).

Petrologically, the Yellowstone Group rocks are relatively uniform. However, some notable mineralogic, textural, and chemical variations are present locally. The basalts are dark, olivine-bearing tholeiites and typically contain phenocrysts of plagioclase and olivine in a matrix containing plagioclase, subcalcic augite or pigeonite, and Fe-Ti oxides.[43] Both high- and low-K varieties are present. Mg numbers fall in the 0.45–0.62 range and $^{87}Sr/^{86}Sr$ range from about 0.703 to about 0.708.[44]

The rhyolites are white to pink to black, hypocrystalline, porphyritic flows and tuffs. Obsidian and pumice are present locally. Phenocrysts are typically quartz and sanidine, but oligoclase, hornblende and biotite also occur.[45] Most rhyolites are relatively high-silica (> 76% SiO_2), low-lime (< 0.8% CaO) rocks. Minor "calcic rhyolite," with about 72% SiO_2 and 1.5% CaO, is also present.[46] Magnesium numbers of the rhyolites are low, but Sr isotopic ratios range from about 0.709 to 0.727. The REE are enriched, with LREE greatly enriched over the HREE (see figure 8.2). Negative europium anomalies are marked. He isotopic ratios in local springs associated with the volcanic rocks are high.[47] The major unit of each of the three volcanic cycles is a rhyolitic tuff. These are the Huckleberry Ridge Tuff, the Mesa Falls Tuff, and the combined Lava Creek Tuff/Central Plateau Member of the Plateau Rhyolite (W. B. Hamilton, 1965; Christiansen and Blank, 1972)(figure 8.7).

Figure 8.5 Maps showing the location and geology of the Yellowstone area. (a) Yellowstone area at the northeast end of the Snake River Plain (after L. A. Morgan, Doherty, and Leeman, 1984). (b) Geology of the Yellowstone area. Heavy lines are faults. Dotted lines encircle calderas numbered I, II, and III. II is the Island Park Caldera *(Modified from Christiansen, 1979).*

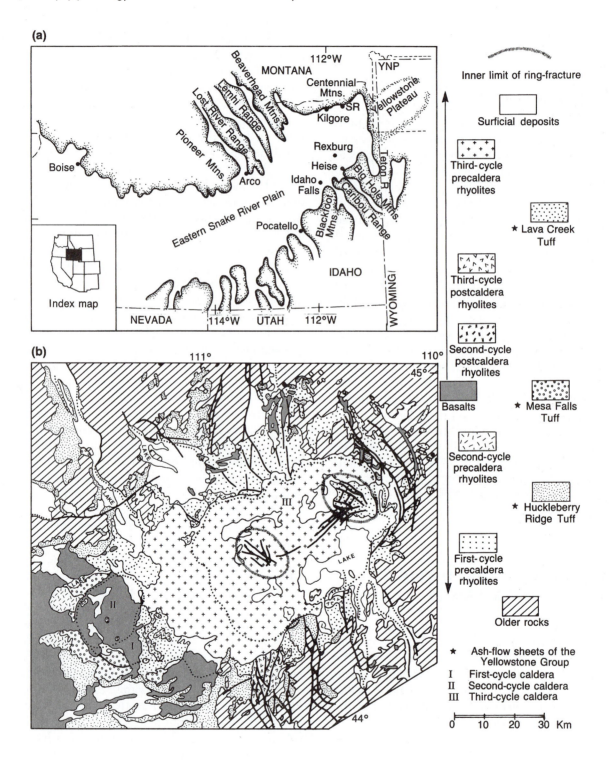

Interpretations

Clearly, the Yellowstone Group volcanic rocks were produced by repetitive explosive volcanic events. The heat and basaltic magmas of the system came from a mantle plume, as suggested by high $^3He/^4He$.[48] The Mg numbers and Sr-isotope ranges of the basalts, however, suggest that they are not simple, primary mantle melts. Whether or how much they have been modified is debated (B. R. Doe et al., 1982).

The rhyolitic magmas probably developed by anatexis of the lower crust, as suggested by Pb- and Sr-isotopic data.[49] The magma rose to form a chamber in the shallow levels of the crust, as indicated by the low-pressure and near-eutectic compositions of the rocks (figure 8.8), and there it underwent modification via some process or processes such as crystal-liquid fractionation, liquid-state thermogravitational diffusion, or wall-rock assimilation and interaction (B. R. Doe et

al., 1982; Hildreth, Christiansen, and O'Neil, 1984). During this time, ring fractures developed, as rhyolite (and local basalt) eruptions occurred (R. L. Christiansen, 1984). Explosive venting to form ash-flow tuffs was followed by caldera formation (collapse) and some postcaldera volcanism, including the formation of domes within the calderas. This cycle of volcanism was repeated three times, with the last eruptions occurring about 75,000 years ago.

OTHER PYROCLASTIC AND VOLCANICLASTIC ROCKS

Not all pyroclastic rocks are rhyolitic.[50] As a consequence, the name *tuff* does not imply rhyolitic compositions, nor does the name *breccia* imply andesitic compositions, as some students think. Pyroclastic deposits of basaltic, andesitic, phonolitic, and other compositions are known. The less siliceous magmas that produce these rocks tend to erupt less explosively and produce pyroclastic deposits that are generally less extensive than the rhyolitic deposits, although phonolitic rocks may locally form extensive pyroclastic sheets.

Volcaniclastic deposits formed by processes other than fragmentation at the vent are also common. Autoclastic breccias,[51] produced by fragmentation during flow, are common among basalts and andesites, but also occur in more siliceous rocks such as flows of rhyolitic obsidian.[51] Epiclastic volcanic deposits, formed by surficial reworking of volcanic materials, may occur in any composition of volcanic rock.

Clearly, volcaniclastic rocks, including pyroclastic rocks, have textures that reflect physical processes of formation that are not exclusive to particular compositions. Yet the physical properties of magmas are partly controlled by magma chemistry. Consequently, specific chemistries favor certain kinds of processes and resulting textures. Thus, rhyolite, which forms from highly viscous magmas that tend to erupt explosively, is the dominant rock type among the tuffs.

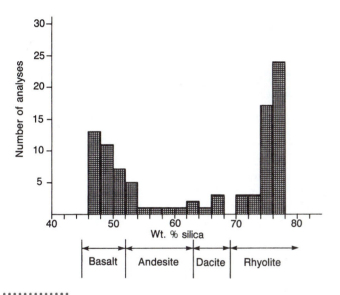

Figure 8.6 Histogram of silica content of rocks from the Yellowstone area displaying bimodal distribution of rock compositions.

(Data from Fenner, 1938; Boyd, 1961; W. B. Hamilton, 1963b, 1965; Hildreth, Christiansen, and O'Neil, 1984; and Hildreth, Halliday, and Christiansen, 1991)

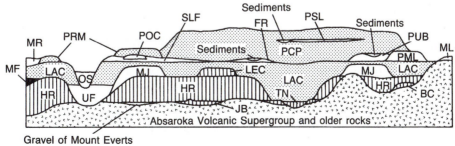

Figure 8.7 Diagrammatic cross section showing the stratigraphy of the Yellowstone National Park area. Note the discontinuous nature of many units. Major cycles are represented by HR-Huckleberry Ridge Tuff; M.F. (at left end)—Mesa Falls Tuff; and LAC and PCP-Lava Creek

Tuff and Central Plateau Member of the Plateau Rhyolite. Other letters represent various local or individual units.

(Source: R. L. Christiansen and H. R. Blank, Jr., "Volcanic stratigraphy of the Quaternary rhyolite plateau in Yellowstone National Park" in U S Geological Survey, Prof. Paper 729–8, 1972.)

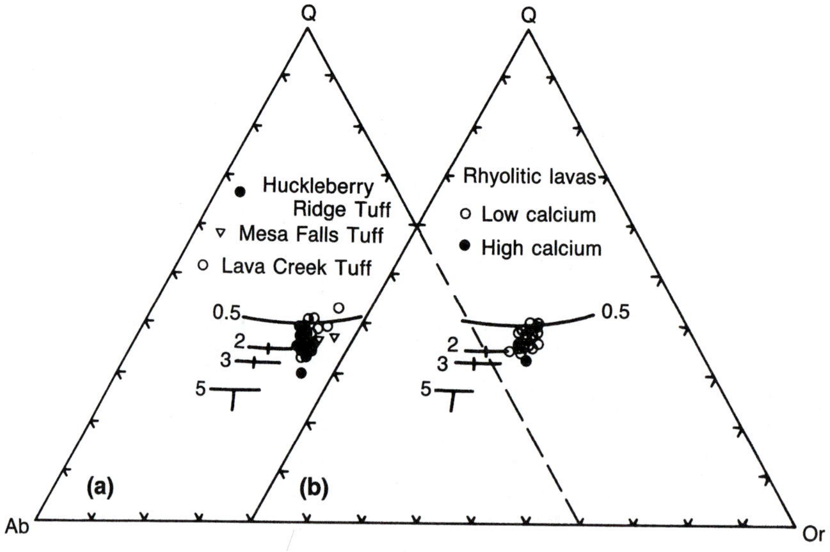

Figure 8.8 Plots of CIPW Or-Ab-Q values of Yellowstone rhyolites on the ternary diagram for silica-alkali feldspar (granite). The ternary minima for 0.5Gpa (5 kb), 0.3 Gpa (3 kb), 0.2 Gpa (2 kb), and 0.05 Gpa (0.5 kb) are shown. (from Tuttle and Bowen, 1958; Luth et al, 1964) (a) Or-Ab-Q plot for rhyolitic ash flow tuffs. (b) Or-Ab-Q for rhyolitic lava flows.
(After B. R. Doe et al., 1982)

SUMMARY

Rhyolites occur at sites in rifts, within plates over hot spots, above subduction zones, and along zones of transform fault/triple junction-continental interaction. They are decidedly more abundant in continental areas, suggesting that large volumes of rhyolite are only generated where continents are involved. Fractional crystallization via crystal-liquid processes is important locally, but anatexis of the base of the continental crust is probably the central process in the origin of many rhyolite magmas. Once generated, other magma-modification processes such as magma mixing, crystal-liquid fractionation, or thermogravitational diffusion may contribute to the production of distinctive rhyolite chemistries. Rhyolites in the Yellowstone National Park area typify rhyolites of siliceous volcanic fields in being porphyritic or tuffaceous and locally welded. As is the case in many other rhyolitic provinces, zoned magma chambers probably developed in the upper crust beneath Yellowstone and repeatedly erupted explosively to form rhyolitic tuffs. Three times these chambers suffered roof collapse to form major calderas.

Pyroclastic and other volcaniclastic rock types are not exclusively rhyolitic. The textures of these rocks reflect fragmentation during eruption, flow, and erosion—physical processes that are only partly controlled by magma chemistry.

EXPLANATORY NOTES

1. R. L. Smith (1960a, 1960b, 1979), Cornwall (1962), Howell Williams (1963), E. F. Cook (1966), McBirney and Weill (1966), McBirney (1968a), Izett (1981), various papers in JGR, v. 89, no. B10 (1984); D. C. Noble et al. (1984). See the definition of tuff below.

2. See papers in R. W. Johnson (1976), JGR, v. 86, No. B11 (1981), and JGR, v. 89, no. B10 (1984). Also see papers such as F. Reid (1983), Clemens and Wall (1984), and Togashi, Shirahase, and Tamanyu (1985).

3. See E. F. Cook (1966) and JGR, v. 89, no. B10 (1984). Also, references to a variety of pyroclastic deposits in these and other locales are given in R. V. Fisher and Schmincke (1984).

4. Raymond (1984, appendix A) provides a summary of common mineral occurrences. Also see Howell, Williams, Turner, and Gilbert (1982), I. S. E. Carmichael (1963), and Moyer and Nealey (1989). Workers who describe specific examples from various parts of the world are B. S. Wood and Carmichael (1973), Hildreth (1979, 1983b), W. I. Rose, Grant, and Easter (1979), F. Reid (1983), Nakada (1983), Clemens and Wall (1984), D. C. Noble et al. (1984), McDonough and Nelson (1984), Worner and Schmincke (1984a, 1984b), Togashi, Shirahase, and Tamanyu (1985), Piccirillo et al. (1987), Spell and Kyle (1989), Vogel et al. (1989), Glazner (1990), and Hazlett (1990). For more information, see papers in Chapin and Elston (1979) and JGR, v. 86, no. B11 (1981); v. 89, no. B10 (1984); and v. 94, no. B5 (1989).

5. See also the sources in note 4.

6. The general term *volcaniclastic* (R. V. Fisher, 1961) encompasses all volcanic clastic rocks. R. V. Fisher and Schmincke (1984) restrict the term pyroclastic rocks to fragmental rocks formed during eruption at a volcanic vent—a practice followed here. Pyroclasts that comprise these rocks have been subdivided into categories on the basis of size (R. V. Fisher, 1960; C. S. Ross and Smith, 1961; W. H. Parsons, 1969; Schmid, 1981; Howell, Williams, Turner, and Gilbert, 1982; Raymond, 1984c; R. V. Fisher and Schmincke, 1984). The names ash, lapillus, and bomb used for the various pyroclast sizes (table 8.1) and the corresponding rock textural names are textural terms that should be combined with a lithic name (e.g., rhyolite tuff or basalt breccia) in naming rocks. Also, some authors use different size categories for the clast types (e.g., Wentworth and Williams, 1932; Raymond, 1984c).

7. Hildreth (1979, 1981), R. L. Smith (1979), Mahood (1981), Fridrich and Mahood (1987), Stix et al. (1988), Warshaw and Smith (1988), Schuraytz, Vogel, and Younker (1989), Vogel, Noble, and Younker (1989).

8. For example, Hildreth (1979), Moll (1981), C. R. Bacon et al. (1981), B. R. Doe et al. (1982), McDonough and Nelson (1984), R. MacDonald et al. (1987), Stix et al. (1988), Moyer and Nealey (1989), Vogel, Noble, and Younker (1989).

9. Contrast Hildreth (1979, 1981, 1983b) with Michael (1983a) and C. F. Miller and Mittlefehldt (1984).

10. I. S. E. Carmichael (1963, 1964), Oskarsson, Sigvaldson, and Steinthorsson (1979, 1982), Sigurdsson and Sparks (1981), P. T. Robinson et al. (1982), Schmincke et al. (1982), R. MacDonald et al. (1987).

11. Wright and Peck (unpublished), in Wright and Fiske (1971).

12. G. A. MacDonald and Katsura (1964).

13. See Chapin (1979) and other papers in Riecker (1979). Also see Spell and Kyle (1989) and the references therein.

14. Dickinson and Snyder (1979), Pilger and Henyey (1979), Donnelly-Nolan et al. (1981), McLaughlin (1981), C. M. Johnson and O'Neil (1984).

15. I. E. M. Smith and Johnson (1981), R. MacDonald et al. (1987), Halliday et al. (1989). Also see Michael (1983). Flood, Vogel, and Schuraytz (1989) use quantitative modelling of mixing and fractionation to assess the likelihood of fractionation in a Nevada tuff.

16. Michael (1983a,b), Miller and Mittlefehldt (1984), Hildreth (1979).

17. Miller and Mittlefehldt (1984).

18. Hildreth (1983a), Miller and Mittlefehldt (1984).

19. Appleton (1972), W. I. Rose, Grant, and Easter (1979), Bickford, Sides, and Cullers (1981), I. E. M. Smith and Johnson (1981), Spulber and Rutherford (1983), Clemens and Wall (1984), F. A. Frey et al. (1984), McDonough and Nelson (1984), Togashi, Shirahase, and Tomanyu (1985). Also see R. MacDonald et al. (1987), Halliday et al. (1989), Flood, Vogel, and Schuraytz 1989) and references in notes 16-18.

20. Wilcox (1979).

21. Chayes (1960, 1964).

22. This has been a subject of debate; see notes 14–16. Also see Mahood (1981), Crecraft, Nash, and Evans (1981), Moll (1981), F. Reid (1983).

23. McBirney (1969), Hildreth (1983a).

24. McBirney and Weill (1966), McBirney (1969), Eichelberger and Gooley (1977), A. S. White and Chappell (1977), Eichelberger (1978a,b), F. Reid (1983), Clemens and Wall (1984), C. M. Johnson and O'Neil (1984), Guilbeau and Kudo (1984), Varne (1985), E. H. Christiansen, Sheridan, and Burt (1986), Glazner and O'Neil (1989).

25. For example, McBirney and Weill (1966), Pichler and Zeil (1972), Speed and Kistler (1980), B. R. Doe et al. (1982), D. C. Noble et al. (1984)

26. McBirney and Weill (1966), Kistler (1978), Leo, Hedge, and Marvin (1980), Hawkesworth (1982), Glazner and O'Neil (1989), but for a different view see McBirney (1969).

27. R. L. Christiansen and Lipman (1972), Eichelberger and Gooley (1977), B. R. Doe et al. (1978), Moyer and Esperanca (1989).

28. Ewart, Taylor, and Capp (1968), Pichler and Ziel (1972), F. Reid (1983), Clemens and Wall (1984), E. H. Christiansen, Sheridan, and Burt (1986).

29. Pichler and Zeil (1972), Glazner (1990).

30. Clemens and Wall (1984), Guilbeau and Kudo (1985).

31. W. K. Conrad, Nicholls, and Wall (1988).

32. Also see A. Rice and Eichelberger (1976), Moyer and Esperanca (1989). For a discussion of the theory of mixing, see Oldenburg et al. (1989).

33. Lipman and Mehnert (1979), R. P. Smith (1979), D. C. Noble et al. (1984), R. MacDonald et al. (1987).

34. The initial suggestion was made by H. R. Shaw, Smith, and Hildreth (1976).

35. For example, Howell Williams (1942), McBirney (1968a), R. L. Smith (1979), Hildreth (1979, 1981), Fridrich and Mahood (1987), Vogel et al. (1987), McCurry (1988), Schuraytz, Vogel, and Younker (1989).

36. Michael (1983), Miller and Mittlefehldt (1984).

37. D. Walker and DeLong (1982), Lesher et al (1982), Lesher (1986).

38. The Yellowstone example is based primarily on the works of Boyd (1961), W. B. Hamilton (1963b, 1965), R. L. Christiansen and Blank (1972), B. R. Doe et al. (1982), and Hildreth, Christiansen, and O'Neil (1984). Also see R. B. Smith and Braile (1984) and older reports by Fenner (1938) and Wilcox (1944). Bimodal means "two peaks of greatest abundance" as is the case in fig. 8.6.

39. Iyer (1979), Schilly et al. (1982), and other papers in JGR, v. 87, no. B4; Morgan, Doherty, and Leeman (1984).

40. R. L. Christiansen and Blank (1972), USGS (1972), Eaton et al. (1975), R. L. Christiansen (1979, and several USGS maps referenced in Hildreth, Christiansen, and O'Neil, 1984). Also see W. B. Hamilton (1965), Smedes and Prostka (1972), and Ruppel (1972).

41. Keefer (1971), Smedes and Prostka (1972), E. T. Ruppel (1972).

42. R. L. Christiansen and Blank (1972), R. L. Christiansen (1979, 1984), Hildreth, Christiansen, and O'Neil (1984).

43. Boyd (1961), B. R. Doe et al. (1982).

44. Boyd (1961), R. L. Christiansen and Blank (1972), B. R. Doe et al. (1982).

45. Boyd (1961), R. L. Christiansen and Blank (1972), B. R. Doe et al. (1982).
46. See B. R. Doe et al. (1982) for these and following chemical data.
47. B. M. Kennedy et al. (1987).
48. B. M. Kennedy et al. (1987).
49. B. R. Doe et al. (1982).
50. Recall that pyroclastic rocks are fragmental rocks erupted from a vent. These rocks are treated in great depth in the text by R. V. Fisher and Schmincke (1984). A variety of pyroclastic rock compositions is described in E. F. Cook (1966), Heiken (1974), R. V. Fisher (1982), R. V. Fisher and Schmincke (1984, pp. 98–99), papers in JGR, 1984, v. 89, no. B10, Worner and Schmincke (1984a, 1984b), and Bourdier, Gourgaud, and Vincent (1985). A. L. Smith and Roobol (1982) discuss andesitic pyroclastic rocks.
51. R. V. Fisher and Schmincke (1984, p. 89).

PROBLEMS

8.1. Calculate the CIPW norm for the following analysis and determine whether it is peralkaline, metaluminous, or peraluminous (values in wt. %):

SiO_2 = 73.91	TiO_2 = 0.36
Al_2O_3 = 13.50	Fe_2O_3 = 2.01
FeO = 0.23	MgO = 0.40
CaO = 1.90	Na_2O = 3.90
K_2O = 3.94	P_2O_5 = 0.10
LOI = 0.43	

8.2. Use the IUGS chemical classification of volcanic rocks to classify all of the rocks in table 8.2.

8.3. Samples (listed below) taken at the indicated levels above the stratigraphic base of a sequence of Recent rhyolitic tuff layers contain the indicated amounts of Cs. Plot the amounts of Cs versus stratigraphic position. Then read Stix and Gorton (1990a, p. 1265) and make a preliminary interpretation of the data.

Distance above Base (meters)	Cs (ppm)
1	12
5	11.5
10	11
15	11
20	7.5
25	7.3
30	6.5
35	6.7
40	5.9
45	4.4
50	4.0
55	3.8
60	3.5
61	3.6

Andesites and Related Rocks

INTRODUCTION

To most people, the image of a volcano is that of a steep-sided composite cone, the typical volcano of the volcanic arcs (figure 9.1). Such volcanoes are the principal domain of the andesites, the most common of the volcanic rocks of intermediate silica content.[1] In these volcanoes and the arcs, andesite occurs with basalts, dacites, and rhyolites, and less commonly with minor amounts of latite and other volcanic rocks.

Defining the term andesite is a problem. Modal, normative, and chemical definitions have been used widely,[2] but the range of rocks encompassed by each type of definition is not the same. For example, examination of Washington's (1917) tables of "superior" chemical analyses of fresh rocks reveals that most of the rocks called andesite are normative dacites, basalts, or

Figure 9.1 Guatemalan composite volcanoes, Agua (right) and Fuego (left). Agua has not erupted in historic times, but Fuego is active and erupts frequently. Note the steep-sided cone shape of these typical, composite basalt-andesite volcanoes, the petrology of which is discussed by Chesner and Rose (1984).

(a)

(b)

....
Figure 9.2   Photomicrographs of **f**-andesites. (a) Hornblende andesite from Black Butte, Shasta County, California. (PL).
(b) Pyroxene andesite from Makushin Volcano, Alaska. (XN). P = plagioclase, H = hornblende, PX = pyroxene. Long dimension of
photos is 3.25 mm.

trachy-andesites. In contrast, of 51 normative andesites in the same table, 67% are called basalt or quartz basalt, whereas only 20% are called andesite.

Reference to the original definition of andesite, applied in the nineteenth century by von Buch[3] to rocks collected from the Andes Mountains of South America, is of little help in clarifying the problem. His definition is vague and reflects incomplete knowledge of the modes of the rocks to which the name was given. These original "andesites" were thought to be hornblende-albite trachytes, but were later found to contain a more calcic plagioclase, plus a pyroxene.

A precise definition of *andesite* depends on the basis of definition and the classification system selected. Petrographically, I recognize that andesites typically are aphanitic-porphyritic rocks containing phenocrysts of sodic plagioclase, pyroxene, and/or hornblende (Raymond, 1984c) and, accordingly, I have defined *f*-andesite on the basis of modal phyric mineralogy (see chapter 4).[4] Because modal mineralogy can be misleading (see chapter 5), I here define andesites as aphanitic to aphanitic-porphyritic (volcanic) rocks with silica values in the range 52–63%, combined alkalis ($Na_2O + K_2O$) of less than 7%, and $Na_2O > K_2O$.[5]

MINERALOGY AND TEXTURES OF ANDESITES

Sodic plagioclase and hornblende are the characteristic phyric and modal phases in andesites (figure 9.2), but the mineralogy of andesites varies considerably. Sodic or calcic plagioclase (An_{30-80}), orthopyroxene, augite, hornblende, and biotite may occur as phenocrysts. Oxide phases, usually magnetite and ilmenite, also occur as phyric phases, but typically dot the matrix. The matrix consists mainly of plagioclase, pyroxene, glass, or combinations of these constituents. Additional phases that may occur in andesites include olivine, quartz, and, rarely, potash feldspars. Xenocrysts, uncommon phases, and alteration products include minerals such as aegerine, cordierite, apatite, sphene, anorthoclase, a wide variety of zeolites, calcite, celadonite, epidote, pumpellyite, chlorite, hematite, and pyrite. In *boninite* (high Mg andesite with Mg number > 0.7), the phyric phases are orthopyroxene, clinopyroxene (pigeonite, augite, or clinoenstatite), and olivine.[6] Glass forms a matrix. In the dacites, which have SiO_2 > 63% and are commonly associated with the andesites, the typical phyric phases are sodic plagioclase, quartz, biotite, and hornblende or augite. Alkali feldspar may be present, especially in the matrix, which consists primarily of glass or of plagioclase and quartz.

Texturally, andesites range from glassy aphyric rocks to holocrystalline aphanitic-porphyritic rocks. Vitrophyric (intersertal) and trachytoidal textures are relatively common, but intergranular-porphyritic textures are typical. Breccias occur commonly among the andesites and related rocks, and some of these rocks occur as tuffs. Plagioclase may be zoned (figure 9.2), with oscillatory zoning occurring locally.

TYPES AND OCCURRENCES OF ANDESITES AND RELATED ROCKS

Arcs are the principal, but not the only, sites of andesite occurrence. From site to site, andesites vary in chemistry. Variations in major element trends and specific chemistries have served as a basis for subdividing andesites into various types. For example, low- and high-silica andesites and low-, medium-, and high-K andesites have been named on the basis of their respective chemistries.[7] G. A. MacDonald (1960) suggested that the term andesite be restricted because of the wide range of major element chemistry that exists within the set of all rocks having silica values in the 52 to 63% range. In particular, he proposed that the term not be used for oceanic (mid-plate), intermediate silica rocks—a suggestion opposed by D. C. Stewart and Thornton (1975). In light of the conflict, arc andesites are commonly distinguished from other intermediate silica rocks as andesites (*sensu stricto*), orogenic andesites, or, as they are called here, *arc andesites*.

A number of special names have been applied to other rocks of intermediate silica content, which are not considered to be "typical" andesites. Boninite is high-Mg andesite. *Mugearite, icelandite, tristanite, benmoreite,* and *shoshonite* are all andesites (*sensu lato*) that do not fit strict definitions of the term.[8] Likewise, the *lamprophyres* are not andesites in the strict sense, but are porphyritic rocks, commonly of intermediate composition, characterized by euhedral phenocrysts of mafic minerals. Mugearite, shoshonite, and lamprophyre are names based on modal characteristics.

Associated with andesites are several, generally more siliceous rock types. *Dacite* and *rhyodacite* are typically quartz phyric volcanic rocks, more siliceous than andesite but less alkali-rich than rhyolite. *Trachyandesite* is a term applied to rocks that have silica contents equal to or lower than andesite, but have a higher alkali content (> 7%). *Trachyte* is a name applied to rocks that have silica contents similar to andesite but in which the alkalis are more abundant. Andesites are commonly accompanied by one or more of these associated rock types, in all types of tectonic sites.

For andesites and related rocks that occur in the arcs, the tectonic setting is clear. Arc andesites form above subduction zones. This is true whether the arc is oceanic or continental.[9] Non-arc andesites occur at sites of transform fault–continent interaction,[10] within continents (e.g., along continental rift zones),[11] above hot spots, and at oceanic spreading centers (mid-ocean ridges).[12] Other rocks of intermediate composition occur in both plate-margin (arc and rift) and intra-plate settings. Boninite occurs in arcs and in some ophiolites.[13]

CHEMISTRY OF ANDESITES

Both major and trace element chemistries are important in assessing the origins of andesites. Unfortunately, analyses of andesite, perhaps more than analyses of rhyolites and basalts, are subject to error. These errors are created by (1) alteration and (2) the presence of abundant impurities (xenocrysts and xenoliths) commonly incorporated in the andesites.

Not atypically, andesites are affected by a form of alteration known as *propylitization* in which hydrothermal solutions alter the rocks to a green color through the replacement of original minerals by such minerals as calcite, chlorite, epidote, and the green mica, celadonite.[14] This alteration adds water and CO_2 to the rock, may alter the oxidation state of the iron, and may remove various ions from the rock, significantly changing its chemistry.

Major Element Chemistry

The major element chemistry of some typical andesitic rocks is given in table 9.1. Analysis 5 in the table is actually the average of 2600 analyses. Notice the medial values of the oxides in this analysis with respect to the other analyses in the table.

Table 9.1 Major Element Analyses and CIPW Norms of Some Andesitic Rocks

	1	2	3	4	5	6	7	8
SiO_2	53.23[a]	54.88	56.20	56.82	57.94	58.96	60.59	61.3
TiO_2	1.48	0.95	0.13	0.92	0.87	1.41	1.25	0.80
Al_2O_3	18.26	18.38	10.57	17.90	17.02	15.50	15.07	16.9
Fe_2O_3	2.68	1.31	2.02	5.09	3.27	4.98	2.31	2.26
FeO	6.45	5.97	6.23	2.43	4.04	1.31	5.73	3.08
MnO	0.11	0.13	0.16	0.10	0.14	0.07	0.19	0.08
MgO	4.32	5.57	11.19	2.72	3.33	3.00	1.73	2.1
CaO	7.52	7.40	7.44	6.19	6.79	4.81	4.94	5.95
Na_2O	4.20	3.88	1.54	4.28	3.48	4.18	4.29	4.36
K_2O	0.70	0.86	0.40	2.12	1.62	3.63	1.59	1.44
P_2O_5	0.28	0.29	0.02	0.37	0.21	0.63	0.43	0.18
Other	0.71	0.18	4.05	0.84	1.22	1.32	1.84	—
Total	99.94	99.80	99.95	99.78	99.93	99.80	99.96	98.45

Sources:

1. Andesite from South Sister (Volcano), Oregon. Analyst: K. Aoki. (McBirney, 1968b, table 2, analysis 1.)*
2. "Olivine basaltic andesite," Paricutin Region, Mexico (H. Williams, 1950).
3. Boninite, Bonin Islands (W. E. Cameron, McCulloch, and Walker, 1983). Analysis includes 0.10 CO2.
4. Calc-alkaline andesite, Southern Peruvian Andes (sample 48, Dostal, Depuy, and Lefevre, 1977).*
5. Average of 2600 andesite analyses (LeMaitre, 1976).
6. Shoshonite, Southern Peruvian Andes (sample 103, Dostal, Depuy, and Lefevre, 1977).*
7. Icelandite, Thingmuli Volcano, Iceland (I. S. E. Carmichael, 1964).*
8. Hornblende andesite, Mount Hood, Oregon (W. S. Wise, 1968).*

[a]Values in weight percent
[b]Norms calculated by the author are noted by an asterisk (*) under Sources.

There are a number of points of note with regard to the major element chemistry revealed in the table. Of course, the range of SiO_2 contents of the andesites is limited by the definition. TiO_2 contents are commonly low, a characteristic common to subduction-zone volcanic rocks (J. B. Gill, 1981, p. 111). Alumina typically falls in the range of 15 to 19%. Thus, in comparison to average basalts and rhyolites, andesites are more aluminous. This is due principally to the high plagioclase content of the andesites. Total iron varies by a factor of two. MgO is typically low in andesites, except in boninites and some low-silica andesites. Mg numbers average about 0.60 in arc andesites (J. B. Gill, 1981, p. 110), but may exceed 0.70 in some non-arc andesites, boninites, and high-Mg andesites (Zielenski and Lipman, 1976; A. J. Crawford, Falloon, and Green, 1989). Alkalis (Na_2O and K_2O) vary, generally by a factor of only two, but much of that variation results from differences in

K_2O. K_2O increases with increasing SiO_2 and also tends to increase across arcs in the direction away from the subduction zone (Dickinson and Hatherton, 1967).[15]

Arcs contain both tholeiitic and calc-alkaline series rocks (Miyashiro, 1974, 1975b; J. B. Gill, 1981). In tholeiitic series, iron undergoes some enrichment at low to moderate levels of SiO_2. In the early stages of magma evolution, little iron is removed from the magma by crystallization, causing the bulk magma chemistry to increase in total iron. Later precipitation of iron-bearing phases causes a decrease in magmatic iron. Together these trends yield a characteristic AFM tholeiite curve (figure 9.3). In calc-alkalic series, magma compositions trend away from the FM side of the diagram, indicating a decrease in iron, as silica increases.[16] The use of AFM and like diagrams represents an attempt to reveal petrogenetically important trends.

	1	2	3	4	5	6	7	8
CIPW Norms								
q	2.41[b]	2.78	12.13	8.61	12.37	9.26	15.42	15.18
or	4.12	5.06	2.34	12.52	9.60	21.43	9.41	8.52
ab	35.56	32.83	13.01	36.19	29.44	35.35	36.29	36.87
an	28.91	30.21	20.78	23.40	26.02	12.82	17.17	22.31
di	5.34	3.78	12.87	3.81	4.84	5.41	3.83	4.86
hy	15.54	20.58	31.55	5.01	9.49	4.96	9.28	5.48
mt	3.89	1.90	2.92	5.49	4.74	0.37	3.36	3.29
il	2.81	1.80	0.24	1.74	1.65	2.67	2.37	1.52
hm	—		—	1.31	—	4.73	—	—
ap	0.67	0.67	0.03	0.87	0.50	1.48	1.01	0.44
Other	0.71	0.18	4.05	0.84	1.33	1.32	1.84	0.00
Total	99.96	99.79	99.92	99.79	99.98	99.80	99.98	98.47

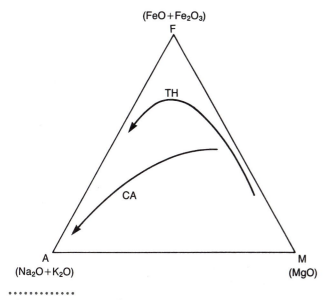

Figure 9.3 AFM diagram showing idealized tholeiitic (TH) and calc-alkaline (CA) trends of magma evolution. Arrows extend from low to high silica compositions. Compare the diagrams and data in figures 9.8 and 9.11 to the trends presented here.

Trace Elements and Isotopes

Trace element trends, like major element trends and values, are used in petrogenetic analyses of andesites.[17] Rare earth element (REE) trends (figure 9.4) vary among andesite types. In general, however, REE enrichment prevails and LREE enrichment increases sympathetically with the K_2O increase. In some rocks, HREE enrichment ranges to 30× chondrite values, whereas LREE ranges from approximately 20× to more than 400× chondrite values in continental rocks (Leeman, Vitaliano, and Prinz, 1976). Only low-K arc andesites, boninites, and ocean-ridge andesites lack a pronounced enrichment in LREE. Atypically, negative Eu anomalies occur, but in most cases they are not marked. Among the other trace elements, U, Pb, Rb, and Ba generally increase with increasing K content of the rocks.

Isotopic ratios also vary among andesite types. Initial $^{87}Sr/^{86}Sr$ values range from about 0.703 to 0.710 in arc andesites and may reach 0.712 in continental, non-arc andesites. ^{10}Be may be significantly enriched (L. Brown et al., 1982a, 1982b). Neodymium isotope ratios also vary, with arc andesites showing both greater and lesser Nd-isotope ratios than the bulk earth.[18] In general, however, Nd isotopic values for andesites are significantly lower than those for MORBs (see figure 6.3).

Chapter 9   Andesites and Related Rocks

PETROGENESIS OF ANDESITES

The available data indicate that andesites (and dacites) are generated by more than a single process. The evidence for this conclusion is of several types. First, the fact that andesites form in markedly different tectonic settings suggests the possibility of more than one process. The varied major element chemistries, especially low versus high Mg numbers, which indicate deriva-

tive and primary mantle sources, respectively, clearly indicate that multiple origins are required. Trace elements and isotopes, too, provide evidence for this view. For example, low Sr-isotopic ratios suggest equilibration with the mantle, whereas andesites with Sr isotopic ratios in the 0.710 to 0.712 range were modified with crustal involvement. Nd isotopes demand the same conclusion (DePaolo and Wasserburg, 1977).

The proposed origins of andesites are many and varied. Non-arc andesites must be created by a process or group of processes that are independent of subduction. In contrast, arc andesites are clearly related to the subduction process. Yet significant similarities exist between many arc and non-arc andesites. Why is this so? Do the variants of "typical" andesite give us a clue to a general model of andesite genesis, or is each andesite type created by a unique process or sequence of events?

In order to address these questions, we will examine (1) the various models of andesite genesis and (2) relevant data, to assess which model or group of models is most able to synthesize and explain the available information. Both source rocks and sites of magma genesis are important in these models (figure 9.5).

A number of kinds of rock have been proposed as source materials for andesitic magmas. The letters in figure 9.5 label materials (and sites) that have been suggested as principal or contributing sources. The letter A marks the subcrustal mantle of the subducting plate. B is the oceanic crust of the subducting plate, composed of MORB and MORB-related plutonic and volcanic rocks or their metamorphic equivalents (eclogite and amphibolite).[19] C designates sediment deposited on the oceanic crust. D is intermediate-depth, upper-mantle peridotite (e.g., garnet lherzolite). E represents shallow-level, upper-mantle peridotite (e.g., garnet or spinel lherzolite). F marks the crust-mantle interface, where ultramafic mantle rock and crustal rocks meet, a site where magmas may pond or mix. G represents deep crustal rocks rich in feldspar or amphibole, and H is a shallow crustal site in sedimentary and/or volcanic rock. Finally, I denotes deep upper-mantle rock comprising the mesosphere, and J is shallow mantle below the continental rift zone. One additional component not shown, but critical to many models, is water (H_2O).

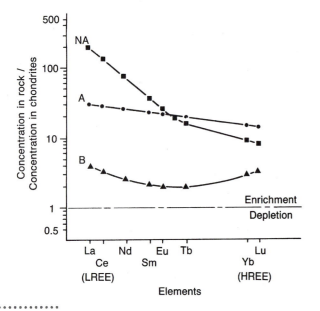

Figure 9.4 Generalized REE curves for arc andesites (A), continental non-arc andesites (NA), and boninites (B). Significant deviations from these generalized curves exist.

A and NA are based on Leeman, Vitaliano, and Prinz (1976), Peccerillo and Taylor (1976), Zielinski and Lipman (1976), Dostal, Dupuy, and Lefevre (1977), Dostal et al. (1977), Hawkesworth and Powell (1980), J. B. Gill (1981), Aoki and Fujimaki (1982), Pearce (1982), T. H. Dixon and Stern (1983), F. A. Frey et al. (1984), Lopez-Escobar (1984), and R. A. Thompson and Dungan (1985). B is based on W. E. Cameron, McCulloch, and Walker (1983)

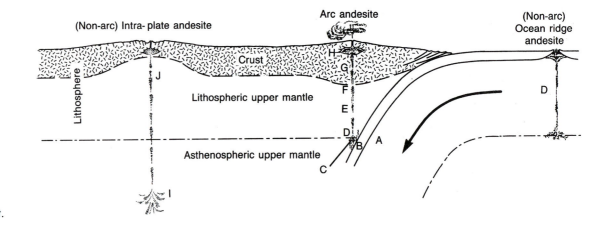

Figure 9.5 Sites (and sources) of andesitic magma origin and modification (A–J). Composition of sources are discussed in the text.

140

The Models and Subduction Effects

As with rhyolitic magmas, several major processes are considered to be responsible for the formation of andesitic magmas. These include fractional crystallization, anatexis, magma mixing, assimilation, and combinations of these processes (e.g., AFC). The various arc and non-arc models that incorporate these processes differ because the tectonic setting, the nature of the magma sources, and the processes affecting the magmas differ. Some models are designed specifically to explain arc-andesite origins. Others only explain the origin of special rock types (e.g., boninite). For discussion here, models for the origin of andesitic magmas and andesites are divided into two groups: (1) those essentially abandoned or of very limited applicability and (2) those currently supported as credible models for arc or non-arc rock origins. Theoretically, no model is solely applicable to non-arc settings, but those that do apply to those settings rely on the relatively simple process of mantle anatexis as a source for parental magma (and heat). In contrast, arc magma genesis, which is linked to subduction of an oceanic plate, is a more complex process.

Because the production of arc magmas relies on subduction, understanding of arc-magma-forming processes requires a rudimenary understanding of the nature of the plate being subducted, the nature of the overlying plate, and the physicochemical processes that occur during subduction. As revealed in chapter 7, oceanic crust is largely tholeiitic basalt and its plutonic equivalents (e.g., gabbro). Importantly, these rocks and the overlying sediments have been in contact with seawater and most have undergone some degree of metamorphic reaction with hydrothermally activated seawater (i.e., seawater circulated through and heated in the oceanic crust). As a result, the rocks of the subducting plate are "wet" in the sense that they contain both uncombined water in pores and fractures (H_2O-) and combined water in zeolites, amphiboles, serpentine, and other hydrous minerals (H_2O+). The plate above the subduction zone consists of a lower part composed of ultramafic mantle rocks, the nature of which is essentially a matter of conjecture, and an upper part composed of either oceanic (simatic) crust or more siliceous continental crust (see chapter 1).

As a plate is subducted, it is heated, dehydrated, and metamorphosed. A number of workers have modelled temperature distributions in subduction zones.[20] Many models predict temperatures at the top of the subducting plate, at the spot below the volcanic front, to be somewhat below 1000° C (figure 9.6a).[21] The temperature contours (isotherms) are subparallel with the plate boundary in this area and show a decrease in temperature towards the interior of the plate. The heat causes dehydration and metamorphism.

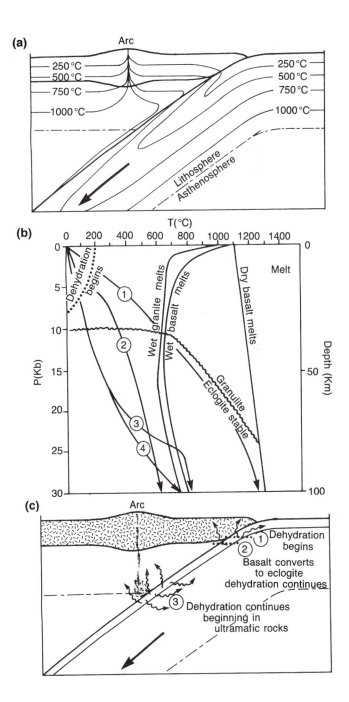

Figure 9.6 *Temperature, dehydration, and melting curves of subduction zones. (a) Schematic distribution of isotherms in a subduction zone. Note the cool center of the subducting plate. (b) Pressure-temperature grid showing P-T paths (1–4) for the top of an oceanic plate being subducted under various conditions. Curve 1 is based on a proposal by B. D. Marsh (1979). Curve 2 follows a path dictated by a model outlined by Wyllie (1983a). Curves 3 and 4 are based on Honda's (1985) models for Japan and the Andes, respectively. Also shown are the melting curves for dry basalt (after Green and Ringwood, 1967), saturated (wet) basalt (after R. E. T. Hill and Boetcher, 1970; Wyllie, 1971b, 1979), and saturated (wet) granite (after W. L. Huang and Wyllie, 1981; C. R. Stern and Wyllie, 1981). The jagged line represents the stability limit of eclogite at high temperature projected to lower temperatures (after Green and Ringwood, 1967). Note the estimated position where dehydration begins. (c) Dehydration zones in a subduction zone. Arrows show fluid migration paths.*

The exact process of metamorphism—the changes in texture, mineralogy, and chemistry—of the oceanic crust, as it descends into the subduction zone, depends on a number of factors. These include the initial temperature of the plate, the temperature of the overlying mantle wedge, the length of time subduction has occurred, the amount of previously subducted lithosphere, the subduction rate, the angle of subduction, the water content of the rocks of the subducting plate, the chemistry of the rocks in the subducting plate, and the amount of frictional heating and shear stress present along the subduction zone boundary.[22] Reference to figure 9.6b and 9.6c will help in visualizing the process. Figure 9.6b shows various phase boundaries, including melting curves for wet and dry basalt and gabbro. Also, projecting from spots very near the origin ($P = 0$ kb, $T = 0°$ C) is a series of numbered curves representing possible P-T paths that might be followed by subducting plates. The steepest curve is based on a model of a northern Japanese subduction zone (Honda, 1985) and corresponds to a geothermal gradient of about 6° C/km. A similar curve has been calculated by Peacock (1990a) for subduction of 200-million-year-old lithosphere. The most gentle curve is based on an idealized model of asthenospheric corner flow developed by B. D. Marsh (1979) and corresponds to a geothermal gradient of about 12.5° C/km. Similar curves are representative of the temperature of young subducting plates in zones where little lithosphere has been previously subducted (Peacock, 1990a). Wyllie's work (1983) suggests intermediate curves. In neither the model of Honda (1985) nor the warm-mantle model of Wyllie (1983b) does the temperature reach the solidus of water-bearing basalt (where melting would begin) at the appropriate 100 km (30 kb = 3 GPa) depth. By design, the B. D. Marsh (1979) geotherm intersects the appropriate solidus at about 100 km.

Many petrologists—among them W. B. Hamilton (1969), Raleigh and Lee (1969), Ringwood (1974), Fyfe and McBirney (1975), J. M. Delaney and Helgeson (1978), Wyllie (1979, 1983), T. J. Lewis et al. (1988), Peacock (1990a, 1990b), and Bebout (1991)—think that the descending plate will undergo progressive dehydration (figure 9.6c). At shallow levels, increasing pressure and temperature will begin to cause reactions in the sediments and volcanic rocks of the crust. These metamorphic reactions include decarbonation and dehydration reactions, which produce CO_2 and H_2O, as well as methane(CH_4)-producing reactions. A fluid phase will evolve and rise into the overlying plate. As the plate approaches a depth perhaps near 30 km, the basalt and gabbro of the descending plate will begin to convert to eclogite, an anhydrous rock.[23] During this time of subduction, the C-O-H fluid phase will continue to be expelled into the overlying plate.[24] As the plate subducts, the upper part becomes dry; however defluidization of the base of the plate may be retarded as a result of some fluid migrating parallel to or down the thermal gradient. At depths of about 80–120km, serpentine and amphiboles in the basal crust and subcrustal mantle will dehydrate, with the resultant fluids moving up the plate or into the overlying mantle wedge (J. M. Delaney and Helgeson, 1978; Tatsumi, 1986; Tatsumi and Nakamura, 1986). Elements transported by these fluids change the chemistry of the invaded rocks.[25] The fluids produced by the various dehydration reactions depress the melting temperature of the subducted oceanic crust, but it is debatable whether this reduction is adequate to cause melting under the range of geothermal gradients predicted on the basis of geophysical studies. In contrast, the fluid entering the mantle wedge will very likely induce metamorphism and anatexis.[26]

Abandoned Models of Andesite Formation

Certain models of andesite formation are no longer considered viable because they are not consistent with experimental data, major and trace element chemistry, or isotopic ratios. In the following descriptions, letters in parentheses represent sources or sites on figure 9.5, where magma genesis occurs.

Anatexis of Subducted Sediment (C) ± Oceanic Crust (B) (Model 1)

Some petrologists have suggested that anatexis of subducted sediment (C) ± oceanic crust (B) produces arc-andesitic magma.[27] Sediments, especially those eroded from continents, have high Sr-isotope ratios and Pb contents. Data from arc andesites on these and other major element, trace element, and isotopic characteristics, especially data on [10]Be, an isotope with a half-life of 1.5 m.y. (Brown et al., 1982b; McHargue and Damon, 1991), favor *some* involvement of subducted sediments in arc-magma genesis.[28] The isotope [10]Be is formed in the atmosphere, transferred to the earth by rain, and absorbed in sediment and soil. The half-life is long enough for the isotope to serve as tracer of sediment moving into a subduction zone and back to the surface via magma transport, but short enough that the [10]Be cannot be concentrated by recycling back into the sediment and again into the arc magmas. Samples from several arcs show elevated [10]Be values that clearly reveal sediment participation in magma genesis (J. D. Morris et al., 1985).

No compelling case has been presented, however, in favor of *large-scale* sediment melting. In fact, much available evidence cited in favor of sediment melting is indistinguishable from that for crustal contamination and fits more reasonably into more complex models of andesite genesis (e.g., R. W. Kay, 1980, 1985; Hildreth and Moorbath, 1988; Morris and Tera, 1989). Various isotopic and trace element studies indicate that the quantity of sediment involved in magma generation may be low (0%–10%)(S. E. Church, 1973b; T. H. Dixon and Batiza, 1979; R. J. Stern and Ito, 1983).[29] For example, [10]Be-isotope data suggest a sedimentary component of less than 4% (Morris and Tera, 1989). Also, experimental melting of sediments has not produced andesitic liquids with appropriate major and trace element chemistries (Winkler, 1965; C. R. Stern and Wyllie, 1973; Wyllie, 1977). Altogether, the data do not provide much support for the model.

Anatexis of Basal Crust (F, G) (Model 2)

A second model proposed by petrologists is based on the notion that anatexis of basal crustal rocks by heat from tectonism or underplated basaltic magmas produces primary andesite magma (F and G).[30] Pichler and Ziel (1969, 1972) and others offer several lines of evidence in support of crustal anatexis.

1. There is a total absence of "true" basalts in the volcanic section of the Chilean Andes.
2. Trace elements (e.g., high $^{87}Sr/^{86}Sr$) support crustal involvement.
3. A close association exists between aluminous rhyolites and andesites in space and time, suggesting a genetic link.

Yet, basalts are not lacking in the Chilean Andes (Deruelle, 1982; F. A. Frey et al., 1984; Lopez-Escobar, 1984) or elsewhere where andesites occur. Further evidence against crustal anatexis as the principle process in andesite genesis is provided by experimental studies of melting in rocks of a range of siliceous to intermediate compositions, which indicate that the melts that form will be rhyolitic rather than andesitic (Wyllie, 1983b).[31] Another major reason that production of andesitic magma by crustal anatexis is unlikely is that the chemistry of natural rocks does not match that predicted on the basis of crustal anatexis models (J. B. Gill, 1981, p. 261–62).

Credible Models of Andesite Formation

Models of andesite petrogenesis that are still generally debated (models of the second group) are predominantly more complex than the abandoned models. Two exceptions are models 3 and 4.

Anatexis of Subducted Crust (B) ± Mantle (A) (Model 3)

Anatexis of subducted crust (B) ± mantle (A) may yield a primary, arc-andesite magma (B. D. Marsh and Carmichael, 1974).[32] In this model, mafic oceanic crust, converted to eclogite during subduction, is melted to yield primary andesitic magmas. Evidence of this process is provided by some experimental melting studies, theoretical calculations, and major and trace element data.[33] Nevertheless, this model has a number of problems. First, the chemistry of most andesites, including REE patterns, just does not match that predicted or produced in experimental studies of melting (T. H. Dixon and Batiza, 1979; J. B. Gill, 1981, p. 235ff.; R. J. Stern and Ito, 1984).[34] Second, as discussed above, most thermal models of

subduction zones do not predict temperatures high enough to melt subducted crustal rocks at sites near the top of the plate below the arc. Third, the model does not allow major fractionation during andesite formation, but abundant evidence indicates that fractionation is a significant process (J. S. Beard, 1986; D. R. Baker and Eggler, 1987).[35] Fourth, the model fails to explain the common association of basalts and andesites in the arcs. Thus, melting of oceanic crust as a model of andesitic magma production encounters major problems.

Anatexis of Mantle Rocks (Model 4)

In this model, anatexis of mantle rocks (A, D, and E; I or J) yields primary andesitic magma that rises to form andesites (M. J. O'Hara, 1965).[36] The idea is that primary andesitic magmas might be generated by partial melting of mantle rock—due to decompression or fluid-induced melting—below a rift zone (I), in the subducting plate (A), or in the mantle wedge above the subducting plate (D–F). Evidence used to support this model is mineralogical, chemical, experimental, and theoretical. Nd isotopic ratios, as well as low Sr-isotopic ratios of some "uncontaminated" arc rocks, are consistent with a mantle source.[37]

Evidence of primary andesitic magmas for certain rock types, is compelling. High-nickel, MgO-rich olivine, and high-Cr_2O_3 chromite in rocks such as high-Mg arc andesite, lamprophyres (spessartites), and boninites suggest that these special types of rocks are representative of primary mantle magmas (Suzuki and Shiraki,1980; Tatsumi and Ishizaka, 1981, 1982; Crawford, Falloon, and Green, 1989).[38] High Mg numbers support this view, as do experimental melting studies (A. J. Crawford, 1989; van der Laan, Flower, and Koster van Groos, 1989; Kushiro, 1990).

Is the model equally applicable to normal arc andesites? Experimental melting studies (T. H. Green and Ringwood, 1968a; Kushiro, 1974a; Mysen and Boettcher, 1975a, 1975b) and theoretical considerations (Mysen, 1983) indicate that andesite-like liquids may be derived from mantle rocks (e.g., eclogite or spinel lherzolite). These (and other) observations led Tatsumi and Ishizaka (1981, 1982) to suggest that *all* calc-alkaline andesites of the Setouchi Volcanic Belt of Japan may be derivatives of primary andesitic magmas. There are, however, a number of problems.[39] Students should realize that experimental studies attempt to approximate real rocks and natural conditions, but are limited by our lack of knowledge about actual mantle conditions and chemistries. They also only indicate what is possible, not necessarily what occurs. There is much room for debate. For example, some experiments suggest that andesitic liquid can only be produced from CO_2-free peridotite, but there is good reason to believe that the upper mantle is not generally CO_2-free (Boettcher, Mysen, and Modreski, 1975; Eggler, 1978). In addition, experimental studies have generally indicated that for andesitic liquids to coexist with peridotites under equilibrium conditions, they must contain 14–25 wt. % H_2O (see reviews in J. B. Gill, 1981; Wyllie, 1978, 1979, 1982). Somewhat lower amounts of water (5–7%) may suffice (Mysen, 1983). Yet andesites generally have less than 5 wt. % H_2O.[40] Also significant is the fact that experimentally produced magmas, albeit generally similar in chemistry to andesites, are not chemically equivalent if examined in detail.[41]

Additional arguments raised in opposition to an anatectic mantle origin for andesitic magmas include the following.[42] The magnesium numbers of arc andesites are generally too low, at about 0.6, to represent primary mantle melts. Also, REE patterns of natural andesites are not like those predicted via modelling of mantle melting (I. A. Nicholls and Harris, 1980). Finally, no explanation of the common association of basalt and andesite is provided by this model. Thus, while the model is still considered viable by some, there are significant reasons to doubt its general applicability.

Anatexis of Pyrolite and Pyroxenite with Fractional Crystallization (Model 5)

The pyrolite-pyroxenite model of I. A. Nicholls and Ringwood (1973) and Ringwood (1974), detailed in chapter 7, appeals to anatexis of mantle rock for the origin of andesitic magmas, but involves a complex set of processes.[43] Recall that these processes include (1) dehydration in the subduction zone, via conversion of meta-igneous rocks of the oceanic crust to eclogite plus fluid (B), (2) hydration and consequent diapiric rise of pyrolite, (3) anatexis and production of tholeiite (E), (4) subsequent anatexis of the subducted, eclogitic crust (B) to form siliceous melts, (5) infiltration of melt and fluids from the subducting slab into the mantle wedge to form garnet pyroxenite, which diapirically rises, (6) melting of the mantle diapir (D, E), and (7) fractional crystallization at higher levels in the lithosphere (E, F, and G?) to produce calc-alkaline basalt, andesite, dacite, and rhyolite magmas that erupt at the surface to form their respective rock types (see figure 7.17d).

This model, like other multiphase models, attempts to account for the variety of chemical characteristics displayed by arc andesites. In such models, complex processes of magma formation and modification can explain the generally low Mg numbers, which do not reflect primary mantle melts. Elevated Sr isotopic ratios, as well as some Nd and oxygen isotopic values, are consistent with models involving continental crust (D. E. James, 1984). Likewise, Pb isotopic ratios reflect crustal or sediment involvement (Barreiro, 1984), and recirculating seawater may increase the relative concentration of K (R. W. Kay, 1980) while adding ^{10}Be to the magma source region (Brown et al., 1982b). Such models also explain low Ni values (characteristic of andesites) and other trace element concentrations in the andesites (S. R. Taylor, 1969). In addition, depleted HREE patterns are consistent with some two-stage pyrolite-pyroxenite-type models that involve fractionation (Lopez-Escobar, 1984).

Experimental tests of several aspects of the pyrolite-pyroxenite model have met with mixed success (I. A. Nicholls and Ringwood, 1972, 1973; Sekine and Wyllie, 1982, 1983; Johnston and Wyllie, 1989). For example, experimental simulation of second-stage hybridization of mantle by siliceous melt may yield pyroxenite (or peridotite), as predicted, but the major and trace element chemistries of the magmas produced from such rocks differ in significant ways from those of andesites (Johnston and Wyllie, 1989). Notably, residual or newly produced garnet in such ultramafic mantle rocks (or in eclogites) will result in the formation of magmas with steeply sloping, HREE-depleted REE patterns, unlike the relatively flat REE patterns of the arc andesite lavas (see figure 9.4). In addition, related experimental results do not suggest that the necessary siliceous melts can be derived from the subducting slab (C. R. Stern and Wyllie, 1981; Huang and Wyllie, 1981). Thus, the model seems inconsistent with some important data that constrain andesite magma genesis.

Fractional Crystallization of Basaltic Magma (Model 6)

Some andesitic magma may form via fractional crystallization of a basaltic magma initially generated by anatexis at depth (D, E, F, H, and I). Since Bowen (1928) proposed the idea, fractional crystallization of basaltic magmas has been a widely accepted process for explaining the origin of andesites.[44] Even where other processes are thought to modify magma chemistry, fractionation is commonly considered to be important.[45] Basically, the process is one in which minerals such as plagioclase, orthopyroxene, olivine, augite, hornblende, and magnetite are crystallized in and separated from a basaltic magma, producing a new magma enriched in silica and alkalis and depleted in iron and magnesium. Considerable debate has centered on whether amphibole, magnetite, or other phases are the key fractionating minerals.[46] In the model, the mineralogical details and the site of fractionation, which may be in the crust (G, H) or mantle (E), may vary.

Where andesitic rocks (*sensu lato*) develop over some hot spots or along mid-ocean ridges, fractional crystallization of olivine, plagioclase, clinopyroxene, and/or Fe-Ti oxides seems capable of explaining major and trace element trends.[47] Negative europium anomalies in such rocks (R. W. Kay, Hubbard, and Bast, 1970) reflect plagioclase fractionation.

In contrast, arc-andesite chemistries are not easily explained by fractional crystallization *alone*, though several arguments favor the operation of this process.[48] First, arc andesites are typically associated with basalts, dacites, and rhyolites in space and time, and each basalt-andesite-dacite-rhyolite association from a specific region typically plots as a smooth line on a variation diagram. Also, basalts and andesites from the same volcano may have identical Sr and Nd isotopic ratios (R. J. Stern, 1979; F. A. Frey et al., 1984). These data suggest that the rocks are **consanguineous,** or of the same parent. Second, agreement exists between (1) the *calculated* compositions of phases removed from basaltic magma via fractional crystallization (to produce andesitic magmas) and (2) the *modal phyric phases* in the rocks themselves (Kuno, 1968). Third, trace element data from some volcanoes provide strong evidence of fractionation (I. A. Nicholls and Harris, 1980; J. B. Gill, 1981, pp. 138ff. and 269).

These arguments are not entirely conclusive. Smooth curves on variation diagrams are also used to support the processes of magma mixing and assimilation, and basalt and andesite are apparently not consanguineous everywhere (Tatsumi and Ishizaka, 1982). Furthermore, the occurrence of identical isotopic ratios may reveal a similar source for magmas, not necessarily a parent-daughter link. Also to be considered is the fact that some experimental and theoretical studies suggest that fractionation *cannot* account for a basalt-to-andesite fractionation path (Tilley, 1950; Eggler and Burnham, 1973) or the abundance of certain elements, such as Rb and Ba, other large ion lithophile (LIL) elements, and certain isotopes.[49] Finally, if andesites derive from basalts via fractionation, they should be equally abundant in plume, rift, and arc environments, which they are not.

In spite of these objections, fractionation is probably central to the formation of many arc andesites (J. B. Gill, 1981). In several arc-andesite provinces, it can explain much of the chemical variation quite adequately, but the trace element chemistry of the rocks requires that other processes must contribute to magma genesis. In continental arcs, isotopic evidence indicates that assimilation is one of these processes, as described in model 7.

Mantle Anatexis plus Crustal Rock Assimilation (Model 7)

Contamination (hybridization) of mantle-derived, anatectic, basaltic parent magma via assimilation of crustal rocks may yield andesitic and related magmas (F, G, H).[50] Assimilation is an old idea, championed by Daly (1933). In modern petrogenetic arguments, it is seldom advocated as the sole modification process in andesite formation. Rather, assimilation is considered to occur either simultaneously with fractional crystallization (AFC) or other processes, *or* to precede or follow those processes (e.g., DePaolo, 1981c; Hildreth and Moorbath, 1988; McMillan and Dungan, 1988).

In the crust (F, G, H), silica, other felsic components, and trace elements could be added to basaltic magma generated in the mantle to form an andesitic magma (Fenner, 1926; A. Holmes, 1932; Daly, 1933; D. E. James, 1982). Evidence supporting this idea includes (1) major element trends on variation diagrams and (2) elevated Sr, Pb, Nd, and O isotopic ratios.[51] Additional evidence is provided by xenoliths of "granulite" (presumed lower crustal rock), granite, and other rocks found both in andesites and in basalts associated with andesites (see chapter 7; Tilley, 1950; Grove et al., 1988; Bacon, Adami, and Lanphere, 1989).

As a general explanation for arc and oceanic andesite genesis, assimilation operating alone fails.[52] First, many andesites occur where there are no known crustal rocks to be assimilated. Second, superheated magmas would be required to melt the assimilated rocks (see chapter 6). Third, assimilation involving crust and basalt endmembers cannot explain the chemistries, especially trace element distributions, of the voluminous arc andesites (J. B. Gill, 1981, p. 264ff.).

In spite of these objections, a case may be made for crustal contamination as one component of multiprocess models. Hildreth and Moorbath (1988) propose such a model, the MASH Model, in which the character of andesitic and dacitic magmas is created by mixing, assimilation, storage, and homogenization of mantle and crustal magmas at the crust-mantle boundary (F). In models such as these, the contribution of continental crust may explain some major element and local isotopic signatures in the andesites.[53]

Mixing of Magmas (Model 8)

Mixing of siliceous and basaltic magmas to form andesitic magma is a major type of andesite-forming process.[54] Like several petrogenetic ideas, the idea that magmas mix to produce compositionally intermediate melts is an old one (Fenner, 1926). Eichelberger (1975b, 1978a, 1978b) resurrected the idea to account for andesites and dacites in the circum-Pacific region. The process is a simple one, requiring only that two magmas (e.g., basalt and rhyolite) in a magma chamber (F, G, H) mix to form a magma of intermediate composition that then erupts and crystallizes to form rocks.

One objection to such a model might be that it is *ad hoc*; that is, it requires the coincidence that two magmas of different compositions arrive in the same area at the same time. However, as we saw in chapter 8, such a circumstance may be far from coincidence, because basaltic magmas rising from the mantle may either fractionate and form rhyolite or they may melt the base of the crust to produce a rhyolite. The problem then becomes one of inducing the two magmas to mix, a process that could be facilitated by the introduction of new, hot, gas-charged basaltic magma into the magma chamber from below.[55]

Evidence favoring magma mixing is observational (field), mineralogical, chemical, and experimental. In the field, basalt, andesite, dacite, and rhyolite may be intimately related spatially and temporally, with alternating rock types in the same volcanic pile or area. These relationships are inferred to represent a genetic link. Additional evidence is provided by xenoliths and xenocrysts in the intermediate rocks, such as basalt inclusions in dacite or Mg-olivine and quartz phenocrysts in andesite.[56] Plagioclase phenocrysts of markedly different compositions and those showing evidence of resorption in their cores are also cited as mineralogical manifestations of mixing (Eichelberger, 1978a). Chemically, straight-line compositional variations on variation diagrams between presumed endmember compositions are consistent with the mixing model.[57] In addition, simulations of mixing, based on various physical parameters (e.g., shape of the magma chamber, viscosity ratio), suggest that mixing will occur, and experimental mixing studies show that mixing is possible and quick (Kouchi and Sunagawa, 1983, 1985; Juster, Grove, and Perfit, 1989; Oldenburg et al., 1989).

In opposition to the magma-mixing model, it has been argued that the physical characteristics of the magmas involved make thorough mixing unlikely (McBirney, 1980). Yet, if magma mixing is an efficient and thorough process, evidence of mixing may be completely destroyed. As a consequence, some petrologists assume that the absence of evidence implies the absence of the process. Evidence of disequilibrium involving xenocryst and phenocryst assemblages and xenoliths can represent assimilation of solid rock rather than magma mixing, as can straight-line variations in chemistry between endmember rock types. Finally, the presence of andesite where rhyolite is absent, rare, or unlikely provides evidence that magma mixing cannot be the only process of andesite formation.

Because no one process or model seems able to account for *all* chemical and mineralogical properties or the origin of all andesites, several petrologists have proposed composite models.[58] These models incorporate combinations of the processes described above and may also involve additional processes, such as *metasomatism*, the alteration of rock by chemically active fluids. Each model seeks to explain the peculiarities of the chemistry of the particular rocks under study.

Conclusions on Petrogenesis

This list of possible origins for andesites may seem overwhelming on first reading. Yet the models are explicit enough to allow carefully directed analyses of their predictions and implications. That fact is reassuring. In contrast, the fact that only a few models may be entirely eliminated from consideration at present signals the complexity of the andesite problem.

Arc andesite magmas are derivative magmas that may result from complex histories of modification. Remembering that andesites are of widely variable chemistry, it is understandable that many combinations of fractional crystallization, anatexis, assimilation, metasomatism, and magma mixing yield these rocks. In contrast, most boninites and high-Mg andesites probably represent primary mantle melts. Other andesites, such as those formed at oceanic spreading centers or hot spots, may result solely from shallow-level crystal-liquid fractionation of tholeiitic basalt.

EXAMPLES OF ANDESITE ORIGINS

As examples of the possible variations in andesite petrogenesis, two arcs are described below. The Aleutian Arc represents an arc resulting from subduction of old crust beneath oceanic and continental crust. The Cascade Arc represents a continental arc resulting from subduction of young, hot oceanic crust.

The Aleutian Arc

The Aleutian Arc is a 2600-km-long volcanic arc that extends westward from south-central Alaska and separates the Pacific Ocean on the south from the Bering Sea on the north (figure 9.7).[59] In the east, the arc is continental, whereas in the west, it consists of a series of islands built on oceanic crust. Because of these differences in basement, the Aleutian Arc is particularly interesting for evaluating andesite origins.

Examination of the geology of several of the islands and the younger history of continent-based volcanoes reveals a broadly similar history for the arc.[60] In the west and east, Paleozoic (east only) and younger volcanic rocks are intruded by granitoid plutons (Coats, 1956a; B. D. Marsh, 1982a; Hein, McLean, and Vallier, 1984).[61] The prominently visible volcanic cones of the arc are of Pliocene to Recent age and are constructed on this basement of older rocks (Hoare et al., 1968).[62] Many volcanoes have gone through a history of construction, caldera formation, and resurgence (reconstruction) of smaller composite cones. Although the focus here is on the young arc, pre-Pliocene arc development may have influenced the contemporary arc.

As is common in well-developed arcs, some Aleutian volcanoes occur *behind* the main volcanic front. In Alaska, the Bogoslov and Amak volcanoes occur along a trend immediately behind the volcanic front. (F. M. Byers, 1959; Arculus et al., 1977; B. D. Marsh and Leitz, 1978). Also the Pribilof Islands (Barth, 1956) and Nunivak Island (Hoare et al., 1968), which may be magmatically unrelated to the arc, occur about 100 km and 200 km, respectively, behind (north of) the arc. The latter two islands have nepheline-normative alkalic basalts, including basanites. The rocks of Bogoslov and Amak are more typical of the main arc but are richer in K_2O.

Subduction controls volcanism in the Aleutian Arc. This is most clearly indicated by the fact that present-day plate motion is such that east of Buldir Island (long. 176° W) subduction and recent volcanism occur, whereas to the west of Buldir Island plate motion dictates strike-slip (transform) faulting and volcanism is absent.[63] Quaternary and older volcanism did occur to the west.

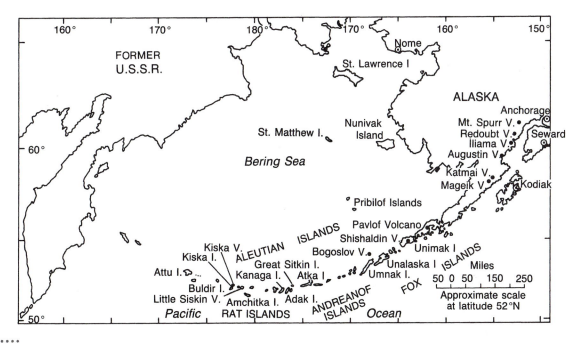

Figure 9.7 Map of the Aleutian Islands showing some of the more well-known volcanoes (V).
(Source: H A. Powers, et al., "Geology and submarine physiography of Amchitka Island, Alaska", U S Geological Survey-B-1028-P, 1960.)

Data bearing on the origin of Aleutian Arc rocks are petrographic, chemical, and structural in nature. The volcanic rocks range modally and chemically from basalts to rhyolites. Petrographically, the rocks are typical porphyritic arc rocks. Phenocrysts and groundmass minerals range from calcic plagioclase, olivine, clinopyroxene, and magnetite in basalts; through calcic to intermediate plagioclase, clinopyroxene, orthopyroxene, olivine, titanomagnetite, and (uncommonly) hornblende[64] in andesites; to quartz, sodic plagioclase, biotite, hornblende, hypersthene, ilmenite, and magneite in rhyolites. Amphiboles are rare in more mafic andesites and basalts (Marsh, 1982a). Chemically, the petrographic range in rock types is reflected by SiO_2 values from 46 to 77.4%.[65] Basalts are dominant (Marsh, 1982a), but andesites are quite common. Siliceous rock types are relatively minor in abundance.

The major element chemistry along the arc shows some variability. Arc rocks comprise both tholeiitic and calc-alkaline trends (figure 9.8) (R. W. Kay, 1977; S. M. Kay, Kay, and Citron, 1982). Though Marsh (1982a) found no obvious correlations between magma chemistry and underlying crustal structure or chemistry, some variations in chemistry are present that do seem to be structurally controlled (S. M. Kay, Kay, and Citron, 1982; Kienle and Swanson,1983; Singer and Myers, 1990). The controlling structural features may be fracture zones in the subducting plate or structural blocks in the overlying plate, either of which make the arc segmented. Crustal thickness variations may also exert some control on chemistry. Tholeiitic rocks form large, basalt-rich volcanoes that occur above arc-segment boundaries overlying oceanic crust. Calc-alkaline rocks form smaller, intra-segment, more andesitic, composite volcanoes on oceanic and continental crustal basements; and they form larger, caldera-forming, more chemically diverse (basaltic to rhyolitic) segment-boundary volcanoes above continental basement (S. M. Kay, Kay, and Citron, 1982; Kienle and Swanson, 1983).

Other major element characteristics of note are as follows. Most andesites have normal K contents, but locally, high-K andesites are present. The latter do not correlate with structural position. Slightly higher total iron and alkalies, thought to distinguish oceanic from continental-based volcanoes (Forbes et al., 1969), apparently characterize the tholeiitic (versus calc-alkaline) rocks (R. W. Kay, 1977; S. M. Kay, Kay, and Citron, 1982). Mg numbers for Aleutian Arc andesites range from at least 0.71 to 0.43, but are typically about 0.55 ± 0.05. Associated Aleutian Arc basalts have Mg numbers that usually fall in the range 0.73 to 0.55, with common values of about 0.60 ± 0.05.

Trace element data have been reported by a number of workers.[66] Sr isotopic ratios from various volcanoes are generally low (< 0.7035) and show little evidence of crustal

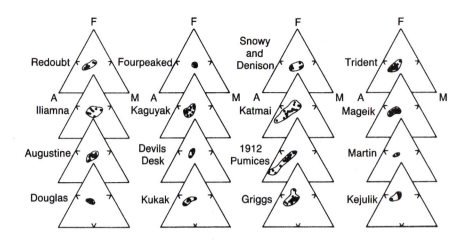

Figure 9.8 AFM diagrams for some volcanic rocks of the eastern Aleutian arc (from Kienle and Swanson, 1983). F calculated as FeO + 0.899 Fe$_2$O$_3$. Pumices of 1912 show the most well-defined calc-alkaline trend (compare to figure 9.3).

contamination. One exception is the Spurr Volcanic Complex, which is founded on continental crust in the eastern arc and has slightly elevated Sr-isotope ratios of between 0.7035 and 0.7041 (Nye and Turner, 1990). Nd-isotope ratios similarly show little crustal contamination (J. D. Morris and Hart, 1983; J. D. Myers, Frost, and Angevine, 1986; von Drach, Marsh, and Wasserburg, 1986). Pb isotopic values, in contrast, are variable and locally exceed those values indicative of uncontaminated, mantle-derived magmas (R. W. Kay, 1977; R. W. Kay, Sun, and Lee-Hu, 1978). [10]Be values are elevated (2.7–6.9 atoms/gram) well above the background value of 0.3 atoms/gram (Brown et al., 1982b). Rare earth elements are generally only slightly enriched, with LREE enriched relative to HREE and more siliceous andesites showing a negative europium anomaly (R. W. Kay, 1977; S. M. Kay, Kay, and Citron, 1982).

What do these data mean? Given that subduction of the Kula Plate earlier in the Tertiary and of the segmented Pacific Plate between the Tertiary and Recent was the direct cause of volcanism in the Aleutian Arc, the data must be analyzed in relation to the subduction process. [10]Be data suggest that water provided by the subducting plate contributes to melting of a source rock, either a metasomatized mantle peridotite or an eclogite.[67] Partial melting produces primary basalt magmas that rise through the overlying oceanic (western) and continental (eastern) crust. The nature of these primary magmas is debatable, but they are relatively dry, containing < 2% water (D. R. Baker and Eggler, 1983). The Pb isotopic data indicate some sediment or crustal contribution to the magma (through melting, assimilation, or contamination), especially in the east; locally, Sr isotopic data support modification; and Nd isotopic data argue for homogenization of the source (von Drach, Marsh, and Wasserburg, 1986; J. D. Myers, Frost, and Angevine, 1986; Nye and Turner, 1990). Most Mg numbers for Aleutian basalts

and andesites (as well as for the more siliceous rocks) fall in the 0.50–.65 range, indicating some magma modification. REE patterns are consistent with mantle derivation of magma and later fractional crystallization. Fractionation of plagioclase is indicated by the negative europium anomalies, especially in calc-alkaline rocks. Major element trends, phase relations, and xenolith compositions also indicate that fractionation has occurred (W. K. Conrad and Kay, 1984; J. S. Beard, 1986; D. R. Baker and Eggler, 1987).

Studies of plutons, xenoliths, mafic phenocrysts and the chemistry of the arc-volcanic rocks support complex models of magma evolution for the andesites.[68] The relative significance of fractional crystallization, magma mixing, and assimilation are unresolved, although assimilation seems to have been generally less important than fractional crystallization and mixing. The tholeiitic rocks develop above arc-segment boundaries, particularly those on oceanic crust. Tholeiitic magmas rise quickly to the crust, perhaps along segment boundary faults, where they undergo shallow-level fractional crystallization, local magma mixing, and subsequent eruption (S. M. Kay, Kay, and Citron, 1982; S. M. Kay and Kay, 1985b). Where such magmas rise into thicker continental crust, they may mix with other less evolved magmas and fractionate over a greater depth range and a longer period of time (Kienle and Swanson, 1983; S. M. Kay and Kay, 1985b; Singer and Myers, 1990). Calc-alkaline magmas may evolve below midsegment regions of the arc, where movement to the surface is presumably more difficult (S. M. Kay, Kay, and Citron, 1982). The calc-alkaline magmas fractionate at various depths and may mix with other primary or evolving magmas (D. R. Baker, 1987; Brophy, 1987, 1990), before erupting at the surface to form the impressive stratovolcanoes of the arc.

In spite of the number of studies on Aleutian volcanic rocks that have been completed, a number of important and fundamental questions remain unanswered or unresolved. An unequivocal case has not been presented in favor of either eclogitic or peridotitic source rocks for parent magmas. The compositions of primary magmas derived from the source are likewise uncertain, and the idea that the primary magmas are either andesitic or high-Al basaltic is debatable (J. B. Gill, 1981, p. 235; Brophy, 1989a, 1989b). Furthermore, the relative importance of fractionation, mixing, and assimilation in controlling major, trace, and isotopic element compositions of the lavas from most of the volcanoes is unresolved. The various conflicting studies do, however, emphasize the difficulty of interpreting the histories of andesites.

The Cascade Range

The Cascade Range of western North America contrasts with the Aleutian Arc in several ways. The range is an exclusively continental arc that extends along an approximately straight trend for about 1000 km, from north of Mount Garibaldi in British Columbia south to Mount Lassen in California (McBirney, 1968b; N. L. Green et al., 1988) (figure 9.9). The Cascade Arc is segmented and the segmentation is reflected by variations in volcanism (S. S. Hughes et al., 1980; Guffanti and Weaver, 1988). Quaternary volcanism in the arc is a direct result of subduction beneath the western North American coast, along the Cascadia Subduction Zone (Silver, 1978).[69]

Like the Aleutian Arc, the Cascade Arc had a multistage history (Howell Williams, 1969).[70] An episode of Eocene to Pliocene volcanism lasting more than 30 million years built a volcanic foundation upon which Quaternary volcanic cones of the High Cascade Range (figure 9.10) were constructed.[71] The older rocks are overlain along the east flank of the High Cascade Range by a series of younger volcanic rocks, assigned by some workers to a high-alumina basalt province, which includes the Medicine Lake Highland Volcano of California, the Newberry Caldera of Oregon, and the Simcoe Volcanic Complex of Washington (Higgins, 1973). Related rocks occur in the central High Cascade Range, where basalts and "basaltic andesites" of Pleistocene age form a platform that overlies the older rocks (S. S. Hughes and Taylor, 1986). Similarly, exposures of Pliocene alkaline basalts along the Klamath River Gorge in southern Oregon (and elsewhere) provide a local foundation for High Cascade volcanoes (Walker and Naslund, 1986).

Cascade basaltic to rhyolitic rocks are decidedly calc-alkaline (figure 9.11 b, c), but chemical diversity is present (S. S. Hughes and Taylor, 1986; Leeman et al., 1990).[72] Over time, the volume percent of mafic rocks has increased such that younger sections are more basaltic. Thus, rhyolite and dacite comprise only about 2% of the volume of the Quaternary-aged rocks in California and Oregon, whereas

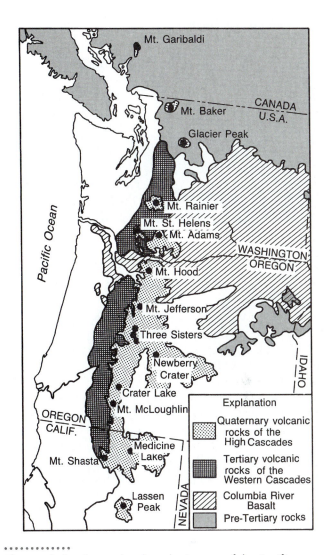

Figure 9.9 Generalized geologic map of the Pacific Northwest of the United States and the southwestern edge of Canada, showing the locations of major Quaternary volcanic centers of the High Cascades and adjoining areas.
(After McBirney, 1968b)

these rock types make up 45 percent (by volume) of the Oligocene-Lower Miocene section (White and McBirney, 1978). Variations with geographic position are also evident, especially among Quaternary sections. Andesite is dominant in the north, whereas basalt is more abundant in the central and southern regions, except in the extreme south, where andesite is abundant.[73]

The general calc-alkaline petrochemistry of Cascade rocks, plus more detailed chemical data, provide the basis for interpreting arc magma genesis. Of particular note are the following. Magnesium numbers among basalts and andesites vary widely (0.82–0.45), with higher values generally occurring in

Figure 9.10 Views of the Cascade Range. (a) Mount Shasta, California (view towards south). (b) Mount McLaughlin, Oregon (view towards northwest). (c) Crater Lake Caldera, Oregon (view east along north rim). (d) Three Sisters volcanoes, Oregon (left view north from Mount Bachelor). (e) Mount Hood, Oregon (view towards northwest). (f) Mount St. Helens, Washington (view towards south).

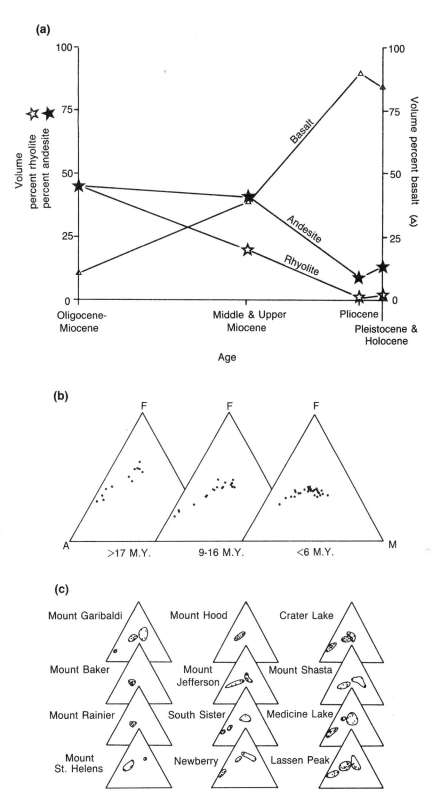

Figure 9.11 Chemical variations in Cascade Range volcanic rocks. (a) Estimated volume percentages of volcanic rock types for various times in the past in central Oregon (values from White and McBirney, 1978). (b) AFM variation diagrams for volcanic rocks of three major age groups from central Oregon. Note strengthening calc-alkaline character with time (from C. M. White and McBirney, 1978). (c) AFM variation diagrams for several High Cascade, Quaternary volcanoes (from McBirney, 1968b).

olivine, pyroxene, and hornblende phyric rocks. Values near .70 and .55 are common. (Inasmuch as most, perhaps all, rocks with higher Mg numbers may have been enriched in Mg through phenocryst accumulation, the liquids represented by the rocks probably were not primary.) Strontium isotopic ratios generally lie between 0.702 and 0.7045, indicating mantle equilibration and little or no sediment or crustal contamination.[74] In contrast, Pb isotopic values require some assimilation or contamination, but in very limited amounts (S. E. Church, 1973a, 1976).[75] The likely Pb values of offshore sediments that might have been subducted are such that a mix of sediment and MORB leads could produce the observed Pb values (S. E. Church, 1976). (Leads from older crust were clearly *not* involved in assimilation or anatexis.) Other elemental clues to the history of Cascade andesites are provided by K, Rb, and the REE. In any given area, K and Rb decrease in abundance with decreasing age of the rocks (White and McBirney, 1978). REE are similar throughout the range, showing LREE enrichment and decreases in abundance over time.[76]

Together, the above data suggest complex histories. These histories probably begin with dehydration and metasomatism in the subduction zone, coupled with progressive partial melting of a mantle peridotite above the subduction zone, followed by magma modification (fractionation, mixing, and assimilation).[77] This general model is consistent with progressive decreases in K, Rb, and REE with time. The nature of the mantle source is indicated by the chemistry of the primary and parental lavas.

The primary magmas derived from the mantle vary in chemistry and include tholeiites, calc-alkaline basalts, high-Al basalts, and low-silica andesites (S. S.Hughes and Taylor, 1986; Grove et al., 1988; Bacon, 1990; Leeman et al., 1990). As these magmas rose from mantle to crust, they were modified, commonly within the crust. Many Cascade andesite compositions require modification. Modification of the magmas is indicated by the derivative nature of the chemistry (Mg number < 0.65) and the silica and alkali enrichments that result in dacites and rhyolites.

Fractionation apparently contributed to the modification of magmas under several Cascade volcanoes (S. S. Hughes and Taylor, 1986; Bacon, 1990; Clynne, 1990; N. L. Green, 1990). Calculations using major and trace element chemistry are used to test whether fractionation is realistic; and chemical data and liquidus lines projected onto ternary phase diagrams allow assessment of fractionation paths.[78] For Medicine Lake andesites, these tests indicate that fractional crystallization contributed to andesite formation, but AFC apparently produced the derivitive magmas (Grove et al., 1988).

Although assimilation did accompany fractional crystallization at Medicine Lake, elsewhere in the arc, assimilation of sediments or crust has been of limited importance in the modification of magmas. For example, isotopic evidence indicates very little sediment involvement in magma genesis,

and, in most cases, low Sr-isotope ratios and Pb-isotope values preclude modification involving assimilation of older crust. In the southern part of the range, however, granite xenoliths at Crater Lake and Medicine Lake plus Th-isotope ratios from Mount Shasta lavas, indicate that locally assimilation of crustal rocks has influenced magma chemistry (S. Newman, Macdougall, and Finkel, 1986; Grove et al., 1988; Bacon, Adami, and Lanphere, 1989).

Mixing also contributed to modification and andesite genesis in the Cascade Arc. C. A. Anderson (1941) and others[79] provided evidence of magma mixing in Medicine Lake volcanic rocks, evidence that includes mafic inclusions with chilled margins, andesite flows with rhyolite inclusions, and phenocrysts clots (Eichelberger, 1975b; Donnelly-Nolan et al., 1990). In addition, Grove, Gerlach, and Sando (1982) have demonstrated through analyses of experimental results and petrochemistry that these particular rocks were formed by a combination of fractional crystallization, assimilation, and magma mixing. Mixing also seems to have been important at Mount Baker, Crater Lake, and Mount Lassen (Bacon, 1990; Bullen and Clynne, 1990; Clynne, 1990).

Together, data from the Cascade rocks, like those from the Aleutian Arc, reveal a complex history both for the arc and for individual volcanoes. Processes vary from place to place. Even within single volcanic systems, for example, at Crater Lake (Bacon, 1990), a number of processes produce andesites, as well as a variety of other rocks. What seems certain is that neither simple melting in or above a subduction zone nor simple fractionation of a basaltic parent, produced the andesitic magmas of the Cascade Range.

SUMMARY

Andesites are typically plagioclase-rich, porphyritic rocks containing hornblende, pyroxene, olivine, and/or Fe-Ti oxide minerals. Chemically, they are intermediate in composition, having SiO_2 values in the range 52–63 wt. %. Within the context of that specific SiO_2 range, the chemistry of andesites varies widely, reflecting a diversity of origins.

Arc andesites, the most common types, may be tholeiitic, calc-alkaline, or boninitic, but calc-alkaline and tholeiitic types are decidedly most abundant. Magma genesis is probably a complex process. It involves some combination of (1) dehydration and/or partial melting in or above a subduction zone, (2) fluid metasomatism and anatexis of the mantle wedge above the subduction zone, (3) assimilation of mantle and crustal rocks along a path between the melting zone and the surface, (4) melting of basal crustal rocks and mixing of the resulting magmas with magmas rising from the mantle, (5) fractional crystallization, and (6) assimilation of crustal rocks. Assessment of the origin of arc andesite at any specific location requires careful assessment of petrographic, chemical, experimental, and tectonic data.

Some arc andesites and those formed along mid-ocean ridge spreading centers, oceanic islands, continental rifts, or transform fault and continental interaction zones may involve less complex processes, such as direct partial melting or melting combined with fractional crystallization. Yet the potential for complications in magma genesis exists at these sites too. Thus, care in petrogenetic analyses is required.

Clearly, the "andesite problem" is not one but many. Different andesite types are probably generated by different processes. Thus, the history of each andesite in each tectonic setting requires careful study.

EXPLANATORY NOTES

1. Le Maitre (1976), J. B. Gill (1981, p. 1).
2. Some texts simply review various definitions without selecting one, others choose specific definitions and ignore others, whereas still other works review and make recommendations. Compare, for example, discussions and definitions of the term in Peccerillo and Taylor (1976), Best (1982, p. 65), D. S. Barker (1983, p. 267ff.), J. B. Gill (1981, p. 2ff.), Cox, Bell, and Pankhurst (1979, p. 14), and P. E. Baker (1982). Illuminating discussions may also be found in Johannsen (1937, pp. 160–176) and Chayes (1969). Also refer to chapter 4 of this text and Streckeisen (1979).
3. See Johannsen (1937, p. 160), who reviews early uses of the term andesite.
4. Olivine phyric andesites occur in the Aleutian Arc and elsewhere. Although they do exist, they are commonly indistinguishable from basalts in handspecimen.
5. J. B. Gill (1981, p. 5) specifies that andesites are hypersthene-normative rocks. Ewart (1976, 1979, 1982) uses the same upper silica limit (63%) but a lower limit of 56%. I. S. E. Carmichael, Turner, and Verhoogen (1974, p. 557) use a silica range of 55–63%, with "basaltic andesite" occupying the range 52–55% SiO_2. Cox, Bell, and Pankhurst (1979) follow I. S. E. Carmichael, Turner, and Verhoogen. (1974).
6. Cameron et al. (1979), Shiraki et al. (1980), W. E. Cameron, McCulloch, and Walker (1983), A. J. Crawford, Falloon, and Green (1989). Note that A. J. Crawford, Falloon, and Green (1989) define boninite as having Mg# > 0.6. Tatsumi and Ishizaka (1981, 1982) refer to relatively aphyric boninites as "sanukitoids." See Johannsen (1937, pp. 172–73) and A. J. Crawford (1989) for derivations and definitions of this and related terms.
7. For example, see S. R. Taylor (1969), Jakes and Gill (1970), Peccerillo and Taylor (1976), Ewart (1979, 1982), and J. B. Gill (1981, ch. 1).
8. These rock types are all defined in R. L. Bates and Jackson (1980) and Johannsen (1937). P. E. Baker (1982) provides a brief overview of the chemistry and occurrences of shoshonites. Peccerillo and Taylor (1976) provide a chemical definition of shoshonite. For definitions of mugaerite and icelandite, see G. A. MacDonald (1960) and I. S. E. Carmichael (1964), respectively.
9. Maaloe and Petersen (1981) consider andesites formed in intraoceanic arcs to be "primitive" and distinguish them as "oceanic andesites."
10. C. M. Johnson and O'Neil (1984).
11. Larsen and Cross (1956), Robyn (1979), C. Zimmerman and Kudo (1979).
12. C. D. Byers, Melson, and Vogt (1983), R. W. Kay, Hubbard, and Gast (1970).
13. See chapter 10 for a definition and discussion of ophiolites. Occurrences of boninite are described by Cameron et al. (1979), Upadhyay (1982), E. N. Cameron, McCulloch, and Walker (1983), and in the papers in A. J. Crawford (1989).
14. Such alterations are not restricted to andesites but are common in these rocks.
15. Also see Dickinson (1968), D. R. Nielson and Stoiber (1973), Whitford and Nicholls (1976).
16. The meaning and significance of iron and magnesium contents are discussed in more detail by Miyashiro (1974), J. B. Gill (1981) and Ewart (1982). Also see Kelemen (1990). M. D. Garcia (1978) indicates problems that arise where altered rocks are used to assign rocks to a series.
17. See Jakes and Gill (1970), Jakes and White (1972), and J. B. Gill (1981, p. 125ff.) for a more complete discussion of trace elements.
18. J. B. Gill (1981, pp. 151–53) and Hawkesworth (1982) review the then available Nd-isotope data. Additional data are provided by Nohda and Wasserburg (1981), F. A. Frey et al. (1984), and D. E. James and Murcia (1984), among others.
19. Amphibolite is a metamorphic rock composed principally of calcic plagioclase and hornblende. Eclogite is composed of garnet and the pyroxene omphacite.
20. For example, see Oxburgh and Turcotte (1970), Toksoz, Minear, and Julian (1971); Turcotte and Schubert (1973), R. N. Anderson, De Long, and Schwarz (1980), R. N. Anderson et al. (1980), Honda and Uyeda (1983), Honda (1985), S. M. Peacock (1990a), and Bebout (1991).
21. B. D. Marsh (1979) and Honda (1985) suggest higher temperatures than most workers (see figure 9.6b).
22. For more information, consult discussions in Wyllie (1971a, 1971b, 1979, 1982, 1983a), Ringwood (1974), Fyfe and McBirney (1975), Ahrens and Schubert (1975a, 1975b), Delaney and Helgeson (1978), J. B. Gill (1981), Peacock (1991a), and Bebout (1991), plus chapters 28 and 29 of this text.
23. The actual depth is variable and depends on the many factors noted previously.
24. See D. H. Green (1990) for a discussion of the roles of oxidation-reduction and C-O-H fluids in mantle melting. At depth, the fluid phase will be a supercritical fluid containing C, O, and H (S. M. Peacock, 1990).

25. Nye and Reid (1986), Tatsumi and Nakamura (1986), J. P. Davidson (1987), A. D. Brandon (1989).

26. S. S. Hughes and Taylor (1986), J. P. Davidson (1987), S. M. Peacock (1990).

27. Oxburgh and Turcotte (1970), Armstrong (1971), Gillully (1971), R. W. Kay, Sun, and Lee-Hu (1978), J. D. Myers, Frost, and Angevine (1986).

28. R. L. Armstrong (1971), S. E. Church and Tilton (1973), S. E. Church (1976), Z. A. Brown et al. (1985), J. D. Morris et al. (1985), Morris and Tera (1989).

29. Also see Stern and Wyllie (1973), J. B. Gill (1974), T. H. Dixon and Stern (1983), Morris and Tera (1989).

30. A. Holmes (1931, 1932), F. J. Turner and Verhoogen (1960, p. 287), Pichler and Zeil (1969, 1972). See Wyllie (1977, 1983b) and J. B. Gill (1981, p. 261ff.) for discussions.

31. Also see Winkler (1965) and Wyllie (1977), and chapters 5 and 6. Less siliceous compositions were produced in experiments by Carroll and Wyllie (1990), but the melts were not calc-alkaline.

32. Also see T. H. Green and Ringwood (1968a), B. D. Marsh (1976a, 1976b); Brophy and Marsh (1986). See Apted (1981) for a discussion of REE calculations related to this model.

33. T. H. Green and Ringwood (1968a), B. D. Marsh and Carmichael(1974), Brophy and Marsh (1986), J. D. Myers and Marsh (1987).

34. Compare the results of Holloway and Burnham (1972) with the arguments of J. B. Gill (1981, p. 235). Also see I. A. Nicholls and Harris (1980) and P. C. Hess (1989, pp. 159–60).

35. Also see W. K. Conrad and Kay (1984), Brophy (1987, 1990), Woodhead (1988), Bacon (1990), Clynne (1990), and Singer and Myers (1990).

36. Boettcher (1973) and J. B. Gill (1981) review this and other models. Some form of mantle anatexis is supported by M. J. O'Hara (1965), T. H. Green and Ringwood (1968a, 1968b), McBirney (1969a), Yoder (1969), Kushiro (1972, 1973a, 1974a, 1990), Colley and Warden (1974), W. A. Morgan (1974), Mysen and Boettcher (1975a,b), B. D. Marsh (1979), Zimmerman and Kudo (1979), Suzuki and Shiraki (1980), Tatsumi and Ishizaka (1981, 1982), Aoki and Fujimaki (1982), T. H. Dixon and Stern (1983), Mysen (1982, 1983), and Nye and Reid (1986). Also see Arculus (1981).

37. DePaolo and Wasserburg (1977), J. I. Matsuda, Zashu, and Ozima (1977), R. J. Stern (1982), Dixon and Stern (1983), R. J. Stern et al. (1991).

38. Also see other papers in A. J. Crawford (1989), including Tatsumi and Maruyama (1989).

39. J. B. Gill (1981, p. 251ff.) provides a more thorough discussion, and opposing views are offered by Wyllie (1978b, 1982). Also see Nicholls and Harris (1980). D. H. Green (1976) suggests, as a result of experimental work, that andesites are not produced by melting of peridotite at depths greater than 30 km.

40. For additional discussion, see reviews in J. B. Gill (1981) and P. C. Hess (1989, p. 160), and see Clemens (1984).

41. For example, see reviews in J. B. Gill (1981, pp. 252–53) and P. C. Hess (1989, pp. 159–60).

42. For additional arguments, see Daly (1933, p. 448) and Longhi (1981, in I. H. Campbell, 1985).

43. This is one of the various multiprocess models that have been proposed and discussed. For other example, see Waters (1955), T. H. Green and Ringwood (1968a), W. Hamilton (1969a,b), S. R. Taylor et al. (1969), I. A. Nicholls and Ringwood (1971, 1972), Ringwood (1974, 1985), Briquen and Lancelot (1979), R. W. Kay (1980,1985), T. H. Green (1980), Deruelle (1982), Sekine and Wyllie (1982, 1983), Feigenson and Carr (1986), Gans, Mahood, and Scherarer (1989), and models 6–8.

44. Fractionation is also discussed by Osborn (1959), Kuno (1968), R. W. Kay, Hubbard, and Gast (1970), Ewart (1976), I. A. Nicholls and Whitford (1976), Byerly, Melson, and Vogt (1976), J. C. Allen and Boettcher (1978, 1983), Yanagi and Ishizaka (1978), T. H. Dixon and Batiza (1979), R. J. Stern (1979), Coulon and Thorpe (1981), Kienle and Swanson (1983), J. A. Walker (1984), S. S. Hughes and Taylor (1986), Woodhead (1988), Brophy (1990), and N. L. Green (1990).

45. For example, J. B. Gill (1981, ch. 12), Grove and Baker (1984), and Brophy (1990).

46. Osborne (1959), Boettcher (1973), Eggler and Burnham (1973), J. C. Allen and Boettcher (1978, 1983), Garcia and Jacobson (1979), J. B. Gill (1981), Grove and Baker (1984), Woodhead (1988).

47. J. P. Eaton and Murata (1960), G. A. MacDonald (1968) R. W. Kay, Hubbard, and Gast (1970), C. D. Byers, Muenow, and Garcia (1983). Also compare Byerly, Melson, and Vogt (1976) and Juster, Grove, and Perfit (1989).

48. Kuno (1968) and J. B. Gill (1981) review the arguments more thoroughly.

49. Dixon and Batiza (1979), Dupuy, Dostal, and Coulon (1979), R. J. Stern (1979), Whitford and Jezek (1979), Aoki and Fujimaki (1982), D. E. James (1982), J. P. Davidson (1985).

50. A. Holmes (1932), R. A. Daly (1933, p. 453), E. S. Larsen, Gonyer, and Larsen (1938), Kuno (1950), Tilley (1950), Waters (1955), Eichelberger (1974), Thorpe and Francis (1979), Coulon and Thorpe (1981), Leeman (1983), D. E. James and Murcia (1984), Harmon et al. (1984), D. E. James (1984), Grove et al. (1988), McMillan and Dungan (1988), Bacon, Adami, and Lanphere (1989), Briot (1990), but see Millhollen (1975) and Klerkx et al. (1977).

51. DePaolo and Wasserburg (1977), Margaritz, Whitford, and James (1978), Thorpe and Francis (1979), D. E. James (1982), D. E. James and Murcia (1984), Harmon et al. (1984), J. P. Davidson (1985). See J. B. Gill (1981, p. 263ff.) for further discussion.

52. McBirney (1979) discusses assimilation problems in more detail, as does J. B. Gill (1981, p. 266).

53. Thorpe and Francis (1979), Coulon and Thorpe (1981), DePaolo (1981c), Leeman (1983), D. E. James and Murcia (1984), Harmon et al. (1984), D. E. James (1984), J. P. Davidson (1985), Hildreth and Moorbath (1988), Bacon, Adami, and Lanphere (1989), Kelemen(1990).

54. Fenner (1926), A. Holmes (1932), E. S. Larsen, Gonyer, and Larsen (1938), Wilcox (1944), Eichelberger (1975a, 1975b, 1978a, 1978b), A. T. Anderson (1976), Eichelberger and Gooley (1977), Sparks, Sigurdsson, and Wilson (1977), Sakuyama (1979), McBirney (1980), Cantagrel et al. (1984), Bourdier, Gourgaud, and Vincent(1985), Kouchi and Sunagawa (1985), Koyaguchi (1986), McMillan and Dungan (1986), N. L. Green (1988), Nixon (1988a, 1988b), Ussler and Glazner (1989), Gourgaud and Thouret (1990), Halser and Rose (1991).

55. A. T. Anderson (1976), Eichelberger and Gooley (1977), Sparks, Sigurdsson, and Wilson (1977), Eichelberger (1978a, 1978b, 1980), Sakuyama (1979), Huppert, Sparks, and Turner (1982), Bourdier, Gourgaud, and Vincent (1985), Grove and Baker (1984), Oldenburg et al. (1989).

56. Eichelberger (1975b, 1978a), Sakuyama (1979).

57. Cantegrel, Didier, and Gourgaud (1984).

58. See note 43.

59. Coats (1950, 1962), B. D. Marsh (1982a); Kienle and Swanson (1983).

60. Many Aleutian islands and volcanoes were studied by USGS geologists under a long-term project begun in 1945. The reports of that project were published as USGS Bulletin 974-B and various parts of USGS Bulletin 1028. Those reports include Coats (1950, 1956a, 1956b, 1956c, 1959a, 1959b), G. C. Kennedy and Waldron (1955), F. S. Simons and Mathewson (1955), F. M. Byers (1959), G. D. Fraser and Barnett (1959), G. D. Fraser and Snyder (1959), W. H. Nelson (1959), G. H. Snyder (1959), H. A. Powers, Coats, and Nelson (1960), R. Q. Lewis, Nelson, and Powers (1960), Coats et al. (1961), Drewes et al. (1961), Waldron (1961), and O. Gates, Powers, and Wilcox (1971). Additional works include those of Fenner (1920, 1926, 1950); Marsh and his coworkers (B. D. Marsh, 1976a, 1976b, 1979, 1982; B. D. Marsh and Leitz, 1979), Meyers et al. (1983); R. W. Kay and S. M. Kay and their coworkers (R. W. Kay 1977, 1980; R. W. Kay, Sun, and Lee-Hu, 1978; S. M. Kay, Kay, and Citron, 1982; S. M. Kay et al., 1983; S. M. Kay and R. W. Kay, 1985a, 1985b; Conrad, Kay, and Kay, 1983; W. K. Conrad and Kay 1984); plus those of Burk (1965, 1966), Curtis (1968), Hoare et al. (1968), Forbes et al. (1969), Marlow et al. (1973), Scheidegger and Kulm (1975), J. R. Hein, Scholl, and Miller (1978), J. R. Hein and McLean (1980), Kienle et al., 1980, Hildreth (1983b), D. R. Baker and Eggler (1983), Kienle and Swanson (1983), Kienle, Swanson, and Pulpan, 1983; J. D. Morris and Hart (1983), Hein et al.(1984), DeLong et al. (1985), Brophy (1986, 1987, 1990), Nye and Reid (1986), D. R. Baker (1987), D. R. Baker and Eggler (1987), Nye and Turner (1990), and Singer and Myers (1990). This discussion of the Aleutian Arc is based on these works.

61. Also see Simons and Mathewson (1955), Marlow et al. (1973), B. D. Marsh (1982a). Reviews and regional syntheses may be found in Burk (1965), Plafker, Jones, and Passagno (1977), J. C. Moore and Connelly (1977), and Raymond (1980). Coney, Jones, and Monger (1980) designate the Alaska Peninsula as the "Peninsular Terrane." See B. L. Reed and Lanphere (1973), J. C. Moore and Connelly (1977, 1979), and T. Hudson (1979) for discussions of the plutonic belts.

62. Additional age data are provided by Scheidegger and Kulm (1975) and J. R. Hein, Scholl, and Miller (1978).

63. Coats (1950).

64. B. D. Marsh (1982a, p. 109) suggests that hydrous phases are absent from the "andesitic basalts" of the Aleutian Arc, but amphiboles are reported in andesitic rocks of the arc by B. D. Marsh (1976a), B. D. Marsh and Leitz (1978), and others including O. Gates, Powers, and Wilcox (1971). Kienle, Swanson, and Pulpan (1983), Kienle and Swanson (1983), S. M. Kay and Kay (1985a), Brophy (1990), and Nye and Turner (1990).

65. For example, see Drewes et al. (1961), Marsh (1976b), Hildreth (1983b), Kienle and Swanson (1983), and S. M. Kay and Kay (1985a).

66. W. H. Nelson (1959), H. A. Powers, Coats, and Nelson (1960), Coats et al. (1961), Drewes et al. (1961), O. Gates, Powers, and Wilcox (1971), R. W. Kay (1977), L. Brown et al. (1982b), J. D. Morris and Hart (1983), DeLong et al. (1985), J. D. Myers, Marsh, and Sinha (1985), Nye and Reid (1986), von Drach, Marsh, and Wasserburg (1986), Brophy (1987, 1990), Nye and Turner (1990), Singer and Myers (1990). Sr and Pb isotopic data were summarized by R. W. Kay, Sun, and Lee-Hu (1978) and J. D. Myers, Frost, and Angevine (1986).

67. Available data do not allow unequivocal distinction between these sources. Some authors argue for eclogitic (garnetiferous) source rocks, either in the subducting slab or the mantle wedge, whereas others suggest ultramafic sources for example, compare B. D. Marsh (1976a,) Brophy (1986), and J. D. Myers, Frost, and Angevine (1986) with J. D. Morris and Hart (1983), DeLong et al. (1985), and Brophy (1989a). Lack of a strong LREE enrichment seems incompatible with a source with garnet, but the complex processes of metasomatism, anatexis, fractionation, mixing, melting, and assimilation that result in production and modification of Aleutian magmas obscure source characteristics (Perfit and Kay, 1986).

68. Perfit et al. (1980), S. M. Kay, Kay, and Citron (1982), W. K. Conrad, Kay, and Kay (1983), Kienle and Swanson (1983), Kienle, Swanson, and Pulpan (1983), S. Kay et al. (1983), W. K. Conrad and Kay (1984), S. M. Kay and Kay (1985a), J. D. Myers, Marsh, and Sinha (1985), D. R. Baker and Eggler (1987), Nye and Turner (1990), Singer and Myers (1990).

155

69. McBirney (1978) and McBirney and White (1982; C. M. White and McBirney, 1978) assert that no significant link exists between volcanism and subduction. Their argument is based on variations in chemistry observed both along strike in the arc and over time, which they suggest cannot be explained by subduction. Guffanti and Weaver (1988) argue otherwise. As noted by Armstrong (1978) and McBirney and White (1982), the petrotectonic history of western North America is complex and resolution of its details awaits additional work. Nevertheless, the exact spatial correlation between Quaternary volcanoes of the arc on the one hand and offshore spreading centers with associated zones of convergent tectonism on the other argue strongly for a genetic link between Quaternary volcanism and subduction (see E. A. Silver, 1978, for references and a review of the evidence for subduction).

70. Much of the foundation of knowledge for contemporary studies of the Cascade volcanoes was provided by Howell Williams (e.g., Williams 1932, 1935, 1942; Williams and Goles, 1968), who summarized the protracted history of the arc (Williams, 1969). Thayer (1937), H. A. Coombs (1939), Peck (1961), Peck et al. (1964), Fiske, Hopson, and Waters (1963, 1964), Greene (1968), and others added to that foundation. Armstrong (1978), McBirney (1978), and McBirney and White (1982) provide summaries of the timing (and other aspects) of Cascade volcanism. Additional works on Cascade volcanic rocks include Crandall et al. (1962), G. A. MacDonald and Katsura (1965), N. V. Peterson and Groh (1965), Dole (1968), McBirney (1968a, 1968b), W. S. Wise (1968), Tabor and Crowder (1969), Higgins (1973), S. E. Church (1973b, 1976), S. E. Church and Tilton (1973), Eichelberger (1975b), Hughes et al. (1980), Lipman and Mullineaux (1981), Gerlach and Grove (1982), Grove et al. (1982), D. L. Williams et al. (1982), C. M. White and McBirney (1982), Bacon (1983, 1990), S. S. Hughes and Taylor (1986), S. Newman, MacDougall and Finkel (1986), Walker and Naslund (1986), N. L. Green (1988, 1990), Grove et al. (1988), Guffanti and Weaver (1988), Bacon, Adami, and Lanphere (1989), Bullen and Clynne (1990), Clynne (1990), Donnelly-Nolan et al. (1990), Goles and Lambert (1990), S. S. Hughes (1990), Leeman et al. (1990), Linneman and Myers (1990), and E. M. Taylor (1990). These works serve as the foundation for this summary.

71. Hay (1963), Greene (1968), Oles and Enlows (1971), E. M. Baldwin (1976), McBirney and White (1982), Evarts (1990).

72. See Walker and Naslund (1986) for an exception.

73. H. A. Coombs (1939), Fiske, Hopson, and Waters (1964), C. M. White and McBirney (1978).

74. S. E. Church and Tilton (1973), McBirney (1978), C. M. White and McBirney (1978), Goles and Lambert (1990).

75. Also see S. E. Church and Tilton (1973).

76. S. E. Church (1973b), C. M. White and McBirney (1978), S. S. Hughes and Taylor (1986).

77. For example, see S. S. Hughes and Taylor (1986) and S. S. Hughes (1990). Leeman et al. (1990) do not agree that subducted fluids were involved, at least in the southern Washington Cascades.

78. See Grove and Donnelly-Nolan (1986), DePaolo (1981d), Grove et al. (1988), and Ragland (1989) for explanations of the method, additional discussions, and further references.

79. Eichelberger (1975b), Gerlach and Grove (1982), Donnelly-Nolan et al. (1990), Linneman and Myers (1990).

PROBLEMS

9.1. Calculate the CIPW norm for analysis 3 in table 9.1 on an anhydrous basis (omit H_2O and recalculate, total the remaining numbers, and recalculate their percentages on the basis of the new total). Compare your results to the norm in table 9.1.

9.2. Calculate the Mg number of each of the andesites in table 9.1.

9.3. What information or experiments are needed to convincingly demonstrate that Aleutian parent magmas are derived from an eclogitic source? from a peridotitic source?

10

Ultramafic-Mafic Complexes and Related Rocks

INTRODUCTION

Ultramafic rocks have received considerable attention in recent years, in part because some of them are considered to represent samples of the normally inaccesible mantle. **Ultramafic rocks** (ferromagnesian rocks) are defined here as those rocks with color indices of 90 to 100 (see chapter 4).[1] Most ultramafic rocks are either plutonic igneous rocks or metamorphic rocks. Those formed in the crust include igneous and metamorphic types, whereas those from the mantle are metamorphic, because they have been recrystallized under directed stresses (see chapters 23 and 31).

Igneous ultramafic rocks are widely distributed in the crust but, volumetrically, are far less abundant in the earth as a whole than are the metamorphic ultramafic rocks. The igneous ultramafic rocks occur in a number of different structures and are commonly associated with other rock types. For example, ultramafic rocks commonly underlie or are interlayered with the more voluminous mafic rocks basalt and gabbro (with color indices less than 90), which comprise the bulk of the oceanic crust.[2] These latter rocks are crystallized from magmas thought to be derived from mantle ultramafic rocks by partial melting.

Rock bodies that contain both ultramafic and mafic rock types are important for a number of reasons. Such plutons typically contain gabbro, peridotite, pyroxenite, and dunite, as well as less abundant rock types ranging in composition from granite to chromitite, some of which have economic importance (H. D. B. Wilson, 1969; UNESCO, 1981). Perhaps most significant from a petrologic perspective is the fact that these bodies of rock provide clear evidence of fractionation in magma bodies. Also important are the clues such bodies provide about the details of fractionation processes, the kinematic processes in magma bodies, and the tectonic history of the regions in which they occur.[3]

The purpose of this chapter is to introduce readers to the fundamental structural, textural, petrographic, and petrologic aspects of crustal ultramafic rocks and ultramafic-mafic complexes. As is the case with other rock types, the features of the ultramafic-mafic rocks provide the raw data for understanding the diverse origins proposed for them. Those origins range from normal plutonic intrusion and differentiation to meteorite impact. Inasmuch as ultramafic-mafic rock bodies are formed both by igneous processes within the crust (the focus of this chapter) and by metamorphic and igneous processes in the mantle, it is important to be able to distinguish crustal and mantle rocks.

KINDS OF ULTRAMAFIC AND ULTRAMAFIC-MAFIC ROCK BODIES

In the crust, ultramafic rocks are present in six major kinds of occurrence.[4] These are (1) layered igneous bodies, (2) zoned to irregularly shaped intrusions, (3) ultramafic lava flows (komatiites) and their differentiates, (4) kimberlite pipes and related structures, (5) alpine-type ultramafic rock bodies (including most crustal metamorphic ultramafic rocks), and (6) nodules (including xenoliths) in igneous rocks. Alpine-type ultramafic rocks, komatiites, kimberlites, and nodules are discussed in chapters 6, 7, and 31.

The major plutonic igneous rock bodies that contain ultramafic rock may be assigned to one of two major groups, the layered bodies (or complexes) and the zoned to irregular bodies (figure 10.1).[5] Layered bodies include dikes and sills, stratiform lopoliths and related bodies, and ophiolites. Zoned to irregular bodies include Alaska-type zoned complexes, alkaline intrusive complexes, and the appinite-type ultramafic bodies. Each of these types has distinctive chemical, structural, and petrographic characteristics (table 10.1). These characteristics distinguish the individual types of igneous body, one from the other, as well as the igneous bodies from metamorphic ultramafic rock masses.

The common metamorphic bodies of the crust are assigned to the category of **alpine-type ultramafic bodies,** ultramafic masses that occur in orogenic belts. Although most alpine-type bodies consist of metamorphosed ultramafic rocks derived either from the mantle or by metamorphism and deformation of igneous bodies, some are igneous bodies of the types listed above that simply have been fragmented and displaced from their sites of crystallization. The metamorphic ultramafic bodies are characterized by **tectonite fabrics,** textures and structures that reflect a history of deformation (see chapters 23 and 31). In addition, they may contain distinctly metamorphic minerals such as the serpentines, talc, tremolite, and anthophyllite, as well as magnesium-rich olivines and pyroxenes. They typically lack contact aureoles, may show either concordant or discordant structural relations with the country rocks, and exhibit post-crystallization structural features, such as folds.

In contrast, the igneous bodies show greater diversity of mineralogy and petrology than metamorphic bodies, typically lack tectonites, commonly have contact aureoles, and have chemistries that reveal coherent differentiation trends. Igneous textures (notably cumulate textures), cumulate layering, and intrusive contacts further distinguish these bodies as igneous.

OCCURRENCES OF IGNEOUS COMPLEXES CONTAINING ULTRAMAFIC ROCKS

Most types of igneous rock masses containing ultramafic and related rocks are crystallized from basaltic (mafic) magmas. As indicated in chapters 6 and 7, basaltic magmas develop at

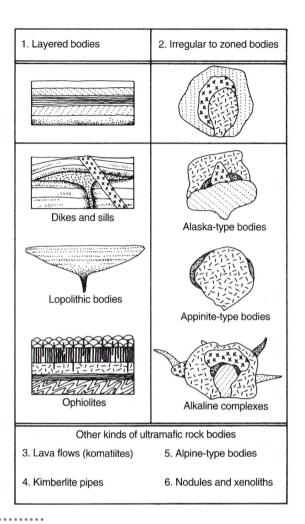

Figure 10.1 Classification of ultramafic rock-bearing rock bodies based on structure.

various levels in the mantle in response to a variety of tectonic conditions. The plutonic rocks considered here are derived from such basaltic magmas and crystallized at depth before or after the magmas were modified. Because the plutonic rocks crystallized from mafic magmas, they should occupy the same kinds of tectonic settings as the basalts. Specifically, they should (and do) occur in the roots of volcanic arcs, underlie volcanic mountain chains formed over mantle plumes, occupy positions at depth in the spreading centers and rift zones, and mark eroded zones of interaction between transform faults and continental crust.

Layered Bodies

As noted above, there are three types of layered bodies: (1) dikes and sills, (2) stratiform lopoliths and related bodies, and (3) ophiolites. Somewhat similar-appearing layered bodies that occur as differentiated, Precambrian komatiitic flows are not discussed here.[6] Dikes and sills may form in any environment in which mafic intrusions and volcanism are occurring, but notable occurrences are present at sites where rifting has taken place. Mesozoic rifting of Africa

Table 10.1 Characteristics of Selected Igneous Complexes Containing Ultramafic Rocks

	Dikes and Sills	Lopoliths	Ophiolites	Alaska-Type	Appinite-Type	Alkaline Complexes
Shape of Body	tabular	plano-convex (dish)	tabular, lenticular	crudely ellipsoidal	lenticular, ellipsoidal, tabular	crudely ellipsoidal
Layering	igneous, locally	igneous, abundant	igneous and MM[1], local	igneous and MM	minor igneous	local MM and igneous
Magma Type	tholeiitic to alkaline	tholeiitic	tholeiitic	tholeiitic to alkaline	tholeiitic to calc-alkalic	alkaline
Major Rock Types	gabbro	2-pyroxene gabbro, norite, harzburgite, websterite, orthopyroxenite, anorthosite	basalt, gabbros, lherzolite, harzburgite, dunite	2-pyroxene gabbro, diorite, quartz diorite, clinopyroxenite, wehrlite, dunite	hornblendite, norite, hornblende quartz diorite, diorite, syenite, gabbros, granodiorite	syenite, trachyte, feldspathoidal gabbro, ijolite, carbonatite, phonolite
Major Textures	diabasic to ophitic	cumulate	cumulate, hypidiomorphic-granular, tectonite	cumulate, hypidiomorphic-granular, tectonite	hypidiomorphic-granular, poikilitic, tectonite	hypidiomorphic-granular, porphyritic, allotriomorphic-granular, cumulate
Common Minerals[2]	plag, cpx, ol	plag, cpx, opx, ol, crt	plag, cpx, opx, ol, hbl, crt	plag, cpx, ol, hbl, mtt	plag, hbl, ol, bio, sp, cpx, opx, akf, qtz	ancl, akf, plag, neph, Na-px, cc, cpx, mtt

Sources: Shand (1942), W. J. French (1966), E. D. Jackson (1971), E. D. Jackson and Thayer (1972), W. S. Pitcher and Berger (1972), Snoke et al. (1982), and Galan and Suarez (1989), and others.

[1] MM = metamorphic.

[2] Mineral abbreviations: ancl = anorthoclase, akf = alkali feldspar, bio = biotite, cc = calcite, cpx = clinopyroxene, crt = chromite, hbl = hornblende, mtt = magnetite, neph = nepheline, Na-px = aegirine, ol = olivine, opx = orthopyroxene, plag = plagioclase, qtz = quartz, sp = spinel.

and Europe from North America, which formed the Atlantic Ocean basin, resulted in the intrusion of numerous dikes and sills along the east coast of North America.[7] Although most of these bodies contain no ultramafic rock *sensu stricto*, fractional crystallization and crystal settling resulted locally in the development of olivine-enriched zones within some intrusions, providing an important indi-

cation of the processes that operate more extensively in larger magma chambers.

Examples of differentiated dikes and sills include the Cenozoic intrusions of Scotland, especially those of the island of Skye, the sills of Antarctica, the Brook Street Complex of New Zealand, the Jimberlana Intrusion of Australia, and the Great Dyke of Rhodesia (Zimbabwe).[8] The latter body may be a faulted part of a lopolith (Worst, 1958). In North America, the Palisades Sill of the New York–New Jersey region represents a differentiated sill with a complex history (J. V. Lewis, 1908a; Shirley, 1987; Husch, 1990).[9] In general, only the larger of these bodies contain ultramafic rocks.

Lopolithic and related layered intrusions are limited in number but are widely distributed.[10] These stratiform bodies range in area of exposure from less than 100 km^2 to more than 65,000 km^2. Most appear to have been emplaced in "anorogenic" tectonic settings similar to those in which the smaller dikes and sills were intruded (e.g., along rift zones), but some may have been associated with mantle-plume sites (Moores, 1973). The most well known of these bodies are the Stillwater Complex of Montana, the Skaergaard Complex of Greenland, and the Bushveld Complex of southern Africa.[11] Other notable occurrences include the Duluth Complex of Minnesota, the Muskox Intrusion of Northwest Territories, Canada, and the Rhum Complex of Scotland.[12]

Ophiolites are structurally distinctive, layered rock bodies composed of mafic and ultramafic rocks (figure 10.2). They include, from bottom to top, ultramafic tectonites (commonly serpentinized metamorphic rocks), cumulate textured ultramafic and mafic rocks, noncumulate gabbros and their siliceous differentiates, a mafic dike and/or sill complex, and mafic to siliceous volcanic rocks (that are locally pillowed).[13] Ophiolites are commonly overlain by cherts or other pelagic sedimentary rocks, turbidites, mafic breccias, or other marine sedimentary rocks. Although for several years ophiolites were considered to represent oceanic crust formed at mid-ocean ridges, many workers now suggest that ophiolites are formed in diverse settings, including back-arc basins and the roots of volcanic arcs.[14] The most well known ophiolites are probably the Bay-of-Islands Complex in Newfoundland and the Troodos Ophiolite on Cyprus.[15] Other ophiolites of note include the Coast Range ophiolites of California, the Canyon Mountain Ophiolite of Oregon, the Semail Ophiolite of Oman, and the ophiolite of Papua New Guinea.[16]

Zoned to Irregular Intrusions

The zoned-to-irregular category of ultramafic and related rock bodies includes Alaska-type zoned bodies and irregular differentiated plutons constituting alkaline igneous complexes and Appinite-type bodies. All of these types of bodies are generally small. The **Alaska-type mafic-ultramafic rock bodies** are intrusions consisting of elliptical, curviplanar, or irregular zones of distinctive rock types such as dunite, peridotites, hornblende pyroxenite, and various gabbros (H. P. Taylor and Noble, 1960; H. P. Taylor, 1967). The bodies occur in linear belts. Together, their various features suggest that they represent subvolcanic intrusions in volcanic arcs (C. G. Murray, 1972). Examples of Alaska-type bodies include the Duke Island Ultramafic Complex, Alaska, and the Emigrant Gap Complex in California.[17]

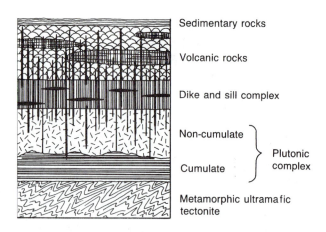

Figure 10.2 Idealized sketch of an ophiolite.
(Based in part on Moores and Vine, 1971; Vine and Moores, 1972; C. A. Hopson et al., 1981).

Appinite is a name that has been applied to quartz diorites characterized by euhedral hornblende.[18] Similarly, Appinite-type ultramafic rocks are olivine-hornblende rock types, typically containing euhedral hornblende. **Appinite-type mafic-ultramafic rock bodies** form small masses associated with granitoid intrusions (W. S. Pitcher and Berger, 1972). As such, they too probably represent subvolcanic arc rocks. Examples of appinite-type ultramafic rocks are found in Scotland, Ireland, and New York.[19]

In contrast to orogenic, zoned Alaska-type and Appinite-type ultramafic bodies, alkaline complexes are considered to be *anorogenic* and probably represent intrusions formed at hot spots or along incipient continental rift zones (Moores, 1973). Most such intrusive complexes are probably derived by differentiation of alkali olivine basalt magmas. Rock types found in these bodies include such unusual rocks as the nepheline-bearing ultramafic rock *jacupirangite* and the *carbonatites*, ultrabasic (but not ultramafic) rocks composed primarily of igneous calcite, dolomite, or ankerite.[20] Examples include the rocks of Magnet Cove, Arkansas, and Iron Hill, Gunnison County, Colorado.[21]

CHEMISTRY, MINERALS, TEXTURES, AND STRUCTURES

Chemistry and Minerals

Typically, ultramafic rocks are ultrabasic, that is, they contain less than 45% silica. Because they are defined on the basis of color index rather than chemistry, however, their chemistries are not exclusively ultrabasic. Where they consist of substantial amounts of pyroxenes, plagioclase, and micas, they may contain 50% or more SiO_2.

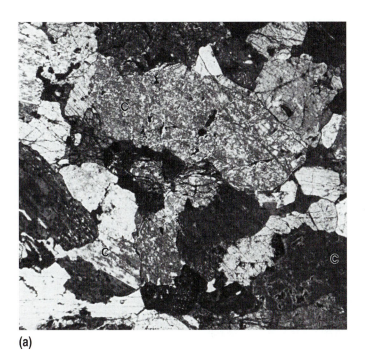

(a)

(b)

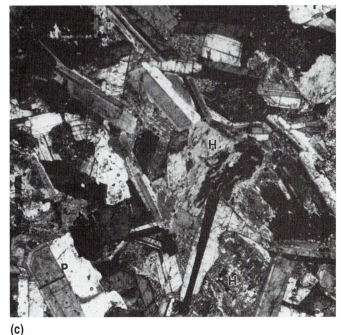

(c)

Figure 10.3 Photomicrographs of ultramafic and related rocks. (a) Clinopyroxenite, Emigrant Gap area, California. (XN). (b) Harzburgite, Stillwater Complex, Montana. (XN). (c) Gabbro, Del Puerto Ophiolite, California. (XN). C = clinopyroxene; O = orthopyroxene; F = forsterite (olivine); P = plagioclase; H = hornblende. Long dimension of photos is 5.5 mm.

Table 10.2 presents analyses of selected ultramafic rocks and table 10.3 gives analyses of other rocks associated with these ultramafic rocks. In the ultramafic rocks, notice that quartz is neither a normative nor, in most cases, a modal mineral. Instead, olivine and, in some cases, nepheline and leucite appear as normative phases. Chemically, alumina is low, except in amphibole-rich rocks, and magnesia is the second most abundant constituent. Magnesia typically comprises 20 to 40% of the rock, whereas total iron makes up about 10%. In associated mafic rocks, alumina is more abundant, as a result of greater amounts of plagioclase, and magnesia is less abundant, because of lesser amounts of pyroxenes and olivines. Mg numbers, which are generally quite high in the ultramafic rocks, especially those containing Mg olivines, are generally lower in associated mafic rocks. Alkalis are low, except in the more differentiated and plagioclase-rich rocks.

Mineralogically, the more common ultramafic and related rocks contain essential olivine, pyroxenes, or hornblende, or combinations of these minerals (figure 10.3). The olivines, which are typically magnesium-rich (Fo_{95-80}) in the

Table 10.2 Major Element Analyses, Modes, and CIPW Norms of Some Ferromagnesian Rocks

	1	2	3	4	5
SiO$_2$	35.39[a]	39.8	44.21	50.8	54.68
TiO$_2$	0.12	1.11	0.11	0.10	0.11
Al$_2$O$_3$	1.34	15.5	0.96	2.4	1.80
Fe$_2$O$_3$	6.57	3.6	1.80	1.2	0.50
FeO	7.72	6.4	9.36	4.5	9.19
MnO	0.11	0.09	0.23	0.15	0.21
MgO	38.16	13.0	32.86	20.7	30.19
CaO	0.80	12.2	8.88	18.7	2.22
Na$_2$O	0.27	2.5	0.11	0.06	0.04
K$_2$O	0.02	0.55	0.01	0.05	0.03
P$_2$O$_5$	tr	—	0.01	0.04	0.02
Other	9.40	—	1.43	1.20	0.98
Total	99.90	94.75+	99.97	99.9	99.97

Sources:

1. Dunite (serpentinized). Duke Island, Alaska, sample H-4-4. Analyst: W. H. Herdsman. Norm adjusted for oxidation and H$_2$O (T. N. Irvine, 1974, table 18A).
2. Hornblendite, Duke Island, Alaska, sample I-31-4. Analyst: Chemistry Section, Geological Survey of Canada (T. N. Irvine, 1974, table 18A).
3. Wehrlite (peridotite), Emigrant Gap area, California, sample CB-0-4 (O. B. James, 1971, table 1). Norm by L. A. Raymond.
4. Clinopyroxenite, Del Puerto, Ophiolite Stanislaus County, California, sample 70-CLE-34. Analysts: G. Chloe, P. Elmore, J. Glenn, J. Kelsey, and H. Smith (E. H. Bailey and Blake, 1974). Mode is an approximation based on qualitative descriptions of Bailey and Blake (1974) and Himmelberg and Coleman (1968) plus a petrographic analysis of a rock from the same locality by L. A. Raymond (unpublished). Norm by L. A. Raymond.
5. Bronzitite (orthopyroxenite), Stillwater Complex, Montana, sample 465-E-3h. Analyst: T. Kameda (H. H. Hess, 1960, table 16). Norm by L. A. Raymond.

[a] Values in weight percent.
[b] Values in volume percent.
[c] Average mode of Stillwater Complex bronzitite (Jackson, 1961, p. 5).
tr= trace

ultramafic rocks, are slightly more iron-rich in the mafic rocks. Olivine may be as iron-rich as fayalite (Fo$_{0-10}$) in silicic differentiates, such as granite, that are associated with the mafic rocks. In a like manner, the pyroxenes are more magnesium rich in ultramafic rocks and more iron-rich in mafic and more siliceous rocks. Clinopyroxene is commonly augite or diopsidic augite (Wo$_{35-50}$En$_{35-50}$Fs$_{0-20}$), but more iron-rich, calcic clinopyroxenes, such as "ferrohedenbergite"

and "ferroaugite" (Wo$_{35-45}$En$_{30-5}$Fs$_{30-60}$), develop during differentiation of some lopolithic layered intrusions. The Ca-poor pyroxenes also exhibit a wide range of chemisty (Wo$_{0-8}$En$_{90-50}$Fs$_{7-50}$). The magnesium-rich orthopyroxenes occur in the ultramafic rocks, but the more iron-rich varieties develop in the differentiates. In addition, pigeonite occurs in some rocks.

	1	2	3	4	5
Modes[b]					
Quartz	—	—	—	—	0.1[c]
Plagioclase	—	—	—	—	12.3
Biotite	—	—	—	—	0.1
Hornblende	—	93.3	1	tr	—
Clinopyroxene	0.2	—	46	98	4.1
Orthopyroxene	—	—	—	< 1	83.2
Olivine	37.2	—	48.5	< 1	—
Serpentine	48.7	—	3.5	tr	—
Fe/Ti oxides	13.2	0.5	1	tr	—
Chromite	0.7	—	—	—	0.2
Other	—	6.2	—	tr	—
Total	100.0	100.0	100	100	100.0
CIPW Norms					
qz	—	—	—	—	—
or	—	—	0.05	0.28	0.17
ab	—	—	0.94	0.52	0.31
an	2.7	31.2	2.11	6.12	4.65
ne	1.4	12.1	—	—	—
lc	0.1	2.7	—	—	—
ol	93.5	28.7	55.67	12.86	3.53
hy	—	—	3.91	8.71	84.10
di	0.04	13.5	32.84	68.21	4.99
he	0.01	5.0	—	—	—
la	0.5	3.1	—	—	—
ap	—	—	0.02	0.10	0.03
mt	1.5	1.5	2.62	1.74	0.72
il	0.3	2.2	0.21	0.18	0.21
cr	—	—	0.45	—	0.69
Other	—	—	1.22	—	0.51
Total	100.05	100.1	100.04	99.9	99.91

Table 10.3 Major Element Analyses, CIPW Norms, and Modes of Some Rocks Associated with (Ultramafic) Ferromagnesi

	1	2	3	4	5
SiO_2	47.0[a]	50.12	51.44	56.43	75.07
TiO_2	0.19	0.19	0.34	0.18	0.15
Al_2O_3	20.7	20.01	12.56	26.10	13.18
Fe_2O_3	0.6	0.80	2.65	0.51	0.76
FeO	3.0	4.29	6.04	0.63	1.15
MnO	0.08	0.09	0.18	0.01	0.03
MgO	7.6	7.91	13.04	0.92	0.23
CaO	17.5	13.97	11.43	8.34	1.10
Na_2O	1.2	1.74	1.68	6.36	4.55
K_2O	0.10	0.05	0.25	0.07	3.27
P_2O_5	—	tr	0.02	tr	0.12
Other	—	0.73	0.32	0.24	0.28
Total	97.97	99.90	99.95	99.79	99.89

Sources:

1. Gabbro, Duke Island, Alaska, sample I-13-2. Norm based on adjusted (anhydrous) analysis. Analyst: Analytical Chemistry Section, Geological Survey of Canada. (T. N. Irvine, 1974). Norm recalculated by L. A. Raymond/Kelsey CIPW Norm Program.
2. Gabbronorite (two-pyroxene gabbro). Banded Series, Stillwater Complex, Montana, sample EB43 (H. H. Hess, 1960). Mode based on text description and point count of 100 grains shown in figure 21a of H. H. Hess (1960).
3. Two pyroxene gabbro, Emigrant Gap area, California, sample LVR-272. (O. B. James, 1971). Norm by L. A. Raymond.
4. Anorthosite, Mid-Indian Ridge sample 97-X (C. G. Engel and Fisher, 1969, table 3). Modal analysis of 1000 points. Norm by L. A. Raymond.
5. Quartz Monzonite, Indian Ocean Ridge, western Indian Ocean, sample 125-4(c). Analysts: C. G. Engel and E. Bingham (C. G. Engel and Fisher, 1975). Norm by L. A. Raymond.

[a]Values in weight percent.
[b]Values in volume percent.
nd = not determined.
tr = trace.

Plagioclase, which comprises less than 10% of the mode in ultramafic rocks, is an important constituent of the associated mafic and felsic rocks (figure 10.3c). Where it occurs in the ultramafic rocks, it is typically anorthite, bytownite, or labradorite (An_{98-50}). In the associated gabbroic-dioritic-granitic rocks, it may range from bytownite to albite (An_{89-0}).

These trends in mineral chemistry are those to be expected in a cooling, crystallizing magma body, as illustrated by both Bowen's and Osborn's reaction series. The more magnesium and calcium-rich phases form at higher temperatures and the more iron- and sodium-rich phases form at lower temperatures.

A number of other minerals occur as accessory and secondary minerals in the ultramafic rocks. These include primary silicates, such as biotite-phlogopite and garnet, and secondary silicates, notably the serpentines. Chromite, magnetite, ilmenite, and ulvospinel are important oxides, the first three of which are locally concentrated to form ores. Pyrite, pyrrhotite, and chalcopyrite are the common sulfides, whereas apatite is the only common phosphate. In the granitoid differentiates, alkali feldspar, quartz, zircon, and sphene may occur.

Structures and Textures

Intrusive ultramafic rocks and the associated mafic to felsic rocks form plutons that are characteristically composite. The irregular to zoned complexes are composed of individual, compositionally distinct intrusions. Structurally, individual intrusions are cylindrical, annular, cone-shaped, or irregular (figure 10.4d), and they commonly cross-cut one another. Each intrusion is composed of a distinct rock type. Some

	1	2	3	4	5
Modes[b]					
Quartz	—	—	—	—	21.5
K-feldspar	—	—	—	—	16.5
Plagioclase	58.3	63	42	94.7	56.0
Biotite	0.2	—	0.5	—	1.0
Hornblende	5.3	—	—	5.2	3.0
Clinopyroxene	30.0	24	32	—	—
Orthopyroxene	3.6	13	23.5	—	—
Olivine	1.9	—	—	—	—
Serpentine	0.2	—	—	—	—
Fe/Ti oxides	0.4	—	2	< 0.1	nd
Other	—	tr	—	—	2.0
Total	99.9	100	100.0	100.0	100.0
CIPW Norms					
qz	—	—	—	—	33.14
or	0.6	0.28	1.45	0.39	19.32
ab	8.7	14.67	14.21	53.81	38.50
an	51.8	46.56	26.01	41.37	4.70
ne	0.9	—	—	—	—
ol	7.4	—	1.28	0.66	—
hy	—	18.29	27.74	1.83	1.85
di	23.9	18.12	24.42	—	—
he	5.3	—	—	—	—
mt	0.9	1.16	3.84	0.74	1.11
il	0.4	0.38	0.64	0.33	0.29
ap	—	—	0.03	tr	0.27
Other	—	—	0.32	0.24	0.28
Total	99.9	99.46	99.94	99.37	99.90

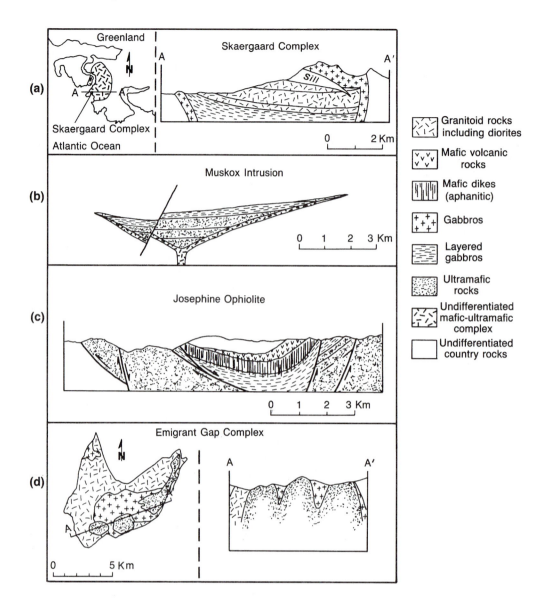

Figure 10.4 Generalized maps and cross sections of selected igneous complexes containing ultramafic rocks. (a) Simplified map of the Skaergaard Complex, Greenland, with section A–A' (simplified from Wager and Brown, 1968). (b) Restored cross section of the Muskox Intrusion, Canada (simplified from Irvine, 1967). (c) Section through the Josephine Ophiolite, California (modified from Harper, 1984). (d) The Emigrant Gap Complex, California. (schematic section by L.A. Raymond.)

(d) From O. B. James, "Origin and emplacement of the ultramafic rocks of the Emigrant Gap Area, California" in Journal of Petrology *12:523-60, 1971. Copyright © 1971 Oxford University Press, Oxford England. Reprinted by permission.*

dikes and apophyses may extend into the country rock from these intrusions.

Layered ultramafic-mafic plutons also contain a variety of rock types, and these form distinct layers(figure 10.4a, b, c). Major groups of layers, locally called series or complexes, develop as approximately horizontal sheets at the time of their formation. These may range from a few to hundreds of meters thick (Wager and Brown, 1968). Within these series or complexes, individual layers range from several meters down to microlayers or laminae of less than one millimeter thick (figure 10.5).[22] The layers are most distinct where they form dark and light bands, each composed of a mineral assemblage with a different ratio of plagioclase to ferromagnesian minerals. Such layering is referred to as **rhythmic layering,** which contrasts with layering lacking conspicuous differences in mineralogy, called **uniform layering** (Wager and Deer, 1939; Wager, 1968). **Cryptic layering** refers to generally invisible layers marked by variations in the chemistries of minerals within the layers (Wager, 1968). For example, variations in the An component of the plagioclase from anorthite (An_{90}) in basal ultramafic layers of a lopolith to labradorite (An_{60}) in the

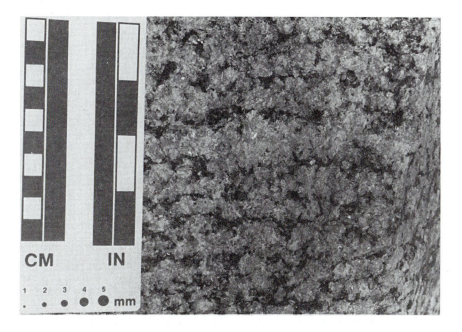

Figure 10.5 Photograph of inch-scale layering in gabbro of the Lower Banded Series, Stillwater Canyon area, Stillwater Complex, Montana. Scale is standard DNAG scale.

gabbros that occur at higher levels of the intrusion, would represent cryptic layering.

Within the various ultramafic-mafic plutons, the rocks exhibit a number of textures. Especially in the layered plutons, the textures commonly are cumulate textures. In such textures, early-formed and -transported crystals, the **cumulus crystals,** are surrounded by **postcumulus crystals** that crystallized from, or recrystallized through interaction with, an **intercumulus liquid** (figure 10.6).[23] Cumulate textures in the ultramafic rocks have intergranular, allotriomorphic-granular, hypidiomorphic-granular, poikilitic, lineated, or foliated appearances, depending on the nature of the crystallization processes involved. Textures in gabbros range from diabasic to ophitic and many are cumulate textures. In the more felsic differentiates, hypidiomorphic-granular and granophyric textures are typical.

The origins of cumulate textures are not fully understood.[24] Accumulation of early crystallizing phases by gravitational settling was an idea advanced by J. V. Lewis (1908a, 1908b) for the Palisades Sill. Similarly, settling and filter pressing were advocated by Bowen (1928, ch. 2, 6, and 9), both as general processes and as specific explanations of the origins of banded gabbros and of the magmas that produced the ultramafic dikes of Skye, Scotland. Recall (from chapter 5) that, simply stated, solid crystals will be heavier than the liquid magma and will sink to the bottom of the magma chamber. There they accumulate. Liquids may be sep-

arated from the crystals either by displacement (the crystals go to the bottom and the remaining magma to the top of the magma chamber) or by tectonic squeezing (filter pressing), in which compression of the magma chamber results in the remaining magma being squeezed out of the chamber (leaving the crystals behind). The liquid that remains behind, crystallizes to complete the cumulate texture. The fact that many layered plutonic bodies have cross-bedded and channeled layers consisting of concentrations of chromite, olivine, pyroxene, or combinations of these or similarly heavy minerals would seem to support the idea of gravitational settling. Experiments confirm that crystal settling can occur in magma chambers (D. Martin, 1990). Yet some textures, such as those with very little postcumulus material between cumulus phases, cannot be explained by simple crystal settling. The lack of postcumulous material may result if convection currents remove intercumulus liquid (Morse, 1986). Alternatively, such textures may result from modification by diffusion, after they develop in layers that form through *in situ* crystallization on the floor and walls of the magma chamber (I. H. Campbell, 1978; McBirney and Noyes, 1979; Morse, 1986; Lesher and Walker, 1988).

The behavior of plagioclase is problematical. As noted in chapter 5 (note 8), because of its lower density, plagioclase should float in certain mafic liquids.[25] If so, how can it form plagioclase-rich layers intercalated with layers of mafic minerals in the layered intrusions, and how could it be a cumulus phase?

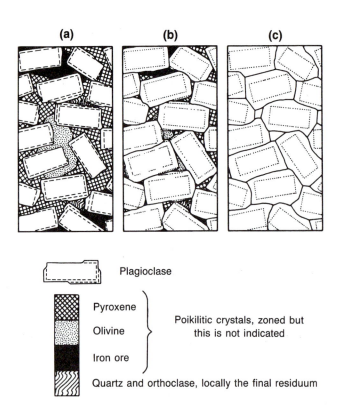

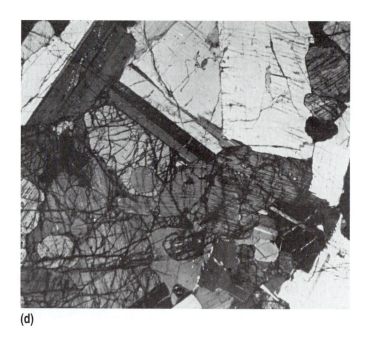

(d)

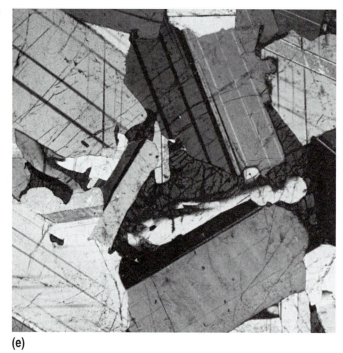

(e)

(f)

Figure 10.6 Cumulate textures. (a) Orthocumulate texture. (b) Mesocumulate texture. (c) Adcumulate texture. Dotted lines show boundaries of cumulate phases, plagioclase in this case. Material on outer edges of cumulate phases is crystallized from intercumulus liquid (from Wager et al., 1960). (d) Orthocumulate texture in norite, Banded Series, Stillwater Complex, Montana. (XN). (e) Mesocumulate texture in gabbro, Banded Series, Stillwater Complex, Montana. (f) Adcumulate texture in anorthosite, Lower Banded Series, Stillwater Complex, Montana. Long dimensions of photomicrographs are (d) 5 mm, (e) 4.9 mm, and (f) 5 mm.

(a)–(c) From L. R. Wager et al "Types of igneous cumulates", in Journal of Petrology, 1:73-85, 1960. Copyright © 1960 Oxford University Press, Oxford England. Reprinted by permission.

The legend accompanying (a)–(c):

Plagioclase

Pyroxene
Olivine   } Poikilitic crystals, zoned but this is not indicated
Iron ore

Quartz and orthoclase, locally the final residuum

The density data are in conflict with many observations. Field observations provide convincing evidence in the form of structures, such as cross-bedding, channeling, grading, and stratification, that many minerals, including plagioclase, were deposited by currents on the floors of magma chambers (Wager and Brown, 1968; T. N. Irvine, 1974; Page et al., 1985). Microscopic petrography reveals that plagioclase *is* a cumulus phase.

T. N. Irvine (1980a) attempted to resolve the conflict via a complex model of cumulate origin in which layer formation is a result of several processes. Some layers are produced by *in situ* crystallization. This may be enhanced by **double-diffusive convection,** a process involving movement of materials by the combined effects of physical movement (convection) and chemical transport (diffusion) (Turner and Chen, 1974).[26] Other layers are formed where the density factor is overcome by movement processes driven by gravity. In such cases, slumping and flow of crystal-rich density currents physically transport plagioclase and other crystalline phases along the bottom of the magma chamber, where they are deposited in layers. They may be kept there by two factors. First, they will have a limited tendency to rise because the differences in density between the crystals and liquid are generally rather small. Second, continuing deposition, crystallization, or both would tend to produce overlying layers that would restrict upward movement of the crystals.

As the crystals develop and are deposited on the base of the magma chamber as cumulus phases, crystallization at the magma-cumulate interface also occurs (Wager, 1963). In the buried part of the cumulate, crystallization of the *intercumulus liquid* modifies the texture. This material may crystallize as separate grains, forming an **orthocumulate texture,** or may crystallize *on* the cumulus crystals, with the result that cumulus crystals and their overgrowths are the only components of the texture, forming an **adcumulate texture** (figure 10.6). **Mesocumulate textures** are intermediate between these two texture types. Recrystallization, through interaction of the cumulus crystals and intercumulus liquid, may also occur.[27]

THE NATURE AND ORIGINS OF ULTRAMAFIC-MAFIC COMPLEXES
• •

Each of the various types of ultramafic-mafic complexes has a distinct origin reflected in its chemistry, mineralogy and petrography, textures, and structure. In some cases, multiple petrogenetic hypotheses have provoked enlightening controversy. The origins of layered dikes and sills provide the least ambiguous petrogenetic models, and serve as a starting point in examining petrogenesis.

Dikes and Sills

General Comments

Dikes and sills of gabbro[28] are relatively common, just like their basaltic volcanic equivalents. They occur widely *beneath* sites of basaltic volcanism. Where they (1) have typical basalt chemistries, (2) cross-cut the structures of the country rock, (3) produce contact metamorphism adjacent to their margins, and (4) have fine-grained (chilled) margins at their contacts, there can be little doubt that they are intrusive features.

Gabbro dikes, sills, and related structures may be simple, multiple, composite, or differentiated (Billings, 1954). Simple dikes and sills result from the intrusion of a single pulse of magma and their only inhomogeneity may be a fine-grained (chilled) border facies. Multiple structures result from more than one intrusion of a particular magma, whereas composite dikes and sills result from multiple intrusions in which the magmas are of different compositions. Differentiated structures result where intrusions undergo differentiation. Differentiated dikes and sills are of interest here, as they are the ones that may contain ultramafic rocks. Features of multiple or composite dikes may be mistaken for differentiation features, as was partly the case for more than half a century, when, following the early works of J. V. Lewis (1908a, 1908b) and F. Walker (1940), the geological community cited the layered Palisades Sill of New York–New Jersey as an example of a differentiated single intrusion. Available evidence now suggests that the Palisades structure is part dike and part sill, is multiple, and is differentiated (K. R. Walker, 1969b; Husch, 1990).

In the simplest cases, differentiation in mafic magma chambers may be thought of as occurring by several complementary processes (S. A. Morse, 1986; Marsh, 1988a). Crystallization alone yields a mafic rock; but together with gravitational settling and convection will lead to segregation and sorting of crystals to yield mafic to ultramafic cumulates. The cumulates consist of a series of layers formed on the bottom, the sides, and, through flotation, the top of the magma chamber. Exchange and diffusion of elements between cumulus phases and intercumulus or residual melts may serve to modifiy cumulus compositions. In the residual magma, enrichment in alkalis, silica, and (in tholeiitic magmas) iron may lead to crystallization of more siliceous differentiates such as granite or quartz monzonite. The magmas that yield these siliceous rocks will tend to move up and form layers beneath the crystallized top of the magma chamber, thus contributing to the layered nature of the body. In dikes and sills, the late differentiates usually are not abundant enough to form complete layers. Instead, they commonly form pods, dikes, or segregations of **granophyre,** porphyritic granite with microscopic intergrowths of quartz and potassic alkali feldspar (figure 10.7).

Figure 10.7 Photomicrograph of granophyre from Mull, Scotland, showing granophyric texture. Long dimension of photo is 6.5 mm.

Mafic dikes and sills subject to differentiation are abundant where rifting results in continental breakup. For example, Mesozoic breakup of the supercontinent of Pangaea to produce most of the present-day continental masses was accompanied by intrusion of numerous mafic dikes. One major suite of such dikes, which formed along the east coast of North America during the Triassic and Jurassic periods, includes the well-known and geologically complex Palisades Sill of New York–New Jersey and a multitude of dikes from South Carolina to New Hampshire (Walker, 1969a; Froelich and Gottfried, 1985). Another such suite in the southern continents is well represented in Tasmania (Hergt et al., 1989) and includes the Red Hill Dike (McDougall, 1961, 1962).

Example: The Red Hill Dike

The Red Hill Dike is one of many mafic intrusions into the Permian and Triassic sedimentary rocks of Tasmania, a large island located off the southeast coast of Australia (A. B. Edwards, 1942; McDougall, 1961, 1962; Hergt et al., 1989). These intrusions, which represent partial melts derived from sediment-contaminated mantle peridotite, intruded the crust during the Jurassic Period. In structure, the Red Hill Dike is a cross-cutting extension of a sill that is more than 20 kilometers long and 800 meters thick (figure 10.8).

The Red Hill Dike clearly reveals effects of the differentiation process. The dike shows fine-grained, chilled margins against the country rock and is assumed to have an Mg-rich cumulate base, but the latter is not exposed. The main mass of the dike is composed of subophitic, medium-grained gabbro composed of plagioclase (commonly zoned, An_{85-40}), pyroxenes (various combinations of one or more of the pyroxenes augite, ferroaugite, orthopyroxene, and pigeonite), and a small amount of interstitial quartz and potassic alkali feldspar. Above the gabbro is a fayalite granophyre layer. The fayalite granophyre is medium-grained and consists of olivine (Fo_{10}), ferroan clinopyroxene, plagioclase (An_{63-27}), quartz, and alkali feldspar. A "silicic" granophyre layer overlies the fayalite granophyre layer. The granophyre is medium-grained and is composed of iron-rich clinopyroxene, plagioclase (An_{54-14}), quartz, alkali feldspar, and iron oxides.

Differentiation in the Red Hill Dike is shown by the changing bulk chemistry of the rocks, the changing proportions of minerals upwards in the dike, the changing compositions of the pyroxenes and plagioclase, and the appearance of fayalite (figure 10.9). Upwards, plagioclase decreases from 40–60% of the rock in the gabbros to 30–50% in the fayalite granophyres and to 15–31% in the "silicic" granophyres.

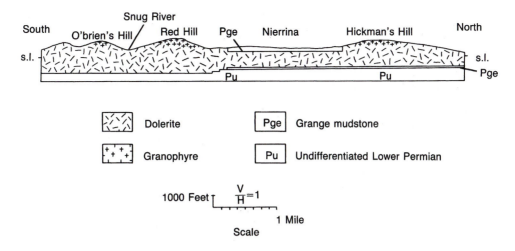

Figure 10.8 Cross section of the Red Hill Dike, Tasmania. (redrawn from McDougall, 1962). Dolerite=gabbro.

Quartz is nearly absent in the most mafic rocks but comprises as much as 30% of some granophyres. Similarly, alkali feldspar ranges from <2% in the most mafic gabbros to over 40% in some granophyres. Orthopyroxenes formed early in the differentiation sequence but were replaced by pigeonite in later formed gabbros. Clinopyroxenes, which comprise 14–37% of the mode of the gabbros and 5–12% of the granophyres, range from more magnesium-rich augites in the gabbros to more iron-rich clinopyroxenes in the granophyres.

These mineralogical changes are reflected by changes in the bulk rock chemistry as plotted on an AFM diagram (figure 10.9c). The initial magma was tholeiitic. The AFM diagram shows that this tholeiitic magma differentiated, producing an early decrease in Mg and increase in Fe and a late increase in alkalis.

Lopolithic Layered Intrusions

Lopolithic and related layered intrusions provide the most extensive evidence and impressive examples of fractional crystallization. These structures represent magma chambers that are larger than dikes and sills and which, consequently, are subject to longer and more extensive differentiation. Some lopoliths, such as the Bushveld Complex of southern Africa, contain the most highly developed fractionated derivatives of tholeiitic basalt present among the ultramafic-mafic complexes.

General Comments

Evidence of fractionation in lopoliths is provided by the regular, usually cyclic, rhythmic layering that characterizes these bodies. Sequences of layered rock thousands of meters thick form the bulk of the larger lopoliths and the similar funnel-shaped intrusions. For example, the Skaergaard Complex of Greenland is more than 2500 meters thick (Wager and Brown, 1968), the exposed part of the Stillwater Complex is 5500 meters thick (W. R. Jones, Peoples, and Howland, 1960), and the Norite Zone of the Bushveld Complex (the Bushveld intrusion of Wager and Brown, 1968) is about 7300 meters thick.

As is the case with dikes and sills, various lines of evidence suggest that the lopolithic bodies are intrusions. Notably the bodies cross-cut the country rocks into which they have intruded, possess finer-grained (chilled) rocks along their margins, and produce contact metamorphism in the adjacent country rocks. Further evidence of magmatic processes is provided by igneous textures and structures, including cumulate textures, in the rocks.

Example: The Stillwater Complex

The Stillwater Complex is a tilted lopolith of Archean (Precambrian) age that crops out for 48 kilometers along the northern edge of the Beartooth Range in southern Montana (figure 10.10a) (Peoples, 1933a, 1933b; N. S. Page and Zientek, 1985).[29] The roof of the body is not exposed, being covered unconformably by Paleozoic and Mesozoic sedimentary rocks. These rocks and the complex dip to the north at 25 to 70 degrees. In the eastern part of the outcrop belt, the complex is intruded by Archean quartz monzonite. It is extensively cut by east-west-trending thrust faults and north-south-trending high-angle faults. Late Cenozoic sediments mask large sections of the body, especially along major river valleys.

The Stillwater Complex is divided into three series. From bottom to top, they are the Basal Series, the Ultramafic Series, and the Banded Series (figure 10.10a). Each series is divided into a number of zones. The Basal Series includes the Basal Contact Zone, the Basal Norite Zone, and the Basal

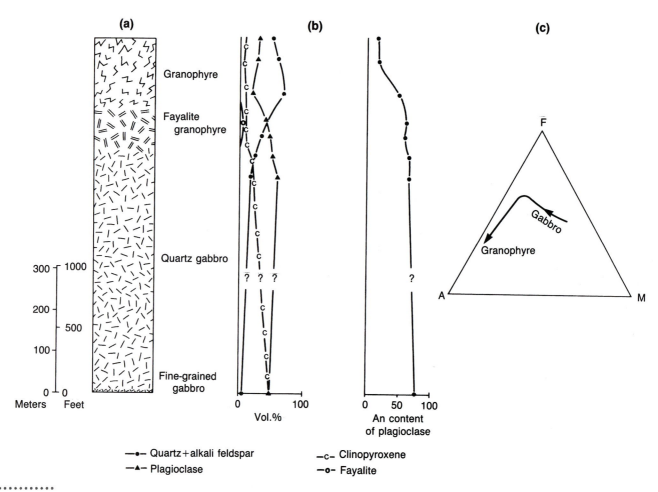

—•— Quartz + alkali feldspar   —c— Clinopyroxene
—▲— Plagioclase   —o— Fayalite

Figure 10.9 Mineralogical and chemical variation in the Red Hill Dike. (a) Columnar section (based on data from McDougall, 1962). (b) Graphs showing variations in mineral abundances and chemistry (based on data from McDougall, 1962). (c) AFM diagram showing chemical variation trend in Red Hill Dike rocks (modified from McDougall, 1962).

Bronzitite Cumulate Zone. The Basal Contact Zone, at the bottom of the Stillwater Complex, is a thin, complicated mass of noncumulate dikes, sills, and other intrusions that are composed of fine- to coarse-grained, diabasic to ophitic, two-pyroxene gabbros (gabbronorite), olivine gabbro, and norites (Helz, 1985).[30] Inclusions and masses of hornfels are common in the zone. The complicated contact zone is overlain by a cumulate unit, the Basal Norite Zone, characterized by norites (orthopyroxene gabbros) with orthopyroxene-rich (bronzite-rich) layers, bronzite-olivine layers, and bronzite-augite layers. Cumulate orthopyroxenites characterize the Basal Bronzite Cumulate Zone. The Basal Series, as described here, includes all the rocks between the base of the complex and the base of the first dunite, which marks the beginning of the Peridotite Zone of the Ultramafic Series.

The Ultramafic Series contains two zones, a lower Peridotite Zone and an upper Bronzitite Zone (E. D. Jackson, 1961; Raedeke and McCallum, 1984).[31] The Peridotite Zone consists of cyclic, rhythmically and cryptically layered units of medium to very coarse-grained dunite, harzburgite (olivine-orthopyroxene rock), and bronzitite (orthopyroxenite), with local layers of chromitite (figures 10.11 and 10.12a). Texturally, the rocks are cumulates, showing every gradation from idiomorphic-poikilitic to allotriomorphic-granular textures. The upper zone, the Bronzitite Zone, consists almost exclusively of orthopyroxenite. The rock is typically coarse-grained and ranges from hypidiomorphic-poikilitic to allotriomorphic-granular. Minor amounts of augite, plagioclase, olivine, and chromite are present within the rock and locally are concentrated to form other rock types (e.g., chromitite).

The Banded Series is the uppermost of the exposed Series.[32] Marking the contact of the Banded Series is the first appearance of cumulus plagioclase (I. S. McCallum, Raedeke, and Mathez, 1980). The series is characterized both by the abundance of plagioclase and by locally distinctive layering in the gabbros (figures 10.5 and 10.13). The Banded Series is subdivided into three parts, the Lower Banded Series, the

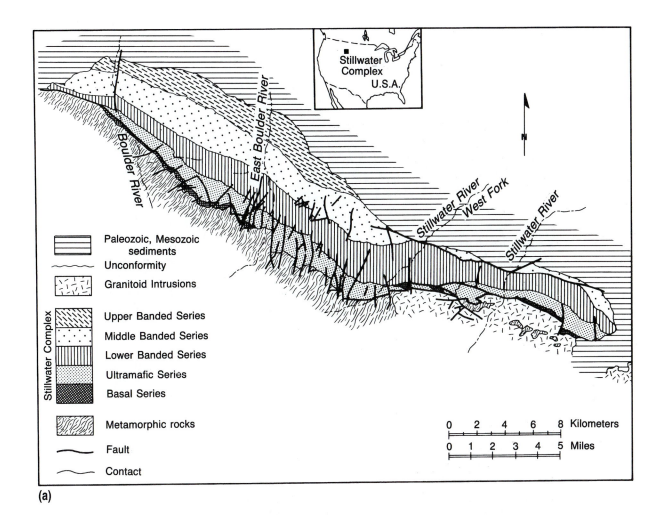

(a)

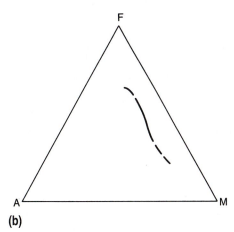

(b)

Figure 10.10 Generalized map (a) and schematic AFM diagram (b) for the Stillwater Complex, Montana
[(a) simplified from N. J. Page and Zientek, (1985), (b) based on data from H. H. Hess (1960) and Helz (1985) and an analysis by Wager and Brown (1968)]

(a)

(b)

Figure 10.11   Rocks of the Ultramafic Series in the Mountain View Lake area of the Stillwater Canyon section, Stillwater Complex, Montana. (a) View south into Ultramafic Series. (b) Harzburgite of the Peridotite Zone. Light patches are weathered, poikilitic orthopyroxenes; dark material is mostly serpentinized olivine.

Middle Banded Series, and the Upper Banded Series, which are subdivided into 12 to 14 zones. Included among the rock types of these various zones are gabbro, norite (orthopyroxene gabbro), two-pyroxene gabbro (gabbronorite), troctolite (olivine gabbro), pigeonite gabbro, and anorthosite (hyperleucocratic gabbro). Hypidiomorphic-granular to allotriomorphic-granular textures are common.

Because of cyclic and rhythmic layering, the cryptic layering (the sequence of chemical changes within the minerals) of the Stillwater Complex is more complicated than that present in the smaller dikes and sills like the Red Hill Dike. The chemistries of minerals fluctuate with the cycles, rather than changing monotonically towards more iron- and alkali-rich compositions. Nevertheless, there are overall variations in

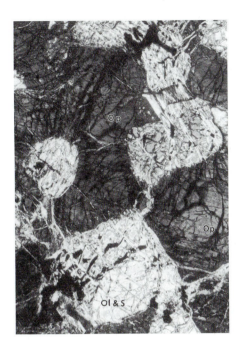

(a)

(b)

Figure 10.12 Photomicrographs of rocks of the Stillwater Complex. (a) Harzburgite (peridotite) from the Peridotite Zone of the Ultramafic Series. (XN). (b) Gabbro from the Lower Banded Series. (XN). Op = orthopyroxene,

Px = clinopyroxene, Ol = olivine, C = chromite, P = plagioclase, M = magnetite-ilmenite, S = serpentine. Long dimension of photos is 6.5 mm.

mineral abundance and chemistry (figure 10.14) (H. H. Hess, 1960). Plagioclase ranges from An_{88} in rocks of the Norite I Zone in the lower part of the Banded Series to An_{62} in gabbros at the top of the Upper Banded Series.[33] Olivine varies from Fo_{90} at the base of the Ultramafic Series to Fo_{64} within the Middle Banded Series. Similarly, Mg decreases upward in orthopyroxenes, which vary from En_{88} at the base of the Ultramafic Series to En_{64} in the Banded Series. Ca-rich clinopyroxenes vary from $Wo_{40}En_{53}Fs_7$ in the Ultramafic Series to $Wo_{40}En_{42}Fs_{18}$ in the Upper Banded Series. These overall trends and the tholeiitic differentiation trend of the Stillwater rocks on the AFM diagram (figure 10.10b) are consistent with an origin by fractional crystallization.

The Origin of the Lopolithic Intrusions

The origin of the Stillwater Complex and similar intrusions is not quite as simple as it may seem from generalized textural, mineralogical, and chemical data. First, although there are general trends of changing mineral abundance and chemistry, there are significant local fluctuations in both the mineralogies and the chemistries of minerals within the various banded rocks. Second, the rock types do not change monotonically upward towards more siliceous, alkali-rich, or

Figure 10.13 Cross-bedded layering in norite of the Lower Banded Series, Stillwater Canyon area, Stillwater Complex, Montana. Scale is standard DNAG scale.

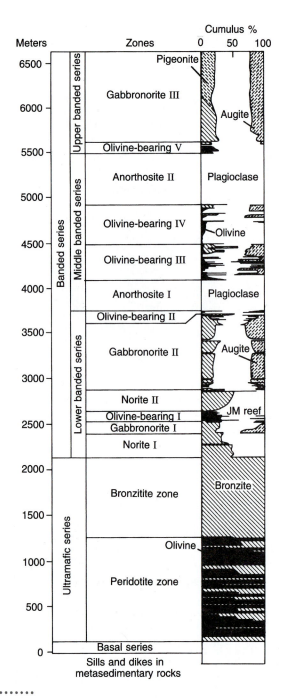

Figure 10.14 Mineralogical variations in the Stillwater Complex, Montana.
(Modified from Boudreau and McCallum, 1989; based on H. H. Hess, 1960; McCallum et al., 1980; Todd et al., 1982; Raedeke and McCallum, 1984)

iron-rich compositions, but instead cyclically repeat compositional types. Third, in the Stillwater Complex, where the base is well exposed and has been carefully examined, the lowest rocks of the complex are not cumulates and do not form a simple chilled gabbro zone; rather they comprise a complicated intrusive dike and sill complex composed of various rock types (N. J. Page, 1979; Helz, 1985; Zientek, Czamanske, and Irvine, 1985).

Attempts to explain the origin of cyclic and rhythmical layering in layered intrusions are numerous.[34] The various hypotheses invoke particular combinations of physical and chemical factors, including variations in magma compositions, density, pressure, temperature, current action, and diffusion, to create particular aspects of these rocks. Although we do not yet fully understand the origins of the layers or the rocks in which they occur, it seems clear that lopolithic intrusions like the Stillwater Complex cannot have evolved by simple fractionation and gravity settling. If they had, the layering would not be nearly so complicated.

For the Stillwater Complex, a complicated history is indicated by a wide variety of available data.[35] The multiple chemistries and various rock types of the dike and sill complex of the Basal Contact Zone represent multiple intrusions of five major magma types. Radiometric dates reveal that these magmas intruded to form the Basal Contact Zone at about the same time as the main magma chamber formed (2704 m.y.b.p.). Major element, REE, and isotopic data, including Mg numbers, Nd- and Os-isotope ratios, and concentrations of platinum group elements, suggest mantle sources for the parent magmas. Two principal parental magma types fed the Stillwater magma chamber; one dominated formation of the Ultramafic Series and the other was the dominant parent for the Banded Series. The parental magma of the Ultramafic Series was similar to one of the dike and sill magma types, a high-MgO two pyroxene gabbro, and probably had a garnet-bearing, LREE (sediment?) contaminated mantle peridotite source. Most Banded Series rocks may have formed from a high-Al_2O_3 basaltic magma, like another dike and sill gabbronorite, that was perhaps derived from the mantle and modified by mixing, contamination, and fractionation when it ponded at the crust-mantle boundary. Abundant injections of this aluminous parent magma into the chamber initiated and sustained the extensive plagioclase crystallization that characterizes the Banded Series (figure 10.14). Repeated injections of the various mantle-derived magmas, which mixed and experienced fractional crystallization (aided by crystal transport and deposition), crystallized to form the diverse layers of the Stillwater Complex.

Ophiolites

Ophiolites are layered igneous bodies but of quite a different kind than the lopoliths, sills, and dikes described above. Although they contain some cumulate-textured rocks, ophiolites have a gross layered structure produced by intrusive and extrusive emplacement of magma in distinct and separate units above a mantle tectonite.

The term ophiolite was originally applied to **serpentinites,** rocks composed primarily of serpentine produced by metamorphism of the magnesium-rich minerals of ultramafic rocks.[36] Later the term was extended to a suite of rocks (Steinmann, 1927), which came to be known as the Steinmann Trinity, because of the repeated association of three rock types throughout the Mediterranean region and elsewhere. The trinity consists of radiolarian chert, diabasic-spilitic rocks,[37] and serpentinite, but gabbro and peridotite were originally included in the suite by Steinmann. The term ophiolite now is used to refer to a structurally distinctive assemblage of ultramafic, mafic, and related rocks consisting, from bottom to top, of four parts: (1) ultramafic tectonites composed of dunite, harzburgite, and/or lherzolite; (2) a plutonic, cumulate to noncumulate hypidiomorphic-granular-textured, layered ultramafic-mafic complex consisting of peridotites, pyroxenites, gabbros, and associated felsic rocks; (3) a mafic "sheeted" dike complex composed primarily of fine-grained gabbros and basalts; and (4) a volcanic complex composed of lenticular to pillow-structured basalts, plus minor andesitic to rhyolitic rocks (figures 10.3 and 10.15).[38] Rocks associated with ophiolites, but not included in the definition, include the radiolarian cherts of the Steinmann Trinity and other sedimentary rocks such as shale, breccia, and limestone, which commonly overlie the ophiolitic rocks.

The Structure and Composition of Ophiolites

Each of the four parts of an ophiolite complex has a different origin. The ultramafic tectonite is essentially metamorphic rock upon which the rest of the ophiolite complex has been built (Moores, 1969).[39] Typically, the rocks of the tectonite unit are younger serpentinite tectonites (e.g., serpentinite schists), with older dunite, peridotite, and pyroxenite tectonite inclusions preserved as unserpentinized layers, blocks, pods, or zones.

Evidence of the metamorphic character of the tectonite unit is provided by structures, textures, and mineral chemistry (Thayer, 1960, Raleigh, 1965; Moores, 1969; Nicolas, 1986).[40] The boundaries of the metamorphic ultramafic rock bodies are commonly faulted and lack both chilled margins and contact aureoles. Internally, the tectonite mass may exhibit folds. Schistose to gneissose textures, including porphyroclastic types, and deformed mineral grains characterize the rocks. Mineralogically, olivines and pyroxenes are essential phases. Olivines are usually magnesium rich (Fo_{94-85}), though they may range to more ferrous types (to Fo_{75}). Pyroxenes are also magnesium-rich. Chemically, the rocks are characterized by high Mg numbers (0.80–0.88), due to the high magnesium content of olivines and pyroxenes. The phase assemblages, textures, and

chemistries reflect equilibration at pressures, temperatures, and conditions of strain characteristic of the mantle (Raleigh, 1965; Moores and MacGregor, 1972; Ave Lallemant, 1976; Nicolas, 1986).[41] Together, these data indicate that the ultramafic tectonite layer of ophiolites is a mantle rock.

The plutonic complex that comprises the second part or major layer of the ophiolite consists of variously textured and structured rocks.[42] At the base is a layered, cumulate-textured mass consisting of both ultramafic and mafic cumulates (Moores, 1969; E. D. Jackson, Shaw, and Barger, 1975).[43] Both rhythmic and cryptic layering are present. The major rock types include dunite, lherzolite, wehrlite, pyroxenites, troctolite, and gabbro. Olivines range from Fo_{91} to Fo_{75}, plagioclase generally has compositions in the range An_{97-40}, and pyroxenes tend to be magnesium-rich ($Cpx = Wo_{51-33}En_{26-65}Fs_{1-20}$; $Opx = Wo_{0-4}En_{92-54}Fs_{7-43}$). Chemically, these rocks do not show the iron enrichment characteristic of lopolithic complexes, as evidenced mineralogically by the absence of iron-rich olivine and pyroxene in silicic rocks.

The upper part of the second layer of the ophiolite consists of noncumulate gabbro, diorite, and more felsic rocks. The latter are interpreted to be differentiates of the basic magma that gave rise to the layered cumulate. The more mafic of these rocks are typically hornblende-bearing and have intermediate plagioclases. The more felsic rocks include quartz diorites (the so-called albite granites, or *plagiogranites;* see R. G. Coleman and Peterman, 1975). Textures are hypidiomorphic-granular to granophyric, but may be pegmatitic locally. Structures include dikes, apophyses, and complex, interfingering intrusives.

The third part or layer of the ophiolite complex consists of a dike and sill complex. The rock types most common are fine-grained gabbro (diabase, dolerite) and basalt, but more silicic rocks also occur.[44] Dikes commonly are distinctly parallel in strike and may have only one chilled margin (Moores and Vine, 1971; Moores, 1982), indicating that successive dikes intruded earlier formed dikes parallel to a regional fracture pattern. Such dike complexes are referred to as **sheeted dike complexes.** In some ophiolites, sills, rather than dikes, comprise a significant part of the third layer of the ophiolite (C. A. Hopson and Frano, 1977). The dikes and sills apparently served as feeders for the overlying volcanic rocks.

The uppermost unit of the ophiolite is composed of extrusive rock. Massive tabular to lenticular and pillowed flows comprise most of the section. Breccias also occur. Interlayered with these volcanic rocks are tuffs, interpillow limestones, and cherts. The mafic volcanic rocks were primarily basalt when they formed, but in many cases they have been altered or metamorphosed to yield *spilites* or *greenstones.*[45] Sodium metasomatism and hydration produced the distinctive albite-chlorite-epidote-calcite-zeolite assemblages that characterize the spilites. More siliceous volcanic rocks and their metamorphic equivalents (e.g., andesite and keratophyre, rhyolite and

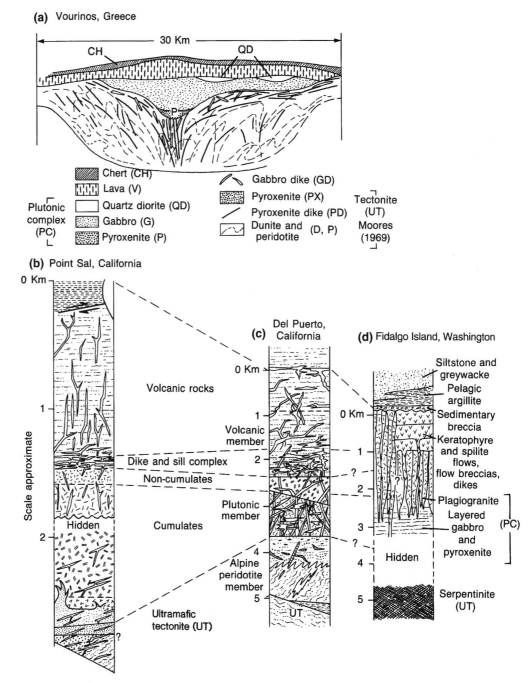

Figure 10.15 Sketches and columnar sections of ophiolite complexes. (a) Vourinos Complex, Greece (from Moores, 1969). (b) Composite columnar section, Point Sal Ophiolite, California (simplified from C. A. Hopson and Frano, 1977). (c) Del Puerto Ophiolite, California (after Evarts, 1977). (d) Fidalgo Island Ophiolite, Washington (from E. H. Brown, 1977). CH = chert, D = dunite, G = gabbro, GD = gabbro dike, P = peridotite, PC = plutonic complex, PX = pyroxenite, QD = quartz diorite, UT = ultramafic tectonite, V = volcanic rocks.

quartz keratophyre) comprise parts of some ophiolites. A source of confusion in understanding some ophiolites, notably the Troodos Ophiolite, is the fact that they contain two volcanic sequences (Moores and Vine, 1971; Beccaluva et al., 1984; P. T. Robinson et al., 1983). In general, the chemistries and origins of these two sequences differ.

The Origin of Ophiolites

The study of ophiolite origins has gone through two major phases. During the early phase, the unique characteristics of ophiolites were not fully realized and a clear distinction was not made between ophiolites and other alpine-type ultramafic rocks (Moores, 1982). During this phase, conflicting views evolved as to (1) whether ophiolites were igneous or metamorphic rocks, and (2) whether they became a part of the crust as intrusive or extrusive magmas or were emplaced as solids, by diapirism or faulting, for example. H. H. Hess (1938, 1955) argued on the basis of field observations that alpine ultramafic rocks were the product of intrusions of ultramafic magmas with the composition of serpentine. Bowen and Tuttle (1949), however, used experimental laboratory analyses to discredit the idea of a serpentine magma.

The view common in Europe was that ophiolites were formed as great basic volcanic extrusions on the ocean floor or as subsea laccoliths. The former idea was clearly described by Aubouin (1965, pp. 151–56) and the latter was described by W. G. H. Maxwell (1968).[46] In the extrusion model (figure 10.16a), a mass of lava, erupts onto the ocean floor, quickly develops chilled margins of basalt or fine-grained gabbro against the surrounding seawater, and slowly cools and differentiates internally. The laccolith model is similar, except that in this model lava intrudes into shallow levels of the subsea crust. At the time these models were developed, the tectonite character of the basal ultramafic rock had not yet been recognized in ophiolites. Now, however, several points argue against such models: (1) the presence of the ultramafic tectonite, (2) the general lack of a basal chilled facies, (3) details of the mineralogy and chemistry of the rocks and minerals in the ophiolites, and (4) the relative proportions of mafic and ultramafic rock in the ophiolites (Bortolotti, Dal Piaz, and Passerini, 1969; Moores, 1982).

Ophiolites were considered to be differentiated crustal intrusions (figure 10.16b) by a number of workers including Steinmann (1927) and McTaggart (1971).[47] In the model advocated by McTaggart (1971), *all* alpine ultramafic bodies (including ophiolites) were envisioned to be high-level, sill-like to lopolithic, crustal intrusions that differentiated in place. The same arguments that contravene an extrusive origin argue against an intrusive origin, with the added problem that most ophiolites are capped by pillow lavas and sedimentary rocks that could not have formed in an intrusion.

Alpine ultramafic rocks in general, and by implication ophiolites, were considered by some petrologists to represent either mantle slabs or hot or cold diapirs tectonically emplaced into the crust (Bowen and Tuttle, 1949; Stoneley, 1975). The nonmagmatic character, tectonite fabrics, and lack of both chilled margins and contact metamorphism around some alpine-type ultramafic rock bodies, coupled with the fact that some ultramafic slabs may be traced geophysically into the mantle, argue for such an origin for ultramafic tectonites.[48] Considering the entire ophiolite sequence (rather than only the tectonite part), however, a number of the same problems posed for the intrusive and extrusive models also contravene models that argue that ophiolites are tectonically emplaced mantle rock. For example, pillow basalts did not form in the mantle. Similarly, although tectonites exist, the ultramafic rocks are not all metamorphic mantle tectonites;

some are cumulates. Thus, such models fail to explain the origin of the ophiolite assemblage as a whole, although they do explain how some ophiolites are emplaced into the crust.

The recognition of ultramafic tectonites and their distinction from the cumulates, along with discovery of the sheeted dike complexes, at a time when the theory of plate tectonics was evolving, spurred the second phase of ophiolite genesis studies. These discoveries further promoted the idea advanced by Brunn (1959) and Dietz (1963) that ophiolites represent mantle and oceanic crust.[49] During this second phase, it was widely accepted that ophiolites represent some kind of mantle-crust sequence, but controversy focused on the question of what tectonic environment that mantle-crust sequence represents.

The suggestion that ophiolites represent mantle plus oceanic crust formed at a mid-ocean ridge is supported by the fact that there is a general correspondence between the layers of the oceanic crust plus the underlying mantle *and* the layers in ophiolites (figure 10.16d) (Moores and Vine, 1971).[50] Furthermore, drilling in the ocean crust near Costa Rica confirms that pillow basalts and flows overlie sheeted dikes in Pacific Ocean crust (Becker et al., 1989). A number of other structural, textural, chemical, and petrographic features are consistent with this interpretation (C. A. Hopson, Mattinson, and Pessagno, 1981). Yet it is also recognized that some problems exist with the comparison between the two types of layered sequence.

The oceanic crust consists of three layers that overlie the mantle.[51] These are defined on the basis of seismic velocities and partly confirmed by dredging and drilling. Layer 1 is a layer of sediments (p-wave velocity = V_p = 1–2 km/sec.). Layer 2 is a layer consisting of various volcanic (basalt) flows, their metamorphic equivalents, and some metamorphosed parts of the sheeted dike complex (V_p = 3–6 km/sec.). Layer 3 is a layer interpreted to consist of mafic intrusive and plutonic rocks and their metamorphic equivalents (V_p = 6–8 km/sec.). The Moho apparently lies within the cumulate sequence at the mafic-ultramafic boundary (Salisbury and Christensen, 1978). Layer 3 overlies the cumulate ultramafic rocks and mantle tectonites (figure 10.17). The total thickness of this normal oceanic crust is about 5–6 kilometers. One problem with the ophiolite = ocean crust + mantle interpretation is that many ophiolites appear thinner than the oceanic crust (R. G. Coleman, 1971; Moores and Jackson, 1974).

The ophiolite = oceanic crust + mantle hypothesis faces other problems. For example, the chemical data from some of the ophiolitic volcanic sequences have been interpreted in different ways. While some petrologists consider the major-element chemical data to support an ocean-ridge origin, to others these data suggest that ophiolites might represent the primitive crust formed beneath an island arc (Miyashiro, 1973c). Still others consider that the various chemical, structural, petrographic, and mineralogical data show that the ophiolites formed in association with arcs, but in marginal basin or

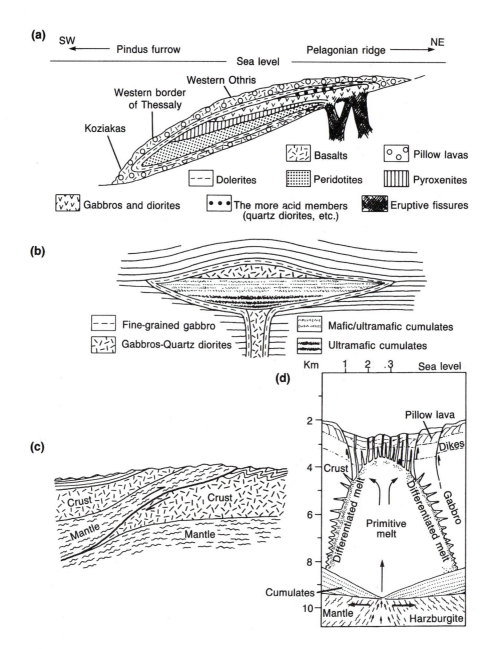

Figure 10.16 Models of ophiolite genesis and emplacement. (a) Submarine extrusion model (from Aubouin, 1965). (b) Crustal intrusion model (based on McTaggart, 1971). (c) Mantle slab model (R. G. Coleman, 1971b).

(d) Oceanic crust and mantle model (based primarily on Bryan and Moore, 1977), and compatible with the ideas of Gass, 1967; Moores and Vine, 1971; Christensen and Salisbury, 1975; Casey and Karson, 1981.

back-arc environments.[52] No consensus has been reached. Even the trace element data seem ambiguous.[53] What does seem clear is that most ophiolitic rocks represent mantle-crust sequences developed in some sort of arc-related or extensional (spreading center) environment. Although mid-ocean ridge sites cannot be excluded for all ophiolites, most likely the more common environment of formation of ophiolites is a back-arc or intra-arc basin. Conceivably, some ophiolites represent one environment, whereas others represent another.

In mid-ocean ridge/spreading center models, tension at the site of seafloor spreading gives rise to melting in the mantle. Mantle diapirs rise up below the ridge and, as they rise, the reduction in pressure induces partial melting to form basaltic magmas. The basaltic magmas intrude into the axial zone of the spreading center to form a magma chamber (see figure 10.16d). Fractionation produces (1) cumulates at the base of the chamber, just above the mantle tectonites, and (2) various gabbroic to felsic differentiates at higher levels in

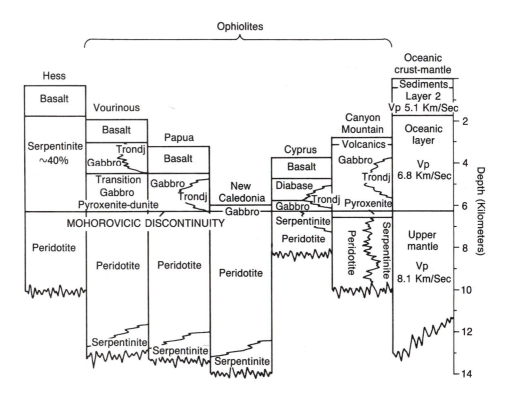

Figure 10.17 Comparison of the thicknesses of oceanic crust and ophiolite complexes. Trondj. = trondjhemite. *(Redrawn from R. G. Coleman, 1971b)*

the chamber. Dikes from the chamber feed sills beneath the surface and lava flows at the surface (some of which develop pillows). The carapace of flows and sills is continually cut by successive dikes that feed new flows at the surface. Subsequent and contemporaneous sedimentation caps the ophiolite with chert, limestone, or other materials.

In arc-related models, the sequence is similar. However, in arc models, rising mantle diapirs yield magmas that pond at the mantle-crust boundary and, in some cases, at higher structural levels. There the magmas fractionate to produce cumulates. Dikes from the fractionating magmas feed eruptive rock units at the surface.

Example: The Del Puerto Ophiolite

The Jurassic Del Puerto Ophiolite is a part of the larger Coast Range Ophiolite of California.[54] It is exposed in the area in and around Del Puerto Canyon on the east side of the Diablo Range, one of the Coast Ranges of central California (figure 10.18). Faulting has fragmented the ophiolite, but parts of all major sections crop out locally, allowing reconstruction of a generalized section through the complex (Evarts, 1977, 1978).

The ophiolite consists of the usual four parts (see figure 10.15). The ultramafic tectonite is composed of serpentinized dunite, wehrlite, and harzburgite, with minor lher-zolite, chromitite, and clinopyroxenite (analysis 4, table 10.2). Olivines are forsteritic (Fo_{93-84}) and pyroxenes are also Mg-rich (Himmelberg and Coleman, 1968; Evarts, 1977, 1978). The ultramafic cumulates are relatively thin and consist of dunite and wehrlite. Gabbro cumulates overlie the ultramafic cumulates. In the cumulates, olivine ranges from Fo_{91} to Fo_{76}. Plagioclase is quite calcic (An_{94-97}) where it first appears but is more sodic in the less mafic rocks (An_{85-42}). The upper part of the plutonic complex consists of hornblende gabbros and hornblende quartz diorite (figure 10.19). Plagioclase compositions in the plutonic complex are variable (An_{85-0}), and are locally quite sodic. The entire tectonite-plutonic section is intruded by dikes of various compositions (see figure 10.15). A sill complex, largely intruded into volcanic rocks, underlies a thick volcanic pile composed of basalts, andesites, and dacites that are altered to spilites, keratophyres, and quartz keratophyres, respectively.

Chemically, the parent magmas were low-K, tholeiitic types. The AFM differentiation trend is neither distinctly tholeiitic nor calc-alkaline (figure 10.20). The data straddle the calc-alkaline-tholeiite boundary and their interpretation is rendered more ambiguous by the fact that many of the rocks are metamorphosed or altered. The trace element data are also ambiguous but reveal some arc and back-arc affinities.

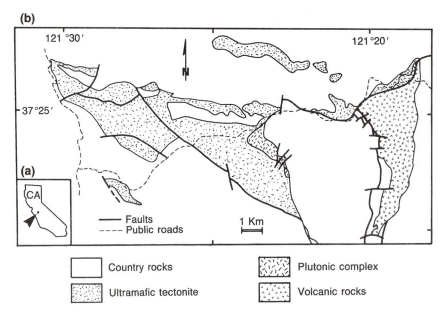

Figure 10.18   Maps showing location (a) and geology (b) of the Del Puerto Ophiolite, California.
(Modified from Evarts, 1977, 1978)

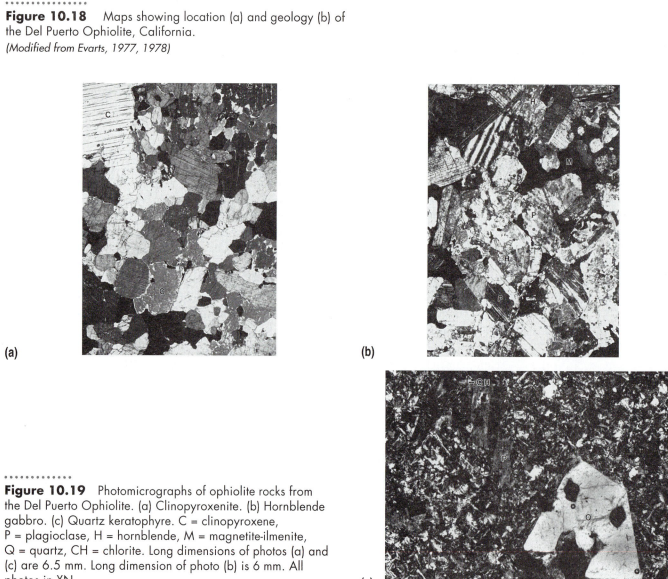

Figure 10.19   Photomicrographs of ophiolite rocks from the Del Puerto Ophiolite. (a) Clinopyroxenite. (b) Hornblende gabbro. (c) Quartz keratophyre. C = clinopyroxene, P = plagioclase, H = hornblende, M = magnetite-ilmenite, Q = quartz, CH = chlorite. Long dimensions of photos (a) and (c) are 6.5 mm. Long dimension of photo (b) is 6 mm. All photos in XN.

182

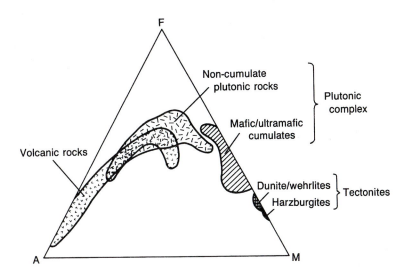

Figure 10.20 AFM diagram showing differentiation trend of Del Puerto Ophiolite, California.
(Modified from Evarts, 1977, 1978)

The Del Puerto Ophiolite formed at a site of spreading, but the exact nature of that site is unresolved. Apparently, low-K tholeiite magmas from the mantle intruded the crust-mantle boundary zone. There, the magmas fractionated and then intruded the overlying crust or erupted onto the surface to form the rocks of the ophiolite. Some of the upper lavas are arclike, but the chemistry of much of the ophiolite has spreading center affinities. Whether that spreading center was a mid-ocean ridge or a back-arc spreading center is still debated (Evarts, 1977, 1978; C. A. Hopson, Mattinson, and Pessagno, 1981; Shervais and Kimbrough, 1986).

Alaska-Type Ultramafic-Mafic Complexes

Alaska-type plutons contain some rocks similar to those of the layered complexes but differ from these in their structure. As defined here, Alaska-type ultramafic-mafic complexes (*sensu lato*)[55] are cylindrical or ellipsoidal to dikelike intrusive complexes composed of a diverse assemblage of mafic and ultramafic rock units that form crudely concentric to irregular intrusive bodies (figure 10.21). Layering is present in the intrusions, but in many cases the layering has been deformed during or soon after crystallization.

Descriptive Petrology and Chemistry

The main rock types that occur in the Alaska-type intrusions are dunite, wehrlite, clinopyroxenite, hornblende pyroxenite, gabbro, and two-pyroxene gabbro (gabbronorite). Locally, other rock types may include harzburgite, troctolite, tonalite, diorite,

and granodiorite. These rocks contain the major minerals olivine (Fo_{95-27}), clinopyroxene ($Wo_{41-49}En_{35-51}Fs_{5-22}$), and plagioclase ($An_{98-25}$). Additional minerals include orthopyroxene, hornblende, biotite, alkali feldspar, quartz, magnetite-ilmenite, chromite, and garnet. Various primary, cumulate to hypidiomorphic-granular textures characterize undeformed rocks, whereas foliated to lineated textures characterize the tectonite units that have experienced post-crystallization deformation.

Chemically, the Alaska-type ultramafic-mafic complexes seem to be derived from tholeiitic to high-Al basaltic magmas, some of which may have an alkaline character (Irvine, 1974; Snoke et al., 1982; Himmelberg, Loney, and Craig, 1986). Many Alaska-type complexes have not been shown to define well-delineated differentiation trends, but early enrichment in iron and other aspects of the petrochemistry suggest that fractional crystallization along a tholeiitic trend has occurred during crystallization of some bodies (Springer, 1980a, 1980b; Himmelberg, Loney, and Craig, 1986).

Example: The Duke Island Ultramafic Complex

The Duke Island Ultramafic Complex, located in southeastern Alaska, is one of a chain of ultramafic-mafic masses from which this type of complex derives its name.[56] As presently exposed, the complex consists of two larger and several smaller related bodies of early Cretaceous mafic and ultramafic rock that intrude larger masses of tholeiitic gabbros and granitoid rock cropping out on Duke and adjacent islands (Irvine, 1974). Hornblende gabbro, hornblende anorthosite, wehrlite, hornblende wehrlite, olivine clinopyroxenite, hornblende clinopyroxenite, magnetite hornblende clinopyroxenite,

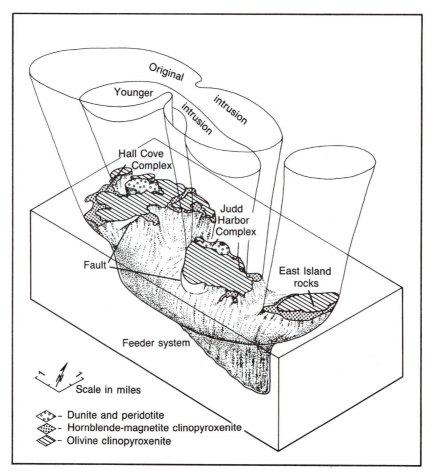

Figure 10.21 Block diagram showing the generally concentric structure and subsurface shape of the Duke Island Ultramafic Complex, Alaska.
(Redrawn from Irvine, 1974)

clinopyroxene hornblendite, and clinopyroxene magnetite hornblendite are the rock types present in this complex. Note that orthopyroxene-bearing rock types are absent. Many of these rocks display rhythmic layering, cross-bedding, and graded bedding, which together with the cumulate textures, indicate a cumulate origin.

Structurally, the ultramafic bodies are deformed, layered, crudely concentric to cylindrical or pluglike intrusions (figure 10.21). In the Duke Island Complex, an early stage of gabbro intrusion was followed later by two phases of emplacement of ultramafic rocks and anorthosite. The pyroxenites, hornblendites, anorthosites, and some peridotites represent the older phase of ultramafic intrusion and dunites and peridotites represent the younger phase.

The main minerals in the Duke Island Complex are plagioclase, olivine, clinopyroxene, hornblende, and magnetite (Irvine, 1974). Additional minerals include chromite, ilmenite, and spinel. Plagioclase ranges from An_{98} in the anorthosites to An_{50} in associated gabbroic rocks. Olivine ranges from Fo_{91} to Fo_{78}. Pyroxenes are generally quite low in iron. The hornblendes, which have lower Mg numbers than the pyroxenes, indicate that some iron enrichment has occurred during fractional crystallization of the magmas.

Chemically, the rocks represent low-K magmas with several distinguishing characteristics (Meen, Ross, and Elthon, 1991). Mg numbers of inferred parent liquids are about 0.60. In the gabbros, the REE patterns are flat to slightly enriched in LREE. In the ultramafic rocks, the REE patterns are slightly more enriched.

The Origin of Alaska-Type Complexes

The magmas from which Alaska-type complexes crystallize are variable in composition, ranging from alkaline to tholeiitic types. C. G. Murray (1972) suggested that Alaska-type bodies are derived through the crystallization of magmas fractionated from a basaltic parent. In contrast, Irvine

(1974) suggests that the Duke Island Complex evolved through fractionation of an ultramafic magma: Yet some new chemical data suggest that low-K basalt or basaltic andesite magmas may have been the parental magmas (Meen, Ross, and Elthon, 1991). In other bodies, tholeiitic parents are likely (Himmelberg, Loney, and Craig, 1986; J. S. Beard and Day, 1988; Springer, 1989). In any case, the magmas were derived from the mantle. Multiple intrusions and complicated, locally diachronous intrusive relations indicate complex histories of subsurface magma emplacement.

Differentiation of the parent magmas through fractional crystallization and crystal settling resulted in the formation of cumulates in most Alaska-type bodies. At Duke Island, the sequence dunite → olivine clinopyroxenite → magnetite clinopyroxenite → hornblende gabbro developed (Irvine, 1963, 1967, 1974). Postcumulate formation of hornblendite and local development of anorthosite followed. At Blashke Island, Alaska, the sequence is dunite → wehrlite → olivine clinopyroxenite → gabbro (Himmelberg, Loney, and Craig, 1986).

The distinctive, concentrically zoned to complex intrusive structure of Alaska-type complexes may develop during magma emplacement, late stages of crystallization, or soon after solidification. For example, the Pine Hill Complex of California (Springer, 1980a, 1980b, 1989) probably experienced early fractionation in a magma chamber, followed by later, presolidification deformation. In contrast, the Emigrant Gap Complex formed via fractional crystallization of a single gabbroic magma and was emplaced as a series of crystal mushes (O. B. James, 1971). In some complexes, diapiric emplacement of ultramafic rocks into higher levels of the crust may be caused by the intrusion of new magma from below or by tectonic or gravity-induced movement. All of these events probably occur in subvolcanic magma chambers, the petrotectonic setting inferred for Alaska-type complexes on the basis of chemistry, petrology, structure, and paleogeography.

Appinite-Type Ultramafic Rocks

The Appinite-type ultramafic rocks are uncommon. As defined here, Appinite-type ultramafic rocks include a variety of hornblende-rich rocks, with euhedral to subhedral hornblende, plus olivine or pyroxene.[57] Hornblende may be poikilitic. The rock bodies form dikelike to irregular intrusions. The category includes both ultramafic hornblendites and the cortlandites.

Most Appinite-type ultramafic rocks have not been analyzed in great detail. The rocks possess petrographic similarities to the hornblendites associated with the Alaska-type ultramafic-mafic complexes and, like the latter, are typically associated with other intrusions (W. S. Pitcher and Berger, 1972). These associated intrusive rocks range from granite to hornblendite in composition. At any particular locality, the ultramafic hornblendites usually comprise a minor percent of the rocks. Appinite-type ultramafic rocks tend to occur in small lenses, dikes, sills, or stocks.

The origins proposed for Appinite-type ultramafic rocks involve (1) fractional crystallization from a magma or (2) assimilation of country rock by an invading magma. Liquid immiscibility as a cause of separation of magmas, with subsequent crystallization, has been suggested for the origin of related rocks (Bender, Hanson, and Bence, 1982). The abundance of hornblende indicates that the parent magmas were water-rich. The proportions and distribution of the ultramafic and related rock types in many intrusions argue for an origin as a differentiate (A. Hall 1967). An origin by assimilation is favored in those cases where hornblende-rich rocks occur at the intrusive boundaries between two plutons (A. K. Wells and Bishop, 1955; Key, 1977).

The histories of most Appinite-type complexes are not well understood. Apparently, mantle anatexis produced mafic magmas that intruded the crust, where the magmas were modified by assimilation and fractional crystallization. The association of some appinitic rocks with shoshonites suggests a possible link with subduction-generated magmatism (Fowler, 1988).

Ultramafic Rocks in Alkaline Complexes

Fractionation of alkali olivine basalts may produce a variety of rocks, including some that are ultramafic. Where these magmas intrude the crust, they form alkaline complexes. In most cases, the included rocks have unusual compositions, such as nepheline + clinopyroxene to make **ijolite,** or nepheline + magnetite + clinopyroxene, forming **jacupirangite.**

The ultramafic rocks in alkaline complexes are associated with a wide variety of other alkaline rocks, including trachytes, phonolites, syenites, and nepheline syenites. These rocks are described and discussed in more detail in chapter 13.

SUMMARY

Ultramafic-mafic complexes occur in a diverse array of forms. Layered dikes and sills, as well as lopolith to funnel-shaped intrusions, are probably developed in tensional ("anorogenic") environments where basaltic magmas from the mantle intrude into various levels of the crust. Small intrusions may represent a single pulse of magma, whereas the larger ones are fed by multiple intrusions. Once the magma chambers are formed, fractionation, commonly along tholeiitic trends, yields progressively more siliceous, alkali-rich, and iron-rich layers and differentiates. Early-formed olivine, pyroxene, and oxide crystals settle or are transported to the bottom of the magma chamber, where they form layered ultramafic cumulates such as dunite, peridotites, and pyroxenites. Plagioclase begins to crystallize later and results in the production of layered gabbros. In the larger intrusions, layering is rhythmic and cryptic. Forsteritic to fayalitic olivines, as well as orthopyroxenes, characterize many of the rocks of these complexes.

Ophiolites also form as layered bodies that develop in tensional environments—the mid-ocean ridges, intra-arc spreading centers, marginal basins, and back-arc basins. Ophiolites represent mantle rock and oceanic crust emplaced on islands or continents *after* the ophiolite formed. The major layers of the ophiolite are (1) an ultramafic tectonite, (2) an ultramafic and mafic plutonic suite, (3) a sheeted dike or sill complex, and (4) a lava flow complex. These are produced by various metamorphic, fractionation, and intrusive processes. The tectonites are metamorphic mantle rocks. The plutonic cumulates are differentiates formed in magma chambers. Cumulates formed in the lower parts of ophiolite complexes consist of peridotites and gabbros that may contain both rhythmic and cryptic layering. Forsteritic olivine and both ortho- and clinopyroxenes occur in ultramafic rocks. The upper parts of ophiolites are composed of coarse- and fine-grained, mafic to silicic intrusive and extrusive rocks.

Alaska-type and Appinite-type complexes may form beneath the volcanoes of volcanic arcs. Layers and cumulate rocks are developed commonly, but the bodies in which they occur are typically stocks, dikes, or sills formed when magmas and crystal mushes are mobilized and emplaced into the crust. Clinopyroxene is the common pyroxene in Alaska-type complexes, and hornblende is present in both Alaska- and Appinite-type ultramafic rock bodies. The presence of euhedral hornblende is important in the recognition of the Appinite-type bodies.

Alkaline intrusive complexes also may contain ultramafic rocks, but these are typically feldspathoidal. Such complexes form in anorogenic settings.

The various ultramafic rock bodies and complexes may become incorporated in mountain belts during orogenesis. There, they may become alpine-type ultramafic rock complexes through metamorphism (including serpentinization) and deformation.

EXPLANATORY NOTES

1. The definition of ultramafic rock used here follows Raymond (1984c) and is similar to that in Streckeisen (1974, 1976). Other petrologists (e.g., Wyllie, 1967c) however, use different CI values, such as CI = 70–100, to define ultramafic rocks. Note that *ultrabasic rocks* are defined as those rocks with $SiO_2 < 45\%$. Thus, *ultramafic* is a petrographic term based on modal mineralogy, whereas *ultrabasic* is a term based on chemistry. Most ultramafic rocks are also ultrabasic and most ultrabasic rocks are ultramafic, but the terms are not synonymous (problem 10.1 reinforces some of the differences). Specific ultramafic rock types are defined in the classification charts in chapter 4.

2. Engel and Fisher (1969, 1975), Bonatti, Honnorez, and Ferrara (1970), Melson and Thompson (1970), N. I. Christensen (1972), Fountain et al. (1975), Hamlyn and Bonatti (1980), R. N. Anderson et al. (1982), Hebert, Bideau, and Hekinian (1983), Shibata and Thompson (1986), and chapter 7.

3. Examples of works that discuss these significant aspects of ultramafic rocks include Wager (1963), Wager and Brown (1967), E. D. Jackson (1971), Irvine (1979, 1980a), and Thy (1983) on crystallization, fractionation, and/or kinematics; and Den Tex (1969), Moores (1970, 1973, 1982), Moores and MacGregor (1972), Coleman (1971b), Upadhyay and Neale (1979); and Gass, Lippard, and Shelton (1984) on tectonic significance.

4. Moores (1973) defines five principal types of ultramafic rock, whereas Naldrett and Cabri (1976) and Wyllie (1967b) define seven basic types. Wyllie (1969a) discusses 11 basic types and Middlemost (1985) recognizes fourteen different types (including nodules and meteorites). Twelve individual types can be defined, but the *major* types of layered igneous complexes and composite intrusions represent just two major groups of igneous ultramafic rock body. Ultramafic meteorites are discussed elsewhere [see the brief reviews in Wyllie (1971a) and Middlemost (1985) and texts, such as Mason (1962) and McSween (1987)].

5. Ophiolites are commonly categorized as alpine-type ultramafic rock bodies (Den Tex, 1969; Moores, 1973), but structurally they are layered bodies of rock that owe their layering to igneous processes. Alpine-type ultramafic rock bodies are bodies found in mountain belts. As such they include a variety of ultramafic rock masses including mantle slabs, mantle diapirs, fragments of igneous bodies, and the metamorphosed (or remetamorphosed) equivalents of these masses.

6. Differentiated, layered komatiite flows, containing phaneritic rocks such as dunite, have been described by Arndt (1977a) and M. J. Donaldson et al., (1986).

7. The Triassic-Jurassic dikes and sills of eastern North America are considered to contain ultramafic rocks by those who define ultramafic rocks as those with CI > 50 or CI > 70. By the definition of *ultramafic* used here, the rocks are mostly, if not entirely, mafic or felsic. For more information on these rocks see F. Walker (1940), Hotz (1953), Hermes (1964), Ragland, Rogers, and Justus, (1968), K. R. Walker (1969a,b), Weigand and Ragland (1970), McHone (1978), Z. A. Brown et al. (1985), Froelich and Gottfried (1985), Gottfried and Froelich (1985), J. A. Philpotts (1985), J. A. Philpotts et al. (1985), Ragland and Arthur (1985), and Husch (1990).

8. The following studies report on the dikes and sills mentioned. *Antarctica*: Gunn (1962, 1963) and W. Hamilton et al. (1965). *Brook Street*: Sivell and Rankin (1983). *The Great Dyke*: H. H. Hess (1950), Worst (1958), and Wager and Brown (1968). *Jimberlana*: Campbell (1987). *Skye*: Drever and Johnston (1967), Simkin (1967), Hatch, Wells and Wells (1973), and Sutherland (1982).

9. The Palisades Sill of New York–New Jersey, a differentiated sheet intruded into Triassic sedimentary rocks during the late Triassic (K. R. Walker, 1969b; Husch, 1990), has long been cited as a differentiated sill with an olivine-rich cumulate zone near the base. It now appears that the sheet formed through a combination of multiple intrusion and fractional crystallization processes, with the olivine-rich zone representing a separate intrusion (Husch, 1990).

10. Hyndman (1985, pp. 229–31) lists several of the major bodies and provides references to descriptive works on them. However, he includes at least one ophiolite in the list, the Bay of Islands Complex, without distinguishing it as such. Among the lopolithic and lopolith-like bodies not discussed here are the La Perouse Intrusion of southeastern Alaska (Loney and Himmelberg, 1983), the Kanichee Complex of Ontario (James and Hawke, 1984), various intrusions in the Nain Anorthosite Complex (Wiebe and Wild, 1983; Wiebe, 1987, 1988, 1990), and the Kiglapait Intrusion of Labrador (S. A. Morse, 1979a, 1979b), the Fongen-Hyllingen Complex of Norway (Thy, 1983), the Freetown Intrusion of Sierra Leone (Welles and Bowles, 1981), the Insizwa Complex of Transkei in southern Africa (Lightfoot and Naldrett, 1984), and the Greenhills Ultramafic Complex of New Zealand (Mossman, 1973).

11. For more information on these bodies see the following. *Bushveld*: Daly and Molengraaf (1924), DuToit (1926), Hall (1932), E. N. Cameron (1963), Wager and Brown (1968), Willemse (1969), Schriffries and Rye (1989). *Skaergaard*: Wager and Deer (1939), Wager and Brown (1968), Maaloe (1976), McBirney and Noyes (1979), R. H. Hunter and Sparks (1987), McBirney (1989), McBirney and Naslund (1990), B. W. Stewart and DePaolo (1990). *Stillwater*: Peoples and Howland (1940), W. R. Jones, Peoples, and Howland (1960), H. H. Hess (1960), E. D. Jackson (1961, 1967), E. Cameron (1963), N. J. Page (1977), McCallum, Raedeke, and Mathez (1980), Raedeke and McCallum (1984), Czamanske and Zientek (1985), Page et al. (1985), S. J. Barnes and Naldrett (1986), Dunn (1986), D. D. Lambert and Simmons (1987, 1988), Boudreau and McCallum (1989), Page and Moring (1990), Premo et al. (1990), Zientek, Fries, and Vian (1990).

12. Information on these bodies may be found in the following and references therein. *Duluth Complex*: Grout (1918), Weiblen and Morey (1980), Chalukwa and Grant (1990), Miller and Weiblen (1990). *Muskox Intrusion*: C. Smith and Kapp (1963), Irvine and Smith (1967), Irvine (1975). *Rhum Complex*: Wager and Brown (1968), P. Henderson (1975).

13. The definition here differs from the definition adopted by the Penrose Conference on Ophiolites (Ophiolites, 1972; also see R. G. Coleman, 1977b). See note 38 for a discussion. *Note*: Mafic dikes in ophiolites are said to be "sheeted."

14. Among those who have considered at least some ophiolites to be arc, back-arc, or marginal basin rocks are Miyashiro (1973c, 1975a,b,c), Evarts (1977, 1978), Upadhyay and Neale (1979), Shervais (1982), P. T. Robinson et al. (1983), Beccaluva et al. (1984), Harper (1984), Leitch (1984), Phipps (1984), Moores et al. (1984), Shervais and Kimbrough (1985), LaGabrielle et al. (1986), and Flower and Levine (1987).

15. Descriptions and discussions of these and other ophiolites are voluminous. Of particular note here are Gass (1967), Moores and Vine (1971), R. G. Coleman (1971b, 1977b), R. G. Coleman and Irwin (1977), Malpas (1977), J. F. Casey and Karson (1981), Elthon, Casey, and Komar (1984), and Gass, Lippard, and Shelton (1984). Also see the references in note 38.

16. For more information, see the references in note 38.

17. The Duke Island Complex was described by Irvine (1963, 1967, 1974). O. B. James (1971) described the Emigrant Gap rocks. Other descriptions of Alaska-type bodies may be found in Guild and Balsley (1942), H. P. Taylor and Noble (1960), H. P. Taylor (1967), Springer (1980a, 1980b), Snoke et al. (1982), and Himmelberg, Loney, and Craig (1986).

18. The name *"appinite,"* coined by Bailey and Maufe (1916), is the kind of petrographic name that results from excessive zeal in assigning each slight variant of a more common rock its own name. Though W. S. Pitcher and Berger (1972) devote an entire chapter to these rocks and Sabine and Sutherland (1982, p. 503) defend the use of the term, I see little benefit in retaining the term. The term *Appinite-type mafic-ultramafic rock* as a descriptor for rock bodies characterized by hornblende-rich ultramafic rocks has some utility. Therefore, I use the latter term.

   Cortlandite, an olivine-bearing hornblende-rich rock, first named for rock exposed near Peekskill, New York (G. H. Williams, 1886), is similar to appinitic ultramafic rock and is here included in the Appinite-type ultramafic rock group. Similar rocks have been described elsewhere, for example, in Spain, where they occur as inclusions (see Galan and Suarez, 1989).

19. G. H. Williams (1886), Bailey and Maufe (1916), Shand (1942), W. J. French (1966), W. S. Pitcher and Berger (1972), Sabine and Sutherland (1982), Middlemost (1985), Fowler (1988), Elsdon and Todd (1989). Also see Galan and Suarez (1989) and the references in note 57.

20. These rocks are discussed in chapter 13.

21. See chapter 13 for references to the Magnet Cove rocks. The Colorado rocks are described by Armbrustmacher (1981).

22. Additional discussion of layering, cumulate textures, and their origins may be found in Wager, Brown, and Wadsworth (1960), E. D. Jackson (1961, 1967, 1971), Wager (1968), I. H. Campbell (1978), Irvine (1979, 1980a), McBirney (1985), S. A. Morse (1986), Campbell (1987), Lesher and Walker (1988), Marsh (1988a), Wiebe (1988), Weinstein, Yuen, and Olson (1988), Conrad and Naslund (1989), and D. Martin (1990).

23. See Wager, Brown, and Wadsworth (1960) and E. D. Jackson (1967). Wager, Brown, and Wadsworth (1960) provide descriptions and figures of cumulate textures, E. D. Jackson (1961) includes good photographs, and Bard (1986) provides excellent line drawings.

24. See Irvine (1979, 1980a), Kerr and Tait (1986), S. A. Morse (1986), Mathison (1987), Lesher and Walker (1988), and Marsh (1988a) for more detailed discussions of cumulus processes in magma chambers. Also see the references in note 22.

25. Refer to Wager and Deer (1939), Bottinga and Weill (1970), McBirney and Noyes (1979), Irvine (1980a), Kushiro (1980, 1982), and Murase (1982).

26. Also see J. S. Turner and Gustafson (1978), J. S. Turner (1978), and I. H. Campbell and Turner (1987). S. A. Morse (1986) questions the role of double-diffusive convection in forming cumulate textures.

27. Additional discussions may be found in Wager and Deer (1939), F. Walker (1940), H. H. Hess (1960), Wager, Brown, and Wadsworth (1960), E. D. Jackson (1961,1967), and Thy (1983). Also compare McBirney and Noyes (1979) and Irvine (1980a).

28. Rocks called *diabase* and *dolerite* in the literature are here called *gabbro*. The somewhat ambiguous term dolerite has been used widely, especially in Europe, to describe aphanitic to phaneritic mafic rocks in dikes. Similarly, the term diabase is widely used, for either fine-grained gabbro or gabbro with diabasic texture, and in some cases even where those gabbros are medium-grained. However, there is no particularly good reason to retain either of these archaic terms, neither of which is a part of contemporary igneous rock classifications. Neither term is used here.

29. The general geology of the Stillwater Complex, as related here, is based primarily on the works of Peoples and Howland (1940), H. H. Hess (1960), W. R. Jones, Peoples, and Howland (1960), N. J. Page (1979), McCallum, Raedeke, and Mathez (1980), Raedeke and McCallum (1984), N. J. Page and Zientek (1985), N. J. Page et al. (1985), S. J. Barnes and Naldrett (1986), D. D. Lambert and Simmons (1987, 1988), D. D. Lambert et al. (1989), C. E. Martin (1989), Premo et al. (1990), and observations by the author. Additional references to the complex include the various papers in Czamanske and Zientek (1985) and the papers listed in P. C. Bennett (1985), Dunn (1986), Boudreau and McCallum (1989), Mathez et al. (1989), Czamanske and Bohlen (1990), Loferski, Lipin, and Cooper (1990), N. J. Page and Moring (1990), N. J. Page and Zientek (1990), Zientek, Fries, and Vian (1990).

30. Earlier, most workers, including H. H. Hess (1960) and N. J. Page (1979), included these rocks in the basal unit of the Stillwater Complex, as do D. D. Lambert and Simmons (1987). For reasons not explained, Zientek, Czamanske, and Irvine (1985) have chosen to exclude these rocks from the Stillwater Complex, though it is clear that they are related to it and represent likely examples of Stillwater source magmas (Helz, 1985).

31. In addition to descriptions in the papers cited in note 29, the Ultramafic Series is discussed by E. D. Jackson (1961) and D. D. Lambert and Simmons (1987).

32. Various aspects of the petrology of the Banded Series are discussed by H. H. Hess (1960), Wager and Brown (1968), McCallum, Raedeke, and Mathez (1980), Raedeke et al. (1985) and N. J. Page et al. (1985). H. H. Hess (1960) first detailed general mineralogical changes in the Banded Series, which were later analyzed in greater detail by others.

33. Compositions of interstitial plagioclases in the ultramafic rocks of the Ultramafic Series have not been analyzed in detail.

34. Hyndman (1985, pp. 227, 237–38) provides lists of explanations and S. A. Morse (1980, ch. 13) presents an overview. Also see D. D. Lambert and Simmons (1988) for data relating to the origin of cyclic units in the Stillwater Complex.

35. The following sources provide the data and most of the interpretations presented in this paragraph on the Stillwater Complex. DePaolo and Wasserburg (1979), Irvine (1980a), McCallum, Raedeke, and Mathez (1980), Raedeke and McCallum (1984), Helz (1985), S. J. Barnes and Naldrett (1986), Dunn (1986), D. D. Lambert and Simmons (1987, 1988), D. D. Lambert et al. (1989), C. E. Martin (1989), Czamanske and Bohlen (1990), Premo et al. (1990).

36. See Brongiart (1813, 1827, in R. G. Coleman, 1977b, and Moores, 1982).

37. *Spilites* are sodic, metamorphosed basalts developed through low pressures and temperatures of metamorphism (Yoder, 1967; Battey, 1974). For more information and contrasting opinions, especially the view that spilites are igneous rocks, see papers in Amstutz (1974).

38. This definition differs from that favored by the majority of the voting participants at a 1972 Penrose Conference on Ophiolites sponsored by the Geological Society of America (Ophiolites, 1972). The Penrose group specifically excluded chromitites, felsic differentiates, and overlying sedimentary rocks from their definition. Inasmuch as the felsic differentiates and chromitites are structurally, texturally, and chemically associated with the ultramafic rocks, I include them as parts of the ophiolite (*sensu stricto*). C. A. Hopson, Mattinson, and Pessagno (1981) and Shervais and Kimbrough (1986), among others, also include felsic differentiates. Moores (1982) provides an alternative to the Penrose definition that lies at an opposite extreme, in that it includes crystalline basement, sedimentary rocks, melange, and metamorphic complexes *below* the ultramafic tectonite

and a variety of sedimentary rocks *above* the upper lavas, even those deposited above overlying unconformities, as part of the "ophiolite association." For descriptions and/or discussions of particular ophiolites, see the following selected papers and the references therein.

Bay of Islands, Newfoundland: C. M. Smith (1958), W. R. Church and Stevens (1971), Harold Williams and Malpas (1972), Malpas (1977), J. F. Casey and Karson (1981), Elthon, Casey, and Komar (1984), J. F. Casey et al. (1985), Komor et al. (1987), S. E. Smith and Elthon (1988), Komar and Elthon (1990).

Coast Range Ophiolite, California: Taliaferro (1943), E. H. Bailey, Irwin, and Jones (1964), M. E. Maddock (1964), Bezore (1969), E. H. Bailey, Blake, and Jones (1970), Loney, Himmelberg, and Coleman (1971), B. M. Page (1972, 1981), E. H. Bailey and Blake (1974), J. Pike (1974), Evarts (1977, 1978), C. A. Hopson and Frano (1977), Menzies et al. (1977), C. A. Hopson et al. (1981), Shervais and Kimbrough (1985), LaGabrielle et al. (1986).

Troodos Ophiolite, Cyprus: Gass (1968), Moores and Vine (1971), Gass and Smewing (1973), Miyashiro (1973c, 1975a,b,c), Gass et al. (1975), Hynes (1975), Moores (1975), Smewing, Simonian, and Gass (1975), Bortolotti et al. (1976), Greenbaum (1977), Malpas and Langdon (1984), Moores et al. (1984), Thy (1984, 1987), Thy, Brooks, and Walsh (1985), Flower and Levine (1987), J. M. Hall, Walls, and Yang (1989), Allerton and Vine (1991).

Other ophiolites: Northern Cordillera—E. H. Brown (1977), Monger (1977), Vance et al. (1980), R. B. Miller (1985). Canyon Mountain, Oregon: Ave Lallement (1976), Thayer (1977), Himmelberg and Loney (1980). Kings-Kaweah Ophiolite belt, Sierra Nevada, California—Saleeby (1977). Nicoya Peninsula, Costa Rica—Kuijpers (1979, 1980); Berrange and Thorpe (1988). Mings Bight, Newfoundland—Norman and Strong (1975), Kidd, Dewey, and Bird (1978). Southern Quebec—Laurent (1975, 1977), Hebert and Laurent (1989). Maine—Coish and Rogers (1987). Northwestern Europe—W. R. Church and Gayer (1973), Sturt, Roberts, and Furnes (1984). Mediterranean region—Aubouin et al. (1964), Bortolotti, Dal Piaz, and Passerini (1969), Moores (1969), Gianelli, Passerini, and Sguazzoni (1972), J. Zimmerman (1972), E. D. Jackson, Shaw, and Barger (1975), Juteau et al. (1977), Pamic (1975a,b, 1977a,b), Ohnenstetter and Ohnenstetter (1976), Bortolotti et al. (1976), Rocci (1976), Beccaluva et al. (1977), Capedri et al. (1980), Beccaluva et al. (1984), Peretti and Koppel (1986), Stating and van Wamel (1989). Semail Ophiolite, Oman—C. A. Hopson et al. (1981), Pallister and Gregory (1983), Browning (1984), Nicolas et al. (1988). Papua, New Guinea—Davies (1968, 1971), Rodgers (1975), Finlayson et al. (1976), H. L. Davies and Jaques (1984). Tasmania—Varne and Brown (1978). New Zealand—Sivell and Waterhouse (1984). Taiwan—Suppe, Liou, and Ernst (1981). Japan—Ozawa (1983), Ishiwatari (1985).

39. E. H. Bailey et al. (1970), Moores and Vine (1971), Gass and Smewing (1973), R. G. Coleman (1977b), J. F. Casey et al. (1981), Ozawa (1983).

40. Also see Loney, Himmelburg, and Coleman (1971), E. D. Jackson and Thayer (1972), R. G. Coleman (1977b), Nicolas and Le Pichon (1980), J. F. Casey et al. (1981), Boudier and Nicolas (1985), Nicolas et al. (1988).

41. In addition, see Medaris (1972) and Nicolas et al. (1988), and consider I. D. MacGregor (1974) and Stroh (1976). Additional data on isolated bodies of alpine ultramafic rock may be found in chapter 31.

42. Descriptions of the second layer of ophiolites are provided by Pallister and Hopson (1981), Smewing (1981), and others listed in note 43.

43. Also see E. D. Jackson (1971), Pamic (1975), Greenbaum (1977), Malpas (1977), Kidd, Dewey, and Bird (1978), Ozawa (1983), Ishiwatari (1985), Komor et al. (1985), Benn and Allard (1989), Hebert and Laurent (1989), and Komor and Elthon (1990).

44. Moores and Vine (1971) and Harold Williams and Malpas (1972) describe sheeted dike complexes with mafic dikes. For examples of dikes other than basalt and gabbro, refer to Hopson and Frano (1977).

45. Both spilites and greenstones are metamorphosed basalts with similar mineralogies. See note 37 for references to spilites and the metamorphic section of this book for details about low-grade metamorphism of mafic rocks, which produces greenstones.

46. Also see J. C. Maxwell and Azzaroli (1962) and papers referred to in Aubouin (1965) for additional information on these hypotheses.

47. Taliaferro (1943) held this view for ultramafic rocks associated with the Franciscan Complex of California, some of which are parts of the Coast Range Ophiolite. Additional discussion of this view is provided by Aubouin (1965) and Moores and Raymond (1972).

48. See Bowen and Tuttle (1949), H. L. Davies (1968, 1971), and Quick (1981a) for evidence and arguments in support of these and related views. E. H. Bailey, Irwin, and Jones (1964) summarize evidence for a cold emplacement hypothesis for ultramafic rocks. Some slabs of lithosphere rooted in the mantle provide clear evidence that thrusting, or obduction, of lithosphere (R. G. Coleman, 1971b; Collot et al. 1987) can provide for the emplacement of hot or cold masses of ultramafic rock and ophiolites into high levels of the crust. In such cases, however, it is important to distinguish the sites of origin (e.g., the oceanic crust and mantle or a back-arc basin) from the site of later emplacement (an orogenic belt).

49. Also, see Gass (1967, 1968), Engel and Fisher (1969, 1975), Thayer (1969), Moores and Vine (1969, 1971), Moores (1970), Coleman (1971b, 1977), Dewey and Bird (1971), Gass and Smewing (1973, 1981), Ferrara et al. (1976), Sinton (1979) and other works in the Reports of the Deep-Sea Drilling Project (DSDP) and the International Project for Ocean Drilling (IPOD), Ogawa et al. (1985), Murton (1986), and Bloomer and Fisher (1987). Beccaluva et al. (1989) and Loucks (1990) discuss the use of pyroxene chemistry for distinguishing ophiolites.

50. H. L. Davies (1968,1971); J. Ewing (1969); Shor and Raitt (1969); Vogt, Schneider, and Johnson (1969); R. G. Coleman (1971b, 1977b); Moores and Vine (1971); Gass and Smewing (1973); Christensen and Salisbury (1975); Fountain et al. (1975); Bonatti and Honnorez (1976); Juteau et al. (1977); Salisbury and Christensen (1978); Pallister and Hopson (1981); Smewing (1981); Anderson et al. (1982); Hebert et al.(1983). Also see Perretti and Koppel (1986).

51. Hill (1957), Raitt (1963), Shor and Raitt (1969), R. E. Coleman (1971b), Ludwig et al. (1971), Moores and Vine (1972), Gass and Smewing (1973), Christensen and Salisbury (1975). Becker et al. (1989) describe the most inclusive section of drilled oceanic crust (through 1988), a section that consists of pillow lavas and their flows interlayered with massive basalts, pyroclastic rocks, and breccias overlying a sheeted dike complex.

52. The controversy produced by Miyashiro's (1973c) paper, in which he suggested that the then preeminent ophiolite, the Troodos Ophiolite, which was taken to be representative of oceanic crust and mantle, is the root of an island arc was both entertaining and productive (Hynes, 1975; Gass et al., 1975; Miyashiro 1975a,b; Moores, 1975). Miyashiro's paper caused many petrologists to take a close look at available data and to carefully seek data that would be definitive in resolving the question raised: Does the Troodos Ophiolite represent ocean crust formed at a mid-ocean ridge or volcanic roots of a primitive arc? Smewing, Simonian, and Gass (1975) decided that the data support an origin in a small marginal ocean basin, whereas Pearce (1975) favored a back-arc or inter-arc-basin origin for the Lower Pillow Lavas and an arc origin for the Upper Pillow Lavas of the ophiolite. An analysis of Ti-V data by Shervais (1982) also supports an arc origin, as do the geochemical analyses of Rautenschlein et al. (1985) and Thy, Brooks, and Walsh (1985). Moores et al. (1984) attempted to harmonize the structural and geochemical data by suggesting an origin in an *intra*-arc spreading zone. Flower and Levine (1987) agree. Similarly, Thy (1987) suggests a complex history involving both spreading and non-spreading stages of development.

53. For example, compare Smewing, Simonian, and Gass (1975) and Rautenschlein et al. (1985) or Menzies et al. (1977), Menzies, Blanchard, and Jacobs (1977), and Shervais and Kimbrough (1985). Also see LaGabrielle et al. (1986) and Kay and Senechal (1976).

54. The Del Puerto Ophiolite has been studied and discussed by a number of workers including Hawkes et al. (1942), Taliaferro (1943), M. E. Maddock (1964), Himmelberg and Coleman (1968), E. H. Bailey and Blake (1974), Raymond (1973b), Evarts (1977, 1978), and C. A. Hopson, Mattinson, and Pessagno (1981). The description here is based on these works and unpublished work by the author.

55. These complexes have been called Alaska-type, concentric complexes, zoned complexes, and peridotitic to dioritic intrusive complexes. Following Irvine (1974, p. 166), I use the name Alaska-type in a broader sense than originally defined (e.g., here, orthopyroxene-bearing rocks are included and the essential elements are the mafic to ultramafic composition and the overall structure of the complex). Description of Alaska-type complexes here is based on Guild and Balsley (1941), Noble and Taylor (1960), H. P. Taylor and Noble (1960), Irvine (1963, 1967, 1974), H. P. Taylor (1967), O. B. James (1971), E. D. Jackson and Thayer (1972), Springer (1980a,b, 1989), Snoke et al. (1982), J. S. Beard and Day (1988), Himmelberg, Loney, and Craig (1986), and Meen, Ross, and Elthon (1991).

56. Description of the Duke Island Complex is based on the extensive work of Irvine (1963, 1967, 1974). Also see Meen, Ross, and Elthon (1991).

57. Appinite-type rocks, as defined here, include both the original appinitic rocks of Bailey and Maufe (1916) and the "cortlandites." See Bailey and Maufe (1916) and W. S. Pitcher and Berger (1972) for descriptions of Appinite-type ultramafic rocks. Also see Iyengar, Pitcher, and Read (1954), A. K. Wells and Bishop (1955), W. J. French (1966), A. Hall (1967), Key (1977), Sabine and Sutherland (1982, pp. 488, 503), Middlemost (1985, p. 194), Fowler (1988), and Elsdon and Todd (1989). Cortlandites are discussed by G. H. Williams (1886), Ratcliffe et al. (1982), and Galan and Suarez (1989). Related rocks are discussed by Bender, Hanson, and Bence (1982).

PROBLEMS

10.1. Classify each of the rocks in tables 10.2 and 10.3 as ultrabasic, basic (see table 4.2), intermediate, or acid, and as ultramafic, mafic (CI = 89–33.3) or felsic (CI = 33–0).

10.2. Carbonatites are igneous rocks composed primarily of a carbonate mineral (e.g., calcite or dolomite). Examine analysis 1 in table 13.1. If this rock were composed of 95% calcite plus minor amounts of pyroxene, apatite, and magnetite, would it be ultrabasic, ultramafic, both, or neither? Explain.

10.3. Calculate the Mg numbers for the odd-numbered samples in tables 10.2 and 10.3.

Granodiorites and Related Rocks

INTRODUCTION

No one has seen them form, but great masses of plutonic rock hundreds of miles long and tens of mile wide—the batholiths—constitute the cores of many Phanerozoic mountain ranges. These batholiths, as well as numerous stocks and dikes, are typically composed of siliceous rocks that, in a general way, are plutonic equivalents of andesites, dacites, rhyolites, and related rocks. Traditionally, such rocks have been called "granitic."[1] Although such a broad use of the term granitic is consistent with its Latin root *granum*, meaning "grain" (granitic rocks have granular textures), it is unfortunate because confusion has resulted from different uses of the term.

The terms granitic and granite must be distinguished. **Granite** as used in this text, refers to phaneritic, crystalline rocks composed of essential quartz and K-rich alkali feldspar. The definition requires that the alkali feldspar comprise two-thirds or more of the total feldspar. Additional minerals typically include sodic plagioclase, micas, and amphiboles. This strict definition of granite is similar to the definitions of Travis (1955) and Peterson (1961, 1965), both widely used for many years in the United States. It is also similar to the definition of Hatch, Wells, and Wells (1973) used by some English petrologists.

Some Europeans have used the term granite for rocks with a wider range of feldspar ratios and silica contents than is included in the definition above.[2] This European definition formed the basis for the IUGS definition (Streckeisen, 1967, 1976, et al., 1973). That definition includes rocks containing quartz to quartz + alkali feldspar + plagioclase ratios of 0.2 to 0.6, alkali feldspar to plagioclase ratios between 9:1 and 35:65, and a color index of less than 90—a range of composition that here includes both granite and quartz monzonite (see figure 4.2). This internationally developed definition is widely accepted.

Alternative definitions of granite of a mineralogical and chemical nature have been proposed. For example, O'Connor (1965) and F. Barker (1979b) proposed normative Ab-An-Or criteria as a basis for the definition.[3]

To emphasize the point that the term *granite* means different things to different petrologists, figure 11.1 shows the various modal limits for rocks called granite in the IUGS, RSO, Nockolds, Knox, and Chinner (1978), and Johannsen (1939) classifications. Notice the variation in quartz content and the wide range of plagioclase percentages allowed in granites by the various classifications. The fact that so many definitions of the term granite exist compounds the problems of communication among those who view granite petrogenesis in different ways.

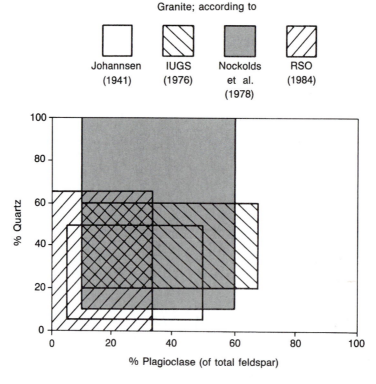

Figure 11.1 Quartz-feldspar plot showing compositional limits of granite in the classifications of plutonic igneous rocks by Johannsen (1939b), IUGS (Streckeisen, 1976), Nockolds, Knox, and Chinner (1978), and Raymond (1984c).

The term granitic has been used, both by geologists and nongeologists in a broad sense (*sensu lato*) to refer to *any* granular, phaneritic feldspar- or quartz + feldspar-bearing rock. The term has also been used by geologists to refer specifically to granite (*sensu stricto*). To avoid the confusion that results from such multiple uses of the term, in this book *granite* and *granitic* are used only in reference to granites *sensu stricto*. The word **granitoid** is used for granular, plutonic, quartz-feldspar rocks of unspecified or a wide range of compositions, in place of granite *sensu lato*.[4]

Many large granitoid batholiths were created by multiple intrusions of magma. In some cases, the various magmas are closely related (consanguineous), having been derived from the same parent through processes of magma modification. In other cases, the magmas are linked only spatially or in a broadly genetic way, having been derived by similar processes occurring in or below the same region of the crust. In either case, the rocks that comprise the batholiths range from granitic to gabbroic compositions. It is these rocks, especially granodioritic to quartz-monzonitic types, that are the subject of this chapter. They comprise many of the Paleozoic to Cenozoic batholiths of the Phanerozoic mountain belts, and also occur in Precambrian terranes. Chapter 12 deals in more detail with granites *sensu stricto*.

CHEMISTRY, MINERALOGY, AND TEXTURES OF GRANITOID ROCKS

Granitoid rocks exhibit a wide range of compositions. Their mineral compositions vary from quartz-rich granites to the (quartz-deficient) olivine gabbros. Similarly, their chemistries are varied, reflecting those various mineralogies. Textures, too, are diverse.

Chemistry

The diverse chemistry of granitoid rocks is reflected by the selected analyses presented in table 11.1. Silica may range from less than 50% in diorites and gabbros to over 77% in granites, especially those of pegmatitic and aplitic texture. As silica increases, alumina, total iron, magnesia, and lime generally decrease. In contrast, total alkalis increase with increasing silica, largely due to an increase in potash. Mg numbers range from values greater than 0.80 in olivine-rich gabbros to those less than 0.10 in some granites.

The range of chemistry exhibited by rocks was subdivided by Shand (1949, pp. 225–29) on the basis of silica and alumina saturation. In terms of silica, rocks are considered to be oversaturated (if they contain quartz), saturated (with

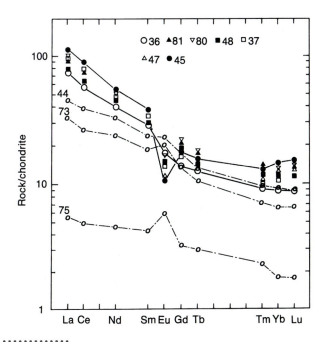

Figure 11.2 Typical REE diagram ("spider diagram") for a granitoid rock suite. Rocks represented are gabbros (samples 44, 73, and 75), diorites (samples 36, 48, 80, 81), granodiorites (samples 37 and 47), and granite (sample 45) of the Linga Group of Peru.
From L. LeBel, et al., "A High-K Mantle derived plutonic suite from "Linga" near Arequipa (Peru)" in Journal of Petrology, *26:124-48, 1985. Copyright © 1985 Oxford University Press, Oxford England. Reprinted by permission.*

neither quartz nor feldspathoids), or undersaturated (with olivine or a feldspathoid). As a method of differentiating various alumina contents, Shand subdivided *molecular proportions* of alumina to alkalis plus lime into four groups (see chapter 3). Three of these groups provide an adequate basis for subdividing granitoid rocks into the following major categories:

Peraluminous rocks, with $Al_2O_3 > CaO + Na_2O + K_2O$
Metaluminous rocks, with $CaO + Na_2O + K_2O > Al_2O_3 > Na_2O + K_2O$
Peralkaline rocks, with $Al_2O_3 < Na_2O + K_2O$

Each analysis in table 11.1 is assigned to one of these groups. Note that metaluminous types are most common in the table, a reflection of rock-type abundances in nature.

Trace elements and isotope ratios, like the major elements, show a range of values.[5] Initial strontium isotopic ratios ($^{87}Sr/^{86}Sr)_i$ range from less than 0.702 to more than 0.760. Similarly, Nd, Pb, O and other isotopes are variable.[6] εNd values are typically in the range 0 to −20 and occur in εND/εSr diagrams on either a mixing curve between MORB and old crystalline crust or in a region characteristic of such crust (see figure 6.3). The variations in these isotopes reflect the involvement of different source materials and processes in magma genesis.

REE patterns for granitoid rocks show some consistency. The two most notable features are a generally concave up LREE versus HREE enrichment curve and, commonly, a distinct europium (Eu) anomaly (figure 11.2).[7] Positive Eu anomalies are present in some analyses of plagioclase-rich rocks, such as gabbro and quartz diorite (tonalite), whereas negative Eu anomalies are characteristic of quartz monzonites and granites.

Minerals and Textures

Quartz and K-rich alkali feldspar are the only two essential minerals that need be present in a rock if it is called granite (according to the definition adopted in this book). Commonly, however, granites also contain discrete grains of sodic plagioclase or consist of quartz plus perthitic alkali feldspar (figure 11.3). The intermediate rocks—quartz monzonite, granodiorite, and some diorites—contain both alkali feldspar and sodic plagioclase, whereas intermediate to calcic plagioclase is typically the only feldspar in quartz diorites, some diorites, and gabbros.[8]

Characterizing accessory minerals are a function of rock chemistry, particularly alumina saturation, and the total abundances of Na_2O, K_2O, and CaO. Peraluminous granites usually contain muscovite or muscovite and biotite, but may also contain garnet, aluminum silicates (e.g., andalusite, sillimanite), topaz, cordierite, or corundum. These accessory minerals also occur in intermediate peraluminous granitoids. Peralkaline granitoids characteristically exhibit sodic pyroxenes (e.g., aegirine) and sodic amphiboles (e.g., riebeckite, arfvedsonite). Hypersthene occurs in the charnokites (the hypersthene granites), and fayalite locally appears as a mafic phase in iron-enriched granitoid rocks. In the metaluminous diorites, granodiorites, quartz diorites, and gabbros, pyroxenes, biotite, and hornblende are typical. Augite is the most common pyroxene in these granitoid rocks, especially in the more calcic rocks, but diopsidic pyroxenes may occur in the alkaline rocks. In gabbros, clinopyroxene is common and is considered to be essential by some petrologists. Also in the gabbros, orthopyroxenes may occur, alone or with augite, and Mg-rich olivine is a common phase.

Minor accessory minerals include apatite, magnetite, ilmenite, hematite, pyrite, and sphene, as well as tourmaline, zircon, rutile, garnet, and fluorite. In addition, where they form less than 5% of the rock, the minerals listed as characterizing accessories become minor accessories. Alteration minerals include such minerals as the clays, calcite, epidote, white mica, chlorite, and hematite. Where pyroxenes and olivines are present, serpentine forms as an alteration product.

The textures of the more siliceous granitoid rocks are typically hypidiomorphic-granular (figure 11.3c). Yet, the quartz monzonites commonly exhibit porphyritic textures, with large poikilitic alkali feldspars as the phenocrysts (figure 11.3b). Seriate and allotriomorphic-granular (aplitic) textures also occur. In the more mafic rocks, especially gabbros, diabasic textures are common. Trachytoidal, subophitic, ophitic, and various cumulate textures are noteworthy in

Table 11.1 Major Element Analyses and CIPW Norms of Some Granitoid Rocks

	1	2	3	4	5	6	7	8
SiO_2	50.08[a]	60.43	63.47	67.40	72.8	73.93	75.07	76.21
TiO_2	1.09	0.72	0.72	0.53	0.21	0.18	0.28	0.06
Al_2O_3	18.79	16.33	15.81	15.80	14.6	12.09	12.85	12.89
Fe_2O_3	5.51	1.39	2.14	1.77	0.43	2.91	0.24	0.29
FeO	3.98	4.30	3.03	1.36	1.7	1.55	1.31	0.84
MnO	0.12	0.08	0.09	0.06	0.08	0.00	0.03	0.07
MgO	5.08	4.06	2.28	1.02	0.39	0.08	0.31	0.05
CaO	8.17	4.86	4.72	3.45	1.5	0.31	1.11	0.69
Na_2O	3.86	3.36	3.32	4.08	3.5	4.66	3.08	3.77
K_2O	0.81	2.32	3.22	3.63	4.0	4.63	4.81	4.72
P_2O_5	0.28	0.39	0.17	0.20	0.10	0.00	0.06	0.01
Other	2.58	1.54	1.32	0.76	0.81	0.71	0.90	0.44
Total	100.35	99.78	100.29	100.06	100	100.95	100.05	100.04

Sources:

1. Diorite, Ingonish, Nova Scotia, sample 46 (Wiebe, 1974).

2. Hornblende-biotite tonalite (quartz diorite), near Ihode, Uusikaupunki, Finland, sample UK-10. Analyst: Elaine L. Brandt (Arth, Peterman, and Friedman, 1978).

3. Hornblende-biotite granodiorite, Half-Dome Granodiorite, Yosemite National Park, California, sample 31 (Bateman and Chappell, 1979).

4. Granodiorite, Cathedral Peak Granodiorite, Yosemite National Park, California, sample 45 (Bateman and Chappell, 1979).

5. Quartz monzonite, Dinkey Dome, Sierra Nevada, California, sample HLd-133. Analysts: H. Smith, J. L. Glenn, J. Kelsey, G. Chloe, and P. L. D. Elmore (Bateman and Wones, 1972).

6. Granite, Quincy Granite, Quincy, Massachusetts, analysis 3. Analyst: H. S. Washington (C. H. Warren, 1913).

7. Granite, Tilmunda Granite, average of four samples considered to be representative of an S-type minimum melt (A. J. R. White and Chappell, 1977). Norm by L. A. Raymond.

8. Adamellite (quartz monzonite), Looanga Adamellite, sample M105, considered to be an I-type minimum melt (A. J. R. White and Chappell, 1977). Norm by L. A. Raymond. *Note:* It is very common for rocks like this that are chemically equivalent to rhyolite to be a modal and normative quartz monzonite.

[a] Values in weight percent.

[b] nr^2= Not reported; nd = not determined.

Al Sat. = Alumina saturation, according to Shand's (1949) criteria: M = metaluminous, PL = peraluminous, PK = peralkaline.

some gabbros. Pegmatitic textures occur in the entire range of granitoid rocks, but are most common in siliceous rocks, especially granites.[9] Graphic-dendritic textures are present in some granitoid pegmatites (figure 11.3a).

OCCURRENCES OF GRANITOID ROCKS, THEIR TECTONIC SIGNIFICANCE, AND GRANITOID ROCK TYPING

Granitoid rocks are widely distributed. They occur on all continents in terranes of the Precambrian shield[10] and in Paleozoic, Mesozoic, and Cenozoic orogenic belts.[11] Earlier

in this century, as well as in the last, petrogenesis of granitoid rocks in Precambrian terranes was attributed to processes quite different from those thought to have operated in Phanerozoic orogens. More recent work suggests that Precambrian tectonic processes were relatively similar to those of the Phanerozoic Eon, with subduction zones and arcs being important tectonic elements (K. Burke et al., 1976; Tarney et al., 1976). Consequently, granitoid rocks in both types of terrane may be viewed in relation to plate tectonic processes, with major granitoid batholiths being developed in the roots of volcanic arcs (Dickinson, 1970b; Dewey and Bird, 1970).

Anorogenic granitoid rocks are found in continental zones that are tectonically passive or that are or were tectonically active, but are not orogenic. Examples include granitoid

	1	2	3	4	5	6	7	8
CIPW Norms								
q	—	14.67	18.24	21.58	32.4	28.83	35.61	34.43
or	4.79	13.74	19.03	21.45	23.6	27.39	28.45	27.89
ab	32.66	28.49	28.09	34.52	29.6	36.40	26.07	31.89
an	31.55	20.36	18.73	14.08	6.8	—	4.26	2.59
ac	—	—	—	—	—	2.68	—	—
en	—	10.13	—	—	1.0	—	—	—
fs	—	5.60	—	—	2.6	—	—	—
di	5.84	—	3.18	1.70	—	1.33	—	—
hy	10.63	—	6.92	2.10	—	0.38	2.57	1.45
fo	5.54	—	—	—	—	—	—	—
fa								
c	—	0.85	—	—	2.0	—	1.01	0.63
mt	3.76	2.02	3.10	2.57	0.6	2.87	0.35	0.42
il	2.07	1.37	1.37	1.01	0.4	0.39	0.53	0.12
ap	0.65	0.93	0.40	0.47	0.2	—	0.13	0.03
cc	—	0.09	nr[b]	nr	nr	nd	0.32	0.27
fr	—	0.28	nr	—	—	—	—	—
pr	—	0.11	—	—	—	—	—	—
Other	—	—	1.06	0.49	—	0.41	0.76	0.32
Total	97.49	98.64	100.12	99.88	99.4	100.97	100.06	100.04
Al Sat.	M	M	M	M	PL	PK	PL	PL

rocks of the Basin and Range Province of the western United States, the Mesozoic Magmatic Province of North America in Central New England (the rocks of the White Mountain Magma Series); the Jurassic Northern Nigerian Magmatic Province of west Africa, and the Upper Devonian Lachlan Fold Belt of Southeastern Australia.[12]

Oceanic areas also contain granitoid rocks, though granitoid rocks are far less abundant in oceanic regions than on the continents. Most granitoid rocks inferred to have formed in oceanic areas are ferromagnesian mineral-poor albite-quartz rocks that have been called *plagiogranite* (R. G. Coleman and Peterman, 1975). Potassium-feldspar-bearing granitoid rocks are rare constituents of spreading center and oceanic arc sequences (e.g., C. G. Engel and Fisher, 1975; S. M. Kay, Brueckner, and Rubenstone 1983).

Given that granitoid rocks form in both orogenic (arc, mountain belt) and anorogenic (spreading center, rift, hot spot) environments, several petrologists have attempted to correlate rock character (e.g., chemistry, mineralogy) with tectonic setting. Attempts have also been made to delineate controls on granite petrogenesis by tectonic position and relative age. Caution must be exercised in attempting such correlations because tectonic environment alone does not control the chemistry.[13]

Numerous subdivisions or classifications of igneous rocks based on such correlations have been developed. A. Harker (1909, p. 90ff.) first divided igneous rocks into Atlantic and Pacific types and these types subsequently assumed a tectonic significance. More recently, Hyndman (1972) divided rocks into (1) preorogenic and early orogenic suites,

Figure 11.3 Granitoid rocks. (a) Photograph of dendritic "snowflake" texture in granite pegmatite, McKinney Mine, Spruce Pine District, North Carolina. Photomicrographs: (b) Porphyritic quartz monzonite, Chaffee County, Colorado. (XN). (c) Sentinal Granodiorite, Yosemite National Park, Sierra Nevada Batholith, California. (XN). (d) Hornblende diorite, Woodleaf, North Carolina. (XN). A = alkali feldspar, B = biotite, H = hornblende, P = plagioclase feldspar, Q = quartz, E = epidote, W = White mica. Long dimension of photomicrographs is 6.5 mm.

(2) synorogenic suites, and (3) postorogenic suites.[14] Each suite consisted of "rocks that characteristically occur together and presumably evolve together." Thus, the rocks were subdivided on the basis of both *observable* criteria (field occurrence) and *inferred* origin and parentage. Avoiding inference, J. J. W. Rogers and Greenberg (1990) use tectonic setting and associated rock types to divide granites into (1) late orogenic, (2) postorogenic, (3) anorthosite/rapakivi, and (4) ring complex types.

Subdivisions based primarily on observed mineralogy and chemistry have also been proposed, but many such subdivisions link observations to inferred origins. For example, Pupin (1980) related zircon morphology (shape) to three main granitoid categories: orogenic crustal-source granitoids, orogenic crust + mantle-source granitoids, and anorogenic mantle-source granitoids. Thus, he used an observable mineralogical criterion to attempt to distinguish and classify granites formed in different structural-tectonic positions.

In an attempt to link chemistry and tectonics, G. C. Brown (1982) and G. C. Brown, Thorpe, and Webb (1984) recognized two fundamental classes of granitoid rock—(1) arc and (2) back arc and anorogenic—on the basis of both major and trace element chemistry.[15] In contrast, Pearce, Harris, and Tindle (1984) subdivided granitoid rocks into four categories—ocean ridge granites, volcanic arc granites, within-plate granites, and collision granites—each clearly associated with a tectonic site. The first and third types are basically anorogenic, whereas the second and fourth are orogenic. Pearce, Harris, and Tindle (1984) suggested that each group exhibits distinctive trace element characteristics. Hussein, Ali, and El Ramly (1982) proposed a similar subdivision based on petrologic, geochemical, age, and field relations. In another such classification, Maniar and Piccoli (1989) define seven types of granitoid rocks, four orogenic and three anorogenic, on the basis of major element chemistry.

Rather than link granitoid rocks to tectonic setting, Didier, Duthon, and Lameyre (1982) divided them on the basis of the inferred site of magma genesis into C-types (crustal) and M-types (mantle or mixed mantle + crust). White (1979) had earlier suggested that M-types exist. Drawing on the seminal work of Chappell and White (1974), the C-types are subdivided by Didier et al. (1986) into those derived from melting of sedimentary sources (CS-types) and those formed by anatexis of igneous materials (CI-types). This classification is based on chemical and petrographic data, and especially on the nature of **enclaves,** inclusions within the granitoid mass (i.e. xenoliths, xenocrysts, and autoliths).

The approach to granitoid rock typing employed by Didier, Duthon, and Lameyre (1982) is essentially genetic, though it is based on observable mineralogical, petrological,

and geochemical data. This kind of genetic approach was initiated by White and Chappell (1977), who distinguished two granitoid types in eastern Australia on the basis of chemical, mineralogical, and field data. They interpreted these types to be derived from two different sources, one igneous (I-type) and one sedimentary (S-type). Thus, they define S-type granite as "one in which geochemical and isotopic characteristics are primarily inherited through partial melting of a crustal sedimentary (or metasedimentary) source" (A. J. R. White et al., 1986). Presumably I-types have an equivalent definition. Both types are considered to be derived from crustal melting (White, 1979; A. J. R. White and Chappell, 1983), a point also emphasized by Didier, Duthon, and Lameyre (1982). To this array of granitoid types, Loiselle and Wones (1979) added the (anorogenic) A-type, Australian examples of which were described by Collins et al. (1982). As noted by Bowden et al. (1984), O-types (orogenic) should be the contrasting type to A-types if the subdivision is based only on tectonic setting.

W. S. Pitcher (1982) also developed a classification of granitoid rocks based on typing. He recognized five types of granitoid rock: M-type (mantle derived), A-type (anorogenic), S-type (granitoid rocks formed from sedimentary crustal melts generated in collision zones), I-type, Cordilleran (granitoid rocks formed by a two-stage process at convergent plate margins), and I-type, Caledonian (postorogenic uplift granitoid rocks). This work, like its predecessors, links granite types to particular tectonic regimes, but it also includes source rock type. The classification emphasizes, particularly in the subdivision of I-types, the fact that granitoid rocks of any given type have a range of characteristics. W. S. Pitcher (1979, 1983) also relates batholiths to various orogenic and anorogenic sites. He suggests that concordant, compositionally restricted batholiths lacking significant quantities of basic rocks, and exhibiting S- and I-type granites, high initial Sr isotopic ratios, and crustal enclaves, are associated with *intracontinental mountain* belts. Within continental *margin* orogens, Pitcher (1979) suggests the occurrence of compositionally diverse, discordant batholiths with associated basic plutonic rocks, andesitic volcanic rocks, and I-type granites (only). These batholithic rocks, he indicates, have low initial Sr isotopic ratios and mantle rock enclaves.

Barbarin (1990a) summarizes many of these attempts to type granites, then offers yet another classification. In his classification, eight types of granitoid rock fit into three main categories of rocks with crustal, mantle, and hybrid (crust + mantle) origins.

What aspects of granite typing are important for students? First, the terminology pervades contemporary literature, so students should be aware of the general meaning of the basic types. Towards that end, the characteristics of some granitoid types are listed in table 11.2. Second, students should realize that because the genesis of each type is *inferred,* the application of the types is subject to debate.[16] For example, although S-type granitoid rocks are strongly peraluminous, considering that

(1) Chappell and White (1974) inferred, but did not demonstrate, that S-type (peraluminous) granitoid rocks are derived from sedimentary source rocks;

(2) Multiple, including nonsedimentary, origins have been proposed for strongly peraluminous rocks (C. F. Miller, 1985);

(3) All granitoid rocks derived from sediments do not qualify as S-types; and

(4) Peraluminous rocks can be derived from metaluminous parents by fractionation (Cawthorn and Brown, 1976; Abbott, 1981),

application of the designation S-type (a genetic name) to any particular peraluminous rock mass may be erroneous.

The difficulty of correctly applying the term S-type (and by implication, other genetic designations such as C-, M-, or I-types) has been emphasized by C. F. Miller (1985) and A. J. R. White et al. (1986). In fact, the restrictions on the use of the S-type designation, placed by the originators of the term, have progressively increased over time (Chappell and White, 1974; White and Chappell, 1977; A. J. R. White et al., 1986) to the point that few granitoid rocks, other than those of the Lachlan Fold Belt of Australia, seem to meet the criteria. Thus, the general utility of S-type and other type designations is questionable. Furthermore, where inevitable debates about terminology arise, in order to clarify their points, authors resort to Shand's (1949) peraluminous-metaluminous-peralkaline classification, which is based solely on observable criteria. Consequently, Shand's terminology is more useful, is referred to as a standard, and is used in this text, except where I, S, or other type designations are used to communicate the points made by other authors.[17]

The reason geologists continue to seek a connecting link between tectonic history and chemistry is simple. If that link can be established, we can evaluate the petrologic and tectonic history of old granitoid rocks whose history is presently obscure.

THE ORIGINS OF GRANITOID ROCKS

The origin of immense bodies of granitoid rock has long been a subject of debate. Competing theories include four types, each designated here by the dominant process involved. Each of these ideas had preeminence at some time or place. **Granitization,** the first of these processes, is the production of granitoid rocks by metamorphism, or the transformation of preexisting rocks into rocks of granitoid composition and texture without melting any significant part of the preexisting material (H. H. Read, 1948). **Fractional crystallization of basaltic magma,** advocated by Bowen (1948), is the second process. **Hybridization,** the third process, ascribes the origin of granitoid rocks to crystallization of magmas modified by assimilation or magma mixing. The fourth process, anatexis, yields granitoid magmas via partial melting of crust and/or mantle rocks (Winkler and van Platen, 1958; Winkler, 1965, p. 178ff.). Crystallization of the magmas yields granitoid rocks. Recent work suggests that combinations of these processes (e.g., AFC) may also yield granitoid magmas and rocks.

As we examine these theories, remember that the word granite meant different things to different workers. G. H. Anderson (1937) and H. H. Read (1948; 1951, in 1957, p. 341) made clear that they were referring to phaneritic, crystalline rocks with quartz, feldspar, and ferromagnesian minerals, including the granodiorites, and, for Anderson, even gabbros. Consequently, the granitization theories they advocated attempted to explain a wide range of granitoid compositions. In contrast, Daly (1933, p. 43) made it clear that he was excluding granodiorites from the category of granite. Bowen (1948) likewise seems to focus on granite *sensu stricto,* as he describes it as having a close similarity to rhyolite and as being the "uppermost differentiate" of a basaltic parent.

Granitization

Granitization, the solid-state transformation of preexisting rocks into rocks of granitoid mineral composition and texture, is a metamorphic rather than an igneous process. It is discussed here because historically it was a major theory of the origin of the granitoid rocks of the batholiths.

H. H. Read (1944, 1948)[18] and Bowen (1948) pointed out that different workers advocate different specific processes of granitization. Some argue that transformations result from ion migrations in dry rocks; others that wet rocks, rocks containing or penetrated by fluids or granitic melts, are required; and still others propose that limited melting produced a partially melted mush or "migma." With some notable exceptions (e.g., D. L. Reynolds, 1946), the evidence offered in support of the efficacy of granitization is principally field and petrographic in nature.[19] H. H. Read (1948, p. 2) states as much where he writes, "I see the granite problem as essentially one of field geology."

Several types of evidence and arguments have been presented in support of the granitization theory (table 11.3).

Table 11.2 Characteristics of Some Granitoid Types

	Type			
	A	0		
Characteristic		S	I	M
SiO_2	60–80%	65–79%	53–76%	54–73%
Mol. Al_2O_3/(Na_2O+K_2O+CaO)	variable	>1.1	<1.1	<1.2
Na_2O in silicic rocks	>2.8%	variable	>3.2%	>3.2%
K_2O/ Na_2O	usually high	high	low	very low
Fe^{3+}/$Fe^{2+}+Fe^{3+}$	medium	low	medium-high	high
Eu	low	low?	low–high	?
Cr, Ni	low	high	low	?
(^{87}Sr/^{86}Sr)$_i$	0.703–0.712	normally >0.708	normally <0.708	0.702–0.714, but <0.704 common
εNd	?	–4 to –17	–4 to –9	?
Main Enclaves	enclaves rare	metasediments, aplites, milky quartz	amphibolite, diorite	hornblende rocks, granites
Key CIPW Minerals	?	>1% C	<1% C	?
Key Modal Minerals	biotite, alkali pyroxene, alkali amphibole, magnetite, fluorite	muscovite, biotite, sillimanite, ilmenite, garnet	hornblende, biotite, sphene, magnetite	hornblende, clinopyroxene, biotite, magnetite
Common Range of Rock types	Alkali granite to anorthosite	Leucogranite to granodiorite	Granite to gabbro	Quartz diorite to gabbro
Associated Volcanic Rocks	Alkalic rhyolites	Siliceous ash flows	Rhyolite ash flows, dacite, andesite	calc-alkalic andesite, dacite, rhyolite; tholeiite

Sources: Chappell and White (1974), White and Chappell (1977, 1983), Loiselle and Wones (1979), White (1979), Collins et al. (1982), Didier, Duthon, and Lameyre (1982), W. S. Pitcher (1982), McHone and Butler (1984), A. J. R. White et al. (1986), J. B. Whalen, Chappell, and Chappell (1987), Barbarin (1990a), Eby (1990).

Among the most important are these: (G1) the association of granitoid rocks and high-grade metamorphic rocks in regional metamorphic terranes reveals a genetic association;[20] (G2a) granitization is revealed by a gradational change, from layered metamorphic rocks (gneisses and schists) to structureless granitoid rocks, in catazonal, high-grade metamorphic terranes (figure 11.4);[21] and (G2b) *ghost stratigraphy*, layers and faint bands of minerals traceable from the country rock into a pluton, reveals that granitoid rocks have formed in place by the transformation of layered metamorphic rocks.[22]

One of the most important, reported examples of a complete gradation from metamorphic rock to granite, on a regional scale, is that of the Pellisier Granite of the Inyo Range of California (figure 11.4) (G. H. Anderson, 1937;

Table 11.3 Some Evidence and Arguments Pro and Con on Granitization

Evidence or Argument	Pro	Con
G1	Association of granitoid rocks and high T-P metamorphic rocks in the catazone reveals a genetic link.[1]	Association in space does not prove a genetic link, only a coincident petrotectonic environment. Granitoid rocks also occur in mesozonal and epizonal settings, a fact not explained by granitization.[2]
G2	Granitoid rocks, some with ghost stratigraphy, grade into layered metamorphic rocks, revealing granitization.[3]	Detailed examination generally shows these gradations to be local contact effects or nonexistent features.[4]
G3	Large feldspar crystals of like composition formed in plutons and country rock (*dent de cheval*; French for "horse's teeth"), represent partial granitization of country rock.[5]	Local growth of alkali feldspar phenocrysts in plutons, xenoliths, and country rock reveal *local* magmatic crystallization and metasomatism.[6] Low ion migration rates make *regional* granitization and ion migration unlikely.[7]
G4	Rapikivi-textured granites in country rock with mantled feldspars reveal incomplete granitization of country rock.[8]	Experimental, field, and phase-chemical data indicate that rapikivi textures are magmatic.[9]
G5	Ferromagnesian-mineral-rich zones on xenolith and pluton margins reveal Fe, Mg, and Al ion migration out of granitizing plutons, while Si migrates in.[10]	Experiments suggest that magmatic crystallization could produce Fe-Mg-rich border zones on plutons.[11] Experiments also show that ion migration patterns are not consistent with the granitization thesis (e.g., Mg and Si migrate in the same, not different, directions).[12] The mafic margins are too small to hold all the Fe-Mg materials that must migrate out of a granite batholith if the process works as suggested.[13]
G6	Heterogeneous and composite plutons reflect ion migration that is incomplete.[14]	Zoned plutons are produced by magmatic processes such as fractional crystallization or multiple intrusion.[15]
G7	Granitization requires no extra room in the crust because ions that migrate in simply replace ions that are migrating out.[16]	Evidence of underplating, fractional crystallization, stoping, forceful intrusion, assimilation, and *lit-par-lit* (layer by layer) injection of magma, plus basement subsidence and continental growth, eliminates the room problem.[17]
G8	No connection exists between granites and rhyolites. They form by different processes.[18]	Field and petrochemical work shows clear connections between rhyolites, other felsic volcanic rocks, and granitoid rocks.[19]

Sources and notes:

[1] H. H. Read (1948), Misch (1949c), Sederholm (1967); also see text note 20.

[2] See note 83.

[3] G. H. Anderson (1937), Jahns (1948); also see notes 21, 22, and 84.

[4] Emerson (1966), W. S. Pitcher (1970); also see note 85.

[5] H. H. Read (1948) and note 86.

[6] Luth and Tuttle (1967), Vernon (1986), and note 87.

[7] Brady (1983) and W. S. Pitcher and Berger (1972, p. 115–16).

[8] Backlund (1938) and H. H. Read (1944).

[9] Tuttle and Bowen (1958), Abbott (1978), Stull (1978, 1979a).

[10] D. L. Reynolds (1946, 1947a), Mehnert (1968, p. 333), and note 88.

[11] See Naney and Swanson (1980), Naney (1982).

[12] See Vidale (1969), Brady (1983).

[13] Bowen (1948).

[14] Goodspeed (1939), Misch (1968; 1969), and note 89.

[15] Taubeneck (1964, 1967), Zorpi et al. (1989), and note 90.

[16] See H. H. Read (1948).

[17] Bowen (1948), Holder (1979), and note 91.

[18] H. H. Read (1948).

[19] See note 92.

H. H. Read, 1944). Devastating to the credibility of the Pellisier example were the revelations of a detailed study in the Inyo area, by Emerson (1966), who showed that the Pellisier Granite studied by Anderson actually consists of several units—two granodiorites and some schist and gneiss units. The granodiorites are separated from nearby sedimentary rocks not by a broad gradational zone, but by both a contact metamorphic aureole and a shear zone. The horizontal ghost stratigraphy shown by G. H. Anderson (1937, pl. 2, fig. 2) in a photograph is apparently an artifact of faulting; it occurs four miles from the nearest outcrops of "Pellisier Granite" and is oriented at a high angle to the nearest exposed sedimentary bedding (Emerson, 1966). Thus, it is actually not ghost stratigraphy but a surface phenomenon; and it does not occur in the "Pellisier Granite," which, in fact, does not exist as a rock unit.

The discovery that this specific example of gradational contacts and ghost stratigraphy is fallacious casts doubt on other less thoroughly examined cases used to support granitization theory. W. S. Pitcher (1970) reviewed many cases of ghost stratigraphy and concluded that most can be explained by magmatic models; for example, they are strings of xenoliths, lenses of country rock, or flow structures (Balk, 1937). Still, a few mineralogical, geophysical, and statistical studies suggest that ghost stratigraphy may exist in some plutons (Jahns, 1948; Whitten, 1959, 1960; Tuominen, 1961).

Granitizationists have also argued that because granitization involves the movement of ions from place to place, no new volume of material is created (G7). Magmatic models apparently have a "room problem," because new materials are theoretically emplaced into the crust where there is no room for them (H. H. Read, 1948). The occurrence of plutons like the Big Oak Flat Pluton, which exhibit no evidence of forceful emplacement (Paterson, 1989), emphasizes this problem. The no-new-volume argument and the other arguments in support of granitization *as a batholith-forming process* can be countered in convincing ways (see table 11.3).

Local magmatic granitization may occur (Grout, 1941, 1948; Goodspeed, 1948), but few petrologists still support the granitization theory for the origin of granitoid batholiths. The theory is fatally flawed by virtue of the fact that no body of experimental data supports it. Some granitizationists have chosen to defend this lack of experimental data by belittling the significance of it, arguing, for example, that laboratory melting experiments yield "entirely inadmissable" conclusions regarding processes that take place in nature (Drescher-Kaden, 1982, p. 9).[23] Yet the experimental data are compelling. For example, Tuttle and Bowen (1958), after experimentally determining the positions of isobaric minima in the granite system (SiO_2-$NaAlSi_3O_8$-$KAlSi_3O_8$-H_2O), compared these positions to the maximum in a contoured plot of the norms of all (571) plutonic

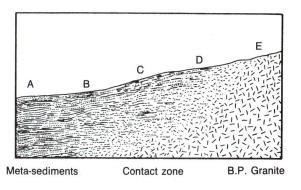

Figure 11.4 Diagrammatic sketch showing a transition from gneiss to massive granitoid rock.
(*From G. H. Anderson, 1937*)

rock analyses that contain 80% or more normative quartz + albite + orthoclase (figure 11.5). Notice that the scatter of analyses is quite small and that most analyses lie near the experimentally determined minima. These data plots strongly suggest a magmatic origin for the rocks.[24] Moreover, the data contravene granitization theory, because by the process it proposes there is no reason to expect a concentration of analyses near the thermal minima, but there is a rationale for expecting the data to be dispersed.

In short, it is unlikely that granitoid batholiths are produced by granitization. Most field evidence is readily explained by other processes and no significant experimental evidence supports the theory.

Fractional Crystallization of Basaltic Magmas

Bowen (1928, 1937, 1948; Tuttle and Bowen, 1958) argued that fractional crystallization of basaltic magma produces granite. This theory of granite formation, which is based in physical chemistry, is one of three magmatic models (the other two are hybridization and anatexis). Evidence offered in support of the theory entails evidence of both magmatism (see table 2.1) and fractional crystallization. Evidence of a magmatic history for granitoid rocks does not *require* fractional crystallization, but it is compatible with it.

In the fractional crystallization and hybridization theories, the source of basaltic magma is not a primary concern. Rather, the focus is on the processes that produce granitoid magmas from a parent magma. Of course, isotopic and other evidence indicate that basaltic parents are derived from the mantle (see chapters 6 and 7).

Arguments in support of fractional crystallization of basaltic magma, as a process that yields granitoid rocks of the batholiths, are numerous. Many workers simply argue for the

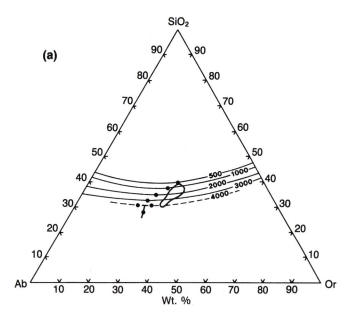

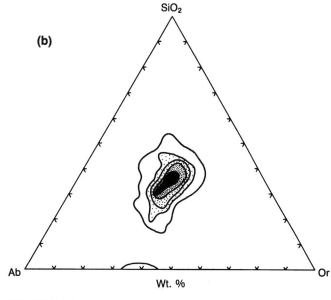

Figure 11.5 Thermal minima and rock analyses. (a) Plotted positions of isobaric thermal minima at various pressures (shown here in kg/cm²) in the "granite" system. The central loop is the maximum from figure 11.5b (from Tuttle and Bowen, 1958). (b) Contour diagram of norms of all plutonic rock analyses in Washington's (1917) tables that yield 80% or more normative q + or + ab. Contours are of 1, 2, 3, 4, 5, and 6–7% per 1% area. More than 25% of the 571 analyses plot in the central black area. Crystallization histories and the addition of Ca to the system may account for the shift of points away from the minima shown here (from Tuttle and Bowen, 1958).

credibility of the fractionation process without addressing the question of whether or not such a process can generate enough volume of granitoid magma. Also, in older literature, workers did not always attempt to discriminate between evidence for magmatism and evidence specifically supporting fractional crystallization. Evidence of a magmatic origin for granites includes (1) sharp, discordant contacts between country rocks and dikes, apophyses, and other plutons; (2) contact metamorphism of country rock; (3) chilled zones along pluton margins; (4) angular xenoliths near the edges of plutons, with shapes that match the shapes of adjoining reentrants in the country rock (thus demonstrating stoping and intrusion of magma) (figure 11.6), and (5) textures indicative of magmatic crystallization.[25]

Specific types of evidence of fractional crystallization, some of which were discussed in chapters 5, 7, and 8, and by Bowen (1928, 1937, 1948), include the following.

F1. Zoned and layered plutons provide field evidence that fractional crystallization of more basic magmas to yield more siliceous differentiates does occur.[26]

F2. The nature of internal layering in some plutons clearly reflects gravitationally induced settling of crystals in a magma.[27]

F3. Experimental studies of phase equilibria demonstrate that fractionation can occur and that the residual minerals in fractionation series (e.g., Bowen's reaction series) are quartz and alkali feldspars, the minerals that comprise granite.[28]

F4. Zoned feldspars are consistent with fractional crystallization.[29]

F5. Glassy groundmass materials or segregations in basaltic rocks have siliceous compositions.[30]

F6. Concentrations and ratios of trace elements in some granitic rocks match those of mantle-derived, ocean island basalts and their presumed differentiates, which can only have formed by fractional crystallization.[31]

F7. Enrichment, depletion, or partitioning of silica, alkalis, and trace elements in the felsic rocks of a group of spatially associated plutons can be explained by fractional crystallization. For example, granitoid rocks with negative Eu anomalies in the REE (spider) diagrams suggest plagioclase fractionation.[32]

F8. Calculations of sequential removal of major and trace elements by crystallization of specific mineral phases indicate fractional crystallization.[33]

F9. Chemical analyses of major and trace elements plot in coherent, curvilinear patterns on Harker, AFM, and other variation diagrams.[34]

Fractional crystallization theory is challenged in three areas. First, will fractional crystallization yield adequate quantities of granitic magma from which to form a batholith? Second, are the compositions of fractionated liquids like the compositions of the granitoid batholiths? Third, do the effects (the evidence) cited actually result from fractional crystallization, or can other processes explain the evidence?

With regard to the volumes of siliceous differentiates derived from basaltic magma bodies, H. H. Read (1948), Presnall (1979), and B. D. Marsh (1988a, 1990) challenged Bowen's (1948) view that adequate amounts of granitic liquid to make a batholith can be generated by fractional crystallization. Marsh's (1988a) argument is based on calculations of crystallization and convection and observations of lava lakes and intrusions. He concludes that neither significant convection nor separation of large quantities of siliceous liquids is possible. Certainly the volumes of granitoid differentiates in known layered intrusions is small. Such is the case at Ingonish and Skaergaard (Wiebe, 1974; Wager and Deer, 1939). Viewed another way, fractional crystallization should produce a residue of 10 to 20 times more mafic-ultramafic rocks than granite, yet the former rock types are rather scarce in granitoid-cored orogenic belts (Bowen, 1948; Barth, 1962). These field observations and the calculations suggest that even if fractional crystallization produces siliceous differentiates, they will not be generated in enough volume to make a batholith.

R. Lambert, Chamberlain, and Holland (1989) challenge that conclusion. They found that siliceous glass residuum in Columbia River basalts ranges up to 24% of the mode (F5). Separation of about 6% glass from the Columbia River Basalt Group and segregation of that glass into a batholith would produce a mass of 10,000 km^3, which is comparable to that of many large batholiths. The mafic residuum remaining after extraction would total about 17 times the amount of siliceous material—just the amount expected. Similar amounts of siliceous glass, about 8%, occur in basaltic dike rocks in Iceland (P. S. Meyer and Sigurdsson, 1978).

Clearly, the arguments and data conflict. Theoretical calculations and studies on convection and magmatic processes in magma chambers yield varied and debatable results (Shaw, 1965; Brandeis and Jaupart, 1986; Marsh, 1988a; Helz, Kirschenbaum, and Marincnko, 1989; Sparks, 1990). Yet, field and petrographic data seem to indicate that siliceous liquids can form in adequate quantities to make a batholith. The next question, then, is are those liquids of the right composition?

Layered plutons, such as the Skaergaard Complex and the Ingonish intrusions, and normally and reversely zoned plutons, such as the Lock Doon Pluton and the Notch Peak

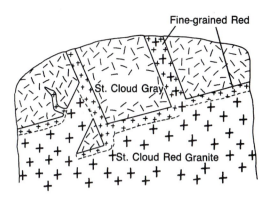

Figure 11.6 Stope blocks of gray granite in red granite ("St. Cloud Granite"), Minnesota, demonstrating a magmatic origin for the red "St. Cloud Granite."
(From Grout, 1948; after M. Skillman)

Granite, provide evidence of fractionation (F1).[35] Where granitoid liquids are produced in the tholeiitic layered intrusions, they may be enriched in iron (Wager and Deer, 1939; Wiebe, 1974; Presnall, 1979) and would not yield the calc-alkaline plutons characteristic of the orogens. In addition, in comparing typical calc-alkaline plutonic rocks of the Sierra Nevada with the late-stage rocks formed as differentiates in the Skaergaard intrusion, Presnall (1979) notes that the Skaergaard rocks are also lower in silica and potash, higher in titania, manganese oxide, and lime, and have from 3-1/2 to 15 times more total iron oxide than the Sierran rocks. Processes have been envisioned to explain how more granitic rocks could evolve from tholeiitic Skaergaard-like magmas (Osborn, 1959; Irvine, 1976; Spulber and Rutherford, 1983), but the possibility that these processes operate at the batholithic scale is debatable. Nevertheless, siliceous calc-alkaline (iron-poorer) differentiates are known (Wiebe, 1974). Furthermore, many glass residues in Icelandic dikes, the Alea Lava Lake of Hawaii, and Columbia River basalts are comparable to Sierran granitoid rocks in silica, total iron, potash, titania, and lime.[36] These results suggest that granitoid liquids and granites *can* form by fractionation, but the chemistries of some liquids may require modification if the liquids are to be comparable to the range of batholithic granitoid rock chemistries. Whether or not those differentiates can be segregated into batholithic masses is an open question.

Of the other evidence cited in support of fractional crystallization, some types can be best explained by this process, but others may be explained equally well by other processes. For example, the presence of zoned plagioclase feldspars (F4) in granitoid rocks generally supports a magmatic origin and is rather clearly explained by fractional crystallization in the system $NaAlSi_3O_8$-$CaAl_2Si_2O_8$ (Bowen, 1913, 1928). Recent students of zoning continue to advocate magmatic methods of generating the zoning (T. P. Loomis and Weber, 1982),[37] but in some cases zoning may be caused, in part, by hybridization (Barbarin, 1990b).

Chemical data (F6, F7, F8, F9) are particularly subject to alternative interpretations (Wall, Clemens, and Clarke, 1987). In particular, curvilinear data patterns on variation diagrams (F9) must be evaluated with caution. Fractionation and mixing may yield similarly continuous curves. Isotopic data may also be interpreted in different ways. As noted above (see table 11.2 and accompanying text), peralkaline to peraluminous granitoid rocks commonly yield initial strontium isotope ratios as high as 0.760. Such high initial Sr-isotope ratios do not support a crystal fractionation origin for batholithic granitoid rocks, since basalts are derived from the mantle, where initial Sr isotopic ratios are low. The basalts *and their derivatives* should have low initial Sr isotopic ratios. Conversely, while low values may result from fractionation, they may also reflect anatexis of mantle rock. Study of other isotopic systems yields similar conclusions. In addition, where isotopic ratios or trace elements have anomalous values or conflicting implications, explanations of the data are complex and involve a variety of processes (Saleeby, Sams, and Kistler, 1987; Speer, Naeem, and Almohandis, 1989).

Because experimental studies of phase equilibria (F3) form the foundation of modern igneous petrology (see chapter 5), they provide an important test of any theory. The works of Bowen (1913, 1928, 1937), Schairer (1950; Schairer and Bowen 1947; 1955; Schairer and Yoder, 1960), Tuttle and Bowen (1958), Winkler, Boese, and Marcopoulos (1975); Winkler and Breitbart, (1978) and their successors have unequivocably demonstrated that equilibrium and fractional crystallization can duplicate the major chemical and mineralogical trends of granitic and other rock systems. What often remains disputed about the formation of magmas is the relative importance of fractional crystallization versus other processes.

To sum up, fractional crystallization of basaltic magmas *can* yield zoned plutons with late-stage, siliceous differentiates containing zoned plagioclase. In some cases, the quantities and compositions of siliceous residues in mafic magmas are appropriate to form batholithic masses of granitoid rock, and they may do so, *if* they can separate from the parent magma and segregate into large bodies. That such processes are possible is suggested where fractional crystallization seems to have produced small masses of granitic rock in oceanic areas (C. G. Engel and Fisher, 1975; Spulber and Rutherford, 1983). In other cases, modification of siliceous differentiates may be required to generate granitoid batholithic rock chemistries.

Hybridization

Hybridization is a process of magma production in which hybrid magma is formed by assimilation or mixing, or a combination of these processes. Daly (1905, 1906, 1914, 1917, 1933), who called hybridization "syntexis," was a principal early advocate of hybridization as a granite-forming process. He built upon the work of his predecessors in developing and clarifying the concept (Daly, 1914, p. 323ff.).[38] For granite petrogenesis, Daly proposed that basaltic magma acts as a superheated solvent that dissolves siliceous sediments, preexisting granitoid and metamorphic rocks, or both, and then crystallizes to form granitoid rocks. Evidence cited by Daly (1914, 1933) and others in support of hybridization includes the following:

H1.   The general distribution of rocks in the crust
H2.   Field and petrographic evidence of assimilation of sediments in the basaltic magmas of sills, dikes, and larger plutons[39]
H3.   Coherent, curvilinear chemical analysis patterns on Harker, AFM, and other variation diagrams, reflecting mixing of endmember compositions[40]
H4.   Isotopic and trace element data[41]

Daly (1914, 1933) argued that the layered nature of the earth, as he perceived it—basalt at depth, sediments on top, and "acid rocks" (granitoids) in between—represents a rock distribution of fundamental importance (H1). Dissolution of siliceous sediments by the basalt, he argued, yields the intermediate granitoid rocks. Local evidence that this process can occur is provided at localities where sediments and other rocks take the form of partially absorbed or melted xenoliths within basaltic magma (H2) (Daly, 1905; W. B. Hamilton, 1956; Dodge, Lockwood, and Calk, 1988).

Chemical data (H3, H4) support the view that significant amounts of lower crustal rocks have been assimilated or melted by large quantities of magmas rising from the mantle.[42] Evidence for assimilation, for example, is provided by Nd isotopic ratios from the Sierra Nevada Batholith. These indicate derivation of the magmas from mixed crust and mantle sources, suggesting that mantle-derived magmas have been contaminated by assimilation of crustal materials (DePaolo, 1980b, 1981a; Domenick et al., 1983). Trace elements and U/Pb, Sr, and O isotopic data support that interpretation (Saleeby, Sams, and Kistler, 1987; Dodge and Kistler, 1990). Similarly, Nd isotopes and/or Pb isotopes suggest assimilation in the petrogenesis of granitoid rocks of the Foyers Complex (Pankhurst, 1979), other Scottish plutons (Halliday, 1983; Harmon, 1983), and the San Nicolas Batholith of Peru (Mukasa and Henry, 1990).

Bowen (1928, ch. 10) and McBirney (1979) discussed the theoretical aspects of assimilation and Bowen found little to recommend the process.[43] Assimilation of solid country rock requires heat. The source of that heat must be the magma. Bowen referred to phase equilibria studies (see chapter 5) and noted that extraction of heat from the magma will cause the magma to crystallize, *unless* the magma is superheated, as proposed by Daly. Bowen (1928) noted that, from experimental and theoretical perspectives, the amounts of heat available in liquid magmas are generally very small compared to the heat (of solution) required to melt mineralogical solids. Consequently, he argued that unrealistically large amounts of heat, amounts of superheat unlikely in nature, are required to melt significant volumes of crustal rock. Bowen (1928, p. 184) also opposed hybridization on the grounds that observations of xenoliths indicate that magmas commonly fail to melt inclusions. In addition, some petrologists doubt that the large volumes of quartz-rich sediments necessary for the generation, by assimilation, of all of the existing quantities of granitoid rocks ever existed (Sederholm, 1907, in 1967, p. 51).

Whether or not batholithic masses of granitoid rocks can be produced by assimilation is the important question here. Although some small to large-scale examples of modification of magma compositions by assimilation have been described,[44] in many cases the granitoid magmas involved already existed prior to emplacement or modification. In such cases, there is no evidence to indicate that the preexisting magma formed by basalt magma assimilating siliceous sediments. For example, Lee and Van Loenen (1971) and Cocherie, Rossi, and Le Bel (1984) describe situations in which magmas were modified at shallower levels of the crust *after* being generated elsewhere.

In cases of hybridization via melting and mixing, siliceous melts mix with mafic melts to form rocks of intermediate composition (J. B. Reid, Evans, and Fates, 1983; Zorpi et al., 1989; L. L. Larsen and Smith, 1990). The volume and composition of granitoid rock thus formed are dependent on a number of factors, including the heat content of the individual magmas and the compositional and viscosity contrasts between them (T. P. Frost and Mahood, 1987). Support for hybridization via mixing is found in the petrochemical data (H3, H4) and field and petrographic evidence (H2). The latter includes (1) unusually zoned plagioclase and (2) partially disaggregated mafic enclaves and mafic pillows in more siliceous rocks (J. B. Reid and Hamilton, 1987; Barbarin, 1990b).[45]

An intriguing example of a hybridization model for the Sierra Nevada Batholith of California, advanced by J. B. Reid, Evans, and Fates (1983), combines aspects of Daly's (1914) ideas and contemporary models of rhyolite petrogenesis. J. B. Reid, Evans, and Fates (1983) propose that basalt both *melts* the lower crust to form magmas that intrude as hyperleucocratic granitic melts (alaskite melt), and *intrudes and mixes* with the granitic materials to form the intermediate

granitoid rocks. Major element and petrographic data support aspects of this model.[46] Barbarin (1990b), for example, reveals that plagioclases from the granitoid rocks have complex zones with three distinct parts; which he attributes to initial crystallization, partial dissolution and crystallization during mixing, and final crystallization from the mixed melt.

Clearly, some data support the hybridization model. Problems of heat are partially resolved if large quantities of somewhat superheated basalt are involved in magma genesis *and* if the rocks to be melted or assimilated are already near their melting temperatures. Intermediate magmas such as quartz diorite or granodiorite have compositions that can be generated more reasonably by hybridization, but few, if any, granitic magmas have been produced by these mixing or assimilation processes. Furthermore, the physical properties of the magmas make it unlikely that granites (*sensu stricto*) can form in this way (Frost and Mahood, 1987). Considering that the same kinds of data used to support hybridization are also used to support alternative models of granitoid rock formation, the importance of hybridization is not entirely resolved.[47] We may conclude, however, that some intermediate magmas do owe their distinctive compositions to hybridization processes.

Anatexis

Anatexis, or partial melting of preexisting rocks, is the final process to be considered in explaining the origin of the magmas that yield granitoid batholiths and other plutons. The theory of anatexis has several advantages: (1) it combines certain aspects of the magmatic and granitization theories, (2) it is consistent with a wide range of chemical data, and (3) it is supported by experimental studies. Some granitizationists consider anatexis to be a process of granitization,[48] yet it differs from granitization in that it involves the generation of a magma.

The anatectic theory for the origin of granitoid batholiths has two parts. First, it addresses the origin of materials that later give rise to the batholiths and other plutons. Specifically, the model suggests that magmas form in the middle to lower crust or upper mantle. One contemporary version of anatectic theory ascribes magma genesis to anatexis of crustal rocks—which yield melt plus a residue of unmelted material, commonly called **restite**—from which the melt is separated (A. J. R. White and Chappell, 1977). In this "restite unmixing model," enclaves are commonly considered to represent the restite. Second, anatexis theory allows for the crystallization of the magmas, either by equilibrium crystallization or by fractional crystallization.

Inasmuch as anatexis yields a magma, any of the evidence cited above that supports the existence of magmas is consistent with the model. In addition, evidence of fractional crystallization (F1–F4) is compatible with the model, as is evidence of *local* assimilation (H2) and granitization (G2, G3), because once the magma is formed by melting, subsequent modifications do not negate the anatectic origin. For the most part, these previously described kinds of evidence are not listed here. Rather, additional evidence and arguments specifically supporting the anatectic origin of melts are given.

A1. Evidence of a magmatic history, coupled with petrotectonic analyses and the geographic restriction of large granitoid masses to continental areas and arcs, suggests that granites are related to the continental crust. These relations suggest that, in many cases, crustal melts are the parents of granitoid rocks (A. Holmes, 1932). Also, the position of Phanerozoic batholithic belts landward of subduction zones (Younker and Vogel, 1976) is consistent with the anatectic generation of granitoid magmas in these areas (Dickinson, 1970b).[49]

A2. The common occurrence of granitoid lenses and pods in metamorphic terranes of high grade, especially those containing amphibolites and mica schists, suggests local anatexis (figure 11.7).[50]

A3. The plot of normative compositions of rocks from Washington's (1917) tables yielding 80% or more normative q + ab + or lies at or near the thermal minimum in the granite system. This is compatible with an anatectic origin of the magmas that produced those rocks.[51]

A4. Studies of major and trace elements, including isotopes, from many granitoid rocks clearly indicate sources other than mafic to ultramafic mantle rocks, and, in some cases, specifically suggest sedimentary or metasedimentary parentage.[52] Where mafic sources are indicated by the data, in some cases these too can serve as parents for granitoid melts.[53]

A5. Experimental melting studies indicate that magmas of appropriate compositions can be generated by anatexis of a variety of crustal rocks over a range of depths (but at deep crustal or subcrustal depths, the source compositions and melting conditions are restricted significantly).[54]

A6. Chemographic and theoretical analyses support the possibility of anatexis (A. B. Thompson and Algor, 1977; Abbott, 1981).[55]

The variety of evidence—field, petrographic, chemical, experimental, and theoretical—assembled in support of the idea of anatexis of mantle or crustal rocks to produce granitoid rocks, gives the theory strength. The data support ways of producing granitoid rocks that are theoretically possible and are consistent with experiment and observation.

What arguments can be raised to refute an origin of granitoid magmas by anatexis? Three general types of argument exist.[56] First among these is the argument of energy. Melting occurs in nature either because there is an influx of heat (energy) or because the solidus temperature is reduced. Experimental melting (A5) is possible because a supply of energy external to the system induces melting to occur, but in nature, Bowen (1948, p. 85) suggests, "no supply is in sight." In fact, an energy source does exist, the internal heat of the earth, which is transferred to the base of the crust by rising diapirs or mafic magmas (Hildreth, 1981; Bergantz, 1989; Barton, 1990). As discussed in chapters 6–9, convection of the mantle drives plates of lithosphere, carries heat energy to the surface, and may focus that energy above subduction zones or beneath spreading centers, especially at the base of the crust or lithosphere. The result is melting. Thus, not only is a source of energy available, a mechanism is evident for continuous addition of that energy both to spreading centers and to sites within the plates above subduction zones, where magmas form.

Some evidence suggests that fluid flux from underthrust or subducted sediments or mantle rocks serves to lower the solidus of crustal rocks, also inducing melting (LeFort, 1981; Hoisch and Hamilton, 1990).[57] If anatexis is caused in this way, less heat (energy) is necessary for melting to progress.

A second argument that may be raised against anatexis, is that evidence and arguments such as those provided in A1, A2, A3, and A7 are permissive, that is, they *allow* anatexis but they do not prove its operation in nature. Viewed another way, while these data and arguments support anatexis, they support other models as well. For example, the coincidence of the thermal minimum in the granite system with the normative maximum in quartz + feldspar-rich granitoid rocks (A3) may also be cited in support of fractional crystallization (F3). Similarly, the close association of granitoid rocks and metamorphic rocks of high grade (A2) has been used as evidence of granitization (G1).

A third argument that may be raised against anatexis is that the chemical data (in specific cases) do not support it. Locally, this may be the case, but the chemical data generally do affirm that anatexis was an important process in the generation of many magmas parental to granitoid plutons.[58] Petrochemical studies reveal that granitoid magmas are derived from anatexis of both metaigneous and metasedimentary rocks.[59] Evaluating magma histories, however, is a complex exercise that relies on various assumptions about the chemistry. For example, one may assume that certain rock compositions or combinations of magmatic rock and enclave compositions reflect parental magmas, but such assumptions can rarely be verified. Making such an assumption, J. L.

Figure 11.7 Pods of granitoid rock (white) in amphibolite facies mica schist, Alligator Back Metamorphic Suite, Highway 421, east of Deep Gap, North Carolina. Drill-hole spacing (for scale) is about 0.7 m.

Anderson and Cullers (1990) use information including (1) modest but elevated initial Sr-isotope ratios (0.707 to 0.709), (2) the absence of Eu anomalies, and (3) the abundance of Ca and Sr to infer that granodiorites and quartz diorites (tonalites) of the Whipple Wash Suite of California crystallized from magmas derived by a high percentage of melting of arc-derived "graywacke" sandstone. In another example involving California rocks, C. F. Miller et al. (1990) use ele-

vated Sr (0.709–0.719), low εNd values, (< −10) Pb and O isotopic data, plus high values of Sr, Rb, Ba, and total REE in the rocks, to infer an ancient crustal source for quartz diorite (tonalite), granodiorite, and granite magmas of the Old Woman Mountain–Piute Range Batholith. In cases such as these, the combination of data cannot be explained as well (or at all) by other theories of magma genesis.

207

Given that anatexis is a primary process in granitoid magma genesis, what more can we learn about the history of anatectic magmas? A. J. R. White and Chappell (1977) note that "ultrametamorphism" (metamorphism with melting) yields melt plus restite. Both restite and melt yield information about the rock history, revealing, for example, whether a rock is an S- or I-type. In particular, restite, *if* it forms enclaves and is available for study, may yield specific information about the parent material melted to yield the magmas.[60] If restite is not available, chemical evidence alone, especially isotopic and trace element data derived from the granitoid rocks, are used to infer both the source of the melt and the processes of magma modification.[61] Experimental melting studies on rocks aid in constraining possible rock origins.[62] These data clearly indicate that *both* crustal and mantle sources yield anatectic melts that later form major types of granitoid rock.

The kinds and amounts of magma generated by anatexis depend on the source rock composition, especially the hydrous mineral content, on the pressure, and on the temperature (Clemens and Vielzeuf, 1987). Isotopic data and experimental melting studies constrain the sources.[63] Basaltic magmas that yield gabbroic rocks develop within the mantle as a result of plate tectonic and mantle convection processes. Magmas that yield more siliceous rocks, from diorite (including tonalite and trondhjemite) to granodiorite, develop from relatively high-temperature melting of mafic rocks, but under H_2O-saturated conditions the common calc-alkaline types only form at depths of less than 40 km (Wyllie, 1984; Huang and Wyllie, 1986; Carroll and Wyllie, 1990). Intermediate magmas may also develop under drier conditions at greater depths. Anatectic granitic to quartz monzonitic melts form at the base of or within the crust at depths of about 30 ± 10 km, by melting of metasediments and metaigneous rocks, but such magmas also develop through fractional crystallization and hybridization of earlier formed anatectic melts.[64]

Granitoid Rock Types: A Summary of Petrogenesis

The various granitoid rocks may develop through simple or complex histories of anatexis, assimilation, mixing, and fractional crystallization. In addition, local metasomatism and metamorphism may yield granitoid-like rocks. Gabbroic (basaltic) magmas form by anatexis of mantle ultramafic rock, but the gabbroic rocks form either by direct crystallization of these gabbroic magmas or by fractional crystallization and crystal settling (see chapters 7 and 10).[65] Diorites, like some

gabbros, may form by equilibrium or fractional crystallization.[66] Quartz diorite (tonalite and trondhjemite) magmas are commonly derived by anatexis of mafic rocks such as basalt, eclogite, or amphibolite.[67] These intermediate rocks may crystallize directly from the primary magmas, may crystallize from hybrid magmas, or, as is the case with "plagiogranites" (albite quartz diorites), form by fractionation of tholeiitic basalt magmas generated in the mantle (R. G. Coleman and Donato, 1979).

Granodiorites may be among the more complex rocks. They are considered to form directly from anatectic melts or by fractional crystallization of melts formed by (1) anatexis of sedimentary rocks and amphibolites (other metaigneous rocks), (2) mixing of mafic and felsic magmas derived from mantle or crust, (3) mixing of anatectic magmas and country rock, or (4) mixing and assimilation in subduction zones and the overlying ultramafic mantle.[68] In contrast, quartz monzonites are typically derived from crystallization of fractionated gabbroic to granodioritic melts or contaminated granitic melts, but they may evolve locally as melts of sediments or preexisting igneous rocks.[69] Some granites, especially peralkaline and peraluminous types, crystallize from anatectic crustal melts, but granites also represent products of fractional crystallization of anatectic crustal or mantle magmas.[70]

EXAMPLE: THE TUOLUMNE INTRUSIVE SERIES, SIERRA NEVADA BATHOLITH

The Sierra Nevada Batholith is a composite plutonic mass of granitoid rocks that underlies over 40,000 square kilometers of eastern California and Nevada (figure 11.8).[71] The main mass of the batholith is exposed in the Sierra Nevada (the mountain range from which it takes its name) as an irregular, ellipsoidal mass nearly 650 kilometers long, 100 kilometers wide, and 30 kilometers thick.[72] The batholith is truncated on the south by the Garlock Fault, is structurally overlain on the west by a wide variety of Paleozoic-to late Cenozoic-age rocks, and has an irregular northern and eastern map boundary marked by isolated plutons, local faults, unconformities, and intrusive contacts (P. C. Bateman, 1983). The rocks of the batholith range in age from about 220 to 75 m.y. (Evernden and Kistler, 1970; Saleeby, 1981; T. W. Stern et al., 1981; J. H. Chen and Moore, 1982).[73]

The Sierra Nevada Batholith is a representative batholith, smaller than the largest batholiths, like the Coast Plutonic Complex of British Columbia,[74] but larger than

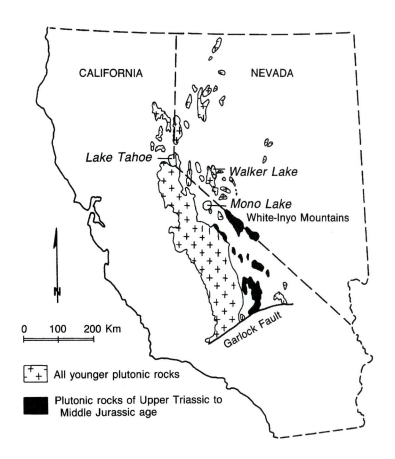

Figure 11.8
Map of California and Nevada showing the general
distribution of Mesozoic batholithic rocks of the Sierra
Nevada Batholith and related plutons. Many small plutons in
Nevada are omitted.
(Modified from Schweickert, 1976)

others like many in New England.[75] Being epizonal to meso-zonal, it is apparently eroded to slightly deeper levels than the batholith of western Peru.[76] Yet it is not eroded as deeply as many mesozonal and all catazonal plutons.[77] By the chemical and petrographic criteria of Pearce, Harris, and Tindle (1984) the batholith is a volcanic arc granite, and by those of G. C. Brown, Thorpe, and Webb (1984) the batholith is representative of a "normal" arc (though some of its oldest parts may be primitive).

Virtually all workers agree that the batholith is magmatic. It is composite, consisting of a large number of plutons belonging to various intrusive suites. For example, in the area of Yosemite Valley alone, there are 10 different plutonic units (Calkins,

Huber, and Roller, 1985). Some of these units belong to the Tuolumne Intrusive Series (Calkins, 1930), a group of "nested" plutons primarily exposed east of the valley (figure 11.9).

Chemically, the Tuolumne Intrusive Series is calc-alkalic.[78] Silica values range between 55 and 78%. Alumina is variable but typically high (11–17.5%). Total iron oxide ranges from less than 1% to more than 7% and Mg numbers are less than 0.45. These data fall on three different trend lines (figure 11.10c). The rare earth elements show variable concentrations, with LREE strongly enriched relative to HREE (F. A. Frey, Chappell, and Roy, 1978) (figure 11.10b). Enrichment is greater in the core plutons (100× chondrite values) than in the marginal bodies (60× chondrite values). Marked Eu anomalies are

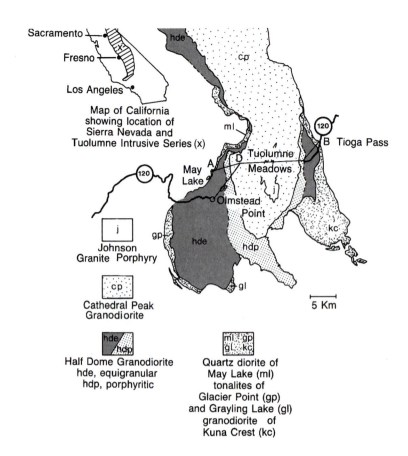

Figure 11.9 Map of the Tuolumne Intrusive Series, Sierra Nevada Batholith, California.

(Modified from P. C. Bateman and Chappell, 1979)

absent. Thus, within the Intrusive Series, the LREE La is relatively depleted in marginal rocks and the HREE Y is relatively depleted in core rocks (P. C. Bateman and Chappell, 1979). Isotopic values also vary, with $^{87}Sr/^{86}Sr$ values ranging from 0.7057 to 0.7067 (Kistler et al., 1986). Nd isotopic ratios vary from 0.51239 to 0.51211, with εNd in the range of –3 to –8.

The Tuolumne Intrusive Series has seven major units, one of which, the Half Dome Granodiorite, consists of both a porphyritic and an equigranular phase (P. C. Bateman and Chappell, 1979) (figures 11.9 and 11.11). The series is zoned, with quartz diorites and granodiorites (the quartz diorite of May Lake, tonalite of Glacier Point, tonalite of Grayling Lake, granodiorite of Kuna Crest, and the Half Dome Granodiorite) forming an outer, partial ring. The

porphyritic phase of the Half Dome Granodiorite occurs as a distinct partial ring within the equigranular phase. The largest unit, occurring in the north and central parts of the series, is the Cathedral Peak Granodiorite. The fine-grained Johnson Granite Porphyry (figure 11.11d), a quartz monzonite in the terminology used here, is exposed in the pluton core. Granitic aplite dikes are present locally throughout the series.

Zoning of the Tuolumne Intrusive Series was originally attributed by P. C. Bateman and Chappell (1979) to fractionation and solidification of magma from the margins of the intrusion inward. Presumably, these magmas were derived by anatexis at depth. Apparently, however, fractional crystallization was only one, and a minor one, of the several processes

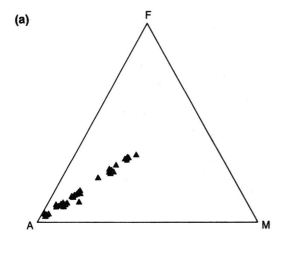

(a)

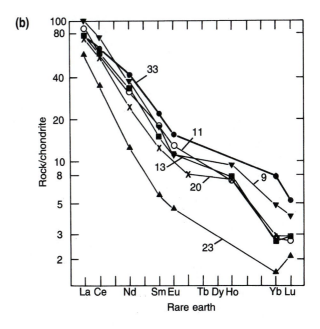

(b)

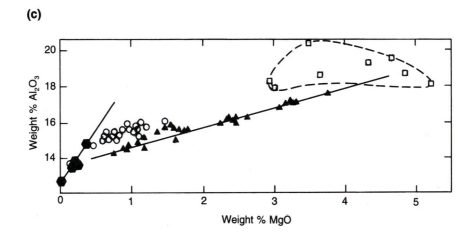

(c)

Figure 11.10 Chemical data from the Tuolumne Intrusive Series. (a) AFM diagram for the series. Note the calc-alkaline trend (data from P. C. Bateman and Chappell, 1979; J. B. Reid, Evans, and Fates, 1983). (b) Selected REE patterns for Tuolumne Intrusive Series rocks: 33 = Granodiorite of Kuna Crest, 9 = equigranular Half Dome Granodiorite, 11 = porphyritic Half Dome Granodiorite, 13, 23 = Cathedral

Peak granodiorites, 20 = Johnson Granite Porphyry (modified from F. A. Frey, Green, and Roy, 1978). (c) Al_2O_3 vs. MgO diagram for the Tuolumne Intrusive Series rocks showing three different chemical trend lines. Similar differences in trend, although in part less pronounced, characterize most of the major and some trace elements.

(*Modified from J. B. Reid, Evans, and Fates, 1983*)

Figure 11.11 Photomicrographs of rocks of the Tuolumne Intrusive Series. (a) Granodiorite of Kuna Crest. (XN). (b) Half Dome Granodiorite. (XN). (c) Cathedral Peak Granodiorite. (XN). (d) Johnson Granite Porphyry. (XN).

Also see figure 11.3c. A = alkali feldspar, B = biotite, H = hornblende, P = plagioclase feldspar, Q = quartz, S = sphene, W = white mica. Long dimension of photos is 3.25 mm.

that contributed to the history of these rocks. J. B. Reid, Evans, and Fates (1983) suggest that the Tuolumne Intrusive Series had a complex history with hybridization as the dominant process. They argue that mafic magmas (presumably generated by mantle anatexis) intruded the base of the crust and caused crustal anatexis that produced granitic magma. The mafic and felsic magmas mixed, with vesiculation and chilling of the intruding mafic magma in the felsic magma chamber resulting in the formation of enclaves. These collected along the domal ceiling of the chamber and were partially assimilated, changing the magma compositions and contributing to the zoned pattern. T. P. Frost and Mahood (1987), in a study of another part of the batholith, found enclaves similar to those described by J. B. Reid, Evans, and Fates (1983), and they support the view that such hybridization can yield rocks of intermediate composition. J. B. Reid and Hamilton (1987) report field, petrographic, and chemical data from Sierran dikes that indicate mixing of rhyolitic and andesitic magmas. The major-element chemical data from these dike rocks compare favorably with those from the Half Dome Granodiorite, suggesting similar processes of formation.

The chemical data from the Tuolumne Intrusive Series reveal that a complex history is likely. The low Mg numbers, elevated Sr-isotope ratios, and negative εNd values indicate that the Tuolumne magmas were not primary mantle melts. The three trends in major element data indicate that the rocks crystallized from more than one magma. Sr, Nd, and Sm data (DePaolo, 1981a; Domenick et al., 1983; Kistler et al., 1986), considered together, suggest that the siliceous rocks of the series (porphyritic Half Dome Granodiorite, Cathedral Peak Granodiorite, and Johnson Granite) were derived from anatectic melts of lower crustal rocks, with little, if any, intermixed mafic melt (Kistler et al., 1986). The strongly enriched LREE patterns are consistent with melting of a crustal source containing garnet (which would retain the HREE). In contrast, some series-margin granodiorites, such as the equigranular Half Dome Granodiorite and the Sentinal Granodiorite (exposed at the west edge of the Tuolumne Intrusive Series), apparently formed from mixed mafic-felsic melts. The isotopic and REE data (e.g., smaller LREE enrichments) are consistent with this interpretation. Mixing is also supported by field relations in nearby intrusions of similar chemistry, which show mafic pillows and disaggregated xenoliths (J. B. Reid, Evans, and Fates, 1983; J. B. Reid and Hamilton, 1987), and by the nature of complex zoning in plagioclase feldspars of the intrusive rocks (Barbarin, 1990b). Fractional crystallization, which apparently made only a minor contribution to the development of the series, as suggested by the absence of Eu anomalies, for example, produced local pegmatite and aplite dikes.

The Tuolumne Intrusive Series is a relatively small part of the larger Sierra Nevada Batholith. As the batholith formed, ultimately, both the heat for magma genesis and some mafic magmas were derived from subcrustal regions. The

mafic melts probably formed by anatexis of mantle rock, but processes of assimilation and melting in the deep crust clearly gave many magmas their dominant character (Kistler et al., 1986; Saleeby, Sams, and Kistler, 1987; Ague and Brimhall, 1987, 1988a, 1988b). At the levels of intrusion and solidification, hybridization and fractionation were important in altering the chemistry and mineralogy of the magmas (J. B. Reid, Evans, and Fates, 1983; Sawka et al., 1990).

Locally and regionally, petrographic and chemical variations occur within the batholith (P. C. Bateman and Dodge, 1970; Ague and Brimhall, 1987, 1988a, 1988b; Sawka, Chappell, and Kistler, 1990). Local chemical variation is reflected by the petrography, typified by the variations in the Tuolumne Intrusive Series.[79] Variations in trace elements, such as HREE, may mimic this petrographically revealed zoning (P. C. Bateman and Chappell, 1979; Noyes, Frey, and Wones, 1983). One major regional chemical variation is an age-independent increase in potassium from west to east across the pluton, reflected by changing rock types.[80] Quartz diorite is common on the west, granodiorite is most abundant in the middle, and quartz monzonite and monzonite typify the eastern batholith. Another major chemical variation is a corresponding increase in initial $^{87}Sr/^{86}Sr$ values (Kistler and Peterman, 1973, 1978; Kistler, 1974, 1978). Ratios of < 0.704 occur in western rocks and the ratios increase across the pluton to >.708 on the east. Nd isotopic ratios on the west suggest that arc magmas were parental to the granitoid rocks (DePaolo, 1980b). F/OH and Mn in amphibole and biotite also increase from west to east (Ague and Brimhall, 1988b). The geochemical variations (P. C. Bateman and Dodge, 1970; DePaolo, 1980b, 1981b; Ague and Brimhall, 1988a) suggest that towards the east old continental crust became increasingly involved in magma modification, but vertical heterogeneity in the subvolcanic crust of the Mesozoic Sierran Arc also contributed to the chemical character of the granitoid rocks (Noyes, Frey, and Wones, 1983; Domenick et al., 1983).

Where younger plutons intrude older ones in the Sierra Nevada Batholith, primary foliations resulting from alignment of mafic minerals and lenticular enclaves may be truncated or deformed (P. C. Bateman, 1983). Similar deformation may occur locally in country rock, but the fact that deformation patterns in separate roof pendants can be correlated across the batholith (Nockleberg and Kistler, 1980) indicates that intrusion was largely passive.

The wallrocks of the batholith, which have been interpreted in a number of ways, appear to be an assembled mass of tectonostratigraphic terranes (fault-bounded regional masses of related rock) (Saleeby, 1981; Nockleberg, 1983). Contacts between the batholith and these terranes are generally discordant. Local contact metamorphism, combined with the cross-cutting relations that exist between plutons and country rock and between individual plutons (figure 11.12), clearly reveals the magmatic nature of the batholith.

(a)

(b)

Figure 11.12 Contacts. (a) Cross-cutting contact, with granodiorite of Kuna Crest (light) cutting country rock hornfels (contact metamorphosed metasedimentary rocks, dark) at Tioga Lake, California. (b) Dikes (light) cutting metamorphic rocks (dark) on the east side of the Sierra Nevada Batholith, east of Yosemite National Park.

The overall picture that evolves from the various studies involves a complex arc history for the formation of the Sierra Nevada Batholith. Transform faulting and rifting in the Triassic Period[81] were followed by subduction and the development of a volcanic arc, extending northward from the continental margin into the oceanic crust (figure 11.13). Magmas generated by partial melting in the subduction zone and mantle wedge area beneath the arc rose diapirically to the base of the crust, where they locally ponded and melted that crust. In some cases, the mafic magmas mixed with the new siliceous melts, and in others they assimilated crustal rocks. The derivative magmas and crustal melts then intruded into the crust to form plutons or through the crust to erupt and form the volcanic arc

(figure 11.14). On the west, the magmas intruded oceanic crust. In the east, where magmas rose through old continental crust, they assimilated parts of and interacted with that crust to become more K- and ^{87}Sr-rich. Locally, fractional crystallization of the magmas in the crust modified their composition, producing zoned plutons, pegmatite bodies, and aplite dikes. Enclaves from the subvolcanic crust and mafic magmas were incorporated in the derivative magmas, giving them distinctive chemical characteristics. Repeated episodes of magma generation occurred as subduction rates, dip of the subduction zone, input of sediment into the subduction zone, and other factors changed, causing variations in the locus and volume of plutonism that produced the overlapping intrusions of the batholith.[82]

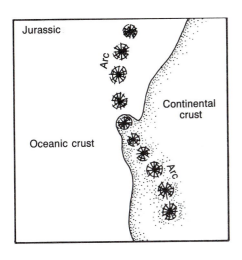

Figure 11.13
Sketch map of the position of the early Jurassic Sierran volcanic-plutonic arc in relation to the continental margin.
(Simplified from Schweickert, 1976)

SUMMARY

Granitoid rocks of all ages were formed in oceanic and continental areas of both orogenic and anorogenic character. They are characterized by the feldspars and quartz, plus pyroxenes, amphiboles, micas, and a host of other minerals that typically crystallize in hypidiomorphic-granular textures. Pegmatitic, porphyritic, and other textures also occur. Chemically, granitoid rocks range in silica from less than 50% to more than 75%. Those that form in orogenic areas tend to be peraluminous to metaluminous, whereas anorogenic rocks are typically peralkaline. Major and trace element characteristics reflect tectonic environments but are also modified by the chemistry of country rocks surrounding sites of melting and intrusion.

Analyses of specific rock types and their chemistries yield more specific information on magma sources and rock history. Several processes may be important in the genesis of any particular magma that produces a pluton of granitoid rock, but there is no question that magmas formed the great batholiths. These

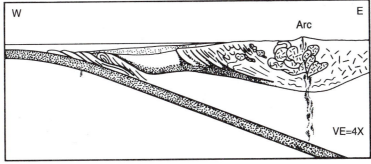

Figure 11.14 Schematic cross section of California and western Nevada for the upper Cretaceous, showing the volcanic-plutonic Sierra Nevada Arc.

magmas apparently developed by anatexis of mantle and crustal rocks beneath volcanic arcs. The evidences of magmatism (cross-cutting relations, contact metamorphism) and the chemistry, particularly isotopic chemistry, demand an anatectic origin for most granitoid magmas. Experimental melting studies in granitoid systems confirm and constrain the possible anatectic processes. Modification of magmas by fractional crystallization, hybridization (assimilation and magma mixing), or combinations of these processes, is locally important in producing granitoid compositions and plutons. In areas of oceanic crust (primitive arcs, back arcs, and spreading centers), fractional crystallization is the principal magma modification process.

The various granitoid rocks may develop through simple or complex histories. Quartz diorite (tonalite and trondhjemite) magmas develop by anatexis of mafic rocks such as basalt, eclogite, or amphibolite. Most of these rocks crystallize directly from primary magmas, but some, like the "plagiogranites" (albite quartz diorites) may be derived by fractionational crystallization. Some gabbros and diorites represent primary, mantle-derived magmas, but these rocks commonly form by accumulation of crystals that settle or crystallize against the floor and walls of a magma chamber during fractional crystallization. Granodiorites form in a variety of ways. Some are considered to form directly from anatectic melts. Others develop by fractional crystallization of melts formed by (1) anatexis of metasedimentary rock or metaigneous rocks, (2) mixing of mafic and felsic magmas, (3) mixing of anatectic magmas and assimilated country rock, or (4) hybridization of magmas in subduction zones and the overlying ultramafic mantle. Quartz monzonites are typically derived from crystallization of either fractionated gabbroic to granodioritic melts or contaminated granitic melts, but they may locally evolve via anatexis of sediments or preexisting igneous rocks. Granites, especially peralkaline and peraluminous types, crystallize from anatectic crustal melts, but some granites also crystallize from fractionated magmas derived from the mantle or lower crust.

No one batholith can represent the spectrum of processes or tectonic history of granitoid batholiths and plutons. The composite Sierra Nevada Batholith, however, typifies the batholithic root of a moderately evolved "normal" arc. Locally, parts of it formed on both oceanic and continental crust, resulting in a variable chemistry. The tectonic history of this and other batholiths is best understood through tectonic, petrographic, petrochemical, and experimental analyses, which characterize and constrain petrogenetic models.

EXPLANATORY NOTES

1. Marmo (1971) gives an excellent review of problems related to the historical use of the term "granitic." He also provides a brief summary of the history of granite studies by European geologists that laid the foundation for twentieth-century work. Among the many other large or collective works of note on granites and related rocks are Gillully (1948), Larsen (1948), E. N. Cameron et al. (1949), H. H. Read (1957), Tuttle and Bowen (1958), Mehnert (1968), W. S. Pitcher and Berger (1972), Didier (1973), *Pacific Geology*, v. 8 (1974), Atherton and Tarney (1979), F. Barker (1979b), Cobbing et al. (1981), *Journal of Geophysical Research*, v. 86, no. B11 (1981), Cerny (1982b), Roddick (1983), *Physics of the Earth and Planetary Interiors*, v. 35 (1984), *American Mineralogist*, v. 71, nos. 3–4 (1986), J. L. Anderson (1990), S. M. Kay and Rapela (1990), H. J. Stein and Hanna (1990), D. B. Clarke (1992), and W. S. Pitcher (1993).

2. For example, see Streckeisen (1967) and Nockolds, Knox, and Chinner (1978).

3. Didier, Duthon, and Lameyre (1982), A. J. R. White and Chappell (1983), Bowden et al. (1984), and Barbarin (1990a) also discuss various granite classifications.

4. *Granitoid* is used in the same general way in the IUGS classification (Streckeisen, 1976), by geologists in the U.S.S.R., and by Smulidowski (1958, in Marmo 1971, p. 6).

5. Allegre and Hart (1978) provide a basic summary. Also see Pearce, Harris, and Tindle (1984).

6. Various papers in Allegre and Hart (1978) provide background on various types of trace element studies, including those on Pb and O isotopes. In addition, site-specific studies such as Pankhurst (1979), Halliday, Stephens, and Harmon (1980), M. D. Hill, Morris, and Whalen (1981), Harmon (1984), Vidal et al. (1984), Saleeby, Sams, and Kistler (1987), R. I. Hill and Silver (1988), and Walawender et al. (1990) employ these and other isotopes in petrogenetic analyses. Nd isotopes, in particular, are used by Farmer and DePaolo (1983), Kistler et al. (1986), Fleck (1990), C. F. Miller et al. (1990), and Stein and Crock (1990).

7. Arth and Hanson (1975), G. N. Hanson (1978), Atherton et al. (1979), Lopez-Escobar, Frey, and Oyarzun (1979), Cullers, Koch, and Bickford (1981), M. D. Hill, Morris, and Whelan (1981), Collins et al. (1982), C. G. Barnes (1983), Noyes et al. (1983a), Cocherie, Rossi, and Le Bel (1984), Lynch and Pride (1984), Allen et al. (1985), Frith and Fryer (1985), LeBel et al. (1985), Nabelek (1986), Capaldi et al. (1987), Speer, Naeem, and Almohandis (1989), D. H. Carlson and Moye (1990), and Sawka, Chappell, and Kistler (1990).

8. The mineralogy of granitic rocks is discussed in greater detail in texts such as Nockolds, Knox, and Chinner (1978), H. Williams et al. (1982), and Raymond (1993). For the mineralogy of specific occurrences, refer to the various papers cited in this chapter and chapter 12 and to T. E. Smith et al. (1983), and Pichavant and Manning (1984).

9. Notable studies of textural development in granitoid rocks include S. E. Swanson (1977), Fenn (1977, 1986), and S. E. Swanson and Fenn (1986).

10. Examples are described in Grout (1925, 1929), Buddington (1939, 1948), Knopf (1955), G. N. Hanson et al. (1971), Goldich et al. (1972), Rankin, Espenshade, and Shaw (1973), Arth and Hanson (1975), Ermanovics, McRitchie, and Houston (1979), Gee et al. (1981), Ludington (1981), E. I. Smith (1983), Frith and Fryer (1985), and Stoeser (1986).

11. For example, Daly (1914), C. R. Williams and Billings (1938), Stark et al. (1949), H. H. Read (1955), Tabor and Crowder (1969), Kistler et al. (1971), Pitcher and Berger (1972), Frisch (1974), Moench and Zartman (1976), Abbott (1978a), Lopez-Escobar, Frey, and Oyarzun (1979), Cobbing et al. (1981), P. C. Bateman (1983), Hudson (1983), Roddick (1983), Vidal et al. (1984), F. Barker, Arth, and Stern (1986), and J. L. Anderson (1990).

12. See the following works and the references cited in them for descriptions and additional information. Bowden and Turner (1974), C. A. Chapman (1976), Kistler, Ghent, and O'Neil (1981), Collins et al. (1982), A. J. R. White and Chappell (1983), Bowden and Kinnaird (1984), McHone and Butler (1984).

13. Kistler and Peterman (1973), Kistler (1974, 1978), and Ague and Brimhall (1988b) have clearly demonstrated that country rock composition can have a significant influence on the chemistry of plutonic rocks, notably on their isotopic and mineral composition. Ague and Brimhall (1988b) and Whalen and Chappell (1988) emphasize the importance of f_{O_2} in controlling mineral composition.

14. Hyndman (1972) based his classification on the work of his predecessors. For earlier discussions of this subject see F. J. Turner and Verhoogen (1960) and Simonen (1969).

15. Similar subdivisions are used by N. J. Jackson (1986b) and Stoesser (1986).

16. For example, see the discussions of C. F. Miller (1985) and A. J. R. White et al. (1986). G. C. Brown, Thorpe, and Webb (1984) suggest that some S-type granites may represent mantle melts. Hensel et al. (1985) note that S- and I-type distinctions are not always clear and they present a discussion of the difficulty of assigning these designations in one Australian case. For example, a sediment derived largely from rapidly eroded igneous rock might yield an apparent I-type magma from a sedimentary source. Arth, Peterman, and Friedman (1978) point out the difficulty of even distinguishing anatectic and fractional crystallization histories for generation of quartz dioritic magmas.

17. The usefulness of genetic classifications of the I- and S-type is quite dubious. Note that Brown, Thorpe, and Webb (1984) suggest that S-type rocks may have crystallized from contaminated mantle melts. C. F. Miller (1985) wisely advocated the use of terminology based on observed mineral composition and chemistry because of the simplicity and lack of ambiguity involved. I concur with C. F. Miller and advocate the abandonment of widespread use of I, S, A, C, M, or other similar genetic designations until such time as the breadth of the definitions and demonstrated links between the classification criteria and the origin allow unambiguous application of the labels.

18. H. H. Read (1944), available in Read's collected works on granite (H. H. Read, 1957). Read (1944) reviews European work on granitization, emphasizing important contributions by the French (Michel-Levy, Fournet, Deville, de Beaumont, Lacroix, and others) and northern European geologists (e.g., Sederholm, Holmquist, Kranck). For somewhat different reviews, see M. Walton (1955, 1960) and Marmo (1967, 1971).

19. For example, D. L. Reynolds (1947b) and H. H. Read (1948).

20. H. H. Read (1948, 1951, in H. H. Read, 1957, 1955), Misch (1949c), and Sederholm (1967). G1 refers to granitization argument 1. Similar terminology (e.g., A1 for anatexis argument 1) is used throughout this chapter.

21. Sederholm (1923; 1926, in Sederholm, 1967), G. H. Anderson (1937), Grout (1948), H. H. Read (1948), Misch (1968), Hutchison (1970), Cela and Yaque (1982).

22. See G. H. Anderson (1937), Backlund (1938), Goodspeed (1939), Buddington (1948), Jahns (1948), Misch (1949a,b; 1968), Whitten (1959, 1960), Tuominen (1961), A. E. J. Engel and Engel (1963), D. J. Gore (1968), and Hutchison (1970). W. S. Pitcher (1970) reviews various types of reported cases of ghost stratigraphy ranging from those of an apparently more static nature to those in which the granitoid rocks have clearly flowed from one place to another. *Note:* In some cases, layers not directly traceable into the country rock, but *interpreted* to be remnants of layering inherited from country rock, are included under the term ghost stratigraphy.

23. H. H. Read (1948) similarly denigrates evidence derived from physical chemistry.

24. Winkler (1976, ch. 18) suggests otherwise.

25. Grout (1925, 1948), Billings (1972), Abbott (1978a), S. E. Swanson (1978). Evidence of the magmatic nature of granitoid rocks summarized here and in table 2.1 is based on G. K. Gilbert (1880), Grout (1932, p. 32; 1948), and Billings (1972, p. 339), and on innumerable papers describing epizonal and mesozonal plutons (e.g., see Allison, 1925; Grout, 1929; Buddington, 1948; Taubeneck, 1964; and Michael, 1984).

Roddick (1982), in a curious paper consisting of numerous anecdotes and *ad hoc* arguments (rather than a synoptic analysis), correctly points out that nonmagmatic intrusions, such as salt diapirs and sandstone dikes, exist. While this is true, it does not follow that dikes composed of other materials are not magmatic. In his paper, Roddick treats geological objects (in this case dikes) as if each feature of the object is unrelated to every other feature. To argue that dikes of sandstone exist and that, therefore, dikes do not *necessarily* indicate magmatic activity may appear to the student as totally logical. It is not. The argument must go deeper. The dikes involved may exhibit *several* features which, *taken together*, make a case for magmatic activity. For example, a given dike may be composed of interlocking crystals of quartz and alkali feldspar, it may show a cross-cutting relationship with the country rock, it may have a fine-grained margin (indicating a large undercooling with a relatively slower growth rate, a point made by Swanson's 1977 paper cited by Roddick, but one he totally ignores in discussing fine-grained margins), and it may have a texture that indicates that crystals nucleated at the dike margin and grew inward. Taken together, these features suggest that a hot fluid intruded the cooler country rock and cooled against it. Experimental studies on the Qtz-Ab-Or system indicate that fluid was probably a magma. Thus, the evidence considered as a whole (rather than as isolated features) indicates magmatic activity.

H. H. Read (1948) and Roddick (1982) both argue that intrusions are not necessarily magmatic. Read (1948, 1955) suggests a "Granite Series," in which granites, formed at depth (in the catazone) during earlier stages in an orogeny, become mobilized and intrude upwards as plastic diapirs during later stages of the orogeny. There they form mesozonal to epizonal plutons. The works of R. Bateman (1985) and Ramsay (1989) contravene this model. These studies, combined with the knowledge that granitoid bodies typically have crystalline textures that can be modelled experimentally by solidification of melts (Fenn, 1972, 1986; Swanson, Whitney, and Luth, 1972; S. E. Swanson, 1977; S. E. Swanson and Fenn, 1986, Petersen and Lofgren, 1986), and not cataclastic to mylonitic textures that should result from plastic, solid-state diapirism, argue for a magmatic origin for the described features. Whether or not deformed granitoids at depth have structures consistent with those in the country rock (Simonen, 1969) is irrelevant to the question of whether or not epizonal to mesozonal granitoids are magmatic.

26. Wager and Deer (1939), Vance (1961). Also see Wager and Brown (1968), Fullagar, Lemon, and Ragland (1971), Ragland and Butler (1972), Bateman and Nokleberg (1978), G. C. Brown et al. (1979), P. C. Bateman and Chappell (1979), C. G. Barnes (1983), and Michael (1984). In addition, note the complexities suggested by Stephens and Halliday (1979).
27. Buddington (1948), Wager and Brown (1967), Wiebe (1974).
28. Refer to chapter 5 for a review of these studies and see Bowen (1928, 1937) and Tuttle and Bowen (1958). See Hildreth (1983a) and C. F. Miller and Mittlefehldt (1984) for related discussions.
29. Bowen (1948), Buddington (1948), Maaloe (1976), Allegre, Provost, and Jaupart (1981), Loomis and Welber (1982), Morse and Nolan (1984).
30. K. R. Walker (1969b), P. S. Meyer and Sigurdsson (1978), T. L. Wright and Peck (1978), D. D. Lambert et al. (1989).
31. For example, Whalen, Currie, and Chappell (1987) and Eby (1990).
32. F. Barker, Arth, and Stern (1986), Nabalek, Papike, and Laul (1986), Boily et al. (1989), Eby (1990), John and Wooden (1990), Sawka, Chappell, and Kistler (1990). Also see Hildreth (1979), Leeman and Phelps (1981), Michael (1983a,b), C. F. Miller and Mittlefehldt (1984) for discussions of fractionation of trace elements.
33. For example, see Ragland (1989 pp. 292 ff., 317–19) for some methods of calculation and Bickford, Sides, and Cullers (1981), Cullers, Koch, and Bickford (1981) and John and Wooden (1990) for examples in which such calculations were used to model fractionation.
34. Barker, Arth, and Stern (1986), Capaldi et al., (1987), Ragland (1989, p. 292 ff.), Hill, Chappell, and Silver (1988), Lira and Kirschbaum (1990), Newberry et al. (1990).
35. Vance (1961). For the Skaergaard Complex, see Wager and Deer (1939), Wager and Brown (1968); for the Ingonish Intrusions, see Wiebe (1974); and for the Loch Doon Pluton, see G. C. Brown et al. (1979). Other examples include the Farrington Igneous Complex (Wagener, 1965), the Galloway Plutons (Stephens and Halliday, 1979), and the Ardara Pluton (Holder, 1979). For reversely zoned plutons see Ayuso (1984) and Nabalek, Papike, and Laul (1986). Also see chapter 10.
36. Compare data in P. S. Meyer and Sigurdsson (1978), T. L. Wright and Peck (1978), and Lambert et al. (1989) to those in P. C. Bateman et al. (1963), P. C. Bateman and Chappell (1979), and T. P. Frost and Mahood (1987). *Note:* Lime is somewhat low in several of the residues.
37. Also see Maaloe (1976), the review in Mehnert (1968, pp. 185–89), Allegre, Provost, and Jaupart (1981), and Morse and Nolan (1984).
38. See references in Daly (1914, 1933) and Sederholm (1907, in Sederholm, 1967) to earlier works on assimilation. Also see Nockolds (1933, 1934) and D. E. Lee and Van Loenen (1971).
39. See Daly's texts (1914, 1933), Daly (1917), Saleeby, Sams, and Kistler (1987), J. B. Reid and Hamilton (1987), Dodge, Lockwood, and Calk (1988), Barbarin (1990b), and Vernon (1990).
40. J. B. Reid, Evans, and Fates (1983), T. P. Frost and Mahood (1987), J. B. Reid and Hamilton (1987), Zorpi et al. (1989), Glazner (1990).
41. For example, DePaolo (1980b; 1981d), J. L. Anderson and Cullers (1987), Glazner (1990), Kistler and Ross (1990), and H. J. Stein and Crock (1990). Also refer back to chapters 6 and 9 of this book for discussions of assimilation and magma mixing.

42. For example, M. D. Hill, Morris, and Whelan (1981), J. L. Anderson and Cullers (1987), Bergantz (1989), L. L. Larsen and Smith (1990), H. J. Stein and Crock (1990). For more references and information, refer to the analogous discussion of the origin of rhyolitic magmas in chapter 8.

43. Also refer back to chapters 6 and 9 of this book for brief discussions of assimilation and magma mixing. References to additional works documenting assimilation may be found in chapter 6, note 56 and chapter 9, note 55.

44. Compton (1955), Lee and Van Loenen (1971), Didier (1973), Wilcox (1979), J. B. Reid, Evans, and Fates (1983), Cocherie, Rossi, and Le Bel (1984), Ague and Brimhall (1987), Saleeby, Sams, and Kistler (1987), Mattinson (1990), and Saleeby et al. (1990).

45. Also see J. B. Reid, Evans, and Fates (1983), L. L. Larsen and Smith (1990), and Vernon (1990). T. P. Frost and Mahood (1987), Hill and Abbott (1989), and C. B. Mitchell and Rhodes (1989) describe situations in which magmas "mingled" without thoroughly mixing.

46. J. B. Reid, Evans, and Fates (1983), Kistler et al. (1986), J. B. Reid and Hamilton (1987), Barbarin (1990b).

47. Heimlich (1965), J. R. Butler and Ragland (1969), Fullagar (1971), Piwinskii (1973a,b), Chappell and White (1974), Arth and Hanson (1975), Arth, Peterman, and Friedman (1978), P. C. Bateman and Chappell (1979), DePaolo (1980b, 1981d), Halliday, Stephens, and Harmon (1980), Czamanske, Ishihara, and Atkin (1981), McCourt (1981), C. G. Barnes (1983), Bussell (1983), Domenick et al. (1983), Noyes, Frey, and Wones (1983), Cocherie, Rossi, and Le Bel (1984), Michael (1984), Frith and Fryer (1985), Hensel, McCullock, and Chappell (1985), Le Bel et al. (1985), Allen et al. (1985), N. J. Jackson (1986b), Cerny et al. (1986).

48. Though H. H. Read (1948) does not give magmas any role in his granitization model, he notes that Sederholm and many French geologists did. C. J. Hughes (1982) apparently equates granitization and anatexis, in part, and notes that the former is "convincingly demonstrated" by the work of W. W. Hutchison (1970) and coworkers (a surprising conclusion, considering that the work reported by Hutchison is largely petrographic and structural, and that Hutchison suggests a somewhat vague combination of mantle and crustal sources for the magmas involved). F. Barker and Arth (1984) have now suggested a magmatic origin for the batholith studied by Hutchison (1970), based on analyses of major and trace element data.

49. See W. B. Hamilton (1969), G. C. Brown (1973), T. H. Green (1980), and Wyllie (1984).

50. Sederholm (1907), van Platen (1965), Mehnert (1968), G. C. Brown (1973), Kenah and Hollister (1983).

51. Tuttle and Bowen (1958), Winkler, Boese, and Marcopoulos (1975). Also see Best (1982, p. 114, figs. 4–10), who cites the work of LeMaitre. Johannes (1983b) applied this kind of analysis to some layered gneisses to show that the light-colored layers in the gneiss are anatectic melts.

52. Chappell and White (1974), Arth and Hanson (1975), A. J. R. White and Chappell (1977), DePaolo (1980b, 1981a), Kistler, Ghent, and O'Neil (1981), Tindle and Pearce (1983), Berg and Baumann (1985), Hensel, McCullock, and Chappell (1985), C. F. Miller (1985), F. Barker, Arth, and Stern (1986), Leake (1990). Roddick (1982) criticizes the use of Sr isotopic data in assessing magmatic history. Here, as elsewhere, his argument is anecdotal and intuitive, rather than synoptic and documented.

53. For example, Noyes, Frey, and Wones (1983), J. L. Anderson and Cullers (1987), Beard and Lofgren (1989), G. B. Morgan and London (1989).

54. Winkler (1957, 1976), Tuttle and Bowen (1958), Winkler and von Platen (1958, 1960, 1961a,b), Wyllie and Tuttle (1961), Luth, Johns, and Tuttle (1964), Boettcher and Wyllie (1968), Piwinskii (1968, 1973a,b), G. C. Brown and Fyfe (1970, 1972), C. R. Stern and Wyllie (1973a,b, 1981), I. B. Lambert and Wyllie (1974), Huang and Wyllie (1975, 1981, 1986), Stern, Juang, and Wyllie (1975), Winkler, Boese, and Marcopoulos (1975), Wyllie et al. (1976), Wyllie (1977, 1983), Winkler and Breitbart (1978), Johannes (1980, 1983a), R. N. Thompson (1983), Clemens, Holloway, and White (1986), W. K. Conrad, Nicholls, and Wall (1988), Vielzeuf and Holloway (1988), Beard and Lofgren (1989), Keppler (1989), J. W. Peterson and Newton (1989, 1990), Carroll and Wyllie (1990), Puziewicz and Johannes (1990), Patino Douce and Johnston (1991). Winkler and his colleagues conducted a number of experimental studies in the area of anatexis of sediments and gneiss. Refer to the fourth edition of Winkler's *Petrogenesis of Metamorphic Rocks* (1976) for a review and references.

55. Additional chemographic or theoretical analyses are reported by Grant (1973, 1983, 1985a,b), A. B. Thompson and Tracy (1979), Abbott and Clarke (1979), MacRae and Nesbitt (1980), Abbott (1985), and DeVore (1981, 1982).

56. Other arguments posed against anatexis are those offered in favor of granitization. These have been summarized by H. H. Read (1948) and Walton (1960). What such arguments represent, in essence, is the assertion that migmatites, *dents de cheval*, inhomogeneous plutons, or other features are *more* compatible with a metamorphic than an anatectic/magmatic origin.

57. Also see the discussion of dehydration of subducting plates in chapter 9.

58. For example, see Kistler, Ghent, and O'Neil (1981), D. E. Lee et al. (1981), Hyndman (1983), Saleeby, Sams, and Kistler (1987), P. J. Sylvester (1989), J. L. Anderson and Cullers (1990), Eby (1990), Leake (1990); C. F. Miller et al. (1990), and Walawender et al. (1990).

59. For discussion and examples, see G. N. Hanson and Goldich (1972), Didier (1973), Chappell and White (1974), Arth and Hanson (1975), J. A. Whitney, Jones, and Walker (1976), A. J. R. White and Chappell (1977), G. C. Brown et al. (1979), DePaolo (1980b, 1981a), Collins et al. (1982), Didier, Duthon, and Lameyre (1982), Tindle and Pearce (1983), Cocherie, Rossi, and La Bel (1984), Hensel, McCullock, and Chappell (1985), Saleeby, Sams, and Kistler (1987), Sylvester (1989), J. L. Anderson and Cullers (1990), C. F. Miller et al. (1990), and Walawender et al. (1990).

60. See A. J. R. White and Chappell (1977), Johannes (1983b), and Chen, Frey, and Garcia (1990), but consider the multiple origins discussed by Didier (1987) and the caveats in Wall, Clemens, and Clanke (1987). Note that many recent studies suggest that enclaves are not restite but, rather, cumulates, wall rocks, or remnants of mixed magmas (Didier, 1987; T. P. Frost and Mahood, 1987; Bedard, 1990; Dodge and Kistler, 1990; Pin et al. 1990).

61. Chappell and White (1974), Arth and Hanson (1975), Hildreth (1981), Michael, (1984), Saleeby, Sams, and Kistler (1987), J. L. Anderson and Cullers (1990).

62. See note 54.

63. See notes 53 and 54 and various topical studies such as DePaolo (1980b) and Kistler et al. (1986).

64. See A. Holmes (1932), Winkler (1965, 1976), Fullagar, Lemon, and Ragland (1971), Presnall and Bateman (1973), Chappell and White (1974), Arth and Hanson (1975), T. H. Green (1976), J. A. Whitney, Jones, and Walker (1976), A. J. R. White and Chappell (1977), P. C. Bateman and Chappell (1979), Halliday, Stephens, and Harmon (1980), Cullers, Koch, and Bickford (1981), C. R. Stern and Wyllie (1981), Huang and Wyllie (1981), Bickford, Sides, and Cullers (1981), Kistler, Ghent, and O'Neil (1981), Collins et al. (1982), C. G. Barnes (1983), Bussell (1983), Duthou et al. (1984), Lynch and Pride (1984), Le Bel et al. (1985), Eby (1990), C. F. Miller et al. (1990), Norton and Redden (1990), and S. E. Swanson et al. (1990). Also see note 54.

65. Wagener (1965), Ragland and Butler (1972), Strong (1979), S. M. Kay et al. (1983), G. I. Smith et al. (1983), Cocherie, Rossi, and Le Bel (1984), Le Bel et al. (1985), N. J. Jackson (1986b).

66. See note 65.

67. See Wagener (1965), G. N. Hanson and Goldich (1972), Holloway and Burnham (1972), Ragland and Butler (1972), Presnall and Bateman (1973), Arth and Hanson (1975), Ishizaka and Yanagi (1975), Gastil (1975), F. Barker and Arth (1976), Wyllie (1977), F. Barker (1979a), Strong (1979), S. M. Kay et al. (1983), G. I. Smith et al. (1983), Cocherie, Rossi, and Le Bel (1984), Le Bel et al. (1985), N. J. Jackson (1986b), J. L. Anderson and Cullers (1987), Saleeby, Sams, and Kistler (1987), Leake (1990), and Walawender et al. (1990).

68. For example, see chapter 6, Nicholls and Ringwood (1973), Wyllie (1977), P. C. Bateman and Chappell (1979), Strong (1979), T. H. Green (1980), Halliday, Stephens, and Harmon (1980), C. G. Barnes (1983), S. M. Kay et al. (1983), J. B. Reid, Evans, and Fates (1983), Duthou et al. (1984), Le Bel et al. (1985), N. J. Jackson (1986b), Kistler et al. (1986), T. P. Frost and Mahood (1987), and J. L. Anderson and Cullers (1990).

69. Fullagar, Lemon, and Ragland (1971), Presnall and Bateman (1973), Arth and Hanson (1975), J. A. Whitney, Jones, and Walker (1976), Le Bel et al. (1985).

70. See A. Holmes (1932), Winkler (1965, 1976), Chappell and White (1974), T. H. Green (1976), A. J. R. White and Chappell (1977), P. C. Bateman and Chappell (1979), Halliday, Stephens, and Harmon (1980), Cullers, Koch, and Bickford (1981), C. R. Stern and Wyllie (1981), Huang and Wyllie (1981), Bickford, Sides, and Cullers (1981), Kistler, Ghent, and O'Neil (1981), Collins et al. (1982), C. G. Barnes (1983), Bussell (1983), Duthou et al. (1984), Lynch and Pride (1984), Le Bel et al. (1985), C. F. Miller et al. (1990), Norton and Redden (1990), and S. E. Swanson et al. (1990). Also see the discussion above on fractional crystallization and see chapter 12.

71. The Sierra Nevada Batholith has been studied for many years by a large number of workers. Early detailed work by Calkins (1930) in the Yosemite Valley area and reconnaissance work by other members of the U.S. Geological Survey were followed by numerous more recent studies. Notable among these is the long-term project to thoroughly evaluate a central east-west strip across the batholith, being carried out by the geologists of the USGS (W. B. Hamilton, 1956, 1969, D. C. Ross, 1962; P. C. Bateman et al., 1963; Eaton, 1963; Rinehart and Ross, 1964; P. C. Bateman and Wahrhaftig, 1966; P. C. Bateman and Eaton, 1967; P. C. Bateman and Lockwood, 1970; Evernden and Kistler, 1970; P. C. Bateman and Wones, 1972; J. G. Moore and Marks, 1972; P. C. Bateman and Clark, 1974; Kistler, 1974, 1978, 1990; Kistler and Peterman, 1978; P. C. Bateman and Nokleberg, 1978; P. C. Bateman and Chappell, 1979; Kistler and Swanson, 1981; Huber, 1982; Domenick et al., 1983; P. C. Bateman, 1981, 1983; Kistler et al., 1986; J. B. Reid and Hamilton, 1987; Dodge and Kistler, 1990; Kistler and Ross, 1990; and Sawka, Chappell, and Kistler, 1990). Additional works include Compton (1955), L. H. Larsen and Poldervaart (1961), Peikert (1965), Piwinskii (1968, 1973b), Schweickert (1976), S. E. Swanson (1978), DePaolo (1980b, 1981a), Raymond and Swanson (1980), Saleeby and Sharp (1980), Saleeby (1981), J. H. Chen and Moore (1982), Noyes, Wones, and Frey (1983), Noyes, Frey, and Wones (1983), J. B. Reid, Evans, and Fates (1983), Krauskopf (1984), Ague and Brimhall (1987, 1988a,b), T. P. Frost and Mahood (1987), Saleeby, Sams, and Kistler (1987), Ishihara and Sasaki (1989), Barbarin (1990b). This review is based on these works.

72. Bateman and Wahrhaftig (1966), Bateman (1981), Ague and Brimhall (1988a).

73. Age groups form a group of intersecting, roughly north-south-trending belts, each of which represents the roots of a volcanic arc, active during the particular magmatic episode involved. Evernden and Kistler (1970) suggested that the ages are divisible into five discrete groups or intrusive episodes, but these groupings are controversial. Saleeby (1981), using 90% of all known unpublished and published dates, including U-Pb dates, divides the rocks into five different time-geographic groupings. T. W. Stern et al. (1981) suggest that magmatism,

although episodic, was more continuous than suggested by Evernden and Kistler. Their summary of the data suggests three major phases of plutonism, one late Triassic, one Jurassic, and one middle to late Cretaceous in age. Suspect, probably exotic, terranes composed of dominantly mafic-ultramafic rocks occur in belts along the west edge of the Sierra Nevada Batholith (Behrman, 1987; Saleeby et al., 1987), but these are not included in the batholith.

74. W. W. Hutchison (1970), F. Barker and Arth (1976), Roddick (1983).

75. C. R. Williams and Billings (1938), R. W. Chapman (1954), C. A. Chapman (1976).

76. Cobbing et al. (1981). The southern end of the Sierra Nevada Batholith is eroded to deeper levels and has some catazonal characteristics (see Saleeby, Sams, and Kistler, 1987).

77. See examples in chapter 2, W. B. Hamilton and Myers (1967), and W. W. Hutchison (1970).

78. Chemical data are taken from F. A. Frey, Green, and Roy (1978), P. C. Bateman and Chappell (1979), J. B. Reid, Evans, and Fates (1983), and Kistler et al. (1986).

79. Also see Schweickert (1976), Bateman and Nokleberg (1978).

80. Lingren (1915, in P. C. Bateman, 1983), J. G. Moore (1959), P. C. Bateman and Dodge (1970), P. C. Bateman (1983), Ague and Brimhall (1987).

81. Schweickert and Cowan (1975), Schweickert (1978), Saleeby et al. (1987).

82. Dickinson (1970b, 1972), Coney (1972), Keith (1978), Raymond and Swanson (1980), P. C. Bateman (1983).

83. Buddington (1959), Tabor and Crowder (1969), Bussell, Pitcher, and Wilson (1976), Cobbing et al. (1981), Kistler and Swanson (1981), Noyes, Frey, and Wones (1983).

84. Sederholm (1923, 1926, 1934), G. H. Anderson (1937), H. H. Read (1944), Misch (1949a, 1968), Whitten (1959, 1960), Touminen (1961), D. J. Gore (1968), Ravich (1968).

85. Grout (1941, 1948), Goodspeed (1948).

86. Michel-Levy (1893–1894, in H. H. Read, 1948), Goodspeed (1939), H. H. Read (1944), Ranguin (1946, in H. H. Read, 1948), Misch (1949a), Goodspeed (1959), Gore (1968), Mehnert (1968, p. 126), Didier (1973, p. 215 ff.), Drescher-Kaden (1982). Mehnert (1968, ch. 6 and 10) reviews ion migration and feldspathization, as does Marmo (1971, ch. 5). Additional papers relating to this subject include A. L. Anderson (1949), M. Walton (1960), and Dietrich (1962), and the papers in note 87.

87. Bowen (1948), Orville (1963), Long and Luth (1986). Also see relevant materials in Gammon, Boresik, and Holland (1969), Luth and Tuttle (1969), Vidale (1969), Vidale and Hewitt (1973), I. Parsons (1978), Brady (1983), Ferry (1983), and G. M. Anderson and Burnham (1983).

88. Goodspeed (1948), Read (1948). Mehnert (1968, ch. 10, p. 332 ff.) provides a good review of ideas relating to this phenomenon, including a summary of the work of D. L. Reynolds. Goodspeed (1955) discusses types of xenoliths and dikes he believes to be relics from preexisting (pregranitization) rocks.

89. G. H. Anderson (1937), Engel and Engel (1963), Buddington (1948), Misch (1949a). Misch's (1968) argument for granitization of the Skagit gneiss is based, in part, on erroneous

P-T estimates, lack of information on melting in tonalitic systems, and false assumptions about the stability fields of minerals such as staurolite (see S. W. Richardson, 1968, for a discussion of staurolite stability). His work does raise important questions about the lack of equilibration evidenced by a wide range of plagioclase compositions. A detailed discussion of this work is beyond the scope of this text.

90. Taubeneck (1964), Emerson (1966), Fullagar, Lemon, and Ragland (1971), Ragland and Butler (1972), Holder (1979), Wones (1980), Noyes, Frey, and Wones (1983), Ayuso (1984), Nabelek, Papike, and Laul (1986), Speer et al. (1989).

91. For example, C. R. Williams and Billings (1938), R. W. Chapman (1954), P. C. Bateman and Eaton (1967), W. B. Hamilton and Myers (1967), Tabor and Crowder (1969), D. E. Lee and Van Loenen (1971), W. S. Pitcher and Berger (1972), Hamilton and Myers (1974), Bussell, Pitcher, and Wilson (1976), Sylvester et al. (1978), P. C. Bateman and Chappell (1979), G. C. Brown et al. (1979), Wones (1980), Cobbing et al. (1981), Sides et al. (1981), C. G. Barnes (1983), P. C. Bateman (1983), Roddick (1983), Fridrich and Mahood (1984), R. Bateman (1985), Bergantz (1989), and Leake (1990).

Lit-par-lit, French for "layer by layer," is a term applied to gneissose rocks with granitoid layers. See the glossary for additional information.

92. Tabor and Crowder (1969), Dickinson (1970), Bussell, Pitcher, and Wilson (1976), Thorpe and Francis (1979), Sides et al. (1981), Bickford, Sides, and Cullers (1981), Kistler and Swanson (1981), Fridrich and Mahood (1984), O'Brient (1986), Hanson, Saleeby, and Schweickert (1988).

PROBLEMS

11.1. Confirm the calc-alkaline character of the Tuolumne Intrusive Series using the data of P. C. Bateman and Chappell (1979) to construct a Peacock diagram (see chapter 4).

11.2. Calculate the Mg number for representative samples of the Tuolumne Intrusive Series (e.g., samples 0, 6, 21, and 26 of J. C. Bateman and Chappell, 1979, and sample F of J. B. Reid, Evans, and Fates, 1983). Do these data indicate any differences in the magmas from which the rocks crystallized?

11.3. Plot Al_2O_3 vs. MgO and TiO_2 vs. MgO for the Palisade Crest Intrusive Suite (Sawka, Chappell, and Kistler, 1990) and compare these graphs to those for the Tuolumne Intrusive Series (see figure 11.10). Explain the differences in light of the origins of the two assemblages of rock outlined by Kistler et al. (1986) and Sawka, Chappell, and Kistler (1990).

11.4. Based on the data presented in Sawka, Chappell, and Kistler (1990), characterize the Palisade Crest Intrusive Suite in terms of depth zone, granite type, and alumina saturation.

11.5. Plot an AFM diagram for the Palisade Crest Intrusive Suite (Sawka, Chappell, and Kistler, 1990) and the Indian Ocean rocks analyzed by C. G. Engel and Fisher (1975, table 2; table 3, analyses 1–4; and table 7). Compare and contrast the nature of the suites of granitoid-rock-bearing samples as revealed by these diagrams.

221

CHAPTER 12

Granites, Aplites, and Pegmatitic Rocks

INTRODUCTION

Granites (*sensu stricto*) are rocks with abundant K-rich alkali feldspar and quartz. Such rocks generally have been considered to be characteristic of Precambrian terranes, as well as both late-stage differentiated and pegmatitic zones of Phanerozoic plutons. More recent studies, however, suggest that rocks of granitic composition are less abundant in these types of occurrences than was implied by early workers (Goldich et al., 1980; Wooden, Goldich, and Suhr, 1980). Yet Precambrian terranes, Phanerozoic orogens, and Phanerozoic anorogenic magmatic belts are all provinces in which granites do occur. We now recognize that the granitoid rocks of the post-Archean Precambrian terranes are chemically much like those of Phanerozoic mountain belts. Many are intermediate in composition, instead of highly silicic.[1] Archean granitoids, though grossly similar to younger rocks, as a group are less silicic and have lower K_2O/Na_2O values (Condie, 1981; H. Martin, 1986). Texturally, many have been deformed and recrystallized into gneisses.

Granitic rocks are typically hypidiomorphic-granular in texture. Yet, in surveying granitic rocks, we encounter two special textural variants of granites, the aplites and the pegmatites. **Aplites** are medium- to fine-grained, allotriomorphic-granular, hyperleucocratic quartz–alkali feldspar rocks. **Pegmatites** are rocks characterized by textures in which the crystals are dominantly 3 cm or more in length. They are not all granitic.[2] Pegmatitic rock bodies, like other granitoid rock masses, may contain large volumes of granodiorite or other intermediate rock types and may also be deformed.

Recall that granitoid rocks are produced by the fractional crystallization of mantle or deep crustal melts, by direct crystallization of anatectic melts of granitoid composition, or by crystallization of hybrid melts (see chapter 11). Thus, different processes may lead to rocks of generally similar composition and texture. Granites, in particular, which may be formed from magmas of different origins, likewise have generally similar characteristics. In detail, however, we find that there are significant variations in both composition and texture among granitic rocks. The purpose of this chapter is to further explore some aspects of these variations and their origins. The mineralogies of granites were discussed in chapter 11 and will not be repeated here, except where additions relating to the special compositions of pegmatitic granites and other rocks are necessary.

(a)

(b)

Figure 12.1 Textures in granites. (a) Hypidiomorphic-granular texture in granite, Concord, N.H. (XN). (b) Allotriomorphic-granular texture in aplite, Sierra Nevada Batholith, California. (XN).

COMPOSITIONS OF GRANITIC ROCKS

The compositions of some granitic rocks are given in tables 11.1 and 12.1. In addition, analyses of some aplites and pegmatites are presented in table 12.1. Note the generally high silica contents for the granites and pegmatitic granites. In the cores of pegmatitic granites, the silica content may range to 100% (though analyses of these parts of pegmatites are rarely reported). Rocks with very high silica content are composed primarily of quartz and have been called **silexites**.[3] Notice also the very low MgO values and the high alkalis that characterize all but the pegmatitic gabbro.

As noted in chapter 11, tectonic position, in part, controls major chemical differences between various evolving granite magmas. The more aluminous, calc-alkaline rocks form in orogenic zones, whereas the alkaline varieties form in anorogenic zones. Crustal melting and hybridization to form granitic liquids are nearly confined to the orogenic zones, whereas fractionation of preexisting, more mafic magmas is important in both types of environment.

TEXTURAL VARIATIONS

Most granitic rocks exhibit one of four major textures. These are the hypidiomorphic-granular, pegmatitic, allotriomorphic-granular (aplitic), and porphyritic textures (figure 12.1). Special variants of pegmatitic textures, such as graphic or dendritic pegmatitic textures, are relatively common.[4] Although other textures occur, they are not common.

Understanding how these various textures form is important to understanding the origin of the rocks in which they occur. There are few experimental studies of textural development in granitoid rocks. S. E. Swanson (1977) reproduced some granitic rock textures experimentally in studies of synthetic compositions in the system $KAlSi_3O_8$-$NaAlSi_3O_8$-$CaAl_2Si_2O_8$-SiO_2.[5] Various weight percentages of water were added to the individual experiments. In systems with 3.5 wt. % H_2O at 8 kb (0.8 Gpa), he found that large undercoolings produced hypidiomorphic-granular textures, because both nucleation density and growth rates are high. In contrast, his studies of both synthetic granite and synthetic granodiorite suggest that small undercoolings yield porphyritic textures, because, although growth rates for alkali feldspars are high, nucleation density is low (figure 2.25).

223

Table 12.1   Chemical Analyses, Modes, and CIPW Norms of Some Granites, Aplite, and Pegmatites

	1[a]	2	3	4	5	6	7
SiO_2	44.32[b]	73.79	73.8	74.80	75.24	77.00	78.0
TiO_2	9.65	0.05	0.12	0.11	0.05	0.15	0.09
Al_2O_3	13.23	15.11	15.2	12.02	14.42	11.83	12.6
Fe_2O_3	1.40	0.26	0.19	1.61	0.14	0.40	0.92
FeO	5.32	0.16	1.1	1.13	0.35	1.05	nr
MnO	0.09	0.05	0.04	0.03	0.18	0.04	nr
MgO	2.04	0.07	0.24	0.12	0.01	0.04	0.00
CaO	18.44	0.97	0.73	0.01	0.20	0.61	0.37
Na_2O	2.77	4.71	3.0	4.41	4.23	3.06	3.48
K_2O	0.45	4.02	4.7	4.43	2.74	4.98	4.90
P_2O_5	1.17	0.01	0.20	0.01	0.13	0.02	0.02
Other	1.22	0.03	0.72	1.33	0.67	0.94	0.00
Total	99.99	99.23	100.	100.01	98.36	100.12	100.4

Modes and Mineral Composition[c]

	1	2	3	4	5	6	7
Qtz		X	X	30.8	X	34.7	X
KFs		X	X	64.3	X	46.4	X
Pl	X	X	X	0.0	X	15.4	X
WM		X	X	0.0	X	0.0	
Bio			X	0.0		3.0	
Other	X	X	?	4.9	X	0.5	—
Total				100.0		100.0	

Sources:

1. Pegmatitic sphene gabbro, Lake Placid Quadrangle, New York. Analyst: R. W. Perlich (from Buddington, 1939, table 7, p. 36).
2. Pegmatitic granitoid rock, Chalk Mountain, Spruce Pine District, North Carolina (Fenn, 1986, table 1).
3. Two-mica granite (Cretaceous), Ruby Mountains, Nevada, sample RM-19-66 (Kistler, Ghent, and O'Neill, 1981, table 1).
4. Fine riebeckite granite, Rattlesnake Pluton, Massachusetts, sample 74. Chemical analysis of sample 74K. Analyst: T. Asari Mode (1000 pts.) of sample 74A.(P. C. Lyons and Krueger, 1976). Norm by L. A. Raymond.
5. Pegmatitic granitoid rock, Harding Pegmatite, Taos County, New Mexico (Fenn, 1986, table 1).
6. Granite (A-type), Mumbulla Pluton, Southeast Australia (Collins et al., 1982, tables 1 and 2). Norm by L. A. Raymond.
7. Aplite, Notch Peak stock, Utah (Nabelek, 1986).

[a] A pegmatitic gabbro is included here to reinforce the idea that pegmatite is a textural, not a compositional, term; that is, any composition of rock may be pegmatitic.
[b] Values in weight percent.
[c] Values in volume percent.
nr=not reported.
X=mineral present.

Extrapolation of S. E. Swanson's (1977) studies to lower pressures, alkali-richer systems, and higher water contents would not seem to alter the results dramatically. S. E. Swanson (1977) suggests that lower pressures will not significantly change the general form of the nucleation and growth rate curves, which means that the same textures formed at higher pressures also form at lower pressures. Of course, at lower pressures, granite systems become hypersolvus, and only a single alkali feldspar (rather than both plagioclase and orthoclase) will crystallize from the magma.

Higher water contents alter the sequence of crystallization (J. A. Whitney, 1975),[6] so that quartz or alkali feldspar may appear before plagioclase, but there is no reason at present to assume that the *form* of the nucleation and growth curves would change. Finally, an increase in alkalis, with a concomitant decrease in lime, results in the occurrence of less calcic plagioclase and might produce a decrease in lime-bearing phases such as sphene, but again these conditions should not radically change the shape of the nucleation and growth curves.

CIPW Norms	1	2	3	4	5	6	7
q	0.90	29.76	36.22	31.74	51.43	38.19	38.4
c	—	—	4.33	—	—	0.32	0.94
or	2.78	24.48	27.78	26.16	18.19	29.45	29.0
ab	23.58	41.07	25.39	37.18	30.38	25.91	29.4
an	21.96	4.69	2.31	—	0.00	2.95	1.7
ac	—	—	—	0.14	—	—	—
di	11.02	—	—	—	—	—	0.26
wo	12.47	—	—	—	—	—	—
en ⎫	—	—	0.60	—	—	—	—
⎬ hy	—	—	—	0.95	—	1.40	—
fs ⎭	—	—	1.74	—	—	—	—
mt	2.09	—	0.28	2.27	—	0.58	—
il	9.73	—	0.23	0.21	—	0.29	0.02
hm	—	—	—	—	—	—	0.92
ap	2.69	—	0.47	0.02	—	0.03	0.05
cc	1.10	—	—	—	—	—	—
other	11.32	—	—	1.33	—	0.94	0.08
Total	99.64	100.00	99.35	100.00	100.00	100.06	100.8

Allotriomorphic-granular textures are most common among granitoid rocks in the rocks called aplites. Aplites are commonly thought to form by a process of *pressure quenching* (Jahns and Tuttle, 1963a, 1963b). This process occurs when granitoid magmas fractionate and become enriched in silica, alkalis, and a water-rich vapor phase. Fracturing of the enclosing rocks may cause a sudden release of this vapor, which lowers the pressure of the system and raises the temperatures of the liquidus and solidus curves. The result is pronounced undercooling and a consequent high nucleation density. A melt stable as a liquid under the initially higher water pressure (P_{H_2O}) becomes unstable under the lower P_{H_2O} and forms a solid, via abundant nucleation and rapid crystal growth. Rapid chilling or depletion of certain chemical species in the melt, as well as other processes, may also initiate the development of allotriomorphic-granular textures (Jahns, 1955; Jahns and Tuttle, 1963a, b).

Recall that pegmatitic textures are those dominated by crystals larger than 3 cm in length (figures 12.2 and 12.3). Although any composition of rock may be pegmatitic,[7]

Figure 12.2 Pegmatitic texture with very large crystals of alkali feldspar (> 1.2 m long) in granite at the Minpro Mine, Spruce Pine District, North Carolina. Dr. Richard Abbott provides scale.
(Photo courtesy of Dr. John Callahan, Applachian State University).

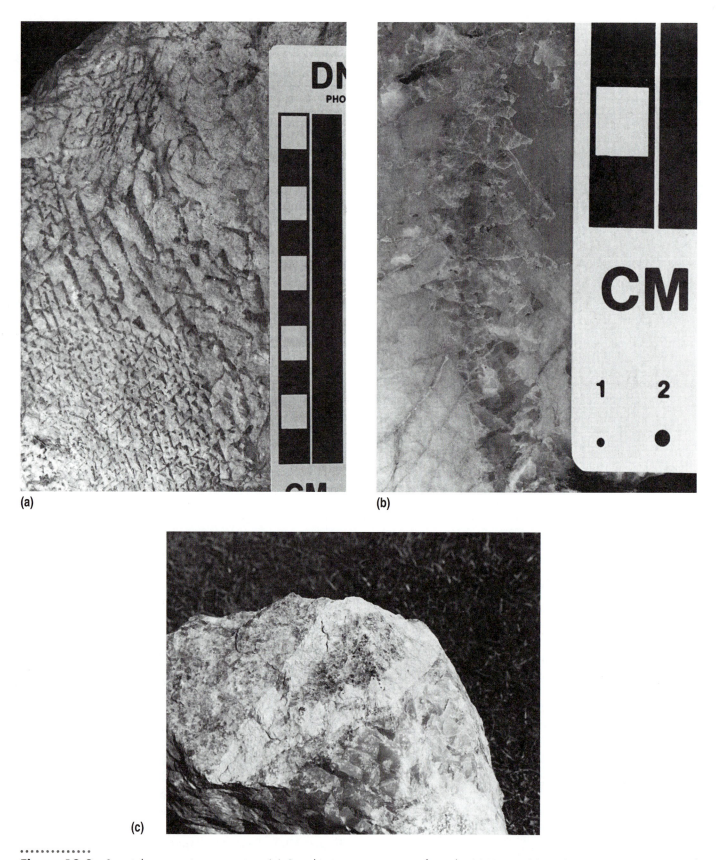

Figure 12.3   Special textures in pegmatites. (a) Graphic texture in granite from the McKinney Mine, Spruce Pine District, North Carolina. (b) Dendritic "snowflake" quartz from the McKinney Mine, Spruce Pine District, North Carolina. (c) Hollow "hopper-shaped" crystal of alkali feldspar in quartz, McKinney Mine, Spruce Pine District, North Carolina.

pegmatitic granites are among the most common types. Consequently, some students and geologists use the word *pegmatite* as a synonym for the more exacting "pegmatitic granite," but *this practice should be avoided.* Pegmatitic textures occur both in small local areas of a few centimeters in length within finer-grained rock masses and as the dominant texture of pegmatitic rock masses that range up to more than 150 m thick and 1500 m long (E. N. Cameron et al., 1949).[8]

Pegmatitic textures have been the subject of somewhat more discussion than have other granitic textures, largely because of the economic importance and unusual minerals associated with the granitoid pegmatites. Within the general category of pegmatitic textures, there are a number of specialized textures. These include, but are not limited to, very coarse hypidiomorphic-granular texture, graphic texture, dendritic ("snowflake") texture, perthitic texture, poikilitic texture, spherulitic texture, and gneissose texture (figure 12.3).[9] Each of these textures reveals details of the crystallization history of the rock. Perthitic textures, for example, result from hypersolvus crystallization and exsolution, as explained in chapter 5. Gneissose or layered textures may result from zonal crystallization (crystallization radially inward from the walls of a dike, "floating layer," or cavity), flow during crystallization, or postcrystallization deformation.[10]

Spherulitic, dendritic, and graphic textures provide important clues to the overall formation of pegmatites. Recall from chapter 2 that spherulitic textures are a nucleation effect, but dendritic forms are favored by silica-rich compositions, rapid rates of cooling, high growth rates, relatively few nucleation sites, and large degrees of undercooling. D. R. Simpson (1962) demonstrated that the geometrical grains of quartz in graphic-textured rocks are actually connected, forming large skeletal crystals. Snowflake dendritic quartz crystals are related to the quartz crystals of graphic-textured rocks, as evidenced in some pegmatites by apparently continuous quartz dendrites extending from snowflake-textured rocks into those that are graphic-textured.[11] Experimental production of graphic textures in granitoid rocks reveals that these textures are developed as a result of simultaneous crystallization of feldspar and quartz under conditions of low undercooling (Fenn, 1979, 1986). In contrast, relatively large undercooling results in dendritic growth (S. E. Swanson and Fenn, 1986). Thus, rapid rates of cooling, high rates of growth, and/or relatively few nucleation sites must be responsible for these textures at either large or small degrees of undercooling.

STRUCTURE OF GRANITIC PLUTONS

Granites occur both in the structural units (plutons) typical of other granitoid rock bodies and in certain distinctive structures. In particular, granites occur in batholiths and stocks, in

Pegmatite type			Sketch
Simple			
Complex	Composite	Zoned	
		Unzoned — With fracture filling	
		Unzoned — With replacement	
		Zoned — With fracture filling	
		Zoned — With replacement	
	Heterogeneous		

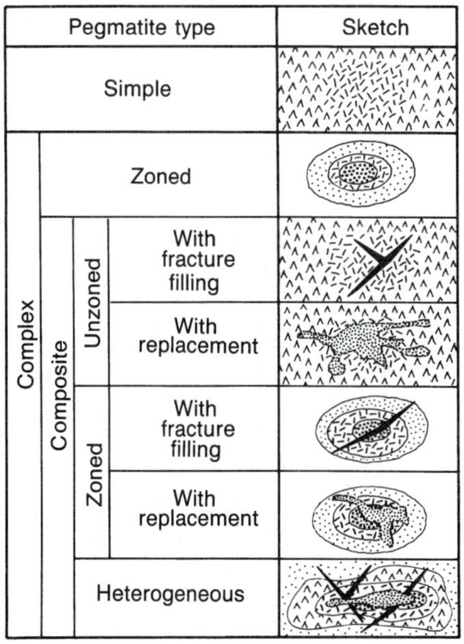

Figure 12.4 Simplified classification of pegmatites with sketches of the structures of pegmatite bodies. See text for a description of each type.

simple and complex pegmatite bodies, and as aplite dikes. These structures, like distinctive textures, provide clues to the origin of the rocks in which they occur.

Granites of porphyritic to hypidiomorphic-granular texture typically occur in the same kind of lenticular, irregular, and zoned plutons in which other granitoid rocks are found. In some plutons, the granite may form a zone or distinct intrusive unit representing a late-stage differentiate of a larger, more mafic magma body. Similarly, if the magma is anatectic in origin, the granite may form a lens-shaped pluton or one that takes the form of another geometrically simple to irregular shape. In any case, the actual shape of the intrusion is controlled by the shape of the magma chamber and may be lenticular and disklike, tabular and dikelike, cylindrical to domical, inverted teardrop-shaped, or irregular (see figure 2.11).

Pegmatite bodies of granitoid composition exhibit a range of shapes and distinctive structures (E. N. Cameron et al., 1949; L. R. Page et al., 1953).[12] In shape, most pegmatite bodies are tabular, elliptical, rod-shaped, or irregular lenticular masses. Internally, pegmatites are of two types, simple and complex (figure 12.4) (Landes, 1933).[13] **Simple pegmatite** bodies typically consist of (1) very coarse-grained areas within finer-grained granitoid plutons or (2) lenses in high-grade (high-T, high-P) metamorphic rocks. Texturally, they commonly contain very coarse hypidiomorphic-granular rocks, but graphic or other inequigranular textures are present in some simple pegmatites. Contacts tend to be diffuse and are marked by a coarsening of grain size, though some contacts are well defined. Mineralogically, the simple pegmatites are usually composed of the minerals characteristic of other granitoid rocks, namely, quartz and feldspars, with or without muscovite, biotite, and a few accessory minerals.

The **complex pegmatites,** though less common than simple pegmatites, are far more interesting because of their structures, textures, and unusual minerals. Six types of complex pegmatites are shown in figure 12.4. **Zoned pegmatites** are those in which layers, lenses, shells, or irregular masses of rock of distinctive composition comprise the body. A core, typically composed almost entirely of quartz, commonly occurs at or near the center of the body (figure 12.5) (Jahns and Burnham, 1969; Norton, 1983). In cases where the pegmatite formed at relatively shallow depths, these cores may contain miarolitic cavities ringed by one or more unusual minerals or may contain "pocket pegmatites," small areas of rock distinguished by euhedral crystals and unusual minerals, such as topaz, beryl, and tourmaline (Jahns, 1955, 1982). **Complex composite pegmatites** show two distinct stages of development, with the later stages consisting of replacements of early-formed minerals, fracture fillings that cross-cut early-formed minerals, or both (figure 12.5d)(E. N. Cameron et al., 1949). Complex composite bodies may be zoned or unzoned. Mineralogically, the complex pegmatites may contain one or more unusual minerals, including lepidolite, spodumene, topaz, beryl, tourmaline, tantalite $[Fe,Mn(Tn,Nb)_2O_6]$, triphyllite $[Li(Fe,Mn)PO_4]$, zeolites like laumontite, clays such as beidelite, and a host of other exotic minerals.[14]

Aplites occur as dikes, layers, lenses, or irregular masses (Jahns and Tuttle, 1963a, b).[15] The dikes cross-cut metamorphic rock masses and various plutonic rock bodies, including pegmatite bodies. The more irregular bodies occur as marginal layers along the edges of pegmatite bodies and as masses within the pegmatites.

PETROGENESIS OF GRANITES, APLITES, AND PEGMATITIC ROCKS

Granites

Granites (*sensu stricto*) occur almost exclusively in continental areas. Thus, the origin of most granites must be related in some way to the continental crust. This origin they share with most of the rhyolites.

Recall that granites are of peraluminous, metaluminous, or peralkaline character. The latter types tend to occur in anorogenic areas, whereas metaluminous granites typify the arcs and orogens. Peraluminous granites occur in both settings.

A Brief Review of Granite Origins

The three main hypotheses for the origin of granitoid magmas—anatexis, hybridization, and fractional crystallization (of a basaltic or other parental magma)—have been employed widely to explain the origin of granitoid rocks in batholiths, stocks, and other plutons (see chapter 11). Granites, in particular, may represent either anatectic melts or differentiates of crustal or mantle melts. Anatectic melts are commonly produced either at the base or in the lower levels of the crust. At the base, heat is transferred to the rocks by basaltic magmas derived from the mantle. In the lower crust, melting results from heating due to pressure release, radioactive decay, injection of superheated magmas, injection of fluids, or combinations of these processes. Fractionation, magma mixing, assimilation, thermogravitational diffusion, or combinations of these processes may further alter the chemistry of siliceous anatectic melts. The chemical and physical-chemical evidence for these explanations was summarized in chapters 6 and 11. The example of the Johnson Peak Granite Porphyry of the Tuolumne Intrusive Series of Yosemite (chapter 11) is representative of the development of a siliceous granitoid rock body within the context of a Phanerozoic batholith developed at a convergent plate margin. There, crustal anatexis was probably caused by heating induced by basaltic mantle melts ponded at the base of the crust.

Hybridization and especially fractional crystallization are important locally in forming granites. Fractional crystallization is a central process in the formation of some quartz monzonites, granites, and alkaline granites, such as those of the Rattlesnake Pluton of Massachusetts.

Example: The Rattlesnake Pluton, Massachusetts

The Rattlesnake Pluton of Massachusetts is one intrusion among many that comprise the belt of Paleozoic plutons extending from Connecticut to Newfoundland.[16] It lies near the southern end of an early Paleozoic belt of plutons, which differ from those of the Cretaceous part of the Sierra Nevada Batholith in chemistry, petrography, and intrusive style (Wones, 1979). The Appalachian rocks tend to be bimodal (gabbro and granite) rather than intermediate in composition, as a group are more siliceous and alkalic than Sierra Nevadan rocks, and commonly contain muscovite or alkali amphiboles.[17]

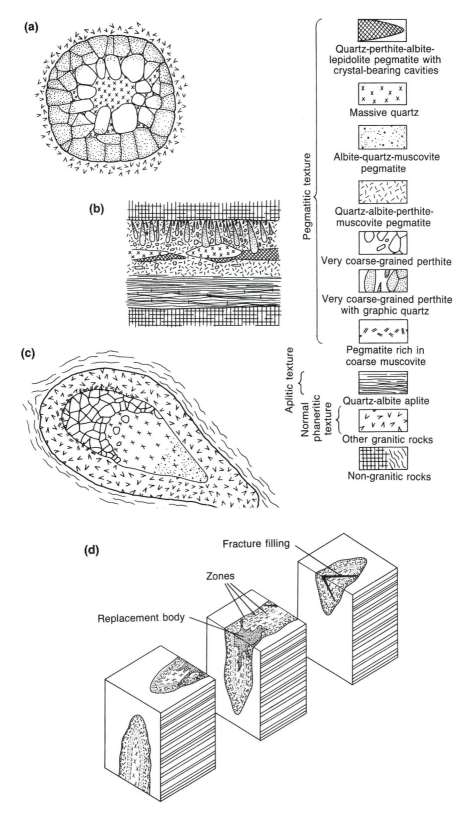

Pegmatitic texture

- Quartz-perthite-albite-lepidolite pegmatite with crystal-bearing cavities
- Massive quartz
- Albite-quartz-muscovite pegmatite
- Quartz-albite-perthite-muscovite pegmatite
- Very coarse-grained perthite
- Very coarse-grained perthite with graphic quartz
- Pegmatite rich in coarse muscovite

Aplitic texture
- Quartz-albite aplite

Normal phaneritic texture
- Other granitic rocks
- Non-granitic rocks

Fracture filling

Zones

Replacement body

Figure 12.5 Sketches of the internal structures of pegmatitic granites. (a) Symmetrically zoned, filled cavity. The diameters of such cavities are commonly several centimeters across. (b) Asymmetrically zoned, tabular pegmatite body with albite aplite base. The usual thickness of such dikes is a few meters. (c) Asymmetrically zoned, podlike pegmatite body with granitoid margin. Such bodies are usually several to tens of meters in thickness. (d) Zoned complex composite pegmatite.

((a), (b), and (c) from Jahns and Burnham, 1969; (d) from E. N. Cameron et al., 1949)

229

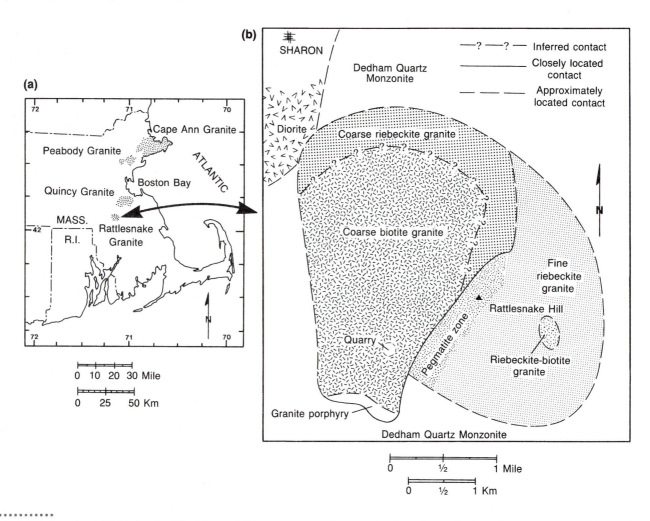

Figure 12.6   Maps of the Rattlesnake Pluton. (a) Location map showing Rattlesnake and other plutons in eastern Massachusetts. (b) Geologic map of the pluton showing various zones.
(From P. C. Lyons and Krueger, 1976)

The Devonian Rattlesnake Pluton consists of six zones, each characterized by a distinct granite. These zones developed in the intrusion as it invaded a larger pluton, the Dedham Quartz Monzonite (figure 12.6). All major rock types of the pluton are hypersolvus. A granite porphyry, with phenocrysts of microperthitic alkali feldspar, plus a small amount of biotite, forms both a marginal zone of the pluton on the south side and xenoliths within the more abundant coarse biotite-granite zone. The coarse biotite granite contains amounts of biotite (< 5%) similar to those of the porphyry, but is coarse-grained and hypidiomorphic-granular. On the north, the coarse biotite granite grades into a coarse riebeckite granite containing amphibole grains up to 15 mm in length. East of both the coarse biotite-granite and coarse riebeckite-granite units, there is a zone of fine riebeckite granite. Along the border between the coarse and fine units and in one area in the east, riebeckite-biotite granite is pres-

ent. Locally, within the riebeckite granite, and especially in the southwest, pegmatitic riebeckite granite occurs. Chemically, the fine riebeckite granite is peralkaline (analysis 4, table 12.1) and the coarse biotite granite is slightly peraluminous.

A magmatic origin for the rocks of the Rattlesnake Pluton is indicated by a variety of evidence. Xenoliths of country rock in the granitic rocks, the hypersolvus nature of the feldspars, the presence of dikes in the country rocks, and phase equilibria studies of riebeckite-granite mineralogies all favor a magmatic history (Lyons and Krueger, 1976). Although the contact relations are unclear, Lyons and Krueger (1976) infer a history involving three major phases of intrusion. The first phase involved early emplacement of a magma that gave rise to the enclosing Dedham Quartz Monzonite. This was followed by the second phase, anatectic development of an alkaline magma and emplacement of this magma into a shallow-level magma chamber. The coarse riebeckite granite developed in this chamber. Fractionation concentration of volatiles containing Na and Li at the top of the magma chamber resulted in the formation of pegmatitic granite. Cauldron subsidence produced ring fractures, along which the alkalic magma intruded and crystallized to form the fine riebeckite granite.[18] Crystal-liquid fractionation enriched the remaining magma in Al, Ca, and K. The final phase, a second cauldron subsidence event, resulted in the emplacement of the remaining magma, which crystallized to form the coarse biotite granite and its porphyritic margin. Fractionation apparently resulted in derivation of a peraluminous magma from a peralkaline parent.

The origins of the magmas that gave rise to the Rattlesnake Pluton are not specifically discussed by Lyons and Krueger (1976). Similar rocks studied elsewhere in the same plutonic belt and elsewhere in the world,[19] combined with theoretical models, such as those of Eichelberger and Gooley (1977) and Hildreth (1981), that explain the origin of siliceous magmas, suggest that basaltic magmas rising from the mantle probably supplied the heat for crustal anatexis. Melting occurred at the base of the crust, where basaltic magmas tend to pond. Siliceous melts thus produced will rise into the crust, where they may crystallize or fractionate or undergo other types of modification before either solidifying to form granitoid rocks or erupting to form rhyolitic volcanic rocks.

North of the Rattlesnake Pluton, the magmas that gave rise to the Ackley Granite of Newfoundland apparently formed via just such a crustal anatexis process (Tuach et al., 1986). There, however, different phases of the pluton have I-, S-, and A-type characteristics and the A-type peralkaline rocks evolved at shallower levels from deeper, older I-type magmas of peraluminous affinity. This sequence, peraluminous →peralkaline, is the opposite of that proposed by Lyons and Krueger (1976) for the Rattlesnake Pluton, but the intrusive sequence is admittedly unclear at Rattlesnake Hill. The important point here is that the chemistry of the magmas, including the high silica and alkali contents, the peraluminous character of some rocks, and the low Mg number, suggest that these rocks developed from magmas *not* derived directly from the mantle. Crustal sources and modification clearly played a role in their development.

Pegmatitic Granites and Aplites

Pegmatitic granites and aplites typically occur together.[20] They are common phenomena in granitoid plutons. This common association, the textures of these rocks, and the

unique minerals of pegmatitic granites provide the raw data for understanding aplite and pegmatite petrogenesis.

Of particular importance in understanding the differences between normal granite, pegmatitic granites, and aplites is knowledge of the textures and their origins (discussed above) and knowledge of the textural implications of the mineral compositions. The mineral compositions of aplites are simple, being composed principally of quartz and alkali feldspar. In contrast, although quartz, alkali feldspar, and plagioclase are principal phases in the pegmatitic granites, these rocks commonly contain additional phases that: (1) are hydrous (e.g., muscovite, biotite, lepidolite, topaz), (2) contain rather large, rather small, and rare ions (e.g., Li, U, Be, La, Nb, Ta), and (3) contain ions that are not abundant in typical rock-forming silicate minerals (e.g., F, Cl, B, P).[21] The presence of the hydrous minerals suggests that a hydrous fluid phase played an important role in the petrogenesis. The presence of unusual ions that do not fit into normally crystallizing rock-forming phases in granitoid rocks suggests that the fluid phase was residual and that it included those elements not taken up during the crystallization of the usual granite minerals. The general absence of the trace-element-bearing phases (such as tourmaline) from the aplites associated with the pegmatites suggests that the fluid phase was not a direct parent or source fluid for the aplite minerals.

Aplite Petrogenesis

As noted above, aplite, in contrast to pegmatitic granite, has a limited mineral composition. Quartz and alkali feldspar are the essential minerals present in most aplites. Locally, however, albite or other plagioclase feldspars; minor biotite, white mica, and garnet; and iron oxide phases do occur. Where albite occurs rather than the normal K-rich alkali feldspar, the rocks are called albite- or Na-aplites (Jahns, 1982). The limited mineral composition and the distinctive allotriomorphic-granular texture of the aplites are the features that must be explained by petrogenetic theories. The origin of the textures, through pressure quenching, was explained above.

Pressure quenching may also explain the limited mineral diversity in aplites. Fracturing of a host rock would allow the vapor phase to escape, carrying with it the unusual ions and H_2O necessary for the formation of the more exotic crystals. Thus, only the normal minerals of granites—quartz and feldspars—would be able to form.

Not all aplites form by pressure quenching, however. For example, irregular to layered aplites that occur *within* pegmatites reflect no structural evidence to indicate that fracturing occurred. Rather, they appear to be integral parts of

Figure 12.7 Photograph of garnet-rich aplite rim (R) on alkali feldspar (A) crystal surrounded by dendritic muscovite-bearing (M) pegmatitic granite from the McKinney Mine, Spruce Pine District, North Carolina.

the pegmatite body. This is especially true of the albite aplites (Jahns and Burnham, 1969; L. A. Stern et al., 1986). Thus, other origins are required. One important process may be selective extraction of a component from a melt via fractional crystallization (Jahns and Tuttle, 1963a, b). The result may be saturation of the melt in the remaining components, which will result in precipitation of phases for which the melt is saturated. Such a "compositional quench" appears to have oc-

curred around some large feldspar crystals in the McKinney Mine Pegmatite of North Carolina, as evidenced by the presence of a garnet albite aplite rim on the feldspar (figure 12.7).

Quenching of granite magmas may also occur as a result of intrusion into substantially colder materials. The quench produces a large undercooling and consequent high density of nucleation sites, resulting in an aplite.

Pegmatite Petrogenesis

Numerous origins have been proposed for pegmatites.[22] Metamorphic (granitization) origins have been advocated by some workers. Yet experimental evidence like that discussed in the previous chapter and described by Jahns and Burnham (1969), London, Morgan, and Hervig (1989), and other petrologists, combined with structural and textural evidence observed in most pegmatites, makes a strong case for a petrogenesis involving a magma (Cerny, 1982d). Like other granitoid rocks, it is likely that most pegmatites owe their character to the crystallization of either an anatectic melt or a differentiated magma (with or without subsequent or associated crystallization from a fluid phase separated from such a magma). Metasomatism and replacement are probably only important locally, notably along the margins of the magma bodies.[23]

Two competing theories of pegmatitic granite formation are outlined here. A definitive model of crystallization of granitic pegmatite bodies was presented by Jahns and Burnham (1969). Though some modifications in detail have been proposed,[24] the general model is widely accepted. In the model, the pegmatite bodies are envisioned to crystallize from a hydrous granitic magma via the following three-stage sequence of events (simplified and slightly modified from the original).

Stage 1. As the magma is emplaced, falling temperature causes crystallization to begin in the hydrous silicate melt. In epizonal and shallower mesozonal bodies, the cool country rocks may induce the development of aplite at the margins of the body. At greater depths, such aplitic margins will be absent. Typical phases characteristic of granite, that is, quartz and feldspars, with or without accompanying hydrous phases such as muscovite, crystallize in (generally) coarse grain sizes to produce granite or pegmatitic granite. The degree of undercooling, of course, will control various aspects of the texture.

One result of crystallization is that the liquid becomes relatively enriched in H_2O as shown in figure 12.8 by the path $a \rightarrow b \rightarrow c$. After about 30% of the melt has crystallized (point b), the percentage of H_2O in that melt will have increased from the initial 5% (at a) to about 7.3%. Continued crystallization will eventually lead to point c, at 55% crystallization, where the remaining melt becomes saturated in H_2O. At this point, further crystallization causes an aqueous fluid of low viscosity to begin to separate from the melt, in a process called *resurgent boiling*.

A number of factors may alter the exact sequence of events. For example, the chemistry of and pressure on the system may have an important influence on (1) the crystallization sequence, (2) the temperature of the liquidus and solidus curves, and (3) the H_2O content required for saturation of the melt.[25] Burnham and Jahns (1962) found that water solubility increases in granitic magmas, from 0% at a pressure of 1 bar to 11% at 5 kbar (0.5Gpa). Thus, increases in pressure will delay resurgent boiling by increasing the solubility of water in the melt. In contrast, higher initial water contents (curve $m \rightarrow n$)

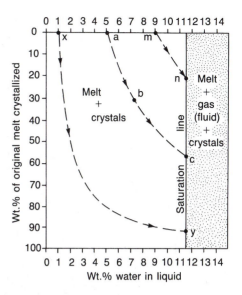

Figure 12.8 Graph showing initial water contents of granitic magmas, changing water contents during crystallization for selected initial water contents (curves $a \rightarrow b \rightarrow c$, $m \rightarrow n$, and $x \rightarrow y$), and the percentage of melt crystallized at the saturation point, at which the fluid (gas) phase separates from the melt. Confining pressure is 0.5Gpa. The 11.2 weight percent water used for the saturation value was determined on the Harding (New Mexico) pegmatite at 0.5Gpa and 650° C by Burnham and Jahns (1962).
(Modified from Jahns and Burnham, 1969)

will result in resurgent boiling after a much smaller percent of the original melt has crystallized. Lower water contents (curve $x \rightarrow y$) will have the opposite effect. Crystallization of hydrous phases (e.g., muscovite) will delay development of the fluid phase, because water will be taken out of the melt and into the structures of those hydrous phases. Finally, the presence of F, Cl, B and other elements may also alter the behavior of the melt and the point of resurgent boiling.[26]

Stage 2. Separation of a fluid phase from the saturated melt gives the system three major phases (melt, crystals, and fluid) and initiates the second stage. During this stage, crystallization from both the melt and the fluid phase occurs simultaneously. The high water content of the fluid phase results in a low viscosity, facilitating rapid ion migration and thus high growth rate. Nucleation density is apparently low. Together, these factors result in the local formation of very large crystals within the developing pegmatite body. Elsewhere within the body, compositional quenching may occur to form aplite. At other localities, the buildup of pressure from resurgent boiling may cause fracturing of previously crystallized material and the resultant formation of aplite dikes. A tendency on the part of the fluids to rise, leads to asymmetric structures, including zones, within the pegmatite bodies (figure 12.5b,c). A quartz core may begin to develop at this stage.[27] When all of the melt is consumed, the system becomes a system consisting of fluid and crystal phases.

Stage 3. Crystallization of the last bit of melt marks the end of stage 2 and the beginning of stage 3. The temperature continues to fall, reaching temperatures as low as 425° C or less (London, 1986a, 1986b). At this stage, in pegmatites developing under high pressure conditions, a mush of crystals and fluids is present in the interior of the pegmatite bodies, where quartz and lepidolite may be the last phases to form (Jahns, 1982; Norton, 1983). In epizonal pegmatite plutons, the interior may consist of a number of individual pockets of fluid containing uncommon ions, from which large crystals of tourmaline, beryl, or other minerals crystallize. High fluid pressures may cause periodic sudden fracturing of the pocket, resulting in the formation of aplites. Fractures not immediately filled by aplite may fill with fluid to produce cross-cutting veins. The fluid may also interact with the previously crystallized materials, metasomatically altering them to or replacing them with new minerals stable at the lower-temperature conditions. Zeolites and clays in the centers of pockets and miarolitic cavities develop at this stage.

Pegmatites may form without the development of stage 3 and may begin with stage 2, but the order of appearance of the stages is not altered. Although the sequence of crystallization described by the Jahns-Burnham model was primarily developed to explain the origin of pegmatitic granite, similar processes, including aplite formation and the development of a fluid phase, may also be important in the origin of pegmatites of other compositions (Jahns, 1955; J. S. Beard and Day, 1986).

The Jahns-Burnham model has some difficulty explaining some of the structural, textural, and mineralogical data (London, 1987, 1990; London, Morgan, and Hervig, 1989). For example, the formation of a large quartz core cannot be a magmatic equilibrium process (Burnham and Nekvasil, 1986), and some evidence indicates that growth rates for feldspars and quartz decrease, rather than increase, as saturation of a melt in H_2O occurs (S. E. Swanson, 1977). A third problem relates to the fact that there is some evidence that certain large ions and HFSE (e.g., Li, Be, Nb, Zr) will *either* concentrate in the melt phase, especially in high alkali melts, rather than in the vapor phase, *or* will be little affected by separation of a vapor phase (London, 1987; London, Hervig, and Morgan, 1988; London, Morgan, and Hervig, 1989).

In response to these problems, an alternative theory of rare element pegmatite formation was developed by London and his coworkers (London, 1987, 1990; London, Morgan, and Hervig 1989). This theory seeks to account for a wide variety of data including (1) zoning, (2) crystal fabrics, (3) crystallization sequences, (4) fluid inclusion data, (5) rare element fractionation trends, and (6) experimental data on the fractionation behavior of rhyolitic (granitic)

magmas. The theory proposes that siliceous, roof-zone granitic magmas, parental to the pegmatites, are separated from the cupolas of granitic magma chambers, that these magmas are injected into their country rock, and that they fractionally crystallize towards alkaline compositions in quasi-closed systems. The London theory differs from the Jahns-Burnham theory (1969) in that vapor saturation does not control crystallization history; that is, the aqueous fluid is of secondary importance in pegmatite petrogenesis. In short, the theory is one of fractional crystallization of *vapor-undersaturated* siliceous magmas.

The primary evidence and argument in favor of the London theory, as with the Jahns-Burnham theory, is experimental (London, 1987, 1990; London, Morgan, and Hervig, 1989). Experimental phase equilibria confine many pegmatites to crystallization at temperatures of less than 700° C (London, 1984; Barton, 1986; Lagache and Sebastian, 1991). Experiments with vapor-undersaturated fractional crystallization on peraluminous rhyolite glass at this and lower temperatures, and pressures of 0.2 Gpa (2 kb), produced structural, textural, and melt fractionation trends like those found in the rocks in siliceous and alkali pegmatites (London, Morgan, and Hervig, 1989). Such features as graphic intergrowths, extreme coarsening of grain size, mineral zoning with quartz cores, and fractionation towards Na-rich residual melts were reproduced in the experiments. In addition, the liquidus field of muscovite was apparently increased. A melt solidus at 450° C in the *dry* experiments suggests that alkalic pegmatite melts can exist to much lower temperatures than normal granitic melts.

The main challenge to the London theory currently is that it focusses on rare element pegmatites rather than dealing with the complete range of pegmatite mineral compositions and structures. The theory depends on experiments on a single composition of magma. In addition, because fractional crystallization was part of the experimental design, equilibrium crystallization was not evaluated. Nevertheless, the theory explains a wide range of features.

Example: The Tin Mountain Pegmatite, Black Hills, South Dakota

The Black Hills area is one of the most well-known pegmatite districts in the world.[28] Here, the 1.7–2.0-billion-year-old Harney Peak Granite forms a domelike structure that intrudes metamorphosed sandstones and shales (Runner, 1943; Redden et al., 1982).[29] The dome consists of a number of large and small sill-like bodies of granite plus numerous dikes. Both simple and complex pegmatites are abundant in the peraluminous granite and surrounding metamorphic terrane and at least 215 zoned pegmatites have been described.[30] The 2.0-billion-year-old Tin Mountain Pegmatite occurs in the metamorphic rocks about 12 km southwest of the main mass of Harney Peak Granite.[31]

The pegmatite body is irregular in plan and cylindrical in cross section (Staatz et al., 1963; R. J. Walker et al., 1986a) (figure 12.9). The complex, zoned pegmatite body contains six zones:

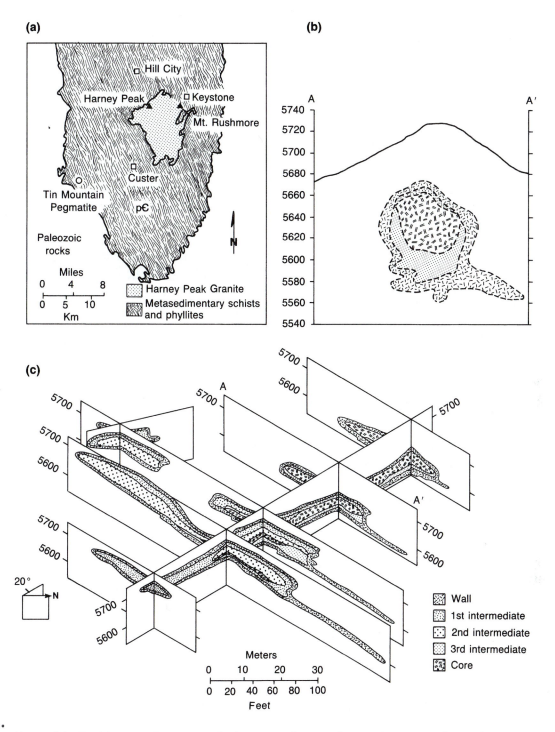

Figure 12.9 Maps of the Tin Mountain Pegmatite, Black Hills, South Dakota. (a) Location map showing the metamorphic core of the Black Hills, the main outcrop of the Harney Peak Granite, and the location of the Tin Mountain Pegmatite. (b) Cross section of the Tin Mountain Pegmatite, showing the zonal structure. The Border Zone is not shown because it is too narrow. (c) Fence diagram of the Tin Mountain Pegmatite. The vertical scale on the fence diagram is reduced 4 ¥, giving the pegmatite a flattened appearance. *(All figures modified from Walker et al., 1986)*

1. A very thin border zone;
2. A relatively thick quartz + albite + muscovite wall zone;
3. A first intermediate zone consisting primarily of perthitic alkali feldspar + albite + quartz;
4. A second intermediate zone of perthitic alkali feldspar;
5. A third intermediate zone of albite + quartz + spodumene + mica; and
6. A core composed of quartz + spodumene + mica (figure 12.9c).

Each of these zones is distinguished by a unique combination of composition, lithology, and texture. For example, the core is distinguished by a high silica (82.90%) and relatively high lithia (1.69%) content; a diverse mineralogy including quartz, albite, spodumene, muscovite, lepidolite, beryl, amblygonite, and pollucite, plus replacement minerals including lepidolite and cookite; and a hypidiomorphic-granular pegmatitic texture consisting of massive, anhedral quartz containing large euhedral to subhedral crystals of spodumene (up to 5 m long), beryl (up to 1 m across), amblygonite (up to 1 m across), micas, and albite. In contrast, the adjacent third intermediate zone consists primarily of subhedral crystals that rarely exceed 5 cm in length, and it is less siliceous (76.54%) but more lithia-rich (1.89%).

The peraluminous Harney Peak magmas were probably derived by anatexis of a sedimentary source, one perhaps similar, but not identical, to the exposed country rocks (R. J. Walker et al., 1986b; Norton and Redden, 1990). The isotopic and trace element data indicate such a source. The Tin Mountain Pegmatite magma was derived from a similar source or from a Harney Peak Granite parent. If the Harney Peak magma was the parent magma, trace element modelling (calculation) suggests that extreme fractional crystallization, perhaps by a process such as progressive equilibrium crystallization, was necessary to produce the chemistry of the Tin Mountain Pegmatite magma (Shearer et al., 1987; R. J. Walker, Hanson, and Papike, 1989).

A combination of chemical characteristics, oxygen isotopic and feldspar geothermometry,[32] and textural and structural data provide evidence useful for understanding the crystallization history of the pegmatite (R. J. Walker et al., 1986a, 1986b). The outer zones (border, wall, and first intermediate zone) crystallized successively as the body cooled. Then the three remaining lower zones (second intermediate, third intermediate, and core) crystallized simultaneously. Although the mechanisms for segregation of elements into various zones are not yet fully understood, an aqueous fluid *may* have developed during the stage of simultaneous crystallization (as in the Jahns-Burnham model). This would allow vertical segregation of the fluid-melt system to take place, promoting a zonal crystallization pattern. The fluid phase, would then be concentrated in the center of the pegmatite as crystallization progressed inward in the second and third intermediate zones, and would aid in redistribution of Si, Na, K, Li, Cs, and Rb. This chemical redistribution would have facilitated mineralogical zoning of the pegmatite.

The history of the Tin Mountain Pegmatite has not been evaluated in detail using the London model. In this model, extreme fractional crystallization of the melt would progressively yield the successsive zones and the Li-rich, quartz-mica core.

In either model, a late-stage fluid phase records the final stage in the pegmatite history. As the pegmatite cooled, these remaining fluids caused minor alteration of previously crystallized phases.

SUMMARY

Granite magmas develop through (1) melting of the base of the crust by mantle derived basaltic magmas, (2) anatexis of crustal rocks during high-temperature metamorphic events, or (3) fractionation of less siliceous magmas. These magmas crystallize directly to form granites, or they may differentiate or undergo other modifications while residing in crustal magma chambers. The original chemistry of the melt, the P-T conditions of fractionation, the nature of the modifications, and the conditions of emplacement will all influence the nature of the resultant magma. Some magmas evolve towards peraluminous compositions, whereas others become peralkaline.

Magmas that fractionate extensively tend to become enriched in fluids and rare elements. If a fluid phase forms and separates, the granitic systems will develop into multiphase systems that include crystalline phases, a melt, and a hydrous fluid phase. From such highly evolved (differentiated) systems or from highly fractionated, B-, F-, and P-enriched magmas, pegmatitic granite, rare element pegmatites, and aplites will crystallize. In nongranitic systems (e.g., gabbroic systems), evolution of a fluid phase can likewise result in the development of pegmatites.

Pegmatites range from mineralogically and structurally simple bodies to complex zoned or complex composite bodies that contain a wide variety of Li, B, F, Ta, and other rare-element-bearing minerals. These minerals may form giant crystals over a meter in length. Texturally, granites and pegmatites are commonly hypidiomorphic-granular, but some are porphyritic. The pegmatites also exhibit other textures, such as graphic and dendritic textures.

The unusual minerals and textures that form in pegmatites, at least in part, form as a result of crystallization from extreme differentiates of more normal granitic magmas. Undercooling varies in these vapor or melt differentiates, but nucleation density is commonly low, whereas growth rates are high. Consequently, very large crystals form. Aplites form where the pressure on a fluid phase is lost due to fracturing and subsequent fluid escape, where large undercoolings occur at contacts, or where compositional quenches occur.

The overall effect of the various processes that influence the development of granite magmas is diversity in texture and composition among the granites. That diversity provides a number of intriguing problems for petrologists, several of which are yet to be solved.

EXPLANATORY NOTES

1. For example, see Lund (1956), G. N. Hanson and Goldich (1972), Frisch (1974), F. Barker and Arth (1976), Arth (1976), Condie (1981), H. Martin (1986), and Robb et al. (1986). For analyses of Precambrian granitoid rocks that range widely within the general limits of granitoid rock composition, see works such as Arth and Hanson (1975), E. Smith (1983), Allen et al. (1986), Du Bray (1985), and Day and Weiblen (1986).

2. Grout (1918a), Buddington (1939), Beard and Day (1986). Also see note 7.

3. W. J. Miller (1919). Also see R. L. Bates and Jackson (1980).

4. For example, see E. N. Cameron et al. (1949), Fenn (1986), S. E. Swanson and Fenn (1986), and P. Keller (1988).

5. See chapter 2 for a review and discussion of this and additional aspects of texture development in igneous rocks.

6. Also see Piwinskii (1973a,b) and Maaloe and Wyllie (1975).

7. Pegmatitic rocks are known in various types of granitoid rocks (E. N. Cameron et al. 1949) and in silica-saturated and -undersaturated rocks (see Hurlbut and Griggs, 1939, pl.5, fig. 2; Jahns, 1955; Kendrick and Edmond, 1981). Beard and Day (1986) discuss the origin of gabbro pegmatites in an ophiolite complex.

8. Other important general works on pegmatites, on which much of this section is based, include Jahns (1955), Jahns and Burnham (1969), Uebel (1977), Cerny (1982a,b,d), the Jahns Memorial Issue of the *American Mineralogist* (1986, v. 71, nos. 3,4), and Modreski et al. (1986).

9. Augustithis (1973) pictures some of these and other textures that occur in granitoid rocks.

10. For example, Jahns and Tuttle (1963) and Gouanvic and Gagny (1987).

11. This phenomenon has been observed at the McKinney Mine in the Spruce Pine District of North Carolina. See Swanson (1978a,b) for a brief comment on North Carolina examples and the cover of Raymond (1993) for a schematic example.

12. Also see Landes (1933), Jahns (1955), Redden (1963), Jahns and Burnham (1969), Norton (1970), Jahns and Ewing (1977), Cerny (1982a,d,e), O'Brient (1986), L. A. Stern et al. (1986), R. J. Walker et al. (1986), Gouanvic and Gagny (1987), and London (1987).

13. The two-fold classification presented here is but one of many classifications. It follows, but differs from, those suggested in the seminal works of Landes (1933), E. N. Cameron et al. (1949), and Jahns (1955, 1982). In the author's view, many classifications are founded too firmly on inferred origin, with descriptive criteria as secondary classification parameters. Additional discussions of classification may be found in Uebel (1977), Cerny (1982a, 1986), and Jahns (1955, 1982).

14. For summaries and discussions of pegmatite mineral compositions that contain more thorough lists of minerals, tables of mineral chemistries, or other comments on occurrence and crystallization sequence, refer to the papers in Cerny (1982b), works such as E. N. Cameron et al. (1949), L. R. Page et al. (1953), Jahns (1955), Norton (1983), V. T. King (1987), and papers on individual pegmatites cited in other notes in this chapter.

15. For a detailed analysis of an aplite origin, see Nabelek (1986). Also see papers on individual pegmatite occurrences cited in this chapter, including O'Brient (1986) and L. A. Stern et al. (1986), for descriptions and discussions of specific aplites.

16. This example is based primarily on the work of Lyons and Krueger (1976).

17. See summaries by Moench and Zartman (1976), Clark et al. (1980), Loiselle and Ayuso (1980), and Wones (1980), for data and additional information, and the specific studies of Abbott (1978a), Nielson et al. (1976), Strong (1979), and Tauch et al. (1986).

18. See C. A. Chapman (1976) and chapter 13, where the process of cauldron subsidence is discussed in somewhat greater detail. See especially figure 13.13, Billings (1972, and references therein) and Bussell, Pitcher, and Wilson (1976), who depict cauldron subsidence in the formation of ring dikes and other plutons.

19. Bonin (1986) reviews many occurrences. See the references in note 17 for sources of additional information.

20. Specific examples and general comments regarding the association of pegmatitic granites and aplites can be found in Jahns (1955, 1982), Jahns and Tuttle (1963a), Foord (1977), B. E. Taylor, Foord, and Friedrichsen (1979), O'Brient (1986), L. A. Stern et al. (1986), and many other reports. Also see these selected works describing well-known pegmatites and their minerals: southern California—Jahns and Wright (1951), Jahns (1955, 1982), Jahns and Tuttle (1963), Foord (1977), Shigley and Brown (1985), L. A. Stern et al. (1986), and Foord, Starkey, and Taggart (1986); New Mexico—E. N. Cameron et al.(1949), Jahns (1955, 1974), Jahns and Ewing (1976, 1977), G. E. Brown and Mills (1986), Lumpkin, Chakoumakos, and Ewing (1986), and Chakoumakos and Lumpkin (1990); South Dakota—E. N. Cameron et al.(1949), L. R. Page et al. (1953), Orville (1960), Redden (1963), Staatz et al. (1963), Norton et al. (1964), Norton (1970), Walker et al. (1986), Jolliff, Papike, and Shearer (1986), and Norton and Redden (1990); Colorado—papers in Modreski et al. (1986); North Carolina—Olson (1944), E. N. Cameron et al. (1949), Brobst (1962), Kunasz (1982), Callahan (1985), and Hanahan (1985); New England—E. N. Cameron et al. (1949), E. N. Cameron et al. (1954), C. A. Francis (1987), and V. T. King (1987).

21. See papers in Cerny (1982b) and the *American Mineralogist* (1986), especially v. 71, nos. 3 and 4, for discussions and descriptions of pegmatite mineralogies. Papers on specific localities also give some details on mineralogy (e.g. L. R. Page et al., 1953). See London (1987, 1990) for discussions of the importance of B, P, and F in pegmatite magmas.

22. Jahns (1955) reviews major petrogenetic theories. Jahns and Burnham (1969) proposed a theory that was widely accepted, Jahns (1982) reviews and updates that theory, and Burnham and Nekvasil (1986) discuss modifications to the theory. London (1987, 1990) and London, Morgan, and Hervig (1989) present an alternative theory. Cerny (1982d) also reviews the various aspects of pegmatite petrogenesis.

23. For an example, see Shearer et al. (1986).

24. Modifications in the details of this model of pegmatite crystallization are suggested by the works of Burnham (1981), Jahns (1982), Burnham and Nekvasil (1986), and Norton (1983), London (1984, 1986b), and Jolliff, Papike, and Shearer (1986).

25. Tuttle and Bowen (1958), Burnham and Jahns (1962), Luth, Jahns, and Tuttle (1964), Jahns and Burnham (1969), Jahns (1982), Burnham and Nekvasil (1986), and S. E. Swanson and Fenn (1986).

26. Burnham (1967), J. C. Bailey (1977), Manning (1981), Pichavant (1981, 1987), Manning and Pichavant (1983), Burnham and Nekvasil (1986), London (1987, 1990) .

27. Luth, Jahns, and Tuttle (1964) provide important experimental data bearing on the formation of quartz cores in granitic pegmatites. Also see Burnham and Nekvasil (1986), but compare London (1987) and London, Morgan, and Hervig (1989).

28. The description of this example is based on the work of Runner (1943), L. R. Page et al. (1953), Stoll (1953), Kupfer (1963), Redden (1963), Staatz et al. (1963), Redden et al. (1982), R. J. Walker et al. (1986a,b), Shearer et al. (1987), E. F. Duke, Redden, and Papike (1988), R. J. Walker, Hanson, and Papike (1989), and Norton and Redden (1990).

29. See Riley (1970) and R. J. Walker et al. (1986b) for radiometric dates on the granite. E. F. Duke, Redden, and Papike (1988) suggest that the domelike structure results from multiple injections of magma into steeply dipping schists (metasediments).

30. Redden et al. (1982), Shearer, Papike, and Laul (1987), Norton and Redden (1990).

31. R. J. Walker et al. (1986 a,b).

32. *Geothermometry* is the use of mineral chemistry to determine temperatures of crystallization of rocks. See the glossary and chapter 24.

PROBLEMS

12.1. Are the pegmatites and aplite in table 12.1 peraluminous, metaluminous, or peralkaline?

12.2. Although the rare element pegmatites contain high alkalis, they are not necessarily peralkaline because they also have high alumina. Using standard mineral formulae, determine whether a pegmatite core rock composed of 50% quartz, 20% spodumene, 20% lepidolite, 5% albite, and 5% muscovite is peraluminous, metaluminous, or peralkaline.

13

Alkaline Igneous Rocks and Carbonatites

INTRODUCTION

Red hot lavas surging forth to form black pahoehoe and aa flows—an exciting scene visited by most geology students via films of Kilauean eruptions—is an image that comes to mind when we hear of volcanoes erupting. Along the East African Rift, the Oldoinyo Lengai Volcano erupted in 1960 and 1961 (Dawson, 1962a, 1962b, 1966). Yet, amazingly, the lavas did not glow when hot material was exposed in cracks. Furthermore, after 24 hours, the flows began to turn white, and within a week they were all colored light gray to white. Close inspection revealed that the flows were coated with the mineral Nahcolite ($NaHCO_3$) and were composed of a hydrated sodium carbonate matrix enclosing phenocrysts of an unknown, complex "sodium calcium potassium carbonate sulphate chloride mineral" (Dawson, 1966).[1]

Although basalts, andesites, dacites, and rhyolites are the most common volcanic rocks present on the Earth's surface, they are clearly not the only ones. The literature reveals that rocks such as trachytes and phonolites (nepheline-bearing volcanic rocks), as well as nepheline syenites, are quite important locally. In some volcanoes, unusual volcanic rocks may comprise the entire volcanic pile. Exotic sodium carbonate lavas, rare calcium carbonate dike rocks, uncommon tuffaceous phonolite, sanidine-rich trachyte flows, and the plutonic equivalents of such rocks—the carbonatites, syenites, and feldspathoidal syenites—are the kinds of rocks that are the subject of this chapter.[2] The syenites and other less common plutonic rocks are typically present where erosion has cut through volcanic piles containing unusual alkalic volcanic rocks. In addition, syenites also may be associated with gabbros and alkali granites. At depth and at the surface, the variety of unusual rock compositions that has formed is large.

An introduction to possible petrogenetic histories for a few of these unusual rocks is presented here to give the student an idea of the kinds of processes that may give rise to unusual volcanic and plutonic rocks. The examples are anecdotal rather than comprehensive.

ROCK TYPES, MINERALS, TEXTURES, STRUCTURES, AND CHEMISTRIES

Chayes (1979) notes that Cenozoic volcanic rocks have been given over 170 different names. Many of these names refer to rocks of one particular chemistry or mineral composition discovered at one location. Others, such as *phonolite*, are names applied widely to rock types less common than the typical basalts, andesites, and rhyolites of the arcs and rifts. Similarly, a large number of names has been applied to odd plutonic rocks. Only a few of these many rock types can be discussed here.

Predictably, the types of alkaline and related rocks and their minerals are interesting and varied. Those minerals reflect a diverse chemistry that is reflected in the examples presented below. Together with the textures and structures, these data reveal interesting histories.

Rock Types and Their Essential Minerals

In the terminology of Shand (1949), the rocks discussed in this chapter are either saturated or undersaturated with respect to silica. They are usually enriched in one or both alkalis, Na_2O and K_2O, and hence are given the name alkaline rocks. Many of these rocks are characterized by the presence of feldspars or feldspathoids and the absence of quartz. Some may lack the feldspar minerals entirely, being composed instead of feldspathoids, alkali pyroxenes, and amphiboles. In some related rocks, there is no quartz, feldspar, or feldspathoid; instead, the principal mineral is calcite, dolomite, or pyroxene.

Essential minerals range from the alkali feldspars (sanidine, anorthoclase, perthitic K-feldspar, microcline, orthoclase, and albite) in the syenites and trachytes (figure 13.1a and b) to feldspathoids (e.g., nepheline, pseudoleucite, cancrinite, sodalite) and inosilicates (e.g., aegirine, titanaugite) in the foidites[3], and carbonates (e.g. calcite, dolomite) in the carbonatites (figure 13.1c and d). Accessory minerals include alkali-bearing amphiboles (e.g., arfvedsonite, barkevikite, hastingsite, kaersutite), calcic and alkali pyroxenes (e.g., diopside, aegirine), biotite, olivine, melanite (garnet), analcite, zircon, magnetite, ilmenite, sphene, perovskite, monticellite, pyrite, pyrrhotite, apatite, fluorite, dolomite, ankerite, and calcite. In less alkaline syenites and monzonites, hornblende and augite are present rather than the sodic inosilicates.[4]

Among the rocks considered here, trachyte, latite, and phonolite are the most common volcanic rocks, and syenite and nepheline syenite are the most common plu-

tonic rocks. Recall from chapter 4 that *trachyte* is a feldspar-rich, aphanitic to aphanitic-porphyritic rock in which more than two-thirds of the feldspar is alkali feldspar.[5] The dominant feldspar is typically sanidine, but anorthoclase, orthoclase, or microcline may occur in some rocks. Characterizing and minor accessory minerals may include biotite, hornblende, augite, aegirine, magnetite, and ilmenite, as well as a number of other minerals. Trachyte is named on the basis of modal characteristics, normative mineralogy, both of these, or chemistry. *Syenite* is the plutonic equivalent of trachyte.

Other volcanic rocks of note include phonolite and latite. *Phonolite* is the feldspathoid-bearing equivalent of trachyte. *Latite* is a name used in some classifications for rocks intermediate between trachyte and andesite, in which the two feldspar types occur in subequal amounts. Accessory minerals are the same as in the trachytes. *Monzonite* is the plutonic equivalent of latite and *feldspathoidal syenite* is the plutonic equivalent of phonolite.

Olivine- and feldspathoid-bearing volcanic rocks (in addition to phonolite) include such types as tephrite, basanite, melilitite, nephelinite, and leucitite. Other unusual volcanic rocks include lamprophyres, kimberlites, komatiites, and the volcanic carbonatites.[6] *Tephrite* is a feldspathoidal basalt lacking olivine, whereas *basanite* is an olivine-bearing feldspathoidal basalt. *Nephelinite* is a nepheline, clinopyroxene rock that does not contain feldspar or olivine. *Leucitite*, like nephelinite, lacks olivine, generally lacks feldspar, and consists mainly of leucite and clinopyroxene. *Melilitite* is a melilite [$(Na, Ca)_2 (Mg, Al) (Si, Al)_2 O_7$]-clinopyroxene rock with minor feldspathoids. Recall that the *lamprophyres* are characterized by euhedral phenocrysts of ferromagnesian minerals. *Carbonatites* are igneous rocks containing 50 percent or more carbonate minerals (calcite, dolomite, and ankerite). Both volcanic and plutonic types occur.

Unusual plutonic rocks include carbonatites, *shonkinite* (dark-colored, augite syenite), and various feldspathoidal syenites. Nepheline clinopyroxenite, an ultramafic rock, is called *jacupirangite*. With more nepheline, the rock is called *melteigite,* and where nepheline is dominant, the clinopyroxene-nepheline rock is called *ijolite. Urtite* is an acmite-nepheline rock.

Textures and Structures

Texturally, the syenites, nepheline syenites, monzonites, ijolites, and related rocks, and the carbonatites usually are either allotriomorphic-granular or hypidiomorphic-granular. In some feldspar-rich rocks (syenites, feldspathoidal syenites, and monzonites), trachytoidal textures are not unusual. In addition, phaneritic-porphyritic, pegmatitic, and

(a)

(b)

(c)

(d)

Figure 13.1 Photomicrographs of alkaline and related rocks. (a) Syenite, Concord Gabbro-Syenite Complex, North Carolina. (XN). (b) Porphyritic trachyte, Christmas Mountains, Texas. (XN). (c) Nepheline syenite, Red Hill, New Hampshire. (XN). (d) Carbonatite, Oka Complex, Quebec. (XN). A = alkali feldspar, B = biotite, C = calcite, H = alkali amphiboles (e.g., hastingsite, arfvedsonite), ME = melilite, N = nepheline, PH = phlogopite, PX = alkali pyroxene (e.g., acmite). Black grains (opaques) are primarily magnetite, ilmenite, and pyrite. Long dimension of photos is 6.5 mm.

Table 13.1 Major Element Chemistry and CIPW Norms of Some Alkaline and Related Volcanic Rocks

	1	2	3	4	5	6
SiO_2	0.0[a,b]	36.75	46.6	51.52	54.5	54.9
TiO_2	0.10	2.41	0.75	0.75	0.80	0.39
Al_2O_3	0.08	11.98	12.9	15.01	17.90	21.1
Fe_2O_3	0.26	6.05	3.95	3.75	4.35	2.40
FeO		7.45	5.10	4.37	2.50	1.71
MnO	0.04	0.08	0.18	0.07	0.13	0.18
MgO	0.49	12.08	11.0	4.72	3.05	1.01
CaO	12.74	13.81	12.3	7.20	5.75	3.15
Na_2O	29.53	4.75	2.70	3.28	6.55	4.85
K_2O	7.58	0.91	1.60	5.46	3.40	6.65
P_2O_5	0.83	1.41	0.38	0.57	0.43	0.19
Other	49.11[c]	2.30	2.43	3.07	0.46	3.24
Total	100.73[d]	99.98	99.88	99.74	99.77	99.77

Sources:

1. Pahoehoe sodium carbonate lava, Oldoinyo Lengai, Tanzania, analysis 1, table IV. Total iron as Fe_2O_3 (Dawson, 1966). Norm by L. A. Raymond.
2. Nepheline melilite "basalt," Kalihi flow, Oahu, Hawaii, analysis 18, table 7 (Winchell, 1947).
3. Potassic basanite, Simberi Island, Tabar Group (R. W. Johnson, Wallace, and Ellis, 1976).
4. Mafic analcime phonolite, Highwood Mountains, Montana, analysis 6, table 2 (E. S. Larsen, 1941).
5. Potassic tephritic phonolite, Ambitle Island, Feni Group, sample 8, table 1 (R. W. Johnson, Wallace, and Ellis, 1976).
6. Potassic *ne*-trachyte, Put Plantation, Malendok Island, Tanga Group, Papua New Guinea (R. W. Johnson, Wallace, and Ellis, 1976).

[a] Values in weight percent.
[b] A trace of silica was reported.
[c] Includes CO_2, BaO, SrO, F, Cl, SO_3, and S.
[d] Adjusted for F, Cl, and S.

seriate textures also occur. The volcanic rocks are typically porphyritic, with phenocrysts of sanidine, analcite, leucite, nepheline, pyroxenes, amphiboles, or biotite, but pyroclastic textures occur.

The structures that the alkaline rocks and carbonatites exhibit are generally like structures found in other rock types. As the lavas at Oldoinyo Lengai reveal, even lavas of unusual compositions form flows (Dawson, 1966; J. Keller and Krafft, 1990). In addition, both the alkaline volcanic rocks and the carbonatites form dikes, sills, cones, and pyroclastic sheets.[7] The alkaline plutonic rocks commonly define composite, locally zoned intrusions, but plutons of a single rock type are known. Carbonatites likewise occur in the alkaline zoned intrusions, but also form dikes, cone sheets, and other shallow-level intrusive structures (Barker, 1989).

Chemistries

The chemistries of the alkaline and related rocks are quite diverse. It is far beyond the scope of this text to attempt to summarize all unusual rock chemistries, but a few general points are worthy of mention. As an introduction and guide to the chemical diversity of these rocks, tables 13.1 and 13.2 present a few analyses.

Silica ranges from zero in some carbonatites to about 65 wt. % in rocks transitional to the more siliceous rocks

	1	2	3	4	5	6
CIPW Norms						
q	—	—	—	—	—	—
or	—	—	9.72	32.35	20.26	40.72
ab	—	—	8.91	17.29	35.31	28.53
an	—	8.90	18.87	10.01	9.49	14.91
lc	—	4.36	—	—	—	—
ne	—	21.58	7.89	5.68	11.14	7.58
di	—	21.06	33.11	19.09	13.42	—
hy	—	—	—	—	—	—
ol	—	18.35	15.79	—	4.49	3.52
mt	—	8.82	3.35	5.57	3.36	2.84
il	—	4.56	1.46	1.37	1.53	0.77
ap	2.59	3.36	0.93	1.34	1.03	0.47
c	0.08	—	—	—	—	—
hm	0.26	—	—	—	—	—
cc	15.67	—	—	—	—	—
ac	—	—	—	4.68	—	—
cs	—	7.40	—	—	—	—
Other	82.13	—	—	—	—	—
Total	100.73	98.39	100.03	97.38	100.03	100.02

such as andesite and granodiorite. Alumina is generally high, because of the feldspar content, except in carbonatites and ultramafic rocks (e.g., see analyses 6, table 13.1; 6 and 7, table 13.2). Mg numbers vary widely. In some rocks, they cluster near 0.70, whereas in others the Mg number is < 0.06 (e.g., in some trachytes).[8] Total iron may be quite low (< 1%), whereas in certain jacupirangites it may exceed 15%. Lime and magnesia also may be quite high or quite low (see table 13.1 and note 8) and in some carbonatites, lime surpasses 50%. Soda and potash each may attain values of 6% or more.[9]

Like the major elements, trace elements vary widely. Some isotopic ratios are like mantle values, whereas others are enriched by crustal isotopes. As a result, the origins of specific rock types or the rocks of specific provinces must be assessed individually.

OCCURRENCES

Alkaline rocks once were thought to be emplaced in nontectonic regions, areas in which normal faulting rather than thrust faulting and folding predominate.[10] Some workers still hold that tectonic quiescence, or the absence of orogeny, is important, because of the wide distribution of alkaline rocks in stable continental areas.[11] Yet Daly (1933) doubted the connection, and the reviews by D. S. Barker (1974) and Woolley (1987) of alkaline rock occurrences in North and South America, and by Borley (1974) of alkaline rocks on oceanic islands, reveal that alkaline rocks (1) occur in continental *and* oceanic areas, and (2) are syntectonic as well as post-tectonic in age.

In North America, alkaline rocks may form as much as 10% of the total volume of igneous rocks.[12] Of all the occurrences, one-third are in the Precambrian shield, a few occur in

Table 13.2 Major Element Chemistry and CIPW Norms of Some Alkaline and Related Plutonic Rocks

	1	2	3	4	5	6	7
SiO_2	1.90[a]	35.42	41.19	45.67	47.87	52.47	58.30
TiO_2	0.10	4.05	1.67	4.05	0.75	0.12	0.10
Al_2O_3	0.33	9.21	17.99	17.68	11.92	22.71	21.38
Fe_2O_3	0.42	8.94	5.63	2.62	3.25	1.96	1.05
FeO	0.32	7.17	1.76	6.67	5.18	1.55	2.04
MnO	0.26	0.29	0.30	0.19	0.15	0.10	tr
MgO	1.05	7.77	3.06	4.89	8.72	0.10	0.22
CaO	53.37	20.83	14.61	10.72	9.92	1.22	0.95
Na_2O	0.00	1.47	6.85	4.25	2.95	8.25	8.66
K_2O	0.16	0.62	2.38	0.84	5.15	6.60	6.06
P_2O_5	2.00	2.23	0.41	1.70	1.11	0.08	0.04
H_2O	1.16	1.16	3.53	0.00	1.95	4.83	0.80
Other[b]	39.67	0.80	1.05	0.00	0.71	0.10	0.45
Total	99.74	99.96	100.43	99.28	99.63	100.09	100.05

Sources:

1. Carbonatite, Magnet Cove, Arkansas, sample L-304. Analyst: M. K. Balazs (Erickson and Blade, 1963). Norm by L. A. Raymond.

2. Jacupirangite, Magnet Cove, Arkansas, sample MC-173. Analyst: S. M. Berthold (Erickson and Blade, 1963).

3. Garnet ijolite, Magnet Cove, Arkansas, sample L-17. Analyst: L. M. Kehl (Erickson and Blade, 1963).

4. Leucogabbro, Mt. St. Hilaire, Quebec, Canada, sample Sa 006 (R. C. Greenwood and Edgar, 1984).

5. Shonkinite (augite syenite), Lower Shonkinite of Shonkin Sag "Laccolith," Montana, sample SS6 (Nash and Wilkinson, 1970).

6. Soda syenite, Shonkin Sag "Laccolith," Montana, sample SS34. (Nash and Wilkinson, 1970).

7. Nepheline-sodalite syenite, Red Hill, New Hampshire, analysis C, table 1. Analyst: H. S. Washington (Quinn, 1937).

[a]Values in weight percent.

[b]Includes CO_2, F, Cl, S, SO_3, BaO, and SrO.

[c]Included in di.

tr-trace

sediment-covered shield areas, and the remainder are found in the Wichita Mountains of Texas, the Black Hills of South Dakota, and in linear belts associated with the Appalachian, Ouachita, and Cordilleran mountain belts. Similarly, carbonatites occur in both stable and orogenic regions, but unlike alkaline rocks, they are confined to continental areas.[13]

A number of regions are known to geologists for their intrusions of silica-undersaturated and alkaline rocks. Among the localities in North America in which these various unusual rock types occur are the Monteregian Hills of Quebec, Canada (east of Montreal);[14] The White Mountain "Magma Series," extending across parts of New England, especially New Hampshire and Maine;[15] the Central Arkansas Belt, including Magnet Cove;[16] the Terlingua–Big Bend, Texas, area;[17] the Black Hills, South Dakota;[18] and the Central

Montana Province, including the Highwood and Bearpaw Mountains.[19] Elsewhere in the world, notable occurrences are found in Norway, Sweden, and the former Soviet Republics in the Baltic Shield; in central Europe; in east Africa, along the rift; and in Brazil.[20]

PETROGENESIS OF ALKALINE AND OTHER ODD MAGMAS AND ROCKS

The processes thought to operate in the petrogenesis of alkaline and other unusual magmas are normal magmatic processes. These include assimilation (usually of silica-poor

	1	2	3	4	5	6	7
CIPW Norms							
q	—	—	—	—	—	—	—
or	—	—	1.11	4.99	30.43	39.00	36.14
ab	—	—	—	28.90	0.41	18.36	36.15
an	0.42	16.96	11.40	26.84	4.07	5.44	2.78
lc	—	2.62	10.46	—	—	—	—
ne	—	6.53	30.96	3.94	13.30	27.87	18.74
th	—	0.14	0.14	—	—	—	—
di	—	37.15	16.42	12.49	31.22	0.27	1.21
wo	—	—	14.50	(6.51)c	(16.35)c	(0.13)c	—
ol	2.18	1.54	—	7.22	9.01	1.01	1.05
cs	0.76	7.40	—	—	—	—	—
mt	0.60	11.14	0.46	3.82	4.71	2.84	1.62
hm	—	1.28	5.28	—	—	—	—
il	0.18	7.60	3.19	7.74	1.42	0.23	0.91
ap	4.74	5.04	1.01	4.05	2.63	0.19	0.00
pr	0.36	0.72	0.75	—	—	—	—
cc	89.63	0.20	1.00	—	0.02	—	—
fr	0.43	—	—	—	—	—	—
Other	0.33	—	—	—	2.06	4.83	0.59
Total	99.63	98.32	96.68	99.99	99.28	100.04	99.19

rocks, such as limestone), fractional crystallization (typically of alkali olivine basalt or basanite magmas), separation of immiscible magmas (developed from a basic or ultrabasic parent), and anatexis.[21] In some cases, parent magmas are clearly considered to be primary mantle melts. In others, the magmas are highly evolved (derivative), having been through several stages of modification.

In the past, many petrologists considered that unusual rock compositions must require unusual origins. Daly (1910) reviewed the then known occurrences of alkaline rocks, found that carbonate rocks occurred near almost all of them, and concluded that alkaline rocks form through assimilation (syntexis) of those carbonate rocks by basaltic magmas. The idea is that limestone, which lacks silica and is Ca-rich, would "desilicate" the magma; that is, it would mix with the magma, reducing its silica content and increasing its Ca con-

tent. Later, Shand (1930, 1945), Daly (1914, 1933), Tilley (1952), and Schuiling (1964b) among others, defended this theory.

The main lines of evidence supporting the assimilation theory are (1) the field association of syenites, basalts, or other igneous rocks, with carbonate rocks, and (2) the fact that alkali- and lime-rich minerals *do* form where basalts encounter limestones.[22] Some experimental evidence supports the second line of evidence, but the volume and kinds of minerals that might be formed in this way are debated.[23] The work of Joesten (1977) suggests that the volume of alkaline rock generated will be very small, even where the intruding magma is gabbroic.

A number of arguments in addition to that provided by Joesten (1977) have been raised against the assimilation theory. Notably, Bowen (1922a, 1928) raised the issue of heat.

Specifically, melting of country rock would extract heat from a magma, causing it to crystallize.[24] Also, CO_2 exerts control over the assimilation process by concentrating in the vapor phase and removing H_2O from the magma (Watkinson and Wyllie, 1964, 1969). A decrease in H_2O results in an increase in the solidus temperature, causing the magma to crystallize. In addition to these experimental and theoretical objections, the field evidence noted by Daly (1910), that is, the alkaline rock–limestone association, has been cast into doubt by later work revealing that many of the carbonate rocks associated with alkaline rocks are, in fact, igneous; that is, they are carbonatites of magmatic origin, not sedimentary or metasedimentary rocks invaded by basaltic or granitic magmas.[25] Furthermore, if carbonate rocks were assimilated by mafic magmas, the Sr isotopic ratios would be high, but these ratios are commonly less than 0.706.[26] Thus, as concluded by many workers, there is little evidence to support the view that significant quantities of alkaline or carbonatite magma evolve by assimilation of limestone by basaltic or granitic magmas.[27] This does not preclude local assimilation as a minor magma modification process.

Assimilation of granite by carbonatite magma has also been proposed to explain the origin of alkaline rocks (A. Holmes, 1950). This model is thermochemically like the limestone assimilation model (Schuiling, 1964a) and fails for similar reasons (Watkinson and Wyllie, 1969; Wyllie, 1974).

The most widely accepted theories of formation of alkaline plutonic rocks are the crystal-liquid fractional crystallization theory, the liquid immiscibility theory, and the fractional fusion theory. Bowen (1922a, 1928, 1937) challenged Daly's (1910) syntexis model, particularly on the basis of phase relations and thermodynamic considerations (the heat problem), and proposed instead that fractional crystallization leads to the variety of alkaline rocks observed. Indeed, fractional crystallization is one of the most straightforward and most commonly invoked ways to explain the generation of syenite, trachyte, feldspathoidal syenite, and phonolite.[28] Reexamination of petrogeny's residua system, the system $NaAlSiO_4$-$KAlSiO_4$-SiO_2 (see figure 13.2a and ch. 5), reveals that three minima are plotted in the system. These are labelled m_o (silica-oversaturated), m_s (silica-saturated), and m_u (silica-undersaturated). These points correspond to the general compositions of rhyolite and granite (m_o), trachyte and syenite (m_s), and phonolite and feldspathoidal syenite (m_u). Inasmuch as each minimum represents a relative thermal low (with respect to some nearby points), Bowen (1937) considered that fractionation would drive liquid compositions down the thermal gradients (figure 13.2b) to those thermal lows, and crystallization of corresponding liquids would yield the respective rocks. For example, fractionation would drive a liquid of composition X (figure 13.2b) from its initial composition of $Ne_{46}Ks_{12}S_{42}$ down the thermal gradient to m_u. Crystallization of the liquid remaining at m_u would yield a feldspar-feldspathoid rock (a phonolite or

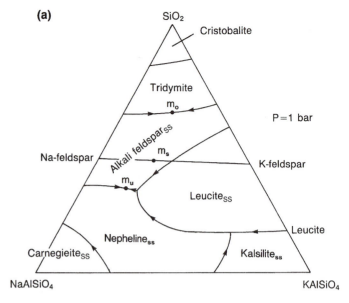

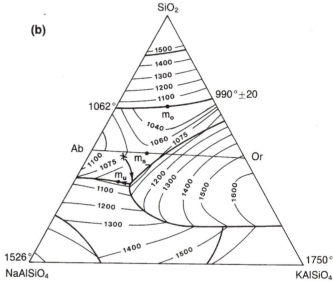

Figure 13.2 Petrogeny's residua system: the system $NaAlSiO_4$-$KAlSiO_4$-SiO_2. (a) Cotectic curves and minima. m_o = silica-oversaturated minimum, m_s = silica saturated minimum, m_u = silica-undersaturated minimum. ss = solid solutions. (b) Thermal contours. X is a sample composition discussed in the text, which cools along a path schematically shown as $X \rightarrow m_u$.
(Diagrams after Schairer, 1950, 1957; Fudali, 1963)

nepheline syenite). Similar processes have been proposed to yield rhyolite and granite, and trachyte and syenite.

Figure 13.3 is a contoured plot of 102 analyses of plutonic rocks with 80% or more albite + orthoclase + nepheline, obtained from Washington's tables (1917). The plot shows that the analyses cluster near m_u and m_s. The fact that a large number of analyses plot at and near those minima

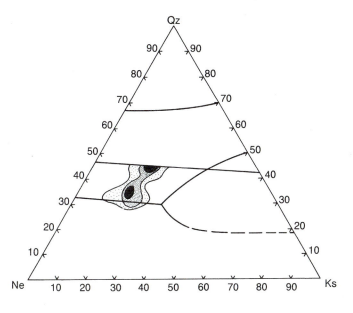

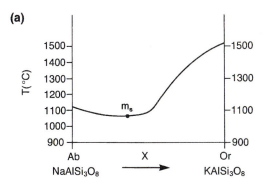

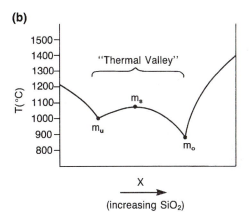

Figure 13.3 Contoured plot of 102 plutonic rock analyses containing 80% or more albite + orthoclase + nepheline, from Washington's (1917) tables.
(From D. L. Hamilton and MacKenzie, 1965)

Figure 13.4 Cross section of the liquidus of petrogeny's residua system along the line Ab–Or (a) and through points m_o–m_s–m_u (the thermal valley)(b), revealing that m_s is a thermal low along Ab–Or, but a high along the thermal valley. Note that m_o–m_s–m_u is not a straight line.
(Based on Schairer, 1957; Fudali, 1963)

suggests that crystal-liquid fractional crystallization (or ana-texis) played an important role in the development of tra-chyte and syenite, and phonolite and feldspathoidal syenite. Similar reasoning leads to the conclusion that fractionation is important to the origin of peralkaline-oversaturated rocks.[29]

In Hawaiian rock suites, the position of trachytes rela-tively near the alkali corner on the AFM diagram (see figure 7.9) suggests that trachyte represents an endmember of a frac-tionation series. Indeed, not only trachyte but hawaiite and mugaerite are attributed to crystal-liquid fractionation (G. A. MacDonald, 1968). These rocks form a common association that may represent the fractionation sequence

alkali olivine basalt → hawaiite → mugaerite → ben-morite → trachyte[30]
Chemically, such a sequence represents both silica and alkali enrichment.

Careful examination of the thermal contours will lead the observant student to the realization that m_s does not actu-ally represent a closed thermal depression on the liquidus of the ternary system. Rather, m_s occupies a high area within the thermal valley that extends from m_u to m_o (figures 13.2 and 13.4). Consequently, one would expect that trachytic liquids would represent metastable compositions, rather than end-member compositions, along a cooling path that would even-tually lead to rhyolite or phonolite. Thus, it seems surprising that trachytes are relatively abundant among the less com-mon rock types.

One cause for fractionation paths ending at m_s could be the development of "critical-plane" liquids (of the basalt tetra-hedron) as primary melts (S. A. Morse, 1980, pp. 278–79). These liquids would tend to fractionate along the feldspar (Ab-Or) line, on which m_s is a thermal low (figure 13.4a). Given such a prerequisite, syenite and trachyte liquids would represent end products. No explanation, however, has been formulated to explain recurring creation of primary critical-plane magmas in the mantle (S. A. Morse, 1980, p. 279).

Some syenites and trachytes may not represent the residue of crystallization. Note that syenites and trachytes typically contain significant amounts of normative plagio-clase (see tables 13.1 and 13.2). Thus, they contain the chemical components required to make plagioclase, but these are incorporated in other minerals, notably in the alkali feldspar. Nekvasil (1990) suggests that resorption of early-formed plagioclase during crystallization may account for the modal dominance of alkali feldspar in these syenites and trachytes. This process can account for some of the alkaline rocks but would require fractionation to explain bulk compositions that lie near m_s.

Phonolite and nepheline syenite compositions (m_u) do plot within a closed thermal depression on the liquidus of petrogeny's residua system. As such, they may represent residual liquids in a fractionation series such as

basanite → nepheline hawaiite → nepheline mugearite → nepheline benmorite → phonolite

or

sanidine basanite → nepheline trachyandesite → nepheline tristanite → phonolite[31]

In all such cases, a parent magma of basic composition is required, perhaps a basalt or basanite generated by partial melting of enriched mantle (Phelps, Gust, and Wooden, 1983; Boettcher, 1984).

Can fractional crystallization also explain the origins of more exotic rock types like ijolite and carbonatite? The common association of these rock types with the syenites, trachytes, and phonolites in alkaline rock complexes is consistent with the possiblity. For example, in the Magnet Cove example discussed below, carbonatite and ijolite form the core of the zoned plutonic-volcanic complex. Their central position, like that occupied by carbonatites, ijolites, jacupirangites, and related rocks in many alkaline complexes, and the fact that they crystallized late (Erickson and Blade, 1963) suggests fractionation. Fractionation is also supported by a continuous chemical gradation from kimberlite to carbonatite compositions in dikes of the Saguenay River Valley of Quebec, Canada (Gittins, Allen, and Cooper, 1975).

The above petrologic and petrochemical data supporting fractionation become more compelling in light of available experimental work. Wyllie (1965) and Watkinson and Wyllie (1969, 1971) examined phase relations in the systems CaO-MgO-CO_2-H_2O and $NaAlSi_3O_8$-CaO-CO_2-H_2O. Their work shows that fractionation of a carbonated, nepheline-rich magma will yield carbonatite from evolved liquids (Watkinson and Wyllie, 1971). Crystal settling and flotation in the feldspathoidal magmas should yield rocks ranging from urtite to ijolite, melteigite, and jacupirangite.[32]

Separation of immiscible liquids—one enriched in Ca, alkalis, and/or volatile components relative to a silicate-richer companion—was suggested long ago as another method of generating certain alkaline magmas (e.g., Weed and Pirsson, 1895). This hypothesis was ignored for many years but has regained some popularity, especially for the generation of carbonatite magmas (Kjarsgaard and Hamilton, 1989).[33]

Evidence favoring liquid immiscibility as a principal process in the formation of alkaline rocks and carbonatites includes textural-structural evidence, chemical evidence, and experimental evidence. The textural-structural evidence is meager. It consists of rare carbonate globules and ocelli in silicate rocks (Bogoch and Magaritz, 1983; Treiman and Essene, 1985; T. D. Peterson, 1989b) and centimeter-scale carbonate diapirs in a silicate layer (Dawson and Hawthorne, 1973). Both major and trace element chemistry contribute additional evidence in support of immiscibility. For example, in Oldoinyo Lengai lavas, a major compositional break exists between presumably related carbonate and silicate lavas; the minimum silica in the silicate lavas is about 37% and alumina is 12% contrasted with 0% silica and <1% alumina in carbonate lavas (Dawson, 1966, 1989). These silicate (nephelinite) and carbonate (natrocarbonatite) lavas have nearly identical Nd and Sr isotopic ratios, suggesting consanguinity (J. Keller and Krafft, 1990).

These data are supported by experimental studies of both natural samples and synthetic systems. Experiments on one nephelinite produced silicate-carbonate liquid immiscibility (T. D. Peterson, 1989a). A number of studies of synthetic systems also demonstrate that under conditions of $T >$ 750° C and $P <$ 0.1–2.5 Gpa (1–25 kb) certain carbonate and silicate magmas are, in fact, immiscible (Koster van Groos and Wyllie, 1963, 1973; Kjarsgaard and Hamilton, 1989; M. B. Baker and Wyllie, 1990).[34]

How important immiscible separation of magmas is to the generation of alkaline rocks remains unclear. Arguments against it being an important process are twofold. First, the rarity of petrographic evidence of immiscibility simply suggests that it is not a common process. Second, the silicate to carbonate rock ratio in carbonatite-bearing complexes is high, a fact that does not support the immiscibility model (Wyllie, 1989).

Anatexis, the final process considered to be important in alkaline magma genesis, has been considered to affect both crustal and mantle rocks. Daly (1914, p. 414ff.; 1933) suggested that alkaline magmas are produced by *crustal* anatexis of basic sediments or igneous rocks. The generally low Sr isotopic ratios of these rocks, along with other major, trace, and isotopic element data, generally deny such an origin.

A number of lines of evidence support the view that various silica-deficient magmas are primary *mantle* melts derived by partial fusion (anatexis) of chemically heterogeneous, enriched or depleted mantle sources.[35] This evidence includes (1) appropriate Pb, Nd, and low Sr isotopic ratios, (2) high Mg numbers of about 0.73 ± 0.05, (3) mantle xenoliths, (4) LREE-enriched rocks, (5) experimental phase equilibria studies, and (6) theoretical considerations. Parent magma types produced by fusion may include alkali olivine basalt, basanite, olivine nephelinite, nephelinite, leucitite, kimberlite, and carbonatite. The type of parent magma produced will depend on a number of factors, including the premetasomatic composition of the mantle, the degree of depletion or metasomatic enrichment of mantle rocks in CO_2,

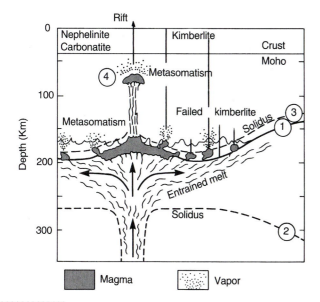

Figure 13.5 The Wyllie model for formation of carbonatite and alkaline magmas. 1 = lithosphere-asthenosphere boundary, 2 = solidus for peridotite + CO_2, 3 = solidus for peridotite + H_2O. 4 = level to which lithosphere-asthenosphere boundary may rise to form nephelinitic magmas. Partial melting occurs between 2 and 3. Alkaline magmas formed at 4 may fractionate to form carbonatites. *(From Wyllie, 1989)*

H_2O, or other components, the depth of melting, and the percentage of partial melting. Crystallization of primary magmas formed by fusion may yield nepheline syenites, nepheline gabbros, or carbonatites. Low Mg numbers and other chemical data suggest that similar-appearing gabbros, syenites, and related rocks are more evolved (B. A. Olsen, McSween, and Sando, 1983).

Fractional fusion may also be important in the origin of some alkaline rocks of the oceanic regions (oceanic trachytes) (Presnall, 1969; Chayes, 1977). As noted in chapter 5, fractional fusion is not the reverse of fractional crystallization. Thus, though the quantities of *residual* magma of composition m_s (see figure 13.2a) formed by fractional crystallization should be small, batches of *new* magma of this composition formed by fractional (or batch) melting could be significant.

None of the data, including isotopic data, allow an unequivocal answer to the question, how do alkaline and associated carbonatite magmas form? Either there are multiple origins or complex processes of magma and rock genesis are involved. Complex models are advocated, for example, by S. C. Eriksson (1989) for the Phalaborwa Complex and by Wyllie (1989; Wyllie, Baker, and White, 1990) for carbon-

atites and related rocks. In the Wyllie model, diapiric rise of mantle rock leads to mantle anatexis that yields an ultramafic (kimberlitic) magma at about 175 km (figure 13.5). Derivative nephelinitic magmas develop at a shallower level of the mantle (about 75 km) above the zone of kimberlite genesis, as these kimberlitic magmas rise towards the surface. At this shallower level, hybridization, immiscibility, and fractionation yield a variety of magmas, which may be modified again by the same processes when the magmas intrude the shallow levels of the crust (Grunenfelder et al., 1986; Sack, Walker, and Carmichael, 1987). Whether or not this model can explain all of the data remains to be judged, but one thing now seems clear from isotopic data, chemical data, and experimental studies: the parent magmas for alkaline and related magmatism are derived from the mantle.[36] Crystallization of these magmas and their derivatives yield the various alkaline and related rocks.

EXAMPLES OF ALKALINE AND RELATED ROCKS

Phonolite, Shonkinite, and Related Rocks of the Highwood Mountains, Montana

The Highwood Mountains, located in the center of Montana, are underlain by a variety of volcanic and plutonic rocks (figure 13.6).[37] By far, the largest surface exposures consist of "quartz latite" and various phonolites. One analysis of the "quartz latite," reported by Pirsson (1905) as a "trachyandesite," contains 59% silica and is chemically similar to typical andesites. Tuffs, breccias, and flows are all represented in the quartz latite unit. Like andesites, the "quartz latites" are typically hornblende and plagioclase phyric (E. S. Larsen, 1941).

The phonolites mimic the quartz latites, occurring as flows, breccias, and tuffs. In addition, they occur as dikes that intrude both the layered volcanic rocks and the surrounding country rocks (figure 13.7a), and they form chilled contact facies rimming syenite and shonkinite[38] cores of laccoliths (Pirsson, 1905; Hurlbut and Griggs, 1939; Buie, 1941; Nash and Wilkinson, 1970). *Tinguaite* (acmite-phyric dike phonolite) and lamprophyre dikes of at least three types (alnoite, minette, and vogesite) are also present (Pirsson, 1905; E. S. Larsen et al. 1935; Nicoll and Nicholls, 1974).

The phonolites (and the lamprophyres) are characteristically porphyritic (figure 13.7b). Some are seriate. In the phonolites, the phenocrysts, which are used to name variants (e.g., biotite phonolite), include augite, olivine, biotite, barium sanidine, pseudoleucite (sanidine and nepheline), pseudoanalcime or analcime, and rare hornblende (Larsen, 1941; Buie, 1941). Additional minerals occurring in the

249

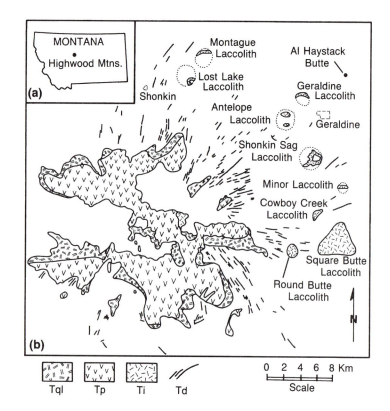

Figure 13.6 Map of the Highwood Mountains, Montana. (a) Location map showing Montana and the site of the Highwood Mountains. (b) Generalized geologic map showing igneous rocks. Al = alnoite, Td = dikes and sills of various compositions, Ti = intrusive rocks (e.g., syenite and shonkinite) excluding dikes, Tp = phonolites, Tql = "quartz latite." Ti stocks occur in the southwest and Ti laccoliths occur in the northeast. Textured areas within the laccoliths show actual exposure; dotted outlines represent the projected limits of the laccoliths.
(After E. S. Larsen, 1941)

(a)

(b)

Figure 13.7
Phonolitic rocks of the Highwood Mountains, Montana.
(a) View east in the Highwood Mountains showing phonolite and related dikes (pronounced exposures along ridge crests).
(b) Photomicrograph of phlogopite phonolite. (PL). A = alkali feldspar, PH = phlogopite, P = plagioclase, PX = clinopyroxene. Long dimension of photo is 3.25 mm.

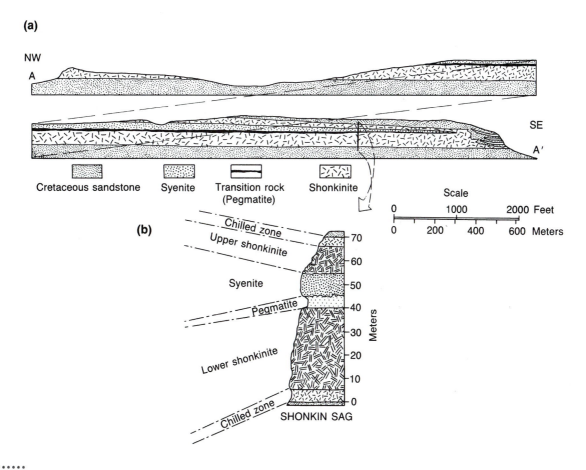

(a)

NW

A

SE

A'

Cretaceous sandstone | Syenite | Transition rock (Pegmatite) | Shonkinite

Scale

0   1000   2000 Feet

0   200   400   600 Meters

(b)

Chilled zone

Upper shonkinite

Syenite

Pegmatite

Lower shonkinite

Chilled zone

SHONKIN SAG

Meters — 70, 60, 50, 40, 30, 20, 10, 0

Figure 13.8 Shonkin Sag Sill. (a) Cross section. (slightly modified from Hurlbut and Griggs, 1939) (b) Columnar section.
(From Kendrick and Edmond, 1981)

groundmass include aegirine, plagioclase, calcite, apatite, and opaque minerals (including magnetite). Alteration minerals include serpentine, iddingsite, chlorite, and white mica, whereas cavities may contain zeolites (analcite, thompsonite, natrolite), calcite, and quartz (Larsen, 1941; Buie, 1941; Woods, 1974).

Among the igneous rocks of the Highwood Mountains is a series of stocks, sills, dikes, and laccoliths. Notable among these is the Shonkin Sag Laccolith, actually a disklike sill, 83 m thick at the center and 4156 m in diameter (228 ft. × 11,400 ft.) (Barksdale, 1937). It is the type locality of the rock *shonkinite* (mafic syenite).

The origin of shonkinite and related rocks exposed at Shonkin Sag and nearby is controversial. At least six stocks and nine large sills and laccoliths occur in the Highwood Mountains area (see figure 13.6). Most stocks are poorly exposed, but they are reported to be relatively homogeneous (Burgess, 1941). The sills and laccoliths, in contrast, show pronounced compositional layering. The Shonkin Sag Sill,

for example, contains six to eight layers (the number depending on the observer). These layers provide one of the clues to the petrogenesis of the rocks.

The layered Shonkin Sag Sill is the most studied of the intrusions of the Highwood Mountains and has served as the focal point of petrogenetic arguments. It intrudes Cretaceous sandstone and shale. A cross section and columnar section through the sill are shown in figure 13.8. At the base and top of the sill are chilled margins of olivine-pseudoleucite-salite–phyric, porphyritic phonolite. The lower zone is about 5.5 m thick, whereas the upper zone is slightly thinner at 4 to 5 m thick. Overlying the basal chilled zone is a thick zone (variously reported as 34–53 m thick), which is called the lower shonkinite. The lower shonkinite is hypidiomorphic-granular in texture and consists of Ca-rich clinopyroxene (salite), olivine, and biotite surrounded by sanidine and lesser amounts of nepheline, calcite, and zeolites. Above the lower shonkinite is a thin (< 1 m), relatively light-colored "hybrid syenite-shonkinite" zone (Barksdale, 1937), commonly included with the lower shonkinite. Pegmatitic syenite with graphic intergrowths and clusters of radiating pyroxene crystals, some greater than 5 cm long, overlies the lower shonkinite. The pegmatite is about 5 m thick. It, in turn, is overlain by 10 to 19 m of syenite, differing primarily from the shonkinite in being dominated by felsic minerals. The upper shonkinite (5.5–16.5 m) directly overlies the syenite, except locally where a very thin zone of coarse aegirine syenite occurs at the contact. Petrographically, the upper and lower shonkinites are similar.

The origin of the various magmas that gave rise to the Highwood Mountain rocks is controversial. E. S. Larsen (1940) suggested that a parent magma of basaltic character formed and differentiated at depth, rose to shallower levels in the crust, and differentiated again to form the various magmas that crystallized into individual rock types. This type of parent does not seem to be consistent with some trace element data (Woods, 1976). A *minette* parent, an alkalic rock with phenocrysts of phlogopite, diopside, and olivine, was proposed by H. E. O'Brien, Irving, and McCallum (1988). In contrast, a monticellite[39] peridotite parent magma was proposed by Nicoll and Nicholls (1974) and a kimberlitic parent is suggested by Woods (1976). Whatever the parent, all workers agree that the magmas were subsequently modified.

Modification of the Highwood parent magma may have involved more than a single process. Local assimilation of granites and limestone is suggested by E. S. Larsen (1940) and E. S. Larsen et al. (1941b) to account for certain rock types, whereas the possibility of liquid immiscibility yielding local variants has been discussed by Weed and Pirsson (1895), Woods (1974), and Kendrick and Edmond (1981). The latter process is typically used to account for variations in dikes and, especially, plutons (laccoliths and stocks). Mixing of minette and phonolite magmas high in the crust is advocated by H. E. O'Brien, Irving, and McCallum (1988). Compositional layering in the plutons is variously attributed to crystal-liquid fractional crystallization, multiple intrusion, and liquid immiscibility.[40]

Four modes of origin have been proposed for the Shonkin Sag Sill, nearby laccoliths, and their layering: (1) intrusion, followed by differentiation in place via crystal-liquid fractionation,[41] (2) intrusion with differentiation by separation of immiscible magmas,[42] (3) multiple intrusions,[43] and (4) intrusion with assimilation.[44] Of these, the assimilation hypothesis is overwhelmingly rejected, because neither the chemistry nor the field relations support the idea (Pirsson, 1905, pp. 182–83; Hurlbut and Griggs, 1939; Woods, 1974). Evidence offered in favor of an origin involving multiple intrusions of magma includes the following. (1) Intrusive, cross-cutting, and sharp contacts exist between various layers in the sill (Barksdale, 1937).

(2) Irregularities on the top of the sill suggest that magma entered the sill at two points (Dockstader, 1982). (3) A slight discordance between bedding in overlying rocks and the upper contact of the sill indicates multiple intrusion (Dockstader, 1982).

These lines of evidence may not necessarily indicate that multiple intrusions of magma produced the principal layering of the laccoliths. The discordance cited by Dockstader (1982) indicates *only* discordance, clearly revealed in outcrop (Barksdale, 1937; Hurlbut and Griggs, 1939), not multiple intrusion. The existence of multiple feeders for the sill (item 2) does not resolve the question of timing (was the sill fed from two sources at the same time or at different times?) nor does it preclude later differentiation. In fact, Hurlbut and Griggs (1939) suggest a second, late intrusion event as part of their overall differentiation model. Thus, some amount of multiple injection is favored by several workers, including those who advocate different models for the development of the principal layering.

The principal question is, then, was the major layering in the sill produced by multiple intrusions at separate times or by differentiation in place? The local intrusive contacts (e.g., local apophyses and dikes) and sharp contacts between major units provide the most significant evidence favoring multiple intrusion. However, Hurlbut and Griggs (1939) argued that local movement of still liquid magma into partly or wholly crystallized areas (autoinjection) is expected in a fractionally crystallizing magma body and that permanent, sharp boundaries will develop where there is a temporary boundary between crystallized and still liquid parts of the body. Kendrick and Edmond (1981) cite sharp contacts as evidence, not of multiple intrusion, but of magma immiscibility developed in place. Thus, at best, the evidence that multiple intrusions produced the major layers seems equivocal.

The alternative to multiple injection of magma is *in situ* differentiation either by separation of immiscible liquids or by crystal-liquid fractionation. Kendrick and Edmond (1981) and Kendrick (1982) argue that the development of immiscible magmas led to the separation of layers of different composition and the consequent gravity-controlled differentiation. Field evidence for immiscible magmas includes (1) streaked, swirled mixtures of light-colored syenite and dark-colored shonkinite, and (2) globular masses of syenite within shonkinite (at the nearby Square Butte Laccolith). Additional evidence includes (3) similar properties of minerals (suggesting similar chemistry) in both syenite and shonkinite, and (4) enrichment of trace and rare earth elements in the lower shonkinite relative to the syenite. Evidences 3 and 4 are debated. Nash and Wilkinson (1970) and Nash (1982) note, and Kendrick (1982) acknowledges, that not all minerals are chemically the same throughout the sill.

Nash (1982) and Nash and Wilkinson (1970) also argue that available data reveal that some trace and rare earth elements *are* enriched in the presumably most differentiated rock, a normative nepheline syenite (rather than in the presumably less differentiated shonkinite). Thus, available data do not compel us to adopt a liquid immiscibility model.

If the data do not unequivocally support liquid immiscibility, do they favor the alternative, crystal-liquid fractionation? The model of crystal-liquid fractional crystallization, in one form or another, was supported by Pirsson (1905), F. F. Osborne and Roberts (1931), Hurlbut and Griggs (1939), and Nash and Wilkinson (1970, 1971), among others. In this model, progressive crystallization of the magma, combined with settling and flotation of heavy and light crystals, respectively, yields a layered intrusion. Evidence offered in support of fractional crystallization includes the following: (1) generally horizontal layering throughout much of the sill, (2) gradational changes in mineralogy within layers, (3) enrichment in Si, Al, and Na and depletion in Fe and Mg from "least differentiated" to "most differentiated" rocks (degree of differentiation is inferred from field relations and chemistry), (4) enrichment of trace elements such as Co and Ni in more mafic, presumably cumulate, shonkinites, and enrichment of REE, such as La, Ce, and Nd, in the "most evolved" syenites, and (5) changes in mineral chemistry—iron enrichment in olivine and biotite and Na enrichment in pyroxenes—from early-formed rocks (phonolites of the chilled margin) to later-formed rocks (the syenites).

These data, too, have been subject to debate. Kendrick and Edmond (1981) suggest that (1) the layering is consistent with immiscibility, not only with fractionation; (2) cryptic and rhythmic layering, typical of cumulate rocks, is not present in the Shonkin Sag Sill; (3) major element partitioning between mafic and felsic rocks is consistent with both fractionation and immiscibility models; (4) trace elements, notably Ce, Sm, and La, are preferentially concentrated in mafic, not felsic rocks; and (5) changes in mineral chemistry are not evident, as olivine, augite, and biotite remain optically "the same throughout the laccolith." Items 4 and 5 clearly contradict the earlier work of Nash and Wilkinson (1970, 1971). The chemical analyses presented in the earlier and later works simply do not agree.

It is safe to conclude that differentiation occurred in place. Available data, however, do not allow discrimination between the immiscibility and fractionation models. As suggested by Nash (1982), a careful chemical study of "critical" rock types and their minerals is needed. Crystal-liquid fractionation was probably the most important *early* differentiation process involved in developing lava chemistries.

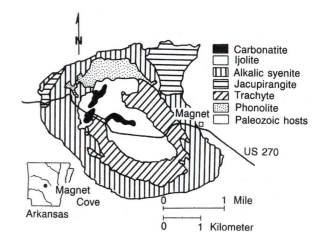

Figure 13.9 Generalized geologic map of the Magnet Cove Igneous Complex.
(Based on Erickson and Blade, 1963)

Alkaline and Related Rocks of Magnet Cove, Arkansas

In Arkansas, during the Cretaceous period, several alkalic igneous complexes intruded folded and faulted Paleozoic sedimentary rocks of the Ouachita Mountain System (Williams and Kemp, 1891, in E. R. Lloyd, 1923). Rocks ranging from aphanitic dike rocks to phaneritic plutonic rocks intruded and crystallized, forming the Magnet Cove Igneous Complex and similar complexes.[45] The histories of the various rock types are intertwined.

The Magnet Cove Igneous Complex consists of a series of ring dikes that constitute a rock mass about 4 km in diameter (figure 13.9). Both volcanic and plutonic rocks are present. The main rock types are trachyte and phonolite, which form a ring-shaped dikelike mass. This ring is encased in "alkalic syenite" (figure 13.10), which forms a second ring. Ijolite fills the core of the complex, but it is intruded by central bodies of carbonatite and lime-silicate rock. Jacupirangite forms two peripheral bodies within the outer syenite ring. Additional dikes that intrude the complex or the surrounding country rocks have varying compositions and consist of such rock types as tinguaite, melteigite, and garnet fourchite and the plutonic rocks melagabbro and pyroxenite (table 13.3). Apparently the complex is a subvolcanic complex intruded at a maximum depth of about 1.5 km (Nesbitt and Kelly, 1977).

As might be expected from the variety of rock types, the chemistry of the complex is diverse (Erickson and Blade, 1963; Flohr and Ross, 1990). The major element chemistries of a carbonatite, jacupirangite, and garnet ijolite are given in

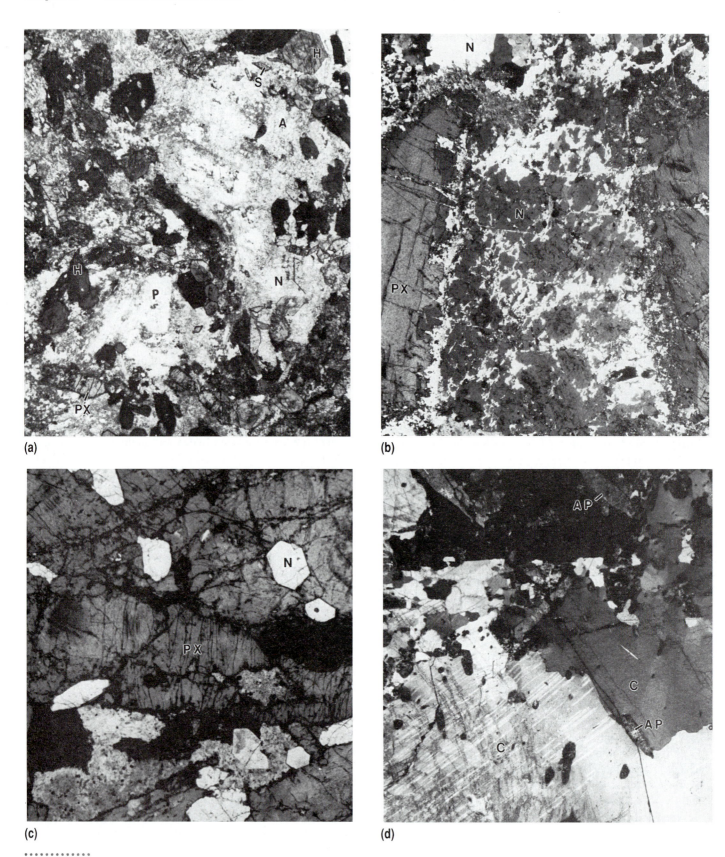

(a)

(b)

(c)

(d)

Figure 13.10 Photomicrographs of Magnet Cove rocks. (a) Nepheline syenite. (b) Ijolite. (c) Jacupirangite. (d) Carbonatite. A = alkali feldspar, C = calcite, H = alkali amphiboles (e.g., hastingsite), N = nepheline,

P = plagioclase, PX = pyroxenes, S = sphene, AP = apatite. Black grains are primarily magnetite, but may be other opaques. Long dimension of photo (a) is 3.25mm; of photo (b) is 6.5 mm, and (c)–(d) is 5mm.

Table 13.3 Brief Descriptions of Some Rock Types from Magnet Cove, Arkansas

Volcanic Rocks (aphanitic to aphanitic-porphyritic)

Trachyte Rock with essential alkali feldspar and plagioclase, phenocrysts of alkali feldspar, and accessory biotite, pyrite, hornblende, aegirine, magnetite, sphene, garnet, flourite, calcite, or melilite.

Phonolite Rock with essential alkali feldspar, plagioclase, and feldspathoids (e.g., nepheline, sodalite) plus accessory biotite, analcite, amphibole, sphene, and others.

Tinguaite Dike rock with phenocrysts of pseudoleucite, nepheline, and aegirine in a matrix with alkali feldspar, sodalite, calcite, magnetite, and other minerals.

Melteigite Nepheline-rich rock with rounded phenocrysts of biotite in a matrix of diopsidic pyroxene, biotite, perovskite, garnet, and other minerals.

Garnet fourchite Aphanitic-porphyritic rock with essential nepheline; augite phyric with a matrix containing garnet, magnetite, sphene, nepheline, and biotite.

Plutonic Rocks (phaneritic)

Syenite Trachytoidal, alkali feldspar rock with biotite, aegirine, acmite, garnet, analcite, sphene, apatite, magnetite, ilmenite, pyrite, and pyrrhotite.

Feldspathoidal syenite Phaneritic rock with essential orthoclase and nepheline, sodalite, cancrinite, and/or pseudoleucite; accessory aegirine, acmite, or hedenbergite; plus hastingsite, magnetite, sphene, garnet, and other minerals.

Ijolite Phaneritic rock with essential nepheline and diopsidic pyroxene plus accessory biotite, magnetite, pyrrhotite, perovskite, garnet, sphene, and calcite.

Jacupirangite Feldspathoidal (nepheline)-bearing magnetite clinopyroxenite.

Carbonatite Calcite-rich rock with accessory amphibole, pyroxenes, magnetite, perovskite, garnet, monticellite, pyrite, phlogopite, apatite, and wavellite.

Sources: Erickson & Blade (1963), McCormick and Heathcote (1987), Flohr and Ross (1990).

Table 13.2. Silica values for these rocks fall between 1 and 42%. The various syenites define linear trends on Harker diagrams and exhibit a range of silica values between 42 and 62%. Alumina ranges from < 1% to more than 22% in the various rocks of the complex, with lowest values in carbonatites and highest values in feldspathoidal syenites. Alkalis likewise range from < 1% to more than 16%. The CaO varies from < 2% in some nepheline syenites to over 50% in carbonatites. Mg numbers range from 5 to 82 in various rocks.

Trace element studies on the syenites show relatively greater uniformity (Flohr and Ross, 1990). LREE are extremely enriched (in some cases more than 200×) over HREE, although in garnet syenites the enrichment is greatly reduced. Modest to pronounced negative Eu anomalies are present in some rocks. Sr behaved as an incompatible element (i.e., it tended to be excluded from fractionating phases) but may be rather abundant. Some HFSE elements, such as Ti and Zr, behaved as compatible elements (i.e., they were taken up preferentially by crystallizing mineral phases).

Isotopic data on certain Magnet Cove rocks indicate mantle sources (Tilton, Kwon, and Frost, 1987). The Sr isotopic ratios for carbonatites average 0.7036. The εNd values are positive, but at about +3.8 are somewhat lower than Hawaiian and MORB values.

As noted earlier in this chapter, the origins of many unusual alkaline rock types, like those found at Magnet Cove, have been the subject of considerable controversy.[46] The eruptions at Oldoinyo Lengai (Dawson, 1962a, 1962b; Dawson et al., 1990) and several experimental studies (Wyllie and Tuttle, 1960; Watkinson and Wyllie, 1971; A. F. Cooper, Gittins, and Tuttle, 1975) verify that magmas of these unusual compositions exist. Which processes produced the Magnet Cove rocks?

Surveys of the Ouachita–Gulf Coast Region (Erickson and Blade, 1963; R. C. Morris, 1987) reveal that alkaline basalt (and related rocks) are widely distributed in the area. Based on major element, trace element, and isotopic data, mantle-derived alkali olivine basalt magma is a likely candidate for the parent magma for Magnet Cove rocks and similar rocks in the region (Erickson and Blade, 1963; R. C. Morris, 1987; Tilton, Kwon, and Frost, 1987).

The intrusive history of the complex, based primarily on the interpretation of Erickson and Blade (1963), is as follows. Fractional crystallization of the olivine basalt magma at depth gave rise to an alkali-, lime-, and volatile-rich "mafic phonolite" magma that intruded to shallow levels of the crust (<10 km). Ring fractures formed and dikes of phonolite and trachyte breccia filled the early fractures. The fractures ultimately filled with enough phonolite and trachyte to comprise 29% of the complex. Presumably, various trachytes and phonolites formed by fractionation of the mafic phonolite. Although some plagioclase fractionation no doubt occurred, Eu anomalies were not produced primarily by plagioclase fractionation because Sr, which is compatible in plagioclase, did not behave as a compatible element. Rather, fractional crystallization of sphene and apatite, common phases in these rocks (see table 13.3), apparently produced the Eu anomalies (Flohr and Ross, 1990).

The next stage in the development of the Magnet Cove Complex was marked by intrusion of two jacupirangite bodies. Jacupirangite formed by the crystallization of a magma created by settling, in the magma chamber, of crystals of clinopyroxene and magnetite (which trapped some nepheline) (figure 13.10b). While clinopyroxene settled, leucite (now represented by pseudoleucite phenocrysts) began to crystallize and float. Fractures on the flanks of the volcano tapped the jacupirangite, which rose toward the surface before crystallizing. Segregations of melteigite and ijolite were caught up in the rising magma and also crystallized.

Fractures that tapped the more felsic magmas at the top of the magma body provided conduits for the rise of feldspathoidal syenite magmas. Early fractionation of sphene and apatite probably removed Eu; crystallization of pyroxenes, sphene, and magnetite-ilmenite fractionated Ti; and garnet removed Zr from the magmas (Flohr and Ross, 1990). These fractionating magmas intruded and crystallized to form the outer syenite ring, which engulfed both the early-formed ring of phonolite and trachyte and one of the jacupirangite bodies. During this stage of intrusion, swarms of nepheline syenite dikes intruded the jacupirangite. At least four different magmas gave rise to the mineralogically different syenites (Flohr and Ross, 1990).

The phases that had crystallized by this point in the evolution of the magma body had diminished the magma in Mg, Fe, Al, Si, Na and K (plus Eu, Ti, Zr, and other trace elements), increasing the relative abundance of Ca, CO_2, H_2O, and rare earth elements. The magma intruded the core of the complex and crystallized ijolite, the very low silica content of the magma resulting in crystallization of a mix of nepheline, cancrinite, sodalite, diopside, garnet, and biotite.

The magma was further enriched in CaO and CO_2 by the crystallization of ijolite. Fluid inclusions suggest that at this point the magma contained 11.4% H_2O and 16.7% CO_2 (Nesbitt and Kelly, 1977). From this Ca- and fluid-rich magma, carbonatite crystallized (figure 13.10c), as did pods of pegmatitic aegirine nepheline syenite and local concentrations of minor to rare minerals such as rutile, garnet, eudialyte, and perovskite. High volatiles and probable low degrees of nucleation, resulting from structural modification of the magma and low degrees of undercooling conditioned by intrusions into a previously heated terrane, allowed the crystallization of pegmatites and coarse-grained veins and pods of late-forming minerals. Throughout the intrusive history, there were probably periods of metasomatism (fluid-induced metamorphism) that altered the chemistry of minerals and spurred the precipitation of additional phases (McCormick and Heathcote, 1987; Flohr and Ross, 1989).

The fractional crystallization scenario for Magnet Cove, summarized above, is generally consistent with available data. Some features of the complex, however, such as carbonatite and various mineral segregations, could be interpreted to reflect separation of immiscible carbonate magmas.

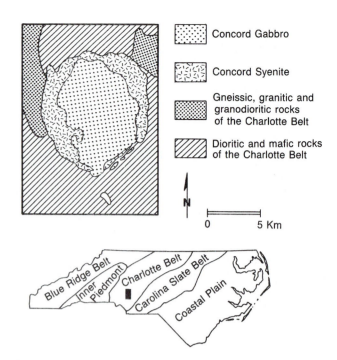

Figure 13.11   Geologic sketch map of the Concord Gabbro-Syenite Complex, North Carolina.
(After Olsen et al., 1983)

Other Examples

Two additional examples of less common syenite occurrences are described briefly below. The first, the Concord Gabbro-Syenite Complex, links the syenites to gabbros. The second, the Mount Monadnock Stock, one of the simpler intrusions of the White Mountain Magma Series of New England, links rock with syenite affinities to granitic rocks.[47]

The Concord Gabbro-Syenite Complex

The early Devonian (405 m.y. b.p.), mesozonal Concord Gabbro-Syenite Complex, North Carolina, consists of the two named rock types.[48] The gabbro forms a central stock, which is almost completely surrounded by a ring dike composed of syenite[49] (figure 13.11). Metamorphic rocks, primarily granitoid gneisses and hornblende schist and gneiss, constitute the country rock.

The gabbros are cumulate rocks. They are composed of cumulus plagioclase ±clinopyroxene ± olivine + magmatite/ilmenite and postcumulus hornblende, biotite, orthopyroxene, and apatite. The syenites are hypidiomorphic-granular to seriate-textured rocks composed of perthitic and antiperthitic alkali feldspar, with minor amounts of albite-oligoclase and microcline, salite and diopsidic augite, quartz, hornblende, biotite, and apatite, and additional minor accessory minerals.

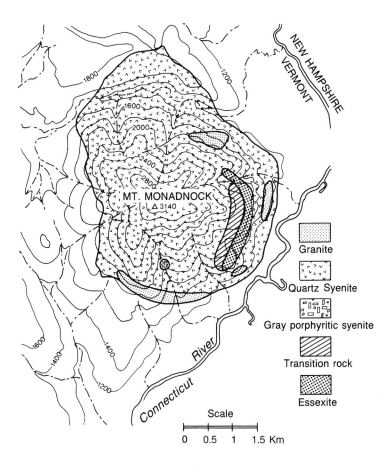

Figure 13.12 Geologic map of the Mount Monadnock Complex, Vermont. *(Redrawn from R. W. Chapman, 1954)*

All syenite samples reported by B. A. Olsen, McSween, and Sando (1983) contain quartz, though one contains less than 1%.

Available evidence indicates that the gabbro and syenite are related and were emplaced at the same time (B. A. Olsen, McSween, and Sando, 1983). Sm and Nd isotopic data show that the gabbro and syenite were intruded contemporaneously. Overlapping Nd- and Sr-isotope data also indicate consanguinity. Apparently, the syenite intruded (after the gabbro was emplaced) along the circular, fractured margins of the gabbro stock. B. A. Olsen, McSween, and Sando (1983) conclude that the syenite developed by fractional crystallization from a basaltic parent magma compositionally similar to the gabbro. Assimilation is excluded as a significant factor in magma evolution.

That bimodal gabbro/syenite-like chemistries (with no intermediate rock compositions) developed in the Concord Complex again raises the question about the origins of significant quantities of syenitic or trachytic magmas. Their compositions lie near m_s (figure 13.2a) in the $KAlSiO_4$-$NaAlSiO_4$-SiO_2 system. B. A. Olsen, McSween, and Sando (1983) follow S. A. Morse (1980) in suggesting that, in the case of the Concord syenite, fractionation drove the liquid to its endpoint near m_s, resulting in a syenitic liquid. Intrusion and crystallization followed. Geophysical modelling, however, suggests that fractionation may have occurred in a deeper (unexposed) magma chamber dominated by rocks of intermediate composition (R. T. Williams and McSween, 1989).

The Mount Monadnock Complex

The crest of Mount Monadnock, Vermont, is underlain by an alkaline rock complex that is part of the Jurassic-Cretaceous White Mountain Magma Series.[50] The complex is a slightly elliptical epizonal stock consisting of five rock units that together intrude Paleozoic schist and metaquartzite of the Gile Mountain Formation (figure 13.12) (R. W. Chapman, 1954).

Most of the complex is composed of *quartz syenite* (granite with ≤ 20% quartz). The other rock types of the complex are *essexite* (feldspathoidal alkali gabbro), *transition*

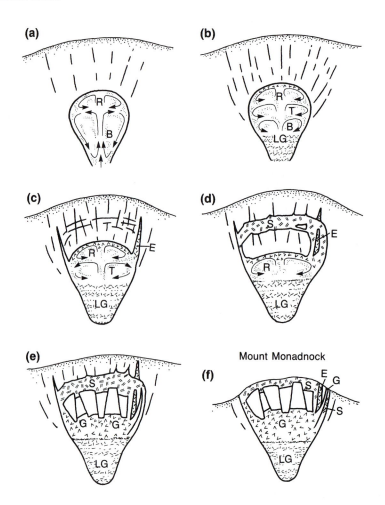

Figure 13.13 Schematic scenario for the emplacement of the Mount Monadnock Complex, Vermont. (a) Emplacement of basalt (B) diapir, which melts roof, forming rhyolitic (granitic) magma (R). (b) Assimilation and differentiation to produce stratified rhyolitic (R), trachytic-syenitic (T), and basaltic (B) magma chamber, with layered gabbro (LG) at base. (c) Marginal fracture is intruded by essexite dike (E).

(d) Stoping of roof block and essexite (E), with intrusion and crystallization of "quartz syenite" (S). (e) Rhyolite magma intrudes ring fractures in stope block and syenite, and crystallizes to form granite (G). (f) Erosion exposes the complex.
(Based on the ideas of C. A. Chapman, 1976, and data of R. W. Chapman, 1954)

rock (modally quartz-poor quartz monzonites and granodiorites), *porphyritic syenite* (porphyritic quartz-poor granite), and granite (≥ 20% quartz). Accessory minerals range from diopsidic pyroxene, Ca-amphibole, biotite, magnetite, apatite, and zircon in the essexite; to biotite, hornblende, magnetite, apatite, zircon, and allanite in the granite. Xenoliths of diorite, metaquartzite, and schist occur in the pluton.

The origin of the Mount Monadnock Complex probably began with a basalt-producing melting event in the upper mantle, perhaps induced by the passage of the continent over a mantle plume (hot spot) (W. J. Morgan, 1971; Crough, 1981a, b) or reactivation of an old, deeply buried transform fault or rift zone (Foland and Faul, 1977; McHone, 1981; McHone and Butler, 1984).[51] As a diapir of

alkali olivine basaltic magma rose to the base of the crust, melting produced a rhyolitic (granitic) cap on the magma chamber (figure 13.13a). Differentiation, accompanied by some assimilation, probably produced intermediate chemistries (D. S. Barker, 1965; C. A. Chapman, 1976) that formed in a layered magma chamber with rhyolite at the top, silica-poor rhyolite magma in an intermediate position, and essexite capping the basalt at depth. Ring fractures may have allowed early intrusion of essexite (figure 13.13b). As a central block of country rock was stoped (dropped downward), intrusion of the main mass of quartz syenite occurred (figure 13.13d). During this event, a section of essexite, emplaced as a dike earlier in the intrusive sequence (figure 13.13c), was stoped into the quartz syenite, and interaction between the still hot essexite and the quartz syenite magma produced the transition rock. Fracturing of the stoped block at depth then allowed rhyolitic magma trapped beneath it to intrude upward along fractures that penetrated the solidified quartz syenite (figure 13.13e). Late fracturing continued to allow a variety of magmas, remaining in pockets at depth, to intrude the complex to form dikes.[52]

SUMMARY

The alkaline and related rocks are chemically, mineralogically, and petrographically diverse. Chemically, the silica contents vary from 0 to 65%. Corresponding variations in other major elements and in trace elements exist, and magnesium numbers range from nearly 0 to 0.88. Distinctive minerals of the trachytes, syenites, and other alkaline rocks include Ca-rich pyroxenes such as diopside, Na-rich pyroxenes such as aegirine, amphiboles such as kaersutite, perthitic alkali feldspar, and the feldspathoids, especially nepheline. In carbonatites, calcite, dolomite, and ankerite occur, as do unusual minerals like perovskite and monticellite.

Alkaline and related rocks are locally important rock types. They develop both in oceanic areas (e.g., Hawaii) and in continental regions (e.g., Montana, Arkansas, East Africa). Syenites and related alkaline plutonic rocks typically are present as shallow intrusions: epizonal stocks, ring dikes, dikes, laccoliths, and sills. They are also found at deeper levels of exposure. In most cases, the alkaline volcanic rocks (and their plutonic equivalents) are associated with tectonically active volcanic sites above mantle plumes, subduction zones, or spreading centers (rifts). Metasomatism of the mantle beneath these sites may play a role in creating a local, fertile, alkali-rich source region, and fractional fusion of that source yields the alkaline parent magmas. That many alkaline magmas are mantle derived is indicated by commonly high Mg numbers (0.70 ± 5), low Sr-isotope ratios (< 0.706), and high ε Nd values (> +2).

After alkaline and related magmas are produced by anatexis of mantle rocks, the melts may either intrude upward and crystallize or they may intrude and differentiate as they crystallize. Basanitic and alkali olivine basalt probably develop primarily through equilibrium crystallization of the mantle melts. Trachyte and syenite, phonolite and nepheline syenite, and other alkaline rocks probably form via fractional crystallization. Oceanic trachyte and some carbonatite magmas, however, may form by crystallization of magmas formed by fractional fusion. Alternatively, some rock types (especially carbonatites) may locally result from liquid immiscibility. Assimilation of limestone is apparently not an important process in alkaline rock origins or carbonatite petrogenesis. The unusual chemistries that we find in alkaline complexes are fundamentally controlled by the initial composition of the melt that is generated in a heterogeneous mantle. Moderate to extreme fractionation simply refines that basic melt into various silica, iron, carbonate, soda, or potash enriched end products.

EXPLANATORY NOTES

1. The phenocrysts were subsequently shown to be $(Na, K)_2 Ca(CO_3)_2$ (Nyererite) and $(Na, K)_2CaCO_3$ (Gregoryite), with P, Cl, F, and SO_4 substituted in the crystal structure (Gittins and McKie, 1980). The volcano erupted again from 1983 to 1988 (Dawson et al. 1990). A beautiful color aerial view of the volcano is presented on the cover of the March 1990 issue of *Geology* (v. 18).

2. Alkaline rocks also include peralkaline granitoids, covered in chapter 12 and below, and peralkaline rhyolites, mentioned in chapter 8, both of which are also widely distributed. The transition from trachyte to peralkaline rhyolite is bridged, for example, in rocks of the Trans-Pecos region in Texas (D. S. Barker, 1977; D. F. Parker, 1983; J. G. Price et al. 1987). *Carbonatite* is an igneous rock with (> 50%) carbonate minerals (Woolley and Kempe, 1989). Several different types of a carbonatite are recognized, based on the texture, mineralogy, and chemistry (Streckeisen, 1980; Woolley and Kempe, 1989). See the discussions below and the glossary for descriptions of this and other unusual rocks mentioned in this chapter.

3. *Foidite* denotes a feldspathoid-rich igneous rock.

4. More complete discussions of minerals and textures typical of these rocks may be found in J. F. G. Wilkinson (1974), Nockolds, Knox, and Chinner (1978), and H. Williams et al. (1982); other descriptions are available in works on individual areas or plutons (R. W. Chapman, 1937; Quinn, 1937; Hurlbut and Griggs, 1939; Buie, 1941; Burgess, 1941; Erikson and Blade, 1963; D. S. Barker, 1977; D. S. Barker and Hodges, 1977; D. S. Barker et al., 1977; B. A. Olsen, McSween, and Sando, 1983; Eby, 1984; R. C. Greenwood and Edgar, 1984; T. Anderson, 1984; Ewart, 1985; Leterrier, 1985; T. D. Peterson, 1989a). Also see various papers in Sorensen (1974a), *Journal of African Earth Sciences*, (1985), v. 3, nos. 1 and 2, *American Mineralogist*, (1985), v. 70, nos. 11 and 12, Fitton and Upton (1987), and E. M. Morris and Pasteris (1987a), as well as the papers referenced therein. See Heinrich (1966), Ouzegane et al. (1988), R. L. Hay (1989), and T. D. Peterson (1989a), and papers in Tuttle and Gittins (1966) and K. Bell (1989) for more information on carbonatite and natrocarbonatite mineralogies.

5. Definitions of this and other related rock types are found in the glossary and in Bates and Jackson (1987). They are discussed more extensively in texts such as Hatch, Wells, and Wells (1973), Nockolds, Knox, and Chinner (1978), Best (1982), D. S. Barker (1983), and A. Hall (1987). Also see Woolley (1987) and papers in Sorenson (1974a), Yoder (1979), and K. Bell (1989). Note that most names are mineralogically based, but Cox, Bell, and Pankhurst (1979) and the IUGS (Streckeisen and LeMaitre, 1979) provide chemically based nomenclature.

6. The komatiites, kimberlites, and lamprophyres are mentioned or discussed briefly elsewhere in this text. See the index to locate additional information on these rocks.

7. See Wolff and Storey (1984); Worner and Schmincke (1984a), C. H. Donaldson et al. (1987), D. S. Barker (1989), D. S. Barker and Nixon (1989), R. L. Hay (1989), and the numerous references to individual localities cited in other notes in this chapter.

8. Dawson (1966), Bultitude and Green (1971), LeBas (1977, 1978), Gest and McBirney (1979), Nicholls and Whitford (1983), Phelps, Gust, and Wooden (1983), Mertes and Schmincke (1985). See table 13.2, analysis 7.

9. For example, see Dawson (1966), J. F. G. Wilkinson (1974), Sahama (1974), R. W. Johnson, Wallace, and Ellis (1976), LeBas (1977), Nochols and Whitford (1983), and Worner and Schmincke (1984a,b).

10. Becke (1896, in Sorensen, 1974a, III.1).

11. Butakova (1974).

12. D. S. Barker (1974).

13. Heinrich (1966, p. 320), Woolley (1989).

14. Gold (1967), A. R. Philpotts (1974), Eby (1984), R. C. Greenwood and Edgar (1984), Grunenfelder et al. (1986).

15. For example, see R. W. Chapman and Williams (1935), R. W. Chapman (1937), Quinn (1937), C. R. Williams and Billings (1938), D.S. Barker (1965), and C. A. Chapman (1976).

16. E. R. Lloyd (1923), Erickson and Blade (1963), Nesbitt and Kelley (1977).

17. Yates and Thompson (1959), R. A. Maxwell et al. (1967), D. S. Barker (1977), Joesten (1977).

18. Darton and Paige (1925).

19. Pirsson (1905), Hurlbut and Griggs (1939), E. S. Larsen (1940), B. C. Hearn, Pecora, and Swadley (1964), Nash and Wilkinson (1970, 1971), D. S. Wood (1974), Kendrick and Edmond (1981).

20. See references and reviews in Sorensen (1974a) and Heinrich (1966).

21. These origins are reviewed in Sorensen (1974a,b,d), Gittins (1979), Fitton and Upton (1987), E. M. Morris and Pasteris (1987b), and K. Bell (1989).

22. Daly (1910, 1933), Shand (1930, 1945), Chayes (1942), Tilley (1952), Schuiling (1964a), Heinrich (1966), Joesten (1977). Wyllie (1974) presents a good historical review of this and other aspects of the assimilation model.

23. See Schuiling (1964a,b), Watkinson and Wyllie (1964, 1969), Joesten (1977), and the reviews in Wyllie (1974) and Gittins (1979). Wyllie (1974) gives a historical review.

24. Also see Watkinson and Wyllie (1969) and reviews in Wyllie (1974) and Gittins (1979).

25. Von Eckerman (1948, in Wyllie, 1974), Heinrich (1966, p. 286), Tuttle and Gittins (1966).

26. Sr isotopic ratios are reported for alkaline rocks by D. S. Barker et al. (1977), Barreiro and Cooper (1987), Bernard-Griffiths et al. (1988), Ellam et al. (1989), Verma and Nelson (1989), and K. Bell and Peterson (1991), among others. D. S. Barker (1989) notes that initial Sr isotopic ratios for carbonatites are similarly < 0.706, a result generally supported by specific studies, but not without exception (K. Bell and Blenkinsop, 1989; S. C. Eriksson, 1989).

27. Wyllie and Tuttle (1960); F. J. Turner and Verhoogen (1960, p. 396); Heinrich (1966, p. 288); I. S. E. Carmichael, Turner, and Verhoogen (1974, e.g., p. 517–519); Wyllie (1974); Gittins (1979); Best (1982, p. 204–206, p. 330).

28. Bowen (1928, 1937), Erickson and Blade (1963), G. A. MacDonald (1968), Edgar (1974), Sorensen (1974a, pp. 421–423), R. Macdonald (1974), D. S. Barker (1977), Fitton and Hughes (1977), LeBas (1978), Gittins (1979), Wolff and Storey (1984), Worner and Schmincke (1984b), Mertes and Schmincke (1985).

29. R. Macdonald (1974) reviews various aspects of this problem.

30. D. S. Coombs and Wilkinson (1969), Irvine and Baragar (1971), R. Macdonald (1974). See *Glossary of Geology* (Bates and Jackson, 1987) for definitions of the unusual rocks named here.

31. D. S. Coombs and Wilkinson (1969), R. Macdonald (1974).

32. Also see L. M. Larsen (1976) and Gittins (1979) for discussions of the layered Ilimaussaq intrusion of Greenland, which may have formed via this process.

33. A. F. Cooper, Gittins, and Tuttle (1975), Roedder (1979), Kendrick and Edmond (1981), LeBas (1977, 1987), Kogarko, Ryabchikov, and Sorenson (1974) and Freestone (1978) review various aspects of liquid immiscibility and related phenomena. Kendrick and Edmond (1981) describe an example in syenitic rocks, and A. F. Cooper, Gittins, and Tuttle (1975), LeBas (1981), and Kjarsgaard and Hamilton (1989) suggest that carbonatite magmas form as immiscible separates from preexisting magmas. Also see J. Ferguson and Currie (1972) for an example involving syenite, carbonatite, and related rocks developed through a combination of immiscibility and fractionation.

34. Also see Koster van Groos and Wyllie (1966, 1968), A. F. Cooper, Gittins, and Tuttle (1975), Koster van Groos (1975), Freestone and Hamilton (1980), Kjarsgaard and Hamilton (1988), and Mattey et al. (1990).

35. J. L. Powell, Hurley, and Fairbairn (1966), P. G. Harris (1974), J. L. Powell and Bell (1974), A. F. Cooper, Gittins, and Tuttle (1975), LeBas (1978), Yagi and Onuma (1978), Gittins (1979), Moller, Marteani, and Schley (1980), Arima and Edgar (1983), Phelps, Gust, and Wooden (1983), Boettcher (1984), Eby (1984), Leterrier (1985), F. E. Lloyd, Arima, and Edgar (1985), Mertes and Schmincke (1985), Grunenfelder et al. (1986), Rock (1986), D. S. Barker, Mitchell, and McKay (1987), Esperanca and Holloway (1987), Eggler (1989), Gittins (1989), P. Wallace and Carmichael (1989), M. B.Baker and Wyllie (1990). Metasomatism as a process that creates mantle heterogeneity or fertile mantle sources for alkaline magmas is discussed, for example, by Boettcher and O'Neil (1980), Menzies and Murthy (1980), Meen (1987), J. E. Nielson and Noller (1987), Wilshire (1987), A. P. Jones (1989), Meen, Ayers, and Fregeau (1989), and Wyllie, Baker, and White (1990).

36. A. F. Cooper, Gittins, and Tuttle (1975), LeBas (1977, 1981), Roedder (1979), Kendrick and Edmond (1981), Grunenfelder et al. (1986), Rock (1986), E. M. Morris (1987), Sial (1987), Tilton, Kwon, and Frost (1987), Ellam et al. (1989), P. Wallace and Carmichael (1989), Wyllie (1989), and K. Bell and Peterson (1991).

37. Early studies of this region include Weed and Pirsson (1895), Pirsson (1905), F. F. Osborne and Roberts (1931), E. S. Larsen et al. (1935), Barksdale (1937), Hurlbut and Griggs (1939), E. S. Larsen (1940, 1941a,b), Buie (1941), Burgess (1941), and Larsen et al. (1941). The Highwood Mountains became known by mid-century as an important locality where alkaline rocks are exposed. Later, works by Nash and Wilkinson (1970, 1971), Nicoll and Nicholls (1974), Woods (1974, 1976), Kendrick and Edmond (1981), Nash (1982), Kendrick (1982), and H. E O'Brien, Irving, and McCallum (1988) added to earlier knowledge. This review is based on the above works. *Note:* Shonkin Sag is a broad valley containing lakes and streams, the latter obviously too small to have carved the valley.

38. See the glossary and table 13.3 for the definition of this and other rock types.

39. Monticellite is an olivine-like mineral with the formula $CaMgSiO_4$ (for more information see Deer, Howie, and Zussman, 1966).

40. Weed and Pirsson (1895), Barksdale (1937), Hurlbut and Griggs (1939), E. S. Larsen (1940), Buie (1941), Nash and Wilkinson (1970, 1971), Nicoll and Nicholls (1974), Kendrick and Edmond (1981), Dockstader (1982), Kendrick (1982), and Nash (1982).

41. Pirsson (1905), F. F. Osborne and Roberts (1931), Hurlbut (1936), Hurlbut and Griggs (1939), Nash and Wilkinson (1970, 1971), Nicoll and Nicholls (1974).

42. Daly (1914), Kendrick and Edmond (1981), Kendrick (1982).

43. This origin is discussed but rejected by Pirsson (1905, p. 182) and favored by E. S. Larsen et al. (1935), Barksdale (1937), and Dockstader (1982).

44. Johnston-Lavis (1896, in Pirsson, 1905, pp. 182–83), Reynolds (1935, in Hurlbut and Griggs, 1939), also discussed in Woods (1974).

45. The geology of Magnet Cove is based primarily on the definitive study of Erickson and Blade (1963) and on subsequent work by Nesbitt and Kelley (1977), McCormick and Heathcote (1987), and Flohr and Ross (1989, 1990).

46. See reviews in Erickson and Blade (1963), Tuttle and Gittins (1966), Heinrich (1966), I. S. E. Carmichael, Turner, and Verhoogen (1974, ch. 10), Sorenson (1974a), and LeBas (1981), and the various papers in E. M. Morris and Pasteris (1987a) and K. Bell (1989).

47. See Quinn (1937), H. Sorensen (1974a), DuBray (1985), O'Halloran (1985a,b), and references to the White Mountain Magma Series in note 52 for other examples of alkaline granitoid complexes.

48. The Concord Gabbro-Syenite Complex has been described briefly in several reports (H. Bell, 1960; J. K. Butler and Ragland, 1969) and its age was determined by Fullagar (1971) and B. A. Olsen, McSween, and Sando (1983). The petrogenesis of the rocks is discussed in J. K. Butler and Ragland (1969), B. A. Olsen, McSween, and Sando (1983), and McSween et al. (1984). R. T. Williams and McSween (1989) discuss the structure of the complex.

49. In the RSO classification (Raymond, 1984c), some of this rock is a granite because it contains minor quartz. In the IUGS classification (Streckeisen, 1976), the "syenites" range from Alkali feldspar syenites to Alkali feldspar granite.

50. The Mount Monadnock Stock has been described by J. E. Wolff (1929) and R. W. Chapman (1954). R. W. Chapman and Williams (1935) and C. A. Chapman (1976) discuss the evolution of the White Mountain Magma Series. Also see McHone and Butler (1984) and C. M. B. Henderson, Pendelbury, and Foland (1989) for related discussions.

51. The scenario described here is based principally on the model of C. A. Chapman (1976), with specific details from R. W. Chapman (1954), and synthesis by the author. Various aspects of this scenario are consistent with evolution of similar complexes in the White Mountain Magma Series (see C. M. B. Henderson, Pendelbury, and Foland 1989; Dorais, 1990).

52. Note that the scenario outlined here for Mount Monadnock is based largely on structural and modal data, rather than mineral-chemical and trace element data. Nevertheless, mineral-chemical, trace element, and isotopic data from other complexes in the White Mountains indicate similar scenarios for the studied complexes (Eby, 1985; C. M. B. Henderson, Pendelbury, and Foland, 1989; Dorais, 1990), lending credence to this model for the Mount Monadnock rocks. Apparently not all plutons in the region have a similar petrogenesis (Foland et al., 1988).

Note that eruption of magmas with the siliceous compositions generated by the anatectic, assimilation, and fractionation processes that formed the Mount Monadnock and other White Mountain alkaline *plutonic* rocks would yield trachytic to peralkaline *rhyolite* compositions. Provinces that contain such volcanic rocks (D. F. Parker, 1983; McCurry, 1988) may represent supracrustal evidence of alkaline plutonic rocks at depth.

PROBLEMS

13.1 (a) Calculate the Mg numbers for the analyses in tables 13.1 and 13.2. (b) Considering the Mg numbers determined for Magnet Cove rocks, what petrogenetic conclusion might you draw from the Mg numbers calculated for these rocks?

13.2 Determine whether the rocks in tables 13.1 and 13.2 are peralkaline, metaluminous, or peraluminous.

13.3 What data are needed to resolve the controversy over the origin of Shonkin Sag shonkinite?

13.4 Using data from Erickson and Blade (1963) and Flohr and Ross (1990), construct an AFM diagram for the Magnet Cove Complex. Is the differentiation trend TH or CA?

Sedimentary Rocks

Sedimentary rocks and sediments are widely distributed over the surface of the earth. Although sedimentary rocks make up only 0.029% of the total volume of the earth, they represent about two-thirds of the exposed rocks of the land surface. Even in terranes of igneous or metamorphic rock, sediment may be found in stream beds and on lake bottoms.

Processes that produce sedimentary rocks are less mysterious than those that form the igneous and metamorphic rocks because many of these processes can be observed. Each process may impart to the rock a distinctive character, as will be shown in this section of the book. Major sedimentary rock types are evaluated individually in the chapters that follow and are assigned to rock associations in the epilogue of the book.

Cross-bedded Jurassic, Navajo Sandstone, Checkerboard Mesa, Zion National Park, Utah.

PART

III

CHAPTER

Sedimentary Rocks: Their Structures, Textures, and Compositions

INTRODUCTION

Sedimentary rocks are rocks formed at the surface of the earth under low-temperature and low-pressure conditions. They result from the accumulation and solidification of **sediments,** materials transported in water, air, or ice. Three main categories of material characterize various types of sedimentary rocks: (1) the silicate fragments and associated grains, (2) the chemical and biochemical precipitates, especially certain carbonate materials, and (3) the **allochems,** fragments of earlier-formed precipitates.

Silicate fragments and associated grains, designated by some petrologists as *detrital* or *terrigenous materials,*[1] include gravel-, sand-, silt-, and clay-size fragments of preexisting silicate minerals or rocks.[2] The category also includes grains of silicates, such as clay minerals (of any size), generated by weathering of preexisting rocks. Minor oxides (e.g., magnetite), sulfides (e.g., pyrite), and other minerals typically associated with the silicates in rocks are also included here. Rocks composed of silicate and related clasts are called **siliciclastic rocks.** Siliciclastic materials are usually derived by erosion of terranes outside the basin of deposition, but the source area or **provenance,** may also include terranes within the basin.

Chemical and **biochemical precipitates** are crystalline-textured, generally fine-grained to aphanitic sediments produced either by inorganic chemical reactions (the chemical precipitates) or by chemical reactions caused by living plants or animals (the biochemical precipitates). These precipitates may form entire rocks by themselves or they may cement other grains together. In both cases, precipitates are usually interpreted to be **authigenic,** that is, they are considered to have been formed *in place,* rather than to have been transported. Not uncommonly, precipitates have been recrystallized.

Allochems are fragments of earlier-formed chemical or biochemical precipitates.[3] Included in this category of materials are whole or fragmented fossils, oolites, organic materials, and fragments of chemically or biochemically precipitated rocks. Allochems, which are abundant in some limestones and cherts, are typically derived from *within* the basin of deposition. Nevertheless, like silicate fragments, these materials also may be derived from a provenance outside of the depositional basin.

The allochems, silicate and related fragments, and chemical and biochemical precipitates occur in various proportions in sedimentary rocks. Raymond (1984c; 1993) has assigned the sedimentary rocks to

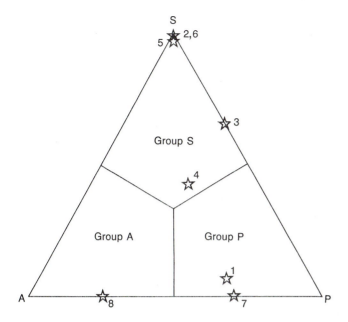

Figure 14.1 Triangular diagram showing subdivision of sedimentary rocks into three major groups. Points represent positions of modes listed in table 14.3.
(Modified from Raymond, 1984c)

Figure 14.2 Beds of sandstone (light-colored), mudrocks (gray), and coal (black) in Pennsylvanian rocks near Pineville, Kentucky. Prominent sandstone (near middle) is about 1 m thick.

Table 14.1 Classification of Stratification

Stratum Thickness	Names
Bed	
> 300 cm (> 3 m)	Massive
100–300 cm	Very thickly bedded
30–100 cm	Thickly bedded
10–30 cm	Mediumly bedded
3–10 cm	Thinly bedded
1–3 cm	Very thinly bedded
Lamina	
0.3–1 cm	Thickly laminated
< 0.3 (< 3 mm)	Thinly laminated

Source: From R. L. Ingram, "Terminology for the thickness of stratification and parting units in sedimentary rocks" in *Geological Society of America Bulletin,* 65:937–38, 1954. Copyright © 1954 Geological Society of America. Reprinted by permission of the author.

groups on the basis of the dominant material characterizing the rock (figure 14.1). Group S, the siliciclastic rocks, encompasses the silicate-rich breccias and conglomerates, the sandstones, and the mudrocks. Group P, the precipitates, includes limestones, dolostones, cherts, evaporites, and other rocks composed predominantly of precipitated materials. Group A rocks are dominated by allochems, which include the fragmental (clastic) limestones, dolostones, and cherts.

STRUCTURES OF SEDIMENTARY ROCKS

Just as specific structures and textures characterize volcanic or plutonic rocks, certain structures and textures also distinguish various types of sedimentary rocks. In addition, such features provide clues to the environment of deposition and origin of each rock type.

Sedimentary structures occur in a wide range of sizes. At the lower size limit (a few millimeters), structures may be confused with textures; the latter, however, involve intergrain relationships and tend to pervade (occur throughout) any given sample. In contrast, structures are larger than individual grains and occur singly or in groups of a few features. Many common sedimentary structures are discussed here, but these and additional sedimentary structures are also described by Pettijohn and Potter (1964), Conybeare and Crook (1968), Collinson and Thompson (1982), and J. R. L. Allen (1984).

Sedimentary structures are subdivided into four main groups: (1) bed and formation shapes, (2) surface structures,

(3) internal structures, and (4) other structures. **Beds,** or layers, are the most characteristic feature of sedimentary rocks (figure 14.2). Surface and internal structures help to distinguish beds, which are recognized on the basis of *differences in color, texture,* and *composition.*

Bedding, Lamination, and Internal Structures

Beds are nearly ubiquitous in sedimentary rocks. Variations in the size and shape of individual beds and their thinner counterparts, the laminations, reflect sedimentation history. In size, beds vary from an arbitrarily selected lower limit of 1 cm to thicknesses of several meters (table 14.1). **Laminations** are layers of less than 1 cm in thickness. Thicker beds may be deposited by vigorous (high-energy) or

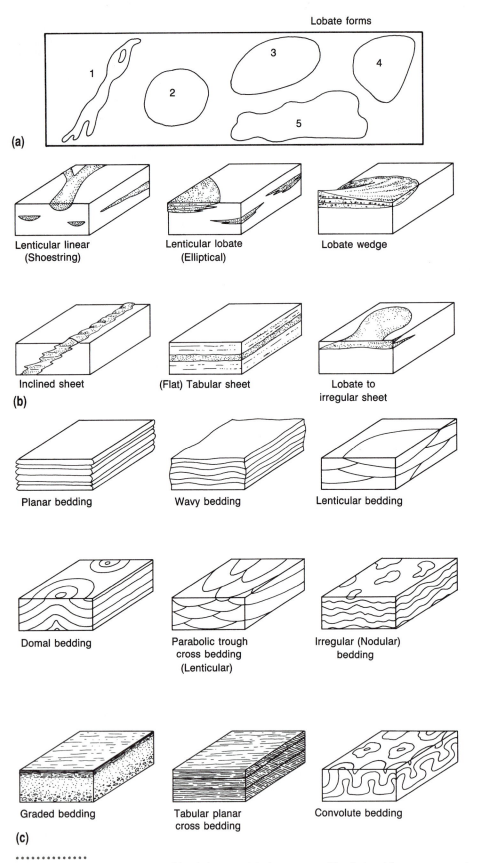

(a)

(b)

(c)

Figure 14.3 Formation and bed shapes. (a) Plan views of beds (and formations and members). 1-linear (shoestring), 2-circular, 3-elliptical, 4-parabolic, and 5-irregular (3, 4, and 5 are lobate forms). (b) Three-dimensional views of formation (and member) shapes. (c) Block diagrams with cross sections of beds showing internal features.

high-density currents, by ice, or under long-term, low-energy, stable conditions.[4] Laminations and thinner beds tend to reflect flow in in-phase waves, laminar flow (water flowing in a planar way), fluctuating conditions, or variable low-energy conditions. Postdepositional destruction of laminations or thin bedding, by bioturbation, burrowing by organisms, may result in thicker beds.

Beds, in plan view (as observed from above or as depicted on a map) may be linear, circular, lobate (elliptical to parabolic), or irregular (figure 14.3a).[5] They are generally sheetlike (tabular) masses. Three dimensionally, beds and laminations may be planar, wavy, lenticular, domal, trough-shaped, or irregular (nodular). Geometrically, each type is distinct (figure 14.3c). For example, in planar beds, the upper and lower contacts are approximately parallel, whereas wavy beds have broadly parallel contacts, but in detail they have local high and low places.

Some beds or laminations are combined in sets (McKee and Weir, 1953; R. G. Thomas et al., 1987)[6] and may contain internal features that distinguish them (figure 14.3c). For example, **cross-beds** or inclined stratification, bed sets that contain regular, local truncations (see the photo on page 263 and figure 14.4b), are of two distinctly different types: tabular planar cross-beds and trough (lenticular) cross-beds. **Convolute beds** are beds or bed sets with contorted internal layers. **Graded bedding** is layering in which there is a decrease in grain size from the bottom to the top of the bed (figure 14.4a). In reversely graded bedding, coarsening occurs from bottom to top. Some **turbidites,** rock units deposited by heavy sediment-water mixtures, may contain all or part of a five-layer sequence of beds and laminations known as the **Bouma sequence.**[7] From bottom to top, a complete Bouma sequence consists of a graded to structureless basal bed (Ta), a parallel laminated bed (Tb), a cross-laminated or convoluted bed (Tc), a second parallel laminated bed (Td), and a cap layer of mudrock (Te) (figures 14.5a, and

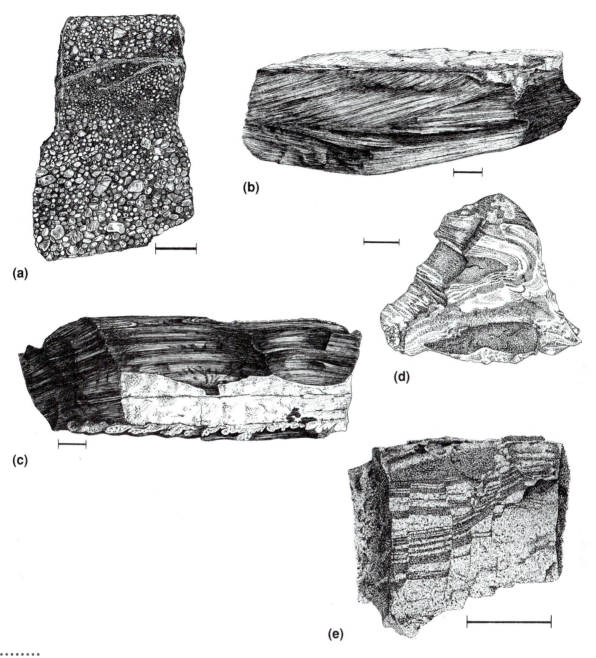

Figure 14.4 Internal structures of sedimentary rocks. (a) Graded bedding in conglomerate, Unicoi Formation (Cambrian), south of Mountain City, Tennessee. (b) Cross-laminated wacke, Knobs Formation (Ordovician), Lodi, Virginia. (c) Flame structures in mudrocks (Tertiary), Shale City, Oregon. (d) Soft-sediment fold in wacke, Knobs Formation (Ordovician), near Abingdon, Virginia. (e) Faults in poorly lithified sandstone-mudrock layers, Franciscan Complex (Cretaceous), near Crescent City, California. All scale bars are 2 cm long. *(From Raymond, 1984c)*

14.6). Other sequences, such as storm sequences (figure 14.5b), are also known.[8] These various bed shapes, sets, and sequences reflect specific conditions of formation. For example, complete Bouma sequences (Tabcde) characterize submarine fan channels, whereas incomplete sequences characterize other fan environments.[9]

Beds and bed sets make up *members*, *formations*, and *lithofacies*.[10] A **formation** is a mappable body of rock of distinctive rock type or rock types and unique stratigraphic position. A **member** is a subdivision of a formation, similarly characterized by distinctive lithologic character and stratigraphic position. Members may be too thin to be mappable,

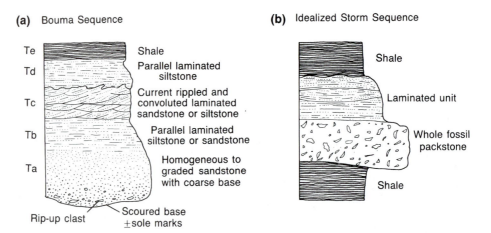

Figure 14.5 Two bed sequences found in sedimentary rocks. (a) Idealized Bouma sequence showing the five turbidite intervals—Ta, Tb, Tc, Td, and Te (modified from Bouma, 1962). (b) Idealized storm sequence with hummocky cross stratification.
(Redrawn from Kreisa, 1981).

Figure 14.6 Bouma sequence in limestone turbidite, Keeler Canyon Formation (Permian), near Keeler, California.

but many are not. Members differ from formations simply in being subunits of the larger formations. Like beds, formations and members occur in a variety of shapes (see figure 14.3b). Each formation or member may represent a particular depositional event, sequence of events, or sedimentary environment.[11] *Facies* are specifically the characteristic deposits of sedimentary environments. A **facies** is defined as a body of sedimentary rocks or sediments with distinct chemical, physical, and biological characteristics (Pickering et al., 1986).[12] A **lithofacies** is a particular type of facies, one with physical and chemical characteristics—rock types, textures, and structures—that represent a particular sedimentary environment. Likewise, **biofacies** are units of sedimentary rock that are distinguished solely by their biota and represent specific environmental conditions.

Within each bed, the structural features, which aid in defining a lithofacies, include features that reflect the environment of formation. Such features as lamination, graded bedding, cross-lamination, and convolute lamination; plus oncolites, stromatolites, reefs, stylolites, flame structure, burrows, bioturbation structure, escape structures, concretions, slump folds, and prelithification faults (described below), are characterized as internal structures.

Structures that occur in carbonate rocks include the following. **Oncolites** are small (generally < 10 cm),[13] concentrically laminated, spherical to irregular bodies formed during deposition by biochemical precipitation and trapping of carbonate mud by algae. **Stromatolites,** larger laminated algal accumulations (>10 cm), are variously shaped, ranging from flat to domal, conical, columnal, or irregular mound-shaped masses (figure 14.7). **Reefs** are domal to elongate, massive to bedded forms built, during carbonate deposition, by organisms that biochemically precipitate carbonate materials. In reefs, the organisms (e.g., corals, bryozoans, algae) are the major component of the reef rock, which typically projects topographically above the different, contemporaneously deposited sediments that surround the reef. Reefs range from less than 1 m to more than 1000 m in height. **Stylolites** are irregular surfaces that commonly appear as dark, jagged lines on exposed surfaces of carbonate rock (and rarely on other sedimentary rock types) (Stockdale, 1943). Their origin is usually attributed to solution that occurs *after* the host rock was formed. The dark layers are insoluble residues.

Additional internal structures are found in a variety of rock types. **Flame structures** are deformed clay or silt laminations with curved, pointed ends that project into the overlying bed and resemble a flame blown in the wind (figure 14.4c). They represent soft-sediment deformation of beds that occurs when more dense, new sediment is deposited above a bed, causing the deformation. **Burrows** are irregular to cylindrical filled depressions or tubes produced

(a)

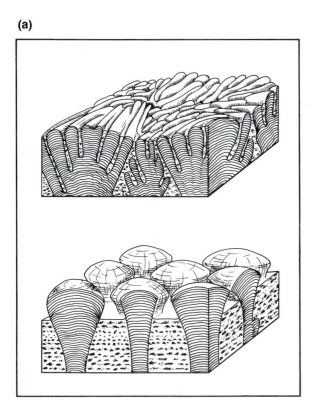

(b)

Figure 14.7 Stromatolites. (a) Some stromatolite forms (from Hoffman, 1974). (b) Stromatolitic layers in dolostone, Helena Formation (Middle Proterozoic), Glacier National Park, Montana.

(a)

(b)

Figure 14.8 Burrows. (a) *Scolithus* burrows (vertical) in Clinch Sandstone (Silurian), Powell Mountain, Highway 58, west of Duffield, Virginia. (b) *Cruziana* (?) trails on bedding surface in Tuscarora Sandstone (Silurian), near Neelyton, Pennsylvania.

by burrowing organisms (figure 14.8). Where burrowing is extensive, it yields highly swirled laminations or complete destruction of lamination and the rock is said to have **bioturbation structure. Escape structures** are columns, holes, or tubes that cut across bedding and are infilled with sediment after the escape of water or a burrowing organism that had been trapped beneath a newly deposited sediment layer.[14]

Concretions are irregular, rounded, or disk-shaped, cemented hard masses of rock that occur within a host rock, and are commonly characterized by different colors or colored rings, called **liesegang rings,** that are produced by oxidation or reduction (figure 14.9). **Soft-sediment folds** and **faults,** bent and broken layers, respectively, form where sediments undergo deformation before they are lithified (R. B. Nelson and Lindsley-Griffin, 1987; C. G. Elliott and Williams, 1988; Shanmugam et al., 1988)(figure 14.4d and e).

Figure 14.9   Liesegang rings in wacke, Lodoga Formation (Lower Cretaceous), Wragg Canyon, near Lake Berryessa, California. Coin is 17mm in diameter.

Figure 14.10   Ripple marks in quartz arenite, Clinch Sandstone (Silurian), Powell Mountain, Highway 58, west of Duffield, Virgina.

Surface Structures and Other Features

A variety of structures define or occur on bedding or lamination surfaces. In some cases, the structures reflect or are reflected by internal features. For example, ripple marks on the surface are characterized internally by cross laminations. Surface structures include mudcracks, sole marks, tracks and trails, and various types of **bed forms** (ripples, sand waves, dunes, and antidunes) (G. F. Jordan, 1962; Bouma et al., 1980; J. C. Harms, Southhard, and Walker 1982).[15] In carbonate rocks, large structures such as reefs, mudmounds, and lithoherms have internal and surface features (W. G. H. Maxwell, 1968; A. C. Neuman, Kofoed and Keller, 1977; A. Lees, 1982).[16]

All surface features may be revealed by impressions on the top of the underlying bed or lamination or on the bottom of the overlying bed or lamination, or both. **Ripple marks** are regularly undulating shapes (figure 14.10) created either by oscillating water or by wind or water currents. Currents produce asymmetrical ripples, whereas oscillating water produces symmetrical ripples. These may be well preserved on the top or the bottom of a bed. Currents carrying pebbles, sticks, shells, or other large objects may make **tool marks** (grooves) on the top of a bed at the sediment-water interface. Small eddies or turbulence may also carve **flutes** (depressions) in the bed surface. Rounded or bulbous protrusions formed during compaction are called **load casts.** Where these grooves and depressions become filled with sediment, especially if the surface is mud, the marks may be preserved. Commonly the weak underlying beds are eroded to expose the base (sole) of the overlying bed with its casts of the filled grooves or flutes. Such casts are called **sole marks** (figure 14.11). Sole marks are useful for determining tops and bottoms (the "facing") of

Figure 14.11   Flute casts (sole marks) in wacke (left of author's head), Knobs Formation (Ordovician), near Lodi, Virginia.

(Courtesy Dr. Fred Webb Jr., Appalachian State University)

beds as well as current direction. The marks generally project into the underlying bed from the overlying bed and their long dimensions indicate direction of current movement (Potter and Pettijohn, 1977). **Mudcracks** develop where fine-grained sediment shrinks and cracks upon drying, to produce polygonal blocks. If filled by sediment different from the blocks, the cracks may be preserved (figure 14.12). **Tracks** and **trails** result from preservation of marks left by various organisms as they walk, crawl, or otherwise move across a sediment surface. **Rain prints** develop where raindrops strike a muddy surface that has enough cohesion to preserve the impact mark. Drying, followed by later filling, will preserve the prints. **Dunes** and **antidunes,** elongate piles of sediment of less than

1 m to tens of meters high, may develop in currents of wind or water and may be preserved under appropriate conditions.

In addition to the surface structures described above, a number of other features appear on bedding surfaces or within rocks. Some of these are transitional between structures and textures. Others are secondary features formed during or after the sediment became rock. Fossils are perhaps the most important of these other features. A **fossil** is any prehistoric evidence of past life.[17] Fossils include leaves, shells, bones, tracks and trails, molds, casts, and several other evidences of past life. An additional feature found in sedimentary rocks is the *vein*, a fracture filling of one or more minerals.

Spherical to elliptical grains that occur in some sediments are transitional between structures and textural elements. Where they are large or isolated, they may be considered structures; whereas, if they pervade the rock, they may be considered to be textural elements. **Oolites** (ooids) are small (1/4–2 mm), concentrically layered, spherical grains composed of primary carbonate minerals or replacement phases (figure 14.13).[18] Oolites form in shoals, where gentle or periodic agitation and warm marine waters combine to allow precipitation on all sides of a grain of sand or a shell fragment. **Pisolites** are similar grains that are larger than 2 mm in diameter. **Peloids** are rounded, structureless, aphanitic calcite or aragonite masses generally less than 1/4 mm in diameter.[19] Some peloids are fecal **pellets** excreted by mud-eating worms, shrimp, and other organisms. Grapestones are cemented clusters of rounded peloids that occur in some modern carbonate sediments (Winland and Matthews, 1974). They apparently form through a sequential history in which forams and other rounded grains are cemented and recrystallized.

Figure 14.12 Mudcracks in mudrock, Catskill Formation (Devonian), Highway 829, near Mill Creek, Pennsylvania.

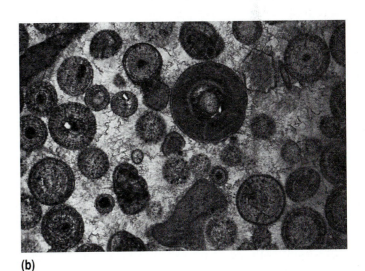

(a)

(b)

Figure 14.13 Oolitic limestone. (a) Photograph of oolitic packstone in outcrop, Ottosee Formation (Ordovician), Maryville, Tennessee. (b) Photomicrograph of rock in (a) under XN. Texture is epiclastic to epiclastic-poikilotopic. Long dimension of photo is 6.5 mm.

TEXTURES IN SEDIMENTARY ROCKS

Sedimentary rocks, like igneous rocks, have either crystalline or clastic textures. Clastic textures, however, are the principal textures in the sedimentary rocks. The textural features of particular importance include grain size, grain shape, sorting, and intergrain relationships. Intergrain relationships allow division of sedimentary textures into the two major categories: *crystalline textures*, with interlocking grains, and *clastic textures*, with rounded to angular grains that are stuck together.

Crystalline Textures and Their Origins

Crystalline textures in sedimentary rocks may be assigned to one of the following three categories:

1. Crystalline textures of primary chemical and biochemical precipitates (Group P rocks)
2. Cystalline textures in cements (Groups S and A)
3. Crystalline textures in recrystallized (diagenetically altered)[20] rocks (Groups P and A)

Textures of categories 1 and 3 are commonly *equigranular-mosaic* (including idiotopic texture) or *equigranular-sutured* (xenotopic texture); that is, the rocks have equant grains with straight or irregular grain boundaries, respectively (figure 14.14a and b). Idiotopic texture is a texture in which the grains with straight boundaries are euhedral or subhedral crystals (J. M. Gregg and Sibley, 1984). Category 2 textures include the two types common in categories 1 and 3 plus poikilotopic, syntaxial, and comb-structured to fibrous-drussy, radial-fibrous, and spherulitic textures.[21] All of the latter textures have elongate grains in different arrangements (figure 14.14c, d, e, and f), except poikilotopic texture which has large crystals of cement that enclose various smaller grains (figure 14.14g). Syntaxial textures are textures in which the new cement grows in the same crystallographic orientation as the preexisting core grain. Other textures, including inequigranular types, may include varying arrangements and proportions of anhedral, subhedral, and euhedral grains. Textures in category 1 tend to be finer-grained than those in category 3. The same size classes used for crystalline igneous textures may be used to describe crystalline sedimentary textures.

Nucleation and growth are important processes in the development of crystalline textures in sedimentary rocks, just as they are in the formation of igneous crystalline textures. Heterogeneous nucleation essentially dominates crystallization in sedimentary rock-forming processes, because preexisting grains that may serve as nucleii are almost always present. In Group P rocks, organisms generate and serve as the nucleii in biochemical precipitation. At sites of chemical precipitation, small grains of sand, clay, shell, or previously formed crystals may serve as nucleii. Cements formed in Groups A and S rocks (textural category 2) develop similarly, with allochems, silicate grains, and other clasts serving as nucleii.[22]

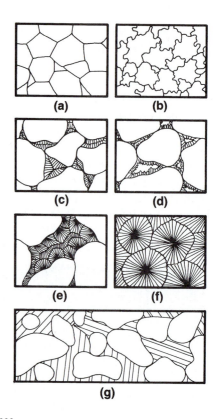

Figure 14.14 Sketches of some textures in sedimentary rocks. (a) Equigranular-mosaic texture. (b) Equigranular-sutured texture. (c) Comb texture in the cement of an epiclastic textured rock. (d) Fibrous-drussy texture in the cement of an epiclastic textured rock. (e) Radial-fibrous texture in the cement of an epiclastic textured rock. (f) Spherulitic texture. (g) Poikilotopic texture.

Crystalline textures of textural category 3 develop in rocks in which preexisting crystals serve as nucleii.

Growth of crystalline material on the nucleii may be influenced by a variety of factors. Among these are (1) the chemical composition of waters involved (seawater, fresh water, mixed seawater and fresh water, pore waters, or other hydrous solutions), (2) temperature, (3) growth activities of organisms, (4) evaporation rate, (5) porosity and permeability, (6) quantities, compositions, and rates of flow of solutions (or water) passing through the sediment, and (7) reactions occurring at crystal-crystal and crystal-solution interfaces. Longman (1980) emphasizes that fluid flow is a particularly important factor in the growth of cements in carbonate rocks, as it may control several of the other factors. Similarly, fluid flow is important in the crystallization process affecting siliciclastic rocks.[23] The factors listed above govern the movement of specific atoms towards (or away from) the growing crystal surface, thus controlling the rate of growth and thereby the crystal shape. Because development of many crystalline textures is a *diagenetic* process, involving both the transformation of sediment into rock and subsequent modification of sedimentary rock (prior to the onset of metamorphism), it will be discussed at greater length in chapter 16.

Clastic Textures and Their Origins

Clastic textures are textures in which clasts are stuck together. **Epiclastic textures** are sedimentary textures, formed at the surface (*epi-*), in which accumulated grains are packed together (figure 14.14c and d, figure 14.15).[24] The descriptors used for epiclastic textures and textural features differ from those used in crystalline rocks. Grain size categories used are usually those of C. K. Wentworth (1922) (table 14.2).[25] It is common practice to convert Wentworth size classes to phi (ø) values, using

$$\varnothing = -\log_2(d/d_o)$$

where d_o = 1mm and d = the grain diameter in millimeters (Krumbein, 1934; McManus, 1963).[26]

Figure 14.15 Epiclastic texture in arenite, Panoche Formation (Upper Cretaceous), Carbona Quadrangle, California. (PL). Long dimension of photo is 3.25 mm.

Table 14.2 Wentworth Size Grades and the ø Scale

ø	Wentworth Scale	Grain Size Names		Group S Rock Names	Texture
−8	256 mm	Boulders			
		Cobbles			
−6	64 mm		Gravel	Conglomerate, Breccia	Epiclastic ruditic
		Pebbles			
−2	4 mm				
		Granules			
−1	2 mm				
		Very coarse sand			
0	1 mm				
		Coarse sand			
1	1/2 mm				
		Medium sand		Sandstone (Arenite, Wacke)	Epiclastic arenitic
2	1/4 mm				
		Fine sand			
3	1/8 mm				
		Very fine sand			
4	1/16 mm				
		Silt		Siltstone, Shale	
			Mud	Mudstone,	Epiclastic lutitic
8	1/256 mm			Claystone	
		Clay			

Sources: Modified from C. K. Wentworth (1922), Krumbein (1934), McManus (1963).

Sorting is an additional size parameter used to describe the textures of clastic rocks. It is a measure of the similarity of grain sizes within a sample. If all the grains are exactly the same size, the sediment is said to be *very well sorted*. In contrast, a wide range of grain sizes indicates *very poorly sorted sediment*. In handspecimen work, qualitative estimates of sorting are made and a degree of sorting is assigned by considering the number of Wentworth size grades in the middle range of grain sizes in the sample. The finest and coarsest grains are ignored. Individual authors define the "middle" differently. For example, Pettijohn (1975) considers the middle 50%, D. W. Lewis (1984) uses the middle 67%, and Compton (1962) specifies the middle 80%. Figure 14.16 depicts the D. W. Lewis (1984) and Compton (1962) divisions with sketches of representative textures.

Sorting may also be defined quantitatively. Pettijohn (1975) defines the sorting as the distribution or spread of values about the average, that is, the standard deviation. In mathematical terms,

$$S_O = \sqrt{Q_3/Q_1}$$

where S_O = sorting, Q_3 = third quartile (the size value below which 75% of the grains in the sample occurs), and Q_1 = first quartile (the size value below which 25% of the grains in the sample occurs). Use of this quantitative measure requires a laboratory analysis of grain size distribution.

Grain shape is described in three ways in the sedimentological literature, by form, sphericity, and roundness. The *form* (general shape) may be classified as equant (subequal length, width, and height), tabular (with two larger dimensions and one substantially smaller dimension), or rod-shaped (two short dimensions and one large one) (figure 14.17).[27] Although some geologists prefer to define four or more general shapes,[28] the three endmembers figured here encompass the shapes defined by others. The *sphericity* of grains is a measure of how closely the overall shape approaches the form of a sphere. Grains that are spherical have high sphericity, whereas those that are rod-shaped or tabular have low

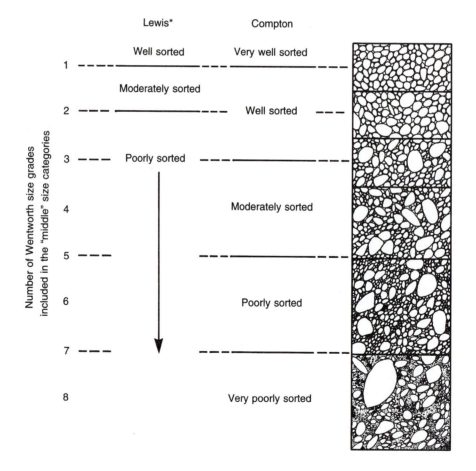

Figure 14.16
Degrees of sorting. D. W. Lewis (1984) defines the "middle" as the middle two-thirds (67%). Compton (1962) designates the "great bulk" as the middle 80%.
(From Raymond, 1984c; after D. W. Lewis, 1984, and Compton, 1962)

sphericity. *Roundness* (vs. angularity) is a measure of the curvature of grain edges. Grains with many sharp, jagged edges are called very angular, whereas those with smooth, rounded edges are called well-rounded (figure 14.18). Visual estimates of grain roundness are commonly made in handspecimen and microscopic petrographic work, but methods of quantifying these parameters are available and are described in standard sedimentary petrology texts.[29]

Grain *surface textures*, examined with the scanning electron microscope (SEM), have been described and used with some success to characterize particular sedimentary environments (Krinsley and Doornkamp, 1973) (figure 14.19). Breaking or chipping of the surfaces of grains and chemical etching of those surfaces produce some distinctive markings and shapes in modern sand grains. In older, lithified deposits, however, those markings may be altered or destroyed by diagenetic processes.

Two other features of note in describing sedimentary rock textures are porosity and permeability. The **porosity** of a rock is the amount of empty space between grains in the rock.

This is usually reported as a percentage and is computed by multiplying the ratio of empty space (void volume) to total sample volume by 100:

$$ps = (V_p/V_T) \times 100$$

where ps = porosity, V_p = the volume of pore space, and V_T = the total volume of the rock sample. Although rocks of all groups (A, P, and S) may have some porosity, the porosities of certain Group A and S rocks may exceed 35%.[30] *Permeability* relates both to the sizes of pores and to the interconnections between pores and is a measure of how well fluid will flow through the rock.

Together, all of the above named measures reported for any given clastic rock will provide an image of its texture. Where necessary, that image may be quantified.

Clastic textures originate through the accumulation of grains produced by weathering and erosion of preexisting rocks. Grains are transported in ice, wind, or water and deposited where those agents lose their carrying capacity, primarily because of changes in velocity. The grain size of any clastic sediment is controlled by (1) the size of clasts present in or produced by weathering from rocks of the source area[31] and (2) the carrying, abrasion and sorting capacity of the transporting medium. Where only sand-size or smaller grains are available in the source area, only sand-size or smaller grains may be transported and deposited. Wind, in particular, sorts materials well, leaving behind the large particles and carrying off, in suspension, the smallest materials. In contrast, ice, a poor agent for sorting, deposits very poorly sorted mixtures of clay-, sand-, and boulder-size materials. Water has an intermediate capacity for sorting, but specific flow regimes or environmental conditions may produce either well-sorted or very poorly sorted sediment.

In the chapters that follow, the origins of specific rock types characterized by particular textures and structures are described individually. The interplay between provenance, environment of deposition, and depositional agents is more fully examined in these chapters.

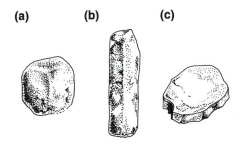

Figure 14.17
Shapes of sedimentary grains: (a) equant, (b) rod-shaped, (c) tabular.
(From Raymond, 1984c)

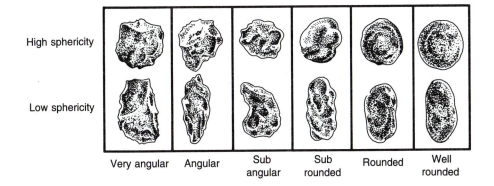

Figure 14.18 Grain roundness and angularity for grains of low and high sphericity.
(From Raymond, 1984c; after Powers, 1953)

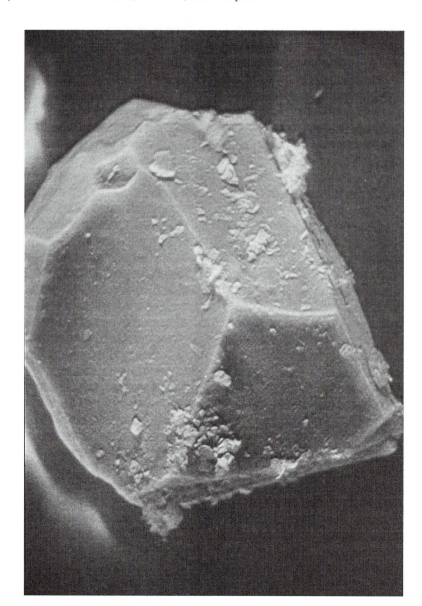

Figure 14.19   SEM photomicrograph of sand grain from a stream deposit (Quaternary), Grandfather Mountain, North Carolina.
(Courtesy Dr. Carol L. Stein, University of Washington)

MINERALOGY OF SEDIMENTARY ROCKS

Sedimentary basins collect whatever material is available. For that reason, the clastic sedimentary rocks (Groups A and S combined) may contain *any* of the common rock-forming minerals, as well as less common to rare mineral types. Nevertheless, certain minerals are notably abundant and others are widely distributed, because they are relatively resistant to weathering and abrasion or are stable under near-surface or surface conditions, or both. Among the most common minerals of the sedimentary rocks are quartz; the clay minerals—kaolinites, montmorillonites (smectites), and illites; and the carbonate minerals—calcite and dolomite.

Other less common minerals that are either resistant to weathering or stable at near-surface conditions are chalcedony, zircon, muscovite, and the oxides of both iron (hematite, goethite) and manganese (todorokite, hollandite). Important, as well, are the feldspars, chlorite, biotite, aragonite, magnetite, and ilmenite. Garnet, sphene, epidote, and a variety of less abundant minerals occur as minor constituents in Group S rocks.

In the Group P rocks, as compared to the epiclastic rocks, the range of minerals is limited, because precipitation is controlled by more restrictive physical-chemical laws (vs. the physical laws that control physical sedimentation). Under specific conditions, minerals such as calcite, aragonite, dolomite, opal, halite, gypsum, and anhydrite crystallize. In addition, under those unusual conditions where they become stable, crystallization of sylvite, ulexite, and a large number of other rare minerals occurs.

The more common sedimentary minerals and their associations are listed in appendix A and mentioned in available petrography texts.[32] Modes of a few selected sedimentary rocks are presented here in table 14.3. Quartz may range from 0 to 100%, and in sandstones typically ranges from 40 to 100%. Other minerals are equally variable in abundance. For example, calcite may constitute 0% of some sandstones and 100% of some limestones. Feldspars are common in the sandstones but rare in most other rocks. Clays are very common as an accessory mineral and constitute 100% of some shales.

CHEMISTRY OF SEDIMENTARY ROCKS

Like the mineralogy of sedimentary rocks, the chemistry of these rocks is diverse. Table 14.4 presents selected major element analyses of a few sedimentary rock types. The intent is neither to show the total possible range of compositions nor to present average values; rather, it is to provide a general impression of the chemical diversity. Note that SiO_2 ranges from 0 to nearly 100%. Alumina is less variable but ranges from very low values in carbonate rocks (< 1%) to moderately high values in shales and in sandstones with abundant feldspar (analysis 3). Magnesia is generally low, except in dolostones, where dolomite is the principal mineral. Similarly, CaO is low, except in limestones and rocks with abundant calcite cement. The alkalis are commonly low as well, but K_2O is abundant in K-feldspar-rich sandstones, shales, and rare evaporites. Soda is moderately abundant in sandstones rich in albite and in evaporites with halite or other sodium-bearing minerals. Iron oxides are similarly quite variable. High loss on ignition (LOI) values in mudrocks and carbonate rocks result from loss of water from clay minerals and loss of CO_2 from carbonate minerals.

Sediments are sometimes analyzed for C, S, and other elements that are abundant in these rocks, but are not measured in normal major element analyses. Organic carbon is usually reported as the total organic carbon, or **TOC**. TOC may range from 0 to more than 15% in muds and mudrocks (Beier and Hayes, 1989; R. Stein, 1990) and can be used to distinguish oxic from anoxic environments. In coals, of course, TOC is very high. The organic carbon to reduced sulfur ratio (C/S) is also used to discriminate between rocks formed in oxic and anoxic, and freshwater and marine, environments (Berner, 1982; Berner and Raiswell, 1984; J. W. Morse and Emeis, 1990). In deposits from oxic environments, the C/S is generally > 2 and may range to 25 or more, whereas in rocks from anoxic environments the C/S is generally < 3.

Trace elements show somewhat less diversity than major elements. Trace element analyses of sediments (Haskin and Gehl, 1963; Wildeman and Haskin, 1965) reveal that the *average* REE distribution in sediments parallels that of the sialic crust (McLennan, 1989).[33] For example, as is the case in the crust, LREE tend to be enriched relative to the chondritic standard and negative Eu anomalies are typical. These data suggest that REE are not generally dissolved and reprecipitated in the oceans, but rather pass through the ocean with the clastic grains, and thus reflect the REE character of the continental source terranes (Bhatia and Taylor, 1981; McLennan, 1989; Basu, Sharma, and DeCelles, 1990). Because of this, trace element variations in *specific samples* reflect broad regional provenances. For example, Bhatia and Taylor (1981) found that where La, Th, and U were high in the provenance rocks, they were high in sediments derived from those rocks. Where they were low in source rocks, they were low in the sediments. Similarly, Feng and Kerrich (1990) distinguished mafic-element (rift) and low mafic-element (arc) source terranes on the basis of major and trace element chemistry. In the former, MgO, FeO, Ni, and Cr are high, various HFSE (e.g., Hf, Zr, U) are low, and REE patterns, which are relatively flat, show only modest enrichment over chondritic values. The arc rocks show more LREE enrichment, higher Sr, and higher HFSE.[34]

Some trace element and isotopic variations do exist, however, among various individual sediment units and rock types.[35] For example, certain elements (e.g., Ni, Cr, Zn, P) have been used in efforts to distinguish marine from nonmarine shales (Degens, Williams, and Keith, 1957; L. J. Walters et al., 1987). Phosphorous, for example, may be twice as abundant in some marine shales as in continental shales (L. J. Walters et al., 1987), but its abundance in marine sediments can be linked to low-latitude zones of high organic productivity (Toyoda and Masuda, 1990). Mn is also high in such zones.[36] Sr is high in Ca-rich rocks,[37] and both Sr and U substitute in appreciable amounts in the aragonite structure (Papekh et al., 1977). Rocks high in TOC, a characteristic of anoxic environments, are enriched in elements such as V, Cd, and Zn (J. D. Vine and Tourtelot, 1970; Leventhal and Hosterman, 1982).[38]

Isotopes also vary in sedimentary rocks, reflecting differences in provenance, depositional environment, or postdepositional modification (diagenesis). In the siliciclastic sediments and rocks, Sr and Nd isotopic ratios typically reflect provenance compositions (Heller et al., 1985; Basu, Sharma, and DeCelles, 1990) and are variable. Sr ratios may

Table 14.3 Modes of Selected Sedimentary Rocks

	1 (Ch)	2 (Ss)	3 (Ss)	4 (Ss)	5 (Ss)	6 (Sh)	7 (Ls)	8 (Ls)
Quartz[a]	84.3[b]	65.3	39.7	41.6	22.2	—	—	—
monocrystalline	(57.6)[c]	(63.0)	(38.7)	(40.6)	(17.8)	—	—	—
polycrystalline	—	(2.3)	(1.0)	(1.0)	(4.4)	—	—	—
fossils[d]	(26.7)	—	—	—	—	—	—	—
Alkali feldspar	—	1.0	4.0	0.3	0.4[e]	—	—	—
Plagioclase	—	2.0	6.0	0.3	23.4	—	—	—
White mica	8.0[f]	—	tr	tr	—	—	—	—
Biotite	—	—	5.7	—	—	—	—	—
Chlorite	0.3	—	5.0	—	—	5	—	—
Clays	—[f]	25.3	—	—	—	95	—	—
Kaolinites	—	(25.3)	—	—	—	(15)	—	—
Montmorillonites	—	—	—	—	—	(70)	—	—
Illites	—	—	—	—	—	(10)	—	—
Other minerals	—	1.3	0.7	7.6[g]	1.0	—	—	—
Rock fragments	—	4.7	4.0	0.3	16.6	—	—	—
Shale and Ss	—	(1.0)	—	(0.3)	(1.6)	—	—	—
Chert	—	(2.0)	(2.0)	(tr)	(8.2)	—	—	—
Siliceous volcanic	—	(1.7)[h]	(1.0)	—	(4.2)	—	—	—
Basic volcanic	—	—[h]	(tr)	—	(0.8)	—	—	—
Metamorphic	—	(tr)	(1.0)	—	(1.6)	—	—	—
Other/Unknown	—	—	—	—	(0.2)	—	—	—
Matrix	(—)[i]	1.7	0.7	—	35.8	na	(—)[j]	(—)[j]
Cement	7.0[k]	—	33.3[l]	33.3[l,m]	0.6	na	na	na
Carbonate Materials								
Mud	—	—	—	—	—	—	58.5	4.5
Spar								
Calcite	—	—	(33.3)	(19.0)	—	—	—	20.0
Dolomite	—	—	—	—	—	—	13.5	—
Allochems[n]								
Oolites	—	—	—	1.3[o]	—	—	—	34.5
Intraclasts	—	—	—	5.0	—	—	—	1.0
Fossils[p]	—	—	—	10.0	—	—	28.0	40.5
Total	100.0	100.0	100.0	100.0	100.0	100	100.0	100.0
Points Counted	300	300	300	300	500	X	200	200

Sources:

1. Red radiolarian chert, Franciscan Complex (Jurassic?), The Geysers, California, sample RF-77A. Point count by L. A. Raymond.

2. Anauxite arenite, Tesla Formation (Eocene), Carbona Quadrangle, California, sample C18a (Raymond, 1969). Point count by L. A. Raymond.

3. Micaceous feldspathic arenite, Panoche Formation (Cretaceous), Carbona Quadrangle, California, sample C1 (Raymond, 1969). Point count by L. A. Raymond.

4. Fossiliferous, hematitic quartz arenite, Rose Hill Formation (Silurian), Clinch Mountain Wildlife Management Area, Virginia. Point count by L. A. Raymond.

5. Feldspathic wacke, Coastal Belt, Franciscan Complex (Cretaceous), west of Willits, California. Point count by L. A. Raymond.

6. Clay shale, Moreno Formation (Cretaceous), Carbona Quandrangle, California. (Raymond, 1969). X-ray analysis by L. A. Raymond.

7. Dolomitic, bioclastic wackestone, Monteagle Formation (Mississippian), near Huntsville, Alabama. Point count by H. Gault.

8. Oolitic fossiliferous grainstone, Monteagle Formation (Mississippian), near Huntsville, Alabama. Point count by H. Gault.

[a]May include chalcedony and opal.

[b]Values in volume percent.

[c]Values in parentheses are not included in the total because they are included in another number on the table (e.g., quartz types are included in the total quartz).

[d]The fossils listed here are radiolaria. Since they are siliceous, they are included as "quartz"; they are not included as fossils below under "Carbonate Materials."

[e]Untwinned feldspar is included with plagioclase.

[f]In column 1, white mica includes all fine grained colorless phyllosilicates (micas, clays). Therefore, clays are not listed as separate minerals.

[g]This value includes 4.3% hematite oolites and 3.3% hematitic fossil fragments.

[h]All volcanic rock fragments in this sample are included under "siliceous volcanic rocks."

[i]Technically, the silica between the radiolaria reported above is matrix.

[j]Calcite mud forms a matrix, but is listed below under "calcite."

[k]The hematite in this rock may be overestimated, as it coats most other grains.

[l]Calcite spar forms cement material and is therefore listed under both cement and spar.

[m]Includes 14.3% hematite cement.

[n]The allochems listed here are only carbonate allochems. Some rocks include siliceous fossils, hematitic fossils, hematitic oolites, and other materials not included here.

[o]The percent of oolites listed here includes only carbonate oolites. The rock also includes 4.3% hematite oolites.

[p]Only carbonate fossils are included. Other fossils are listed in other compositional categories.

Ch = chert, Ls = limestone, Sh = shale, Ss = sandstone.

tr = trace amounts.

na = not applicable.

X = analysis by X-ray diffraction.

be low (< 0.706) where sediments are derived from mantle sources, but high if the provenance is an ancient sialic source. Similarly, εNd values that are strongly negative (−10 to −20) indicate older crustal components, whereas positive values reflect mantle affinities for the sediments. Sr and Nd isotopes are also used to assess the histories of cherts and carbonate rocks (Weis and Wasserburg, 1987). In addition, O, C, and S isotopes are used to evaluate environmental conditions and diagenetic changes that affect sedimentary rocks (J. D. Hudson, 1977; Longstaffe, 1983; C. P. Rao, 1990).[39]

The chemical diversity suggested here is great. Chemical parameters, in addition to textural, structural, and mineralogical features, are used to characterize environments of deposition and provenance in an effort to understand ancient sedimentary rocks.[40] In the chapters that follow, such integrated assessments are provided, where possible, in the examples.

Table 14.4 Chemical Analyses of Selected Sedimentary Rocks

	1 (Ss)	2 (Ch)	3 (Ss)	4 (Sh)	5 (Fe-st)	6 (Ls)	7 (Dlst)
SiO_2	96.65[a]	94.7	67.2	61.84	4.21	1.15	0.28
TiO_2	0.17	0.06	0.05	0.83	0.12	nd	nd
Al_2O_3	1.96	1.1	14.6	13.40	4.38	0.45	0.11
Fe_2O_3	0.58	2.7	1.9	3.83	37.72	nd	0.12
FeO	nd	0.22	2.3	1.15	7.27	0.26	nd
MnO	nd	0.05	0.1	0.05	0.18	nd	nd
MgO	0.05	0.14	2.3	2.69	1.68	0.56	21.30
CaO	0.08	0.06	1.8	2.68	22.49	53.80	30.68
Na_2O	0.05	0.01	3.7	0.97	0.01 ⎫	0.07	0.03
K_2O	0.27	0.37	1.9	2.8	0.00 ⎭		0.03
P_2O_5	nd	0.03	0.1	0.44	1.00	nd	0.00
LOI[b]	0.59	0.79	3.4	9.74	20.81	43.61	47.42
Total	100.40	100.2	99.4	100.42	99.87	99.90	99.97

Sources:

1. Quartz arenite (?), Abbott Formation (Pennsylvanian), Illinois, sample B19. Analyst: L. D. McVicker (Bradbury et al., 1962).
2. Thin-bedded, red chert, Franciscan Complex (Jurassic-Cretaceous?), California, sample 4, table 9. Analysts: P. L. D. Elmore, I. H. Barlow, S. D. Botts, and G. Chloe (E. H. Bailey, Irwin, and Jones, 1964).
3. "Graywacke" (sandstone), Coastal Belt, Franciscan Complex (Cretaceous?), California, sample 10, table 1. Analysts: P. L. D. Elmore, I. H. Barlow, S. D. Botts, and G. Chloe (E. H. Bailey, Irwin, and Jones, 1964).
4. Shale, Cody Shale (Cretaceous), Wyoming, sample SCo-1. Analyst: S. M. Berthold (L. G. Schultz, Tourtelot, and Flanagan, 1976).
5. Hematitic ironstone, Keefer Formation (Silurian), Pennsylvania, sample D, table 9. Analyst: P. M. Buschman (H. L. James, 1966).
6. Limestone, Solenhofen Formation (Jurassic), Bavaria. Analyst: G. Steiger (F. W. Clarke, 1924).
7. Dolostone, Royer Dolomite (Cambrian), Oklahoma, sample 9294, table 5. Analyst: A. C. Snead (W. E. Ham, 1949).

[a]Values is weight percent.
[b]Loss on ignition and other.
Ch = chert, Dlst = dolostone, Fe-st = ironstone, Ls = limestone, Sh = shale, Ss = sandstone.
nd = not determined or not reported. In the case of iron, total iron may be reported as Fe_2O_3.

SUMMARY

Sedimentary rocks are rocks formed at the surface of the Earth from accumulated silicate fragments and allochems that have been transported by wind, water, or ice and from precipitates formed from aqueous solutions. They are characterized by a wide variety of structures, the most common of which are beds. Internal and external structures such as stromatolites, oncolites, cross-bedding, sole marks, burrows, ripple marks, mudcracks, tracks and trails, dunes, and fossils characterize lithofacies and biofacies and reflect particular environments of deposition.

The grains that comprise sedimentary rocks may be diverse in size, shape, and composition. The most common texture in sedimentary rocks is the epiclastic texture, a texture consisting of an aggregate of grains accumulated at the Earth's surface. Epiclastic textures range from very well sorted types with well rounded grains to very poorly sorted types with angular clasts up to the size of boulders. Crystalline textures also occur in sedimentary rocks, in precipitates as well as in cements and in recrystallized rocks. The mineralogical composition of sedimentary grains is controlled by the source region, transportation, and depositional agent, with calcite predominant among the precipitates, and the quartz and clay minerals dominant in the siliciclastic rocks. The chemical compositions of sedimentary rocks reflect the diversity of sedimentary grain types, with silica values ranging from 0 to 100%. REE, HFSE, and isotopes commonly reveal the character of the sediment provenance.

EXPLANATORY NOTES

1. For example, see Spock (1962), Ehlers and Blatt (1982), or Nockolds, Knox, and Chinner (1978). Terrigenous materials are generally considered to be derived from a continent or landmass.

2. These size categories are defined in the text below.

3. The use of *allochem* here follows a similar but genetic use by Folk (1974). Folk specified that allochems must be derived from *within* the basin of deposition. This restriction is not applied here because in most cases the discrimination between intrabasinal and extrabasinal sources is interpretive and problematical.

4. Selley (1976, p. 212) notes the low-energy origin of thick beds. See Campbell (1967) for comparison of the features of beds, laminae, and other units used in stratigraphy. Carey and Roy (1985) and Cheel and Middleton (1986) examine the nature and origin of laminations. Varves, distinctive laminae formed in lakes, are discussed by R. Y. Anderson and Dean (1988). Other papers dealing with aspects of bedding and lamination include McKee and Weir (1953), Kreisa (1981), Clemmensen and Blakey (1989), G. V. Middleton and Neal (1989), and Cheel (1990). See these works, the references therein, and sedimentology texts for additional information.

5. See Krynine (1948) and P. E. Potter (1963) for descriptions of rock body shapes, and see papers in Peterson and Osmond (1961).

6. McKee and Weir (1953) introduced "set" and "coset" terminology. R. G. Thomas et al. (1987) suggest replacement of the traditional "cross-bedding" terminology with the new, nongenetic terms *inclined stratification* and *inclined heterolithic stratification*.

7. The Bouma Sequence, named after Arnold H. Bouma, who described these sequences (Bouma, 1962, pp. 48–49), is widely described and used in articles and texts on sedimentary rocks (D. J. Stanley, 1963; Reineck and Singh, 1975; Selley, 1976; Mutti and Ricci Lucchi, 1978; Ingersoll, 1978c; Friedman and Sanders, 1978; Blatt, Middleton, and Murray, 1980; Boggs, 1987).

8. The idealized storm sequence illustrated by Kreisa (1981) contains "hummocky cross stratification," named by Harms et al. (1975). Further discussion of this kind of stratification may be found in Dott and Bourgeois (1982), B. Greenwood and Sherman (1986), Nottvedt and Kreisa (1987), Leckie (1988), Sherman and Greenwood (1989), and W. L. Duke, Arnott, and Cheel (1991).

9. Submarine fans and their characteristics are discussed in chapter 19.

10. See the North American Commission on Stratigraphic Nomenclature (NACSN, 1983) for the exact definitions adopted by that group and a more extensive discussion of the terminology. See Campbell (1967) and von Wagoner et al. (1990) for definitions of beds, laminae, and other units used in sequence stratigraphy.

11. Particular sedimentary units are discussed and illustrated by authors of general survey works (Spearing, 1974; Reineck and Singh, 1975; Selley, 1976; Blatt, Middleton, and Murray, 1980; Frazier and Schwimmer, 1987; Baars et al., 1988; Milici and DeWitt, 1988), as well as by those who treat more specialized topics (Cloud, 1952; Lugn, 1960; H. N. Fisk, 1961; J. R. L. Allen, 1965; Aubouin et al., 1970; N. D. Smith, 1970; Bull, 1972; Picard and High, 1972a; Reineck, 1972; Visher, 1972; Kulm et al.,1974; Bay, 1977; Miall, 1977a; Ingersoll, 1978c; Bentor, 1980; N. Eyles, Eyles, and Miall, 1983; Walters et al.,1987; M. T. Whalen, 1988; Clemmensen and Blakey, 1989; W. Morris and Busby-Spera, 1990).

12. The history of the use of the term *facies* is reviewed by Selley (1988, p. 165ff.) and is discussed by R. G. Walker (1984b). See these works and R. C. Moore (1949), Krumbein and Sloss (1953), and Pickering et al. (1986), and the references cited therein, for additional information and alternative definitions of the terms facies, lithofacies, and biofacies. Pickering et al. (1986) present a facies classsification.

13. For example, see Ginsburg (1960, 1967) and Gebelen (1969); also see Collinson and Thompson (1982), who suggest that oncolites are typically < 5 cm in diameter. Leinfelder and Hartkopf-Froder (1990) argue for the *in situ* formation of oncolites. See that discussion and the included references for more information. Stromatolites are discussed by Hofman (1973), Hoffman (1974), and Braithwaite et al. (1989), among others.

14. For a description of water-escape structures, see Lowe (1975), H. D. Johnson (1977), and Postma (1981b).

15. Bed forms (e.g., ripples, dunes), developed during sedimentation, are discussed by Blatt, Middleton, and Murray (1980), Harms, Southard, and Walker (1982), and Boggs (1987). Other works on bed forms include Shipp (1984), J. Nielson and Kocurek (1986), Leckie (1988), and Arnott and Hand (1989).

16. These features are discussed in chapter 21.

17. Definitions of the term fossil vary. The exact point in history at which things are no longer fossils is debated and a discussion of it is beyond the scope of this text.

18. Newell, Purdy, and Imbrie (1960), Loreau and Purser (1973), and Land, Behrens, and Frishman (1979) describe the internal structure of Holocene oolites. Milliman (1974) and Simone (1981) review their nature and origin.

19. See Folk (1974), Pettijohn (1975, pp. 83–84), Blatt, Middleton, and Murray (1980, pp. 453–57), Boggs (1987, p. 225), or other sedimentary petrology texts for further discussion of these features. B. Jones and Squair (1989) describe the formation of some peloids in a modern environment.

20. Diagenesis is discussed in chapter 16.

21. For example, see Folk and Weaver (1952), Pettijohn (1975, pp. 80–81), Nockolds, Knox, and Chinner, (1978, pp. 230–32), Scholle (1978), Blatt, Middleton, and Murray, (1980, p. 498), R. Thomas (1983b), and A. E. Adams, MacKenzie, and Guilford, (1984).

22. Longman (1980) provides a summary of cements and diagenetic histories in carbonate rocks. Cements in clastic rocks are discussed by R. Thomas (1983b). Chapter 16 provides a further discussion of cementation.

23. Hutcheon (1983).

24. Epiclastic textures are distinguished from the pyroclastic textures common in volcanic rocks and cataclastic textures found in metamorphic rocks. Prince and Ehrlich (1990) discuss a quantitative method for assessing the epiclastic textures of sandstones.

25. Numerous authors discuss various competing size scales for grains. Blatt, Middleton, and Murray (1980) give a good review in which different scales in use in North America and Europe are noted. They and others (Krumbein and Pettijohn, 1938; Pettijohn, 1975) describe the size scales that preceeded Wentworth's scale.

26. Krumbein (1934) introduced the phi (ø) scale, the basis of which was changed by McManus (1963).

27. Raymond (1984c), Howell Williams Turner, and Gilbert (1982), Spock (1962). Zingg (1935, in Blatt, Middleton, and Murray, 1980) divides the tabular grains into disk and bladed shapes. Disks are special cases in which the two larger dimensions are approximately equal. Boggs (1987, p. 123ff.) gives a more thorough review and discussion of this subject. Also see references in note 28.

28. For example, see Folk (1974), Pettijohn (1975), and D. W. Lewis (1984).

29. For example, Folk (1974), Pettijohn (1975), Friedman and Sanders (1978), Blatt, Middleton, and Murray (1980), D. W. Lewis (1984), and Boggs (1987).

30. For some examples of porosities see Archie (1952), G. E. Thomas and Glaister (1960), and Pittman (1979). Also see Choquette and Pray (1970).

31. McKee, Crosby, and Berryhill (1967).

32. Mineralogy is discussed in a number of texts including Folk (1974), Pettijohn (1975), and Raymond (1993).

33. Also see Bhatia (1983, 1985b), S. R. Taylor and McLennan (1985, ch. 2), and Bhatia and Crook (1986).

34. Larue and Sampayo (1990) also show a trace element link between sediment and probable source rocks. R. W. Murray et al. (1990) use Ce anomalies and REE to distinguish between tectonic environments of deposition of chert and shale, and Huebner and Flohr (1990) use similar data to evaluate the genesis of manganiferous cherts.

35. Van Weering and Klaver (1985), Dabard and Paris (1986), Amajor (1987), Brumsack (1989), Beier and Hayes (1989), S. J. Carpenter and Lohmann (1989), W. J. Meyers (1989), C. P. Rao (1990).

36. Toyoda and Masuda (1990).

37. Toyoda and Masuda (1990).

38. Also see Dabard and Paris (1986) and Brumsack (1989).

39. Also see Land (1980), Leeder (1982, p. 260ff.), and N. P. James and Choquette (1983) for reviews. Examples of specific studies include O'Neil and Hay (1973), Knauth and Epstein (1976), and Gautier (1986). Variations in Sr isotopic ratios may also reflect diagenesis (Clauer, Chaudhuri, and Subramanium, 1989).

40. See J. N. Weber (1960), Bhatia (1983), Bhatia and Crook (1986), and Rao (1990) for examples.

PROBLEMS

14.1 Plot the positions of rocks with the modes listed in table 14.5 on a copy of figure 14.1 and assign each to a group.

Table 14.5 Modes of Sedimentary Rocks for Problem 14.1

	Samples				
	1	*2*	*3*	*4*	*5*
Quartz	94	52	30	2	0
Alkali feldspar	0	4	7	0	0
Plagioclase	0	4	9	0	0
Chert	2	3	4	0	0
Other silicate rock fragments	0	6	5	0	0
Carbonate rock fragments, etc.	0	0	1	58	7
Biotite	0	0	1	0	0
Muscovite	1	0	1	0	0
Chlorite	0	0	1	0	0
Clays	2	20	2	1	1
Calcite mud	0	0	0	32	0
Calcite spar	0	10	34	2	69
Dolomite spar	0	0	0	0	10
Fossils	0	0	4	5	13
Other	1	1	1	0	0
Total	100	100	100	100	100

14.2 Calculate the sorting coefficient (S_O) for each of the curves shown in the Figure 14.20. You will need to estimate Q_1 and Q_3 values, the sizes corresponding to the points at which the curves cross the 25% and 75% lines. Note that the size scale is logarithmic. Q_1 and Q_3 for the curve on the far right are given.

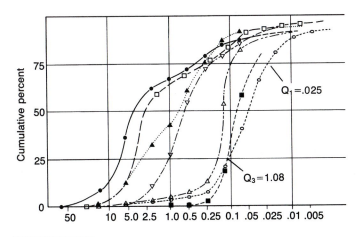

Figure 14.20 Cumulative (weight percent) curves for sorting of individual alluvial fan beds from the Carbona Formation.
(Modified from Raymond, 1969)

14.3 Given a wedge-shaped aquifer that is entirely filled with water, with a maximum thickness of 200 m, a length (radius) of 2 km, a semicircular plan, and a porosity of 20%, calculate (a) the total volume of the rock body, and (b) the volume of water that could be obtained if 25% of the available water were removed by pumping from the aquifer.

14.4 Calculate an *igneous rock* norm for analysis 3 in table 14.4. (a) If this rock had been produced by erosion of a plutonic rock terrane, what plutonic rock name would you apply to the average composition of that terrane? (b) Suppose this rock were totally melted and recrystallized as a plutonic rock, what kind of rock would it be?

14.5 Consider analysis 6 on table 14.4. Calculate the weight percent dolomite and weight percent calcite in the sample. (a) Is the LOI value reported consistent with these values? Explain. (b) What other minerals may have been in this sample as suggested by the remaining chemistry?

15

Classification of Sedimentary Rocks

INTRODUCTION

Sedimentary rocks are diverse, abundant at the Earth's surface, and economically important. For these, and other reasons, much attention has been given to their classification. Nevertheless, no single classification scheme or set of schemes is accepted by all geologists. The variety of classification schemes in use has led to the same rock being assigned different names by different geologists. As frustrating as this may be for students and professionals, it is a fact of geological life. In this book, several classifications are presented, including those that are the most widely used.

In chapter 14, the sedimentary rocks were divided into three major groups—the siliciclastic rocks (Group S), the allochem-rich clastic rocks (Group A), and the precipitate-rich rocks (Group P) (see figure 14.1). In contrast, some geologists recognize only two subdivisions. For example, Pettijohn (1975) divides sedimentary rocks into "exogenetic" and "endogenetic" rocks. Exogenic rocks are grossly equivalent to the allochthonous, clastic, fragmental, or terrigenous rocks defined by other geologists, whereas endogenic rocks are more or less equivalent to the autochthonous, chemical and biochemical, crystalline and precipitate rocks of other geologists.[1] In contrast, Folk (1974)

advocates a five-fold subdivision based on the three kinds of materials (S, A, and P) recognized here.

Each of the three sedimentary rock groups may be subdivided. Group S, in which the percentage of silicate and associated rock and mineral fragments exceeds that of either the allochems or the precipitates, includes the siliciclastic conglomerates, breccias, and diamictites; the sandstones; and the mudrocks. Group A, in which the percentage of allochems is greater than either the percentage of silicate and associated fragments or the precipitates, includes fragmental limestones and dolostones, fragmental cherts, and other fragmental allochem-rich rocks. Included in Group P, in which chemically precipitated minerals exceed either allochems or silicates and associated fragments, are the crystalline limestones and dolostones, the cherts, the evaporites, and other chemical and biochemical precipitates.

CLASSIFICATION OF GROUP S ROCKS

Siliciclastic Conglomerates, Breccias, and Diamictites

Conglomerate, breccia, and diamictite (as defined here) are all rocks containing 25% or more clasts larger than 2 mm in

Classification of conglomerates, breccias, & diamictites			
Matrix/support	Clast shape	Clast composition	Name*
Gravelly or sandy (Generally clast-supported)	Rounded	• Single compostion Quartz±chert Calcareous • Varied composition	Name of clast type & "conglomerate" Quartzitic conglomerate Calcirudite or limestone conglomerate Polymict (=Petromict) conglomerate
	Angular	• Single composition Quartz±chert Calcareous • Varied composition	Name of clast types & "breccia" Quartzitic breccia Limestone breccia Polymict breccia
Muddy (clay±silt) and mud-supported	Rounded, angular or both	• Single composition • Varied composition	Oligomict diamictite ** Polymict diamictite**
*Prefix the word conglomerate, breccia or diamictite with the clast size designation (e.g. pebble conglomerate; boulder breccia). **Where the rock is known to be of glacial origin, the term tillite is substituted for diamictite			

Figure 15.1
Classification of coarse silici-clastic sedimentary rocks (conglomerates, breccias, and diamictites), with > 25% of the clasts larger than 2 mm in diameter. Oligomict refers to rocks of a single clast type. Polymict refers to rocks with many clast types. *(Modified from Raymond, 1984c).*

diameter (or length) enclosed in a finer-grained matrix.[2] **Conglomerates** have rounded clasts in a sandy matrix, **breccias** have angular fragments in a sandy matrix, and **diamictites** have rounded clasts, angular clasts, or both in a matrix dominated by mud (figure 15.1). Modifiers for individual rock names are based on the clast composition, the dominant clast size, and the root name (e.g., quartz pebble conglomerate).

Alternative criteria are used by some authors to define and name coarse-grained, siliciclastic rocks. Compton (1962) uses a definition similar to the one offered here (figure 15.2a) but applies the names conglomerate and breccia to rocks with either sandy or muddy matrix material. Moncrieff (1989) does likewise with the conglomerates. Inasmuch as diamictites have unique textures and origins, this seems undesirable. Pettijohn (1975) discusses varied uses of the terms gravel and conglomerate, noting that 30% and 50% gravel-size clasts are also possible lower-limit values for applying the terms conglomerate and breccia. Folk (1974) favors a value of 30% (figure 15.2b). In a modified version of the Folk classification, Moncrieff (1989) uses a value of 20% and designates all rocks having between 20% and 1% clasts and a sand: mud ratio of <9 as diamictites, regardless of the small amount of mud they may contain (figure 15.2c).

As used here, diamictite, a rock originally named by Flint et al. (1960), may have mud (silt or clay) or rock flour as a matrix. The term is nongenetic; that is, it does not imply

a specific origin. Diamictites are deposited by glaciers and landslides (including debris flows and mudflows), both of which do a poor job of sorting sediment. Where specific origins are known, more exact names (e.g., tillite, for a glacial diamictite) or names with modifiers (e.g., mudflow diamictite) should be applied.

There are no compelling genetic reasons to favor one of the classifications over another. The 20, 25, 30, or 50% gravel value used as boundaries for distinguishing conglomerates, breccias, and diamictites from sandstones and mudstones are arbitrary. In the field, one tends to notice larger clasts preferentially over smaller clasts; hence the lower values are more commonly adopted. The classification of figure 15.1 is favored here because diamictites, which are distinguished on the basis of observable criteria (poor sorting, dominantly fine-grained matrix, and clast size) and do seem to have unique origins, are distinguished from other coarse-grained siliciclastic rocks.

Sandstone Classification

Sandstones have been named and classified in a large variety of ways.[3] Scholle (1979) illustrated many of these classifications. Although the numerous classifications differ, they have many similarities, with most of the more recent classifications representing refinements of those published earlier. Fortunately, in recent years, only three classifications—those of

McBride (1963), Folk (1974), and Dott (1964)—have become preeminent. The Dott classification (or modified forms of it), based on an earlier classification by Gilbert (Howell Williams, Turner, and Gilbert, 1954), is used widely. A modified version of the Dott classification is adopted in this book (figure 15.3).[4]

The Dott (1964) classification of sandstones consists of two triangles, each representing one of the two major categories of sandstone—**arenite** and **wacke.** The two basic types of sandstone are distinguished on the basis of matrix. **Matrix** is *clastic* material that occupies the spaces between grains and usually consists of clay with silt-size particles of quartz and other minerals.[5] Matrix should not be confused with **cement,** which is *precipitated* material that fills the interstices between grains. Arenites contain less matrix than wackes. Various petrologists use different amounts of matrix to define the arenite-wacke boundary.[6] Values typically selected are 0, 5, 10, and 15% matrix. In this text a value of 5% is used; in other words, arenites have less than 5% matrix, whereas wackes have 5% or more matrix.[7]

Because they are well known, two additional classifications, those of Folk (1974) and McBride (1963), are presented here in figure 15.4. Both allow more discrimination of sandstone types than the Dott classification, but both ignore the matrix. Folk subdivides sandstones that plot at the feldspar and rock fragment corners of his diagram. This yields a large number of sandstone names. Note, however, that in both the McBride and the Folk classifications, only two basic root names—arenite and arkose—are employed. Rocks relatively rich in feldspar contain the root name **arkose.** Other rocks have the root name arenite.

A nongenetic scheme for naming sandstones that requires no triangular plot and could be used for classification is presented by Brewer, Bolton, and Driese (1990). Sandstones of any composition, including those dominated by glauconite or other less common minerals, may be named using the scheme. The root names include sandstone (matrix undetermined), arenite (matrix < 10%), and wacke (matrix ≥ 10%). The detailed names (e.g., lithic quartz arenite; subglauconitic quartz wacke) are based on the three most abundant components and are derived through a five-step process.

The merits and deficiencies of the various sandstone classifications are, in part, a reflection of personal preference. The Dott (1964) and Brewer et al. (1990) classification schemes are relatively simple, easy to remember, and relatively easy to use. The primary basis for classification, the distinction between matix-poor ("clean") and matrix-rich ("dirty") sandstones, is useful and has some genetic significance. Some consider the fact that the Dott classification provides for only six rock names (some modifications include several more) to be a flaw.

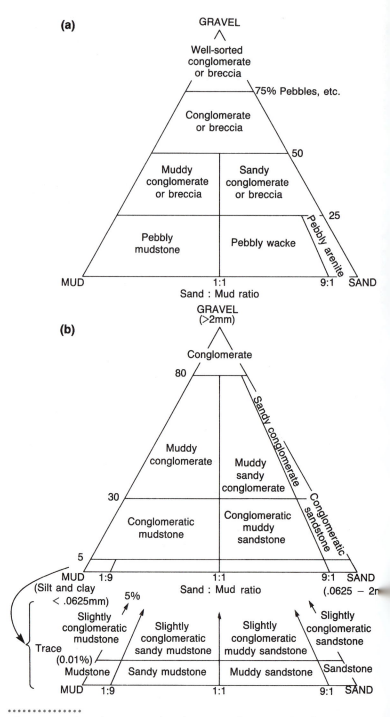

Figure 15.2 Alternative classifications of coarse siliciclastic sedimentary rocks. (a) Compton's (1962) subdivisions of clastic rocks. (b) Folk's classification of conglomerates (drawing modified from Folk, 1974). (c) Moncrieff's (1989) classification of coarse siliciclastic rocks.

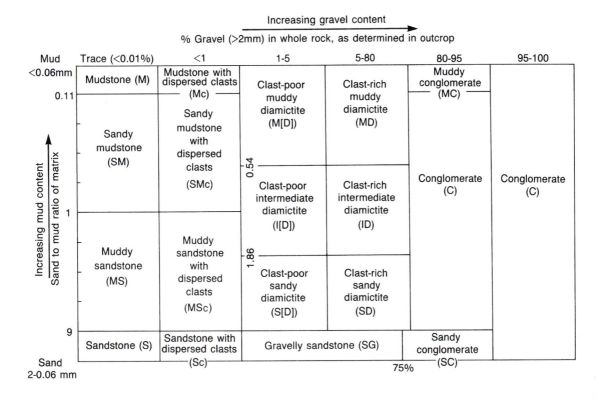

Figure 15.2 con't.

The McBride (1963) and Folk (1974) classifications ignore matrix and emphasize, instead, **framework** (grain) composition. Thus, the distinction between "dirty" and "clean" sandstones is not possible. Both classifications, especially Folk's, provide for more discrimination than does Dott's. The careful study of sandstones required in using the Folk classification may reveal significant information about provenance, and the rock name will reflect that knowledge. Such is not the case with the names from the Dott or McBride classifications. The relatively greater complexity of the McBride and Folk classifications, however, makes them somewhat more difficult to remember and use in handspecimen identification during fieldwork. In addition, they do not discriminate between clean and dirty sandstones.

Classification of Mudrocks

Mud is siliciclastic silt, clay, or mixtures of these materials. Mixtures may occur in any proportion. Where mud becomes lithified, it becomes **mudrock** (mudstone to some authors). Mudrocks are quite abundant among sedimentary rocks, as they are commonly interbedded with carbonate rocks, cherts, and sandstones, and they comprise thick sections of rock by themselves.

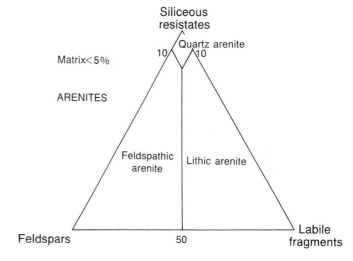
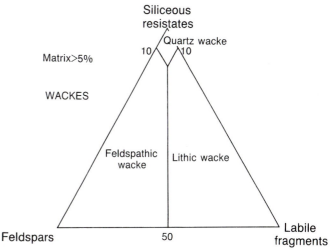

Figure 15.3 The modified Dott (Gilbert) classification of sandstones. Labile fragments are easily weathered rock fragments. *(Modified from Dott, 1964)*

287

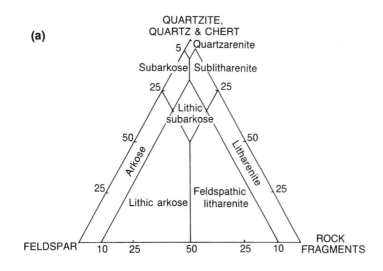

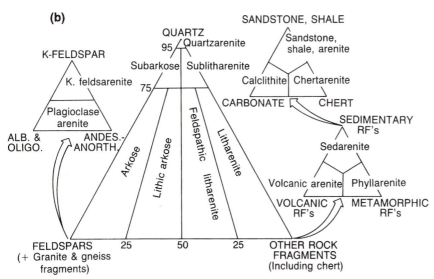

Figure 15.4   Sandstone classifications of (a) McBride (1963) and (b) Folk (1974).

In recent years, the classification of mudrocks has received more attention, perhaps in part because of the application of new techniques of study. X-radiography, SEM, and other techniques now may be used in conjunction with X-ray diffraction and optical petrography to gain a greater understanding of these aphanitic, difficult-to-study rocks.

Of the several classifications proposed for mudrocks,[8] four are shown in figure 15.5. Those of Pettijohn (1975), Lundegard and Samuels (1980), and Raymond (1993) have the merit of being easily used in the field. That of Spears (1980) is more difficult to use, as it requires the preparation of standards of various proportions of quartz and clay, and discrimination between quartz and other silt-size silicates.[9] The

classification of Lundegard and Samuels (1980) is a modification of a classification proposed by Folk (1968, 1974). The scheme of P. E. Potter, Maynard, and Pryor (1980) is quite similar.

The classifications generally concur in assigning the names mudstone and shale but differ on the names of silt-rich sedimentary rocks. **Shale** is a laminated or fissile mudrock.[10] **Mudstone** is nonlaminated or nonfissile mudrock. The name **siltstone,** assigned to rocks containing substantial, but variable, amounts of silt-size grains, has been assigned to different rocks, including laminated, fissile, nonlaminated, and nonfissile rocks. A modified version of the Lundegard-Samuels (Folk) classification is used here (figure 15.5d). Modification is needed because mudrocks may contain sand or larger grains. The rock name should be based on the *mud fraction*, which must account for 50% or more of the rock for the rock to qualify as a mudrock. Using the Lundegard-Samuels classification, in which the name is based on the "silt fraction" of the rock, a nonlaminated rock composed of 45% sand, 30% silt, and 25% clay would qualify as a claystone. This is unacceptable because the name does not reflect the true character of the rock, which is dominated by grains coarser than claysize. The classification proposed here bases the name on the relative proportion of silt- and clay-size materials in the mud fraction.

CLASSIFICATION OF GROUP P ROCKS

Group P is dominated by the carbonate rocks—the limestones and dolostones—but also includes such diverse rocks as cherts, evaporites, some iron-rich rocks, phosphatic rocks, and siliceous sinter. Many of these rocks occur in close association with Group A rocks. For that reason and because of their similar compositions, both Group P and Group A rocks are included by most geologists in the same classification scheme.

A general classification of Group P rocks is presented in table 15.1. Major categories of rock (e.g., limestones) are defined on the basis of composition. Specific rock names are then determined by consulting either the classification of rocks for that major category or the detailed descriptions provided in table 15.1.

Figure 15.5 Classifications of mudrocks:
(a) Pettijohn (1975), (b) Lundegard and Samuels (1980), (c) From D. A. Spears, "Towards a classification of shales" in *Journal of Geological Society of London,* 137:125–129, 1980. Copyright © 1980 Geological Society of London. Reprinted by permission. (d) Raymond (1993).

CLASSIFICATION OF MUDROCKS

(a) Pettijohn (1975)

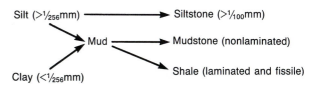

Silt (>1/256mm) ⟶ Siltstone (>1/100mm)

Mud ⟶ Mudstone (nonlaminated)

Clay (<1/256mm) ⟶ Shale (laminated and fissile)

(b) Lundegard and Samuels (1980)

INDURATED		SILT FRACTION 2/3	1/3
	Non-laminated	Mudstone	Claystone
		Siltstone	
	Laminated	Mudshale	Clayshale

(c) Spears (1980)

	FISSILE	NON-FISSILE
>40% Quartz	Flaggy siltstone	Massive siltstone
30–40% Quartz	Very coarse shale	Very coarse mudstone
20–30% Quartz	Coarse shale	Coarse mudstone
10–20% Quartz	Fine shale	Fine mudstone
<10% Quartz	Very fine shale	Very fine mudstone

(d)

	MUDROCKS Rocks containing >50% mud			Rocks with <50% mud
	Silt dominant (>2/3 of mud)	Clay and silt	Clay dominant (>2/3 of mud)	Sand-sized or larger grains dominant
Non-laminated	Siltstone	Mudstone	Claystone	Conglomerates breccias diamictites and sandstones
Laminated	Laminated siltstone	Mudshale	Clayshale	

Table 15.1   Classification of Sedimentary Precipitate Rocks (Group P)

Composition {major mineral(s)}	Rock Families	Description and Special Names
Calcite	Limestones	**Limestone** Aphanitic to phaneritic rock composed dominantly of calcite (see figures 15.6, 15.7, 15.8 for specific names in addition to those listed below).
		Travertine Aphanitic to phaneritic, layered rock, usually light-colored and concretionary. Deposited by ground and surface waters emanating from springs.
		Tufa Cellular travertine (i.e., with holes).
		Caliche White, chalky, travertine-like rock developed in soil layers.
		Chalk Soft, earthy, friable, porous, light-colored limestone.
		Sparite Crystalline limestone (see figure 15.7).
		Micrite Lime mudstone (see figures 15.6, 15.7).
Dolomite	Dolostones	**Dolostone** Aphanitic to phaneritic rock composed dominantly of dolomite (see figures 15.7, 15.8 for additional names).
		Dolomudstone Aphanitic dolostone.
		Microdolostone Cryptocrystalline to aphanitic dolostone.
		Dolosparite Phaneritic crystalline dolostone.
Halite	Evaporites	**Rock salt,** or **Halite evaporite** Aphanitic to phaneritic, usually light-colored, salty-tasting, soft rock dominated by halite.
Gypsum		**Gypsum evaporite** Aphanitic to phaneritic, soft, commonly layered, gypsum-rich rock.
Anhydrite		**Anhydrite evaporite** Aphanitic to phaneritic, soft, commonly layered, anhydrite-rich rock.
Other evaporite minerals		Mineral name **evaporite.**
Quartz, Chalcedony,	Siliceous rocks	**Chert** Aphanitic to fine-grained phaneritic, multi- or variously colored, waxy to grainy, hard, siliceous rock.
Cristobalite,		**Radiolarian chert** Chert containing significant numbers of radiolaria.
Opal		**Diatomite** Aphanitic, light-colored, soft, friable, siliceous rock composed of diatoms; where well cemented the rock is called **diatomaceous chert.**
		Siliceous sinter Aphanitic to fine-grained, typically porous and layered, variously colored, hard, siliceous rock deposited by groundwater or surface water at or near a hot spring or geyser.
Goethite ± hematite ± calcite± dolomite	Iron-rich rocks	**Ironstone** Aphanitic to phaneritic, massive to bedded, commonly oolitic, yellow to maroon, silver, or black, iron-rich rock.
± magnetite ± hematite ± greenalite		**Iron formation** Aphanitic to phaneritic, thin-bedded and interbedded with chert, typically red to black, iron-rich rock.
Manganese minerals	Mn-rich rocks	**Manganese nodule** Aphanitic, black to brown, nodular mass composed of various manganese oxides.
		Manganolite Aphanitic, black to brownish-black rock composed of various manganese oxides.
Apatites	Phosphorous-rich rocks	**Phosphorites** Aphanitic to phaneritic; typically brown to black; oolitic, laminated, nodular, or fossiliferous rock, with > 50% apatite.
		Phosphatic rock Conglomerate to mudrock, commonly dark, variously colored, with apatite cement (the root rock name is inserted with *phosphatic* as a modifier, e.g., phosphatic shale).

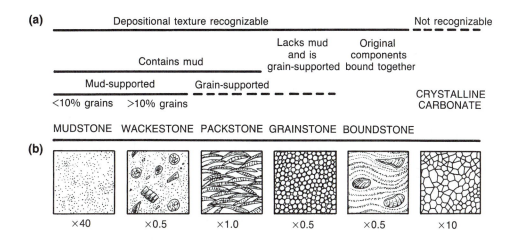

(b)

MUDSTONE WACKESTONE PACKSTONE GRAINSTONE BOUNDSTONE

×40 ×0.5 ×1.0 ×0.5 ×0.5 ×10

(c)

Predominantly calcite (Cc >95%)	Dominantly calcite (95%>Cc>50%) with dolomite	Dominantly dolomite (Do >50%)	Thoroughly recrystallized rocks with some relict structures	
			Dominantly dolomite	Dominantly calcite
Lime mudstone	Dolomitic lime mudstone	Dolomudstone	Crystalline dolostone	Crystalline limestone
Wackestone	Dolomitic wackestone	Dolowackestone		
Packstone	Dolomitic packstone	Dolopackstone		
Grainstone	Dolomitic grainstone	Dolograinstone		
Boundstone	Dolomitic boundstone	Doloboundstone		

Figure 15.6 Dunham-type (1962) classifications of carbonate rocks. (a) Definitions of limestone types (after Dunham, 1962). (b) Sketches of typical textures. (Different types of grains are used in the sketches, but grain types are not restricted to specific rock types and have no bearing on the root names.) (c) Classification of carbonate rocks based on (a).
(From Raymond, 1993)

Carbonate Rock Classifications

Carbonate rocks are of two main types: the limestones, rocks dominated by calcite (and aragonite), and the dolostones, rocks dominated by dolomite. Two major classifications of carbonate rocks are widely used today.[11] Both incorporate Group P and Group A rocks and both use the percentage of allochems as a classification parameter. The Dunham (1962) classification (figure 15.6) is based primarily on depositional textures, specifically the ratio of grains to mud. In this classification, only two rock types belong to Group P, the boundstones and *crystalline carbonates* (crystalline limestone or crystalline dolo-

stone). To distinguish a dolomitic rock, the prefix *dolomitic* is used before the root name (e.g., dolomitic boundstone). Boundstones are limestones and dolostones in which the original components of the rock (the organisms and their precipitated materials) were bound together during life. Crystalline limestone and crystalline dolostone are rocks with a crystalline texture and no recognizable depositional texture.

The Folk (1962) classification (figure 15.7) is more complex than the Dunham classification. It too is based largely on depositional textures. Folk recognized four basic materials: the allochems; microcrystalline ooze, called

Volumetric allochem composition		Limestones, partially dolomitized limestones, and primary dolomites (see Notes 1 to 6)					Replacement dolomites (V) (see note 7)	
		>10% Allochems Allochemical rocks (I and II)		<10% Allochems Microcrystalline rocks (III)		Undisturbed bioherm rocks (IV)		
		Sparry calcite cement > microcrystalline ooze matrix	Microcrystalline ooze matrix > sparry calcite cement	1-10% allochems	<1% allochems		Allochem ghosts	No allochem ghosts
		Sparry allochemical rocks (I)	Microcrystalline allochemical rocks (II)	Most abundant allochems			Evident allochem	
<25% Intraclasts (i)		Intrasparrudite (Ii:Lr) Intrasparite (Ii:La)	Intramicrudite* (IIi:Lr) Intramicrite* (IIi:La)	Intraclasts: intraclast-bearing micrite* (IIIi:Lr or La)	Micrite (IIIm:L); if disturbed, dismicrite (IIImX:L); if primary dolomite, dolomicrite (IIIm:D)	Biolithite (IV:L)	Finely crystalline intraclastic dolomite (Vi:D3) etc.	Medium crystalline dolomite (V:D4)
<25% Intraclasts	>25% Oolites (O)	Oosparrudite (Io:Lr) Oosparite (Io:La)	Oomicrudite* (IIo:Lr) Oomicrite* (IIo:La)	Oolites: oolite-bearing micrite* (IIIo:Lr or La)			Coarsely crystalline oolitic dolomite (Vo:D5), etc.	
<25% Intraclasts; <25% Oolites	Volume ratio of fossils to pellets >3:1 (b)	Biosparrudite (Ib:Lr) Biosparite (Ib:La)	Biomicrudite (IIb:Lr) Biomicrite (IIb:La)	Fossils: fossiliferous micrite (IIIb:Lr, La,or L1)			Aphanocrystalline biogenic dolomite (vb:D1), etc.	Finely crystalline dolomite (V:D3)
<25% Intraclasts; <25% Oolites	Volume ratio of fossils to pellets 3:1 to 1:3 (bp)	Biopelsparite (Ibp:La)	Biopelmicrite (IIbp:La)	Pellets: pelletiferous micrite (IIIp:La)			Very finely Crystalline pellet dolomite (Vp:D2), etc.	etc.
<25% Intraclasts; <25% Oolites	Volume ratio of fossils to pellets <1:3 (p)	Pelsparite (Ip:La)	Pelmicrite (IIp:La)					

Figure 15.7 The Folk classification of carbonate rocks. Notes: * designates rare rock types. (1) Names and symbols in the body of the table refer to limestones. If the rock contains over 10% replacement dolomite, add the prefix *dolomitized* to the rock name, and use *DLr* or *DLa* for the symbol (e.g., dolomitized intrasparite, Ii:DLa). If the rock contains over 10% dolomite of uncertain origin, add *dolomitic* before the rock name, and use *dLr* or *dLa* for the symbol (e.g., dolomitic biomicrite, IIb:dLa). If the rock is a primary dolomite, add *primary dolomite* to the rock name, and use *Dr* or *Da* for the symbol (e.g., primary dolomite intramicrudite, IIk:Dr). Instead of *primary dolomite micrite* the term *dolomicrite* may be used. (2) Upper name in each box (I and II) refers to calcirudites (median allochem size larger than 1.0 mm), and lower name refers to all rocks with median allochem size smaller than 1.0 mm. Grain size and quantity of ooze matrix, cements, or terrigenous grains are ignored. (3) If the rock contains over 10% terrigenous material, add *sandy, silty,* or *clayey* to the rock name, and *Ts, Tz,* or *Tc* to the symbol depending on which is dominant (e.g., sandy biosparite, TsIb:La, or silty dolomitized pelmicrite, TzIIp:DLa). Glauconite, collophane, chert, pyrite, or other modifiers may also be added. (4) If the rock contains other allochems in significant quantities that are not mentioned in the main rock name, these should be added immediately before the main rock name (e.g., fossiliferous intrasparite, oolitic pelmicrite; peletiferous oosparite; or intraclastic biomicrudite). These can be shown symbolically as Ik(b), Io(p), IIb(i), respectively. (5) If one or two types of fossils are dominant, this fact should be shown in the rock name (e.g., pelecypod biosparrudite, crinoid biomicrite, brachiopod-bryozoan biosparite). (6) If the rock was originally microcrystalline and can be shown to have recrystallized to microspar (5–10 microns, clear calcite) the terms *microsparite, biomicrosparite,* and so on, can be used instead of *micrite* or *biomicrite.* (7) Specify crystal size as shown in the examples.
(From Folk, 1962)

micrite; crystalline calcite, called **spar** or **sparite;** and intergrown, fossiliferous reef rock, called **biolithite.** In addition, **dismicrite,** which is micrite containing local areas of sparry calcite, and crystalline carbonate is recognized. Of these, sparite, biolithite, dismicrite, crystalline carbonate, and some micrite belong to Group P. The various other rock names in the classification are based on the percentages of allochems of specific types (e.g., oolites, pellets), the overall abundance of allochems, and the presence or absence of spar versus micrite. Specific names consist of a group of prefixes attached to a root (e.g., biopelsparite). The prefixes dolomitic and primary dolomite precede the root name where appropriate. Although these Group P rocks form a significant part of the geologic record, Group A carbonate rocks are more abundant.

Dunham's classification has the merits of being both easy to use and easy to remember. It is largely descriptive, but it may be used to draw tentative conclusions about the energy of the environment of deposition. For example, the grainstones presumably represent a higher-energy environment where mud does not accumulate. Lime mudstones, in contrast, may represent low-energy environments. The addition of modifiers to root names (e.g., oolitic, skeletal, pelletal) gives more information, but this, of course, adds to the apparent complexity. Geologists who prefer genetic classifications or classifications with larger numbers of names may find Dunham's classification wanting. Dunham's classification is also lacking in that it does not include the less common carbonate rocks—travertine, tufa, and caliche (table 15.1). Travertine and tufa could be called boundstone and caliche could be called mudstone, yet their unique textures and origins would be obscured by doing so.

Folk's classification provides more discrimination (more rock names) than does Dunham's. Because of that discrimination, more genetic information about a rock is conveyed by the name. A disadvantage is that the Folk classification is much more difficult for the beginner to remember and use. It too overlooks travertine, tufa, and caliche, rocks that clearly have no niche in the classification scheme.

Classification of Other Precipitates

Classification of noncarbonate Group P rocks is largely based on mineral or chemical composition. In addition, few names exist for any given composition. For example, for silica-rich rocks, three types—diatomite, chert, and siliceous sinter—are defined (table 15.1).

Evaporites provide an exception to the rule of generally limited variability in rock type shown by noncarbonate rock groups. A large number of minerals may develop as a result of evaporation of salt, fresh, or saline waters (see chapter 22). Yet few names have been proposed for the various evaporites. Rock salt, a rock composed primarily of halite, is an exception. In this book, *evaporite* is used as a root name. A complete name consists of (1) a textural term, (2) a list of major minerals, in order of increasing abundance, and (3) the root name evaporite (e.g., fine-grained, gypsum-anhydrite evaporite).

CLASSIFICATION OF GROUP A ROCKS

Group A rocks, the rocks dominated by allochems, are generally classified jointly with their precipitated relatives. As noted above, the two most commonly used classifications of limestones include both allochemical and crystalline varieties of limestone. Further, these classifications use the percentage of allochems as a classification parameter. Nevertheless, there are both empirical (observational) and genetic reasons to distinguish Group A and Group P rocks.

Recall that Folk (1962) distinguishes two main intraclastic materials, lime mud, or micrite, and crystalline lime cement (spar). As defined by Folk, micrite consists of grains less than or equal to 0.004 mm (4 microns) in diameter. Spar grains are greater than 0.004 mm. Discrimination between particles above and below this size limit is difficult, even under the microscope. For that reason, some geologists, in an attempt to make the classifications more usable, suggest that the boundary be drawn between sizes of grains that can be more easily distinguished, for example, at 0.06 mm, the lower limit of sand-size grains.[12] Here, that convention is adopted for micrite (micrite = grains less than 0.06 mm in diameter). Micrite forms the matrix in many allochemical rocks.

In the Dunham classification (1962), the relative amounts of mud (micrite) versus carbonate framework grains is used to define four major allochemical limestone types (figure 15.6). **Lime mudstone** contains less than 10% grains. **Wackestone** contains more than 10% grains but is micrite-supported (the larger grains do not generally touch one another). In **packstone,** micrite fills the spaces between framework grains (the grains generally touch one another and the rock is grain-supported). **Grainstones** are grain-supported and lack micrite matrix.

In the Folk classification (1962), the names are based on relative amounts of micrite versus sparite, plus the abundances of specific types of allochems. For example, a rock with sparite cement in excess of micrite matrix, more than 10% allochems, and greater than 25% oolites is called an *oosparite*. Folk accounted for the fact that in many allochemical rocks the allochems are larger than 1.0 mm. For such rocks, he used the root name rudite. Thus, the rock described above but in which the oolites are larger than 1.0 mm would be called an *oosparrudite*.

Embry and Klovan (1971) and Cuffey (1985) expanded the Dunham classification (1962) to accommodate large bioclastic grains, giving it a similar capacity for size discrimination to that of the Folk classification. In this expanded classification, rocks with sand-size bioclasts in abundances of less than 10% are assigned the original Dunham names, but those with more than 10% bioclasts, especially those of gravel size, are assigned one of eleven additional names (e.g., rudstone, shellstone, lettucestone)(figure 15.8).

In some detailed work, the specific types of fossils present among the biogenic allochem clasts are distinguished and their names are used as modifiers for the root names (e.g., bryozoan packstone). Recognition of specific bioclasts is useful in refining the facies analysis because each organism has a limited range of tolerance for salinity, depth, temperature, or other factors. Such analyses require expertise in recognizing the petrographic characters of different skeletal framework components, which change in abundance and character over time (M. Pitcher, 1964; Wilkinson, 1979).

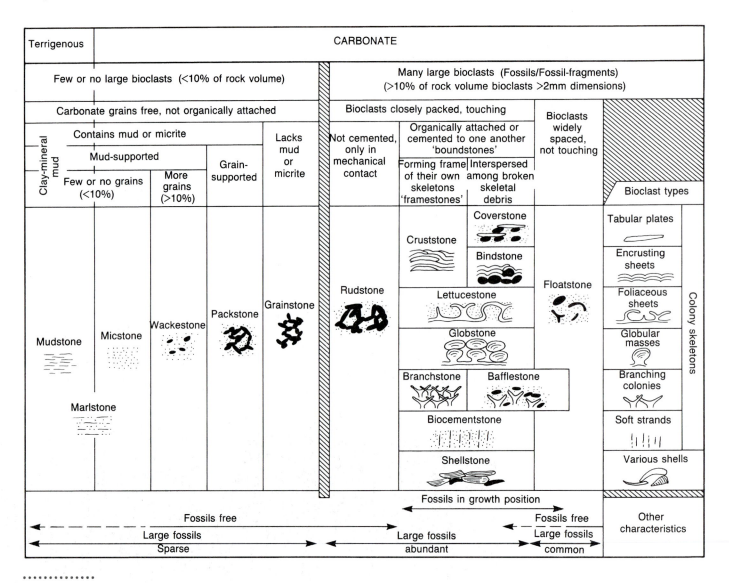

Figure 15.8
Cuffey (1985) classification of clastic limestones.

Figure 15.9 is an alternative classification of Group A rocks based on the ideas of Grabau (1924). Like the general classification of Group P rocks, it divides rock types into major categories. Except for the carbonate rocks, which are commonly assigned the special names discussed above, the textural names conglomerate, breccia, or rudite are used to designate allochemical rocks composed of clasts larger than 2 mm in diameter (e.g., chert breccia). Similarly, the textural names arenite and wacke may be applied, as they are to siliciclastic rocks, to allochemical rocks with sand-size grains. Allochemical mudrocks are assigned the textural root *lutite* (e.g., calcilutite). Modifiers such as pelletal, oolitic, fenestral, or radiolarian may be used before the root names, as is the practice for Group P rocks.

SUMMARY

A wide range of sedimentary rock classifications exists. Because of the variety of available sedimentary rock classifications, the same rock may be assigned different names by different geologists. Although this is confusing to students, it is a reality. The student who accepts and understands this diversity of nomenclature will be able to read, understand, compare, and contrast the professional literature on sedimentary rocks.

Chapters 18 through 22 describe and discuss specific rock types and associations. The coarse clastic-rock and mudrock classifications of Raymond (1984; 1993), the sandstone classification of Dott (1964), the Dunham (1962) and Cuffey (1985) classifications of carbonate rocks, and the classifications of noncarbonate Group A and Group P rocks presented in this chapter serve as a basis for those discussions.

Texture	Principle clast type(s)	Names	
Grains ≥ 2 mm in 25% of rock	Fossils	Skeletal rudite*	
	Limestone	Calcirudite* (porous skeletal calcarenite is called coquina)	
	Dolostone	Dolorudite*	
	Evaporites	Evaporite rudite (e.g. Gypsum rudite)	
	Chert	Chert rudite	
	Other	Main clast type name followed by the root rudite, conglomerate, or breccia (e.g., Chert breccia)	
		Matrix > 5%	Matrix < 5%
Grains 1/16 to 2 mm dominate rock	Fossils	Skeletal wacke ‡	Skeletal arenite ‡ (Porous skeletal calcarenite is called coquina)
	Limestone or calcite	Calcwacke**▼	Calcarenite**▼
	Dolostone or dolomite	Dolowacke**▼	Doloarenite**▼
	Chert	Chert wacke	Chert arenite
	Other	Prefix root name wacke with name(s) of main clast type(s)▼	Prefix root name arenite with name(s) of main clast type(s)▼
Grains mainly 1/16 – 1/256 mm	Calcite or limestone	Calcisiltite	
	Dolomite or dolostone	Dolosiltite	
	Chert	Chert-siltite	
	Other	Prefix siltite root with major clast type	
Grains mainly < 1/256 mm	Calcite	Calcilutite	
	Dolomite	Dololutite	
	Siliceous	Chert	

Figure 15.9 Grabau-type classification of Group A (allochemical) rocks. *Conglomerate or breccia, following the use in figure 15.1, may be substituted for the root name *rudite*. See other carbonate rock classifications for alternative names. ▼Use modifiers such as *oolitic* where appropriate. ‡Specific fossil names may be substituted for the word *skeletal* (e.g., fenestral rudite).

EXPLANATORY NOTES

1. For further discussion of general classification problems, see standard texts such as Folk (1974), Pettijohn (1975), Selley (1976), Greensmith (1978), Nockolds, Knox, and Chinner (1978), Blatt, Middleton, and Murray (1980), Tucker (1981), and Boggs (1987).
2. This classification is essentially that of Raymond (1984c).
3. For example, see Krynine (1948), Williams, Turner, and Gilbert (1953), Packham (1954), McBride (1963), Dott (1964), Okada (1971), Folk (1974), Scholle (1979), Zuffa (1980), Pettijohn, Potter, and Siever (1987), and Brewer, Bolton, and Driese (1990).
4. Pettijohn (1975), Greensmith (1978), Nockolds, Knox, and Chinner (1978), and Tucker (1981) depict or use Dott-like classifications. Blatt, Middleton, and Murray (1980), curiously, do not depict the Dott classification; rather, they depict classifications of Gilbert (Howell Williams, Turner, and Gilbert, 1954) on which Dott's was based; Pettijohn (1957), who later (1975) abandoned his for a Dott-like classification; McBride (1963); and Folk (1968, 1974). Packham (1954) subdivides sandstones on the basis of bedding types, that is, on structural characteristics. Brewer, Bolton, and Driese (1990) attempt to provide a rational classification that avoids the bias towards the most common types of sandstones that is a part of many other classifications.
5. The origin of matrix is addressed in chapter 19.
6. Dott (1964) designated 10%, but subsequently other workers have opted for different values.
7. The rationale for this choice is as follows. Few rocks have absolutely no matrix materials; thus, the choice of 0% would be so restrictive as to make the classification useless. Turbidite and other sands in modern environments usually have less than 10% matrix (Shepard, 1961, 1964; C. D. Hollister and Heezen, 1964; Kuenen, 1966). In rocks with >10% matrix, much of it is generated by postdepositional processes. Thus, use of 10% would, in effect, serve to primarily distinguish sandstones with postdepositional matrix from those without such matrix and would have no meaning in terms of the sedimentology of the sandstone. The choice of 5% is arbitrary and may also have a weak sedimentological basis, but it does separate sands deposited by agents that do not sort the sediment well from sands deposited by agents that do sort well or sands with a history of reworking that eliminated the fine material.
8. For example, Folk (1974), Pettijohn (1975), Selley (1976), Lundegard and Samuels (1980), P. E. Potter, Maynard, and Pryor (1980), and Spears (1980).
9. The latter point was noted by Lundegard and Samuels (1981).
10. *Fissility* is the tendency in shales to break or split into flat pieces.
11. For other classifications, see papers in Ham (1962); also see T. W. Todd (1966) and Bissell and Chilingar (1967).
12. For example, see Greensmith (1978, p. 131).

PROBLEMS

15.1. Place each of the rocks represented by the modes in table 15.2 into as many of the classifications presented in chapter 15 as you can.

Table 15.2 Modes of Selected Rocks for Problem 1 (values in volume percentages)

	1	2	3	4	5	6
Clasts >2mm*	(28a)	(25r)	(20a)	(0)	(0)	(0)
Quartz	6	22	—	—	—	—
Chert	—	3	1	—	—	—
Sandstones	—	—	—	—	—	—
Shale	1	—	—	—	—	—
Limestones	—	—	14	—	—	—
Fossils	—	—	5	—	—	—
Granitic Rocks	17	—	—	—	—	—
Siliceous Volcanic Rocks	2	—	—	—	—	—
Mafic Volcanic Rocks	1	—	—	—	—	—
Schist and Gneiss	1	—	—	—	—	—
Sand-size Clasts	(13)	(66)	(19)	(84)	(81)	(37)
Quartz	8	64	—	10	50	—
Chert	—	2	1	—	3	—
Feldspars	4	—	—	35	19	—
Fossils	—	—	3	—	2	5
Oolites	—	—	—	—	—	22
Pellets	—	—	7	—	—	6
Limestones	—	—	8	—	—	4
Shale	—	—	—	4	5	—
Volc. Rocks	1	—	—	34	2	—
Meta. Rocks	—	—	—	1	—	—
Mud	(59)	(1)	(63)	(16)	(2)	(63)
Silt (siliceous)	29	—	—	10	2	—
Clay	30	1	—	6	—	—
Micrite	—	—	63	—	—	63
Cement and Spar	(0)	(8)	(8)	(0)	(17)	(0)
Calcite	—	—	6	—	17	—
Dolomite	—	—	2	—	—	—
Siliceous	—	8	—	—	—	—
Total	100	100	100	100	100	100

*a = angular, r = rounded.

16

Sedimentary Provenance, Processes, and Diagenesis

INTRODUCTION

Sediment is derived from the weathering and erosion of preexisting rocks. Once sediment is available for transportation, various agents—including gravity, running water and aqueous currents, the wind, and moving ice—will move it from its site of formation to various sites of deposition. There, it is deposited in the various lithofacies that characterize the particular environments of deposition (chapter 17). While at the surface, following deposition, and when buried under subsequent sedimentary accumulations, the sediment will experience **diagenesis,** the physical, chemical, and biological processes that collectively result in (1) transformation of sediment into sedimentary rock and (2) modification of the texture and mineralogy of a rock. Diagenesis contrasts with **weathering,** the processes that transform rock into soil. The direction of the two types of reactions is opposite. Weathering is degradational and includes processes in which rock is changed into less lithified material composed of minerals stable at the Earth's surface. In contrast, during diagenesis, sediments are generally changed into more lithified materials. Diagenesis is replaced at elevated temperatures and pressures by the processes of metamorphism.

WEATHERING AND PROVENANCE

The nature of the sediment deposited in any particular environment is a function of many factors. These include (1) the provenance, or the place of origin of the sediment, which controls the kinds of materials that will be available as sediment; (2) weathering and transportation, which control modification of the materials derived from the source terrane; and (3) the nature of the depositional sedimentary environment, which is a function of such factors as velocity of transport just prior to deposition and the chemistry of the environment of deposition. Of these factors, weathering and provenance exert primary, initial control on the compositions of sediments.

Weathering

Weathering includes two general categories of processes—physical processes, collectively referred to as **disintegration,** and chemical processes, collectively referred to as **decomposition.** The principal role of disintegration during the formation of a soil or sediment is to reduce the grain size of the materials. This is accomplished by the physical breaking of rock and mineral materials through *abrasion,* the grinding away of rock materials by transporting agents;

frost action, particularly frost wedging, in which ice freezes and expands in the cracks in rock and mineral materials, breaking them apart; and *biological activity*, such as the fracturing of rocks due to the growth of roots in cracks.[1] Reduction in grain size increases the surface area of particles, resulting in an increase in the rates of chemical reactions that occur during decomposition.

Decomposition processes include oxidation, reduction, solution, hydration, hydrolysis, chelation, and colloid formation with cation exchange. Oxidation and reduction are opposite processes. **Oxidation** is the process in which the oxidation number (valence) of an ion is increased and **reduction** is the process in which the valence is decreased. One of the most common of the oxidation reactions in weathering involves the oxidation of iron. For example, magnetite, a very common mineral in igneous, sedimentary, and metamorphic rocks experiences oxidation of iron as it is transformed into the common weathering product, hematite:

$$4Fe_2O_3 \cdot FeO + O_2 \rightarrow 6Fe_2O_3$$

$$(Fe^{3+})\ (Fe^{2+}) \rightarrow (Fe^{3+})$$

magnetite + oxygen → hematite

In this reaction, the ferrous iron in magnetite is oxidized yielding ferric iron. Similar reactions involving ferrous iron cations released by other decomposition reactions also occur. Reduction is simply the reverse of oxidation. For example, the production of pyrite in anaerobic (oxygen-deficient), sulfidic environments may involve the reduction of ferric iron to the ferrous state.

Water is important in many decomposition reactions as a solvent or a reactant. For example, water and acids in aqueous solution are the two principal agents of solution. **Solution** is the process in which soluble materials are dissolved, or broken down to release ions. A representative solution reaction involves the decomposition of pyroxene, as follows (Koster van Groos, 1988):

$$(Mg,Fe,Ca)SiO_3 + 2H^+ + H_2O \rightarrow Mg^{2+} + Fe^{2+} + Ca^{2+} + H_4SiO_4$$

pyroxene + hydrogen ion + water → Mg, Fe, and Ca ions + silicic acid molecule

Similar reactions may be written for other ferromagnesian silicates.[2] The Ca and Mg ions and the silicic acid produced in this reaction may be transported away in aqueous solution, whereas the iron may oxidize or hydrate, or both, and precipitate as hematite or goethite. Similarly, carbonate minerals dissolve yielding calcium ions, magnesium ions, and bicarbonate molecules, all of which are transported in aqueous solution. Solution, in particular, may generate increased porosities of up to 40% in weathered bedrock (Velbel, 1988). This allows for greater fluid flow and enhanced decomposition.

Water is also critical to hydration and hydrolysis. **Hydration** reactions are those in which water combines with another component to yield a new phase. For example, goethite is produced from hematite by the following hydration reaction (Krauskopf, 1979, p. 86):

$$Fe_2O_3 + H_2O \rightarrow 2FeOOH$$

Hydrolysis, in contrast, refers to reactions in which an excess of H^+ or OH^- is produced in the associated solution (Krauskopf, 1979, p. 36). Although the reactions may be written in a number of ways, hydrolysis reactions may be viewed as reactions in which hydrogen replaces another cation in a mineral structure. Thus, olivine weathers to silicic acid plus iron and magnesium ions via the reaction

$$(Mg,Fe)_2SiO_4 + 4H_2O \rightarrow xMg^{2+} + 2\text{-}xFe^{2+} + H_4SiO_4 + 4(OH)^-$$

in which hydrogen replaces the Mg and Fe. Similarly, feldspars hydrolyze via reactions such as

$$KAlSi_3O_8 + H_2O \rightarrow HAlSi_3O_8 + K^+ + OH^-$$

and immediately hydrate to form the clay mineral kaolinite (Huang and Kiang, 1972; Krauskopf, 1979, pp. 91–92):[3]

$$2HAlSi_3O_8 + 9H_2O \rightarrow Al_2Si_2O_5(OH)_4 + 4H_4SiO_4$$

In geological environments, colloid formation and cation exchange also depend on water. **Colloids** are finely divided materials in suspension. The surfaces of colloidal particles typically carry a negative charge, which attracts hydrogen ions from the surrounding aqueous solution. When colloids come in contact with other materials, they may *exchange* the weakly held hydrogen with ions that constitute part of the contacted material. In so doing, colloids induce decomposition.

Chelates are compounds in which a metal cation is bonded with and connects organic ring structures. **Chelation** (the formation of chelates)[4] promotes decomposition, where chelates form by extraction of a metal cation (e.g., silicon, copper) from a mineral. The chelate also serves to protect the metal ion from reaction, so that it is more likely to be transported away in solution than to be reprecipitated at the site of weathering.

Each of the decomposition processes is one in which minerals that are not in equilibrium at the Earth's surface react to form new minerals, molecules, or ions that are more stable under surface conditions. Among the most important of the products of such processes are quartz, clay minerals, iron oxides, and ions such as Ca^{2+} and Mg^{2+}. Three principal products of weathering—*carbonate minerals* formed from Ca^{2+} and Mg^{2+}, the *clay minerals*, and *quartz and opal*—were produced in approximately equal amounts over the past 4.5 billion years of Earth history (Koster van Groos, 1988). Other weathering products are generally subordinate to these three kinds of materials.

The relative stability of common minerals during weathering was recognized by Goldich (1938). He found that olivines, pyroxenes, and Ca-rich plagioclase feldspars are among the most readily weathered phases, whereas quartz and muscovite are among the last minerals to weather. This sequence of weathering replicates Bowen's reaction series. The first minerals to weather are those that form first during fractional crystallization and the last to weather are those that form last. The first-formed minerals, which form at the highest temperatures, are farthest from their fields of stability when they are present at or near the Earth's surface. Furthermore, these minerals contain one or more of the elements sodium, calcium, potassium, and magnesium, all of which are lost early during weathering of rocks, due to the easily broken ionic bonds they form with oxygen in various mineral structures (Goldich, 1938; Ehlers and Blatt, 1982). The residual elements in weathered rock—silicon, aluminum, and titanium—form covalent bonds with oxygen that are much more difficult to break.

Provenance

The provenance, the source terrain for sediments, exerts primary control over the sediment composition. Provenance factors control weathering processes and the nature of the sediments that can be supplied to any transporting agent. Among these factors are the relief and elevation, the climate and associated vegetation, and composition of the bedrock. In terms of the composition of the bedrock, to use a simple example, if only quartz sandstone is exposed in the source area, the sediments derived from that source region must be quartz-rich. If the rocks in the provenance are feldspar-rich, either feldspars or clays will be abundant in the sediment, depending on the degree of weathering of the feldspar. Thus, the nature and abundance of rock types in the source region limit the kinds of materials that will be present in the sediment derived from that region.

The relief and elevation of the provenance will affect both (1) the relative roles of disintegration and decomposition and (2) the nature of transportation (Gibbs, 1967; Pettijohn, Potter, and Siever, 1987, p. 37).[5] The relief, the difference between elevations within an erosional basin, controls the rate of erosion. In general, areas of high relief, especially those in which uplift is active, undergo rapid erosion. In contrast, those that are generally flat have low rates of erosion. Flat areas serve as local base levels, metastable sites at which potential energy is at a minimum. As a consequence, in flat areas, both mass wasting and the downward component of flow in streams are diminished, reducing the degree to which downward erosion and disintegration of a landscape will occur. The result is that the more gradual processes of weathering, especially decomposition processes, have the opportunity to operate over longer periods of time (Velbel, 1988; Johnsson and Stallard, 1989; Johnsson, 1990).

In conjunction with relief, the elevation of the provenance is important. Elevation affects the climate, which in turn affects the relative roles of decomposition and disintegration. At high elevations, especially at medium to high latitudes, where freezing and thawing are important, frost wedging is a major weathering (disintegration) process. Mass wasting and abrasion also are important in such environments. Thus, where relief is high, disintegration is facilitated. In contrast, at low elevations, particularly in tropical latitudes in areas of low relief, decomposition is the principal kind of weathering that occurs.

The climate and the vegetation are also important. Cold climates (i.e., cold temperatures) reduce decomposition rates and enhance disintegration. Warm climates produce the opposite effect. Similarly, dry climates reduce the role of decomposition, whereas in wet climates decomposition is enhanced (Mack and Jerzykiewicz, 1989). In addition, vegetation is more abundant in warm, moist climates than in cold, dry climates. Vegetation produces organic acids and other compounds that promote decomposition. For example, young lava flows in Hawaii that have a cover of vegetation (lichen), have an oxide-rich, weathering crust thicker by a factor of 10 or more than the weathering crust on similarly aged but bare rock (T. A. Jackson and Keller, 1970). This pronounced impact of vegetation on decomposition raises questions about the relative roles of disintegration and decomposition in pre-Devonian time, when no significant vegetation covered continental rocks (Schumm, 1968). The likely effect was one of increasing the role of disintegration, and, as a consequence, decreasing the amount of clay-rich sediment supplied to depositional basins.

The Products of Weathering

Given the limits imposed by provenance and climate, weathering will yield a variety of products. These products are summarized in table 16.1. Note that relatively insoluble quartz, plus the stable clays, and the iron oxides and hydrates are the main residual materials left in soil derived from highly decomposed rock. Silicic acid and various metal cations—including Ca, Mg, Fe, Mn, Na, and K—and P will be transported away from the weathering rock in ground and surface water solutions.

In areas of high relief, rapid erosion, and significant disintegration, relatively unstable (labile) rock fragments and mineral grains will be a part of the soil and will be eroded from bedrock to provide additional sediment to be transported and deposited. In areas of low relief and slow erosion, where prolonged decomposition is the principal kind of weathering that has taken place, only the most insoluble and chemically stable materials will remain in the soil to become a part of the sediment derived from the provenance. This relationship between the nature of the provenance and the sediment type is, of course, extremely valuable in interpreting the histories of sedimentary rocks.

Table 16.1 Weathering Products of Common Minerals

Common Minerals in Rocks	Weathering Products
Quartz	Quartz, dissolved silica[1]
Feldspars	Clays; Ca, Na, K ions; dissolved silica
White micas	Clays; Na, K ions; dissolved silica; gibbsite
Biotite	Clays; iron oxides; K, Mg, Fe ions; dissolved silica
Amphiboles	Iron oxides; Na, Ca, Fe, Mg ions; clays; dissolved silica
Pyroxenes	Iron oxides; Ca, Fe, Mg, Mn ions; clays; dissolved silica
Olivines	Iron oxides; Fe, Mg ions; dissolved silica; clays
Garnets	Ca, Mg, Fe ions; clays; iron oxides; dissolved silica
Al-silicates	Clay; silica; gibbsite
Magnetite	Hematite, goethite, limonites
Calcite	Ca ions, HCO_3^- ions
Dolomite	Ca, Mg ions, HCO_3^- ions
Iron carbonates	Ca, Mg, Fe ions; iron oxides, HCO_3^- ions

Sources: Huang and Kiang (1972), Krauskopf (1979), Koster van Groos (1988), and Velbel (1988), and many others.
[1]Generally considered to be a silicic acid molecule.

TRANSPORTATION OF SEDIMENTS

Transportation of sediments begins when (1) particles or clasts of fresh or weathered material and (2) the dissolved ions and molecules are removed from a surface or bed and moved. Dissolved materials are moved *in solution.* Such transportation is here called **solute transportation.** Solid materials are moved via one or more of several other processes and are said to be **entrained** when they are lifted from the surface and begin to move. The transportation of solid materials is here called **mechanical transportation.** The processes of mechanical transportation include falling, sliding, rolling, bouncing (saltation), flowing, and **suspension transport.** Subaqueous transport via a combination of sliding, rolling, and saltation is called **bed-load transportation.**

The nature of all sediment transportation depends on the physical properties of the transportation agent, the nature of the transported material, the physical properties of the mixture of transportation agent and transported material, and the forces that have caused transportation to occur. The agents that transport or cause the transportation of sediment include gravity, flowing water, wind, and moving ice. Gravity not only causes movement of material by itself, it drives many aqueous currents and induces ice to move downhill. Thermal perturbations and pressure differences in the atmosphere result in air currents (wind), and thermal perturbations are locally quite important in the oceans.

Fluid flow on the surface of the Earth is a response to the gravitational force pulling water down (figure 16.1).[6] The magnitude of the gravitational stress[7] is a function of the fluid density (ρ)—in the case of water, 0.998 g/ml at 20° C. The density of water, which is about 770 × greater than that of air, is one of the physical properties of water that allows it to transport larger particles than can the wind. The gravitational stress can be resolved into two components, a normal stress (σ_n) that is oriented perpendicular to the bed over which the fluid is flowing, and a shearing stress in the fluid (τ_f) oriented parallel to the bed (figure 16.1). This shearing stress is resisted by a frictional stress, the boundary shear stress (τ_b), which is related to flow velocity (Boggs, 1987, pp. 42–43). The flow velocity generally increases upwards from the bed, reaching a maximum somewhere between a point two-thirds of the way up from the bed and the water surface itself (Morisawa, 1968).

The shear stress in a fluid (τ_f) is a function of the change in velocity (dv) relative to the varying height above the bed (dh):

$$\tau_f = \mu(dv/dh)$$

The value μ is the *dynamic viscosity.* Dynamic viscosity may be thought of as the resistance of a substance to flow, in particular, the resistance to a change in shape taking place during flow. Different materials and mixtures of materials, such as water, sand in water, or water in mud, have different viscosities. Specific processes of mechanical transportation depend, in part, on those different viscosities. Other

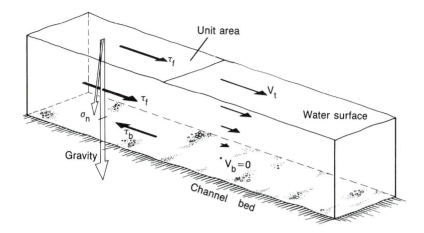

Figure 16.1 Diagram showing the forces acting on a body of water (e.g., a stream) on a slope. See text for description and discussion of parameters.

physical aspects of the sediment mix that influence transportation and deposition include its fluid-plastic character, cohesiveness (materials with significant clay are cohesive), density, and flow character (laminar vs. turbulent)(figure 16.2; 16.4). All of these characteristics influence the nature of the deposited sediment.

Fluids and plastic materials may be assigned to one of four categories—Newtonian fluid, non-Newtonian fluid, Bingham (ideal plastic), or pseudoplastic (Blatt et al., 1972, p. 160)(figure 16.3). Water, a typical **Newtonian fluid,** has no strength and does not change in viscosity as the shear rate increases. Addition of sand or similar material to water, in volume concentrations of 30% or more, *does* result in a mixture of variable viscosity, but one that still lacks strength. Such mixtures are **non-Newtonian fluids.** Mixtures in which the sediment-to-water ratio is quite high have an initial strength, called a yield strength, that must be overcome before flow will occur. If the viscosity is constant after the yield strength is exceeded, the material is a **Bingham plastic,** whereas if the viscosity varies during flow, the material is **pseudoplastic.**

Mechanical Transportation and Deposition

Materials are mechanically transported via several mechanisms, including low-viscosity turbulent suspension, non-Newtonian viscous-incohesive flow, plastic high-viscosity cohesive flow, and elastic cohesive-laminar flow (figure 16.2). Each transportation mechanism results in deposits of different character, such that the sedimentological characteristics reveal the transportation mechanism (Postma, 1986).

Gravity Transportation and Deposition

Gravity is the principal agent causing transportation in landslides and mass flows. In subaerial mass movements—falls, slides, slumps, avalanches, mudflows, and subaerial debris flows (**landslides** *sensu lato*)—and submarine mass movements (**olistostromal flows**), transportation begins when a yield stress is exceeded. Landslides *sensu stricto* move by falling, sliding, or rolling. Olistostromes, mudflows, and debris flows move by viscous plastic flow. In both cases, movement is triggered by (1) overloading of a slope, (2) removal of slope support, (3) earthquakes, or (4) mud diapirism.[8] Overloading occurs when water or sediment are added to the mass on the slope, adding weight and shear stress. Mud diapirism loads slopes by adding extra mass. Earthquakes, however, provide extra yield stress necessary to initiate movement. In contrast, removal of support through erosion at the base of the slope decreases the magnitude of the yield stress needed for failure. Some submarine slides are reactivated by wave-induced stresses (D. B. Prior et al., 1989).

In falls, slides, slumps, and avalanches, brittle failure results in a general loss of cohesion between the substrate and the moving mass (and commonly between the particles or clasts within the mass), but as frictional forces remove energy from the moving mass, some cohesion is reestablished and deposition occurs. As deposition begins, movement stops, generally rather abruptly. The product of such deposition is usually a breccia or a diamictite. The resulting rocks have very little sorting and no bedding and are characterized by both a high clast to matrix ratio and by angularity of clasts.

Transport process	Physical character of mix				Transport mechanism	Sediment character
Suspension transportation	Newtonian fluid	Low viscosity	Low density	Incohesive (= non-cohesive = cohesionless)	Turbulent flow	Suspension	Massive to bedded and laminated sediments
Turbidity current	Newtonian to Non-Newtonian fluid						Bouma sequences with laminated, cross laminated, and graded strata
Bed-load transportation	Non-Newtonian fluid to Bingham plastic		High density		Laminar flow	Temporary suspension, rolling and saltation	Laminated, crossed to structureless beds of well to moderately sorted sediment
Grain flow (*sensu lato*)						Sediment supported by dispersive pressure	Laminated to structureless, thin to massive beds of well sorted sand with dish structure and pebbles
Mass flow (debris flow, mudflow, olistostromal flow)	Bingham plastic to pseudo – plastic	High viscosity		Cohesive		Flow with shear on penetrative surfaces	Medium to massive beds of diamictite
Landslide (*sensu stricto*) (slump, debris slide, rock slide)	Elastic/brittle			Incohesive	Turbulent flow	Rotation and/or sliding with shear on spaced planes and surfaces	Thick to massive beds; typically matrix poor; commonly with slickensided clasts

(Left-margin bracket spanning the lower rows: Landslide (sensu lato))

Figure 16.2   Summary of mechanical transportion processes, mechanisms, and sediment types.
(Modified from Nardin et al., 1979; and based in part on Lowe, 1976; Postma, 1986)

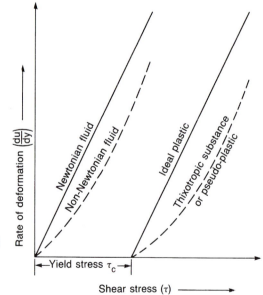

Figure 16.3
Qualitative graph showing differing parameters for fluid and plastic materials in terms of rate of deformation versus shear stress.
(After Blatt et al., 1972, p. 160)

302

In debris flows, mudflows, and olistostromes, a muddy or similarly fine-grained matrix (e.g., serpentinite) is present and is typically abundant. This matrix gives enough strength to the mass that it behaves either as a Bingham plastic or as a pseudoplastic flow. Once the yield strength is exceeded, flow may occur on rather gentle slopes of a few degrees (Curray, 1966).[9] The matrix typically is saturated with water during flow, yet apparently gives enough strength to support very large boulders or slabs of rock (Lowe, 1972).

In debris flows, mudflows, and olistostromes, the entire mass of material is deposited at once. When the shear stress component of the gravitational stress no longer exceeds the frictional shear stress at the base of the mass, flow stops quickly. In the case of some subaqueous flows, however, mixing of water with the flow mass during downslope movement results in a loss of strength and the conversion of the flow into a non-Newtonian or Newtonian fluid, such as a fluidized flow or a turbidity current (G. V. Middleton and Hampton, 1973; Lowe, 1982; G. V. Middleton and Southard, 1984). In such cases, deposition occurs in accordance with the physical properties of such fluids.

Mudflows produce poorly sorted, clast-poor, mud-rich diamictites with beds of medium to massive dimensions. In debris flow and olistostromal diamictites, the clast-to-matrix ratio is relatively higher than that in mudflows, and beds are nearly always thick to massive.

Grain flows are gravity-induced flows of incohesive, grainy sediments that occur on steep slopes (Bagnold, 1956; Blatt et al., 1972; Lowe, 1976). Flow begins either (1) when sediment accumulations exceed the angle of repose, with the result that the gravitational shear stress component exceeds the resistive shear stress, or (2) when an earthquake disturbs a nearly unstable mass, adding enough additional shear stress to the sediment mass to initiate movement. Grain flows occur both subaerially, on dunes, or under subaqueous conditions, notably in submarine canyons. In water, the grain flow may behave as a plastic flow or as a viscous non-Newtonian fluid, depending on the concentration of sand. During flow, grain-grain collisions and near collisions create dispersive pressures that keep the grains apart and keep the flow moving. Deposition of sediment occurs suddenly, when the grain flow stops moving due to a decrease in both the slope and the accompanying shear stress component of the gravitation stress.

Grain flow deposits typically are well-sorted sands that occur in structureless to locally laminated beds.[10] The thickness of the beds is commonly medium to massive, but these beds may be amalgamated units consisting of a composite of several individual grain flow deposits that each do not exceed a few centimeters in thickness. Pebble to cobble-size clasts, dish structures, and reverse grading may be present locally in grain flow sands.

Fluidized and liquidized flows are cohesionless, concentrated sediment-water mixtures that result from sudden shock (e.g., from an earthquake) or injection of fluids into a satu-rated mass of sediment (Lowe, 1976; Middleton and Hampton, 1976).[11] These flows may move down gentle slopes for significant distances, but the sediment is deposited progressively from the bottom of the flow up, as the grains reestablish grain-to-grain contact. The deposits of such flows are thick-bedded sands characterized by abundant fluid escape structures, such as dish structures, convolute beds, and fluid escape pipes.

Turbidity currents are an important type of density current characterized by a difference in density between the fluid in the current and that of the surrounding fluid.[12] Turbidity currents owe their increased density to suspended sediment. The currents arise from sudden or steady flows of sediment-rich waters into relatively clear waters (Middleton and Hampton, 1976; Kersey and Hsu, 1976). Sudden injection of muddy waters into clear waters (surges) result from storm activity, including flooding, or from earthquake-induced slope failure. Turbidity currents arise from steady flows, for example, where muddy stream waters enter a lake or the sea on a continuous basis. Sediment becomes entrained in the turbidity current by the initial event, the introduction of the muddy water into the clear water by a stream or shock-induced slope failure, and will continue to be transported as long as turbulence and flow velocity allow. Once started, the flow moves in response to gravity.

In contrast to most of the types of mass flows discussed above, deposition from a turbidity current is gradual. Loss of velocity, due to decrease in slope or mixing of the current with the invaded water, results in deposition. Coarser sediment is deposited either nearer the source, that is, in a **proximal** environment, or farther offshore along the channels of submarine fans. Finer sediment is deposited at a considerable distance, in a **distal** environment, as well as in proximal overbank areas and, as the current wanes, within the channel environment (Nilsen, 1980). Turbidites typically exhibit part or all of the Bouma sequence (figure 14.5a). Turbidites resulting from high-density currents tend to be thick-bedded, coarse-grained, and poorly graded, whereas those deposited by low-density currents are thin-bedded, graded, and have well-developed laminations.[13]

Glacial Transportation and Deposition

Glacial transportation results from gravity-induced fluid flow, but the flow rate is very low. Ice behaves as a high-viscosity, non-Newtonian pseudoplastic (Boggs, 1987, p. 55). Glaciers entrain particles simply by physically picking them up. Sediment is transported by being dragged along the base and sides of the glacier, by being suspended within the ice, and by being carried atop the glacier, where it accumulates as the ice melts.

Glacial deposition occurs, not from a decrease in velocity, but as a result of melting and evaporation (sublimation) of the ice. Warming causes some ice to sublimate, leaving its sediment load behind. Most of the ice melts, with the result that large clasts (in a matrix) are left behind and many smaller ones are transported away by glaciofluvial streams. The more sorted, finer-grained sediment is deposited as *outwash*; the very poorly sorted sediment with coarse clasts is deposited as *till*.

Transportation in and Deposition from Air and Water

Flow of materials may be laminar or turbulent (figure 16.4). Glacial ice and some landslides move via laminar flow. In contrast, materials are entrained and transported in air and water primarily in turbulent flows. When water and air flow, shear occurs between the moving fluid and the surroundings (e.g., the bed of the stream over which the current is moving).[14] Turbulence is initiated near the boundary with the surroundings (for example, near the bed of a stream) as a result of the interplay of forces at that site.

The turbulent water will begin to move particles when a critical threshold value for velocity is attained. Factors that are important in determining whether or not particles will begin to move include the size, density, and shape of the particles, the velocity of flow, the fluid viscosity; and the boundary (bed) shear stress. At the threshold value for given grain sizes, the flow of water over a grain lying on the surface will produce a force that lifts the particle from the surface.[15] Once it is lifted or moved, the particle may roll forward, rise higher into the current before falling back to the surface (saltate), or may rise into the current and remain suspended. Suspended particles remain in suspension due to the support provided by the flowing, turbulent water. Particles with masses too great to be continuously supported are moved as part of the bed load.[16] Other grains within the bed load are moved forward as a result of being struck by saltating or rolling grains.

Sedimentation from fluid flows occurs when the current slows. Individual sizes of particles fall successively out of the suspension and appropriate parts of the bed load cease movement incrementally. The lower the current velocity, the smaller is the maximum dimension of particles held in suspension. Thus, diminishing currents produce beds characterized by fining upward sequences. In units deposited from suspension, tabular laminations are common and bed thicknesses are variable, but the beds tend to be thin. Beds derived from the bed load, that is, those deposited from traction currents, may be thin, but tend to be medium- to thick-bedded and may exhibit cross-bedding, clast imbrication, and current ripple marks.[17]

Chemical Transportation and Deposition

The ions and molecules that are produced by weathering either react to yield less soluble compounds that remain in the soil or they become a part of the ground or surface water solution. Migration of these aqueous solutions results in trans-

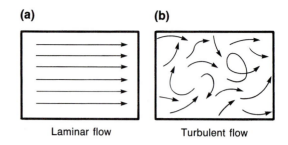

Figure 16.4
Sketches showing the nature of laminar (a) and turbulent (b) flow.

portation of the dissolved sediment load. As they migrate, particularly if they do so as groundwater, the solutions may be diluted, concentrated, or altered in chemistry due to reaction with the rocks through which they pass. Alternatively, they may mix with other waters, changing their physical-chemical character. If they react with rocks or sediments, the rocks or sediments undergo diagenetic changes (see below). Precipitation of chemical constituents during diagenesis constitutes one form of chemical deposition. The waters with a dissolved load that make it to the seas and lakes, mix there with the resident waters, yielding their chemical load through (1) diagenetic sediment-water reactions, (2) inorganic precipitation, or (3) biochemical precipitation.

Among the principal factors controlling precipitation—whether it occurs during diagenesis in the subsurface, at the sediment-water interface, or as direct inorganic precipitation—are the Eh and pH. The **Eh,** or **redox potential,** is a measure of the ability of a solution to produce oxidation or reduction. Positive values indicate oxidation, whereas negative values indicate reduction. The **pH** is the negative logarithm of the hydrogen ion concentration of a solution. Values above 7 reflect basic solutions, whereas values below 7 indicate acid solutions. A solution with a pH of 7 is neutral.

The Eh-pH fence diagram of Krumbein and Garrels (1956) provides a semiquantitative framework for understanding deposition of chemical sediments from the aqueous solutions that occur in various sedimentary environments (figure 16.5). Under basic, oxidizing conditions, iron oxides and hydrates such as hematite and goethite precipitate, as do manganese oxides. Under basic, reducing conditions, minerals such as apatite, siderite, and pyrite will precipitate. Pyrite is the major phase stable under acidic, reducing conditions. Calcite, dolomite, gypsum, and anhydrite precipitate from solutions with pH values above about 7.8. To some degree, biologically induced precipitation, the presence of CO_2, and other factors may alter the stability fields shown on figure 16.5, but these fields provide a general framework for understanding chemical deposition, a subject examined in more detail in chapters 21 and 22.

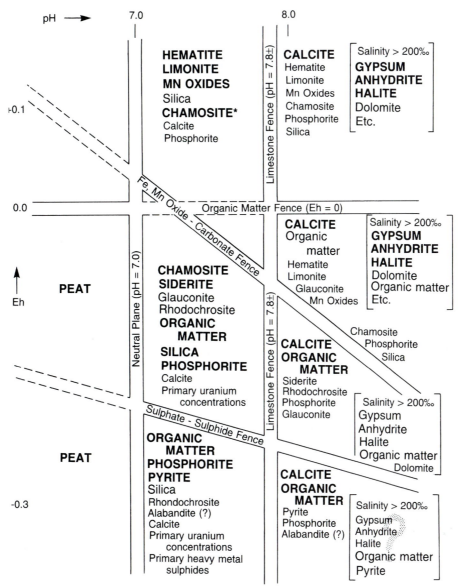

Figure 16.5 Eh-pH fence diagram showing the generalized stability fields of various minerals. *(After Krumbein and Garrels, 1956)*

DIAGENESIS

Once sediment has been deposited, it is subject to diagenesis. Recall that diagenesis includes those physical, chemical, and biological processes that collectively result in transformation of sediment into sedimentary rock. Diagenesis may continue to operate after the sediment has become rock, altering the rock texture and mineralogy. Diagenetic processes give sedimentary rocks many of the characteristics observed in outcrop, handspecimen, and thin section.

Diagenesis occurs where some change in conditions or chemistry renders the mineralogy of the sediment (or rock) unstable. Typically, instability occurs at contact zones between the grains and the surrounding or enclosed fresh or marine waters, the air, or both (Folk, 1974, p. 176; Bathurst, 1975, ch. 8, 9). At such contacts, the chemistry of the fluids is changed. Changes in P and T also drive diagenetic reactions. New minerals form or modifications of preexisting minerals take place in response to the changes, as the chemical system of the sediment or rock adjusts towards a new equilibrium condition.

Types of Diagenetic Processes and Diagenetic Environments

Seven processes belong to the general category of diagenesis. These are

1. Compaction,
2. Recrystallization,
3. Solution (including pressure solution),
4. Cementation,
5. Authigenesis (neocrystallization),
6. Replacement (including neomorphism), and
7. Bioturbation (Krumbein, 1942; Folk, 1974).[18]

The degree to which each operates on any given sediment depends on one or more of several factors, including the composition of the sediment, the pressure (which results from burial), the temperature, the composition and nature of the intergranular fluids that are the principal agents of diagenesis—including their pH and Eh, and the amount of fluid flow through the sediment or rock (Blatt, 1966; Longman, 1982; Scoffin, 1987, pt. 4).[19]

Compaction is the process by which the volume of a sediment is reduced as grains are squeezed together. It results primarily from overburden pressure, the stress provided by the weight of the overlying sediment and rock. This stress causes a reorganization of the packing of grains and the expulsion of intergranular fluids, yielding an overall reduction in the porosity of the sediment or rock. The amount of compaction possible is a function of the grain size, the grain shape, the sorting, the original porosity, and the amount of pore fluid present in the sediment (Kuenen, 1942; Chilingarian, 1983; Bjorlykke, Ramm, and Saigal, 1989).[20] Very well sorted, well-rounded grains compact less than poorly sorted, angular grains, as the former are limited to a cubic closest packing arrangement. In poorly sorted sediments, the small grains can fill in between the larger grains. Angular grains fit more tightly together than well-rounded grains. In *sands*, the original porosity is typically 25–50%, whereas in carbonate *sediments*, it may be as high as 50–75% (Choquette and Pray, 1970; Pryor, 1973; Choquette and James, 1987).[21] In sedimentary *rocks*, the porosity may be reduced to as little as 0–2%, in part through compaction and in part through the operation of the other diagenetic processes, especially cementation (Textoris, 1984; Cavazza and Dahl, 1990).[22] Time and temperature of exposure are also related to porosity reduction (van de Kamp, 1976; Schmoker and Gautier, 1988, 1989).

Recrystallization is a process in which physical or chemical conditions induce a reorientation of the crystal lattices of mineral grains. Sediments become lithified through associated textural changes. The process is one that occurs in response to such factors as pressure, temperature, and fluid phase changes. Increases in grain size and regularity of grain outline yield reduced surface and, therefore, reduced *surface free energy*, both of which are effects of recrystallization.

Recrystallization is accomplished, in part, through solution and reprecipitation of mineral phases already present in

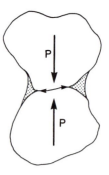

Figure 16.6 Sketch showing pressure solution in the contact region of two quartz sand grains. Large arrows (P) show stress (i.e., pressure) acting on zone of contact. Small arrows show direction of migration of silica and shaded areas show areas of precipitation of silica.

the rock. *Solution* refers to the process in which a mineral is dissolved. As fluids pass through sediment or rock, those constituents of the sediment or rock that are not stable—in terms of either the pressure, or the pH, Eh, and temperature of the fluid phase—will dissolve (Longman, 1981b). They are either transported away in the fluid or they are reprecipitated in adjacent pores or in nearby parts of the sediment, where conditions are different. Of particular importance in some rocks and sediments is **pressure solution**, a process in which pressure is concentrated at the point of contact between two grains, causing solution and subsequent migration (diffusion) of ions or molecules away from the point of contact (figure 16.6).[23] Via this process mass is transferred from the point of contact to a lower-pressure site, usually in the adjoining pore space, where reprecipitation of the dissolved phase occurs. Obviously, pressure solution and reprecipitation reduce the porosity of a sediment and will facilitate textural recrystallization.

Cementation is the process in which chemical precipitates, in the form of new crystals, form in the pores of a sediment or rock, binding the grains together.[24] Common cements include quartz, calcite, and hematite, but a wide variety of cements are known, including aragonite, Mg calcite, dolomite, gypsum, celestite, goethite, and todorokite. Mg calcite, aragonite, and protodolomite are particularly important in certain carbonate sediments and rocks, especially those of young age. Pressure solution produces locally derived cement, but many cements consist of new materials (allochemical materials) introduced in solution by the fluid phase. Clearly, cementation reduces porosity and contributes to lithification. New textures, including spherulitic, comb texture, and poikilotopic texture result from cementation.

Authigenesis (neocrystallization) is the process in which new mineral phases are crystallized in the sediment or rock during diagenesis. These new minerals may be produced via reactions involving phases already present in the sediment or rock, they may arise through precipitation of materials introduced in the fluid phase, or they may result from a combination of primary sedimentary and introduced components. Authigenesis overlaps with weathering and cementation, usually involves recrystallization, and may result in replacement. The diversity of authigenic phases is even greater than is that of cement minerals. Authigenic phases include silicates such as quartz, alkali feldspar, clays, and zeolites; carbonates such as calcite, dolomite, and ferroan calcite; evaporite minerals, including halite, sylvite, gypsum, and anhydrite; oxides such as hematite, goethite, todorokite; and a host of other minerals, including sulfates, sulfides, and phosphates.[25]

Where a new mineral replaces an originally sedimented phase *in situ*, the process is known as **replacement.** Replacements may be **neomorphic,** in which either the new grain is the same phase as the original phase[26] or is a polymorph of that phase (Folk, 1965); **pseudomorphic,** in which the new phase mimics the external crystal form of the replaced phase but is a different phase; or **allomorphic,** defined here as a replacement in which a new phase, usually of different crystal form, replaces an original sedimentary phase.[27] Replacement phases are as diverse as authigenic phases, but among the most important replacement phases are dolomite, opal, quartz, and illite.

Bioturbation (*sensu lato*) refers to those physical-biological activities that occur at or near the sediment surface, including burrowing, boring, and mixing of sediment by organisms. Bioturbation results in mixing of sediment, locally to a depth of a meter below the sediment surface (Shinn, 1968). In some cases, it also yields increased compaction and it usually destroys laminations and bedding (Purdy, 1965). During bioturbation, some organisms precipitate materials that act as cements.

Diagenetic environments, where the various processes of diagenesis operate, include surficial environments, shallow burial environments, and deep burial environments. Surficial and shallow environments may be marine or nonmarine; in deep burial environments, this distinction is irrelevant. In nonmarine and shallow burial environments, meteoric water is commonly the main fluid phase. At depth, connate water, water trapped in the sediment pores at the time of deposition, and waters of dehydration, derived from chemical reactions involving hydrous or hydrated mineral phases, provide the foundation for the fluid phase. Reactions add new chemical species to that foundation. Diagenesis is also subdivided temporally and spatially into **eogenesis,** early diagenesis that occurs between deposition and burial, at or near the surface; **mesogenesis,** middle-stage diagenesis that occurs after burial; and **telogenesis,** late stage diagenesis that occurs after reexposure of formerly buried rocks (Choquette and Pray, 1970; Textoris, 1984; also see Purdy, 1965).

The Fluid Phase

Fluids make many diagenetic changes possible. The abundance and movement of fluids are controlled both by original depositional factors and by the very processes the fluids control. The development of porosity through solution, for example, is locally a rather important diagenetic process, as it controls the amount of fluid that can be present in a rock at any given time. More importantly, it controls the amount of throughflow of fluid that is possible (a function of permeability).

Fluids control precipitation of cements, the development of authigenic and replacement minerals, and the rates of solution.[28] In addition, they facilitate recrystallization. The fluids arise as original sedimentary pore fluids (depositional waters), meteoric waters, or seawater, but they evolve as they mix, as dehydration reactions and compaction of mineral grains add new water, as diagenetic reactions of hydrocarbons produce and contribute methane, and as alterations of various minerals add additional gas phases such as CO_2 (Hutcheon, 1983; Galloway, 1984; J. E. Andrews, 1988).[29] Furthermore, as the fluids pass through the rocks, the diagenetic reactions that they produce or catalyze change the chemical composition of the fluids. As a consequence, the kinds of diagenetic reactions that occur will depend on the composition and volume of this evolving fluid phase, as well as on the temperature, Eh, pH, and pressure conditions extant during the diagenetic reactions.

A changing composition of the fluid phase means that diagenetic reactions produced by a migrating fluid may be different at different places and at different times. Over time, the same rock or sediment may experience different diagenetic changes, either as a result of the changing composition of the fluid or as a result of the passage of different fluids through the rock. Consequently, diagenetic histories may be complex and include several distinct stages.

Diagenesis of Sediments and Rocks

The relative importance of the particular processes of diagenesis varies and depends both on the sediment or rock type affected and on the environment in which the diagenesis occurs. In particular, composition, grain size, porosity, permeability, and the presence (or absence) of diagenetic changes developed at an earlier time have a major influence on what kinds of diagenetic changes can and will take place. Furthermore, the pressure and temperature, plus the composition, nature, and flow rate of the fluid phase, control the nature and rate of diagenetic change.

Because of the overriding influence of rock composition on diagenetic change, each sediment and its equivalent rock type is addressed separately below. Most studies are conducted in the context of single compositions; but lithologies are usually interlayered in nature. This fact has significance in terms of influencing changes in the composition and nature of the migrating fluid phase, so essential to diagenesis.

Carbonate Rocks

The diagenesis of carbonate rocks has been studied in considerable detail.[30] In order to recognize and describe the diagenetic changes that occur, it is necessary to recall the nature of unlithified carbonate sediment. As emphasized by Folk's

(1962) classification, carbonate rocks are composed of sparite and two main types of detrital materials—carbonate mud and allochems. *Carbonate muds* are dominated by calcite, Mg calcite, aragonite or, in rare cases, protodolomite or dolomite.[31] These muds are produced both by inorganic precipitation and by organic precipitation, primarily by algae. Allochems include a wide variety of biochemically and inorganically precipitated materials that have been moved, reworked, and deposited within the basin of deposition. Biochemical carbonate allochems that occur in Mesozoic to Recent rocks and sediments include the shells (tests) of coccolithophores, foraminifers, pelecypods, gastropods, cephalopods, corals, various echinoids, and to a minor degree bryozoa, crinoids, and brachiopods (Wilkinson, 1979). In Paleozoic rocks, the tests of corals, brachiopods, bryozoa, crinoids, and blastoids are important contributors of carbonate grains, but forams, gastropods, pelycepods, cephalopods, trilobites, miscellaneous echinoids, algae, and sponges contributed biochemical allochems to some sediments. Nonskeletal biochemical allochems include fecal pellets excreted by various marine organisms. Inorganic allochems include oolites, pelletoids of inorganic origin (MacIntyre, 1985), grapestones, and clasts of preexisting carbonate rocks. These various carbonate grains are deposited together with a mud matrix to form the sediment precursors of packstone, grainstone, shellstone, and related rocks. In many cases, the porosities of these carbonate sediments are from 40 to 75% (Pray and Choquette, 1966; Halley, 1983; Choquette and James, 1987), values that locally may result from a multitude of tiny holes (microfenestrae) within the mud (Lasemi, Sandberg, and Boardman, 1990).

General Processes. Diagenesis of carbonate rocks involves all of the processes of diagenesis—recrystallization, solution, cementation, replacement, bioturbation, compaction, and authigenesis. Different processes operate in different diagenetic zones, both above and below the water table and at and below the sediment-water interface.[32] As a result, the compositions of the fluids that pass through the sediment or rock will vary from meteoric to mixed fresh water–seawater to brines trapped during sedimentation and altered during diagenesis.[33]

The relative importance of individual processes varies in carbonate diagenesis. The role of compaction is debated,[34] yet some experimental studies show that significant compaction (up to 30% or more) can occur in carbonate materials (Robertson, 1965; Shinn and Robbin, 1983).[35] Bioturbation is common and reflects the abundance of organisms in shallow marine environments. The other processes—solution, recrystallization, cementation, authigenesis, and replacement—occur during eogenesis and mesogenesis, in or beneath subtidal and supratidal settings; during mesogenseis at depth; and during telogenesis in a variety of settings.

Solution is locally quite important (Friedman, 1964, 1975).[36] At shallow levels in the ocean, calcite is a stable phase. At depth, however, the solubility decreases as a function of temperature and pressure, and at a depth called the carbonate compensation depth (CCD), calcite becomes unstable and will dissolve.[37] The CCD varies in depth but seldom occurs below 4500 m.[38] Carbonate sediments experiencing eogenesis below the CCD will undergo solution. Similarly, during mesogenesis circulating groundwaters will dissolve carbonate minerals if the appropriate partial pressure of CO_2 exists. Meteoric waters also may facilitate dissolution, especially during telogenesis. Because the solubilities of high-Mg calcite, calcite, aragonite, and dolomite differ, a typical carbonate rock composed of mixtures of two or more of these minerals will undergo differential dissolution in which the most soluble phase is dissolved in preference to less soluble phases. Aragonite and high-Mg calcite are particularly soluble under many diagenetic conditions (Purdy, 1965; Longman, 1980).[39] Which phase is the most soluble depends on several of the controlling factors in the environment (e.g., P, T, P_{CO_2}), as well as on the microstructural details of grain surfaces, such as surface roughness (Walter, 1985). The rate of dissolution will also be a function of a number of factors, including the mineralogy, the grain size, deviatoric stresses, the ambient temperature, the pressure, the pH, the Eh, the rate of throughflow of fluid, the volume and chemistry of the fluid, and the partial pressure of CO_2 in that fluid (Fyfe and Bischoff, 1965; Longman, 1980, 1982; Sanford and Konikow, 1987).

Pressure solution in carbonate rocks occurs both along surfaces and on grain boundaries.[40] The product of pressure solution along a surface is a stylolite. The product of pressure solution along grain boundaries is usually a cement.

Generally, solution occurs where waters move *through* carbonate rocks, dissolving carbonate minerals as they go. As they do so, they become saturated with various carbonate phases. This is particularly true, if (1) the waters are changed in chemistry, (2) the waters enter a different lithology, or (3) the variables controlling solubility (e.g., P, T, Eh, P_{CO_2}) undergo change. Precipitation of cements results from such changes.[41] The kinds and amounts of organic matter and biogenic silica also influence dissolution and precipitation (Mitterer and Cunningham, 1985; Hobert and Wetzel, 1989).

Any of the four major carbonate phases (calcite, Mg calcite, aragonite, dolomite) may become a cement.[42] Mg calcite and aragonite typically develop in early stages of eogenesis and mesogenesis, whereas calcite and dolomite are common as mesogenetic phases. There is a tendency over time for calcite and dolomite to become the stable phases in carbonate rocks, although aragonite is reported in the cements of some Paleozoic rocks (Sandberg, 1985). Additional cements include such minerals as ferroan calcite, ferroan dolomite, anhydrite, gypsum, halite, sphalerite, celestite, and quartz (including quartzine) (Woronick and Land, 1985).[43]

Texturally and structurally, carbonate cements occur as fibers, blades, and equant grains that form localized to pervasive intergranular material, layered void fillings, and surficial crusts (see figure 14.14)(Folk, 1965; P. M. Harris, Moore, and Wilson, 1985; Coniglio, 1989).[44] Zoned calcite cements, which are rather common, reveal the changing chemistry of the cement over time (W. J. Meyers, 1978; M. R. Lee and Harwood, 1989).[45] The growth of new phases to form cement in the pores of a rock is essentially a process of authigenesis, as well as one of cementation.

Recrystallization generally involves the coarsening of grain size in cement and framework grains. For example, calcitic micrite and microspar recrystallize to microspar and spar, respectively, over time (Prezbindowski, 1985). Similarly, dolomites may become coarser through recrystallization.

In most carbonate rocks, textural changes are accompanied by the formation of new minerals. Because the stabilities of the individual carbonate phases are different, as conditions change, the identity of the phase that is stable will change.[46] As a consequence, authigenesis and replacement are generally more important than recrystallization. Aragonite is replaced by calcite, Mg calcite is replaced by calcite, and calcite is replaced by dolomite during the diagenesis of many carbonate rocks. Similarly, quartz (as quartzine and chalcedony in chert) replaces calcite and dolomite, as well as associated anhydrite, gypsum, or halite. Calcite is also replaced by hematite and apatite (see Marlowe, 1971). Neomorphic replacement of Mg calcite by calcite is a common process. In contrast, the frequency of neomorphic replacement of aragonite by calcite is a matter that is open to debate.[47] Locally, calcite replaces dolomite in a process called *dedolomitization* (Woronick and Land, 1985).

Carbonate sediments deposited both in shallow and in deep water undergo diagenesis (N. P. James and Choquette, 1983). Most carbonate rocks in the geologic record, however, are shallow-water types. Thus, a general model for progressive diagenesis developed for these types of rock is applicable to many occurrences. Such an idealized sequence of events begins with eogenesis, following deposition of the sediment in the shallow marine environment, and includes the following stages (Longman, 1981)[48]

1. Micritization (development of small grains) and pore cementation under marine phreatic (saturated) conditions,
2. Intergranular cementation under marine phreatic conditions,
3. Precipitation of sparry calcite cement as marine water is replaced by fresh water,
4. Leaching of aragonite and Mg calcite and conversion of Mg calcite to calcite in the freshwater phreatic zone,
5. Filling of molds with sparry calcite, conversion of unstable phases to calcite, and recrystallization of micrite to microspar or spar,
6. Solution to form vuggy porosity, plus continued recrystallization of micrite in the freshwater vadose (unsaturated) zone,
7. Precipitation of sparry calcite in vugs, in the freshwater phreatic zone.

These stages may then be followed by additional events *if* mesogenesis is initiated by deeper burial (Choquette and Pray, 1970; Longman, 1981b).[49] The ensuing events might include

8. Compaction and expulsion of fluids,
9. Changes in organic matter as temperature increases,
10. Formation of stylolites by pressure solution,
11. Formation of dolomite and chert (dolomitization and chertification),
12. Cementation by calcite,
13. Fracturing followed by additional dolomitization and chertification, and
14. Solution to form secondary porosity and stylolites.

The secondary pores may later be filled by calcite. If telogenesis follows mesogenesis, additional replacement relationships, brecciation, development of biogenic borings, and the formation of infillings and crusts will further complicate the diagenetic history of the rock (Hurley and Lohman, 1988; B. Jones, 1988). Deciphering these complex diagenetic histories is clearly a challenge and rarely, if ever, does any single rock unit reveal evidence of all of these events.

The Origin of Dolostone. The origin of dolostone[50] has been a subject of considerable interest and debate over the past century, especially because dolostones are common in the geologic record but do not form in typical contemporary marine environments of sedimentation. Most dolostone is the product of the diagenetic processes called dolomitization. Nevertheless, dolomitic sediment does form rarely as a primary precipitate in lakes, and may form under some deep marine conditions.[51] Evidence that dolomite in these settings is primary includes textural, structural, and temporal data (the sediment ages range from a few hundred to a few thousand years).

Why dolomite does not precipitate from seawater remains somewhat of an enigma. Seawater is supersaturated with respect to dolomite and thus should be precipitating dolomite (K. J. Hsu, 1966). It does not precipitate, probably because of factors relating to nucleation and growth of the dolomite structure, which is a highly ordered atomic structure (K. J. Hsu, 1966; Sibley, 1990).

The processes of **dolomitization** are primarily ones of replacement. These processes effectively replace primary sedimentary structures, preserving detail, including fossil structures, allochems, and fine laminations (Dietrich, Hobbs, and Lowry, 1963). Dolomitization occurs where large volumes of Mg-bearing fluid pass through a preexisting sediment or carbonate rock, the original carbonate phases are dissolved and/or replaced by dolomite, and the exiting fluid removes dissolved calcium.

Dolostones apparently form in a variety of ways; that is, dolomitization includes several different processes. Four principal hypotheses (models) of dolomitization, linked to

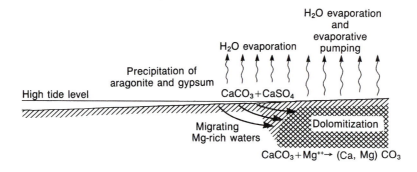

Figure 16.7
Sketch of Evaporite Brine Model of dolomitization. See text for a description of the process.

specific sites of dolomitization, are currently considered as possible explanations for the origin of dolostones. These models are the Evaporite Brine Model, the Groundwater Mixing Model (Dorag Model), the Convection Flow Model, and the Formation Water Model (Illing, Wells, and Taylor, 1965; Badiozamani, 1973; Leeder, 1982, p. 298; E. N. Wilson, Hardie, and Phillips, 1990). Variants of these models are sometimes invoked to explain special circumstances.[52]

The *Evaporite Brine Model* is based on studies of *sabkhas*, the supratidal salt flats that occur along arid coastlines, and similar supratidal and lagoonal areas (J. E. Adams and Rhodes, 1960; Deffeyes, Lucia, and Weyl, 1965; Illing, Wells, and Taylor, 1965).[53] In Qatar, along the coast of the Persian Gulf, evaporation of waters in algal tidal flats that separate the sabkha from the main intertidal zone, increases the salinity and facilitates precipitation of aragonite and gypsum (Illing, Wells, and Taylor, 1965). Extraction of Ca via the precipitation of these minerals increases the Mg concentration of the waters. These waters are drawn in and through the sabkha sediments by hydraulic gradients resulting from evaporation (i.e., water is driven by "evaporative pumping")(K. J. Hsu and Siegenthaler, 1969). In these sabkha sediments, previously deposited aragonite is replaced by dolomite (and gypsum)(figure 16.7).[54] In more humid climates, a similar eogenetic dolomitization process operates in supratidal regions (Deffeyes, Lucia, and Weyl, 1965). Evidence that dolomitization occurred in sabkhas and related supratidal zones includes (1) associated marine lithofacies indicative of such environments, (2) in the sabkhas, significant development of evaporites, and (3) isotopic evidence (D. W. Muller, McKenzie, and Mueller, 1990). Textural evidence of early dolomitization (Dietrich, Hobbs, and Lowry, 1963) also favors this model.

The *Groundwater Mixing Model* is primarily one of mesogenetic dolomitization. In this process, a zone of mixing develops in buried carbonate sediments or rocks beneath a coastal

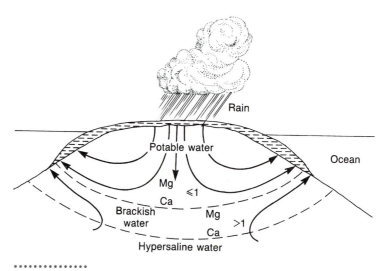

Figure 16.8   Sketch of the Groundwater Mixing Model of dolomitization.
(Modified from Hanshaw, Back, and Deike, 1971)

region, where fresh groundwaters moving seaward from the land and salt waters moving shoreward mix (figure 16.8)(Hanshaw, Back, and Deike, 1971; Badiozamani, 1973; Land, 1973).[55] If the mixed water is dominated by the freshwater component (50–95% fresh), so that the ratio Mg/Ca is in the range 1:1 to 1:4, the water will be supersaturated with respect to dolomite but undersaturated with respect to calcite (figure 16.9) (Badiozamani, 1973; Leeder, 1982, p. 299). Under such conditions, dolomitization will proceed. Under higher salinities, with seawater comprising up to 95% of the mix, some evidence suggests that under certain conditions the same solubility relationships may apply (Stoessell et al., 1989; E. N. Wilson, Hardie, and Phillips, 1990). If so, dolomitization is possible through most of the range of mixed-water compositions in the mixing zone. Evidence of dolomitization by groundwater mixing is provided by (1) large volumes of dolostone made possible by this mesogenetic process, (2) the purity of the dolomite (which would result from slow crystallization under mixing model conditions), and (3) isotopic evidence (Leeder, 1982, p. 300; Swart, Ruiz, and Holmes, 1987; Humphrey, 1988).

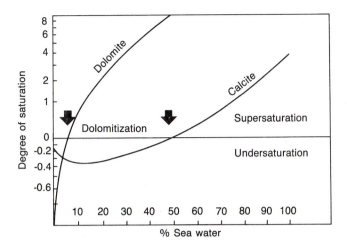

Figure 16.9 Graph showing the degrees of saturation of calcite and dolomite in mixtures of fresh water and seawater with varying percentages of seawater.
(Modified from Badiozamani, 1973)

The *Convection Flow Model* invokes thermally driven, upward flow of seawater through limestones, as a causative agent of dolomitization (Aharon, Socki, and Chan, 1987; E. N. Wilson, Hardie, and Phillips, 1990). Dolomite *is* a stable phase in hot seawater; thus, hot seawater becomes an effective dolomitizing medium. Subsurface volcanic activity provides one plausible thermal source for the heat. Although this model was designed to explain special cases in which the pattern of dolomitization is not consistent with other models, more generalized fluid flow patterns and a similar process may explain the origin of other dolostones.[56]

The *Formation Water Model* is a model of mesogenetic dolomitization that involves deeper burial than the Groundwater Mixing Model. Dolomitization in this model results from reactions between rocks and a fluid phase derived from pore waters in associated buried sediments (Leeder, 1982, p. 300; Gawthorpe, 1987). The magnesium needed for dolomitization is probably derived from diagenetic alteration of Mg calcite in carbonate rocks and the conversion of smectites to illite in associated or interbedded mudrocks. Because iron is also released by diagenetic reactions in mudrocks, Formation Water dolostones may contain ferroan dolomite or ankerite. The presence of these minerals and their distribution are primary evidences for this type of dolomitization.

More than one of these processes can affect the same rock (figure 16.10). For example, Shukla and Friedman (1983) ascribe dolomitization of the Silurian Lockport Formation of New York to a two-stage history with early supratidal dolomitization followed by a second-stage dolomitization induced by mixing of groundwater and seawater in the phreatic zone.

Finally, it should be noted that dolomite is a minor accessory mineral in some marine sediments and may occur in minor to dominant amounts in some lake deposits. For example, minor amounts of dolomite have been found in some deep marine carbonate sediments and rocks (Friedman, 1965b; Lumsden, 1988). Such dolomites are thought to be eogenetic materials precipitated from pore solutions. In general, they do not form in enough abundance to yield dolostone. In contrast, in some lakes, dolomite does form in large amounts, yielding so-called oil shales, which are, in fact, dolostones with high organic carbon contents. For example, dolomitic "oil shales" of this type formed in great abundance in the Eocene lakes that produced the Green River Formation of the western United States.[57]

Cherts

Chert is a crystalline sedimentary rock composed predominantly of fine-grained silica minerals. Cherts are formed by a variety of processes. Some result from primary biogenic precipitation of silica, some are clastic deposits of biogenic materials, and others are essentially diagenetic rocks. Diagenesis is important in the petrogenesis of most chert, regardless of its origin.

The origins of cherts are discussed in some detail in chapter 22. Here it will suffice to note that cherts form in a diverse group of environments, including marine and nonmarine types. In both types of environment, either clastic or precipitated constituents may dominate the rock. Where clastic constituents are dominant, they are typically bioclastic; that is, they are sponge spicules (microscopic siliceous parts of sponges) or fragments or whole shells of single-celled animals (the radiolaria) or algae (the diatoms). Nonclastic sources of silica in cherts include silica-enriched fluids derived from weathering, diagenesis, or volcanic-hydrothermal sources (Hesse, 1988).

Bioclastic (allochemical) cherts may experience all of the processes of diagenesis. Compaction of the bioclasts reduces porosity by as much as 30%, just as it does in carbonate rocks (Isaacs, Pisciotto, and Garrison, 1983). Authigenesis yields a variety of new phases, rare bioturbation results in mixing of sediment and disruption of laminations, and replacement results in the development of new phases. For example, allomorphic replacement of silica minerals by dolomite is common in the cherts that occur in dolostones.

Solution, recrystallization, and cementation are probably the most important processes in the diagenesis of cherts. Under pressure, siliceous sediment experiences the same pressure solution processes that occur in carbonate rocks. Primary bioclasts are dissolved at pressure points, the silica migrates, and it reprecipitates at intergranular sites of lower pressure, cementing the remaining bioclastic grains. This process is particularly aided by the fact that biogenically precipitated silica occurs as an easily destabilized, amorphous form of opal called **opal-A.** Opal-A readily converts to quartz or to a metastable form of silica, **opal-CT** (Calvert, 1971; Jones and Segnit, 1971; Kastner, 1979).[58] Pressure solution and recrystallization convert opal-A to opal-CT, which then converts to fibrous quartz (chalcedony) or quartz. The resulting rocks have crystalline textures and reduced porosities.

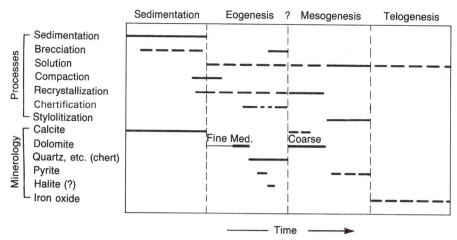

Figure 16.10 Diagram showing the diagenetic history of Upper Knox Group rocks (Ordovician) from the Nebo Quadrangle, southwestern Virginia.

Cherts also result from the diagenesis of siliciclastic sediments such as siltstone and quartz arenite. For example, local chertified zones are present in nearly pure quartz arenites of the Silurian Keefer Sandstone and in siltstones of the Cambrian Rome Formation of the southern Appalachian Valley and Ridge Province. Presumably, in such rocks, the solution and reprecipitation of quartz and replacement of quartz by chalcedony or opal have produced the cherts. Similarly, partial silicification of quartz arenites, with chalcedonic quartz cements, occurred in the eastern Alps (Hesse, 1987). These diagenetic processes are aided locally by hot spring activity on the sea floor.

Diagenetic processes form the **replacement cherts.** Such cherts are produced where silica minerals replace preexisting minerals of various types that exist in other (nonchert) rocks. The most common types of replacement cherts are those that form from limestones and dolostones. Silicified fossils, silicified oolites, and silicified pellets reflect the calcareous parentage of the cherts (Folk and Pittman, 1971; Namy, 1974)(figure 16.11a). Structurally, replacement cherts commonly form nodules, lenses, or thin beds in carbonate rock sequences.

The diagenetic origin of chert requires that the fluid phase involved in the diagenetic changes be supersaturated with regard to the precipitating silica phase and undersaturated with regard to the dissolving phase. In the case of replacement of limestones or dolostones, the fluid phase is supersaturated in quartz but undersaturated in calcite or dolomite (Knauth, 1979). As a consequence, when the fluid phase moves through the rock, the carbonate minerals are dissolved and silica is precipitated. Where this occurs on an atom for atom basis, the fine details of allochems are preserved.

An example of diagenesis involving both chert and carbonate rocks is provided by shelf and possibly sabkha limestones of the Ordovician Upper Knox Group in southwestern Virginia (C. R. B. Hobbs, 1957; Dietrich, Hobbs, and Lowry, 1963; Webb and Raymond, 1989). These rocks experienced eogenetic compaction and recrystallization followed by dolomitization,

replacement by silica minerals (chertification) with local pyrite authigenesis, local brecciation and authigenetic crystallization of halite, additional chertification, a second phase of dolomitization that produced larger dolomite crystals, and later mesogenetic solution, producing stylolites (figures 16.10, 16.11). Telogenesis and weathering produced additional solution and local replacement of pyrite by iron oxides. Evidently, the chemistry of the fluid phase and the conditions of diagenesis varied over time, producing successive stages in the evolution of the textures and mineral compositions of the rocks.

Mudrocks

The diagenesis of mudrocks is becoming more clearly understood. The fine grain size of these rocks has heretofore hampered studies of diagenetic change, but new techniques in microscopy combined with older X-ray techniques have allowed expanded investigations of mudrock diagenesis.

Mudrocks are distinctive both in texture and mineralogy. The textures are uniformly fine-grained, epiclastic textures that are locally weakly foliated due to the alignment of the phyllosilicate grains that dominate the mineralogy. The sheet structures of the phyllosilicates create textures with locally high porosities but very low permeabilities. Mineralogically, the rocks consist of clays (the smectites, illites, and kaolinites), chlorites, and mixed-layer phyllosilicates, with varying amounts of quartz, feldspar, calcite, and other minerals.

The major changes that occur during the diagenesis of mudrocks result from bioturbation, authigenesis, and replacement, with lesser amounts of solution, compaction, recrystallization, and cementation. Bioturbation is a common eogenetic process that affects the structure of the mudrocks. In as much as many shallow marine environments are characterized by muddy bottoms and muddy bottoms in shallow marine environments typically are populated by abundant infaunal (burrowing) organisms and epifaunal (surface) detritus feeders, mudrocks are commonly bioturbated, reflecting the activity of these organisms.

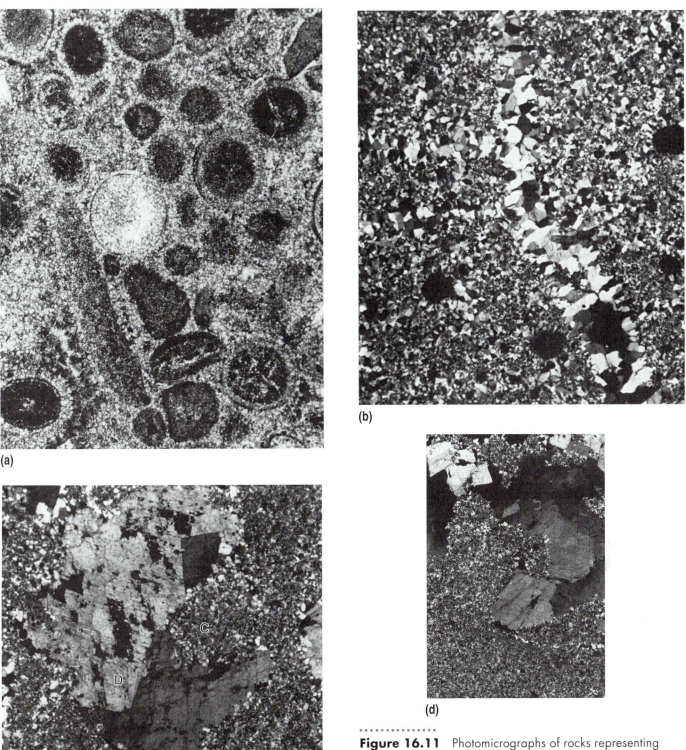

(a)

(b)

(c)

(d)

Figure 16.11 Photomicrographs of rocks representing various stages in the diagenetic history of the Upper Knox Group rocks (Ordovician) of southwestern Virginia.
(a) Chertified, oolitic packstone. (b) Brecciated, silicified, oolitic packstone with veins of quartz with chevron textures (replacements of halite?). Entire rock is chertified. (c)Dolomite replacing chert. C = chert, D = dolomite. (d) Fine-grained dolostone containing coarse-grained dolomite crystals, cut by chert. All figures have long dimensions of 3.25 mm and all views are in XN.

During authigenesis, replacement, solution, recrystallization, and cementation, the mineralogy of the mudrocks is altered. The major mineralogical changes that occur include (1) the appearance of kaolinite and later replacement of kaolinite by dickite, pyrophyllite, or other phases; (2) reduction in the amounts of smectite, through replacement by illite and mixed-layer clays; (3) solution or replacement of feldspar; (4) solution or replacement of calcite and dolomite; (5) reduction in the amounts of organic carbon; (6) an increase in the crystallinity of illite; (7) increases in abundances of mixed-layer clays; (8) the crystallization of chlorite, as an authigenic or replacement phase; and (9) the precipitation of zeolites as cements, authigenic phases, or replacement minerals (Grim, 1958; C. E. Weaver, 1961; Dunoyer de Segonzac, 1970).[59] Because of the abundance of smectites in muds, the genesis of illite-smectite (I/S) mixed-layer clays and the formation of illite are the dominant processes in mudrock diagenesis.

Sandstones and Conglomerates

Sandstones and conglomerates experience significant diagenetic changes of both academic and commercial interest. Because sandstones are major reservoir rocks for both hydrocarbon accumulations and groundwater, changes in porosity and permeability resulting from diagenesis have been studied in considerable detail.

Recall that sandstones and conglomerates are composed of a variety of mineral grains and rock fragments. Although there are exceptions, the majority of noncarbonate conglomerates consist of clasts of rock dominated by quartz and/or feldspar. Similarly, the majority of sandstones are composed of quartz, feldspar, siliceous rock fragments, or various combinations of these grain types. Micas and clays are abundant to minor constituents.

All of the processes of diagenesis affect sandstones.[60] Bioturbation is not uncommon and serves both to mix sediment and to destroy primary structures. New minerals form through authigenesis and replacement, and some are precipitated during cementation. Compaction, cementation, recrystallization, authigenesis, and replacement reduce porosity. Porosity reduction also seems to be related to the age and temperatures attained by a sand (Schmoker and Gautier, 1988, 1989), perhaps because as time passes, under the elevated temperatures (and pressures) that accompany burial, the likelihood that diagenetic processes such as compaction, cementation, authigenesis, and recrystallization will occur is greatly increased. Solution, too, especially pressure solution, facilitates porosity reduction, but late-stage solution (e.g., during mesogenesis or telogenesis) commonly results in porosity *increases*. Such porosity, developed via post-sedimentation processes, is called **secondary porosity** (V. Schmidt and McDonald, 1979).[61]

Compaction is an important, but not a dominant, process in diagenetic alteration of sand (J. M. Taylor, 1950; Blatt, 1966). Upon deposition, *sand* has a porosity of 15 to 60%, depending on the grain size, the grain shape, and the sorting (Gaither, 1953; Pryor, 1973).[62] A porosity value of about 40% is characteristic and is the maximum porosity recognized in *sandstones*. Most sandstones, however, have porosities that are much lower than this. Some yield measured values of < 3%.[63] The evident reduction in porosity that occurs during lithification results from one or more of the several diagenetic processes that reduce porosity, but the amount of porosity reduction that can be attributed directly to compaction is variable and generally small.

Recall that the amount of compaction possible is a function of the grain size, the grain shape, the sorting, the original porosity, and the amount of pore fluid present in the sediment (Krumbein, 1942; Blatt, 1966).[64] Compaction alone simply serves to reduce pore spaces that result from irregular packing of grains *and* from intergranular fluids that keep the grains separated. In poorly sorted sands, both before and after compaction, the smaller grains fill in the interstices between the larger grains, reducing the porosity.[65] In well-sorted sands with well-rounded spherical grains, compaction reduces the grain arrangement to one that approximates closest packing, with retained porosities of 25 to 45%.[66] In the former case, some significant reductions in porosity may be realized through compaction, whereas in the latter, porosity reductions are generally small.

In addition to the reductions in porosity resulting directly from compaction-induced reorganization of grains, other effects of compaction may also result in porosity reductions. For example, the deformation of ductile lithic fragments and phyllosilicate grains during compaction causes these grains to be squeezed into the pores, reducing the porosity and creating a pseudomatrix (Dickinson, 1970a). In lithic wackes, such a process is quite significant, being one of the principal causes of porosity reduction. Also during compaction, stylolite formation yields reduced porosity via pressure solution and pore space reduction.[67] Pressure solution is particularly important in the porosity reduction of some arenites, especially quartz arenites.[68]

Recrystallization of grains further facilitates lithification and porosity reduction. Like compaction, recrystallization is induced primarily by lithostatic pressures, but chemical instability also can cause this process to occur. Under pressure, the high stresses at grain contact points facilitate structural reorganization of the crystal lattices, causing solution of the mineral grains, and resultant reprecipitation to yield interlocking, sutured grain boundaries (figure 16.12). Recrystallization of opaline cement (Pettijohn, Potter, and Siever, 1987, p. 487) yields similar results. How much sutured texture results from recrystallization *sensu stricto* is moot.

Cementation, authigenesis, pressure solution, and replacement are the most significant processes of porosity reduction. For example, in the Ordovician Bromide Sandstone of Oklahoma, an estimated 38% of the original porosity was removed by cementation and a similar amount was eliminated by pressure solution related phenomena (Housknecht, 1987).

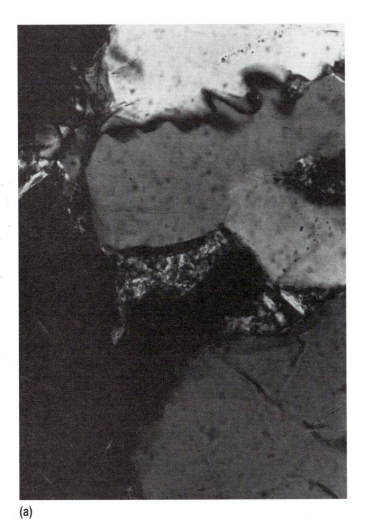

(a)

(b)

Figure 16.12 Photomicrographs showing diagenetic textures in sandstones. (a) Sutured grain boundaries in quartz arenite. (XN). Long dimension of photo is 0.33 mm. (b) Quartz overgrowths on detrital quartz grains, quartz arenite, Potsdam Sandstone, Cambrian, Potsdam, New York. (XN). Long dimension of photo is 1.27 mm. (c) Poikilotopic texture in calcite-cemented feldspathic arenite, Moreno Formation, Great Valley Group, Cretaceous, northeastern Diablo Range, California. (XN). Note dark calcite grain on left part of photo and light calcite grain in upper right part of photo containing enclosed grains. Long dimension of photo is 1.15 mm.

(c)

These processes are also important for other reasons. Cement is a major contributor to lithification and most commonly consists of quartz or calcite, but feldspars, dolomite, illite, kaolinite, hematite, analcite, anhydrite, and other minerals also serve as cements.[69] Quartz typically forms overgrowths (in crystallographic continuity) on detrital quartz grains (figure 16.12b), but chalcedonic quartz and cristobalite also fill in the pores between quartz grains (Blatt, 1966). The chalcedony forms radial fibrous textures. Phyllosilicates, such as chlorite, illite, and kaolinite, form authigenic cements that also form radial fibrous textures or occur as interclastic meshworks of grains (Hutcheon, 1983). Calcite is a common cement that typically creates poikilotopic textures by enclosing clastic grains in large crystals (figure 16.12c), but it also occurs as intragranular micrite and spar.[70] In some rocks, Mg calcite, dolomite, and aragonite cements form as a result of chemical stability conditioned by methane-enriched pore waters.[71]

Solution is important in the diagenesis of sandstones for a number of reasons. Most importantly, it affects the amount of porosity and the permeability of the rocks. These aspects of the rock, particularly permeability created by secondary porosity development, control the volumes and flow rates of the fluids that make many diagenetic reactions possible. Another effect is a change in the bulk composition of the rock. Solution can remove detrital framework grains, as well as diagenetic minerals formed as cements, authigenic minerals, or replacement minerals (Siebert et. al., 1984; Shanmugam and Higgins, 1988). For example, in the Jurassic Brent Group of the North Sea oil fields, an increase in quartz content with depth is attributed to increasing dissolution of feldspar (Harris, 1989). The shallow-level sandstones are feldspathic, whereas more deeply buried sandstones are quartz arenites.

Authigenesis commonly produces the same minerals as cementation and replacement. Mixed-layer phyllosilicates, chlorites, feldspars, and zeolites are among the many minerals that develop as authigenic phases. In combination with solution and replacement, authigenesis is particularly important in the production of matrix in sandstones via the alteration and replacement of lithic rock fragments (Siever, 1986; also see chapter 19).

Over time, sandstones, like limestones, may undergo progressive diagenesis. A generalized sequence of stages for sandstones is as follows.[72]

1. Sedimentation and bioturbation accompanied by minor compaction.
2. Cementation involving such phases as calcite, dolomite, quartz, pyrite, and clays; accompanied by compaction and minor recrystallization and solution.
3. Compaction with authigenesis and cementation by calcite, quartz, zeolites, and chlorite; accompanied by replacement of smectites by I/S mixed-layer clays, pressure solution, and recrystallization.

4. Authigenesis yielding new quartz, chlorite, I/S, calcite, and zeolites; accompanied by recrystallization and replacement (especially of lithic fragments), and minor compaction and solution.
5. Solution to form secondary porosity; authigenic development of chlorites, zeolites, and quartz; recrystallization; and minor compaction.
6. Late authigenic precipitation and cementation yielding kaolinite, illite, pyrite, quartz, calcite, and/or zeolites, with continued solution and replacement.
7. Late solution with minor recrystallization, replacement, and cementation, including the formation of hematite, replacement of pyrite by goethite, and precipitation of calcite.

The first of these stages is clearly eogenetic and the last is telogenetic. Stage 2 is eogenetic to mesogenetic, whereas stages 3 to 6 are mesogenetic.[73] As is the case with carbonate rocks, rarely does any one sandstone exhibit evidence of all of these stages of diagenesis.

An example of sandstone diagenesis is provided by the diagenetic history of feldspathic sandstones of the Santa Ynez Mountains of California. A three-stage history (figure 16.13), described by Helmold and van de Kamp (1984), includes

1. An eogenetic stage, in which minor compaction, solution, calcite cementation, and pyrite formation were accompanied by significant authigenesis and cementation by pore-filling clays (particularly mixed-layer illite/smectite);
2. A mesogenetic stage, in which authigenic development of a number of phases (e.g., sphene, laumontite, and chlorite) was accompanied or followed by development of authigenetic quartz and feldspar overgrowths, the albitization of plagioclase, solution of heavy minerals and plagioclase, the replacement of mixed-layer phyllosilicates by chlorite, additional compaction, and the development of laumontite, barite, and calcite cements; and
3. A telogenetic *sensu lato* (outcrop vadose) stage characterized by the formation of iron oxide coatings.

Other Sedimentary Rocks

Other sedimentary rocks, including evaporites, phosphorites, and ironstones, also undergo diagenetic changes.[74] In general, the processes of diagenesis affecting these rocks are the same as those that act on the more common rock types. Yet, because of the unique compositions of these rocks, their mineral compositions are distinctive. For example, in phosphatic rocks, carbonate fluorapatite may crystallize, whereas hydroapatite is recrystallized or replaced during diagenesis. In iron-rich sediments, authigenetic hematite, hydrated iron oxides, or glauconite may form, as carbonate minerals are replaced by the iron oxides (Alling, 1947). In both Fe- and P-rich rocks,

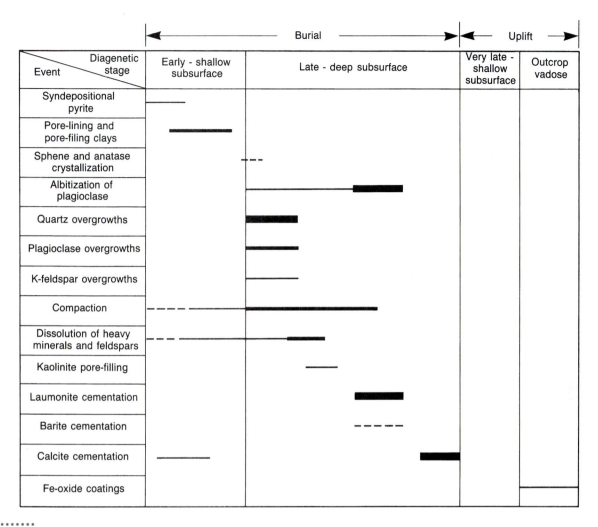

	Diagenetic stage	Burial		Uplift	
Event		Early - shallow subsurface	Late - deep subsurface	Very late - shallow subsurface	Outcrop vadose
Syndepositional pyrite					
Pore-lining and pore-filing clays					
Sphene and anatase crystallization					
Albitization of plagioclase					
Quartz overgrowths					
Plagioclase overgrowths					
K-feldspar overgrowths					
Compaction					
Dissolution of heavy minerals and feldspars					
Kaolinite pore-filling					
Laumonite cementation					
Barite cementation					
Calcite cementation					
Fe-oxide coatings					

Figure 16.13 Diagram showing diagenetic stages and events in Paleogene sandstones of the Santa Ynez Mountains, California. Width of bars indicates relative abundance of diagenetic phases, whereas length indicates duration of diagenetic change.
(From Helmold and van de Kamp, 1984)

recrystallization of quartz and carbonate minerals is an important diagenetic process.

In evaporites, recrystallization and replacement are the significant processes. As is typical of recrystallization phenomena, recrystallization may result in an early reduction in grain size, but ultimately enlargement of grains results. Evaporite minerals (e.g., halite, anhydrite) are also replaced by gypsum, dolomite, quartz, or other phases during diagenetic changes.

SUMMARY

Sedimentary rocks form from sediment that is weathered and eroded from various rock types, transported by a variety of processes—either in solution or by physical movement of clasts—and deposited in a sedimentary basin or other sedimentary environment. The provenance, weathering processes, transportation history, and sedimentary environment all control the nature of the resulting sediment. Provenance ultimately controls composition. Weathering and transportation can modify that composition and influence the texture of the sediment. Depositional processes and environmental conditions control the textural development of the sediment and may further modify the composition.

Weathering, the transformation of rock into soil, involves various disintegration (physical) and decomposition (chemical) processes. Important disintegration processes are frost wedging, abrasion, and animal activity. Decomposition processes include solution, oxidation, reduction, colloid formation and cation exchange, chelation, hydration, and hydrolysis.

Weathered and eroded materials are transported in moving air, ice, and water. By far the most important agent or facilitator of transportation is water. Materials are transported via solute transportation or by mechanical transportation. In addition, under the influence of gravity, various types of gravity and density flows (including landslides) move materials towards sites where they are deposited. The depositional processes affect the textural and structural character of the resulting sediments.

After deposition, the various processes of diagenesis—compaction, recrystallization, solution, cementation, authigenesis, replacement, and bioturbation—transform sediment into sedimentary rock. Eogenesis occurs at or near the sediment-water interface, whereas mesogenesis occurs after burial, and telogenesis occurs after burial and exhumation. The degree to which each of the diagenetic processes affects any given sediment or rock depends on the composition of the sediment or rock, on the physical characteristics of the sediment or rock, on the volume and nature of the fluid phase present during diagenesis, and on the conditions existing during the diagenetic history. Because of changing conditions over time, diagenetic histories are commonly complex.

EXPLANATORY NOTES

1. For additional discussions of the processes of weathering, see texts such as Rieche (1950), Verhoogen et. al. (1970, ch. 7, 8), Krauskopf (1979), Ritter (1986, ch. 3, 4), Boggs (1987, ch. 2), Pettijohn, Potter, and Siever (1987), and Richardson and McSween (1989, ch. 6).
2. For example, see Velbel (1988, 1989b).
3. Also see Koster van Groos (1988) and Velbel (1989a). Velbel (1988) describes a solution model for the formation of kaolinite from feldspar.
4. Boggs (1989, pp. 29–30). For additional information, consult standard chemistry texts.
5. Also see Grantham and Velbel (1988) and Johnsson and Stallard (1989).
6. For a more complete discussion of this topic, see standard sedimentology texts such as Blatt et al. (1972, ch. 4); Leeder (1982, ch. 5); and Boggs (1987, ch. 3), and sources such as G. V. Middleton (1966a, b), Lowe (1982), G. V. Middleton and Hampton (1973, 1976), and G. V. Middleton and Southard (1984), all of which serve as a basis for the discussion here.
7. Stress = force/unit area = ma/A.
8. Mud diapirism was suggested by melange studies (Cloos, 1984; Barber et al. 1986) and described by Prior et al. (1989).
9. For additional descriptions of slides see R. P. Sharp and Nobles (1953), Embly (1976), and Jacobi (1984), and the references therein.
10. Examples are depicted and described in Aalto (1978) and Nilsen and Abbott (1981). Also see Middleton and Hampton (1976).
11. Also see Nardin et al. (1979) and Boggs (1987). Lowe (1976) distinguishes liquidized flows in which solids settle down through the liquid from fluidized flows in which fluid moves up through the solids.
12. Turbidity currents and their deposits have been the subject of extensive study. Selected works on these currents and their deposits include Bouma (1962), Mutti and Ricci Lucchi (1978), and R. G. Walker (1984d). Also refer to chapter 19 and the numerous references therein.
13. Boggs (1987, p. 64)

14. See Leeder (1982, ch. 5, 6) and Middleton and Southard (1984) for detailed summaries of fluid flow, which in conjuction with Blatt et al. (1972, ch. 4) and Boggs (1987, ch. 3) provide the foundation for this review.
15. Leeder (1982, ch. 5) presents a discussion of the Bernoulli equation, which provides the theoretical explanation for the lifting force.
16. Additional and more detailed discussions of bed-load transport are provided by Bagnold (1973), Abbott and Francis (1977), Middleton and Southard (1984), I. Reid and Frostick (1987), and Whiting et al. (1988).
17. For example, see D. B. Simons, Richardson, and Nordin (1965), V. R. Baker (1984), and Nemec and Steele (1984). Refer to Chapter 20 and to the references therein relating to alluvial deposits.
18. Also see Blatt (1966), Bathurst (1975, ch. 8–13), Textoris (1984), and Scoffin (1987, ch. 4).
19. Bathurst (1958, 1975, 1983), T. R. Walker (1962), Blatt (1979), Hayes (1979), Bjorlykke (1983), Hutcheon (1983), R. Thomas (1983a), Velde (1983), Edman and Surdam (1984), Gautier and Claypool (1984), Kaiser (1984), Loucks et al. (1984), Textoris (1984), Wood and Hewett (1984), Kantorowicz (1985), Choquette and James (1987).
20. Also see Choquette and Pray (1970), Pryor (1973), and R. Thomas (1983b).
21. Also see Graton and Fraser (1935), T. W. Doe and Dott (1980), and N. P. James and Bone (1989).
22. Also see J. M. Taylor (1950), Longman (1981b), and Harbour and Mathis (1984).
23. See Kerrich (1977) for a historical review of pressure solution studies and the additional references in note 40.
24. See Schneidermann and Harris (1985) for discussions of carbonate cements. Cementation and cements are also discussed in Ginsburg and Schroeder (1969), Ginsberg et al. (1971), Meyer (1974, 1978), Bathurst (1975, ch. 10), Hanor (1978), Longman (1981b), R. Thomas (1983b), Textoris (1984), C. H. Moore (1985), Walls and Burrowes (1985), James et al. (1986), Choquette and James (1987), Dorobek (1987), McBride and Marsh (1987), Mitchell et al. (1987), and Dutton and Land (1988), Barnaby and Rimstidt (1989), and many other articles.
25. Sibley (1978), Mankiewicz and Steidtmann (1979), Scholle (1979), Oldershaw (1983), Edman and Surdam (1984), Lamando et al. (1984), Markert and Al-Shaieb (1984), Reinson and Foscolos (1986), Burton et al. (1987), McBride and Marsh (1987), Molenar and de Jong (1987), Dutton and Land (1988).
26. Neomorphic replacement is basically a process of recrystallization.
27. Other crystallographic and paleontologic definitions of the term allomorphic exist (see R. L. Bates and Jackson, 1987) and should not be confused with the definition presented here.
28. For example, see Surdam and Boles (1979) and Dutton and Land (1988).
29. Also see Dorobek (1987), Ritger, Carson, and Suess (1987), Budd (1988), Shanmugam and Higgins (1988, fig. 9), M. R. Lee and Harwood (1989), and Vavra (1989).

30. Steidtmann (1917), J. E. Adams and Rhodes (1960), Pray and Murray (1965), Badiozamani (1973), P. A. Harris (1979), Longman (1981b), Friedman and Ali (1981), Bathurst (1983), Halley (1983), N. P. James and Choquette (1983a, b, 1984), Wanless (1983), Textoris (1984), Schneidermann and Harris (1985), Choquette and James (1987), Halley and Matthews (1987), and Stoessell et al. (1989).

31. For example, see Friedman (1964), Lasemi and Sandberg (1983), von der Borch and Lock (1979), Muir et al. (1980).

32. Longman (1980), Halley (1983), N. P. James (1983), Choquette and James (1987), N. P. James and Choquette (1984).

33. For example, see Dorobek (1987) and Budd (1988).

34. Compare Friedman (1975) and Shinn et al. (1977). See Beales (1965). Schmoker (1984) argues that loss of porosity is a function of time and temperature, although he does not explain what processes occur to produce that loss.

35. Choquette and Pray (1970); Shinn et al. (1977).

36. Solution may produce *secondary porosity*, important in some petroleum fields and aquifers. See, for example, Matthews (1967), V. Schmidt and McDonald (1979a,b), P. M. Harris (1984a), Lomando, Birdsall, and Goll (1984), several of the papers in McDonald and Surdam (1984), Bjorlykke, Ramin, and Saigal (1989), W. E. Sanford and Konikow (1989), and Howlander (1990).

37. The term *lysocline* is used as a synonym for CCD by some geologists, whereas others use it to denote the depth at which dissolution features are detectable in sediments (compare R. L. Bates and Jackson, 1987, and Richardson and McSween, 1989, p. 102).

38. Scoffin (1987, p. 91). Also see Bathurst (1975, p. 267ff.) for a review and additional references.

39. Also see Walter (1985) and Boggs (1987, p. 282).

40. Various aspects of pressure solution or its effects in carbonate rocks are discussed by Weyl (1959), Renton, Heald, and Cecil (1969), Kerrich (1977), Robin (1978a,b), Wanless (1979), Mitra and Beard (1980), James, Wilmar, and Davidson (1986), Porter and James (1986), Micke and Wenk (1988), Tapp and Cook (1988), and N. P. James and Bone (1989).

41. Folk (1974, p. 176), MacIntyre (1977), Halley and Harris (1979), Halley (1983), Barnaby and Rimstidt (1989), Braithwaite and Heath (1989), Coniglio (1989), D. Emery and Dickson (1989). See Feazel and Schatzinger (1985) and Porter and James (1986) for discussions of the prevention of cementation.

42. For example, see Ginsburg and Schroeder (1969), Chafetz, Wilkinson, and Love (1985), Friedman (1985), P. M. Harris, Moore, and Wilson (1985), Kendall (1985), Lighty (1985), C. H. Moore (1985), Pierson and Shinn (1985), Prezbindowski (1985), Sandberg (1985), Walls and Burrows (1985), Wilkinson, Smith, and Lohmann (1985), Ritger, Carson, and Suess (1987), and Woronick and Land (1985). Also see papers in Schneidermann and Harris (1985).

43. Also see P. M. Harris, Dodman, and Bliefnick (1984), Heydari and Moore (1989), and M. R. Lee and Harwood (1989).

44. Also see M. Pitcher (1964), Ginsburg and Schroeder (1969), P. A. Harris (1979), Given and Wilkinson (1985), P. M. Harris, Dodman, and Bliefnick (1985), Lighty (1985), MacIntyre (1985), C. H. Moore (1985), Pierson and Shinn (1985), Sandberg (1985), and Searl (1989). See Kendall (1985) for a discussion of the distinction between radiaxial, radial fibrous, and fasicular-optical calcite. Walls and Burrowes (1985), C. H. Moore (1985), and Wilkinson et al. (1985) describe equant-mosaic calcite cements. Given and Wilkinson (1985) describe environments of formation.

45. Dorobek (1987), D. Emery and Dickson (1989), D. Emery and Marshall (1989), Horbury and Harms (1989).

46. See Woronick and Land (1985), Budd (1988), Hurley and Lohman (1989), and Noble and Stempvoort (1989), and see Land (1967) for experimental studies of diagenetic phase changes.

47. Compare Friedman (1964) and Sandberg (1985), for example.

48. Also see Bathurst (1958), S. J. Mazzulo et al. (1978), Shlager and James (1978), and Shukla and Friedman (1983). N. P. James and Choquette (1984) provide a comprehensive review and a bibliography. N. P. James and Bone (1989) describe the stages of diagenesis of a "cool water" limestone and suggest that these are more common than is realized.

49. Hurley and Lohman (1989), Webb and Raymond (1989).

50. Most authors use the word dolomite for both the rock and the mineral. Here, I use *dolostone* for the rock, reserving *dolomite* for the mineral.

51. Mawson (1929), Alderman et al. (1957), M. N. A. Peterson, Bien, and Berner (1963), Illing, Wells, and Taylor (1965), von der Borch (1976), von der Borch and Lock (1979), Muir et al. (1980), Friedman (1964, 1989), Lumsden (1988, 1989). Direct precipitation of dolomite from organic-rich pore water in surface sediments is described by Middelburg et al. (1990).

52. For example, see S. J. Mazzulo, Reid, and Gregg (1987) and J. E. Andrews, Hamilton, and Fallick (1987).

53. Mawson (1929) had previously suggested a similar process for dolomitization of arid Coorong Salt Lake sediments in Australia.

54. Also refer to Deffeyes, Lucia, and Weyl (1965), Shinn, Ginsburg, and Lloyd (1965), Shukla and Friedman (1983), Budai, Lohmann, and Wilson (1987), and M. W. Wallace (1990). S. J. Mazzulo, Reid, and Gregg (1987) describe similar dolomitization by "near-normal" marine waters. See the questions raised about sublagoonal dolomitization by K. J. Hsu and Siegenthaler (1969).

55. Also see Budai, Lohmann, and Wilson (1987), Swart, Ruiz, and Holmes (1987), Humphrey (1988), Humphrey and Quinn (1989), and W. A. Nelson and Read (1990).

56. Graber and Lohmann (1989), Machel and Anderson (1989), Cervato (1990), E. N. Wilson, Hardie, and Phillips (1990).

57. For more information, refer to the discussions of oil shales and the origin of the Green River Formation in chapter 22.

58. Also see Ernst and Calvert (1969), R. Greenwood (1973), Kastner, Keene, and Gieskes (1977), Isaacs, Pisciotto, and Garrison (1983), Kastner and Gieskes (1983), L. A. Williams, Parks, and Crerar (1985), and L. A. Williams and Crerar (1985). A more detailed review of this transformation and other aspects of chert diagenesis than is possible here is provided by Hesse (1988).

59. See Dapples (1962), Blatt (1966), Weaver and Beck (1971), Wallace (1976), Bjorlykke (1983), McBride (1984), Chan (1985), Kantorowicz (1985), and B. F. Jones (1986). Examples of mudrock diagenesis are described by Bjorkum and Gjelsvik (1988), Bohrmann, Stein, and Faugeres (1989), Palastro (1989), and Shaw and Primmer (1989).

Discussion of the mixed-layer clay phyllosilicate formation is found in Hower et al. (1976), Braide (1987) Glasmann et. al. (1989), and Alt and Jiang (1991). Illite formation is discussed in Eberl (1986), Colton-Bradley (1987), Sass, Rosenberg, and Kittrick (1987), G. Whitney (1990), and Aja, Rosenberg, and Kittrick (1991). Reviews of these and related subjects are presented by Eslinger and Pevear (1988, ch. 5) and C. E. Weaver (1989).

60. Weaver and Beck (1971), Hiltabrand et al. (1973), Blatt (1979), Hayes (1979), Hoffman and Hower (1979), Aoyagi and Kazama (1980), Oldershaw (1983), Moncure et al. (1984), Siever (1986), Braide (1987), Burton et al. (1987), Kisch (1987), Raiswell and Berner (1987), Sass, Rosenberg, and Kittrick (1987). Several additional examples of the diagenesis of sandstones are found in McDonald and Surdam (1984), including several papers on the Frio Sandstone of the Texas Gulf Coast.

61. For more information, refer to Matthews (1967), Moncure et al. (1984), Siebert et al. (1984), Surdam et al. (1984), Shanmugam and Higgins (1989), Shanmugam (1990), and several of the papers in McDonald and Surdam (1984).

62. Also see Choquette and Pray (1970), D. C. Beard and Weyl (1973), and Houseknecht (1987).

63. Blatt (1966), Textoris (1984), Houseknecht (1987).

64. Also see Shinn et al. (1977).

65. D. C. Beard and Weyl (1973).

66. For example, Graton and Fraser (1935), Gaither (1953), Rogers and Head (1961).

67. Heald (1955), Renton et al. (1969), Mitra and Beard, (1980), Hutcheon (1983), James et al. (1986), Houseknecht (1987).

68. For example, W. C. James et al. (1986), but see Sibley and Blatt (1976).

69. Sibley (1978), Blatt (1979), Klein and Lee (1984), Imam and Shaw (1985), James (1985), Reinson and Foscolos (1986), Lee (1987), McBride et al. (1987), Pettijohn, Potter, and Siever (1987, p. 434; 447ff.), Dutton and Land (1988), Duffin et al. (1989), Vavra (1989), Cavazza and Dahl (1990), Hansley (1990).

70. Additional illustrations of textures may be seen in Scholle (1979), A. E. Adams, MacKenzie, and Guilford (1984), W. C. James (1985), and Reinson and Foscolos (1986). Also see the various papers in McDonald and Surdam (1984).

71. Ritger, Carson, and Suess (1987), Friedman (1988), Andrews (1988).

72. This sequence is idealized. Variations in rock type (e.g., quartz arenite vs. volcanic lithic wacke), depth and P-T conditions of burial, and the chemistry and quantity of the fluid phase will have a significant impact on the exact minerals that develop during the various stages.

73. Examples of diagenesis of sandstones may be found in Galloway (1979), Hoffman and Hower (1979), Helmold and van de Kamp (1984), Fishman and Reynolds (1986), Shanmugam and Higgins (1988), Pollastro (1989), Cavazzi and Dahl (1990), Hansley (1990), and Howlander (1990).

74. For examples of diagenesis of the less common rocks, see Alling (1947), Froelich et al. (1988), Casas and Lowenstein (1989), and Chipley and Kyser (1989).

PROBLEMS

16.1. Given a cubic piece of rock, 1 cm on a side, (a) calculate the surface area of the cube. (b) If the cube is broken down into smaller grains, each 1 mm on a side, what will be the combined surface area of the material (i.e., of all the smaller cubes)?

16.2. Write a balanced equation for the solution of andradite garnet ($Ca_3Fe_2Si_3O_{12}$).

16.3. Compare the three stages of diagenesis described by Helmold and van de Kamp (1984)(figure 16.13) to the seven idealized stages of sandstone diagenesis listed in the text on page 316. What if any correspondence is there and what differences exist?

CHAPTER 17

Sedimentary Environments

INTRODUCTION

Sedimentary rocks form in sedimentary environments. A **sedimentary environment** is a surface region of the lithosphere, either above or below sea level, that is distinguished by a particular set of chemical, physical, and biological characteristics. These characteristics control development of sediments and rocks that have unique combinations of textures, structures, compositions, and fossils; that is, they control the formation of *sedimentary facies*. The textures, structures, and compositions characterize *lithofacies* that represent specific environments.

By studying both the processes in present-day environments and the lithofacies formed by those processes, sedimentologists are able to infer how similar, ancient rocks were formed. Stratigraphers study layered sequences of sediments as a key to stratigraphic sequences in the rock record. Together with paleontologists, these geologists discover information necessary for interpretation of the veneer of sedimentary rocks that blankets the Earth's surface.

In this chapter, various sedimentary environments and their distinctive stratigraphic sequences and lithofacies are de-scribed. The petrogeneses of particular rock types are to be found in succeeding chapters in this part of the book.

TYPES OF SEDIMENTARY ENVIRONMENTS AND THEIR DEPOSITS

Traditionally, geologists have subdivided sedimentary environments into three categories—continental, transitional, and marine.[1] **Continental environments** are totally above sea level at high tide and are not influenced by marine processes. **Marine environments,** in contrast, are totally below sea level at low tide. **Transitional environments,** as the name implies, lie between the levels of marine and continental environments and are influenced by marine and continental agents (e.g., fresh and salt water, wind and wave action). The subdivision of environments adopted in this text, shown in table 17.1, is similarly tripartite. Students should note that great volumes of sedimentary rock in the geologic record were produced in certain of these environments (e.g., in reef, shelf, and submarine fan environments), whereas in other environments (e.g., spelean (cave) environments), the volume of rock in the stratigraphic record is rather small.

Table 17.1 Classification of Sedimentary Environments

Major Categories	General Environment	Specific Environments
Continental	Fluvial (river)	Channel and bar
		Overbank, high-energy (e.g., levee)
		Overbank, low-energy (e.g., swamp)
	Desert	Alluvial fan
		Playa
		Erg
	Glacial	Subglacial
		Englacial
		Supraglacial
		Cryolacustrine
		Proglacial fluvial
		Proglacial aeolian
	Lacustrine	Cryolacustrine
		Playa lake (salina)
		Freshwater lacustrine (each of the above may have associated deltaic and shoreface environments)
	Paludal (swamp)	Intrapaludal
		Deltaic paludal
	Landslide	—
	Spelean (cave)	—
Transitional	Coastal deltaic	Channel bar
		Overbank-crevasse splay
		Deltaic paludal
		Deltaic lacustrine
		Prodelta
		Delta front
	Estuarine-lagoonal	Estuarine

CONTINENTAL ENVIRONMENTS

Continental sedimentary rocks are less abundant in the geologic record than are marine rocks. The continental environments that yield the most common rocks are the fluvial (stream), glacial, and lacustrine (lake) environments.

Fluvial Environments

Streams, viewed from above, have one of five general patterns—straight, sinuous, meandering, braided, and anastomosing.[2] Of these, meandering and braided streams have created

significant deposits recognized in the geologic record.[3] The areas associated with streams that serve as depositional sites include the stream channel and the surrounding plains.

In a *meandering stream* (figure 17.1), water flows in one main channel that moves back and forth across the floodplain in a looping, snakelike fashion. During floods, sands and gravels in the channel bottom (the streambed) are moved and redeposited, either in the channel or along the channel margins. Erosion results in channel migration. Through time, as the channel migrates, gravel and sand are deposited locally to form cross-bedded *channel bars* within the channel, *bank benches* along the stream banks or margins, and

Major Categories	General Environment	Specific Environments
	Estuarine-lagoonal	Lagoonal
		Salt marsh
	Littoral-beach	Beach foreshore
		Beach backshore
		Beach dune (and berm)
		Tidal channel
		Tidal flat
Marine	Shelf–shallow sea	Low-energy open
		Low-energy restricted
		High-energy
		Glaciomarine
	Reef	Reefal
		Forereef
		Reef lagoon
	Submarine canyon	—
	Slope and rise	Open slope-rise
		Slope basin (submarine fan may occur in either of the above)
	Pelagic	Basinal or Abyssal plain
		Oceanic plateau
	Trench	Trench slope
		Trench slope basin
		Trench floor (submarine fan may occur in the latter two environments)
	Rift-fracture zone	—

point bars along the channel edge on the concave side of meander bends.[4] Upon lithification, the sand and gravel become sandstone and conglomerate.

Flooding occurs when water leaves the channel and flows over the banks. Floods produce high-energy overbank deposits that consist of (1) linear natural-levee deposits of sand and silt dropped by the water that pours over the channel banks, and (2) lobate, crevasse splay deposits of sand and silt developed as a result of breaching of the levee and escape of water onto the floodplain.[5] On the floodplain, silt and clay are deposited. Organic-rich muds and layers of decayed vegetation (peat) form in swamp (paludal) areas. Both the flood-plain and swamp areas are overbank low-energy environments. Lakes that form in abandoned meanders are also low-energy environments.

Various depositional processes operate in the fluvial environment and produce a complexly interfingered stratigraphy consisting of lenticular to tabular masses of conglomerate, sandstone, and mudrock (figure 17.1a). This stratigraphy is characterized by *fining upward sequences*, sequences of rock with coarser-grained (conglomeratic) rock at the base, intermediate-grained rock (sandstone) in the middle, and mudrock at the top (figure 17.1b) (J. R. L. Allen, 1970b).[6] Cross-bedded, parabolic-lenticular,

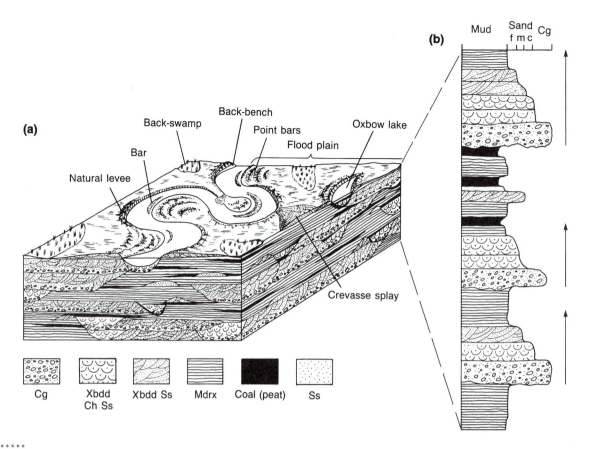

Figure 17.1 Simplified sketch of a meandering stream environment and cross section. (a) Block diagram showing various subenvironments and the stratigraphy developed in them. Cg = conglomerate, Xbdd = cross-bedded, Ss = sandstone, Xbdd Ch Ss = cross-bedded channel sandstone, Mdrx = mudrocks. (b) Columnar section showing stratigraphy of a meandering stream environment. Note fining upward sequences indicated by arrows. f = fine sand, m = medium sand, c = coarse sand, cg = conglomerate. *(Based in part on Selley, 1976, and R. G. Walker and Cant, 1984).*

lenticular-lobate, and lenticular-linear sandstone units and nonmarine fossils (in middle Paleozoic and younger rocks) are indicative of this environment.

Braided streams generally are dominated by coarse-grained sediment (sand and gravel)(N. D. Smith, 1970; Miall, 1978b; Cant, 1982). Because the channels carry a large load of coarse sediments, overbank conditions result in deposition of these sediments, which upon lithification become interfingering linear-, parabolic-, to lenticular-lobate masses of cross-bedded sandstone and conglomerate. Channels and cross-stratification are common structures in these rocks and nonmarine fossils may be present in rocks younger than middle Paleozoic age.

Desert Environments and Rocks

The desert environment consists of three main subenvironments—the erg (or aeolian environment), the playa, and the alluvial fan (figures 17.2, 17.3). Each has distinguishing features.

Ergs are large, sand-covered desert terrains. Aeolian dunes, interdune areas, and sand sheets comprise the erg. Aeolian **dunes** are piles of sand deposited by the wind. The main feature characteristic of the dune environment and lithofacies is well-sorted sand (and sandstone) comprising simply to complexly cross-stratified units (McKee, 1966).[7] Between the dunes are relatively thin sheets of sand, the interdune sand flats, which cover small to large areas. These interdune sand flats are characterized by thin, horizontally laminated and low-angle, wind-ripple cross-stratified sandstone.[8] Sand sheets, large flat areas of sand generally not closely associated with dunes, form where there is a high water table, periodic flooding, cementation, and vegetation (Kocurek and Nielson, 1986). Like the interdune areas, sand sheet deposits are characterized by parallel-bedded to wind ripple, low-angle cross-stratified layers (Kocurek and Nielson, 1986; Chan, 1989). Aeolian sand grains show pitted surfaces ("frosting"). Vegetation is usually lacking in the erg, but may be present locally. Fossils typically consist of footprints and, less commonly, of burrows.[9]

Ergs may be adjoined by **playas,** flat, vegetation-free desert basins that occasionally contain ephemeral lakes called **playa lakes.** Playa salt flats called **continental sabkhas** (Blatt, Middleton, and Murray, 1980; A. C. Kendall, 1984) are characterized by thin-bedded or laminated mudrocks and evaporites. Sandstones may form, especially near the margins of the playa. Bedded evaporites, however, are the most distinctive rock type of playa stratigraphies.

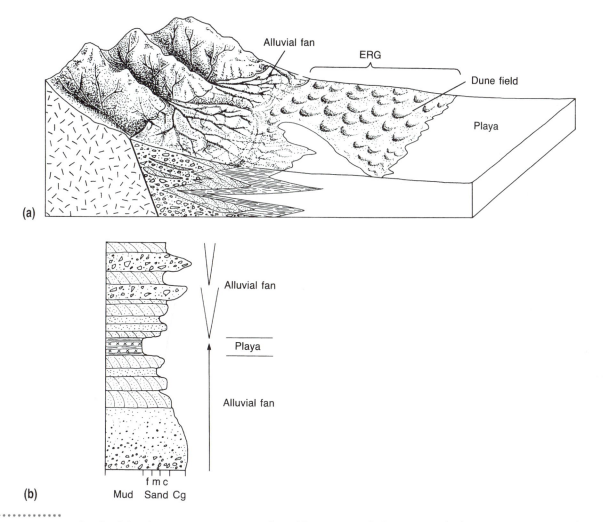

Figure 17.2 Sketch of the desert environments—alluvial fan, erg, and playa (a), with diagrammatic section through the fan-playa section (b). Angular fragments show diamictite and fanglomerate. Long Vs at right show coarsening upward sections. Arrow shows fining upward sequence.

The alluvial fan environment is perhaps the most studied of the desert subenvironments.[10] Alluvial fans typically form lobate wedges where stream courses exit mountain canyons and enter valleys (figure 17.2). A decrease in water velocity and carrying capacity caused by the decrease in gradient and increase in channel widths results in voluminous sedimentation.

Most sediments of alluvial fans are coarse fluvial in character (gravel, sand), but bouldery muds produced by mudflows or debris flows are commonly interlayered with the fluvial deposits. Upon lithification, these sediments become conglomerate, sandstone, and diamictite, respectively. The structures are those of stream deposits. Locally, however, on the proximal areas of fans, breccias and conglomerates with a large percentage of angular clasts, called "fanglomerates," are present (Lawson, 1913; Krynine, 1948). Fossils are rare in alluvial fan deposits. Nevertheless, tree trunks and insect and vertebrate burrows may be preserved locally (Gostin and Rust, 1983, in Rust and Koster, 1984).

Figure 17.3 View northeast from west edge of Death Valley, California, showing sand dunes overlapping alluvial fan sediments that grade out into a playa.

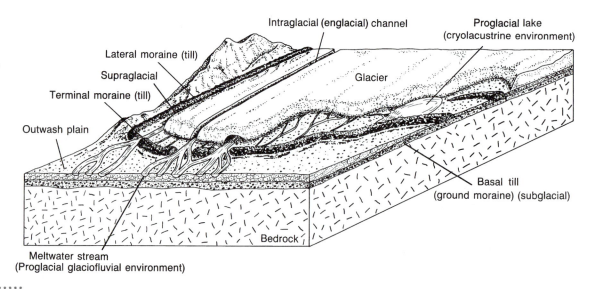

Lateral moraine (till)
Supraglacial
Terminal moraine (till)
Outwash plain
Intraglacial (englacial) channel
Proglacial lake (cryolacustrine environment)
Glacier
Basal till (ground moraine) (subglacial)
Bedrock
Meltwater stream (Proglacial glaciofluvial environment)

Figure 17.4   Sketch showing various glacial subenvironments surrounding a mass of glacial ice.

Glacial Environments and Deposits

Glaciers are moving masses of ice that are quite effective as agents of erosion. As the glacier advances and retreats, it deposits materials at its base in **subglacial environments,** within tunnels and channels within the ice in **englacial environments,** and along the sides and at the end of the glacier in **supraglacial environments** (figure 17.4). In front of the glacier, lakes may develop, forming **cryolacustrine environments.** Also in front of the glacial terminus, streams of meltwater form **proglacial fluvial environments,** especially where lakes are absent. Winds deposit fine sediment in **proglacial aeolian environments.**

Till is glacially deposited, very poorly sorted sediment (a diamict) composed of coarse, angular to rounded, locally faceted and striated clasts in a finer-grained matrix that typically consists of clay- to sand-size rock and mineral fragments. The equivalent rock, called **tillite,** is a glacial diamictite. Till is the sediment of subglacial environments and is characteristic of some supraglacial environments as well.[11]

In supraglacial, glaciofluvial, and proglacial environments, and in englacial tunnels, debris flow diamicts, stream-deposited, cross-bedded and channelled sands and gravels, lacustrine sediments, and till may be intimately interlayered (Flint, 1971; M. B. Edwards, 1978).[12] Because the meltwater streams are usually braided, sediment and structures (including channels and crossbeds) are typically fluvial. Yet striated clasts contained in the conglomerates indicate the glacial source of the sediment.[13] Cryolacustrine environments contain (1) laminated sands and muds (in part deposited by turbidity currents) that contain stones dropped from melting ice (*dropstones*), (2) diamict dropped from melting ice and icebergs, and (3) coarsening upward sequences of cross-bedded sand and mud deposited in stream-generated deltas that build out into the lake (N. D. Smith and Ashley, 1985). Laminites,

or **varvites,** containing dropstones are diagnostic of these cryolacustrine deposits (figure 17.5). Varves (rock = *varvite*) are rhythmically repeated sediment couplets, each representing a year, consisting of silty, light summer layers and dark, clay-rich, organic winter layers.[14]

Winds sweeping across proglacial fluvial deposits pick up, transport, and deposit finer sediment downwind in proglacial aeolian deposits. Many **loess** deposits—tabular deposits of light-colored, nonstratified, commonly calcareous, dominantly silt-size, quartz-rich sediment—are assigned an origin involving aeolian deposition of glacial sediment (Flint, 1971; Smalley, 1975a).[15] Such deposits blanket large areas in several parts of the world, including midwestern North America.

Lacustrine and Other Continental Environments and Deposits

Lacustrine environments develop in a variety of locations (Collinson, 1978c), including near glaciers, in deserts, on floodplains, on deltas, in intermountain areas, and in rifts. In small lakes, on lake margins, and where relief is high, lake deposits may consist of interlayered coarse and fine sediment. Where relief is low or where the central parts of lakes are remote from the shorelines, the sediment formed is usually fine-grained.[16] Mud and organic matter deposited from suspension and chemical precipitates (e.g., biogenic silica, carbonate minerals, and evaporite minerals) are dominant. The layers formed are usually parallel and thinly laminated, commonly comprising diagnostic varves (McLeroy and Anderson, 1966; Picard and High, 1972a; Dean and Fouch, 1983).[17] Even in mud-dominated lakes, lake margin facies and local turbidites within the lakes consist of coarser sediments. Fossils include vertebrates (fish), mollusks, plants (including algal stromatolites), and insects.

Figure 17.5 Varved laminite with dropstone layer (arrows, upper left), Mount Rogers Formation (Neoproterozoic), north of Mount Rogers, Virginia.

Swamps, like lakes, have central and marginal zones, the intrapaludal and deltaic paludal subenvironments, respectively. The intrapaludal subenvironment is dominated by quiet water. Here clay and peat, which become mudrock and coal, are deposited. The deltaic paludal environment is characterized both by cross-bedded sandstones and by mudrocks.

Landslides occur in continental areas. These range from small to very large features, and typical deposits are diamicts or *megabreccias* (Krieger, 1977).[18] The compositions of landslide deposits depend entirely on the source terrain, as the landslide itself, with few exceptions, does little sorting or other modification of the landslide debris. Clasts are typically angular. Where fluvial deposits are involved in landsliding, large, well-rounded clasts may occur in the deposit. In contrast, many of the landslides generated by the 1976 Guatemala earthquake occurred in a volcanic terrane and were composed exclusively of ash. Fossils are uncommon in landslide deposits, but locally, tree trunks and other vegetation, plus vertebrates, may become trapped and fossilized.

Landslide deposits are closely associated with other rocks in temperate to tropical regions, where active tectonism creates valleys that serve as small to large basins of sedimentation. These valleys are similar in many ways to those in arid regions. Differences are due to the controlling effects of the more humid conditions on weathering, erosion, and

sedimentation. The sedimentary rocks formed in these basins include breccias and diamictites deposited by landslides along steep mountain fronts, meandering and braided stream conglomerates and sandstones, mudrocks formed on meandering stream floodplains and in lakes, and sandstone and mudrock turbidites deposited by turbidity currents (R. J. McLaughlin and Nilsen, 1982). Fossils, where present, consist of plant fragments, vertebrate remains, and freshwater mollusks.

TRANSITIONAL ENVIRONMENTS

Transitional environments include coastal deltaic, estuarine-lagoonal, and littoral and related types. Rocks formed in these environments have been influenced by a variety of agents, including wind, waves, and water currents. Because the agents of sedimentation are varied and because sedimentation and biologic activity are subject to markedly variable chemical conditions, the sedimentary rocks formed in transitional environments are diverse.

Coastal Deltaic Environment

Deltas are protruding geomorphic forms along shorelines that form where streams enter oceans, lakes, or other bodies of water (figure 17.6).[19] In three dimensions, deltas are irregular

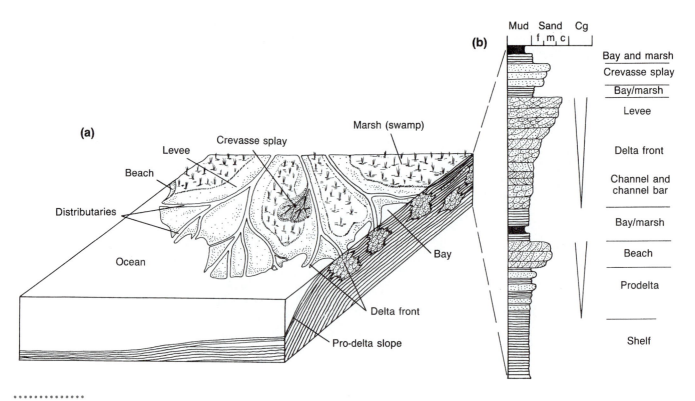

Figure 17.6   Sketch showing simple deltaic environment (a) and columnar section (b). Note crevasse splay, distributaries, delta front, and prodelta slope. In (b), large *Vs* show coarsening upward sequences.

to arcuate, lenticular to tabular masses of sediment. The sediment is deposited because the carrying capacity of the stream is reduced as its water velocity decreases when it enters the water body of the receiving basin.

The largest deltas, such as the Nile, Mississippi, and Mekong deltas, occur where major rivers enter the seas or oceans. Such deltas owe their character primarily to deposition by the rivers, but also to wave or tidal action (W. E. Galloway, 1975). Waves and tides alter the character of sediment and the bed geometries within the delta.

The deltaic environment is most easily subdivided into two major parts, the delta plain and the delta front. The delta plain is a low-lying area on the upstream part of the delta that may be above or below sea level, or parts may lie both above and below sealevel. **Distributaries,** the branch channels of a river that diverge from the main channel and spread like fingers across the delta (figure 17.6), deposit fluvial sediment, such as cross-bedded sand and gravel, within channel bar subenvironments (H. N. Fisk, 1961; J. R. L. Allen, 1970a, 1970b).[20] Between channels are overbank-crevasse subenvironments characterized by interbedded sand, silt, and mud. Lakes may develop within these interchannel areas and smaller lacustrine deltas form and fill the lake basin (Tye and Coleman, 1989). In many deltas, large tracts of the interchannel delta plain become swamps (deltaic paludal subenvironments) in which mud and organic debris are deposited to later form coal and associated mudrocks. Such was the origin

of Mississippian and Pennsylvanian coals of the eastern United States.[21] Locally, bays develop where interchannel areas lie below sea level. The subenvironments of the delta plain migrate over time, producing a complexly interfingered stratigraphy.

The delta front environment consists of the delta front (*sensu stricto*) and the prodelta subenvironments. The delta front (*sensu stricto*) is characterized by cross-bedded sand and mud (and locally gravel) deposited by the distributaries. Beyond that zone, farther into the receiving basin, mudrocks with local sand lenses are deposited. Slumps and associated mass flows develop above or at the base of the prodelta slope and extend out onto the shelf, slope, or basin plain (J. M. Coleman, 1982; Bjerstedt and Kammer, 1988; Postma, et al., 1988). Overall, deltaic deposits form coarsening and thickening upward sequences (figure 17.6b). Fossils in deltaic rocks range from profuse in coal-bearing sequences to sparse in sandstones of the deltaic environment and include trace fossils (e.g., burrows, trails), marine invertebrates, and roots, leaves, and other plant debris.

Estuarine-Lagoonal Environments

Estuaries and lagoons are transitional shoreline environments in which seawater and fresh water mix in varying proportions. **Estuaries** are irregular to triangle-shaped, widened river mouths connected to the open ocean. **Lagoons,** shallow water bodies that generally parallel the coastline, develop

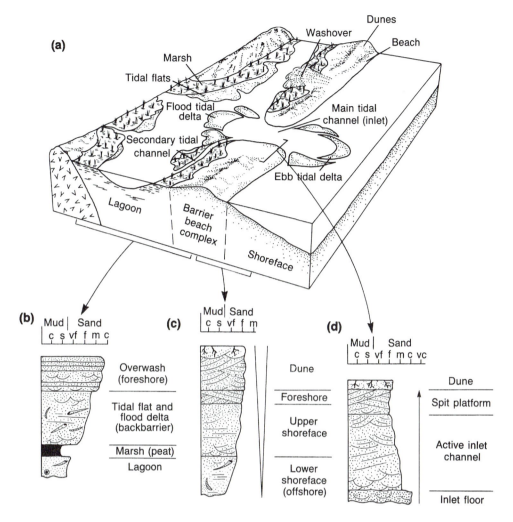

Figure 17.7 Sketch showing littoral and related environments, and representative stratigraphic sections of a coastline with a barrier beach complex. (a) Block diagram showing various subenvironments (modified from Reinson, 1984). (b) Stratigraphic section of the back-barrier region. (c) Stratigraphic section of dune-foreshore-shoreface area. (d) Stratigraphic section of the tidal inlet area
[(b)–(d) from Heron et al., 1984]

where bars block the mouths of estuaries or where barrier islands or spits separate low-lying (intertidal to subtidal) areas from wave action (figure 17.7). **Salt marshes** are flat vegetated areas, periodically flooded by salt water, that adjoin estuaries and lagoons.

Estuaries and lagoons may develop interfingered stratigraphies because of the varying salinity of the water, variable influence of currents, variable rates and compositions of sediment influx, changing sea level, and variety of adjoining subenvironments. Estuaries are characterized by well-sorted, cross-bedded sand that gives way locally either to parallel-laminated sand and mud, wavy interbedded sand and mud, or sandy muddy peat.[22] In some estuaries, there is a tendency for the sediment to be finer upstream, but along the channels and bars, the sediment is typically sandy.

In lagoons, the sediment is dominantly bioturbated mud. Nevertheless, parallel-laminated to cross-bedded sand interlayers are also typical. Sand may dominate locally (Rusnak,

1960; R. H. Parker, 1960). Where tidal flats and salt marshes adjoin the estuarine or lagoonal environments, the deposits of the latter grade into mudcracked, wavy-bedded mud and sand with roots or root molds. In low-latitude, carbonate-rich lagoonal environments, the carbonate equivalents of mud and sand—lime mud, pelletal lime mud, and carbonate sands that yield lime mudstone, wackestone, packstone, and grainstone—replace the siliciclastic sediments.[23] Lagoonal sediments associated with reef environments are discussed below.

Littoral and Related Environments

The land-sea interface, including the **littoral** zone *sensu stricto*, between high and low tide, may take a number of different forms. Simply defined, these interfaces are the beach, barrier, and delta. Deltas have been discussed above. A continuum exists between beach- and barrier-type interfaces. A **beach** is an accumulation of sediment formed along the margin of a water body. Wide beaches are called **strand-plains.**

Linear beaches, isolated from the land by a lagoon, are called **barrier islands** or **barrier beach complexes** (figure 17.7). Submerged linear sediment accumulations are called **bars.**

Coastlines with barrier islands and lagoons have the largest number of interrelated subenvironments. The nature of the subenvironments and the character of the sediment deposited will vary widely, depending on climate and vegetation, tidal range, rising (transgressing) or falling (regressing) sea level, coastal type or form, tectonic setting, sediment supply, and prior sedimentation history.[24]

The general form of a barrier-type transitional environment is shown in figure 17.7. Along the edge of the continental environments are **tidal flats,** including marshes and sabkhas, where fine sediment is typically deposited.[25] The marshes may have tidal creeks (small channels) cutting through them, but commonly the marshes lie above the normal high tide level (i.e., they are supratidal) (Shinn, Lloyd, and Ginsburg, 1969; Reineck and Singh, 1975).[26] Lower parts of the tidal flats, the parts affected by daily tides (i.e., intertidal areas), may also have tidal channels in which coarser sediments are deposited (G. R. Davies, 1970b).

The lagoon lies seaward of the tidal flats. Where main tidal channels (inlets) cut through the barrier island, daily tides flow and deposit sandy flood tidal deltas that project into the lagoon. The seaward side of the lagoon will have a shoreline that may have marsh and tidal flat zones. In addition, washovers, low areas along the barrier where storms wash sand across the island to the lagoon, form fan- or delta-like lobes extending across the marsh and tidal flat and into the lagoon.

The spine of the barrier island is formed by dunes. Together with the *berm*, a sandy, somewhat flat area above the normal high tide level, the dune forms a supratidal zone called the **beach backshore.** The **beach foreshore** includes the zone between high and low tide and the swash zone where breaking waves "run" up the beach face. Seaward of the foreshore is the *shoreface*, the zone above effective wave base and below low tide level. This zone is not strictly a part of the transitional environment, but is usually included in discussions of the sea-land interface because it is closely related to beach environments. Extending across the shoreface are ebb tidal deltas formed by outward-flowing tides.

Because the diversity of environments associated with the sea-land interface is large, the rocks and stratigraphic sequences produced are varied. For example, along coastlines that are rocky or tectonically active or that have high wave energy, shoreface and foreshore conglomerates and conglomeratic sandstones may be the principal sedimentary rock (Bourgeois and Leithold, 1984; Leithold and Bourgeois, 1984; W. Duffy, Belknap, and Kelley, 1989). Along passive, low-relief coastlines, either well-sorted siliciclastic sand or carbonate sand and mud are the principal sediments (Shinn, Lloyd, and Ginsburg, 1969; Kent, Van Wyck, and Williams, 1976; Duc and Tye, 1987). Parallel laminations, ripple cross-stratification, and meter-scale low-angle cross-beds are typical structures (Komar, 1976a).

The salt marsh is characterized by thin, wavy beds of fine-grained sediment such as silty clay (which when lithified becomes mudshale)(Reineck, 1967). Fossiliferous mudstones form local interlayers. Plant roots or thin molds are common, and abundant vegetation may result in peat layers that become coal. In carbonate environments, thinly laminated algal muds (G. R. Davies, 1970a; Hardie and Ginsburg, 1977) are precursors to stromatolitic lime mudstones. Thus, in both carbonate and siliciclastic settings, fine-grained sediments are typical. In contrast, tidal creeks cutting the marsh may have shell lag deposits[27] and a muddy to sandy fill that yields sandstones, mudrocks, packstones, or grainstones.

Tidal flats tend to be muddy at higher levels and more sandy at lower levels (Weimer, Howard, and Lindsay, et al., 1982; R. W. Frey, 1989).[28] Bedding occurs in a range of types, including rhythmically laminated, lenticular bedding with mud layers in sand and bioturbated beds of mud, pelletal lime mud, and sand. Ripples, mudcracks, and burrows are present locally. Tidal channels within the tidal flats may have a basal lag gravel (with clasts of mud, shells, or beachrock) overlain by cross-bedded sandy (siliciclastic, packstone, or pelletal lime mudstone) channel fill.[29] Washover fans on the seaward side of the lagoon, that is, the landward side of the barrier island, are characterized by parallel-laminated and foreset cross-bedded sands that yield, upon lithification, arenites, grainstones, and packstones.[30]

The barrier dunes, like other aeolian dunes, are characterized by high-angle, cross-bedded, well-sorted, fine to coarse sands.[31] Seaward of the dunes, the berm, foreshore, and shoreface sediments are represented by arenites, grainstones, or packstones deposited in subhorizontal, parallel-laminated to low-angle, cross-bedded units.[32] Local fossiliferous or conglomeratic layers may result from storm deposition. In general, these sediments fine and show increasing bioturbation seaward. Tidal channels (inlets) that cut the barrier are characterized by basal pebbly or fossiliferous sand, overlain by poorly sorted, parallel-laminated to cross-bedded, coarse, shelly sand with local clay layers (Moslow and Tye, 1985; Israel, Ethridge, and Estes, 1987; Hennessy and Zarillo, 1987). The overall sequence for the shoreline zones may either coarsen or fine upward, depending on history. Fossils, particularly invertebrate fossils, are generally common in the marine shoreline environments (R. H. Parker, 1960). In tidal flats and marshes, algal mats generate and trap mud.

MARINE ENVIRONMENTS

Large masses of sedimentary rock were deposited in marine environments, notably in shelf and reef environments. The transition from the shoreface to the shelf, in lithologic terms, is commonly almost imperceptible. The shelf, with local reefs, gives way to the slope (figure 17.8), and that, in turn, gives way to the rise or a trench. Pelagic environments in deep marine

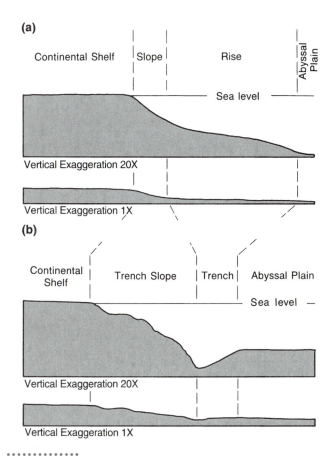

Figure 17.8
Generalized profiles of major marine environments of deposition. (a) Passive margin (modified from C. L. Drake and Burk, 1974; Cook, Field, and Gardner et al., 1982).
(b) Active margin.

basins and on abyssal plains represent environments that generally are remote or otherwise protected from deposition of coarse clastic sediment. Exceptions occur where submarine canyons that cut the shelf, slope, and rise allow flows and currents to transport fine to coarse sediment into basins and trenches that otherwise might receive only fine-grained sediment (Thornburg and Kulm, 1987; Kuehl, Hariu, and Moore, 1989).

Marine fossils, notably corals, mollusks, foraminifers, and bryozoans in Mesozoic or Cenozoic rocks, and corals, stromatoporoids, brachiopods, echinoderms, and bryozoans in Paleozoic rocks, are locally quite abundant in shallow marine rocks. Algae are represented by stromatolites and oncolites.

Shelf–Shallow Sea Environments

The shoreface is succeeded seaward by the shelf, the part of the ocean floor that extends from normal wave base to the shelf-slope break, a break in the slope of the seafloor that occurs at an average depth of 124 m (Bouma et al., 1982). Like the environments of the transitional zones, shelf environ-

ments are quite diverse. Variations result from differences in (1) climate, (2) sediment input, (3) biological activity, (4) waves, currents, and storm activity, (5) previous sedimentary and tectonic history, (6) present tectonic setting, and (7) sea-level changes.[33] For example, carbonate sediments develop particularly in tropical to subtropical climates, whereas siliciclastic sediments occur at moderate to high latitudes. In addition, siliciclastic sediments are deposited anywhere rivers, because of their large size or their occurrence in areas of local tectonism, provide large amounts of sediment to the coastal zones (e.g., along the northwestern edge of the Gulf of Mexico or along the western North American coast).

Shelf environments include low-energy types, such as large bays, platforms, ramps,[34] and deep shelf areas, and high-energy types in shallow areas and glaciomarine settings. The distribution patterns of sediments along the shelves reflect variable conditions and diverse environments of deposition (figure 17.9).

High- and low-energy shelf environments have some similarities, because occasional lulls in the high-energy environments result in low-energy sedimentation. In both environments, shoreface sands locally grade laterally into moderately to well-sorted shelf sands, which grade into muds and, locally, into gravels (figure 17.9).[35] The sands, which may be carbonate (and upon lithification yield skeletal wackestones or packstones and oolitic grainstones) or siliciclastic (yielding quartz arenites and wackes), are characterized by irregular bedding, tabular bedding, subparallel laminations, ripple cross-laminations, and high-angle cross-beds. The muds consist of lime muds (e.g., pelletal lime mud) or phyllosilicate muds. Bioturbation is typical, especially in muds, and commonly completely destroys the bedding. Glauconitic sands and phosphatic rocks are also important locally.[36]

Low-energy environments are dominated by muds. Where siliciclastic sedimentation has dominated, mudstone and wacke derived from mud and clay-rich sand (Shepard and Moore, 1955) are the main rock types. Lime mudstone is the equivalent rock developed in carbonate settings (Enos, 1983). Organic material may give either the carbonate or the siliciclastic sediments a dark color. This is especially true where circulation is restricted (e.g., in fjords), and anaerobic conditions yield dark, organic, laminated sulfidic sediments. Structurally, the rocks derived from low-energy shelf sediments form parallel-laminated, wavy to tabular, thin or medium to bioturbated-massive beds containing both marine invertebrate and minor vertebrate fossils (figure 17.10).

High-energy environments, in contrast, are characterized by relatively coarse sediments. Locally cross-bedded gravels, skeletal or oolitic grainstones and packstones, and arenites and wackes reveal higher energies of transportation. Currents produce sand waves and sandbars (submarine dunelike forms) that have high-angle cross-beds. Currents, like waves, deposit the sandy precursors of sandstone or grainstones and packstones, notably oolitic, pelletal, and skeletal types (figure 17.11).

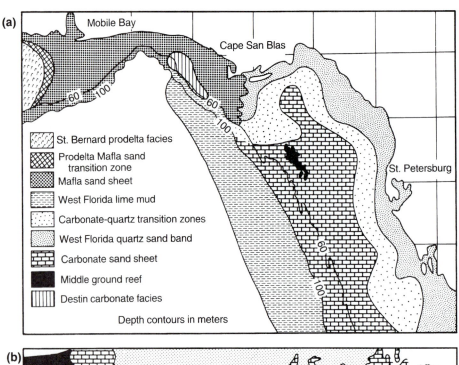

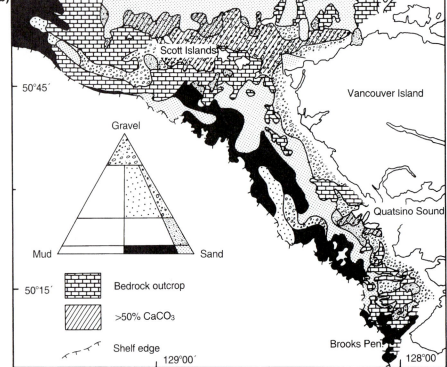

Figure 17.9 Examples of sediment distribution patterns in shelf environments. (a) Mississippi, Alabama, Florida (MAFLA)–Gulf of Mexico shelf. Destin carbonate facies is a shell, algal, foram sand. (Redrawn from Doyle and Sparks, 1980). (b) Shelf along part of the West Coast of North America, northern Vancouver Island, British Columbia. (after Bornhold and Yorath, 1984)

Storms create sediments that yield conglomerate and skeletal packstones, and form hummocky cross-stratification (figure 17.11b)(Harms et al., 1975; Kreisa, 1981)[37].

The advance of a glacier into the ocean modifies the depositional processes, both of the glacier and of the marine environment. Here, glacial, glaciomarine, and marine sediments are intimately related (R. D. Powell and Molnia, 1989).[38] Notably, glacial diamicts occur with turbidites. The glaciomarine sediments locally include a varvelike, laminated lithofacies consisting of coarse-fine couplets deposited from glacial sediment-enriched water (E. A. Cowan and Powell, 1990). Marine invertebrate fossils, in conjunction with diamicts, erratics, and polished and scratched clasts, reveal the glaciomarine character of the stratigraphic section. Thus, this environment differs from other shelf and glacial environments but has characteristics of both.[39]

Figure 17.10 Interbedded shelf rocks—packstone and mudshale with pelmatazoan stems; Dennis Formation(?), Pennsylvanian, Kansas City Group, I-470 at Raytown Road, Kansas City, Missouri.

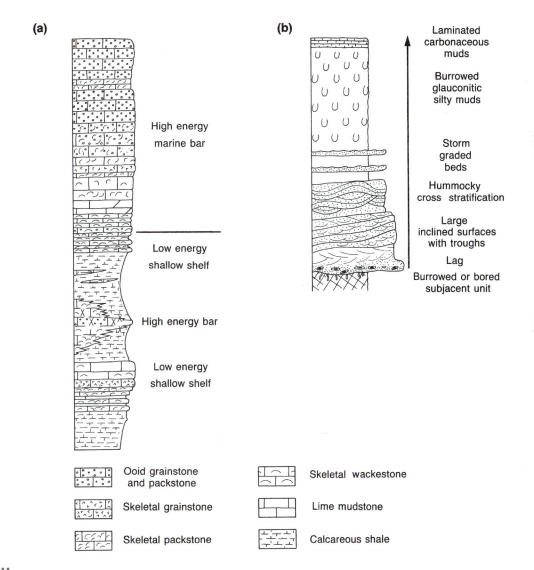

(a)

High energy marine bar

Low energy shallow shelf

High energy bar

Low energy shallow shelf

(b)

Laminated carbonaceous muds

Burrowed glauconitic silty muds

Storm graded beds

Hummocky cross stratification

Large inclined surfaces with troughs

Lag

Burrowed or bored subjacent unit

Ooid grainstone and packstone

Skeletal grainstone

Skeletal packstone

Skeletal wackestone

Lime mudstone

Calcareous shale

Figure 17.11 Shelf stratigraphies. (a) Low- and high-energy carbonate shelf rocks, Bangor Limestone, Mississippian, Reid Gap, Alabama (modified from McKinney and Gault, 1980). (b) Schematic stratigraphy of a storm-dominated (high-energy) transgressive shelf (modified from W. E. Galloway and Hobday, 1983).

333

(a)

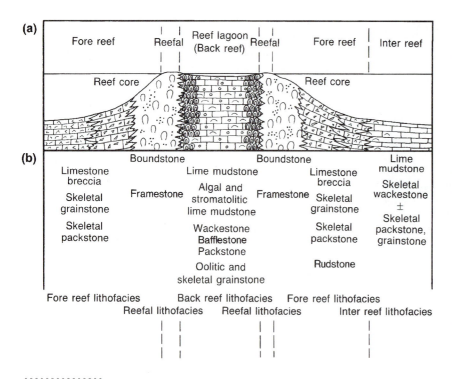

Figure 17.12 Reef environments and facies. (a) Schematic reef model showing positions of subenvironments. (b) Rock types and lithofacies common in reef subenvironments. Sketch assumes vertical growth of reef. Migration to right or left would superpose various rock types, one upon another. Rock symbols as in figure 17.11. Boundstone and stromatolite symbols added here. *(Based in part on Enos and Moore, 1983; N. P. James, 1983, 1984a, 1984b, 1984c)*

Reefs

Reefs (sensu stricto) are carbonate masses that (1) differ in nature from surrounding rocks, (2) are thicker than carbonate deposits formed nearby at the same time, (3) stand or stood higher than surrounding rocks during deposition, (4) display evidence suggesting that they are or were wave-resistant, (5) contain evidence of control over the surrounding environment, and (6) typically have a significant organic component (see Heckel, 1974).[40] Similar deposits, called mounds, banks, bars, lithoherms,[41] and Waulsortian reefs differ in their composition, environment, and structure from reefs *sensu stricto*, but similarly involve accumulations of carbonate materials thicker than those in the surrounding rocks.[42] Together, all such depositionally thickened, fossiliferous, carbonate bodies, including reefs, are assigned to the general category of **carbonate buildups.**

Reefs form in the open ocean in shallow water, high-energy areas, as fringes or rings developed on volcanoes or other prominences, and as circular or elongate forms on shelves (W. G. H. Maxwell, 1968). Over the span of geologic time, the principal reef-building animals have varied, resulting in varied reef forms.[43] Mounds, including Waulsortian "reefs" (A. Lees, 1982), developed in deeper, quieter water.

A simplified view of a reef is shown in figure 17.12. The reef environment is divided into three subenvironments—the forereef, the reefal, and the reef lagoon. Each subenvironment is characterized by different sediments and lithofacies.

The reefal subenvironment includes three parts. The subtidal to intertidal *reef front* is composed of organisms in growth position (N. P. James, 1983) bound together with space-filling shell debris, pellets, and mud to form boundstone.[44] The intertidal to supratidal *reef crest* produces boundstone and skeletal and oolitic grainstone and packstone from whole organisms and their reworked parts. Ancient reefs may include crystalline limestone here (V. Schmidt, 1977). In the geologic record, reef front and reef crest rocks typically form thick to massive beds. The *reef flat* is a subtidal to intertidal, quiet-water to high-energy wave and surf zone that yields laminated lime mudstones, grainstones, packstones, limestone breccias, rudstones, and floatstones.

Reef lagoons are similar to other lagoonal environments except that the sediments commonly are entirely carbonate. Where reefs fringe island arcs or other continental areas, however, siliciclastic components are present (Polsak, 1981). The lagoonal waters are quiet, resulting in laminated stromatolitic lime mudstones to thin-bedded wackestones, but local waves and currents may generate sands that yield packstones and grainstones. Bioturbation locally disrupts the bedding (R. W. Scott, 1979).

The *forereef* subenvironment forms the slope seaward of the reef.[45] This subenvironment may interfinger with carbonate slope or basinal environments (Enos and Moore, 1983; N. P. James, 1983; Pomar, 1991). Nearest the reef, talus (carbonate boulder breccia) may blanket the base of the slope. Where breccias are absent, lime mudstone and packstone blanket the lower slope (Wilber, Milliman, and Halley, 1990; Pomar, 1991). Seaward from the talus, carbonate-mass flow deposits and turbidites yield various conglomerates, breccias, grainstones, and packstones. These become finer grained seaward and are replaced in the stratigraphic column by wackestones and mudstones of the slopes and basin floors.

Slope-Rise Environments

The **submarine slope** is the steep part of the seafloor that extends from the shelfbreak to the **rise,** a zone of gentler gradient at the base of the slope (1500–4000 m)(figure 17.8). Slopes are commonly inclined from 1 to 10°, but locally reach inclinations of 15° or more (H. E. Cook and Mullins, 1983; W. E. Galloway and Hobday, 1983). The rise is a broad, gently sloping transition zone between the slope and a trench, a basin, or an abyssal plain.[46]

Subenvironments recognized within the slope-rise environment, include the *open slope-rise* subenvironments that occur along uniformly inclined ocean margins and *slope basin* subenvironments, developed where folding, faulting, or both processes have created a basin on the slope. Included in the latter category are the forearc basins of island arc systems.

Three main types of processes lead to the deposition of sediment in both slope-rise subenvironments. Pelagic **sediment rain** is simply the falling of sediment, including siliciclastic, allochemical, and precipitated sediment, from the water column onto the seafloor. **Current deposition** is bedload or suspended load deposition by currents, as they lose their ability to transport the sediment. **Mass deposition** is deposition that results from grain flows or mass flows. Interlayering of sediments produced by each of the three depositional processes occurs in various ways that depend on details of the specific conditions in the subenvironment.[47] In the open slope-rise subenvironment, fine-grained, parallel, thin-bedded or laminated sediments (mud, hemipelagic mud, or lime mud) are typical.[48] These may contain a large component from sediment rain and additional material moved onto the slope from the shelf. Geostrophic or *contour currents* (currents that flow parallel to the slope), as well as various bottom currents, may rework older sediment or deposit additional lime sand or mud, cross-bedded sand, or siliciclastic mud (Heezen, Hollister, and Ruddiman, 1966; Rona, 1969; Bouma and Hollister, 1973; Stow and Lovell, 1979). Turbidity currents deposit both typical Bouma sequence turbidites and laminated mud turbidites (Stow and Shanmugam, 1980; Heller and Dickinson, 1985; Coniglio and James, 1990). Mass deposition creates olistostromes and related rocks (Abbate, Bartolotti, and Passerini, 1970; Jacobi, 1976; and Cook, 1979).[49] Idealized sections of slope-rise sedimentary rocks contain (1) thin-bedded, laminated mudrocks or lime mudstones with some interlayered wackes or wackestones, (2) thin- to thick-bedded, locally graded, turbidites, and (3) a few olistostromal diamictites. Sedimentation in slope basins locally involves a greater contribution from turbidity currents and mass deposition, but mud dominates in many basins.[50] Cherts, tuffs, and other rocks also form locally within slope basins (e.g., M. Earle, 1983).

Submarine Canyons and Trenches

Submarine canyons are valleys eroded into the shelf and slope. They generally serve as channels along which sediment moves from shallower to deeper levels in the marine environment. Consequently, submarine fans develop where canyons open out onto rises, into slope basins, into trenches, or onto the abyssal plains. Currents in the canyons flow both up and down canyon (Shepard and Marshall, 1978), but turbidity currents and mass movements move down canyon.

Submarine canyon sediments are varied.[51] They include olistostromes; thick-bedded, pebbly, coarse to medium sandstones; medium-bedded to laminated, rippled, and convoluted wackes and siltstones; cross-bedded arenites and wackes; and mudrocks (figure 17.13b).

Trenches are deep valleys that lie on the ocean floor adjacent to continental margins or island arcs. Tectonically, they represent convergent plate boundaries along which crust is subducted. In terms of sedimentary environments, trenches include trench floor and trench slope environments. Submarine fans occur in both.

The Submarine Fan Model

Submarine fans vary tremendously in size. Many mature modern fans are mud-rich and contrast with ancient sandstone-rich fans (N. E. Barnes and Normark, 1985; Shanmugam and Moiola, 1988).[52] Nevertheless, general submarine fan and fan facies models (figure 17.13), developed in the 1970s by Normark, Ricci-Lucchi, Mutti, and Walker,[53] are widely adopted (though they have been questioned).[54] In the fan model, the fan is subdivided into inner-fan, mid-fan, and outer-fan regions (figure 17.13a).

Although the fan itself is subdivided into three parts, six environments of deposition are present. The first of these is the channel or canyon floor environment, an environment largely restricted to inner- and mid-fan regions. Between channels, there are interchannel or overbank areas. At the channel ends, channels merge into fan lobes, features present in the mid-fan to outer-fan regions.[55] Beyond the lobes is the fan fringe, which is bounded more distally by the basin plain. The sixth environment is the slope-base environment, where mass flows and submarine landslides interfinger with a variety of sediments deposited in the various fan environments. Each environment is characterized by particular lithofacies.

Submarine Fan Facies

Mutti and Ricci-Lucchi linked lithofacies to submarine fan environments in their fan facies model (Mutti and Ricci-Lucchi, 1972).[56] Of course, the *grain size* and the actual *composition* of the sediments in any environment depend on the material in the source area and the control exercised by the transportation agent (G. V. Middleton and Hampton, 1973; R. G. Walker, 1975; Aalto, 1976). For example, along the slope and the canyon that feeds the fan, submarine landslides and mass flows develop, but inasmuch as the slope may be underlain by any type of rock, the composition of the landslides will be variable and dependent on the local geology. The slope, however, is generally coated with a layer of fine sand, silt, or hemipelagic mud, and these materials will be incorporated into or form the matrix of mass flows. Similarly, Bouma sequence turbidites found in fan channels and lobes have compositions controlled by the provenance and typically consist of siliciclastic mudrocks, sandstones, and conglomerates composed of sediment derived from orogenic zones landward of the canyon. Yet mudrock, carbonate and chert turbidites with nonorogenic sources are known as well. Other fan rocks include olistostromal diamictites, massive sandstones and conglomerates, and mudrocks, the latter deposited by sediment rain on the slope, over the fan, and on the

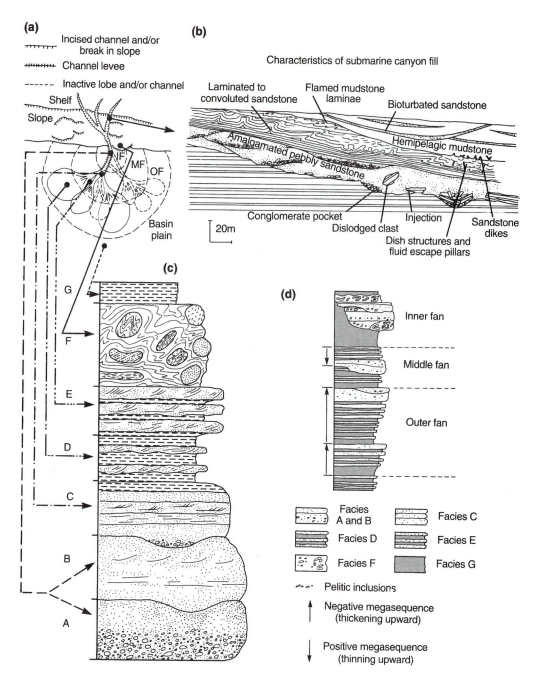

Figure 17.13 Submarine fan with channel and fan lithofacies. (a) Map view of model submarine fan (after Ricci-Lucchi, 1975; modified from Ingersoll, 1978c). (b) Submarine canyon fill (from May, Warme, and Slater 1983). (c) Section showing Mutti and Ricci-Lucchi submarine fan facies A through G (based on Mutti and Ricci-Lucchi, 1972, 1975; Ingersoll, 1978c). (d) Idealized stratigraphic section through a submarine fan (redrawn from Mutti and Ricci-Lucchi, 1972).

basin plain. A simplified version of idealized submarine fan facies composed of these various rocks is presented in figure 17.13c. Following Mutti and Ricci-Lucchi (1972), the facies are assigned letter designations A through G.

Facies A typically consists of coarse conglomerate, conglomerate, pebbly sandstone, or sandstone. These rocks occur in beds that range from 0.3 to 15 m thick that lack any mudrock laminations (figure 17.14). Shale chips and crude laminations comprise the main structures of the beds. Facies A rocks were deposited in inner-fan channels by grain flows, mass flows, and turbidity currents.[57]

Medium- to coarse-grained sandstones usually comprise Facies B. Beds are 30 to 200 cm thick and contain shale chips, laminations, and dish structures (figures 17.13, 17.14). Very little mudrock occurs in this Facies, which is considered to be deposited by grain flows or mass flows. Facies B rocks occur in inner-fan and mid-fan channels.

Although neither Facies A nor B contains the turbidite features of the Bouma sequence, Facies C is characterized by complete Bouma sequences. Sand, silt, and mud interbeds together form layers 10 to 300 cm thick. Sole marks, shale chips, parallel laminations, convolute laminations, and ripple cross-laminations are characteristic structures. Facies C sediments are deposited by turbidity currents, primarily on the fan lobes of the midfan and outer fan.

Facies D and E both contain partial Bouma sequences. The rocks are typically sandstone, siltstone, and shale that form layers 3 to 20 cm thick. Facies D is generally dominated by mudrock. It occurs mainly on the fan fringe of the midfan and outer fan, but locally is developed in interchannel areas. Sand (and silt) dominate in Facies E. Beds of this facies are thin, usually falling in the 3–20 cm range (figure 17.14e). Facies E rocks are basically overbank and interchannel deposits in many areas of the fan, including the outer fan.

Facies F occurs in fan sequences and elsewhere. It may be associated with the fan sequence in channels or may locally interfinger with other fan facies along the slope edge or inner fan (Normark and Gutmacher, 1988). Facies F consists of olistostromal diamictites (with characteristic block-in-matrix structure) that form masses of up to 300 m thick. Soft-sediment folds and faults, boudinage, and sedimentary dikes are also typical structures. Facies F rocks are produced by gravity sliding and mass flows, and occur with and grade into Facies A and B rocks in the fan channels.

Normal marine muds comprise Facies G. Mudstone and lime mudstone beds of this facies are a few millimeters to 60 cm thick and have parallel layering (figure 17.14g). Sediment rain of fine-grained suspended particles produces the layering, which will develop anywhere (slope, channel, lobe, basin plain) that other, more voluminous sedimentation does not flood the environment.

The occurrence of these various facies in the trench and trench slope is shown in figure 17.15. Note that turbidites, as well as Facies G and F rocks, also occur in the slope-rise environments, particularly in slope basins. In addition, fans are found on the abyssal plains. Thus, these facies are restricted neither to fans *sensu stricto* nor to the trench environment.

Trench Slope and Trench Floor Environments

Sedimentation on trench slopes is basically like that in slope-rise environments (Scholl and Marlow, 1974; Underwood and Bachman, 1982). Trench slope basins, like other submarine basins, contain pelagic or hemipelagic mudrocks, turbidites, grain flow deposits, and olistostromes.[58] The compositions and structures of these rocks are like those from similar environments except that siliciclastic rocks are commonly volcanic-rich (e.g., volcanic lithic wacke).

Trench floors are dominated by sands and muds of submarine fan facies turbidites, sands deposited or reworked by bottom currents, and diamicts produced by mass flows (figure 17.15) (Speed and Larue, 1982; Thornburg and Kulm, 1987).[59] Pelagic and hemipelagic, structureless to laminated muds and volcanic ash layers form local interbeds.[60] At high latitudes, glaciomarine deposits may be present. Some trench floors are "starved" of sediment because the trench is cut off from major clastic sediment sources by structural barriers such as folds, fault blocks, or slope basins, which intercept sediment moving downslope. Like rises, such trenches dominantly contain muds that become mudrocks, plus very minor sands and olistostromes.[61]

Pelagic Environments

Pelagic environments are tectonically passive environments of deposition that lie below the deep sea or ocean.[62] Two major types of pelagic environment are recognized—pelagic basin/abyssal plain and oceanic plateau. *Pelagic basins* are environments on the sea or ocean floor that are elongate to circular areas essentially surrounded by hills or ridges. *Abyssal plains* occur as broad, flat pelagic environments found seaward of the continental rise and lacking enclosing hills. Both are low-lying areas. *Oceanic plateaus* are large elevated areas in the deep oceans that rise above adjoining abyssal plains.

The sediments of pelagic environments are, by definition, dominantly fine-grained.[63] The main types include variously colored muds, clays, and oozes (W. H. Berger, 1874; Stow and Piper, 1984; Dean, Leinen, and Stow, 1985). **Hemipelagic muds** are silty sediments with some biogenic material (dominantly calcareous and siliceous microfossils) and more than 25% material of greater than 5-micron size derived from silicate continental sources, volcanoes, or shallow marine settings. **Pelagic clays** are clay-rich sediments containing less than 50% biogenic material and less than 25% silicate fragments of greater than 5-micron size. **Oozes** are generally clay-bearing and have less than 25% siliciclastic materials of greater than 5-micron size, but are dominated by more than 50% biogenic materials. Pelagic sediments are typically laminated or thin-bedded, but in some cases bioturbation has totally destroyed the stratification, which forms via sediment rain.

The factors that control sedimentation and the resulting stratigraphy in pelagic environments include tectonic history, temperature and fertility of near-surface water, the history of the calcium carbonate compensation depth (CCD), and the paleobiologic history. Tectonic history controls both the rate and amount of siliciclastic sediment input into the pelagic environment and the basin location.[64] The temperature and fertility of near-surface waters (where most of the organisms live that generate microfossils) control the volume and type of organisms that fall to the bottom. The CCD is the level below which the solution rate of calcite exceeds the rate of supply. Where the ocean floor lies above the CCD, calcareous oozes and fossils may accumulate. Where the ocean floor is at a greater depth, siliceous fossils (e.g., radiolaria or diatoms) and other silicate materials dominate the sediment produced by sediment rain. Paleobiologic history controls the type of organism that contributes to pelagic sedimentation. For example, radiolarian cherts, common in Mesozoic rocks, may have present-day analogues in diatomaceous sediments (Jenkyns and Winterer, 1982).

The sediments of both basinal/abyssal plain and plateau subenvironments are essentially the same. Stratigraphic sections consist of sequences of interbedded mudrock, chert, chalk, and lime mudstone or their unlithified equivalents, mud, clay, and ooze.

Rift-Transform Environments

Trenches represent convergent plate boundaries. The other two types of plate boundary—spreading centers and transform faults—also serve as marine environments of sedimentation.

(a)

(b)

(c)

(d)

Figure 17.14   Submarine fan facies rocks. (a) Facies A conglomerate channel fill, Knobs Formation, Ordovician, Avens Ford Bridge area, south of Abingdon, Virginia. (b) Facies B sandstones in the Sandstone of West Fork, Franciscan Complex, Jurassic/Cretaceous, northeastern Diablo Range, California. (c) Facies C turbidites, Knobs Formation, Ordovician, near Lodi, Virginia. (d) Facies D siltstone and shale, Paperville Formation, Avens Ford Bridge, southwestern Virginia. (e) Facies E sandstone and shale, Panoche Formation, Cretaceous, Ingram Creek, Diablo Range California. (f) Facies F olistostrome, Venado Formation, Cretaceous, Putah Creek at Monticello Dam, Lake Berryessa, California. (g) Facies G shales, Paperville Formation, Holston Reservoir, northeast Tennessee.

Both environments mark sites where active faulting produces cliffs of mafic igneous rocks (see Chapter 10) that will yield volcanic epiclastic sediment.[65] These sediments, which become volcanic breccia and sandstone, are probably most abundant at the base of the stratigraphic sections. As seafloor spreading and erosion of the cliffs progress, sand deposition becomes less common, so that sandstones are replaced upward in the stratigraphic section by mudrocks and bioclastic sediment formed by sediment rain. Where either a spreading center rift or trench environment lies near a continent, the sections may contain continental siliciclastic turbidites throughout (Vallier, Harold, and Girdley 1973; Einsele, 1985). Nevertheless, mud is likely to be the dominant sediment at most localities (see Faugeres, Gonthier, and Pontiers, 1983).

In both spreading center and transform fault environments, sediment rain, mass flows, landslides, turbidity currents, and bottom currents operate as depositional agents (Faugeres and Pontiers, 1982; Faugeres, Gonthier, and Pontiers, 1983). As a result, sediments may be laminated, cross-bedded, or massive. Along rift zones of the spreading centers, mafic lava flows and intrusions occur locally within the section.

SUMMARY

Most sedimentary subenvironments are present somewhere on the Earth today and associated facies are forming in them. Consequently, we find that at any specific geologic time (such as today, the late Silurian, or the middle Cretaceous) there are

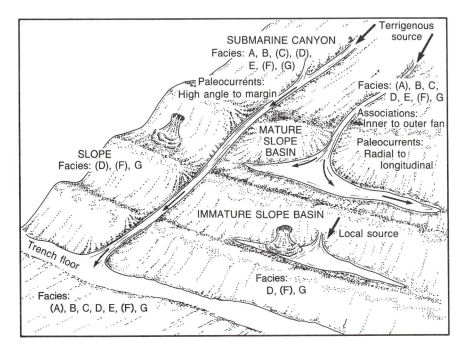

(e)

(f)

(g)

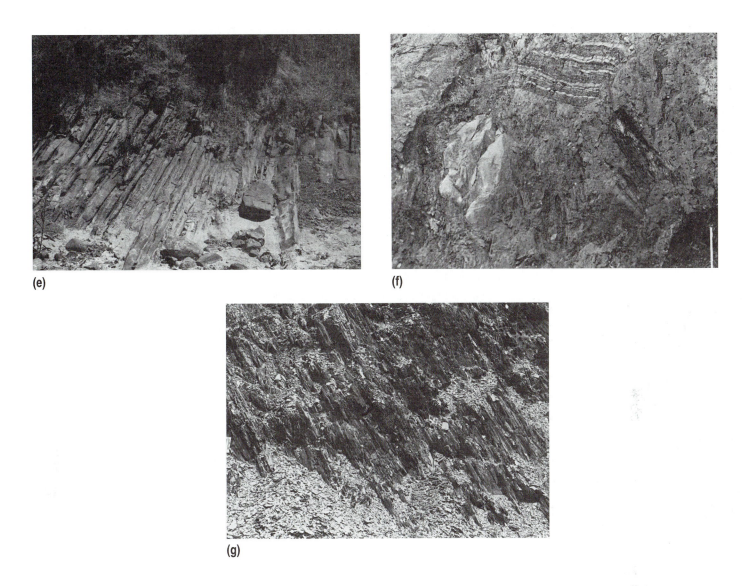

SUBMARINE CANYON
Facies: A, B, (C), (D), E, (F), (G)

Terrigenous source

Paleocurrents: High angle to margin

Facies: (A), B, C, D, E, (F), G

Associations: Inner to outer fan

Paleocurrents: Radial to longitudinal

SLOPE
Facies: (D), (F), G

MATURE SLOPE BASIN

IMMATURE SLOPE BASIN

Local source

Trench floor

Facies: D, (F), G

Facies: (A), B, C, D, E, (F), G

Figure 17.15
Trench slope and trench environments showing a submarine canyon, slope basins, trench floor, and the submarine fan facies associated with various trench slope environments. (after Underwood, Bachman, and Schweller 1980; Underwood and Bachman, 1982)

Table 17.2 Summary of Sedimentary Environments and Their Facies Characteristics

Environment	Facies	
	Common Rock Types	**Notable Characteristics**
Continental		Fossils of land plants and animals
Meandering river	cg, ss, sltst, sh	Fining upward cycles, xbdd ss and cg, channels, lenticular-linear units
Braided river	cg, ss, sh, sltst	xbdd cg and ss, channels, lenticular deposits, slight fining upward cycles
Glaciofluvial	cg, ss, sh, sltst	xbdd ss and cg with striated clasts
Glacial	tillite	striated clasts; poor sorting
Landslide	breccia	lacks sorting; angular clasts
Alluvial fan	fgl, cg, ss, sltst, sh, dm	xbdd cg and ss with dm interbeds; coarsening upward sequence with fining upward cycles
Erg	ss	xbdd ss with tabular laminated ss bds; well-sorted, frosted sand grains
Playa/sabkha	evaporites, sh, sltst, ss, ls, dlst, ch	thin-laminated, tabular bds; mudcracks
Lacustrine	sh, sltst; local ss, cg, ls, coal	varves ± turbidites
Cryolacustrine	sh, sltst, ss, cg, dm	varves, striated dropstones, dm interbeds ± turbidites
Inland basin	fgl, br, cg, ss, sh, sltst ± dm	interbed river, lake and landslide deposits
Spelean (cave)	ls	dripstones
Deltaic	ss, sltst, sh, coal, and mdst; local cg, and dm	xbdd ss; coal; coarsening upward sequences
Transitional		
Deltaic	ss, sltst, sh, coal, and mdst; local cg, and dm	xbdd ss; coal; coarsening upward sequences
Estuarine	ss, sltst, sh ± cl	xbdd ss; wavy to parallel laminated mudrocks; limited faunal diversity

Abbreviations: bddg = bedding, bds = beds, br = breccia, bst = boundstone, cg = conglomerate, ch = chert, cl = coal, dlst = dolostone, dm = diamictite, evap = evaporites, fgl = fanglomerate, gst = grainstone, lmst = lime mudstone, ls = limestone, mdst = mudstone, pkst = packstone, SFF = Submarine Fan Facies, sh = shale, sltst = siltstone, ss = sandstone, xbdd = cross-bedded, xbds = cross-beds.

lateral facies equivalents. Today, meandering river facies are forming in Mississippi, while various deltaic facies form in Louisiana, clastic shelf facies form along the west coast of Florida, and carbonate shelf facies form on the southern Florida shelf. These facies interfinger and grade into one another. Similarly, in the past, lateral facies changes have existed and are mappable, both on the surface and in the subsurface.[66] In any specific location, either facies are developing and leaving a geologic record *or* erosion is occurring, leaving a gap in that record. Because the Earth is dynamic, the environments and facies, as controlled by tectonism,[67] will change over time leaving a vertical record of facies changes in the stratigraphy.

Thinking back over the rock associations described in this chapter, one can realize the tremendous range of sedimentary rocks and rock sequences that develop over the gamut of continental, transitional, and marine environments. These environments grade into each other. Also, each environment may be subdivided into subenvironments, each of which is characterized by a distinct sedimentary facies. As we study the rock record, we identify rocks, fossils, and structures and use them to reconstruct the facies. By comparing the reconstructed facies with facies of known environments, we are able to interpret ancient sedimentary environments (table 17.2).[68]

Environment	Facies	
	Common Rock Types	Notable Characteristics
Transitional		
Lagoonal	sh, mdst, sltst, ss, coal, lmst, pkst, gst	bioturbated mdst; wavy bddg ± xbdd or laminated ss; limited faunal diversity; stromatolites
Littoral	ss, stlst, gst, pkst, dlst, local cg	low and high angle xbdd ss ± sltst with marine fossils
Tidal flat	ss, sltst, mdst, sh, lmst, pkst, dlst, ch, flat-pebble cg	laminated to lenticular bddg; bioturbation; limited biological diversity; stromatolites; mudcracks; evaporites; xbds in tidal channels
Marine		marine fossils
Glaciomarine shelf–shallow sea	ss, dm, sltst	interlayered ss and dm with marine fossils; ± turbidites and laminites; dropstones
Open low- to high-energy (shelf)	sh, lmst, ss, gst, dlst	abular-parallel bds; xbdd ss; bioturbation; diverse fauna
Restricted	sh, sltst, ss	laminated and thin bdd, ± pyrite
Reefs	bst, ls br, with pkst, gst, lmst, sh	fossils in growth position; locally thick ls; stromatolites
Slope-rise	sh, sltst, ss, ± dm	SFF G, laminated ss-sh beds or SFF B-E with local SFF F
Submarine canyon	dm, ss, sltst ± sh	SFF A, B; xbdd ss; linear-lenticular units
Trench	ss, sltst, sh, cg, ± dm	SFF A–G, lenticular to tabular units
Rift-fracture zone	mafic volcanic breccia, ss, sltst, sh, ch, ls, dm	SFF A–G with basalt interbeds
Pelagic (abyssal)	mdst, sh, ch, lmst, ss, sltst	tabular units; laminated SFF G and other fine-grained rocks, ± SFF B–E

EXPLANATORY NOTES

1. See for example, Pettijohn (1957), Krumbein and Sloss (1953), and Blatt, Middleton, and Murray (1980). Selley (1976) uses a somewhat different form of threefold classification because (1) many types of present-day environments have not produced large volumes of rock in the past, and (2) water depths commonly used to subdivide the marine environments are not easily determined in ancient rocks. Pettijohn (1975) uses five major categories of environment.

2. See W. E. Galloway and Hobday (1983). Miall (1977a,b) defines four categories, lumping those here called straight and sinuous into one "straight" category. Brierley (1989) cautions against the use of channel patterns for facies analysis.

3. For example, see papers in Collinson and Lewin (1983) and Ethridge, Flores, and Harvey (1987), and the references therein.

4. For example, see Lattman (1960), J. C. Harms, MacKenzie, and McCubbin (1963), J. R. L. Allen (1965, 1970b), Selley (1976), plus G. Taylor and Woodyer (1978), R. G. Jackson (1978), Nijman and Puigdefabregas (1978), and Puigdefabregas and van Vliet (1978) in Miall (1978a), and Haszeldine (1983), Nanson and Page (1983), T. E. Moore and Nilsen (1984), R. G. Walker and Cant (1984), and Collinson (1986a).

5. Crevasse splay deposits also form on deltas. See figure 17.6 for an illustration.

6. For examples, see Plint (1983), Lorenz (1987), and D. G. Smith (1987).

7. Also see R. E. Hunter (1977), McKee and Bigarella (1979), Ahlbrandt and Fryberger (1982), McKee (1982), Collinson (1986b), Clemmensen (1987), and Chan (1989). Brookfield (1984) presents a summary of aeolian facies models.

341

8. Glennie (1970), Reineck and Singh (1975), McKee and Bigarella (1979), Ahlbrandt and Fryberger (1980, 1982), Clemmensen and Abrahamsen (1983), Collinson (1986b), Lancaster (1988), Clemmensen, Olsen, and Blakey (1989).

9. For example, Ahlbrandt and Fryberger (1980, 1982), Clemmensen and Abrahamsen (1983), and Brookfield (1984).

10. See Bull (1963, 1964, 1972), Hooke (1967), Miall (1977a, 1978b), Heward (1978), Rust (1978), Rachocki (1981), Nilsen (1982), W. E. Galloway and Hobday (1983), Rust and Koster (1984), Shultz (1984), Nilsen (1985), Collinson (1986a), Lorenz (1987), and Went, Andrews, and Williams (1988).

11. Flint (1957, 1971), Goldthwait (1971), Boulton (1972), M. B. Edwards (1978, 1986), D. E. Lawson (1981), Easterbrook (1982), N. Eyles, Eyles, and Miall (1983), J. Shaw (1985), Mullins and Hinchey (1989).

12. For example, see Augustinus and Terwindt (1972) in Reineck and Singh (1975), Banerjee and McDonald (1975), Rust and Romanelli (1975), Saunderson (1975), and N. Eyles and Miall (1984).

13. Flint (1971), Boothroyd and Ashley (1975), Boothroyd and Nummedal (1978), J. J. Clague (1975b), Rust and Romanelli (1975), T. Zielinski (1980), Easterbrook (1982), Gustavson and Boothroyd (1987), and Mustard and Donaldson (1987) describe glaciofluvial sediments and structures in more detail. Also see articles in Jopling and McDonald (1975).

14. See Flint (1971), Ashley (1975), Gustavson (1975), Gustavson et al. (1975), J. Shaw (1975), and Leonard (1986) for additional discussion of sediment and structures in cryolacustrine environments.

15. Several other origins have been assigned to Loess (see Smalley, 1975b, and Lugn, 1960, 1962).

16. This tendency is widely recognized. See Fouch and Dean (1982), Dean and Fouch (1983), P. J. W. Gore (1989), and the references cited therein.

17. For additional discussions of lacustrine sediments see Fouch and Dean (1982), Hazeldine (1984), P. A. Allen and Collinson (1986), and P. J. W. Gore (1989).

18. Megabreccias are breccias containing clasts dominantly >1 m in length. Clasts can range from one to hundreds of meters.

19. Because of the occurrence of coal and oil in deltaic deposits, deltas have been the object of numerous studies. Among the many works that focus on deltas or review their characteristic deposits are Scruton (1960), M. L. Shirley and Ragsdale (1966), J. R. L. Allen (1970a), J. P. Morgan and Shaver (1970), G. Briggs (1974), Weimer (1975), T. Elliott (1978b, 1986a), J. M. Coleman (1982), J. M. Coleman and Prior (1982), W. E. Galloway and Hobday (1983), Miall (1984), Bjerstedt and Kammer (1988), and Tye and Coleman (1989).

20. Also see Scruton (1960), A. C. Donaldson, Martin, and Kanes (1970), Tankard and Barwis (1982), and Tye and Coleman (1989), as well as reviews by Reineck and Singh (1975), Elliott (1978b, 1986a), J. M. Coleman and Prior (1982), W. E. Galloway and Hobday (1983), and Miall (1984). The stratigraphy of the several subenvironments of the delta plain and delta front, as reviewed here, is discussed in these works.

21. Ferm et al. (1967), Ferm (1974), Edmunds et al. (1979), Ettensohn (1980).

22. For further discussion of estuarine environments and sediments, see Land and Hoyt (1966), J. R. Schubel and Pritchard (1972a,b), and summaries of the work of Terwindt (1971) and Howard et al. (1975) in Elliott (1978a), plus Kulm and Byrne (1967), Nichols (1972), Buller, Green, and McManus (1975), Cronin (1975), Clifton and Phillips (1980), Clifton (1982), W. Miller (1982), Darmoian and Lindqvist (1988), and Horne and Patton (1989). Reineck and Singh (1975) and Elliott (1978a) review the characteristics of lagoonal environments. Rusnak (1960) and R. H. Parker (1960) report original work on northwest Gulf of Mexico localities.

23. For example, see Colby and Boardman (1989) and chapter 21.

24. Shepard (1963), Logan, Read, and Davies (1970), Ginsburg (1975), Reineck and Singh (1975), Elliott (1978a, 1986b), Till (1978), Kraft et al. (1979), Bouma et al. (1982), Weimer, Howard, and Lindsay (1982), Reinson (1984), A. D. Short (1984), V. P. Wright (1984), Hennessy and Zarrillo (1987), Duffy, Belknap, and Kelley (1989).

25. For example, see Shinn (1983) and Gunatilaka (1986). Warren and Kendall (1985) describe the differences between marine sabkha and marine salina sediments.

26. Also see Ginsburg (1975), Hardie (1977), and G. deV. Klein (1977).

27. Lag deposits are accumulations of coarser materials that are left behind (lag), where currents remove fine-grained materials.

28. Tidal flat sediment and stratification are discussed or reviewed in Reineck (1967), Hardie and Ginsburg (1977), G. deV. Klein (1977), Weimer, Howard, and Lindsay (1982), Shinn (1983), N. P. James (1984c), V. P. Wright (1984), Duc and Tye (1987), Frey et al. (1989), and Cloyd, Demmico, and Spencer (1990).

29. For example, Duc and Tye (1987) and Cloyd, Demmico, and Spencer (1990).

30. Logan and Cebulski (1970), Weimer, Howard, and Lindsay (1982), and Shinn (1983).

31. See Barwis (1978), Moslow and Heron (1979), McCubbin (1982), Inden and Moore (1983), Reinson (1984), and the references therein.

32. See W. O. Thompson (1937), Davidson-Arnott and Greenwood (1976), Kent, Van Wyck, and Williams (1976), Barwis (1978), Moslow and Heron (1979), and Kraft et al. (1979). For discussions of particular subenvironments and progressive changes in sediments and their structures from backshore to offshore, see Brenner and Davies (1973), Kent, Van Wyck, and Williams (1976), Barwis (1978), Elliott (1978a, 1986b), Moslow and Heron (1979), Kreisa (1981), McCubbin (1982), Shinn (1983), Reinson (1984), and Short (1984).

33. Although shelf sedimentation has been reviewed by several workers—including H. D. Johnson (1978), Sellwood (1978), Bouma et al. (1982), Bridges (1982), Stride et al. (1982), Enos (1983), Galloway and Hobday (1983), J. L. Wilson and Jordan (1983), R. G. Walker (1984c), and Johnson and Baldwin (1986); and Sellwood (1986)—models of shelf sedimentation and stratigraphy are not adequately developed (R. G. Walker, 1984c; Johnson and Baldwin (1986); and Sellwood, 1986). Subdivisions of environments differ, as do interpretations. Nevertheless, the works listed above serve as a basis, both for the discussion presented here and for focusing future studies. For some specific, studies, see Kulm et al. (1975), Doyle and Sparks (1980), the papers in Nittrouer (1981), Bouma et al. (1982), Enos (1983), J. L. Wilson and Jordan (1983), Saito (1989a,b,c), and Saito, Nishimura, and Matsumoto (1989).

34. See chapter 21 for a definition and discussion.

35. For discussions of sediment types, structures, and lateral variations, see Shepard and Moore (1955), R. H. Parker (1960), Reineck (1967), Kent, Van Wyck, and Williams (1976), Moslow and Heron (1979), Doyle and Sparks (1980), Bouma et al. (1982), J. C. Harms, Southard, and Walker (1982), Otvos (1982), Stride et al. (1982), Enos (1983), Galloway and Hobday (1983), Halley, Harris, and Hine (1983), J. L. Wilson and Jordan (1983), and J. M. Hurst, Sheehan, Pandolfi (1985). M. L. Irwin (1965) and J. L. Wilson (1974) fit sediment character into general models for shallow marine sedimentation. J. F. Read (1980a) provides a description of equivalent Ordovician rocks, and J. M. Hurst, Sheehan, and Pandolfi (1985) describe a Silurian example. Siliciclastic bay sediments are described by Shepard and Moore (1955) and Rusnak (1960). Also see the brief description of subtidal lagoonal sediments in Reineck and Singh (1975). Enos (1983) and J. L. Wilson and Jordan (1983) briefly mention sedimentation in restricted carbonate environments. Restricted environments also occur in association with reefs as discussed in the next section.

36. See Rooney and Kerr (1967) and the several papers in Bentor (1980) for more information on phosphatic rocks. Triplehorn (1966) discusses the origin of glauconite.

37. J. Simpson (1987) describes unusual storm generated wackestones.

38. Also see Mustard and Donaldson (1987) and Josenhans and Fader (1989).

39. See Easterbrook (1982), N. Eyles and Miall (1984), and M. B. Edwards (1986) for reviews and Blondeau and Lowe (1972), Schwab (1976), and Eyles and Eyles (1984), for discussions of some ancient examples. Also see R. D. Powell (1984), Elverhoi (1984), E. A. Cowan et al. (1988), R. D. Powell and Molnia (1989), and E. A. Cowan and Powell (1990), and see Rust and Romanelli (1975) for a description of similar subaqueous, cryolacustrine deposition.

40. This definition is slightly modified from that of Heckel (1974). There are considerable differences in opinion about how reefs should be defined, as reviewed by Heckel (1974). Furthermore, the varied nature of reeflike deposits and their distinct differences that are dependent on geological age (Heckel, 1974; Longman, 1981a; and N. P. James, 1983, 1984b) make the subject of reefs a complicated one. The brief review presented here is based primarily on the studies and reviews by W. G. H. Maxwell (1968), Krebs and Mountjoy (1972), Heckel (1974), Krebs (1974), Purdy (1974), J. L. Wilson (1974, 1975), R. W. Scott (1979), Burchette (1981), Longman (1981a), Polsak (1981), Hine and Mullins (1983), N. P. James (1983, 1984b), Enos and Moore (1983), Sellwood (1986), Ausich and Meyer (1990), and Pomar (1991). The interested student should also refer to other important works and collections including Textoris and Carozzi (1964), Laporte (1974), AAPG (1975), Hileman and Mazzulo (1977), S. H. Frost, Weiss, and Saunders (1977), Lees and Conil (1980), Toomey (1981), and Bolton, Lane, and LeMone (1982).

41. Newmann et al. (1977).

42. Also see the discussion of clinothems in DeVaney et al. (1986).

43. Heckel (1974), Longman (1981a), N. P. James (1983, 1984b).

44. Recall that recently, because some workers feel that the single term boundstone does not adequately reflect the diverse sediments of the reef, biogenic carbonate rocks have been given additional names such as floatstone, bafflestone, and lettucestone (Embry and Klovan, 1971; N. P. James, 1983; Cuffey, 1985).

45. Enos and Moore (1983) give a good review of this particular subenvironment. Also see McIlreath and James (1984) and Pomar (1991).

46. The slope-rise environment and sedimentation are discussed by Bouma and Hollister (1973), K. O. Emery (1977), Cremer, Faugeres, and Poutiers (1982), F. J. Hein (1985), Surlyk (1987), and Pickering, Hiscott, and Hine (1989).

47. For additional details on open slope-rise sediments, consult the reviews and studies of H. E. Cook, Field, and Gardner (1982), H. E. Cook and Mullins (1983), Galloway and Hobday (1983), McIlreath and James (1984), Hein (1985), Heller and Dickinson (1985), and Surlyk (1987) which, in part, serve as a basis for this review. Also consult Damuth (1977), Sheridan et al. (1982), Barrett and Fralick (1989), and various Initial Reports of the Deep-Sea Drilling Project (DSDP) and its successor, the International Project for Ocean Drilling (IPOD).

48. For example, see Cremer, Faugeres, and Poutiers (1982), J. C. Moore et al. (1982a,b), Surlyk (1987), and Von Huene et al. (1988). Hemipelagic mud is marine mud composed of both biogenic and siliciclastic materials in which at least 25% of the material larger than 5 microns in size is siliciclastic (R. L. Bates and Jackson, 1980; Stow and Piper, 1984).

49. An *olistostrome* is "a mappable sedimentary slide deposit, within a normal geologic sequence, which is characterized by lithologically or petrographically heterogeneous bodies of harder rock mixed and dispersed in a matrix of prevalently pelitic, heterogeneous material" (Raymond, 1984a). For additional discussions of submarine slides and olistostromes see Flores (1955), Elter and Trevisan (1973), Embly (1976), Jacobi (1984), and Raymond (1984a). Aalto (1976) describes some of the variations found in grain flow sandstones. Slope deposits are described by Doyle and Pilkey (1979).

50. For example, see Dickinson (1974), Karig and Moore (1975), Gorsline (1978), Ingersoll (1978c, 1982), Dickinson and Seely (1979), Damuth (1980), G. F. Moore et al. (1980), Underwood et al. (1980), G. F. Moore et al. (1982), J. C. Moore et al. (1982), and McMillen et al. (1982).

51. For example, see Almgren and Schlax (1957), Heezen et al. (1964), C. H. Nelson and Kulm (1973), Martini and Sagri (1977), Cacchione, Rowe, and Malahoff (1978), D. E. Drake, Hatcher, and Keller (1978), R. M. Scott and Birdsall (1978), R. M. Carter (1979), Fischer (1979), May, Warme, and Slater (1983), and Valentine, Cooper, and Usmann (1984). Dingler and Anima (1989) discuss grain flows and their deposits in the head of Carmel Canyon. Pickering, Hiscott, and Hine (1989) review canyon characteristics and sedimentation.

52. See papers in Bouma, Normark, and Barnes (1985).

53. Normark (1970), Mutti and Ricci-Lucchi (1972, 1975, 1978), R. G. Walker and Mutti (1973), Ricci-Lucchi (1975). Pickering, Hiscott, and Hine (1989) suggest an expanded version of the fan facies classification to incorporate all marine sediments.

54. Shanmugam, Moiola, and Damuth (1985). Pickering, Hiscott, and Hine (1989) argue that the universal model is not justified, because there are so many variations in modern fans.

55. Shanmugam and Moiola (1991) discuss different kinds of lobes and their properties.

56. For additional information on submarine fans and their rocks see Bouma and Hollister (1973), C. H. Nelson and Nilsen (1974), Normark (1974), R. G. Walker (1975, 1984d), Whitaker (1976), Martini and Sagri (1977), Ingersoll (1978c, 1979), several papers in D. J. Stanley and Kelling (1978), Siemers, Tillman, and Williamson (1981), Howell and Normark (1982), the collection of papers in Bouma, Normark, and Barnes (1985), Droz and Mougenot (1987), S. Reynolds (1987), Shanmugam et al. (1988), Shanmugam and Moiola (1988), Fergusson, Cas, and Steward (1989), and W. Morris and Busby-Spera (1990).

57. Although most such deposits have generally been attributed to grain flow, mass flow, and related agents of transportation and deposition (D. J. Stanley and Unrug, 1972; G. V. Middleton and Hampton, 1973; Lowe, 1976; Aalto, 1976; Pickering, 1984), Thornburg and Kulm (1987) ascribe some relatively thick, contemporary channel fill sands in the Chile Trench to turbidity current deposition or reworking.

58. Kulm, von Huene, et al. (1973), J. C. Moore and Karig (1976), G. W. Smith, Howell, and Ingersoll (1979), G. F. Moore et al. (1980), Underwood and Bachman (1982), Taira et al. (1982), von Huene and Arthur (1982), von Huene et al. (1982), J. C. Moore et al. (1982a), S. H. Stevens and Moore (1985). Also see note 50 for other related references.

59. The literature on trench sedimentation is moderately abundant. Representative works include those of Scholl, von Huene, and Ridlon (1968), Scholl et al. (1970), von Huene (1972), Scholl and Marlow (1974), Schweller and Kulm (1978), G. F. Moore et al. (1980), Underwood, Bachman, and Schweller (1980), G. F. Moore, Curray, and Eurmel (1982), J. C. Moore et al. (1982a), McMillen et al. (1982), Erba, Parisi, and Cita (1987), Thornburg and Kulm (1987), and various papers in the Initial Reports of the Deep-Sea Drilling Project.

60. For example, McMillen et al. (1982) and Thornburg and Kulm (1987).

61. See, for example, Scholl and Marlow (1974), Underwood and Bachman (1982).

62. The word *pelagic* has been given different meanings by different authors. The definition adopted here is similar to that adopted in the *Glossary of Geology*, 2nd ed. (Bates and Jackson, 1980). Scholle, Arthur, and Ekdale (1983), however, follow Jenkyns (1978) in defining *pelagic* in a broader way as any open marine deposit, either shallow or deep.

63. The Deep-Sea Drilling Project (DSDP) and the International Project for Ocean Drilling (IPOD) have revealed considerable information about oceanic sediments. The results of those projects, as reported in *Geotimes* and the Initial Reports of the Deep-Sea Drilling Project, are a veritable storehouse of information on oceanic sediments. These, plus various summaries, reports, and reviews including M. Pratt (1968), D. K. Davies (1972), Bouma and Hollister (1973), Rothe (1973), W. H. Berger (1974), Damuth (1977), van Andel et al. (1977), Jenkyns (1978, 1986), Sheridan et al. (1982), Scholle, Bebout, and Moore (1983), D. A. Johnson and Rasmussen (1984), Dean et al. (1985), and Pickering, Hiscott, and Hein (1989) provide the foundation for this review. Students wishing more information should see these reports plus works on individual rock types such as Garrison (1972) on limestones, Calvert (1972) and Jenkyns and Winterer (1982) on cherts, Cleary and Conolly (1974) on sands, and W. H. Berger (1974) on ooze and clay.

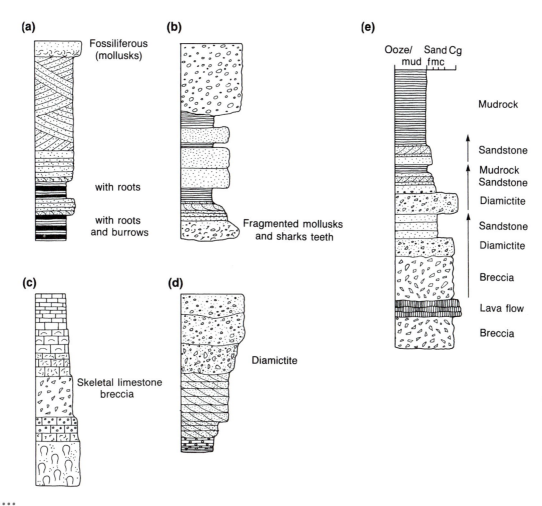

Figure 17.16
Stratigraphic sections for use in problem 1.

64. Contemporary basinal sedimentation is common in ensimatic basins (such as Lau Basin, Mozambique Basin, Shikoku Basin; see Karig and Moore, 1975), but less deformed basinal sediments in the geologic record are typically found in ensialic basins, such as the Middle Ordovician Foredeep Basin of the Southern Appalachian Orogen (Read, 1980; Shanmugam and Walker, 1980). See Pickering (1984) for a discussion of sedimentation in a fault-produced basin.

65. Sedimentation patterns and sediment in spreading center rifts and transform fault zones were revealed by several DSDP holes (e.g., site 26, Bader et al., 1970; site 35, McManus et al., 1970), but available studies are limited (see comments by Faugeres and Poutiers, 1982). See Faugeres, Gonthier, and Poutiers (1983) for a definitive study of sedimentation in the Charlie-Gibbs fracture zone and Vallier, Harold, and Girdley (1973) on the Escanaba Trough deposits. Saleeby (1979, 1982) discusses an ancient example of a probable fracture zone now preserved in an orogenic belt. Phipps (1984) contrasts theoretical stratigraphic sequences for forearc basins, rifts, and transform fault zones.

66. For example, see McKee (1949), R. C. Moore (1949), S. W. Muller (1949), and Spieker (1949) in Longwell (1949), P. Hoffman (1974), Cant (1984); and G. H. Davis (1984).

67. Dapples, Krumbein, and Sloss (1948), G. deV. Klein (1982).

68. R. C. Moore (1949); Selley (1970, 1978), Hallam (1981), Cant (1984), R. G. Walker (1984b).

PROBLEMS

17.1 Examine each of the stratigraphic columns in figure 17.16. (a) What general environment does each represent? Explain your reasoning. (b) Where possible, subdivide the column into parts representing subenvironments and label each part.

18

Mudrocks

INTRODUCTION

Mudrocks are the most abundant of the sedimentary rocks. They constitute in excess of 50% of the sedimentary record.[1] Although they are less obvious than other rock types because generally they do not form prominent outcrops, mudrocks constitute major stratigraphic units. Mudrocks occur as interbeds between other rock types in a large variety of continental stratigraphic sections, are a major constituent of estuarine and lagoonal deposits, and make up a large part of the stratigraphic record of marine origin.

Recall that mudrocks are siliciclastic sediments in which 50% or more of the particles are <1/16 mm in diameter or length. Because of the small grain size, the mudrocks are difficult to study. Yet their general mineralogical composition, consisting of clays and silt-size grains of other materials, is well known. This general mineral composition and lamination, one of the principal primary structures of mudrocks (figure 18.1), are used as the basis for classification.

TYPES AND OCCURRENCES OF MUDROCKS

The main types of mudrocks are siltstone, mudstone, mudshale, claystone, and clayshale.[2] Shales are distinguished from the mudstones and claystones by their lam-

inations. Two special types of mudrocks are worthy of mention. **Bentonite** is claystone or clayshale composed essentially of smectite (montmorillonite) group clays.[3] *Loess* is a porous, friable, commonly calcareous siltstone that forms blanket deposits.[4]

Mudrocks are widely distributed in time, space, and environments of deposition. Mudrocks are known to occur in Archean greenstone belts, which are more than 3 billion years old[5] and in Pleistocene to Holocene stratigraphies of a few thousand years in age. Chapter 17 revealed the diversity of environments in which mudrocks occur. On the continents, these include fluvial floodplains, alluvial fans, playa lakes, playas and sabkhas, swamps, caves, and lakes representing a variety of climatic, topographic, and tectonic environments. In transitional environments, mudrocks are commonly formed in deltas, estuaries, lagoons, and marshes. Mudrocks are widely distributed in marine shelf and slope deposits, as well as in those of the rise, trench, and basin plains. Considering this diversity, it is not surprising that mudrocks are abundant on all continents and are presently forming in the ocean basins.

Bentonites and loess have distinct origins. Bentonites are produced by the alteration of volcanic ash. They occur where volcanism, particularly siliceous volcanism, is or was common. For example, in parts of North America, such as in the Cretaceous Hell Creek Formation in southwestern

(a)

(b)

Figure 18.1 Outcrop views of laminated mudrocks. (a) Laminations in mudrock turbidites of the Paperville Formation, Ordovician, Highway 421 at Holston Reservoir, northeast Tennessee. (b) Laminations in red mudrocks of the Snowslip Formation, Belt Supergroup, Mesoproterozoic, Glacier National Park, Montana.

(a)

(b)

Figure 18.2 Photomicrographs of mudrocks. (a) Laminated mudrock from the Paperville Formation, Ordovician, U.S. Highway 421 at Holston Reservoir, northeast Tennessee. (XN). Q = quartz, P = plagioclase, Cl = clay minerals, C = calcite. Long dimension of photo is 1.3 mm. (b) Electron microscope photomicrograph of mudrock from Nanfen Formation, Mesoproterozoic, Waizi Commune near Benxi City, Liaoning Province, Peoples Republic of China. Long dimension of photo is 30 microns.
(Courtesy of Dr. Fred Webb Jr., Appalachian State University)

North Dakota, bentonites are common. Loess, which is wind-deposited, forms important deposits at several localities around the globe. Significant occurrences are present in the United States, China, and the former Soviet Union.[6] In the United States, loess occurs in the Great Plains, especially in Nebraska (Schultz and Stout, 1945; Lugn, 1960), is widely distributed in the Mississippi Valley, as far south as Mississippi and Louisiana (H. N. Fisk, 1951; Glass, Frye, and Willman, 1968; Matalucci et al., 1969), and is present in Alaska (Pewe, 1955). Loess is arguably the product of wind transportation and deposition of fine-grained, glacially derived sediment.[7]

MINERALOGY, CHEMISTRY, AND COLORS OF MUDROCKS

Mineralogy

The mineralogy of mudrocks is dominated by clay minerals, micas, and chlorites; quartz and feldspars; and carbonate minerals (figure 18.2, table 18.1).[8] Other important minerals include iron and aluminum oxides and hydroxides, zeolites, sulfates and sulfides, apatite, and "heavy minerals," such as

Table 18.1   Mineral Composition of Selected Mud and Mudrocks

	1	2	3	4	5
Quartz	70–80[a]	26–35	4	—	—
Alkali feldspar	—	—	tr	—	—
Plagioclase	—	26–35	5	—	—
Kaolinites	—	11–15	1	18	15
Montmorillonite	—	11–15	4	—	70
Mixed-layer phyllosilicates	—	—	—	61[b]	—
Illite	—	2–5	—	21	10
Micas	—	—	3	—	—
Chlorite	—	6–10	tr	—	5
Palygorskite	—	—	tr	—	—
Clinoptilolite	—	—	tr	—	—
Phillipsite	—	—	1	—	—
Calcite	5–10	6–10	—	—	—
Gypsum	—	—	—	—	p
Limonite	—	—	—	—	p
Goethite	—	—	—	p	—
Hematite	—	—	—	p	—
Pyrite	1–3	—	—	—	—
Microfossils	—	—	—	—	p
Other organic debris	5–10	—	—	—	p
Amorphous materials[c]	—	—	81	—	—
Other	tr	—	tr	—	—
Total	100	100	100	100	100

Sources:
1. Siliceous "siltite" (siltstone)(Permian), Dollarhide Formation, Idaho (Wavra, Isaacson, and Hall, 1986).
2. Calcareous, sandy, clayey siltstone, (Mio-Pliocene), New Zealand, sample 33401 (Ballance et al., 1984).
3. Volcanic glass-rich mud (Plio-Pleistocene), Peru Basin, Pacific Ocean (Zemmels and Cook, 1976).
4. Shale (Paleocene), Scaglia Rossa, Gubbio, Italy. Pelagic marine shale, interbedded with limestone, sample 347S (Johnsson and Reynolds, 1986).
5. Clayshale (Cretaceous), Moreno Shale, Diablo Range, California. Shelf shale, with locally interbedded arenites (L. A. Raymond, unpublished analysis).

[a] Percents estimated from XRD analysis or petrographic observations.
[b] Illite-smectite.
[c] Includes volcanic glass, allophane, biogenic silica, and organic material.
p = present, but not included in the quantitative analysis.
tr = trace.

hornblende. In addition, mudrocks contain nonminerals such as volcanic glass and the organic materials (P. E. Potter, Maynard, and Prior, 1980). In table 18.1, which lists the minerals of selected mudrocks, note the range of quartz contents (0–80%), with high values in siltstone and low values in clayshale. Amorphous materials (probably glass, but perhaps very poorly crystallized or cryptocrystalline minerals) are abundant in volcanic mud. Diagenesis renders these materials into clays, quartz, and feldspar; minerals that characterize older mudrocks (e.g., analyses 4 and 5). Note also that the clays and chlorites, although occurring in various proportions, are major constituents in all but the quartz-rich siltstone and the volcanic mud.

Clay minerals in mudrocks include kaolinite; the smectites, including montmorillonite, beidellite (an aluminous smectite), and nontronite (an iron smectite); chamosite;

illite; mixed-layer (I/S) clays; and Mg-bearing clays, including corrensite, sepiolite, and attapulgite (palygorskite). Each individual clay forms and persists under particular conditions.[9] Among these clay minerals are those that alter to illite, chlorite, and muscovite—minerals that characterize diagenetically altered and weakly metamorphosed mudrocks. For example, montmorillonite converts to illite, which may recrystallize to muscovite with increasing diagenesis or metamorphism. Corrensite alters to chlorite.

Carbonate minerals are those typical of other sedimentary rocks. They include calcite, dolomite, siderite, and ankerite. Like the micas, each of these crystallizes and persists under specific conditions.

Numerous other minerals and nonminerals are rare to abundant in mudrocks. Zeolites found in mudrocks include minerals such as phillipsite, clinoptilolite, and analcite. Iron and aluminum oxides or hydroxides include hematite, goethite, limonite, and gibbsite. Pyrite and marcasite are the common sulfides, and gypsum and anhydrite are typical sulfates. Minerals such as apatite, plus hornblende, zircon, and other heavy minerals, constitute a very minor fraction of the mineral composition of many mudrocks. Volcanic glass is common in muds, but is less so in mudrocks, as it is readily converted to zeolites, clays, and other minerals.[10] Organic materials include *kerogens* and other compounds discussed below.

In general, provenance exerts primary control over the mineralogical composition of a mudrock (Droste, 1961; Hume and Nelson, 1986).[11] For example, G. Muller and Stoffers (1974) showed that in the Black Sea, the clay mineral ratios in the sediments are related to provenance, with an illite-rich area to the north and a montmorillonite-rich area to the south. Similarly, on the Alaska shelf, Naidu and Mowatt (1983) were able to relate the compositions of sediments in various regions to particular continental sources. Also, Windom et al. (1971) were able to show that illite and montmorillonite are transported into Georgia estuaries from offshore during flood tide, and are mixed with kaolinite and montmorillonite, plus minor talc, mixed-layer clay, and vermiculite, derived from continental sources.

Transportation and the environment of deposition also influence sediment composition. In general, kaolinite tends to dominate in shoreward regions of basins and is successively replaced basinward by illite, chlorite, and palygorskite (Parham, 1966). Oxygenated waters yield oxidized minerals, such as hematite, whereas anoxic waters yield sulfidic phases, such as pyrite. Particularly in saline environments, zeolites such as erionite, chabazite, and phillipsite form (Sheppard and Gude, 1968; Surdam and Eugster, 1976).

There is some evidence that mudrock compositions have changed over geologic time.[12] Smectites are abundant in many Quaternary muds and shales, but pre-Cenozoic shales commonly contain abundant illite (> 50%) and only small amounts of smectites (tables 18.1 and 18.2). Chemical variations resulting from these mineralogical differences include an abundance of K_2O in older, as compared to younger, rocks and a corresponding dearth of CaO. These differences correlate with the diagenetic changes discussed in chapter 16, in which smectite containing calcium is converted to illite, a potassium-bearing mineral. Other explanations have also been suggested as causes of the chemical differences between younger and older mudrocks, including a change in the biologic controls of weathering over time and changes in tectonism, volcanism, and climate (E. G. Ehlers and Blatt, 1982, p. 291ff.; C. E. Weaver, 1989, p. 563 ff.).

Chemistry

The chemistry of mudrocks is a function of their mineral composition.[13] Inasmuch as mudrocks commonly contain significant quantities of both silicates and organic debris, their chemistry can be discussed in terms of both inorganic and organic aspects. Neither aspect has received adequate study.

Inorganic Chemistry

Mudrocks are siliciclastic. Consequently, their inorganic chemistry is dominated by SiO_2 (table 18.3). Because of the abundance of phyllosilicates, which contain less SiO_2 than does quartz (the dominant component of most sandstones), mudrocks are typically lower in silica than are sandstones (compare analyses 1 and 3 with analysis 4 in table 14.4, and the analyses in table 18.3 with those in table 19.2). The facts that mudrocks form in a diverse array of environments and may include a wide range of materials other than the phyllosilicates—including abundant quartz, organic debris, carbonate minerals, evaporite minerals such as halite and gypsum, phosphatic materials, and iron oxides and hydroxides—suggest that mudrock chemistry may vary widely. Such is the case.

The large chemical variability of mudrocks is suggested by the analyses presented in table 18.3. Silica may range from less than 40% to more than 80%. Similarly, alumina varies widely, from 1% to more than 22%. Either ferrous or ferric iron may dominate, but total iron oxides range from less than 1% to nearly 30%. MnO and Na_2O are commonly less than 1%, but may range to 5% or more in particular rocks. Potash generally exceeds 1%, as it is a constituent of the illites, micas, and alkali feldspars. CaO and MgO vary widely, depending primarily on the abundance of the carbonate component of the mineralogy. Where calcite or dolomite are abundant, the CaO and/or MgO contents are correspondingly high. In the case of MgO, high values may also be a function of the chlorite content. Phosphorus is high in the "phosphatic shales" (Heckel, 1977; Giresse, 1980).

Trace element analyses of shales reveal that a number of trace elements are present in significant amounts. Several sedimentary petrologists and geochemists have attempted to link specific trace element chemistries of mudrocks to particular types of shale, associated environments of deposition, or source areas.[14] For example, Vine and Tourtelot (1970) sought to characterize black shales by their trace element

Table 18.2 Comparison of Generalized Mineral Compositions of Selected Muds and Ancient Mudrocks

| | Muds | | | | | Mudrocks | | | |
	1	2	3	4	5	6	7	8[b]	9
Quartz	4[a]	A	10	11	45	22	20	10	22
Alkali Feldspar	tr	} m	5	7	5	} 9	—	} tr	} tr
Plagioclase	5		11	12	10		—		
Kaolinite	1	m	4	6	5	tr	—	—	tr
Smectites	4	A	10	6	—	—	—	—	—
Illite	—	C	26	22	15	31	36	48	48
Mixed-layer clays	—	—	21	10	5	—	9[c]	21	12
Chlorite	tr	?	6	6	—	2	15	17	6
Vermiculite	—	—	5	2	—	—	—	—	—
Biotite	—	—	—	—	—	tr	—	tr	—
Palygorskite	tr	—	—	—	—	—	—	—	—
Calcite	—	—	—	11	10	tr	15	tr	3
Dolomite	—	—	—	7	5	tr	—	tr	7
Zeolites	tr	—	—	—	—	—	—	—	—
Pyrite/marcasite	—	—	—	—	—	11	5	tr	tr
Iron oxides	—	—	—	—	?	2	—	—	—
Total organic carbon	?	?	?	—	—	22	?	<1	?
Amorphous	81	—	—	—	—	—	—	—	—
Other	tr	—	tr	tr	—	1	—	tr	—

Sources:
1. Volcanic glass-rich mud (Pleistocene-Holocene), Peru Basin, Pacific Ocean (Zemmels and Cook, 1976).
2. Texas Gulf Coast muds (Pleistocene-Holocene) (generalized from several sources; see text).
3. Gray clay, (Pleistocene-Holocene), Lake Superior (Lineback et al., 1979, with modifications from Mothersill and Fung, 1972).
4. Red varved clay, (Pleistocene), Lake Superior (Lineback et al., 1979, with modifications from Mothersill and Fung, 1972).
5. "Shale," (Eocene), Garden Gulch Member, Green River Formation, Piceance Creek Basin, Colorado, sample 28 (Hosterman and Dyni, 1972).
6. Black shale, (Devonian), composition generalized, Chattanooga Shale, (T. E. Bates and Strahl, 1957, with additions from Conant and Swanson, 1961; Ettensohn and Barron, 1982).
7. Shale (Devonian), Genessee Formation and equivalents, average of 64 samples (Hosterman and Whitlow, 1983).
8. Green to gray Devonian shale (Ettensohn and Barron, 1982; Leventhal and Hosterman, 1982).
9. Shale (Cincinnatian, Ordovician), Clarkesville Member of the Waynesville Formation (Big Bull Formation), average (Scotford, 1965).

[a]Values in volume percent.
[b]Values in this column in weight percent.
[c]Illite-smectite mixed-layer minerals.
? = probably present, but not reported.
A = abundant; C = common; m = minor.
tr = trace.

content. Bhatia (1985b) uses REE patterns to distinguish the tectonic settings of sedimentary basins. Leventhal (1983, 1987) distinguishes normal (oxygenated) marine environments from euxinic (anoxic) depositional environments on the basis of C/S (ratios) in mudrocks.[15] Walters et al. (1987) use major and trace elements, including phosphorous, iron, vanadium, chromium, nickel, and zinc, in conjunction with mineralogy to distinguish between marine and nonmarine mudrocks. These attempts have met with only modest success, although trace elements *do* seem to be useful for distinguishing between adjacent units, distinguishing facies groups, or correlating stratigraphic units within a basin (Amajor, 1987; Stow and Atkin, 1987; Walters et al., 1987).

Table 18.3 Chemical Analyses of Selected Mudrocks

	1	2	3	4	5	6	7	8	9
SiO_2	40.1[a]	46.30	58.32	60.0	61.84	61.99	66.00	74.8	76.44
TiO_2	0.76	0.48	0.48	0.73	0.83	0.89	0.11	0.38	0.63
Al_2O_3	10.9	16.11	8.59	18.1	13.40	22.25	1.30	9.1	11.25
Fe_2O_3	—	—	2.04	1.3	3.83	1.25	—	2.6	—
FeO	27.6[b]	6.33[c]	—	—	—	—	0.65[b]	—	4.45[c]
FeO	—	—	0.18	5.0	1.15	0.42	—	0.4	—
MnO	4.8	0.06	0.07	0.11	0.05	0.01	0.01	nr	0.04
MgO	3.5	3.01	3.65	2.9	2.69	1.34	2.60	1.2	0.99
CaO	0.53	16.20	8.45	1.1	2.68	0.02	16.00	1.4	3.03
Na_2O	0.12	0.35	0.72	1.8	0.97	0.10	0.20	0.34	0.24
K_2O	4.5	1.25	2.71	3.2	2.8	6.32	0.70	1.4	1.14
P_2O_5	0.14	0.50	0.05	0.17	0.44	0.03	0.16	0.5	0.14
LOI[d]	—	—	—	—	—	—	11.20	—	—
H_2O^-	2.2	—	0.52	0.64	2.45	0.10	nr	3.8	—
H_2O^+	4.8	0.58[e]	1.40	4.4	3.85	4.57	nr	3.0	0.43[e]
CO_2	0.08	7.55	12.08	0.10	2.55	0.45	nr	nr	2.11
Other	1.37	1.44	0.43	0.34	1.05	0.26	tr	0.60	0.00
Total	100.0	100.16	99.69	99.4	100.44	100.00	98.93	99.52	100.89

Sources:
1. Red Shale interlayered with chert, Franciscan Complex (Jurassic-Cretaceous), California, (E. H. Bailey, Irwin, and Jones, 1964). Analysts: P. L. D. Elmore, S. D. Botts, I. H. Barlow, and G. Chloe. Deep sea.
2. Shale, Ezeaku Formation (Cretaceous), Nigeria, sample 8 (Amajor, 1987). Passive margin shelf or slope.
3. Red sandy shale, Spearfish Formation (Triassic), South Dakota (G. B. Richardson, 1903). Transitional to shallow epicontinental sea (?).
4. Shale accompanying graywacke, Franciscan Complex (Jurassic-Cretaceous), California, sample SF-2126 (E. H. Bailey, Irwin, and Jones, 1964). Active margin slope or trench.
5. Shale, Cody Shale (Cretaceous), Wyoming, sample Sco-1. (L. G. Schultz, Tortelot, and Flanagan, 1976). Analyst: S. M. Berthold, Epicontinental marine seafloor.
6. Mudrock, Cookman Suite (Silurian-Devonian), Australia, sample MK55 (Bhatia, 1985a). Passive margin.
7. Calcareous, carbonaceous, siliceous siltite, Dollarhide Formation, (Permian), Idaho, sample Ex-1 (Wavra, Isaacson, and Hall, 1986). Active margin slope (?).
8. Siliceous shale or claystone, Pierre Shale (Cretaceous), Potter County, South Dakota, sample 259536 (L. G. Schultz et al., 1980). Epicontinental marine seafloor.
9. Shale, Asu River Group (Cretaceous), Nigeria, sample 2 (Amajor, 1987). Passive margin shelf or slope.

[a]Values in weight percent.
[b]Total iron as Fe_2O_3.
[c]Sum of FeO + Fe_2O_3.
[d]Loss on ignition.
[e]Total water reported only.
nr = not reported.

Isotopic studies have also been useful to some degree for characterizing sediments and their environments of deposition. Maynard (1980) and Gautier (1986) found that sulfur isotope ratios are related to sedimentation rate. Gautier (1986) also relates isotopic composition to specific environmental conditions. He shows that disseminated pyrite that contains "light" sulfur occurs in rocks with high organic carbon contents. These rocks are laminated (not bioturbated) and the muds apparently were deposited in anoxic environments. Bioturbated (nonlaminated) mudrocks deposited as muds in oxygenated waters are low in organic carbon and exhibit a range of sulfur isotope values.

Organic Chemistry

The organic carbon, used in conjunction with sulfur by Gautier (1986) to characterize mudrocks, is a common constituent of these rocks. In shales, organic matter averages 2.1% (Degens, 1965, p. 202), but may range to 35% or more.[16] It is organic carbon that gives rise to coal (where plant matter is concentrated in large amounts) and petroleum (where amorphous organic matter is abundant).[17] Whatever the original source of organic debris, there is some evidence that microbial action by bacteria and fungi alters the organic chemistry of that material (Ourrison, Albrecht, and Rohmer, 1984).

The organic constituents of sediments are numerous. These include amino compounds, carbohydrates and their derivatives, lipids, isoprenoids, steroids, heterocyclic compounds, phenols, quinones, humic compounds, hydrocarbons, and asphalts (Degens, 1965, p. 2). **Lipids,** long-chain carboxylic acids, particularly plant and animal fats, and **kerogen,** a fine, brown to black, insoluble powder composed principally of carbon, hydrogen, oxygen, and nitrogen plus or minus sulfur, are two of the more important types of organic materials (Forsman and Hunt, 1958a; Degens, 1965).[18]

Particular types of organic compounds and ratios of compounds are useful for characterizing sediment sources and the thermal history of mudrock. Within the bulk organic matter of a mudrock (i.e., among the lipids and kerogens), indicator molecules may be identified that signal a source. For example, the lipid perylene may be an indicator of "terrigenous" origin (Aizenshtat, 1973; Simoneit, 1986). Similarly, high pristane/phytane ratios characterize laminated mudrocks of continental derivation (L. M. Pratt, Claypool, and King, 1986). Simoneit (1986) summarizes such inferred origins for organic debris in Cretaceous rocks of the oceans, origins that are both "terrestrial" and marine. These origins become more difficult to decipher with increasing diagenesis, because of induced changes in elemental ratios and because of the addition of bacterial lipids (Ourrison, Albrecht, and Rohmer, 1984).

Distinctive among organic carbon-rich mudrocks are the black shales.[19] They are widely distributed in the geologic record, and were a particularly significant product of both worldwide Cretaceous marine deposition[20] and Devonian marine deposition in eastern North America.[21] The hypothesis that such shales form from organic-rich sediment in anoxic bottom water is commonly accepted.[22] Yet among the Atlantic Jurassic and Cretaceous black shales, Tissot et al. (1980) and B. J. Katz and Pheifer (1986) identified *three different types* of organic matter:

1. Marine organic matter that is well preserved, with relatively high H/C values (of approximately 1.2) and O/C values in the kerogen of less than 0.15,
2. Moderately well preserved organic matter of "terrestrial" origin, with H/C values of about 0.85 and O/C values in the kerogen of up to 0.3, and
3. Highly altered and/or recycled organic material, with H/C values of less than 0.7 and O/C values, in the kerogen, that vary widely.

The fact that the relative proportions of these types varies from site to site and between stratigraphic levels, indicates that the origins of the sediment constituting the mudrocks vary. These data, combined with sedimentological data, also suggested to B. J. Katz and Pheifer (1986) that anoxic conditions are not requisite to the formation of black shale. This view is becoming more widely accepted. Rapid burial, high organic productivity, reducing conditions, and metastable persistence of organic compounds may all be important locally in contributing to the presence of a large organic component in a black shale.[23]

Colors

The colors of mudrocks reflect their mineralogical and organic content. The colors span the range of rock colors, from nearly white to gray and black, and includes a range of hues including violets, blues, greens, yellows, oranges, browns, and reds.

Three factors seem to be important in producing colors. To some extent, the color of detrital organic material, may influence the color of mudrocks. Brown, gray, and black rocks commonly owe their colors to these organic materials, especially where the organic content is high (figure 18.3) (P. E. Potter, Maynard, and Pryor, 1980, p. 55). For example, Sheu and Presley (1986) found that highly laminated dark gray and black shales from the Orca Basin of the Gulf of Mexico contained > 1.0% organic carbon, whereas those with < 1.0% organic carbon were light gray.

The processes of deposition and diagenesis may be even more important than the color of the detrital material in influencing the colors of mudrocks. In either process, oxidation or reduction may occur. Oxidation results in higher ferric to ferrous ratios (Fe^{3+}/Fe^{2+}) and the formation of red hematitic rocks. Reduction results in an increase in ferrous iron and a concomitant reduction in the ratio. Reduction characterizes rocks with high organic carbon. In such rocks, pyrite and marcasite are the common iron-bearing phases.

In a study of *slates*, Tomlinson (1916) showed that colors are a function of the ratio of ferric to ferrous iron. High ratios (> 2:1), in rocks with 3–6% total iron, yield red slates. Purple slates have generally lower values, but ratios exceed 1:1. Green and black slates have ratios of 1:2 or less. In a qualitative way, A. M. Thompson (1970) confirmed this result in a study of the mudrock-bearing Juniata and Bald Eagle formations of Pennsylvania, in which he found that hematite characterizes red rocks, whereas pyrite and chlorite are the important phases in gray-green rocks. McBride (1974) extended these observations to multicolored shales derived from a common source terrane. He showed that decreasing values of ferric/ferrous yield a progression of colors from red to yellow to green, and he found that (1) red and brown shales contain iron oxide grain coatings; (2) green shales are characterized by chlorite and illite, but lack significant hematite, organic matter, and sulfides; (3) olive and yellow shales contain mixtures of chlorite, illite, organic materials, and iron sulfides; and (4) gray shales owe their

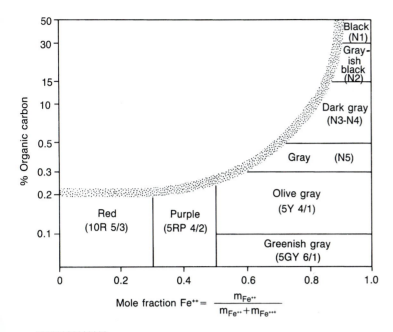

Figure 18.3
Diagram showing the relationship between color, organic content, and oxidation state of iron in mudrocks.
(From P. E. Potter, Maynard, and Pryor, 1980)

color to organic material and sulfides. Although there is evidence that, in some cases, green and gray patches in red rocks contain less total iron than the adjoining red areas, this is not requisite to the development of the green coloration.[24]

Distinctively colored clay minerals and micas may also give a rock color. For example, blue-gray mudrocks of the Pliocene Neroly Formation of the California Coast Ranges owe their color to iron-rich beidellite or iron-poor nontronite grain coatings derived from alteration of andesitic source materials (Louderback, 1924; Lerbekmo, 1956, 1957). In a like manner, authigenic development of glauconite in mudrocks gives them a distinctive green color.

STRUCTURES AND TEXTURES OF MUDROCKS

The principal structures of mudrocks are bedding and lamination (figure 18.1). These features may be parallel, wavy, or lenticular.[25] Parallel stratification in mudrocks results from sediment rain from suspension in the case of normal deposition, storm- and flood-generated deposition, contour-current-deposition, and varve deposition (D. L. Reed et al., 1987; R. Y. Anderson and Dean, 1988; Leithold, 1989; E. A. Cowan and Powell, 1990). Wavy laminations, including flaser laminations, result from (1) deposition (from suspension) from a combination of traction currents, with local scour, and sediment rain, (2) slight soft-sediment deformation, or (3) postde-

positional crystal growth.[26] Lenticular laminations occur in the Tc intervals of mudrock turbidites (figure 18.4b), where they represent ripple marks, and in rocks in which other traction currents have been involved in deposition.

Several additional structures are found in mudrocks. Primary structures include sole marks, ripple marks, flame structures, salt crystal casts, and graded bedding. Secondary structures, including diagenetic and deformational structures, consist of concretions, crystal casts, load casts, rain prints, mudcracks, color banding (including liesegang rings) burrows, bioturbated, disrupted, and convoluted bedding, escape structures, soft sediment folds, soft sediment faults, clastic dikes, folds, joints, faults, and slikensides (figure 18.4). **Fissility,** the tendency of shales to split into flat pieces, is a common feature of enigmatic origin that may result from a combination of sedimentation, compaction, and weathering effects. Typically, it is present in mudrocks exposed at the surface, but at depth it is absent (Lundegard and Samuels, 1980).

The textures of mudrocks are essentially epiclastic. Clay-rich rocks, however, may have a weak foliation resulting from the alignment, during deposition and compaction, of the tabular clay grains. Because of the fine grain size, nearly all mudrocks are well sorted to very well sorted. Some siltstones containing a significant sand component are moderately sorted and pebbly mudstones are very poorly sorted. Grains are dominantly angular, except for silt, sand, or coarser-size clasts, which locally are well rounded (e.g., Mazzulo and Peterson, 1989). Recrystallization during diagenesis imparts crystalline textures to the mudrocks, ranging from equigranular sutured to poikilotopic types.

CONTEMPORARY SETTINGS OF MUDROCK DEPOSITION

Mud is currently being deposited in diverse environments worldwide. These environments range from continental to deep marine types. From studies of these depositional settings and the materials formed in them, we gain a perspective on the *prelithification history* of the mudrocks.

Contemporary muds are derived from weathering and erosion of rocks on the continents and from neocrystallization of minerals within the various depositional environments. Muds derived by weathering and erosion are transported to depositional basins by streams and by winds. For example, sediment transported by the Mississippi River between Davenport, Iowa, and Cairo, Illinois, includes significant amounts of both mud and sand (Lugn, 1927). In contrast, abyssal marine sediments commonly are composed almost entirely of mud, a small fraction of which was airborne (Rosato and Kulm, 1981). The clay component of these muds may experience recrystallization, and new clays and other minerals form during deposition and diagenesis, altering the final composition of the muds and the resulting mudrocks.

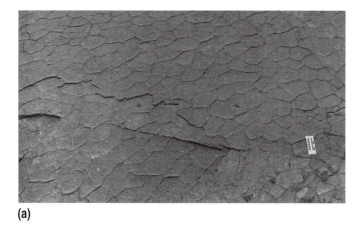

(a)

(b)

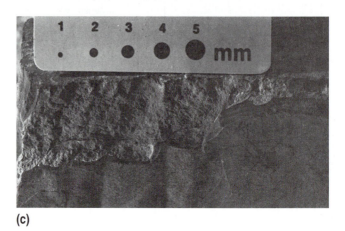

(c)

Figure 18.4 Structures in mudrocks. (a) Mudcracks in calcareous shale and argillaceous lime mudstone, Bowen Formation, Ordovician, Gate City, Virginia. (b) Cross-laminations in mudrock turbidites of the Sevier (Paperville) Formation, Ordovician, Avens Ford Bridge at Holston Reservoir, northeast Tennessee. (c) Ripplemarks and flame structures in oil shale, Tertiary, Oil City, Cascade Range, Oregon.

Deposition of Mudrock Precursors on the Nazca Plate

Clays, oozes, and muds of various types blanket the seafloor (figure 18.5).[27] One of many deep marine seafloor sites at which incipient mudrocks are forming is on the Nazca Plate. The Nazca Plate lies between the East Pacific Rise on the west, the Peru-Chile Trench (off the west coast of South America) on the east, the Galapagos Rift on the north, and the Chile Rise on the south (figure 18.6).

Numerous samples of seafloor sediment from the Nazca Plate, obtained during the Oregon State University/Hawaii Institute of Geophysics Nazca Plate Project (Dymond, 1981; Rosato and Kulm, 1981), and DSDP data from sites 319, 320, and 321 provide the primary sources of information on Nazca Plate sediments.[28] Surface sediments range in composition from site to site on the plate (figure 18.6). Carbonate and other biogenically generated sediments are dominant on topographically high areas and at low latitudes, where dissolution has not been effective and/or organic productivity is high. Siliceous ooze is particularly abundant in the north, at equatorial latitudes. Hydrothermal sediments are abundant along the ridge crests. In the Bauer Deep, an intraplate basin, the surface sediment is iron-rich brown clay that locally is rich in biogenic components (Dymond, 1981; Yeats et al., 1976a). As would be expected, detrital sediments, derived from weathering and erosion of rocks on the South American continent, are abundant near the continent margin, especially in deeper basins. It is the detrital sediments and the clays of the intraplate basins (like the Bauer Deep) that will yield mudrocks upon lithification.

The muds of the Bauer Deep contain components derived from a variety of sources (G. R. Heath and Dymond, 1977). These include *biogenic* sources (precipitated by organisms living in the water column), *hydrogenous* sources (precipitates from seawater), *hydrothermal* sources (precipitates from hot-water solutions that interact with ocean crust), and *detrital* sources (materials weathered and eroded from continental areas, primarily South America). Si, Al, and Fe are detrital and hydrothermal in origin. Part of the Si is biogenic as well. These various components combine with others to generate the constituents of the clays deposited on the seafloor.

Near-surface clays are dusky yellowish brown to dark reddish brown or gray. They are composed primarily of *amorphous materials*, including poorly crystallized ferromanganese oxides and hydroxides; *clay minerals*, notably an iron-rich smectite and kaolinite; *iron and manganese oxides*, including goethite and todorokite $[(Mn,Ca,Mg)Mn_3O_7 \cdot H_2O]$;[29] the *zeolite* phillipsite; and a variety of minor constituents, including calcareous nannofossils, quartz, and feldspar (see table 18.2,

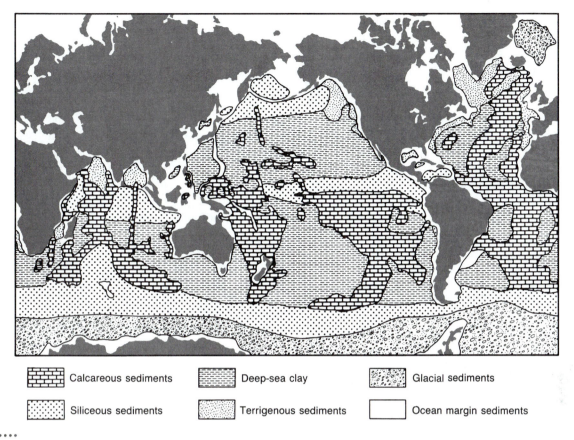

Calcareous sediments	Deep-sea clay	Glacial sediments
Siliceous sediments	Terrigenous sediments	Ocean margin sediments

Figure 18.5 Distribution of dominant sediment types on the sea floor.
(After T. A. Davies and Gorsline, 1976)

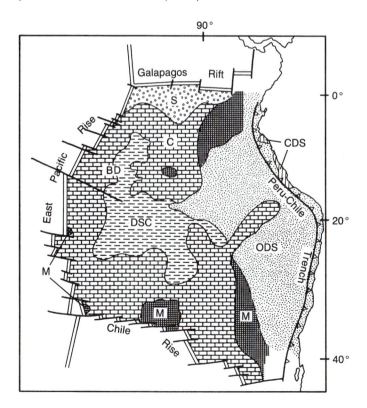

Figure 18.6
Generalized map of sediment distribution on the Nazca Plate.
BD = Bauer deep, C = carbonate sediment, CDS = continental detrital sediment, DSC = deep-sea clay, M = mixed compositions, ODS = oceanic detrital sediment, S = siliceous sediment.
(Modified from T. A. Davies and Gorsline, 1976; Dymond, 1981; Rosato and Kulm, 1981)

column 1)(Dymond et al., 1973; Yeats et al., 1976a; G. R. Heath and Dymond, 1977). Some contain significant quantities of calcite (Bisschof and Sayles, 1972). Organic carbon content is low, at < 0.3% (D. H. Cameron, 1976).

Typical examples of the detrital sediments deposited nearer the continent on the eastern part of the plate were sampled at DSDP sites 320 and 321 (Yeats et al., 1976b,c; Zemmels and Cook, 1976). The near-surface sediment is olive gray to dark gray, siliceous fossil-rich, detrital, silty clay and clay. It is dominated by a clay-size fraction, but contains 15–20% silt and a trace of sand. Laminations and beds are present locally. Amorphous material, including volcanic glass and siliceous microfossils; 'mica' and the clay minerals montmorillonite and kaolinite; chlorite; quartz; and plagioclase feldspar are the major constituents of the sediment (in approximate order of decreasing abundance). In the northeastern part of the plate, the dominant clays vary from smectites and mixed-layer clays in the north and south, to illite in the central area (Rosato and Kulm, 1981). Organic carbon is somewhat higher than in the Bauer Deep sediments, ranging from < 1.0 to 3.0% (D. H. Cameron, 1976; Rosato and Kulm, 1981).

Shelf Muds Along the South Texas Gulf Coast

The Gulf of Mexico contains a variety of depositional environments (figure 18.7).[30] These include such environments as the anoxic Sigsbee Deep, various rise and slope environments, the deltaic-shelf-slope environment of the Mississippi River Delta, and shelf environments such as the Eastern Gulf/Florida Shelf, which is blanketed with carbonate sands, and the Western Gulf/South Texas Shelf, which is blanketed by sand and mud. Muds occur in the deep basin/abyssal plain, on the rise and slope, and on the shelf.

Along the South Texas Shelf, muds are the dominant sediment type (figure 18.8) (Curray, 1960; Shideler, 1976).[31] A central "Interdelta Province," southeast of Corpus Christi Bay, is bounded on the north, in the area offshore from Matagorda and San Antonio bays, by the Ancestral Brazos-Colorado Delta Province, and on the south, offshore from Padre Island and Laguna Madre, by the Ancestral Rio Grande Delta Province (Shideler, 1977). In the northern and southern provinces, significant amounts of sand occur with the mud in the surface or near-surface sediment. Sediment in the Interdelta Province is dark greenish gray to gray clayey silt.

The mineralogical composition of the mud of the South Texas Shelf is relatively uniform, even though the abundances of clay, silt, and sand-size particles vary (Pinsak and Murray, 1960; C. W. Holmes, 1982; Mazzullo and Crisp, 1985). The dominant minerals are quartz, smectites, illite, kaolinite, feldspar, and chlorite (?), generally in that approximate order of decreasing abundance (see table 18.2, column 2). Locally, illite predominates over smectite, but more commonly smectites are the most abundant clay minerals.

The present-day surface sediment distribution is a function of three factors: (1) supply from adjacent areas, (2) redistribution by contemporary shelf currents, and (3) profound influence by Late Pleistocene sediment distribution, paleogeography, and sea-level fluctuation associated with climatic change (Shideler, 1978; Mazzullo and Crisp, 1985; Mazzulo and Peterson, 1989). Sediment is derived primarily from relict Pleistocene deposits underlying the shelf and from turbid lagoonal-estuarine waters that effect offshore transport as a result of ebb-tide discharge from the coastal inlets. Much of the sediment in the northern and southern delta areas is

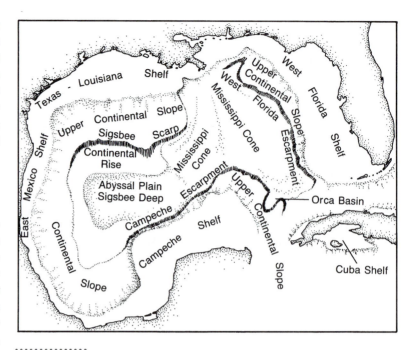

Figure 18.7
Map of physiographic provinces (depositional environments) in the Gulf of Mexico.
(Modified from Ewing, Ericson, and Heezen, 1958)

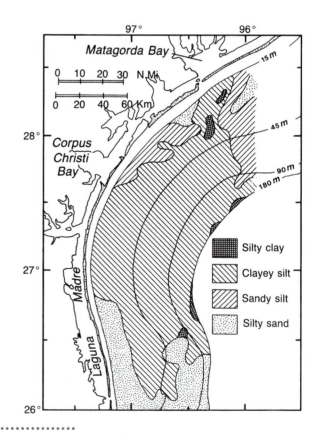

Figure 18.8   Map of sediment distribution on the South Texas shelf.
(Modified from C. W. Holmes, 1982; and based in part on Curray, 1960, and Shideler, 1977, 1978)

residual Pleistocene sediment, as is some in the Interdelta Province. Significant fractions of the sediment—notably some quartz, presumably some feldspar, and the smectite—are inherited from these Pleistocene deposits formed at a time when sea level was lower. Most modern fluvial sediment is trapped in the bays and estuaries along the coast; thus the modern component of the shelf sediment is small. Distribution of the silt-size fraction, for example, reflects Pleistocene depocenters related to Gulf Coast rivers (Mazzulo and Peterson, 1989). Variations in the quartz-grain roundness and surface textures in this fraction allow subdivision of the Interdelta Province into a southern Guadalupe Province and a northern Brazos-Colorado Province. The former is dominated by rounded and fractured quartz, whereas the latter is dominated by more angular, fractured quartz, with subordinate rounded and "crystalline" quartz (Mazzulo and Peterson, 1989). Each population of grains presumably reflects the quartz in the source areas of the respective river systems.

The major type of sediment that is *currently* being transported across the shelf is mud consisting of the very finest silts and clays. The mud is suspended in a turbid bottom layer called the **nepheloid layer.** Sediment transport is accomplished primarily by *advection,* that is, transportation by currents. The currents are largely generated by the wind and include nearshore littoral currents, shelf bottom currents, and storm produced currents. The shelf bottom currents, which seem to be the most important in long-term mud transport and from which sediment of the nepheloid layer is deposited, achieve an overall southerly transport of the sediment.

The muds of the Interdelta Province are well laminated locally, with only scattered burrows (Shideler, 1977). Overall, however, muddy sediments range from laminated to mottled to homogeneous, and may contain significant numbers of foraminifera (Shepard and Moore, 1955; Curray, 1960).[32] The laminated sediments of the Interdelta Province contrast with more bioturbated, sandy, and locally shelly sediments in the delta and nearshore areas (Shideler, 1977).

Muddy Sediments in Western Lake Superior

The Great Lakes provide an environment in which one can examine contemporary sedimentation in large lakes and examine the effects of recently changed climatic conditions (and mining activities) on lacustrine sedimentation. Lake Superior approaches the size of a small ocean basin and experiences some of the same sedimentary processes, such as sediment rain, contour current deposition, and turbidity current deposition (Normark and Dickson, 1976; T. Johnson, Carlson, and Evans, 1980; Klump et al., 1989).[33] Of the Great Lakes, it is one of the least affected by human activity. As is the case with all of the Great Lakes, geologically Lake Superior is a very young feature, having formed as a result of Pleistocene continental glaciation. Thus, its early history is glacial, but its later history is that of a climatically more temperate lake.

The late Pleistocene and Holocene history of Lake Superior is summarized by Farrand and Drexler (1985). Advance and readvance of continental glacial ice across the region, with a lobe projecting out from the ice margin into the western Lake Superior region, facilitated erosion of the Lake Superior basin. Deposition followed recession of the ice, with red, late Pleistocene till, outwash, and lacustrine sediments forming the basal sedimentary unit (Farrand, 1969; Dell, 1975; T. Johnson, Carlson, and Evans, 1980).[34] In the western part of the lake, these basal sediments are up to 400 m thick and cover an irregular topography characterized by troughs. The present lake floor is rather smooth. In contrast, in the east the glacial deposits are thin, so that irregular topography characterizes the present lake floor.

In the offshore parts of western Lake Superior, the thick, red glacially derived sediments are overlain successively by an Upper Pleistocene to Holocene sequence of (1) grayish red silty clays, (2) red varved clay, and (3) gray, olive, and brown varved clay, with locally interbedded, nonvarved clays (figure 18.9)(Farrand, 1969; Lineback, Dell, and Gross, 1979).[35] Locally, sandy turbidites are interlayered with the varved clays (Teller and Mahnic, 1988). At least 1300 varve years of sediment are represented by the varved clay section (Farrand and Drexler, 1985). The gray to brown varved muds, which represent glacial-lacustrine deposition, consist of light-shaded, thicker, calcareous summer layers alternating with dark, thinner, slightly calcareous to noncalcareous winter layers (Dell, 1972; Wold, Hutchinson, and Johnson, 1982).

The minerals of the varved sediments include quartz, plagioclase, and calcite, with alkali feldspar, chlorite, dolomite, kaolinite, illite, and smectite mixed-layer minerals (see table 18.2)(Mothersill and Fung, 1972; Dell, 1973; Lineback, Dell, and Gross, 1979).[36] Calcite and clays are more abundant in the upper gray varved mud, whereas quartz and feldspars are more abundant in the lower red varved sediment. The carbonate minerals are concentrated in the light-colored, thicker, summer silty layers and may constitute up to 30% of the rock. They were probably precipitated from surface waters, the **epilimnion,** during the summer, and rapid sedimentation may have precluded total dissolution during its transit through the water column. In the winter, carbonate minerals were probably not precipitated, and suspended carbonates in the water column were largely dissolved before being deposited. Organic carbon is quite low in the varved mud.

The Upper Pleistocene to Holocene varved section of sediments is overlain by a Holocene to Recent layer of gray and brown muds (Swain and Prokopovich, 1957; Farrand, 1969; Lineback, Dell, and Gross, 1979).[37] These range from silts (with sands) in nearshore, higher-energy environments, to silty clays and clays, which dominate in the deeper, offshore environments. Structurally, these youngest sediments are massive to laminated, with laminations being concentrated in shallower water sediments. Laminations also occur in the deeper-water contourites present north of the Keweenaw Peninsula of Michigan. These gray and brown muds are interpreted to be postglacial sediments.

Mineralogically, the youngest muds are generally similar to the varved sediments, but they lack carbonate minerals (table 18.4)(Mothersill and Fung, 1972; Lineback et al., 1979). Clays, especially illite, smectite, and mixed-layer illite-

smectite, are more abundant than in the underlying varved sediment. Smectite appears to be forming within the clays, a few centimeters below the sediment-water interface, in part from biogenic (diatom generated) silica (T. Johnson and Eisenreich, 1979).[38] The yellowish brown color of some Lake Superior sediments is due to the presence of limonite ± hematite in the upper, oxidized layers (Mothersill and Fung, 1972). The underlying gray sediment is reduced. Total carbon content is variable and ranges from rather low values (< 1%) to moderately high values (> 6%) (Callender, 1969; Mothersill and Fung, 1972; Klump et al., 1989). Most of this carbon is organic.

ANCIENT MUDROCKS

Having examined some examples of incipient mudrock formation in modern environments, we now have a better perspective for the study of ancient mudrocks. The ancient mudrocks have undergone diagenesis and are lithified, but their structures, minerals, and associated facies provide information about their origins and paleoenvironments of deposition.

Upper Ordovician Shales of the Cincinnati Arch Area

During the Ordovician period, the western Ohio–eastern Indiana–northern Kentucky area was a marine shelf. Upper Ordovician (Cincinnatian) mudrocks, interbedded with limestones, now are exposed in the area of the Cincinnati Arch in this region (figure 18.10).[39] Here, clayshales and mudshales are the most common types of mudrock. The interbedded limestones are argillaceous, fossiliferous packstones, wackestones, and lime mudstones that contain a fauna indicating the shallow marine conditions of deposition. The fossiliferous character and the **benthic** (bottom-dwelling) faunal assemblages indicate that deposition occurred in a normal, aerobic marine environment. Bedding typically is thin and beds of mudrock alternate with layers of limestone (figure 18.11). Massive limestone and dolomitic limestone occur locally (W. T. Fox, 1962).

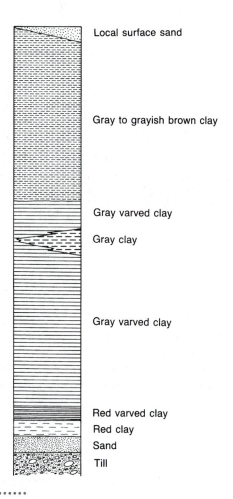

Local surface sand

Gray to grayish brown clay

Gray varved clay

Gray clay

Gray varved clay

Red varved clay
Red clay
Sand
Till

Figure 18.9 Stratigraphic column showing the sediment types in western Lake Superior.
(Modified from Lineback, Dell and Gross, 1979)

Table 18.4 Minerals of the Fine Fraction of Lake Superior Sediments

	Qtz	Afs	Plg	Cc	Dol	Kao	Ill	Vrm	Chl	MxL	Smt	Amp
Gray Clay	10[a]	5	11	0	0	4	26	5	6	21	10	tr
Gray Varved Clay	9	6	11	9	4	5	26	2	5	15	8	tr
Red Varved Clay	11	7	12	11	7	6	22	2	6	10	6	tr

Sources: Lineback, Dell, and Gross (1979), with modifications from Mothersill and Fung (1972).
[a]Values are percents, averages of several samples.
Qtz = quartz, Afs = alkali feldspar, Plg = plagioclase feldspar, Cc = calcite, Dol = dolomite, Kao = kaolinite, Ill = illite, Vrm = vermiculite, Chl = chlorite,
   MxL = mixed-layer phyllosilicates, Smt = smectites, Amp = amphibole.

The mineral composition of the Cincinnatian shales is dominated by illite (see table 18.2, column 9) (Scotford, 1965). The carbonate minerals calcite and dolomite, plus quartz, chlorite, and mixed-layer clays, each exceed 10% in individual rocks. In addition, trace amounts of kaolinite, plagioclase, alkali feldspar, and pyrite, with hematite, garnet, zircon, and additional heavy minerals, constitute the remainder of the minerals. The various minerals are distributed differently among the clay-, silt-, and sand-size fractions. Quartz and calcite are more abundant in the sand- and silt-size fractions, whereas illite is more abundant in the silt and clay fractions. Illite is the most abundant mineral in the silt fraction. Smectites are absent. In general, organic carbon is low, as indicated by the gray color of the shales. One black, carbonaceous shale bed does occur in the Whitewater Formation, near the top of the Upper Ordovician section, in Indiana (W. T. Fox, 1962).

The generally high content of illite and significant quantities of carbonate minerals have a predictable impact on the chemistry (Scotford, 1965). Alumina is high and CaO, CO_2, and H_2O occur in major amounts.

The chemistry, textures, and mineral composition of the Cincinnatian shales are relatively uniform, indicating relatively constant conditions of deposition through time. Subsidence must have kept pace with deposition, in order to maintain a relatively uniform, shallow depth. In the early part of the late Ordovician period, broad shallow bars may have existed in the Cincinnati area (H. J. Hoffman, 1966). Later, local shoaling occurred along an east-west line, as indicated by an increase in carbonate sediment and a dominance of sand and silt in the southern part of the basin (Scotford, 1965; Borella and Osborne, 1978). Some paleocurrent data suggest that current directions varied through time, as well (Hofmann, 1966). Although the variations in sediment influx were subtle, storms and currents, coupled with precipitation of biogenic material and a flux of clastic material, probably controlled deposition of the alternating compositions of sediments that resulted in the interbedded mudrock and limestone (Scotford, 1965; Shrake, Schumacher, and Swinford, 1988).

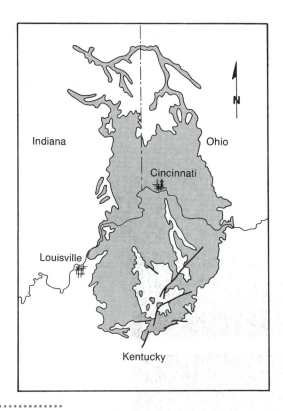

Figure 18.10 Map of the Cincinnati Arch region showing the distribution of exposures of Cincinnatian Ordovician rocks.

(From P. B. King and Beikman, 1974)

(a)

(b)

Figure 18.11 Shales interbedded with limestones of Upper Ordovician age, (a) exposed in highway cuts (b) in Covington, Kentucky. View north along I-75 towards Cincinnati, Ohio.

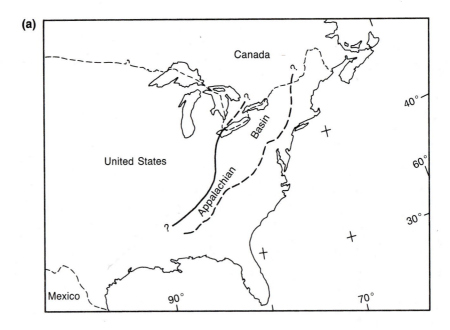

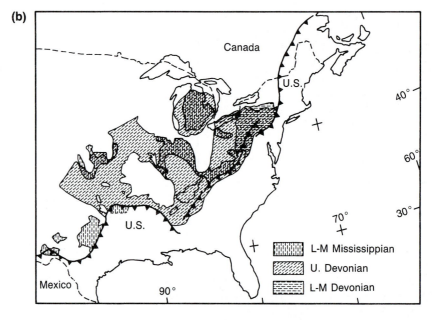

Figure 18.12 Maps of Devonian rocks and the Appalachian Basin in eastern North America. (a) Minimum limits of the Appalachian Basin. The eastern edge of the basin has been shortened by thrust faulting. The northern and southern limits are unknown. (b) Distribution of Lower-Middle Devonian, Upper Devonian, and Lower-Middle Mississippian black, gray, and green mudrocks in the Appalachian Basin and adjoining regions in eastern North America. *(Modified from Cook et al., 1975)*

Devonian Shales of the Appalachian Basin and Adjacent Areas

During the Devonian Period, shales were deposited in the Appalachian Basin of eastern North America, but in contrast to the Ordovician case described above, the prevailing conditions, including a deeper basin, resulted in the formation of black shales.[40] The Appalachian Basin is a region of eastern North America in which thick Paleozoic stratigraphic sections mark the locus of extensive Paleozoic deposition (figure 18.12a). It extends from Quebec and Ontario south to Alabama (Colton, 1970). On the east, the Appalachian Basin is bounded by the highly deformed and crystalline terranes that now make up the Blue-Green-Long axis (Rankin, 1976), including the Blue Ridge Province of the southern and central Appalachian Orogen and the Taconic Klippe, Green Mountain Anticlinorium, and associated structures in New England and southern Quebec. The Appalachian Basin is bounded on the west by ancestral to present-day structural arches. This western boundary extends from the Nashville Dome in the south, north through the Cincinnati Arch and the Findlay Arch in western Ohio, to the Algonquin Axis in southern Ontario. The western boundary essentially marks a hinge zone between the Appalachian Basin, which subsided to

(a)

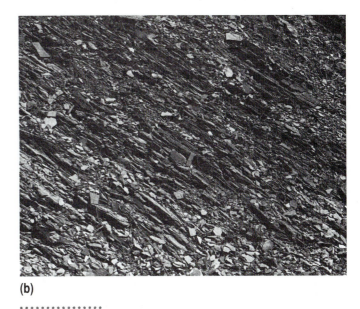

(b)

....

Figure 18.13
Photographs of outcrops of Devonian shales in the Appalachian Basin. (a) Chattanooga Shale, I-24 and Browns Ferry Road, Chattanooga, Tennessee. (b) Brallier Shale, Allison Gap, near Saltville, Virginia.

receive great thicknesses of sediment (> 10,000 m in the east), and the platform and basins of the continental interior, which did not subside to as great an extent.[41] Rocks in the eastern part of the Basin were deformed during the late Paleozoic mountain-building events that deformed the Appalachian Orogen, whereas those in the western part of the basin were not. Thus, the basin overlaps the boundary between the Appalachian Orogen (on the east) and the stable continental interior (on the west).

Devonian-Mississippian shales are widely distributed in the Appalachian Basin. Rocks underlying the Devonian-Mississippian shales of the Appalachian Basin are generally Silurian and early Devonian in age. These Silurian and Lower Devonian rocks are dominantly sandstones, limestones, and dolostones, with associated shales and conglomerates (Colton, 1970). In general, they exhibit features indicative of shallow marine conditions of deposition.

The conditions that gave rise to deposition of the shallow marine sands and carbonate sediments of the Silurian and early Devonian periods were progressively replaced during early Devonian, middle Devonian, or late Devonian time by shallow to deep, oxygen-poor marine conditions that resulted in the deposition of dark muds. The latter conditions extended across most of the Appalachian Basin and affected major regions of the continental interior (figure 18.12b). In the central part of the Appalachian Basin, subsidence brought about deeper-water sedimentation, but by middle Mississippian time, the mud deposition had been succeeded by deposition of sands, some red muds, and carbonate sediments (Pepper, DeWitt, and Demarest, 1954; Colton, 1970; B. R. Moore and Clarke, 1970).

Black, gray, and green shales, with interbedded sandstones and carbonate rocks, formed during the Devonian to early Mississippian phase of deposition in the Appalachian Basin and adjoining regions (figure 18.12). In the basin, major mudrock units that developed include the Chattanooga Shale, the Millboro Shale, and the Brallier Formation of the southern and central Appalachian region, one or more of which is exposed in Alabama, Tennessee, Virginia, West Virginia, Maryland, and Pennsylvania; the Ohio Shale of Ohio and eastern Kentucky; and the Marcellus Formation and other shales of the Hamilton Group in New York and Pennsylvania (figure 18.13).[42] In addition, mudrocks make up significant parts of Devonian sandstone-rich sections, such as those of the Genesee and West Falls groups in central New York and northern Pennsylvania (Woodrow and Isley, 1983). To the west of the Appalachian Basin, the Antrim Shale of the Michigan Basin and the New Albany Shale (Group) of the Illinois Basin in Kentucky, Indiana, and Illinois are representative units (G. Campbell, 1946; Lineback, 1968; Beier and Hayes, 1989).[43] Important stratigraphic markers, referred to as "ash beds" or "bentonites," extend for long distances across the region and allow correlation of the various units.[44]

The mineral compositions of the Devonian-Mississippian mudrocks of eastern North America vary somewhat, as might be expected considering the diversity of ages and environments of deposition represented. Yet these compositions have a general similarity. The mudrocks are composed of varying amounts of clays, principally illite; plus chlorite, quartz and feldspars, white mica, organic materials, calcite and dolomite, and minor minerals such as biotite, apatite, barite, and gypsum (see table 18.2, columns 6, 7, and 8).[45] The mudrocks of the middle Devonian Hamilton Group rocks of New York, for example, consist of quartz, illite, and some chlorite, with locally abundant calcite (Towe and Grim, 1963). Kaolinite, smectite, and mixed-layer minerals were not observed. Similarly, illite is the dominant clay mineral of the Chattanooga Shale (Leventhal and Hosterman, 1982), a batch sample of which was found by T. F. Bates and Strahl (1957) to consist of 31% illite (including muscovite and minor kaolinite), 22% organic matter, 22% quartz, 11% pyrite and marcasite, 9% feldspar, 2% chlorite, 2% iron oxides, and 1% minor minerals. More recent studies have revealed significant amounts of mixed-layer minerals and some kaolinite, apparently not recognized in older studies, but they verify the dominance of illite in the clay mineral assemblage (Leventhal and Hosterman, 1982; Hosterman and Whitlow, 1983).

Locally, some less common minerals are abundant in Devonian-Mississippian shales. Pyrite is a significant component of many shales, including the Chattanooga Shale (described above), some New Albany Group shales in the Illinois Basin, and the Millboro Shale in southwestern Virginia. Some Devonian shales are phosphatic or glauconitic. For example, the Mississippian Grainger Formation of eastern Tennessee contains a glauconitic unit with a glauconitic shale bed (Hasson, 1972).

Devonian Appalachian shales with more than 2% (by volume) of organic materials tend to be black, whereas those with less may be gray or green. The black shales may contain more than 20 volume percent organic material, as they do in the western parts of the Appalachian Basin, and the organic carbon may total up to 20 weight percent of the rocks (Conant and Swanson, 1961; Schmoker, 1980; Ettensohn and Barron, 1982).[46] The black shales tend to be laminated and unfossiliferous. In contrast, gray and green shales are commonly bioturbated and may be fossiliferous.

The fossils, structures, minerals, and chemistry, combined with regional structural and geophysical data, indicate that the Devonian-Mississippian mudrocks of eastern North America were deposited in an equatorial, continental marine basin that shallowed towards both the east and the west. Erosion of uplifted crystalline and other rocks to the east provided most of the sediment. Siliciclastic sediments derived from these eastern source regions formed deltas, notably the Catskill Delta, along the eastern edge of the basin. Towards the west, the basin shallowed to form a shallow marine platform.

Within the basin, the water column was apparently stratified, with (1) an upper, oxygenated **aerobic layer,** (2) an intermediate **dysaerobic layer,** or **pycnocline,** of variable oxygen content, salinity, and density, and (3) a cold, oxygen-deficient **anaerobic layer** on the bottom (figure 18.14)(Rhoads and Morse, 1970; C. W. Byers, 1977; Ettensohn and Barron, 1982; Ettensohn and Elam, 1985). During at least part of the Devonian Period, the anaerobic bottom layer formed a euxinic basin-floor environment[47] in which sulfur and organic carbon were abundant, contributing to the formation of organic, carbon-rich laminated shales. The deepest part of this euxinic basin extended from eastern Pennsylvania, through western West Virginia and Virginia, to northeastern Tennessee (Leventhal, 1987).

The Green River Formation

The Green River Formation, famous for its "oil shale," is exposed in six structural basins in southwestern Wyoming, northeastern Colorado, and northern Utah.[48] These basins are separated by uplifts, folds, and/or faults, but during the Eocene Epoch, the area contained three depositional basins occupied by ancient lakes—Lake Gosiute, Lake Uinta, and a small, unnamed lake near the western Wyoming–northeastern Utah border (W. H. Bradley, 1931, 1963, 1964).[49] Rivers from adjacent uplifted mountain ranges fed the lakes and transported sediment to the lake basins.[50] Like all lakes, these Eocene lakes were ephemeral, but the larger lakes lasted at least 13 million years (Picard, 1963).[51]

The Green River Formation is enclosed within fluviatile deposits of the Wasatch, Colton, Uinta, and Bridger formations. It forms a series of lenses and "tongues" that interfinger with the fluvial sedimentary rocks of the latter formations. The ancient Lake Gosiute clearly fluctuated in size, so that it was larger in earlier and later stages than in the intervening stage. In the large basins—the Piceance Creek Basin in northwestern Colorado; the Great Divide, Washakie, and Green River basins in southwestern Wyoming; and the Uinta Basin of northeastern Utah—the stratigraphy has been studied thoroughly enough to allow subdivision of the Green River Formation into members and beds of different lithology that reflect various environments of deposition.[52] These environments include fluvial-deltaic, mudflat lake margin, shore and near shore lacustrine, and offshore lacustrine types.

Although they are commonly referred to as "shales" or "oil shales," the rocks of the Green River Formation are lithologically diverse. The dominant rock types of the oil shales, in fact, are kerogen-rich dolostones and limestones. The less abundant mudrocks are but one class of rock in the formation, which includes a varied assemblage of laminated

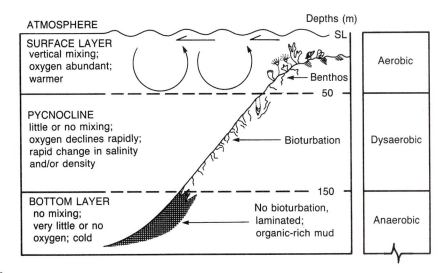

Depths (m)

ATMOSPHERE

SURFACE LAYER
vertical mixing;
oxygen abundant;
warmer

SL

Benthos

Aerobic

50

PYCNOCLINE
little or no mixing;
oxygen declines rapidly;
rapid change in salinity
and/or density

Bioturbation

Dysaerobic

150

BOTTOM LAYER
no mixing;
very little or no
oxygen; cold

No bioturbation,
laminated;
organic-rich mud

Anaerobic

Figure 18.14 Model of stratified basin waters showing sediment structures associated with the aerobic zone, pycnocline (dysaerobic zone), and anaerobic zone. Devonian shales of the Appalachian Basin formed in stratified waters like those indicated here, with the black shales forming below the pycnocline.
(Modified from Ettensohn and Elam, 1985)

rock types including limestones {e.g., stromatolitic algal lime mudstones, hydrocarbon-rich lime mudstone, oolitic packstone (?), and ostracodal and molluscan grainstones and/or packstones}, dolostones, evaporites, siltstone, shale, tuff, and sandstone (figure 18.15).[53] Siliciclastic mudrocks are most common in the fluvial and near-shore environments, but clay-bearing rocks also occur in offshore sections (Eugster and Hardie, 1975; Dyni, 1976; Surdam and Stanley, 1979).

The mudrocks and oil shales are various shades of gray, green, and brown and weather to lighter shades of the same colors, as well as to blue-gray, red, orange, or yellow colors. The last three colors result from oxidation of iron-bearing minerals, particularly pyrite, which is a common phase in some of the rocks.

The major minerals in the mudrocks include quartz, feldspars, calcite, dolomite, and illite (table 18.5).[54] Additional minerals include kaolinite, mixed-layer clays, chlorite, calcite, dolomite, pyrite, analcite, and zeolites such as clinoptilolite and mordenite. Some quartz and feldspar, kaolinite, chlorite, some smectites (montmorillonite), and some illite are detrital. The other minerals and some quartz, feldspar, illite, and smectite are authigenic, having formed via chemical reactions in the lake waters and bottom sediment (W. H. Bradley and Eugster, 1969; Hosterman and Dyni, 1972).[55] The distribution of the clay minerals—with kaolinite plus smectite dominating in near-shore sedimentary rocks, but illite and some mixed-layer minerals dominating in many offshore rocks—suggests that kaolinite reacted to form illite, whereas montmorillonite altered to form illite and mixed-layer minerals (Hosterman and Dyni, 1972; Dyni, 1976).

Figure 18.15
Outcrop of mudrocks and "oil shales" of the Green River Formation, Highway 6, Tucker, Utah.

A variety of models have been proposed to explain the origin of the Green River rocks (see chapter 22). In all of the models, clay and silt are introduced from the surrounding region and some clays are produced as authigenic precipitates, that is, they are neocrystalline phases formed in the lake. Mudrocks associated with the lake margins and deltas, and those formed in freshwater phases of lacustrine deposition, originated from these materials.

Table 18.5   Mineral Compositions of Some Green River Mudrocks and Associated Rock Types

	Mudrocks			Associated Rocks		
	1	**2**	**3**	**4**	**5**	**6**
Quartz	45[a]	A	A	10[a]	20	6[b]
Alkali Feldspar	5	m	m	5	5	12
Plagioclase	10	—	m	5	10	—
Kaolinite	5	—	—	—	—	—
Smectite	—	—	A	—	—	—
Illite	15	m	m	5	5	—
Mixed-layer clays	5	—	—	tr	tr	—
Chlorite	—	—	m	—	—	—
Calcite	10	m	p	45	—	2
Dolomite	5	m	m	25	55	34
Gypsum	—	m	—	—	—	—
Analcite	—	m	m	5	5	—
Shortite	—	—	—	—	—	26
Pyrite	—	—	—	—	—	2
Anatase(?)	—	—	—	—	—	tr
Organic matter	—	—	—	—	—	16

Sources:

1. "Shale," sample 28, Garden Gulch Member, Green River Formation, Piceance Creek Basin, Colorado (Hosterman and Dyni, 1972).
2. Siltstone-claystone (composite), Parachute Creek Member, Green River Formation, Uinta Basin, Utah (Picard and High, 1972b).
3. Claystone, sample 9, Parachute Creek Member, Green River Formation, Uinta Basin; Duchesne County, Utah (Dyni, 1976).
4. "Marlstone," sample 3, Evacuation Creek Member, Green River Formation, Piceance Creek Basin, Colorado (Hosterman and Dyni, 1972).
5. Dolomitic "oil shale," sample 4, Parachute Creek Member, Green River Formation, Piceance Creek Basin, Colorado (Hosterman and Dyni, 1972).
6. "Oil shale," Wilkins Peak Member, Green River Formation, Green River Basin, Sweetwater County, Wyoming (W. H. Bradley and Eugster, 1969).

[a]Percentages estimated from (1) weight loss due to carbonate dissolution and (2) X-ray peak height ratios.

[b]Percentages based on a combination of optical, X-ray, and chemical tests.

A = abundant, m = moderate to minor amounts, p = present in undetermined amount, tr = trace.

SUMMARY

Mudrocks, which include siltstones, mudstones, mudshales, claystones, and clayshales, are the most abundant of the sedimentary rocks. They are deposited in a wide range of continental and marine environments, where they constitute entire formations or are interbedded with other rock types, such as sandstone and limestone.

Because mudrocks are typically dominated by quartz, feldspars, and clay minerals, their chemistries are usually characterized by abundant silica, alumina, and H_2O. However, where mudrocks are interlayered with carbonate rocks, as they are in the Cincinnatian Series and the Green River Formation, they contain substantial amounts of carbonate minerals. As a consequence, the chemical analyses of such

mudrocks contain relatively large amounts of CaO, MgO, or both, and substantial amounts of CO_2. Among the clay minerals, smectites and kaolinite tend to occur in younger mudrocks and in smaller quantities. Bentonites, altered volcanic ash primarily composed of smectites, are an exception. Illites dominate in older mudrocks, at least in part because they are produced by diagenetic reactions in which smectites are converted to illite. A variety of additional minerals occurs in mudrocks, including chlorite, zeolites, gypsum and anhydrite, pyrite and marcasite, iron oxides, apatite, and a wide range of heavy minerals such as garnet and hornblende.

The colors of mudrocks are related to their chemistry and mineralogy. Mudrocks with large amounts of organic carbon (notably kerogen and lipids) are dark gray to black. Organic carbon may total as much as 80% of the volume of some rocks, but mudrocks commonly have between 0.1 and

6% organic carbon. Pyrite or marcasite typically occurs in the high-carbon mudrocks and these rocks are laminated, because more complex life-forms, which burrow in and bioturbate the sediment, are unable to live in the anaerobic to dysaerobic reducing environments in which these minerals develop. Green and light gray mudrocks typically contain one or more of the minerals pyrite, chlorite, and glauconite. Mudrocks with iron oxides are red to yellow and reflect either aerobic environments of deposition or postdepositional oxidation.

Contemporary muds, the precursors of mudrocks, contain a wide variety of materials. These may include amorphous materials produced by hydrogenous, hydrothermal, or biological precipitation; and they may include detrital minerals derived by weathering and erosion of preexisting continental or marine rocks or sediments. The source terrane may control the mineralogy of the mud, but authigenic minerals are dominant in some mudrocks. High carbon contents in mudrocks are preserved in deep, basinal anaerobic environments, but may also result from high rates of organic production or rapid burial.

Most ancient mudrocks have experienced significant diagenesis. Illite is commonly the dominant mineral, but chlorites and mixed-layer clays may be abundant, and quartz, feldspar, calcite, and dolomite are typical minerals that constitute most of the remainder of the mineralogy. In spite of diagenetic changes in mineralogy that may obscure the early rock history; the associated rocks and fossils, the chemistry, the structures, and the regional structural and facies relationships allow an evaluation of the petrogenesis of the mudrocks.

EXPLANATORY NOTES

1. Kuenen (1941) estimates that shale makes up about 56% of the stratigraphic record; Blatt (1970) estimates that, of the continental sedimentary rocks, 65% is shale; and Boggs (1987, p. 215) estimates that "shales" constitute 50–60% of the total sedimentary rock record. Ehlers and Blatt (1982, p. 283) suggest that mudrock constitutes approximately two-thirds of the sedimentary record.
2. Refer to chapter 15 for a discussion of classification.
3. Knight (1898 in Bates and Jackson, 1987).
4. K. Bryan (1945), Obruchev (1945), C. B. Schultz and Stout (1945), Swineford and Frye (1945), H. N. Fisk (1951), Swineford and Frey (1951), Pewe (1955), Lugn (1960, 1962, 1968), Glass, Frye, and Willman (1968), E. C. Reed (1968), C. B. Schultz and Frey (1968), Matalucci et al. (1969), Smalley (1975b), Bates and Jackson (1987).
5. For example, see Windley (1977, p. 25), K. A. Eriksson (1980), and Taylor and McLennan (1985, p. 151ff.).
6. See the sources in note 4.
7. For example, see papers in Smalley, (1975b).
8. See P. E. Potter, Maynard, and Pryor (1980) for a review of the general mineralogy of shales, C. E. Weaver (1989, ch. 9) for a review focussed on phyllosilicates, Degens (1965) for a review of the mineralogy of sediments in general, and the papers in A. C. D. Newman (1987), especially A. C. D. Newman and Brown (1987) and Velde and Meunier (1987), for discussions of the details of clay mineral chemistry and phase equilibria. These papers provide the foundation for the review here. Additional papers that list minerals of muds and mudrocks are numerous. A few examples include T. F. Bates and Strahl (1957), Scotford (1965), Parham (1966), T. P. Hill, Werner, and Horton (1967), Heling (1969), J. C. Moore (1974), G. Muller and Stoffers (1974), D. A. Ross and Degens (1974), D. J. Stanley et al. (1981), Bhatia (1985), R. D. Cole (1985), Lenotre, Chamley, and Hoffert (1985), Schoonmaker et al. (1985), Johnsson and Reynolds (1986), G. D. Williams and Bayliss (1988), and Remy and Ferrell (1989).
9. Refer to chapter 16 for additional details.
10. The voluminous reports of the Deep-Sea Drilling Project (DSDP) and the International Phase of Ocean Drilling (IPOD), published as a set of volumes and briefly summarized in a series of reports in *Geotimes*, provide extensive coverage of the compositions of modern deep-sea muds and oozes.
11. Also see Windom, Neal, and Beck (1971), G. Muller and Stoffers (1974), P. E. Potter, Maynard, and Pryor (1980), and Naidu and Mowatt (1983).
12. Garrels and MacKenzie (1971, ch. 9), van Moort (1972), Ehlers and Blatt (1982, pp. 291–294), C. E. Weaver (1989, ch. 9).
13. See Erba, Parisi, and Cita (1987) and Stow and Atkin (1987), plus other typical studies cited in the explanatory notes below.
14. For example, J. D. Vine (1969), J. D. Vine and Tourtelot (1970), Bhatia (1985), van Weering and Klaver (1985), Amajor (1987), Leventhal (1987), Stow and Atkin (1987), and Walters *et al.* (1987). Also see J. N. Weber (1960).
15. Leventhal follows the method of Berner (1982) and Berner and Raiswell (1984). See chapter 14 for a brief discussion of C/S.
16. W. H. Bradley (1948), Simoneit (1974), Claypool, Love, and Maugham (1978), Deroo et al. (1978), Tissot et al. (1980), P. A. Meyers and Mitterer (1986), Parisi, Erba, and Cita (1987).
17. For example, see Potter, Maynard, and Pryor (1980, p. 52).
18. Forsman and Hunt (1958b); Simoneit (1986). A detailed discussion of these and other individual types of compounds is beyond the scope of this text, but an introduction to their characteristics may be found in Degens (1965).
19. For discussion of black shales, see *Marine Geology*, v. 70, nos. 1, 2, including B. J. Katz and Pheifer (1986), Meyers and Mitterer (1986), and Simoneit (1986), plus other papers such as Rich (1951), Conant and Swanson (1961), Heckel (1977), Dean et al. (1984), Ettensohn and Elam (1985), Dabard and Paris (1986), and Leventhal (1987).
20. See Arthur and Schlanger (1979), Tissot et al. (1980), Summerhayes (1981), Dean, Arthur, and Stow (1984), and Simoneit (1986).
21. For example, Rich (1951), Conant and Swanson (1961), B. N. Cooper (1968), Schmoker (1980), Maynard (1981), Ettensohn and Elam (1985), and Levanthal (1987).

22. For example, Pettijohn (1957, 1975), Heckel (1977), Arthur and Schlanger (1979), Ettensohn and Elam (1985).

23. Arthur, Dean, and Stow (1984), Meyers and Mitterer (1986). Refer to Summerhayes (1981) and Waples (1983) for discussions of high productivity as a factor in the formation of black shales, and Habib (1983), and D. J. Stanley (1986) for an explanation involving rapid burial. Summerhayes notes that layers rich in organic carbon are deposited under reducing conditions. Also see Pettijohn (1975, p. 275) and Potter, Maynard, and Pryor (1980).

24. See the discussions in G. B. Richardson (1903), T. R.Walker (1967), A. M. Thompson (1970), Pettijohn (1975, pp. 274–275), Schluger and Roberson (1975), Durrance et al. (1978), Potter, Maynard, and Pryor (1980, p. 55), and Morad and AlDahan (1986). G. B. Richardson (1903) in a study of rocks from the Black Hills of South Dakota and Durrance et al. (1978) in a study of red mudrocks (lutites) in southeast Devon, England, demonstrated that pale patches contain about half the total iron content of the surrounding red mudrocks. In an analogous study of chert, J. O. Berkland (pers. comm., 1964) found that, in some Franciscan cherts, where green patches in red chert are common, the ferric iron content is less in the green cherts.

25. These and other features are reviewed by Potter, Maynard, and Pryor (1980), who summarize their own observations and those of others.

26. For example, see R. D. Cole and Picard (1975, in Potter, Maynard, and Pryor, 1980) and see Leithold (1989), who shows wavy laminations resulting from climbing ripples.

27. Lisitzin (1972), W. H. Berger (1974), and T. A. Davies and Gorsline (1976) may be consulted for general discussions of oceanic sediments and their distribution.

28. Also see Bischoff and Sayles (1972), Dymond et al. (1973), Sayles, Ku, and Bowker (1975), Yeats et al. (1976a,b,c), G. R. Heath and Dymond (1977), and Scheidegger and Krissek (1982). (DSDP = Deep-Sea Drilling Project.)

29. Formula from Bates and Jackson (1987).

30. For descriptions of Gulf of Mexico depositional environments and facies, see the various papers in Shepard, Phleger, and van Andel (1960), and see Goldstein (1942), Shepard and Moore (1955), Greenman and LeBlanc (1956), R. H. Parker (1956), M. Ewing, Ericson, and Heezen (1958), J. I. Ewing, Worzel, and Ewing (1962), W. R. Bryant, et al. (1968), Uchupi and Emery (1968), D. K. Davies and Moore (1970), A. C. Donaldson, Martin, and Kanes (1970), Gould (1970), Kanes (1970), H. F. Nelson and Bray (1970), Davies (1972a,b), Devine, Ferrell, and Billings (1973), Berryhill et al. (1976, in J. Mazzullo and Crisp, 1985), Shideler (1976, 1977, 1978), Doyle and Sparks (1980), C. W. Holmes (1982), J. Mazzullo and Withers (1984), Joyce, Kennicutt, and Brooks (1985), J. Mazzullo and Crisp (1985), J. Mazzullo (1986), Sheu and Presley (1986), Bouma et al. (1986, Initial Reports of the Deep-Sea Drilling Project, v. 96), Israel, Ethridge, and Estes (1987), and J. Mazzulo and Peterson (1989).

31. This review is based on the work of Curray (1960), Shideler (1976, 1977, 1978), C. W. Holmes (1982), and J. Mazzullo and Crisp (1985). The NTIS report of Berryhill et al. (1976, in J. Mazzullo and Crisp, 1985) clearly provided background for several of these published reports. Additional information was contributed by Shepard and Moore (1955), Greenman and LeBlanc (1956), and M. Ewing, Ericson, and Ewing (1958).

32. Also see Greenman and LeBlanc (1956) and M. Ewing, Ericson, and Ewing (1958).

33. This review of Lake Superior sedimentation and history is based on the works of Swain and Prokopovich (1957), Callender (1969), Farrand (1969), Mothersill (1971), Dell (1972, 1973, 1976), Mothersill and Fung (1972), Normark and Dickson (1976), T. Johnson and Eisenreich (1979), Lineback, Dell, and Gross (1979), T. Johnson (1980), T. Johnson, Carlson, and Evans (1980), Wold, Hutchinson, and Johnson (1982), Farrand and Drexler (1985), Teller and Mahnic (1988), and Klump et al. (1989). Hough (1958) gives an overview (now dated) of the geology of the Great Lakes. For an introduction to the study of ancient lakes, see Reeves (1968).

34. Also see Farrand (1969b, in Wold, Hutchinson, and Johnson, 1982), Wold, Hutchinson, and Johnson (1982), and Teller and Mahnic (1988).

35. In addition, refer to Mothersill (1971), Dell (1972), Mothersill and Fung (1972), T. Johnson (1980), and Teller and Mahnic (1988).

36. Also see Mothersill (1971), who did work on Thunder Bay, a large embayment in Ontario on the northeastern part of western Lake Superior. The sedimentation history of the bay mimics that of the lake as a whole.

37. Also see Dell (1972) and Mothersill and Fung (1972).

38. A point of interest: In contrast to some surface sediments of Lake Superior, the fine fraction of Lake Ontario surface sediments lacks smectites and is dominated by illite (R. L. Thomas, Kemp, and Lewis, 1972).

39. This review is based on the works of a few authors. Scotford (1965), Weiss et al. (1965), and Bassarab and Huff (1969) have examined the mineralogy, petrology, and chemistry of the mudrocks. W. T. Fox (1962) has studied the stratigraphy and paleoecology of the uppermost unit, the Richmond Group, and H. J. Hofmann (1966) reports on paleocurrent analyses of lower Upper Ordovician rocks. The middle to late Ordovician history of the Cincinnati Arch was examined by Borella and Osborne (1978). Cincinnatian rocks described by the geologists listed above constitute a standard section of the Upper Ordovician Series, which, combined with their fossiliferous character, has made them the subject of considerable study (Shrake, Schumacher, and Swinford, 1988). Also see Sweet and Bergstrom (1971) and Weir, Peterson, and Swadley (1984) for discussions of the stratigraphy. See the references in Shrake, Schumacher, and Swinford (1988) and the above papers for citations of earlier reports.

40. The general geology and stratigraphy of the Appalachian region are reviewed by Eardley (1962) and P. B. King (1977), and the stratigraphy is discussed in detail by Frazier and Schwimmer (1987). This review of the mudrocks of the Appalachian Basin and adjoining regions is based on Willard (1939), Butts (1940), G. A. Cooper et al. (1942), G. Campbell (1946), Rich (1951), Pepper, DeWitt, and

Demarest (1954), Hass (1956), T. F. Bates and Strahl (1957), K. V. Hoover (1960), Conant and Swanson (1961), Dennison (1961), Dennison and Naegele (1963), Towe and Grim (1963), W. A. Oliver et al. (1967), Lineback (1968), North (1969), Colton (1970), R. G. Sutton, Bowen, and McAlester (1970), Hasson (1972), Droste and Vitaliano (1973), T. D. Cook et al. (1975), C. W. Byers (1977), Schmoker (1980), Maynard (1981), Ettensohn and Barron (1982), Leventhal and Hosterman (1982), Hosterman and Whitlow (1983), Woodrow and Isley (1983), Roen (1984), Dennison (1985), Ettensohn (1985a,b), Ettensohn and Elam (1985), Sevon and Woodrow (1985), and Leventhal (1987). Also see C. R. Stauffer (1909), M. Kay (1951), B. N. Cooper (1961), G. M. Ehlers and Kesling (1970), R. G. Walker and Harms (1971), Liebling and Sherp (1976), Roen and Hosterman (1982), and Beier and Hayes (1989).

41. M. Kay (1951), Eardley (1962), Colton (1970), P. B. King (1977), Ettensohn and Barron (1982).

42. Willard (1939), Butts (1940), G. A. Cooper et al. (1942), K. V. Hoover (1960), Colton (1970), Kepferle et al. (1981), Roen (1984), Ettensohn and Elam (1985), Sevon and Woodrow (1985).

43. Also see G. A. Cooper et al. (1942), North (1969), and G. M. Ehlers and Kesling (1970).

44. For example see Dennison (1961), Dennison and Naegele (1963), Droste and Vitaliano (1973), Roen and Hosterman (1982), Hosterman and Whitlow (1983), N. F. Forsman (1984), and Frazier and Schwimmer (1987, fig. 6.1).

45. The data reported here are based on Bates and Strahl (1957), Conant and Swanson (1961), Towe and Grim (1963), Lineback (1968), North (1969), Hasson (1972), Liebling and Sherp (1976), Kepferle et al. (1981), Ettensohn and Barron (1982), Leventhal and Hosterman (1982), Hosterman and Whitlow (1983), on miscellaneous comments by the authors of many of the papers cited in note 32, and on limited observations by the author.

46. Also see Lineback (1968), Maynard (1981), Leventhal (1987), and Beier and Hayes (1989).

47. *Euxinic basins* are those with stagnant, oxygen-deficient bottom waters.

48. Because of the economic importance of the oil shales in the Green River Formation, the abundance of other economically important minerals, such as trona, and the presence of unique "evaporite" minerals, the literature on the Green River Formation is voluminous. Papers that provided background and information for this report include W. H. Bradley (1926, 1931, 1948, 1973, 1974), Picard (1955, 1985), Milton and Eugster (1959), Culbertson (1961, 1966), Cashion (1967), W. H. Bradley and Eugster (1969), Picard and High (1972b), Eugster and Surdam (1973), Brobst and Tucker (1973), Desborough and Pitman (1974), Roehler (1974), Wolfbauer and Surdam (1974), Dyni (1974, 1976, 1987), Eugster and Hardie (1975), Surdam and Wolfbauer (1975), Bucheim and Surdam (1977), R. D. Cole and Picard (1978), Desborough (1978), Surdam and Stanley (1979, 1980a,b), Moncure and Surdam (1980), R. C. Johnson (1981, 1984), R. C. Johnson and Nuccio (1984), Sullivan (1985), Baer (1987), MacLachlan (1987), and Remy and Ferrell (1989), and others listed in chapter 22.

49. The Green River rocks are described in greater detail in chapter 22. See the discussion there for maps and sections.

50. For example, see W. H. Bradley (1963, 1964), Eugster and Hardie (1975), K. D. Stanley and Surdam (1978, 1979), Moncure and Surdam (1980), Surdam and Stanley (1980a), and Dyni and Hawkins (1981).

51. Also refer to Mauger (1977) and Surdam and Stanley (1980a).

52. For example, see Cashion (1967), Picard and High (1972), Cashion and Donnell (1974), D. C. Duncan et al. (1974), Eugster and Hardie (1975), R. C. Johnson (1984), and MacLachlan (1987).

53. Many authors report marl or marlstone as a major rock type. *Marl* is a name that is applied to friable argillaceous limestones and calcareous mudrocks, rocks intermediate between limestone and shale in composition. *Marlstones* are more lithified marls. Inasmuch as mudrocks are defined as those rocks in which the mud sized components make up 50% or more of the rock and limestones are those rocks dominated by calcite, the terms marl and marlstone are unnecessary. Argillaceous lime mudstone, calcareous shale, etc. are used where appropriate.

54. Mineralogy of Green River Formation rocks is discussed by W. H. Bradley (1948), Milton and Eugster (1959), Culbertson (1966), Hosterman and Dyni (1972), Picard and High (1972), Roehler (1972), Tank (1972), Brobst and Tucker (1973), Desborough and Pitman (1974), Dyni (1974, 1976, 1985), Milton (1977), R. D. Cole and Picard (1978), R. D. Cole (1985), and Remy and Ferrell (1989). Desborough and Pitman (1974) indicate that illite makes up less than 5% of the mineralogy of the rocks they studied.

55. Also refer to Roehler (1972), Brobst and Tucker (1973), and Dyni (1976).

PROBLEMS

18.1. Using a geologic time scale and the mineral percentages tabulated in this chapter, plot the percent illite, smectite, and kaolinite in muds and mudrocks against geologic time. Does your graph verify the general conclusions stated in this chapter with regard to the abundances of these minerals over time?

18.2. Plot the major element variation diagrams (e.g., MgO vs. K_2O, MgO vs. Na_2O) using different symbols for each tectonic setting of the mudrock samples tabulated in table 18.3. Do you detect any relationship between chemistry and tectonic setting in this limited data set?

18.3. Average the three Devonian shale analyses in table 18.2 and compare and contrast the Devonian composite with the Ordovician shale (associated with carbonate rocks) (analysis 9). What, if any, mineralogical differences are significant?

CHAPTER 19

Sandstones

INTRODUCTION

Sandstones, like mudrocks, are widely distributed. Though they constitute 5 to 15% of all sediments, they may make up 25% of the continental stratigraphic record (Kuenen, 1941; Blatt, 1970). Sandstones form in a wide range of environments, from alpine glacio-fluvial to deep submarine fan types. Shallow marine, transgressive conditions result in deposition of extensive sand sheets covering thousands of square kilometers. Such conditions likely produced both the Ordovician St. Peter Sandstone, a major unit in the upper Mississippi Valley, and the Silurian Clinch/Tuscarora Sandstone of the central and southern Appalachian Region.[1] Similarly, submarine fan complexes are sediment accumulations with surface areas of thousands of square kilometers, but these are up to 15 km thick and contain tabular to wedge-shaped sand units. Submarine fan complexes are represented by many thick sandstone-shale sequences in the stratigraphic record, including the Great Valley Group of California.

Contemporary sand textures, structures, and lithofacies provide the basis for interpreting ancient sandstones. Included fossils allow discrimination between continental and marine environments, but the specific environments of deposition are determined principally on the basis of structure, lithofacies, and facies associations. Few detailed studies of the compositions of modern sands have been published. Inasmuch as diagenesis has modified ancient deposits, an understanding of diagenesis is also important to a full understanding of sandstone petrogenesis.

Sandstones, in contrast to sands, have received a tremendous amount of study. This is because of the economic importance of sandstone beds. Pure quartz sands and sandstones are valuable as a source of silica in glass manufacturing; sandstones are major reservoir rocks for gas, oil, and water; and sand (and gravel) are used extensively in construction.

SANDSTONE CLASSIFICATIONS AND TEXTURES

Sandstones are rocks dominated by sand particles ranging in size from 1/16 to 2 mm.[2] The term is customarily reserved for siliciclastic and related sedimentary rocks, rather than clastic carbonate rocks, because a special set of names is used to designate the latter. The compositions of sand grains are diverse, but quartz, feldspars, and various types of rock fragments are dominant.

Classifications and the Nature of Matrix

Recall that a large number of sandstone classifications have been proposed. These use either the framework grains or the matrix as a primary disciminator. Cement, which is diagenetic (postdepositional), is ignored.

Also recall that, in particular, the classifications of McBride (1963) and Folk (1968) rely on the composition reflected by the ratios of the framework grains of the rock and do *not* use matrix as a classification criterion. In contrast, the classification of Dott (1964), adopted here, uses matrix materials to distinguish between *arenites*, which have 5% or less matrix, and *wackes*, which have more than 5% matrix.[3] Dott's classification is founded on the suggestion of Pettijohn (1954), that matrix should be considered as a distinct component of sandstones.

In some older classifications, such as those of Krynine (1948), Packham (1954), and Pettijohn (1954), the term **graywacke** was used to designate sandstones in which matrix material, composed of micas, chlorite, or clays, typically constitutes more than 10 or 15% of the rock.[4] This name is still widely used for matrix-rich sandstones, in spite of the fact that it has no place in the commonly adopted classifications of Dott, Folk, and McBride. The continued use of the terms graywacke and wacke (by Dott and others) calls attention to the importance of the matrix in sandstones.

Matrix

Study of modern to Pleistocene marine sands, especially turbidites, that presumably become wackes in the geologic record, reveals that these sands commonly have only modest amounts of clay matrix (<10%) (Shepard, 1961, C. D. Hollister and Heezen, 1964; Kuenen, 1966).[5] Yet many wackes have 10 to 30% matrix. In some rocks and sediments, large amounts of matrix material appears to be primary detrital sediment (J. C. Moore, 1974), but in many others, matrix appears to be secondary. If that is the case, what are the sources of matrix materials and what significance do they have, if any, for petrogenetic studies?

Careful analyses reveal that some sandstone matrices may contain both detrital and diagenetic components. The distinction is important in petrogenetic studies, because detrital matrix is primary and reveals information about the provenance and transportation of the sediment, whereas diagenetic matrix is secondary and reveals information about postdepositional conditions. Matrix materials owe their character to the interaction of three formative processes (Pettijohn, Potter, and Siever, 1987, ch. 2, 5). First, the *weathering and erosion* of the rocks of the provenance provide the primary materials from which some matrix is derived. Two principal types of detrital material are known to contribute to the matrix. These include phyllosilicates—the clays, micas, and chlorites that form primary matrix material—and "labile" rock fragments, those rock fragments easily altered by diagenesis and low-grade metamorphism. Several studies have demonstrated that the provenance exerts the primary control on phyllosilicate composition in sediments (Biscayne, 1965).[6] Similarly, the provenance controls the compositions of lithic fragments.[7]

The second group of processes that influence the nature and amounts of matrix in a rock is the combination of *physical and chemical processes in the environment of deposition*. For example, current velocity and density control the amount of fine-grained matrix materials that may be transported and deposited with sand.[8] In addition, chemical controls, such as the acidity (pH) and oxidation potential (Eh), regulate the stability of various mineral phases during and immediately after deposition. The stabilities of phyllosilicates, in particular, are controlled by the chemistry of the bottom and interstitial waters.[9]

Diagenetic processes comprise the third and final group of processes that influence the nature of the matrix. Diagenesis also results in changes in the composition and amount of matrix materials. Recrystallization, neocrystallization, and deformation of soft, clay-rich rock fragments all play a role in the production of matrix from preexisting detrital materials. Feldspars alter to or are replaced by clay minerals or micas; new chlorite and clays precipitate from intergranular solutions; and other minerals, even quartz, are replaced by clays (W. F. Galloway, 1974; Morad, 1984).

Because most sands containing 10 to 30% matrix do not form directly through deposition (C. D. Hollister and Heezen, 1964; J. F. Hubert, 1964), either depositional or diagenetic processes must be responsible for the origin of additional matrix materials found in wackes.[10] Instructive in this regard are the works of Whetten (1966b) and J. W. Hawkins and Whetten (1969), who showed experimentally that in sediments with a bulk chemistry typical of "graywackes," collected from the Columbia River, matrix can form by alteration of detrital components. The presence of lithic fragments showing varying degrees of recrystallization in wackes suggests that alteration of such fragments contributes to the formation of matrix in these rocks (figure 19.1).

Six different kinds of matrix and cement materials are recognized in sandstones (Dickinson, 1970a). These include (1) detrital, clay-rich mud, or **protomatrix**, (2) recrystallized protomatrix, or **orthomatrix**, (3) deformed and recrystallized lithic fragments, called **pseudomatrix**, (4) polymineralic, diagenetic matrix produced by neocrystallization and alteration of framework grains, called **epimatrix**, (5) homogeneous **phyllosilicate cement**, including smectites, chlorites, chlorite-vermiculite, kaolinites, celadonite, illite, and muscovite, and (6) **nonphyllosilicate cement**, consisting of minerals such as calcite, quartz, dolomite, hematite, phosphate minerals, manganese oxides, and zeolites.[11] Distinguishing the various types of matrix and cement in any given rock may be

Figure 19.1 Photomicrograph showing matrix developing from lithic clasts (arrows) that are somewhat deformed and blend to a limited degree with the protomatrix. Width of photo is 3.25 mm. (PL)

difficult, but detailed petrographic, chemical, and textural analyses can be used to discriminate between detrital and nondetrital constituents (Almon, Fullerton, and Davies, 1976; S. C. Meyer, Textoris, and Dennison, 1987).

Because matrix may result both from deposition (protomatrix) and from diagenesis (orthomatrix, epimatrix, and pseudomatrix), the ideal situation would be to distinguish those rocks with protomatrix matrix from those with diagenetic matrix. Thus, if possible, a rock with a kaolinite epimatrix derived by diagenetic alteration of feldspar could actually be recognized as an arenite, whereas a rock with a detrital illite + chlorite protomatrix would be recognized as a wacke. Such distinctions are often difficult to make in routine practice and, in many cases, they are not possible.

In this text, I follow Dott (1964) in recognizing arenites and wackes, but I distinguish them on the basis of the presence or absence of matrix in amounts of more or less than 5% (figure 19.2). If all matrix materials are unequivocally diagenetic in origin, the rock is called an arenite. In all other cases, sandstones with 5% or less matrix are called arenites, whereas those with more than 5% matrix are referred to as wackes. The term graywacke is not used, except in the context of relating the work of others who have used the term.

Textures

The textures of sandstones are primarily epiclastic, but some are volcaniclastic.[12] Framework grains—dominantly quartz, feldspars, and rock fragments—are joined together by recrystallization of their grain boundaries, either in grain-to-grain or in grain-to-matrix relationships; by crystallization of cement in the rock pores; or by combinations of these processes. The grains may be rounded to angular.

There is no widely accepted terminology for *specific* types of epiclastic textures. Yet textures in which the grains are well sorted and well rounded and in which there is little or no matrix are sometimes designated as "mature" or "supermature" (Folk, 1974, p. 103). In contrast, rocks with moderate to poor sorting and little or no matrix are considered to be "submature" texturally, whereas those with significant matrix are considered to be "immature." The underlying assumption, for which there is some evidence, is that extensive working of sediments will yield matrix-free sediment with rounded grains of only the most resistant types, such as quartz (Johnsson, 1990).

As an alternative to using a maturity index (which is interpretive), some petrographers use descriptive terms relating to roundness and sorting of grains to describe the texture. For example, such a textural designation might read: "poorly sorted, coarse-grained, epiclastic texture with angular grains in a phyllosilicate matrix (7%)." The complete textural term combines a basic term, either *epiclastic* or *volcaniclastic*, with terms indicating the sorting and grain size, the latter based on the C. K. Wentworth (1922) size grades (see table 19.1). Where appropriate, terms describing the matrix or cement (as in "weakly foliated matrix" or "poikilotopic cement") are added. The root names for the rocks—wacke or arenite—indicate the presence or absence of matrix in excess of 5%. Typical sandstone textures are shown in figure 19.2. Where special textural elements are present, additional terms (given in table 19.1) are used to designate those elements.

SANDSTONE COMPOSITIONS

The compositions of sandstones can be described both mineralogically and chemically. As noted above, the main mineralogical constituents of sandstones are quartz (Q), feldspars (F), and lithic or rock fragments (L or R, respectively). The quartz may be single grains or polycrystalline aggregates (i.e., clusters of grains with intergrown grain boundaries), and feldspars include both alkali feldspars and plagioclases (Graham et al., 1976; Dickinson and Suczek, 1979). Rock fragments (R) include all polycrystalline aggregates, including chert.[13] Lithic fragments (L) include only "labile" clasts, such as shale fragments, volcanic rock clasts, plutonic rock clasts, and the metamorphic equivalents of these, including schist, phyllite, slate, greenstone, and serpentinite grains.[14] Various additional minerals, particularly those that are heavier than average or are resistant to abrasion or solution, also occur in sandstones. These include such minerals as hornblende, glaucophane, muscovite, biotite, garnet, epidote, tourmaline, zircon, sillimanite, kyanite, magnetite, and ilmenite. Some minerals, such as alkali feldspar, rutile, and tourmaline, may form diagenetically (Folk, 1974, p. 99; Sibley, 1978). Any rock-forming mineral may occur in a sandstone, given the appropriate conditions of erosion and deposition.

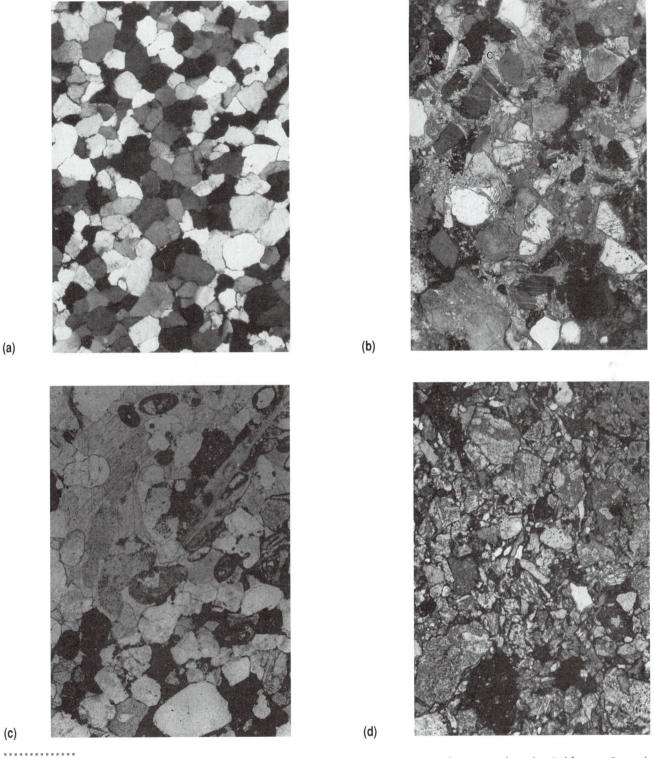

(a)

(b)

(c)

(d)

Figure 19.2 Photomicrographs of typical arenite and wacke showing some sandstone textures. (a) Very well sorted, medium-grained, epiclastic to sutured-textured, quartz arenite, Clinch Sandstone, Silurian, Clinch Mountain Wildlife Management area, Virginia. (XN). (b) Well-sorted, medium-grained, poikilotopic, epiclastic texture in calcareous, feldspathic arenite, Panoche Formation, Great Valley Group, Cretaceous, Carbona Quadrangle, California. C = calcite cement. (XN). (c) Moderately well-sorted, medium- to coarse-grained, oolitic, hematitic quartz arenite, Rose Hill Formation, Silurian, Laurel Bed Lake area, Virginia. (PL). (d) Poorly sorted, medium-grained, epiclastic, lithic wacke, Dothan Formation (Franciscan Complex), Jurassic, Dutchman's Butte Quadrangle, Oregon. Long dimension of all photos is 3.25 mm.

Table 19.1 Textures of Sandstones

Epiclastic Texture A texture consisting of rounded to angular grains, derived by normal processes of surficial weathering, erosion, and abrasion, and bound together through recrystallization of grain boundaries, cementation, or matrix-grain amalgamation.

Very coarse-grained	2–1 mm (2.0–1.0 mm)
Coarse-grained	1–1/2 mm (1.0–0.5 mm)
Medium-grained	1/2–1/4 mm (0.5–0.25 mm)
Fine-grained	1/4–1/8 mm (0.25–0.125 mm)
Very fine-grained	1/8–1/16 mm (0.125–0.0625 mm)

Volcaniclastic Texture A texture consisting of angular grains of volcanic rock fragments, feldspar, quartz, and/or other minerals generated by volcanic processes and deposited as sediment.

Very coarse-grained	2–1 mm (2.0–1.0 mm)
Coarse-grained	1–1/2 mm (1.0–0.5 mm)
Medium-grained	1/2–1/4 mm (0.5–0.25 mm)
Fine-grained	1/4–1/8 mm (0.25–0.125 mm)
Very fine-grained	1/8–1/16 mm (0.125–0.0625 mm)

Special Textural Terms Used as Modifiers

Bouldery A texture in which boulders enclosed in a sand matrix comprise less than 25% of the rock.

Cobbly A texture in which cobbles enclosed in a sandy matrix comprise less than 25% of the rock.

Fossiliferous A texture characterized by included fossil fragments (bioclasts) or other types of fossils.

Granular A texture in which granules enclosed in a sand matrix comprise less than 25% of the rock.

Oolitic A texture characterized by small round, ellipsoidal, to irregularly rounded grains consisting of concentric layers of precipitated material.

Pebbly A texture in which pebbles enclosed in a sand matrix comprise less than 25% of the rock.

Poikilotopic A texture in which framework grains are enclosed by crystals of cement materials (e.g., calcite).

Weakly foliated A texture in which some alignment of grains (usually phyllosilicate grains) gives a faint layered quality to the rock.

Source: From C. K. Wentworth, "A scale of grade and class terms for clastic sediments" in *Journal of Geology,* 30:377–392, 1922. Copyright © The University of Chicago Press. Reprinted by permission.

The modes of several sandstones are presented in table 19.2. Note the extreme variations in mineralogy. Quartz ranges from 2 to 90% and may be higher or lower than these values. Feldspar ranges from a trace to more than 25%, whereas lithic fragments may exceed 50%. Other detrital components typically comprise less than 15% of most sandstones, but matrices may exceed this value and cements may constitute as much as 30% or more of the volume of a sedimentary rock.

Detrital modes of sands and sandstones can provide important information about provenance and depositional setting (Dickinson, 1970a, 1982; Graham et al., 1976; Ingersoll et al., 1984).[15] Modes rich in both feldspar and igneous rock fragments (volcanic and plutonic), and which also contain a metamorphic component, suggest volcanic-plutonic arc sources; whereas those rich in monocrystalline quartz characterize many craton interior-continental block terranes (figure 19.3, table 19.3). Those rich in quartz and lithic fragments typically represent recycled sediment from mountain belts, especially orogens with abundant sedimentary rocks (Critelli et al., 1990). Modes rich in quartz plus feldspar but low in lithic fragments suggest a cratonic provenance (Valloni and Mezzadri, 1984).

Several factors may complicate these interpretations of sediment modes, because compositions depend on transportation, depositional environment, and diagenesis, as well as on provenance (Suttner, 1974). For example, weathering and abrasion may alter Q:F:L values (Johnsson and Stallard, 1989; G. H. Mack and Jerzykiewicz, 1989).[16] Also, depositional sites are not always in the same local tectonic setting as the provenance. Thus, while sandstone compositions may reflect the provenance, they do not necessarily reflect the specific tectonic setting (e.g., transform rift vs. forearc) of the depositional basin (Velbel, 1985; Schwab, 1991). Furthermore, basins may receive sediment contributions from multiple sources, complicating their modal signatures.[17]

Table 19.2 Modes of Selected Sandstones

Grain type	1	2	3	4	5	6
quartz	(90)[a]	(88)	(40)	42	51	(2)
monocrystalline	89	81	39	—	—	1
polycrystalline	tr	7	1	—	—	1
alkali feldspar	tr	1	4	5	<1	4
plagioclase	—	4	6	7	<1	22
muscovite	—	—	tr	—	—	—
biotite	—	—	6	—	—	—
other minerals	tr	—	6	—	—	12
rock fragments						
chert/metachert	tr	1	2	6	—	—
sedimentary	—	tr	—	—	6	—
felsic volcanic	—	—	1	—	—	38
mafic volcanic	—	—	tr	—	—	6
metamorphic	tr	tr	1	—	3	tr
other/undiff.	—	—	—	8	—	tr
cement						
calcite	—	—	33	—	30	tr
other[b]	7	—	—	1	4	—
matrix[c]	2	5	1	31	5	16
Total	100	100	100	100	100	100
Number of points counted	400	650	300	500	400	600
QFL	100:0:0	95:5:0	77:19:4	68:19:13	84:1:15	3:36:61

Sources:

1. Quartz arenite, Clinch Formation (Silurian), Clinch Mountain in Clinch Mountain Wildlife Management Area, Virginia (L. A. Raymond, unpublished data).
2. Quartz arenite, Navajo Sandstone (Lower Jurassic), east entrance to Zion National Park, Utah (L. A. Raymond, unpublished data).
3. Feldspathic arenite, sample C1, Panoche Formation (Upper Cretaceous), Carbona Quadrangle, California (L. A. Raymond, unpublished data).
4. Feldspathic wacke, sample 430.1 of 30-1 Anderson core, Eagle Sandstone (Upper Cretaceous), Bearpaw Mountains, Montana (Gautier, 1981, table 3).
5. Lithic arenite, sample C4A, Trivoli Sandstone (Pennsylvanian), Wayne County, Illinois (Andresen, 1961, table 2).
6. Lithic wacke, sample 10 (Archean), Vermilion District, Minnesota (Ojakangas, 1972, table 2).

aVolume percents based on number of points counted, as indicated.

bCements other than calcite include manganese oxides (column 1), iron carbonates, silica, and clays.

cMatrix materials may include clays, chlorite, and silt- to clay-size quartz, feldspar, or other minerals.

tr = trace.

The bulk rock chemical compositions of sandstones also may reflect the composition of the provenance. Inasmuch as sandstone compositions are commonly modified by diagenesis, however, it is generally not possible to determine the provenance on the basis of chemistry alone. Nearly pure quartz arenites, derived either from preexisting quartz sandstone terranes or from extensive working of sediment are predictably silica-rich (analysis 1, table 19.4). Such rocks characterize cratonic and passive margin environments and have relatively high K_2O/Na_2O (Crook, 1974; Roser and Korsch, 1986). Those sandstones derived from volcanic arcs or terranes composed of a diversity of rock types are generally lower in silica and richer in aluminum, the alkalis, calcium, magnesium, and iron (see tables 19.3 and 19.4, analyses 4 and

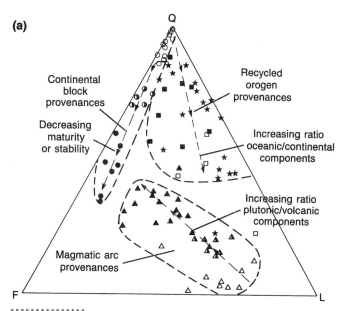

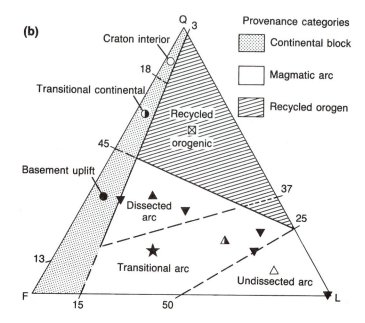

Figure 19.3   QFL diagrams showing the provenances for sandstone with various framework modes. (a) Simplified version of QFL diagram of Dickinson and Suczek (1979), showing fields of the continental block provenance, magmatic arc provenance, and recycled orogen provenance. (b) QFL diagram of Dickinson et al. (1983) showing fields of the three main provenance types and selected subfields. Inverted triangles in (b) represent compositions of some deep-sea sands (see Harrold and Moore, 1975; discussion in text). The star is the average subduction zone sand of Valloni and Maynard (1981).

Table 19.3   Characteristics of Sandstones from Various Provenances

Characteristic	Oceanic Arc	Continental Arc	Orogenic Belt	Craton Interior	Craton Basement
QFL ratio	Q<F≤L	Q≤≥F≤L	Q>L≤F	Q>>F>L	Q≥F>L
Chemistry	Low to intermediate SiO_2	Low to intermediate SiO_2	Intermediate SiO_2	High SiO_2	High SiO_2
	Low K_2O/Na_2O	Low to intermediate K_2O/Na_2O	Low to intermediate K_2O/Na_2O	Intermediate to high K_2O/Na_2O	Intermediate to high K_2O/Na_2O
	Low REE	Moderate REE	LREE>HREE	LREE>HREE	LREE>HREE
Characteristic Sandstone Types	Volcanic lithic wacke	Feldspathic-lithic wacke or arenite	Feldspathic-lithic arenite or wacke	Quartz arenite	Quartz or feldspathic arenite or wacke

Sources: Based in part on Dickinson and Suczek (1979), Bhatia (1983, 1985), and Roser and Korsch (1986).

Table 19.4 Chemical Compositions of Representative Sandstones

	1	2	3	4	5	6
SiO_2	98.91[a]	88.7	76.6	67.3	65.0	60.9
TiO_2	0.05	0.21	0.6	0.6	—	0.6
Al_2O_3	0.62	5.03	12.4	15.5	9.57	16.4
Fe_2O_3	0.09	—	0.7	0.4	1.59	1.4
$FeO*$[b]	—	3.60	—	—	—	—
FeO	—	—	0.2	3.8	1.08	4.4
MnO	—	0.04	—	0.1	—	0.1
MgO	0.02	0.29	0.3	1.9	0.4	3.1
CaO	—	0.43	0.4	0.6	10.1	3.9
Na_2O	0.01	1.14	0.3	4.2	2.14	4.2
K_2O	0.02	0.92	3.8	3.2	1.43	0.6
P_2O_5	—	0.03	—	0.1	—	0.1
H_2O+	—	—	2.7	1.8	0.82	3.7
H_2O-	—	—	—	0.2	0.23	0.5
CO_2	—	—	—	—	6.9	0.1
LOI	0.27	1.02	—	—	—	—
Other	—	tr	—	—	0.31	—
Total	99.99	99.60	100.6	99.7	99.54	100.0

Sources:

1. Quartz arenite, St. Peter Sandstone (Ordovician), Mendota, Minnesota. Analyst: A. William (Thiel, 1935).
2. Quartz wacke, metamorphosed, Pinal Schist (Paleoproterozoic), near Globe, Arizona. (Condie and DeMalas, 1985).
3. Arkose (Oligocene), Auvergne, France (Huckenhotz, 1963).
4. Feldspathic wacke, Franciscan Complex (Jurassic-Cretaceous), San Bruno Mountain, San Francisco Peninsula, California (E. H. Bailey, Irwin, and Jones, 1964).
5. Calcareous lithic arenite, "Frio" (Oligocene), Texas, composite of ten samples from Wells and Kleberg counties (Nanz, 1954).
6. Lithic (volcanic) wacke, Franciscan Complex (Jurassic-Cretaceous), San Francisco, California (E. H. Bailey, Irwin, and Jones 1964).

[a]Values in weight percent.

[b]Total iron as Fe_2O_3.

6). Oceanic magmatic arcs typically yield low-silica wackes with low K_2O/Na_2O (Crook, 1974; Roser and Korsch, 1986). Where depositional or diagenetic processes have introduced precipitated components such as hematite, calcite, or apatite, the bulk chemistry is correspondingly modified and may be substantially reduced in silica.

Trace element studies also reveal information about provenance and depositional setting (J. N. Weber, 1960; Bhatia, 1985).[18] For example, Bhatia (1985) showed that oceanic arc wackes from Paleozoic terranes in eastern Australia have relatively low REE abundances and generally lack a negative europium anomaly. In contrast, continental arc wackes show relatively higher REE abundances and a small negative Eu anomaly. In wackes from passive margin to cratonic basins, LREE are enriched over HREE and there is a pronounced negative Eu anomaly. Discrimination diagrams based on specific trace elements (e.g., La-Th-Sc diagrams) also show differences between suites of sandstones and reflect sand provenance (Bhatia and Crook, 1986; Larue and Sampayo, 1990).

SANDSTONE STRUCTURES

Sandstones exhibit a large variety of structures ranging from macroscopic to mesoscopic scales (Pettijohn and Potter, 1964). These structures are listed in table 19.5, and several were discussed and illustrated in chapter 14.[19] Primary structures, at the largest scale, consist of beds of sandstone in tabular, lenticular, wedge, or shoestring shapes. The largest of these sandstone bodies may constitute formations and members of formations. Beds of various shapes also range in size down to small-scale mesoscopic features (figure 19.4). Filled channels, which form lenticular beds, range from macroscopic to mesoscopic in scale. At the mesoscopic scale, inorganic features within beds, such as crossbeds, laminations, cross-laminations, and grading, are characteristic primary structures. Primary surface structures include sole marks, trace fossils, and ripple marks (see figures 14.10, 14.11, and 19.4). Such primary structures are syndepositional; that is, they form at the time of deposition.

Soft sediment and postdepositional structures are also numerous. These include such bedding features and discontinuities as mudcracks, soft-sediment folds, convolute beds (figure 19.4), bioturbated beds, ball-and-pillow structures, soft-sediment faults, and sandstone dikes. Biogenic structures, such as burrows and escape structures also are postdepositional.

OCCURRENCES AND ORIGINS OF WACKES

Wackes occur in continental, transitional, and marine sequences. In continental environments, they form in alluvial fans, in fluvial channels and on floodplains, and in lacustrine deltas. Transitional environments in which wackes develop include estuarine, deltaic, and, to a lesser degree, tidal flat–strandline types. In marine environments, wackes may form on the shelf, but are characteristic of slope contourites and basin plain turbidites.

Turbidites and Related Rocks

Turbidites form from turbidity currents that develop wherever sediment accumulations become unstable and move downslope in turbid flows. Such currents have been recognized or inferred to have occurred in environments ranging from continental lakes to deep marine basins.[20] Nevertheless, most turbidites in the geologic record probably represent submarine deposits formed in trenches, in slope basins, or on continental rise/basin plain sites.[21]

Turbidites are characterized by complete or partial Bouma sequences (see figures 14.5 and 14.6). These sequences, which contain beds that range from very thin to

Table 19.5 Sedimentary Structures Found in Sandstones

Depositional Structures

Bedding	Laminations
Cross bedding	Cross-laminations
Ripple marks	Graded bedding
Salt crystal casts	

Erosional Structures

Channels	Flute casts
Tool marks	Load casts
Mudcracks	Burrows
Tracks and trails	Rip-ups

Deformational Structures

Soft-sediment folds	Penecontemporaneous faults
Slump or slide scars	Slump or slide casts
Breccias	Sand or mud volcanoes
Sandstone dikes and sills	Convolute laminations
Flame structures	Dish structures
Fluid escape channels	Organic escape structures
Ice wedge casts	Root casts and molds
Ball-and-pillow structure	

Diagenetic Structures

Concretions	Sand crystals
Stylolites	Liesegang bands or rings

Sources: Conybeare and Crook (1968), Collinson and Thompson (1982), Pettijohn, Potter, and Siever (1987), and observations by the author. For descriptions of these structures and discussion of their origins, see chapter 14 of this text and the sources listed above.

thick, form the basis for recognizing the submarine fan lithofacies C, D, and E of Mutti and Ricci-Lucchi (1972). In general, these lithofacies occur in channel, fan lobe, interchannel, and basin plain areas (Mutti and Ricci-Lucchi, 1972; Shanmugam and Moiola, 1985), but they may also occur on submarine ramps (Postma, 1981a; Chan and Dott, 1983; Heller and Dickinson, 1985; Surlyk, 1987). Related sand (grain) flow deposits, constituting submarine fan lithofacies A and B of Mutti and Ricci-Lucchi, are massive- to medium-bedded. These rocks form in submarine canyons and fan channels. In the geologic record, both turbidites and sand flow deposits commonly are composed of wackes.

(a)

(b)

(c)

Figure 19.4 Photographs of selected structures in sandstones. (a) Bedding in a sandstone-mudrock sequence of the Hinton Formation, Mississippian, Highway 460 west of Princeton, West Virginia, showing penecontemporaneous faulting (A) in bedding (B), with a sandstone-filled channel (C). Stake at lower right is about one meter in length. (b) Overturned convolute bedding and laminations in arenite, Panoche Formation, Great Valley Group, Upper Cretaceous, Hospital Canyon, northeastern Diablo Range, California. (c) Rippled, mudcracked, burrowed quartz arenite (circular to elliptical dark spots are the ends of *Scolithus* burrows), Clinch Sandstone, Silurian, Powell Mountain, near Duffield, Virginia.

Two curious aspects of contemporary submarine fans confound their use as analogues of ancient turbidite sequences. First, the largest contemporary fans are much larger than the largest ancient fans of the geologic record (Normark, Mutti, and Bouma, 1983; N. E. Barnes and Normark, 1985). Notably, the area of the modern-day Bengal Fan in the northeast Indian Ocean is larger than that of most of the well-known modern and ancient fans *combined* (Curray et al., 1982; N. E. Barnes and Normark, 1985). Second, the dominant grain size of ancient turbidite sequences is sand size, whereas that reported for most modern fans is clay or silt size (N. E. Barnes and Normark, 1985).

Contourites and Shelf Wackes

Wackes also form on the shelves and as contourites on the slopes. **Contourites** are rocks composed of sediment deposited by currents that flow *along* submarine slope contours rather than down the slope. They form where bottom currents flowing parallel to the slope rework and transport previously deposited sediment (Heezen, Hollister, and Ruddiman, 1966; Bouma and Hollister, 1973).[22] Sandstone contourites are characterized by thin beds and cross-laminations or bioturbation structure (Lovell and Stow, 1981).

Contourites may develop in any large body of water in which bottom currents are able to develop. Typically, they occur in marine sequences, but they also occur in lake deposits (T. Johnson, Carlson, and Evans, 1980).

On the shelf, offshore bars composed of muddy sands have produced wackes. For example, wackes of the Upper Cretaceous Woodbine-Eagleford Formations of Texas are interpreted to have developed in such a setting (Turner and Conger, 1984). These rocks formed from sediments deposited by longshore currents and storms.

Transitional Environment and Continental Wackes

Transitional environments along the continent-ocean boundary, especially deltaic environments, serve as depositional sites for some wacke-type sands.[23] Although arenites are typical of transitional environments such as strandlines, the wackes are commonly associated with mudrocks on deltas, on mudflats, and in tidal channels. An example of a deltaic wacke is the Davis Sand of the Yegua Formation of the Texas Gulf Coast, which contains 9.5% mud (S. R. Casey and Cantrell, 1941, in Shelton, 1973).

Alluvial wackes and their parental sands are also relatively common. Some Mississippi River sands collected in Iowa, for example, contain significant amounts of mud (Lugn, 1927). Equivalent sandstones are included in the Pennsylvanian Anvil Points Sandstone of the Illinois Basin (Hopkins, 1958).[24] These alluvial sandstones typically form cross-bedded, shoestring units characterized in outcrop by channels.

Thorough studies of contemporary environments that include detailed petrographic analyses of sand-size sediment are remarkably few in number. Nevertheless, a few studies of Pleistocene to Recent sands and studied examples of ancient wackes are instructive in characterizing the nature of these rocks.

Example: Deep-Sea Sands

Many of the thick sections of wacke found in the geologic record are interpreted to represent deep-sea sands deposited on submarine fans in forearc basins and trenches. The tectonic, structural, and stratigraphic framework of such settings have been described in detail (J. C. Moore and Karig, 1976; Dickinson and Seely, 1979; Thornburg and Kulm, 1987).[25] Yet, few studies have detailed the compositions of deep-sea sands and extensive studies from specific sites have not been published.

In general, the lithofacies of deep-sea sands are those of the submarine fan facies of Mutti and Ricci-Lucchi (1972), particularly Facies C, D, and E marked by the Bouma sequence. A summary of recent deep-sea sand compositions from these lithofacies, including some from trench-slope sites in the western Pacific Ocean (Harrold and Moore, 1975), reveals an overall average composition for such (subduction zone) settings of $Q_{16}F_{53}L_{31}$ (Valloni and Maynard, 1981) (figure 19.3). Either lithic-volcanic (Lv) or lithic-sedimentary (Ls) clasts may be dominant in a given region. For example, at DSDP site 298 on the inner wall of the Nankai Trough (a trench southeast of Japan), seven samples of Pleistocene to Recent sand, six of which are Ls-rich, average $Q_{16}F_{17}L_{67}$. Sands from the Middle America Trench (DSDP site 570), however, are L_v-rich, with estimated QFL values of 1:1:98 to 22:13:65, and in these sands the lithic clasts are almost entirely volcanic glass (von Huene et al., 1985). These data are comparable with more voluminous data from Jurassic-Cretaceous sedimentary rocks (presented below) that are considered to have formed in similar settings.

Example: Wackes of the Franciscan Complex, California

The Franciscan Complex is a lithologically diverse, structurally complicated rock body that forms the basement for a major part of the Coast Ranges of California, as well as parts of the Klamath Mountains of California and Oregon, the southern Coast Ranges of Oregon,[26] and parts of Baja California (figure 19.5) (E. H. Bailey, Irwin, and Jones, 1964; Berkland et

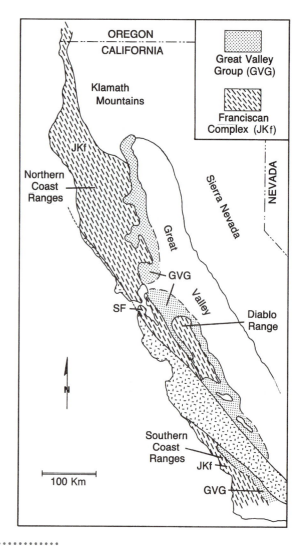

Figure 19.5 Generalized map showing the distribution of the Franciscan Complex (JKf) and Great Valley Group (GVG) in California. SF = San Francisco.

al., 1972; E. M. Baldwin, 1976). Major rock types in the Franciscan Complex include wacke ("graywacke"), shale, conglomerate, breccia, chert, and basalt, plus the metamorphic equivalents of these rock types. In addition, there are serpentinites and related ultramafic rocks, gabbros and metagabbros, arenites, marbles and limestones, eclogites, and various glaucophane schists (most of which are metabasalts and metagabbros). By far the most abundant rock types are wacke and metawacke (E. H. Bailey, Irwin, and Jones, 1964). An estimate of the volume of the sedimentary and metasedimentary rocks indicates that they represent over 600,000 km[3] of rock.

The wackes have been variously attributed to sand deposition in continental (E. F. Davis, 1918), shallow-marine (Taliaferro, 1943), and deep-marine (E. H. Bailey, Irwin, and Jones, 1964) environments. The latter is favored by the general lack of fossils,[27] the presence of deep-marine trace

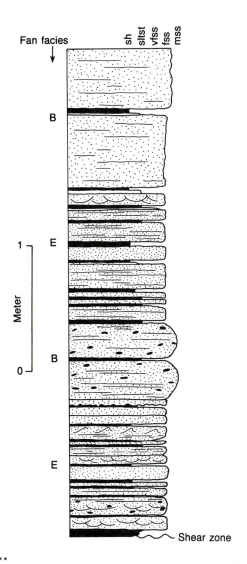

Fan facies

sh sltst vfss fss mss

B

E

1

Meter

0

B

E

Shear zone

Figure 19.6 A measured section showing submarine fan facies in the Franciscan Complex, West Fork of Hospital Creek, Carbona 15′ Quadrangle, northeastern Diablo Range, California.

(a)

(b)

(c)

Figure 19.7 Photographs of submarine fan facies in the Franciscan Complex. (a) Facies F olistostrome containing pillow-basalt block (at right near hammer) near Helsinger Canyon, northeastern Diablo Range, California. Hammer is 1/3 m long. (b) Facies B wackes, West Fork of Hospital Canyon, Carbona Quadrangle, northeastern Diablo Range, California. Scale bar is 30 cm. (c) Turbidites (Facies C and E), Sulphur Gulch Broken Formation (of Raymond, 1973a), northeastern Diablo Range, California. Middle sandstone bed at right center is approximately 8cm thick.

fossil assemblages, the general absence of associated carbonate beds, the lack of large-scale crossbeds and ripple marks (expected in shallow marine and continental deposits), and the presence of interbedded and associated radiolarian cherts (see chapter 22). Except for radiolaria, which are pelagic and therefore do not directly indicate water depth, the only abundant fossil remains appear to be specimens of a rich assemblage of trace fossils of the *Nereites* ichnofacies, a group of trace fossils indicative of bathyl to abyssal environments (W. Miller, 1986, 1991). The presence of submarine fan lithofacies A–F, including Facies C graded beds, indicates that turbidity currents, grain flows, and debris flows deposited the Franciscan wackes and associated olistostromes and conglomerates (figures 19.6 and 19.7).

Table 19.6 Chemical Compositions of Wackes

	Franciscan Wackes			Average "Graywackes"		
	1	2	3	4	5	6
SiO_2	58.4[a]	67.3	71.72	57.2	69.0	85.0
TiO_2	0.5	1.8	0.35	1.0	0.6	0.3
Al_2O_3	14.2	12.4	13.23	16.0	11.7	7.2
Fe_2O_3	2.4	0.6	0.30	—	—	—
$FeO*$[b]	—	—	—	8.8	6.2	4.2
FeO	1.4	4.0	3.58	—	—	—
MnO	0.2	0.1	—	0.1	tr	tr
MgO	1.2	2.3	1.81	3.4	nr	0.1
CaO	8.2	3.3	1.80	5.4	2.6	0.1
Na_2O	3.3	3.0	2.72	5.0	2.0	0.8
K_2O	2.0	1.2	1.29	0.7	1.5	1.7
P_2O_5	0.1	0.1	0.09	0.2	nr	tr
H_2O+	—	2.5	2.53	—	—	nr
H_2O-	3.1	0.3	0.15	—	nr	—
CO_2	4.8	0.6	0.32	—	nr	—
LOI	—	—	—	2.0	—	1.0
Other	—	—	0.04	—	tr	—
Total	100	99.5[c]	99.93	99.8	93.6	100.4

Sources:

1. "Graywacke," Franciscan Complex (Cretaceous?), about 2 km north of Reese Gap, Sonoma County, California. Analysts: P. L. D. Elmore, I. H. Barlow, S. D. Botts, and G. Chloe (E. H. Bailey, Irwin, and Jones, 1964, table 1, analysis 11).

2. "Graywacke," Franciscan Complex (Jurassic-Cretaceous), New Almaden District, Santa Clara County, California. Analyst: A. C. Vlisidis (E. H. Bailey, Irwin, and Jones, 1964, table 1, analysis 2).

3. "Sandstone" (metawacke), Franciscan Complex (Jurassic?), junction of Buckeye Gulch and Hospital Canyon, Carbona Quadrangle, California. Analyst: Herdsman Laboratory, Glasgow (Taliaferro, 1943, table 3, analysis 3, p. 136).

4. Graywackes, average of 10 analyses, Sierra Madre Mountains, Wyoming (J. R. Reed and Condie, 1987, table 2).

5. Graywackes, average of 6 analyses, Gazelle Formation (Silurian), near Yreka, California (Condie and Snansieng, 1971, table 2, columns 2 and 3).

6. High-quartz graywackes, average of 2 analyses, (Precambrian), Mazatzal Mountains, Arizona (J. R. Reed and Condie, 1987, table 2).

[a]Values in weight percent.

[b]Total iron as Fe_2O_3.

[c]Original analysis, reported to hundredths place, totalled 99.41

nr = not reported.

tr = trace.

Mineralogically and chemically, Franciscan wackes are typical of "graywackes" from around the world (tables 19.6 and 19.7). Both lithic and feldspathic varieties are present, but quartz wackes are absent (E. H. Bailey, Irwin, and Jones, 1964; Jacobson, 1978; Dickinson et al., 1982; Underwood and Bachman, 1986).[28] Franciscan wackes contain abundant plagioclase and volcanic lithic fragments. Alkali feldspar is minor to absent and chert is generally moderately abundant. QFL plots of Franciscan wackes show that they are dominantly the product of erosion of magmatic arc sources, but the youngest wackes (Upper Cretaceous to Eocene) have QFL values suggesting a mixed cratonal–orogenic-arc provenance. The most common heavy minerals are apatite, biotite, chlorite, clinozoisite and epidote, garnet, hornblende, ilmenite, magnetite, muscovite, sphene, and zircon (E. H. Bailey, Irwin, and Jones, 1964). Uncommon oxyhornblende, augite, and hypersthene among the heavy minerals, combined with the abundance of volcanic lithic fragments and plagioclase feldspar, indicate a volcanic arc source for much of the sediment. A wide variety of additional *detrital* heavy

Table 19.7 Modes of Franciscan and Other Wackes

	Franciscan Wackes				Other Wackes		
Grain Type	**1**	**2**	**3**	**4**	**5**	**6**	**7**
quartz	(33.7)[a]	(20.3)	(22.2)	38	5.9	(36.8)	61
monocrystalline	15.7	13.5	17.8	—	—	22.4	—
polycrystalline	18.0	6.8	4.4	—	—	14.4	—
alkali feldspar	—	0.2	0.4[b]	2.7	5.9	16.9	4
plagioclase	18.6	10.5[c]	23.4[c]	33	6.1	5.5	
chlorite/serpentine	0.3	0.5	—	3	3.0	—	—
biotite	0.6	—	—	1	—	—	—
white mica	0.9	—	—	—	—	—	—
other minerals	0.3	2.0	1.0	0.2	4.0	2.9[d]	1
rock fragments	—	—	—	6.1		18.0	14
chert	5.4	8.2	8.2	5	26.9[e]	—	—
sedimentary	4.5	0.5	1.6	—	14.6	—	—
felsic volcanic	4.9	9.3	4.2	—	1.7	—	—
mafic volcanic	0.3	7.8	0.8	—	11.6	—	—
metamorphic	4.5	8.2	1.6	—	—[f]	—	—
other/undiff.	0.6	0.2	0.2	—	1.0	—	—
Cement	—	2.8	0.6	—	—	4.5	—
Matrix	25.4	29.5	35.8	11.0	19.3	15.4	20
Total	100	100	100	100	100	100	100
Number of points counted	350	400	500	nr	1000	500	nr
Q/F/L	54:26:20	50:19:31	50:39:11	48:45:7	49:9:42	48:29:23	77:5:18

Sources:

1. "Graywacke" (lithic-feldspathic wacke), Buckeye-Grummett Dismembered Formation, Franciscan Complex (Jurassic?), Northeastern Diablo Range, California (unpublished data L. A. Raymond).

2. "Graywacke" (lithic wacke), Franciscan Complex (Jurassic-Cretaceous), Phelan Beach, San Francisco, California (unpublished data, L. A. Raymond).

3. "Graywacke" (feldspathic wacke), Coastal Belt, Franciscan Complex (Cretaceous), west of Willits, California (unpublished data, L. A. Raymond).

4. "Graywackes," average of 27 samples, Bohemian Grove unit, Coastal Belt (?), Franciscan Complex (Cretaceous?), Occidental-Guerneville area, California (W. P. Christiansen, 1973, table 1, column 3).

5. "Graywackes," average of six samples, Gazelle Formation (Silurian), near Yreka, California (data from Condie and Snansieng, 1971, table 1, columns 3 and 5).

6. "Graywacke" (lithic-feldspathic wacke), Visingso Group (Neoproterozoic), Lake Vattern area, Sweden (Morad, 1984, sample 1, table 1).

7. "Greywacke" (lithic wacke), Oaklands Formation (Cambro-Ordovician), southeast Ireland (Shannon, 1978, sample 1, table 1).

[a]Volume percents based on number of points counted, as indicated.

[b]Untwinned feldspars included with plagioclase.

[c]Zeolites and other minerals replacing plagioclase feldspar included with plagioclase.

[d]Includes all others not specifically listed, including micas.

[e]Includes quartzite.

[f]Phyllite and slate included with sedimentary fragments.

nr = not reported.

Q = monocrystalline and polycrystalline quartz + chert,

F = total feldspar, L = all labile lithic fragments.

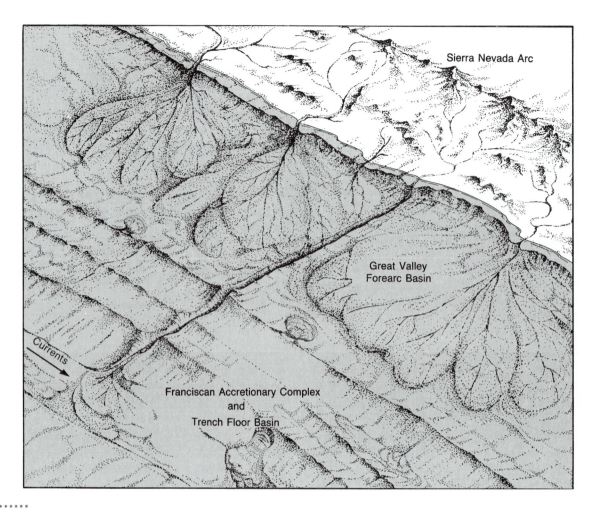

Figure 19.8   Paleogeographic model for Franciscan Complex and Great Valley Group deposition during the Tithonian (late Jurassic) to Eocene epochs along the west coast of North America.

minerals is present, including tourmaline, kyanite, andalusite, actinolite, and glaucophane, suggesting derivation of portions of the sediment from both high pressure–low temperature and high temperature–low pressure metamorphic terranes.

The matrix of Franciscan wackes is a combination of protomatrix, orthomatrix, epimatrix, and pseudomatrix. In the western part of the Franciscan terrane, where diagenesis and metamorphism are less severe than in the east, epimatrix and orthomatrix predominate. Chlorite, white mica, and mixed-layer phyllosilicates are important matrix materials. In the east, psuedomatrix is an important constituent of the matrix, and typically the minerals chlorite, chlorite-vermiculite mixed-layer minerals, white mica, and local stilpnomelane combine with jadeitic pyroxene, lawsonite, glaucophane, or pumpellyite to form a metamorphic matrix in the rocks.

Chemically, the wackes belong to the intermediate category of Crook (1974), in which silica, which *averages* 68–74%, reflects quartz contents of 15–65%. The K_2O/Na_2O values of <1, combined with the silica contents, indicate an active continental margin/Andean-like arc setting (E. H. Bailey, Irwin, and Jones, 1964; Crook, 1974; Roser and Korsch, 1986).

Together, the data indicate that Franciscan wackes were deposited along the continental margin seaward of a volcanic-plutonic arc (see table 19.3, figure 19.8). Coeval sandstones of the Great Valley Group (see below) were deposited in an arc-trench gap, whereas Franciscan rocks were deposited in trench-slope basins, in a trench, and on the adjoining ocean floor (W. B. Hamilton, 1969; Bachman, 1978; Ingersoll, 1982).[29] Much of the sediment was derived by erosion from the arc, but some sediments apparently were derived from other sources including uplifted trench sediments previously deposited in the trench and then accreted to the continental margin to form the trench slope. Turbidity currents and grain flows transported sands from the arc and the eroding trench-slope break into the basins, where they formed submarine channel fan complexes (e.g. Aalto, 1982; Aalto and Murphy, 1984).

OCCURRENCES AND ORIGINS OF ARENITES

Arenites form from sands that are transported and deposited by agents that sort the sediment well, separating mud and silt from the sand grains. During this process, fine-grained sediment is winnowed out and washed away, whereas coarse-grained sediment is not transported to the depositional site. Agents capable of these particular conditions of transportation and deposition include submarine currents of relatively uniform flow velocity, longshore currents, waves, and the wind. Alternatively, if the *only* sediment available in a source terrane is well-sorted sand, fluvial currents and turbidity currents may transport and deposit sand that yields an arenite. Arenites may also develop where weathering, transportation, and diagenesis remove fine-grained and soluble components of a sediment.

Arenites may contain up to 5% matrix. Matrix materials represent (1) minor protomatrix deposited with the sand or infiltrated into the sand immediately following deposition, (2) epimatrix derived from modification of detrital grains, especially the feldspars, and (3) orthomatrix and pseudomatrix.

Most arenites have epiclastic textures, equigranular sutured textures, or equigranular mosaic textures. Various types of cement and matrix materials may bind the grains, but grains may also be joined by interlocking grain boundaries produced by recrystallization during diagenesis. Some equigranular mosaic textures result from overgrowths, especially of quartz on quartz framework grains. Poikilotopic textures are common in carbonate-cemented arenites, whereas cements composed of silica minerals, zeolites, and phyllosilicates form radial fibrous, comb-textured, fibrous drussy, or spherulitic textures (Hoholick, Metarko, and Potter, 1984).

The most distinctive arenites are the pure quartz arenites, which have very few grains that are not quartz (figure 19.9, and see figure 19.2a). Pure quartz sands may develop from (1) extensive working of sediment from a quartz-bearing source terrane, (2) reworking of previously well-sorted, mature sands, (3) deep weathering of quartz-bearing rocks in the source terrane, or (4) combinations of these processes. Such sands form dune sands, aeolian sand sheets, strandline stringer sands, and the blanket sand deposits of marine shelves and epicontinental seas. Thus, both continental and marine environments are represented. Upon lithification, such sands become quartz arenites. Examples of quartz arenites include the St. Peter Sandstone of the mid-continent region, the Navajo Sandstone of the Rocky Mountain Region, and the Clinch-Tuscarora and Keefer formations of the central and southern Appalachian Valley and Ridge Province.[30]

Quartz arenites also develop where weathering and diagenetic destruction of rock fragments, feldspars, and other grains, leaves a quartz-rich residue (D. W. Lewis, 1984;

McBride, 1984, in Chandler, 1988; M. J. Johnsson, 1990). The former process occurs before or during transportation, whereas the latter may occur during or after lithification.

Lithic arenites and feldspathic arenites (i.e., arkoses) are rather abundant locally. Both contain abundant quartz, but lithic arenite is distinguished by its abundant rock fragments, whereas feldspathic arenite is characterized by abundant feldspar (figure 19.9). The particular conditions necessary to form feldspathic arenites have been a matter of some debate.[31] Clearly, a feldspar-bearing or feldspar-rich provenance is required. Beyond that, it is essential that the feldspar neither weather away before erosion, transportation, and deposition nor be abraded away during transportation. Conditions also must be such that diagenesis does not destroy the feldspar during lithification. It might seem that both arid conditions, where decomposition is limited, and short, single-cycle transportation histories are required for the preservation of feldspar, yet the presence of second-cycle feldspars in a tropical Mexican climate (Krynine, 1935) indicates that such conditions are not the only ones that can yield feldspathic sands. High relief and rapid erosion also contribute to the preservation of feldspars. Thus, Pettijohn, Potter, and Siever (1987, p. 155) conclude that feldspathic sands may result *either* from a "rigorous climate," in which decomposition is inhibited, *or* from accelerated erosion in an area of high relief.

Lithic arenites reflect a provenance with sedimentary rock, metamorphic rock, volcanic rock, or combinations of these rock types exposed at the surface. Fluvial transport commonly provides the necessary transportation and sorting of grains, but turbidity currents may also deposit lithic sands. Examples of alluvial lithic arenites include some Pennsylvanian sandstones associated with coal beds in the eastern United States, such as some sandstones of the Gizzard Formation of Lookout Mountain, in Georgia and Alabama, and those of the Pottsville Formation of Pennsylvania.[32] Some sandstones of the Great Valley Group of California (described below) are good examples of lithic arenites deposited by turbidity currents and grain flows.

The Great Sand Sheets

The geologic record contains a number of large blanket sandstone bodies that cover thousands of square kilometers. Essentially all such bodies are composed of arenites. It is generally believed that these sandstones resulted from deposition in broad **epeiric seas**, great shallow-water shelf areas that form when the oceans flood low-lying continental margins. Alternatively, such sands may have formed along the transgressive shorelines of such seas (Driese, Byers, and Dott, 1981). Arenites also form in desert or shoreline dune fields. Unfortunately, there are no contemporary analogs for many epeiric-sea depositional environments (M. L. Irwin, 1965), though some modern shelves, such as the northwest Florida shelf (Hine et al., 1988), do have some similarities.[33]

383

Figure 19.9 Photomicrographs of arenites. (a) Quartz arenite, St. Peter Sandstone, Ordovician, Missouri. (XN). (b) Quartz arenite, Navajo Sandstone, Jurassic, east of Zion National Park, Utah (see mode 2 in table 19.2). (XN). (c) Feldspathic arenite, Panoche Formation, sample C-1 (see mode 3 in table 19.2), Cretaceous, Hospital Canyon, northeastern Diablo Range, California. (XN). (d) Lithic arenite, Myrtle Group (Jurassic-Cretaceous), Canyonville area, Oregon. (PL) Long dimension in all photos is 1.27 mm.

The shallow-marine depositional environments of epeiric-sea blanket sandstones is indicated by associated carbonate rocks with shallow-marine benthic fossils, extensive bioturbation and trace fossils, nearly ubiquitous cross-bedding in the sandstones, ripple marks, and common channeling. The types of cross-beds and associated sedimentary structures make possible the discrimination between such shallow marine rocks and similar-appearing nonmarine rocks (Driese, Byers, and Dott, 1981; Cudzil and Driese, 1987).

Aeolian arenites are formed in dune fields and in associated interdune sand flats. The rocks are typically cross-bedded and composed of very well sorted sand. Commonly the sand is well rounded, but this is not true in every case.[34] Associated evaporites and alluvial fan deposits indicate desert environments for aeolian arenites, whereas associated shallow marine facies reflect coastal dune environments.

Shoestring Sandstones

Shoestring arenites represent fluvial environments, offshore bars, or beaches.[35] Along shorelines, wave action sorts the sand well.

Distinguishing between fluvial and shoreline arenites is easily accomplished on the basis of the associated lithofacies, the sedimentary structures, and the enclosed fauna. Both types of environments yield cross-bedded sands, but those of the shoreline are less commonly associated with major channels and are generally lower angle. Associated facies, such as coal-bearing floodplain mudrocks and fluvial conglomerates, also indicate the fluvial character of the continental sandstones, whereas associated marine mudrocks and fossiliferous carbonate rocks indicate shoreline environments.

Examples of Sands and Sandstones

The St. Peter Sandstone, North-Central North America

The Ordovician St. Peter Sandstone forms a blanket quartz arenite in the midwestern upper Mississippi Valley and Michigan Basin areas of North America.[36] The unit is commonly less than 50 m thick, reaches a maximum of only 140 m thick, but originally covered well over 500,000 km^2 (Dake, 1921; Dapples, 1955; A. H. Bell et al., 1964).

A marine origin is indicated for much of the St. Peter Sandstone by the structures, fossils, and associated marine rocks. Fossils include the trace fossil *Scolithus* sp. and local occurrences of pelecypods, gastropods, cephalopods, and bryozoans (Sardeson, 1910, in Dake, 1921, p. 195; Dake, 1921; Lamar, 1927). Structures within the formation include thin to massive tabular beds, large- and small-scale cross-strata, and ripplemarks. The larger sets of cross-strata

are 10–15 m thick. Convex upward and convex downward types of cross-beds are present and the foreset beds dip toward the southwest. This direction parallels the regional paleocurrent pattern; and together the data suggest a northeast-southwest trending shoreline.[37] The large scale of the cross-beds suggests that they represent the internal stratification of beach deposits, littoral to neritic migrating dunes, and a barrier island complex. The beds overlying the St. Peter Sandstone represent a muddy, back-barrier lagoon sequence and an overlying marine carbonate unit (G. S. Fraser, 1976).

The sandstones are rather pure, fine- to coarse-grained quartz arenites. Monocrystalline quartz exceeds 99% of the grains in most rocks, the remainder being composed of traces of heavy minerals, such as zircon or tourmaline, and clays and iron oxides (Lamar, 1927; Thiel, 1935). This mineralogy is reflected in a chemistry that typically consists of 98 to 99.5% SiO_2 and only very minor amounts of other oxides (e.g., see table 19.4, column 1).

Cements include dolomite, calcite, chalcedony, quartz, anhydrite, and minor chlorite (Hoholick, Metarko, and Potter, 1984). Calcite and dolomite are the most common. The cements occur in depth-controlled zones, except for quartz overgrowths, which occur everywhere. Chalcedony and calcite, calcite and dolomite, and anhydrite form the cements in shallowly, moderately, and deeply buried rocks, respectively.

Texturally, the St. Peter rocks are epiclastic and moderately well to very well sorted (figure 19.9a). Grains are commonly very well rounded, and, in many cases, frosted, but fine grains tend to be subangular (Dake, 1921; A. H. Bell et al., 1964). The well-rounded grains have surface textures indicative of a phase of aeolian transport (Mazzullo, 1987). Common irregular grains have surface textures and compositions that suggest both aqueous transport and composite sources (with sandstone, igneous rocks, and metamorphic rocks in the source terrane) (Mazzullo, 1987). Grain sizes are uniformly fine, medium, or coarse, depending on the sample. Cements give the rocks poikilotopic–, radial fibrous–, comb–, fibrous drussy–, or spherulitic – epiclastic textures.

Together, the data suggest that the St. Peter Sandstone is a blanket sandstone initially formed as beach to near-shore sands deposited in a transgressing epeiric sea. The shoreline extended from northeast to southwest and perhaps migrated northwestward. Sand derived from preexisting sandstones as well as from some igneous and metamorphic rocks, probably exposed on the craton to the north and west, were mixed with wind transported sands before and during deposition. Longshore currents and waves worked the sands, ultimately depositing them on beaches and in bars and submarine dunes. Diagenetic precipitation of cements resulted in the final textures of the sandstones.

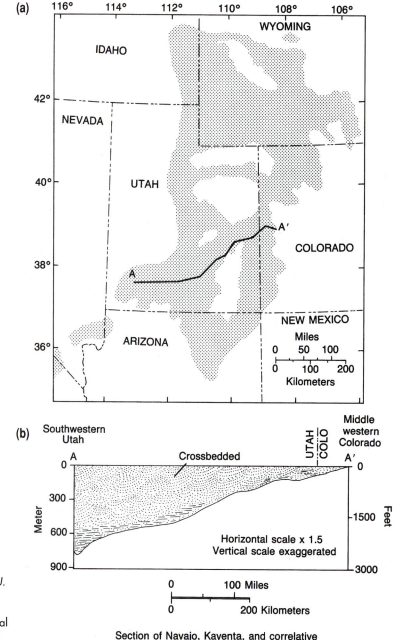

Figure 19.10 Map (a) and section (b) of the distribution of Navajo Formation and related quartz sandstones in the western United States.

(Source: E. D. McKee and J. J. Bigarella, "Sedimentary structures in dunes" in E. D. McKee, ed., A Study of Global Sand Seas in U S Geological Survey pp. 1052, 1979.)

The Navajo Sandstone of the Rocky Mountains–Colorado Plateau Region

Spectacular cross-beds characterize the nonmarine, Lower Jurassic Navajo Sandstone.[38] This formation and its equivalents, including the Aztec Sandstone of southern Nevada and California and the Nugget Sandstone of northwestern Colorado, northern Utah, Idaho, and Wyoming, form a west-ward-thickening wedge that reaches thicknesses in excess of 2500 m (figure 19.10) (Gregory and Moore, 1931; McKee, 1979; F. Peterson and Pipiringos, 1979).[39] In the four corners region, the Navajo Sandstone ranges from zero to 500 meters thick. To the west, the large red, orange, and white cliffs of Zion National Park are composed of cross-bedded sandstones of the Navajo Sandstone (figure 19.11, and see page 263).

Figure 19.11 Photographs of cross-bedding in Navajo Sandstone, Zion National Park, Utah. Note Matt Raymond (1.6 m in height) at left for scale.

Most Navajo and correlative sandstones are quartz arenites. Unlike the St. Peter sandstones, however, Navajo sandstones contain significant but minor amounts of feldspar, minor to trace amounts of chert and polycrystalline quartz, and minor to rare accessory minerals, such as biotite, muscovite, magnetite, staurolite, zircon, and garnet. The feldspar is commonly alkali feldspar, but plagioclase is abundant locally. Framework quartz and feldspar grains range from angular to well rounded in shape and are generally well to very well sorted, with an average grain size of about 0.15 mm (see figure 19.9b). Many grains are frosted. The framework grains are cemented locally by quartz overgrowths, but calcite, dolomite, kaolinite, and limonite cements also occur. In some rocks neither cement nor matrix is present. Where matrix occurs (it is absent from most rocks), fine quartz and clay particles are the matrix materials (Gregory and Moore, 1931).

Sedimentary textures, structures, and rare fossils, as well as associated facies, suggest that most of the Navajo sandstones are continental, aeolian deposits. The cross-strata are typically large-scale, wedge-planar to tabular-planar structures in sets 2–33 m thick (Hatchell, 1967, in McKee, 1979). Paleo-slip faces may exceed 20°. Angles of intersection between steeply dipping sets of cross-strata commonly exceed 30°. Rare wind-faceted pebbles and terrestrial dinosaur fossils indicate a continental origin for the enclosing sediments. Thin, mudcrack-bearing, interdune limestones and dolomitic limestones attest to local evaporation and dessication (Gregory and Moore, 1931; McKee, 1979). These features, the excellent sorting, and the complete absence from much of the Navajo section of sedimentary structures and fossils that are exclusively marine in origin, plus the fact that channels and other fluvial features are generally lacking from the rocks, strongly suggest a continental, aeolian origin for much of the Navajo Sandstone (McKee, 1979). The aeolian environment may have been marginal to a marine basin, as suggested by smaller-scale cross-bedding, rare burrows, and interlayered marine sediments at the top of the Navajo Sandstone (K. O. Stanley, Jordan, and Dott, 1971; T. W. Doe and Dott, 1980), but no body of evidence exists that would indicate that most Navajo rocks were deposited in a deep marine environment as large submarine sand waves, a possibility suggested by a few workers.[40]

SEQUENCES WITH MIXED SANDSTONE LITHOLOGIES

Environments and Occurrences

In some environments, sorting and other processes affecting matrix abundance vary, resulting in deposition of both arenite and wacke sands. Diverse sources, fluctuating transportational and depositional conditions, and differing degrees of diagenesis are among the factors that result in such stratigraphic sequences with interlayered matrix-rich and matrix-poor sandstones (Benchley, 1969).

Marine, transitional, and continental environments may all be sites of coeval arenite and wacke formation.[41] For example, on submarine fans, variations in sediment composition fed to the fans, variations in sorting as a function of the transporting agent (debris flow, grain flow, or turbidity current), and variations in matrix abundance resulting from differences in diagenetic history may lead to sequences containing both wackes and arenites. Similar factors yield like sequences in shallow marine environments, bays, estuaries, and continental basins. Alternatively, organisms may rework and mix sands and muds to produce wackes from interbedded arenite sand–mudrock sequences (Kulm et al., 1975; Hine et al., 1988). Thus, one cannot expect a priori to find only one kind of sandstone within a given stratigraphic sequence. Sequences with mixed lithologies are common. The Great Valley Group of California is an example of one such sequence.

Example: The Great Valley Group, California

The Great Valley Group of California is a Middle Mesozoic–Lower Cenozoic forearc basin unit deposited between a volcanic arc on the east (now the Sierra Nevada) and a trench subduction zone on the west (at the site of the present Coast Ranges) (see figure 19.5) (E. H. Bailey, Irwin, and Jones, 1964; Berkland et al., 1972; B. M. Page, 1981).[42] The group is exposed both northeast and west of the Great Valley and crops out almost continuously from southern Oregon to south-central California. Additional sections of Great Valley Group rocks occur as outliers to the west of the main belt of outcrops and along the south-central California coast

(in the Naciamento block). Equivalent rocks occur in southern and Baja California. The thickest sections of Great Valley rocks, most of which are sandstones and mudrocks, are about 12 km thick.

Both arenites and wackes occur in the Great Valley Group (table 19.8). Older sandstones are typically volcaniclastic to epiclastic, volcanic lithic wackes. Younger sandstones are feldspathic wackes and arenites. The arenites are locally cemented with iron oxides and other minerals, but calcite cements are common.

The Great Valley Group has been subdivided into both lithostratigraphic units, that is, formations and members (F. A. Schilling, 1962; Ojakangas, 1968), and petrofacies (W. G. Gilbert and Dickinson, 1970; Dickinson and Rich, 1972; Ingersoll, 1978; Mansfield, 1979). The various formations and members are based on ratios of sandstone, mudrock, and conglomerate in any given section. Formations and members are useful for mapping and local subdivision of the stratigraphy. Lithofacies are useful in determining environments of deposition. Many of the Great Valley units, however, are lenticular and grade into other Great Valley units, and as such, cannot be mapped continuously through the outcrop belt. Therefore, **petrofacies,** stratigraphic subdivisions based on similarity of sandstone petrology, have been used to subdivide the stratigraphic section. Because petrofacies are a function of the compositions of individual sandstones, which in turn depend on the provenance, rather than being dependent on stratigraphic features such as bed thickness and rock type abundances, the petrofacies may be used for long-distance correlation and for provenance studies.

Lithofacies and stratigraphic studies reveal that the Great Valley Group consists of a thick accumulation of deltaic, shelf, slope, submarine fan, and basinal clastic sedimentary rocks.[43] All of the Mutti and Ricci-Lucchi (1972) submarine fan and shelf facies are represented (see figures 19.12 and 19.13). Paleocurrent indicators such as sole marks and ripple cross-laminations suggest that sediment was fed into a north-south-trending forearc basin from the north, east, and south. Turbidity currents and mass flows flowed downslope towards the west and along the basin axis towards the north and south.[44] Locally, intrabasin areas of high relief, including the trench-slope break, may have shed sediment towards the east. Clearly, the dominant sources were in the Sierra Nevada to the east and the Klamath Mountains to the north.

The Great Valley Group was first subdivided into nine petrofacies in the north (Ingersoll, 1978b) and seven petrofacies in the south (Mansfield, 1979), but these were later consolidated into eight petrofacies (Ingersoll, 1983). The petrofacies are

Table 19.8 Modes of Selected Great Valley Group Sandstones

Grain Type	1	2	3	4	5	6	7
quartz	{7.6}[a]	{39.0}	14.3	25.7	{34.0}	23.7	{24.2}
monocrystalline	7.2	16.8			33.6		22.2
polycrystalline	0.4	22.2			0.4		2.0
alkali feldspar	—	3.5	—	14.7	13.0	14.0	1.6
plagioclase	61.4[b]	24.2	7.6	17.7	18.4	16.2	16.0[c]
micas	—	—	6.2	2.7	4.0	7.4	
muscovite	—	—					0.6
biotite	—	—					2.0
glauconite	—	—					10.4
other minerals[d]	1.4	3.0	2.2	1	0.6	2.4	0.2
rock fragments							
chert/metachert	0.4	3.8	3.2	<1	0.2	1.9	1.6
sedimentary	1.0	0.5	17.1	1.9	1.2	1.7	—
felsic volcanic	3.6	1.5	} 32.4	} 22.2	} 10.5	} 26.3	3.2
mafic volcanic	2.0	0.2					0.2
metamorphic	0.2	4.5	4.1	1.5	1.3	1.7	0.6
other/undiff.	0.2	0.8	—	—	—	—	0.2
Cement		0.2	—	0.5	6.3	—	36.8
Matrix	22.2	18.8	12.9	11.6	10.5	4.7	1.2
Total	100	100	100	100	100	100	100
Number of points counted	500	400	400	400	400+	400	500
Q/F/L	10:81:9	55:36:9	22:10:68	31:39:30	44:40:16	30:35:35	55:37:8

Sources:

1. Volcaniclastic, Feldspathic wacke, Lotta Creek Formation (Upper Jurassic), Great Valley Group, Hospital Canyon, Northeastern Diablo Range, California (Raymond, 1969, p. 21ff., and L. A. Raymond, unpublished data).
2. Feldspathic wacke (lower Upper Cretaceous), Great Valley Group, Hospital Canyon, northeastern Diablo Range, California (L. A. Raymond, unpublished data).
3. Lithic wacke, "lower unit," Great Valley Group (Upper Jurassic?), southern Santa Lucia Range, California (sample PR-43 of Gilbert and Dickinson, 1970, table 2).
4. Feldspathic wacke, "middle unit," Great Valley Group (Cretaceous?), southern Santa Lucia Range, California (sample PR-83 of Gilbert and Dickinson, 1970, table 2).
5. Lithic feldspathic wacke, Great Valley Group, (Campanian, Upper Cretaceous), Mono Creek/Agua Caliente Canyon area, Transverse Ranges, California (sample 35 of MacKinnon, 1978, table 1).
6. Lithic feldspathic arenite, "middle unit," Great Valley Group (Cretaceous?), southern Santa Lucia Range, California (sample PR-37 of Gilbert and Dickinson, 1970, table 2).
7. Glauconitic, calcareous, feldspathic arenite, Moreno Formation, Great Valley Group, Rumsey Petrofacies (Upper Cretaceous), Hospital Canyon, northeastern Diablo Range, California (L. A. Raymond, unpublished data).

[a]Volume percents based on number of points counted, as indicated.

[b]Zeolites and other metamorphic minerals replacing plagioclase are counted as plagioclase.

[c]Includes some untwinned feldspar.

[d]Includes all others not specifically listed.

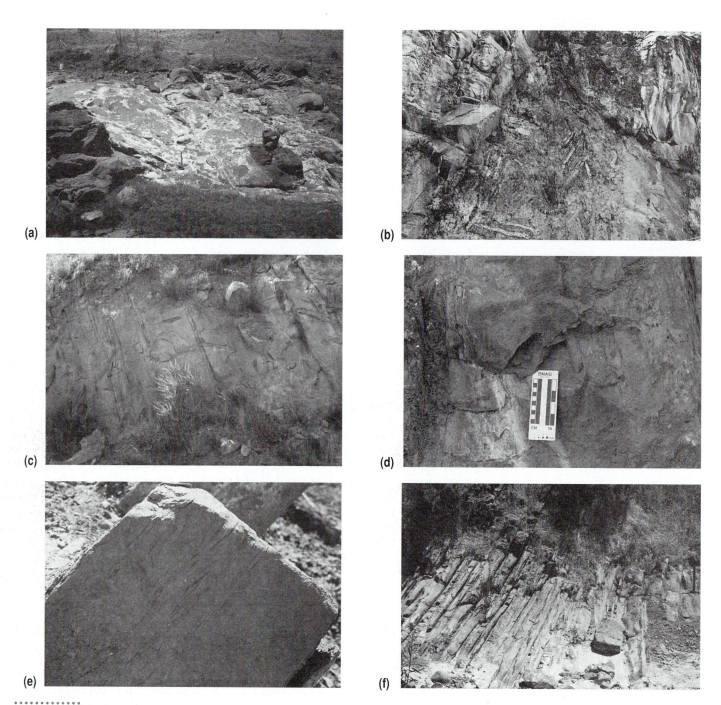

Figure 19.12 Photographs of submarine fan facies in the Great Valley Group. (a) Massive, concretionary, channel sandstones (Facies A) and associated thinner-bedded Facies B sandstones, Panoche Formation, Great Valley Group, Upper Cretaceous, Ingram Canyon, northeastern Diablo Range, California. Bedding dips steeply to right. Hammer handle, left of center, is 1/3 m long. (b) Thick-bedded Facies A and B sandstones above and below a Facies F olistostrome, Cortina Formation, Great Valley Group, Upper Cretaceous, Monticello Dam, Lake Berryessa, California. (c) Facies B and C sandstones, Panoche Formation, Great Valley Group, Upper Cretaceous, Lone Tree Creek, northeastern Diablo Range, California. (d) Facies C turbidite (note pebbles at base and grading), Cortina Formation, Great Valley Group, Upper Cretaceous, Monticello Dam, Lake Berryessa, California. (e) The T_{bc} interval in shelf-sandstone turbidite, Moreno Formation, Great Valley Group, Upper Cretaceous, Hospital Canyon, northeastern Diablo Range, California. (f) Facies E sandstones and shales, Panoche Formation, Great Valley Group, Upper Cretaceous, Ingram Canyon, northeastern Diablo Range, California. (g) Close-up view of Tcde and Tce intervals in Facies E turbidites. (h) Facies G mudrocks and Facies D mudrocks with minor sandstones, Lodoga Formation, Great Valley Group, Lower Cretaceous, Wragg Canyon, near Lake Berryessa, California.

(g)

Figure 19.12 Con't.

(h)

(a)

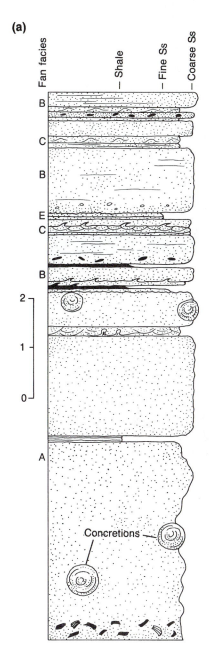

(b)

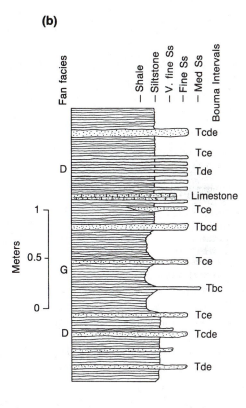

Figure 19.13 Measured sections of submarine fan facies rocks in the Great Valley Group. (a) Panoche Formation (UK), Ingram Canyon; NW 1/4 Sec. 10, T.5S., R.6E. M.D.B.M. (b) Hospital Formation (MK), NW 1/4 Sec. 34, T.4S., R.5E., M.D.B.M., Hospital Canyon; Carbona 15′ Quadrangle, northeastern Diablo Range, California.

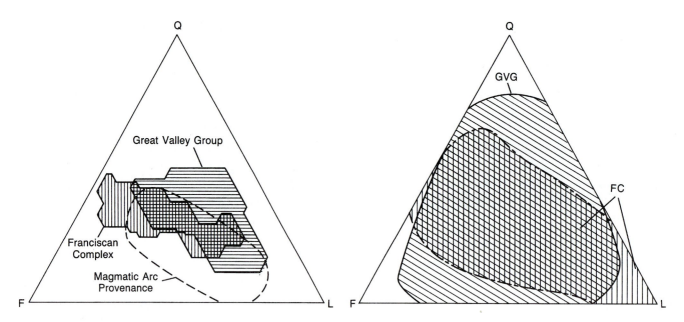

Figure 19.14 QFL diagrams showing the fields encompassing the framework QFL values for Franciscan Complex (FC) and Great Valley Group (GVG) sandstones. (a) Simplified version of QFL diagram of Dickinson et al. (1982; based in part on Ingersoll, 1978b) showing fields defined by *average values* and standard deviations. (b) Generalized QFL diagram showing sandstone compositional ranges, including additional data not included in Dickinson et al. (1982) (includes data from W. G. Gilbert and Dickinson, 1970;

Mansfield, 1979; Dickinson et al., 1982; Bertucci, 1983; Ingersoll, 1983; Golia and Nilsen, 1984; Jayko and Blake, 1984; Underwood and Bachman, 1986; Aalto, 1989b; Larue and Sampayo, 1990; P. F. Short and Ingersoll, 1990; L. A. Raymond, unpublished data). *Note:* Because different workers have used different grain counting techniques, the data from various studies are not strictly comparable. The fields shown in (b), however, encompass all of the *individual sample* values from all workers cited.

recognized on the basis of QFL ratios and four additional petrographic parameters: plagioclase to total feldspar ratios (P/T), volcanic lithic grains as a percentage of total unstable lithic grains (Lv/L), percentage of monocrystalline phyllosilicate grains (M), and the ratio of polycrystalline quartz grains to total quartz grains (Qp/Q). The QFL field for the Great Valley sandstones is shown in figure 19.14, as is the field for magmatic arc sandstones. Comparison of figure 19.14 with figure 19.3 reveals that the Great Valley rocks include a recycled orogenic source in addition to an arc source. Biotite and white mica are common accessory minerals, but a wide range of minor accessory minerals is present including garnet, magnetite, ilmenite, epidote, chlorite, and hornblende. These minerals likewise suggest composite sources, consistent with the Sierra Nevadan and Klamath terranes.

Additional evidence supporting the theory that the Sierra Nevada and Klamath Mountains served as the provenances for Great Valley Group sediments is provided by trace element data. The relationship of Nd-isotope ratios to Sr-isotope and trace element data is the same in Great Valley sandstones and plutonic rocks of the arc, indicating that the latter served as the source for sediments of the former (Linn, De Paolo, and Ingersoll, 1991).

SUMMARY

Sandstones may be subdivided and classified in a number of ways, but here they are assigned to two main categories—arenites with 5% or less matrix and wackes with more than 5% matrix. Both wackes and arenites occur in a wide variety of continental to deep marine environments, and locally occur together.

Matrix materials consist of four types, primary matrix, or protomatrix; recrystallized protomatrix, called orthomatrix; deformed and recrystallized lithic fragments, called pseudomatrix; and polymineralic, diagenetic matrix produced by neocrystallization and alteration of framework grains, called epimatrix. Protomatrix seldom exceeds 10%. In wackes with large amounts of matrix, the matrix results from diagenetic processes. Cements, which commonly bind the grains together in arenites, consist of quartz, calcite, and dolomite, as well as iron oxides, manganese oxides, gypsum, and a variety of other minerals.

The main framework grains in siliciclastic sandstones are quartz, feldspars, and lithic fragments. The ratios of these constituents (QFL) or subdivisions of these constituents may be used to define petrofacies that reflect the provenance of the sediment. Additional factors that influence the amounts of various framework grains (as well as the percentage of matrix) include physical and chemical processes in the environment of deposition and diagenetic processes. The mineral compositions of the sandstones are clearly reflected in their chemistries, and these in turn reflect the provenance.

Texturally, sandstones are relatively simple, but structurally they contain a wide variety of features. Most sandstones have epiclastic textures. A few are volcaniclastic. Sedimentary structures include numerous types of depositional, erosional, chemical, and deformational structures, including bedding and lamination, channels, folds, and concretions. Of these, bedding and lamination are nearly ubiquitous. All textures and structures reveal something of the history of the sandstone.

EXPLANATORY NOTES

1. For example, see the works of Dake (1921), Lamar (1927), and Dapples (1955) on the St. Peter Sandstone, and Butts (1940), B. N. Cooper (1961), Yeakel (1962), Shelton (1973), M. O. Hayes (1974), and Whisonant (1977) for discussions of the Clinch/Tuscarora Sandstone.
2. Pettijohn, Potter, and Siever (1987, ch. 1) discuss the differing views on the size range of sand particles. In engineering practice, for example, the size grades differ from those commonly used by sedimentologists.
3. Recall that various geologists use different percentages of matrix as the boundary between arenites and wackes. Here, I adopt a matrix value of 5% as the boundary, considering only those sandstones with 5% or less matrix as arenites. See note 7 in chapter 15 for the rationale.
4. See Dott (1964) for a discussion of the history and uses of this term.
5. Also refer to the numerous reports of the Deep-Sea Drilling Project (DSDP), especially Bode (1973). Note that the clay-plus-silt (mud) component of sands commonly does exceed 10%.
6. See the reports of the Deep-Sea Drilling Project and Pettijohn, Potter, and Siever (1987, p. 40).
7. For example, see Dickinson (1970a, 1982), G. F. Moore (1979), Gergen and Ingersoll (1986), and Packer and Ingersoll (1986).
8. See Brush (1965), J. E. Sanders (1965), Leliavsky (1966, ch. 9), and G. V. Middleton and Hampton (1973).
9. Refer to chapter 16 and to Garrels and Christ (1965, pp. 352–362).
10. Kuenen (1966), Benchley (1969), Pettijohn, Potter, and Siever (1987, p. 172ff., 431).
11. Cements are described in Scholle (1979) and Meyer, Textoris, and Dennison (1987), for example.

12. The textures of sandstones are illustrated with color photomicrographs in Scholle (1979) and A. E. Adams, Mac Kenzie, and Guilford (1984). Also see petrography texts such as Howell Williams et al. (1982) or Raymond (1993). Volcaniclastic sandstones are discussed in Boggs (1992) and R. V. Fisher and Smith (1991).
13. Some authors include chert with quartz, as polycrystalline quartz (Dickinson and Suczek, 1979), whereas others include it with lithic fragments (Folk, 1974). This is one of the principal distinctions between L and R.
14. In the widely used Gazzi-Dickinson method of counting grains, plutonic clasts are not counted as such and are not included in L (Ingersoll et al., 1984). Rather the individual crystals of quartz, feldspar, and other minerals in plutonic clasts are counted as quartz, feldspar, and so on. In this method, L = sum of fine-grained clasts of sedimentary + volcanic/hypabyssal + metamorphic rock.
15. Also see Ingersoll (1978), Mansfield (1979), G. F. Moore (1979), Dickinson and Suczek (1979), Dickinson and Valloni (1980), Dickinson et al. (1982), McLennan (1984), Valloni and Mezzadri (1984), Gergen and Ingersoll (1986), Packer and Ingersoll (1986), and Underwood and Bachman (1986).
16. For additional discussion, see DeCelles and Hertel (1989, 1990), Dickinson and Ingersoll (1990), Johnsson (1990), Johnsson and Stallard (1990), and Johnsson, Stallard, and Lundberg (1990).
17. For other discussions of these complicating factors see Cleary and Connolly (1974), G. F. Moore et al. (1982), M. Ito (1985), Lash (1987) and Girty and Armitage (1989).
18. Additional examples of the use of trace elements in studying sand and sandstones are J. N. Weber and Middleton (1961a,b), Wildeman and Haskin (1965), Bhatia and Taylor (1981), Z. E. Peterman, Coleman, and Bunker (1981), Knedler and Glasby (1985), Bhatia and Crook (1986), and Larue and Sampayo (1990).
19. Sedimentary structures, in general, including those characteristic of sandstones, are also described by Krynine (1948), G. V. Middleton (1965), Conybeare and Crook (1968), and Collinson and Thompson (1982). Additional works such as Dzulynski and Walton (1965), Harms, Southard, and Walker (1982), and Pettijohn, Potter, and Siever (1987) deal specifically with the structures of sandstones.
20. Lacustrine examples of turbidity current or turbidite formation are described by Nelson (1967) and Normark and Dicksen (1976). Shallow-water turbidites are described by Fenton and Wilson (1985). For information on submarine turbidites, refer to Kuenen and Migliorini (1950), Shepard (1961), Bouma and Brouwer (1964), J. M. Coleman, Gagliano, and Smith (1970), Hampton (1972), Bouma and Hollister (1973), Middleton and Hampton (1973), Vallier, Harold, and Girdley (1973), R. G. Walker and Mutti (1973), Bouma, Norwark, and Barnes (1985), Jansen et al. (1987), S. Reynolds (1987), and the references cited therein.
21. See the sources in note 20.

22. Also see Hollister and Heezen (1972), Lovell and Stow (1981), and D. L. Reed et al. (1987).

23. For example, see Hantzschel (1939), Krumbein (1939), Paine (1964), Kanes (1970), Busch (1971, 1974), Shelton (1973), and J. M. Coleman and Prior (1982).

24. The Anvil Points Sandstone is also discussed by P. E. Potter and Simon (1961).

25. Also see Underwood, Bachman, and Schweller (1980) and chapter 17 of this text and the references cited therein.

26. In Oregon, Franciscan rocks are given individual formation names including the Whitsett Limestone, Dothan Formation, and others (F. G. Wells and Peck, 1961; E. H. Bailey, Irwin, and Jones, 1964; Koch, 1966; Dott, 1971; E. M. Baldwin, 1976; Worley and Raymond, 1978).

27. See E. H. Bailey, Irwin, and Jones (1964) for a review of megafossils, other than ichnofossils, found in Franciscan rocks. Radiolaria are described by Pessagno (1977a,b) and Murchey (1984). W. Miller (1986, 1988, 1991) discusses ichnofossils and some foraminifers and Sliter (1984) discusses foraminifers.

28. Additional petrographic data on Franciscan sandstones are provided by several other geologists, including Soliman (1965), Swe and Dickinson (1970), W. P. Christensen (1973), D. S. Cowan (1974), O'Day (1974), Kleist (1974), Jordan (1978), Jayko and Blake (1984), Underwood and Bachman (1986), Aalto (1989), and Larue and Sampayo (1990).

29. Also see Ernst (1970), Raymond (1974a), Underwood (1977), Ingersoll (1978a), G. W. Smith, Howell, and Ingersoll (1979), Aalto (1982), Dickinson et al. (1982), and Murchey (1984) . A controversy exists with regard to whether or not the Franciscan rocks accumulated seaward of the Great Valley arc-trench gap. Some have argued that Franciscan rocks, were formed elsewhere, and represent suspect terranes moved northward to a position now lying west and seaward of the Great Valley rocks (see D. L. Jones et al., 1978; Coney, Jones, and Monger, 1980; Blake and Jones, 1981; Blake, Howell, and Jayko, 1984). Petrotectonic considerations and modal averages of Franciscan wackes that are more feldspathic than those of Great Valley Group sandstones are considered to support this interpretation (Blake and Jones, 1981; Jayko and Blake, 1984). The same modal differences between Franciscan and Great Valley rocks were considered by Dickinson et al. (1982) to reflect variations in composition within the source terrane and transportational working of sediments, supporting the alternative view that Franciscan Complex and Great Valley Group rocks represent contiguous parts of the same arc-trench terrane. The latter interpretation has also been supported by data from conglomerates in both rock sequences (Seiders, 1988; Seiders and Blome, 1984, 1988) and by petrotectonic and metamorphic arguments (Ernst, 1981b, 1984).

30. See B. N. Cooper (1961), Yeakel (1962), A. W. Hayes (1974), Whisonant (1977), and S. C. Meyer, Textoris, and Dennison (1987).

31. See Krynine (1935) and the reviews of Blatt, Middleton, and Murray (1972, p. 279ff.) and Pettijohn, Potter, and Siever (1987, p. 155).

32. For information on the Gizzard Formation, see Cramer (1986). Pettijohn, Potter, and Siever (1987, figs. 5–8) show photomicrographs of Pottsville sandstone. Additional examples of lithic arenites are found in units such as the Mesoproterozoic Snowslip Formation of Glacier National Park, Montana (Whipple and Johnson, 1988) and the Pennsylvanian Trivoli Sandstone of Illinois (Andresen, 1961).

33. For example, see papers in Bouma et al. (1980), Nittrouer (1981), and Tillman and Siemers (1984).

34. The aeolian Navajo Sandstone described in this chapter, for example, contains many angular grains. See Figure 19.14.

35. For example, see S. R. Casey and Cantrell (1941), Nanz (1954), Busch (1974), and K. M. Scott (1982).

36. The description of the St. Peter Sandstone here is based on the descriptions of Dake (1921), D. J. Fisher (1925), Lamar (1927), Thiel (1935), Dapples (1955), Templeton and Willman (1963), A. H. Bell et al. (1964), Pryor and Amaral (1971), Amaral and Pryor (1976), G. S. Fraser (1976), Hoholick, Metarko, and Potter (1984), Visocky, Sherrill, and Cartwright et al. (1985), Frazier and Schwimmer (1987), and J. Mazzullo (1987), among others. Also see Dapples, Krumbein, and Sloss (1948).

37. Data on bedding and cross-bedding may be found in Dake (1921), Lamar (1927), Dapples (1955), and Pryor and Amaral (1971).

38. This summary of the Navajo Sandstone and correlative units is based on the works of Darton (1925), Longwell et al. (1925), Gregory and Moore (1931), A. A. Baker, Dane, and Reeside (1936), Gregory and Williams (1947), Gregory (1950), Kiersch (1950), Averitt et al. (1955), K. O. Stanley, Jordan, and Dott (1971), Novitsky and Burchfiel (1973), E. L. Miller and Carr (1978), McKee (1979), F. Peterson and Pipiringos (1979), T. W. Doe and Dott (1980), and Kocurek and Dott (1983).

39. For references and discussions of units correlative to the Navajo Sandstone see K. O. Stanley, Jordan, and Dott (1971), Novitsky and Burchfiel (1973), E. L. Miller and Carr (1978), McKee (1979), and Kocurek and Dott (1983).

40. Whether the Navajo Sandstone represents an inland or coastal desert deposit has been a matter of some debate (Frazier and Schwimmer, 1987, p. 323). It has even been suggested that much of the unit may represent marine sand deposited in sand waves (K. O. Stanley, Jordan, and Dott, 1971). However, the bulk of both negative evidence (e.g., the absence of marine fossils) and positive evidence (e.g., the presence of ventifacts and dinosaur fossils) favors an aeolian origin.

41. T. W. Todd (1968), K. O. Stanley, Jordan, and Dott (1971), Rupke (1977), MacKinnon (1978), G. F. Moore (1979), Kocurek and Dott (1983), Coch (1986, 1987).

42. The Mesozoic strata of the Great Valley area of California have been referred to for many years as the "Great Valley sequence" (E. H. Bailey, Irwin, and Jones, 1964; Page, 1966; Ingersoll, 1978c). In accord with modern stratigraphic nomenclature, these strata have been renamed the Great Valley Group (Ingersoll and Dickinson, 1981; Ingersoll, 1982) and are referred to here as such. Works on the Great Valley Group are numerous, but among the historically and petrologically important works are the studies of F. M. Anderson (1905, 1958), R. Anderson and Pack (1915), Taliaferro (1943), C. E. Weaver (1949), M. B. Payne (1951, 1962), W. P. Irwin (1957), F. A. Schilling (1962), E. H. Bailey, Irwin, and Jones (1964), B. M. Page (1966, 1981), Ojakangas (1968), W. G. Gilbert and Dickinson (1970), Swe and Dickinson (1970), Dott (1971), Berkland et al. (1972), Dickinson and Rich (1972), Ingersoll, Rich, and Dickinson (1977), Lee-Wong and Howell (1977), Ingersoll (1978a,b,c, 1979, 1982, 1983, 1988), MacKinnon (1978), Mansfield (1979), Bertucci (1983), Golia and Nilsen (1984), Underwood and Bachman (1986), S. A. Reid (1988), McGuire (1988), Imperato, Nilsen, and Moore (1990), and P. F. Short and Ingersoll (1990).

43. For example, F. A. Schilling (1962), Charles Bishop (pers. commun., 1968), Ingersoll (1978c), R. Garcia (1981), Cherven (1983), Nilsen (1984a,b), L. A. Raymond (unpublished data).

44. Paleocurrent data and interpretations are given by several workers, including Shawa (1966), Colburn (1968), Ojakangas (1968), Mansfield (1979), Cherven (1983), and Nilsen (1984a, 1990).

PROBLEMS

19.1. Plot the compositions of the sandstones in table 19.2 on a copy of the QFL diagram in figure 19.3b. Using that plot and table 19.3, suggest a provenance for each of the sandstones.

19.2. (a) Use the data in table 19.7 to construct a Lv:Ls:Lm (lithic volcanic : lithic sedimentary : lithic metamorphic) plot for the tabulated Franciscan sandstones. (b) Using different symbols and the data in table 19.8, plot the Lv:Ls:Lm data for the Great Valley Group rocks tabulated here. (c) Compare the two plots. Do the provenances for the two sets of rocks appear to be different on the basis of these few data?

20

Conglomerates, Diamictites, And Breccias

INTRODUCTION

Conglomerates—the lithified equivalents of gravels—plus breccias and diamictites are widely distributed but not voluminous components of the stratigraphic record. Conglomerates comprise no more than 1% of the sedimentary rocks on the Earth's surface (Blatt, 1970). Yet, they form in a wide variety of environments ranging from alpine glacial to deep marine.[1] On the continents, conglomerates and related rocks are deposited by glaciers, by landslides, by lake shore waves, and by streams and rivers. Rivers deposit gravels in channels, on floodplains, and in alluvial fans and deltas. Marine conglomerates and related rocks form along beaches, on the shallow shelves, on the slope and rise, along submarine scarps at spreading centers, and in the channels that lead to the lobes of submarine fans. Notable examples of conglomerates and related rocks include the fanglomerates of the Big Horn Mountains of Wyoming (R. P. Sharp, 1948), conglomerates and breccias of the Beaverhead Formation of Montana (R. P. Lowell and Klepper, 1953), the Collings Ranch and related conglomerates of Oklahoma (Ham, 1954), the Sharon Conglomerate of Ohio (J. O. Fuller, 1955), deposits of the Blackhawk landslide of California (Shreve, 1968), the Miocene landslide breccias of southeastern Arizona (Krieger, 1977), deep-marine Otter Point conglomerates of Oregon (R. G. Walker, 1977), the Dunnage Melange of Newfoundland (Horne, 1969; Kay, 1970, 1976; Jacobi, 1984), the Cambrian carbonate conglomerates of the Nolichucky Formation of Virginia (Markello and Read, 1981), the subaerial debris flows of the Cutler Formation of Colorado (Schultz, 1984), and the shallow marine conglomerates of Wapiabi Formation of Alberta (Rosenthal and Walker, 1987).

Conglomerates and similar rocks are important for both petrologic and practical reasons. From a petrologic viewpoint, conglomerates contain clasts that are large enough to be analyzed independently as rock types, allowing the petrologist to gain a unique perspective on the compositions of rocks of the provenance from which the sediment was derived. Counts of pebble types provide information on the lithologic diversity of the provenance. Such information, combined with facies relationships, facilitates paleogeographic reconstructions. From a practical perspective, conglomerates may be porous and permeable, allowing them to serve as reservoir rocks for economically important fluids such as groundwater. In addition, gravelly or conglomeratic fluvial facies may contain heavy minerals, including gold, that constitute economically important placer deposits (Eric, Stromquist, and Swinney, 1955; Kingsley, 1987).

DISTINCTIONS BETWEEN MAJOR TYPES OF COARSE CLASTIC ROCKS

The major types of coarse clastic sedimentary rocks are conglomerates, breccias, and diamictites. All of these clastic rocks have certain features in common, notably their overall *clast-in-a-matrix* character (figure 20.1). In some, the interclast material is largely cement.

Clast- or *block-in-matrix* structure is characteristic not only of coarse clastic sedimentary rocks, but of igneous breccias and xenolithic plutonic rocks, of some metamorphic rocks, and of melange[2] rocks of both sedimentary and igneous origin (table 20.1) (Raymond, Yurkovich, and McKinney, 1989). Because rocks formed by one process may grade into or show very similar features to those formed by another process, it is important to recognize the distinctions between the various coarse clastic rocks that exhibit block-in-matrix structure.

Rocks of igneous origin consist of both matrix and clast materials that are entirely or predominantly igneous. Igneous textures, structures, and minerals reveal that igneous character. Rocks with clasts of plutonic rock in a phaneritic plutonic rock matrix are unequivocally igneous in origin. The origins of rocks with volcanic clasts tend to be more ambiguous. Rocks with matrix and clasts that are entirely volcanic are likely to be of volcanic origin. Volcanic materials, however, may be reworked by sedimentary agents, so it is important to examine such rocks for definitive structures of either igneous or sedimentary origin (e.g., welding or fluvial cross-bedding).[3] In addition, some coarse clastic rocks with volcanic clasts have a complex history. Pyroclastic materials erupted from a volcano may be transported by streams before deposition. Such a situation developed during the 1980 eruption of Mount St. Helens in Washington, when pyroclastic and previously formed volcanic materials were erupted and transported subaerially by the initial blast and avalanche, but were mixed with lake and river water before ultimately being deposited (Cummans, 1981; Foxworthy and Hill, 1982).[4] Particularly in ancient deposits of volcanic or volcani-fluvial origin, where the glassy matrix has been recrystallized, altered, or weathered to clay or other minerals, the exact origin of the rock may be difficult to decipher.

Similarities and gradations also exist between coarse clastic rocks of sedimentary origin and those with a metamorphic history. In particular, sedimentary diamictites formed by debris flows and submarine landslides, called **olistostromes,**[5] are similar to diamictites found in diapirs and tectonic melanges (compare columns 3 and 4 to column 5 in figure 20.2) (Raymond, 1984a).[6] The similarities are especially marked where soft-sediment deformation has affected the diamictite or where the clasts in the diamictite were formed either by an early phase of *tectonic* fragmentation and mixing or by *diapiric transport* of subducted sediments or rocks (Sarwar and DeJong, 1984; Cloos, 1982, 1984).

In any diamictite, the clasts may be of various origins and compositions, and the matrix may consist of a variety of

Table 20.1 Origins of Rocks with Block-in-Matrix Structure

Origin	Product
Sedimentary	Conglomerates
	Breccias
	Diamictites
Igneous	
Extrusive	Xenolithic flow rocks
	Agglomerates
	Pyroclastic rocks
Intrusive	Xenolithic plutonic rocks
Metamorphic (Structural/Tectonic)	
Brittle deformation	Cataclasites,[2] especially breccias
Brittle and ductile deformation	Protomylonites, quasimylonites[2]
Ductile deformation	Mylonites[1,2]
	Migmatites[1,2]
Diapiric	Diamictites
	Cataclasites[2]
	Quasimylonites[2]

Source: From L. A. Raymond, et al., "Block-in-matrix structures in the North Carolina Blue Ridge belt and their significance for Orogen" in J. W. Horton, Jr. and N. Rast, Editors, "Melanges and Olistostromes of the U.S. Appalachians" in *Geological Society of America Special Paper* 228, 195–215, 1989. Copyright © 1989 Geological Society of America. Reprinted by permission of the author.

[1] These rocks do not have clastic textures.
[2] See glossary for a definition.

rock types. In determining the origins of such rocks, the structures and regional relationships are of critical importance. Olistostromes, for example, have depositional contacts at their bases and tops and are interstratified with other marine sedimentary rocks (see figure 20.1c). Alluvial fan (mudflow) diamictites have similar depositional contacts and are interlayered with continental fluvial deposits. Tillites contain a matrix of rock flour and may be associated with varved lacustrine sediments containing dropstones or with glacio-fluvial deposits. They too have depositional contacts above and below. In contrast, tectonic melanges and diapirs have sheared and cross-cutting contacts.

Knowledge of the major types of coarse clastic sediments, their corresponding sedimentary rocks, and associated structures and textures may also aid in distinguishing between sedimentary, igneous, and metamorphic origins for a rock body with block-in-matrix structure. Recall that *conglomerate* is composed of 25% or more *rounded* clasts greater than 2 mm in length in a matrix that is sandy or gravelly. Diamictite contains the same type of clasts in a muddy matrix.[7] Thus, the terms are essentially texturally based. In both conglomerates and diamictites, the presence of well-rounded clasts in

(a)

(b)

(c)

(d)

(e)

Figure 20.1 Photos of coarse clastic sediments and sedimentary rocks. (a) Glacial till, south moraine of North St. Vrain Creek, Allenspark, Colorado. (b) Polymict, marine, cobble conglomerate, Cortina Formation, Great Valley Group, Cretaceous, Lake Berryessa, California. (c) Olistostrome with boulders (large blocks) of sandstone and shale, Venado Member, Cortina Formation, Great Valley Group, Cretaceous, Lake Berryessa, California. (d) "Flat pebble" dolostone conglomerate overlying laminated dolostones, Conococheague Formation, Knox Group, Cambrian, Little River Dam, Radford, Virginia.(e) Lenticular beds (2–30 cm thick) of polymict, clast-supported, alluvial fan breccia, Pleistocene-Recent, La Hood, Montana. (Also see figure 20.6.)

	Sedimentary rocks				Melange rocks	
	Diamictites				Diapiric	Tectonic
Texture	Epiclastic with blocks, fragments, and/or grains in a finer grained matrix		Epiclastic to foliated with slabs, fragments, and/or grains in a finer grained matrix		Clastic to foliated with blocks and fragments in a finer grained matrix	Foliated with slabs, blocks, fragments, and/or grains in a finer grained matrix
Clast size	Fragments to small blocks <1-15m	Fragments to small blocks <1-15m	Fragments to small slabs <1-1500m	Fragments to large slabs <1->1500m	Fragments to small blocks <1-15+ m	Fragments to large slabs <1->1500m
Block/matrix	Low to moderate	Generally low	Generally high	Low to high generally low	Generally low	Low to high
Contacts	Depositional		Depositional to sheared		Discordant to sheared	Sheared
Associated rocks and features	Rhythmites, dropstones, striated clasts, polished pavements boreal fossils	Crossbedded and channelled fluvial sandstone, shale and conglomerate non-marine fossils	Colluvium, alluvium, fluvial sedimentary rocks, continental organic debris	Marine sedimentary rocks (including turbidites); marine fossils	Marine sedimentary rocks Continental sedimentary rocks	Any kind of rock (including gouge; breccia); slickensides
Rock types	Tillite	Tilloid or pebbly mudstone	Tilloid, breccia, or megabreccia, SF-tectonite	Tilloid, megabreccia, pebbly or bouldery mudstone or sandstone, SF-tectonite	Bouldery mudstone, SF-tectonite	Breccia, megabreccia, mylonite (sensu lato), SF-tectonite, Quasimylonite
Rock unit	Formation or member	Formation or member	Formation, member, broken formation, landslide	Formation, member, broken formation, olistostrome	Dismembered formation, diapiric melange	Broken formation, dismembered formation, tectonic melange
Origin	Glacial deposition	Mudflow	Subaerial landslide (sensu lato)	Submarine landslide	Mud diapirs Mud volcanoes	Brittle or ductile deformation in a fault or shear zone

Figure 20.2 Comparison of diamictites and melange rocks. The third and fourth columns represent rocks with characteristics of both diamictites and sedimentary melanges.
[Modified from Raymond (1984a) with additions based in part on Barber et al. (1986) and Orange (1990)]

the matrix is an important, although not unequivocal, indicator of the sedimentary origin for the rock. A fossiliferous matrix is, similarly, a strong indicator of a sedimentary origin. **Talus, rubble,** and **skree** are names commonly assigned to *angular* sediments dominated by clasts of coarse size (> 2mm). Recall that *breccia* is sedimentary rock derived from talus and similar sediments in which 25% or more of the angular clasts are larger than 2 mm and in which the matrix is sandy or coarser-grained material. If the matrix is muddy, the rock is a *diamictite*. *Tillite* is diamictite of glacial origin, with a muddy or rock-flour matrix.

Because angular clasts are produced by brittle failure (breaking)—a process that can occur in igneous, metamorphic, and sedimentary environments—a careful compositional and textural analysis of clasts and matrix, coupled with a paleoenvironmental analysis, may be necessary to distinguish rocks with angular clasts formed by sedimentary processes from those formed by other processes. Mixed clast and matrix compositions is a property consistent with a sedimentary origin, but is not one that is exclusive to sedimentary rocks. Metamorphic

clastic rocks may also show mixed rock types, but they exhibit significant signs of grain deformation (see chapter 30). Depositional contacts are another indicator of a sedimentary origin for breccias and diamictites. Likewise, gradational facies relations, in which breccias grade laterally into sandstones or limestones, suggest sedimentary origins. Finally, fossils in the matrix of breccias and diamictites are strongly suggestive of a sedimentary origin, regardless of the matrix composition (e.g., Moiseyev, 1970; C. Carlson, 1984a, 1984b, 1984c).

TEXTURES, STRUCTURES, AND COMPOSITIONS OF COARSE CLASTIC SEDIMENTARY ROCKS

Textures and Structures

The textures of coarse clastic sedimentary rocks are *epiclastic*; that is, they are clastic textures of surficial origin. In most coarse clastic rocks, the larger clasts, which dominate the rock

(a)

(b)

···············

Figure 20.3   Matrix-supported versus clast-supported conglomerates. (a) Matrix-supported, polymict pebble conglomerate, Knobs Formation (Ordovician), Avens Ford Bridge area, South Holston Reservoir, Tennessee.

(b) Clast-supported, polymict conglomerate, Citico Conglomerate, Wilhite Formation, Neoproterozoic, Chilhowee Dam, Little Tennessee River, Tennessee.

visually, are surrounded by a finer-grained matrix. Examined in more detail, coarse clastic rocks exhibit variations (1) in both texture and amount of cement and matrix and (2) in the relationship between and distribution of clasts and matrix. For example, calcareous cements may have crystalline textures of the mosaic, comb, or drussy types. Alternatively, calcareous as well as siliceous cements commonly give the rocks a poikilotopic texture. The cements may be coarse-grained to aphanitic. Matrix materials may also be coarse-grained to aphanitic, and are epiclastic to weakly foliated.[8] Foliation is caused by the alignment of tabular clay and mica grains.

Rocks with abundant matrix and with scattered clasts that are rarely touching, are referred to as **matrix-supported** (figure 20.3a). Those rocks in which the clasts are abundant enough to be touching several neighboring clasts are called **clast-supported** (figure 20.3b). The large clasts are referred to as *framework*. In both clast- and matrix-supported rocks, grain-size distributions range from unimodal (one dominant size) to polymodal (two or more dominant sizes)(figure 20.4).[9] For example, sample C50a in figure 20.4 is bimodal (two dominant sizes), with the most common clast size being about 8 mm and the most abundant matrix grains being about 1/2 mm in diameter. In matrix-supported rocks, the most abundant grains are of sand, silt, or clay size, whereas in clast-supported rocks, the framework is the major component of the sediment. In general, because of the disparity in grain size between clast and matrix materials; conglomerates, breccias, and especially diamictites are very poorly sorted. Some granule and pebble conglomerates, however, are moderately to well sorted.

The causes of the polymodal and bimodal distributions in clastic sediments are debatable (Schlee, 1957; Pettijohn, 1975, ch. 3). Possible origins include (1) control by the environment, nature, and weathering of materials in the source region (e.g., weathering in the source region may produce two sizes of clast), (2) transportational and depositional control (e.g., bed load and suspended load may be mixed during deposition, or variations in grain size may develop as flow discharges change),[10] or (3) postdepositional modification (e.g., fine sediment may infiltrate between larger grains *after* deposition of the framework). In many cases, postdepositional processes may be eliminated as a cause on the basis of texture or composition, but the roles of source, transportation, and deposition in generating the bimodal character of the sediment are less easily discriminated.

Clast sizes in coarse clastic sedimentary rocks vary tremendously. In olistostromes, individual clasts may measure more than a kilometer in length (J. C. Maxwell, 1964; Hsu, 1967). Yet, the matrix of the olistostrome may be dominated by mud. In contrast, in many granule conglomerates the maximum clast size is four millimeters. In conglomerates and gravels, there is a direct, straight-line relationship between maximum clast size and geometric mean clast size; *the larger the clast of maximum size, the larger the average clast size of the sample* (figure 20.5a) (Schlee, 1957, figs. 9, 12; Pettijohn, 1975, p. 159). Similarly, at least in some formations, there is a linear relationship between maximum particle size and bed thickness (figure 20.5b) (Bluck, 1967; Nemec and Steele, 1984; A. W. Walton and Palmer, 1988).

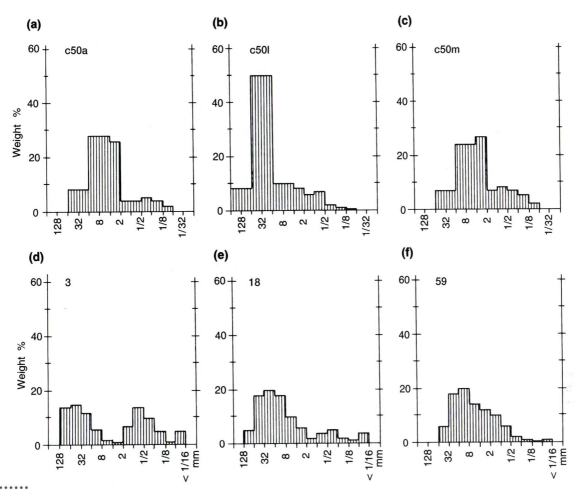

Figure 20.4 Histograms showing the distribution of grain (clast) sizes in coarse clastic sediments and sedimentary rocks. (a), (b), (c) "Carbona" conglomerates, northeastern Diablo Range, California (L. A. Raymond, unpublished data). (d), (e), (f) Upland gravels, southern Maryland (from Schlee, 1957).

Additional factors that should be considered in evaluating the textures of the coarse-grained rocks are (1) roundness of clasts, (2) sphericity and shape of clasts, and (3) the relative proportions of matrix and cement. Roundness is generally a function of weathering and transport. Because the surface energy and therefore the rate of chemical reaction is greater at corners and edges of clasts, these areas weather more rapidly than adjoining flat surfaces. Consequently, weathering tends to render sharper edges and corners into more rounded forms. Transportation abrasion, however, is the major process of rounding, and it may be aided by weathering (Wentworth, 1919; Krumbein, 1942; W. C. Bradley, 1970). The amount of rounding (as well as grain-size reduction) depends on a number of factors, including the distance of transport, the composition of the clast, the compositions of associated clasts, the sizes of associated clasts and grains, and the nature of the transporting medium (Wentworth, 1919; Plumley, 1948). Obviously, softer rocks such as limestone round more readily than

harder rocks such as metaquartzite. Similarly, rounding is favored by greater distances of transport; hence as a rule, rounding increases downstream. This suggests that most sedimentary breccias result from deposition after little or no significant transportation of the clasts. Blocks may retain substantial angularity, even after being transported several kilometers, *if* they are transported in the mud matrix of a debris flow.

The structures of conglomerates, breccias, and diamictites include bedding, cross-bedding, pebble imbrication, grading of various types, and channel fills. Because of the coarse grain size, beds commonly are medium bedded to massive. In shape, they are typically tabular-lenticular or linear-lenticular, but other shapes such as wedge-shaped and lobate beds also occur. Cross-beds of both planar and trough shape are known (figure 20.6a). Low-angle cross-beds occur in shallow marine conglomerates and fluvial deposits, whereas high-angle cross-beds occur in fluvially deposited channel deposits and alluvial fans.

(a)

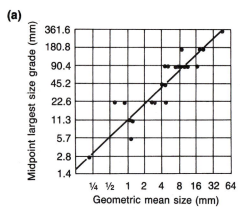

(b)

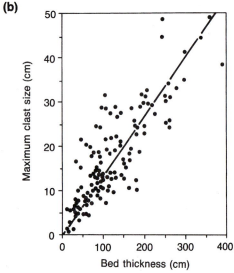

(a)

(b)

⋯⋯⋯⋯⋯⋯

Figure 20.5 Graphs showing the relationships between (a) maximum clast size and average clast size (*from Pettijohn, 1975*) and (b) maximum clast size and bed thickness in clast-supported, submarine fan complex conglomerates, Ksiaz Formation, Devonian, Poland.
(Modified from Nemec and Steel, 1984)

⋯⋯⋯⋯⋯⋯

Figure 20.6 Some structures in coarse clastic sediments and sedimentary rocks. (a) Cross-bedded alluvial fan breccia, Pleistocene-Recent, La Hood, Montana. Matt Raymond is about 1.4 m tall (for scale). See figure 20.1b for a close-up view of the layers behind and to the left of Matt. (b) Channel fill of conglomerate in sandstone, Knobs Formation, Ordovician, near Avens Ford Bridge, Holston Reservoir, northeastern Tennessee. Scale is meter stick.

Grading is also an important indicator of the transportational and depositional conditions. In terms of grading, six types of structures are recognized. These are: (1) non-graded or *disorganized* beds, (2) *normally graded* beds, (3) *inversely graded* beds, (4) *inversely-normally graded* beds, (5) *inversely graded-disorganized* beds, and (6) *graded-stratified* beds (figure 20.7) (R. G. Walker, 1975; 1977; F. J. Hein, 1982; Lowe, 1982). In **normally graded beds,** the clasts or grains gradually get smaller at higher levels in the bed. Inverse grading is the opposite. In **inversely-normally graded beds,** the clasts (or grains) gradually get larger towards the middle of the bed and then gradually get smaller towards the top. **Inversely graded disorganized beds** show coarsening of grains upward from the base to the middle of the bed, where grading is replaced by a random distribution of grains. In

graded stratified beds, a lower graded part of the bed is overlain by a layered or stratified part. In a given bed, if the grading is defined *only* by a change in the sizes of the framework or coarsest clasts, the grading is called **coarse-tail grading.** If all of the grain-size fractions of the sediment change, the grading is referred to as **distribution grading.** Graded beds should not be confused with bedded, fining upward or bedded, coarsening upward sequences (figure 20.7).

The various types of graded bed indicate specific flow regimes and transporting agents (see figure 16.2) and thus reflect environmental conditions. For example, inversely

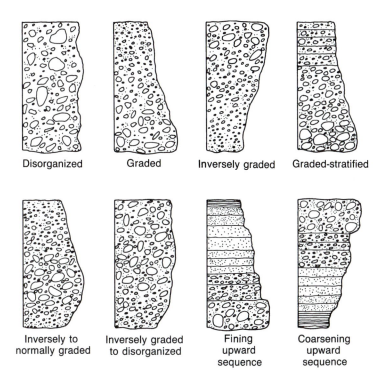

Disorganized	Graded	Inversely graded	Graded-stratified

Inversely to normally graded	Inversely graded to disorganized	Fining upward sequence	Coarsening upward sequence

Figure 20.7 Columnar sections showing graded and disorganized bed types in conglomerates compared to fining and coarsening upward sequences.
(In part, after R. G. Walker, 1975; F. J. Hein, 1982)

Figure 20.8 Photo of imbricated gravel in Boone Fork stream bed, Grandfather Mountain, North Carolina. Downstream to left.

graded beds may develop from grain flows, common in submarine slope, channel, and fan environments. Graded beds are deposited by turbidity currents in the same environments. Disorganized beds, however, are deposited by subaerial and submarine slides and debris flows.

Clast orientation, particularly imbrication, has been studied in some detail in both fluvial and glacial deposits, as well as in marine conglomerates.[11] In general, unless clasts are equant, they may become oriented during deposition.

Elongated clasts may lie with their long axes *either* parallel to or perpendicular to the direction of flow of the transportation agent. In tillites, the long axes are commonly parallel to the direction of ice flow (Richter, 1932, in Pettijohn, 1975; Krumbein, 1939; but see Visser, 1989, for caveats). A similar effect is generated during deposition of some matrix-supported conglomerates (Collinson and Thompson, 1982, ch. 7; R. G. Walker, 1984d). In clast-supported conglomerates, the long axes may lie perpendicular to the direction of flow (Boggs, 1987, p. 132). Fluvial clast-supported conglomerates, in addition to exhibiting clast orientation, show **imbrication** of clasts, a kind of tilted stacking of clasts in which the dip of the stack is upstream (figure 20.8). Imbrication also occurs in beach gravels and conglomerates (Bluck, 1967; Bourgeois and Leithold, 1984).

Compositions

The compositions of conglomerates may be described in terms of bulk chemistry or in terms of clast and matrix compositions. Few analyses of the bulk rock chemistry of conglomerates and related rocks are reported in the literature, probably because the large clast size makes obtaining a *representative* sample a laborious task. Nevertheless, the analyses of a few coarse clastic sediments and sedimentary rocks are reported in table 20.2. These analyses reveal the wide range of compositions that would be expected considering the compositional variability of dominant clasts (e.g., quartz in sample 1 and limestone in sample 4). The close association of sandstones and conglomerates,

Table 20.2 Chemical Compositions of Some Coarse Clastic Sedimentary Rocks

	1	2	3	4
SiO_2	98.95[a]	64.59	64.49	2.64
TiO_2	0.04	0.47	0.58	nr
Al_2O_3	0.52	14.66	26.50	0.67
Fe_2O_3	0.17	2.89	1.20	0.36
FeO	nr	3.55	nr	nr
MnO	0.01	0.08	nr	nr
MgO	0.02	3.83	nr	0.36
CaO	0.00	0.46	nr	52.82
Na_2O	0.01	1.60	nr	nr
K_2O	0.04	5.86	nr	nr
P_2O_5	nr	0.22	nr	tr
H_2O+	nr	1.66	—	nr
H_2O-	nr	0.08	—	nr
CO_2	nr	0.31	—	41.89
LOI[b]	nr	—	7.23[c]	—
Other	0.01	0.08	nr	tr
Total	99.77	100.35	100.00	98.74

Sources:

1. Quartz conglomerate, Sharon Conglomerate (Pennsylvanian), Geauga County, Ohio (J. O. Fuller, 1955).
2. Fern Creek Tillite (Precambrian), Dickinson County, Michigan. Analyst: B. Bruun (Pettijohn, 1975, table 6.3).
3. Clay-rich, quartz conglomerate (Eocene), Ione Formation, Amador County, California (Parker and Turner, 1952).
4. Limestone conglomerate, Siskiyou County, California (Heyl and Walker, 1949; adjusted by Hill, 1981).

[a]Values in weight percent.
[b]LOI = loss on ignition.
[c]Includes H_2O.
nr = not reported.
tr = trace.

in combination with the similar mineralogies of these two rock classes, suggests that the chemistries of conglomerates (and breccias) may mimic those of related sandstones. The same is probably true of carbonate breccias and related limestones and dolostones. The overall chemistry of the few coarse clastic rocks analyzed to date suggests that this is the case. Thus, oligomict quartz conglomerate and quartz arenite have similar compositions (compare analysis 1 in table 20.2 with analysis 1 in table 19.4). Likewise, lithic wackes and polymict volcanic lithic conglomerates and breccias would be expected to have similar compositions. Limestone conglomerates (e.g., analysis 4, table 20.2) are chemically like limestones (e.g., analysis 1, table 21.1).

The compositions of conglomerates, diamictites, and breccias are commonly reported as clast or block counts of the included rock types. Table 20.3 presents some representative analyses of this type. Notice the wide range of clast compositions, but also note that there is a complete gradation from nearly oligomict conglomerates (those with one clast type) to polymict conglomerates with highly diverse clast assemblages. Oligomict quartz conglomerates may result either from extensive transportation and abrasion (**working**) of sediment that was derived from a quartz-rich source terrane or from reworking of sedimentary materials already enriched in quartz through previous working. In contrast, oligomict carbonate breccias and conglomerates, in most cases, probably are locally derived and have been worked little. Oligomict chert breccias may also be locally derived. Polymict volcanic lithic conglomerates are derived from volcanic arcs or plate boundary spreading centers and transform fault scarps. Polymict conglomerates with an even greater clast diversity, those that include granitoid, volcanic, sedimentary, and metamorphic rock types, are generally derived from orogenic belts.

ORIGINS OF COARSE CLASTIC SEDIMENTARY ROCKS

The environments in which we can witness the formation of coarse clastic sediments, plus the facies and fossils that we find associated with coarse clastic sedimentary rocks, give a clear indication of the general origins of these rocks. They are deposited as a result of transportation and deposition by flowing or oscillating water, by gravity-induced flows, slides, and falls, and by moving ice.

Most conglomerates and some breccias are transported and deposited in water. On the continents, the major agents that form conglomerates and some breccias are streams and lacustrine waves. In streams, gravel is transported primarily as bed load.[12] In some cases, however, high flow velocities and turbulence may lift gravels into suspension (Krumbein, 1942; K. M. Scott and Gravlee, 1968; V. R. Baker, 1984). Whenever shear stress acting on the streambed exceeds the forces that hold particles in place, especially during periods of high flow (floods), major bed-load transport begins. Decrease in the current velocity and associated shear stress occurs either (1) when the flood subsides, or (2) when the velocity of the water decreases, as on floodplains, fans, or deltas, where single-channel, channelized flow is replaced by sheet flow or multichannel distributary flow.[13] Deposition follows when the shear stress decreases to a value perhaps on the order of one to six times less than that initially required to begin movement (Reid and Frostick, 1984). The conglomerates resulting from fluvial transport are typically

Table 20.3 Clast Counts of Selected Conglomerates and Related Materials

	1	2	3	4	5	6	7
Igneous							
Felsic plutonic	26[a]	2	0	1	0	0	0
Mafic plutonic	12	0	1	0	0	0	0
Felsic volcanic	11	23	0	3	0	0	0
Mafic volcanic	21	1	1	1	0	0	0
Sedimentary							
Conglomerate	0	0	0	0	tr	0	0
Sandstone and siltstone	20	7	32	<1	1	12	0
Shale and mudstone	0	15	0	50	0	0	0
Limestone	0	2	0	0	0	78	<1
Dolostone	0	0	0	0	0	1	61
Chert	3	41	64	38	2	3	38
Fossils	0	0	0	0	0	0	tr
Metamorphic							0
Metaquartzite	0	1	0	<1	tr	4	0
Slate and argillite	6	0	0	2	1	0	0
Serpentine	0	0	2	0	0	0	0
Quartz	0	6	0	3	92	2	0
Feldspar	0	0	0	0	3	0	0
Unknown and miscellaneous	0	2	0	<1	tr	0	0
Number of pebbles counted	422	494	102	213	564	848	600

Sources:

1. Polymict conglomerate, Herring Head Formation, Toogood Sequence (Silurian), Herring Neck Canal, New World Island, Newfoundland (Helwig and Sarpi, 1969).
2. Polymict conglomerate, sample 39, Franciscan Complex (Jurassic/Cretaceous), Pickett Peak Quadrangle, northern California (Seiders and Blome, 1988).
3. Poorly consolidated, polymict conglomerate, sample C51c (Miocene), Carbona Quadrangle (Raymond, 1969).
4. Polymict shale chip, granule breccia, Franciscan Complex (Jurassic/Cretaceous), Mount Oso area, northeastern Diablo Range, California (unpublished data, L. A. Raymond).
5. Quartz conglomerate, Citico Conglomerate of the Wilhite Formation (Proterozoic), Chilhowee Dam, Little Tennessee River, Tennessee (unpublished data, L. A. Raymond).
6. Polymict conglomerate, Tellico Sandstone (Formation), Ordovician, Cisco, Georgia (Kellberg and Grant, 1956).
7. Chert-dolostone breccia, Mosheim Limestone (Formation), Ordovician, Rich Valley, near Saltville, Virginia (unpublished data, L. A. Raymond).

[a]Values in percent.

tr = trace.

clast-supported (figure 20.9a), although those from flash floods may contain considerable amounts of matrix (Boothroyd and Ashley, 1975; Nemec and Steel, 1984; L. T. Middleton and Trujillo, 1984). Clast imbrication is common, as is cross-bedding, but grading is present only locally.

Fluvial conglomerates and breccias formed on alluvial fans (the "fanglomerates") include matrix- and clast-supported types. These are associated with diamictites produced by mudflows and debris flows.[14] Transportation and deposition mechanisms of fluvial fanglomerates and breccias are the same as those operating in stream channels and on floodplains. In contrast, matrix-supported units result from deposition of sediment from flows in which the sediment to water ratio is high (R. P. Sharp and Nobles, 1953). Mudflows and debris flows are

405

Figure 20.9 Fluvial, clast-supported, cobble-to-boulder polymict gravel overlying sands. Note clast imbrication to the right and towards the viewer. Terrace gravel (Pleistocene?) along the south side of Arroyo Hondo, north of Taos, New Mexico, on Highway 3. Gravel escarpment is about 5 m high.

Figure 20.10 Beach conglomerate (polymict, clast-supported pebble type) interlayered with sandstone, Pleistocene marine terrace, Portuguese Beach, Sonoma County, California.

typically initiated where gravitational instability is created in accumulated weathering products by large amounts of rainfall. These flows move downslope or down channels until resistance to flow produced by cohesion between particles and the friction exceed the shear stresses produced by gravity acting on the sediment mass. Then, the flow stops abruptly and the sediment is deposited. The resulting subaerial mudflow and debris-flow fanglomerates and diamictites typically are poorly sorted, disorganized, and matrix-supported. Some reverse grading is present at the base of individual units, but grading is uncommon. Clasts tend to be angular, especially in the diamictites.

Quaternary to Recent debris flows and landslides are common (Pomeroy, 1980; S. G. Wells and Harvey, 1987), as are rockfalls, rockslides, and other forms of landslides.[15] Yet, because these processes occur in orogenic regions that are undergoing erosion, most diamictites, breccias, and conglomerates formed by them are eroded before being preserved in the long-term stratigraphic record. Tills are locally common in high-latitude and high-elevation areas, but their lithified equivalents, the tillites, likewise do not comprise much of the stratigraphic record.

Diamictites produced by mass flows are similar enough in character to tills to be confused with them (R. P. Sharp and Nobles, 1953; D. E. Lawson, 1981; Madole, 1982). Tills result from glacial deposition, where melting ice drops its load of sediment. Distinctive features such as striated clasts, chatter-marked grains, unimodal clast-orientation fabrics, and associated dropstone-bearing, laminated mudstones and varvites are important in the recognition of ancient tillites

(Schwab, 1976).[16] Although tills appear disorganized, they commonly have a fabric resulting from orientation of clasts, with their long axes parallel to ice flow directions (Krumbein, 1939; Clague, 1975a).

Marine and lakeshore beach deposits are similar in nature.[17] They are characterized by relatively good sorting, imbrication of clasts, and commonly, discoidal shaped pebbles (Bluck, 1967; Bourgeois and Leithold, 1984). In addition, they are cross-bedded and are locally, but uncommonly, fossiliferous. Both clast- and matrix-supported conglomerates form in beach to shallow marine environments (R. L. Phillips, 1984), but clast-supported conglomerates are the norm. They are commonly interbedded with or grade laterally into sandstones (figure 20.10). Beach gravels (and the resulting conglomerates) owe their characteristics to the breaking of waves and the ebb and flow of waters caused by the wave action.

Waves also produce carbonate breccias. For example, storm waves will fragment mudcracks and laminated lime muds on tidal flats and redeposit them as breccias or "flat pebble conglomerate" (figure 20.1d).[18] In somewhat deeper water, carbonate breccias or conglomerates may develop where wave-fragmented reefal limestones accumulate on the reef flank (P. Hoffman, 1974).[19]

In offshore areas, mass flow and submarine landsliding are the dominant means of transportation and deposition of coarse clastic materials. Subaqueous debris flows apparently occur on submarine deltas, develop on the slope, and are common in submarine canyons and channels (Postma, 1984; J. M. Coleman, 1988). Similarly, submarine landslides occur widely on submarine slopes (Embley, 1976; Jacobi, 1984).

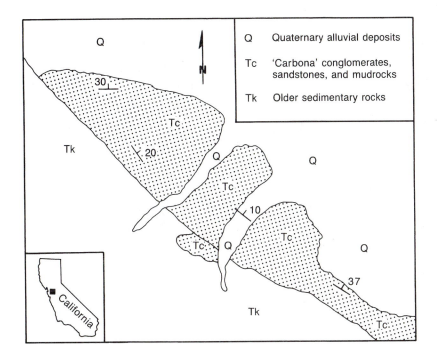

Figure 20.11 Map of "Carbona" conglomerate (Tc), Carbona 15' Quadrangle, California. *(Modified from Raymond, 1969)*

Like subaerial mass movements, the submarine mass movements begin to move when shear stresses exceed the yield strength of the sediment. Deposition occurs as that stress diminishes, generally due to a decrease in gradient.

EXAMPLES

The examples presented here illustrate the diversity of the occurrences, compositions, and origins of coarse clastic rock. They also show how sedimentary structures, rock compositions, and paleogeographic data can be used to understand the history of a rock body and the region in which it occurs.

Upper Miocene Conglomerates of the Carbona Quadrangle, California

Upper Miocene gravels and poorly lithified conglomerates, breccias, and diamictites, interbedded with sandstones and mudrocks, crop out along the northern and eastern edge of the Diablo Range, one of the Coast Ranges of central California (figure 20.11). These greenish gray to reddish brown rocks have been variously referred to as the Livermore Gravels, the Carbona Formation, and the Oro Loma Formation (Huey, 1948; L. I. Briggs, 1953; Pelletier, 1961; Raymond, 1969). In the Carbona Quadrangle, south of Tracy, California, and in the Tesla Quadrangle to the west, the rocks reach their maximum thickness of about 1200 m. In the Carbona area, the principal exposures form a semicircular map unit representing a northeast-dipping sedimentary pile that thins to the west and east.

Exposures are poor in some areas, but the unit is well exposed along gullies and canyon walls. Individual beds are lenticular and typically, in exposed sections, the lenses are one to several meters across. They range from about 5 cm to 1.5 m in thickness (figure 20.12). Locally they may exceed these limits. The maximum clast size recognized is 60 cm. Sedimentary structures in the beds include cross-bedding, channels, some grading, and a tubular feature of unknown origin. Minor inverse grading is present locally, but is not common. Some clast imbrication occurs as well.

Texturally, the individual beds of coarse clastic debris are usually bimodal (see figure 20.4 a, b, and c). Sorting is poor. Both clast- and matrix-supported lithofacies are present in the section. The clasts range from well rounded to angular.

The provenance of the sediment is indicated both by the position of the unit on the flank of the mountain range and by clast compositions. Pebble counts of Carbona conglomerates (table 20.4) and heavy mineral analyses of conglomerates and interbedded sandstones (table 20.5) indicate that the dominant source terrane was the structurally underlying Mesozoic System, which forms the core of the Diablo Range and is exposed to the south (Raymond, 1969). The Jurassic-Cretaceous Franciscan Complex was the principal contributor of sediment, as is indicated by the abundance of wacke sandstone and chert clasts and by the presence of the metamorphic minerals glaucophane, jadeite, and lawsonite, among the heavy minerals (see table 20.5, especially samples C50L and C50r). Minor amounts of sediment were probably derived from underlying Great Valley Group and Cenozoic volcanic conglomerates, sandstones, and mudstones, which

Table 20.4 Clast Counts for Individual Beds of "Carbona" Conglomerate

	C51a	C51b	C51c	C50a	C50L	C50m	C50n
Chert	**55**[a]	**63**	**64**	**80**	**43**	**63**	**74**
green	28	26	33	18	18	33	27
red-pink	12	22	17	19	12	14	21
yellow-orange	5	10	5	2	3	0	4
black	1	0	0	3	0	3	1
white	9	6	9	37	10	13	20
breccia	0	0	0	0	0	0	1
Sandstones	**39**	**35**	**32**	**10**	**54**	**29**	**21**
"graywacke"	—	—	—	8	—	25	17
other	—	—	—	2	—	4	4
Conglomerate	**0**	**0**	**0**	**2**	**3**	**2**	**0**
Mafic metavolcanic rocks	**3**	**1**	**1**	**0**	**0**	**3**	**2**
Felsic volcanic rocks	**0**	**0**	**0**	**0**	**0**	**1**	**0**
Mafic plutonic rocks	**1**	**0**	**1**	**0**	**0**	**0**	**0**
Serpentinite	**1**	**0**	**2**	**0**	**0**	**0**	**0**
Quartz	**0**	**0**	**0**	**2**	**0**	**0**	**0**
Calcite	**0**	**0**	**0**	**0**	**0**	**1**	**0**
Caliche	**0**	**0**	**0**	**6**	**0**	**0**	**2**
Unidentified	**1**	**1**	**0**	**2**	**0**	**2**	**2**
Clasts counted	136	126	102	124	61	117	128

Source: Raymond (1969).
[a]Values in percent. Totals may exceed 100% due to rounding errors. Only boldface numbers total to about 100%.

had an eastern (Sierra Nevada) provenance. This secondary source is indicated by the presence of hypersthene and oxy-hornblende (e.g., sample C22, table 20.5), phases absent in the underlying basement rocks.

The evidence, considered together, indicates that the origin of the "Carbona" conglomerates, breccias, and diamictites is alluvial. Terrestrial vertebrate fossils in the rocks, including those of horses, support a nonmarine origin (Pelletier, 1951; 1961). Trough-shaped cross-beds, channels, lenticular beds, and clast imbrication, combined with the size distribution of the sediment (see cumulative size curves for Carbona conglomerates in problem 14.2 page 283), suggest that most beds represent sediment transported as bed load and graded suspensions in streams and deposited in channels and floodplains on the surface of an alluvial fan or alluvial plain. The presence of a few diamictites and some inversely graded beds

Figure 20.12 Bedded "Carbona" conglomerate, sandstone, and diamictite. The lowermost complete bed here is approximately 1.5m thick.

Table 20.5 Heavy Minerals from "Carbona" Conglomerates

	C22	C22b	C50L	C50q	C50r
Actinolite	—	—	—	—	r
Amphibole, colorless	—	—	C	C	m
Apatite	—	r	—	r	r
Biotite	—	—	r	r	m
Chromite	—	—	A	A	C
Epidotes	r	m	m	m	m
Garnets	r	C	m	m	m
Glauconite	—	—	r	—	—
Glaucophane	m	m	C	m	m
Hornblende	C	A	m	m	m
Oxyhornblende	C	A	C	m	r
Hypersthene	A	m	—	—	r
Jadeite	—	—	m	C	C
Lawsonite	—	—	m	m	C
Limonite	—	—	A	C	C
Magnetite	C	C	C	C	m
Sphene	—	r	m	m	m
Zircon	—	m	m	r	m

Source: L. A. Raymond unpublished data

A = abundant, C = common, m = minor, r = rare.

suggests that some deposition from debris flows also occurred. Sediment shed from the rising mountain range to the south and southwest was transported to and deposited on a developing bajada along the flank of the mountains.

Allochemical Dolostone-Chert Breccias of the Middle Ordovician Mosheim Limestone, Virginia

In the southern Appalachian Orogen, deposition of a thick sequence of Cambrian- to Ordovician-aged carbonate rocks, including the Cambro-Ordovician Knox Group, occurred along the passive eastern margin of North America (Colton, 1970; Markello and Read, 1981). Carbonate deposition was interrupted by the Ordovician Taconic Orogeny, which uplifted the margin, resulting in subaerial exposure, development of karst topography, and erosion during Middle Ordovician time. Weathering and erosion resulted in the formation of hills and valleys, creating a paleotopography with up to 150 m of relief (Webb, 1959; Mussman and Read, 1986; Webb and Raymond, 1979, 1989).

The Middle Ordovician transgression over the eroded Knox Group resulted in deposition of various carbonate rocks and shales at the base of the Tippecanoe Sequence. Thus, the Knox–Middle Ordovician unconformity was formed. The post-unconformity units include sinkhole-filling breccias, limey and dolomitic red beds, basal carbonate (± chert) conglomerates and breccias, and fossiliferous limestones and dolostones (Mussman and Read, 1986). Breccia-filled depressions suggest that solution and associated collapse of roofs over solution cavities contributed to breccia formation. Basal breccias are exposed in the Rich Valley area, near Saltville, Virginia, where paleotopography on the Knox Group reaches 140 m (Webb, 1959).

The Rich Valley exposures consist of small lenses or wedges of breccia, which are assigned to the Mosheim Limestone (Formation). These breccias flank topographic highs in the Knox basement (figure 20.13). Although several of the breccia beds occur at the base of the formation, forming the lower 2/3 m of the section, others are present as interlayers a few meters above the base. Beds typically are less than 1 m in

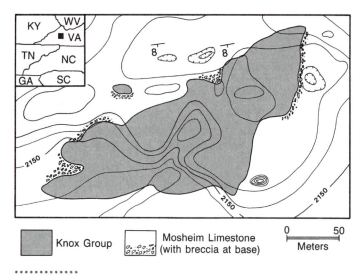

Knox Group   Mosheim Limestone (with breccia at base)

0   50
Meters

Figure 20.13 Map of the Ben Clark Farm area, showing Knox Group carbonate rocks overlain by Mosheim breccias and Mosheim Limestone. *(Modified from Webb, 1959)*

Figure 20.14 Photograph of Mosheim breccia. Clasts include chert, dolostone, and a cephalopod.

thickness and the maximum clast size is about 45 cm. Dolostone forms the largest clasts, whereas chert clasts, which range from 1 to 60 mm, average about 9 mm in diameter (Webb, 1959, p. 18). Above the breccia beds, Mosheim limestones, primarily fenestral lime mudstones, indicate shallow marine conditions of deposition.[20]

The Mosheim breccias vary markedly from place to place. Most are matrix-supported, but clast-supported types are present as well (figure 20.14). The matrix is dominantly dolomite. In some areas, the breccias consist of oligomict dolostone breccia, whereas in others polymict chert-dolostone breccia is dominant. Locally, the polymict breccia is sparsely fossiliferous, containing marine mollusks .

Pebble counts reflect the variable composition of the breccias. Some counts yield 100% chert or 100% dolostone and others reveal a polymict composition. Analysis 7 in table 20.3 is a composite clast count of several slabs of Mosheim breccia that reflects the average composition of the breccias at the Ben Clark Farm locality of Webb (1959). Dolostone is dominant, but chert is common. Rare limestone clasts occur and calcite forms veins and "eyes" that represent cavity fillings. All of the clasts, except the fossils, are derived from the underlying Knox Group.

The available data indicate that the Mosheim breccias were formed by erosive wave action and subsequent deposition of sediment in the intertidal zone that developed as the Middle Ordovician sea transgressed the eroded North American margin. Solution-induced brecciation (sinkhole collapse) may have locally contributed to the fragmentation of Knox Group carbonate rocks and chert. Nevertheless, whether the fragmentation resulted in part from doline collapse or was caused entirely by wave action, transportation of the clasts was quite limited prior to deposition. Limited transportation,

indicated by the angularity of the clasts, is supported by the fact that the breccias occur immediately adjacent to Knox Group basement (see figure 20.13). Because most breccias are matrix- supported, winnowing of finer materials from the sediment by currents or waves to produce "lag deposits," apparently was not important in the formation of Mosheim breccias.

Conglomerates of the Cape Enragé Formation, Quebec

The Cambro-Ordovician Cape Enragé Formation is exposed along the south side of the St. Lawrence River in Quebec (C. Hubert, Lajoie, and Leonard, 1970).[21] Up to 270 m of conglomeratic rocks and sandstones form a complex of layers that is divisible into six lithofacies (F. J. Hein, 1982; F. J. Hein and Walker, 1982).

The six lithofacies constitute a deep submarine channel complex that was deposited over basin-plain mudrocks of the Orignal Formation. These lithofacies are (1) coarse conglomerate (clasts >16mm), (2) graded-stratified and cross-bedded fine conglomerate, (3) ungraded, cross-bedded, fine conglomerate, pebbly sandstone, and sandstone, (4) coarse-tail-graded, matrix-supported, pebbly sandstone and fine conglomerate, (5) graded fine conglomerate, pebbly sandstone, and sandstone with fluid escape structures, and (6) massive pebbly sandstone and sandstone (F. J. Hein, 1982; F. J. Hein and Walker, 1982). Normal grading occurs in all units except unit 3, and the grading is distribution grading in all units except 3 and 4. Other grading types—graded stratified, inverse-normally graded, inversely

St. Simon Sur Mer, Est

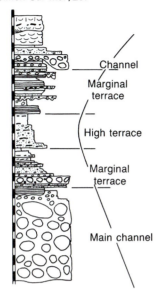

Channel

Marginal terrace

High terrace

Marginal terrace

Main channel

Figure 20.15
Stratigraphic sequence at St. Simon Sur Mer Est, Cape Enragé Formation, showing channel and terrace conglomerates interlayered with sandstones and mudrocks.
(Simplified from F. J. Hein and Walker, 1982)

graded disorganized, and inversely graded—as well as disorganized beds, occur in the formation. The basal contacts of the units are flat or scoured, except for unit 3, which has loaded to scoured basal contacts. The various lithofacies are complexly interlayered with and channel into one another (figure 20.15).

The conglomeratic rocks are polymict and are dominated by limestone clasts (C. Hubert, Lajoie, and Leonard, 1970; B. A. Johnson and Walker, 1979). Quartz is the other dominant clast type. At least one unit contains boulders of carbonate rock up to 4 m in maximum dimension, but clast dimensions are generally less than 10 mm, falling in the granule and pebble size categories.

The conglomerates are channelized and facies are complexly interrelated. This fact and the paleocurrent and paleogeographic considerations suggest that the Cape Enragé Formation was deposited at the base of the continental rise, in a southwest-trending complex of conglomerate-filled submarine channels *and* on the adjoining channel terraces (F. J. Hein and Walker, 1982). With the exception of lithofacies 3, the sediments themselves were deposited from concentrated sediment masses transported at the bottom of large turbidity currents (F. J. Hein, 1982). The ungraded units (facies 3), which represent previously deposited materials, may be sediments reworked by traction currents formed where spillover from major channels established currents outside of the main depositional site of a given turbidity current. Clasts of shelf rocks indicate reworking of shallow marine rocks. Thus, the various deposits represent interlayered, resedimented conglomerates and sandstones, composed of materials first deposited on the shelf and later redeposited in deeper water.

SUMMARY

Conglomerates, breccias, and similar rocks are formed in a diverse array of environments. These include glacial, fluvial, alluvial fan, lacustrine shore, deltaic, beach, tidal flat, slope, and submarine fan/basin plain environments. Each environment affects the particular characteristics of the coarse clastic sediment formed. Those characteristics aid in discriminating between nonsedimentary and sedimentary origins for coarse clastic rocks. Fossils associated with the rocks reflect the specific environments of formation.

Framework clasts of coarse clastic sedimentary rocks may be rounded or angular. The matrix in conglomerates and breccias is limey, dolomitic, sandy, or gravelly in nature, whereas that of the diamictites is mud. Both clast- and matrix-supported coarse clastic rocks may show grading. Additional structures, present locally, include cross-bedding and channelling (cut and fill). These structures and the associated fossils indicate specific depositional settings.

Conglomerates are particularly useful in making paleogeographic reconstructions. The clasts may contain fossils and the clast lithologies give specific information about the provenance. Thus, although they are a volumetrically minor component of the stratigraphic record, coarse clastic sediments are extremely valuable sources of information.

EXPLANATORY NOTES

1. Examples of papers describing various gravels, conglomerates, breccias, and diamictites or their origins include the following: Lawson (1913), R. P. Sharp (1948), W. S. White (1952), Chase (1954), Ham (1954), J. O. Fuller (1955), Kellberg and Grant (1956), R. C. Anderson (1957), Crowell (1957), E. H. Bailey, Irwin, and Jones (1964), W. B. Hamilton and Krinsley (1967), Horne (1969), Helwig and Sarpi (1969), Abbate, Bartolotti, and Passerini (1970), Nordstrom (1970), J. L. Wilson (1970), Lowe (1972), Mutti and Ricci-Lucchi (1972), Andersen and Picard (1974), D. S. Cowan and Page (1975), J. F. Hubert, Suchecki, and Callahan (1977), R. G. Walker (1975, 1977), Krieger (1977), Stanley (1980a), Daniels (1982), F. J. Hein (1982), Kalliokoski (1982), Madole (1982), McLaughlin and Nilsen (1982), Merk and Jirsa (1982), Seiders (1983), Jacobi (1984), Leithold and Bourgeois (1984), T. E. Moore and Nilsen (1984), Phipps (1984), Pickering (1984), Postma (1984), Schultz (1984), Seiders and Blome (1984, 1988), Nadon and Middleton (1985), Postma and Roep (1985), Rosenthal and Walker (1987), Went, Andrews, and Williams (1988), Barany and Karson (1989), C. H. Stevens, Lico, and Stone (1989), N. Eyles (1990), Mustard and Donaldson (1990), A. H. F. Robertson (1990), and L. H. Tanner and Hubert (1991).

2. A *melange* is "a body of rock mappable at a scale of 1:24000 or smaller and characterized both by the lack of internal continuity of contacts or strata and by the inclusion of fragments and blocks of all sizes, both exotic and native, embedded in a fragmented matrix of finer-grained material" (Raymond, 1984a).

3. The boundary between volcaniclastic igneous rocks and volcaniclastic sedimentary rocks is gradational and some rocks cannot easily be assigned to one class or the other.

4. See additional papers in Lipman and Mullineaux (1981) for descriptions of the Mount St. Helens eruption and its consequent deposits.

5. The term *olistostrome* is used to designate a mesoscopic-scale to mappable-scale rock body of submarine slide origin (Flores, 1955; Abbate, Bartolotti, and Passerini, 1970; Hsu, 1974; Raymond, 1984a). Mappable olistostromes represent one type of melange. Olistostromes are described and discussed by various workers including Horne (1969), Abbate, Bartolotti, and Passerini (1970), B. M. Page and Suppe (1981), Naylor (1982), M. T. Brandon (1989), P. D. Muller, Candela, and Wylie (1989), and Eyles (1990). Also see the discussions of related "megaturbidites" in Bouma (1987), Souquet, Eschard, and Lods (1987), and Labaume, Mutti, and Seguret (1987).

6. Related aspects of this problem are discussed by Raymond, Yurkovich, and McKinney (1989).

7. *Pebbly mudstone* is another name used for rocks with rounded clasts in a muddy matrix. For a discussion, see Crowell (1957).

8. The term foliation is used here in a descriptive rather than a genetic way and refers to the tabular alignment of platy to acicular grains.

9. For example, Lugn (1927), Krumbein (1942), Schlee (1957), Raymond (1969), Pettijohn (1975, p. 158), Nemec and Steele (1984).

10. Shih and Komar (1990) discuss the impact of variation in flow discharge.

11. For more information, see Krumbein (1939a), and the reviews in Pettijohn (1975, ch. 6) and Collinson and Thompson (1982, ch. 7). For examples, see W. S. White (1952), Byrne (1963), Bluck (1967), Pessl (1971), J. J. Clague (1975a), and Visser (1989).

12. The mechanics and processes of stream erosion, transportation, and deposition are discussed in a large number of papers, such as Krumbein (1942), Plumley (1948), Kuenen (1956), Wolman and Miller (1960), Byrne (1963), Bagnold (1966; 1973), D. B. Simons and Richardson (1966), K. M. Scott and Gravlee (1968), Keller (1971), Shroba et al. (1979), Faye et al. (1980), Kochel and Baker (1982), V. R. Baker (1984), I. Reid and Frostick (1984, 1987), S. G. Wells and Harvey (1987), and Shih and Komar (1990). Refer to these works and the references therein for more information.

13. Bull (1963, 1968), Nilson (1982), Rust and Koster (1984).

14. The origins or nature of subaerial debris flows and mudflows are discussed by R. P. Sharp and Nobles (1953), Middleton and Hampton (1976), Nilson (1982), Schultz (1984), and Osterkamp, Hupp, and Blodgett (1986). Also see Dott (1963). Alluvial fan sediments are discussed by A. C. Lawson (1913), Bull (1963), Raymond (1969), Nilsen (1982), Harvey (1984), Kochel and Johnson (1984), Wells (1984), and Nadon and Middleton (1985), among others.

15. Landslide deposits are discussed by Shreve (1968), B. Johnson (1978), Voight (1978a,b), Plafker and Ericksen (1978), McLaughlin and Nilsen (1982), Harvey (1984), and Osterkamp, Hupp, and Blodgett (1986). Miocene landslide breccias in Arizona are described by Krieger (1977). Subaerially formed diamictites in the Upper Paleozoic Cutler Formation of Colorado are described by Schultz (1984).

16. See also W. B. Hamilton and Krinsley (1967), Schwab (1976), Easterbrook (1982), Collinson (1986b), and J. Mazzullo and Anderson (1987).

17. Literature on lacustrine pebble or cobble beaches is rare. See Krumbein and Griffith (1938) for an example. Marine beaches are discussed by Clifton (1973), other authors cited in this text, and A. T. Williams and Caldwell (1988).

18. Flat pebble conglomerates are discussed by Matter (1967), Shinn (1983), and Whisonant (1987, 1988).

19. Reef-flank breccias and conglomerates are also described, for example, in H. E. Cook et al. (1972) and in Beard (1985, in McKnight 1988).

20. See chapter 21 for a discussion of carbonate facies and their significance.

21. This review is based on the work of C. Hubert, Lajoie, and Leonard (1970), I. C. Davies and Walker (1974), B. A. Johnson and Walker (1979); F. J. Hein (1982), and F. J. Hein and Walker (1982).

PROBLEMS

20.1. Convert the histogram in figure 20.4a into a CM diagram.

20.2. Describe the provenance for each of the conglomerates listed in table 20.3.

CHAPTER 21

Carbonate Rocks

INTRODUCTION

Tropical waters offshore of palm-shaded beaches in areas such as the west Florida coast, Cancun, Mexico, or the Bahamas, represent the environments in which many carbonate rocks form. Carbonate rocks are interesting for several reasons. They are excellent indicators of specific depositional environments and thus are instrumental in reconstructing geologic history; they provide reservoirs for hydrocarbons and are hosts to certain types of ore deposits; and they contain large numbers of fossils, which are interesting as objects of nature, are essential to the documentation of biological evolution, and are instructive in reconstructing depositional environments.

The carbonate sedimentary rocks—the limestones and dolostones—comprise only about 4% of the volume of rocks covering the total surface of the Earth, but on the continents they make up between 10 and 35% of the rocks (Blatt, 1970; Folk, 1974).[1] Carbonate sediments blanket large areas in the Atlantic Ocean, the Indian Ocean, and the southern Pacific Ocean, and are notable on banks and shallow shelves in tropical to subtropical regions. On the continents, carbonate sediments form in lakes, in soils and dunes of arid and semiarid regions, and in areas of hydrothermal activity and groundwater springs.

CHARACTERISTICS OF CARBONATE ROCKS

Like all rocks, carbonate rocks are classified on the basis of their minerals and textures (see chapter 15). Yet, because their mineral compositions are restricted, the textures of carbonate rocks assume added importance. Some carbonate rocks are crystalline, whereas others are clastic. Many contain both crystalline and clastic elements.

By definition, **carbonate rocks** are those rocks with 50% or more carbonate minerals. In terms of their primary classification, they include precipitates and clastic rocks, including allochemical types. Mineralogically, the carbonate rocks include primary phases, secondary phases, and reworked grains. Most of the primary materials are precipitated by organisms. The origins of the most notable secondary phase, dolomite, and other diagenetic carbonate minerals were discussed in chapter 16.

The clastic grains in carbonate rocks consist of whole or broken shells and fragments of other chemical and biochemical precipitates, including pieces of preexisting rocks. Where these grains (fragments) have been transported away from their sites of origin they are called *allochems*.[2] Recall that allochemical rocks are rocks composed predominantly of allochems.

Folk (1974) made an important distinction, in separating allochemical rocks from primary precipitates, between rocks with single and multistage sedimentary histories. Primary precipitates form in response to chemical controls provided by the physical environment or biota, or both. Allochemical rocks owe their character not only to the chemical conditions that contribute to their initial precipitation, but to the physical conditions that led to erosion, transportation, and redeposition.

MINERAL COMPOSITIONS AND CHEMISTRIES OF CARBONATE ROCKS

Limestones, composed predominantly of calcite, Mg calcite, and aragonite, and *dolostones*, consisting primarily of dolomite, are the two main types of carbonate rock. Mg calcite and aragonite are particularly subject to diagenetic change; thus calcite and dolomite are the most common phases in older carbonate rocks.

Chemically the carbonate rocks are dominated by CaO, MgO, and CO_2. The isotopic ratios of the oxygen and carbon are significant in environmental analyses, as are some trace elements.

Carbonate Minerals

Several carbonate minerals occur in carbonate sediments and rocks, including two varieties of calcite. The calcite ($CaCO_3$) may be a relatively pure form called *calcite* or *low-Mg calcite* or it may contain significant amounts of magnesium ion (up to 30 mol %), in which case it is referred to as *magnesian calcite, Mg calcite, high-magnesium calcite,* or *high-Mg calcite* (Friedman, 1964, 1965b; Scoffin, 1987, sec. 1.2).[3] It also may contain small amounts of elements such as iron, manganese, or sodium. Calcite is precipitated both organically and inorganically. Mg calcite is produced mainly by benthic organisms in shallow tropical marine environments. It may also form as a result of the combination of Ca and Mg ions with chemically or biologically oxidized methane gas in the sediments of non-tropical regions.[4] In addition, high-Mg calcite is reported from Lake Tanganyika, a large tropical lake in east Africa (Cohen and Thouin, 1987). Low-Mg calcite is produced by pelagic and planktonic organisms in the open ocean and is the stable phase at depth (Friedman, 1965b; Mullins, 1986).

Aragonite (orthorhombic $CaCO_3$) is also precipitated via organic and inorganic processes. Much of that present in carbonate rocks is of organic origin, as aragonite is the main constituent of many invertebrate skeletons and the principal precipitate of certain algae (Lowenstam, 1955; Halley, 1983).[5] Like Mg calcite, aragonite is typically produced by benthic

organisms.[6] Inorganic aragonite precipitates from springs such as Mammoth Hot Springs in Yellowstone National Park, Wyoming (Pentecost, 1990). In contrast to the trigonal calcite structure, the aragonite lattice, which is orthorhombic, does not incorporate significant amounts of Mg ion. Sr and Na, among other elements, do occur in trace amounts in aragonite.[7] Because it is metastable at the surface of the Earth and readily converts to calcite, pre-Pleistocene aragonite is relatively rare.

Dolomite [$CaMg(CO_3)_2$] occurs both as a primary precipitate and as a replacement mineral, but most is secondary.[8] Dolomite has a structure in which one-half the cation sites are occupied by the relatively small Mg atoms; thus, larger Ca ions (0.99 Å) do not readily substitute for Mg (0.66 Å) in dolomite.[9] Ferrous iron (0.74 Å) does substitute, however, and dolomites with significant iron are called ferroan dolomite.

Other Minerals

A wide range of other minerals occurs in carbonate rocks. The most common of these are quartz and the clays. Typically these minerals are present only in minor to trace amounts. Yet exceptions are not uncommon. For example, in carbonate-clastic transition zones, calcite or aragonite is mixed with abundant quartz, clays, and even feldspars (Driese and Dott, 1984; Mount, 1984).[10] The most abundant of the phyllosilicate minerals found in the carbonate rocks include the clays—illites, smectites, and kaolinites—plus glauconite and the chlorites.[11]

Examples of the additional phases that occur in carbonate rocks, some of which are formed during diagenesis, include opal, chalcedony, ankerite, rhodochrosite, braunite, pyrite, graphite, hematite, magnetite, alkali feldspars, talc, apatites, gypsum, anhydrite, and halite.[12] Particularly where replacement has been extensive, the occurrence of these minerals may only be indicated by pseudomorphs.

Chemistries of Carbonate Rocks

Because they are composed largely of calcium and magnesium carbonate minerals, the carbonate rocks exhibit chemistries high in these major elements and CO_2. Table 21.1 presents the chemistries of some carbonate rocks and sediments revealing the abundance of these elements. Note that silica is rather low, as is the case in most carbonate rocks. Where clays and other minerals are a major constituent of the rocks, the amounts of silica, alumina, and iron are greater.

Some trace element studies have also been completed on carbonate rocks (Parekh et al., 1977; Tlig and M'Rabet, 1985). In general, trace element values are low, because many trace elements do not readily substitute in carbonate minerals. For example, dolostones and associated (parental) limestones in Tunisia have low abundances of REE (Tlig and M'Rabet, 1985). Furthermore, the work of Tlig and M'Rabet

Table 21.1 Chemistry of Selected Carbonate Rocks and Sediment

	1	2	3	4
SiO_2	1.15[a]	13.9	7.96	0.28
TiO_2	nd	0.26	0.12	nd
Al_2O_3	0.45	4.57	1.97	0.11
Fe_2O_3	nd	3.8[b]	0.14	0.12
FeO	0.26	—	0.56	nd
MnO	nd	0.53	0.07	nd
MgO	0.56	1.56	19.46	21.30
CaO	53.80	38.2	26.72	30.68
Na_2O	0.07	0.20	0.42	0.03
K_2O	—	<0.02	0.12	0.03
P_2O_5	nd	0.19	0.91	0.00
H_2O+	—	—	0.33	—
H_2O-	—	—	0.30	—
CO_2	—	nd	41.13	—
LOI	43.61	35.8	—	47.42
Other	—	—	0.19	0.00
Total	99.90	98.7	100.40	99.97

Sources:

1. Limestone, Solenhofen Formation (Jurasssic), Bavaria, Germany. Analyst: G. Steiger (F. W. Clarke, 1924).
2. Foraminiferal ooze with quartz, feldspar, and minor clay, sample FJ-1, (Recent?), Braemer Ridge, north of Fiji, Pacific Ocean (Knedler and Glasby, 1985).
3. Dolostone (Silurian), Joliet, Illinois. Analyst, D. F. Higgins (Fisher, 1925, in Pettijohn, 1975, p. 362).
4. Dolostone, Royer Dolomite (Cambrian), Oklahoma. Analyst: A. C. Shead, (Ham, 1949, Table V, sample 9294).

[a]Values in weight percent.

[b]Total iron as Fe_2O_3.

nd = not determined.

suggests that dolomitization does not produce radical changes in the shape of REE patterns but does lower the overall REE values. Thus, *if* particular REE patterns reflect provenance or environmental conditions, those patterns may be preserved during diagenesis.

Isotopic analyses of carbonate materials are more common than other trace element studies. The former are important because they can reveal the nature and relative amounts of the waters present during deposition or diagenesis (Land, 1980).[13] The stable isotopes of hydrogen, carbon, and oxygen are the isotopes typically used for such analyses.

TEXTURAL ELEMENTS AND TEXTURES OF CARBONATE ROCKS

Carbonate rocks show a wide variety of textures. Because most sedimentary carbonate rocks are composed of one or more of the three main carbonate minerals, the character of the rock is largely controlled by the nature and distribution of textural elements.[14]

Textural Elements of Carbonate Rocks

The main textural elements of carbonate rocks are micrite, sparite, allochems, and biolithic elements. *Micrite* is microcrystalline carbonate (diameter < 0.004mm). Here, micrite is commonly considered together with other components of *lime mud*, which encompasses all materials less than 0.06 mm in diameter. Lime mud consists largely of fragmented algal remains (Stockman et al., 1967).[15] *Sparite* is crystalline carbonate material of greater than 0.004 mm size, and includes macrosparite (> 0.06 mm) and microsparite (0.06–0.004 mm). Both micrite and sparite form by precipitation. Sparite also develops via recrystallization of finer-grained carbonate material and may be transported and redeposited as epiclastic grains.

The two remaining categories of textural element, allochems and biolithic elements, include a variety of materials. Allochems include any *transported* chemical or biochemical precipitate.[16] These include fragments of preexisting carbonate rock or mineral, chert, or other precipitate rock fragments (including intraclasts); oolites; pellets and pelloids; grapestones;[17] and fossils or fossil fragments (bioclasts) (table 21.2). The biolithic elements are materials that are formed by organisms *in situ* and are bound together by precipitated material or mud.[18] These include various types of fossils, such as bryozoans, corals, stromatoporoids, pelecypods, and algal mats and lumps (stromatolites and oncolites), but do not include any transported fossil fragments.

Textures and Classifications of Carbonate Rocks

Because carbonate materials occur both as precipitates and as clastic materials, it is obvious that the textures of carbonate rocks are both epiclastic and crystalline in nature. Epiclastic textures may be given the same size designations as those used for the textures of siliciclastic epiclastic rocks (i.e., the Wentworth size grades). Most commonly, however, grain size is not the major criterion used in the description of carbonate rock textures, or in their classification.[19]

The *abundance* (Dunham classification) or the *type* of grain (Folk classification) is now commonly used for subdividing textural and rock types. Recall that in Folk's

Table 21.2   Major Textural Elements of Carbonate Rocks and Sediments

Lime Mud-Material consisting of carbonate grains smaller than sand size.

  Micrite-Microcrystalline carbonate grains < 0.004 mm in diameter.

  Microsparite-Crystalline carbonate grains between 0.004 and 0.06 mm in diameter.

Sparite-Crystalline carbonate grains larger than 0.004 mm.

  Microsparite-Crystalline carbonate grains between 0.004 and 0.06 mm in diameter.

  Macrosparite-Crystalline carbonate grains larger than 0.06 mm in diameter.

Allochems-Transported fragments of precipitated materials.

  Intraclasts-Fragments of preexisting precipitate rocks.

  Ooids (oolith, oolites)-Spherical to oval, concentrically layered particles of carbonate material between 0.25 and 2.0 mm in diameter.

  Pellets-Small, rounded, structureless grains less than 0.25 mm diameter.

  Grapestones-Rounded clusters of carbonate grains cemented together.

  Skeletal fragments-Rounded to angular fragments of carbonate (or other) material originally precipitated as skeletal material for an organism (for specific types see the list under "Biolithic Elements").

  Oncolites-Small (< 10 cm), rounded, concentrically laminated carbonate bodies formed by algal precipitation.

Biolithic Elements-Carbonate materials formed in place by precipitation by organisms.

  Stromatolites-Laminated, micritic carbonate material in the form of layers, columns, or semispherical masses, formed by organic precipitation and trapping of mud.

  Oncolites-Small (< 10 cm), rounded, concentrically laminated carbonate bodies formed by algal precipitation.

  Tests-The precipitated shells of organisms, especially microorganisms (e.g., coccolithophores and foraminifera).

  Skeletons-The precipitated exoskeletons or endoskeletons of organisms.

  Typical organisms producing skeletons and providing skeletal fragments to the carbonate rocks include corals (coelenterates), stromatoporoids, sponges, bryozoans, archeocyathids, annelids, brachiopods, mollusks (including gastropods, pelecypods, and cephalopods), arthropods (including trilobites), and the echinoderms (especially crinoids, blastoids, and echinoids).

classification (1962) grains distinguished as micrite, sparite, intraclasts, oolites, pellets or pelloids, and fossils (bioclasts), provide the textural basis for the subdivision of rocks. In this classification, the textures are incorporated into the name (e.g., oomicrite). In contrast, the root names in the Dunham classification (1962) are based on the *percentages* of the two main size categories of material—mud and grains—rather than on the specific size or character of the grains. The grain types recognized by Folk, as well as other clast types, such as other allochems and exotic clasts (including lithic granules, pebbles, and cobbles), may serve as modifiers for the root names of Dunham, but are not incorporated into the textural subdivision or the classification. Thus, textures such as intraclastic textures, oolitic textures, pelletal textures, and bioclastic or fossiliferous textures (figure 21.1) are recognized, but do not provide a basis for classification (figure 15.6).

Recall that Embry and Klovan (1971) and Cuffey (1985) provide additions to the Dunham classification that essentially give it the same level of size discrimination afforded by the Folk classification. In this modification, the textural arrangements and types of bioclasts determine the name. These names are partly a function of age, because the specific types of bioclasts and biogenic elements have changed in abundance and character over time (figure 21.2) (M. Pitcher, 1964; Wilkinson, 1979). For example, reef formers were dominantly sponges and algae early in the Ordovician Period, but that niche is filled by corals in the Cenozoic Era (figure 21.2b).

Crystalline textures typically are simply given the designation *crystalline* and the name may be modified by the prefixes *fine-, medium-,* or *coarse-.* Most primary crystalline textures are fine-crystalline, but some medium- to coarse-crystalline textures may develop in continental carbonate rocks. Typically, these textures are equigranular-mosaic or equigranular-sutured textures (see figures 21.1 and 14.14). During diagenesis, crystalline textures usually become coarser.

The textures of the biolithites or boundstones (names based largely on interpretations of the textures and origins rather than on strictly observable criteria) may be clastic, crystalline, or both. The essential characteristic of such rocks

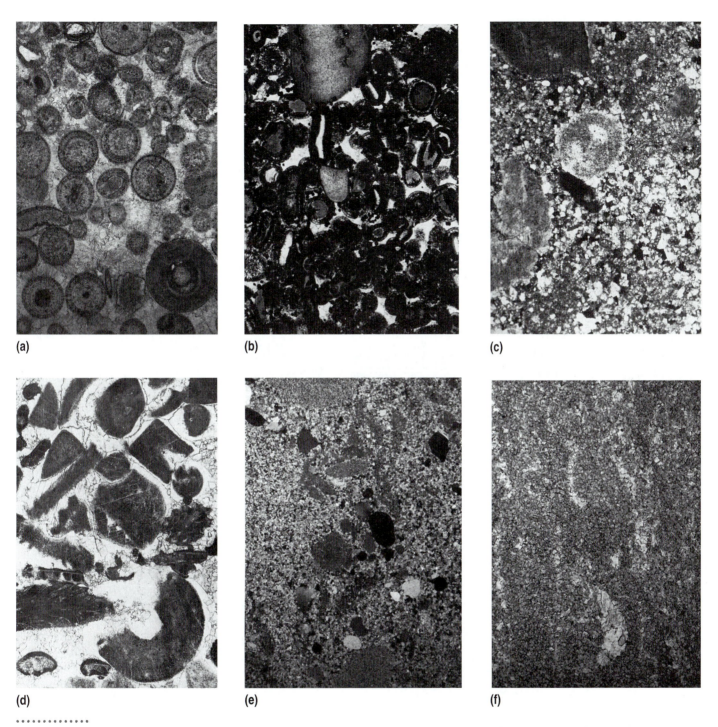

(a)

(b)

(c)

(d)

(e)

(f)

Figure 21.1   Photomicrographs of textures in carbonate rocks. (a) Oolitic texture, oolitic packstone, Ottosee Formation, Ordovician, Alcoa, Tennessee. (XN). (b) Pelletal texture in oolitic-pelletal packstone, Pence Springs Limestone, Mississippian, Pence Springs, West Virginia. (XN). (c) Skeletal grainstone, Kaibab Formation, Permian, Grand Canyon, Arizona. (XN). (d) Kerogen-stained, crystalline skeletal packstone, Pennington Formation, Mississippian, Woodson Bend, Kentucky, (PL). (e) Sandy intraclastic dolopackstone, Knox Group, Ordovician, Nebo Quadrangle, Virginia. (XN) (f) Crystalline dolowackestone, Ordovician Knox Group, Nebo Quadrangle, Virginia. (PL). Long dimension of photos in (a), (b), (e), and (f) is 6.5 mm. Long dimension of photos in (c) and (d) is 3.25 mm. Top is to the right in (f).

(d) Courtesy Dr. F. K. McKinney, Appalachian State University

(a)

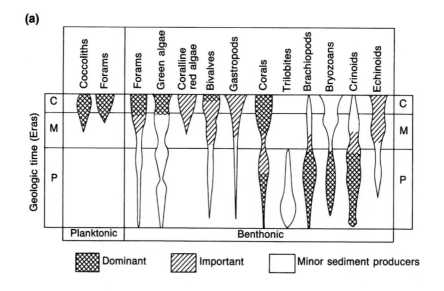

(b)

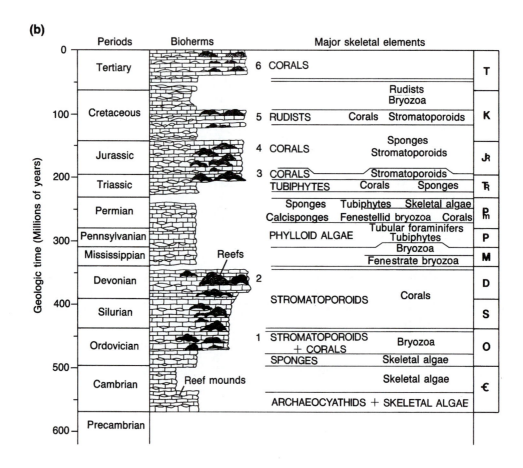

····················
Figure 21.2 Illustrations of changes in abundances and dominance of marine, calcareous skeletal-producing organisms over time. (a) Variations in abundance and dominance of various groups of organisms (after Wilkinson, 1979). (b) Dominant reef-forming organisms during various Phanerozoic periods.
(After James, N. P., 1983.)

is that the biogenic elements occur in orientations interpreted to be growth positions; that is, the orientations do not indicate that transportation of the biogenic element has occurred. Subdivision of such rocks may be made, as is the case for other carbonate rocks, on the basis of the type of fabric element that is present, yielding names such as bafflestone or lettucestone (Cuffey, 1985).

Recall that in the Dunham (1962) and Folk (1962) classifications, the modifiers *dolo-* and *dolomitic* are used to designate rocks of that composition. Calcitic and aragonitic rocks are designated by the various root names for limestones (e.g., pelsparite or grainstone) and no distinction is made between rocks composed of calcite and those composed of aragonite.

Structures in Carbonate Rocks

Various types of bedding and laminations are the dominant sedimentary structures of carbonate rocks. Bedding ranges from massive to very thinly bedded types and laminations are common (figure 21.3). The thickest bed types are relatively uncommon and generally result from (1) deposition in a buildup, (2) postdepositional bioturbation, or (3) amalgamation during diagenesis. The more common thin beds are typically separated by shale partings or interbeds. Bedding may be tabular, nodular, wavy, lenticular, crossed, convoluted, or graded.

Additional structures present in carbonate rocks include stromatolites, stromatactis, mounds, reefs, fenestrae, vugs, concretions, nodules, and tepees (see table 21.3). Several of these were described in chapter 14. **Vugs** are cavities lined with crystals. *Concretions* are irregular to round masses of more resistant rock formed as a result of cement (chalcedony, iron oxides) precipitating around a core material, often a fossil or grain of different composition. In diameter, concretions are typically less than 30 cm, but may occasionally exceed 2 m. **Nodules** are small, rounded to irregular concretions with a knobby surface. **Stromatactis** is a layered structure formed after deposition. It results from the development of lenticular, water-filled cavities in which layers of sparry calcite precipitate within micritic lime mudstone (Scoffin, 1987, p. 87). Replacement may also contribute to stromatactis formation (Walker and Ferrigno, 1973). **Tepees** are small (centimeter to meter scale) angular anticlines produced by compression of a layer towards a fracture in such a way as to produce two opposing, concave-up layers that come to a point like a Native American tepee (J. E. Adams and Frenzel, 1950).[20] The compression results from a lateral force produced by expansion that accompanies the crystallization of minerals within the layer. Tepees form in arid environments in evaporite-bearing sequences. **Hardgrounds** are hard, lithified layers of sediment a few centimeters thick that form on the seafloor during periods of nondeposition. They may be bored, burrowed, fossiliferous (with *in situ* communities), and

Table 21.3 Sedimentary Structures Found in Carbonate Rocks

Depositional Structures

Bedding	Laminations
Cross bedding	Cross laminations
Stromatolites	Oncolites
Pellets[1]	Grapestones[1]
Ripple marks	Graded bedding
Reefs	Mounds

Erosional Structures

Channels	Flute casts
Tool marks	Load casts
Slump or slide scars	Rip-ups
Burrows	Tracks and trails

Deformational Structures

Soft-sediment folds	Penecontemporaneous faults
Dessication cracks	Convolute laminations
Breccias	Flame structures

Diagenetic Structures

Concretions	Nodules
Veins	Vugs
Stylolites	Stromatactis
Tepees	Hardgrounds
Breccia	Fenestrae
Liesegang bands or rings	Pisolites

Sources: Conybeare and Crook (1968), Collinson and Thompson (1982), Reijers and Hsu (1986), Pettijohn, Potter, and Siever (1987), and observations by the author. For further description of these structures and discussions of their origins, see these sources as well as chapter 14.
[1]May be considered as either structural or textural elements.

of a different color than the underlying or enclosing rocks.[21] **Fenestrae** are holes ("birdseyes"), occurring in carbonate sediments, that may later be filled with precipitated calcite, as they usually are in ancient rocks (Tebbutt, Conley, and Boyd, 1965; Shinn, 1983). They tend to be elongate and they form in intertidal to subtidal environments by algal mat decay and shrinkage during drying of the sediment or trapping of air in loose sediments. Each of these structures reveals something of the history or depositional environment of the host carbonate rock or the subsequent diagenesis of that rock.

(a)

(b)

(c)

(d)

Figure 21.3 Photos of some structures in carbonate rocks. (a) Wavy, lenticular, irregular, and tabular beds in subtidal (shelf) skeletal grainstone, packstone, wackestone, and clay shale of the Dennis Formation, Kansas City Group, Pennsylvanian, I-470, Kansas City, Missouri. (b) Cross-bedding in grainstone, Monteagle Formation, Mississippian, Lookout Mountain, Georgia. (c) Chert nodules in dolostone, Ordovician Knox Group, Nebo Quadrangle, Virginia. (d) Stromatolitic laminations in stromatolitic doloboundstone, Helena Formation, Proterozoic, Glacier National Park, Montana.

OCCURRENCES AND ORIGINS OF CARBONATE ROCKS

Most carbonate rocks form from sediment deposited in shallow marine environments. Nevertheless, carbonate rocks also form in deep marine, transitional, and continental environments. In transitional environments, such as tidal flats, carbonate sedimentation is quite common locally. Continental carbonate sediments, in contrast, tend to be limited in abundance and distribution, but in rare circumstances (e.g., in the case of the Green River Formation of Wyoming, Colorado, and Utah), large lacustrine deposits of carbonate rock may develop. Continental carbonate sediments form as lacustrine, alluvial, dune, spring, hydrothermal vent, or soil deposits.

Most of the carbonate material that is deposited in the various environments is initially precipitated from solution by organisms. The shells of invertebrate animals contribute substantial carbonate material to shelf, shoal, ramp, reef, and mound facies, whereas algae are important contributors of lime mud and precipitate to a wide range of lithofacies, including sabkha, tidal marsh, platform, reef, and deep marine types. In beaches, dunes, and shoals, the organic remains are reworked by wind or waves. Ooids in shoals probably form primarily via inorganic precipitation on grain cores of organic or inorganic origin, but algal precipitation may be of secondary importance in the formation of some ooids (N. D. Newell, Purdy, and Imbric, 1960; Simone, 1980; B. Jones and Goodbody, 1984b). In the deep marine basins, pelagic microorganisms, notably coccolithophores and foraminifera (especially *Globogerina* in younger rocks), are the principal contributors of carbonate sediment (McIntyre and Be, 1967; Taft, 1967).[22] In continental environments, both algal and inorganic precipitates occur. The caliches and evaporites are largely inorganic, but algae do play a role in carbonate sedimentation in lakes and some springs (Dean and Fouch, 1983; Pentecost, 1990). These various carbonate materials and their characteristic textures and structures provide the basis for distinguishing carbonate rocks formed in different environments.

Marine Carbonate Rocks

Nine major depositional environments are represented by marine carbonate rocks in the geologic record (Bathurst, 1975; J. L. Wilson, 1975; J. F. Read, 1980a).[23] These include basins, slopes, ramps, shelf margins, foreslopes, reefs and other carbonate buildups, open shelves, shoals, and platform marine environments. Each is characterized by particular lithofacies, structures, and fossils (figure 21.4).

Basinal Rocks

Basinal carbonate rocks typically are thin bedded, laminated, dark-colored lime mudstones and wackestones (Enos, 1974; J. L. Wilson, 1974, 1975; Yurowicz, 1977). They are characterized by even laminations and mudrock interbeds. Locally, however, some beds range to massive thicknesses (Enos, 1974b). In other cases, laminated Facies C and D carbonate and noncarbonate turbidites with T_{abe} and T_{bde} sequences—as well as Facies F olistostromal mud- and grain-supported diamictites and breccias—may occur with the basinal carbonate rocks, especially near the basin margins (R. E. Garrison and Fisher, 1969; Bornhold and Pilkey, 1971).[24] Chert interbeds occur at sites far removed from sources of clastic sediment (J. F. Hubert, Suchecki, and Callahan, 1977). Fossils in basinal rocks are predominantly pelagic types, but some benthic types may be transported into the basin by turbidity currents.

Thin sections of light to dark, gray to dusky red basinal carbonate rocks also form on volcanic topographic highs, along spreading centers and on seamounts. Here, the carbonate sediments form infillings between basaltic pillows (as interpillow limestones) or thin-bedded layers overlying the mafic rocks (figure 21.5) (R. E. Garrison, 1972).

Slope Rocks

Slopes mark the deep-water transitions from basins to shelves, or platforms.[25] Carbonate slope sedimentation yields lime mudstone and wackestone, with local grainstones (McIlreath, 1977; Cook and Taylor, 1977).[26] These rocks are characterized by dark colors, thin beds, and laminations. Carbonate contourites consisting of thin-bedded, well-sorted grainstones and lime mudstones, locally capped by hardgrounds, also occur along the slopes (Mullins, 1983; Cook, 1983).

Contrasting rocks formed at the base of the slopes, where submarine fan or carbonate slope-apron sediment bodies develop, are characteristically carbonate breccias and the carbonate equivalents of the Mutti and Ricci-Lucchi (1972) fan Facies A, C, F, and G (J. E. Sanders and Friedman, 1967; Enos, 1974; Bennett and Pilkey, 1976).[27] The Facies A rocks are generally thick-bedded to massive, grain-supported conglomerates and grainstones. Grainstones in Bouma sequences form the thinner (medium- to thick-bedded) beds of Facies C, and these are interbedded with lime mudstones of Facies G. The Facies F rocks are breccias and diamictites. The clasts, blocks, or slabs in these rocks are cobble- to boulder-size pieces of reef or shelf rock that range up to several hundred meters in length (Davies, 1977; J. F. Hubert, Suchecki, and Callahan, 1977). The fragments are supported by lime mud (Mullins and Cook, 1986). Examples of slope rocks occur in the Devonian Ancient Wall Carbonate Complex of Alberta and the Silurian-Devonian Roberts Mountain Formation of Nevada (H. E. Cook et al., 1972; Cook, 1983).

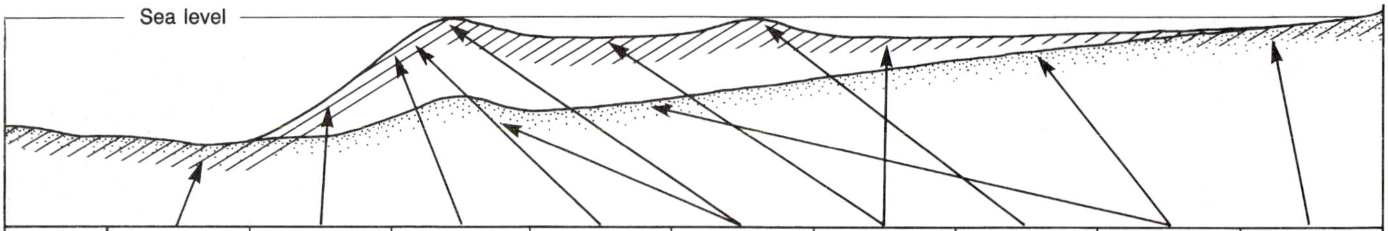

Environment	Deep basin	Slope and rise	Foreslope	Shelf margin	Build-up (Reefs)	Shelf	Shoal	Ramp	Platform (Epeiric seas)
Rock types	Lime Mudstone, Wackestone, ±Packstone	Lime Mudstone, Wackestone, ±local Grainstone, +base of Slope Breccia, Diamictite and Packstone	Skeletal Packstone, Skeletal Grainstone ±Wackestone Rudstone, Breccia and Diamictite	Skeletal Grainstone ±oolites and pellets, and local Packstone	Boundstone with Lime Mudstone, Wackestone, Packstone, and Grainstone containing skeletal clasts, oolites, and pellets Bafflestone, Floatstone	Wackestone, Packstone, ±Lime Mudstone and Grainstone with skeletal clasts and pellets	Grainstone composed of skeletal clasts and oolites ±pellets, local Packstone	Grainstone, Packstone, Wackestone, and Lime Mudstone with skeletal clasts and pellets	Lime Mudstone, Wackestone, ±Packstone with oolites, pellets, and skeletal clasts
Structures and associated facies	Laminated beds, Mudrock interbeds, laminated to thin bedded Chert, Facies C and D (fan facies) near margins	Thin to massive beds, hardgrounds, fan facies at base of slope, transported skeletal clasts, local grading, Mudrocks	Local cross bedding, local chaotic beds, thin to massive beds	Thin to thick bedded, cross-bedded units	Thin to massive beds, local stromatolites, whole fossils, local Sandstone and Mudrock	Laminated thin to medium beds, Mudrock interbeds, burrows, hummocky cross strata, whole fossils, local Sandstone and Mudrock	Thin to thick bedded, cross bedded units	Thin to thick beds, local grading, burrows, whole fossils, local interbedded Shale and Sandstone	Burrows, local stromatolites, Dolostone, Evaporites (landward), hardgrounds, restricted fauna with whole fossils, thin to thick beds, Quartz arenite

Figure 21.4 Diagram showing some characteristics of the nine major marine environments of carbonate rock formation. (Based on J. L. Wilson, 1975, and J. F. Read, 1980a)

Figure 21.5 Laminated, thin-bedded, pelagic, crystalline lime mudstone overlying mafic igneous rocks (oceanic crust?), Franciscan Complex, Jurassic-Cretaceous, northeastern Diablo Range, California.

Platform, Shoal, Shelf, and Ramp Rocks

Carbonate ramps are very gently sloping surfaces that extend from shore to basin without a pronounced change in slope (Ahr, 1973; J. F. Read, 1980a).[28] Ramp slopes are typically less than 1° (J. F. Read, 1980a, 1982a). The rocks formed on ramps are light to dark, thin- to thick-bedded, skeletal grainstones, packstones, wackestones, and lime mudstones (J. F. Read, 1980a). Coarser, thick-bedded, less muddy rocks (grainstones) develop in shallower parts of the ramp, nearer shore and around carbonate buildups, where wave action creates high-energy environments. Burrowing is extensive in ramp lithofacies. Thin to nodular beds with shale interbeds form on deeper parts of the carbonate ramp. Some of the latter may be burrowed and others are graded or laminated (J. F. Read, 1980a).

The important petrologic distinction between ramp and shelf rocks is that, in ramp environments, grainstone and packstone facies occur shoreward of muddy lithofacies (Ahr, 1973). On shelves and platforms, the reverse is generally true; that is, coarse-grained rocks occur seaward of fine-grained rocks (Enos and Perkins, 1977), especially where reefs occur at the shelf break.

Shelves are distinguished from ramps in having a seaward shelf margin characterized by an abrupt steepening of slope. On the shelf, shoals and buildups yield coarse sediment. The shelf environments differ from platform environments in being open-marine, neritic, and subtidal (J. L. Wilson, 1975). Shelf rocks are characteristically thin- to medium-bedded, skeletal wackestone and packstone, plus pelloidal packstone and wackestone, but lime mudstones and conglomeratic sediments (rudstones) form in some areas (LaPorte, 1969; Doyle and Sparks, 1980; P. M. Harris, 1985; Dix, 1989).[29] Interbedded clastic rocks, particularly mudrocks, occur between the carbonate layers. The beds of carbonate rock are wavy to irregular (figure 21.3a). Burrowing is common to abundant, as are whole and fragmented fossils of typical organisms (J. L. Wilson, 1975; Enos and Perkins, 1977; Driese and Dott, 1984). The rocks tend to be dark and commonly occur in shades of gray, green, or brown. Much of the deposition on the shelves occurs below wave base, where currents transport nutrients to marine organisms and also transport sediment that is trapped between or deposited with these organisms. Occasional storms do affect the bottom in the shallower shelf areas, and hummocky cross-stratified, fossiliferous packstone wave deposits result from the storm activity (Harms et al., 1975; Kreisa, 1981). The contemporary Mississippi-Alabama-Florida (MAFLA) shelf sediments (Doyle and Sparks, 1979), those of the south Florida shelf (Enos and Perkins, 1977), and those on the Belize shelf (Purdy, Pusey, and Wantland, 1975; Pusey, 1975) are good examples of shelf sediments.[30] Upper Ordovician limestones, interbedded with mudrocks of the Cincinnati Arch region (described in chapter 18), are an example of shelf

limestones. Storm deposits make up common shelf rocks that comprise certain calcareous parts of the Martinsburg Formation of the southern Appalachian Basin (Kreisa, 1981; Kreisa and Springer, 1987).

Platforms are essentially shelf or epicontinental areas, inundated by the sea, that serve as environments of deposition for blanket sandstones and extensive sheets of carbonate rocks (M. L. Irwin, 1965).[31] Platform waters are essentially marine, but because of restricted nutrients and/or restricted circulation, as well as variations in salinity, the fauna and flora commonly are limited in variety. Burrows (trace fossils) are common, however, and algae are abundant (J. L. Wilson, 1975). Platform subenvironments include sand flats, shoals, and carbonate buildups. Dark- to light- shaded, locally skeletal, grainstones, packstones, wackestones, and mudstones all are formed on the platforms (Krebs and Mountjoy, 1972; Steiger and Cousin, 1984; Handford, 1988).[32] The grainstones and packstones form in shoal and intershoal (sand flat) areas through wave reworking of previously formed carbonate materials. Wackestone-mudstone lithofacies form in various carbonate buildups and in intershoal areas. The rocks may be pelletal, oolitic, or skeletal. In the carbonate buildups, the skeletons of *in situ* marine organisms form a framework that traps carbonate sand and mud derived from the breakdown of algae and other marine organisms. As a result, a variety of coarse carbonate rocks characterize the buildups, including boundstones—such as floatstones, bafflestones, and cruststones—and rudstones. Some modern environments on the Great Bahamas Bank (described below) are like those of the ancient platforms.

Shoals occur on platforms, shelves, and ramps.[33] In the shoals, indigenous faunas may be even more restricted than is the case on the platform sand flats. Where fossils occur, for example in fossiliferous grainstones (coquinas), they are primarily transported and reworked invertebrate remains derived from elsewhere in the marine environment. By definition, the shoals are shallow places, and as a result, wave action is important in working the sediment. The resulting rocks are light-colored, fossiliferous (skeletal) grainstones and oolitic grainstones, plus local packstones (Newell, Purdic, and Imbric, 1960; P. M. Harris, 1985).[34] Cross-bedding is a common feature in these wave-worked sediments (J. L. Wilson, 1975; Driese and Dott, 1984); thus, the lithofacies of the shoal is a sequence of cross-bedded grainstones. Typical contemporary carbonate shoal deposits are found on the Great Bahamas Bank.[35] Ancient examples occur in the Cambrian Hoyt Formation of New York, the Mississippian Bangor and Monteagle Limestones of Alabama and Georgia, and the Pennsylvanian Morgan Formation of Colorado and Utah.[36]

In shallow parts of shelves and ramps, tidal channels may cut across other environments. In these channels, intraclastic packstones form (P. M. Harris, 1985).

Rocks, Reefs, and Buildups on Bank, Shelf, and Platform Margins

The larger shallow marine environments—the platforms, shelves, and ramps—are commonly bounded by or include zones of sediment accumulation that form a distinct buildup or marginal environment, or both. These marginal and carbonate buildup environments include bank margin and ramp-bar belts and shoals, reefs, and mounds.[37] The carbonate buildups typically form in association with other carbonate sediments, but they also occur with siliciclastic sediments on deltas, fan deltas, beaches, and tidal flats (J. L. Wilson, 1975; Multer, 1977; Santisteban and Taberner, 1988).[38]

Marine sand accumulations, variously referred to as bank margin sands, platform edge sands, tidal bar sands, and marine sand belts, commonly rim banks, shelves, and platforms and form within ramps (Ball, 1967; J. L. Wilson, 1974, 1975).[39] The sands, worked by storm and tide-generated currents, locally project above sea level to form islands on which wind-formed dunes develop (Ebanks, 1975). The marine carbonate (and siliciclastic) sands form thin to thick cross-bedded units. Carbonate rocks representative of such accumulations include oolitic, skeletal, and peloidal grainstones and packstones. Indigenous organisms are only locally abundant in these marine sands, but rounded or abraided skeletal fragments and some algal precipitates are common.[40] The shoal sands of the Conasauga Formation of Alabama represent platform edge sands (Sternbach and Friedman, 1984).

Recall that reefs are a type of carbonate buildup characterized by (1) lithologies different from those of the surrounding area, (2) thicker sections of rock than those formed in the contemporaneous surroundings, (3) a significant organic component, and (4) evidence of wave activity (Heckel, 1974). Mounds are similar carbonate buildups formed in quiet water, but they are characterized by having no significant large skeletons (Lees, 1982; N. P. James and Macintyre, 1985, p. 28). Both structures represent localized accumulations of carbonate sediment, but reefs are dominated by skeletons, whereas mounds are dominated by mud.

Reef environments are complex. They contain both high-energy and low-energy areas, and, as a result, contain a diversity of sediments and rock types (Krebs and Mountjoy, 1972; N. P. James, 1983, 1984b; Adjas, Masse, and Montaggioni, 1990).[41] Hence, the various reef environments, including the reef core, lagoonal areas, and reef flanks (foreslopes), yield lithofacies consisting of a variety of rock types, bed types, and bed thicknesses. The core of the reef consists of light-colored boundstones of one or more types (e.g., shellstone, bafflestone, bindstone), depending on the organisms responsible for reef construction and the location of the reef (beach, platform, ramp). The beds are thick to massive. Lime mudstones, packstones, and grainstones, including peloidal and oolitic types, occur in layers and lenses between boundstone masses. In the lagoon or reef flat, lithofacies and rock types are similar to those of the shoals. Local stromatolitic layers, however, may form interbeds in the lagoonal lithofacies. Floatstones and rudstones with skeletal grainstone matrices occur where reef flat sediments are worked by currents or waves (Viau, 1983). On the reef flank (foreslope), the rocks are light- to dark-shaded and of varied types, including lime mudstones, wackestones, skeletal grainstones, skeletal packstones, rudstones, boundstones, breccias, and carbonate olistostromal diamictites. Beds are thin to thick. Some of the grainstones and packstones exhibit foreset bedding (J. L. Wilson, 1975; N. P. James and Macintyre, 1985). Examples of reef rocks include those of the famous Permian Reef Complex of the Guadalupe Mountains of New Mexico and west Texas, the Devonian Swan Hills and Golden Spike Reefs of Alberta, the Silurian Pinnacle Reefs of the Michigan Basin, and the Jurassic Smackover Reefs of Arkansas (King, 1948; Huh, Briggs, and Gill, 1977; Baria et al., 1982; Harris and Crevello, 1983; Toomey and Babcock, 1983; Viau, 1983; Walls, 1983).

Inasmuch as mounds contain more mud than the reefs, they are characterized more by mudstones, wackestones, and bafflestones than by more abundantly fossiliferous rock types (D. T. King, 1986; Ausich and Meyer, 1990). Floatstones are present locally, as are packstones (e.g. Toomey and Babcock, 1983, p. 21). Bedding is thin to massive. On the flanks of the mounds, working of the sediments yields packstones and grainstones (Toomey and Babcock, 1983, p. 74; D. T. King, 1986). The rocks range from dark to light shades and are commonly grey or brown in color.

Carbonate Rocks Formed in Transitional Environments

Carbonate rocks formed in transitional environments include sedimentary rocks formed in intertidal to subtidal environments on beaches, in bays and lagoons, and on sabkhas. The rocks include beachrock, various tidal flat lithologies, mudstones to grainstones, and evaporites. In several of these environments, significant evaporation is an important environmental condition that affects the nature of the rocks. Specifically, inorganic carbonate precipitation becomes possible in areas of high evaporation rate. The commonly high and variable salinities of transitional environments restrict the amount of direct organic contribution to the sediment, except for contributions from algae, which thrive there. Locally, mollusks contribute significant amounts of carbonate material, and this, plus bioclasts derived from erosion in adjoining environments, provides the bulk of the coarse clastic component of the carbonate sediment (High, 1975; Wigley, 1977). Examples of rock sequences formed in transitional environments include some sections of units such as the Middle Ordovician Black River Group of New York,[42] similarly aged rocks in the southern Appalachian Valley and Ridge Province,[43] and the Devonian Manilaus Formation of the Helderberg Group of New York.[44]

Tidal Flat, Bay, and Lagoonal Rocks

The rocks formed in tidal flats, bays, and lagoons vary in character as a function of the energy of the specific environment.[45] Tidal flats are typically low- to moderate-energy environments that are protected from the open ocean and, therefore, from incoming waves, except during periods of storm activity. Because tidal flats contain supratidal (above mean high tide), intertidal, and subtidal (below mean low tide) zones, the rocks that form in them display a wide variety of colors, sedimentary structures, and rock types (LaPorte,1967; Ebanks, 1975; Shinn, 1986).[46] The structures include thin, alternating beds of carbonate rock and shale ("ribbon rock"); thin, irregular, and cross-laminae; flat and round pebble conglomeratic beds; mudcracks; burrows and irregular burrowed beds; evaporite mineral casts and molds; stromatolitic laminae; and tepee structures (Hardie, 1977). The supratidal rocks are distinctive, being characterized by mudcracks, stromatolitic layering,[47] and fenestrae (which in ancient rocks are typically filled with calcite; see Textoris, 1968). The rocks range from cryptalgal boundstone to flat pebble conglomerates and breccias—the latter being formed where laminated muds are ripped up and redeposited by storms—and from skeletal lime mudstones to skeletal packstones. Many of these rocks are dolomitized and contain chert (LaPorte, 1967, 1971; Cluff, 1984). The dolostone-chert association results from diagenetic alteration of sediments in surface to subsurface zones of the tidal flat region. A variety of colors including red, orange, yellow, brown, gray, and white distinguish the rocks of the supratidal zone from less colorful lithofacies formed in adjacent environments.

The intertidal and subtidal rocks contrast in shade and structure. The intertidal rocks are typically medium- to thick-bedded wackestones and grainstones of light shade that lack well-developed laminations. The laminations are destroyed by extensive burrowing activity by organisms (LaPorte, 1971; N. P. James, 1984c; Shinn, 1986). Locally, stromatolitic laminations are present, and hardgrounds and fenestrae are common. Some intraclastic conglomerates and breccias also form in the intertidal zone (N. P. James, 1984c).

Subtidal rocks tend to be darker gray shades, perhaps due to a higher organic carbon content, and they are almost devoid of sedimentary structures. Skeletal lime mudstones, packstones, and grainstones occur, but whole fossils are not characteristic. In subtidal-tidal channels, energetic currents produce lag deposits that give rise to fossiliferous (skeletal) floatstones, wackestones, packstones, and grainstones (Shinn, 1983; Cloyd, Demmico, and Spencer, 1990). These rocks are of light shade but contain local patches of mixed dark and light grains ("salt-and-pepper sands"). Cross-bedding is common and subtidal point-bar lithofacies are present in some tidal channel sections (J. L. Wilson, 1975; Shinn, 1986).

In bays and lagoons, water is generally quiet and consequently the sediments are fine-grained. Skeletal and pelletal lime mudstones, wackestones, packstones, and grapestones typify these environments (Gebelein, 1973; D'Aluisio-Guerrieri and Davis, 1988; Colby and Boardman, 1989). Where currents and wave activity are more common, however, skeletal and oolitic grainstones, skeletal packstones, and skeletal wackestones also form from skeletal fragments, which are washed in from adjacent environments or are derived from nearby faunas (Howard, Kissling, and Lineback, 1970).[48] The rocks may be well laminated or nonlaminated and may be bioturbated, but the fauna is typically limited in diversity due to variations in salinity or the high level of salinity of the water (Ginsburg, 1956; Ebanks, 1975). Dolostones and stromatolites develop locally (P. M. Harris, 1985).

Sabkha Rocks

Sabkhas are simply special types of supratidal zones that form in arid to semiarid regions. They are characterized by the same types of rock found in supratidal mudflats, including stromatolitic cryptalgal boundstones and flat pebble breccias and conglomerates, but they also contain evaporites (gypsum evaporite, anhydrite evaporite, dolostone). Dolostones are common and gypsum nodules occur locally within the dolomitic rocks (Hardie, 1986a). Red beds also occur here, just as they do in tidal flats. Structurally, the beds are laminated, irregular, or nodular.

Beach Rocks and Beachrock

Beaches range in style from gentle strandline types to rocky cliff-backed shores.[49] Beach environments may be divided into two parts, the foreshore and the backshore. The *foreshore* includes the zone between high and low tide. The *backshore*, a supratidal zone, is above the normal high tide level. On rocky, cliff-backed shores, the backshore area may be entirely missing. Seaward of the foreshore, in the zone between low tide level and wave base, is the *shoreface*, a marine zone equivalent to the landward edge of the shelf.

Rocks of the foreshore are characteristically grainstones. These exhibit parallel lamination, low- to high-angle cross-beds, and vertical burrows (Multer, 1977, ch. 2; Inden and Moore, 1983; Strasser and Davaud, 1986). Whole shell fossils may occur scattered throughout the grainstone, but the grains themselves typically consist of oolites and well-rounded intraclasts and fossil fragments, including algal grains and bioclasts such as foraminifera, mollusk, coral, bryozoan, and echinoderm fragments. On rocky, cliff-backed shorelines, the foreshore sediments may include, in addition to grainstones, reefal boundstones, bioclastic or skeletal conglomerates, or rudstones, (Multer, 1977, ch. 2; Harland and Pickerill, 1984). Large intraclasts of beachrock and "keystone vugs" (holes with a protective shield of sand grains that are wedged together) are good indicators of a foreshore petrogenesis for a grainstone (Inden and Moore, 1983; Strasser and Davaud, 1986). Together with these particular structures, the facies association—cross-bedded grainstone flanked by shoreface or supratidal facies—indicates a foreshore unit.

425

Beachrock is a foreshore grainstone or conglomerate cemented by calcite *on the beach,* soon after deposition (Ginsburg, 1953).[50] The rock commonly contains typical features of foreshore grainstones, notably cross-bedding. Although fragments and slabs of beachrock are present in some foreshore formations,[51] they are relatively scarce in the geologic record, perhaps because of wave erosion and biological destruction of the distinctive character of the rock (B. Jones and Goodbody, 1984a).

The supratidal part of the beach consists of berm and dune grainstones. The former contain laminations and local (wind) ripple cross-beds. Where these rocks are cemented, they may contain cubic (halite) molds in the cement (Strasser and Davaud, 1986). Local shell layers, representing storm-deposited shell lags on the berm, occur in the berm grainstones. The dune grainstones show typical, large-scale aeolian cross-beds, are commonly bioturbated, and may contain large burrows, roots, or root molds.[52] Local, planar-bedded, overwash grainstones occur as interbeds in the dune rocks (Strasser and Davaud, 1986).

Continental Carbonate Rocks

Carbonate rocks deposited in continental environments include carbonate grainstones formed in wind-generated dunes, alluvial carbonate rocks, lacustrine carbonate rocks, spring deposits, and caliche.[53] None of the continentally derived carbonate rocks are particularly common, but lacustrine carbonate rocks locally form large deposits.

Carbonate rocks formed in lakes include inorganically precipitated carbonate, biogenic carbonate generated in the lake by resident organisms, and detrital carbonate transported into the lake by incoming streams.[54] Lakes in each climatic regime yield unique lithofacies. In many cases, the carbonate rocks are "marls" (argillaceous carbonate rocks) formed in the central, deeper parts of lakes. Varved or laminated rocks are common. Stromatolitic and other algal boundstones develop in shallow, temperate to tropical lakes, especially along lake margins (Eardley, 1938; Cohen and Thouin, 1987). Dolostones form in playas and in tropical freshwater lakes (R. L. Hay et al., 1986; Demicco, Bridge, and Cloyd, 1987). Shore facies include lime mudstones, grainstones, and packstones, which are locally oolitic or bioclastic. In spite of the apparently distinct lithofacies formed in each environment, discrimination of the paleoenvironment of deposition of ancient lake deposits may be difficult. The lacustrine dolostones and limestones ("marls") of the Green River Formation, described in chapters 18, are good examples of lacustrine carbonate rocks, yet the origin of these rocks is debated.

Alluvial carbonate rocks are carbonate-rich equivalents of the shales, sandstones, and conglomerates formed by typical fluvial and alluvial processes. Such rocks form under arid conditions, where both extrabasinal carbonate clasts, transported by wind and water, and biochemical and inorganic carbonate precipitates are deposited in alluvial fans and associated lakes (E. Nickel, 1985).

Carbonate dune sands, which yield cross-bedded, laminated, and locally ripple-marked grainstones, occur primarily in coastal areas (Sayles, 1929; McKee and Ward, 1983). The dune fields may represent either expanded backshore areas or coastal desert sand fields. Because of the similarity between aeolian grainstones and other grainstones (e.g., those of the foreshore), however, recognition of ancient aeolian grainstones is difficult and may depend on recognition of the grainstone as part of a continental facies association. Among the known examples of carbonate dune rocks are Pleistocene rocks present near Cancun, along the coast of the Yucatan Peninsula of Mexico (Ward, 1975).

Travertine, a layered to concretionary, variously colored rock, and *tufa* (vuggy travertine) are rocks deposited by springs, lakes, and rivers (figure 21.6) (J. E. Sanders and Friedman, 1967; Julia, 1983; Goff and Shevenell, 1987).[55] Where groundwater or thermal waters emerge at the surface, the temperature of the water may change, the water may mix with surface water, or both, resulting in saturation and inorganic precipitation of the carbonate minerals calcite and aragonite. The role of organic precipitation from similar waters is moot.[56] The commonly vuggy, layered character of the travertines and the cellular nature of tufa are distinctive features indicative of a spring origin. In addition, high initial dips on the layers and the laterally restricted distribution of carbonate rock are useful in recognizing such deposits (Steinen, Gray, and Mooney, 1987). Travertines of Mammoth Hot Springs in Yellowstone National Park, Wyoming, and those of Saratoga Spa State Park, New York (Friedman, 1972), are good examples of contemporary travertines. Ancient examples are rare, but have been described e.g., in the Mesozoic rocks of the Hartford Basin of Connecticut (Steinen, Gray, and Mooney, 1987).

Caliche is chalky, light-colored, micritic to microsparitic rock formed by evaporation and precipitation of calcite in the soil, in sediments, or in preexisting rock (Bretz and Horberg, 1949; Multer and Hoffmeister, 1968; Esteban and Klappa, 1983).[57] It is characterized by its general light color, micritic character, and association with present or former surface areas (Esteban and Klappa, 1983). Both low- and high-Mg calcites are known to occur in caliche, but low-Mg calcite is considered to be typical (Coniglio and Harrison, 1983; Esteban and Klappa, 1983). Caliche commonly contains sparry veins and may contain pisolites, rhizoliths, ooliths, pelloids, or nodules.

EXAMPLES

Examples of modern carbonate sedimentation in well-studied areas like the Bahamas provide the foundation for understanding ancient carbonate rocks. Towards that end, both modern examples of carbonate sediments and examples of ancient limestones are provided here.

(a)

(b)

Figure 21.6 Travertine. (a) Travertine forming from hot springs, Mammoth Hot Springs, Yellowstone National Park, Wyoming. (b) Layered travertine, tufa, and siliceous sinter, Yellowstone National Park, Wyoming.

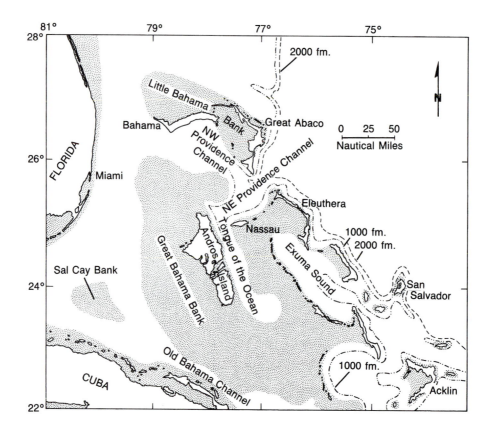

Figure 21.7 Geographic map of the Bahamas region, southeast of Florida in the Atlantic Ocean. *(Modified from Newell, 1955)*

Shallow Marine Carbonate Sediments of the Andros Platform and Adjoining Areas, Great Bahama Bank

The Great Bahama Bank is an elevated, shallow-marine platform-shelf area east and southeast of the south coast of Florida (figure 21.7).[58] It is bounded on the seaward margins to the east and southeast by escarpments and steep slopes, has several deep embayments and basins that project into the margin, and is separated from the Florida lowland by the Florida Straight. As a consequence of these factors, the Great Bahama Bank has long been isolated from any significant sources of siliciclastic sediment (Newell, 1955). The central and eastern part of the bank are marked by several islands (the Bahamas). From the islands, the bank slopes gently westward from depths of less than 2 m to depths of about 5 m, before dropping off abruptly along the western slope.

The Great Bahama Bank, also referred to as the Andros Platform, is located in the west central part of the Bahama Bank. The Platform and Andros Island have been the particular focus of many studies. Six major facies are recognized on the Andros Platform, including coral reef facies, coralgal (carbonate sand) facies, oolitic (sand) facies, grapestone facies, pelletal (lime) mud facies, and lime mud facies (Imbrie and Purdy, 1962; Purdy, 1963b). The distribution of these facies is shown in figure 21.8. Coral reefs flank Andros

Island on the east. Note that the coralgal sand facies is a shelf margin sand that extends around the east, north, and northwestern margins of the bank. Grapestone and oolitic sand facies locally rim the shelf and also constitute sand-flat blanket deposits in areas northwest and southwest of Andros Island. In the lee of the islands, protected from the effects of prevailing easterly winds, lime muds were deposited. Additional facies (mentioned only briefly here) are present in the transitional areas of Andros Island and along the reef foreslopes, bank slopes, and basins of the Bahamas region.[59] On the island, for example, beach ridges and levees composed of carbonate sand grade laterally into marshes in which dark fenestral and bioturbated lime muds are being deposited (Shinn, Lloyd, and Ginsburg, 1969; Multer, 1977). Locally the subtidal zones are replaced shoreward by tidal flats in which dolomitic sediments are forming (Gebelein, 1974). Foreslopes, slopes, and basins contain typical bedded carbonate sediments representative of those environments.

In the area around Andros Island, the six major facies have developed in response to various environmental conditions, such as water temperature, water depth, and prevailing wind and wave directions. Bahamian reefs are relatively small and thin, occurring as patch reefs and fringing reefs developed over various bedrock lithologies generally east (on the windward side) of Andros Island. Stony corals and algae are the dominant reef formers in the Bahamian reefs, as is typical

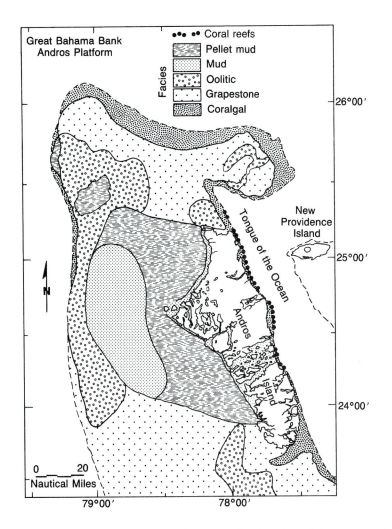

Figure 21.8 Facies map of the Andros Platform, Great Bahama Bank. The edge of the patterned area marks the 100-fathom depth contour.
(Modified from Purdy, 1963b)

of Cenozoic reefs elsewhere, but in some patch reefs, bryozoans are a major reef-forming organism (Newell and Rigby, 1957; Cuffey and Gebelein, 1975). The principal reef rocks are various boundstones, such as framestones and bafflestones. Wave action converts much of the reefal material into coralgal sands. Burrowing by organisms and infilling of burrows by lime mud and sand alter the nature of the reef rocks that become a part of the geologic record.[60]

The coralgal facies is skeletal sand consisting of fragments of coral, algae, mollusks, and other organisms, plus oolites, grapestones, pellets, mud fragments, and other allochems (table 21.4, column 6) (Purdy, 1963b). Locally, a significant fraction of the sediment is mud. Upon lithification, such sediments would become skeletal grainstones or packstones.

The oolitic (sand) facies occurs in shoals formed as marine sand belts and tidal bar belts, in sand flats, in channels, and in lobate sheets (P. A. Harris, 1979). It contrasts with the coralgal facies in being dominated by oolites (figure 21.9 and

Figure 21.9 Photo of oolitic sand from the Andros Platform.

Table 21.4 Modes of Selected Carbonate Rocks and Sediments

	1	2	3	4	5	6	7	8
Mud	100[a]	42.8	4.5	4.5	5.0	10.8	10.5[b]	0.0
Spar	—	0.0	0.0		0.0	0.0	—	—
calcite	—	—	—	20.0	—	—		0.0
dolomite	—	—	—	—	—	—	—	99.9
Cement	—	—	—	—	—	—	18.8[c]	—
Intraclasts	—	1.9[d]	26.9[d]	1.0	7.5	22.8	<1	0.0
Ooids	—	6.3	14.9	34.5	66.6	6.4	—	0.0
Pellets	—	32.6	4.9	—	7.2	4.7	tr	0.0
Grapestones	—	0.3	32.0	—	4.5	5.4	—	0.0
Skeletal grains				40.5				0.0
Corals	—	0.0	0.1	—	0.1	5.9	nr	—
Mollusks	—	2.7	4.1	—	1.4	7.1	<1	—
Brachiopods	—	—	—	—	—	—	<1	—
Echinoderms	—	—	—	—	—	—	33.8	—
Algae	—	1.3	2.8	—	1.9	11.6	4.5	—
Forams	—	3.9	2.6	—	1.0	7.1	nr	—
Bryozoans	—	—	—	—	—	—	17.2	—
Trilobites	—	—	—	—	—	—	2.2	—
Ostracods	tr	—	—	—	—	—	<1	—
Other	—	2.9	3.2	—	1.9	9.5	3.8	—
Other	—	5.3	4.0	—	3.0	8.8	0.0	tr[e]
Total	100	100	100	100	100.1	100.1	100	100
Points Counted	nr	11500[f]	32000[f]	200	35500[f]	16500[f]	nr	200

Sources:

1. Lime mudstone (basinal), average of 9 samples, Liberty Hall and Rich Valley formations (Middle Ordovician), western Virginia (J. F. Read, 1980a).
2. Pelletal mud facies, average of 23 samples, Great Bahama Bank (Recent) (Purdy, 1963b).
3. Grapestone facies, average of 64 samples, Great Bahama Bank (Recent) (Purdy, 1963b).
4. Oolitic fossiliferous grainstone, Monteagle Formation (Mississippian), near Huntsville, Alabama (unpublished data, H. Gault).
5. Oolitic facies, average of 71 samples, Great Bahama Bank (Recent) (Purdy, 1963b).
6. Coralgal facies, average of 33 samples, Great Bahama Bank (Recent) (Purdy, 1963b).
7. Skeletal grainstone, average of 25 samples, Effna, Murat, Rockdell, and Ward Cove limestones (Ordovician), western Virginia (J. F. Read, 1980a; see note b below).
8. Fine-grained crystalline dolostone, upper Knox Group (Ordovician), Nebo Quadrangle, Virginia (unpublished data, L. A. Raymond).

[a]Values in volume percent, based on point counts as indicated.

[b]Values are converted from those reported, assuming clasts + cement = total.

[c]Reported as blocky and fibrous cement.

[d]In these sediments, grains reported as crytocrystalline grains are included as intraclasts.

[e]Consists of a few scattered quartz grains and cryptocrystalline iron oxides.

[f]Number of samples × 500 pts./sample.

nr = not reported.

tr = trace.

table 21.4, column 5) (Purdy, 1963b). In some samples, the ooids constitute 98% of the sediment. Additional constituents, though varied, are not particularly abundant in many sediments of this facies, but the most abundant of those that are present are grapestones, pellets, skeletal fragments, and micritic (?) allochems. Locally, mud constitutes up to 33% of the sediment (P. A. Harris, 1979; F. K. McKinney, pers. commun., 1989). Sedimentary structures characteristic of this facies include tabular cross-strata, ripple marks, and tabular beds (Imbrie and Buchanan, 1965; P. A. Harris, 1979). Lithification of most of these sediments would yield light-colored, cross-bedded, medium-grained oolitic grainstones. Where mud is more abundant, the resultant rock type would be an oolitic packstone. Cores from the subsurface beneath the shoal contain both of these rock types (P. A. Harris, 1979).

The grapestone facies is also a sand facies, but in this case, the sands are dominated by grapestones and cryptocrystalline (micritic?) allochems (table 21.4, column 3) (Purdy, 1963b). Other major constituents of the sand include pellets, oolites, and skeletal fragments, especially mollusks. Locally, mud is present in significant amounts. The principal sedimentary structure of the grapestone facies is cross-bedding, but in most areas bioturbation has destroyed the bedding (Imbrie and Buchanan, 1965; Multer, 1977, p. 130). Inasmuch as grapestone-rich limestones are not generally recognized in the geologic record, upon lithification, such sediments would probably become rocks that would be identified as lime mudstones, wackestones, or cross-bedded, intraclastic to pelletal grainstones or packstones.

The pelletal mud facies is characterized by fecal pellets and sediment finer than 1/8 mm in size (table 21.4, column 2) (Purdy, 1963b). Pellets may constitute up to 50% of the sediment. Other allochems that occur in abundances greater than 5% include skeletal fragments, oolites, and mud clasts. Bioturbation, common in the muddy facies, destroys laminations and causes amalgamation of bedding. Upon lithification, the pelletal muds would yield pelletal packstones and wackestones.

The mud facies of the Andros Platform is richer in mud and poorer in pellets than the pelletal mud facies (Purdy, 1963b). Allochems locally constitute as much as 70% of the sediment, but typically they comprise less than 50%. Among the most abundant of the allochems are skeletal fragments and pellets. The mud itself consists mostly of small rods of aragonite (Bathurst, 1975, p. 137). Bioturbation features are the dominant structures (Imbrie and Buchanan, 1965; Multer, 1977). Lithification of these sediments would yield mottled or poorly stratified, skeletal or pelletal lime mudstones, wackestones, or packstones, like those discovered in the subsurface by core drilling (Beach and Ginsburg, 1980).

In general, the various carbonate sediment facies of the Andros Platform grade from one into another and interfinger in ways typical of lithofacies observed in the geologic record. Thus, understanding their distribution and origins provides insight into carbonate stratigraphic sequences of that record.

The Ordovician Ramp-to-Basin Facies of Southwestern Virginia

Middle Ordovician rocks of the Appalachian Valley and Ridge Province in southwestern Virginia constitute an excellent example of a carbonate ramp to basin sequence in the geologic record (J. F. Read, 1980a).[61] Shallower parts of the ramp contain lithofacies similar to some of the sediments of the Great Bahama Bank. During the Ordovician period, an ensialic, foreland basin developed from a Cambrian-Ordovician, passive-margin shelf that had marked the eastern margin of the continent, while uplift produced a "tectonic highland" to the southeast. Siliciclastic and, locally, calcareous turbidites of the Knobs Formation, shed from the highlands to the southeast, filled the southeastern part of the basin (F. B. Keller, 1977; Raymond et al., 1979).[62] On the northwestern edge of the basin, a carbonate ramp and adjoining supratidal mudflat served as a depositional site for carbonate sediments (figure 21.10) (J. F. Read, 1980a).

The sediments that formed the middle Ordovician carbonate rocks were deposited over an eroded landscape that became the Knox (Group)–Middle Ordovician unconformity (Webb, 1959; Webb and Raymond, 1979; Mussman and Read, 1986). These rocks, which formed in various environments on the tidal flat, ramp, and basin, are each assigned to a formation based on their lithofacies (figure 21.11). Supratidal, intratidal, and restricted subtidal units, including the New Market, Blackford, and Mosheim limestones and the Bowen and Moccasin formations, contain both limestone and dolostone lithofacies. These include carbonate lime mudstone red beds, thin-bedded to massive lime mudstone, algal laminated lime mudstone, wackestones, pelletal packstones, intraclastic dolostones, and lesser amounts of breccia[63] and shale (figures 21.12 and 21.13). Mudcracks, bioturbated beds, and fenestrae are characteristic structures. In some units, such as the Bowen Formation and Lincolnshire-Wardell Limestones, hardgrounds are present. Rare chevron texture preserved in Knox chert indicates sabkha conditions for pre-unconformity tidal flats (Webb and Raymond, 1989) that may have persisted locally into the Middle Ordovician Period.

Seaward of the tidal flats, lime sand flats are represented by rocks of the Whistle Creek, Witten, and parts of the Elway, Wardell, and Wassum Limestone (formations) (J. F. Read, 1980a). These rocks are characterized by thin- and

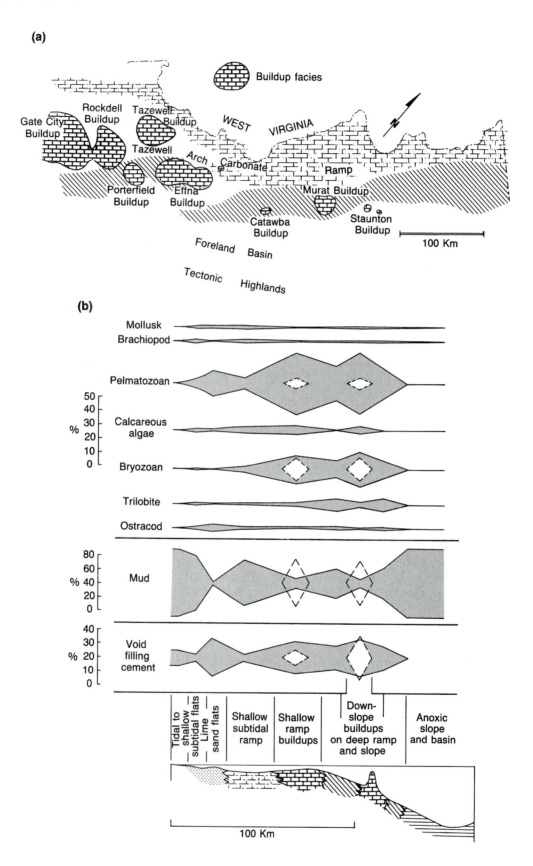

Figure 21.10 Ramp-to-basin model for southwestern Virginia. (a) Map showing location of ramp and basin. (b) Schematic section showing the variations in the composition of each lithofacies. Compositions of the buildup lithofacies are shown by dashed triangles.
(Modified from J. F. Read, 1980a)

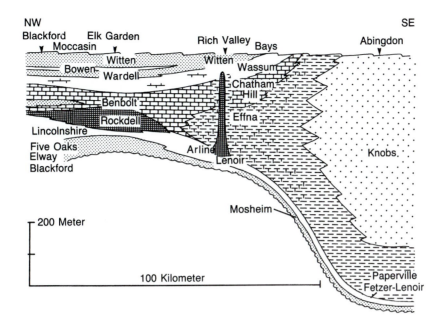

Figure 21.11 Ramp-to-basin section from northwest to southeast across southwestern Virginia. The full thicknesses of the basinal facies are not shown. Names are those of individual formations.
(Modified from J. F. Read, 1980a)

medium-bedded, locally cross-laminated, gray, skeletal-pelletal grainstones with interbedded chert and some intraclastic and fenestral limestones (figures 21.12 and 21.13).

At various sites on the ramp, both on the shallow and deeper parts, carbonate buildups developed (J. F. Read, 1980a; Grover and Read, 1983). These buildups are marked by light gray boundstone or lime mudstone-wackestone cores, with local bafflestones, and they are flanked by similarly colored, medium- to thick-bedded, skeletal grainstones and packstones (figures 21.12 and 21.13; table 21.4, column

The ramp was seaward of the sand flats. The shallowest, quiet-water, subtidal part of the ramp sequence contains cherty limestones, including dark gray, skeletal lime mudstone, packstone, and wackestone (figures 21.12 and 21.13) (J. F. Read, 1980a). The rocks are thin- to thick-bedded and burrowed, and they contain hardgrounds and oncolites. Formations such as the Lincolnshire Limestone and parts of the Elway, Wardell, and Wassum limestones represent this inner part of the ramp. Deep ramp rocks are represented by formations such as the Benbolt and Chatham Hill formations. These rocks are gray to black, thin-bedded skeletal packstone, wackestone, and lime mudstone (figures 21.12 and 21.13) with interbedded shale. Hardgrounds are locally abundant in these units (Markello, Tillman, and Read, 1979).

4) (J. F. Read, 1980a, 1982b; Grover and Read, 1983). Bryozoans, with various combinations of pelmatazoan, sponge, and algal fragments, plus local trilobites, corals, and stromatoporoids, form the reef frameworks and framework grains (Grover and Read, 1983). The flank deposits are locally cherty or shaly. They are cross-bedded, ripple cross-laminated, or thin-bedded, whereas the cores are massive. The Rockdell, Ward Cove, Effna, and Murat limestones represent carbonate buildups.

The basin and adjoining slope sections consist of lime mudstones and shales with local, fine-grained lime mudstone turbidites (table 21.4, column 1). Bedding is generally thin, but in some turbidites is medium to thick. Laminations are common, and grading and soft-sediment deformation features (e.g., folds) are present locally. Except for some gray turbidites, the rocks are black. The Paperville Shale, Rich Valley Formation, and Liberty Hall Formation are basinal units.

Clearly, most of the Middle Ordovician units described here are directly analogous to modern units on carbonate shelves and platforms. Knowledge of such direct analogues, combined with paleontologic studies, allows us to accurately reconstruct the paleogeography and history of rock sequences in the stratigraphic record. If we wish to understand the processes that produced specific features, however, we must take a closer look.

(a)

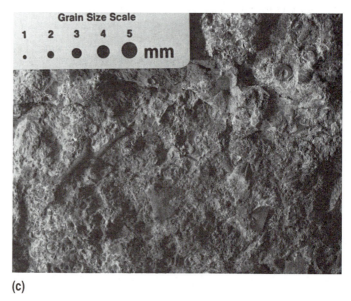

(c)

(b)

Figure 21.12 Photographs of selected lithofacies from the Middle Ordovician ramp-to-basin sequence in southwestern Virginia. (a) Moccasin Formation, intertidal lithofacies, Highway 16, Walker Mountain, Virginia. (b) Mosheim Limestone, subtidal lithofacies, north of Walker Mountain, Highway 16, Virginia. (c) Benbolt Formation, ramp lithofacies, Rye Cove, Virginia. (d) and (e) Effna Formation, carbonate buildup lithofacies, Highway 16, north of Walker Mountain, Tazewell County, Virginia. (f) Rich Valley Formation, basinal lithofacies, Rich Valley, near Saltville, Virginia.

(d)

(e)

(f)

435

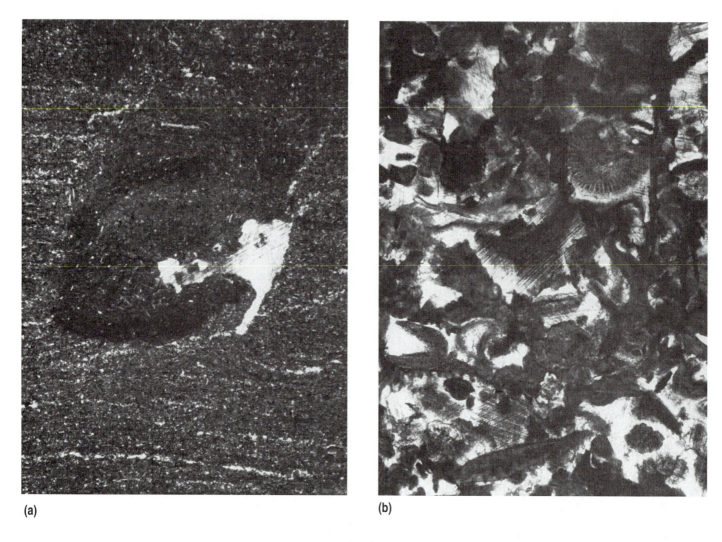

(a)

(b)

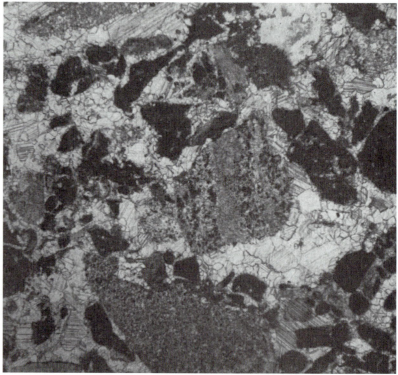

(c)

Figure 21.13 Photomicrographs of selected lithofacies from the Middle Ordovician ramp-to-basin sequence in southwestern Virginia. (a) Moccasin Formation; hematitic, argillaceous, red lime mudstone, with burrow, intertidal lithofacies. (PL). (b) Wassum Limestone; pelletal skeletal grainstone, sand flat lithofacies, Highway 16, Walker Mountain, Virginia. (XN). (c) Mosheim Limestone; intraclastic packstone, subtidal lithofacies, Nebo Quadrangle, Virginia. (XN). (d) Bowen Formation; skeletal wackestone, subtidal lithofacies, Gate City, Virginia. (PL). (e) Effna Formation; bafflestone, carbonate buildup lithofacies, Highway 16, north of Walker Mountain, Virginia. (XN). (f) Rich Valley Formation, lime mudstone, basinal lithofacies, Rich Valley, near Saltville, Virginia. (PL). Long dimension of (a), (b) and (e) is 6.5 mm. Long dimension of (d) and (f) is 3.25 mm. Long dimension of (c) is about 2 mm.

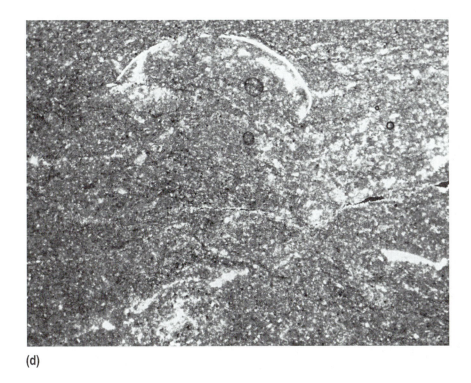

(d)

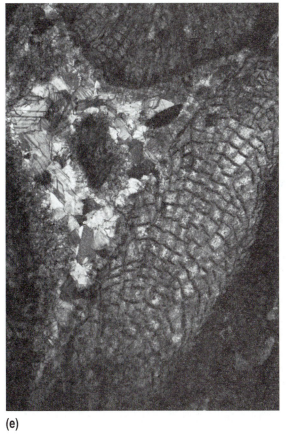

(e)

(f)

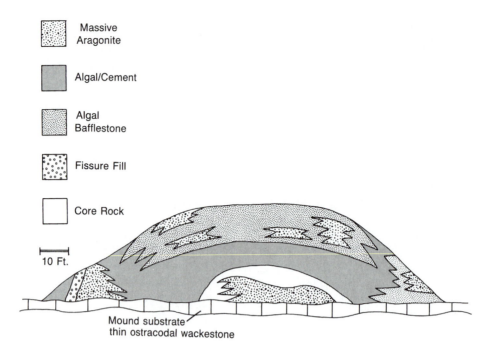

Massive
Aragonite

Algal/Cement

Algal
Bafflestone

Fissure Fill

Core Rock

10 Ft.

Mound substrate
thin ostracodal wackestone

Figure 21.14 Scorpion Mound, Laborcita Formation, Permian, New Mexico, showing various lithofacies. *(Modified from S. J. Mazzullo and Cys, 1979; see Toomey and Babcock, 1983)*

A Closer Look at a Buildup: Algal Mounds of the Permian Laborcita Formation, New Mexico

In the Sacramento Mountains of New Mexico, the Permian Laborcita Formation contains numerous carbonate buildups (Toomey and Cys, 1977, 1979; Mazzulo and Cys, 1979; Toomey and Babcock, 1983). Among these bioherms is the Scorpion Mound, an algal mound. The mound was built upon an ostracodal wackestone base and consists of five major lithologic units (figure 21.14). These include "aragonite massive rock," algal/cement rock, algal bafflestone, fissure fill, and core rock (S. J. Mazzullo and Cys, 1979; Toomey and Babcock, 1983). The core rock and the aragonite massive rock are actually black crystalline calcite (formerly aragonite?) in a matrix of gray, skeletal lime mud. Overlying the aragonite massive/core facies is a distinctive light gray, whole-pelecypod packstone-wackestone unit, with black bands and pockets of cement, called the algal/cement facies (the rock was originally thought to consist primarily of algal plates). The top of the mound consists of an algal bafflestone facies. This facies is also light-colored, contains local patches of ferroan dolomite, and includes algal wackestone. Fissures cutting the mound contain a fissure-fill facies consisting of rudstones and breccia with skeletal packstone-wackestone and algal packstone and wackestone clasts in a sparry calcite cement.

The mounds have a multistage history (Toomey and Babcock, 1983). Early "pioneer" communities of pelecypods and other invertebrates started the buildup, but were overgrown by algae. This algal reef surface was then populated by brachiopods, bryozoans, mollusks, and ostracods. Finally, after periods of exposure and cementation, the reef was drowned and covered with a fusilinid skeletal wackestone.

SUMMARY

Carbonate rocks are composed predominantly of one or more of the minerals calcite, high-Mg calcite, dolomite, and aragonite. Quartz, clays, and other minerals make up the remainder. The chemistry of the rocks reflects that dominant carbonate chemistry and the isotopic ratios indicate conditions of formation.

Various textures and structures characterize carbonate rocks formed in each of the major marine, transitional, and continental environments. The two principal classifications of carbonate rocks in use currently rely either on the dominance of the main textural elements—micrite, sparite, oolites, pellets, intraclasts, and skeletal fragments (the Folk classification)—or on the relative abundance of the two size categories of carbonate material—lime mud and grains (the Dunham classification).

Carbonate rocks form in nine major marine environments, as well as in transitional and continental environments. Most carbonate rocks are marine. The major marine environments are the basin, slope, foreslope, carbonate buildup, ramp, open shelf, shoal, and platform environments. Transitional environments include intertidal, supratidal (including mudflat and lagoon), and bay types. Lakes, alluvial deposits, spring, dunes, and the soil are sites in which continental carbonate rocks are formed. Basin, slope, shelf, and lagoon lithofacies contain abundant lime mudstones and wackestones. Packstones form on slopes, foreslopes, buildups, ramps, shelves, platforms, bays, and lagoons. Grainstones are characteristic of the shoals, intertidal zones (beaches), platforms, some alluvial deposits, and dunes. Springs yield travertine and tufa, whereas caliche is formed in soil and alluvial deposits. Boundstones of various types (e.g., floatstone, bafflestone, shellstone, branchstone) and rudstones form in reefs and other buildups. Stromatolitic boundstones characterize some lacustrine settings and the intertidal to supratidal environments, especially tidal flats and sabkhas. Dolomites form primarily through diagenesis of the rocks formed in these same environments. Together with distinctive structures and associated facies, lithofacies serve as keys to the recognition of the various environments of carbonate rock formation in the stratigraphic record.

EXPLANATORY NOTES

1. Estimates of carbonate rock abundance vary widely. Ronov and Yaroshevsky (1969) estimated that carbonates comprise only 2% of the volume of crustal rocks, whereas Halley (1983) estimated 8%. Of the continental rocks, Blatt's estimate of 10% is low compared to those of Kuenen (1941), who estimated 29%, and Folk (1974), who estimated 25–35%.

2. Allochems or allochemical constituents are defined by Folk (1974, p. 1) as "those substances precipitated from solution *within* the basin of deposition but which are 'abnormal' chemical precipitates because in general they have been later moved as solids within the basin." As examples, Folk (1974) cites oolites, whole or fragmented shells, fecal pellets, and clasts of penecontemporaneous carbonate sedimentary material reworked to form pebbles. He distinguishes between clasts composed of materials derived from erosion, transportation, and deposition *within* the basin and those derived from erosion of terranes *outside* the depositional basin, designating the former as *allochems* and the latter as *terrigenous components*. The intra-/extrabasinal distinction requires a second-order genetic interpretation, and in many cases is either very difficult to apply or requires a subjective judgement. Consequently, Raymond (1984c) redefined *allochems* as "fragments of chemically and biochemically formed materials," including oolites, fossils, and pellets. As defined by Raymond, the term, though still requiring a first-order genetic interpretation (as to whether or not the materials are precipitates), is based on textural and mineralogical criteria that can be observed in rock fragments. Also see note 3, chapter 14.

3. Bathurst (1975, p. 235ff.) also presents a summary of carbonate mineralogy and chemistry and see Wray (1971).

4. Friedman (1965b), Ebanks (1975), Multer (1977, ch. 2), C. S. Nelson and Lawrence (1984). The latter paper discusses the role of methane gas.

5. See Friedman (1965b) for a discussion of the occurrence of aragonite in the oceans. Also see A. C. Newman and Land (1975). See Wolf (1965), W. B. Lyons et al. (1984), and Lasemi and Sandberg (1984) for discussions of the synsedimentary and early diagenetic transformation of algal-generated carbonate.

6. For example, see Mullins (1986).

7. Scoffin (1987, pp. 5–6) and Wray (1971).

8. Refer to the discussion of dolomitization in chapter 16 and see Friedman and Sanders (1967) and Scoffin (1987, p. 134).

9. Ionic radii from Mason and Berry (1968, p. 78). Also see Pauling (1927), J. Green (1959), and Clark (1989, in Carmichael, 1989, p. 35). Bloss (1971, p. 201ff.) gives a good review of ionic radii data.

10. Also see P. G. Flood and Orme (1988) and the papers in Doyle and Roberts (1988).

11. For example, see Spock (1953), Bathurst (1975, p. 137), M. R. Scott (1975), D. L. Williams et al. (1982).

12. Spock (1953), Friedman (1965a), D. L. Williams et al. (1982), A. C. Kendall (1984), Huebner et al. (1986a), and observations by the author.

13. See M. G. Gross and Tracey (1966), Milliman (1974, p. 30ff.), J. D. Hudson (1977), Boggs (1987, p. 688ff.), and Richardson and McSween (1989, pp. 208–232) for brief introductions to stable isotope studies. Also see M. L. Keith and Weber (1964), J. D. Hudson (1975, 1977), R. L. Hay et al. (1986), T. R. Taylor and Sibley (1986), and Ditchfield and Marshall (1989) for more specific discussions and examples.

14. The textures of carbonate rocks are illustrated in color photographs in Scholle (1978) and A. E. Adams, MacKenzie, and Guilford (1984).

15. Also see Land (1970), Ginsburg (1971), and Halley (1983). Howard, Kissling, and Lineback (1970) suggest that locally, fine-grained carbonate sediment may form by disintegration of coarser shell fragments.

16. Particles are considered to be transported if they have been moved away from their source area or, in the case of organic clasts, if they have been moved from the local environment in which the organism lived. In practice, it is likely that allochems in rocks are only recognized if the particles show evidence of transport (e.g., rounding) or if they are exotic with respect to the matrix, fauna, or lithofacies in which they are found.

17. *Grapestones* are sedimentary grains consisting of clusters of sand grains cemented together by micrite in the form of clusters of bunches of grapes (Illing, 1954; Bathurst, 1975, pp. 89 and 316ff.). For color photomicrographs of these and other allochems see A. E. Adams, MacKenzie, and Guilford (1984). Pellets are generally too small to be recognized in handspecimen. Consequently, thin sections or acetate peels are generally required to distinguish pelletal wackestone and packstone from lime mudstone.

18. *Biolithic elements* are parts of a biolithite or boundstone that are formed by organisms. Biolithic elements, by definition, form in place. *Biogenic elements,* as the term is used here, are any textural components of a rock formed by biologic activity. Biogenic elements include biolithic elements, bioclasts (transported fossil fragments), and pellets.

19. Carbonate grains form by fragmentation of preexisting materials (in which case they get smaller) or by cementation processes, as in the formation of oolites and grapestones (in which case they get larger). Either of these processes may occur very near the site of deposition or at some distance from it. Consequently, the grain sizes of carbonate particles do not reflect hydrodynamic conditions as clearly as do siliciclastic grains. For this reason grain sizes are less important in the classification of carbonate rocks than in the classification of siliciclastic rocks.

20. See D. B. Smith (1974), Assereto and Kendall (1977), Worley (1979), and C. Kendall and Warren (1987) for descriptions or examples.

21. Hardgrounds are described by Scoffin (1987, pp. 70, 96–98) and depicted in color by H. E. Cook and Mullins (1983, pp. 603–606).

22. Also see R. E. Garrison (1972), Milliman (1974, p. 54ff. and ch. 8), and the reports of the Deep-Sea Drilling Project.

23. Additional general discussions of carbonate environments are provided by Halley (1983), Halley, Harris, and Hine (1983), N. P. James (1984a), and Mullins and Cook (1986).

24. Also see Yurewicz (1977), Mullins (1986), and C. W. Holmes (1988).

25. Carbonate slope and basin margin sediments are now commonly called *peri-platform sediments* (Schlager and James, 1978; Cook, 1983).

26. Also see Mullins (1983), McIlreath and James (1984), S. C. Ruppel (1984), and C. W. Holmes (1988).

27. Carrasco-V (1977), Cook and Taylor (1977), Davies (1977), Evans and Kendall (1977), J. F. Hubert, Suchecki, and Callahan (1977), Yurewicz (1977), Davis (1983), Mullins (1983), Coniglio and James (1985), Mullins and Cook (1986).

28. The terms shelf and platform are commonly used as synonyms. For example, see P. M. Harris, Moore, and Wilson (1985), who, like many other authors, use the term platform for open marine, as well as epicontinental marine, environments. See note 31.

29. Additional relevant discussions may be found in Enos (1983), J. L. Wilson and Jordan (1983), Cluff (1984), Driese and Dott (1984), and C. W. Holmes (1988).

30. C. W. Holmes (1988) also describes shelf and peri-platform sediments offshore of southern Florida. Dix (1989) discusses high-energy shelf facies in Australia.

31. The term platform has been used in different ways by different geologists. J. L. Wilson (1975, p. 21) defines *platforms* as "carbonate bodies . . . with a more or less horizontal top and abrupt shelf margins." Ahr (1973) uses the term platform to include shallow lagoonal or marsh areas that are flat. I use the term here to refer to generally flat, epicontinental areas of regional extent, with somewhat restricted circulation or tidal activity.

32. Also see Enos (1974b), Hagan and Logan (1974), Read (1974b), J. L. Wilson (1975), and Driese and Dott (1984).

33. See Ball (1967), J. L. Wilson (1975), Driese and Dott (1984), Smosna (1984), Sternbach and Friedman (1984), and Handford (1988).

34. Also see LaPorte (1969) and J. L. Wilson (1975).

35. See note 58 for references.

36. S. J. Mazzulo et al. (1978), McKinney and Gault (1979), Driese and Dott (1984), Sternbach and Friedman (1984), Cramer (1986).

37. For example, see Halley, Harris, and Hine (1983), Handford (1988), Holmes (1988), J. H. Anderson and Machel (1989), and Poppe, Circe, and Vuletich (1990).

38. Refer, in addition, to Friedman (1988) and other papers in Doyle and Roberts (1988) and to Poppe, Circe, and Vuletich (1990).

39. Also see P. A. Harris (1979), Halley, Harris, and Hine (1983), and Hine (1983).

40. Wilson (1975, p. 27), P. A. Harris (1979), Halley, Harris, and Hine (1983).

41. Milliman (1974, ch. 6), Pusey (1975), Hopkins (1977), Multer (1977, chs. 2, 6, 7), R. W. Scott (1979), Enos and Moore (1983), Viau (1983), Toomey and Babcock (1983), N. P. James and Macintyre (1985), Shaver and Sunderman (1989).

42. Friedman and Sanders (1967), Textoris (1968), LaPorte (1971), Friedman (1972).

43. See the example of the Ordovician ramp to basin facies, below.

44. LaPorte (1967).

45. This section is based on the works of LaPorte (1967, 1969, 1971), Read (1974b), Ebanks (1975), High (1975), Hardie (1977a,b, 1986a,b), Hardie and Ginsburg (1977), Cluff (1984), P. M. Harris (1985), Shinn (1983b), A. C. Kendall (1984), Steiger and Cousin (1984), Hardie and Shinn (1986), Colby and Boardman (1989), and Cloyd, Demicco, and Spencer (1990).

46. Also see LaPorte (1971), Read (1974b), High (1975), P. M. Harris (1985), and Hardie (1986a,b).

47. For discussions and descriptions see Ginsburg (1960, 1967), P. Hoffman (1974), Logan, Hoffman, and Gebelein (1974), and Multer (1977).

48. Also see Hagan and Logan (1974), P. M. Harris (1985), and D'Aluisio-Guerrieri and Davis (1988).

49. The environments and sediments of a modern tropical beach-barrier-lagoon-deltaic shoreline are described in Purdey, Pusey, and Wantland (1975), Pusey (1975), and High (1975).

50. Stoddart and Cann (1965), Multer (1977, pp. 33–34), Strasser and Davaud (1986); H. H. Roberts and Murray (1988).

51. Inden and Moore (1983), Strasser and Davaud (1986).

52. F. Webb (pers. commun., 1989) reports significant root penetration in carbonate aeolianites of the Tulum area of the Yucatan Peninsula, Mexico.

53. Bretz and Horberg (1949), Dean and Fouch (1983), Esteban and Klappa (1983), McKee and Ward (1983), E. Nickel (1985), R. L. Hay et al. (1986), Pentecost (1990).

54. Dean and Fouch (1983) provide a good review of lacustrine sedimentation. Also see Friend and Moody-Stuart (1970).

55. Also see Friedman (1972) and Pentecost (1990).

56. See Folk et al. (1985), Pentecost (1990), and the references therein.

57. Also see J. E. Sanders and Friedman (1967), Read (1974a), Folk and McBride (1976), J. F. Hubert (1978), and Coniglio and Harrison (1983).

58. A large body of literature exists on the sediments of the Bahamas. This example is based on the mapping of Imbrie and Purdy (1962) and Purdy (1963a,b); additional studies by Illing (1954), Newell (1955), Newell and Rigby (1957), Newell, Purdy, and Imbrie (1960), Cloud (1962), Imbrie and Buchanan (1965), Shinn (1968), Ball (1969), Shinn, Lloyd, and Ginsburg (1969), Bornhold and Pilkey (1971), Gebelein (1973, 1974), Bathurst (1975, ch. 3), Cuffey and Gebelein (1975), Hardie (1977), Hardie and Garrett (1977b), Hardie and Ginsburg (1977), Multer (1977), P. A. Harris (1979), Beach and Ginsburg (1980), N. P. James (1983); and the reviews of Milliman (1974, ch. 6), Multer (1977), Leeder (1982, pp. 219–226), and Halley, Harris, and Hine (1983). Enos (1974b) also presents a map of sediments in the Bahamas region. Related sedimentation on the south Florida shelf is described by Enos and Perkins (1977).

59. The facies of Andros Island and the surrounding deeper marine environments are not the primary focus of this review, but information on those facies may be found in Newell and Rigby (1957), Cloud (1962), Shinn, Lloyd, and Ginsburg (1969), Bornhold and Pilkey (1971), and Multer (1977).

60. See the works of Shinn (1968) and the discussion of the work of Ginsburg et al. (1971) in Milliman (1974, p. 168).

61. This example is based largely on the studies published by Read and Tillman. (1977), Markello, Tillman, and Read (1979), J. F. Read (1980a,b), and Grover and Read (1983), with additions and background provided by the works of Read (1982a,b), Butts (1940), B. N. Cooper (1961, 1964, 1968), Shanmugam and Walker (1978, 1980), additional references cited in this section, and observations by the author and F. Webb (pers. communications, 1972–1989). Related works include R. B. Neuman (1951), Pitcher (1964), Walker and Benedict (1980), S. C. Ruppel and Walker (1982, 1984), Walker et al. (1983), Harland and Pickerill (1984), R. E. Johnson (1985), Simonson (1985), and Wedekind (1985). Shanmugam and Lash (1982) depict the details of the foundering of a shelf and the development of a foreland basin, and they suggest a modern analog in the Timor region. The foreland basin described here extended south into Tennessee; equivalent rocks there are described by F. B. Keller (1977), Shanmugam and Walker (1978), Walker et al. (1983), and S. C. Ruppel and Walker (1984), among others. In contrast to the ramp in Virginia, the shelf in Tennessee appears to have had a slope break with a steepened lower slope (S. C. Ruppel and Walker, 1984). Also see the discussion in Sternbach and Friedman (1984).

62. F. B. Keller (1977) and Shanmugam and Walker (1978, 1980) more thoroughly describe similar rocks exposed in eastern Tennessee.

63. Recall the discussion of the Mosheim breccias in chapter 20.

PROBLEMS

21.1. Suggest an environment of formation for each of the following carbonate rock sequences. Refer to this chapter and chapter 17.
   a. Interlayered stromatolitic limestone (laminated) and oolitic grainstone with local patches of thick-bedded boundstone
   b. Thin-bedded, locally coccolithic, light gray lime mudstone with interbeds of clay shale
   c. Flaser-bedded, thin- to medium-bedded, gray, skeletal packstone interbedded with skeletal (bryozoan) calcareous clay shale.

21.2. a. Draw a columnar section with the following units (from top to bottom):
   Medium- to thick-bedded, cross-bedded skeletal grainstone (6.3 m)
   Flat pebble dolostone breccia with fine-grained skeletal grainstone matrix (0.2 m)
   Iron-oxide-stained, mud-cracked laminated dolostone (0.5 m)
   Laminated dolostone with white chert nodules and gastropods (3.5 m)
   Dolomitized, medium-bedded, cross-bedded skeletal grainstone (7.5 m)
   b. Explain the geologic history suggested by this section of carbonate rocks.

C H A P T E R 22

Cherts, Evaporites, and Other Precipitated Rocks

INTRODUCTION

Cherts, siliceous sinter, evaporites, salinites, phosphorites, ironstones, and iron-formations are the major types of noncarbonate rock produced by precipitation of minerals from solutions. The solutions may be marine or nonmarine, dilute or concentrated, and are generally of moderate pH and Eh.[1] Minerals that precipitate from such solutions are precipitated by both organic and inorganic processes and include both primary and diagenetic phases.

Chert is hard, waxy to grainy, aphanitic sedimentary rock composed predominantly of silica minerals. These minerals include opal, chalcedony, and quartz, which are precipitated inorganically, and in the case of opal, organically. The solutions from which cherts are precipitated include seawater, fresh to saline lacustrine waters, diagenetic fluids, and hydrothermal solutions. Cherty rocks are currently forming in bathyl to abyssal marine environments—for example at the northern edge of the Nazca Plate and in the Gulf of California (T. A. Davies and Gorsline, 1976; Calvert, 1966)—and in ephemeral lakes associated with the Coorong Lagoon in South Australia (Peterson and von der Borch, 1965). **Siliceous sinter** is also a siliceous rock, but is porous and is precipitated from hot springs and geysers. Siliceous sinter is formed, for example, around geysers in Yellowstone National Park, Wyoming.

Salinastones are rocks composed of saline minerals.[2] Two major types of salinastones are recognized—evaporites and salinites. **Evaporites** (*sensu stricto*) are sedimentary rocks formed by the crystallization of salts from concentrated solutions produced by evaporation of a watery solvent. Inasmuch as significant evaporation is required for their formation, evaporites form in arid regions, such as in the playas (salinas) of desert basins, in sabkhas along hot coastal regions, and in extremely restricted to isolated marine basins (marine salinas) along dry coastlines. Examples of such evaporite-forming environments include Death Valley and Searles Lake in southeastern California, the Great Salt Lake in Utah, the sabkhas of the Persian Gulf and Baja California, and the Lake of Lanarca on the island of Cyprus in the Mediterranean Sea.[3] Carbonate sedimentation via chemical precipitation occurs in these same environments, and the carbonate rocks thus formed are true evaporites.

In addition to forming as a result of evaporation, evaporite-like rocks or **salinites** form via other processes. Such nonevaporite salinastones may crystallize from brines developed in continental to marine basins, in polar region groundwaters and lakes, and in the pores of subsurface rocks and sediments, especially in supratidal environments

(Craig et al., 1975; A. C. Kendall, 1984; Sonnenfeld and Perthuisot, 1989). Minerals of the former two environments are simply precipitates. Those of the subsurface are commonly diagenetic alteration or replacement phases, but may also be evaporite precipitates. Rarely, salinastones are also recognized to have been deposited in a wide range of additional environments, including in "deep water" and on submarine fans (A. C. Kendall, 1984; Sonnenfeld, 1984; J. K. Warren and Kendall, 1985).

In meromictic lakes and extremely isolated marine basins, the bottom waters may yield rocks that include (1) evaporites, (2) salinites, (3) laminated, kerogen-rich carbonate rocks, (4) cherts, and (5) mudrocks. Some of the carbonates and salinites and all of the mudrocks are *not* evaporites, yet they are genetically related to the evaporites and therefore are mentioned in discussions in this chapter.

Phosphorite is a sedimentary rock containing more than 19.5% P_2O_5 (Cressman and Swanson, 1964; Pettijohn, 1975, p. 427). Such a P_2O_5 content is equivalent to a mode with about 50% apatite. Most phosphorites of significant volume are marine, but continental phosphatic gravel and guano (bird manure) deposits are known. Perhaps the best-known phosphorite-bearing formation is the Permian Phosphoria Formation of Colorado, Idaho, Montana, Nevada, Utah, and Wyoming (McKelvey et al., 1956, 1959; Cressman and Swanson, 1964).

Ironstone and **iron-formation** are iron-rich sedimentary rocks containing 20% or more total iron oxides (FeO + Fe_2O_3). Ironstone is noncherty, whereas iron-formation is cherty. The term "taconite" is used in the Great Lakes region to designate unweathered iron-formation. Although iron-rich rocks are widely distributed in space and time in the geologic record, their origins are poorly understood. Many seem to occur in shallow marine and continental swamp or lake deposits. Good contemporary examples of ironstone or iron-formation development were unknown until their recent discovery in the southeast Asian region and the southern Caribbean Sea (G. P. Allen et al., 1979; and Kimberly, 1989). The general absence of contemporary analogues for iron-rich sediments is, perhaps, an indication that diagenesis and replacement are important in the formation of these rocks. Whatever the case, iron-formation is largely restricted to Precambrian rocks, particularly Proterozoic rocks such as the various iron-formation-bearing units of the Great Lakes region and the Labrador Trough.[4] Ironstones are widely distributed in early Phanerozoic rocks, such as the Silurian "Clinton" formations of New York and adjoining states—including the Rose Hill Formation of New York, Pennsylvania, Maryland, and Virginia—and the Silurian Red Mountain Formation of Alabama (Van Houten, 1990).[5]

CHERTS

Cherts form in several ways: (1) as surficial deposits in deep to shallow marine environments, (2) as similar deposits in various types of lakes, (3) as vein or cavity fillings precipitated

from hydrothermal solutions in continental to marine rocks, especially volcanic rocks, and (4) as diagenetic replacement deposits in preexisting rocks, such as limestone. Cherts may be chemical precipitates or clastic allochemical accumulations. In the case of chemical precipitation, silica minerals crystallize directly from aqueous solutions, commonly as replacements of earlier formed minerals. Clastic allochemical cherts form by a two-stage process in which biochemical precipitation results in the accumulation of organic (usually opaline) tests, which are subsequently transported and redeposited by currents. Because of the fine grain size of cherts and the fact that organisms precipitate amorphous opal; recrystallization, which occurs readily, tends to obscure or entirely eliminate evidence of allochemical and biogenic origins.

Among the well-known examples of cherts are the radiolarian cherts of the Jurassic-Cretaceous Franciscan Complex of California, the Devonian-Mississippian Arkansas "Novaculite," and the cherts associated with iron-formation of the Precambrian Gunflint and Biwabik formations of Minnesota and Ontario. Other examples include the Miocene Monterey Formation cherts of California, the Mississippian Fort Payne Chert of Georgia and Tennessee, and the Silurian-Devonian Caballeros Novaculite of Texas, to name just a few.[6] Some of these have partly allochemical histories. Cherts that form by replacement commonly form smaller masses associated with other rock types. In contrast, extensive regions of the ocean floor, especially at low and high latitudes, are covered with biogenically precipitated silica (see figure 18.5).[7] Similarly, the Precambrian iron-bearing formations form large volumes of chert-rich rock.[8]

The Compositions, Textures, and Structures of Cherts

As noted above, cherts are composed primarily of one or more of the silica minerals—opal, chalcedony, and quartz. Biogenically precipitated silica is amorphous opal, or *opal-A*.[9] Opal-A sediment consists of radiolaria, diatoms, or sponge spicules (figure 22.1a). Upon diagenesis, opal-A converts either to *opal-CT* (a metastably crystallized, interlayered cristobalite-tridymite form of silica) or directly to quartz (R. Greenwood, 1973). Replacement cherts may form either from direct precipitation of quartz *or* through a multistage process of precipitation and recrystallization in which opal is first precipitated and later replaced by granular or chalcedonic quartz (chalcedony).[10]

Additional minerals in cherts are numerous. The nature and variety of accessory minerals is largely a function of the associated rocks. Iron oxides, especially hematite, and clays are common accessory minerals. Where cherts occur as nodular masses in carbonate rocks, calcite and dolomite are the typical accessories. In addition, other minerals—including graphite, pyrite, magnetite, manganese oxides (e.g., hausmannite, braunite), alkali feldspars (albite, microcline), clays (e.g., illite, smectites, sepiolite), micas (biotite, stilpnomelane, muscovite), chlorites (e.g., chamosite), serpentines (e.g., greenalite), talcs (e.g., minnesotaite), zeolites

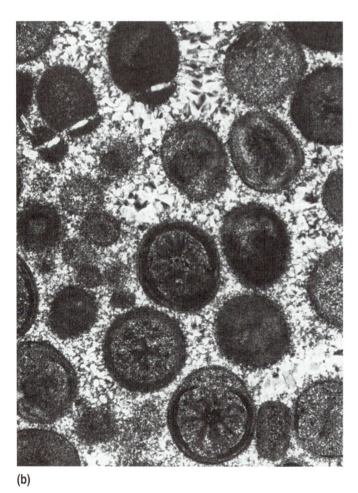

(a)

(b)

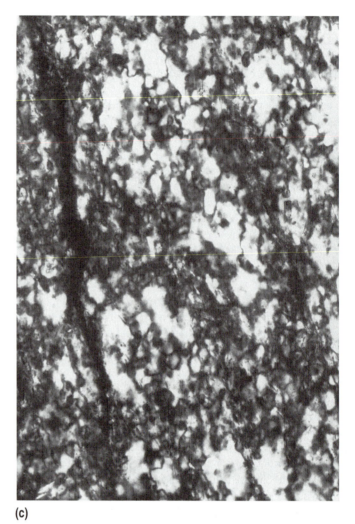

(c)

(d)

Figure 22.1   Textures in cherts. (a) Bioclastic, diatomite; "Val Monte Formation," Miocene, Lompoc, California. (PL) (b) Radial fibrous texture in oolitic chert, Mines Member, Gatesburg Formation, Cambrian, State College, Pennsylvania. Here chert has apparently replaced oolitic limestone. (XN) (c) Equigranular sutured texture in bedded chert, Falcon Formation, Jurassic, Franciscan Complex, northeastern Diablo Range, California. (XN) (d) "Hopper" to chevron texture in chert; Upper Knox Group, Ordovician, Smyth County, Virginia. Because hopper and chevron textures characterize halite in evaporites, the texture suggests that chert has replaced an evaporite. (PL) Long dimension of photos in (a), (c), and (d) is 0.33 mm and in (b) is 3.25 mm.

Table 22.1 Chemical Analyses of Chert and Siliceous Sinter

	1	2	3	4	5	6
SiO_2	99.10[a]	97.4	91.7	83.66	82.2	69.00
TiO_2	0.06	0.03	0.17	—	nr	0.10
Al_2O_3	0.19[b]	0.47	3.31	1.94	0.67	1.5
Fe_2O_3	0.06	1.3	—	—	—	—
$FeOx_{total}$	—	—	0.93[c]	0.22[d]	0.38[d]	3.2[d]
FeO	nr	<0.26	—	—	—	—
MnO	—	tr	0.42	tr	nr	nr
MgO	0.64	0.05	0.92	0.21	0.16	0.39
CaO	0.10	0.05	0.06	0.76	5.12	13.20
Na_2O	0.02	0.01	0.23	1.20	0.08	0.32
K_2O	0.06	0.55	0.77	0.42	0.09	0.42
P_2O_5	nr	0.04	0.04	nr	0.06	4.71
LOI	nr[e]	0.62	1.39	11.60	10.8	6.0
Other	tr	—	tr	tr	—	0.78
TOTAL	100.23	100.2	99.9	100.34	99.7	99.6

Sources:

1. "Soft opal" in silty and sandy limestone (?), Ogallala Formation (Miocene), Scott County, Kansas (Franks and Swineford, 1959).
2. Red chert (massively bedded). Franciscan Complex (Jurassic-Cretaceous), near Ortega Street, San Francisco, California (E. H. Bailey, Irwin, and Jones, 1964).
3. Chert, sample BJ14, Franciscan Complex (Jurassic-Cretaceous), Blue Jay Mine, Trinity County, California (Chyi et al., 1984).
4. Siliceous sinter (Recent), near Daisy Geyser, Yellowstone National Park, Wyoming (Allen and Day, 1935, *in* T. P. Hill, Werner, and Horton, 1967, p. 13).
5. "Porcellanite" nodule, Monterey Formation (Miocene), Lompoc Quarry, California (Weis and Wasserburg, 1987).
6. Chert (phosphatic), Phosphoria Formation (Permian), Lincoln County, Wyoming (McKelvey et al., 1953, in T. P. Hill, Werner, and Horton, 1967, p. 70).

[a]Values in weight percent.
[b]Includes MnO and Ga_2O_3 if present.
[c]Total iron as FeO.
[d]Total iron as Fe_2O_3.
[e]Analysis recalculated as water free.
nr = not reported.
tr = trace.

(e.g., clinoptilolite), detrital silicates (e.g., augite or hornblende), other carbonates (ankerite, siderite, and rhodochrosite), apatites, and sulfates (gypsum, anhydrite, and barite)—occur in various cherts.[11]

Predictably, many chert analyses show cherts to be very high in silica (table 22.1). Those, however, that are impure (of which there are many) contain significant amounts of iron, alumina, lime, or other components.

The textures and structures of cherts show some variety. Those dominated by opal are amorphous. Chalcedonic varieties of chert have fibrous, spherulitic, or comb textures (figure 22.1b). Perhaps the most common texture in chert is the equigranular-sutured texture (figure 22.1c), but equigranular-mosaic textures are developed in thoroughly recrystallized cherts (Folk and Weaver, 1952). Breccia textures are relatively

common as well. In the allochemical cherts, clastic textures are typical.

Many varieties of cherts have been given special names. These include radiolarite (radiolarian-rich chert), ribbon chert (rhythmically bedded chert), jasper, jasperite, and diasporiti (red chert), lydite (black chert), flint (homogeneous, black or dark gray chert), and porcelanite (hard, fine-grained, clay and calcite-bearing chert) (e.g., D. L. Jones and Murchey, 1986). Diatomite is a soft, lightweight, light-colored rock composed of the siliceous tests of algae (aquatic plants). Diagenesis transforms diatomaceous sediments into chert (Ernst and Calvert, 1969; J. R. Hein, Yeh, and Barron, 1990).

Structurally, cherts are laminated and bedded, nodular, or podiform to veinlike (figure 22.2). In addition, some chert is dispersed and gives a mottled appearance to the rock

(a)

(b)

(c)

Figure 22.2   Chert structures. (a) Bedded to nodular chert in a limestone-shale sequence, Admire Group, Permian, I-70, Exit 311, south of Manhattan, Kansas. (b) Locally nodular, very thin-to medium-bedded ("ribbon") chert and siliceous shale, dismembered formation of Grummett Creek, Jurassic (?), Franciscan Complex, northeastern Diablo Range, California. Note the local folds and pinch-and-swell structures in the beds. Hammer handle is 46 cm in length. (From Raymond, 1973a.) (c) Thin- to thick-bedded "novaculite" (chert) of the Arkansas Novaculite, Devonian-Mississippian, Highway 70, two miles east of Magnet Cove, Arkansas.

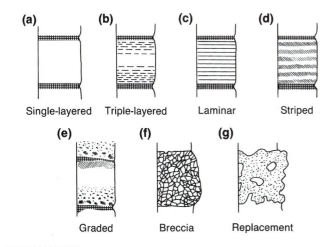

Figure 22.3 Types of beds in chert-bearing formations:
(a) single-layered bed, (b) triple-layered bed, (c) laminar
bed, (d) striped bed, (e) graded bed, (f) breccia bed,
(g) replacement bed.
[(a)–(e) After Iijima, Matsumoto, and Tada (1985)]

(Bustillo and Ruiz-Ortiz, 1987). The podiform and veinlike
cherts that particularly characterize volcanic rocks, are volumetrically far less important than the bedded and nodular
types. Although some nodules occur in bedded chert-shale
sequences, such as the widely distributed Mesozoic "ribbon
cherts," nodules are particularly characteristic of cherts in
carbonate rocks (figure 22.2a). The nodules range from well-
formed ellipses, typically flattened in the plane of the bedding, to very irregular masses with numerous bulbous
protrusions.

Bedded cherts occur (1) in sequences of regularly interlayered chert and shale, (2) as beds in evaporite or salinite sequences, (3) in carbonate-rich formations, (4) as a major
component of banded iron-formation, and (5) with bedded
phosphorites. The layers in bedded cherts vary in thickness
from microlaminae to massive beds several meters in thickness. Iijima and Utada (1983) and Iijima et al. (1985) describe five types of layered sequences (figure 22.3). These
include homogeneous *single-layered bed* types, bounded above
and below by shale; *triple-layered bed* types, with a clay-poor
central zone that grades up and down into more argillaceous
parts of the bed; *laminar bed* types, which have parallel laminations; *striped beds*, characterized by millimeter-wide laminations of clay-rich material; and *graded bed* types, with
associated, small-scale, cherty Bouma sequences. To these five
types, two additional types may be added—*breccia beds* and
nodular beds. (Nodular to irregular beds are characterized by
numerous pseudomorphs, spherulites, or other diagenetically
developed features interpreted to be of a replacement origin.)

Origins of Cherts

Cherts form through one or more of the following processes:
(1) biochemical precipitation, (2) hydrogenous precipitation,
(3) hydrothermal precipitation, (4) replacement, and (5) erosion, transportation, and deposition of previously formed
siliceous materials. The silica content of materials formed by
one or more of these processes may be increased by (6) diagenetic processes; that is, the silica may be further concentrated
in chert beds. Process 5 is one of deposition of allochems.
Processes 1 to 4 involve precipitation of silica from a solution, whereas process 6 involves remobilization of previously
deposited silica.

Biochemical precipitation encompasses all those
processes by which living organisms induce the crystallization of solids from solutions. In ordinary streams and
groundwater, the concentration of silica is about 10 to 60
ppm, and in seawater it is even less (1 to 2 ppm)(Krauskopf,
1967, pp. 168–169).[12] These values are well below the equilibrium solubility of amorphous silica, which is about
60–150 ppm (Krauskopf, 1967; L. A. Williams, Parks, and
Crerar, 1985). Thus, silica should not precipitate from these
waters under equilibrium conditions. Nevertheless, silica-
secreting organisms are able to extract and precipitate silica
from these undersaturated waters and protect the biogenically precipitated opal from dissolution (perhaps with some
form of organic coating), as long as the organisms are alive.
After the organisms die, solution and diagenetic recrystallization will act on the opaline material, altering it to opal-
CT, chalcedony, or quartz (Bramlette, 1946; Thurston,
1972; R. E. Garrison et al., 1975).[13]

Hydrogenous precipitation—precipitation from low-
temperature, water-based solutions—is a second process involved in the origin of chert. Only where solutions become
supersaturated with silica will hydrogenous precipitation
occur. Such conditions exist principally where solutions become concentrated, for example, in the saline bottom waters
of meromictic lakes, in ephemeral lakes where pH values vary
periodically, or in warm, shallow marine environments where
evaporation rates are high (e.g., Peterson and von der Borch,
1965).[14] As the fluids become concentrated, the silica content exceeds the solubility limit of the water (especially if the
pH is lowered below about 9), and silica precipitates.[15] Inasmuch as the present oceans are generally undersaturated with
silica and may have been so since the Precambrian, inorganic
hydrogenous precipitation of silica from the open ocean during the Phanerozoic Eon seems unlikely (G. R. Heath, 1974).

Hydrothermal precipitation of silica has been invoked
widely to explain the origin of cherts. In thermal waters,
the solubility of silica increases by more than an order of
magnitude (Morey et al., 1964; Krauskopf, 1967, p. 169).

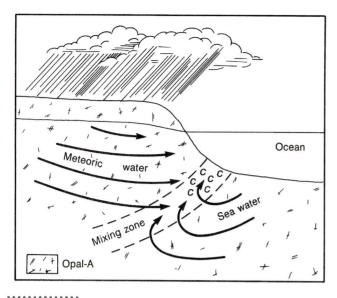

....
Figure 22.4
Schematic diagram showing the mixing model of chertification
for shoreline areas. The zone of chertification is shown by Cs.
(Modified from Knauth, 1979)

Waters enriched in silica as a result of dissolution of silica
under high-temperature conditions become saturated and
precipitate silica upon cooling. Siliceous sinter forms
around hot springs and geysers as a result of this process.
More importantly, recent discoveries of extensive hydrothermal activity on the seafloor along mid-ocean ridges
provide support for the long advocated hypothesis that hydrothermal fluids, either released from submarine volcanoes
or produced by the interaction of hot volcanic rocks and
seawater, are the source of the silica that is precipitated
as chert (E. F. Davis, 1918; Taliaferro, 1943; E. H. Bailey,
Irwin, and Jones, 1964; Crerar et al., 1982). Major element,
trace element, and isotopic data from some types of chert
provide support for this view.[16]

Replacement is a process in which silica substitutes for
the chemical components of other types present in a preexisting rock, notably for calcite in limestones. Evidence that this
process occurs is provided by originally calcareous fossil forms
preserved in detail as siliceous fossils in chert nodules and
beds. Similarly, other textures and structures (e.g., oolites)
may also be replaced by chert, with full preservation of detail
(Folk and Pittman, 1971; Namy, 1974). In limestones, the
process requires that solutions facilitating the replacement be
supersaturated with respect to silica, but undersaturated with
respect to calcite (Knauth, 1979).[17] Such a situation may
exist in coastal regions, where groundwater from the landward side of the shore passes through silica-bearing rock, dissolving silica as it moves seaward. (figure 22.4) Recall that
such water is likely to contain substantially more dissolved
silica (under equilibrium conditions) than does seawater. This

water would continue to move seaward and mix with seawater, whereupon it would become supersaturated with respect
to silica, but undersaturated with calcite, *if* the partial pressures of CO_2, the temperatures, or the pHs of the two waters
are different. Under these conditions, silica will precipitate,
while calcite dissolves.

Diagenesis is known to convert opaline silica to quartz
(Ernst and Calvert, 1969). In silica-rich sediments, ionic migration of silica during diagenesis may result in the additional
concentration of silica in some beds and its depletion in others. Through such a process, silica-enriched layers become
cherts, whereas clay-bearing, silica-depleted, intervening beds
form shales (E. F. Davis, 1918; Jenkyns and Winterer, 1982).
The resulting chert is layered or rythmically bedded and is, in
part, of diagenetic origin.

A special type of diagenetic chert, Magadi-type chert,
forms in evaporite sequences (Eugster, 1967; 1969; Surdam,
Eugster, and Mariuer, 1972; Sheppard and Gude, 1986).[18]
Magadi-type chert develops by a two-stage process. Magadiite
$[NaSi_7O_{13}(OH)_3 \cdot 3H_2O]$ or a similar mineral precipitates
from a high-pH, silica-rich brine, when pH reduction of the
brine occurs (caused, for example, by mixing of the brine
with fresh water, during flooding). The precipitated magadiite
is then converted to chert via diagenetic processes that cause
the removal of sodium ions and water. Magadi-type cherts
may be either nodular or bedded.

Clearly, cherts vary in chemical composition, mineralogy, and lithologic association, and it is equally clear that
they form in a variety of environments via one or more of the
processes described here. In some cases, a unique series of
events may lead to abundant silica precipitation (McGowran,
1989). In others, sediments derive from several sources and
diagenesis further alters the chert (Chyi et al., 1984). In still
others, cherts originate via a single-stage replacement process
(N. A. Wells, 1983). The presence of primary features, including fossils and oolites in the nodular to bedded cherts
that occur in limestones and dolostones, indicates such a replacement origin. The origins of many cherts with complex
constituents or histories is usually open to debate.

The origin of the bedded, radiolarian ("ribbon") cherts,
with their characteristic interlayered chert and siliceous shale
sequences, has been particularly controversial for many
years.[19] Such rocks are common in Mesozoic sections of orogenic belts, particularly in the Alpine-Himalayan and western Cordilleran orogens. Deep-sea drilling suggests that there
are few analogues for these cherts (Jenkyns and Winterer,
1982; J. R. Hein and Parrish, 1987), as drilling in the open
ocean has revealed only rare *thick* sequences of similar rock
(Tucholke et al., 1979). Yet the details of the chert mineralogies, textures, and structures from both the oceanic and on-
land occurrences (Garrison et al., 1975) along with the chert
chemistries (K. Yamamoto, 1987; R. W. Murray et al., 1990,
1991) suggest oceanic depositional environments for many
ribbon cherts. Deep ocean floor, ocean ridge, marginal basin,

forearc basin, and shallow intertidal mudflat environments of deposition have been suggested.[20]

Several models of origin have been proposed for the ribbon cherts. These include the following:

1. Ribbon cherts result from alternating conditions of clay and radiolarian (biogenic) deposition in zones of upwelling, where plankton productivity is high. Periodic influx of clastic sediment (for example, as a result of turbidity current deposition) or periodic increase in radiolarian production (e.g., due to climatic effects) produces the alternating layers of sediment.
2. Ribbon cherts result from alternating deposition of clay and radiolarian-rich sediment, with the increase in biogenic component resulting from radiolarian blooms caused by volcanic additions of silica to the water column.
3. Ribbon cherts result from turbiditic redeposition of radiolarian-rich sediment in tabular layers, with intervening periods of pelagic or hemipelagic mud deposition.
4. Ribbon cherts result from the diagenetic transformation of clay-bearing, siliceous sediment into interlayered sequences of chert and siliceous shale.

Radiolarian turbidites containing partial Bouma-like sequences are present locally in the Franciscan Complex, and some chemical, textural, and stratigraphic evidence supports the view that, locally, ribbon cherts exist that owe their character to processes 1–3. Which of these processes is most important in ribbon chert formation is, as yet, unresolved. Consequently, for any given chert occurrence, careful analyses of the various characteristics of the chert and the associated rocks are generally required before a definitive history for that chert can be reconstructed. Even given those kinds of analyses, the origins of some cherts, such as the Northern Apennine radiolarian cherts, remain enigmatic (Bosellini and Winterer, 1975; Folk and McBride, 1978; McBride and Folk, 1979; Barrett, 1982).

Examples

Chert of the Kaibab Formation, Grand Canyon, Arizona

The Kaibab Formation consists of clastic and carbonate rocks exposed along the rim of the Grand Canyon. Sediments that gave rise to these rocks were deposited during the middle Permian Period on a marine shelf and the adjoining tidal flat and dune field (McKee, 1938, 1969; J. W. Brown, 1969; R. A. Clark, 1980). The formation is divided into three members, the upper two of which are exposed near Grand Canyon village.

The Kaibab Formation around and to the west of the village consists of a series of cherty dolostones, cherts, gypsum evaporites, and red beds (figures 22.5 and 22.6). To the east,

cross-bedded aeolian sandstones crop out. Petrography and facies changes, mapped in the Grand Canyon area by McKee (1938), J. W. Brown (1969), and R. A. Clark (1980), indicate that the Kaibab environments represent a migrating shelf, associated lagoons and supratidal mudflats, and an adjoining desert—in short, a shelf and adjoining coastal sabkha (R. A. Clark, 1980). The occurrence of mudcracks, halite pseudomorphs, algal stromatolites, and gypsum nodules attests to the mudflat origin of some of the chert-bearing dolostones. Marine fossils in other dolostones indicate that they were derived from shelf lime mudstones, packstones, and grainstones. Repetitive cycles in the chert-dolostone sequences suggest cyclic processes of formation (McKee, 1969; F. K. McKinney, pers. comm., 1988).

The cherts range from nodular to bedded masses. The nodules are elliptical to irregular and are typically flattened in the plane of the bedding (figure 22.5). Although commonly 5 to 15 cm in length, the nodules may range up to 1/2 m long (J. W. Brown, 1969; R. A. Clark, 1980). Some composite masses are over a meter in length. The bedded cherts are typically a few centimeters thick, but some thicker beds, including some chert breccias, also occur in the formation.

The minerals of the cherts include quartzine, chalcedony, and microcrystalline to phaneritic quartz, but additional minerals are common (J. W. Brown, 1969). These include carbonate grains and detrital quartz grains. Sand grains and recrystallized lime mud, in textural relations like those in the surrounding carbonate rocks, suggest postsedimentation chertification (J. W. Brown, 1969).

Silicified fossils, preserved in detail, suggest that chertification occurred before dolomitization, which generally destroys the detailed features of fossils. Silicified fossils and evaporite crystal pseudomorphs (gypsum and halite) clearly indicate a replacement origin for the chert, but sponge spicule-like structures (R. A. Clark, 1980) suggest that some silica is biogenic and locally derived. Together, the features are consistent with a model in which a flux of silica-bearing waters passed through the already deposited but incompletely lithified limestones, promoting replacement of carbonate minerals by silica. Replacement apparently was centered on fossils, detrital accumulations of sponge spicules, or bedding irregularities. These waters may have been fluvial waters (on their way to the sea) that had passed through the quartz-rich sands present to the east. In the sands, silica would be dissolved. Upon reaching the sabkha, mixing of fresh water and seawater (R. L. Nielson, 1983) would change the pH and result in inorganic precipitation of silica. How much silica was introduced and how much is locally derived (from sponge spicules) is unknown.

McKee (1969) argues that bedded cherts in the Kaibab Formation are inorganic precipitates. He suggests that precipitation occurred where silica-bearing, fluvial fresh waters were mixed with marine waters. As evidence McKee (1969) cites the following. (1) The bedded cherts occur in a region that lies between pure quartz sandstones on the east and pure limestones on the west. (2) There is a complete change in fauna, from a nearshore molluscan fauna on the east to a shallow-marine brachiopod-bryozoan-echinoid fauna on the west. (3) The occurrence of chert is cyclic; that is, it appears repeatedly in the section. Observation (3) suggests that influxes of siliceous fresh waters were periodic.

Chert of the Franciscan Complex, California

The Franciscan Complex forms the basement rock of much of the Coast Ranges from southern Oregon to Baja California (E. H. Bailey, Irwin, and Jones, 1964; Berkland et al., 1972).[21] The complex consists of a diverse assemblage of rocks and rock bodies, including submarine fan facies rocks, olistostromal melanges (in part, submarine debris flows), chert-shale sequences, ophiolite fragments, tectonic melanges, and the metamorphosed equivalents of all of these. Radiolarian chert occurs in locally thick sections that are interbedded with siliceous shale, interbedded with lithic wacke, overlying ophi-

olite fragments, and in minor occurrences as interbeds with and as nodules in pelagic limestone (figure 22.7, figure 22.2b) (E. F. Davis, 1918; Taliaferro, 1943; E. H. Bailey, Irwin, and Jones, 1964; Raymond, 1973a, 1974a). The thickest sections of chert range in age from early Jurassic to late Cretaceous (Murchey, 1984; Karl, 1984; K. Yamamoto, 1987). Typically the cherts are folded (figure 22.8).

Like many Mesozoic cherts, the most abundant Franciscan cherts are rhythmically bedded and contain interlayers of siliceous shale. The beds are typically 1 to 5 cm thick, but thinner and thicker beds are present. Bedding types include thin single-layer beds, triple-layer beds, graded beds, breccia beds, and massive beds. Many cherts are red in color, but

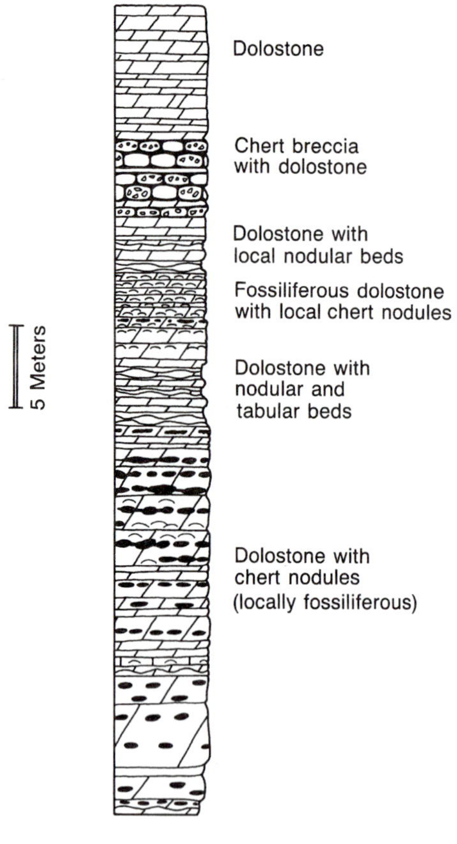

Dolostone

Chert breccia
with dolostone

Dolostone with
local nodular beds

Fossiliferous dolostone
with local chert nodules

Dolostone with
nodular and
tabular beds

Dolostone with
chert nodules
(locally fossiliferous)

5 Meters

Figure 22.5   Measured section showing the various lithologies in the upper 50 m of section in the Permian Kaibab Formation along the Hermit Trail, west of Grand Canyon Village, Arizona.
(*Unpublished data courtesy of F. K. McKinney*)

Figure 22.6   Nodular chert in the Kaibab Formation, Permian, about 500 m west of Grand Canyon Village, Arizona.

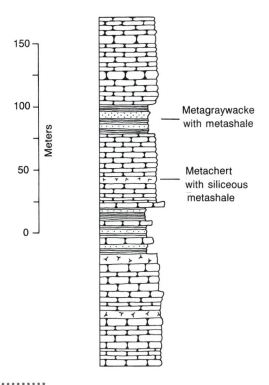

green, gray, black, pink, yellow, orange, and white cherts also occur. The white cherts are veined, commonly thick to massively bedded, and coarser grained than the red and green cherts, suggesting extensive recrystallization.

The dominant mineral of the cherts is quartz. Although opal-CT was reported by J. R. Hein, Koski, and Yeh (1987) and J. R. Hein and Koski (1987), its presence is questioned by Huebner and Flohr (1990), who sought, but did not find it in any of their samples collected from the same locality. Additional minerals include hematite, a wide variety of manganese minerals (Huebner and Flohr, 1990), calcite, feldspar, chlorite, and clay, as well as Prehnite-Pumpellyite and Blueschist Facies metamorphic minerals, such as acmite, pumpellyite, crossite, and aragonite. Many cherts are radiolarian-rich (figure 22.9).

The origin of the Franciscan cherts has been a matter of interest and debate for nearly a century. E. F. Davis (1918) summarized earlier work, demonstrated through laboratory experiment that rhythmically bedded chert-shale sequences could result from diagenetic changes, and suggested that the Franciscan cherts were precipitated from siliceous hot springs

Figure 22.7 Measured section showing interbedded chert, shale, and wacke sandstone (all metamorphosed), Falcon Formation, Franciscan Complex, Jurassic-Cretaceous, northeastern Diablo Range, California. (Modified from Raymond, 1973a)

Figure 22.8 Folded, interlayered, very thin-bedded to thin-bedded chert and siliceous shale; Franciscan Complex, Jurassic-Cretaceous, near Goat Rock, Sonoma County coast, California.

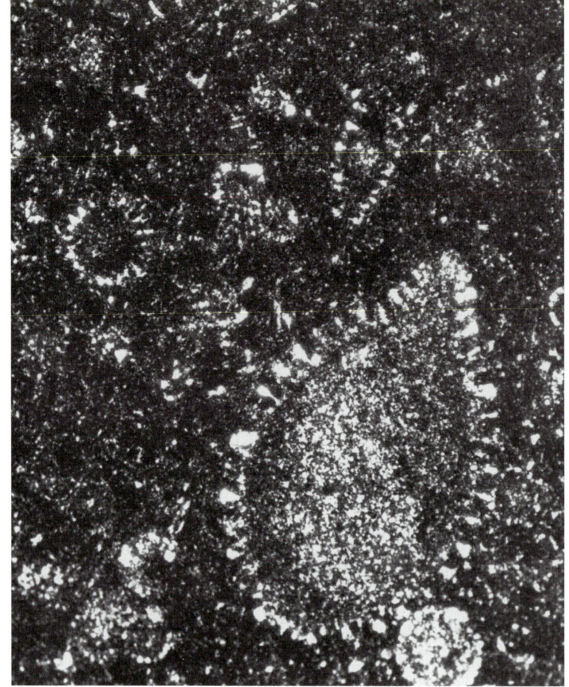

Figure 22.9 Photomicrograph showing radiolaria in red, bedded radiolarian chert from the Franciscan Complex, Jurassic, near The Geysers, Sonoma County, California. (XN.) Long dimension of photo is 1.27 mm.
(See Raymond and Berkland, 1973; McLaughlin and Ohlin, 1984)

associated with mafic volcanic rocks.[22] Taliaferro (1943), E. H. Bailey, Irwin, and Jones (1964), and Granau (1965) also emphasized the association of cherts and mafic volcanic rocks, but Taliaferro (1943) and E. H. Bailey, Irwin, and Jones (1964) suggested an origin for the Franciscan cherts involving interaction of seawater with volcanic rocks or magma. E. H. Bailey, Irwin, and Jones (1964) detailed this *inorganic precipitation— diagenetic alteration hypothesis*, in which magma-seawater interaction produces heated seawater saturated with silica. In this hypothesis, upon rising, cooling, and mixing with other seawater, the heated seawater becomes supersaturated and precipitates a silica gel that settles on the seafloor. The abundance of

silica in the water column allows radiolaria to multiply and these accumulate as incidental bioclastic components of the sediment. In this model, rhythmic layering is produced during diagenesis, when opal-A converts to quartz.

The inorganic precipitation–diagenetic alteration hypothesis is favored by (1) the common association of chert and volcanic rocks in the Franciscan Complex, (2) the experimental evidence for diagenetic development of rhythmic layering, produced by E. F. Davis (1918), and (3) some chemical evidence from manganiferous cherts that suggests a hydrothermal source for various elements in the manganiferous cherts (Crerar et al., 1982; Chyi et al., 1984). It is also consistent with (1) the common occurrence of single-layer chert beds, (2) the presence of triple-layer chert beds, and (3) the available data on silica solubilities. The environment of deposition for the Franciscan rocks, including the cherts, suggested by E. H. Bailey, Irwin, and Jones (in a 1964, pre–plate tectonics discussion), is that of a large-scale, continental margin, ensimatic basin.

A second hypothesis of chert formation that has been applied to Franciscan cherts is that the chert represents diagenetically altered, biogenic siliceous sediment (Ransome, 1894; Chipping, 1971; Thurston, 1972).[23] In this *biogenic sedimentation–diagenesis hypothesis*, the chert layers are considered to be composed principally of recrystallized radiolarian tests, diatoms, and/or sponge spicules, and the intervening shale layers consist of terrigenous material, including volcanic ash, mixed with varying proportions of biogenic siliceous and authigenic materials. Areas of upwelling and high silica-organism productivity at high and low latitudes, in part protected from clastic sedimentation, provide sites for accumulation of the large quantities of biogenic siliceous sediment necessary to yield thick sections of chert. In this model, diagenesis alters opal-A to quartz, coarsens the chert textures, and may concentrate silica in the chert layers.

Versions of the biogenic sedimentation–diagenesis hypothesis for Franciscan cherts are advocated by Chipping (1971), K. J. Hsu (1971), Raymond (1974a), Karl (1984), and Murchey (1984), who draw comparisons with Mesozoic to contemporary analogs from deep-sea drilling data (DSDP).[24] Chipping (1971) proposed that the association of chert with volcanic rocks is a topographic association rather than a genetic one. The cherts, he suggests, develop on the volcanic rocks because the latter form high areas protected from the diluting effects of turbidity currents and other clastic sedimentation processes. Murchey (1984) suggests that only large distances from sediment sources are needed to produce the same effect. Neither Chipping (1971) nor Raymond (1973, 1974) address the origins of the rhythmic layering in any detail, but Chipping's remarks seem to suggest that he considers that episodes of clastic sedimentation intervened during periods of more or less constant radiolarian production, producing the alternating beds. Murchey (1984) suggests that the layering is original, but was enhanced by diagenesis. Chipping (1971), Raymond (1974a), Karl (1984), and Murchey (1984) all favor open-ocean depositional sites, and Murchey and Karl argue that the cherts were deposited beneath a low latitude zone of high productivity. The models of B. M. Page (1970), K. J. Hsu (1971), and Murchey (1984) suggest that clastic rocks may have been added when the abyssal ocean floor, with its underlying mafic volcanic rocks, approached the subducting western margin of North America.

A biogenic model is favored by the abundance of radiolarian tests in less recrystallized cherts and by thick sections of chert that lack any volcanic or terrigenous interlayers. The biogenic sedimentation–diagenesis hypothesis is consistent with the occurrence of triple-layer beds, the presence of locally interlayered turbidite sandstones (fig. 22.7), the frequent association of chert and mafic volcanic rocks, and the presence of oxidized manganese and iron in the cherts—a condition characteristic of open ocean, but not continental margin, sediments (Karl, 1984). The model is also compatible with Ce anomaly and REE data that indicate deposition on oceanic crust near and progressively more distant from an oceanic spreading center (R. W. Murray et al., 1990, 1991).

A third model of chert formation, in which the cherts represent resedimented allochemical accumulations, is proposed by J. R. Hein and Karl (1983), J. R. Hein, Koski, and Yeh (1987), and J. R. Hein and Koski (1987, 1988). They too suggest that the silica is biogenic, rather than inorganic, but they argue that the textures, lithlogic associations, and origins of the cherts differ from those that characterize ocean floor, radiolarian, siliceous oozes of contemporary equatorial regions. They argue that the sediments were deposited in an incipient ocean basin, back-arc basin, forearc basin, or rifted continental margin basin, a site closely linked to an adjoining continent. In this third model, the chert is the product of (1) upwelling-driven, high radiolarian productivity and deposition, (2) cyclic redeposition of biogenic siliceous sediment by bottom and turbidity currents, and (3) deposition of hemipelagic clay, which forms the intervening siliceous shale layers. In short, rhythmic bedding is produced by mechanisms of rapid, periodic current redeposition of radiolarian sediment during periods of constant mud deposition. Hein and Karl (1983) suggest that the depositional sites are anoxic. Subsequent diagenesis results in the present character of the chert.

This *biogenic resedimentation* model is favored primarily by the occurrence of radiolarian turbidites containing Bouma-like sequences (Raymond, unpublished data) and graded beds (J. R. Hein, Koski, and Yeh, 1987; Murchey, 1984; Chipping, 1971), *provided that* the graded beds are depositional features.[25] The hypothesis is consistent with the observation that, locally, the cherts are interbedded with turbidite sandstone and shale in sections deficient in volcanic rocks (Raymond, 1974a; 1988; J. R. Hein and Koski, 1987; J. R. Hein, Koski, and Yeh, 1987).

A variation of the biogenic resedimentation model, developed in conformity with a model of manganese ore genesis, has been suggested by Huebner, Flohr, and Jones (1986a), Huebner, Flohr, and Matzko (1986b), and Huebner and Flohr (1990). They argue that the Mn-bearing, chert-shale sequences represent chemically precipitated mud and gel formed near the sediment-seawater interface, as a result of an interaction between Mn-bearing hydrothermal fluids and biogenic silica deposited as radiolarian turbidites. Huebner, Flohr, and Matzko (1986b) suggest that the mudrocks are chemical precipitates and that the compositional layering may be climatically controlled. Sedimentation occurred in the open ocean, perhaps near a spreading center, where hydrothermal activity would serve as the source of Mn and Fe in the water column (Crerar et al., 1982). Evidence supporting this *biogenic-resedimentation, hydrothermal, gel-precipitate hypothesis* is provided by (1) relic, gel-like materials preserved in Mn-ore-bearing sections of chert and (2) the presence of radiolarian turbidites containing Bouma-like sequences and graded beds.[26]

Considering that the Franciscan Complex is a very large unit and is known to contain a diverse array of rocks deposited in a variety of environments (E. H. Bailey, Irwin, and Jones, 1964; Wachs and Hein, 1975; R. W. Murray et al., 1990), it is conceivable that several models of Franciscan chert formation are valid (Murchey and Jones, 1984). The inorganic precipitation–diagenetic alteration hypothesis has been discredited for most Franciscan cherts, because the cherts have a major biogenic component and at least some lack any chemical evidence of a volcanic contribution (Chyi et al., 1984; Karl, 1984; Murchey, 1984). The biogenic resedimentation models may be valid locally, where chert turbidites can be documented. As general models, however, they fail because they require (1) a reducing or anoxic environment of sedimentation for the cherts (J. R. Hein and Karl, 1983), (2) a clastic component in most cherts, and (3) dominance of chert turbidites, contourites, and bottom current deposits among Franciscan cherts—none of which are indicated by the compositions and textures of most Franciscan cherts. The gel-precipitate variant of this model, although interesting, is, as yet, only supported by data from one unusual locality. Thus, in spite of the arguments to the contrary by Jenkyns and Winterer (1982), J. R. Hein and Karl (1983), and J. R. Hein and Koski (1987),[27] the biogenic sedimentation–diagenesis model, which is supported by several lines of evidence, seems to be the best general petrogenetic model currently available for most Franciscan cherts.

EVAPORITES AND RELATED ROCKS

Mineralogy, Petrography, and Structures of Evaporites

Evaporites may contain common minerals, but many are composed of uncommon to rare chlorides, sulfates, and borates. Table 22.2 lists the names and formulas of several evaporite minerals. The most common minerals, in bold print, include the *carbonates* calcite, dolomite, and magnesite; the *chlorides* halite, sylvite, and carnallite; and the *sulfates* anhydrite, gypsum, polyhalite, kainite, and langbeinite (F. H. Stewart, 1963; Borchert and Muir, 1964).

There is no widely used classification of evaporite and related rocks. In general, rock names are based solely on mineral composition. Textural terms are not used. Shrock (1948) referred to all rocks composed of saline (salt) minerals as sali-

Table 22.2   Selected Evaporite and Associated Minerals and Their Formulas

Carbonates and Bicarbonates

Nahcolite	$NaHCO_3$
Aragonite[1]	$CaCO_3$
Calcite	$CaCO_3$
Magnesite	$MgCO_3$
Dolomite	$CaMg(CO_3)_2$
Ankerite	$(Ca,Mg,Fe)CO_3$
Trona	$NaCO_3(HCO_3) \cdot 2H_2O$
Pirssonite	$CaCO_3 \cdot Na_2CO_3 \cdot 2H_2O$
Dawsonite	$Na_2AlCO_3(OH)_2$

Chlorides

Sylvite	KCl
Halite	$NaCl$
Bischoffite	$MgCl_2 \cdot 6H_2O$
Carnallite	$KMgCl_3 \cdot 6H_2O$
Tachyhydrite	$CaMg_2Cl_6 \cdot 12H_2O$

Sulfates

Picromerite	$K_2SO_4 \cdot 6H_2O$
Thenardite	Na_2SO_4
Mirabilite	$Na_2SO_4 \cdot 10H_2O$
Glauberite	$Na_2SO_4 \cdot CaSO_4$
Anhydrite	$CaSO_4$
Gypsum	$CaSO_4 \cdot 2H_2O$
Kieserite	$MgSO_4 \cdot H_2O$
Hexahydrite	$MgSO_4 \cdot 6H_2O$
Epsomite	$MgSO_4 \cdot 7H_2O$
Celestite	$SrSO_4$
Aphthitalite	$K_3Na(SO_4)_2$
Glauberite	$Na_2Ca(SO_4)_2$
Bloedite	$Na_2Mg(SO_4)_2$
Schoenite	$K_2Mg(SO_4)_2$
Langbeinite	$K_2Mg_2(SO_4)_2$
Polyhalite	$K_2MgCa_2(SO_4)_4 \cdot 2H_2O$
Kainite	$KMg(SO_4)Cl \cdot 3H_2O$

Table 22.2 Continued

Borates

Kernite	$Na_2B_4O_7 \cdot 4H_2O$
Tincalconite	$Na_2B_4O_7 \cdot 5H_2O$
Borax	$Na_2B_4O_7 \cdot 10H_2O$
Colemanite	$Ca_2B_6O_{11} \cdot 5H_2O$
Ulexite	$NaCaB_5O_9 \cdot 8H_2O$

Others (including those with combinations of anions)

Burkeite	$2Na_2SO_4 \cdot Na_2CO_3$
Galeite	$Na_2SO_4 \cdot Na(F,Cl)$
Hanksite	$9Na_2SO_4 \cdot 2Na_2CO_3 \cdot KCl$
Northupite	$Na_2CO_3 \cdot MgCO_3 \cdot NaCl$
Teepleite	$Na_2B_2O_4 \cdot 2NaCl \cdot 4H_2O$
Pyrite	FeS_2
Realgar	AsS
Orpiment	As_2S_3

Associated Silicates of Authigenic Origin

Quartz	SiO_2
Adularia	$KAlSi_3O_8$
Albite	$NaAlSi_3O_8$
Analcite	$NaAlSi_2O_6 \cdot H_2O$
Searlesite	$NaBSi_2O_6 \cdot H_2O$
Magadiite	$NaSi_7O_{13}(OH)_3 \cdot 3H_2O$
Phillipsite	$KCaAl_3Si_5O_{16} \cdot 6H_2O$
Heulandite	$CaAl_2Si_6O_{16} \cdot 5H_2O$
Illite	$KAl_4Fe_4Mg_{10}(Si,Al)_8O_{20}(OH)_4$
Smectites	$(K,Na,Ca,Mg)_{0.33}Al_2Si_4O_{10}(OH)_2 \cdot nH_2O$

Sources: Many sources, including Gale (1915), J. E. Adams (1944), Scruton, (1953), Murdoch and Webb (1956, 1960), G. I. Smith (1962, 1979), F. H. Stewart (1963), Borchert and Muir (1964), G. I. Smith and Haines (1964), V. Morgan and Erd (1969), Hosterman and Dyni (1972), Roehler (1972), Dyni (1976), Holser (1979), G. I. Smith et al. (1983), Donahoe and Liou (1984), Sheppard and Gude (1986), and observations by the author. For lists of additional minerals see Borchert and Muir (1964). Sonnenfeld (1984), and Perthuisot (1989).

[1]Major evaporite minerals are shown in boldface.

nastones. The term has not been widely adopted by geologists but is useful as a general term to designate sedimentary rocks composed of halides, sulfates, borates, and associated minerals. In common practice, halite-rich rocks are called "rock salt" and most other salinastones are named by the dominant mineral name (e.g., anhydrite or gypsum). Some geologists have used terms such as anhydrock, andhydrite rock, gypsite,

gyprock, and gypsum rock. Raymond (1984c) appends mineral names to the root name evaporite. Thus, where the evaporite origin is known, the genetic rock names are names such as anhydrite evaporite, halite evaporite, ulexite evaporite, or anhydrite-gypsum evaporite. Where salt-rich rocks *do not form as a result of evaporation*, the equivalent names are anhydrite salinite, anhydrite-gypsum salinite, or halite salinite. These conventions are adopted here.

Evaporites and salinites may have microcrystalline-sutured texture or hypidiomorphic-granular texture, but many other textures occur as well. Spherulitic texture, comb texture, porphyroblastic texture, poikilotopic texture, allotriomorphic-granular texture, clastic texture,[28] and various replacement textures, are among those found in evaporites and salinites (see Lowenstein and Spencer, 1990). Chevron and "hopper" textures in halite and halite aggregates (figure 22.10) and pseudomorphs after halite are particularly distinctive of an evaporite or salinite origin.

The structures of evaporite and salinite sequences are diverse (Lucia, 1972; Lowenstein and Hardie, 1985; J. K. Warren and Kendall, 1985).[29] These include in supratidal and subaerial sequences irregular to laminated bedding, ripple cross-laminations, stromatolites, gypsum and anhydrite nodules, enterolithic anhydrite beds,[30] burrows, and flat pebble conglomerates and breccias. Tepees form along salina margins.[31] Intertidal zones are characterized by burrowed, bioturbated, and pelletal, irregular to flaser bedding; burrows; and stromatolites. Saline lagoonal, sabkha, and offshore deposits are typically well-laminated, bedded units; but irregular to flaser bedding, comb-textured crystalline beds, cross-laminated beds, and chevron-textured beds are also common. In sabkhas, anhydrite diapirs form in gypsum muds.

Origins of Evaporites and Related Rocks

Evaporites and related rocks may be grouped into two main chemical groups—a $MgSO_4$-rich group and a $MgSO_4$-poor group (Hardie, 1990). The former is represented by mineral suites with phases such as gypsum, anhydrite, halite, and kainite, whereas the latter is characterized by minerals such as halite, sylvite, and carnallite. Large deposits of the former are derived from marine brines, whereas deposits of the latter involve "continental brines" (Sonnenfeld, 1989; Hardie, 1990). Small-scale occurrences of evaporite minerals in veins, cavities, and even atoll cliff roofs[32] are also known. The veins and cavity fillings are precipitated from various groundwater, diagenetic, or hydrothermal solutions.

The major occurrences of salinastones form by the crystallization of salts from concentrated, hydrous solutions (brines). A number of factors control the resulting mineral assemblages, textures, and structures. These include the climate, the hydrographic conditions (e.g., currents and density stratification of brines and waters), the chemistry of contributing solutions, and the basin geometry (Logan, 1987; Lowenstein, Spencer, and Pengxi, 1989; Sonnenfeld, 1989; Hardie, 1990). Evaporites form in arid regions, such as

"Hopper" textured crystals   Chevron texture

Figure 22.10   Sketches of chevron and hopper textures in halite evaporite (rock salt).

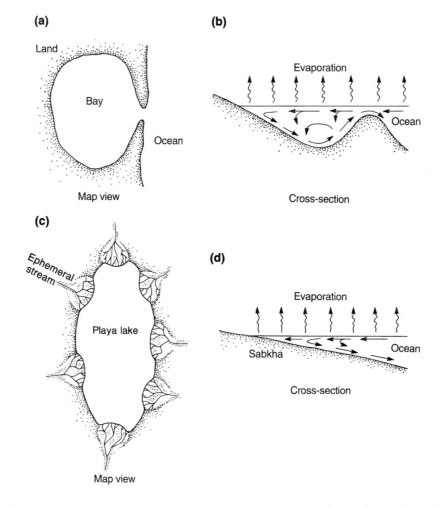

Figure 22.11   Sketches of evaporite depositional settings. (a) A basinal bay isolated by a constricted entrance from the open ocean (cf. Sonnenfeld, 1989, figs. 13–16). (b) Cross section of an isolated (barred) basin with a topographic obstruction at the basin entrance (cf. Schmalz, 1969). (c) Map view of a landlocked, arid basin with a playa lake (cf. Kendall, 1984, fig. 1). (d) Cross section of a sabkha along an arid coastline (vertical scale is exaggerated to allow depiction of water and currents).

in the playas (salinas) of desert basins, in sabkhas along hot coastal regions, and in extremely restricted to isolated marine basins (marine salinas) along dry coastlines. The main condition for evaporite formation is that there be a *water-balance deficit*; that is, losses due to evaporation must exceed, at least periodically, gains from inflow (Logan, 1987). Nonevaporite salinastones may crystallize from brines developed in conti-

nental to marine basins, in polar region groundwaters and lakes, and in the pores of subsurface rocks and sediments, especially in supratidal environments.[33]

Several basin models have been suggested for various evaporite deposits (figure 22.11).[34] No single model can explain all occurrences, but all models meet the requirement that water loss exceed inflow. This condition develops where the entrance to a basin is topographically or geographically restricted (figure 22.11a and b) or where evaporation exceeds water inflow due to a combination of topographic and climatological or hydrographic conditions (figure 22.11c and d). In the latter case, (e.g., along shallow arid shelves), evaporation may exceed inflow from the open ocean, due to high evaporation rates. Likewise, where prolonged evaporation follows or accompanies precipitation events in or around closed, arid, continental basins, evaporation may similarly exceed inflow. Whatever the basin shape, less saline, lower-density water that enters the basin will begin to spread across the top of the resident water mass. Because it is exposed at the surface, it will undergo evaporation and consequent density increase. With salinity and density increases, the waters precipitate evaporite minerals and, in barred basins, sink to levels that preclude their escape from the basin. In models involving sabkhas or closed continental basins, high evaporation rates alone maintain the high salinity of water from which evaporite minerals precipitate. In sabkhas, evaporation occurs away from the open ocean. In continental basins, evaporation occurs at a distance from the mouths of streams that contribute less saline water to the basin. Where marine waters feed evaporite basins, $MgSO_4$-rich evaporite mineral sequences develop. In contrast, continental brines that yield potash evaporites apparently form where hydrothermal fluids containing $CaCl_2$ mix with basin lake waters to produce a brine that yields $MgSO_4$-poor evaporite mineral sequences (Lowenstein, Spencer, and Pengxi, 1989; Hardie, 1990).

Examples

Evaporites of the Silurian Salina Group, Michigan Basin

The upper Silurian Salina Group forms a major stratigraphic unit in the Michigan Basin and adjoining regions of the United States and Canada (figure 22.12).[35] The unit *overlies* a carbonate bank, reef, shelf, and basin sequence of the Niagara Group, which contains dolomitized stomotoporoid-algal reefs ("pinnacle reefs") scattered along the shelf zone (Huh, Briggs, and Gill, 1977). As many as twelve units, some formally named and others informally designated, make up the Salina Group (figure 22.13). These include a basal (A0) carbonate unit of middle Silurian age, overlain successively by the A1 evaporite, the A1 carbonate or Ruff Formation, the A2 evaporite, the A2 carbonate, and a series of salt, mudrock, and carbonate units (units B to G, and the Bass Islands Formation) (Alling and Briggs, 1961; Budros and Briggs, 1977; Huh, Briggs, and Gill, 1977).

At the base, overlying the pinnacle reefs, a supratidal, algal stromatolitic limestone forms the basal carbonate (A0) unit. In the inter-reefal and basinal areas, locally laminated, micritic limestone that grades upward into enterolithic

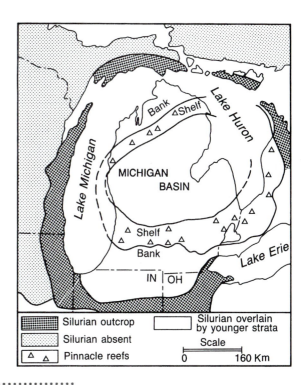

Figure 22.12 Map of the Michigan Basin area, showing the distribution of Silurian outcrops containing the Salina Group and the subsurface position of bank, shelf, and "pinnacle reefs."
(Modified from Huh, Briggs, and Gill, 1977)

	SERIES	GROUP	
Upper Silurian	Cayugan	Salina	Bass Islands Formation
			G Unit
			F Salt
			E Unit
			D Salt
			C Shale
			B Unit / B Salt
			A2 Carbonate
			A2 Evaporite
			Ruff Formation
			A1 Evaporite
			A0 Carbonate
Middle Silurian	Niagaran	Niagara	

Figure 22.13
Stratigraphic subdivision of the Salina Group.
(Modified from Budros and Briggs, 1977)

457

anhydrite comprises the basal A0 unit. The A1 evaporite is characterized by nodular and enterolithic anhydrite evaporite with halite evaporite and sylvinite evaporite, but it also includes micritic dolostone. The Ruff Formation is composed primarily of unfossiliferous carbonate mudstones that range from limestone to dolostone. Nodular anhydrite evaporite also occurs in the A2 evaporite, which is overlain by another dolomitic carbonate unit, the A2 carbonate. The latter units are succeeded by locally thick halite evaporite with minor dolostone (units B and D), shale and dolostone with anhydrite evaporite (units C, E, F, and the Bass Island Formation), and dolostone with red and green shale (unit G). The latter units, together, comprise the upper part of the Salina Group. The sequence of units has a generally cyclic character, with dolostone- and shale-richer sections alternating with evaporite-rich sections.

Detailed analysis of the various units reveals that they consist of several lithofacies (Dellwig and Evans, 1969; Budros and Briggs, 1977; D. Gill, 1977; Nurmi and Friedman, 1977). For example, in the *nodular anhydrite lithofacies* of the A1 evaporite, among the most common rock types are "laminated massive" anhydrite, consisting of alternating laminae of anhydrite nodules, carbonaceous material, and micritic dolostone; and "distorted mosaic" anhydrite, composed of closely packed, coalescing, distorted nodules. These two rock types are interbedded on a centimeter scale with less abundant rock types, such as enterolithic anhydrite. Some small chalcedonic chert nodules up to 6 mm in diameter occur in the anhydrite (Gill, 1977). Dessication cracks and brecciated horizons with laminated dolostone micrite clasts and an anhydrite matrix also occur. A second lithofacies, a *gypsum mold–laminated anhydrite lithofacies*, consists of alternating layers of laminated anhydrite and gypsum molds infilled with halite and anhydrite (Nurmi and Friedman, 1977). The *salt* or *halite lithofacies* consists of layered halite evaporite, with typical beds 10 to 50 cm thick, and thin laminae and interbeds of gypsum, anhydrite, and dolostone (Dellwig and Evans, 1969; Nurmi and Friedman, 1977). Locally, the halite evaporites exhibit ripple marks (Kaufmann and Slawson, 1950). Additional lithofacies include microlaminated carbonate mudstone lithofacies, pelletal wackestone lithofacies, laminated algal (stromatolitic) lithofacies, intraformational breccia and flat-pebble conglomerate lithofacies, peloid-oolite lithofacies, and sylvinite lithofacies.

The interbedded rock types and their structures suggest that the various lithofacies represent subtidal, intertidal, and supratidal mudflat (sabkha) environments of deposition (Nurmi and Friedman, 1977; Gill, 1977). Particularly important in indicating a shallow-water to subaerial origin for these evaporites are the oolites and stromatolites.[36]

During the late Silurian Period, the Michigan Basin apparently harbored an arm of the sea. High evaporation rates and few inlets from the larger, surrounding epicontinental sea caused sea level to drop and the Michigan Basin sea to become hypersaline (Cercone, 1988).[37] As a consequence, evaporites formed. Subaerial exposure resulted in the development of flat pebble conglomerates and facilitated the development of the evaporites. Periodic influx of large amounts of marine water accounts for the cyclic nature of the sequence (Mesollela et al., 1974; Bay, 1983). In addition, subsurface flow from the adjoining open sea may have facilitated dolomitization of the lime muds and pseudomorphic replacement of gypsum by halite or anhydrite.

The Green River Formation of Colorado, Utah, and Wyoming

Study of the Green River Formation provides some insight into the difficulty of determining the paleoenvironmental conditions under which some salinastones and related rocks are formed.[38] There is no exact contemporary analogue for these rocks, although W. H. Bradley (1966) finds similarities in some Floridian and African lakes.

Recall that the Green River Formation, famous for its "oil shale," is exposed in six structural basins in southwestern Wyoming, northeastern Colorado, and northern Utah (see ch. 18). The rocks of these structural basins formed during the Eocene epoch in three depositional basins that were rimmed by alluvial fans and fluvial floodplains. The basins were occupied by lakes—Lake Gosiute, Lake Uinta, and a small, unnamed lake near the western Wyoming–northeastern Utah border (figure 22.14) (W. H. Bradley, 1931, 1948, 1963, 1964). Rivers from adjacent uplifted mountain ranges fed the lakes and transported sediment to the lake basins.[39] These Eocene lakes, like all lakes, were ephemeral, but the larger lakes lasted at least 13 million years and accumulated sediments that account for stratigraphic sections that locally exceed 5000 m (Picard, 1963; R. C. Johnson and Nuccio, 1984).

The lacustrine nature of the sedimentary rocks of the Green River Formation is indicated by (1) the included fossils, (2) the laterally extensive and continuous nature of beds and laminations in the rocks, (3) the minerals of the rocks, and (4) the paleogeographic and stratigraphic relationships between beds of the Green River Formation and those of adjacent fluvial facies.[40] Fluvial sediments were transported far into the lake basin at times, as indicated by extensive tongues of fluvial sandstones that project into the lacustrine section from the basin margin (W. H. Bradley, 1964; Cashion, 1967; Roehler, 1991). As a consequence of the paleogeography, the Green River Formation forms a set of lenses of lacustrine beds enclosed by fluviatile strata of the Wasatch, Uinta, and Bridger formations (figure 22.15). The

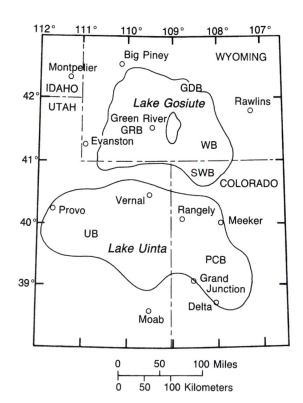

Figure 22.14 Map of the Colorado, Utah, Wyoming area showing the location of the two large Eocene lakes, Lake Gosiute and Lake Uinta, in which the Green River Formation was deposited. The locations of some present-day structural basins are indicated by letter symbols as follows: GDB = Great Divide Basin (≈Bridger Basin), GRB = Green River Basin, PCB = Piceance Creek Basin, SWB = Sand Wash Basin, UB = Uinta Basin, WB = Washakie Basin
(Source: J. R. Dyni, "The origin of oil shale and associated minerals" in U S Geological Survey PP 1310, 1987.)

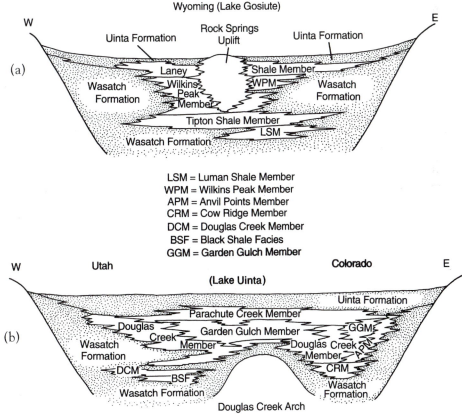

LSM = Luman Shale Member
WPM = Wilkins Peak Member
APM = Anvil Points Member
CRM = Cow Ridge Member
DCM = Douglas Creek Member
BSF = Black Shale Facies
GGM = Garden Gulch Member

Figure 22.15 Schematic cross sections showing relationships between various members of the Green River Formation.
(a) Stratigraphy formed in Lake Gosiute, Wyoming (based on W. H. Bradley, 1964; with modifications based on the data of W. H. Bradley and Eugster, 1969; Surdam and Wolfbauer, 1975; and Surdam and Stanley, 1980a). (b) Stratigraphy of Lake Uinta, Colorado and Utah.
(Based on data from Picard, 1955; Cashion, 1967; Cashion and Donnell, 1974; Johnson, 1984; R. D. Cole, 1985; and MacLachlan, 1987)

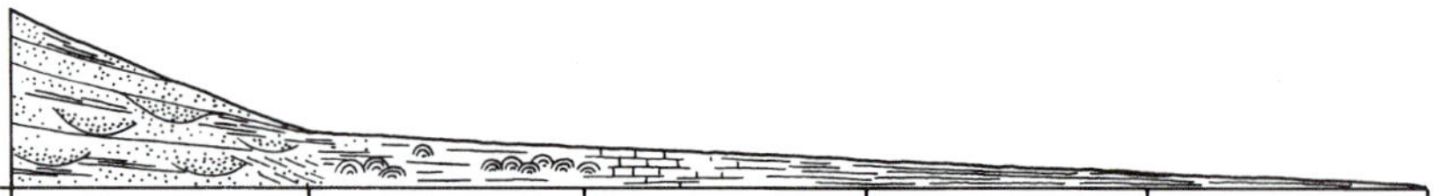

Environment	Alluvial fan/ Fluvial/ Deltaic	Carbonate mudflat Swamp	Nearshore lacustrine	Offshore lacustrine	Basin center
Features	Cross beds Channels Gilbert-type deltas Sandstone and shale	Mudcracks Stromatolites Carbonate conglomerate Oolitic carbonate Ripple marks laminations cross laminae Mudrock (minor)	Laminated carbonate and mudrock Minor "oil shale"	Laminated "oil shale" and argillaceous carbonate rock	"Oil shale" and salinastones (Laminated and bedded)
Water Salinity Eh	Fresh Oxidizing	Fresh to saline Oxidizing	Saline Mildly reducing	Saline Moderately to strongly reducing	Hypersaline Strongly reducing

Figure 22.16 Schematic illustration of the various depositional environments of the Green River and adjacent formations, showing some characteristics of each environment.
(Based on many sources, especially Dyni, 1976; Cole and Picard, 1978; and Roehler, 1990)

lakes that produced the Green River Formation fluctuated in size.[41] The environments represented by Green River and associated rocks include fluvial deltaic, mudflat/swamp/lake margin, strandline, near-shore lacustrine, and offshore lacustrine types (figure 22.16).

The dominant and distinctive rocks of the Green River Formation, the so-called "oil shales," although lithologically diverse, are predominantly kerogen-rich carbonate rocks. Both limestones and dolostones are abundant. These and other rocks are distinctively laminated and bedded (figure 22.17). In addition to the carbonate rocks, there are various salinastones, with minerals such as gypsum, anhydrite, halite, and trona. Rocks with these minerals, including some "oil shales," "trona beds," and "halite beds," contain interbeds of siltstone, shale, tuff, and sandstone.[42] The rocks are typically brown, gray, or green, but they weather to various colors, including red, orange, and yellow, from oxidation of iron-bearing minerals such as pyrite.

The mineral compositions of the various rocks of the Green River Formation are diverse (W. H. Bradley, 1948; Milton and Eugster, 1959; W. H. Bradley and Eugster, 1969).[43] Dolomite, calcite, sodium-bearing carbonates and bicarbonates (e.g., shortite, trona, dawsonite, and nahcolite), quartz, and feldspars are the most abundant minerals. Halite, gypsum, various clay minerals, and zeolites are less common but locally abundant. In addition, sodium pyroxene (acmite) and amphibole (magnesioriebeckite) occur, as do the organic phases gilsonite, utahite, and uintahite. Sodium-bearing carbonates and bicarbonates, plus halite and gypsum, in association with dolomite and calcite, are good indicators of extremely saline conditions.

Figure 22.17 Photomicrograph of laminated "oil shale" of the Green River Formation, Eocene, Piceance Creek Basin, Garfield County, Colorado. (PL). Long dimension of photo is 3.25 mm.

The various sedimentary rocks of the Green River Formation contain a wide range of structures. Marginal facies, those that grade into the surrounding fluvial deposits, exhibit typical fluvial structures, especially cross-bedding and ripple marks. Deltas, indicated by fine-grained lacustrine and bottomset beds overlain by coarsening upward sequences of foreset and topset sandstone beds, are well represented in the Laney Member (K. O. Stanley and Surdam, 1978). Fluvial or deltaic marginal facies are replaced locally by mudflat facies with interbedded sequences of stromatolitic lime mudstone, grainstones containing algal-encrusted logs, and mudcracked mudstone and carbonate beds (W. H. Bradley, 1964; Eugster and Hardie, 1975; Surdam and Wolfbauer, 1975). Flat pebble conglomerates are interpreted by Eugster and Hardie (1975) to reveal lacustrine transgression over mudflat environments.

The marginal facies grade laterally into lacustrine facies. These lake beds vary lithologically but have distinctive rhythmic sequences including

marlstone–mudstone,
marlstone–oil shale,
dolostone–oil shale–trona salinastone,
dolostone–oil shale–claystone,
ostracodal grainstone–oil shale–dolomitic mudstone, and
algal boundstone–oil shale–dolomitic mudstone.[44]

In detail, marlstone–oil shale consists of alternating laminae of kerogen-bearing dolomite and analcite + quartz mudrock, but local pyritic oil shale contains dolomite interlaminated with dawsonite + quartz + dolomite mudrock (Brobst and Tucker, 1973). Beds of trona and halite associated with oil shale, plus shortite- and eitelite-bearing oil shale (W. H. Bradley, 1964; Dyni, 1974; Dyni, Milton, and Cashion, 1985), clearly indicate that salinastones formed with oil shale.

In addition to the distinctive rocks and minerals, the Green River Formation contains an important lacustrine fauna and terrestrial and lacustrine flora represented by fossils (MacGinitie, 1969; Grande, 1980). Fossil fish, including catfish; crocodile, bat, and bird fossils; numerous insects; mollusks; ostracods; microfossils of algae and fungi; and pollens from spruce, pine, and various angiosperms are known from the shales (D. E. Winchester, 1923; W. H. Bradley, 1931; Bucheim and Surdam, 1977). All of these organisms indicate a continental environment and several specifically indicate a freshwater lake.

The origins of the Green River Formation salinastones, including the oil shales, are controversial. The fossils, laminations, minerals, and facies clearly indicate a lacustrine origin. Most workers agree that sand and mud were fed into the lakes by rivers draining the surrounding highland areas. The fluvial and near-shore mudrocks are predominantly products of that detrital deposition. Yet, while virtually all workers agree that the offshore sediments were deposited in lakes, the nature of the lakes, the nature and distribution of lake waters, and the causes and processes of deposition of oil shale, trona beds, and associated rocks are matters of considerable debate. The debate is complicated by the fact that Lakes Gosiute and Uinta had different histories (Picard, 1985). Thus, the history and conditions in one lake do not define those of the other lake.

Three lacustrine models have been proposed as frameworks for explaining the development of the Green River Formation. The first is that of a *permanently stratified (meromictic) lake*, probably greater than 25 m deep, in which a layer of oxygen-deficient, saline bottom water, that is an **hypolimnion,** occupies the deeper, central parts of the lake

basin (W. H. Bradley, 1948; W. H. Bradley and Eugster, 1969; Desborough, 1978; Picard, 1985).[45] A less dense, freshwater **epilimnion** overlying the saline water provides both a locale for some carbonate precipitation and a habitat for aerobic organisms. The oil shale results primarily from deposition of algal and fungal materials generated in the photic (lighted) zone (the epilimnion), with additions of (1) minor airborne silt and clay, (2) minor suspended clay added by fluvial input, and (3) lake-bottom precipitates, including trona, halite, gypsum, and clay minerals. The algal and fungal materials are generally agreed to be the sources of the kerogen and other organic hydrocarbons (W. H. Bradley, 1973), and the algal material may have served as a source of high-Mg carbonate (Desborough, 1978).

Evidence offered in support of the meromictic lake theory (and in opposition to the playa lake theory, described below) includes the following.

1. Stratigraphic evidence suggests that Lake Uinta, one of the ancient lakes, was at least 300 m deep (R. C. Johnson, 1981).
2. The laminated beds of "oil shale" are continuous over significant distances (Desborough, 1978).
3. Extensive oil shale turbidites suggest a deeper, less transient lake (Dyni and Hawkins, 1981).
4. Turbidites in the Green River Formation favor a deeper lake model (Dyni and Hawkins, 1981).
5. Micrite laminae alternate with organic-rich laminae that grade into micrite—a feature of contemporary, meromictic lake sediments (B. W. Boyer, 1982).
6. Few disconformities exist; yet if the lake periodically dried out, they would be expected between halite or trona beds and overlying lake beds (W. H. Bradley and Eugster, 1969).
7. Higher Mg contents occur in lake-center sediments, as indicated by the abundance of dolomite (and ankerite) there; and they do not occur in lake margin sediments, where calcite exhibits its greatest abundance (R. D. Cole and Picard, 1978).
8. There is a general lack of oxidized (trivalent) iron-bearing minerals in the oil shale. Instead, pyrite, a phase indicating reducing conditions, is abundant in these rocks (Desborough, 1978; Dyni and Hawkins, 1981).
9. There is an absence of thicker limestone and dolomite sequences along the lake margin, as might be expected if these minerals were initially precipitated there, as predicted by the playa lake model (Desborough, 1978).
10. A meromictic lake is necessary for the growth of the large volumes of algae indicated by the volumes of oil shale (W. H. Bradley and Eugster, 1969).
11. The Mg-concentrating ability of blue-green algae, which can thrive in saline waters, means that algal growth provides a mechanism by which both high Mg and high kerogen can be generated to produce an oil shale sediment (Desborough, 1978).

461

12. There is a general lack of benthic organisms in the oil shale, even of those that might be tolerant of high salinity (Boyer, 1982).

13. The oil shales lack bioturbation (Boyer, 1982), but bioturbation is expected in bottom sediments under oxygenated waters in which bottom-dwelling organisms churn the sediment.

14. Whole catfish skeletons suggest an absence of scavengers, a characteristic of anaerobic bottom waters (Boyer, 1982).

A second model for the origin of the Green River rocks is the *playa lake* model (W. H. Bradley, 1973; Eugster and Surdam, 1973).[46] In the playa lake theory, the lake is interpreted to have been aerobic, broad, and shallow, and to have experienced repeated periods of evaporation and desiccation. Large playa flats surround the lake, separating it from basin-margin fluvial regimes. The playa flats provide a site in which alkaline brines evolved in the capillary zone above the groundwater table via a process involving evaporation of water and precipitation of calcite and protodolomite. Periodic floods wash calcite and dolomite into the lake, and the fresh water dilutes the salinity of the lake, providing an environment for algal blooms. Thus, the oil shale is a combined product of deposition of organic material generated in the lake and detrital carbonate derived from the playa flats. During dry periods, the lake shrinks and the flow of saline brines downslope towards the lake is accompanied by their depletion in calcium, brought about by the precipitation of calcite and protodolomite. By the time brines reach and enter the lake, they are enriched in Na. As the waters evaporate, precipitation of trona, halite, and other evaporite minerals occurs.

Evidence offered in support of the playa lake theory includes the following:

1. Sedimentary structures and dessication features indicating alternate wet and dry and extremely shallow water conditions are widely distributed in the Green River Formation. These include mudcracks, flat pebble conglomerates and breccias, flattened crests on ripple marks, saline mineral casts, and lenticular silt laminae (Eugster and Surdam, 1973; Eugster and Hardie, 1975; Lundell and Surdam, 1975; Surdam and Wolfbauer, 1975).

2. Ripple marks of the interference and oscillation types are widespread in Green River rocks (Lundell and Surdam, 1975).

3. Oolitic and pisolitic carbonate rocks are widely distributed (Lundell and Surdam, 1975).

4. Dolomite and quartz grains of similar size in carbonate silt laminae suggest transport of detrital dolomite by currents, rather than precipitation of dolomite as a cement (Eugster and Hardie, 1975).

5. Dolomite peloids, presumably derived from erosion of preexisting dolomitic mud, are present in some of the sedimentary rocks (Eugster and Hardie, 1975; Smoot, 1978).

6. Repetition of the dolomite–trona bed sequence within the lacustrine sedimentary section suggests fluctuating conditions (Eugster and Surdam, 1973).

7. Carbonate layers interpreted to be caliche, tufa, and surface dolomite crusts are present in the Green River Formation (Smoot, 1978).

8. High Mg/Ca ratios and high Mg content of oil shales is consistent with the saline lake theory (Eugster and Surdam, 1973).

9. There is a large volume of Na-salts and dolomitic mudstone in the Wilkins Peak Member, the member associated with a reduced lake size (Eugster and Surdam, 1973).

10. The presence of numerous and widely distributed catfish fossils in the oil shale units of the Laney Member (catfish are a bottom-dwelling fish) indicate that a meromictic Lake Gosiute was not required for the deposition of oil shale (Buchheim and Surdam, 1977).

11. Extraordinary preservation of delicate fossils suggests that they were deposited and protected in a flocculent algal ooze that experienced periodic drying (W. H. Bradley, 1973).

12. In the Tipton Member, there is a major decrease in the benthic fauna and a replacement of that fauna by algae, as indicated by algal structures such as stromatolites, suggesting a change to hypersaline conditions (Surdam and Wolfbauer, 1975).

A third model for the origin of Green River rocks is a hybrid *polygenetic lake* model, in which the lake histories include both playa lake and meromictic phases (Boyer, 1982).[47] In this model, playa phases producing desiccation features and evaporites are replaced periodically (in time) or laterally (in space) by meromictic conditions that yield oil shale.

Evidence favoring a polygenetic theory would include some of those data cited above for the playa lake and meromictic lake theories. In addition, the theory is supported by the following arguments.

1. The Wilkins Peak Member is interdigitated with extensive fluvial, clastic units of the Wasatch Formation, indicating significant freshwater influx during the time when Wilkins Peak rocks were deposited (minimum lake dimension) (Sullivan, 1985).

2. Evidence of subtropical climatic cycles suggests a variable climate and consequent variations in influx of fresh water (Sullivan, 1985).

3. Evaporite layers overlain by benthic fossil-free, microlaminated sediments indicate that the lake waters were saline and subsequently became stratified with saline bottom waters (Boyer, 1982).

Which of the theories best explains the data? The answer is debatable. In part, what is needed to resolve the controversy are detailed maps of the Green River Formation showing coeval sedimentary features formed at specific times.

Surdam and Wolfbauer (1975) have provided initial maps of this type. These data would provide a detailed paleogeography that would resolve questions related to lake size and facies relations. Presently available data, including that for mudcracks and flat pebble conglomerates and breccias, indicate that dessicated playa flats did develop in some areas. Yet other factors—the distribution of Mg in the lake sediments, the presence of articulated vertebrate fossils in laminated (nonbioturbated) beds, the apparent link between magnesium concentration and organic carbon formation by algae, the continuity and nature of "oil shale" laminae, and the paleotopography—suggest that the "oil shales," and at least some of the interlayered salinastones, were deposited in a meromictic lake. Thus, the evidence favors the existence of both playa and meromictic conditions over time and space, with meromictic conditions predominating. In short, the available evidence supports a polygenetic model for the origin of the Green River Formation.

IRONSTONES AND IRON-FORMATIONS

Iron and oxygen are among the most abundant elements in the Earth and, as such, might be expected to form iron-oxide-rich rocks. Much of the iron in the crust, however, is combined with other elements to form silicate minerals that are not generally abundant enough to constitute iron-rich rock types. Nevertheless, under certain conditions, iron oxides, iron silicates, iron sulfides, or combinations of these, do form iron-rich sedimentary rocks—the ironstones and the banded iron-formations.

Petrography and Structures of Ironstones and Iron-Formations

The mineralogy of the iron-rich sedimentary rocks is surprisingly diverse.[48] Undoubtedly the most common iron-bearing minerals in these rocks are magnetite and hematite, both of which occur with abundant quartz. These rocks also contain the iron oxides kenomagnetite, maghemite, goethite, and "limonite"; the iron carbonates siderite and ankerite; the iron sulfides pyrite and marcasite; and iron silicates, such as stilpnomelane and minnesotaite, plus chamosite, glauconite, and greenalite (iron-rich chlorite, mica, and serpentine, respectively). Associated minerals include feldspars, calcite, dolomite, clay minerals, white mica, chlorite, talc, graphite, and apatite. Where altered or metamorphosed, additional minerals such as fayalite, ortho- and clinopyroxenes, grunerite-cummingtonite, hornblende, actinolite, biotite, and garnets also occur (S. M. Richards, 1966; C. Klein, 1973; Dymek and Klein, 1988).[49]

The textures of iron-rich rocks are also varied. Like other chemical precipitates, the rocks may contain clastic components, including allochems, plus cement and matrix.

Allochems include oolites, pisolites, pellets, "granules," intraclasts, and fossils (Dimroth and Chauvel, 1973). Other clastic components consist of typical rounded to angular materials found in other sedimentary rocks, including quartz and feldspar grains, lithic fragments, and various minor minerals. Cements are present in comb textures,[50] radial fibrous textures, and poikilotopic textures. These components typically consist of microcrystalline quartz, chalcedony, calcite, siderite, iron silicates, and hematite. Matrix may also include hematite, but clay, quartz, and calcite occur too. The mud-size, iron-mineral cement and matrix materials have been called *femicrite*, and the silica cements, *matrix chert*.[51] Rocks with significant amounts of matrix have epiclastic textures, and some have breccia textures. In rocks with little matrix or clastic materials, equigranular-mosaic, equigranular-sutured, or spherulitic textures are present.[52] A very typical texture of ironstones and one that appears in some iron-formation is an epiclastic texture consisting of oolites in matrix or cement (figure 22.18a). In ironstones, the oolites, and commonly associated sand grains, are bound by hematitic clay matrix. In iron-formation, quartz cement is typical.

Structurally, iron-formation is so commonly laminated to thin-bedded that it is often called *banded iron-formation*, or BIF (figure 22.18c). Banded iron-formation is characteristically composed of chert beds and laminae that alternate with beds or laminae of siderite packstone, siderite grainstone, recrystallized sideritestone, hematitite, or magnetitite (LaBerge, 1964; Gole, 1981; Ewers and Morris, 1981).[53] Other structures that occur in iron-formation include stylolites, raft-structured chert beds, graded laminae, cross-beds, nodules, microfossils, and soft-sediment folds.[54] Ironstones, including glauconitic sandstones and siltstones, are typically laminated to bedded rocks that locally contain cross-beds and cross-laminations, ripple marks, and fossils, including trace fossils (figure 22.18d).

Origins of Ironstones and Iron-Formations

The origins of ironstones and iron-formations have been debated for many years.[55] A principal reason for the debate is that contemporary analogues for most types of iron-rich rock are unknown. Although glauconite-rich rocks are known to form in contemporary oceans (Galliher, 1935; Ehlmann, Hulings, and Glover, 1963), hematitic sediments of the types and volumes necessary to yield the major deposits of the geologic record are largely unknown. Two contemporary analogs of iron-rich sedimentation have been recognized in southeastern Asia and the Caribbean Sea (G. P. Allen, Laurier, and Thouvenin, 1979, p. 92ff.; Kimberly, 1989), and some iron-formations have chemical similarities with Red Sea and other basinal marine sediments (Barrett, Fralick, and Jarvis, 1988). The diversity of iron-rich rocks precludes the possibility that these analogs can be representative of all of iron-rich rocks.

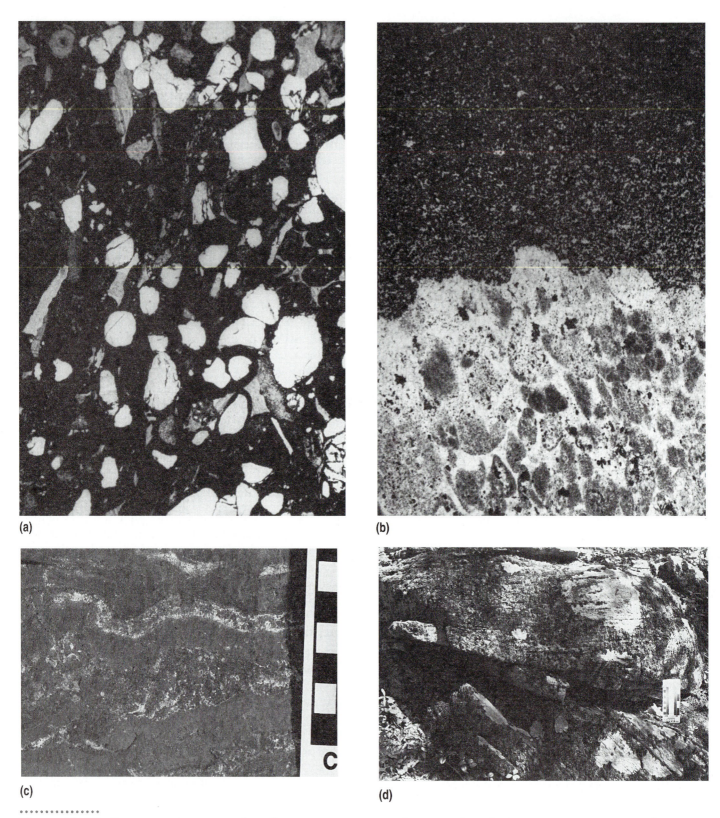

(a)

(b)

(c)

(d)

· · · · · · · · · · · · · · · · · ·
Figure 22.18 Textures and structures of iron-formations and ironstones. (a) Photomicrograph of oolitic ironstone of the Red Mountain Formation, Silurian, Highway 31, Birmingham, Alabama. (PL) (b) Photomicrograph of banded iron-formation; Riverton Formation, Precambrian, Florence County, Wisconsin. (PL) (c) Photo of bedded and laminated iron-formation, from the Riverton Formation, Precambrian, Florence County, Wisconsin. (d) Cross-bedded ironstones and ferruginous sandstone of the Rose Hill Formation, Silurian, Laurel Bed Lake area, Saltville Quadrangle, southwestern Virginia.

In discussing the origin of the iron-rich rocks, two major questions are pertinent: What was the source of the iron and what conditions and processes prevailed during deposition? In terms of the source, one view is that the iron was derived from volcanism, volcanic rocks, volcanic hot springs, or seafloor hydrothermal springs (Goodwin, 1956; G. A. Gross, 1980; Dymek and Klein, 1988). REE patterns and La, Eu, and Ce anomalies in some BIFs support this view.[56] The general absence of a one-to-one association of volcanic rocks and iron-rich sedimentary rocks has been used as an argument against it. Alternatively, it has been proposed that the iron may be derived from normal subaerial weathering and erosion (H. L. James, 1951; and Govett, 1966). Certainly, stream waters in tropical and subtropical regions carry adequate iron to account for major iron deposits (Gruner, 1922; H. L. James, 1954). Yet, because it is bound up and transported as colloidal ferric hydroxide, as a component in Fe-bearing silicate clastic particles, and as oxide surface films on clay particles, this iron is generally unavailable, and therefore, an unlikely source for large volumes of iron in sedimentary units being deposited over large regions.[57] A variation on the idea that the source of the iron is the weathering of surface rocks is central to the *supergene enrichment hypothesis,* which suggests that the original banded iron-formations were enriched in iron *after* they initially formed (Gair, 1975; R. C. Morris, 1980, 1987). In this model, weathering of surface rocks provides iron to solutions that penetrate downward, where they diagenetically or hydrothermally alter subsurface rocks, increasing their iron contents.

Many ironstones show evidence of replacement. Fossils, known to be composed of carbonate minerals, and calcareous oolites partially replaced by hematite, testify to the occurrence of this process. Burchard, Butts, and Eckel (1910) and Alling (1947) note these and several additional replacement textures in the "Clinton" ironstones of Alabama and New York, respectively.

Replacement textures have also been recognized in the iron-formations. For example, Lougheed (1983) describes a number of replacement textures, including replaced oolites, and magnetite pseudomorphs after gypsum, in Precambrian iron-formations of the Lake Superior region. He argued that the rocks were originally supratidal, intertidal, and subtidal *carbonate* sediments that were subsequently altered to iron-formations.

Transportation and deposition of iron is controlled by the Eh and pH of the hydrous solvent (figure 22.19). Solution of iron is favored by low pH (acidic conditions), whereas precipitation of hematite is favored by combined high pH and Eh (oxidizing to slightly reducing-basic conditions) (Garrels and Christ, 1965, ch. 7). Precipitation of magnetite and iron silicates is favored by low Eh and high pH (reducing-basic conditions), with magnetite seemingly unstable in the presence of fluids saturated in silica. The coexistence of magnetite and silica, which would seem to be mutually exclusive, is enigmatic.

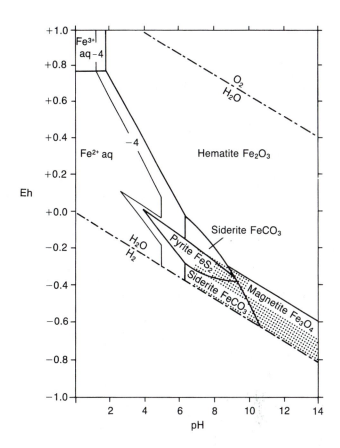

Figure 22.19 Eh-pH diagram showing the stability of iron-bearing minerals at one atmosphere pressure under conditions of high dissolved carbonate and very low reduced sulfur. Under conditions of very low dissolved carbonate and high silica, iron silicates may be stable in the region shown by the stipple pattern.
(From R. M. Garrels and C. L. Christ, Solutions, Minerals, and Equilibrium. © 1965 Boston: Jones and Bartlett Publishers. Reprinted by permission.)

Garrels (1987) summarizes the controversy relating to the origin of banded iron-formation and notes that a wide variety of initial environments of deposition have been proposed. The environments include playa lake (Eugster and Chou, 1973), supratidal to subtidal (Lougheed, 1983), shelf (Ewers and Morris, 1981), and marine basin (H. D. Holland, 1973) types. Garrels (1987) points out that the majority view is that these rocks form in marine environments of a restricted nature. Yet an analysis of geochemical and mass balance data leads him to advocate a model involving primary precipitation of iron minerals and chert layers, *as evaporites,* from stream waters that enter a restricted basin.

Clearly, the problems are not yet resolved. Replacement and alteration have clearly conditioned the *present state* of many iron-rich rocks, but the depositional environments and conditions are debatable. Some ironstone formations, such as the Rose Hill Formation of southwestern Virginia, contain marine fossils, oolites, and cross-beds that

465

clearly indicate a subtidal, open-marine environment of deposition for the primary sediments. Those sediments were clastic and at least part of the iron replaces or cements originally deposited materials. Modern-day ironstones are apparently forming in subtidal marine and interdistributary environments (G. P. Allen, Laurier, and Thouvenin, 1979, p. 92ff.; Kimberly, 1989). In contrast, both marine and hydrothermal waters clearly contributed to the formation of some BIFs (Beukes and Klein, 1990). For such units as the Biwabik Formation of the Great Lakes region, the environment of deposition, the original compositions of the sediments, and the processes involved in precipitation of iron are at least partially unresolved. What is clear is that there are different kinds of iron-rich rocks, and thus a single petrogenetic theory cannot explain the origin of all of them.

PHOSPHORITES

Phosphorites, like iron-formations, have been the subject of petrogenetic controversy. Phosphorous is nearly ubiquitous in rocks; it occurs in apatite in igneous and metamorphic rocks, and as organic skeletal, amorphous phosphate, and in apatite and other minerals in sediments and sedimentary rocks. In most rocks, however, the concentration of P_2O_5 is well below 1 wt. %. How then do sedimentary rocks form that have P_2O_5 in excess of 20%? In seeking an answer to that question, we first examine the mineral and textural evidence provided by the nature of phosphorites.

The Petrography of Phosphatic Rocks

Phosphorites occur as allochemical clastic rocks, phosphatic shales, nodular rocks, and diagenetically altered clastic rocks.[58] The allochemical forms consist of resedimented phosphatic materials that have been reworked by currents. The nodular, shaly, and diagenetically altered rocks result from primary and secondary precipitation of phosphate minerals from solutions. Oolites and pellets are the most common objects replaced by phosphorous minerals, and calcareous fossils are also replaced.

The principal phosphatic materials in most phosphorites are collophane, a cryptocrystalline form of apatite, and calcium fluorapatite, or *francolite* [$Ca_5(PO_2)_3F$] (Heinrich, 1956; Manheim and Gulbrandsen, 1979; Slansky, 1980). Other phosphate minerals are less common and most are rare. Among these are calcium hydroxyapatite or dahllite [$Ca_5(PO_2)_3(OH)_2$], metavariscite ($AlPO_4 \cdot 2H_2O$), and wavellite [$Al_3(OH)_3(PO_4)_2 \cdot 5H_2O$](D. McConnell, 1950; Heinrich, 1956).[59] Other precipitates in phosphorites include calcite, chalcedony, and quartz, which are particularly common as cements, and gypsum, anhydrite, and pyrite. Quartz, feldspar, micas, and clays are common detrital constituents of phosphorites.

Phosphorite textures are typically epiclastic-cryptocrystalline to epiclastic-microcrystalline (mud), epiclastic sand-size textures (packstone and wackestone types), epiclastic-conglomeratic to breccia types; or they are epiclastic oolitic, fossiliferous, or pelletal types in which phosphorous minerals have replaced allochems, fossils, or both.[60] In the epiclastic sand-size and coarser textures, collophane or other phosphatic materials may serve as cements. Collophane, francolite, and the nonphosphate minerals quartz and calcite are common as cements. Goethite may also occur as a cement (G. F. Birch, 1980). Clay and silt are typical matrix materials. Additional textures observed in some phosphorites include microlaminated mosaic and equigranular-mosaic textures (Cressman and Swanson, 1964).

The principal structures of phosphorites include beds, laminae, and nodules. The beds range from a few centimeters to tens of meters thick (P. J. Cook, 1984) and are commonly interlayered with black shale, chert, or carbonate rocks. Nodules are typically small, but may range to several tens of centimeters. Bones and other fossils are also major constituents of some phosphorites.

Origins of Phosphorites

The origin of phosphorites was particularly problematical before the discovery of phosphatic deposits on the ocean floor. A number of hypotheses were advanced in the nineteenth and early twentieth centuries, including the following: (1) phosphorites are the recrystallized accumulations of bones and other vertebrate remains; (2) phosphorites are marl nodules and beds replaced by phosphatic materials in solution, derived from organic remains and fecal materials; and (3) phosphorites are residual crusts or soils left after dissolution or alteration of phosphorous-bearing marls or other sedimentary rocks (G. S. Rogers, 1915). Clearly, both organic and inorganic origins were considered. Both may have been important locally.

The origin of phosphorites remains somewhat enigmatic. Modern assessments of phosphorite petrogenesis include a primary precipitation hypothesis, diagenetic replacement and precipitation hypotheses, and a mechanical concentration hypothesis. The model of *primary phosphorite precipitation* was proposed in preliminary form by Kazakov (1937, in Cressman and Swanson, 1964). He and others (McKelvey et al., 1959; Heckel, 1977) suggested that upwelling of phosphorous-bearing deep-ocean waters, at low latitudes in offshore areas, could provide the conditions for *inorganic* deposition of phosphorites. The shelf break and shelf are likely sites for such deposition. Contemporary phosphorite deposition *appears* to be occurring off the coasts of Africa and Peru, as suggested by the Recent radiometric sediment ages and the occurrence of phosphatic sediment at the sediment-water interface (Baturin Merkulova and Chalov, 1972; Manheim, Rowe, and Jipa, 1975).[61] Paleomagnetic data suggest that ancient phosphorites formed at similar sites.[62] In the contemporary lagoonal, estuarine, and shelf environments of deposition, whether phosphate precipitation is inorganic or organic seems to be arguable and there is increasing evidence that microbial action plays some (unknown) role in the precipitation process (P. J. Cook, 1976; Sheldon, 1980, 1987).[63] Precipitation is favored by slightly basic, reducing conditions (Krumbein and Garrels, 1952; Manheim, Rowe, and Jipa, 1975).

The development of phosphorite is a several-stage process (P. J. Cook, 1976; G. F. Birch, 1980; Sheldon, 1980, 1987). Phosphate is added to ocean water via weathering of rocks and fluvial transport to the sea. Additional phosphorous is contributed from volcanic sources. Planktonic organisms in the surface waters take up the phosphorous in their bodies, contribute some phosphorous to the deeper levels of the sea and to the seafloor as fecal pelletal material, and, upon dying, fall to the bottom, transferring additional phosphorous to the depths. Dissolved phosphorous is then transfered back to the surface through **upwelling,** a process of vertical circulation in which deep waters, moving as currents, flow up to the surface in equatorial areas and along coastlines. The upwelling waters are nutrient-rich and supply the means by which plankton can flourish, again concentrating phosphorous and transferring it, in abundance, to the shallow bottom as a primary biochemical precipitate. In some cases, direct fluvial transportation of phosphorus to shallow depositional sites may also occur (Glenn and Arthur, 1990).

On the shelf, the abundance of organic phosphatic material on the seafloor may create phosphorous-bearing water that passes downwards through the upper few centimeters of surficial sediment (Glenn and Arthur, 1988). These pore fluids facilitate precipitation of collophane as cement and phosphatic ooids and promote diagenetic replacement of carbonate materials, including ooids, pellets, and fossils. The end product is bedded to nodular phosphorite.

Reworking of phosphatic sediments by currents and waves produces second-generation, resedimented, *mechanically concentrated* phosphorite (Soudry, 1987; Glenn and Arthur, 1990). Lenses of phosphatic sand, mud, and conglomerate may develop in zones where transportation agents act along the shelf to concentrate previously formed phosphatic sediment.

The cherts, composed primarily of opal-A, opal-CT, or various forms of quartz, are bedded and laminated to nodular and podiform in structure. Over time, opal-A converts to opal-CT and then to quartz. The nodular cherts typically form through replacement of carbonate rocks. Bedded types have more diverse origins, including inorganic precipitation, biochemical precipitation and sedimentation, diagenetic segregation of layers, and resedimentation of bioclasts. The ribbon cherts may have formed via several of these processes.

The salinastones, including evaporites and salinites, form primarily as a result of crystallization from aqueous brines. Mineralogically, these rocks contain a wide array of unusual minerals, including carbonates, bicarbonates, chlorides, sulfates, and borates. Evaporites form in arid regions—in the playas of desert basins, on sabkhas along hot coastal regions, and in extremely restricted to isolated marine basins (marine salinas) along dry coastlines. Salinites crystallize from brines in arctic or alpine lakes, from the hypolimnion of merimictic (stratified) lakes, or from marine basinal brines. Apparently, high-$MgSO_4$ evaporites precipitate from marine brines, whereas low-$MgSO_4$ evaporites are derived from fluids generated by the mixing of $CaCl_2$-bearing hydrothermal fluids and lake or basin waters.

The ironstones and iron-formations are rocks rich in magnetite, hematite, or other iron-bearing minerals. Iron-formations are distinguished by their cherty character and commonly have a banded structure. Ironstones are commonly oolitic and may form where hematite diagenetically replaces and cements oolitic shoal sediments. The origin of iron-formations is more problematical, but trace element data suggest that mixing of marine and hydrothermal fluids may create the waters from which iron-rich sediments are deposited.

Phosphorites develop through the biochemical and inorganic precipitation of phosphorous in shallow marine environments. Upwelling marine waters transport phosphorous up from deeper levels of the sea onto the shelf or into coastal lagoons. The high concentrations of phosphate in many phosphorites may result from diagenetic concentration of phosphorous through cementation and replacement in the topmost layers of the shelf or lagoonal sediments. Alternatively, some phosphorites show evidence of physical concentration of clastic phosphate by waves or currents.

SUMMARY

Noncarbonate precipitated rocks include cherts, salinastones, ironstones and iron-formations, and phosphorites. Both biochemical and inorganic precipitation contribute to the formation of cherts and phosphorites. In contrast, the salinastones and iron-bearing sediments are generally inorganic precipitates. Reworking of any of these precipitated materials will yield epiclastic rocks of like composition.

EXPLANATORY NOTES

1. For definitions of Eh and pH, see chapter 16 or Krauskopf (1967, p. 34 for pH, p. 243 for Eh). Also see Garrels and Christ (1965) and Krumbein and Garrels (1952).
2. *Salinite* is a term proposed here to refer to nonevaporite saline rocks, as contrasted with evaporites. Both salinites and evaporites are *salinastones*, as defined by Shrock (1948), which are rocks composed of saline minerals of unspecified origin. Salinastone is a nongenetic term, whereas evaporite is genetic.

3. More information on the rocks of these particular occurrences is available in Bellamy (1900), Gale (1915), Hicks (1916), Eardley (1938), G. I. Smith (1962), G. Evans et al. (1964), G. I. Smith and Haines (1964), Holser (1966), Kirkland, Bradbury, and Dean (1966), J. R. Butler (1969), C. Kendall and Skipwith (1969), Kinsman (1969), Pettijohn (1975, ch. 11), G. I. Smith (1979), G. I. Smith et al. (1983), J. K. Warren and Kendall (1985), and the references cited therein. Other evaporite sequences and environments are described in the following works: *Permian Castile and Salado Formations of Texas and New Mexico*—R. H. King (1942, 1947), J. E. Adams (1944), R. Y. Anderson et al. (1972), W. E. Dean (1975), Presley (1987), Lowenstein (1988); *Permian Zechstein Formation of the North Sea Basin*— Brunstrom and Walmsley (1969); *middle Paleozoic formations of Alberta, Saskatchewan, Montana, and North Dakota*— Andrichuck (1965), J. G. C. M. Fuller and Porter (1969); *MacLeod Evaporite Basin of Australia*—Klingspor (1969), J. A. Peterson and Hite (1969), Bosellini and Hardie (1973), G. R. Davies and Nassichuk (1975), Cita et al. (1985a,b), Corselli and Aghib (1987), Hovorka (1987), Logan (1987); *western China*—Presley (1987), Lowenstein (1988), Lowenstein, Spencer, and Pengxi, (1989); Southgate et al. (1989), and Lowenstein and Spencer (1990).

4. Iron-formations of the Great Lakes region are described by many workers including Leith (1903), Leith et al. (1935), H. L. James (1951), Goodwin (1956), LaBerge (1964), Lepp (1966), G. B. Morey et al. (1972), Bayley and James (1973), E. C. Perry, Tan, and Morey (1973), Lougheed (1983), and Gair (1975). Similar rocks in the Labrador Trough are described by Dimroth and Chauvel (1973).

5. Ironstones of the Appalachians are described in Alling (1947), B. N. Cooper (1961), Simpson (1965), R. E. Hunter (1970), and Thomas and Bearce (1986).

6. Selected references to the occurrences mentioned include the following: *Franciscan cherts*—E. F. Davis (1918a), Taliaferro (1943), E. H. Bailey, Irwin, and Jones (1964), R. W. Murray et al. (1990, 1991); *Arkansas novaculite*—W. D. Keller, Stone, and Hoersh (1985), J. Zimmerman and Ford (1988); *Gunflint and Biwabik cherts*—Leith (1903), Goodwin (1956), W. J. Perry et al. (1973), Floran and Papike (1975), J. D. Miller, Morey, and Weiblen (1987); *Monterey Formation*—Bramlette (1946), Garrison and Douglas (1981), L. A. Williams and Graham (1982); *the Fort Payne Chert*—Hurst (1953), Cramer (1986), Hasson (1986); *Caballeros Novaculite*—McBride and Thompson (1970), E. F. McBride (1988).

7. For the distribution of ocean-floor siliceous sediments, see W. H. Berger (1974), T. A. Davies and Gorsline (1976), and Leinen et al. (1986).

8. For discussions of Precambrian iron-formation, see Trendall and Morris (1983), Garrels (1987), R. C. Morris (1987), Dymek and Klein (1988), Beukes and Klein (1990), and the discussion later in this chapter.

9. Calvert (1971), Jones and Segnit (1971), R. Greenwood (1973), Kastner et al. (1977), Kastner (1979), Hesse (1988), Maliva and Siever (1988).

10. Chalcedony occurs in two forms, an optically length-fast form and an optically length-slow form, commonly referred to as *quartzine*. The occurrence of quartzine was at first considered to be indicative of cherts of evaporitic origin (Folk and Pittman, 1971), but is now known to occur elsewhere. Nevertheless, quartzine does seem to indicate diagenetic crystallization in sulfate- and Mg-rich environments (Keene, 1983).

11. For example, see H. Williams et al. (1954, 1982), Heinrich (1956), Dietrich, Hobbs, and Lowry (1963), Calvert (1971), Siedlecka (1972), Chowns and Elkins (1974), Floran and Papike (1975), McBride and Folk (1977), Steinitz (1977); Folk and McBride (1978), Crerar et al. (1982), Jenkyns and Winterer (1982), M. Earle (1983), Huebner et al. (1986a), J. R. Hein, Koski, and Yeh (1987), and Pollock (1987).

12. Additional information on silica behavior and solubility may be found in Siever (1957), Davis (1964), G. W. Morey, Fournier, and Rowe (1964), Aston (1983), L. A. Williams, Parks, and Crerar (1985), Maliva and Siever (1988), Isshiki, Sohrin, and Nakayama (1991), and the review of Hesse (1988).

13. Also see Ernst and Calvert (1969), Lancelot (1973), Folk and McBride (1978), Kastner (1979), Reich and von Rad (1979), L. A. Williams and Crerar (1985), Maliva and Siever (1988), and the review of Hesse (1988).

14. Also see Eugster (1967), McKee (1969), and Surdam, Eugster, and Mariner (1972).

15. Hesse (1988) and L. A. Williams and Crerar (1985) provide reviews.

16. Hydrothermal activity at mid-ocean ridges is described in Wolery and Sleep (1976), Corliss et al. (1979), and in Rona et al. (1983). Chemical evidence supporting a hydrothermal origin or component for some cherts and cherty Mn ores is provided by Crerar et al. (1982), Chyi et al. (1984), and K. Yamamoto (1987). Chyi et al. (1984) do suggest that some silica is initially of biogenic origin and that some components are detrital.

17. See Weis and Wasserburg (1987) for isotopic data consistent with this model.

18. Magadi-type cherts are named for Lake Magadi in Kenya, where magadiite, the precursor of this type of chert was first recognized by Eugster (1967). Also see B. F. Jones, Rettig, and Eugster (1967), Eugster and Surdam (1971), O'Neil and Hay (1973), Muraishi (1989), and Schubel and Simonsen (1990).

19. These problems are reviewed by Jenkyns and Winterer (1982), D. L. Jones and Murchey (1986), and J. R. Hein and Parrish (1987). Also see Karl (1984), Iijima, Matsumoto, and Tada (1985), and the various papers in Iijima, Matsumoto, and Utada (1983), including Iijima and Utada (1983).

20. Chipping (1971), Raymond (1974a), Garrison et al. (1975), Folk and McBride (1978), Crerar et al. (1982), Sugisaki, Yamamoto, and Adachi (1982), M. Earle (1983), Chyi et al. (1984), Karl (1984), Iijima, Matsumoto, and Tada (1985), Bustillo and Ruiz-Ortiz (1987), Hein and Koski (1987), Hein, Koski, and Yeh (1987), R. W. Murray et al. (1990), Sedlock and Isozaki (1990), T. Matsuda and Isozaki (1991).

21. This review is based primarily on the works of E. F. Davis (1918), Taliaferro (1943), E. H. Bailey, Irwin, and Jones (1964), Raymond (1973a, 1974), Karl (1984), Murchey (1984), D. L. Jones and Murchey (1986), Huebner and Flohr, (1990), R. W. Murray et al. (1990), and on the unpublished observations of the author.

22. A. C. Lawson (1895, pp. 423–426) and Soliman (1965) shared this view.

23. See additional and related discussions in Fairbanks (1895, p. 82), Calvert (1971), Garrison (1974), Raymond (1974a), Garrison et al. (1975), McBride and Folk (1979), Murchey (1984), and D. L. Jones and Murchey (1986).

24. For example see, M. N. A. Peterson et al. (1970), Pimm, Garrison, and Boyce (1971), Winterer and Riedel (1971), Winterer et al. (1971), Heath (1973), Lancelot (1973), Garrison et al. (1975), Riech and von Rad (1979), and Tucholke et al. (1979).

25. See Murchey (1984) for a discussion of the graded beds.

26. Chipping (1971), Murchey (1984), Hein, Koski, and Yeh (1987), Raymond (unpublished data).

27. These arguments are rebutted by Murchey (1984), who argues that chert sections equivalent to those in the Franciscan Complex have developed in the oceans (Tucholke et al., 1979).

28. Logan (1987, p. 28 & 31) discusses clastic gypsum and halite.

29. For additional descriptions of structures in evaporite sequences, see Dellwig and Evans (1969), J. G. C. M. Fuller and Porter (1969), Comite des Techniciens (1980), Collinson and Thompson (1982, Ch. 8), Shinn (1983), A. C. Kendall (1984), Logan (1987, p. 78ff.), and Lowenstein (1988).

30. Enterolithic beds are irregularly folded, intestine-like beds, produced through diagenetic expansion of bedding resulting from replacement.

31. Tepees are deformed sediment layers of mesoscopic scale (from a few centimeters wide and high to structures $1 m \times 1 m \times 10 + m$ or larger) shaped in cross section like a Native American tepee. The layers result from compression of a layer towards a fracture in such a way as to produce two opposing, concave-up sediment layers that come to a point like a tepee. See J. E. Adams and Frenzel (1950), D. B. Smith (1974), Assereto and Kendall (1977), P. L. H. Worley (1979), and C. Kendall and Warren (1987) for descriptions or examples.

32. See Braithwaite and Whitton (1987).

33. Craig et al. (1974), A. C. Kendall (1984), Sonnenfeld and Perthuisot (1989).

34. See Woolnough (1937), Scruton (1953), Schmalz (1969), and the reviews in Logan (1987) and Sonnenfeld (1989).

35. This review is based on the works of Kaufmann and Slawson (1950), Dellwig (1955), Alling and Briggs (1961), Dellwig and Evans (1969), Mesolella et al. (1974, 1975), Budros and Briggs (1977), D. Gill (1975, 1977), Huh, Briggs, and Gill (1977), Nurmi and Friedman (1977), and Cercone (1988).

36. In spite of the abundance of anhydrite in the Salina Group, Hardie (1990) suggests a hydrothermal brine source for the rocks. The principal evidence supporting this interpretation is the composition of fluids within inclusions in halite, which chemically are significantly different from modern seawater. The presence of sylvite with halite in the center of the basin is also cited as evidence.

37. As noted by Cercone (1988), some workers have also suggested that the salinastones of the Michigan Basin are salinites formed in a stratified water body (e. g., as described by Sloss, 1969), not evaporites resulting from an evaporative drawdown of the Michigan Basin sea (see Droste and Shaver, 1977). However, the evidence for periodic subaerial exposure and very shallow water deposition seems compelling.

38. Papers that provided background and information for this summary include D. E. Winchester (1923), W. H. Bradley (1931, 1948, 1963, 1964, 1970, 1973, 1974), Picard (1955, 1985), Milton and Eugster (1959), Culbertson (1966), Cashion (1967), Bradley and Eugster (1969), Hosterman and Dyni (1972), Picard and High (1972b), Roehler (1972, 1974, 1990), Brobst and Tucker (1973), Eugster and Surdam (1973, 1974), Cashion and Donnell (1974), Desborough and Pitman (1974), D. C. Duncan et al. (1974), Dyni (1974, 1976, 1987), D. B. Smith (1974), Wolfbauer and Surdam (1974), Eugster and Hardie (1975), Lundell and Surdam (1975), Surdam and Wolfbauer (1975), Buchheim and Surdam (1977), Mauger (1977), R. D. Cole and Picard (1978), Desborough (1978), Smoot (1978), K. O. Stanley and Surdam (1978), Surdam and Stanley (1979, 1980a,b), Kornegay and Surdam (1980), Moncure and Surdam (1980), Dyni and Hawkins (1981), B. W. Boyer (1982), Sullivan (1985), R. C. Johnson (1981, 1984), R. C. Johnson and Nuccio (1984), R. D. Cole (1985), Dyni, Milton, and Cashion (1985), Baer (1987), and MacLachlan (1987). For a proposal for a revision of the stratigraphic nomenclature, see Roehler (1991).

39. For example, see W. H. Bradley (1926, 1963, 1964), Roehler (1972, 1974), Picard and High (1972), and Surdam and Stanley (1980a).

40. W. H. Bradley (1926, 1948, 1964, 1973), Cashion (1967), W. H. Bradley and Eugster (1969), Picard and High (1972), Eugster and Surdam (1973), Roehler (1974), D. B. Smith (1974), Desborough and Pitman (1974), D. C. Duncan et al. (1974), Lundell and Surdam (1975), Eugster and Hardie (1975), Buchheim and Surdam (1977), Desborough (1978), Surdam and Stanley (1980a), B. W. Boyer (1982), Picard (1985), and Dyni (1987).

41. For example, see W. H. Bradley (1964), W. H. Bradley and Eugster (1969), Eugster and Surdam (1973), Roehler (1974), Eugster and Hardie (1975), Surdam and Wolfbauer (1975), Surdam and Stanley (1979), Surdam and Stanley (1980a), R. D. Cole (1985), and Picard (1985).

42. W. H. Bradley (1948, 1964), Picard (1955, 1985), Culbertson (1966), Cashion (1967), W. H. Bradley and Eugster (1969), Picard and High (1972), Brobst and Tucker (1973), Dyni (1974), Roehler (1974), Eugster and Hardie (1975), and Lundell and Surdam (1975).

43. Also see Culbertson (1966), Roehler (1972), Hosterman and Dyni (1972), Brobst and Tucker (1973), Desborough and Pitman (1974), Dyni (1974, 1976), Wolfbauer and Surdam (1974), Milton (1977), and R. D. Cole and Picard (1978).

44. W. H. Bradley (1964), Picard and High (1972), Brobst and Tucker (1973), Surdam and Wolfbauer (1975), Dyni (1976), Surdam and Stanley (1979), R. D. Cole (1985).

45. This model is also supported by D. B. Smith (1974), Desborough (1978), Dyni and Hawkins (1981), R. C. Johnson (1981), and R. D. Cole (1985). Bradley and Eugster, who originally advocated this model, switched to the playa lake model. Picard (1985) provides a summary of the controversy between the meromictic lake and playa lake models.

46. The playa lake model is also advocated in Wolfbauer and Surdam (1974), Lundell and Surdam (1975), Surdam and Wolfbauer (1975), and Eugster and Hardie (1975).

47. The data and arguments of B. W. Boyer (1982) and Sullivan (1985) support such a model. Picard (1985), though favoring the meromictic lake model, summarizes observations consistent with a lake history involving both meromictic lake and playa lake aspects.

48. See H. L. James (1951, 1954, 1966), Heinrich (1956), G. B. Morey et al. (1972), Floran and Papike (1975), Gair (1975), A. E. Adams, MacKenzie, and Guilford (1984), and E. M. Morris (1987).

49. Metamorphic mineralogy is also discussed by Lepp (1972), G. B. Morey et al. (1972) B. M. French (1973), Floran and Papike (1975, 1978), and Gole (1981).

50. Dimroth and Chauvel (1973) refer to this as "columnar impingement texture."

51. Dimroth (1968), Dimroth and Chauvel (1973).

52. For descriptions and illustrations of various textures, see Alling (1947), H. L. James (1951, 1954), Dimroth and Chauvel (1973), R. W. Bayley and James (1973), Lougheed (1983), Dymek and Klein (1988), C. Klein and Beukes (1989), and Buekes and Klein (1990).

53. Also see H. L. James (1954), Alexandrov (1973), R. W. Bayley and James (1973), Trendall (1973a,b), Gair (1975), and Lougheed (1983). The terms sideritestone, magnetitite, and hematitite as used here are defined as follows. *Sideritestone* is a sedimentary rock, the major component of which is siderite. The term is equivalent to limestone and dolostone. Specific types of sideritestone include siderite mudstone, siderite wackestone, siderite packstone, and siderite grainstone. *Hematitite* is any rock that has hematite as its principal mineral (most abundant and > 33% or > 50%). *Magnetitite* refers to any rock in which magnetite is the principal mineral (most abundant and > 33% or > 50%). Sideritestone, magnetitite, and hematitite may form iron-formation *if* they have a silica cement or contain chert laminae.

54. Goodwin (1956), LaBerge (1967), Eugster and Chou (1973), Beukes and Klein (1990).

55. The treatment of this subject here is brief. For additional discussions, see Pettijohn (1975, p. 420ff.), Trendall and Morris (1983), Boggs (1987, p. 94ff.), Garrels (1987), and E. M. Morris (1987)(the principal works on which this summary is based), the references therein, and more recent works such as Barrett et al. (1988), Dymek and Klein (1988), and Beukes and Klein (1990).

56. For example, see Barrett, Fralick, and Jarvis (1988), Dymek and Klein (1988), and Beukes and Klein (1990).

57. Contrast the views of H. L. James (1966) and G. P. Allen, Laurier, and Thouvenin (1979, p. 97) on this subject.

58. For example, see Heinrich (1956, p. 147), McKelvey et al. (1956, 1959), Cressman and Swanson (1964), Pettijohn (1975, p. 427ff.), Heckel (1977), and Riggs (1979a,b). R. P. Sheldon (1987) gives a bibliography of phosphatic rock occurrences.

59. These and other phosphate minerals are listed, described, or discussed in D. McConnell (1950), Heinrich (1956), Manheim and Gulbrandsen (1979), Slansky (1980), and Nriagu (1984).

60. McKelvey et al. (1959), Cressman and Swanson (1964), R. J. Parker and Siesser (1972) G. F. Birch (1980), Slansky (1980), Adams, MacKenzie, and Guilford (1984), Soudry (1987).

61. Also see Burnett and Veeh (1977), A. O. Fuller (1979), G. F. Birch (1980), Burnett, Veeh, and Soutar (1980), Baturin, Merkulova, and Chalov (1972), Burnett et al. (1988), and P. N. Froelich et al. (1988).

62. P. J. Cook and McElhinney (1979).

63. G. F. Birch (1980), Porter and Robbins (1981), Soudry (1987), P. N. Froelich et al. (1988), V. P. Rao and Nair (1988).

PROBLEMS

22.1 (a) Recalculate the chemistry of the porcellanite (sample 5, table 22.1) on a volatile-free basis (subtract LOI from the total, assume that total equals 100%, and recalculate each of the other percentages as fractions of that total). (b) Compare the analysis to analyses 2 and 3. How does it differ? What mineralogical factors might explain the major chemical difference? Considering that the Monterey Formation, from which sample 5 was taken, is a continental margin unit, whereas the Franciscan cherts were probably deposited on the ocean-basin floor, what is a possible explanation for this mineralogical difference?

22.2 Consider an evaporite system. (a) If gypsum has a density of 2.32, what would be the mass of a bed of gypsum evaporite that is 10 cm thick and covers 5 km^2? (b) If this gypsum was formed by evaporation of a marine brine with the ionic equivalent 1.7 kg of gypsum per m^3, what would have been the original volume of brine (V_o) that produced the bed of gypsum? (c) The volume reduction ratio (V_{er}) for evaporite systems is given by the expression: $V_{er} = (V_o - V_e)/V_o$, where V_o is the original volume of fluid and V_e is the volume lost to evaporation (Logan, 1987). If the average V_{er} for gypsum precipitation is 0.15, what volume of brine evaporated to yield the gypsum bed?

Metamorphic Rocks

Metamorphic rocks characterize mountain belts and pervade the continental shields, marking the roots of long-eroded mountain belts. In the mantle, they are ubiquitous. These rocks form through transformations of pre-existing igneous, sedimentary, and metamorphic rocks brought on by changes in the prevailing intensive variables and fluids. The processes of transformation are not generally visible at the Earth 's surface and must be reproduced in the lab. In nature, the transformation is not always complete. That incompleteness is a fortunate circumstance in that it allows the petrologist to see beyond the most recent events and into the past history of the rock.

PART IV

Swirled patterns in metamorphic rocks, Indian Peaks Wilderness Area, Colorado.

23

Metamorphism and Metamorphic Rock Textures and Structures

INTRODUCTION

Banded and swirled patterns typify the metamorphic rocks (see page 471). These are rocks that have undergone a change (*meta-*) in their form (morph) from preexisting igneous, sedimentary, or metamorphic progenitors. Metamorphic rocks, which are common in the cores of mountain ranges, bear the imprints of heat, pressure, chemically active fluids, and deformation. Like the igneous and sedimentary rocks from which they are derived, metamorphic rocks have histories reflected in their textures and mineral compositions.

The purposes of this chapter are to define metamorphism, to discuss the agents that cause it, and to describe the structures, textures, and mineral compositions of metamorphic rocks. Through these features, the conditions of metamorphism are recognized. Yet, if the history of the metamorphic rock is to be fully understood, one must also understand the chemistry of the rock and recognize any minerals, structure, and texture remaining from the premetamorphic parent rock or **protolith** (literally, first stone). Together, all of these data enable petrologists to decipher metamorphic rock histories, which may be complex and incompletely recorded.

DEFINITIONS: METAMORPHISM AND METAMORPHIC ROCKS

Metamorphism is a process or set of processes that affect rocks in such a way as to produce textural changes, mineralogical changes, or both, under conditions in the Earth between those of diagenesis and weathering (at the lower limit) and melting (at the upper limit).[1] Processes of textural change that may occur *without accompanying mineralogical change* are of two types: recrystallization and cataclasis. **Cataclasis** is the crushing and breaking of grains in rocks. It may affect any type of rock. **Recrystallization,** however, is a process of reorganization of crystal lattices and intergrain relationships through ion migration and lattice deformation, without accompanying breaking of grains. Recrystallization occurs most commonly in monomineralic rocks, such as pure limestone, quartz arenite, or dunite. It may also occur where a directed stress acts on a rock under conditions of pressure (P), temperature (T), and composition (X) for which existing minerals in the rock are stable. **Neocrystallization** is the process that results in the formation of new minerals that did not previously exist in the metamorphic rock. Recall that equivalent processes occur during diagenesis. Thus, metamorphism is similar to diagenesis, but encompasses only those processes that occur *beyond* the near-surface (low-*P*), low-temperature, and low-stress limits of diagenesis.

Metamorphic rocks are rocks with textures, minerals, or both, that reflect cataclasis, recrystallization, or neocrystallization in response to conditions that differ from those under which the rock formed and that lie between those of diagenesis and anatexis. Like igneous rocks and sedimentary precipitates, metamorphic rocks have features that are adjusted to or are adjusting to specific physical and chemical conditions. As a consequence, they are amenable to study using principles of physics and physical chemistry, such as the Phase Rule.

AGENTS AND TYPES OF METAMORPHISM

The *agents* of metamorphism are pressure, temperature, directed stress, and chemically active fluids. In general, when a rock is transferred to a new set of conditions from the set of conditions under which it formed, it will have been transferred from a condition of stability to one of instability.[2] Thus, the minerals, the texture, or both, are out of equilibrium. Provided adequate energy is available, changes will occur in the rock to bring the minerals and texture into equilibrium under the new set of conditions. Metamorphic rocks exist at the surface of the Earth, even though they are not at equilibrium there, because there has been insufficient energy to transform the minerals and textures of the metamorphic rocks back into those stable at the surface.

Pressure

Stress (σ) is defined as force per unit area ($\sigma = F/A$). Pressure is a uniform stress.[3] It confronts the rock equally in all directions. Trapped fluid phases, such as H_2O and CO_2, may create pressure and such pressures are referred to as P_{fluid}, P_{H_2O}, or P_{CO_2}, whichever is appropriate. Alternatively, pressure is created by the load of the overlying rock and is referred to as P_{load} or lithostatic stress.

Recall that, as a "rule of thumb," P_{load} in the crust increases by about 0.1 Gpa for every 3.3 km of burial (0.1 Gpa = 1 kb = 3.3 km). Pressures of metamorphism range from less than 0.1 Gpa up to the enormous pressures of tens of Gigapascals present in the deep mantle and core.[4] Because rock masses exposed at the surface are derived only from the crust and uppermost mantle, however, most metamorphic petrologists are primarily concerned with pressures in the range of 0.1 Gpa (1 kb) to about 1.5 Gpa (15 kb).

Deviatoric Stress

In addition to pressure, directed or **deviatoric stresses,** stresses acting in particular directions and exceeding the (mean hydrostatic) stress, commonly affect rocks during metamorphism. Such stresses may act (1) along a line in opposite directions away from a point, producing tension (figure 23.1a), (2) along a line in opposite directions towards a point, producing compression (figure 23.1c), or (3) in opposite directions along different lines, yielding a force couple and producing attendant compression, tension, and shear (figure 23.1e). In each case, distinctive textures and structures may develop (figures 23.1b, d, and f).

Foliation, the most typical characteristic of metamorphic rocks, is a planar feature.[5] It results primarily from the parallel to subparallel alignment of inequant mineral grains, such as mica or amphibole grains. Though the details of the development of foliation in polymineralic rocks have been investigated experimentally only in recent years[6] and are subject to debate,[7] it is generally agreed that deviatoric stress induces the alignment of minerals that gives rocks their foliation. Thus, deviatoric stress plays a critical, yet commonly overlooked, role in the formation of many metamorphic rocks.

Temperature

The temperature conditions under which metamorphism occurs are relatively well defined, but variable. At the absolute upper limit, temperatures of metamorphism are bounded by those of the solidus of dry ultramafic and ultrabasic rocks (figure 23.2). This lies between about 1200° C and 2000° C, depending on the pressure and composition of the rock.[8]

Because melting occurs at markedly different temperatures in rocks of different composition, the upper limit of metamorphism is different for each different bulk composition. For granitic or quartz-feldspar-mica rocks, the minimum upper limit of metamorphism is the solidus for wet granite, which may be as low as 600° C (figure 23.2) (Huang and Wyllie, 1981; C. R. Stern and Wyllie, 1981). Between the melting curves of wet granite and dry ultramafic rock is a zone in which rocks will undergo either metamorphism, partial melting, or complete melting—depending on the rock composition, the pressure, the temperature, and the composition and quantity of the fluid phase present.

The lower temperature limit of metamorphism is not quite as well defined. Diagenesis and weathering end and metamorphism begins where new minerals form that are not stable at or very near the surface. The reactions that give rise to such minerals are commonly considered to begin to occur at about 100° C, though it is possible that some might occur at slightly lower temperatures. Alt et al. (1986) showed that zeolites and prehnite form in the oceanic crust

473

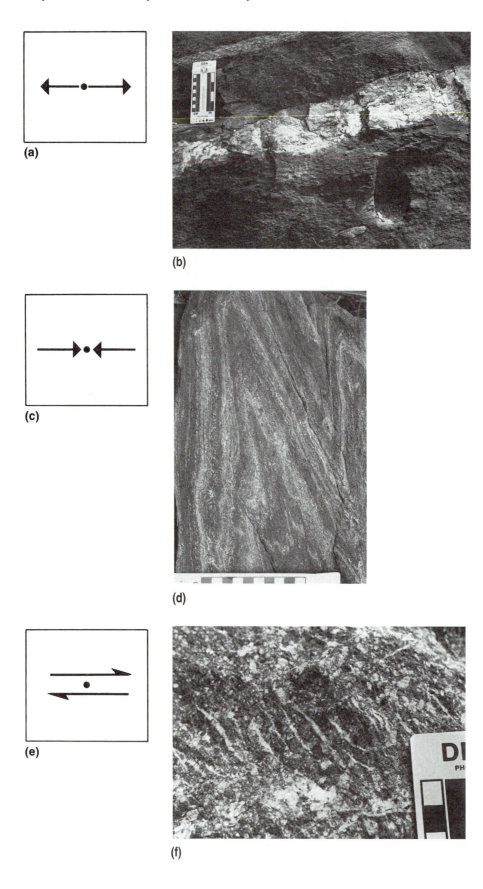

(a)

(b)

(c)

(d)

(e)

(f)

Figure 23.1 Stress and metamorphic structures. (a) Force vectors (arrows) showing configuration of stress for conditions of tension. (b) Boudinage produced by tension (extension) acting on a tabular granitoid dike, Highway 321, between Boone and Blowing Rock, North Carolina. (c) Force vectors (arrows) showing configuration of stress for compression. (d) Folds in amphibole gneiss, produced by compression, Ashe Metamorphic Suite, northwestern North Carolina. (e) Force vectors (arrows) showing configuration of stress for a shear couple. (f) Tension fractures (vein filled) produced by a shear couple acting on meta-arkose of the Grandfather Mountain Formation, Boone, North Carolina. Scale in (b), (d), and (f) is DNAG scale.

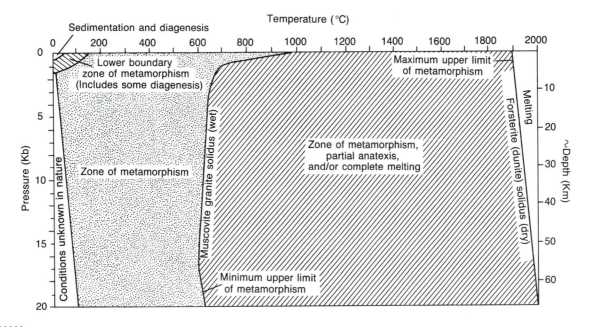

Figure 23.2 Pressure-temperature grid showing the P-T limits of metamorphism. The forsterite solidus is based on Bowen and Anderson (1914), B. T. C. Davis and England (1964), and Ohtani and Kumazawa (1981). The muscovite granite solidus is from Huang and Wyllie (1981).

at temperatures as low as 100° C, whereas weathering occurred in the same rocks at temperatures below about 50° C. Most common metamorphic rocks apparently were formed under temperatures between 100° C and 750° C. Nevertheless, metamorphism is known to have occurred, in unusual cases, at temperatures outside these limits.[9]

The main sources of the heat that results in temperature changes in metamorphic rocks are (1) increases in pressure with depth, (2) radioactive decay, (3) deformation, and (4) migrating magmas. It is well known that temperature increases with depth in the Earth (see chapter 1). In general, this increase results both from the temperature increase that accompanies increased pressure and from radioactive decay.[10] In addition, small to large regions may be heated by the heat escaping from magmas that have migrated into the rocks. Such is obviously the case where rocks have been metamorphosed adjacent to an intrusion. It is less obvious, though equally true, that the numerous intrusions that penetrate into actively forming mountain belts bring with them the heat that pervades the mountain root, causing regional heating and associated metamorphism (Lux et al., 1986). Locally, thermal effects also result from frictional shear heating along fault zones.[11]

Chemically Active Fluids

Except for certain localized areas in which very high temperature conditions or impermeable rocks prevail, rock masses contain a fluid phase. The volatile-rich phase is called fluid because at nearly all metamorphic P-T conditions, the volatile phase is in a supercritical state. In such a state, a distinction between gas and liquid cannot be made.[12]

Several kinds of evidence suggest that, in fact, a fluid phase does exist in rocks during metamorphism.

1. Fluid inclusions occur in metamorphic minerals (Touret, 1971; Touret and Dietvorst, 1983; Craw, 1988; Nwe and Grundmann, 1990).
2. Formation of metamorphic phases that include CO_2 (e.g., calcite), H_2O (e.g., muscovite), or other components (S, N_2, F, Cl, B) requires the presence of a fluid phase (Ferry and Burt, 1982).
3. Whole-rock analyses show that metamorphic rocks developed under conditions of high P and T are depleted in volatile components relative to rocks formed under lower-grade conditions (Ferry and Burt, 1982).
4. Isotopic studies indicate the involvement of fluids during metamorphism (Losh, 1989; Nesbitt and Muehlenbachs, 1989).
5. The presence of veins in metamorphic rocks suggests that fluids were present during metamorphism (Bucher-Nurminen, 1981; Walther and Orville, 1982; Yardley, 1983; Nishiyama, 1989).
6. Active metamorphism is occurring today in active geothermal regions (Muffler and White, 1969; Rona et al., 1983; Charles, Buden, and Goff, 1986).[13]
7. Metamorphic reactions commonly involve dehydration and decarbonation reactions that yield a fluid phase (Bowen, 1940; Walther and Orville, 1982; Mohr, 1985).[14]

Usually the fluid phase is dominated by H_2O, but CO_2, CH_4, N_2, Cl, S, B, Na, K and other components may be present. In some cases, species other than H_2O are dominant.[15]

In a given rock body, the fluid phase may be in equilibrium with the solid phases that comprise the rock. If, however, that fluid phase subsequently changes in composition, disequilibrium between the rock and fluid results. The rock will adjust through mineralogical changes, textural changes, or both, in order to re-equilibrate. Fluid phases that have changed become "active" with respect to the rock and will interact with it.

A number of events may result in the activation of a fluid. Fluids may be induced to migrate from one rock mass to another as a result of temperature, pressure, or stress changes. Fluid migration of this kind activates the fluids *if* the new rock into which the fluid migrates has a composition different from that from which the fluid came. Such activated-fluid migrations result in alteration zones associated with ore-bearing intrusions and in other metamorphic changes (Ferry, 1983a).[16]

Activation may also occur *in situ*. Igneous intrusions in a region may add *new fluid or components* to the fluid phase, changing its composition and thereby activating it (Burnham, 1959). Some alteration zones around plutons are produced by such a process. Finally, metamorphic reactions themselves, induced by regional or local changes in pressure or temperature, may result in changes in the chemistry of the fluid phase. Such a process may activate the fluid by producing chemical potential gradients, i.e. gradational differences in chemistry between different locations in the rocks (Ferry, 1983a; Grambling, 1986). Altered fluids produced in this way may also migrate into other rocks and effect metamorphism in them.

Types of Metamorphism

Metamorphism is usually subdivided into a variety of types on the basis of the chemical nature of the metamorphism, the dominant agent of metamorphism, and/or the area or volume of rock affected.[17] In the last case, metamorphism is subdivided into local and regional types. **Local metamorphism** is metamorphism that affects relatively small volumes of rock (less than 100 km^3). **Regional metamorphism** typically affects thousands of cubic kilometers of rock.

Subdivision of the types of metamorphism on the basis of the dominant metamorphic agent yields several subtypes (table 23.1). Where temperature is dominant, metamorphism is local and is called **contact metamorphism,** because such metamorphism occurs in country rocks at and near their contact with an igneous rock mass. At shallow levels in the crust, at low pressures (LP), contact metamorphism may be referred to as LP-contact metamorphism. At deeper levels, where pressure becomes a factor the metamorphism may be referred to as MP-contact metamorphism at medium pressures or HP-contact metamorphism at high pressures. HP-contact metamorphism is rare,

Table 23.1 Types of Regional and Local Metamorphism

Dominant Agent	Local	Regional
Pressure	—	Static
Deviatoric stress	Local dynamic (includes impact)	Regional dynamic
Temperature	LP-contact	—
Chemically active fluids	Local metasomatic (including alteration)	Regional metasomatic
Temperature + pressure ±deviatoric stress ±chemically active fluids	MP-contact HP-contact	Dynamothermal

because high temperatures usually accompany high pressures, so that at depth, katazonal intrusions cannot supply enough heat to effect noticeable changes in the already hot country rocks. Where rare intrusions occur within accretionary complexes in subduction zones (Carden et al., 1977), HP-contact metamorphism may develop.

Metamorphism induced primarily by deviatoric stress is called **dynamic metamorphism.** In this book, local dynamic and regional dynamic types are distinguished. Local dynamic metamorphism develops along discrete fault zones, in metamorphic core complexes, and in the areas associated with meteorite impacts (e.g., at Meteor Crater, Arizona).[18] Regional dynamic metamorphism occurs in the mantle and in developing mountain belts—particularly in accretionary complexes at convergent plate margins—where deviatoric stress is distributed over large regions. Such regional distributions of deviatoric stress at moderate to high temperatures result in regional mylonite belts, and at low temperatures yield tectonic melanges.[19]

Pressure may also be the principal agent of metamorphism at the regional scale. Such metamorphism is called **static metamorphism.**[20] Static metamorphism is considered to occur at depth, in thick piles of sedimentary rock in continental and forearc basins, in trenches, and in sedimentary prisms along passive continental margins. Structural burial, produced where thick sections of rock are thrust upon other rock masses, also results in static metamorphism.[21] Local metamorphism caused primarily by pressure does not occur under natural conditions.

Chemically active fluids produce a kind of metamorphism, called **metasomatism,** that is dominated by chemical changes. Near pluton contacts, local metasomatism may be important.[22] Metasomatism is usually referred to as *alteration*, where it is associated with ore deposits. On the regional scale, the importance of metasomatic processes in the production of metamorphic rocks is debatable (see chapter 11). Some geologists *do* consider some types of regional metasomatism to be significant, referring to them as regional alteration, granitization, or basification.[23]

Dynamothermal metamorphism, the most widely distributed type of metamorphism, is metamorphism induced primarily by a combination of pressure and temperature. In some texts, the terms *regional* and *dynamothermal* are used interchangeably. This is an outdated practice derived from earlier works in which *only two* major kinds of metamorphism were recognized, thermal (contact) and regional (dynamothermal) (Harker, 1932).

In all forms of regional metamorphism—dynamothermal, regional dynamic, regional metasomatic, and static metamorphism—fluids are important. In some cases of dynamothermal metamorphism, fluids may play a role that is essentially equal to that of pressure and temperature (Grambling, 1986). In contrast, it may be the absence or scarcity of fluid in some static to regional dynamic terranes that controls the distribution of phases (B. A. Morgan, 1970; Raymond, 1973b).

Metamorphism may also be subdivided on the basis of the nature of the chemical processes that occur. **Isochemical metamorphism** is metamorphism in which there is no change in the bulk chemistry of the **domain,** or rock volume, being considered. Changes in water content, however, are commonly ignored. **Allochemical metamorphism,** in contrast, is metamorphism in which there *is* a change in the bulk chemistry of the domain being considered. Metasomatism is an allochemical metamorphic process.

In the above definitions, the phrase "of the domain being considered" is critical. The **domain** refers to the volume of rock under consideration. Definition of the domain, which may range from less than one cubic millimeter to tens of cubic kilometers or more, is an essential first step in evaluating whether or not a particular metamorphic event is isochemical or allochemical. For example, in figure 23.3, metamorphism of domain A, which consists of all of the material within the limits of the outer ellipse (including that in domain B), is isochemical, because the total content of chemical species (represented by the dots) in the domain is unchanged by the metamorphic process. In contrast, if we consider domain B, the volume of the small circle within domain A, metamorphism is allochemical. Chemical species have migrated within the larger domain to concentrate in the smaller one. In the case of domain B, the total content of the chemical species within the domain changes as a result of the metamorphic process; that is, the chemistry of domain B is changed. Clearly, then, definition of the domain is prerequisite to considerations of allochemical metamorphism in general and metasomatism in particular.

Two additional terms used to describe the metamorphic history of terranes are prograde and retrograde metamorphism. Traditionally, *prograde metamorphism* has referred to metamorphism that progressed from lower to higher temperatures. In practice, such a change is recognized where minerals stable at

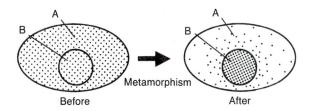

....
Figure 23.3 Diagram showing two domains (A and B) before and after metamorphism. Chemical species are indicated by dots. The total chemistry of domain A (which includes B) does not change during metamorphism; thus, the metamorphism is isochemical. During metamorphism the total chemistry of domain B changes; thus, the metamorphism of domain B is allochemical. Metasomatism has occurred in domain B.
(Modified from Bayly, 1968, p. 225)

lower temperatures are only partially replaced by those stable at higher temperatures. Alternatively, an increase in temperature across a region may be reflected by a series of mineral zones characterized by minerals that are stable at progressively higher temperatures. *Retrograde metamorphism* has traditionally been considered to be re-metamorphism that progressed from higher to lower temperatures (Harker, 1932, p. 193). Usually, retrograde metamorphism is reflected by the incomplete or pseudomorphic replacement of minerals stable at higher temperatures by those stable at lower temperatures.

The view that prograde versus retrograde metamorphism is temperature-dependent is supported by the fact that many mineral reaction curves have steep slopes (figure 23.4a). Changes in pressure, even of large magnitude, result in relatively few mineralogical changes in some systems. For example, path $a{\rightarrow}a'$ in figure 23.4a is a classical prograde path and crosses several univariant reaction curves, whereas curve $b{\rightarrow}b'$ crosses few reaction curves.

Experimental study of many systems over the past half century has led to the delineation of large numbers of mineral reaction curves (see appendix C). The slopes of these curves are variable. Consequently, we now recognize that curves in P-T-X space—representing important metamorphic reactions—slope in such a way that similar appearing mineralogical changes may result from all of the following:

1. An increase in temperature at constant pressure (path $a{\rightarrow}x$; figure 23.4b),
2. A decrease in pressure at constant temperature (path $b{\rightarrow}x$),
3. Combined decreases in P and T (path $c{\rightarrow}x$),
4. An increase in T with a concomitant decrease in P (path $d{\rightarrow}x$),
5. An increase in T with a concomitant increase in P (path $e{\rightarrow}x$), and
6. A change in the fluid phase (where fluid is involved in the reaction).

477

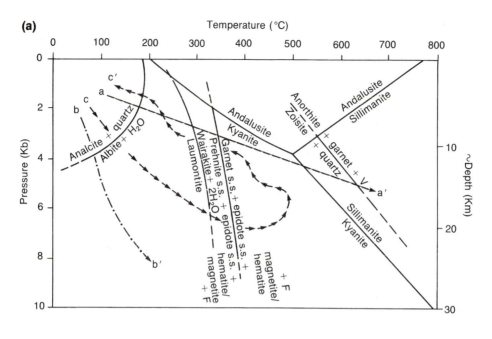

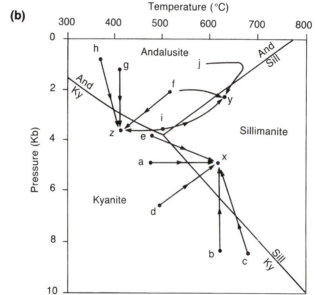

Figure 23.4 Pressure-temperature plots showing the locations of various experimentally determined univariant curves and selected metamorphic paths. Metamorphic paths (P-T-t paths) are curves that represent the set of all successive positions in pressure-temperature (P-T) space occupied by a rock as it undergoes metamorphism in the earth over time (t). (a) P-T-t paths a→a' and b→b' are prograde paths that cross many and few mineral reaction curves, respectively. Path c→c' is a representative path that includes both prograde and retrograde parts. (b) P-T plot of part of A showing three sets of prograde and retrograde paths that yield the same mineralogical change.

[Analcite-albite curve from A. S. Campbell and Fyfe (1965), laumontite-wairakite curve from Liou (1971b), prehnite-garnet curve from Liou, Kim, and Maruyama (1983), aluminum silicate curves from Holdaway (1971), zoisite + quartz = anorthite + garnet curve from Boettcher (1970).]

Thus, a change from one mineral assemblage to another (e.g., from an andalusite-bearing assemblage to a sillimanite-bearing assemblage, in figure 23.4b), thought earlier to represent a simple increase in temperature, may reflect any number of path histories (e.g., f→y, i→y, or j→y). In some cases, decreases in temperature with concomitant increases in pressure (path f→z) may result in the *same* mineralogical change as that produced by either an increase in P at constant T (path g→z) or an increase in both temperature and pressure (path h→z). Consequently, the terms prograde and retrograde must be used with caution.

In this book, **prograde metamorphism** is defined as metamorphism that follows a path generally *away* from surface temperature and pressure conditions (STP). Where fluid phases induce the metamorphic changes, the metamorphism is considered prograde *if* the new assemblages mimic those that are produced by a P-T path that progresses away from STP conditions. In figure 23.4b, paths such as $a{\rightarrow}x$, $e{\rightarrow}x$, and $h{\rightarrow}z$ are prograde. **Retrograde metamorphism** is metamorphism that follows a path generally towards STP conditions. Again, where a fluid phase is the primary agent of metamorphism, the metamorphism is considered retrograde *if* the path history mimics one that approaches STP conditions. Paths such as $c{\rightarrow}x$ and $i{\rightarrow}z$ are retrograde. In nature, rocks probably follow irregular prograde-retrograde paths through P-T space like path $c{\rightarrow}c'$ in figure 23.4a. Dated paths of this kind are called **P-T-t paths** (pressure-temperature-time paths) (England and Thompson, 1984; Spear and Peacock, 1989).

STRUCTURES AND TEXTURES OF METAMORPHIC ROCKS

In many cases metamorphic rocks are easily recognized in the field on the basis of their distinctive structures and textures. Many are foliated. Like most sedimentary rocks and some igneous rocks, metamorphic rocks may have layers. Metamorphic layering is usually distinguished by the fact that the layers are the result of, or are enhanced by, the alignment of phyllosilicates, inosilicates, or other inequant mineral grains.

Metamorphic rocks are also distinctive because of the presence of associated **mesoscopic structures** (handspecimen-to outcrop-scale structures), such as folds, veins, and rock cleavage.[24] Ductile shear zones, boudin, and other structures that are common in outcrops of metamorphic rock, further aid in distinguishing between metamorphic rocks and sedimentary or igneous rocks.

Structures

Structures are those features that characterize handspecimen or larger masses of rock. They differ in scale from *textures*, which are microscopic to small-scale mesoscopic features of the rock produced by grain shapes, grain sizes, grain orientations, grain distributions, and intergrain relationships. Many textures are *penetrative* (i.e., they pervade all parts of a handspecimen), whereas most structures are not penetrative. The distinction between structures and textures is not clear-cut, and one of the most important metamorphic structures, cleavage, bridges the boundary between the two types of feature.

In metamorphic rocks, metamorphic structures result from deformation. Rocks with a fabric that reflects a history of deformation are called **tectonites** (Turner and Weiss, 1963). Here, the word **fabric** refers to *all* of the structural and textural features of a rock that together define its geometrical character.[25] Inasmuch as most metamorphic rocks exhibit structures and textures that are the product of deformation, most are tectonites. Some contact-metamorphic rocks, as well as other undeformed rocks produced by metasomatism and nontectonic processes, are not tectonites.

Tectonites are of two major types. Those with fabrics produced predominantly by flow in the solid state (i.e., by ductile deformation with recrystallization) and characterized by *oriented* mineral grains are called S-tectonites or L-tectonites (Sander, 1970; Turner and Weiss, 1963).[26] S-tectonites have planar fabrics, whereas L-tectonites have lineated fabrics. Tectonites with fabrics that were produced primarily by brittle fracture or shear along a pervasive set of anastamosing, subparallel, or parallel surfaces have been called SF-tectonites (Raymond, 1975). These tectonites are characterized by commonly slickensided, mesoscopic, parallel to subparallel fractures that are dominantly *independent* of the overall arrangement and orientation of mineral grains or rock fragments within the rock.[27]

Not all structures in metamorphic rocks are the product of either deformation or metamorphic processes. Because metamorphic rocks had sedimentary or igneous protoliths, they may retain structures from those parent rocks. Such remnant primary structures are called **relict structures** or palimpsest features.[28] The relict structures that are commonly retained in metamorphic rocks are those less intricate structures that are visible because of differences in composition or grain size. Bedding, cross-bedding, and grading are the typical relict structures in metasedimentary rocks. Primary layering, amygdaloidal structure, and pillow structure are among the most frequently encountered relict igneous structures.

Cleavage

Rock **cleavage,** the tendency of rocks to break along parallel to subparallel surfaces, is the most widespread metamorphic structure (figure 23.5). Cleavage reflects either the textural alignment of mineral grains, called **preferred orientation,** or the subparallel arrangement of discontinuities (abrupt changes in physical properties) in a rock. Traditionally, cleavage was classified into a variety of types that are assignable to two main categories—fracture cleavage and flow cleavage.[29] The two types correspond to the two main types of tectonites. Thus, SF-tectonites are characterized by fracture cleavage and S-tectonites are characterized by flow cleavage.

In an effort to avoid use of terms that imply origin, C. M. Powell (1979) proposed that cleavage types be classified according to the physical character of the cleavage. Accordingly, cleavage is divided into two main types—continuous cleavage and spaced cleavage—on the basis of the spacing of cleavage domains or surfaces (figure 23.6a). *Cleavage domains*, consisting of zones of minerals with preferred orientation, separate *microlithons*, consisting of less cleavable material.

Figure 23.5
Continuous cleavage (approximately vertical) cutting bedding (inclined) in metamorphosed Precambrian metasediments, Wallace, Idaho. DNAG scale is in lower left (see arrow).

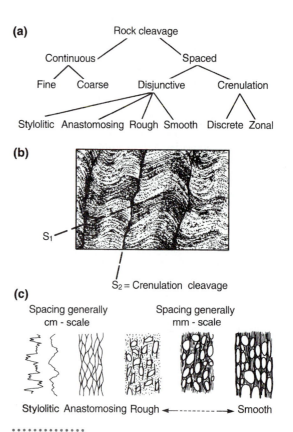

Figure 23.6 Morphology and morphological classification of rock cleavage. (a) Morphological classification of rock cleavage (after C. M. Powell, 1979). (b) Sketch of microscopic view of crenulation cleavage in phyllite. Crenulation cleavages are marked by both microfolds in preexisting foliation and the concentrations of opaque minerals (black). (c) Shape and spacing of cleavage in the four types of disjunctive spaced cleavage.
(After C. M. Powell, 1979)

Cleavage that has a spacing of <0.01 mm *or* that results from a penetrative preferred orientation of phyllosilicates or other minerals is called **continuous cleavage.** Continuous cleavage is further subdivided into fine and coarse types based on the grain size (fine = <0.1 mm, coarse = >0.1 mm). Cleavage with domain spacings of >0.01 mm and which lacks a microscopic penetrative fabric is **spaced cleavage.** Spaced cleavage is subdivided into disjunctive and crenulation types. Crenulation cleavage is cleavage imposed on rocks that already possess a planar fabric, such as a preexisting cleavage (figure 23.6b). Disjunctive cleavages, of which there are four types (figure 23.6c), are cleavages in rocks lacking a preexisting planar fabric. Anastamosing and some smooth disjunctive cleavages correspond, to some degree, with fracture cleavage, as defined above, whereas rough and some smooth disjunctive cleavages, crenulation cleavage, and continuous cleavage are approximately equivalent to "flow cleavage".

The origin of the various types of cleavage is a subject about which much has been written.[30] Briefly, the origins are as follows. Stylolitic disjunctive- and crenulation-spaced cleavages result from solution under pressure (pressure solution) and movement of materials (diffusion) out of the cleavage domain (Gray, 1979a; Wanless, 1979).[31] Residues of less soluble materials left in the domain mark the cleavage surface.

Relatively little work has been done on the formation of anastamosing, disjunctive-spaced cleavage, in spite of the fact that such cleavages occur widely in scaly clays and other matrix materials of melanges. Some anastamosing cleavage appears to have developed as a result of gravity-induced shearing in submarine and subaerial slide deposits. Alternatively, such cleavage may be induced by tectonic stresses. Detailed studies suggest that mechanical rotation of phyllosilicate grains into the plane of the cleavage and

cataclasis of grains are important processes responsible for the development of this type of cleavage (J. C. Moore et al., 1986; Lucas and Moore, 1986, Lash, 1989).

The origins of rough and smooth disjunctive-spaced cleavages and continuous cleavages have been the subject of the most debate and intensive study.[32] The origin of these cleavages, in part referred to as "slaty cleavage," has been variously attributed to one or more of the following processes:

1. Mechanical rotation of phyllosilicate grains into the cleavage plane,
2. Pressure solution, accompanied by rotation of residual phyllosilicate grains into the plane of the cleavage, and
3. **Syntectonic recrystallization,** that is, dissolution of detrital phyllosilicates and other minerals and recrystallization or neocrystallization of phyllosilicate and other grains, parallel to the cleavage direction, during deformation (J. C. Maxwell, 1962; Kanagawa, 1991).

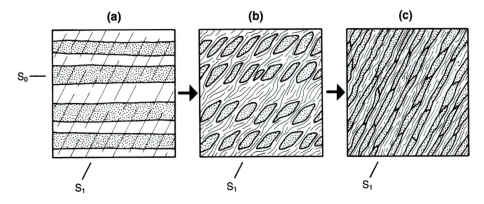

Figure 23.7 Schematic sequential diagrams showing transposition of bedding (S_0) along a cleavage (S_1). (a) Bedding transected by incipient cleavage.

(b) Separation, rotation, and deformation of bedding segments. (c) New compositional bands parallel to cleavage.

All may play some part in the development of cleavage in rocks at specific localities. Yet, a growing body of evidence indicates that syntectonic recrystallization may be the dominant process of cleavage formation in rocks with well-developed cleavage, especially those of higher metamorphic grade (W. J. Gregg, 1985; J. H. Lee et al., 1986).

Layers and Transposition of Bedding

Metamorphic rocks commonly have mesoscopic layers. The layering is compositional banding defined by differences in mineral composition and texture (e.g., grain size). Physically, metamorphic layers may be meters to millimeters thick. Layers that are less than 1 mm thick are called microlayers, whereas layers larger than 1 m in thickness are designated megalayers. Layers that have thicknesses between these limits define the structure called **gneissic structure.**[33] Gneisses are banded rocks in which alternate layers are composed of different minerals.[34]

The compositional layers observed in metamorphic rock may have a variety of origins. Some bands represent primary layering (i.e., relict bedding) in the sedimentary rock from which the metamorphic rock was derived. In contrast, faulting may interleave layers of rock of different compositions that are later metamorphosed to produce a banded rock. In other cases, banding may represent dikes or veins that invaded a host rock and were later metamorphosed along with it. In still other cases, banding results from a process of chemical migration called metamorphic differentiation (discussed below).

New compositional banding in metamorphic rocks may also result during cleavage formation through **transposition** of bedding or other planar features. In this process, the layering is transected at a significant angle by a cleavage (labelled S_1 in figure 23.7a).[35] As the cleavage (S_1) develops, the bedding is separated into segments that are physically rotated or deformed by flow, or both, so that elongate lenses that parallel the cleavage direction are created (figure 23.7b). Continued

cleavage development may cause the lenses to merge yielding new compositional bands that strike and dip in a direction different from that of the original layering (figure 23.7c).

Other Structures

A number of other structures are important locally in metamorphic rocks. Among these are folds, kink bands, boudins, mullions, rods, faults (including ductile shear zones), joints, and veins. Folds, kink bands, some boudins, mullions, rods, and ductile shear zones are produced by ductile deformation.[36] Some boudins, brittle faults, and joints result from brittle deformation. Veins and (commonly) compositional banding represent the effects of chemical migration of materials.

Folds are bends in planar structures of the rock (figure 23.8a). Folds may occur in bedding, foliation, veins, or other features, including previously existing folds. In terms of their scale, they range from microscopic structures to macroscopic structures covering tens or hundreds of square kilometers. Folds are formed as a result of compressional or shearing stresses acting parallel or at an angle to rock layers.

Kink bands are small, abrupt folds developed in rocks that already have a fabric (figure 23.9a). They develop from compressive stresses directed at an angle to the preexisting fabric (Dewey, 1965; Cobbold, Cosgrove, and Summers, 1971; P. F. Williams and Price, 1990). Most commonly, kink bands are found in fine-grained rocks such as phyllite, but they also occur in schists and gneisses.

Boudin (French for sausage) are cylindrical masses of rock, originally part of a single bed or layer that has been stretched and pulled apart.[37] The cylinders in true boudins lie side by side. Three-dimensional exposures of boudin are uncommon and typically one sees cross sections that appear as a series of crude ellipses or rectangles with rounded corners.

Mullions and rods are similar to boudins in that they are long, cylindrical structures (G. Wilson, 1953, 1982). **Mullions** are columns, 2 cm to 2 or more meters in diameter,

(a)

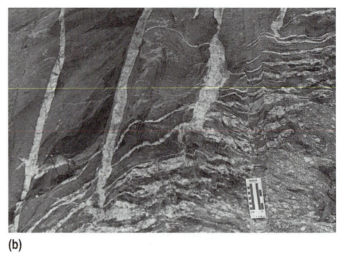

(b)

(c)

(d)

Figure 23.8 Photographs of structures in metamorphic rocks. (a) Mesoscopic folds in gneiss and schist, Franciscan Complex, Jurassic-Cretaceous, Laytonville Quarry, south of Laytonville, California. (b) Quartz veins in meta-quartz arenite of the Grandfather Mountain Formation, Neoproterozoic, Highway 321, south of Boone, North Carolina. (c) Melange Franciscan Complex Jurassic-Cretaceous, exposed at Goat Rock, south of Jenner, California. Notice the football-shaped boudin at the lower left, the folds in the lower center, and the boudins at the right center. Blocks in the melange include serpentine quasimylonite, metachert, metawacke, and metabasalt. The beach is at the lower right. (d) Laminated mylonite of the Linville Falls Fault Zone, Linville Falls, North Carolina. Here, Mesoproterozoic semischistose granitoid rocks of the Cranberry Gneiss are thrust over younger, Neoproterozoic metaquartz arenite and phyllite of the Chilhowee Group.

composed of the country rock of the metamorphic terranes in which they occur. The exterior of the columns are angular to rounded and are commonly polished or striated parallel to their length. In cross section, they may exhibit folded internal layering. The origins of mullions are varied, as there are several types of mullions.[38] **Rods** are similar to mullions, but they are composed of segregated or introduced material, (i.e., dike or vein material), such as quartz.

Joints are fractures along which there has been no significant movement parallel to the plane of the structure. In metamorphic rocks, joints commonly occur in sets that have been filled by veins (figure 23.8b). Joints commonly develop from primary or secondary tensional stresses resulting from stored stress in the rocks (N. J. Price, 1966), but may also develop from active shear or tensional stresses.

(a)

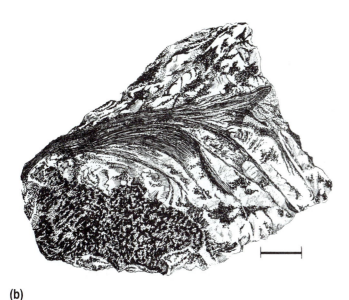

(b)

············
Figure 23.9 Sketches of two structures found in metamorphic rocks. (a) Kink bands in ultramylonite from the Cranberry Gneiss, Mesoproterozoic from the banks of the New River east of Mouth-of-Wilson, Virginia. Scale bar is 2 cm. (b) Sketch of mylonite zone forming "horse-tail shear" in Striped Rock Granite, Proterozoic, near Independence, Virginia. Scale bar is 2 cm.
(Both sketches from Raymond, 1984c)

Faults, in contrast, are fractures, along which there has been significant movement parallel to the plane of the structure. Brittle, brittle-ductile, and ductile faults occur in metamorphic rocks (J. G. Ramsay, 1980). The former tend to be relatively clean breaks characterized by polished and scratched surfaces (slickensides) *or* cataclastic zones marked by breccia, tectonic melanges, or related rock masses. Tectonic melanges are tectonically deformed *bodies* of rock "mappable at a scale of 1:24,000 or smaller and characterized both

by the lack of internal continuity of contacts or strata and by the inclusion of fragments and blocks of all sizes, both exotic and native, embedded in a fragmented matrix of finer-grained material" (figure 23.8c) (Raymond, 1984a). Native blocks are a part of the original unit that was fragmented, whereas exotic blocks were formed elsewhere and were incorporated into the melange. Ductile shear zones are zones of dislocation a few millimeters to hundreds of meters wide, composed of rocks called **mylonites** that are recrystallized and/or neocrystallized under the influence of a shearing stress (figure 23.8d and 23.9b).[39]

Veins are tabular joint or fault fillings composed of one or more minerals (figure 23.8b). The distinction between veins and dikes is somewhat arbitrary, but the term vein tends to be applied to tabular bodies that are monomineralic, bimineralic, or contain ore minerals, whereas the term dike is used for igneous rock types crystallized from magmas or magmatic vapor phases. Veins form where fluids penetrate a fracture and precipitate minerals.

Textures

Textures are a function of grain size, grain shape, intergrain relationships, grain distribution, and grain orientation (Spry, 1969, p. 5). Somewhat different sets of textural terms are commonly used in field and handspecimen work than are used in thin-section petrography. In this text, those sets of terms are integrated.

Texture Types

Metamorphic textures may be divided into five major types—foliated textures, granoblastic textures, diablastic textures, cataclastic textures, and relict textures.[40] The word relict assumes the same meaning here that it has with regard to structures; it refers to textures retained from a protolith. In general, the names of relict textures are preceded by the prefix *blast-* or *blasto-* (e.g., blastoporphyritic) to indicate that the texture is metamorphic, but retains the relict character.[41] "Blast" as a suffix indicates a new metamorphic texture (e.g., porphyroblastic).

Each of the major texture types is characterized by distinctive grain shapes or grain orientations, or both. **Foliated textures** are textures characterized by an alignment of mineral grains in such a way as to give the rock the appearance of or the tendency for splitting into layers or flat pieces (figure 23.10a). Commonly, the minerals in foliated rocks are predominantly acicular or tabular. **Granoblastic textures** are those characterized by more or less equidimensional mineral grains (figure 23.10b). As used here, the term granoblastic refers to a general category of texture and implies nothing about grain boundary shape, grain size variation, or preferred orientation or the lack thereof.[42] **Diablastic textures** are those in which tabular or acicular minerals are intergrown in a nonfoliated, interlocking, locally radiating manner (figure 23.10c) (Harker, 1932).[43]

483

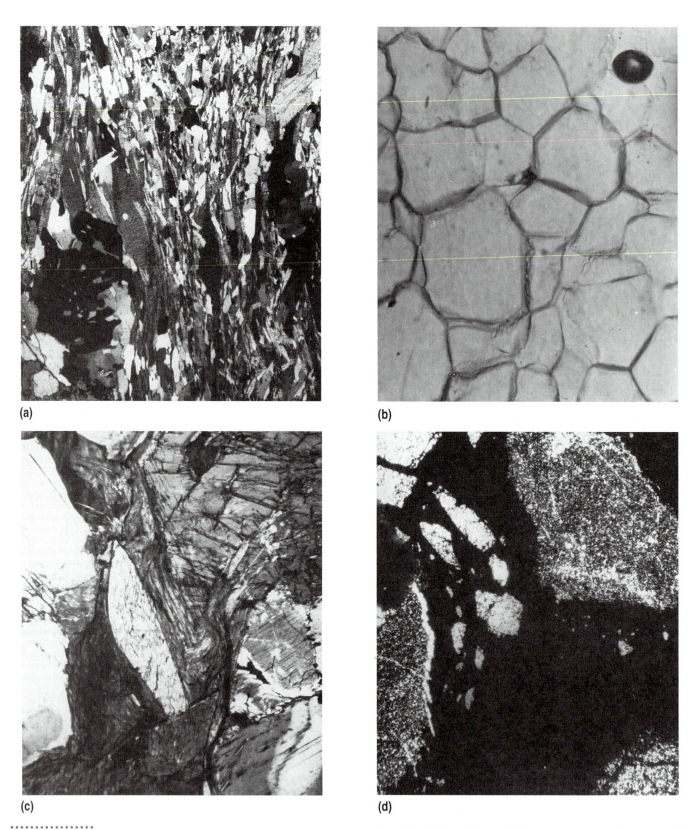

Figure 23.10 Photomicrographs showing major types of metamorphic textures. (a) Foliated (lepidoblastic) texture in pelitic schist of the Ashe Metamorphic Suite, Neoproterozoic (?), northwest North Carolina. (XN). (b) Granoblastic (equigranular-mosaic) texture in metadunite, Corundrum Hill, western North Carolina. (PL). (c) Diablastic texture in talc-anthophyllite-chlorite diablastite, Greer Hollow ultramafic body, Todd, North Carolina. (XN). (d) Cataclastic texture in fault breccia in Rome Formation, Cambrian, Mountain City, Tennessee. (XN). Long dimension of photos is 6.5 mm in (a), (c), and (d), 1.27 mm in (b).

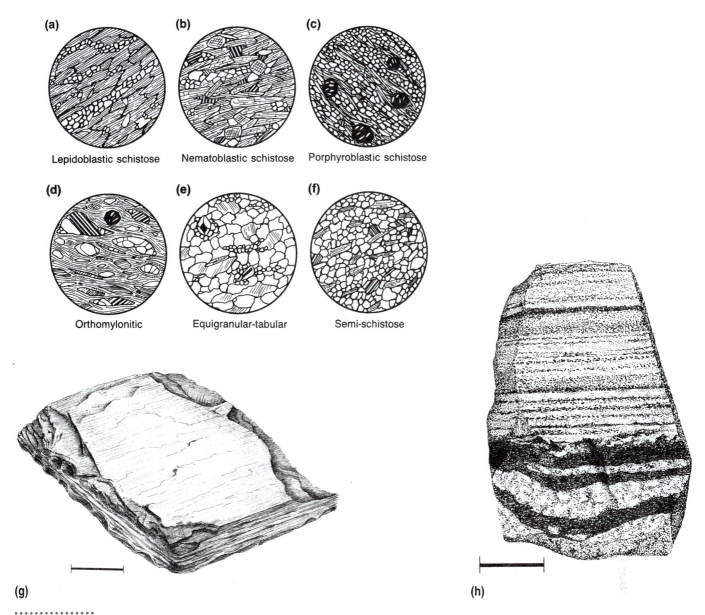

(a) Lepidoblastic schistose

(b) Nematoblastic schistose

(c) Porphyroblastic schistose

(d) Orthomylonitic

(e) Equigranular-tabular

(f) Semi-schistose

(g)

(h)

....
Figure 23.11 Sketches of selected types of foliated metamorphic textures. (a)–(f) are thin-section views; (g) and (h) are handspecimen views (scale bars = 2 cm).
(a) Lepidoblastic schistose texture in chlorite-albite-quartz-white mica schist. (b) Nematoblastic schistose texture in plagioclase-hornblende schist. (c) Porphyroblastic schistose texture in garnet-biotite-plagioclase-white mica-quartz schist. (d) Orthomylonitic texture in plagioclase-quartz-white mica-chlorite orthomylonite. (e) Equigranular tabular texture in metadunite. (f) Semischistose texture in white mica-biotite-plagioclase-quartz semi-schist. (g) Slaty texture (and continuous "slaty" cleavage) in slate, from the Mettawee Formation, Pawlet, Vermont. (h) Gneissose texture in magnetite-quartz-feldspar gneiss, from the Valley Springs Group, Llano Co., Texas.
[(g) and (h) from Raymond, 1984c]

Cataclastic textures are nonfoliated textures characterized by fractured rock materials and mineral grains (figure 23.10d). Each of these major textural types may be subdivided into two or more individual types (see table 23.2 and figures 23.11 and 23.12).

Foliated textures characterize metamorphic rocks and therefore assume the most importance. The two major categories of foliated textures are strongly foliated and weakly foliated. Strongly foliated rocks include (1) rocks dominated mineralogically by platy, bladed, or acicular minerals, (2) rocks deformed to the extent that minerals, such as quartz, which typically show little tendency to be elongate parallel to foliation, are very elongate and define a well-developed cleavage, and (3) rocks characterized by mineral segregations, microlithons, or spaced cleavages. In strongly foliated rocks, continuous cleavage, spaced crenulation cleavage, or smooth,

485

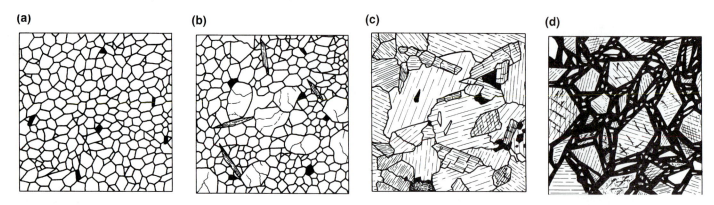

(a)   **(b)**   **(c)**   **(d)**

Figure 23.12   Sketches of selected nonfoliated metamorphic textures. (a)–(d) are thin-section views; (e) and (f) are handspecimen views. Scale bars are 2 cm. (a) Granoblastic-polygonal texture in metadunite (olivine granoblastite). (b) Heterogranoblastic texture in metadunite (olivine granoblastite). (c) Diablastic texture in tremolite-chlorite diablastite, Ashe Metamorphic Suite, northwestern North Carolina. (d) Cataclastic texture (schematic) in fault breccia; with Mn-oxide cement. (e) Hypidioblastic-heterogranular texture in quartz-plagioclase-hornblende granoblastite, Winding Stair Gap, southwestern North Carolina. (f) Porphyroblastic, allotrioblastic-heterogranoblastic texture in metadunite (olivine granoblastite) from Day Book Mine, Spruce Pine District, North Carolina.

[(e) and (f) from Raymond, 1984c]

rough, or anastamosing disjunctive-spaced cleavage are well developed. Weakly foliated rocks are those in which (1) linear, but not planar, arrangements of bladed to acicular grains dominate the texture, (2) equant to subequant grains, such as quartz and feldspar, comprise most of the rock, or (3) platy to bladed minerals, such as phyllosilicates, are present but are only weakly aligned. In such rocks, cleavage is stylolitic, anastamosing, or rough disjunctive-spaced cleavage and the microlithons between the cleavage surfaces lack a preferred orientation of grains.

Recrystallization, Neocrystallization, Nucleation, and Crystal Growth: An Overview

The development of metamorphic textures is similar in many ways to the development of igneous textures. *Neocrystallization* involves nucleation and crystal growth, as is the case with igneous textures. In metamorphic rocks, however, diffusion (migration) of chemical species towards the growing crystal occurs in a crystalline rock, rather than in a melt. In addition, *recrystallization* of the phases that are already present may occur, either simultaneously with neocrystallization or independent of it. In any case, recrystallization also involves nucleation and growth of crystals.

The metamorphic reactions and processes (e.g., neocrystallization and crystal growth) that give rise to new or reoriented crystals, that is, to new textures, are attempts by the rock system to attain equilibrium under the new metamorphic conditions. Those conditions are generally different from the conditions under which the rock formed. At equilibrium, the rock will have attained the lowest possible energy state under the extant conditions.

All metamorphic reactions and processes, including nucleation, diffusion, and growth, are activated by energy, or more specifically by differences in energy states (Vernon, 1976, p. 75). Chemical reactions (neocrystallization processes) are attempts to reduce the thermodynamic energy (called the Gibbs Free Energy) of the system.[44] Similarly, recrystallization involves progress towards reduction of the structural *free energy* of the system, especially that energy associated with the boundaries of each phase, the *surface free energy*. Surface free energy results from the fact that atoms at the surface of a crystal, while bonded internally to other atoms, are not bonded on the side of the atom facing the surface. Thus, that side *and* the surface have a higher energy. The more surface there is, the more surface energy there will be (all other factors being equal). Because a few large grains have less surface area than numerous small grains (in an equal volume of rock), larger grains are favored at equilibrium, because they have less surface free energy. Both nucleation and recrystallization aid the rock (the system) in its progress towards equilibrium, that is, towards a lower state of free energy.

Nucleation and Growth

Recall that nucleation is the process in which a relatively few atoms of the right type cluster together to form the rudiments of a new crystal structure. Most nucleation in metamorphic

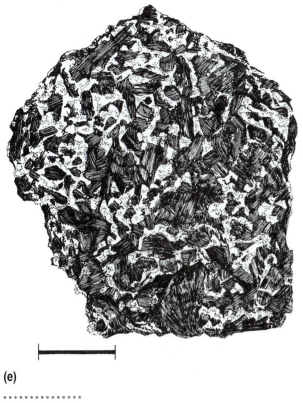

(e)

Figure 23.12 Continued

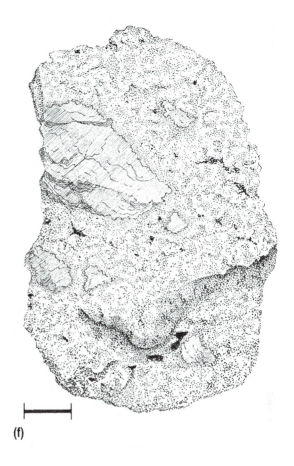

(f)

rocks is heterogeneous, because crystal lattices already exist upon which new minerals may be constructed. Nucleation occurs where (1) the appropriate atoms are available to form the nucleus, and (2) the appropriate conditions exist for the persistence of the nucleii. The latter condition prevails where the energy state, based on a given set of P-T conditions, is such that the free energy of the nucleus is favored over that of unbonded atoms.[45] The availability of ions depends either on their initial presence or on their ability to migrate to the nucleation site (i.e., on the process of diffusion).

Once the nucleus is formed, growth may occur. Growth will occur as long as it is favored by the energy state, that is, by large differences in free energy, and by the availability of ions. Grains that are strained or have relatively large surface areas (i.e., small grains) tend to recrystallize to reduce their free energy. New minerals will form where a new phase is more stable than a preexisting mineral that formed under different thermal conditions. The availability of ions depends on their presence in the surrounding rock and their ability to migrate to the growth site. Again, diffusion is important.

Diffusion

Clearly, diffusion is a critical process in the development of metamorphic textures. Both nucleation and crystal growth depend on it. **Diffusion** may be defined as a process in which chemical species migrate, in a solvent phase, under the influence of a chemical potential gradient between two sites (Jensen, 1965).[46] The chemical potential gradient represents a difference in chemistry, as well as temperature and/or pressure, between two parts of the system. In small volumes of rock, differences in chemistry exist where different mineral phases exist side by side or near one another. Differences in pressure exist because pressures are high where grains are in contact, but are generally lower in adjoining voids. In general, differences in temperature do not exist at the small mesoscopic scale. The solvent phase in metamorphic rocks may be a crystal or a pore fluid.

Within crystals, diffusion occurs where atoms jump from one lattice position to another (Condit, 1985; Nicolas and Poirier, 1976, p. 52). In rocks, diffusion occurs along crystal boundaries and along linear and planar defects within crystals (e.g. Spry, 1969, p. 14). The presence of a fluid phase on crystal boundaries greatly facilitates the diffusion process, because ions migrate more easily through a fluid.

Recrystallization

Recrystallization, which depends on diffusion, is more easily understood than neocrystallization, because in recrystallization, chemical changes do not accompany the textural changes. The rocks involved are generally monomineralic. Considering recrystallization only, a rock undergoing metamorphism may experience changes in grain size, grain shape, and grain orientation.

In situations of LP-contact metamorphism, stress will not be a significant factor, and recrystallization may be considered to result primarily from the influence of temperature.

Table 23.2 Metamorphic Rock Textures

A. Textural Terms Relating to Intergrain Relationships, Grain Shapes and Sizes, Grain Orientations, and Grain Distributions

Foliated Texture Aligned minerals (i.e., a preferred orientation of minerals) give a layered, flaky, or lineated appearance.

Strongly Foliated Texture Rock exhibits well-developed continuous or finely spaced cleavage.

Slaty Texture Very fine-grained texture (grains < 0.1 mm) characterized by abundant phyllosilicates strongly aligned in a planar or subplanar orientation. In handspecimen, this texture is aphanitic and specimens break into flat, smooth pieces.

Phyllitic Texture Very fine-grained to fine-grained texture (grains < 0.5 mm) characterized by the presence of crenulation cleavage, microfolds, or kink bands.

Schistose Texture Fine-grained to very coarse-grained texture (grains > 0.1 mm) characterized by subparallel arrangement of acicular, bladed, and/or tabular minerals (especially phyllosilicates and amphiboles).

Lepidoblastic texture Platy or sheet-structured minerals predominate.

Nematoblastic texture Acicular or elongate prismatic minerals predominate.

Gneissose Texture Fine-grained to very coarse-grained texture with more or less continuous bands (at the handspecimen scale) of contrasting mineralogy.

Porphyroblastic-Foliated Texture Very fine-grained to very coarse-grained rock with larger crystals (porphyroblasts) in a finer-grained, foliated matrix. (The term *foliated* should generally be replaced with the appropriate specific term, e.g., *porphyroblastic schistose* or *porphyroblastic gneissose texture*.) The porphyroblasts are commonly mineralogically distinct from the matrix.

Nodularblastic-Schistose Texture A texture consisting of nodular clusters of small grains of one or two minerals in a matrix of a different composition.

Mylonitic Texture (sensu lato) Foliated to porphyroclastic-foliated texture with fine-grained to very fine-grained matrix or fabric materials that typically show evidence of crystal-plastic deformation and syntectonic recrystallization or recovery features (i.e., unstrained, sutured grains).

Porphyroclastic texture A weakly to strongly foliated texture characterized by a bimodal distribution of grain sizes; with larger, typically deformed, protolith grains (the porphyroclast) in a finer-grained matrix of materials derived, in total or in part, from the porphyroclast. Matrix may be strain free, but typically is not.

Protomylonitic texture Mylonitic texture with > 50% porphyroclasts.

Orthomylonitic texture Mylonitic texture with 10–50% porphyroclasts.

Ultramylonitic texture Mylonitic texture with < 10% porphyroclasts. (Phyllonitic texture, as used in older literature, is more or less equivalent to ultramylonitic texture and is used especially where a crenulation cleavage is present.)

Foliated Cataclastic Texture (see *Cataclastic Textures* below).

Weakly Foliated Texture Texture characterized by lineations or weakly developed or widely spaced cleavages.

Semi-slaty Texture Very fine-grained to fine-grained weakly foliated texture.

Semi-schistose Texture Fine-grained to very coarse-grained weakly foliated texture.

Lepidoblastic-semi-schistose texture Fabric dominated by poorly aligned phyllosilicates or other platy minerals.

Nematoblastic-semi-schistose texture Fabric dominated by poorly aligned elongate prismatic or acicular minerals.

Sheaf texture Texture containing grain clusters consisting of arrays of diverging platy to acicular grains.

Comb texture Texture characterized by parallel to subparallel grains arranged perpendicular to a surface or set of surfaces upon which the grains nucleated.

Equigranular-Tabular Texture (granoblastic elongate) Texture characterized by aligned, elongate to tabular, subequant grains.

Porphyroclastic Texture A weakly to strongly foliated texture characterized by a bimodal distribution of grain sizes, with larger, deformed grains in a finer-grained matrix of materials derived, in total or in part, from the porphyroclast. Matrix may be strain free, but typically is not.

Diablastic Texture (*decussate texture*) Composed of non-aligned, radiating to randomly oriented acicular to platy grains.

SHEAF TEXTURE Texture characterized by grain clusters consisting of arrays of diverging platy to acicular grains.

SPHERULOBLASTIC TEXTURE (spherulitic texture, rosette texture) Texture characterized by clusters of radiating acicular to bladed minerals.

Table 23.2 Continued

FIBROBLASTIC TEXTURE Diablastic texture characterized by acicular minerals of approximately the same size.

Granoblastic Texture Equant to subequant grains dominate a granular aggregate.

HOMOGRANULAR TEXTURE Grains are all approximately the same size.

Granoblastic-Polygonal Texture (mosaic texture, equigranular-mosaic texture, equigranular texture, granulitic texture, granular texture) A texture characterized by polygonally shaped crystals of equal or nearly equal size with straight to slightly curved grain boundaries. Rocks with this texture are typically almost monomineralic. In monomineralic rocks, grain boundaries typically meet at 120° angles and grains are anhedral.

Protogranular texture (granular texture) Medium- to coarse-grained, granoblastic-polygonal texture typical of some ultramafic rocks. In ultramafic rocks, spinel forms vermicular intergrowths in pyroxenes. Note that this name has a genetic implication; i.e. that the texture is an original (versus second-stage) metamorphic texture.

Equigranular-mosaic texture (equigranular texture, granulitic texture) Fine-to coarse-grained, granoblastic-polygonal texture typical of some ultramafic rocks, granulites, and hornfelses.

Granoblastic-Polysutured Texture Texture characterized by polygonally shaped crystals of equal or nearly equal size with lobate or serrate margins.

HETEROGRANULAR TEXTURE Grains are of notably different sizes.

Heterogranoblastic Texture Texture characterized by equant or subequant minerals of varied sizes.

Nodularblastic Texture A texture consisting of nodular clusters of small grains of one or two minerals in a matrix of a different composition.

Clastic Texture Nonfoliated to foliated texture in which broken grains or rock fragments are enclosed in a matrix of more finely broken materials.

MORTAR TEXTURE (cataclastic texture) Nonfoliated texture with larger broken grains in a matrix of smaller, broken grain fragments of the same composition.

FOLIATED CATACLASTIC TEXTURE Foliated texture consisting of broken grains of rock or mineral materials in elongate segregations (meso-to microlithons) arranged in preferred orientations and separated by anastamosing to subparallel shear surfaces or microfoliated laminae.

VITRICLASTOBLASTIC TEXTURE Nonfoliated texture with broken fragments of rock or mineral in a matrix of frictionally generated glass.

Relict Textures Primary epiclastic, pyroclastic, or crystalline textures preserved in a metamorphic rock.

B. Textural Terms Relating to Nonpenetrative, Nonplanar, and Intragrain Textures

Porphyroblastic Texture A texture in which there is a bimodal distribution of grain sizes (equivalent to porphyritic texture in igneous rocks; large grains are called porphyroblasts). In general, the smaller grains surrounding the porphyroblast are of varied minerals, some or all of which are different from the composition of the porphyroblast.

Allotrioblastic (Texture) This term, which should be used as a prefix to other textural terms, indicates that grains are dominantly anhedral (e.g., allotrioblastic-granoblastic-polygonal texture).

Hypidioblastic (Texture) This term, which should be used as a prefix to other textural terms, indicates that grains are dominantly subhedral.

Idioblastic (Texture) This term, which should be used as a prefix to other textural terms, indicates that grains are dominantly euhedral.

Augen Texture A texture characterized by eye-shaped, larger grains or grain clusters enclosed in a finer-grained matrix.

Coronitic Texture A texture in which larger grains have rims of a generally finer-grained mineral, presumed to be produced by a reaction between the core mineral and the surrounding material.

Poikiloblastic Texture A texture in which a larger grain encloses many smaller grains.

Symplectic Texture A texture in which there is an intimate, commonly wormlike, intergrowth between two minerals.

Helicitic Texture A texture consisting of bands of small relict inclusions, typically folded or arranged in spiral patterns, enclosed in poikilitic grains or porphyroblasts.

Sources: Spry (1969), Bell and Etheridge (1973), Collerson (1974), Mercier and Nicolas (1975), J. E. N. Pike and Schwarzman (1976), Basu (1977b), Bates and Jackson (1980, 1987), Raymond (1984c), D. U. Wise et al. (1984), Chester, Friedman, and Logan (1985), and Bard (1986).

Under such conditions, major reorientation of grains is unlikely, so grain size and grain shape changes are the expected metamorphic effects. Metamorphism will involve breaking and reforming of bonds, heterogeneous nucleation, diffusion of ions to the newly nucleated sites, and grain growth. These changes allow the rock system to progress towards a state of lower free energy. Rocks with large amounts of grain surface area are the most susceptible to recrystallization. These would include rocks with fine grain size, irregular grain shapes, or large porosities. Recrystallization promotes a reduction in surface free energy by reducing the amount of surface.

We can observe the expected effects of recrystallization in rocks. The two effects anticipated, an increase in grain size and a smoothing of grain boundaries, are reported by Joesten (1983) from an ideal natural occurrence of LP-contact metamorphism in the Christmas Mountains of Texas. Here, a gabbro intrudes a chert-nodule-bearing limestone. At positions progressively closer to the contact, where temperatures increase correspondingly, the texture of the chert changes from a "mosaic texture" (granoblastic-polysutured texture) to a "granoblastic polygonal microstructure" (equigranular-mosaic texture). The average grain size increases from 0.0075 mm at 101.7 m from the contact to 1.06 mm at 1.8 m from the contact (figure 23.13).

These same effects are reproduced in experiments and predicted on the basis of theoretical considerations.[47] In monomineralic rocks composed of equant grains, the grain boundary between any two grains will be expected to have the same relative surface energy as the boundary between any other two grains. As a result, under equilibrium conditions, grains will be equidistant from one another, will meet at a triple point, and will have interfacial angles of approximately 120° (figure 23.14a). This is the lowest-energy configuration for the grain boundaries. The resulting texture is a granoblastic polygonal texture (figure 23.14b). Once such a configuration is developed, further reduction of surface free energy is attained by an increase in grain size.

Where a second mineral is introduced into the aggregate, surface free energies will vary, as will the resulting interfacial angles (figure 23.14c) (Vernon, 1976). The resulting textures will also be equigranular if equant grains are dominant (figure 23.14d). In rocks in which acicular or platy minerals are dominant, the textures will be diablastic.

The addition of deviatoric stress to the variables affecting the system creates additional variations in the textures. In general, the result will be the development of preferred orientations.[48] Two types of preferred orientations exist (Vernon, 1976, p. 24). **Dimensional preferred orientations** or **shape preferred orientations** (SPO), are those in which inequant grains have a tendency towards parallel alignment. Such preferred orientations are generally quite visible to the unaided eye and are referred to as foliation or lineation. **Lattice preferred orientations** (LPO) are those that result from special orientations of optical and crystallographic axes. Commonly,

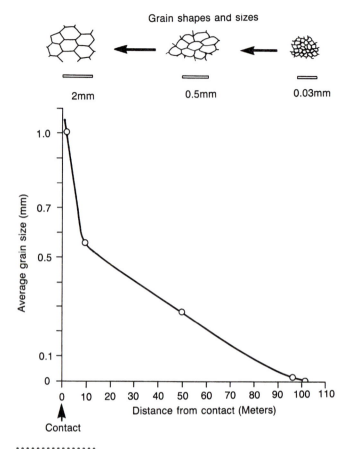

Figure 23.13
Graph showing grain size in metachert plotted against distance from contact with gabbro.
(Data from Joesten, 1983)

such preferred orientations are not visible to the unaided eye and occur in apparently randomly oriented grain assemblages.

In monomineralic or nearly monomineralic rocks, recrystallization under the influence of a deviatoric stress results in the following progressive sequence of textural changes and characteristics.[49]

1. Grains develop deformation features, such as deformation bands, and, in some cases, serrated boundaries (figure 23.15b).
2. Polygonization of grains occurs—a process in which larger, strained grains are reorganized into a number of smaller, strain-free grains, typically with serrated or irregular boundaries—beginning along grain boundaries and deformed regions in the host grain (figure 23.15c). This results in a porphyroclastic texture.
3. Coarsening of grain size and the straightening of grain boundaries follows the elimination of older, deformed grains, resulting in a granoblastic-polysutured texture.
4. A granoblastic polygonal texture develops, as enlarged grains form interfacial angles of 120° (figure 23.15d).

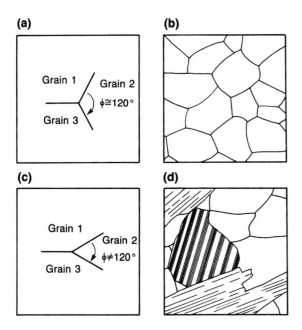

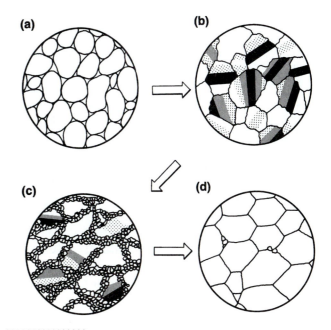

Figure 23.14 Equilibrium configurations for grain boundaries in rocks with granoblastic polygonal texture. (a) Grain boundaries and interfacial angles for three grain boundaries at equilibrium in a monomineralic rock. (b) Sketch of granoblastic polygonal texture with 120° grain-boundary interfacial angles. (c) Interfacial angles for three grain boundaries at equilibrium in a polymineralic rock, ∅ ≠ 120°. (d) Equilibrium texture in polymineralic rock in which equant minerals predominate.

Figure 23.15 Sketches showing idealized sequence of textural changes in a monomineralic rock under the influence of a deviatoric stress during metamorphism. (a) Original grains (in this case, of quartz sand). (b) Stage 1, development of deformation bands. (c) Stage 2, polygonization of grain margins. (d) Stage 3, coarsening of grain size with accompanying straightening of some grain boundaries, is followed by stage 4 in which granoblastic polygonal texture results from continued grain size increases and straightening of grain boundaries.

The textural changes may vary, depending on the pressure, temperature, strain rate, and the nature of the stress.[50] In particular, lattice preferred orientations will vary substantially under different conditions. In some cases, stage (4) is followed by the development of a porphyroblastic-granoblastic-polygonal texture, with porphyroblasts 1 cm or more in length (C. J. L. Wilson, 1973). Further grain growth yields a *coarse* granoblastic-polysutured texture. In cases of high strain rate, mylonitic, rather than granoblastic, textures are developed (C. Simpson, 1983).

The reduction in grain size during the polygonization stage, as well as during mylonitization, is a result of the high strain energies imposed by the deviatoric stress. These energies overcome the tendency to reduce surface area, because the smaller unstrained grains represent a *lower* free energy state than the larger strained grains.

All of the processes described above may be affected by phase changes and the presence of additional phases. C. J. L. Wilson (1973) found, for example, that the presence of additional phases inhibited the development of larger grains. In contrast, the presence of only a few nuclei of a certain phase may result in the development of a porphyroblastic texture with porphyroblasts of phases different from those of the matrix.

Metamorphic Differentiation

The development of gneissic structure and gneissose texture, like polygonization, is a process that seems at first glance to operate in opposition to the direction of lower free energy. In general, one would expect that metamorphism would tend to homogenize rocks, i.e., to make them uniform in composition and mechanical properties. This is so because homogenization eliminates chemical potential gradients resulting from differences in composition. Thus, a homogeneous rock should be one of lower free energy. In spite of this logic, compositional layering is very common in metamorphic rocks (figure 23.16). In many cases, the layering may represent relict primary features, but not in all.

Metamorphic differentiation is the process or processes that lead to the development of banded or lenticular segregations of minerals from an initially homogeneous rock. The process of metamorphic differentiation has been discussed by a number of petrologists.[51] To date, several explanations have been proposed. (1) The bands represent reaction zones developed between chemically incompatible rock types. (2) The bands represent synmetamorphic dikes or veins, in some cases formed by anatexis (Sawyer and Robin, 1986). (3) The bands develop due to preferential nucleation of

Figure 23.16 Banded (gneissic) structure in Cranberry Gneiss near Roan Mountain, Tennessee.

particular phases in preexisting structural zones (Bramwell, 1985). (4) The bands form via some combination of shearing, solution, and precipitation (M. B. Stephens, Glasson, and Keays, 1979). All of the above models may be possible, with each pertaining to particular cases. In the first two, continued metamorphism would presumably homogenize the rocks. In (3) and (4) (which may represent the more widely applicable models and the only models of metamorphic differentiation *sensu stricto*), the bands are produced and *enhanced* by the metamorphic process. In such cases, high strain may overcome the influence of the chemical potential gradient resulting from chemical differences. In other cases, the rates of solution and precipitation may vary and be enhanced in phyllosilicate layers by the growing physical differences (e.g., porosity differences) between layers (M. B. Stephens, Glasson, and Keays, 1979). In still other cases, physical separation of minerals with various mechanical properties may accompany deformation at high strain rates. Because all of these processes seem viable, each case must be assessed on its own merits.

SUMMARY

Metamorphism is the textural, structural, and mineralogical response of rocks subjected to the agents of pressure, temperature, deviatoric stress, and fluid phase under conditions between those of diagenesis and those of anatexis. Those conditions generally fall in the range $T = 100$–$750°$ C and $P = 0.1$ Gpa to 1.5 Gpa (1–15 kb). A fluid phase is commonly a critical part of the system during metamorphism and, though typically dominated by H_2O, may contain significant quantities of or be dominated by CO_2, CH_4, or other components. Each of the agents of metamorphism (pressure, temperature, chemically active fluid, deviatoric stress) may dominate during a given metamorphic event, but most commonly, two or more of these agents act in concert. Either local or regional forms of metamorphism may result.

Metamorphic rocks are commonly distinctive because of their structures and textures. In outcrop, the association of folds, faults, veins, boudins, and bands, with foliation, the dominant fabric element, is characteristic. Rock cleavage—which straddles the boundary between nonpenetrative structures and generally penetrative fabric elements (i.e. the textures)—is the most common feature of metamorphic rocks. Continuous cleavages are closely spaced (<0.01 mm), pervasive at the handspecimen scale, or both, and are distinguished from spaced cleavages that are characterized by a wider spacing (>0.01 mm). The dominant process in cleavage formation is syntectonic recrystallization, but mechanical rotation of phyllosilicates and pressure solution occur locally. Metamorphic layering may be a relict structure or it may result from intrusion, veining, transposition of bedding, or metamorphic differentiation.

Textures are a function of grain size, grain shape, grain distribution, grain orientation, and intergrain relationships. The major texture categories are (1) foliated textures, which are characterized by mineral grains showing a dimensional preferred orientation, (2) granoblastic textures, characterized by subequant to equant grains, (3) diablastic textures, in which elongate to tabular grains are radially to randomly arranged, (4) cataclastic textures, generally nonfoliated textures characterized by broken grains or rock fragments, and (5) relict textures, those textures inherited from the protoliths. The development of these various textures through recrystallization, neocrystallization, or both processes, involves nucleation and crystal growth. Both of the latter depend on diffusion, the migration of chemical species through grains or along their boundaries.

EXPLANATORY NOTES

1. This definition follows that of F. J. Turner (1948, p. 3) and is also generally consistent with that of Barth (1962, p. 232), A. R. Philpotts (1990), and other petrologists.
2. Refer to figure 5.1 and the accompanying discussion for a description of these states of a system.
3. See standard texts in structural geology such as G. H. Davis (1984), Suppe (1985), Hatcher (1990), or Twiss and Moores (1992) for more detailed discussions of stress here and in the text that follows.
4. See Cho, Liou, and Bird (1988) and Becker et al. (1989) for a description of very low P metamorphism (at < 0.05 Gpa).
5. In some cases, the term foliation has been applied to layered rocks that lack grain alignment. Such uses of the term should be avoided unless it can be shown that the layering is a metamorphic feature, rather than one inherited from an igneous or sedimentary protolith.
6. For example, Etheridge and Hobbs (1974), P. F. Williams, Means, and Hobbs (1977), and Fernandez (1987).
7. Compare J. C. Maxwell (1962), P. F. Williams, Collins, and Wiltshire (1969), and J. C. Moore and Geigle (1974) with Geiser (1975), Groshong (1976), T. H. Bell (1978), Beutner (1978), D. R. Gray (1978), W. J. Gregg (1985), J. H. Lee et al. (1986), Ishii (1988), and S. J. Sutton (1989). See D. S. Wood (1974) for a review. Also see Lundberg and J. C. Moore (1986).

8. For more information on the melting of ultramafic compositions, see Bowen and Shairer (1935) and Wyllie (1971c), as well as reviews in Wyllie (1971a), Yoder (1976), S. A. Morse (1980), and chapters 5 and 6 of this text.

9. Examples of unusual thermal conditions of metamorphism may be found in contact metamorphic environments discussed in chapter 26. For additional references to unusual conditions see references in note 16. See Bayly (1968), Winkler (1979), and F. J. Turner (1981) for additional discussions of the limits of metamorphism.

10. Wyllie (1971a) and standard geophysics texts such as A. H. Cook (1973) and Stacey (1977) provide discussions of these phenomena.

11. Frictional heating is discussed by Scholz (1980), Turcotte and Schubert (1982, pp. 189–190), and Molnar, Chen, and Padovani (1983).

12. Note that the same kind of fluid (i.e. a supercritical fluid phase) evolves during the origin of granitoid pegmatites (see chapter 12).

13. Also see Mariner and Wiley (1976), Sorey (1985), Alt et al. (1986), Hulen and Nielson (1986), and A. J. R. White (1986).

14. For additional information on fluids in metamorphism see Trommsdorff and Skippen (1986), Bickle and McKenzie (1987), Brady (1988), Ferry (1987; 1988), S. M. Peacock (1989), Ferry and Dipple (1991), the reviews of Rumble (1989) and Torgerson (1990), and the references therein.

15. Touret (1970), Roedder (1972, 1984, ch. 12, 13), Greenwood (1976, p. 198ff.), Fyfe, Price, and Thompson (1978, ch. 6), Ferry and Burt (1982), H. P. Taylor (1983), Touret and Dietvorst (1983), M. L. Crawford and Hollister (1986), Newton (1986).

16. J. D. Lowell and Guilbert (1970), Beane and Titley (1981), and Guilbert and Park (1986, ch. 5) review and discuss alteration around ore bodies associated with plutons. P. B. Larsen and Taylor (1986) discuss alteration surrounding a caldera.

17. Harker (1932), F. J. Turner (1948, ch. 1; 1981, pp. 3–5), Miyashiro (1973a, p. 22ff.), Mason (1978, p. 4), Winkler (1979, p. 1ff.), Best (1982, p. 348ff.), Suk (1983, ch. 1), Hyndman (1985, ch.13).

18. Shock metamorphism is discussed by Boon and Albritton (1936), Shoemaker (1960), Chao, Shoemaker, and Madsen (1960), Chao et al. (1962), Bunch and Cohen (1964), N. M. Short (1966), Currie (1967), B. M. French and Short (1968), and Spry (1969, pp. 247–249). Metamorphic core complexes are discussed by Coney (1979) and G. H. Davis (1980) and in the various papers in Crittenden, Coney, and Davis (1980). Examples of local dynamic metamorphism in high-temperature peridotites and related rocks are described by Nicolas and Poirier (1976, ch. 10). Dynamic metamorphism associated with fault zones is described by Beach (1980) and R. Kerrich et al. (1980). Associated metamorphic and/or deformational effects along discrete fault zones and small-scale shear zones are described by Brabb, Maddock, and Wallace (1966), Mosher (1980), and M. J. Watts and Williams (1983). Also refer to Spry (1969) and A. J. Barker (1990) for discussions of these various types of metamorphism and the resulting textures.

19. Mylonites, rocks produced by ductile deformation, are defined and described elsewhere in this text (ch. 30). Examples of regional shear zones, typically characterized by mylonites, are described by Theodore (1970), Roper and Justus (1973), Bak, Korstgard, and Sorenson (1975), Hatcher et al. (1979), Beach (1980), and C. Simpson (1985). Also see J. G. Ramsay and Graham (1970), Nicolas et al. (1977), and J. G. Ramsay (1980) . Melanges and associated broken and dismembered formations are discussed by K. J. Hsu (1968, 1969, 1974), Berkland et al. (1972), and Raymond (1975, 1984a), as well as in the various papers in Raymond (1984c) and Horton and Rast (1989). Also see papers in J. C. Moore (1986).

20. Bates and Jackson (1980).

21. Discussions of static metamorphism (typically under the heading "burial metamorphism") are provided by Coombs (1960, 1961), Dickinson et al. (1969), and Ernst (1971a,b,c), and in standard metamorphic texts such as F. J. Turner (1981) and Yardley (1989).

22. H. Ramberg (1952, p. 229). For examples, see Holser (1950), Floyd (1975), Joesten (1977), S. E. Swanson (1981), and Bowman, O'Neil, and Essene (1985).

23. For discussions of these processes see H. H. Read (1944, 1948), Ramberg (1952), Sederholm (1967), and Beloussov (1980, pp. 306–314). Guilbert and Park (1986, ch. 5) discuss alteration. Granitization was discussed in chapter 11. For more information on that process, see that chapter and the references therein.

24. The origins of the structures mentioned here are described in the text below and are discussed in more detail in various books on structural geology, such as J. G. Ramsay (1967), Davis (1982), G. Wilson (1982), Suppe (1985), Dennis (1987), and Twiss and Moores (1992).

25. Sander (1930, in Turner and Weiss, 1963; Sander, 1970) introduced the terminology now used to define the structural character of metamorphic rocks. He used the German word *Gefuge*, translated as "fabric," to refer to the geometrical character of rocks discussed here (Paterson and Weiss, 1961; Turner and Weiss, 1963; Sander, 1970; Bates and Jackson, 1987).

26. Additional tectonite types have been defined. See Turner and Weiss (1963), Sander (1970), Schwerdtner, Bennett, and Janes (1977), and Shelly (1989), and see Raymond (1987) for a review.

27. Scanning electron microscopy (SEM) has now shown that even those cleavages dominated by slickensided shear surfaces consist of small domains of aligned phyllosilicates (J. C. Moore et al., 1986). Rocks with such cleavages, nevertheless, are characterized by brittle shear fractures and are fundamentally different in structure than mylonitic S-tectonites. The SF-tectonites exhibiting such structures are foliated cataclasites (Chester, Friedman, and Logan, 1985; Kano and Sato, 1988).

28. The definition of *relict* or *palimpsest* offered here is essentially the same as that of Bates and Jackson (1987). Their definition follows that used in older works (Grout, 1932).

29. Some geologists (G. Wilson, 1982) continue to use this type of binary classification of secondary cleavage types.

30. The volume of literature and the diversity of ideas on cleavage formation is considerable. A detailed treatment is beyond the scope of this text. The interested reader should consult J. C. Maxwell (1962), J. G. Ramsay (1967, p. 177ff.),

P. F. Williams, Collins, and Wiltshire (1969), J. C. Moore and Geigle (1974), D. S. Wood (1974), Geiser (1975), Means (1975), Alvarez, Engelder, and Lowrie (1976), D. R. Gray (1976, 1978, 1979b, 1981), Groshong (1976), T. E. Tullis (1976), Beutner, Jancin, and Simon (1977), Tobisch et al. (1977), P. F. Williams, Means, and Hobbs (1977), Alvarez, Engelder, and Geiser (1978), T. H. Bell (1978a), Beutner (1978), Etheridge and Wilkie (1979), Lebedeva (1979), M. B. Stephens, Glasson, and Keays (1979), Borradaile, Bayly, and Powell (1982), D. S. Cowan (1982), Onasch (1983), W. J. Gregg (1985), B. E. Hobbs (1985), Ogawa and Miyata (1985), Rosenfeld (1985); J. H. Lee et al. (1986), Ishii (1988), S. M. Agar, Prior, and Behrmann (1989), Lash (1989), S. J. Sutton (1989), Bhagat and Marshak (1990), Kanagawa (1991), the papers on cleavage in Stauffer (1983), and the papers referred to in the reference lists in these various reports, for more information on conflicting views on cleavage origins. The work of Borradaile, Bayley, and Powell (1982) provides a good introduction to cleavage studies for those who wish to expand their understanding.

31. The origin of stylolitic cleavage is also discussed by Stockdale (1943), Heald (1955), Logan and Semeniuk (1976), Alvarez, Engelder, and Lowrie (1976), Alvarez and Engelder (1982), Borradaile (1982), and Engelder and Alvarez (1982). The origin of crenulation cleavage is also discussed by Cosgrove (1976), D. R. Gray (1976, 1978, 1979b) and T. H. Bell and Rubenach (1980). For more information and excellent photographs, see Borradaile, Bayley, and Powell (1982b). Pressure solution, an important process in the formation of stylolitic and other cleavages is discussed by Renton et al. (1969), A. Beach (1974, 1979), A. Beach and King (1978), and P.-Y. F. Robin (1978).

32. See the references to papers on the development of slaty cleavage in note 30, such as Geiser (1975), Maxwell (1962), and Lee et al. (1986).

33. Phaneritic metamorphic rocks with handspecimen-scale layers are called gneiss and those with megalayers are here called megagneiss.

34. Gneisses, as defined here, are characterized by a layered or banded structure and like texture. The two main types of phaneritic, strongly foliated textures are the schistose and gneissose textures. In much of the literature and as used by many geologists, definitions of gneissose texture and the corresponding rock type, gneiss, and the distinctions between schistose and gneissose textures and between schists and gneisses, are at best ambiguous. Harker (1932, p. 61) acknowledged the problem in describing the term gneiss as "vague and unsatisfactory."

Three uses of the term gneiss are currently applied by geologists. In one use, gneiss is a phaneritic, metamorphic rock of any composition characterized by alternating bands of contrasting mineral composition, at least some of which are characterized by the preferred orientation of the included minerals (Raymond, 1984c; R. S. Mitchell, 1985; cf. Bergman, 1784, in Tomkeieff, 1983). Here, both bands and preferred orientation are considered to be definitive characteristics. A second definition, based on both common practice and the definition of Werner (1787, in Tomkeieff, 1983), is that gneiss is a phaneritic, metamorphic rock of

granitoid composition with a preferred orientation of minerals. In this definition, the granitoid composition and preferred orientation are central, but banding is not required. A third definition is that a gneiss is a phaneritic, metamorphic rock of any composition that has a preferred orientation of some minerals, but that lacks continuous cleavage (L. Acker, pers. comm., 1987). In the latter case, gneiss is contrasted with schist, which has continuous cleavage. The word foliation has been judiciously avoided in these definitions because some geologists consider *foliation* to be synonymous with continuous cleavage, whereas others use it, as I do in this book, to refer to a visible, dimensional, preferred orientation.

Note that two of the contrasting definitions are based on definitions first presented in the eighteenth century. The controversy is long-standing. The fact is, there is a continuum of rock types extending from phyllosilicate-rich rocks with continuous cleavage to quartz- and feldspar-rich, banded rocks with spaced cleavage. The boundary between the two rock types, schist and gneiss, and between their corresponding textures, is arbitrarily chosen. Some definitions proffered in the last half century have added to the confusion. For example, C. M. Rice (1941) defines gneiss as "a foliated or banded crystalline rock in which granular minerals, or lenticles and bands in which they predominate, alternate with schistose minerals, or lenticles and bands in which they predominate. . . . It is most commonly of the same composition as granite." A similar definition is given by Bates and Jackson (1987). There are several problems with this definition: (1) linking the composition to granite supports the definition of Werner; (2) the suggestion that banding may occur supports the definition of Bergman; and (3) the suggestion that lenticles can occur in gneiss or schistose materials does not provide for a distinction between gneiss and schist, inasmuch as the latter is typically composed of lenticles of granular minerals separated by layers or lenticles of platy and/or acicular minerals.

In the past, many geologists have applied the terms schist and gneiss loosely. Rocks with abundant mica were called schist. Those not dominated by mica, especially those with considerable amounts of quartz and feldspar or those containing augen, were called gneiss—regardless of the presence or absence of continuous cleavage, banding, or granitoid composition. Precision of language and clarity of communication require that such applications of these terms be abandoned.

35. Metamorphic layers, cleavage, and relict bedding are all crudely planar features found in metamorphic rocks. Following the lead of Sander (1930, in Turner and Weiss, 1973), petrologists and structural geologists refer to all planar features in metamorphic rocks as S-surfaces. Bedding is designated S_0 and successively younger S-surfaces are designated $S_1, S_2, \ldots S_n$.

36. The interested reader should consult modern structural geology texts (see note 3) for discussion of the differences between brittle and ductile deformation. In a very general sense, ductile deformation involves flow and recrystallization, whereas brittle deformation involves breaking of materials.

37. G. Wilson (1982, ch. 9) reviews the history of the use and misuse of the term boudin. See J. G. Ramsay (1967) for a detailed discussion and analysis of boudins, related "chocolate tablet" structure, and the process of boudinage.

38. See G. Wilson (1982) for a discussion of the origin of mullions.

39. J. G. Ramsay (1980) and J. G. Ramsay and Huber (1987, Session 26) discuss brittle and ductile shear zones. Also see Nicolas et al. (1977), Sibson (1980), L. Anderson et al. (1983), C. Simpson (1983), and Blenkinsop and Rutter (1986). Chapter 30 focuses on mylonites. Refer to that chapter for details and references to the subjects of ductile deformation, cataclasis, and mylonite formation.

40. Additional information on textures is available in Spry (1969), Nicolas and Poirier (1976), Wenk (1985), Bard (1986), A. J. Barker (1990), and Yardley, MacKenzie, and Gilbert (1990).

41. Becke (1903, in Grout, 1932, p. 353) introduced the use of the root *blast* ("to sprout") to refer to metamorphic textures.

42. The term granoblastic has been used in several ways. As defined here, the use is much like Harker's (1932) original use. He did, however, specify that granoblastic textures were mosaic textures, a term that likewise has several meanings (Spry, 1969, p. 186). Heinrich (1956) also refers to granoblastic texture as a mosaic texture. Spry (1969) specifies that the texture be equidimensional-xenoblastic (composed of equant grains lacking crystal faces). Bard (1986, p. 182) defines the texture as one in which there is no preferred orientation of minerals. Because determination of the presence of preferred orientation (or demonstration of its absence) in rocks with equant minerals like quartz or olivine typically requires time-consuming petrofabric analysis, such a stricture renders the term useless for normal petrographic work and is rejected here.

43. Such textures have been loosely assigned to the foliated category by many petrologists simply because they are composed of inequant mineral grains. In spite of that, the textural arrangement of grains neither promotes the breaking of rocks with this texture into flakes, tabular pieces, or folia; nor does it give the rock the appearance of having aligned grains. Therefore, the use of the term foliated for such textures and the use of the word schist for rocks exhibiting such textures should be abandoned.

44. The Free Energy of a system can be considered in a general way to be the capacity of the system to do work. Additional discussion and quantitative treatment of this topic may be found in standard texts on thermodynamics, physical chemistry, or geochemistry (e.g. Krauskopf, 1967). Also see Ehrlich et al. (1972).

45. See Fyfe et al. (1958), Spry (1969), Vernon (1976, ch. 3), Gottstein and Mecking (1985), Rubie and Thompson (1985), Bard (1986), and Joesten and Fisher (1988) for more thorough discussions of nucleation and crystal growth.

46. Diffusion is discussed in more detail by Fyfe et al. (1958, p. 60ff.), Spry (1969), H. W. Green (1970), G. W. Fisher (1973, 1978), Vernon (1976), Walther and Wood (1984), Joesten (1985), Lasaga (1986), Wheeler (1987), and Joesten and Fisher (1988).

47. See Spry (1969), Nicolas and Poirier (1976, p. 163ff.), Vernon (1976, ch. 5), Ricoult (1979), and Mercier (1985).

48. The works of Nicolas and Poirier (1976) and in H. R. Wenk (1985) focus on problems related to the development of preferred orientations. Also see N. L. Carter, Christie, and Griggs (1964), H. W. Green (1967), Ave Lallemant and Carter (1970), J. Tullis, Christie, and Griggs (1973), C. J. L. Wilson (1973), Poirier and Nicolas (1975), T. H. Bell (1979), Etheridge and Wilkie (1979), Lister and Hobbs (1980), Zeuch (1983), Zeuch and Green (1984a,b), Toriumi and Karato (1985), Dell'Angelo and Tullis (1986), Shelley (1989), Ji and Mainprice (1990), Ildefonse, Lardeaux, and Cason (1990), H. R. Wenk and Pannetier (1990), and the references therein.

49. A. G. Sylvester and Christie (1968) describe such a sequence in deformed quartz arenites in the Inyo Mountains of California, C. J. L. Wilson (1973) describes a similar sequence in the same kind of rocks from the Mount Isa area of Australia, and Masuda (1982) details like changes in quartz schists from Shikoku, Japan. A detailed discussion of the causes of these changes is beyond the scope of this text. For more information, see the works listed in note 22 and in Means (1980).

50. For example, see Raleigh (1968), Ave Lallemant and Carter (1970), Lister and Hobbs (1980), Zeuch and Green (1984a), Ave Lallemant (1985), Mercier (1985), G. P. Price (1985), and Blumenfeld, Mainprice, and Bouchez (1986).

51. Suk (1983) and Hyndman (1985) summarize explanations and examples of metamorphic differentiation, and F. J. Turner (1948) and H. Ramberg (1952) devote chapters to the subject. Also see Dietrich (1963), Vidale (1974), Vernon (1976), Gray (1977), Dick and Sinton (1979), M. B. Stephens, Glasson, and Keays (1979), C. Simpson (1983), Bramwell (1985), and Sawyer and Robin (1986).

PROBLEMS

23.1 Examine figure 23.4. Consider a rock with the assemblage quartz + biotite + andalusite + muscovite + garnet + plagioclase, with veins of laumontite cutting the grain boundaries between andalusite and quartz. Describe the metamorphic history and explain whether the metmorphism indicated is prograde or retrograde.

23.2 Compare the texture in figures 23.12b with that in figure 23.11e. Note that the two rocks mineralogically are almost identical. Provide a general explanation, based on your present knowledge, of how a dunite like that in figure 23.12b might be converted to one with a texture like that in figure 23.11e. Comment on processes and their effects.

23.3 Joesten and Fisher (1988) provide information on wollastonite grain sizes in the Christmas Mountains contact aureole for which chert grain sizes are plotted in figure 23.13. The values for matrix wollastonite are 0.011 mm at 101.7 m, 0.053 mm at 46.5 m, and about 0.25 mm at 20 m from the contact. Plot these data on a copy of figure 23.13 and compare and contrast the plotted curve with the chert grain size data. (*Note:* The contact aureole also contains large wollastonite porphyroblasts with diameters of more than 20 mm. See Joesten (1983) and Joesten and Fisher (1988) for a discussion of how the various textures develop.)

24

Metamorphic Conditions, Mineralogies, Protoliths, Facies, and Facies Series

INTRODUCTION

Metamorphic rocks may exhibit a wide range of chemistry and minerals. The mineral composition reflects the conditions of metamorphism. Rock chemistry controls the mineral composition and, as mentioned in chapter 23, it also controls the upper and lower limits for metamorphic conditions. This chapter summarizes the common minerals of metamorphic rocks, notes the possible range of mineralogies and chemistries, discusses the significance of minerals and chemistry, and links minerals and rock chemistry through the concept of metamorphic facies.

The concept of metamorphic facies connects the observed mineralogical composition of metamorphic rocks to their conditions of metamorphism. A series of facies in a mountain belt reflects the tectonic setting of the metamorphism.

MINERALOGY, PROTOLITHS, AND ROCK CHEMISTRY

Petrographic observations are a powerful tool for deciphering the basic aspects of metamorphic rock history. In particular, the minerals present in a metamorphic rock can be used to identify the P-T conditions under which the rock formed. Although many of the minerals present in metamorphic rocks also occur in igneous and sedimentary rocks, additional minerals particularly characteristic of metamorphic rocks comprise a significant part of some rocks and distinguish the metamorphic rocks from rocks of other classes.

Minerals are also important, in part, because they reveal something of the chemistry of the protolith. Any kind of rock can be metamorphosed. Consequently, the range of protoliths for metamorphic rocks encompasses all rock types and rock chemistries that exist in nature. In order to make the discussion of metamorphic rocks manageable, the range of rock chemistries is divided into several groups—ultrabasic (silicate), basic, carbonate/nonsilicate, aluminous, siliceous-alkali-calcic, and silicic rocks. Ultrabasic (silicate) rocks are those with silica contents of less than 45%. Basic rocks have silica contents between 45 and 52%. The terms carbonate and nonsilicate are self-explanatory. Here, aluminous rocks are those siliceous rocks that are relatively rich in alumina, the term being akin to the term peraluminous used for igneous rocks. Siliceous-alkali-calcic rocks are rocks with moderate to abundant amounts of lime *and* the alkalis. Siliceous rocks are those that are very rich in silica (> 90% SiO_2). Each chemical group is characterized by a particular set of minerals (table 24.1). Differences in mineral content among the various

Table 24.1 Selected Common Minerals of Metamorphic Rock Chemical Groups

Ultrabasic Rocks	Basic Rocks	Carbonate Rocks	Aluminous Rocks	Siliceous-Alkali-Calcic Rocks	Siliceous Rocks
Olivines	Augite	Calcite	Quartz	Quartz	Quartz
Augite	Omphacite	Dolomite	White micas	Plagioclases	Plagioclases
Diopside	Jadeites	Aragonite	Biotite	Alkali feldspar	Alkali feldspar
Orthopyroxenes	Orthopyroxenes	Olivine	Chlorites	Chlorites	Biotite
Tremolite	Glaucophane	Diopside	Plagioclases	Biotite	White micas
Anthophyllite	Hornblende	Tremolite	Alkali feldspar	White micas	Chlorites
Serpentines	Actinolite	Wollastonite	Pyrophyllite	Sillimanite	Garnets
Chlorites	Epidotes	Talc	Sillimanite	Kyanite	Sillimanite
Talc	Lawsonite	Phlogopite	Kyanite	Andalusite	Kyanite
Phlogopite	Plagioclases	Periclase	Andalusite	Garnets	Andalusite
Chromite	Biotite	Idocrase	Staurolite	Cordierite	Cordierite
Magnetite	Zeolites	Graphite	Garnets	Jadeites	Aegirine
	Quartz	Garnets	Calcite	Lawsonite	Crossite
	Calcite	Pyrite	Chloritoid	Epidotes	Stilpnomelane
	Sphene	Pyrrhotite	Cordierite	Pumpellyite	Hematite
	Garnets		Tourmaline	Zeolites	Magnetite
	Magnetite		Kaolinite	Glaucophane	
	Ilmenite		Magnetite	Calcite	
			Ilmenite	Magnetite	

groups of metamorphic rocks emphasize the point made by Es-kola (1915, in Turner, 1958) that the mineralogy is *controlled* by the chemistry, an indication that metamorphic rocks form in response to processes that are controlled by physical-chemical laws.

Each chemical group includes major metamorphic rock types that represent various protoliths. The ultrabasic group encompasses the metamorphosed silicate-ferromagnesian mineral-rich rocks, including peridotites, pyroxenites, and dunites. The basic group includes metamorphosed basalts and gabbros and related rocks. The carbonate/nonsilicate group includes the marbles, most of which have sedimentary protoliths (e.g., limestone or dolostone), and various other metamorphic rocks derived from evaporites or other less common rocks. The aluminous category refers to aluminum-rich rocks typically characterized by abundant micas and containing aluminum silicates such as kyanite or andalusite. These rocks are principally derived from shales and are referred to as

pelitic (e.g., pelitic schists), but may also be derived from per-aluminous igneous rocks. Most intermediate to siliceous igneous rocks give rise to quartz-feldspar rocks that may be assigned to the siliceous-alkali-calcic group. Feldspathic and lithic sandstones are also protoliths for the metamorphic rocks of this group. Silicic rocks include metamorphosed quartz arenites, quartz wackes, cherts, siliceous sinter, and "silexites."

The chemistries of the various protoliths are represented by the chemistries presented in chapters 3 through 22. Although metamorphic rock chemistries may have been altered by metasomatism, they nevertheless fall in the general range provided by the limits of sedimentary and igneous rock compositions. The general fields for the chemistries of the various protoliths are presented in a triangular plot in figure 24.1. Note that the boundaries of some fields overlap, in part because alumina is not accounted for by the compositions of the corners.

Figure 24.1 Triangular diagram showing fields representing the chemistry of the six general chemical groups of metamorphic rocks. Values are in weight percent. A = aluminous rocks; B = basic rocks; C = carbonate and nonsilicate rocks; SAC = siliceous-alkali-calcic rocks; S = siliceous rocks; U = ultrabasic (ferromagnesian) silicate rocks.

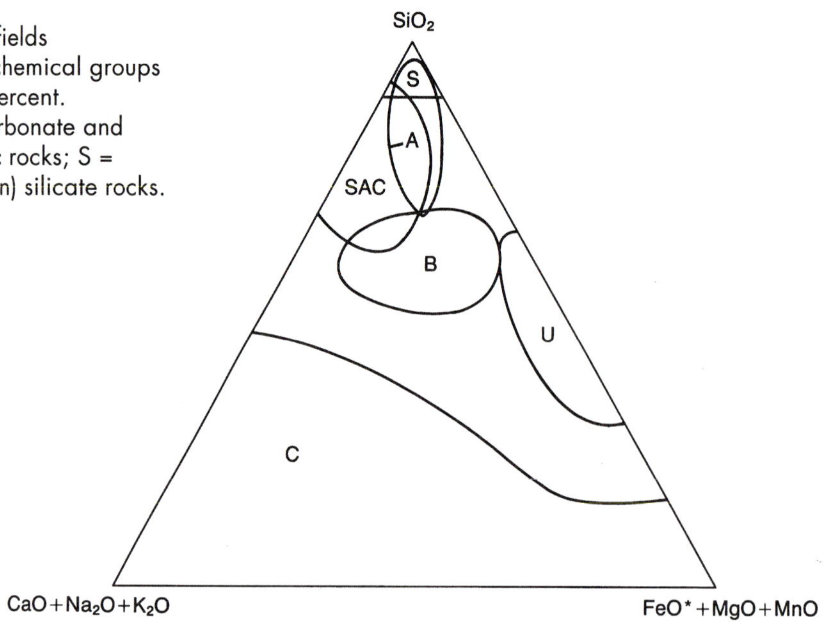

CLASSIFICATIONS OF METAMORPHIC ROCKS

Unlike igneous and sedimentary rock names, many metamorphic rock names are based entirely on rock texture. Others are based only on mineral content. This makes metamorphic rock classification somewhat different from igneous and sedimentary rock classification. In several texts, metamorphic rock names are simply presented on a list.[1]

Published classifications are based on texture, mineralogy, chemistry, or combinations of these parameters.[2] It is not uncommon to find classifications in which some names are based on mineralogy and others are based on texture, because that is the *common practice* among geologists.[3] Thus, schist, a name based solely on the texture of the rock appears in the same classification with marble, a name based primarily on composition.

Textural Classifications

Several petrologists have adopted metamorphic rock classifications based primarily on textural criteria. These are designated here as textural classifications. Spock (1962), W. T. Huang (1962), Best (1982), and Raymond (1984c) have adopted this approach. Spock (1962, p. 241) divides metamorphic rocks into two main categories—those with parallel structure and those lacking parallel structure (figure 24.2a).

(a)

Rocks with visible parallel structure (foliates and banded rocks)	
Slate	Schist (continued)
Mylonite (in part)	Tremolite
Phyllite	Actinolite
Schist	Staurolite
Muscovite	Graphitic
Chlorite	Gneiss
Talc	Granitic, diorite, etc.
Biotite	Hornblende
Quartz-mica	Biotite
Garnetiferous	Banded
Hornblende	Augen
Rocks apparently lacking parallel structure	
Quartzite	Soapstone
Marble	Amphibolite
Dolomitic	Granulite
Serpentine, etc.	Eclogite
Hornfels	

(b)

Strongly foliated rocks	Weakly foliated rocks	Non-foliated to weakly foliated rocks
Slate	Gneiss	Granofels
Phyllite	Migmatite	Amphibolite
Schist	Mylonite	Serpentinite
		Greenstone
		Greisen
		Hornfels
		Quartzite
		Marble
		Argillite
		Skarn

Figure 24.2 Two classifications of metamorphic rocks based on texture. a) Binary classification of Spock (slightly modified from Spock, 1962). b) Threefold classification of Best (1982).

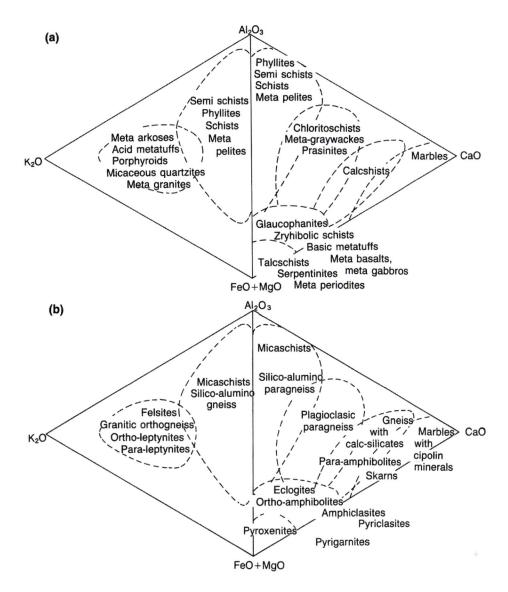

(a)

Al$_2$O$_3$

Phyllites
Semi schists
Schists
Meta pelites

Semi schists
Phyllites
Schists
Meta
pelites

Chloritoschists
Meta-graywackes
Prasinites

Meta arkoses
Acid metatuffs
Porphyroids
Micaceous quartzites
Meta granites

K$_2$O

Marbles CaO

Calcshists

Glaucophanites
Zryhibolic schists
Basic metatuffs

Talcschists
Serpentinites
Meta basalts,
meta gabbros
Meta periodites

FeO + MgO

(b)

Al$_2$O$_3$

Micaschists

Micaschists
Silico-alumino
gneiss

Silico-alumino
paragneiss

Felsites
Granitic orthogneiss
Ortho-leptynites
Para-leptynites

K$_2$O

Plagioclasic
paragneiss

Gneiss
with
calc-silicates

Marbles
with
cipolin
minerals

CaO

Para-amphibolites

Skarns

Eclogites
Ortho-amphibolites

Amphiclasites
Pyriclasites

Pyroxenites

Pyrigarnites

FeO + MgO

Figure 24.3 Bard's chemical classification of metamorphic rocks: a) for low-grade rocks, b) for medium- to high-grade rocks.
(From Bard, 1986)

Best (1982, p. 390) adopts three categories—strongly foliated rocks, weakly foliated rocks, and nonfoliated to weakly foliated rocks (figure 24.2b). W. T. Huang (1962, pp. 384–85) also uses three categories—cataclastic, nonfoliated, and foliated. He includes mylonites as cataclastic rocks, a practice that has been abandoned, because of the current understanding that mylonitic rocks are rocks with textures resulting primarily from recrystallization.

Raymond (1984c, pp. 128–29) adopts two main categories—crystalline and clastic rocks—which are further subdivided. Clastic rocks are subdivided into foliated and nonfoliated types. Crystalline rocks are subdivided into strongly foliated, weakly foliated, granoblastic, and diablastic types. The classification of Raymond (1984c) has been adopted here, but has been modified in an attempt to produce a more rational (albiet less conventional) classification (table 24.2).

Other Classifications

Two other classifications, one based primarily on chemistry and one based primarily on mineralogy, are presented here for the purpose of demonstrating that nontextural classifications are possible and have been proposed. These are the classifications presented in Winkler (1979) and Bard (1986). Bard's (1986) classification is a double triangle, chemical classification in which four endmember components are used as parameters of classification (figure 24.3). These are K$_2$O, Al$_2$O$_3$, CaO, and FeO + MgO. Two versions are presented, one for low-grade rocks and one for higher-grade rocks. Many of the names used by Bard are unusual (e.g., amphiclasite, zryhibolic schist) or archaic (e.g., prasinite, leptynite) and no definitions or explanations are provided.

Table 24.2 Classification of Metamorphic Rocks

Texture and Composition	Root Name	Examples of Names[1]
Crystalline Rocks		
STRONGLY FOLIATED		
Slaty	Slate	Black Slate
Phyllitic	Phyllite	Quartz-chlorite Phyllite
Schistose	Schist	Biotite-quartz-white mica Schist
Serpentine-rich		Serpentine Schist (Serpentinite)
Hornblende-rich		Hornblende Schist (Amphibolite)
Calcite-rich		Calc-schist
Gneissose	Gneiss	Biotite-quartz-plagioclase Gneiss
Mylonitic	Mylonite	Quartz-chlorite Mylonite
Protomylonitic	Protomylonite	Biotite-quartz Protomylonite
Orthomylonitic	Orthomylonite	Chlorite-quartz Orthomylonite
Ultramylonitic	Ultramylonite	Muscovite-quartz Ultramylonite
WEAKLY FOLIATED		
Semi-slaty	Semi-slate *or*	Maroon Semi-slate
	Argillite	Black Argillite
Semi-schistose	Semischist	Biotite-garnet-sillimanite-
	or	plagioclase-quartz Semischist
	prefix *meta-*	Porphyroclastic Meta-dunite
	followed by a	Muscovite Meta-arkose
	protolith name	Epidote Metabasalt
		Serpentine Semi-schist
		(Serpentinite)
		Hornblende Semi-schist
		(Amphibolite)
		Calcite Marble Semi-schist

The classification presented by Winkler (1979, pp. 340–344) was developed by Austrian petrographers. It, too, has a double triangle shape (figure 24.4). The classification is quantitative and the classification parameters are the minerals quartz, carbonate, micas, and feldspars. Textural terms are used as a secondary parameter for naming rocks. Among the drawbacks of this classification are (1) it employs commonly used textural names for compositional categories, and (2) it is ambiguous in specifying that textures may override the names based on the mineralogical categories.

CONDITIONS OF METAMORPHISM AND PETROGENETIC GRIDS

Metamorphism is usually described in terms of the general field of P-T space in which it occurs (figure 23.2). The limits of this P-T space are based on experimentally determined conditions, such as the melting of wet granite or the P-T conditions for crystallization of certain phases. Within these limits, it is possible and desirable to divide P-T space into

Table 24.2 Classification of Metamorphic Rocks

Texture and Composition	Root Name	Examples of Names[1]
DIABLASTIC	Diablastite	Biotite Diablastite
		Serpentine Diablastite
		(Serpentinite)
		Hornblende Diablastite
		(Amphibolite)
GRANOBLASTIC	Granoblastite	Garnet-quartz-plagioclase
		Granoblastite
Varieties include:	Marble-granoblastite	Calcite Marble-granoblastite
		Dolomite Marble-granoblastite
	Tactite (Skarn)	Garnet-epidote Tactite
	(Ca-silicate-rich rock)	
	Metaquartzitic	Muscovite Metaquartzitic
	Granoblastite	Granoblastite
	Eclogite	Phlogopite Eclogite
	(Pyrope-omphacite rock)	
Clastic Rocks		
FOLIATED		
Foliated cataclastic	Quasimylonite	Polymict Quasimylonite
		Illite-quartz Quasimylonite
NON-FOLIATED		
Cataclastic	Cataclasite	Plagioclase-quartz Cataclasite
Mortar	Breccia	Rhyolite Breccia
	or	
	Cataclasite	Rhyolite Cataclasite
Vitriclastoblastic	Pseudotachylite	Quartz-plagioclase Pseudotachylite

[1]Minerals are placed before the root name in order of increasing abundance, with the most abundant mineral adjacent to the root name.

subregions that reflect more restricted or specific conditions of metamorphism. Similarly, subdivision of P-X (X = composition) or T-X space is possible.

In practice, a metamorphic petrologist observes mineral assemblages in rocks, not P-T conditions. In order to link those assemblages to specific conditions of metamorphism, it is essential to know the P-T conditions of stability (and metastability) of any given assemblage that has been observed. Experimental laboratory analyses of phase relations provide that link. Once the stability of a mineral or assemblage is known. The mineral or assemblage can be assigned properly to a region in P-T space. Numerous phase studies

have been completed (see appendix C). Thus, P-T space has been subdivided into regions characterized by particular minerals or mineral assemblages, each corresponding to a particular, limited set of P-T conditions. Bowen (1940) referred to graphs of subdivided P-T space, with regions characterized by particular phases or phase assemblages, as **petrogenetic grids.**

Conditions of Metamorphism

In order to determine the limits of metamorphic conditions, it is necessary to specify exactly what physical-chemical indicators mark those limits. It was noted in chapter 23 that the

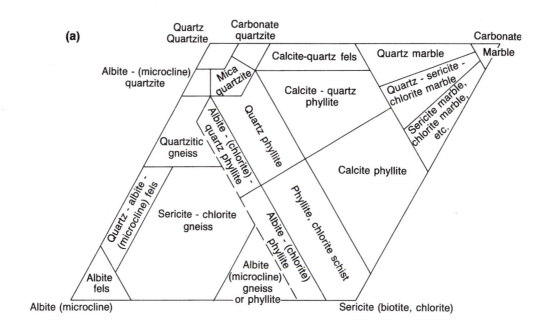

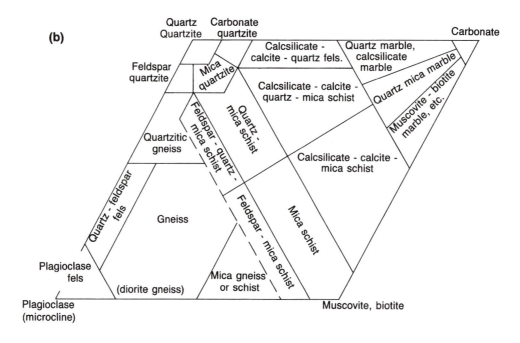

Figure 24.4 The Austrian mineralogical classification of metamorphic rocks: a) for low-grade rocks and b) for high-grade rocks.
(From Winkler, 1979)

temperature limits of metamorphism are approximately 100° C and 600–2000°C. At the lower limit, the beginning of metamorphism is marked by the first appearance of a mineral, mineral assemblage, or texture that is not stable under the surface to near-surface conditions of weathering and diagenesis. Knowledge of the stability limits of that mineral, as-

semblage, or texture is based on experimental analysis and field measurements. At the upper limit, the end of metamorphism, marked by the beginning of melting, is also determined by experiment. Similarly, pressure and fluid phase limits require experimental or theoretical delineation.

The Beginnings of Metamorphism

To recognize the beginnings of metamorphism, one needs to know the nature of weathering and diagenesis (see chapter 16) and to recognize the occurrence of minerals and textures not associated with those processes. Recall that weathering commonly produces such minerals as calcite, the clays, and metallic oxides like hematite, goethite, or todorokite.[4] Similar minerals, plus additional phases like dolomite and phillipsite, are diagenetic products.

What minerals mark the inception of metamorphism? Some zeolites, such as phillipsite, form in sediments on the ocean floor[5] and are, therefore, diagenetic; but other zeolites such as wairakite, heulandite, and laumontite do not appear under diagenetic conditions. Therefore, we reason that the appearance of these minerals, as well as white mica, certain chlorites, and other minerals that appear *only* in rocks buried beyond the levels of weathering or heated beyond the ambient temperature of the surface zones of the Earth, must mark the beginning of metamorphism.

By definition, the conditions at the beginning of metamorphism must exceed the surface conditions, which lie in the range $T = < 0–60°C$ and $10^5 Pa–0.1$ Gpa (0.001–1 kb).[6] Though some studies have been completed, experimental syntheses of the minerals that form under the lowest P-T conditions of metamorphism are fraught with difficulties because of the slow reaction rates under such low-energy conditions.[7] For example, Liou (1971c) showed that laumontite formed from stilbite at about 150°C and 0.1 Gpa (1 kb) and Thompson (1970) demonstrated that kaolinite and quartz react to form pyrophyllite at 0.1 Gpa (1 kb) and 325°C. Yet, because experimental data are inadequate, we rely considerably on observations of *where* the minerals in question occur in nature as a basis for inferring *what* the conditions of formation were.[8]

Such observations have been made in a number of tectonic settings. For example, Alt et al. (1986) and Becker et al. (1989) discovered that Na-zeolite, laumontite, talc, and mixed-layer chlorite-smectite formed at shallow levels in the oceanic crust, representing conditions with lower limits of about 100°C and 0.02 Gpa (0.2 kb). Similar data are derived from sedimentary basins (C. E. Weaver, 1960, 1984, ch. 8; B. M. French, 1973; Hower et al., 1976) and the Salton Sea geothermal field (Muffler and White, 1969; Cho, Liou, and Bird, 1988). Together, experiments and observations all suggest lower limits of metamorphism of about 50–150°C and 0.02–0.15 Gpa (0.2–1.5 kb).

Diagenesis occurs over a range of low temperatures and low pressures (see chapter 16). Review of the P-T conditions of diagenesis reveals that they overlap with those of the lower limits of metamorphism. This is possible because of variations in rock and fluid phase compositions and in temperature. Diagenetic minerals can form at depths of burial that, given different bulk rock or fluid phase compositions or temperature, would yield metamorphic minerals. Consequently, the lower limit of metamorphism, as shown in figures 24.5a and 23.2, is a broad boundary zone.

The Upper Limit of Metamorphism

The upper limit of metamorphism is more diffuse than the lower limit, but for different reasons. Recall that the upper limit is marked by the beginning of melting. Experimental analyses of the conditions of beginning melting for various bulk rock compositions provide well-defined P-T values for that anatexis.[9] Yet, because the bulk rock chemistry has such a profound influence on the temperature of melting, the temperature range for the upper limit of metamorphism is 600–2000°C. The solidus for wet granite and the liquidus for dry dunite bracket that range of conditions. At a specific temperature between those bracketing values, a quartz-feldspar rock might melt entirely, a pelitic or amphibole schist might melt partially, and a basalt might recrystallize into an olivine-pyroxene-plagioclase granoblastite without melting at all.

In terms of pressure, upper mantle pressures of about 1.2–1.5 Gpa (12–15 kb) provide the upper limit *normally considered* in discussions of metamorphism (figure 24.5). (Considering that the mantle and core rocks are, in fact, metamorphic, the extreme pressure in the center of the core provides the actual upper limit.) As is the case with temperature, pressure-controlled melting is a function of bulk rock and fluid phase compositions.

The presence or absence of a fluid phase provides an important control on melting and hence on the upper limit of metamorphism. For example, the melting temperature of granitic rock may be depressed by as much as 450°C by the addition of a fluid.[10] Similarly, the presence of a fluid phase depresses the melting temperature of ultrabasic rocks by values of up to 450°C.[11]

Clearly, then, the upper limit, like the lower limit of metamorphism, cannot be defined as a single, fixed value. Rather, the limit may be specifically defined for particular bulk rock compositions at specified P-T conditions, and a fluid phase composition. Consequently, that boundary too, as shown on figure 24.5a, is a broad boundary zone.

Petrogenetic Grids, Geothermometry, and Geobarometry

Once the P-T limits of metamorphism are established, either in a general way for rocks of variable bulk composition or more specifically for rocks of particular bulk compositions, the petrogenetic grid is outlined and may be subdivided (figure 24.5a). It may also be expanded into the third dimension by adding an additional axis representing the composition of the fluid phase. In any case, subdivision is based on experimentally determined fields of mineral stability.

A number of mineral reaction curves are particularly important in examining the metamorphism of various terranes of metamorphic rock. For example, in terranes metamorphosed under conditions of static metamorphism, the reaction

$$NaAlSi_3O_8 \quad <==> \quad NaAlSi_2O_6 + SiO_2$$

albite   jadeite+ quartz   (24.1)

503

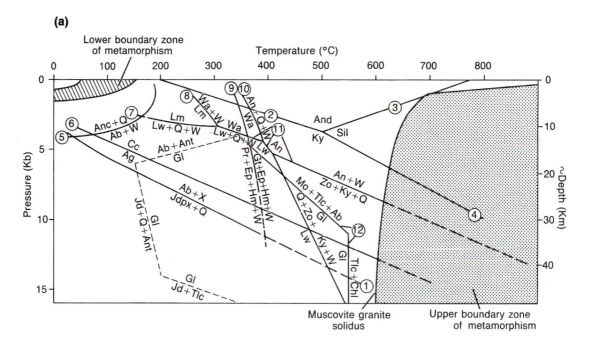

(a)

....
Figure 24.5 Partial petrogenetic grid for metaclastic rocks. (a) Grid showing the locations of various experimentally determined (solid lines) and calculated (short dashed lines) univariant mineral reaction curves. Long dashes show projected extensions. Curves represent the following reactions and are from the noted works. (1) Albite (Ab) + X (necessary elements) = Jadeitic pyroxene (Jdpx) + quartz (Q). Jdpx = $Jd_{82}Ac_{14}Di_4$ (Newton and Smith, 1967). (2) Andalusite (And) = kyanite (Ky) (Holdaway, 1971). (3) Andalusite = sillimanite (Sil) (Holdaway, 1971). (4) Kyanite = sillimanite (Holdaway, 1971). (5) Analcite (Anc) + quartz = albite + water (W) (Campbell and Fyfe, 1965). (6) Calcite (Cc) = aragonite (Ag) (Jamieson, 1953; Clark, 1957). (7) Laumontite (Lm) = lawsonite (Lw) + 2 quartz + 2 water (Liou, 1971a). (8) Laumontite = wairakite (Wa) + 2 water, and wairakite = lawsonite + 2 quartz (Liou, 1971b). (9) Prehnite (solid solution)(Pr) + epidote (solid solution)(Ep) + hematite-magnetite buffer (HM) + water = garnet (solid solution)(Gt) + epidote (solid solution) + HM + water

explains the appearance of jadeitic pyroxene in metawackes and marks the boundary between moderate and high pressure conditions. The curve for this reaction, which was studied by F. Birch and LeCompte (1961) and Newton and Smith (1967), is shown in figure 24.5b. If only this curve were depicted, the diagram would be a phase diagram showing that albite is stable at low pressures, whereas the combination of jadeite plus quartz is stable at high pressures. This phase diagram becomes a part of the petrogenetic grid.

The petrogenetic grid in figure 24.5a also shows curves representing the three reactions

$$
\begin{array}{lcl}
\underset{\text{andalusite}}{Al_2SiO_5} & <==> & \underset{\text{kyanite}}{Al_2SiO_5} \qquad (24.2)
\end{array}
$$

$$
\begin{array}{lcl}
\underset{\text{andalusite}}{Al_2SiO_5} & <==> & \underset{\text{sillimanite}}{Al_2SiO_5} \qquad (24.3)
\end{array}
$$

$$
\begin{array}{lcl}
\underset{\text{kyanite}}{Al_2SiO_5} & <==> & \underset{\text{sillimanite}}{Al_2SiO_5} \qquad (24.4)
\end{array}
$$

These reactions are particularly important in pelitic schists (metamorphosed shales) and other aluminous rocks. Because of several peculiarities involving these apparently simple reactions, the exact positions of the reaction curves in the petrogenetic grid have been somewhat difficult to establish.[12] A reasonably

satisfying set of curves is now available and they meet at an invariant point that lies at 0.387 Gpa (3.87 kb) and 511°C (Hemingway et al., 1991). P-T space is divided into three distinct phase fields by the three reaction curves. At low pressures, andalusite is the stable aluminum silicate. At relatively low temperatures and relatively higher pressures, kyanite is stable. Sillimanite is the stable phase at high temperature.

The experimental studies on which petrogenetic grids are based and the petrogenetic grids themselves provide a framework for estimating *approximate* conditions of metamorphism for particular rocks. These P-T estimates are based on distinctive mineral assemblages known as *critical mineral assemblages* (discussed below). Because reaction curves are generally bivariant, assessments of the P-T-X conditions of metamorphism based on petrogenetic grids are often quite general. For example, if a rock found in a study area contained the assemblage

kyanite + jadeitic pyroxene + lawsonite + glaucophane + white mica + quartz

the P-T conditions of formation indicated on a petrogenetic grid such as that shown in figure 24.5a would be those shown in the shaded area in figure 24.5b. The area of stability for the rock, within the petrogenetic grid, is determined by finding that part of the grid in figure 24.5a in which *all* of the listed phases are stable. Albeit this limits the possible P-T conditions, the range of

(b)

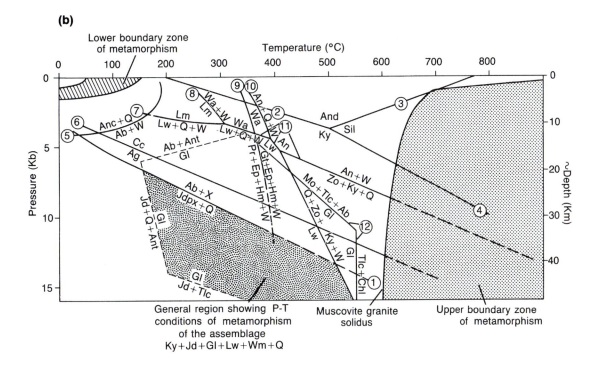

Figure 24.5 Continued
(Liou, Kim, and Marayama 1983). (10) Wairakite = anorthite (An) + quartz + water (Liou, 1970). (11) Lawsonite = anorthite + water, zoisite + kyanite + quartz = anorthite + water, and lawsonite = zoisite (Zo) + kyanite + quartz + water (Newton and Kennedy, 1963). (12) Upper stability limit of glaucophane; glaucophane (Gl) = montmorillonoid clay (Mo) + talc (Tlc)? + albite, and glaucophane = talc? + Chlorite (Chl) (Maresch, 1977). Lower stability limits of glaucophane (short dashes) calculated by Muir Wood (1980) Jd = jadeite; Ant = antigorite. b. Petrogenetic grid (P,T space identical to (a)) showing general region of stability for the assemblage Ky + Jdpx + glaucophane + lawsonite + white mica + quartz, shaded for emphasis. Note the wide range of conditions (T > 300°C and P > 10 kb) under which this six-phase assemblage will crystallize.

conditions under which the rock may have formed is still considerable (T = 150°–550°C, P = 0.6–2 Gpa, or 6–20 kb).

Because one of the goals of studying metamorphic terranes is to determine the *specific* P-T-X conditions of metamorphism, a method of further refining P, T, and X estimates is needed. The composition of the fluid phase (X_{fluid}) may be restricted by the mineralogy present in the rock. Yet, with regard to pressure and temperature, many minerals are stable over a wide range of conditions. Furthermore, where diagnostic (critical) mineral assemblages are absent, that is, where the phase assemblages in the rocks are stable over temperature ranges of more than 200°C and pressure ranges of several kilobars, petrogenetic grids provide little help in determining the exact P-T conditions. Quantitative studies used to determine more exact T and P conditions than those provided by the petrogenetic grid are referred to as **geothermometry** and **geobarometry,** respectively.[13] Obviously, what is needed for geothermometry and geobarometry are indicators of very specific conditions. Such specificity is realized via study of the P-T conditions associated with the particular chemistries of certain minerals that comprise mineral assemblages.

The assemblage biotite + garnet, which occurs over a wide range of conditions in a variety of metamorphic rocks, has been widely used as a geothermometer (A. B. Thompson, 1976a; Ferry and Spear, 1978; Indares and Martignole,

1985a,b).[14] The basis for the geothermometer is the idea that, if pressure is accounted for, the temperature of metamorphism will control the ratios of Fe to Mg in the coexisting mineral phases. The ratios will be different in the individual minerals. Thus, considering the endmember reaction

$$Fe_3Al_2Si_3O_{12} + KMg_3AlSi_3O_{10}(OH)_2 <==>$$
garnet + biotite

$$Mg_3Al_2Si_3O_{12} + KFe_3AlSi_3O_{10}(OH)_2$$
garnet + biotite

the distribution of iron and magnesium in the actual minerals can be expressed by a constant distribution coefficient K_D, given by

$$K_D = \frac{(Fe/Mg)_{biotite}}{(Fe/Mg)_{garnet}}$$

Chemical analysis of the respective Fe and Mg values for garnets and biotites presumed to be in equilibrium in a rock are then used to determine the specific temperature, either via a thermodynamic calculation (using this equilibrium constant) or by a comparison with experimentally studied phases.

Other geothermometers have been developed for a range of rock compositions and conditions of metamorphism.[15] Thus, a wide variety of rocks may now be used for

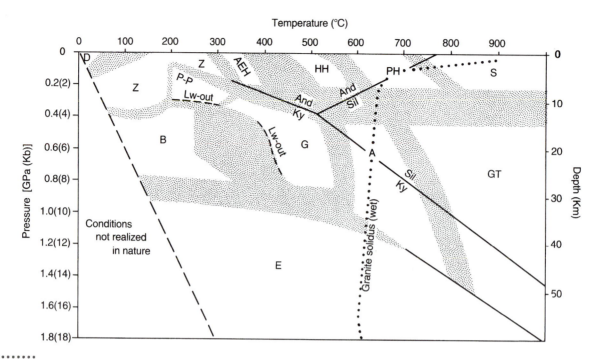

Figure 24.6 Metamorphic facies diagram, showing the general stability fields of the eleven metamorphic facies used in this text. Facies are as follows: AEH = albite-epidote hornfels, A = amphibolite, B = blueschist, E = eclogite, G = greenschist, GT = granulite, HH = hornblende hornfels, PP = prehnite-pumpellyite, PH = pyroxene hornfels, S = sanidinite, and Z = zeolite. D = zone of diagenesis. See appendix C for a compilation of specific reactions on which the boundaries are based. The facies names are those used by F. J. Turner (1981).

analyses of temperatures of metamorphism. In a like manner, several mineral assemblages may be used for geobarometry, to evaluate the pressures of metamorphism.[16]

THE FACIES CONCEPT

Facies and the Petrogenetic Grid

A **metamorphic facies** is defined as a set of rocks representing the full range of possible rock chemistries, with each rock characterized by an equilibrium assemblage of minerals that reflects a specific, but limited, range of metamorphic conditions (cf. Eskola, 1915, 1920, in Turner, 1958).[17] Again, the conditions of metamorphism and the chemistry of the rock control the mineralogy of each metamorphic rock. To illustrate this concept and clarify its meaning, consider the following subset of three rocks, all metamorphosed under the same conditions (3.5 kb, 300°C). The three rocks—a *marble*, a crystalline metamorphic rock composed of calcite, dolomite, and minor quartz; a *metabasite* (a metabasalt) composed of chlorite, albite, actinolite, epidote, and minor quartz; and a *pelitic schist* (metamorphosed shale) composed of quartz, chlorite, muscovite, and albite—are quite different mineralogically. Yet, because they were metamorphosed under the *same conditions*, their respective mineral assemblages each represent the same region on a petrogenetic grid. Similarly, rocks of any other composition metamorphosed

under these same conditions also exhibit mineral assemblages that represent the same area on the petrogenetic grid. Thus, although the mineral assemblages are not the same, each represents the same general conditions of metamorphism. All of these rocks belong to the same metamorphic facies.

Three rocks of the same chemical composition as the three listed earlier but metamorphosed under *different conditions* (e.g., 4 kb and 550°C) are found to have different minerals. Under these conditions, a metabasite would consist of andesine + hornblende ± quartz, the pelitic schist might consist of quartz + sillimanite + white micas + biotite + garnet + oligoclase, and the marble might contain calcite + dolomite + tremolite + talc. Again, the minerals in the three rocks are quite different, yet they represent the same general P-T conditions and, therefore, the same facies.

Eskola,[18] having noted these relationships between mineralogy, bulk chemistry, and metamorphic conditions, advanced the idea of the *metamorphic facies*. He emphasized that under a given set of conditions, the minerals of a metamorphic rock are solely a function of its chemistry. To each facies, Eskola assigned a name. Over the years, his facies scheme has been modified by many petrologists, in the interest of developing a more usable or more complete subdivision of P-T space.[19]

Here, P-T space, the petrogenetic grid, is divided into 11 regions, each of which corresponds to a facies (figure 24.6).[20] The low-P, high-T facies—the Albite-Epidote Hornfels Facies, Hornblende Hornfels Facies, Pyroxene

Table 24.3 Selected Rock Types Typical of Various Facies

Facies	Protoliths			
	Shale	**Basalt**	**Dunite**	**Limestone**
Zeolite	Zeolitic shale	Zeolitic greenstone	Lizardite serpentinite	Calcite marble-granoblastite
Prehnite-Pumpellyite	Chlorite phyllite	Pumpellyite greenstone	Lizardite serpentinite	Calcite marble-granoblastite
Blueschist	Glaucophane phyllite	Glaucophane schist	Lizardite serpentinite	Aragonite marble-granoblastite
Eclogite	Kyanite granoblastite	Eclogite	Dunite	Wollastonite granoblastite
Greenschist	Slate	Actinolite greenstone	Antigorite serpentinite	Calcite marble-granoblastite
Amphibolite	Kyanite-mica schist	Amphibolite	Anthophyllite-talc schist	Tremolite-calcite marble-granoblastite
Granulite	Sillimanite-orthoclase granoblastite	Garnet-pyroxene granoblastite	Dunite	Wollastonite-diopside marble-granoblastite
Albite-Epidote Hornfels	Chloritoid hornfels	Epidote hornfels	Antigorite diablastite	Calcite marble-granoblastite
Hornblende Hornfels	Andalusite hornfels	Hornblende diablastite	Talc dunite	Diopside marble-granoblastite
Pyroxene Hornfels	Cordierite hornfels	Augite hornfels	Dunite	Wollastonite-diopside marble granoblastite
Sanidinite	Cordierite-mullite hornfels	Augite-hypersthene hornfels	Dunite	Akermanite-spurrite marble-granoblastite

Hornfels Facies, and Sanidinite Facies—characterize LP-contact metamorphism. The Greenschist, Amphibolite, and Granulite Facies are typical of dynamothermal metamorphism. The Zeolite, Prehnite-Pumpellyite, and Blueschist Facies are found in regions of static metamorphism. Note that facies names are just that, names, they do not indicate minerals that are always present in every rock of the facies. Also note that the boundaries between the facies shown in figure 24.6 are diffuse. This is because, for each of the various chemical groups of rocks, specific reactions are used to define each boundary between two facies, but the locations of the individual reaction curves do not coincide. For example, the reaction marking the boundary between the Greenschist and Amphibolite facies for metabasites will not be exactly the same as that marking the same facies boundary for aluminous rocks (see appendix C).

With the petrogenetic grid subdivided into named areas, petrographic observations may be used to assign rocks to a specific facies. For example, the first subset of rocks described, metamorphosed under the lower-grade conditions, belongs to the Greenschist Facies. The higher-grade subset belongs to the Amphibolite Facies. Such assignments facilitate communication about specific metamorphic conditions. Every protolith yields a rock composed of a particular mineral assemblage that may be assigned to a facies. That assemblage will vary, depending on the facies, and will contrast with the assemblages indicative of other protoliths under similar or different conditions (table 24.3).

Critical Minerals

Observed minerals and mineral assemblages are useful for mapping and distinguishing rocks of one facies from those of another. Minerals and mineral assemblages for which the

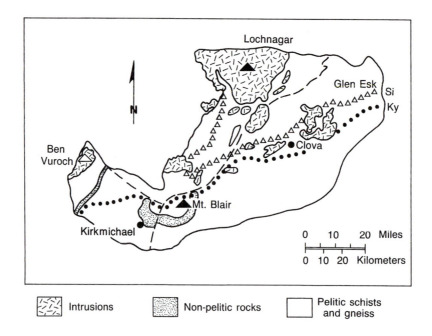

Figure 24.7   Generalized geologic and metamorphic map of a part of the Grampian Highlands of the type area of Barrovian metamorphism, Scottish Highlands. Ky (shown by the dotted line) is the kyanite isograd and marks the first appearance of kyanite in the pelitic rocks along south to north traverses. Si (shown by the line of triangles) marks the southwestern boundary of the regional zone of sillimanite-bearing rocks.

(From E. L. McLellan, "Metamorphic reactions in the kyanite and sillimanite zones of the Barrovian type area" in Journal of Petrology, *26:789–818, 1985. Copyright © 1985 Oxford University Press, Oxford England. Reprinted by permission.)*

reaction curves and stability fields are generally known and which may be used to distinguish between one facies or zone and another are called **critical minerals** and **critical mineral assemblages,** respectively. For example, in terranes containing quartz-mica schists, assemblages with the critical mineral andalusite are recognized as lower-pressure assemblages than are similar schists bearing the critical mineral kyanite. In practice, petrologists map the occurrences of such minerals or mineral assemblages in order to show changing conditions of metamorphism across a region (figure 24.7). In mapping or subsequent petrographic study, the locations of the first appearance of a critical mineral are marked by a line. This map line was originally considered to correspond to a reaction curve in the petrogenetic grid. Thus, the line marked "Si" in figure 24.7 would have been considered to correspond to a reaction curve such as curve 4 in figure 24.5a. Because each point along the map line was considered to represent the *same* reaction, and therefore the *same* P-T condition or "grade" of metamorphism, the line was called an **isograd** (*iso* = same, *grad* = grade)(Tilley, 1924).

Winkler (1974, 1976, ch. 7) rightly pointed out that, in fact, the first appearance of a mineral at different locations in an area may result from different reactions. For example, kyanite might form via the reaction

$$\text{Mg-chlorite} + \text{staurolite} + \text{quartz} + \text{muscovite} \longleftrightarrow \text{kyanite} + \text{biotite} + \text{water}$$

or via the reaction

$$\text{Mg-chlorite} + \text{quartz} + \text{muscovite} \longleftrightarrow \text{kyanite} + \text{biotite} + \text{water.}[21]$$

It is likely that these two reactions occur at somewhat different P-T conditions, and the absence of staurolite from the second reaction suggests a different bulk composition for the two parent rocks represented by the assemblages on the left. Thus, the first appearance of a mineral along an isograd may actually represent somewhat different P-T-X conditions from place to place. To account for this fact and to distinguish more well-defined isograds, Winkler proposed that isograds based on specific reactions be called *isoreaction-grads* (Winkler, 1976, p. 66). He later changed this term to *reaction isograds* (Winkler, 1979, p. 66).

The minerals (and mineral assemblages) most useful for mapping isograds—the critical minerals (or critical mineral assemblages)—are those that are distinctive and that have a limited range of conditions over which they are stable. Minerals like quartz and biotite, which are present in a wide variety of rocks metamorphosed over a wide range of conditions, have limited use in this regard. Nevertheless, the *first appearance* of biotite has been used as an isograd. Minerals such as andalusite, sillimanite, kyanite, jadeitic pyroxene, and laumontite and mineral assemblages such as lawsonite + jadeitic pyroxene + glaucophane, which have a more restricted range of stability, assume more importance in mapping metamorphic changes across a region.

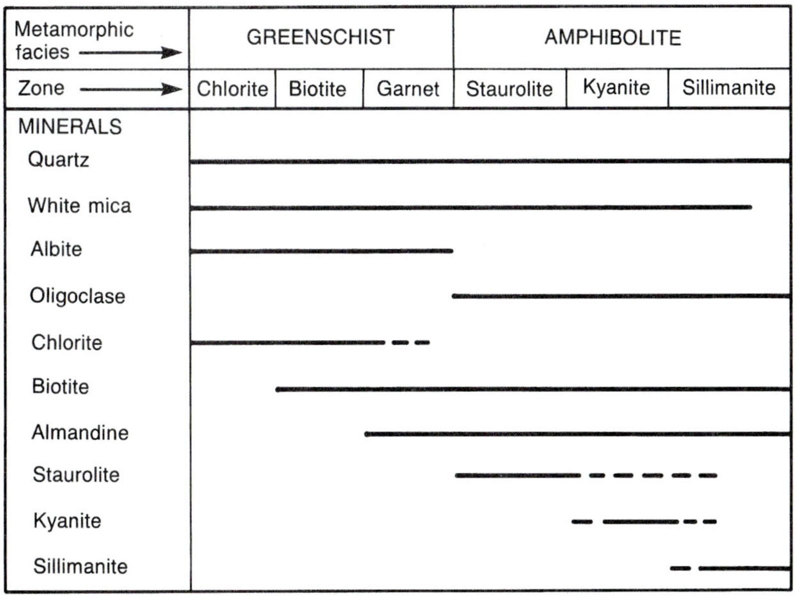

Metamorphic facies →	GREENSCHIST			AMPHIBOLITE		
Zone →	Chlorite	Biotite	Garnet	Staurolite	Kyanite	Sillimanite
MINERALS						
Quartz	————————————————————————————————					
White mica	———————————————————————————————					
Albite	————————————————————					
Oligoclase	————————————————					
Chlorite	——————————————— – –					
Biotite	————————————————————————					
Almandine	———————————————————————					
Staurolite	———————— – – – – –					
Kyanite	– – ——————— – –					
Sillimanite	– ———					

Figure 24.8 Mineral-zone diagram showing the occurrence of various mineral phases in pelitic (aluminous) rocks across the zones of the classical Barrovian Facies Series in the Highlands of Scotland. Solid lines show the common occurrence of the phase; dashed lines indicate sporadic occurrences of the phase in rocks of restricted bulk composition.
(Based on data from Barrow, 1893; Harker, 1932; Chinner, 1960; 1978; Harte and Hudson, 1979; McLellan, 1985a, 1985b)

FACIES SERIES

Because early studies in metamorphism were done in northwestern Europe and the eastern United States, two areas that were actually linked as one prior to the plate tectonically driven separation of the continents,[22] the particular pattern of metamorphic zones present in the above areas was considered to be normal. That pattern of zones, each of which is bounded by isograds, was first discovered by Barrow (1893, 1912) in a study of the Scottish Highlands and was later described by Harker (1932).[23] From lower to higher grade, it consists of chlorite, biotite, and garnet zones (of the Greenschist Facies) followed by staurolite, kyanite, and sillimanite zones (of the Amphibolite Facies)(figures 24.7 and 24.8). These zones were recognized in pelitic rocks on the basis of the first appearance of the named critical minerals. The rocks types and major mineral assemblages of each zone are as follows.

Chlorite Zone (slates, phyllites, and schists)
　quartz + albite + white mica + chlorite
Biotite Zone (phyllites and schists)
　quartz + albite + white mica + chlorite + biotite
Almandine (Garnet) Zone (phyllites and schists)
　quartz + albite + white mica ± chlorite + biotite + garnet
Staurolite Zone (schists)
　quartz + oligoclase + white mica + biotite + garnet + staurolite
Kyanite Zone (schists)

　quartz + oligoclase + white mica + biotite + garnet ± staurolite + kyanite
Sillimanite Zone (schists, gneisses, and granoblastites)
　quartz + oligoclase ± K-rich alkali feldspar ± white mica + biotite ± kyanite + sillimanite

Notice that once a mineral appears at an isograd, it may persist through several zones. Chlorite, albite, staurolite, kyanite, and white mica eventually disappear. Notice also that the rock types change from slates and phyllites, which are fine-grained rocks, to schists and gneisses, which are coarser-grained rocks.

Miyashiro (1961), in a classic work on regional metamorphic belts, pointed out that different metamorphic belts (1) have different mineralogies, (2) are characterized by different isograds, and (3) have had different histories. These differences are reflected by the different metamorphic facies present in the belts. In metamorphic belts in which there were high temperatures (e.g., in northern New England, of North America), the series of facies present across the belt is Zeolite, Prehnite-Pumpellyite, Greenschist, Amphibolite (curve BFS, figure 24.9). Here, andalusite occurs (rather than kyanite), as the aluminum silicate of medium-temperature rocks. Cordierite, sillimanite, and biotite are also present.[24] In contrast, in the Franciscan Complex of California, the geothermal gradient was low (curve FFS, figure 24.9). The facies represented there are the Zeolite, Prehnite-Pumpellyite, and Blueschist Facies. Pelitic rocks contain minerals such as stilpnomelane, glaucophane, lawsonite, quartz, and white mica.[25] In summary, metamorphic belts in different places

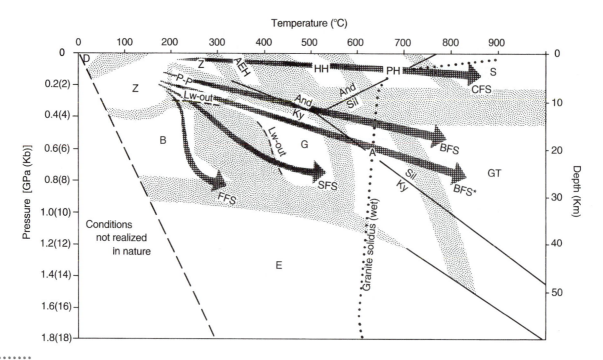

Figure 24.9 Metamorphic facies diagram showing geothermal gradient paths for the five facies series discussed in the text. CFS = Contact Facies Series, BFS = Buchan Facies Series, BFS* = Barrovian Facies Series, SFS = Sanbagawa Facies Series FFS = Franciscan Facies Series. Facies are as follows: AEH = albite-epidote hornfels, A = amphibolite, B = blueschist, E = eclogite, G = greenschist, GT = granulite, HH = hornblende hornfels, PP = prehnite-pumpellyite, PH = pyroxene hornfels, S = sanidinite, and Z = Zeolite. Note that the Blueschist Facies and the Greenschist Facies overlap in a wide zone at pressures above about 3.8 kb. Aluminum-silicate triple point and reaction curves are from Holdaway (1971; modified by Hemingway et al., 1991). The minimum melting curve (solidus) for muscovite granite is from Huang and Wyllie (1981).

were formed under different conditions and are characterized by different minerals and sequences of facies. Miyashiro (1961) called the progression of facies across a metamorphic belt a **metamorphic facies series.**

Five types of facies series were initially proposed by Miyashiro (1961). These were the andalusite-sillimanite type, the low-pressure intermediate group, the kyanite-sillimanite type, the high-pressure intermediate group, and the jadeite-glaucophane type. Later, Miyashiro (1973a, pp. 71–86) dropped the intermediate groups and adopted the names low-pressure baric type, medium-pressure baric type, and high-pressure baric type. In this book, following Miyashiro's original study (1961), five types are recognized.

1. A very low-pressure, andalusite-sillimanite type, represented by the rocks of contact metamorphic zones, is called the **Contact Facies Series.**
2. A low-pressure andalusite-sillimanite type, represented by the rocks of the Buchan area of northeastern Scotland, is designated the **Buchan Facies Series.**
3. A medium-pressure, high-temperature, kyanite-sillimanite type, represented by rocks of the Scottish Highland described by Barrow (1893), is named the **Barrovian Facies Series.**

4. A high-pressure, moderate-temperature type, represented by rocks of the Sanbagawa Metamorphic Belt of Japan, is here designated the **Sanbagawa Facies Series.**[26]
5. A high-pressure, very low temperature, jadeite-glaucophane type, represented by the rocks of the Franciscan Complex of California, is here named the **Franciscan Facies Series.**

Each of the series is characterized by a particular sequence of metamorphic facies (table 24.4). Just as facies are characterized by particular minerals, each facies series is represented by a specific sequence of mineral assemblages that occurs in low- to high-grade rocks. As is always the case, the mineral composition of any particular rock is a function of the bulk chemistry of that rock.

Each facies series represents a particular line through the petrogenetic grid and this line mimics a geothermal gradient (figure 24.9). Based on the experimentally determined stabilities of the phase assemblages present in the rocks, these *apparent geothermal gradients*[27] are approximately as follows:

Contact-type Facies Series	> 80°C/km
Buchan-type Facies Series	40–80°C/km
Barrovian-type Facies Series	20–40°C/km

Table 24.4 Facies and Critical Minerals in the Five Facies Series

Facies Series	Facies Included	Critical Minerals[1,2]
Contact	Zeolite → Albite-Epidote Hornfels → Hornblende Hornfels → Pyroxene Hornfels → Sanidinite	Analcite, *Wairakite*, Albite + epidote + chlorite + chloritoid, Hornblende + tremolite + diopside, Andalusite, Cordierite, Sillimanite, *Sanidine, Mullite, Monticellite, Tillyite, Spurrite*
Buchan	Zeolite → Prehnite-Pumpellyite → Greenschist → Amphibolite → Granulite	Analcite, Laumontite, Prehnite + pumpellyite, Albite + epidote + chlorite, *Andalusite + chloritoid, Cordierite* + garnet + biotite, Sillimanite + Orthoclase, *Hypersthene + orthoclase +* garnet + quartz + plagioclase
Barrovian	Zeolite → Prehnite-Pumpellyite → Greenschist → Amphibolite → Granulite	Analcite, Laumontite, Prehnite + pumpellyite, Epidote + albite + chlorite ± actinolite, Epidote + hornblende + garnet, *Kyanite*, Sillimanite, *Hypersthene + orthoclase* + garnet + quartz + plagioclase
Sanbagawa	Zeolite → Prehnite-Pumpellyite → Blueschist → Greenschist → Amphibolite	Analcite, Laumontite, *Heulandite*, Prehnite + pumpellyite, *Lawsonite +* glaucophane ± stilpnomelane, *Epidote + glaucophane* or crossite, Epidote + hornblende + garnet, Kyanite
Franciscan	Zeolite → Prehnite-pumpellyite → Blueschist → Eclogite	Analcite, *Laumontite, Heulandite*, Prehnite + pumpellyite, Ferrocarpholite, *Lawsonite* + glaucophane ± stilpnomelane, *Lawsonite* ± aragonite, *Jadeitic pyroxene, Omphacite + pyrope*

Sources: Zen (1960), Miyashiro (1961, 1973a), Hietenan (1967), Zwart (1969), Ernst et al. (1970), Chopin and Schreyer (1983), Liou, Maruyama, and Cho (1985) and observations of the author.

[1] Not all critical minerals will appear in every metamorphic belt.

[2] Particularly distinctive minerals of a facies series are italicized.

Sanbagawa-type Facies Series	10–20°C/km
Franciscan-type Facies Series	< 10°C/km

Facies series with high apparent geothermal gradients typify volcanically active regions. Those with low apparent geothermal gradients represent areas of subduction and rapid burial, where heat flow from the interior of the Earth is neither aided by movement of magmas nor enhanced by convection.

Complete facies series are rarely present in an orogenic belt. For example, in the Grampian region of the Scottish Highlands, the Zeolite Facies is apparently missing, the first sign of metamorphism being the development of phase assemblages of the Greenschist Facies.[28] England and Thompson (1984) and D. Robinson (1987) suggest that, in many cases, the incompleteness of a facies series is a function of the tectonically controlled thermal conditions. For example, in extensional settings (e.g., back-arc spreading centers), high heat flow results from magmatic and associated hydrothermal activity at low pressures (i.e., there is a high geothermal gradient). However, spreading removes the rocks relatively rapidly from the region of high heat flow, restricting the complete development of a facies series (Robinson, 1987). In such regions, upper Amphibolite and Granulite

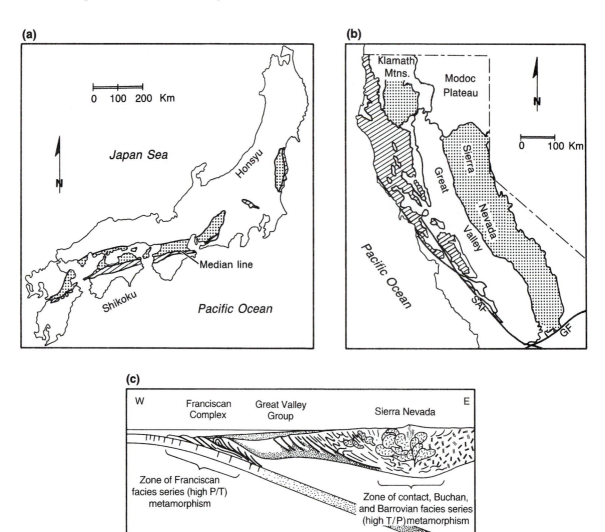

Figure 24.10   Schematic maps and a section showing paired metamorphic belts in two parts of the circum-Pacific region. a) Map showing Mesozoic low- and high-temperature belts in southeastern Japan. Diagonal lines represent Sanbagawa Metamorphic Belt (low to high *P,* low *T*) and stippled pattern marks location of Ryoke-Abukuma Metamorphic Belt (high *T,* low *P*). b) Map showing late Mesozoic metamorphic belts in California. Diagonal lines mark the extent of the low temperature, low to high pressure Franciscan belt. Stipples mark the inferred extent of the high-temperature belt (based on data from Durrell, 1940; Hietenan, 1951, 1973, 1976, 1977; L. D. Clark, 1954, 1970; Eric, Stromquist, and Swinney, 1955; E. H. Bailey,

Irwin, and Jones, 1964; Creely, 1965; Davis et al., 1965; Bateman and Wahrhaftig, 1966; G. A. Davis, 1966; Kistler, 1966; Blake, Irwin, and Coleman, 1967, 1969; Lanphere, Irwin, and Hotz, 1968; Ernst, 1971b; Kistler et al., 1971; B. M. Page, 1981; Dickinson, 1981; Irwin, 1985; Mortimer, 1985; Donato, 1987; and observations by the author). c) Diagrammatic section across California for the late Mesozoic Period, showing paired metamorphic belts developing in association with an east-dipping subduction zone.

((a) From A. Miyashiro, "Evolution of metamorphic belts" in Journal of Petrology, *2:277-311, 1961. Copyright © 1961 Oxford University Press, Oxford England. Reprinted by permission.)*

Facies rocks are usually absent. Slow reaction rates, the effects of various fluid phases, or the absence of appropriate P-T conditions may also result in the absence of a particular facies from a series.

In the circum-Pacific region, mountain systems commonly exhibit pairs of metamorphic belts (figure 24.10) (Miyashiro, 1961). Miyashiro named such zones **paired metamorphic belts.** On the continent side of a pair, the facies series of an "inner metamorphic belt" is typically of the

Buchan or Barrovian type. Within the inner belt, granitoid igneous rocks, reflecting *in situ* igneous activity, are common. The inner metamorphic belt represents a volcanic arc, plus its basement and its roots. The "outer metamorphic belt," on the ocean side of the pair, is characterized by facies series of the Sanbagawa or Franciscan type. Mafic and ultramafic rocks, including ophiolitic rocks, are the usual igneous rocks found in these outer metamorphic belts. In many cases, these igneous rocks formed elsewhere (e.g., at a mid-ocean ridge) and have been tectonically emplaced into the belt. The outer metamorphic belt represents an accretionary complex, a crustal expression of subduction. Together, the paired metamorphic belts mark the site and direction of subduction, with the high-T, lower-P belt marking the position of the volcanic arc, the low-T, high-P belt marking the subduction zone and plate boundary; and the pair indicating subduction in the direction from outer towards the inner metamorphic belt (figure 24.10).

SUMMARY

Metamorphic rocks may be grouped into six broad chemical groups—the silicic, siliceous-alkali-calcic, aluminous, carbonate/nonsilicate, basic, and ultrabasic groups—each of which encompasses the chemistry of various sedimentary and/or igneous protoliths. Mineralogically, each of the groups is distinctive. Because the chemistry of the protolith and the conditions of metamorphism control the equilibrium mineral assemblage, the observed assemblages can be used as indicators of the chemistry of the protolith.

Metamorphic rock classification is commonly founded on a textural subdivision of the rocks. Nevertheless, mineralogical and chemical classifications have been proposed. In this text, the classification adopted is based on texture and has two major categories—the crystalline textures, including foliated, weakly foliated, diablastic, and granoblastic types, and the clastic textures, including foliated and nonfoliated types.

One of the major goals of metamorphic petrology is to discover the general and specific conditions under which metamorphism has occurred. Laboratory analyses provide data for a composite phase diagram referred to as a petrogenetic grid. Petrogenetic grids are used for comparative purposes and to estimate the general conditions under which particular metamorphic mineral assemblages formed. Specific P-T conditions are assessed using geothermometry and geobarometry, analyses based on chemical variations within the individual minerals of a metamorphic mineral assemblage.

A set of all rocks, each with a distinctive mineral assemblage, formed under a particular set of conditions is called a metamorphic facies. Eleven metamorphic facies are recognized here. Linear groups of adjacent facies comprise metamorphic facies series and each facies series represents an apparent geothermal gradient. The Franciscan Facies Series represents the lowest apparent geothermal gradient, whereas these gradients are increasingly higher in the Sanbagawa, Barrovian, Buchan, and Contact Facies series. Orogenic belts, formed via subduction-related events, may exhibit paired metamorphic belts, with outer, high-pressure Franciscan or Sanbagawa facies series oceanward of inner, high-temperature Barrovian or Buchan facies series.

EXPLANATORY NOTES

1. Among those who have simply listed metamorphic rock names are Grout (1932), Miyashiro (1973a), and Ehlers and Blatt (1982).
2. Winkler (1979, p. 340ff.) adopts a quantitative mineralogical classification suggested by Austrian petrographers, whereas Bard (1986) adopts a chemical classification. Most other authors have used texture as a primary basis of classification (W. T. Huang, 1962; Spock, 1962; Best, 1982; D. L. Williams et al., 1982; and Raymond, 1984c) or have mixed texture, genesis, mineralogy, and/or chemistry in some combination (J. F. Kemp, 1929; Harker, 1932, 1939; Mason, 1978; F. J. Turner, 1981; Hyndman, 1985). Fry (1984) promotes the use of mineralogically based names first, then texturally based names where mineralogically based names are inadequate or inappropriate.
3. For example, F. J. Turner (1981) and Raymond (1984c).
4. For more information on weathering, see Reiche (1950), Ollier (1969b), and Ritter (1986, chs. 3,4). The stabilities of various oxides are discussed in texts such as Garrels and Christ (1965) and Krauskopf (1967, 1979). The common manganese oxide stains and veins in rocks are composed of minerals such as birnessite, romanechite, todorokite, and hollandite (R. M. Potter and Rossman, 1979).
5. For example, see reviews in R. L. Hay (1966) and Boles (1981) and see Alt et al. (1986).
6. This range of conditions is based on typical surface conditions, both on the continents and beneath the water in the ocean basins. Different petrologists suggest various criteria for defining the beginning of metamorphism. For example, B. M. French (1973) indicates that the appearance of minerals that consistently show a secondary textural relationship and are associated with axial plane cleavage and fractures marks the beginning of metamorphism. The definition of Mason (1978) is like that used in this text. Winkler (1979, p. 11) suggests that the beginning of metamorphism is marked by the first appearance of a mineral assemblage that cannot form in a sedimentary environment. One result of the use of these various criteria is that the P-T conditions at the *beginning* of metamorphism cited by various authors differ, ranging from those cited here up to temperatures of 250–350°C and pressures of 0.6 Gpa (6 kb) (see B. Bayly, 1968; B. M. French, 1973; Winkler, 1979; Weaver, 1984).
7. For example, see A. S. Campbell and Fyfe (1965), A. B. Thompson (1970b), Liou (1971a,b), Nitsch (1971), Velde (1973), Ivanov and Gurevich (1975), and Donahoe and Liou (1985). Also see the review of Zen and Thompson (1974).

513

8. Examples include those reported by Muffler and White (1969), Blake, Irwin, and Coleman (1967, 1969), Dickinson et al. (1969), Ernst et al. (1970), B. M. French (1973), Floran and Papike (1978), M. Frey (1978), S. D. Weaver et al. (1984), Alt et al. (1986), and Becker et al. (1989). Also see R. L. Hay (1966, 1977) and Surdam (1977).

9. Refer back to chapter 6 for information and references on anatexis. See Bowen and Anderson (1914), S. A. Morse (1980, p. 130), and Ohtani and Kumazawa (1981) for information on the melting of forsterite (T = 1890°C at 1 bar to 2000°C at 20 kb).

10. Luth (1969), Huang and Wyllie (1975, 1981), C. R. Stern and Wyllie (1981).

11. For example, see D. H. Green (1973a), Kushiro (1973a), Eggler (1978), and Ribe (1985).

12. Richardson et al. (1969), Holdaway (1971), Robie and Hemingway (1984), Grambling and Williams (1985), Salje (1986), Kerrick (1990), Hemingway et al. (1991).

13. Discussions of geothermometry and geobarometry may be found in Essene (1982) and Newton (1983). For additional examples and discussions, see the papers cited in the text. D. M. Carmichael (1978) earlier proposed that pressure zones ("bathozones") be defined on the basis of mineral assemblages, the first appearance of which marks a surface called a *bathograd*. The development of geobarometry enhances the capability of metamorphic petrologists to define depth zones specifically, thereby making the delineation of bathograds more widely applicable.

14. Also see Goldman and Albee (1977), Baltatzis (1979), Ghent and Stout (1981), Schreurs (1985), and Hoisch (1990).

15. Geothermometers include such mineral assemblages as garnet-clinopyroxene (Banno, 1970; Raheim and Green, 1974; Ellis and Green, 1979; Krogh, 1988; T. H. Green and Adam, 1991), garnet-hornblende (P. R. A. Wells, 1979; C. M. Graham and Powell, 1984; Ghent and Stout, 1986), hornblende-plagioclase-bearing assemblages (Plyusnina, 1982; Blundy and Holland, 1990), garnet-staurolite-Al_2SiO_5-quartz-H_2O (Hodges and Spear, 1982), two feldspars (Barth, 1951; Stormer, 1975; Fuhrman and Lindsley, 1988; but see W. L. Brown and Parsons, 1985, who argue that this geothermometer is based on an erroneous assumption), plagioclase-muscovite- Al_2SiO_5-quartz-H_2O (Cheney and Guidotti, 1979), muscovite-biotite (Hoisch, 1989), and calcite-dolomite (R. Powell, Condliffe, and Condliffe, 1984). For other geothermometers and additional discussion, see Ghent (1976), Ghent and Stout (1981), N. L. Green and Usdansky (1986), Kawasaki (1987), Pownceby, Wall, and O'Neill (1987), Triboulet and Bassias (1988), Sen and Jones (1989), Grambling (1990), Spear et al. (1990), Berman and Koziol (1991), and Witt-Eickschen and Seck (1991).

16. Geobarometric analyses have been attempted on a range of mineral assemblages, including sphalerite-pyrrhotite-pyrite (Hutcheon, 1978), olivine-clinopyroxene (G. E. Adams and Bishop, 1986), garnet-rutile-ilmenite-plagioclase-quartz (Bohlen and Liotta, 1986), garnet-rutile-ilmenite-Al_2SiO_5-quartz (Bohlen, Wall, and Boettcher, 1983), garnet-plagioclase-biotite-muscovite (Ghent and Stout, 1981), and biotite-muscovite-chlorite-quartz (R. Powell and Evans, 1983). For additional discussions of geobarometry, see Anovitz and Essene (1987), Bucher-Nurminen (1987),

Holdaway, Dutrow, and Hinton, (1988), Koziol and Newton (1988), McKenna and Hodges (1988), Kohn and Spear (1989; 1990), Hoisch (1990), and Mukhopadhyay (1991).

17. Although the wording is different, this definition follows the original definitions of Eskola (1915; 1920, in Turner, 1958) in part. The definition embraces two stipulations, noted by Turner (1958, p. 18): (1) rocks of each facies form in response to the same physical conditions, (2) equilibrium is attained at the time of formation of the constituent mineral assemblage of each facies. Eskola (1920, 1939; in Turner, 1958) also stipulated that for each specific rock composition, *the same set of minerals* is produced by the physical conditions for each facies (i.e., each set of minerals in a bulk composition corresponds to *one* facies). This means that a basalt metamorphosed to the assemblage pumpelleyite + actinolite + chlorite + albite + quartz in one place, but to the assemblage epidote + actinolite + chlorite + albite + quartz in another, should be assigned to different facies in each of the two places. Some petrologists have moved toward this (Liou, Maruyama, and Cho, 1985). Petrographic studies show that numerous mineralogical assemblages exist in each bulk composition over the range of possible metamorphic conditions. Thus, following Eskola, the number of facies should be quite large, and it will increase as knowledge of mineralogical variability increases. Winkler (1974) correctly pointed out that, as a result, the number of facies (and "subfacies") has increased over time to the point that the facies concept is losing its utility. He chose to abandon the use of the concept (Winkler, 1974, 1976, 1979). As have others, I choose to lump like assemblages under single facies names.

18. The views of Eskola reported here are based on quotations in Turner (1958) and on the summaries by numerous authors, including Winkler (1979).

19. Published facies schemes vary from author to author. The reaction curves, on which facies boundaries are based, are selected on the basis of criteria each author thinks are important. Consequently, the facies boundaries of one author may lie in the middle of a facies field of another (e.g., compare the facies fields of F. J. Turner, 1981, p. 420, and those of Dobretsov and Sobolev, 1972, p. 199).

As noted above, Winkler (1979, ch. 6) reasoned that because the facies concept has been made obsolete by the wealth of petrographic information now available, it should be abandoned. That information, he argues, shows that within areas like those designated here as individual facies on the P-T grid, there are, in fact, numerous groups of equilibrium mineral assemblages (i.e., numerous subsets of rocks), each of which represents a restricted range of P-T conditions and chemistry. Thus, *sensu stricto*, there are numerous metamorphic facies. For certain bulk compositions, this may be the case. For others it is not. However, I concur with Winkler (1974) in the belief that the proliferation of facies and subfacies titles is cumbersome. It inhibits and confuses communication because the facies names are established in the literature and reflect general sets of P-T conditions to most geologists. If the concept of metamorphic facies is to remain a useful tool for the metamorphic petrologist, Eskola's original idea (1920, in Turner, 1958) that each specific mineral assemblage must represent a new facies must be abandoned in favor of the kind of broader definition adopted in this text, in which certain *similar assemblages* that represent a limited range of metamorphic conditions are lumped together under the title of a facies. I here retain the facies concept, as modified, because of its utility in designating areas on the petrogenetic grid and because its use is widespread. As Winkler (1979) did, I consider specific mineral assemblages within each facies to mark *zones* (e.g., the staurolite zone).

20. The facies names used here are those used by F. J. Turner (1981, pp. 202–210). F. J. Turner (1981, p. 202) and others apply the designation Eclogite Facies to rocks thought to have been recrystallized under conditions of metamorphism of high pressure and moderate to high temperature. The validity of the concept of an Eclogite Facies is discussed in chapter 29.

21. See J. B. Thompson and Norton (1968), D. M. Carmichael (1970), and Winkler (1979, p. 227).

22. See P. M. Hurley (1968), Harold Williams and Max (1980), and standard texts on tectonics (Condie, 1982; Windley, 1977, p. 148ff.) for discussions of this tectonic history and references to the important body of literature supporting this contention.

23. Additional works on this area are numerous. For more information, see Chinner (1960, 1966, 1978, 1980), Harte and Hudson (1979), McLellan (1985), and the works cited therein.

24. Miyashiro (1961, 1973a) emphasizes that andalusite characterizes low-pressure facies series, notes that sillimanite occurs at higher temperatures in the facies series, and indicates that cordierite is a common phase in both the higher-temperature parts of the andalusite zone and in the sillimanite zone.

25. Work on the pelitic rocks of the Franciscan Complex is limited. The Franciscan terrain and several other metamorphic belts developed in areas of low temperature are characterized by metasandstones (siliceous-alkali-calcic rocks) and metabasites (see Coombs, 1960; E. H. Bailey, Irwin, and Jones, 1964; McKee, 1962; Ernst, 1964, 1965, 1971b; Ernst et al., 1970; Brothers, 1974; Roeske, 1986). In the Franciscan Complex, siliceous rocks, carbonate rocks, ultramafic rocks, and pelitic rocks also occur, but in some low-T, high-P terranes, these rocks are characteristic (Seki, 1958; Ernst et al., 1970; Black, 1977; Liou, 1981a; Cloos, 1983; Matthews and Schliestedt, 1984; E. H. Brown, 1986; Mottana, 1986; Okay, 1986).

26. Miyashiro (1961) initially cited areas of the Sanbagawa Metamorphic Belt as examples of his jadeite-glaucophane-type facies series. However, the definitive assemblage jadeite + lawsonite + quartz apparently does not occur in Sanbagawa schists (Hirajima, 1983, in Banno, 1986). Instead, assemblages containing epidote ± Na-amphibole ± actinolite, which form at somewhat higher temperatures, occur (Ernst et al., 1970; Miyashiro, 1973a; Toriumi, 1975; E. H. Brown, 1977; Liou, Maruyama, and Cho, 1985; Maruyama, Cho, and Liou, 1986). Thus, the Sanbagawa terrane represents a high-pressure intermediate type of facies series, rather than the high-pressure (jadeite-glaucophane) type.

27. Continuous geothermal gradients did not exist in most mountain systems, the thermal histories of individual belts within the mountain range reflect the disparate individual tectonic histories of those belts. Consequently, the apparent geothermal gradients, in many cases, are an artifact of the tectonic history, rather than an actual record of the thermal history of the mountain system as a whole. Within individual belts, it may be possible to assess the actual geothermal gradient.

28. See Barrow (1893, 1912), Harker (1932), Chinner (1960, 1966), and Harte and Hudson (1979).

PROBLEMS

24.1. Assign each of the following rock groups to a facies series based on its mineral assemblages.
 (a) Group A
  Rock 1 Quartz-albite-laumontite-muscovite-chlorite
  Rock 2 Quartz-albite-epidote-muscovite-chlorite
  Rock 3 Quartz-plagioclase-muscovite-biotite-andalusite-staurolite
  Rock 4 Quartz-plagioclase-alkali feldspar-biotite sillimanite-garnet
 (b) Group B
  Rock 1 Quartz-albite-laumontite-muscovite-chlorite
  Rock 2 Quartz-albite-prehnite-muscovite-chlorite
  Rock 3 Quartz-albite-muscovite-chlorite-stilpnomelane-lawsonite
  Rock 4 Quartz-white mica-lawsonite-jadeitic pyroxene glaucophane
 (c) Group C
  Rock 1 Quartz-albite-wairakite-muscovite-chlorite
  Rock 2 Quartz-albite-epidote-muscovite-chlorite
  Rock 3 Quartz-plagioclase-muscovite-biotite-cordierite-garnet
  Rock 4 Quartz-plagioclase-cordierite-orthopyroxene (hypersthene)

25

Metamorphic Phase Diagrams

INTRODUCTION

Phase diagrams are as useful for the study of metamorphic rocks as they are for the study of igneous rocks. The typical diagram used in metamorphic studies differs from that used in igneous studies, however, because melts are not present in most metamorphic rocks and are not included on the diagrams. Inasmuch as metamorphic rocks, by definition, remain solid as they undergo changes, metamorphic petrologists are concerned with changes of one mineral assemblage to another. The nature of these reactions, the progress of the reactions, and the controls on the reactions are important in understanding the metamorphic history. At the highest grades of metamorphism, where melting occurs, igneous and metamorphic processes overlap.

Recall that the Phase Rule is

$$P = C - F + 2$$

where P is the number of phases, F is the number of degrees of freedom, and C is the number of components. As a generalization, minerals are considered to be phases in natural metamorphic systems at equilibrium.[1] Because there are commonly at least two degrees of freedom (P and T), the number of phases (minerals) will be less than or equal to the number of components. In short, P = C. This relationship was first noted by Goldschmidt (1911, in Mason,

1978) and is referred to as **Goldschmidt's Mineralogical Phase Rule.**

Metamorphic rocks, like other rocks, may be characterized chemically by 10 or 11 major oxides. Substitutions (e.g., Mn for Fe, Ti for Al) reduce from 10 or 11 to 6 or 7, the number of components that must be considered. Thus, metamorphic rocks can be treated as if they have six or seven components and, following Goldschmidt's Mineralogical Phase Rule, metamorphic phase assemblages will consist of about six or seven phases. In plotting such metamorphic phase assemblages, one encounters the same difficulty as in plotting multiphase igneous assemblages. A maximum of only four coexisting phases may be plotted conveniently in three-dimensional space. To deal with this problem, in metamorphic petrology, systems are considered to be saturated in certain phases; that is, *certain phases are assumed to be present in all assemblages.* For example, in pelitic rocks quartz is always present,[2] so that SiO_2 is in excess and does not affect the number of degrees of freedom (F). Consequently, in plotting the phase assemblages, we can consider the system saturated in SiO_2 and we can delete one phase (quartz) and its corresponding component (SiO_2) from the diagram without affecting the system or the Phase Rule calculations. In general, three-phase *partial assemblages* are plotted on triangular phase diagrams, with the additional phases (those present in all assemblages) listed to the side of the triangle.

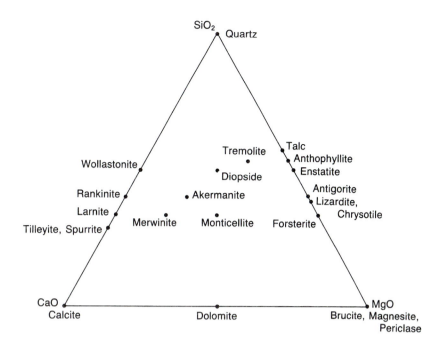

Figure 25.1 Diagram representing the system CaO-MgO-SiO_2-H_2O-CO_2. Here only the CaO-MgO-SiO_2 plane of the system is presented, with dots indicating the positions of minerals in the system that plot on or project onto the plane. Table 25.1 lists the chemical formulae for each of the phases shown.
(Based on Bowen, 1940, and B. W. Evans, 1977)

THE SYSTEM SiO_2-CaO-MgO-H_2O-CO_2 AS A MODEL

One of the easiest phase diagrams to understand is that for a simplified form of the system SiO_2-CaO-MgO-H_2O-CO_2.[3] In order to reduce the complexity of the initial analysis, we temporarily ignore the components of the fluid phase—water and carbon dioxide. This can be done *if* these components are externally controlled, that is, if they are **externally buffered** by a large influx of fluid from outside of the system (outside of the rock being considered in the analysis).[4] Ignoring the fluid phase allows one to plot the remaining three components at the corners of a composition triangle (figure 25.1). Each corner of the triangle represents 100 mole percent of the component at that corner. Points located on lines or within the triangle represent combinations of the components at the corners, the position of the point reflecting the molecular percentage of the various components. Reading the triangle is directly analogous to reading a triangular rock classification chart, except that instead of percentages of minerals, we read moles of components.

Rock, mineral, or chemical compositions may be plotted on triangular phase diagrams as molecular ratios. Thus, in figure 25.1, wollastonite, which has the formula $CaSiO_3$ (table 25.1) and consists of one mole of CaO and one mole of SiO_2, plots exactly one-half of the way between CaO and SiO_2. It contains no MgO and plots on the 0% MgO line. Similarly, diopside plots at a point above the center of the tri-angle, as it consists of one mole of CaO, one mole of MgO, and two moles of SiO_2 (25% CaO, 25% MgO, 50% SiO_2). Dolomite is plotted exactly halfway between the CaO and MgO corners. It contains one mole of MgO per mole of CaO. Of course, dolomite also contains CO_2, but here one may consider that CO_2 is available in abundance; it is externally buffered and is available wherever it is needed to form a mineral. Actually, CO_2 can be plotted at the apex of a tetrahedron (figure 25.2). The position of dolomite, $CaMg(CO_3)_2$, is projected from the CO_2 apex onto the CaO-MgO-SiO_2 (CMS) plane of the tetrahedron, and lies at a point exactly halfway between CaO and MgO. Note that silica is not present in dolomite, with the result that dolomite is plotted on the 0% SiO_2 line.

To plot a rock analysis on the three-component CaO-MgO-SiO_2 (CMS) diagram, the number of moles (in the rock) of each of the three endmember components of the diagram is determined by dividing the three respective oxides in the chemical analysis by their corresponding molecular weights. For example, the weight percent of silica in the analysis is divided by 60.09 g/mol. Similarly, the lime and magnesia are divided by their molecular weights. The three values are then summed to 100% (moles SiO_2 + moles CaO + moles MgO = 100%) and mole percentages are calculated. The analysis is plotted by locating the point that corresponds to the molecular *percentages* of the three components. The procedure is identical to that used in plotting points on triangular variation diagrams in igneous petrology.

Table 25.1 Chemical Formulae of Minerals Plotted in the CMS System

Mineral	Formula	Mineral	Formula
Quartz	SiO_2	Wollastonite	$CaSiO_3$
Calcite	$CaCO_3$	Rankinite	$Ca_3Si_2O_7$
Dolomite	$(Ca,Mg)CO_3$	Larnite	Ca_2SiO_4
Magnesite	$MgCO_3$	Tillyite	$Ca_3Si_2O_7 \cdot 2CaCO_3$
Brucite	$Mg(OH)_2$	Spurrite	$2Ca_2SiO_4 \cdot CaCO_3$
Periclase	MgO	Merwinite	$Ca_3Mg[SiO_4]_2$
Forsterite	Mg_2SiO_4	Akermanite	$Ca_2MgSi_2O_7$
Lizardite	$Mg_3Si_2O_5(OH)_4$	Diopside	$CaMgSi_2O_6$
Chrysotile	$Mg_3Si_2O_5(OH)_4$	Tremolite	$Ca_2Mg_5Si_8O_{22}(OH)_2$
Antigorite	$Mg_3Si_2O_5(OH)_4$	Anthophyllite	$Mg_7Si_8O_{22}(OH)_2$
Talc	$Mg_6Si_8O_{20}(OH)_4$	Enstatite	$MgSiO_3$

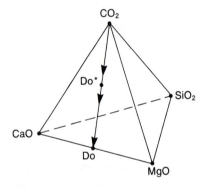

Figure 25.2 Sketch of the system CaO-MgO-SiO_2-CO_2. The point marked Do* is a plot of the composition of dolomite. For consideration in the CMS diagram, Do* is projected to the point marked Do on the basal plane of the system CaO-MgO-SiO_2-CO_2.

The minerals commonly found in the rocks that plot in the system CaO-MgO-SiO_2-H_2O-CO_2 are shown in figure 25.1 and their formulae are listed in table 25.1. Minerals such as tremolite, talc, and dolomite do not plot directly on the CMS plane of the diagram because they contain H_2O or CO_2. Therefore, they are *projected* onto the CMS plane (as was demonstrated for dolomite in figure 25.2). The listed minerals are those that are common in carbonate rocks (marbles) and in metamorphosed ultramafic rocks (e.g., metadunite, metaperidotite), the rock types for which this diagram is used.

Mineral Assemblages, Reactions, and Facies

Equilibrium mineral assemblages representing a facies or subfacies may be plotted on the triangular phase diagram. The stable coexistence of any mineral pair is indicated on

the diagram by a *tie line*, the line connecting two stable phases. Each three-phase assemblage is plotted as a triangle, the corners of which are connected by tie lines. Because the diagram has only three corners (three components), only three phases from any one assemblage can be plotted as an equilibrium assemblage.[5] Each point within the diagram represents a system.

A stable assemblage in limestones and diagenetically altered limestones (at low P-T conditions), for example, is calcite + dolomite + quartz. This assemblage is shown by the three corners of the triangle (Qz, Cc, Do) on the phase diagram in figure 25.3. The bulk composition of a representative rock is shown by a star (this is the system under consideration). The mineral assemblage calcite + dolomite + quartz represents a stable, three-phase assemblage for the divariant field of the petrogenetic grid in which the phase diagram is plotted. That this field is divariant is evident from application of the Phase Rule ($F = C - P + 2 => F = 3 - 3 + 2 => F = 2$).

The univariant curve in the petrogenetic grid represents the reaction

3 dolomite + 4 quartz + H_2O = talc + 3 calcite + 3 CO_2

Dolomite and quartz are stable together to the left of the curve. Under the higher-temperature conditions on the right, talc and calcite are stable together. Note that in the phase diagram to the right of the curve, dolomite and quartz are no longer connected by a tie line, indicating that they are no longer stable together. Note also that the triangular phase diagram is subdivided into smaller triangles, the sides of which are tie lines connecting stable phases. The small triangles indicate stable three-phase assemblages (quartz + calcite + talc, calcite + dolomite + talc, dolomite + magnesite + talc) in the Amphibolite Facies.

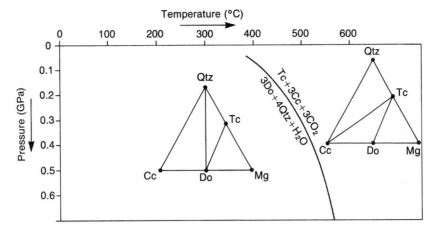

Figure 25.3 Part of the petrogenetic grid for carbonate rocks showing the *nonterminal* reaction curve and phase diagrams for the reaction 3 dolomite + 4 quartz + H$_2$O = talc + 3 calcite + 3 CO$_2$ (see appendix C for the complete grid). The reaction curve here, studied by T. M. Gordon and Greenwood (1970), Metz and Puhan (1971), and Eggert and Kerrick (1981), is located on the basis of the work of Metz and Puhan (1971). The star represents one possible bulk composition for which the reaction shown would result in a nonterminal reaction.

Reactions and the Use of Triangular Phase Diagrams

Three-phase assemblages represented by the smaller triangles within a triangular phase diagram, in general, represent mineral assemblages stable over some range of P-T-X conditions. That range is represented by the area between reaction curves in the petrogenetic grid. If the assemblage changes as a result of a change in pressure, temperature, or composition, the change will be represented in one of two ways, as a terminal or a nonterminal reaction. In **terminal reactions,** one or more phases cease to be possible on one or the other side of the reaction. In such cases, minerals appear or disappear from the phase diagram. In **nonterminal reactions,** some mineral pairs become unstable, while others become stable. On the phase diagram, there is a *tie line flip*, but mineral phases neither appear nor disappear from the diagram.

The reaction examined above (see figure 25.3) is representative of a nonterminal reaction. The first phase diagram shows that quartz, calcite, and dolomite are stable together. They are connected by tie lines. The second phase diagram shows that—for the bulk composition shown—calcite, dolomite, and talc are stable together. As a result of the reaction, the tie line connecting dolomite and quartz is replaced by one connecting calcite and talc. Because one of the rules of using triangular phase diagrams is that under equilibrium conditions *tie lines do not cross,* the dolomite-quartz line is removed and replaced by the calcite-talc tie line. In considering the change from one phase diagram to the other, it appears that the tie line has "flipped" in position. (Note also that the two solid phases on the left side of the reaction are the two that are connected by a tie line and are stable together at low T, whereas the two solid phases on the right side of the equation are the two that are connected by a tie line and are stable together at higher T.)

As may be induced from the above discussion, in working with triangular phase diagrams, two general rules are maintained. First, the phase diagrams must be subdivided into three-phase fields (i.e., divided into triangular subareas). Second, tie lines should not cross. If they do, the assemblage involved represents (1) a disequilibrium assemblage, (2) an equilibrium assemblage stable over a range of P-T conditions, (3) a condition resulting from projection of phases onto a plane (in which case the tie lines only appear to cross), or (4) an assemblage in which one or more phases partition the components used to define the diagram (F. J. Turner, 1968, p. 178).

As noted, terminal reactions involve the appearance or disappearance of a phase. In diagramming such reactions, tie lines are added or subtracted to maintain triangular subdivision of the interior of the phase diagram, but no tie line flip occurs. As an example, consider the reaction

$$\text{calcite} + \text{quartz} = \text{wollastonite} + CO_2$$

In this reaction, two phases connected by a tie line (calcite and quartz) react to form a new phase that plots *between* them on the same line (figure 25.4). Note that to retain a triangular subdivision within the phase diagram, an additional tie line must be added to the diagram after wollastonite is plotted. Terminal reactions involving the appearance or disappearance of a phase *within* one of the small triangles of the phase diagram can also occur. Both terminal and nonterminal reactions that involve univariant curves are **discontinuous reactions.**

Triangular phase diagrams containing solid solution phases may contain assemblages in which only two phases, at least one of which is a solid solution phase, coexist stably. In these cases, the systems are trivariant ($F = C - P + 2 => F = 3 - 2 + 2 => F = 3$). Such is the case in the AFM diagrams discussed

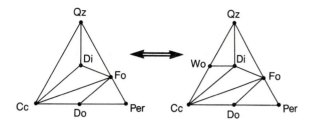

Figure 25.4 CMS phase diagrams for the *terminal* reaction quartz + calcite = wollastonite + CO_2. Note that the only changes are the addition of Wo and a tie line in the triangle on the right. Cc = calcite, Di = diopside, Do = dolomite, Fo = forsterite, Mg = magnesite, Qz = quartz, and Wo = wollastonite.
(After Bowen, 1940)

below, where two-phase assemblages are represented by groups of tie lines rather than by a single line and the solid solution phases are shown as areas or bars rather than points (figure 25.5). Both terminal and nonterminal reactions, including discontinuous reactions, may be depicted on such diagrams. However, some reactions involving solid solutions are continuous reactions. To illustrate such a reaction, recall that iron and magnesium may be partitioned between phases such as garnet and biotite (see chapter 24). During metamorphism, as the temperature changes, the ratios of iron to magnesium in the two phases change; that is, the compositions of the garnet and biotite vary. As a consequence, the position, but not the general configuration, of the tie lines will vary as the reactants and products change composition (see figure 25.5). Reactions such as this, in which the compositions of coexisting phases change gradually over the course of the reaction, are called **continuous reactions.**

Continuous reactions and metamorphic rock histories involving a number of solid solution phases stable over a wide range of conditions may be examined using the concept of **reaction space** (J. B. Thompson, 1982a, 1982b, 1991; Schneiderman, 1990).[6] This method of analysis involves defining, for a given bulk composition, an imaginary volume, the axes of which are reactions and the faces of which are the mineralogical limits beyond which a reaction cannot proceed (figure 25.6). The limits result from the depletion of a reacting phase or some similar chemical control. By examining the chemical data, minerals, and textures of a rock, it is possible to define one *specific reaction path* within this volume that was followed by the rock during metamorphism. Quantification of the progress of the reactions involved, including changing modal mineralogy, chemistry, or fluid volume during metamorphism, is possible using a function called the *reaction progress variable* (Ferry, 1983a,b; Ridley, 1986).[7]

Facies and Phase Diagrams

Triangular phase diagrams can be used to depict stable mineral assemblages representing a facies. Observation of the rocks from a region *covering a wide range of compositions* reveals a set of mineral assemblages, each corresponding to one of the bulk rock compositions. Using these mineral assemblages, a complete phase diagram or a set of diagrams can be constructed that will show the possible, stable, three-phase assemblages for the region in which the rocks occur. Particular metamorphic grades (i.e., zones, subfacies, and facies) are represented by the specific arrangement of tie lines that plot on each diagram.[8] Different assemblages in rocks of the same bulk composition from adjoining regions represent different conditions of metamorphism and hence, different facies, subfacies, zones, or grades of metamorphism.

A series of triangular phase diagrams can be used to depict the mineralogical and, therefore, the facies changes in a metamorphic belt. Such a series of diagrams may show graphically the facies changes over an entire facies series. Recall that the P-T conditions indicated by facies series reflect the tectonic setting and may reflect the geothermal gradient.

THE ACF DIAGRAM

The ACF diagram (figure 25.7) is a triangular diagram used to plot mineral assemblages in metabasites and impure carbonate rocks (Eskola, 1939, in Winkler, 1976). It is not a true phase diagram in a Phase Rule sense; rather, it is a pseudophase diagram, because the diagram combines components, notably FeO and MgO, that are not completely interchangeable (Guidotti, 1982).[9] Nevertheless, for some purposes, the ACF diagram can be and is used as a phase diagram. Each of the three corners of the diagram represents moles of a particular major element oxide or group of oxides. The oxides SiO_2, Al_2O_3, Fe_2O_3, FeO, MnO, MgO, and CaO are accounted for in the diagram, but in order to plot seven oxides on a three-component diagram, adjustments must be made. First, all assemblages are assumed to be saturated with silica, that is, quartz must be present in all assemblages for which the diagram is used and it is not plotted. Next, Fe^{2+}, Mg, and Mn are assumed to substitute freely for one another, because they occupy the same sites in mineral structures.[10] The F-corner represents moles of the oxides of these three cations. Similarly, Al and Fe^{3+} are assumed to substitute freely for one another and they are lumped together. The sum $Na_2O + K_2O$ is subtracted from the sum $Al_2O_3 + Fe_2O_3$, as the former combine with Al to make feldspar (this manipulation allows the diagram to function as a projection from feldspar). The molecular value remaining after this subtraction is plotted at the A corner. The C corner represents the moles of CaO. In summary, the three corners of the diagram are represented by the molecular values

$$A = Al_2O_3 + Fe_2O_3 - (Na_2O + K_2O)$$

$$C = CaO$$

$$F = FeO + MgO + MnO$$

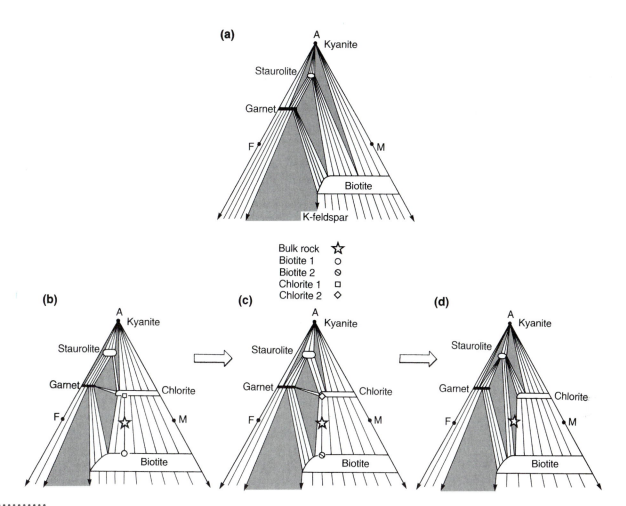

Figure 25.5 Phase diagrams showing two-phase assemblages plotted on a triangular diagram and tie line shifts due to continuous reactions. (a) AFM diagram showing areas in which two, rather than three phases form assemblages plotted on the diagram. The groups of subparallel tie lines show two-phase fields. The ends of each tie line connect a mineral of one particular chemistry (within the range of compositions possible for that phase) with another mineral of a particular chemistry (within the range of compositions possible for the second phase). The shaded areas represent three-phase fields. (b)–(d) Hypothetical AFM diagrams showing a change in chemistry of two phases during a continuous reaction [(b)→(c)] and a subsequent change from a two-phase assemblage on the diagram to a three-phase assemblage on the diagram [(c)→(d)]. The bulk composition of the rock is shown as a star. During the continuous reaction, the two phases, chlorite 1 and biotite 1, initially in equilibrium and connected by a tie line, change composition to become chlorite 2 and biotite 2, connected by a different tie line. Continued metamorphism [(c)→(d)] causes a shift in tie lines that is large enough that the bulk composition, which at lower grades was within a two-phase region, is encompassed by a three-phase region.
(Based in part on J. B. Thompson, 1957; Guidotti, 1974; and Abbott, 1979b).

Figure 25.6 Schematic reaction-space polyhedron showing the direction of a reaction during the metamorphism of a pelitic rock . The axes, labelled A, C, and D, represent particular reactions (not shown). The arrow represents the reaction path followed by the rock, which is a path away from phases that are increasing (B, K, S, M) and towards those being depleted (C, G, P). B = biotite, C = chlorite, G = garnet, K = kyanite, M = muscovite, P = plagioclase, and S = staurolite.

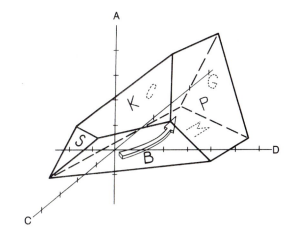

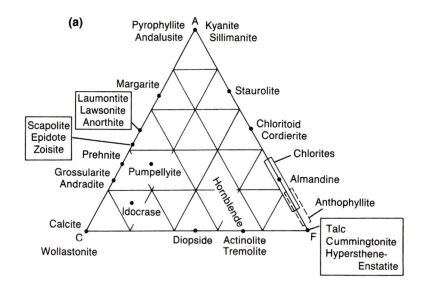

(a)

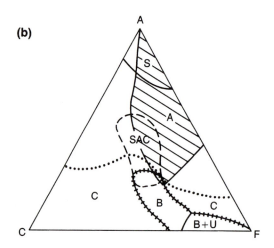

(b)

Figure 25.7 ACF diagram. (a) Diagram showing composition plots of minerals that occur in the ACF compositional triangle (from Winkler, 1979). (b) ACF diagram showing the approximate compositional limits of the six chemical groups of metamorphic rocks defined in chapter

24. A = aluminous rocks, B = basic rocks, B + U = basic and ultrabasic rocks, C = carbonate rocks and other chemical precipitates, S = siliceous rocks, and SAC = siliceous alkali-calcic rocks.

Before plotting data on the diagram, these values must be normalized by making A + C + F = 100%. To plot the analysis of a rock, further adjustments must be made to the analysis to account for minor or additional minerals that may appear in the assemblage and affect the plot, but that cannot be plotted on the diagram.[11]

Some typical minerals that plot on the ACF diagram are shown in figure 25.7a. Minerals that have a substantial range of composition (solid solution phases) plot as areas rather than points. Figure 25.7b shows where the bulk rock chemistries of various common rock types will plot. From

the positions of the bulk rock chemistries, one can easily determine which minerals might be expected in the various metamorphic rocks. For example, calcite, dolomite, wollastonite, anorthite, and grossular are among the minerals expected in metamorphosed limestones (carbonate rocks). In contrast, metabasites might contain diopside, hornblende, actinolite, glaucophane, hypersthene, olivine, epidote, laumontite, pumpellyite, or anorthite, whereas andalusite, kyanite, sillimanite, cordierite, and anorthite are minerals that might appear in an aluminous rock such as a metapelite.

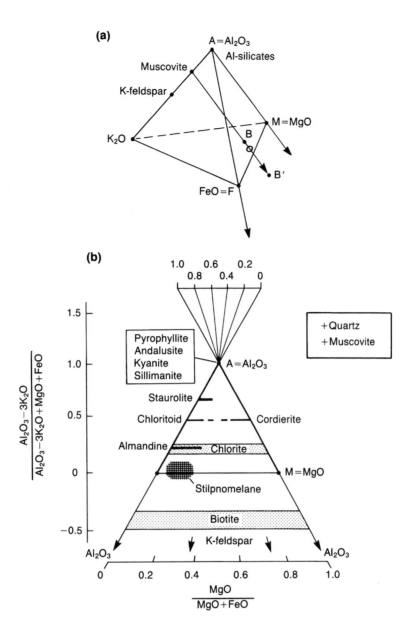

Figure 25.8 AFM diagram. (a) AKFM tetrahedron showing positions of muscovite and K-feldspar. Mineral positions on the AFM plane are projected from muscovite (+ quartz). For example, a biotite of composition B that plots within the tetrahedron is projected through the side (circled point of intersection) onto the AFM plane at B'. Minerals lacking K$_2$O plot directly on the AFM plane. (b) The AFM plane and its extension showing the compositional points, bars, or fields of minerals that plot in the AFM diagram. Scales for plotting molecular ratios of compositional data are shown at the top, bottom, and left side of the diagram.
(Modified from J. B. Thompson, 1957)

THE AFM DIAGRAM

The AFM diagram is a true phase diagram. Developed by J. B. Thompson (1957), it is used to depict the mineral assemblages present in metapelites, some metasandstones, metaigneous rocks of generally similar chemistry, and any other rock containing both quartz and muscovite.[12] In this diagram, FeO and MgO are treated as separate components (as they should be). The diagram is derived from a four-component plot, the AKFM tetrahedron (figure 25.8a).

Six major element components are accounted for in the system SiO$_2$-Al$_2$O$_3$-FeO-MgO-K$_2$O-H$_2$O, which serves as the foundation for the AFM diagram. Adjustments may be made for the additional components TiO$_2$, Fe$_2$O$_3$, Na$_2$O, and P$_2$O$_5$.[13] The diagram is used only for systems saturated in silica; i.e. quartz is always present and need not be depicted other than to be listed adjacent to the diagram. Similarly, H$_2$O is considered to be available where needed, either because the system will be saturated with it or because the external environment can provide adequate water, if it is needed to form a particular phase. Adjustments made to

523

account for the presence of the "additional" components listed leaves only the four remaining components—Al_2O_3, FeO, MgO, and K_2O—to be plotted.

Any aluminous or ferromagnesian mineral or rock composition may be plotted in the AFM diagram. Mineral or rock compositions that lack K_2O (e.g., garnet) plot directly on the basal AFM plane. Projecting from the composition of muscovite (point M, figure 25.8a), one may plot any point lying within the volume of the tetrahedron onto the basal AFM plane or its projection to infinity. Minerals like biotite (point B) that are K-rich project beyond the FeO-MgO line (outside of the AFM triangle). Inasmuch as muscovite is used as a projection point, the system is considered to contain excess muscovite, that is, it can only be used for assemblages containing muscovite. Muscovite is listed with quartz adjacent to the diagram (figure 25.8b). Actual plotting of compositions in the AFM diagram is done using calculated coordinate values derived from the chemical analysis. The coordinate axes, shown in Figure 25.8b, are the molecular values MgO/(MgO+FeO) and $(Al_2O_3 - 3K_2O)/(Al_2O_3 - 3K_2O + MgO + FeO)$.

The positions at which common mineral compositions are plotted in the AFM diagram are shown in figure 25.8b. Pelitic rock compositions may plot anywhere within the diagram, but most commonly fall in the lower center. Similarly, sandstone compositions are variable. Quartz arenites typically plot near the A corner, because they consist of quartz and aluminous clays, whereas lithic wackes and arenites have compositions that plot nearer the base (the M–F line), because they contain Fe-Mg-bearing minerals and volcanic rock fragments.

THE CFM DIAGRAM

The CFM phase diagram is useful for plotting the phase changes in metabasites (Abbott, 1982, 1984). In this diagram (figure 25.9), a number of major components are taken into account, including SiO_2, Al_2O_3, Fe_2O_3, FeO, MgO, CaO, Na_2O, and K_2O, but the two-dimensional plane used for graphical display of phase assemblages is the CFM plane. That plane has corners with the following molecular values:

$$C = CaO + Na_2O + K_2O - Al_2O_3$$

$$F = FeO - Fe_2O_3$$

$$M = MgO$$

Additions and subtractions are made to adjust for the presence of the additional oxides in various minerals. The system is considered to be saturated with quartz, plagioclase, water, and magnetite, and may also contain alkali feldspar.

The compositions of common minerals found in metabasites are plotted on the CFM diagram in figure 25.9. Bulk rock compositions for basalts, andesites, gabbros, and diorites plot near, but commonly above, the F–M line.

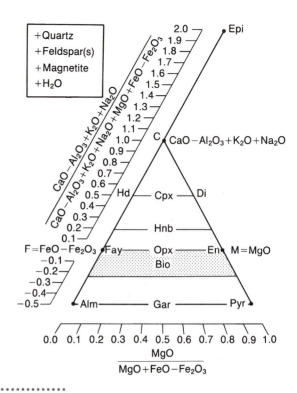

Figure 25.9 CFM diagram showing the endmember compositions for C, F, and M, the plotting scale, and the compositional ranges (lines and shaded area) of various minerals, as projected from quartz, H_2O, the feldspars, and magnetite onto the CFM plane. Epi = epidote, Cpx = clinopyroxene (Hd = hedenbergite = $CaFeSi_2O_6$, Di = diopside = $CaMgSi_2O_6$), Hnb = hornblende, Opx = orthopyroxene (En = enstatite = $Mg_2Si_2O_6$), Fay = fayalite, Bio = biotite (shaded area), Gar = garnet (Alm = almandine, Pyr = pyrope).
(Modified from Abbott, 1982)

OTHER DIAGRAMS

A number of additional diagrams have been used to present the phase assemblages of particular types of rock for which the diagrams above were either not appropriate or not yet available. These include the A'KF, AKN, and ACF^3 diagrams (figure 25.10).[14] The A'KF diagram allows the plotting of muscovite and biotite, as well as K-rich alkali feldspar, but has the same flaw as the ACF diagram in having FeO and MgO combined at the F corner. The AKN diagram, based on the system Na_2O-K_2O-Al_2O_3-SiO_2-H_2O, like the AFM diagram, is useful for showing phase changes in metashales and metasandstones, as well as in felsic metaigneous rocks. The rocks plotted on this diagram should be poor in ferromagnesian minerals. The ACF^3 diagram was developed for use with low- to moderate-temperature, low- to high-pressure metamorphosed shales, sandstones, and mafic igneous rocks.

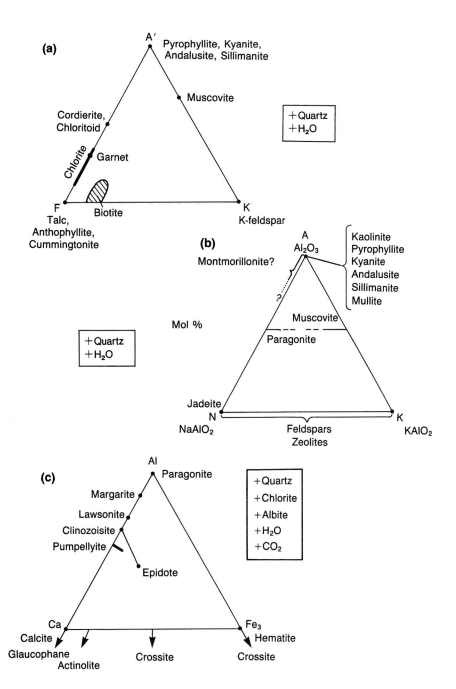

Figure 25.10 A'KF, AKN, ACF³ diagrams showing compositional ranges of minerals that plot in the diagrams. (a) A'KF diagram (after Eskola, 1939, in Winkler, 1965). (b) AKN diagram (after J. B. Thompson and A. B. Thompson, 1976). (c) ACF³ diagram (after E. H. Brown, 1977). Minerals and components in box represent phases from which projection is made. For details on corner compositions, refer to the cited references.

The A'KF, AKN, and ACF³ diagrams, as well as those representing other systems, provide some versatility in showing phase changes for different kinds of rocks. Each diagram is used in the same way, with terminal, nonterminal, and continuous reactions marking the changes from one metamorphic grade to another. In the chapters that follow, where various facies series are described, selected diagrams are used to show distinctive phase assemblages for various grades of metamorphism.

A method of constructing other (geometrical) phase diagrams from data on phase compositions was developed by Shreinemaker (in Zen, 1966; Yardley, 1989). This method allows the development of diagrammatic petrogenetic grids for specific systems in which univariant reaction curves meet at one or more invariant points.[15]

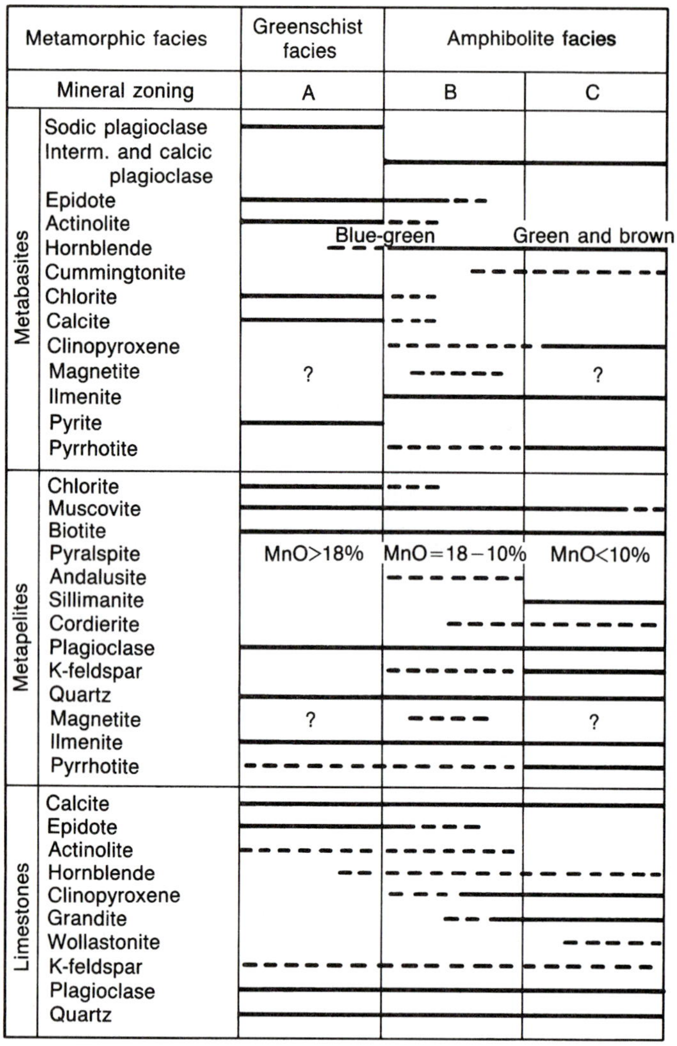

Metamorphic facies		Greenschist facies	Amphibolite facies	
Mineral zoning		A	B	C
Metabasites	Sodic plagioclase				
	Interm. and calcic plagioclase				
	Epidote				
	Actinolite				
	Hornblende		Blue-green	Green and brown	
	Cummingtonite				
	Chlorite				
	Calcite				
	Clinopyroxene				
	Magnetite		?		?
	Ilmenite				
	Pyrite				
	Pyrrhotite				
Metapelites	Chlorite				
	Muscovite				
	Biotite				
	Pyralspite		MnO>18%	MnO=18−10%	MnO<10%
	Andalusite				
	Sillimanite				
	Cordierite				
	Plagioclase				
	K-feldspar				
	Quartz				
	Magnetite		?		?
	Ilmenite				
	Pyrrhotite				
Limestones	Calcite				
	Epidote				
	Actinolite				
	Hornblende				
	Clinopyroxene				
	Grandite				
	Wollastonite				
	K-feldspar				
	Plagioclase				
	Quartz				

Figure 25.11 Mineral-facies chart for the Central Abukuma Plateau of Japan.
(From Miyashiro, 1973a)

MINERAL-FACIES CHARTS

The changes in mineral assemblages through a facies series or across several grades of metamorphism are clearly revealed by mineral-facies charts. Ernst (1965; 1971a, 1971b, 1971c, 1973b, 1977a) and Miyashiro (1973) have popularized the use of such charts, which consist of a rectangular plot with facies or grades of metamorphism listed on the horizontal axis and individual minerals, lumped by major bulk rock composition, listed on the vertical axis (figure 25.11). Lines extending parallel to the horizontal axis show the range of facies or grades over which each mineral is stable. Dashed lines reveal limited conditions of stability.

In the example in figure 25.11, note that two aspects of the changing mineralogy are easily observed. First, the points of appearance or disappearance (by terminal reactions) of critical minerals and mineral assemblages are obvious. Second, the correlation between specific mineralogical changes and the facies or grade boundaries based upon them is evident. In addition, the mineralogical variations between the various bulk rock chemistries may be compared with ease. The principal detriment of such diagrams is that stable phase assemblages are not shown. Thus, such diagrams cannot be used as a substitute for phase diagrams, but rather should be used in conjuction with them.

SUMMARY

Several types of phase diagrams are used in metamorphic petrology. In each case, the diagram is selected on the basis of its usefulness in portraying the mineralogical changes in a particular rock type. All of these diagrams show assemblages of solid phases, rather than melt compositions. Changes from one mineral assemblage to another are depicted on the diagrams as terminal reactions (addition or subtraction of a phase from the diagram) or nonterminal reactions (tie-line flips on the diagram). Univariant curves depict discontinuous reactions, whereas some reactions involving solid solutions are continuous. Reaction paths involving solid solutions may be depicted in reaction space and quantitatively defined using a reaction progress variable.

The most commonly used phase diagrams include the AFM and the CMS diagrams. The former is used for pelitic (aluminous) and quartz-feldspar (siliceous-alkalic-calcic) rocks and the latter is useful in the study of carbonate and ultramafic rocks. The CFM diagram is used for basic rocks. In addition, the pseudo-phase diagrams, A'KF and ACF, are used for pelitic and mafic plus carbonate rocks, respectively.

Mineral-facies charts are useful for portraying general changes in mineralogy across a facies or facies series. Such charts are not phase diagrams, however, and cannot be used in place of phase diagrams, because they do not depict stable phase assemblages.

EXPLANATORY NOTES

1. Mineral grains are not strictly phases in many cases, as they contain inclusions and exsolution lamellae, which, in fact, are mechanically separable.
2. Quartz may be replaced in pelitic rocks by other phases under extreme conditions. In the Sanidinite Facies, the silica phase may be tridymite, and in the Eclogite Facies, it may be coesite.

3. The metamorphic relations in the system SiO_2-CaO-MgO-H_2O-CO_2, related systems, and particular examples have been studied by several workers. Important works include those of N. L. Bowen (1940), R. I. Harker and Tuttle (1956), H. J. Greenwood (1967), Turner (1968), Metz and Trommsdorf (1968), Trommsdorf and Evans (1972), B. W. Evans and Trommsdorf (1974), Skippen (1974), Slaughter, Kerrick, and Wall (1975), Joesten (1976), B. W. Evans (1977), Kase and Metz (1980), G. Franz and Spear (1983), and S. B. Tanner, Kerrick, and Lasaga (1985). Winkler (1967, 1974, 1976, 1979) and F. J. Turner (1981) provide good reviews of the subject. Also see chapters 26 and 31 of this text for additional discussions.

4. Systems are *internally buffered* if the mineral assemblage controls the fluid phase composition. This occurs if reactions produce or consume CO_2 or H_2O in amounts that significantly affect the fluid phase composition. Under such circumstances, externally generated fluids are relatively insignificant in volume.

5. Any number of phases may be plotted by projection onto the plane, but information is lost in doing so and the diagram cannot be as easily used to represent equilibrium assemblages.

6. A detailed treatment of this subject is beyond the scope of this text. For additional examples, see J. B. Thompson, Laird, and Thompson (1982) and Poli (1991).

7. Detailed treatment of this topic is beyond the scope of this text.

8. To construct a complete phase diagram from field and petrographic data, it is important to collect samples of rock covering a wide range of bulk compositions. By doing so, one is able to obtain a wide range of three-phase mineralogies that will allow complete definition of the tie lines of the phase diagram. This point has been emphasized by J. B. Thompson (1957) and Guidotti (1982).

9. In this chapter, the term *pseudo-phase diagram* refers to phase diagrams that look like typical triangular metamorphic phase diagrams but do not meet the requirements of the Phase Rule, because of the way that oxides are combined to make the components used as definitive endmembers for the apices of the triangle. Guidotti (1982) discusses this problem.

10. Though it was assumed in older works that Fe^{2+} and Mg^{2+} substitute freely for one another in various minerals in rocks, more recent studies reveal that there is an uneven distribution of these two ions between coexisting ferromagnesian minerals. As noted in chapter 24, uneven distributions are referred to as partitioning of the element. See chapter 24 for a brief discussion and for references to important works on the subject.

11. For details on the specific procedures involved in ACF calculations and more thorough discussions of the use of this and other diagrams, see the quotations of Eskola's (1939) work and other discussions in older metamorphic petrology texts such as Winkler (1965, 1967, 1974, 1976, 1979), F. J. Turner (1968, 1981), Miyashiro (1973a), and Mason (1978).

12. More detailed discussions of the nature and use of AFM diagrams are provided in metamorphic petrology texts such as those cited in note 11. Also see Best (1982, p. 403ff.) and Yardley (1989, ch.3).

13. See J. B. Thompson (1957) and Mason (1978).

14. The A'KF diagram was introduced by Eskola (see F. J. Turner, 1968, p. 175 ff.) and is described in metamorphic petrology texts such as Turner (1968), Miyashiro (1973a), and Winkler (1976). The AKN diagram (a name applied here) was introduced by J. B. Thompson (1961, in A. B. Thompson, 1974). The AKN diagram is derived from the system $NaAlO_2$-$KAlO_2$-Al_2O_3-SiO_2-H_2O, with points projected from SiO_2 and H_2O onto the Al_2SiO_5-$KAlSi_3O_8$-$NaAlSi_3O_8$ plane. A. B. Thompson (1974) and J. B. Thompson and Thompson (1976) discuss its use. The ACF^3 diagram was designed by E. H. Brown (1977a). See E. H. Brown (1974) and E. H. Brown et al. (1981) for additional discussions relating to the sorts of rocks for which this diagram was devised. H. J. Greenwood (1975) provides a calculation for the production of any desired projection of phase relations in any system.

15. See Zen (1966) and Yardley (1989, appendix) for details on how to construct Schreinemaker diagrams.

PROBLEMS

25.1. Using the chemical analyses of ultramafic rock 3 in table 10.2 and carbonate rock 3 in table 21.1, (a) calculate the coordinates for and plot the positions of these bulk compositions on a CMS diagram. (b) Using the low-T phase diagram shown in figure 25.3, determine what mineralogy would be present in these rocks. (c) Similarly, determine the stable minerals for these rocks for the higher-T phase diagram in figure 25.3.

25.2. Using the chemical analysis of pelitic rock 6 in table 18.3, (a) calculate the coordinates for and plot the position of this bulk composition on an AFM diagram. (b) Using a chemical analysis of an almandine garnet from a mineralogy book, calculate the coordinates for and plot the position of its composition on the AFM diagram.

CHAPTER 26

Contact Metamorphism

INTRODUCTION

Contact metamorphism, as the name implies, occurs locally, at and near the contacts between intrusions and the surrounding country rock. As might be expected in such a setting, the metamorphism is dominantly controlled by the heat introduced by the intrusion. The effects of increased temperature are most pronounced where intrusions occur at shallow levels in the crust. There, contrasts in temperature between country rock and intrusion are at a maximum. With increasing depth of intrusion, the temperature contrast generally decreases, as do the contact effects, except in relatively rare cases where intrusion occurs in higher-P/T facies series.

As will become evident, the fluid phase is also an important agent of contact metamorphism. It transports heat and has a profound influence on the chemistry and mineral composition of the rocks with which it comes in contact. Fluids are particularly important (1) in the metamorphism of carbonate rocks, where metamorphism yields CO_2, (2) in rocks undergoing metamorphism in hydrothermally active areas, and (3) in rocks that yield H_2O upon metamorphism, such as those with abundant phyllosilicates. Especially along the mid-ocean ridges and in other active hydrothermal areas, H_2O-rich fluids are very important agents of metamorphism.

Contact metamorphism produces fine-grained granoblastites and diablastites, commonly called **hornfelses** (figure 26.1).[1] In addition to a variety of common minerals, such as quartz, feldspars, and epidote, hornfelses locally contain unique phases. Such minerals as spurrite and tilleyite (calcium silicates) may form in carbonate rocks; mullite, an aluminum silicate, forms in pelitic rocks; and minerals more commonly associated with igneous rocks, such as sanidine, form in rocks of appropriate chemistry. The high temperatures of metamorphism, in their extreme, may even cause local melting in the contact zone, yielding glass.

Examination of the effects of contact metamorphism allows us to gain insight into metamorphic processes without the complicating influences of high pressure and deviatoric stress. Typically, contact metamorphism occurs at shallower levels of the crust, where the pressure is relatively low (< 0.4 Gpa = 4 kb). At those shallow levels, the deviatoric stresses characteristic of the deeper levels of mountain belts are generally absent and the contact metamorphic rocks lack foliation. Locally, however, high-pressure contact rocks do exist, for example, in forearc regions, shields, and root zones of mountain belts.

Contact Facies Series rocks are mineralogically similar and in some cases identical to those of the Buchan Facies Series (Pattison and Tracy, 1991), but contact

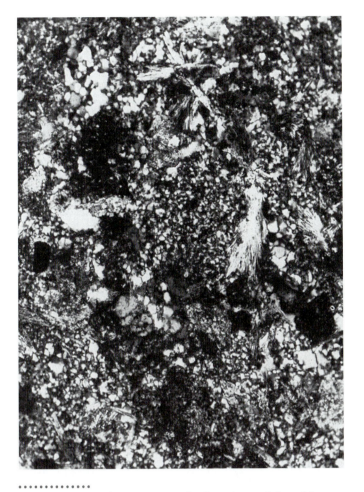

Figure 26.1 Photomicrograph of epidote-rich hornfels. Tioga Lake, Sierra Nevada, California. Note dominant granoblastic texture. (PL). Long dimension of photo is 0.33 mm.

rocks are distinguished by their general lack of foliation. They differ from rocks of the other facies series in their mineralogy, textures, and local distribution.

FACIES AND FACIES SERIES

Contact metamorphic rocks are found in **aureoles**, zones of metamorphic rock surrounding and associated with plutons (figure 26.2), as well as in roof pendants within plutons and in xenoliths in plutons and lava flows. In this text, the hydrothermal metamorphism associated with igneous activity along the mid-ocean ridges is discussed with contact metamorphism because (1) the metamorphic effects are restricted to distances of a few kilometers perpendicular to strike and vertically within the rock column, (2) the effects are produced primarily as a result of the igneous and associated hydrothermal activity, and (3) deviatoric stress does not play a dominant role in fabric development in this environment.

Observation of the occurrences of contact metamorphic rocks and examination of petrogenetic grids (see appendix C) reveals that Zeolite, Prehnite-Pumpellyite, Albite-Epidote Hornfels, Hornblende Hornfels, Pyroxene Hornfels, and Sanidinite facies constitute the *Contact Metamorphic Facies Series*. At progressively higher temperatures, in pelitic and siliceous-alkali-calcic rocks, minerals indicative of these facies include analcite, stilbite, wairakite, pyrophyllite, cordierite, andalusite, sillimanite, K-feldspar,

Figure 26.2 Map of the contact aureole of the Onawa Pluton, Maine, showing crudely concentric metamorphic zones. *(Modified from Philbrick, 1936; J. M. Moore, 1960)*

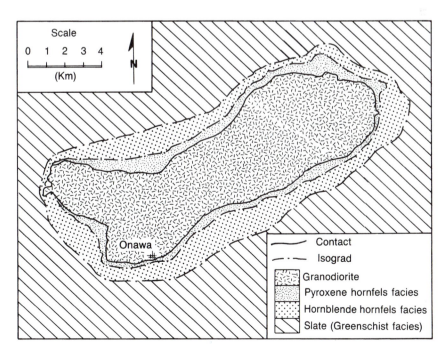

Scale

0 1 2 3 4

(Km)

N

Onawa

——— Contact

—·—·— Isograd

Granodiorite

Pyroxene hornfels facies

Hornblende hornfels facies

Slate (Greenschist facies)

orthopyroxene, sanidine, and mullite. Wairakite, albite, actinolite, epidote, hornblende, the pyroxenes, and olivine occur in corresponding basic rocks. In carbonate rocks, minerals such as talc, tremolite, diopside, forsterite, grossularite, wollastonite, and spurrite may develop.

In any given aureole, a complete sequence of facies will not likely occur. Both low-temperature facies and high-temperature facies are often missing. Because many intrusions occur in previously metamorphosed rocks, occurrence of the lower-grade facies is precluded by the resistance of preexisting, partially dehydrated, metamorphic country rocks to retrograde metamorphism. At the highest grades, Sanidinite Facies rocks can only develop where very low pressures, particular bulk rock compositions, or the absence of a fluid phase inhibit melting. The upper parts of both the Pyroxene Hornfels Facies and the Sanidinite Facies overlap the zone of melting, where P-T conditions and a fluid phase cause melting in many bulk compositions (see figure 24.6 and appendix C).

A typical example of a partial facies series is provided by the contact aureole of the Devonian Onawa pluton of Maine (figure 26.2). (Philbrick, 1936; J. M. Moore, 1960)[2] The pluton is an elongate, composite mass of granitoid rock that was intruded into slate country rock previously metamorphosed under conditions of the lower Greenschist Facies. The country rocks contain the assemblage Fe-Ti oxide + white mica + chlorite + quartz. The first evidence of contact metamorphism is the appearance of spots in the slates as far as 2 km from the pluton margin. The spots were cordierite porphyroblasts (now largely replaced by phyllosilicates) and are part of the assemblage biotite + andalusite + cordierite + white mica + quartz + albite (figure 26.3). This assemblage is representative of the Hornblende Hornfels Facies.

The outer zone of spotted slates surrounds a second zone. This second zone is composed of porphyroblastic granoblastites (hornfelses) with the same mineral assemblage as the rocks of the outer zone, but lacking their continuous (slaty) cleavage. The second zone surrounds a third zone, adjacent to the pluton, composed of coarser-grained granoblastites with the assemblage biotite + sillimanite + cordierite + alkali feldspar + quartz (figure 26.3). This assemblage indicates the Pyroxene Hornfels Facies. Local, small-scale dikes and fine-mesoscopic, millimeter-size patches of quartz-alkali feldspar rock indicate that partial melting has occurred locally. In summary, only two facies are recognized in the Onawa aureole, the Hornblende Hornfels Facies and the Pyroxene Hornfels Facies.

CONDITIONS OF CONTACT METAMORPHISM

The conditions of contact metamorphism, with rare exception, are those of low to moderate pressure and low to high temperature. Both isotopic studies and comparisons of

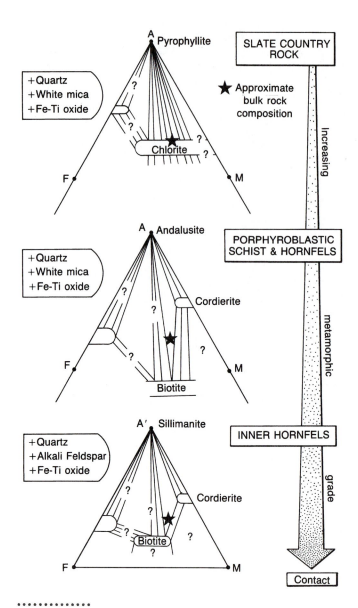

Figure 26.3 AFM diagrams showing the metamorphic mineral assemblages of the contact aureole and metamorphic terrane surrounding the Onawa Pluton, Maine. Sodic plagioclase is present in some lower-grade assemblages.

observed mineral assemblages with experimentally determined phase equilibria yield information about the P-T conditions. Table 26.1 lists the maximum P-T conditions estimated for several contact aureoles. Note that pressures are generally less than 0.4 Gpa (4 kb). The common presence of andalusite in contact aureoles is a good mineralogical indicator that the pressure in the middle grade of most contact aureoles was lower than that of the aluminum silicate triple point, which occurs at 0.387 Gpa (3.87 kb) (Hemingway et al., 1991).

Temperatures of metamorphism vary widely. Among the controlling factors are (1) the temperature of the

Table 26.1 Estimated Maximum P-T Conditions for Selected Contact Aureoles

Locality	T (°C)	P (kb)	References
DSDP Hole 504B	380	0.4	Alt et al. (1986)
Brewster Co., Tex.	470	0.3	Droddy and Butler (1979)
Mid-Atlantic Ridge 6° N	>500	0.5	Bonatti et al. (1975)
Blakes Ferry, Ala.	510	7.0	R. G. Gibson and Speer (1986)
Hope Valley, Calif.	540	2.0	Kerrick, Crawford, and Randazzo (1973), Ferry (1989)
Marysville, Mont.	600	1.0	J. M. Rice (1977), Lattanzi, Rye, and Rice (1980)
Notch Peak, Utah	600	2.0	Hover Granath, Papike, and Labotka (1983)
Duluth Complex, Minn.	620	1.5	Labotka (1983), Labotka, White, and Papike (1984)
Tioga Pass, Calif.	670	2.0	Kerrick (1970)
Bergell Intrusion, Italy	700	3.5	Trommsdorf and Evans (1972, 1977a), Wenk, Wenk, and Wallace (1974)
Mathematician Ridge	700	0.6	Stakes and Vanko (1986)
Liberty Hill, S. C.	725	4.5	Speer (1981, 1987)
Onawa, Maine	725?	3.6	J. M. Moore (1960), Pattison and Tracy (1991)
Connecticut Valley	745	<0.1	April (1980)
Lilesville, N. C.	750	3.5	N. H. Evans and Speer (1984)
Mount Royal, Quebec	750	0.5	Williams-Jones (1981)
King Island, Tasmania	800	0.7	Hing and Kwak (1979)
Ronda, Spain	800	4.3	Loomis (1972a)
Morton Pass, Wyo.	933	3.0	Russ-Nabelek (1989)
Christmas Mtns., Tex.	1035	0.3	Joesten (1974, 1976)
Bushveld Complex, S. Africa	>1200	1–2	Wallmach, Hatton, and Droop (1989)

magma, which is a function of its chemistry, (2) the temperature of the country rock at the time of intrusion, (3) the conductivities of the solidifying magma and the country rock, (4) a factor called the diffusivity (of both the country rock and the intrusion), (5) the heat of crystallization of the magma, (6) the heat capacity (the rate of change in the energy of reaction with change in temperature), (7) fluid transport, the heating or cooling by influx of water, and (8) contributions from other sources, such as radioactive decay (Jaeger, 1957, 1959; F. J. Turner, 1981; Ferry, 1983b).[3] In a simplified case, considering only the temperatures of the intrusion (T_i) and the country rock (T_{cr}), the temperature at the contact (T_c) will be $T_c = 1/2(T_i + T_{cr})$. In fact, however, the effects of the additional factors will increase or reduce that value. These factors and the temperature of the contact will also affect the *width* of the contact aureole.

PROCESSES IN CONTACT METAMORPHISM

The brief description of the Onawa aureole (above) provides a glimpse of some of the kinds of processes that operate during contact metamorphism. In particular, the growth of porphyroblasts and progressive recrystallization and neocrystallization were indicated. Local melting accompanied metamorphism at the highest grades.

In addition to the above processes, processes involving a fluid phase may be quite important. The fluid phase may result from decarbonation and dehydration reactions, which are common in metamorphism, and may cause melting or metasomatism in rocks invaded by it.

The Fluid Phase

Evidence that a fluid phase exists in metamorphic rocks was presented in chapter 23. The fluid phase may begin as (1) meteoric water, (2) interstitial water or brine in sedimentary and igneous protoliths, (3) juvenile water or fluids derived from intrusions, or (4) as water or other volatile species chemically bound in volatile-bearing mineral phases, such as clays or carbonate minerals.[4] Particularly important in the evolution of the fluid phase are decarbonation and dehydration reactions.

If we examine the phase assemblages present in the Onawa aureole, we find that dehydration is clearly revealed. The Fe-Ti oxide + white mica + chlorite + quartz assemblage of the country rocks is replaced in the outer zone of the aureole by one containing andalusite and cordierite. Chlorite [$(Mg,Fe,Al)_6(Al,Si)_4O_{10}(OH)_8$], a hydrous phase, disappears from the assemblage, and is replaced by the assemblage biotite + cordierite + andalusite, two minerals of which are anhydrous phases [andalusite = Al_2SiO_5 and cordierite = $(Mg,Fe)_2Al_4Si_5O_{18}$]. Closer to the pluton, white mica [$(K, Na)Al_2(Si_3Al O_{10})(OH)_2$], a hydrous phase, is replaced by alkali feldspar [$(K,Na)AlSi_3O_8$], an anhydrous phase. This pattern is typical of prograde metamorphism, in which the rocks become increasingly anhydrous at progressively higher grades of metamorphism as fluids are driven off. Thus, progressive metamorphism generates a fluid phase that apparently migrates away from the zones of highest-grade metamorphism.

A similar situation pertains in rocks involving carbonate phases. For example, a reaction that occurs during the contact metamorphism of carbonate rocks is

$$\text{calcite + quartz} <=> \text{wollastonite} + CO_2 \qquad (26.1)$$

$$CaCO_3 + SiO_2 <=> CaSiO_3 + \quad CO_2$$

In this reaction, the carbonate phase calcite combines with quartz to yield wollastonite and the volatile species CO_2.[5] As is the case with dehydration, progressively higher grades of metamorphism yield increasingly CO_2-poor rocks. CO_2 is driven from the minerals during progressive metamorphism and becomes a part of a fluid phase.

In some cases, fluids evolved from the rocks themselves are overwhelmed by fluid derived from an intrusion or that involved in hydrothermal activity. Below we will examine a case where an intrusion induces Si-Al-Fe metasomatism by introducing these elements via a magmatic fluid phase. In contrast, along mid-ocean ridges, heated ocean water in great volumes moves through newly formed oceanic crust, altering the rock chemistry (Rona et al., 1983).

The *composition* of the fluid phase controls the mineral assemblage developed under given sets of metamorphic conditions and vice versa (Ferry and Burt, 1982, A. B. Thompson, 1983).[6] Thus, it is important to know the nature of the fluid phase present during metamorphism. That fluid phase

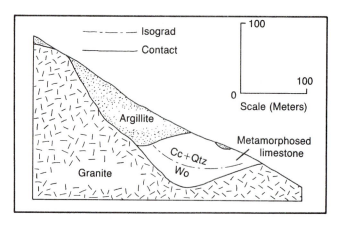

....
Figure 26.4   Cross section of the Victory Mine contact aureole near Salmo, British Columbia. The contact geology is based on drilling and surface mapping. Only the wollastonite isograd in the carbonate rock is shown.
(Simplified from H. J. Greenwood, 1967)

may be dominated by H_2O, but in carbonate rocks and at higher grades of metamorphism, the mole fraction of H_2O (X_{H_2O}) may decrease to values of 0.1 or less.[7] In carbonate rocks, the ratio of H_2O to CO_2 is especially important. Other components of the fluid phase may include Cl, F, B, S, Na, C and H (CH_4) (Russ-Nabelek, 1989; Sisson and Hollister, 1990; Labotka, 1991).

The influence of varying mole fractions of CO_2 and H_2O in metamorphism of carbonate rocks was demonstrated by H. J. Greenwood (1967) in his work on a contact aureole near Salmo, British Columbia (figure 26.4). Here, during metamorphism, water and carbon dioxide formed a fluid phase dominated by these two components, and the fluid phase controlled the development of wollastonite in the contact aureole. The important phase relations are best depicted on a T-X diagram constructed for a fixed pressure (figure 26.5). At low mole fractions of CO_2, wollastonite is stable at lower temperatures. At higher values of X_{CO_2}, wollastonite requires higher temperatures to form. A change in pressure simply raises or lowers the temperature of the reaction.

In general, a series of such $T\text{-}X_{CO_2\text{-}H_2O}$ curves are applicable to metamorphism of carbonate and ultramafic rocks. Where more than one curve exists, the curves will tend to intersect at invariant points (figure 26.6), beyond which the stable phase assemblage will differ. More generally, $T\text{-}X_{fluid}$ curves may be used for any fluid phase in any composition of rock.

Recrystallization and Neocrystallization

Recrystallization and neocrystallization are evident in the Onawa aureole, just as was dehydration. In terms of recrystallization, it was noted that there was an increase in grain size in the rocks from the outer to the inner zones of the aureole.

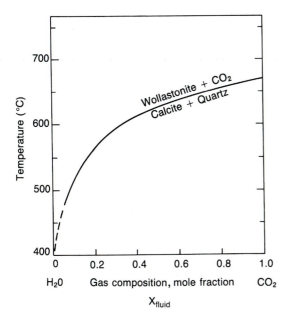

Recall that the same effect, quantified by Joesten (1983), was described in chapter 23. In general, rocks become progressively more coarse-grained with increasing grades of metamorphism, though polygonization and grain-size reduction may occur as an intermediate stage in the coarsening process.

Neocrystallization in the Onawa aureole involved the development of biotite, andalusite, cordierite, sillimanite, and alkali feldspar. The appearance of each new phase or phase assemblage marks a reaction in which there is a chemical and structural readjustment in the rocks. In general, such readjustments occur as the rock equilibrates with new P-T conditions and responds geochemically to the fluid phase present at the time.

The first indication of contact metamorphism in the Onawa contact aureole, the development of porphyroblasts, is a common form of neocrystallization in the outer zones of contact aureoles. Of course, porphyroblasts also develop in regionally metamorphosed rocks, wherever conditions are appropriate. In general, porphyroblast development involves the formation of a few nucleii and the subsequent growth of those nucleii into large crystals.[8] Nucleation is possible where the necessary phases are present to yield chemical species needed to form the nucleus of a new phase stabilized by the

Figure 26.5 T-X_{fluid} diagram showing the stability relations of calcite, quartz, and wollastonite at one kilobar (P_f = 1 kb). Note the strong downward bend of the reaction curve at high values of X_{H_2O}. An increase in pressure raises the position of the curve to higher temperatures but does not change its general shape.

(Simplified from H. J. Greenwood, 1967)

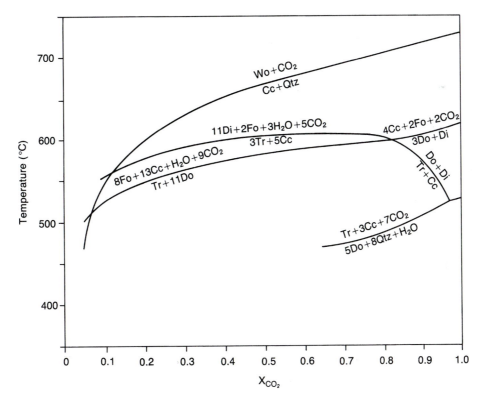

Figure 26.6 T-X phase diagram showing intersecting stability curves for some reactions important in the metamorphism of carbonate rocks. Reactions involve the phases calcite (Cc), diopside (Di), dolomite (Do), forsterite (Fo), quartz (Qtz), tremolite (Tr), wollastonite (Wo), and the components of the fluid phase P = 2 kb (0.2 Gpa).

(Simplified from Hover Granath, Papike, and Labotka, 1983)

533

imposition of different P, T, or X_{fluid} conditions. Diffusion of species through the surrounding rock allows both nucleation and growth.

In zones of high shear stress, nucleation and growth of porphyroblasts are prohibited by the effects of that stress, especially effects such as dissolution and solution transfer that result from chemical potential gradients created by deformation of crystals (T. H. Bell, Rubenach, and Fleming, 1986). If shear stress is moderate, porphyroblast nucleii may form at local sites of lower stress. Once nucleated, porphyroblasts grow. The increased size of the porphyroblast may then create **pressure shadows,** zones of lower stress and equant grain growth adjacent to the porphyroblast that facilitate further porphyroblast growth (see figure 23.11c) (D. J. Prior, 1987). In contact zones, shearing stress is typically moderate to nil and porphyroblasts commonly lack pressure shadows. Where they are present, they indicate that shearing stresses accompanied contact metamorphism.

MINERALOGICAL CHANGES DURING CONTACT METAMORPHISM

Mineralogical changes during contact metamorphism are a function of bulk rock chemistry, P-T conditions, and the nature of the fluid phase. Below, each of the major categories of bulk rock is discussed in terms of the mineralogical changes that occur during contact metamorphism.

Aluminous Rocks (Pelitic Rocks)

The Onawa aureole is representative of contact aureoles in pelitic rocks. Yet, as is typical, it does not exhibit all of the facies of contact metamorphism. To see the range of mineralogies that may develop, a composite array of facies diagrams (figure 26.7) must be compiled from observed occurrences of the Contact Metamorphic Facies Series or calculated using model systems such as the so-called KFMASH system (Pattison and Tracy, 1991).[9] The phase diagram most useful for portraying assemblages in aluminous rocks is the AFM diagram. *Note:* Only a few representative diagrams are presented here, and the addition or subtraction of each phase during the progressive metamorphism of any assemblage of rocks yields a new topology.

At the lowest grades of metamorphism, in the Zeolite Facies, various combinations of quartz, chlorite, alkali feldspar, calcite, kaolinite and various mixed-layer clays (smectite/illite, vermiculite/chlorite), and illite, as well as a number of zeolites and iron oxides, characterize the rocks.[10]

With progressive metamorphism, the clays are replaced by white mica, chloritoid may appear in iron-rich rocks, kaolinite is replaced by pyrophyllite, and biotite appears (see figure 26.7). The reaction

$$\text{kaolinite} + 4\ \text{quartz} \Longleftrightarrow 2\ \text{pyrophyllite} + 2\ H_2O \quad (26.2)$$

(Velde, 1969) is one of the important reactions marking the boundary between the Zeolite and Albite-Epidote Hornfels Facies.[11] Assemblages such as

quartz–white mica–chlorite–biotite–chloritoid–albite–epidote

and

quartz–chlorite–white mica–alkali feldspar–albite–epidote

are typical of the Albite-Epidote Hornfels Facies. The presence of biotite in this facies indicates that a reaction such as

$$\text{chlorite} + \text{alkali feldspar} \Longleftrightarrow 2\ \text{biotite} + 2H_2O \quad (26.3)$$

has occurred. The assemblages listed above are essentially identical to those of the adjacent, higher-pressure Greenschist Facies. This is no doubt one of the reasons that the Albite-Epidote Hornfels Facies is less distinct and less commonly recognized than the higher-temperature Hornblende Hornfels Facies, which is characterized by more distinctive mineral assemblages. The similarity also raises the question of whether or not the Albite-Epidote Hornfels Facies should be recognized as a separate and distinct facies.

The Albite-Epidote Hornfels Facies is succeeded at higher temperatures by the Hornblende Hornfels Facies. This facies, at the middle grades of contact metamorphism, exhibits some of the most characteristic contact metamorphic phase assemblages (see figure 26.7). These include

quartz–white mica–andalusite–cordierite–biotite

and

quartz–alkali feldspar–cordierite–plagioclase–magnetite.[12]

Staurolite and garnet may also occur in this facies, as may Mg-amphiboles.[13] Important reactions leading to Hornblende Hornfels Facies assemblages include

$$\text{pyrophyllite} \Longleftrightarrow \text{andalusite} + 3\ \text{quartz} + H_2O \quad (26.4)$$

(Kerrick, 1968: Haas and Holdaway, 1973),

$$\text{Fe-chlorite} + \text{quartz} + \text{magnetite} \Longleftrightarrow \text{garnet (almandine)} + H_2O \quad (26.5)$$

(L. C. Hsu, 1968), and

$$2\ \text{muscovite} + 2\ \text{chlorite} + 4\ \text{quartz} \Longleftrightarrow 2\ \text{cordierite} + 2\ \text{biotite} + 7\ H_2O \quad (26.6)$$

(Fawcett and Yoder, 1966; N. H. Evans and Speer, 1984; Pattison, 1987).

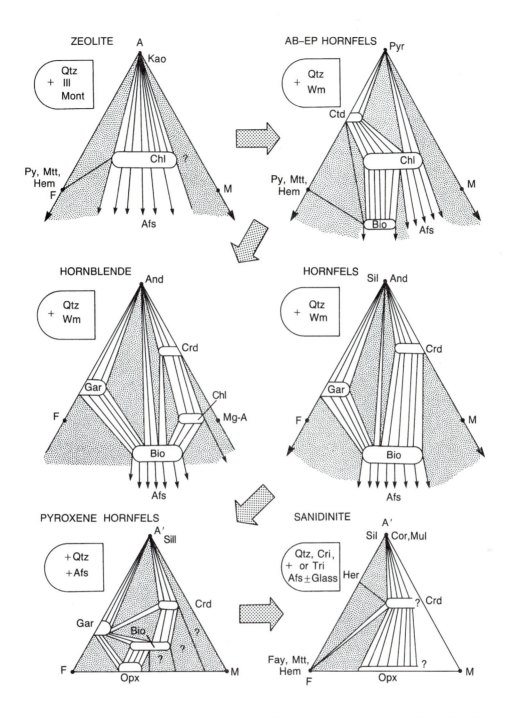

Figure 26.7 AFM and A'FM phase diagrams, showing typical phase assemblages for the facies of contact metamorphism. These diagrams are applicable to pelitic rocks and aluminous siliceous-alkali-calcic (quartz-feldspar) rocks. Afs = alkali feldspar, And = andalusite, Bio = biotite, Chl = chlorite, Cor = corundum, Crd = cordierite, Ctd = chloritoid, Cum = cummingtonite, Fay = fayalite, Gar = garnet, Hem = hematite, Her = hercinite, Ill = illite, Kao = kaolinite, Mg-A = magnesium amphibole, Mont = montmorillonite, Mtt = magnetite, Mul = mullite, Opx = orthopyroxene, Py = pyrite, Pyr = pyrophyllite, Qtz = quartz, Sil = sillimanite, Tri = tridymite, Wm = white mica.

(Based on sources cited in the text. A'FM plot after E. W. Reinhardt, 1968)

Where intermediate to mafic magmas intrude aluminous country rocks and incorporate fragments of those rocks as xenoliths, Pyroxene Hornfels and Sanidinite facies rocks are developed. Occurrences of the former are not uncommon in contact aureoles, but those of the latter are rare. In these two facies, the hydrous phases white mica and biotite may yield to the anhydrous phases alkali feldspar and orthopyroxene. Muscovite breaks down via the reaction

$$\text{muscovite} + \text{quartz} <==> \text{andalusite (or sillimanite)} + $$
$$\text{alkali feldspar} + H_2O \qquad (26.7)$$

(Evans, 1965).[14] This reaction is one of the important reactions marking the boundary between the Hornblende Hornfels and Pyroxene Hornfels facies. White mica has also been found to break down, under disequilibrium conditions that are probably typical of xenolithic Sanidinite Facies metamorphism, via the coupled reactions

$$\text{white mica} <==> \text{alkali feldspar} + \text{biotite} + \text{mullite} \qquad (26.8)$$

and

$$\text{white mica} <==> \text{alkali feldspar} + \text{biotite} + \text{corundum} + $$
$$\text{hercynite} \qquad (26.9)$$

(Brearley, 1986).
Biotite may break down via reactions such as

$$4 \text{ biotite} + 18 \text{ quartz} + 12 \text{ andalusite} <==> $$

$$6 \text{ cordierite} + 4 \text{ alkali feldspar} + H_2O \qquad (26.10)$$

(Holdaway and Lee, 1977; Pattison, 1987) or

$$\text{biotite} + \text{quartz} <==> \text{sanidine} + \text{magnetite} \pm \text{hematite}$$
$$(26.11)$$

(Eugster and Wones, 1962; Wones and Eugster, 1965). Both reactions occur generally within the P-T conditions of the Pyroxene Hornfels Facies, but the latter reaction is important at low pressures in defining the boundary between Pyroxene Hornfels and Sanidinite facies. Typical assemblages of the Pyroxene Hornfels Facies[15] include

quartz–K-feldspar–biotite–cordierite–garnet–plagioclase,

quartz–K-feldspar–andalusite–cordierite–biotite , and

quartz–alkali feldspar–plagioclase–biotite–garnet–
cordierite–hypersthene.

Note the absence of white mica in these assemblages.

In the Sanidinite Facies,[16] garnet may break down to form Fe-cordierite, fayalite, and hercynite, via the reaction

$$5 \text{ almandine} <==> 2 \text{ Fe-cordierite} + 5 \text{ fayalite} + \text{hercynite}$$
$$(26.12)$$

(L. C. Hsu, 1968). Other reactions important within this facies are

$$\text{orthoclase} <==> \text{sanidine} \qquad (26.13)$$

(J. V. Smith, P. H. Ribbe, and D. B. Stewart, 1965, in Hyndman, 1972) and

$$\text{quartz} <==> \text{tridymite} \qquad (26.14)$$

(Tuttle and England, 1955). Each yields phases characteristic of the Sanidinite Facies. Thus, assemblages such as

tridymite–sanidine–Fe-cordierite–hercynite–mullite

are particularly definitive of the facies (see figure 26.7). Other assemblages include

sillimanite–mullite–cordierite–glass,

mullite–cordierite–cristobalite, and

corundum–hercynite–ilmenite–magnetite.[17]

Reviewing all of the reactions listed above, we can see that most represent either dehydration reactions or polymorphic changes. The parageneses of the Zeolite Facies are quite hydrous, whereas those of the Sanidinite Facies are rather dry. Clearly, dehydration is the norm in progressive contact metamorphism. In some cases, the fluid phase generated by the dehydration (and related) reactions migrates away from the region of metamorphism, whereas in others it promotes melting of the less refractory part of the rock, yielding a glass phase.

Silicic and Siliceous-Alkalic-Calcic Rocks

The silica-rich rocks typically have a restricted bulk composition, with high silica, moderate to low alumina, and minor amounts of other components. Because of the limited range of chemistry, mineral assemblages are also limited. Where either alumina is relatively abundant or calcium and the alkalis are significant, the mineralogical changes will be like those of the pelitic rocks and siliceous-alkali-calcic rocks, respectively.

The siliceous-alkali-calcic (SAC) rocks are quartz- and feldspar-rich and are sometimes referred to as quartzofeldspathic. Protoliths include sandstones and felsic to intermediate igneous rocks. Some of these rocks are aluminous and those that are bear mineralogical similarities to the pelitic rocks. The major differences are that the SAC rocks are characterized by calcium-rich phases and generally lack the aluminum silicates (i.e., andalusite, sillimanite). Those that are more calcium-, iron-, and magnesium-rich are gradational in composition with the basic rocks. In general, the SAC rocks differ from the basic rocks in containing abundant quartz and lacking abundant Ca-amphiboles (and clinopyroxenes).

SAC rocks may be represented on either AFM or ACF diagrams. The AFM diagrams in figure 26.7 are applicable to *aluminous* SAC rocks. Mineral assemblages of those with appreciable calcium may be shown more clearly on the ACF pseudo-phase diagram (figure 26.8). The dotted polygon outlined on the Zeolite Facies diagram in figure 26.8 shows the general range of compositions for SAC rocks. A few unusual

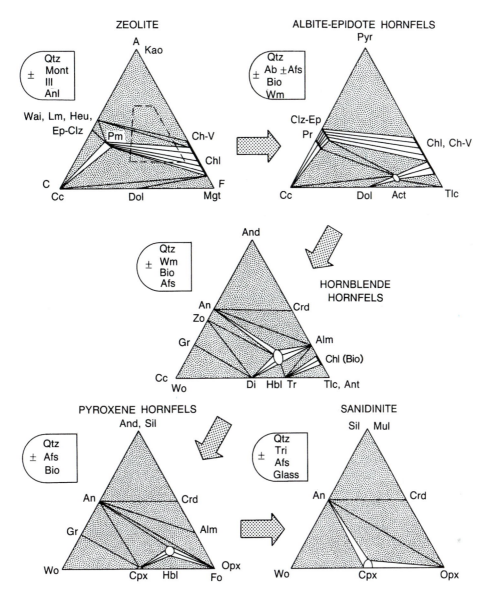

Figure 26.8 ACF diagrams depicting typical mineral assemblages for siliceous-alkali-calcic rocks, intermediate igneous rocks, and basic igneous rocks in the various facies of contact metamorphism. Abbreviations as in figure 26.7 and as follows: Ab = albite, Act = actinolite, Alm = almandine (garnet), An = anorthite, Anl = analcite, Ant = anthophyllite, Ch-V = chlorite/vermiculite, Clz = clinozoisite,

Cpx = clinopyroxene, Di = diopside, Dol = dolomite, Ep = epidote, Fo = forsterite, Gr = grossularite (garnet), Hbl = hornblende, Heu = heulandite, Lm = laumontite, Mgt = magnesite, Pr = prehnite, Pm = pumpellyite, Tlc = talc, Tr = tremolite, Wai = wairakite, Wo = wollastonite, Zo = zoisite. *(Based on sources cited in text)*

compositions may lie outside of this field, but sandstones and siliceous igneous rocks generally plot on the left and upper parts of the polygon, whereas intermediate igneous rocks plot near the bottom and bottom right. Basic rocks (discussed below) plot in the bottom center and below the bottom line of the polygon. Ultrabasic rocks (also discussed below) plot in the lower right corner of the ACF triangle. *Note:* Only a few representative diagrams are presented here; numerous additional topologies exist between the diagrams shown.

At the lowest grades of metamorphism, in the Zeolite Facies, a number of phase assemblages characterize rocks metamorphosed under various conditions.[18] Figure 26.8 shows a common group of assemblages. Metasandstones typically contain an assemblage that includes quartz, analcite, clay minerals, heulandite, and chlorite or chlorite mixed-layer minerals (chlorite/vermiculite, chlorite/smectite). More calcic sandstones may also include pumpellyite, calcite, or both. At lower temperatures, xanthophyllite may occur (Droddy and Butler, 1979), whereas higher temperatures may yield laumontite, prehnite, or wairakite (Liou, Maruyama, and Cho, 1987; Inoue and Utada, 1991).

Marking the transition from the Zeolite Facies to the Albite-Epidote Hornfels Facies are the conversion of kaolinite to pyrophyllite (see equation 26.2), the conversion of analcite to albite, and the disappearance of Ca-zeolites. Albite forms via the reaction

$$\text{analcite} + \text{quartz} <==> \text{albite} + H_2O \qquad (26.15)$$

(Liou, 1971a; A. B. Thompson, 1971).

Typical assemblages in SAC rocks of the Albite-Epidote Hornfels Facies include

quartz–albite–epidote–chlorite–white mica–biotite,
quartz–albite–chlorite–white mica–
epidote–prehnite–sphene, and
albite–quartz–biotite–calcite–chlorite –white mica

(Loney et al., 1975).[19] Talc and actinolite may form in more mafic sandstones (S. D. McDowell and Elders, 1980). In the lower-temperature part of the facies, pumpellyite is present. In the highest-temperature part of the facies, pumpellyite, prehnite, or both minerals may be absent. The Prehnite-Pumpellyite Facies is indicated where prehnite and pumpellyite occur together.

The appearance of andalusite (see equation 26.4), garnet, and hornblende mark the transition to the Hornblende Hornfels Facies (figure 26.8).[20] Plagioclase is more calcic than in lower-grade facies and zoisite may be present. In the higher-temperature part of the facies, cordierite occurs and calcite is replaced by wollastonite (see equation 26.1). Typical mineral assemblages mimic those of pelitic rocks, except that plagioclase and quartz are more abundant than they are in pelitic rocks. A characteristic assemblage is

quartz–plagioclase–cordierite–andalusite–biotite.

In rocks of intermediate composition, the assemblage

plagioclase–quartz–hornblende–garnet–epidote

is representative.

The Pyroxene Hornfels Facies, in its lower-temperature part, contains aluminous SAC rocks that are mineralogically like their pelitic counterparts, except that quartz and the feldspars are more abundant.[21] White mica is absent. A typical assemblage is

quartz–plagioclase–alkali feldspar–
cordierite–andalusite–biotite.

In rocks of intermediate composition, epidote-group minerals are absent and clinopyroxene is present. At higher temperatures, andalusite rather than sillimanite is present in aluminous rocks, orthopyroxene is an important phase in more basic assemblages, and hornblende does not occur. A typical assemblage might consist of

quartz–plagioclase–alkali feldspar–cordierite–
orthopyroxene.

SAC rocks may melt under Pyroxene Hornfels Facies conditions. For example, in the Sierra Nevada of California, a trachybasalt intruded a biotite-hornblende granodiorite and partially melted it (Dodge and Calk, 1978).[22] The resulting rocks contain the assemblage

quartz–plagioclase–alkali feldspar–
glass–orthopyroxene–sphene.

The phase assemblages of the Sanidinite Facies are less complex than those of the lower-temperature facies (figure 26.8).[23] SAC rocks either contain an aluminum silicate (sillimanite or mullite) or they contain a pyroxene. In either case, they contain quartz (reverted from tridymite), plagioclase, and may contain cordierite. In addition, these rocks may have alkali feldspar and glass. The glass-bearing rocks are sometimes called *buchites*. Characteristic assemblages include

mullite–cordierite–glass,
quartz–alkali feldspar–plagioclase–titanomagnetite–glass,
and
sanidine–clinopyroxene–hematite–glass.

Basic Rocks

Basic rocks—the metamorphic equivalents of gabbros, basalts, and related rocks—are characterized by the phase assemblages depicted by tie lines that cross the lower, right-hand third (the F-rich corner) of the ACF triangle (figure 26.8).[24] Contact/hydrothermally metamorphosed basic rocks are widely distributed among oceanic crustal rocks[25] and contact-metamorphosed mafic rocks exist elsewhere as well.

In the Zeolite Facies, typical assemblages include

chlorite–chlorite/smectite–analcite–heulandite–quartz–calcite–
smectite,
chlorite–albite–laumontite–epidote–pumpellyite–quartz–sphene,
and
chlorite–wairakite–albite–pumpellyite –quartz– calcite.

A wide variety of additional minerals may occur, including K-feldspar, thomsonite, chabazite, stilbite, prehnite, mixed-layer minerals (e.g., chlorite/vermiculite), various smectites, talc, anhydrite, pyrite, hematite, and magnetite. Reactions marking the upper boundary region of the Zeolite Facies, as defined here, include the following:

$$5 \text{ prehnite} + \text{chlorite} + 2 \text{ quartz} <==> 4 \text{ epidote} + \\ \text{actinolite} + 6 H_2O \qquad (26.16)$$
and

$$\text{wairakite} <==> \text{anorthite} + 2 \text{ quartz} + 2 H_2O \qquad (26.17)$$

These reactions mark the appearance of actinolite and the upper stability limit of the common zeolites, respectively (Liou, Maruyama, and Cho, 1985; Liou, 1970).[26]

The typical assemblage of the Albite-Epidote Hornfels Facies is

chlorite–albite–epidote–actinolite–quartz–sphene,

but assemblages containing prehnite *or* pumpellyite do occur within this facies. Calcite, iron oxides, and iron sulfides are also common. The upper boundary of the facies and corresponding lower boundary of the Hornblende Hornfels Facies is a broad zone, the lower-temperature limit of which is marked mineralogically by the appearance of two plagioclases (albite coexisting with oligoclase) and the coexistence of two Ca-amphiboles (an actinolite and a hornblende) (Maruyama, Liou, and Suzuki, 1982; Maruyama, Suzuki, and Liou, 1983).

The Hornblende Hornfels Facies thus begins with reactions such as

$$\text{actinolite} + \text{albite} <==> \text{hornblende} + \text{oligoclase} + \text{quartz} \tag{26.18}$$

and

$$\text{epidote} + \text{actinolite} + \text{chlorite} <==> \text{hornblende} + H_2O \tag{26.19}$$

that yield hornblende and oligoclase at temperatures just below 400°C (Spear, 1981; Maruyama, Suzuki, and Liou, 1983). At slightly higher temperatures, albite disappears. The appearance of garnets is defined by reactions such as

$$\text{prehnite} <==> 2\ \text{zoisite} + 2\ \text{grossularite} + 3\ \text{quartz} + 4\ H_2O \tag{26.20}$$

and

$$\text{Fe-chlorite} + \text{quartz} + \text{magnetite} <==> \text{almandine} + H_2O \tag{26.21}$$

(L. C. Hsu, 1968; Liou, 1971; Helgeson et al. 1978). In the middle of the facies, diopside is produced via the reaction

$$\text{dolomite} + 2\ \text{quartz} <==> \text{diopside} + 2\ CO_2 \tag{26.22}$$

(F. J. Turner, 1981, pp. 163–64). As calcium is consumed in the reactions above, sphene is converted to ilmenite (Moody, Meyer, and Jenkins, 1983). Together these various reactions yield parageneses such as

actinolite–hornblende–albite–oligoclase–chlorite–epidote–sphene

in lower-grade Hornblende Hornfels Facies metabasites, and

hornblende–andesine–clinopyroxene–quartz –ilmenite, or

anthophyllite–cummingtonite–cordierite–plagioclase–quartz–ilmenite

at higher grades. Biotite and garnet occur in some assemblages. Note that the latter assemblages contain no actinolite and only one plagioclase.

The transition from the Hornblende Hornfels Facies to the Pyroxene Hornfels Facies is marked by the disappearance of zoisite and Mg-chlorites. A zoisite-out reaction is

$$6\ \text{zoisite} <==> 6\ \text{anorthite} + 2\ \text{grossular} + \text{corundum} + 3\ H_2O \tag{26.23}$$

Boettcher (1970). Mg-chlorite may be eliminated by reactions such as

$$5\ \text{chlorite} <==> \text{cordierite} + 3\ \text{spinel} + 10\ \text{forsterite} + 20\ H_2O \tag{26.24}$$

(Helgeson et al., 1978) or

$$\text{chlorite} + 2\ \text{quartz} <==> \text{garnet} + \text{orthopyroxene} + 4\ H_2O. \tag{26.25}$$

Representative assemblages in metabasites of the Pyroxene Hornfels Facies are

orthopyroxene–clinopyroxene–hornblende–biotite–Fe-Ti oxide, and orthopyroxene–clinopyroxene–Ca-plagioclase–hornblende–ilmenite

(Russ-Nabelek, 1989). Quartz or olivine may also occur.

The Sanidinite Facies and the higher-pressure parts of the Pyroxene Hornfels Facies are characterized by the absence of amphibole. A representative assemblage in a metabasite at this grade is

Ca-plagioclase–augite–olivine–ilmenite–magnetite

(I. D. Muir and Tilley, 1957). The pyroxenes are more Fe-rich and the clinopyroxenes are more aluminous. Garnet is absent from basic rocks of the Sanidinite Facies.

The Origin of Spilites

Spilite is a term used to refer to sodic basic rocks composed of assemblages such as albite–chlorite–clinopyroxene–epidote–quartz–calcite–sphene. **Keratophyre** and **quartz keratophyre** are related rocks of more siliceous character. The origin of these rocks has long been controversial, with some workers arguing that they represent the products of direct crystallization from a magma (e.g. Amstutz and Patwardhan, 1974) and others favoring some form of low-grade metamorphic origin (e.g. Battey, 1974). The former interpretation is based primarily on textural data, whereas the latter is based primarily on phase equilibria studies. The compositions of the clinopyroxenes and their textures appear to be igneous, but it has been argued that many of the other phases have a secondary appearance.

The question of the origin of spilites is raised here because the mineralogy of the spilites (and keratophyres) is much like that of metabasites of low metamorphic grade (compare the above assemblage to those listed on page 538 for the Zeolite Facies). In fact, spilite assemblages lacking pyroxene and containing zeolites or pumpellyite are identical to the Zeolite Facies assemblages. Do identical assemblages form both by low-grade metamorphism at low P and T and via crystallization of magma at high T and low P?

A number of lines of evidence suggest that the answer to the above question is no. First, we have seen that at higher T and low P the phase assemblages that are stable are those containing hornblende and a plagioclase more calcic than albite; the albite–chlorite–epidote assemblage is not stable. Second, experimental phase-stability studies of rocks of spilite-like composition indicate that the assemblage clinopyroxene–chlorite (as well as other assemblages containing chlorite) is not stable under magmatic conditions (Yoder, 1967; Liou, Kuniyoshi, and Ito, 1974; Moody, Meyer, and Jenkins, 1983). Furthermore, no spilite is known to have been erupted in historic times and the only contemporary volcanic rocks that are similar are hydrothermally metamorphosed oceanic basalts (Humphris and Thompson, 1978). Experimental studies of seawater-basalt interaction indicate that spilitization is a viable hydrothermal-metamorphic process (Wedepohl, 1988). Together these data make a strong case for a metamorphic origin for the spilites. Inasmuch as many spilites contain > 4% Na_2O, it is likely that sodium metasomatism, probably on the seafloor, was an important process in the development of the spilites and keratophyres.

Carbonate Rocks

Because they commonly develop into marbles and skarns containing interesting and valuable minerals, limestones and dolostones in contact aureoles have received considerable study.[27] The formation of the interesting minerals, however, requires that the carbonate rock be impure. Calcite, dolomite, or magnesite alone are stable, at slightly elevated pressures and in the absence of H_2O, to temperatures in excess of 700° C.[28] Consequently, contact metamorphism of a pure limestone or dolostone may yield no more than a coarsening of grain size in the carbonate minerals of the rock.

Most carbonate rocks are not pure calcite or dolomite. Many contain silica in the form of chert or sand grains, and calcite-dolomite-quartz mixtures are also common. Clay minerals in carbonate rocks provide aluminum to the system. Organic carbon, another common impurity, is converted to graphite upon metamorphism. Iron, potassium, and sodium are among the other elements that may occur, become involved in reactions, or control the stability of phases during metamorphism (see Skippen and Trommsdorff, 1986).

Bowen (1940) presented the first comprehensive analysis of progressive metamorphism of siliceous limestones and dolostones, which focussed on the anhydrous system $CaO-MgO-SiO_2-CO_2$. He proposed that through a series of decarbonation reactions, siliceous dolostones and limestones yield successive assemblages marked by ten index minerals: tremolite (which, of course, is a hydrous phase), forsterite, diopside, periclase, wollastonite, monticellite, akermanite, spurrite, merwinite, and larnite.[29] The compositions of these phases and their positions on the $CaO-MgO-SiO_2$ (CMS) plane are shown in figure 26.9. Figure 26.10 shows a petrogenetic grid with coordinates of T and $P \approx P_{CO_2}$ on which the positions of selected, experimentally determined and calculated curves relevant to such decarbonation reactions are plotted. Between the curves, phase diagrams depict the assemblages (tie lines connect the various phases, the compositions and plotted positions of which are shown in figure 26.9). Note that in each of the reactions, CO_2 is produced (and may become a part of the fluid phase).

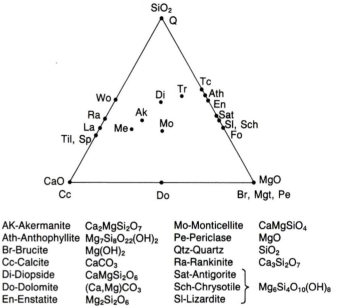

AK-Akermanite	$Ca_2MgSi_2O_7$	Mo-Monticellite	$CaMgSiO_4$
Ath-Anthophyllite	$Mg_7Si_8O_{22}(OH)_2$	Pe-Periclase	MgO
Br-Brucite	$Mg(OH)_2$	Qtz-Quartz	SiO_2
Cc-Calcite	$CaCO_3$	Ra-Rankinite	$Ca_3Si_2O_7$
Di-Diopside	$CaMgSi_2O_6$	Sat-Antigorite	
Do-Dolomite	$(Ca,Mg)CO_3$	Sch-Chrysotile	$Mg_6Si_4O_{10}(OH)_8$
En-Enstatite	$Mg_2Si_2O_6$	Sl-Lizardite	
Fo-Forsterite	Mg_2SiO_4	Sp-Spurrite	$Ca_5(SiO_4)_2CO_3$
La-Larnite	Ca_2SiO_4	Tc-Talc	$Mg_3Si_4O_{10}(OH)_2$
Me-Merwinite	$Ca_3Mg(SiO_4)_2$	Til-Tilleyite	$Ca_5Si_2O_7(CO_3)_2$
Mgt-Magnesite	$MgCO_3$	Tr-Tremolite	$Ca_2Mg_5Si_8O_{22}(OH)_2$
		Wo-Wollastonite	$CaSiO_3$

Figure 26.9 The $CaO-MgO-SiO_2$ plane of the system $CaO-MgO-SiO_2-H_2O-CO_2$ with plotted positions of various calc-silicate minerals.

The Importance of the Fluid Phase

In many aureoles, tremolite and talc are phases that develop at the lowest grades of metamorphism.[30] Both are hydrous phases. Neither is produced by the reactions shown in figure 26.10. During progressive metamorphism, these minerals develop through interaction with H_2O and yield H_2O to the fluid phase, which commonly also derives H_2O from the intrusion that causes the contact metamorphism. Given the reactions depicted in figure 26.10 and those involving talc and tremolite, it is clear that progressive metamorphism yields both CO_2 and H_2O. These components are completely miscible and form a single fluid phase. Thus, it is necessary to consider phases in the system $CaO-MgO-SiO_2-CO_2-H_2O$ if we are to correctly discuss phase assemblages of carbonate rocks in the facies of contact metamorphism.

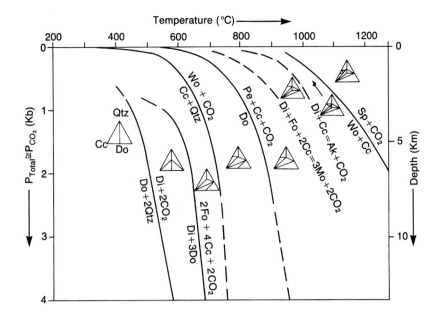

Figure 26.10 Petrogenetic grid for carbonate rocks in the system $CaO\text{-}MgO\text{-}SiO_2\text{-}CO_2$. $P_{total} \approx P_{CO2}$. Selected reaction curves are shown; phase diagrams show the phase assemblages that are stable between the curves. Positions of minerals in the $CaO\text{-}MgO\text{-}SiO_2$ plane are as shown in figure

26.9. All curves are experimentally determined, with slight modifications based on thermodynamic calculations.
(Sources as in table 26.2 and appendix C; also see Tracy and Frost, 1991)

The inclusion of two (or more) fluid components in the system means that (1) by the Phase Rule, most reactions will be bivariant (at the least), and (2) the effect of various ratios of CO_2 to H_2O (and other components) in the fluid phase, that is, the effect of the mole fraction of CO_2 (X_{CO_2}) or H_2O (X_{H_2O}) on the stability fields of minerals in carbonate rocks must be considered.[31] That effect, discussed briefly in chapter 25, is shown schematically in figure 26.11. In this T-X phase diagram, pressure is fixed. At very low values of X_{CO_2}, the sequence of key minerals that will develop with increasing temperature during the metamorphism of a siliceous dolomitic marble is talc → tremolite → antigorite → wollastonite → brucite → periclase (path a-a, figure 26.11). At intermediate values of X_{CO_2}, the sequence will be talc → tremolite → forsterite → wollastonite (b-b′); and at very high values of X_{CO_2}, the sequence will be diopside → forsterite → wollastonite (c-c′). The pressure and the sequence of increasing temperatures are the same in each case. Only the composition of the fluid phase varies. Thus, a particular key mineral or phase assemblage cannot be used as a P-T indicator *unless* the composition of the fluid phase is known and is taken into

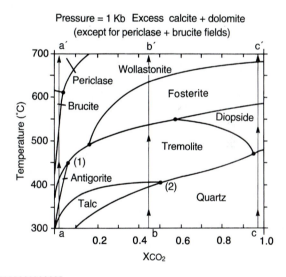

Figure 26.11 T-X_{fluid} phase diagram showing three equivalent thermal paths (a–a′, b–b′, c–c′), at different X_{CO_2} values, which result in three different mineral sequences.
(Modified from Bucher-Nurminen, 1982)

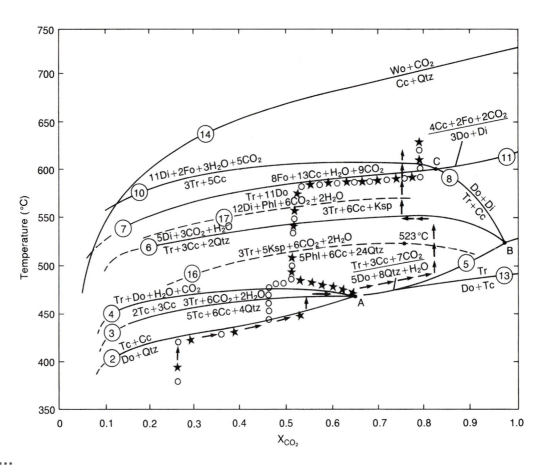

Figure 26.12
T-X$_{fluid}$ phase diagram showing suggested paths of
metamorphism in T-X space (stars and arrows) for the Notch
Peak Aureole, Utah. Notice the changing composition of the
fluid phase (as indicated by left-to-right shifts in paths). A

simplified version of the top part of this diagram is shown in
figure 26.6. Abbreviations not listed in previous figures: Ksp
= K-rich alkali feldspar, Phl = phlogopite. P = 2 kb
(0.2 Gpa).
(From Hover Granath, Papike, and Labotka, 1983)

account. One additional complexity that exists is that the
fluid phase may change composition during the progress of
the metamorphic event, because of additions or subtractions
of CO_2 and H_2O (figure 26.12).[32] For example, decarbonation reactions may increase the mole fraction of CO_2 in the
fluid phase: In figure 26.11, this would force the reaction path
to the right.

Because the fluid phase is transient, we must seek information about its former composition through direct or indirect methods. One approach is through fluid inclusion studies,
which can give direct information about the fluid phase.[33] A
second, but indirect, method is based on mass balance calculations of the volume of fluid generated during the length of the
metamorphic event (Labotka, White, and Papike, 1984; Ferry,
1989; Labotka, 1991). A third method involves examining
the phase assemblages in the aureole in order to discover
phase assemblages that indicate the locations of invariant
points in the T-X diagrams (where the various reaction curves
intersect). For example, at a given pressure, if the sequence of
key minerals in the aureole was identical to the sequence a–a´
of figure 26.11, it would be possible to conclude that the

composition of the fluid phase was $X_{CO_2} < 0.1$ and $X_{H_2O} > 0.9$
i.e. to the left of the invariant point (1). Because the positions
of invariant points and the configurations of tie lines change
with changing pressure, the pressure must be known if we are
to use this technique. If the pressure can be determined independently, however, discovery of the sequence of key minerals
and assemblages that represent invariant assemblages allows us
to restrict the possible fluid phase composition to a particular
range. As another example, an equilibrium assemblage of
talc–tremolite–calcite–dolomite–quartz would indicate conditions at invariant point 2, indicating a fluid phase composition of X_{CO_2} = X_{H_2O} = 0.5. Here again, an independent
determination of pressure may be necessary, although in some
instances the invariant assemblages are stable only over a restricted range of pressures.

Metamorphism of Dolomitic, Argillaceous, and Siliceous Carbonate Rocks

Taking into account X_{H_2O} and X_{CO_2}, we can now define
phase assemblages for the various facies of contact metamorphism of impure carbonate rocks. Typical examples are

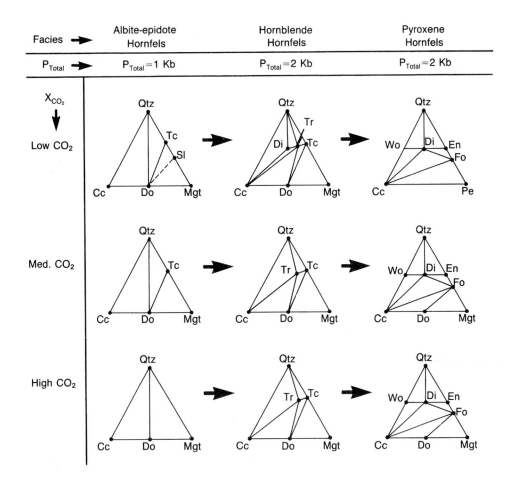

Figure 26.13 Phase assemblages in carbonate rocks (plotted on the CMS phase diagram) for various facies of contact metamorphism at low, medium, and high values of X_{CO_2} (based on various sources cited in the text). Refer to figure 26.9 for mineral compositions and plotted positions.

depicted in figure 26.13. Assemblages for low, medium, and high values of X_{H_2O} and X_{CO_2} are presented. Several important reactions are listed in table 26.2. As was the case in the figures for other bulk compositions of rock, only representative phase diagrams are shown. Additional diagrams between those shown are necessary to develop a complete sequence of diagrams for the range of possible conditions.

The important phases that develop from these dolomite- and calcite-rich rocks include talc, tremolite, diopside, forsterite, wollastonite, grossularite, phlogopite, and a host of other, uncommon minerals. The sequence in which they appear depends on the chemistry of both the rock and the fluid phase, as well as on the P-T conditions. Not all carbonate rocks containing aluminous, siliceous, or magnesian phases were initially impure. Some relatively pure limestones or other carbonate rocks have undergone *metasomatism*, resulting in the formation of Mg-, Al-, and Si-bearing phases.

Because the chemistry of carbonate rocks differs significantly from that of most other rocks (see figure 25.1), we can readily observe the effects of metasomatism in carbonate rocks

that have experienced this process. For example, consider the metamorphic aureole at Crestmore, California (figure 26.14) described by Burnham, (1954, 1959). Here, quartz diorite and porphyritic quartz monzonite have intruded a relatively pure Mg-bearing limestone unit, causing recrystallization, isochemical neocrystallization, and metasomatically induced neocrystallization. The igneous rocks (which locally exhibit a contaminated contact unit that resulted from assimilation of carbonate country rocks) are surrounded by an aureole of variable width (< 3 cm–> 15 m) consisting of four parts. The outermost zone, here referred to as the *marble zone*, consists of calcite marble and brucite-calcite marble.[34] In this zone, isochemical recrystallization, resulting in a coarsening of grain size in calcite, and neocrystallization, yielding brucite, produced these two main rock types. The marble zone is succeeded inwardly by the *monticellite zone*, consisting of rocks composed of calcite and monticellite in association with one or more of the various minerals clinohumite, forsterite, melilite, spurrite, tilleyite, and merwinite. An *idocrase zone* occurs interior to the monticellite zone. The idocrase zone contains rocks composed

Table 26.2   Selected Reactions Important in the Metamorphism of Carbonate Rocks

$$3 \text{ magnesite} + 4 \text{ quartz} + H_2O = \text{talc} + 3 CO_2$$
$$3 MgCO_3 + 4 SiO_2 + H_2O = Mg_3Si_4O_{10}(OH)_2 + 3 CO_2$$

$$3 \text{ dolomite} + 4 \text{ quartz} + H_2O = \text{talc} + 3 \text{ calcite} + 3 CO_2$$
$$3 CaMg(CO_3)_2 + 4 SiO_2 + H_2O = Mg_3Si_4O_{10}(OH)_2 + 3 CaCO_3 + 3 CO_2$$

$$2 \text{ dolomite} + \text{talc} + 4 \text{ quartz} = \text{tremolite} + 4 CO_2$$
$$2 CaMg(CO_3)_2 + Mg_3Si_4O_{10}(OH)_2 + 4 SiO_2 = Ca_2Mg_5Si_8O_{22}(OH)_2 + 4 CO_2$$

$$5 \text{ talc} + 6 \text{ calcite} + 4 \text{ quartz} = 3 \text{ tremolite} + 6 CO_2 + 2 H_2O$$
$$5 Mg_3Si_4O_{10}(OH)_2 + 6 CaCO_3 + 4 SiO_2 = 3 Ca_2Mg_5Si_8O_{22}(OH)_2 + 6 CO_2 + 2 H_2O$$

$$\text{tremolite} + 3 \text{ calcite} + 2 \text{ quartz} = 5 \text{ diopside} + 3 CO_2 + H_2O$$
$$Mg_5Si_8O_{22}(OH)_2 + 3 CaCO_3 + 2 SiO_2 = 5 CaMgSi_2O_6 + 3 CO_2 + H_2O$$

$$\text{dolomite} + 2 \text{ quartz} = \text{diopside} + 2 CO_2$$
$$CaMg(CO_3)_2 + 2 SiO_2 = CaMgSi_2O_6 + 2 CO_2$$

$$\text{diopside} + 3 \text{ dolomite} = 2 \text{ forsterite} + 4 \text{ calcite} + 2 CO_2$$
$$CaMgSi_2O_6 + 3 CaMg(CO_3)_2 = 2 Mg_2SiO_4 + 4 CaCO_3 + 2 CO_2$$

$$3 \text{ tremolite} + 5 \text{ calcite} = 2 \text{ forsterite} + 11 \text{ diopside} + 5 CO_2 + 3 H_2O$$
$$3 Ca_2Mg_2Si_8O_{22}(OH)_2 + 5 CaCO_3 = 2 Mg_2SiO_4 + 11 CaMgSi_2O_6 + 5 CO_2 + 3 H_2O$$

$$\text{calcite} + \text{quartz} = \text{wollastonite} + CO_2$$
$$CaCO_3 + SiO_2 = CaSiO_3 + CO_2$$

$$\text{dolomite} = \text{periclase} + \text{calcite} + CO_2$$
$$CaMg(CO_3)_2 = MgO + CaCO_3 + CO_2$$

$$2 \text{ calcite} + \text{forsterite} + \text{diopside} = 3 \text{ monticellite} + 2 CO_2$$
$$2 CaCO_3 + Mg_2SiO_4 + CaMgSi_2O_6 = 3 CaMgSiO_4 + 2 CO_2$$

$$\text{diopside} + \text{calcite} = \text{akermanite} + CO_2$$
$$CaMgSi_2O_6 + CaCO_3 = Ca_2MgSi_2O_7 + CO_2$$

$$3 \text{ calcite} + 2 \text{ wollastonite} = \text{tilleyite} + CO_2$$
$$3 CaCO_3 + 2 CaSiO_3 = Ca_5Si_2O_7(CO_3)_2 + CO_2$$

$$3 \text{ calcite} + 2 \text{ wollastonite} = \text{spurrite} + 2 CO_2$$
$$3 CaCO_3 + 2 CaSiO_3 = Ca_5(SiO_4)_2CO_3 + 2 CO_2$$

$$2 \text{ monticellite} + \text{spurrite} = 2 \text{ merwinite} + \text{calcite}$$
$$2 CaMgSiO_4 + Ca_5(SiO_4)_2CO_3 = 2 Ca_3Mg(SiO_4)_2 + CaCO_3$$

Table 26.2 Continued

tilleyite	+	4 wollastonite	=	3 rankinite	+	2 CO_2	
$Ca_5Si_2O_7(CO_3)_2$	+	4 $CaSiO_3$	=	3 $Ca_3Si_2O_7$	+	2 CO_2	
spurrite	+	rankinite	=	4 larnite	+	CO_2	
$Ca_5(SiO_4)_2CO_3$	+	$Ca_3Si_2O_7$	=	4 Ca_2SiO_4	+	CO_2	
spurrite	=	calcite	+	2 larnite			
$Ca_5(SiO_4)_2CO_3$	=	$CaCO_3$	+	2 Ca_2SiO_4			
rankinite	=	larnite	+	wollastonite			
$Ca_3Si_2O_7$	=	Ca_2SiO_4	+	$CaSiO_3$			

Sources: Bowen (1940), R. I. Harker and Tuttle (1955), Burnham (1959), Metz and Winkler (1963), Metz and Trommsdorff (1968), Metz (1970), Metz and Puhan (1970), Skippen (1971, 1974), Puhan and Hoffer (1973), Joesten (1974), Slaughter, Kerrick, and Wall (1975), Zharikov, Schmulovich, and Vulatov (1977), Winkler (1979), Eggert and Kerrick (1981).

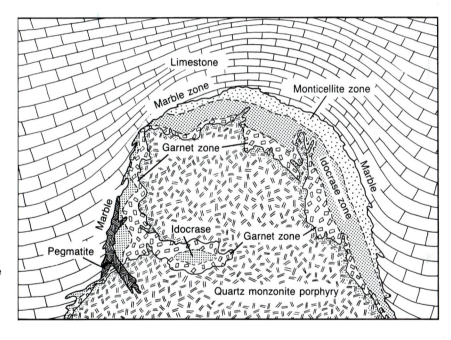

Figure 26.14 Idealized cross section through the quartz monzonite and contact aureole at Crestmore, California, showing the various zones of contact metamorphism.
(Slightly modified from Burnham, 1959)

of idocrase in association with such minerals as calcite, diopside, wollastonite, phlogopite, monticellite, and xanthophyllite. Closest to the intrusion is the *garnet zone*, where diopside-wollastonite-grossularite rocks, containing minor calcite and quartz, are the dominant rock types.

Examination of the key minerals indicates that metasomatism has occurred. The progressive sequence of key minerals and their chemistries is as follows:

calcite $CaCO_3$
calcite + brucite $CaCO_3 + Mg(OH)_2$
monticellite $CaMgSiO_4$
idocrase $Ca_{10}Mg_2Al_4Si_9O_{34}(OH)_4$

grossularite - wollastonite - diopside $Ca_3Al_2Si_3O_{12}$ - $CaSiO_3$
$CaMgSi_2O_6$.

Notice that there is a progressive increase in the ratio Si/Ca towards the contact and a similar increase in Al. Chemical analyses of the rocks confirm these trends and also indicate a slight enrichment in Fe^{3+} (figure 26.15). Inasmuch as the original rock was a Mg-bearing limestone, the first two assemblages indicate isochemical metamorphism (the water was probably present initially or is a retrograde addition). The latter three assemblages reflect introduction of silica and alumina (i.e., they reflect metasomatism). Although less obvious in less pure rocks, similar metasomatism may occur during contact metamorphism in rocks of any composition.

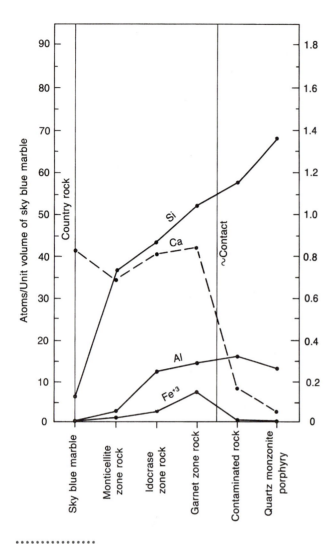

Figure 26.15 Chemical plot showing the variation in abundance of selected elements across the contact aureole at Crestmore, California.

(Simplified from Burnham, 1959)

Ultramafic Rocks

In considering contact metamorphism of ultramafic rocks, the system $CaO\text{-}MgO\text{-}SiO_2\text{-}CO_2\text{-}H_2O$ is pertinent, but for these rocks, we must focus on the Mg-rich side of the CMS triangle (B. W. Evans, 1977). The following idealized sequence of phase assemblages, based principally on assemblages reported from several localities,[35] represents a progression from lower to higher grade.

Zeolite Facies

talc–chrysotile–lizardite–dolomite–chlorite–magnetite
talc–chrysotile–lizardite–tremolite–chlorite–magnetite

Albite-Epidote-Hornfels Facies

talc–chrysotile–lizardite–tremolite–chlorite–magnetite
talc–antigorite–tremolite–chlorite–magnetite

Hornblende Hornfels Facies

talc–antigorite–tremolite–chlorite–magnetite
talc–forsterite–tremolite–chlorite–chromite

Pyroxene Hornfels Facies

forsterite–tremolite–anthophyllite–chlorite–chromite
forsterite–tremolite–enstatite–chlorite–chromite
forsterite–hornblende–enstatite–spinel–chromite
forsterite–diopside–enstatite–spinel.

These assemblages represent a limited range of bulk composition. Other bulk compositions will yield somewhat different assemblages in which, for example, diopside and/or forsterite will appear at lower-temperature conditions. Note that hornblende does not represent the Hornblende Hornfels Facies in these bulk compositions, but rather forms in the Pyroxene Hornfels Facies. In the absence of H_2O (i.e., at very low activites of water) or at very low pressures, the temperatures at which the neocrystallization reactions occur are substantially lower (see figure 26.10).

Carbonate phases are generally absent from the assemblages listed above. In those assemblages in which they are absent, the composition of the fluid phase may be essentially $X_{H_2O} = 1.0$. Under such circumstances, brucite is a common phase in Mg-rich rocks. Carbonate phases *are* present in some rocks (S. E. Swanson, 1981), and in such rocks the phase relations will be controlled by the composition of the fluid phase (i.e., by the ratio X_{CO_2}/X_{H_2O})(B. W. Evans and Trommsdorff, 1974).

Metamorphic reactions in ultramafic rocks are discussed more fully in chapter 31. Nevertheless, reactions of note include

$$17 \text{ chrysotile} \Longleftrightarrow \text{antigorite} + 3 \text{ brucite} \qquad (26.26)$$

$$\text{antigorite} + 20 \text{ brucite} \Longleftrightarrow 34 \text{ forsterite} + 51\, H_2O \qquad (26.27)$$

and

$$9 \text{ talc} + 4 \text{ forsterite} \Longleftrightarrow 5 \text{ anthophyllite} + 4\, H_2O \qquad (26.28)$$

(H. J. Greenwood, 1963; 1971; B. W. Evans, 1977).[36] These define the approximate lower limits of the Albite-Epidote Hornfels Facies, the Hornblende Hornfels Facies, and the Pyroxene Hornfels Facies, respectively.

SUMMARY

Contact metamorphic rocks are typically nonfoliated granoblastic rocks called hornfelses. They form from elevated temperatures in aureoles that surround igneous intrusions. The temperatures at some contacts may have exceeded 1000° C, but in all aureoles, they diminished to regional values over relatively short distances. Because of this decline in temperature, the aureoles are typically narrow, seldom measuring more than 2 km across and commonly measuring only a few to hundreds of meters across. Pressures of metamorphism seldom exceeded 0.4 Gpa (4 kb) and were typically less than 0.2 Gpa (2 kb), indicating that contact metamorphic rocks are characteristically associated with epizonal intrusions. Hydrothermal metamorphism in active geothermal regions, including the mid-ocean ridges, is a form of contact metamorphism characterized by large fluxes of H_2O-rich fluids.

The facies recognized in contact metamorphic terranes include the Zeolite Facies, the Prehnite-Pumpellyite Facies, the Albite-Epidote Hornfels Facies, the Hornblende Hornfels Facies, the Pyroxene Hornfels Facies, and the Sanidinite Facies. Together, these form a Contact Metamorphic Facies Series. Across this facies series, aluminum silicates change from kaolinite to pyrophyllite to andalusite to sillimanite to mullite or corundum. Partial melting may yield glass in the Sanidinite Facies. Chloritoid and cordierite are also characteristic phases in pelitic rocks. In metabasites, actinolite is the stable inosilicate at the lower grades of metamorphism, whereas hornblende and the pyroxenes (diopside plus orthopyroxenes) characterize the medium and higher grades, respectively. In impure carbonate rocks, a variety of minerals is produced, including talc and tremolite at the lower grades; diopside, forsterite, periclase, and wollastonite at the intermediate grades; and monticellite, merwinite, spurrite, tilleyite, akermanite, rankinite, and larnite at the highest grades (high T, low P). In all of these rock types, but especially in the carbonate rocks, careful attention must be given to understanding the fluid phase that existed during metamorphism, if the conditions of metamorphism are to be understood. The fluid phase may, in some cases, promote metasomatism, and in many cases controls the mineral composition of the rocks.

EXPLANATORY NOTES

1. The term hornfels, a genetic term for granoblastic to diablastic rocks formed in contact metamorphism, is well ingrained in the literature and practice of geology. It is a different kind of term (being based on origin) than schist or slate, which are based on texture, or marble, which is based on composition. The term usually implies a fine-grained, nonfoliated texture, but is also used for coarse-grained rocks or those that may have a foliation. Furthermore, while some geologists envision hornfelses as dark-shaded rocks, others apply the term to any contact rocks, whether dark or light. Consequently, I recommend that this term be abandoned over time. **Contact granoblastite** and **contact diablastite** are suitable substitutes for *hornfels*, where there is a need to imply genesis with the rock name. As descriptive terms, granoblastite and diablastite are superior.

2. Additional descriptions of this contact aureole, more readily accessible, may be found in F. J. Turner (1968, pp. 6–8) and Best (1982, pp. 415–417).

3. Also see Ferry (1980a) and Nisbet and Fowler (1982), Carslaw and Jaeger (1959) give a quantitative treatment of the conduction of heat in solids. A treatment of the thermodynamics of metamorphic reactions is beyond the scope of this text, but the interested student may find a good introduction in Krauskopf (1967) and in basic texts on thermodynamics for geologists, such as R. Kern and Weisbrod (1967).

4. For example, see Burnham (1959), Lachenbruch et al. (1976), Fyfe, Price, and Thompson (1978, p. 33, chs. 3, 6), Kanaris-Sotiriou and Angus (1979), Ferry and Burt (1982), J. M. Rice and Ferry (1982), Tracy et al. (1983), Walther (1983), Labotka, White, and Papike (1984), Bowman, O'Neil, and Essene (1985), Truesdell and Janik (1986), Labotka et al. (1988), McKibben, Andes, and Williams (1988), Wedepohl (1988), Novick and Labotka (1990), and Sisson and Hollister (1990). Also see various papers in Walther and Wood (1986) for detailed treatments of subjects related to metamorphic fluid compositions, analyses, and reactions.

5. Kridelbaugh (1973) discusses this reaction and its kinetics.

6. H. J. Greenwood (1962, 1967, 1975a), J. M. Rice and Ferry (1982), Speer (1982), Hover Granath, Papike, and Labotka (1983), Ferry (1983b,c), Trommsdorf and Skippen (1986).

7. Greenwood (1962, 1967), Metz (1970), Touret and Dietvorst (1983).

8. Additional discussions of porphyroblast development may be found in Rosenfeld (1968), Spry (1969), Misch (1971), van den Eeckhout and Konert (1983), T. H. Bell, Rubenach, and Fleming (1986), D. J. Prior (1987), Vernon (1988), T. H. Bell and Johnson (1989), W. D. Carlson (1989), J. Reinhardt and Rubenach (1989), and S. E. Johnson (1990).

9. KFMASH is an abbreviation for the system K_2O-FeO-MgO-Al_2O_3-SiO_2-H_2O, which can generally be represented on AFM diagrams. Calculations using Shreinemakers' technique (see ch. 25, note 15) may be used to construct schematic petrogenetic grids for this system. Pattison and Tracy (1991) review contact metamorphism of pelitic rocks using the KFMASH system and divide the contact rocks into four contact facies series, corresponding with Contact- Buchan- (2 series), and Barrovian facies series. They note that low- to high-P conditions exist in various contact metamorphic aureoles, but their discussion emphasizes the fact that most contact metamorphism is low pressure metamorphism.

10. Zeolite Facies rocks are not common in metamorphic aureoles because intrusions commonly occur in previously metamorphosed rocks. Where pelitic sediments occur along mid-ocean ridges and on the cratons in hydrothermal areas, however, Zeolite Facies rocks are present. The phase diagrams presented here are based on data obtained from Muffler and White (1969), S. D. McDowell and Elders (1980), Barger and Beeson (1981), Evarts and Schiffman (1983), C. E.Weaver et al. (1984), Alt et al. (1986), and Liou, Maruyama, and Cho (1987). Additional data are from M. Frey (1978). Also see Shearer et al. (1988), Yau et al. (1988), Inoue and Utada (1991), and the summaries in Winkler (1979) and Hyndman (1985).

11. Helgeson et al. (1978) discuss this and many of the other reactions mentioned here and those important in subsequent discussions of metamorphism. Phase assemblages important in the Albite-Epidote Hornfels Facies are described by Muffler and White (1969), Verhoogen et al. (1970, p. 549ff.), Loney et al. (1975), Evarts and Schiffman (1983), C. E. Weaver et al. (1984), Liou, Maruyama, and Cho (1987), and Loney and Brew (1987) and are discussed by Winkler (1979), F. J. Turner (1981), and J. B. Thompson and Thompson (1976), among others. These works form the foundation for the discussion here. Also see M. Frey (1978, 1987c), Ashworth and Evirgen (1984), and Miyashiro and Shido (1985) for additional discussions of phase assemblages and mineralogical changes relevant to this facies.

12. Hornblende Hornfels Facies rocks are discussed widely in the literature. This discussion is based primarily on examples described in Philbrick (1936), Chapman (1950), Compton (1960), J. M. Moore (1960), Verhoogen et al. (1970, p. 549ff.), T. P. Loomis (1972a, 1976), Loney et al. (1975), Vaniman, Papike, and Labotka (1980), Speer (1982), Labotka (1983), Evans and Speer (1984), Labotka, White, and Papike (1984), Pattison and Harte (1985), Loney and Brew (1987), and Pattison (1987). Also see Reverdatto (1970), Miyashiro (1973a), Mason (1978), Winkler (1979), F. J. Turner (1981), and Best (1982) for general discussions, as well as E. W. Reinhardt (1968), J. B. Thompson and Thompson (1976), Holdaway (1978), April (1980), and Kerrick (1987) for treatments of particular relevant topics.

13. Mg-amphiboles are reported by Seki (1957, 1961 in Miyashiro, 1973a). Staurolite is reported, for example, by Compton (1960), T. P. Loomis (1972a), and Loney et al. (1975).

14. Also see Kerrick (1972), Chatterjee and Johannes (1974), and Schramke et al. (1987).

15. Rocks of the Pyroxene Hornfels Facies have been described by Tilley (1924), T. P. Loomis (1972a, 1976), Simmons, Lindsley, and Papike (1974), Dodge and Calk (1978), Vaniman, Papike, and Labotka (1980), Benimoff and Sclar (1984), N. H. Evans and Speer (1984), Droop and Charnley (1985), Pattison and Harte (1985), and Pattison (1987). The phase relations are reviewed by Miyashiro (1973a) and F. J. Turner (1981).

16. Sanidinite Facies pelitic rocks are described by Brauns (1911, in F. J. Turner, 1981), H. H. Thomas (1922), Agrell and Langley (1958), Searle (1962), and Brearley (1986), and their parageneses are reviewed by F. J. Turner (1981). These works provide the basis for the discussion here.

17. See Agrell and Langley (1958), Searle (1962), and F. J. Turner (1981, p. 283ff.). Also see Cosca et al. (1989) for some extremely unusual mineral assemblages.

18. The mineral assemblages of low T metamorphic rocks are reviewed by M. Frey (1987b) and Liou, Maruyama, and Cho (1987). For descriptions of specific Zeolite Facies siliceous-alkali-calcic rocks, see Muffler and White (1969), Droddy and Butler (1979), S. D. McDowell and Elders (1980), Barger and Beeson (1981), Brauckmann and Fuchtbauer (1983), and Evarts and Schiffman (1983), which together with the reviews in M. Frey (1987b) and Liou, Maruyama, and Cho (1987) provide the basis for the mineral associations and reactions discussed here. Some experimental temperature limits are reported by Barth-Wirsching and Höller (1989).

19. Additional works reporting Albite-Epidote Hornfels Facies assemblages include Verhoogen et al. (1970, p. 553ff.), Droddy and Butler (1979), S. D. McDowell and Elders (1980), Brauckmann and Fuchtbauer (1983), Evarts and Schiffman (1983), Cho and Liou (1986), and Inoue and Utada (1991). Also see the review in F. J. Turner (1981).

20. The character of Hornblende Hornfels Facies SAC rocks is reviewed by Miyashiro (1973a) and Turner (1981). See those works and studies of specific areas, including Harker and Marr (1893), Compton (1960), Verhoogen et al. (1970, p. 549ff.), Loney et al. (1975), Droddy and Butler (1979), R. G. Gibson and Speer (1986), and Loney and Brew (1987).

21. The mineral assemblages of Pyroxene Hornfels Facies SAC rocks are like those of the pelitic rocks. Few separate descriptions have been published, but see Le Maitre (1974), Loney et al. (1975), and Loney and Brew (1987). Also, refer to F. J. Turner (1981).

22. A similar case is described by Kitchen (1989).

23. H. H. Thomas (1922), Searle (1962), Turner (1981), Brearley (1986).

24. The phase assemblages reported and the discussion here are based on several works including H. H. Thomas (1922), Thomson (1935), G. A. MacDonald (1944), Muir and Tilley (1957, 1958), Agrell and Langley (1958), Seki, Ernst, and Onuki (1969, in Liou, 1971b), Spooner and Fyfe (1973), Chinner and Fox (1974), Hietanen (1974), Jolly (1974), Bonatti (1975), Floyd (1975), Kuniyoshi and Liou (1976b), Dimroth and Lichtblau (1979), S. E. Swanson and Schiffman (1979), Ernst, Liou, and Moore (1981), Spear (1981a,b), F. J. Turner (1981, ch. 8), Abbott (1982, 1984), Hynes (1982), J. B. Thompson, Laird, and Thompson (1982), Evarts and Schiffman (1983), Kristmannsdottir (1983), Maruyama, Suzuki, and Liou (1983), J. B. Moody, Meyer, and Jenkins (1983), Andrew (1984), Sivell and Waterhouse (1984a), Ishizuka (1985), Liou, Maruyama, and Cho (1985), Alt et al. (1986), Cho, Liou, and Maruyama (1986), Cho and Liou (1987), Ferry, Mutti, and Zuccala (1987), Liou, Maruyama, and Cho (1987), Becker et al. (1989), and Russ-Nabelek (1989).

25. For example, see Spooner and Fyfe (1973), Bonatti et al. (1975), Humphris and Thomson (1978), Dimroth and Lichtblau (1979), Evarts and Schiffman (1983), Kristmannsdottir (1983), Sivell and Waterhouse (1984a), Ishizuka (1985), Alt et al. (1986), JGR, v. 93, no. B5 (1988), Becker et al. (1989), Goodge (1989), Gillis and Robinson (1990), and the numerous works cited therein.

26. Liou, Maruyama, and Cho (1987) do not consider wairakite to be indicative of the Zeolite Facies, as they have adopted, for metabasites, the facies scheme of Liou, Maruyama, and Cho (1985). In that scheme, the Contact Metamorphic Facies Series would consist of the successive facies, Zeolite Facies–Prehnite-Actinolite Facies–Greenschist Facies–(Epidote Amphibolite Facies)–Amphibolite Facies. Liou, Maruyama, and Cho (1985) choose the reaction prehnite + chlorite + quartz = tremolite + zoisite + H_2O, which falls in the same general region of P-T space as the wairakite reaction, for the upper limit of their Prehnite-Actinolite Facies. The lower limit of that facies (i.e., the upper limit of their Zeolite Facies) occurs at a temperature that is about 150° C lower (see Appendix C). Thus, the reaction involving wairakite is not a critical reaction for their Zeolite Facies because it occurs at temperatures above the upper limit of the Zeolite Facies, *as they define it.*

27. References pertinent to the discussion here and on which it is based include A. Harker and Marr (1893), Bowen (1940), Tilley (1948, 1951a), Holser (1950), Burnham(1954, 1959), R. I. Harker and Tuttle (1955, 1956), R. L. Rose (1958), L. S. Walter (1963a, 1963b), H. J. Greenwood (1963, 1967a, 1967b, 1975a), Johannes and Metz (1968), Metz and Trommsdorff (1968), Johannes (1969), T. M. Gordon and Greenwood (1970), Metz (1970, 1976), Metz and Puhan (1970), Reverdatto (1970), Skippen (1971, 1974), Hoschek (1973), Kerrick, Crawford, and Randazzo (1973), Puhan and Hoffer (1973), Joesten (1974, 1976), B. W. Evans and Trommsdorff (1974), Slaughter, Kerrick, and Wall (1975), J. M. Rice (1977), Zharikov, Shmulovich, and Bulatov (1977), Winkler (1979), Lattanzi, Rye, and Rice (1980), Kase and Metz (1980), Bucher-Nurminen (1981, 1982), Eggert and Kerrick (1981), Hoersch (1981), F. J. Turner (1981), Williams-Jones (1981), Flowers and Helgeson (1983), Hover Granath, Papike, and Labotka (1983), Andrew (1984), S. B. Tanner, Kerrick, and Lasaga (1985), Skippen and Trommsdorff (1986), Labotka et al. (1988), Ferry (1989), Wallmach, Hatton, and Droop (1989), Jenkins and Clare (1990), Trommsdorf and Connolly (1990), and Tracy and Frost (1991).

28. R. I. Harker and Tuttle (1955, 1956), Johannes and Metz (1968), Winkler (1979).

29. Bowen suggested a jingle as an aid to remembering the sequence he proposed (though the jingle may be more difficult to remember than the list of mineral names): Tremble, for dire peril walks, / Monstrous acrimony spurning mercy's laws. Note that Bowen lists the appearance of periclase before wollastonite. Experimental work (R. I. Harker and Tuttle, 1955, 1956; H. J. Greenwood, 1967) reveals that wollastonite precedes periclase (under anhydrous conditions).

30. For example, Tilley (1948), Puhan and Hoffer (1973), Hoersch (1981), and Williams-Jones (1981).

31. The works of Korzhinskii (1959, in Skippen, 1974), Wyllie (1962), and H. J. Greenwood (1967) emphasized the necessity of using T-X_{fluid} plots in analyzing carbonate phase assemblages in metamorphic rocks.

32. For example, see Puhan and Hoffer (1973), Williams-Jones (1981), Bucher-Nurminen (1982), and Flowers and Helgeson (1983). For an analogous situation in noncarbonate rocks see Labotka, White, and Papike (1984).

33. For example see Touret and Dietvorst (1983), Craw (1988), Nwe and Grundmann (1990), Sisson and Hollister (1990), Labotka (1991), and the references therein.

34. In Burnham (1954, 1959), the brucite-calcite marbles are called "predazzites."

35. See Pinsent and Hirst (1977), Trommsdorff and Evans (1972, 1977), B. W. Evans and Trommsdorff (1974), Springer (1974, 1980a, 1980b), B. R. Frost (1975), S. E. Swanson (1981), and the review of B. W. Evans (1977).

36. Also see Johannes (1969), B. W. Evans et al. (1976), Trommsdorf (1983).

PROBLEMS

26.1. Construct a series of ACF diagrams representing a theoretically possible step-by-step progression from the Zeolite Facies diagram to the Albite-Epidote Hornfels Facies diagram shown in figure 26.8.

26.2. Construct a series of ACF diagrams representing a theoretically possible step-by-step progression from the Pyroxene Hornfels Facies diagram to the Sanidinite Facies diagram shown in figure 26.8.

26.3. Write an equation for each of the topology changes shown in your answer to problem 26.2.

26.4. Select reaction curves for the appearance of talc and tremolite from appendix C, and using a copy of figure 26.10, plot the curves. Construct phase diagrams showing a consistent set of topologies for a contact-metamorphic sequence in which the minerals talc, tremolite, diopside, and forsterite appear, in that order.

26.5. Construct topologies in the system CaO-MgO-SiO_2(-CO_2-H_2O) for each of the fields shown in figure 26.12.

26.6. Using the sequence of phase assemblages (representing low CO_2) listed in the section on ultramafic rocks (page 546), construct a consistent set of CaO-MgO-SiO_2 topologies representing a sequence of progressive metamorphism in which those assemblages were produced.

27

Regional Metamorphism Under Low to Medium P/T Conditions: Buchan and Barrovian Facies Series

INTRODUCTION

Mountain systems typically contain large belts of regionally metamorphosed rock. The more obvious of these regional belts, from which the term regional metamorphism was originally derived, are characterized by foliated metamorphic rocks developed under medium to high temperatures. The accompanying pressures vary from low to high values; thus, the ratio of P to T (P/T) is low to moderate and T/P is moderate to high. Geothermal gradients, which are likewise moderate to high, produce Buchan and Barrovian Facies series.

Because the pressures of Buchan and Barrovian Facies series are commonly higher than are those of Contact Facies Series, they may contain different sequences of critical minerals (see table 24.4).[1] Recall that Buchan Facies Series form under pressures, which, in the middle grades of metamorphism, are *lower* than that of the aluminum silicate triple point (i.e., pressures in the lower Amphibolite Facies are less than 0.387 Gpa or 3.87 kb). Consequently, the critical sequence of aluminum silicates is kaolinite → pyrophyllite → andalusite → sillimanite. The Barrovian Facies Series, in contrast, develops where pressures in the middle grades of metamorphism are *higher* than that of the aluminum silicate triple point (i.e., at intermediate temperatures, $P > 3.87$ kb). The resulting aluminum silicate mineral sequence is

kaolinite → pyrophyllite → kyanite → sillimanite. The presence in pelitic rocks of either andalusite or kyanite at the middle grades of metamorphism is one feature that distinguishes these facies series from one another.

In chapter 24, it was noted that Barrow (1893, 1912) described the sequence of mineral zones in the Scottish Highlands that later became known as the Barrovian Facies Series. Recall that this sequence was considered to be normal. Although the idea that Barrovian metamorphism is normal was no doubt overemphasized in the past, there is some tendency for metamorphic belts to evolve towards a Barrovian kind of sequence. Such a sequence represents a geothermal gradient that is somewhat average for the continents (about 20° C/km).[2] As colder metamorphic belts are heated up and hotter belts cool down, they tend to develop mineral assemblages characteristic of Barrovian Facies Series. The tendency to approach the average is one of the factors responsible for the paucity of low-temperature metamorphic belts in older mountain systems.[3] High-temperature, low-pressure belts may be more readily preserved, because high-temperature metamorphism induces dehydration, and later retrogression is inhibited by a lack of fluids. In this chapter, we first examine high T/P regional metamorphism—metamorphism that yields Buchan Facies Series—and then contrast that series with the Barrovian Facies Series.

BUCHAN FACIES SERIES

Buchan Facies Series, like Barrovian Facies Series, take their name from a region in the Scottish Highlands (H. H. Read, 1952).[4] The Buchan rocks lie to the east of the Barrovian rocks and were metamorphosed under lower pressures. Recall that the geothermal gradient of Buchan Facies Series rocks is 40°C–80°C/km. Thus, at depths of 10 km, temperatures are about 400–800° C and the pressure is 0.3 Gpa (3 kb); that is, pressures at the middle grades of metamorphism are *lower* than those of the aluminum silicate triple point.

In general, the geothermal gradients that give rise to the low pressures and high temperatures of Buchan Facies Series may be attributed to regional heating. *Where* that heating occurs may vary. In some cases, heating results from intrusion of groups of plutons at shallow to moderate depths (Lux, DeYoreo, and Guidotti, 1986; DeYoreo et al., 1989; R. B. Hanson and Barton, 1989). In northeastern North America, a thermal event generated by such a circumstance was caused by magmatic or orogenic events that resulted from plate collisions at a convergent margin.[5] Alternatively, heating associated with thinning of crust or lithospheric mantle may cause this type of metamorphism (Wickham and Oxburgh, 1985; Golberg and Leyleroup, 1990; Loosveld and Etheridge, 1990). Hence, low-pressure regional metamorphism may also develop in a rift zone at a divergent plate margin or along a major strike-slip fault. The possibilities require that a petrotectonic analysis of each Buchan-type metamorphic belt be made to assess the paleotectonic environment of metamorphism.

The appropriate paleotectonic environments for Buchan metamorphism apparently are common, and metamorphic belts with Buchan Facies Series are widely distributed. Miyashiro (1973a, ch. 7) describes a number of Buchan belts from various parts of the world, notably Spain and Japan. Other localities containing Buchan Facies Series include Maine, New Hampshire, Colorado, Oregon, Alaska, Australia, India, and Ireland.[6] Buchan Facies Series are also common in Precambrian shields, for example, in Canada and Finland (see Schreurs, 1985; Schreurs and Westra, 1986), as well as in Phanerozoic gneissic core complexes (Blumel and Schreyer, 1977). In the Precambrian, especially the Archean Eon, heat flow from the interior of the Earth may have been higher, resulting in steeper geothermal gradients and high-temperature, low-pressure metamorphism in the crust (Fyfe, 1974).

As noted above, the low pressures of metamorphism that produced the various Buchan Facies Series metamorphic belts yield the idealized critical mineral sequence in aluminous rocks: kaolinite → pyrophyllite → andalusite → sillimanite. In the type area in the Scottish Highlands, kaolinite has been reported as a possible retrograde phase, but it is andalusite that is distinctive of the series.[7] Here, as well as elsewhere, cordierite is also an important phase in the middle to high grades of Buchan Facies Series metamorphism.

The general sequence of phase assemblages in pelitic rocks at the type locality is as follows.

Hydrous Zone

muscovite–chlorite–quartz–albite–ilmenite±chloritoid

muscovite–chlorite–biotite–quartz–albite–ilmenite

Andalusite Zone

muscovite–biotite–quartz–oligoclase–garnet–andalusite–ilmenite

muscovite–biotite–quartz–oligoclase–andalusite–staurolite–cordierite–ilmenite

Sillimanite Zone

muscovite–biotite–quartz–oligoclase–andalusite–staurolite–cordierite–garnet–sillimanite–K-rich alkali feldspar

microcline–biotite–quartz–oligoclase–garnet–sillimanite–cordierite–magnetite[8]

The hydrous zone represents the Greenschist Facies, the Andalusite and lower Sillimanite zones represent the Amphibolite Facies, and the upper Sillimanite Zone represents the Granulite Facies (figure 27. 1).

Buchan Phase Assemblages and Reactions

The overall progression of mineralogical changes in the Buchan Facies Series at the type locality defines only the facies sequence Greenschist → Amphibolite → Granulite. As is typical, the Zeolite and Prehnite-Pumpellyite Facies are not represented. The complete facies series is revealed by the sequence of representative mineral assemblages listed in table 27.1. The low-grade assemblages are virtually identical to those of the Barrovian Facies Series described below. Similarly, Greenschist Facies rocks are mineralogically similar to their equivalents in Barrovian Facies Series. It is in the Amphibolite Facies, where andalusite and cordierite appear, that the Buchan Facies Series is distinguished from the higher-pressure Barrovian rocks. Although garnet and staurolite may appear in addition to cordierite in the Buchan Facies Series, the *relative order* of appearance of these minerals differs.

The various phase assemblages developed in each metamorphic zone of the Buchan Facies Series indicate various reactions. In pelitic rocks, at the lowest grade, the Zeolite Facies contains assemblages such as

kaolinite–illite–illite/smectite–chlorite–quartz–analcite

which is characterized by the clay minerals.[9] Alkali feldspar, smectites, iron oxides, calcite, dolomite, Ca-zeolites, and organic carbon may also occur.

The perceptive student will have noted the similarity between the above phase assemblage and those of low-grade contact and hydrothermal metamorphism. Because very

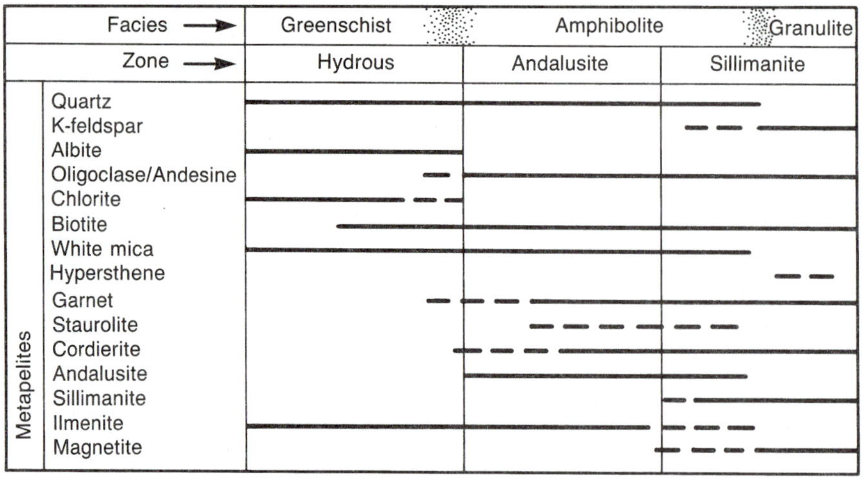

Facies →	Greenschist		Amphibolite		Granulite
Zone →	Hydrous		Andalusite	Sillimanite	
Quartz					
K-feldspar					
Albite					
Oligoclase/Andesine					
Chlorite					
Biotite					
White mica					
Hypersthene					
Garnet					
Staurolite					
Cordierite					
Andalusite					
Sillimanite					
Ilmenite					
Magnetite					

Figure 27.1 Mineral-facies chart for metapelites of the classic Buchan area of Buchan Facies Series metamorphism in the Scottish Highlands.

[Data sources include A. J. Baker (1985), A. J. Baker and Droop (1983), Chinner (1966), Harker (1932), Harte and Hudson (1979), N. F. C. Hudson (1980, 1985), Porteous (1973) and H. H. Read (1952).]

low-grade conditions for *all* facies series overlap, they contain identical or nearly identical phase assemblages. Thus, low-grade phase assemblages and reactions described in any of these facies series are generally pertinent to others.

At slightly higher-grade conditions, where assemblages of the Zeolite Facies are replaced by those of the Prehnite-Pumpellyite Facies, some minerals, such as K-rich alkali feldspar, are absent from many rocks, and new phases appear, such as "phengite" (greenish white mica), albite, and illite/chlorite or illite/smectite/chlorite mixed-layer minerals. These mineral changes indicate that metamorphic reactions have occurred. Smectites and K-rich alkali feldspar are among the first minerals that may disappear from aluminous rocks, and they do so via discontinuous reactions[10] such as

$$\text{smectite} + \text{K-rich alkali feldspar} \Longleftrightarrow \text{illite} + \text{quartz} \pm \text{chlorite} \quad (27.1)$$

The amount of chlorite produced by such a reaction is small, as chlorite development is controlled by the limited number of Mg ions available from the smectite and pore fluid. Kaolinite also commonly disappears from pelitic assemblages before or during development of Prehnite-Pumpellyite Facies assemblages. The disappearance of kaolinite apparently does not result from the (perhaps anticipated) reaction

$$\text{kaolinite} + 4 \text{ quartz} \Longleftrightarrow 2 \text{ pyrophyllite} + 2 \text{ H}_2\text{O} \quad (26.2)$$

which occurs at about 300° C. This is clear because (1) the reaction temperature is higher than that of the Zeolite/Prehnite-Pumpellyite Facies boundary (see figure 24.6); (2) in restricted bulk compositions, kaolinite is stable into the Prehnite-Pumpellyite Facies or lowermost Green-

schist Facies, where it is replaced by pyrophyllite;[11] and (3) kaolinite typically disappears from pelitic rocks of metamorphic sequences within or at the top of the Zeolite Facies.[12] A reaction that would yield the appropriate change[13] is

$$\text{kaolinite} + \text{quartz} + \text{alkali feldspar} + \text{magnetite} + \text{calcite} + \text{Mg}^{++} + (\text{OH})^{-} \Longleftrightarrow$$

$$\text{illite/montmorillonite/chlorite (mixed layer)} + \text{illite} + \text{CO}_2 + \text{H}_2\text{O}. \quad (27.2)$$

Such a reaction involves most of the phases present in the rock, except organic carbon, which yields methane (CH_4) to the fluid phase.

Greenschist Facies assemblages are perhaps the most widely distributed (or at least recognized) metamorphic assemblages on Earth. This is probably the case, because (1) the Greenschist Facies occurs in the lower grades of Buchan and Barrovian Facies series, and in the middle to upper grades of the Sanbagawa Facies Series, and (2) the P-T conditions of Greenschist Facies metamorphism are high enough to readily promote reactions, but low enough to be attained in many environments of metamorphism.

Typical assemblages in Greenschist Facies pelitic rocks include

white mica–chlorite–quartz–albite–magnetite,

white mica–chlorite–quartz–epidote–albite, and

white mica–chlorite–biotite–albite–quartz–ilmenite.

Stilpnomelane, alkali feldspar, pyrophyllite, chloritoid, sphene, and calcite are among the other phases that may appear in this facies. The appearance of the various new

Table 27.1   Typical Mineral Assemblages of Buchan Facies Series Rocks

Zeolite Facies

illite/smectite–kaolinite–chlorite/smectite–quartz–hematite (pelitic rock)

quartz–analcite–heulandite–chlorite–illite–hematite (SAC rock)

chlorite–prehnite–analcite–quartz–sphene (metabasite)

Prehnite-Pumpellyite Facies

illite/smectite–chlorite–quartz–albite–hematite (pelitic rock)

quartz–albite–white mica–chlorite–stilpnomelane–prehnite–pumpellyite–hematite (SAC rock)

chlorite–quartz–albite–white mica–pumpellyite–prehnite–stilpnomelane –sphene (metabasite)

Greenschist Facies

muscovite–chlorite–biotite–quartz–albite–magnetite (pelitic rock)

quartz–albite–muscovite–chlorite–K-feldspar–calcite–pyrite (SAC rock)

chlorite–actinolite–epidote–albite–quartz–sphene (metabasite)

Amphibolite Facies

muscovite–biotite–quartz–andalusite–cordierite–staurolite–oligoclase–ilmenite (pelitic rock)

quartz–oligoclase–muscovite–biotite–garnet–andalusite–ilmenite (SAC rock)

hornblende–plagioclase–garnet–biotite–quartz–sphene (metabasite)

Granulite Facies

K-feldspar–sillimanite–garnet–andesine–quartz–biotite–magnetite (pelite)

quartz–andesine–K-feldspar–biotite–garnet–magnetite (SAC rock)

hornblende–garnet–clinopyroxene–plagioclase–quartz–ilmenite–apatite–sphene (metabasite)

Sources: See notes 4 and 15; unpublished data from the author.

minerals signals several reactions. Changes in chlorite and white mica compositions involve continuous reactions. For example, biotite is produced by the reaction

$$\text{celadonite (Fe,Mg-bearing white mica)} + \text{chlorite} <==> $$
$$\text{muscovite} + \text{biotite} + \text{quartz} + H_2O \qquad (27.3)$$

in which the composition of the white mica changes (Brown, 1971).[14] Biotite may also form from alkali feldspar and chlorite via the discontinuous reaction equation 26.3. Pyrophyllite is produced from kaolinite via reaction equation 26.2 or from the breakdown of smectites (Helgeson and others, 1978). Epidote is derived from Ca-zeolites, prehnite, or detrital plagioclase. Continuous reactions of the form

$$\text{chlorite} + \text{biotite}_1 + \text{quartz} <==> \text{garnet} + \text{biotite}_2 + H_2O$$
$$(27.4)$$

give rise to garnet at higher grades.

The Greenschist-Amphibolite Facies boundary is a broad zone. The disappearance of albite marks the maximum upper limit of the Greenschist Facies. Both albite and pyrophyllite are absent from Amphibolite Facies rocks, whereas cordierite and the aluminum silicates andalusite (at lower grades) and sillimanite (at higher grades) characterize aluminous bulk compositions. Additional phases that may occur in pelitic rocks include, but are not restricted to, chloritoid, alkali feldspar, tourmaline, apatite, and sphene.

Reactions distinctive of Buchan Facies Series are those defining the appearance of andalusite and cordierite, which combined with the disappearance of albite, mark the transition to the Amphibolite Facies. Although the conversion of pyrophyllite to andalusite via reaction equation 26.4 may occur, it is probable that alternative reactions produce the andalusite. Reactions such as paragonite + quartz <==>

andalusite + albite (component in plagioclase) + H_2O
(27.5)

and

cordierite$_1$ + biotite$_1$ + muscovite <==>

cordierite$_2$ + biotite$_2$ + andalusite + quartz   (27.6)

may be important (Chatterjee, 1972; N. F. C. Hudson, 1980). Cordierite may form from muscovite and associated phases via reactions such as

chlorite$_1$ + biotite$_1$ + muscovite + quartz <==>

cordierite + chlorite$_2$ + biotite$_2$ (27.7)

(N. F. C. Hudson, 1980) or from assemblages that include chloritoid.

Pelitic rocks in the Granulite Facies are distinguished by the general absence of white mica, by the presence of alkali feldspar + sillimanite or orthopyroxene, and by the occurrence of the assemblage cordierite + orthopyroxene (figure 27.1, table 27.1). The characteristic assemblages develop through reactions such as

muscovite + quartz <==> sillimanite + orthoclase + H_2O (27.8)

which describes the breakdown of muscovite (Evans, 1965; Althaus et al., 1970) and

biotite + quartz <==>

hypersthene + almandine + K-rich alkali feldspar + H_2O (27.9)

which describes the breakdown of some biotite to form orthopyroxene (Winkler, 1979, p. 265). Additional cordierite may arise via reactions involving garnet or biotite, plus sillimanite and quartz (Holdaway and Lee, 1977; S. M. Lee and Holdaway, 1977).

The mineralogical changes described here for the pelitic rocks are paralleled by mineralogical changes in all other bulk rock compositions. Representative phase assemblages are presented for a variety of bulk compositions in table 27.1. In carbonate rocks, tremolite occurs with calcite, dolomite, and quartz in the Greenschist Facies, and is joined by diopside, forsterite, phlogopite, and wollastonite in the Amphibolite Facies. In the Granulite Facies, tremolite is absent, and phlogopite, quartz, and calcite may disappear. In metabasites, the lowest-grade rocks contain combinations of albite, epidote, prehnite, pumpellyite, heulandite or laumontite. At higher grades, chlorite, albite, and actinolite of the Greenschist Facies are replaced by biotite, oligoclase, hornblende, cummingtonite, and augite in the Amphibolite Facies. Granulite Facies rocks contain two pyroxenes, brown hornblendes, and intermediate plagioclase (e.g., Bard, 1969).

Example: Buchan Metamorphism, Northern New England, U.S.A.

Perhaps the best-known Buchan Facies Series is that of northern New England, in the United States. As a result of detailed work by a number of petrologists, the petrology is rather well known.[15] A line representing the aluminum silicate triple point extends through New England—from Rhode Island, through central Massachusetts, across western New Hampshire, and into northeastern Vermont—marking a change from a Barrovian Facies Series on the southwest to a Buchan Facies Series on the northeast (figure 27.2) (J. B. Thompson and Norton, 1968).

In the Buchan Facies Series of northeastern New England, several Acadian (mid-Paleozoic) isograds have been mapped in the widely distributed pelitic rocks, including biotite, garnet, andalusite-staurolite, cordierite-staurolite, sillimanite, and K-feldspar–sillimanite isograds. Locally, muscovite coexists with sillimanite and K-feldspar in pelitic rocks of the uppermost zone; thus, the rocks containing these minerals belong to the Amphibolite Facies. Granulite Facies rocks are present only to the south, in New Hampshire, Massachusetts, and northern Connecticut (Osberg et al., 1989; Schumacher et al., 1989).[16] In carbonate rocks, biotite, amphibole, zoisite, and diopside isograds have been mapped (Ferry, 1976, 1982, 1983b,c). In northernmost Maine, Quebec, and New Brunswick, the Zeolite and Prehnite-Pumpellyite Facies are represented by analcite, prehnite-pumpellyite, and pumpellyite-epidote-actinolite zones in metaclastic and metavolcanic rocks (Pavlides, 1973).[17] In summary, Zeolite, Prehnite-Pumpellyite, Greenschist, Amphibolite, and Granulite Facies rocks are represented in central to northern New England. At least one late Carboniferous intrusion, the Sebago Batholith of western Maine, caused overprinting of earlier Acadian metamorphic rocks (Alienikoff, 1984; Hayward and Gaudette, 1984; Lux and Guidotti, 1985). Elsewhere, local areas containing low- to high-grade Taconic and Alleghenian metamorphic rocks adjoin the dominant Acadian regional metamorphic terrane (figure 27.2) (Murray and Skehan, 1979; Guidotti, 1985).

Radiometric dating, geothermometry, and geobarometry indicate that a diachronous, four-stage Acadian metamorphism (405–350 m.y.b.p.)[18] produced the Buchan rocks at variable temperatures and maximum pressures of 0.24–0.45Gpa (2.4–4.5 kb).[19] Temperatures ranged from 400° C at the biotite isograd (Ferry, 1984) to 670° C in the

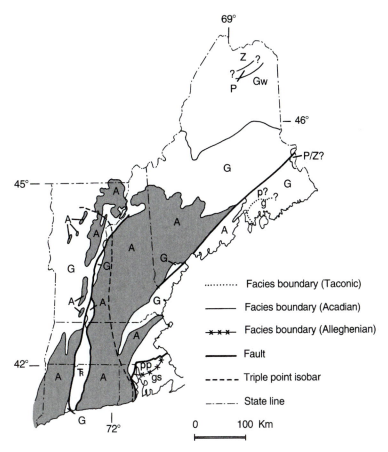

Figure 27.2 Generalized metamorphic map of New England showing zones of Paleozoic regional metamorphism, with Buchan Facies Series northeast of the triple-point isobar and Barrovian Facies Series southwest of the isobar. Acadian metamorphic facies are represented as follows: Z = Zeolite Facies, P = Prehnite-Pumpellyite Facies, Gw = weakly recrystallized Greenschist Facies, G = Greenschist Facies, and A = Amphibolite Facies. Taconic facies are

g = Greenschist Facies and p = Prehnite-Pumpellyite Facies. Alleghenian-Hercynian metamorphic facies are pp = Prehnite-Pumpellyite Facies and gs = Greenschist Facies. Tr = unmetamorphosed Triassic rocks.

[Based principally on the syntheses of J. B. Thompson and Norton (1968) and Guidotti (1985); also see B. A. Morgan (1972), Richter and Roy (1974), and Murray and Skehan (1979).]

K-feldspar–sillimanite zone (Lux et al., 1986; DeYoreo et al., 1989).[20] Locally, contact-like geothermal gradients of 80–100° C/km may have been attained, especially near the numerous intrusions associated with the regional metamorphic event.[21]

A generalized phase-facies diagram for the northeastern New England part of the Northern Appalachian Orogen is presented as figure 27.3. Note the presence of andalusite and cordierite and the absence of kyanite in the middle grades of metamorphism, as compared to the Barrovian Facies Series rocks of the southern Appalachian Orogen described below. Local migmatites, formed by partial melting, are developed at the highest grades of metamorphism.[22]

In south-central Maine, an exceptional opportunity exists to examine mineralogical changes with increasing grade of metamorphism within single formations. The Wa-

terville and Vassalboro/Songerville Formations, which trend northeast-southwest, are crossed by isograds that trend approximately east-west (figure 27.4). The Waterville Formation is composed of metamorphosed shale, argillaceous sandstone, and argillaceous limestone, whereas the Vassalboro/Songerville Formation consists of metamorphic equivalents of argillaceous carbonate rocks, calcareous quartz wackes, and shales (Osberg, 1968, 1979; Ferry, 1983c; Ferry and Osberg, 1989). Metamorphic grade increases from chlorite-zone pelitic rocks and ankerite-zone carbonate rocks in the northeast to sillimanite-zone metapelites and diopside-zone metacarbonate rocks in the southwest (figures 27.4 and 27.5). As is typical in zones of increasing grade, there is an overall increase in grain size towards the southwest.

555

Figure 27.3 Mineral-facies chart for the middle Paleozoic, Buchan Facies Series of northeastern New England. (a) Metapelites, metawackes, and metabasites. (b) Carbonate rocks. Amph = amphibole, An = anorthite (component of plagioclase), And = andalusite, Ank = ankerite, Bio = biotite, Chl = chlorite, Gar = garnet, P/P = prehnite-pumpellyite (facies), Pr-Anl = prehnite-analcite, Pr-P = prehnite-pumpellyite (zone), S-K = sillimanite-alkali feldspar, Sil = sillimanite, Zo = zoisite.

[Sources are listed in explanatory note 15 and in the paragraph following that footnote number in the text. Unpublished observations by the author are also included.]

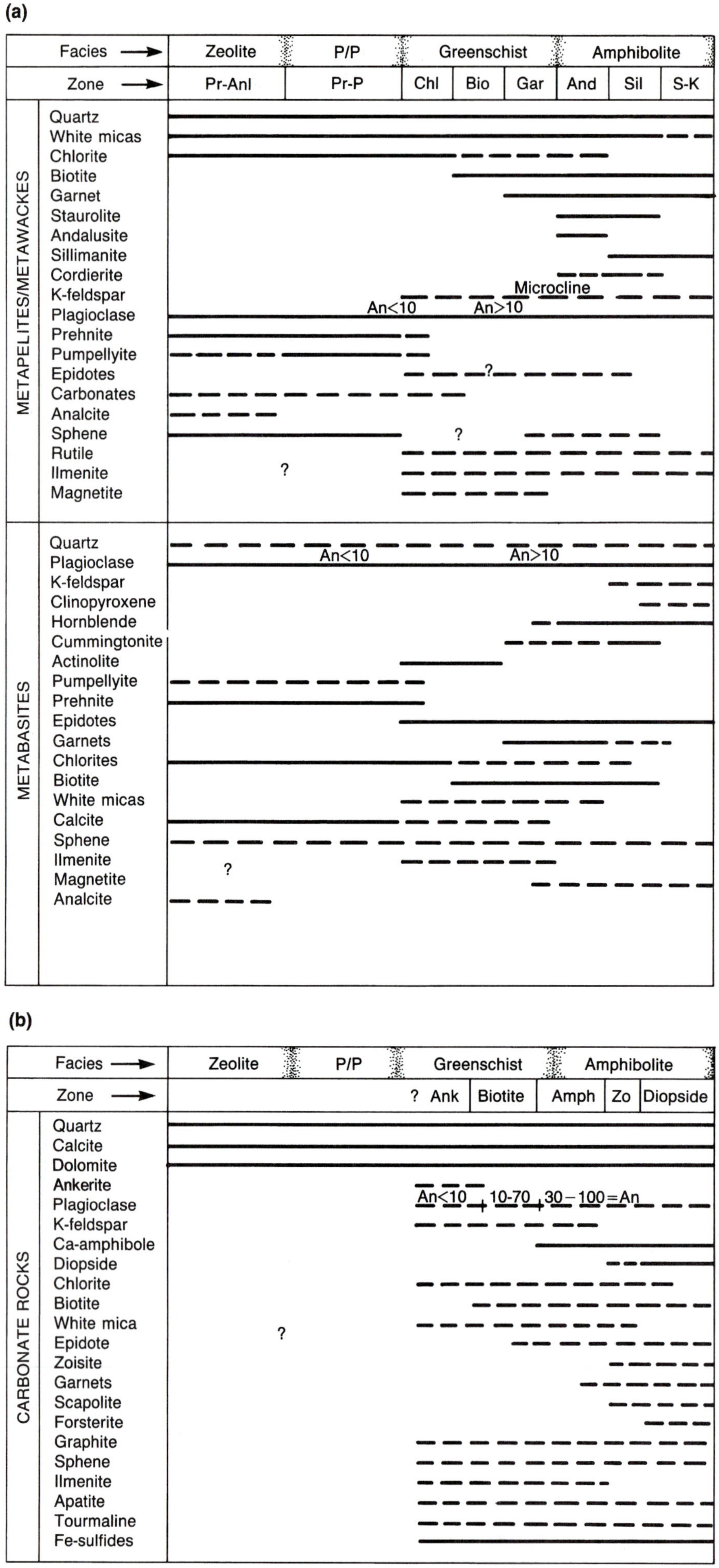

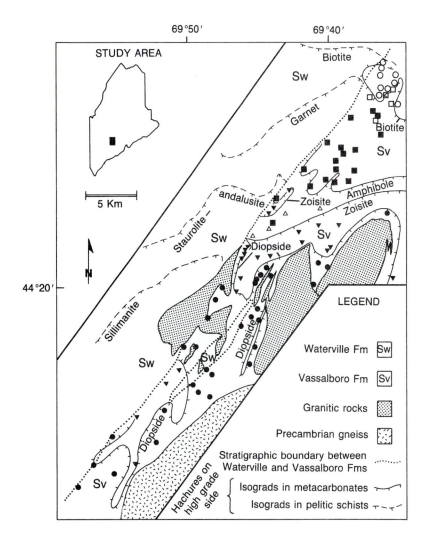

Figure 27.4 Geological sketch map of part of southern Maine showing metamorphic isograds in the Vassalboro and Waterville formations.

(From J. M. Ferry, "Regional metamorphism of the Vassalboro Formation, south-central Maine, USA: A case study of the role of fluid in metamorphic petrogenesis" in Journal of the Geological Society of London, 140:551-76, 1983. Copyright © 1983 Geological Society of London. Reprinted by permission.)

In the Vassalboro/Songerville Formation, the carbonate rocks are characterized by assemblages such as the following (after Ferry, 1983c):

Ankerite Zone

ankerite–quartz–albite–muscovite–calcite–chlorite–pyrite

Biotite Zone

ankerite–quartz–albite–muscovite–biotite–calcite–chlorite–pyrite

biotite–quartz–oligoclase/labradorite–muscovite–calcite–chlorite–pyrite

Amphibole Zone

Ca-amphibole–biotite–quartz–andesine/anorthite–calcite–chlorite–pyrrhotite

Zoisite Zone

zoisite–Ca-amphibole–quartz–andesine/anorthite–calcite–biotite–microcline–pyrrhotite

Diopside Zone

diopside–zoisite–Ca-amphibole–quartz–andesine/anorthite–calcite–biotite–pyrrhotite.

The minerals garnet, graphite, sphene, tourmaline, apatite, and scapolite are accessory minerals at various grades of metamorphism.

In the pelitic rocks, the mineral assemblages are typical of Buchan Facies Series rocks (figures 27.3 and 27.5). The following assemblages are representative (Osberg, 1968; Ferry, 1980a, 1982):

Chlorite Zone

muscovite–quartz–albite–chlorite–ankerite–calcite–magnetite

Biotite Zone

muscovite–quartz–plagioclase–biotite–chlorite–calcite–ilmenite

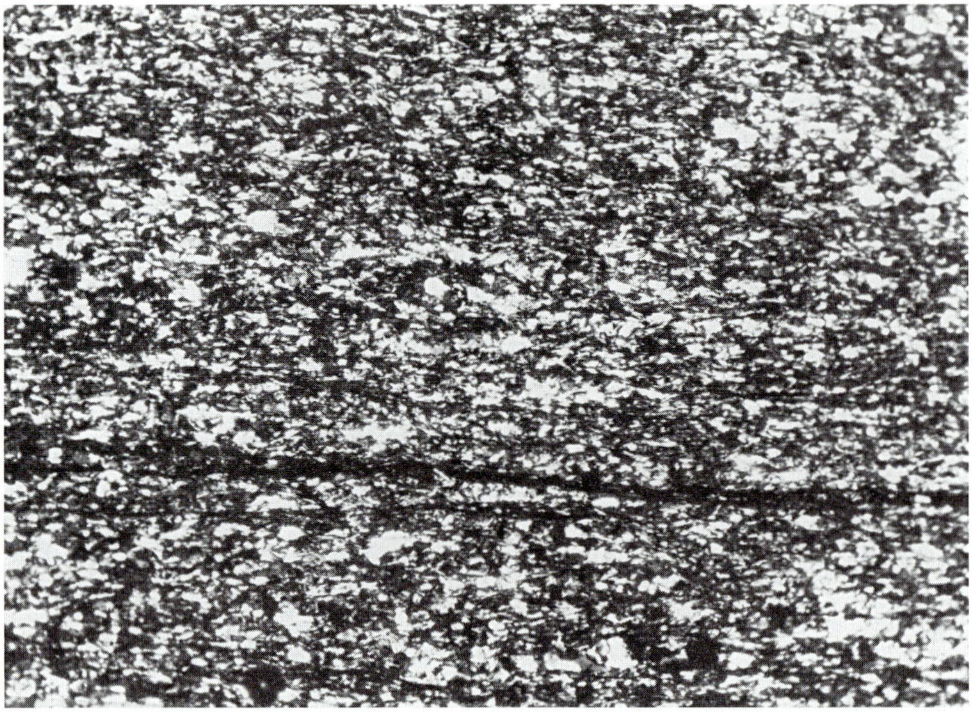

(a)

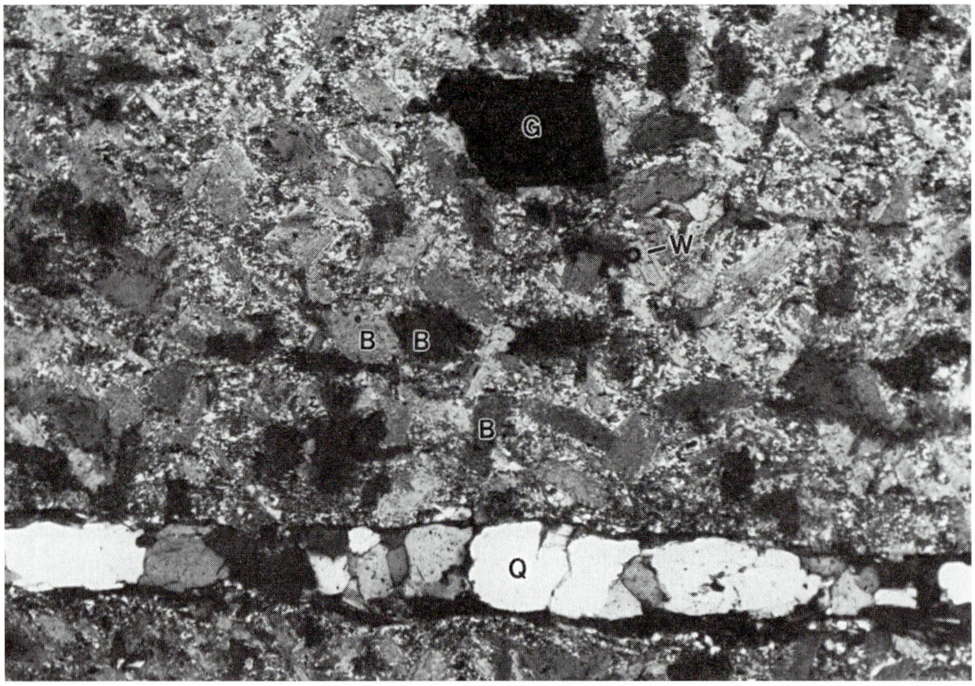

(b)

·················
Figure 27.5   Photomicrographs of pelitic schists from the Waterville Formation. (a) Chlorite Zone schist. (XN). Long dimension of photo is 1.27 mm. (b) Garnet Zone schist with quartz lens. (XN). (c) Andalusite Zone schist. (XN). (d) Sillimanite Zone schist. (XN). A = K-rich alkali feldspar, B = biotite, CH = chlorite, CD = cordierite, G = garnet, O = opaque minerals, P = plagioclase feldspar, Q = quartz, SI = sillimanite, ST = staurolite, and W = white micas. In (b), (c), and (d) long dimension of photo is 3.25 mm.

(c)

(d)

Figure 27.5 Continued

Garnet Zone

muscovite–quartz–plagioclase–biotite–garnet–chlorite–calcite–ilmenite

Staurolite-Andalusite Zone

muscovite–quartz–plagioclase–biotite–garnet–staurolite–andalusite–ilmenite

Sillimanite Zone

muscovite–quartz–plagioclase–biotite–garnet–sillimanite–cordierite–microcline–ilmenite.

Pyrrhotite, pyrite, tourmaline, rutile, clinozoisite, and sphene are among the accessory minerals that occur at various grades of metamorphism.

The fact that all of the metamorphic rocks of each type are derived from a single formation means that compositional variations are minimized as a factor in controlling the mineral assemblages. Similarly, the pressure was relatively uniform during the metamorphic event. As a consequence, temperature, fluid phase composition, and element migration were responsible for the mineralogical differences that developed. The temperature was a primary controlling factor, but fluids were extremely influential, as they transport heat, transport ions (changing the rock chemistry), and drive mineral reactions (Ferry, 1980b, 1982, 1983b,c). Fluid/rock ratios are estimated to have been between 0.7 and 2.0 during metamorphism, and fluid flow rates were about 10^{-1} mm/year (Ferry and Osberg, 1989).

Differences and Similarities Between Contact and Buchan Facies Series

A complete gradation exists between Contact and Buchan Facies series and the conditions of metamorphism overlap. The major differences that exist between rocks of Contact Facies Series and those of the Buchan Facies Series are that the latter are regional in distribution (rather than local), are characterized by foliated fabrics, and lack Sanidinite Facies at the highest grades. Buchan Facies Series are associated with orogenic belts that have experienced a regional influx of heat, rather than being locally associated with particular plutons. In contrast, in contact metamorphism, the heat producing the metamorphism is local and may clearly be related to an intrusion or a set of closely associated intrusions.

The second major difference—that Buchan rocks are characteristically foliated, whereas contact rocks are typically granoblastic to diablastic—is related to the tectonic regime in which such rocks form. In contact metamorphism, intrusions cause local metamorphism and may intrude a passive (nondeforming) country rock. As a consequence, many contact metamorphic rocks lack foliation altogether. Fabrics do develop where movements during intrusion and metamorphism generate deviatoric stresses that produce foliations or lineations in the rocks, but foliation is not a definitive characteristic of contact metamorphic rocks. In contrast, Buchan rocks develop in orogenic belts associated with zones of plate convergence, rifting, or transcurrent faulting. As a consequence, metamorphism may be either synkinematic (occurring at the same time as deformational movements) or slightly pre- or postkinematic. In general, either synkinematic metamorphism or prekinematic metamorphism followed by regional deformation will result in regionally developed foliations. It is these regionally developed foliations that characterize Buchan rocks.

TRIPLE-POINT ROCKS

Recall that Buchan Facies Series are metamorphosed at pressures lower than that of the aluminum silicate triple point (3.87 kb = 0.387 Gpa). Andalusite is a critical indicator of that kind of facies series. Barrovian Facies Series occur in rocks metamorphosed at pressures that, in the middle grades of metamorphism, are greater than the 0.387 Gpa aluminum silicate triple point. Kyanite is a critical mineral indicator of that kind of facies series.

As might be expected, in some occurrences of transitional facies series, rocks containing all three aluminum silicates (andalusite, kyanite, and sillimanite) are present. Such occurrences are reported in New England, Idaho, New Mexico, and Alaska.[23]

BARROVIAN FACIES SERIES

Occurrences

Barrovian Facies Series occur in a number of Phanerozoic orogenic belts, as well as in some of Precambrian age. Notable among the Phanerozoic belts are the Caledonides of northwestern Europe, including the classic region in the Scottish Highlands, and parts of the Appalachian Mountain System of eastern North America.[24] Other Phanerozoic belts with Barrovian rocks occur in western North America (e.g., Idaho, Colorado, British Columbia, Alaska), Venezuela, Spain, southern Europe and Asia (the Alpine-Himalayan Orogen), central Asia (the Ural Mountains), and Japan.[25]

Precambrian belts of Barrovian rocks occur in the Black Hills of South Dakota, the Rocky Mountains, and Labrador, Quebec, and Ontario (Canada).[26] Outside of North America, such belts are found in the former Soviet Union, India, and Sri Lanka.[27]

The Phanerozoic orogenic belts are clearly associated with convergent plate margins. Both Barrovian and Buchan Facies series develop at such margins. Those of Precambrian age may represent convergent margins as well, but their origins are debated (Windley, 1977; P. F. Hoffman, 1989). In convergent zones, regional heating due to the rise of plutons into the overlying plate (the plate above the subduction zone) is the general cause of metamorphism (Dewey and Bird, 1970; Miyashiro, 1982, p. 153), but migrating fluids may also transport heat.

The Barrovian Facies Series Revisited

The zones of metamorphism in the Scottish Highlands originally described by Barrow (1893, 1912) include six distinct mineral assemblages that occur in the rock types listed below:

Chlorite Zone

(slates, phyllites, and schists)
quartz–albite–white mica–chlorite–microcline ± calcite

Biotite Zone

(phyllites and schists)
quartz–albite–white mica–chlorite–biotite ± microcline ± calcite ± epidote

Almandine (Garnet) Zone

(phyllites and schists)
quartz–albite–white mica–biotite–garnet ± chlorite

Staurolite Zone

(schists)

quartz–oligoclase–white mica–biotite–garnet–staurolite

Kyanite Zone

(schists)

quartz–oligoclase–white mica–biotite–garnet–kyanite ± staurolite

Sillimanite Zone

(schists, gneisses, and granoblastites)

quartz–oligoclase–biotite–sillimanite ± kyanite ± K-rich alkali feldspar ± white mica.

The additional granulite facies assemblage

K-rich alkali feldspar–quartz–sillimanite–garnet–biotite–plagioclase–rutile–Fe-Ti oxide (?)

has been reported more recently (A. J. Baker and Droop, 1983; A. J. Baker, 1985). These assemblages in the pelitic rocks are paralleled by others in mafic rocks, in which there are increases in the calcium content of plagioclase, a change in the Ca-amphibole composition, and other mineralogical changes (figure 27.6).[28]

As was the case in occurrences of Contact and Buchan Facies series rocks, metamorphic belts exhibiting Barrovian Facies Series do not always exhibit *all* of the facies possible within the series. Clearly this is the case in the Scottish Highlands, where Zeolite and Prehnite-Pumpellyite Facies are not described as part of the sequence. To completely characterize the phase assemblages of a facies series, it is necessary to depict the phase assemblages of the entire facies series by constructing a composite series of phase diagrams and an associated set of petrogenetic grids (see appendix C). Such an illustration, showing phase topologies for *all* conditions of P T, and X_{fluid} in the facies series, can appear overwhelming due to the large number of phase diagrams required to present such a complete sequence.[29] A representative sequence of topologies, however, can reveal the general trends of phase changes that occur during progressive metamorphism, as shown by the selected phase diagrams for Barrovian Facies Series in figure 27.7.

Phase Assemblages and Reactions in Barrovian Facies Series

The representative phase diagrams for rocks of Barrovian Facies Series shown in the topologies in figure 27.7 serve four purposes. First, they allow us to obtain an impression of the overall mineralogical trends that characterize the facies series. Second, each phase diagram reveals typical assemblages for a part of the facies it represents. Third, the composite nature of

the illustration allows us to compare phase assemblages in rocks of different composition metamorphosed at the same general grade. For example, assemblages for aluminous and aluminum-rich siliceous and siliceous-alkali-calcic (SAC) rocks are shown on AFM diagrams, whereas more mafic SAC rocks plus intermediate to basic igneous rocks have mineral assemblages that are plotted on ACF pseudo-phase diagrams or CFM diagrams. Mineral assemblages of carbonate and ultramafic rocks are plotted on the CSM diagram. The diagrams for the carbonate rocks represent intermediate X_{CO_2} values, whereas those for ultramafic rocks represent low values of X_{CO_2} (high X_{H_2O}). Fourth, the sequences imply several reactions.

The reactions implied by the topologies in figure 27.7 are numerous. In most cases, several reactions must have occurred in order for adjoining topologies to develop. Experimentally determined and calculated reaction curves for many of the more important reactions are presented in appendix C.

Assemblages and Reactions in Pelitic Rocks

A survey of occurrences of pelitic rocks in Barrovian Facies Series reveals the important phase assemblages at various grades of metamorphism. These, in turn, suggest the kinds of reactions that are important during prograde metamorphism. At the lowest grade, in the Zeolite Facies, which forms under conditions just above those of diagenesis, assemblages are characterized by clay minerals (C. E. Weaver et al., 1984).[30] Assemblages may include

kaolinite–illite–illite/smectite–chlorite–quartz–analcite,

illite–illite/smectite–kaolinite–chlorite, and

illite–chlorite–kaolinite–K-rich alkali feldspar.[31]

In addition, smectites, iron oxides, calcite, and dolomite may occur, as may Ca-zeolites (in appropriate bulk compositions). Organic carbon is also generally present.

At slightly higher-grade conditions, assemblages of the Zeolite Facies are replaced by those of the Prehnite-Pumpellyite Facies. In this facies, some minerals present in lower-grade rocks, such as kaolinite and K-rich alkali feldspar, are absent from rocks of certain bulk compositions. New phases appear, including albite, phengitic white mica, and illite/chlorite or illite/smectite/chlorite mixed-layer minerals (figures 27.7 and 27.8). In addition, changes occur in the structures of some minerals, for example, in chlorite and illite (J. Hoffman and Hower, 1979). Notably, illite changes in crystallinity (C. E. Weaver, 1960; Kubler, 1967; Kisch, 1990).

The mineralogical changes imply several reactions. As was the case in Buchan Facies Series, K-rich alkali feldspar and smectites are among the first minerals to disappear from aluminous rocks. Discontinuous reactions such as equation 27.1 account for the replacement of smectite and feldspar by illite, chlorite, and quartz.[32] Kaolinite also is

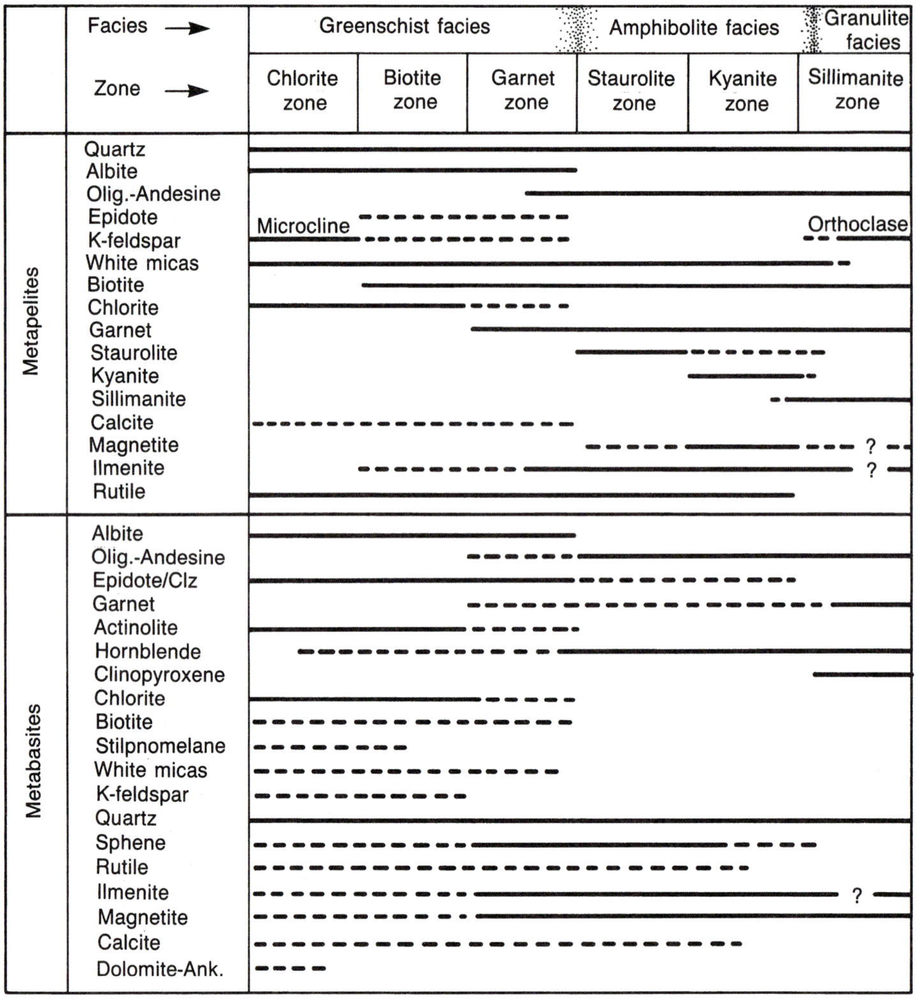

Facies →		Greenschist facies			Amphibolite facies		Granulite facies
Zone →		Chlorite zone	Biotite zone	Garnet zone	Staurolite zone	Kyanite zone	Sillimanite zone
Metapelites	Quartz						
	Albite						
	Olig.-Andesine						
	Epidote						
	K-feldspar	Microcline					Orthoclase
	White micas						
	Biotite						
	Chlorite						
	Garnet						
	Staurolite						
	Kyanite						
	Sillimanite						
	Calcite						
	Magnetite						?
	Ilmenite						?
	Rutile						
Metabasites	Albite						
	Olig.-Andesine						
	Epidote/Clz						
	Garnet						
	Actinolite						
	Hornblende						
	Clinopyroxene						
	Chlorite						
	Biotite						
	Stilpnomelane						
	White micas						
	K-feldspar						
	Quartz						
	Sphene						
	Rutile						
	Ilmenite						?
	Magnetite						
	Calcite						
	Dolomite-Ank.						

Figure 27.6 Mineral-facies chart for the type area of Barrovian Facies Series metamorphism in the Scottish Highlands. Ank = ankerite, Clz = clinozoisite, Olig = oligoclase.

[Data from Harker (1932), Chinner (1960, 1966), C. M. Graham (1974, 1985), Harte and Hudson (1979), Graham et al. (1983), A. J. Baker (1985), McLellan (1985a, 1985b), Moles (1985), and K. P. Watkins (1987).]

commonly absent from Prehnite-Pumpellyite Facies rocks. Although the reactions involved are unknown, kaolinite commonly diminishes or disappears as illite appears or increases in abundance, as chlorite increases in abundance, and as 2M white micas appear.[33] Illite/smectite mixed-layer minerals may also appear where kaolinite disappears from the rocks. The chemistry of the new phases suggests that Si, K, Na, Ca, Mg, and Fe must be available for these phases to form. Such elements may be derived from calcite, Mg-calcite, and dolomite; quartz, chalcedony, or opal; K-rich alkali feldspar; smectites; iron oxides; minor amounts of ferromagnesian minerals, especially detrital biotite and chlorite; and *pore fluids.* C. E. Weaver, Beck, and Pollard (1971) suggest that some of the necessary ions may be transported into the rock by migrating solutions, and several other workers suggest that Eh, pH, and the chemistry of the pore solutions may control the nature of the reactions that occur (Garrels and Christ, 1965; Grim, 1968; B. M. Sass, Rosenberg, and Kittrick, 1987).[34] A reaction such as equation 27.2, involving most of the phases in a typical rock, would account for the disappearance of kaolinite and would yield appropriate Prehnite-Pumpellyite Facies phases. Inasmuch as chlorite and illite may already exist in low-grade rocks or sediments, reactions such as equations 27.1 and 27.2 merely add to the abundance of these phases in the rock. Of course, as the pressure and temperature increase, continuous reactions may produce compositional and structural changes in these phases.[35]

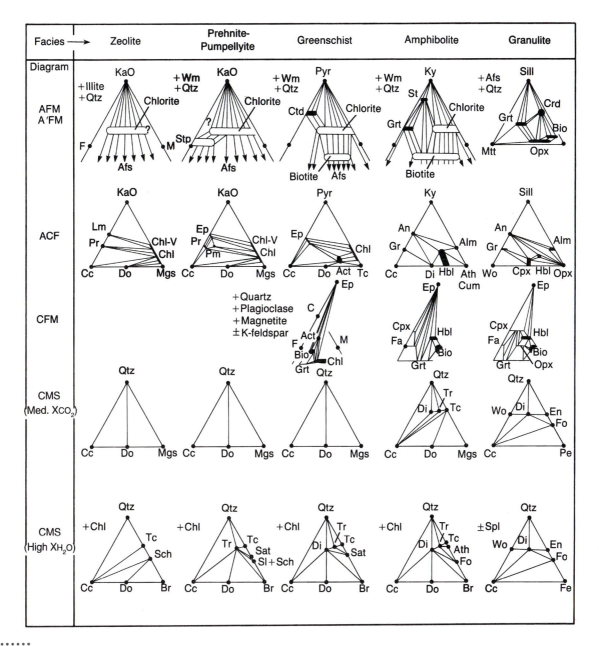

Figure 27.7 Representative phase diagrams for the various facies of the Barrovian Facies Series. AFM diagrams are useful for depicting phase assemblages in pelitic rocks and aluminous SAC rocks. ACF diagrams are useful for metabasites, metaigneous rocks of intermediate composition, and more iron- and aluminum-rich SAC rocks. CFM diagrams depict the phase assemblages of metabasic rocks. The CMS diagram is useful in representing phase assemblages in carbonate and ultramafic rocks.

[CFM diagrams are from Abbott (1982, 1984). Additional sources include Miyashiro (1973a), B. W. Evans (1977), Winkler (1979), F. J. Turner (1981), Evarts and Schiffman (1983), Absher and McSween (1985), various sources listed in the text and notes 24–66, and the observations of the author.]

An additional phase that may appear in pelitic rocks of the Prehnite-Pumpellyite Facies is stilpnomelane. Stilpnomelane may form either by modification of detrital biotite or via a reaction such as

chlorite + K-rich alkali feldspar + magnetite <==> stilpnomelane   (27.10)

Stilpnomelane is generally restricted to iron-rich bulk rock compositions (e.g., see Floran and Papike, 1978).

As the P-T conditions increase, Greenschist Facies assemblages with new minerals form[36]. Typical assemblages in pelitic rocks (figure 27.9a) include

white mica (muscovite and paragonite)–chlorite–quartz–albite–magnetite,

white mica–chlorite–biotite–quartz–albite–magnetite,
white mica–chlorite–chloritoid–epidote–magnetite–
hematite, and
white mica–chlorite–garnet–biotite–quartz–albite–
oligoclase–magnetite.

Stilpnomelane, pyrophyllite, alkali feldspar, calcite, and
sphene also occur in some rocks of this facies.

Reactions that produce the various new minerals include both discontinuous and continuous types. Changes in
chlorite and white mica compositions involve continuous reactions, such as reaction equation 27.3, in which the composition of the white mica changes and biotite is produced
(E. H. Brown, 1971).[37] Biotite may also form from chlorite
and alkali feldspar via the discontinuous reaction equation
26.3 or via the continuous reaction

$$\text{microcline} + \text{chlorite} + \text{white mica}_1 \Longleftrightarrow$$

$$\text{biotite} + \text{white mica}_2 + \text{quartz} + H_2O \quad (27.11)$$

(Mather, 1970). Pyrophyllite forms from the breakdown of
smectites or is produced from kaolinite via reaction equation
26.2 (Helgeson et al., 1978). Reactions such as

$$\text{chlorite} + 5 \text{ pyrophyllite} \Longleftrightarrow 7 \text{ chloritoid} + 17 \text{ quartz} + 4 H_2O \quad (27.12)$$

may yield chloritoid (J. B. Thompson and Norton, 1968) and,
at higher-grade conditions, continuous reactions such as

$$\text{chlorite} + \text{biotite}_1 + \text{quartz} \Longleftrightarrow \text{garnet} + \text{biotite}_2 + H_2O \quad (27.4)$$

generate garnet. Because of the variable composition of garnet, its first appearance in a regional metamorphic terrane
probably occurs over a range of P-T conditions, depending on
the composition of the rock and the fluid phase. In some
cases, it may not appear until the lower part of the Amphibolite Facies.[38]

As is the case in the Buchan Facies Series, the
Greenschist-Amphibolite Facies boundary is a broad zone.
The disappearance of albite marks the maximum upper limit
of the Greenschist Facies. Thus, albite, like pyrophyllite, is
absent from Amphibolite Facies rocks. Staurolite, rather
than chloritoid, occurs in the lower part of the Amphibolite
Facies and the aluminum silicates kyanite (at lower grades)
and sillimanite (at higher grades) characterize aluminous
bulk compositions. Typical assemblages (figures 27.7 and
27.9b) include

white mica–biotite–quartz–plagioclase (oligoclase)–
garnet–magnetite,
white mica–biotite–chlorite–quartz–oligoclase–garnet–
staurolite–ilmenite,
white mica–biotite–quartz–oligoclase–garnet–kyanite–
ilmenite, and
white mica–biotite–quartz–oligoclase–garnet–sillimanite–
ilmenite.[39]

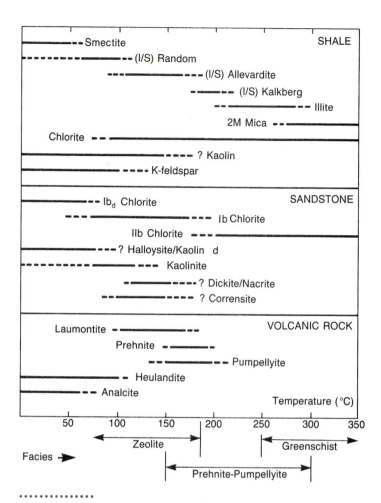

Figure 27.8 Progressive mineralogical changes in weakly
metamorphosed rocks of the Montana disturbed belt. In the
shale, note the changes from one mixed-layer illite-smectite
(I/S) mineral to another. In the sandstones, different structural
varieties of chlorite are stable under different conditions.
Facies designations are assigned here on the basis of the
mineralogical relations discussed in this chapter and P-T limits
defined in chapter 24.
(Modified from J. Hoffman and Hower, 1979).

Additional phases that occur locally include, but are not restricted to, alkali feldspar, chloritoid, tourmaline, epidote,
apatite, and sphene. Note that the first assemblage above is
one that occurs over a wide range of conditions, the specific
nature of which can only be determined through geothermobarometry.

Possible reactions leading to phase assemblages in
pelitic rocks of the Amphibolite Facies are numerous.[40] A
number of reactions have been proposed to yield the first appearance of staurolite. These involve various combinations of
the phases chloritoid, biotite, chlorite, muscovite, quartz,

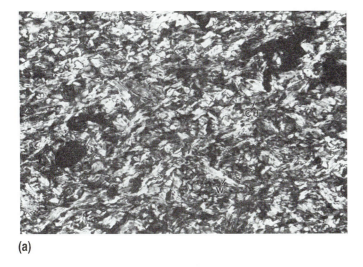

(a)

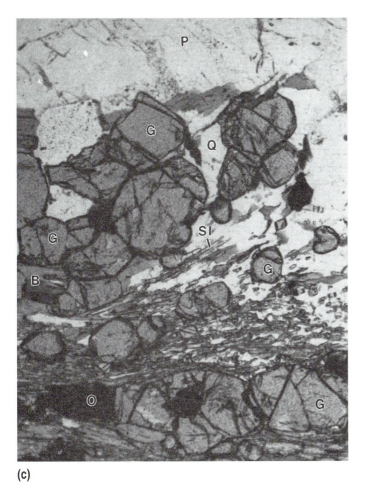

(c)

(b)

Figure 27.9 Photomicrographs of Barrovian Facies Series pelitic rocks. (a) Greenschist Facies metashale composed of white mica, chlorite, quartz, and other minerals, Grandfather Mountain Formation, North Carolina. (XN). (b) Amphibolite Facies pelitic schist, Ashe Metamorphic Suite, Ashe County, North Carolina. (PL). (c) Granulite Facies gneiss, Winding Stair Gap, North Carolina. (PL). B = biotite, CH = chlorite, G = garnet, K = kyanite, O = opaque minerals, P = plagioclase, Q = quartz, SI = sillimanite, ST = staurolite, W = white mica. Long dimension of photo (a) is 1.27 mm, of (b) is 3.25 mm and of (c) is 1.19 mm.

kyanite, ilmenite, and garnet. For rocks of the Snow Peak, Idaho, area, one such reaction is

$$\text{chlorite} + \text{garnet} + \text{muscovite} + \text{ilmenite} \Longleftrightarrow$$

$$\text{staurolite} + \text{biotite} + \text{plagioclase} + \text{quartz} + H_2O \quad (27.13)$$

(Lang and Rice, 1985b). Kyanite may arise by the conversion of pyrophyllite via the reaction

$$\text{pyrophyllite} \Longleftrightarrow \text{kyanite} + 3\ SiO_2 + H_2O \quad (27.14)$$

(Kerrick, 1968; Haas and Holdaway, 1973) or by more complex reactions involving other phases, such as

$$\text{staurolite} + \text{muscovite} + \text{quartz} \Longleftrightarrow \text{biotite} + \text{kyanite} + H_2O \quad (27.15)$$

(McLellan, 1985b). The particular reaction depends on the bulk composition and other factors. Sillimanite may form by the simple polymorphic transformation

$$kyanite \Longleftrightarrow sillimanite \qquad (27.16)$$

as it appears to have done in western Labrador (Rivers, 1983) and in the Himalayan Orogen (Lal, Mukerji, and Ackermand, 1981), but reactions involving muscovite or biotite, such as

$$6\ staurolite + 4\ muscovite + 7\ quartz \Longleftrightarrow 4\ biotite +$$
$$31\ sillimanite + 3\ H_2O \qquad (27.17)$$

(J. B. Thompson and Norton, 1968; McLellan, 1985a) may be more common.

The upper part of the Amphibolite Facies and the Granulite Facies are characterized locally by zones of *lit-par-lit* (literally, bed-by-bed) "injection" gneisses—banded rocks with alternating dark and light (granitoid) layers. Although such rocks may have begun as pelitic schists, partial melting has altered their character. They consist of a light-colored, quartz-feldspar-rich **leucosome**, produced by partial melting and/or the local injection and crystallization of a lower-temperature (eutectic) melt between layers or foliation planes, and a **melanosome**, a dark-colored, ferromagnesian mineral-rich, refractory rock. Leucosomes typically have granitoid mineral assemblages, whereas melanosomes are characterized by biotite- or amphibole-rich assemblages.[41]

The Granulite Facies is distinguished by the general absence of white mica, the presence of orthopyroxene in pelitic rocks, and by the occurrence of the assemblage cordierite + orthopyroxene.[42] Rocks of this facies are more common in Buchan Facies Series, but some occur in Barrovian Facies Series, for example, in Ontario (E. W. Reinhardt, 1968), the Adirondack Mountains of New York (A. E. J. Engel and Engel, 1958, 1960; Stoddard, 1980; Bohlen, Valley, and Essene, 1985; Metzger, 1987), the southern Appalachian Orogen (Force, 1976; Absher and McSween, 1985; Gulley, 1985), Antarctica (Sheraton, 1984/1985; Harley, 1986, 1987), and Calabria, Italy (Schenk, 1984). Granulite assemblages are also found in xenoliths in mafic volcanic rocks (Rudnick and Taylor, 1987). Pelitic assemblages (figure 27.9c) include

biotite–garnet–sillimanite–K-rich alkali feldspar–andesine–quartz–magnetite–ilmeno-hematite,

biotite–K-rich alkali feldspar–andesine–quartz–cordierite–garnet–iron oxides, and

garnet–andesine–quartz–rutile–ilmenite–orthopyroxene–sillimanite–zircon.

The alkali feldspar is typically perthitic. Additional phases include apatite, spinel, and graphite.

These assemblages develop through reactions such as equation 27.8, which describes the breakdown of muscovite, and equation 27.9, which describes the breakdown of biotite to form orthopyroxene. Cordierite may arise in Granulite Facies rocks of the Barrovian Facies Series via reactions involving garnet or biotite, plus sillimanite and quartz (Holdaway and Lee, 1977; S. M. Lee and Holdaway, 1977).

Siliceous-Alkali-Calcic (SAC) Rocks

Metamorphosed sandstones, felsic to intermediate volcanic rocks, and granitoid rocks exhibit mineral assemblages that range from those identical to the assemblages of pelitic schists to those like assemblages in mafic rocks. The SAC rock modes, however, differ from those of pelitic rocks. Quartz and feldspar are the dominant phases, rather than the phyllosilicates, and calcium-bearing phases are common. Additional minerals that may occur in various assemblages include stilbite, epistilbite, calcite, pyrophyllite, stilpnomelane, garnet, Fe-Ti oxides, actinolite, and hornblende.

Typical mineral assemblages in SAC rocks for various facies of metamorphism are shown in table 27.2.[43] Compare the assemblages listed to those depicted in the AFM and ACF diagrams in figure 27.7. Notice that these rocks have an intermediate character, with compositions that are between those for which these diagrams are best suited. Figure 27.10 shows photomicrographs of SAC rocks metamorphosed under a variety of conditions.

Reactions involving the various phases present in SAC rocks include some of those described above for the aluminous rocks, as well as some applicable to the mafic rocks described below. Important reactions include equation 27.18, which describes the disappearance of laumontite and the upper limit of the Zeolite Facies, equation 27.21, which marks the appearance of actinolite and the Prehnite-Pumpellyite to Greenschist Facies transition, and equation 27.10 marking the appearance of stilpnomelane. At higher grades of metamorphism, reactions such as equations 27.4 and 27.13 mark the broad transition from the Greenschist Facies to the Amphibolite Facies, and reaction equation 27.8, defining the conversion of white mica to alkali feldspar, delimits the lower boundary of the Granulite Facies.

Assemblages and Reactions in Mafic Rocks

Mafic rocks of the Barrovian Facies Series have not been studied as extensively as have aluminous rocks, but their compositions are generally well known. The work of Liou, Maruyama, and Cho (1985, 1987) has been definitive in outlining the phase relations in metabasic rocks at low grade.[44] These rocks contain assemblages in the Zeolite Facies such as:

albite–chlorite/smectite–prehnite–pumpellyite–heulandite, and laumontite–analcite–albite–pumpellyite–epidote–chlorite–quartz–sphene.[45]

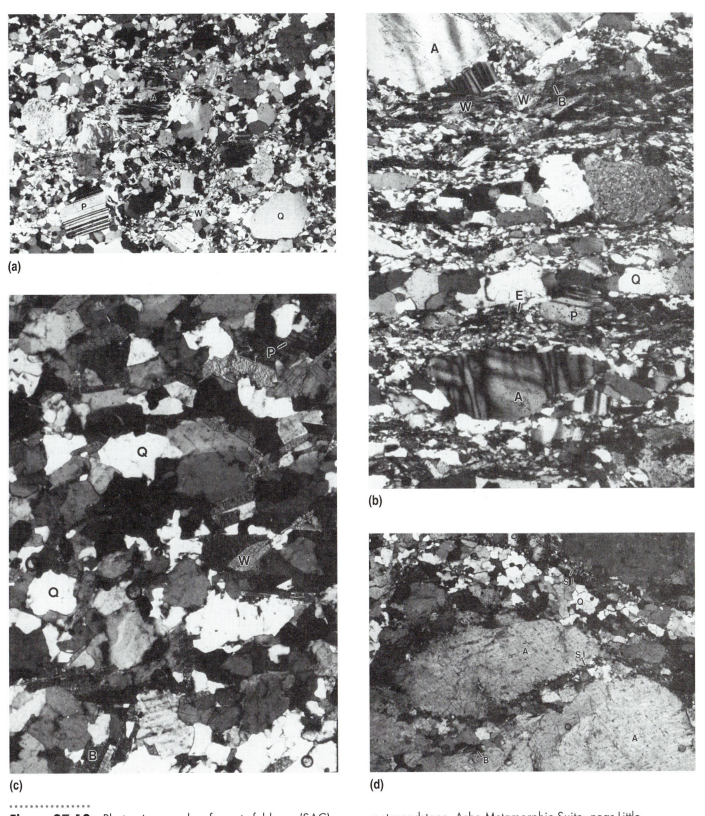

(a)

(b)

(c)

(d)

Figure 27.10 Photomicrographs of quartz-feldspar (SAC) rocks of the Barrovian Facies Series, metamorphosed under a variety of conditions. (a) Greenschist Facies metaquartz arenite, Grandfather Mountain Formation, Boone, North Carolina. (XN). (b) Greenschist Facies Gneiss, Caldwell County, North Carolina. (XN). (c) Amphibolite Facies metasandstone, Ashe Metamorphic Suite, near Little Switzerland, North Carolina. (XN). (d) Granulite Facies quartz-feldspar gneiss, Cloudland Gneiss, Roan Mountain, Tennessee. (XN). A = K-rich alkali feldspar, C = calcite, E = epidote. Other symbols as in figure 27.9. Long dimension of photos (a) – (c) is 3.25 mm and of (d) is 1.27 mm.

Notice that epidote, prehnite, and pumpellyite all may occur in the Zeolite Facies. The Zeolite Facies is distinguished, not by the presence or absence of these minerals, but by the presence of a zeolite in association with them. The Prehnite-Pumpellyite Facies is distinguished by associations containing these minerals but *lacking zeolites and actinolite*.

Important reactions that mark the transition from the Zeolite Facies to the Prehnite-Pumpellyite Facies include equation 26.15, for the conversion of analcite to albite, and the reactions

$$\text{laumontite} + \text{prehnite} <==> \text{clinozoisite} + \text{quartz} + H_2O \tag{27.18}$$

and

$$\text{laumontite} + \text{pumpellyite} <==> \text{clinozoisite} + \text{chlorite} + \text{quartz} + H_2O \tag{27.19}$$

which define the disappearance of laumontite (Liou, Maruyama, and Cho, 1985, 1987). These and similar reactions involving other zeolites occur at temperatures of about 200° C ± 75° C.

Phase assemblages of the Prehnite-Pumpellyite Facies are similar to those of the Zeolite Facies, but, as noted, lack the zeolites and analcite.[46] No new Ca-Al silicate minerals appear in the Prehnite-Pumpellyite Facies. Stilpnomelane does appear as a new phyllosilicate phase in some rocks of this facies, and sphene appears as a titanium-bearing phase.[47] Additional minerals that may occur include calcite, white mica, and alkali feldspar.

Phase assemblages of the Greenschist Facies are different. They are marked by the occurrence of actinolite with chlorite and either pumpellyite or epidote-clinozoisite (figure 27.11a). Pumpellyite-bearing Greenschist Facies rocks form at temperatures below about 350° C (except at high pressures) and thus characterize only the lowest-grade part of the facies. The pumpellyite-bearing zone is equivalent to the Pumpellyite-Actinolite Facies of Liou, Maruyama, and Cho (1985). Representative Greenschist Facies assemblages include

albite–actinolite–epidote–pumpellyite–calcite,

chlorite–epidote–quartz–calcite–garnet–prehnite–albite–actinolite,

actinolite–chlorite–albite–quartz–magnetite, and

epidote–albite–actinolite–calcite–chlorite–quartz–sphene.[48]

Additional phases reported include stilpnomelane, white mica, biotite, K-rich alkali feldspar, and pyrite.[49]

Reactions marking the transition from the Prehnite-Pumpellyite Facies to the Greenschist Facies, or the *lower boundary* of the Greenschist Facies, are those defining the appearance of actinolite. Such reactions as

$$10 \text{ calcite} + 3 \text{ chlorite} + 21 \text{ quartz} <==> 2 \text{ epidote} + 3 \text{ actinolite} + 8 H_2O + 10 CO_2 \tag{27.20}$$

and

$$25 \text{ pumpellyite} + 2 \text{ chlorite} + 29 \text{ quartz} <==> 43 \text{ epidote} + 7 \text{ actinolite} + 67 H_2O \tag{27.21}$$

are important (Nakajima, Banno, and Suzuki, 1977; Liou, Maruyama, and Cho, 1985, 1987; Cho and Liou, 1987).

The *upper (high-T) boundary* of the Greenschist Facies is marked by a transition zone in which (1) oligoclase appears, (2) chlorite disappears as a common phase, (3) hornblende appears, (4) garnet appears, (5) albite disappears, and (6) actinolite disappears.[50] The reactions involved are complex and continuous, involving changing compositions of amphibole, chlorite, garnet, and plagioclase (Laird, 1980; Spear, 1980, 1982; J. B. Thompson, Laird, and Thompson, 1982).[51] An example of such a reaction is

$$\text{amphibole}_1 + \text{chlorite}_1 + \text{epidote} + \text{albite} <==> \text{amphibole}_2 + \text{chlorite}_2 + \text{oligoclase} + \text{quartz} + H_2O \tag{27.22}$$

(Laird, 1980). In this text, the disappearance of albite is considered to be a principal indicator of the upper limit of Greenschist Facies conditions in mafic rocks.

Because of these various reactions, rocks of the Amphibolite Facies are characterized by hornblende and plagioclase (oligoclase, andesine, bytownite, and/or anorthite) (figure 27.11b). Amphibolite Facies metabasites may also include garnet, quartz, epidote, cummingtonite, gedrite, anthophyllite, staurolite, clinopyroxene, biotite, paragonite, K-rich alkali feldspar, magnetite, ilmenite, rutile, calcite, dolomite, chlorite, and pyrite. Representative assemblages include

hornblende–epidote–andesine–biotite–quartz–sphene–ilmenite–apatite,

hornblende–garnet–plagioclase (bytownite)–quartz–ilmenite,

hornblende–garnet–cummingtonite–gedrite–anthophyllite–andesine–ilmenite–magnetite, and

hornblende–Ca-plagioclase–clinopyroxene–ilmenite–sphene.[52]

Table 27.2 Representative Mineral Assemblages in Metamorphosed Siliceous-Alkali-Calcic (SAC) Rocks of the Barrovian Facies Series

Zeolite Facies

quartz–kaolinite

quartz–analcite–heulandite–chlorite–illite–hematite

quartz–albite–laumontite–chlorite–illite–illite/chlorite–hematite

Prehnite-Pumpellyite Facies

quartz–albite–white mica–chlorite–stilpnomelane–prehnite–pumpellyite–hematite

quartz–albite–white mica–chlorite–chlorite/vermiculite–pumpellyite–stilpnomelane–sphene

quartz–albite–chlorite–epidote–prehnite–pumpellyite–calcite

Greenschist Facies

quartz–alkali feldspar–white mica–chlorite–magnetite

quartz–albite–epidote–white mica–chlorite–sphene

quartz–alkali feldspar–plagioclase–white mica–chlorite–biotite–magnetite

Amphibolite Facies

quartz–alkali feldspar–plagioclase–white mica–biotite–garnet–ilmenite

quartz–plagioclase–biotite–garnet–gedrite–ilmenite

quartz–plagioclase–biotite–garnet–sillimanite

Granulite Facies

plagioclase–quartz–alkali feldspar–biotite–garnet–sillimanite–ilmenite

quartz–alkali feldspar–plagioclase–orthopyroxene–ilmenite

quartz–alkali feldspar–plagioclase–orthopyroxene–clinopyroxene–ilmenite

Sources: See note 43.

Note both the absence of albite and the common (but not ubiquitous) occurrence of garnet.

Hornblende persists into the Granulite Facies, as does biotite. In metabasites, the Granulite Facies is characterized by the appearance of orthopyroxene, the coexistence of plagioclase with *both* orthopyroxene and clinopyroxene, and the occurrence of biotite *only* in magnesium-richer bulk compositions. In most metabasites, biotite and epidote are absent. Typical assemblages (figure 27.11c) include

hornblende–plagioclase–orthopyroxene–clinopyroxene–garnet–ilmenite and

orthopyroxene–labradorite–clinopyroxene–ilmenite–apatite–zircon.[53]

The appearance of orthopyroxene in these assemblages may arise from a reaction such as

$$hornblende_1 + plagioclase_1 <==> hornblende_2 + plagioclase_2 + clinopyroxene + orthopyroxene + ilmenite + H_2O$$

$$(27.23)$$

(Spear, 1981b). Notice that this is a dehydration reaction. The generation of a fluid phase via such dehydration reactions may cause local anatexis and the formation of migmatites. (Recall that the rocks of the Granulite Facies develop above the minimum melting curve for wet granite.)

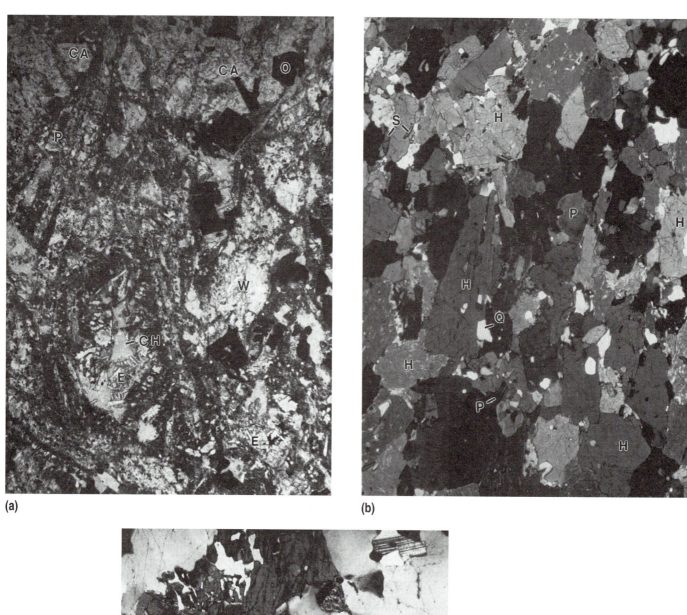

(a)

(b)

(c)

Figure 27.11 Photomicrographs of metabasites. (a) Metadiabase, Greenschist Facies, Linville Metadiabase, Linville, North Carolina. (PL). (b) Hornblende schist (amphibolite), Amphibolite Facies, Ashe Metamorphic Suite, Todd, North Carolina. (XN). (c) Hornblende-plagioclase granoblastite, Granulite Facies, Winding Stair Gap, southwestern North Carolina. (XN). CA = Ca-amphibole, CH = chlorite, E = epidote, G = garnet, H = hornblende, O = opaques (magnetite, ilmenite, etc.), OP = orthopyroxene, P = plagioclase, Q = quartz, S = sphene, W = white mica. Long dimension of photos (a) and (b) is 6.5 mm and of (c) is 4.9 mm..

Calcareous Rocks and Ultramafic Rocks

Regional metamorphism of limestones, dolostones, and calcareous shales yields marbles and calcareous schists that are mineralogically like contact-metamorphosed carbonate rocks.[54] Comparison of the phase assemblages in figures 27.7 and 26.13 reveals that similarity. At low grades of metamorphism, that is, within the Zeolite and Prehnite-Pumpellyite Facies, as well as within much of the Greenschist Facies, at moderate to high values of X_{CO_2}, the phase assemblage in marbles is essentially the same as the primary assemblage (calcite–dolomite–quartz). At low values of X_{CO_2}, Greenschist Facies assemblages may contain phases such as ankerite, talc, and chlorite (compare the Albite-Epidote Hornfels Facies, figure 26.13). At higher metamorphic grades, phases such as tremolite, phlogopite, diopside, idocrase, scapolite, garnet, hornblende, spinel, and forsterite will appear, depending on the pressure temperature, and the compositions of both the rock and the fluid phase.

In a similar way, the phase assemblages of ultramafic rocks parallel those of equivalent contact metamorphic assemblages.[55] Representative topologies for the various facies of metamorphism are shown in figure 27.7 and representative assemblages are listed in table 27.3. At low grades of metamorphism, serpentines (lizardite and chrysotile) typically are the dominant phases, with variable amounts of talc, carbonate minerals, and magnetite comprising the remainder of the mineralogy. Likewise, serpentine (antigorite) is commonly the principal phase within ultramafic rocks of the Greenschist Facies, but talc and chlorite may predominate in some schists. Talc, tremolite, chlorite, magnetite, brucite, and magnesite are important accessory minerals in Greenschist Facies ultramafic rocks (B. W. Evans, 1977).

Several reactions clearly mark the facies boundaries in terranes with ultramafic rocks. The reaction that marks the transition from the Zeolite to the Prehnite-Pumpellyite Facies is reaction equation 26.26:

$$17 \text{ chrysotile} \Longleftrightarrow \text{antigorite} + 3 \text{ brucite.}$$

Serpentine may convert to forsterite in the upper part of the Greenschist Facies, near the beginning of the Amphibolite Facies, via reaction 26.27 (page 546). The upper limit of the Greenschist Facies is marked by a terminal reaction for antigorite:

$$\text{antigorite} \Longleftrightarrow 18 \text{ forsterite} + 4 \text{ talc} + 27 \text{ H}_2\text{O} \quad (27.24)$$

(Johannes, 1975). Within the Amphibolite Facies, anthophyllite and enstatite appear. Anthophyllite forms via reaction equation 26.28 and enstatite forms through either the reaction

$$\text{anthophyllite} + \text{forsterite} \Longleftrightarrow 9 \text{ enstatite} + \text{H}_2\text{O} \quad (27.25)$$

Table 27.3 Selected Metamorphic Mineral Assemblages in Barrovian Facies Series Ultramafic Rocks

Zeolite Facies

chrysotile–magnetite

chrysotile–talc–calcite–magnetite

chrysotile–lizardite–magnetite–brucite–dolomite

lizardite–chrysotile–chlorite–magnetite–magnesite–dolomite

Prehnite-Pumpellyite Facies

chrysotile–lizardite–brucite–chlorite–magnetite

antigorite–lizardite–tremolite–chlorite–magnetite

Greenschist Facies

antigorite–magnesite–magnetite

antigorite–chlorite–talc–tremolite–magnetite

antigorite–forsterite–diopside–chlorite–magnetite

Amphibolite Facies

forsterite–anthophyllite–talc–chlorite–magnetite

anthophyllite–tremolite–talc–chlorite–magnetite

forsterite–enstatite–chlorite–tremolite–chromite

Granulite Facies

forsterite–enstatite–diopside–spinel–chromite

Sources: See note 55.

at low pressures (< 6.5 kb) or

$$\text{talc} + \text{forsterite} \Longleftrightarrow 5 \text{ enstatite} + \text{H}_2\text{O} \quad (27.26)$$

at high pressures (H. J. Greenwood, 1963).[56] In the Granulite Facies, anthophyllite and talc are absent, and assemblages containing coexisting enstatite, diopside, and forsterite characterize the rocks (B. W. Evans, 1977).

Example: Barrovian Metamorphism in the Southern Appalachian Orogen

The southern Appalachian Orogen extends from central Virginia to Alabama (figure 27.12a). It is a complex orogenic belt, parts of which have experienced regional metamorphism during each of four orogenic events. The ages of these events are Mesoproterozoic, Ordovician (the Taconic Orogeny), Devonian-Mississippian (the Acadian Orogeny), and Pennsylvanian-Permian (the Alleghanian/Appalachian

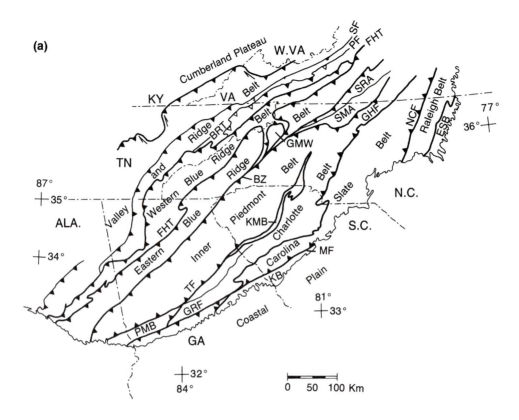

Figure 27.12 Maps of the Southern Appalachian Orogen.
(a) Structural map of the Southern Appalachian Orogen
showing major lithotectonic belts (based on many sources,
including Hatcher et al., 1979). From northwest to southeast,
abbreviations are as follows: SF = Saltville Fault, PF = Pulaski
Fault, BRT = various faults of the Blue Ridge Thrust Fault
system, FHT = Fries-Hayesville Fault, GMW = Grandfather
Mountain Window, BZ = Brevard Zone, SRA = Smith River
Allochthon, SMA = Sauratown Mountains Anticlinorium, TF
= Towaliga Fault, PMB = Pine Mountain Belt, KMB = Kings

Mountain Belt, GHF = Gold Hill Fault, GRF = Goat Rock Fault,
MF = Modoc Fault, KB = Kiokee Belt, NCF = Nutbush Creek
Fault Zone, and ESB = Eastern Slate Belt.
(b) Paleozoic metamorphic facies map for the Southern
Appalachian Orogen. The ages of the metamorphic events
that produced the various facies are not the same throughout
the orogen.
*(Sources include Abbott and Raymond (1984), Absher and McSween
(1985), Bartholomew (1983), Bearce (1973, 1982), Black and Fullagar
(1976), S. E. Boyer (1978), S. E. Boyer and Elliott (1982), B. Bryant*

Orogeny).[57] R. H. Carpenter (1970) and others[58] have delineated various zones of metamorphic rock and L. Glover et al. (1983), Hatcher (1987), and J. R. Butler (1991) have attempted to discriminate zones associated with the individual orogenies. A new map based on these and other works is shown in figure 27.12b.

While the Southern Appalachian Orogen is one of the major regions of Barrovian Facies Series rocks in North America, analysis of the metamorphism there has been confounded by several factors. First, the various tectonic belts (terranes) in the southern Appalachian Orogen have been juxtaposed by significant movements of various types along major faults—in several cases, *after* metamorphism had oc-

curred (Hatcher, 1978, 1987; Abbott and Raymond, 1984; Secor, Snoke, and Dallmeyer, 1986; Vauchez, 1987a, 1987b). This problem is particularly significant in the central and eastern parts of the Orogen. Second, the thermal significance of various metamorphic zones is open to question.

Within individual belts, various Paleozoic metamorphic events have been attributed either to a series of thermal maxima associated with the individual orogenic events[59] (figure 27.13a) or to cooling and uplift associated with a single event[60] (figure 27.13b). The first of these models requires that the metamorphic zones, as they now exist, are diachronous and polymetamorphic. In the west, peak metamorphism occurred during the Taconic Orogeny. Yet

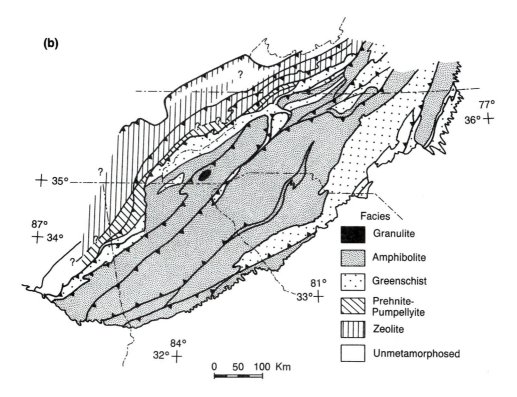

(b)

Facies

- ■ Granulite
- ░ Amphibolite
- ∷ Greenschist
- ▨ Prehnite-Pumpellyite
- ▥ Zeolite
- □ Unmetamorphosed

0   50   100 Km

....

Figure 27.12 Continued

and Reed (1970a, 1970b), J. R. Butler (1972, 1973; 1984 on Brown et al., 1985, 1991), R. H. Carpenter (1970), Conley (1987), Dallmeyer (1975a, 1975b, 1988), Dallmeyer et al. (1986), Dietrich, Fullagar, and Bottino (1969), Drake et al. (1989), Eckert et al. (1989), Espenshade et al. (1975), Farrar (1984), Force (1976), Fullagar and Dietrich (1976), Fullagar et al. (1980), Gillon (1989); Glover et al. (1983), Gulley (1985), Hadley and Nelson (1971), Hatcher (1976, 1987), Hatcher et al. (1979, 1980), Hatcher et al. (1989), Hopson, Hatcher, and Stieve (1989); Horton and Stearn (1983), V. J. Hurst (1970, 1973), Kish (1990), L. E. Long, Kulp, and Eckelmann (1959), McConnell and Costello (1984), Mohr and Newton (1983), W. J. Morgan (1972), Mose and Nagel (1984), Nesbitt and Essene (1982), Noel, Spariosu, and Dallmeyer (1988), Osberg et al. (1989), Rankin, Espenshade, and Neuman (1972), Rankin, Espenshade, and Shaw (1973), G. S. Russell, Russell, and Farrar (1985), J. W. Sears and Cook (1984), Secor et al. (1986), Secor, Snoke, and Dallmeyer (1986), Sinha and Glover (1978), Snoke, Kish, and Secor (1980), Sundelius (1970), Tull (1980, 1982), C. E. Weaver et al. (1984), and unpublished observations of the author. See Morgan (1972), Hatcher (1987), and Drake et al.(1989) for additional relevant sources.)

most zones of Acadian- and Alleghenian-age metamorphism in the west and many in the central part of the orogen are of Greenschist or lower grade (L. Glover et al., 1983). This evidence is compatible with the hypothesis that post-Taconic metamorphism in the west-central orogen may have resulted from monotonic cooling. Such a cooling history, resulting from uplift punctuated by periods of deformation and local heating along fault zones, also suggests that the relatively regular zones of Barrovian Facies Series metamorphism present today are diachronous, but not polymetamorphic (in the sense that thermal maxima developed repeatedly). In the east and locally in the central and western parts of the orogen, thermal maxima

did yield Amphibolite Facies conditions during the Acadian deformation (Dallmeyer et al., 1986; Dallmeyer, 1988; Kish, 1989). Throughout the orogen, analysis of the metamorphic history is complicated where such younger metamorphic events have overprinted the mineral assemblages and textures generated by older events (J. R. Butler, 1973b, 1991; Abbott and Raymond, 1984).

Barrovian metamorphism in the *western* part of the orogen is the focus of this example.[61] In this region, strike-slip faulting is insignificant and the isograds are primarily cut by thrust faults that have telescoped, but not laterally offset, the metamorphic zones. A map of the orogen, showing the approximate positions of metamorphic facies of Paleozoic

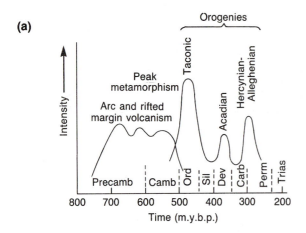

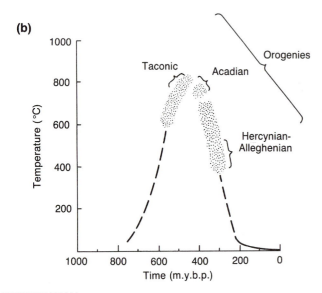

Figure 27.13 Graphs showing the timing of Paleozoic metamorphism and orogeny in the Southern Appalachian Orogen. (a) Time vs. intensity plot showing multiple peaks of metamorphism and orogenic activity (modified from Hatcher et al., 1979). (b) Time vs. temperature plot showing a single metamorphic peak and monotonic cooling of the orogen punctuated by orogenically induced recrystallization. *(Modified from Abbott and Raymond, 1984)*

age, is presented in figure 27.12b. A broad range of rock types exists in the region, but carbonate rocks, especially impure carbonate rocks, are relatively rare in the higher-grade parts of the metamorphic belt, whereas mafic and ultramafic rocks are rare to nonexistent in the low-grade zones.

Rocks of the Zeolite and Prehnite-Pumpellyite Facies occur primarily in the Valley and Ridge Belt, but some occur along the northwestern edge of the Blue Ridge Belt. At these lowest grades of metamorphism, the pelites are characterized by clays and the carbonate rocks by calcite and/or dolomite ± quartz (figures 27.14a and 27.15). Rare basic rocks contain chlorite, calcite, and locally may contain rare pumpellyite. Greenschist Facies assemblages are distributed in the western Blue Ridge Belt and in structural windows in that belt. Rocks of this grade consist of younger (Neoproterozoic to Cambrian) sedimentary and igneous rocks and older (Mesoproterozoic) polymetamorphic rocks. Quartz-rich metaclastic rocks typically contain the assemblage quartz–white mica–chlorite–alkali feldspar–magnetite in this facies (figure 27.14b). Quartz-feldspar gneisses, in part probably products of retrograde metamorphism of Precambrian Amphibolite and Granulite facies rocks, contain similar assemblages, plus assemblages such as biotite–chlorite–quartz–albite–K-rich alkali feldspar–calcite–pyrite. Pelitic rocks, at the lower grades, typically are composed of the assemblage chlorite–white mica–quartz–albite–magnetite (figure 27.14c). In higher-grade assemblages, garnet is present. Metabasites contain assemblages such as chlorite–epidote–albite–quartz–hematite and actinolite–chlorite–albite–quartz–magnetite. Iron-rich carbonate rock and calcareous metasiltstones contain ferroan dolomite–calcite–quartz–plagioclase-biotite ± white mica (Bryant and Reed, 1970a; Raymond, Neton, and Cook, 1991).

Much of the eastern Blue Ridge Belt is composed of rocks of the Amphibolite Facies. Migmatites are common.[62] Pelitic mica schists consist of various assemblages containing staurolite, kyanite, and sillimanite (figures 27.14d and 27.15). Cordierite is present locally, in the central Blue Ridge,[63] but it is rare and its significance is unknown. Quartzo-feldpathic (SAC) rocks are composed predominantly of the assemblage plagioclase–quartz–biotite–white mica–garnet, and may contain magnetite, ilmenite, and K-rich alkali feldspar (figure 27.15). Some feldspar-rich rocks contain the assemblage plagioclase–gedrite–biotite–epidote–apatite–quartz–ilmenite ± garnet. Mafic rocks are typical amphibole schists and gneisses, with hornblende and plagioclase as the dominant phases. Garnet, epidote, biotite, quartz, ilmenite, magnetite, pyrite, pentlandite,[64] and chalcopyrite are accessory minerals in these rocks (figures 27.14f and 27.15). Diopside-calcite marble is rare. Granite to quartz diorite (trondhjemite) dikes are common throughout the Amphibolite Facies terrane, suggesting that local anatexis was a widespread event.[65] Geothermometry and geobarometry indicate that the Amphibolite Facies rocks of the central core of the Blue Ridge were metamorphosed at temperatures between 500° C and 850° C at pressures of 0.5–1.1 Gpa (5–11 kb).[66]

Paleozoic Granulite Facies rocks have been recognized at only a few localities (Force, 1976; Absher and McSween, 1985; Eckert, Hatcher, and Mohr, 1989). Absher and McSween (1985) give complete descriptions of a full range of

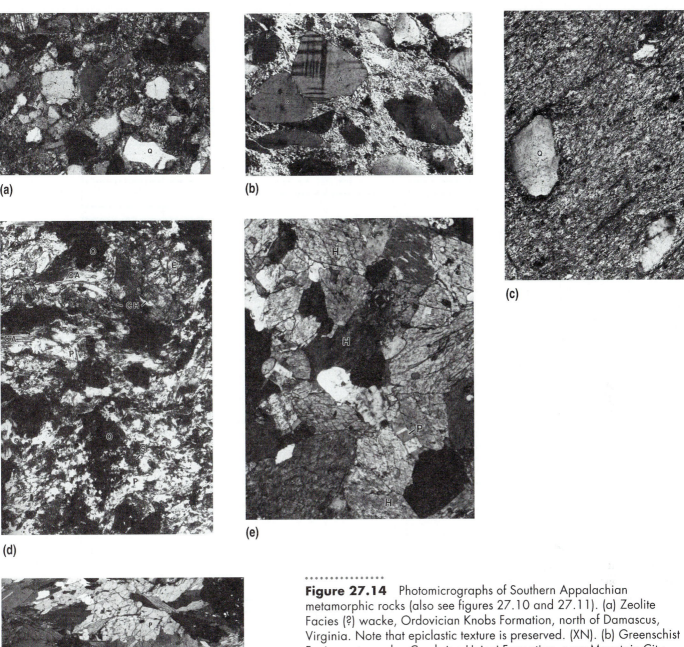

(a)

(b)

(c)

(d)

(e)

(f)

Figure 27.14 Photomicrographs of Southern Appalachian metamorphic rocks (also see figures 27.10 and 27.11). (a) Zeolite Facies (?) wacke, Ordovician Knobs Formation, north of Damascus, Virginia. Note that epiclastic texture is preserved. (XN). (b) Greenschist Facies metawacke, Cambrian Unicoi Formation, near Mountain City, Tennessee. (XN). (c) Greenschist Facies metasiltstone, Cambrian Unicoi Formation, near Mountain City, Tennessee. (XN). (d) Greenschist Facies metabasalt, Grandfather Mountain Formation, near Valle Crucis, North Carolina. (PL). (e) Amphibolite Facies metabasite (amphibolite), near Celo, North Carolina. (XN). (f) Amphibolite Facies pelitic schist, near Todd, Ashe County, North Carolina. (PL). C = calcite, K = kyanite, St = staurolite. Other abbreviations the same as those on earlier figures. Long dimension of photos is 0.33 mm in (a) and (c), 3.25 mm in (b), 2.7 mm in (e), and 1.27 mm in (d) and (f).

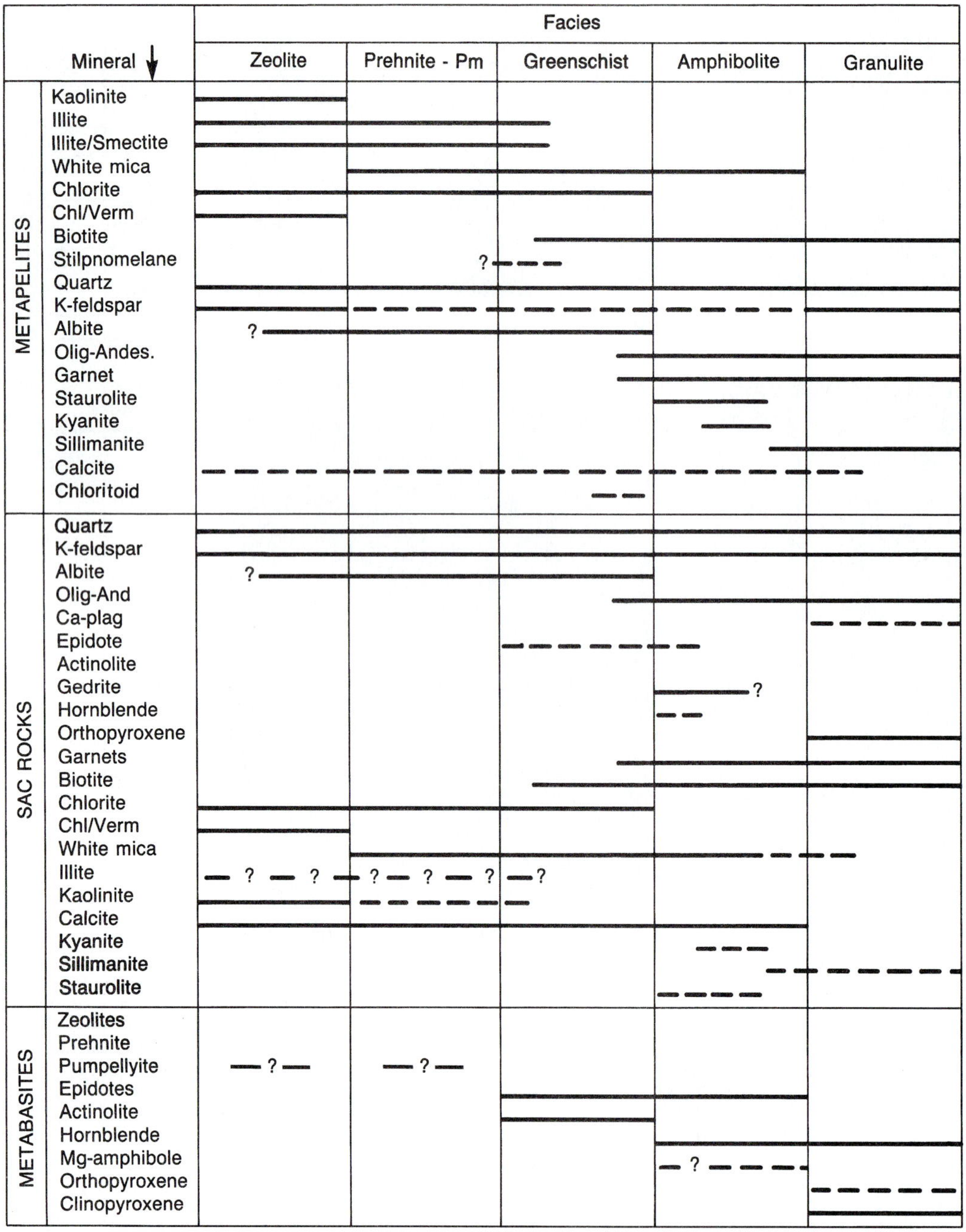

Figure 27.15 Mineral-facies chart for Paleozoic metamorphic rocks of the western edge of the Southern Appalachian Orogen.

[Based on Abbott and Raymond (1984), Absher and McSween (1985), Bearce (1973), Brewer (1986), Bryant and Reed (1970a, 1970b), J. R.Butler (1972, 1973b), R. H. Carpenter (1970), Espenshade et al. (1975), Gillon (1989), Gulley (1985), Hadley and Nelson (1971), Hatcher (1976), Hatcher et al. (1979), Helms et al. (1987), V. J. Hurst (1973), McElhaney and McSween (1983), McSween and Hatcher (1985), Mohr and Newton (1983), Nesbitt and Essene (1982, 1983), Rankin, Espenshade, and Neuman (1972), Rankin et al. (1973), G. S. Russell, C. W. Russell, and Farrar (1985), S. E. Swanson (1980, 1981), Weaver et al. (1984), and unpublished observations of the author.]

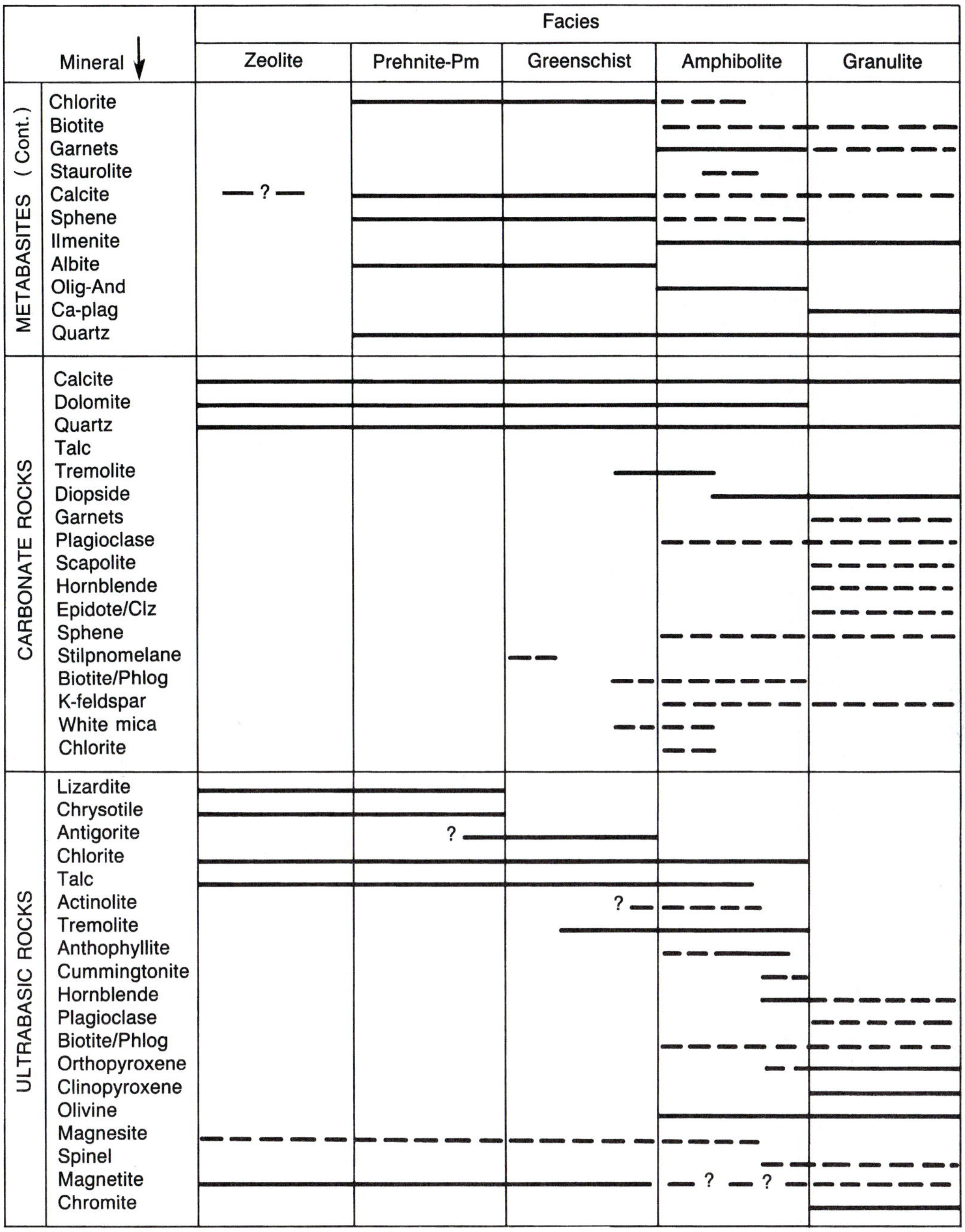

	Mineral ↓	Facies				
		Zeolite	Prehnite-Pm	Greenschist	Amphibolite	Granulite
METABASITES (Cont.)	Chlorite					
	Biotite					
	Garnets					
	Staurolite					
	Calcite					
	Sphene					
	Ilmenite					
	Albite					
	Olig-And					
	Ca-plag					
	Quartz					
CARBONATE ROCKS	Calcite					
	Dolomite					
	Quartz					
	Talc					
	Tremolite					
	Diopside					
	Garnets					
	Plagioclase					
	Scapolite					
	Hornblende					
	Epidote/Clz					
	Sphene					
	Stilpnomelane					
	Biotite/Phlog					
	K-feldspar					
	White mica					
	Chlorite					
ULTRABASIC ROCKS	Lizardite					
	Chrysotile					
	Antigorite					
	Chlorite					
	Talc					
	Actinolite					
	Tremolite					
	Anthophyllite					
	Cummingtonite					
	Hornblende					
	Plagioclase					
	Biotite/Phlog					
	Orthopyroxene					
	Clinopyroxene					
	Olivine					
	Magnesite					
	Spinel					
	Magnetite					
	Chromite					

Figure 27.15 Continued

rock types present in what is arguably a melange metamorphosed to this grade. Aluminous schist consists of biotite–garnet–sillimanite–K-feldspar–andesine–quartz–magnetite–ilmeno-hematite (figure 27.15). SAC rocks contain assemblages such as andesine–quartz–K-feldspar–biotite–garnet–magnetite–ilmenite. A typical metabasite assemblage is hornblende–bytownite–biotite–orthopyroxene–quartz–magnetite–ilmenite. Calc-silicate gneisses consist of calcite–quartz–scapolite–bytownite, with accessory minerals that include hornblende, garnet, diopside, clinozoisite, sphene, and apatite. Ultramafic rocks contain such assemblages as orthopyroxene–andesine–biotite–hornblende–cummingtonite–quartz–pyrrhotite–pyrite–chalcopyrite. It is likely that the quartz, andesine, biotite, hornblende, and cummingtonite are retrograde phases. Granitoid veins in many of the rocks suggest that, where fluids were available, anatexis occurred. Given that the estimated P-T conditions ($T = 750$–$775°$ C, $P = 0.65$–0.7 Gpa, or 6.5–7 kb) do not differ significantly from those for Amphibolite Facies metamorphism in the area, the zones of Granulite Facies

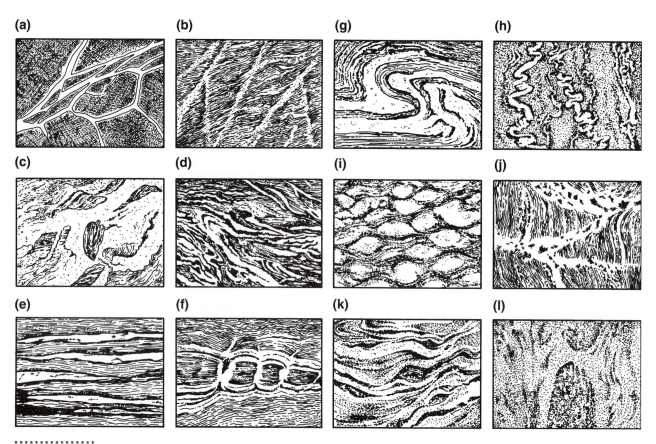

(a)   **(b)**   **(g)**   **(h)**

(c)   **(d)**   **(i)**   **(j)**

(e)   **(f)**   **(k)**   **(l)**

Figure 27.16   Sketches showing the various structural types of migmatites: (a) breccia (agmatic) structure, (b) diktyonitic structure, (c) raft (schollen) structure, (d) vein (phlebitic) structure, (e) layered (stromatic) structure, (f) boudinage (surreitic) structure, (g) folded structure, (h) ptygmatic structure, (i) augen (opthalmitic) structure, (j) stictolithic (fleck) structure, (k) Schlieren structure, (l) nebulitic structure.
(After Mehnert, 1968)

metamorphism probably represent local areas in which the rocks were dehydrated by previous or prograde metamorphic events.

Because the overall metamorphic pattern in the Southern Appalachian Orogen developed over a long period of time (i.e., it is diachronous), it is difficult, especially in both low-grade and thoroughly recrystallized metamorphic rocks, to discern the *complete* patterns of metamorphism *associated with each orogenic event*. In the western part of the orogen, that problem is increased where thrust faults have shortened the width of the orogen, concealing sections of the metamorphic belt. Nevertheless, the elongate metamorphic zones are typical of orogenic Barrovian Facies Series metamorphic belts.

MIGMATITES

Migmatites are quite common in the highest-grade terranes of Buchan and Barrovian Facies series. **Migmatites** are masses of crystalline, mixed rocks, consisting of various pro-

portions of dark, ferromagnesian mineral-rich rock and light quartz- or feldspar-rich rock, that occur in medium- to high-grade metamorphic terranes (Sederholm, 1967).[67] The light-shaded rock is referred to as the *leucosome*, whereas the dark-shaded rock is referred to as the *melanosome*. Together, the leucosome and melanosome comprise new rock formed by migmatization processes, rock referred to collectively as the **neosome.** Leucosomes commonly have a nonfoliated, igneous-like appearance, whereas melanosomes are usually foliated. The **mesosome** is rock typically of intermediate shade that has the appearance of ordinary metamorphic rock (Ashworth, 1985a). Some mesosomes may represent the metamorphic protolith of the migmatite. **Restite** is a term applied to the residual rock remaining after leucosome has been removed from a protolith.

Twelve types of migmatite structures are recognized (figure 27.16) (Mehnert, 1968, ch. 2). Among the more common of these are breccia (agmatic) structure, raft (schollen) structure, layered (stromatic) structure, vein (phlebitic) structure, folded structure, augen (opthalmitic) structure, and

Table 27.4 Migmatite-Forming Processes

Process	Open System Required	Open System Not Required
Magmatic Processes		
Magmatic injection	X	
Anatexis		X
Metamorphic Processes		
Metasomatism	X	
Metamorphic differentiation		X
Combination Processes		
Structural-metamorphic processes		
Melange formation + metamorphism	X	
Ductile deformation yielding boudinage, folding, or layering during metamorphism		X
Other combinations (many combinations are possible, including the following.)		
Melange formation + metamorphism + magmatic injection	X	
Ductile deformation followed by anatexis		X

Source: Based on Ashworth (1985a) and Raymond, Yurkovich, and McKinney (1989).

boudinage (surreitic) structure. In breccia structure, angular blocks of melanosome, mesosome, or protolith, the edges of which correspond, are separated by thin veins of leucosome. In contrast, in raft structure, commonly rounded and/or rotated blocks of darker material are enclosed in larger masses of leucosome. Augen, folded, boudinage, and layered structures have their usual meanings. In some cases, many of these structures may be found within a single outcrop.[68]

The textures of migmatites are typical of textures in other kinds of metamorphic and igneous rocks (Yardley, 1978; McLellan, 1983a; Ashworth and McLellan, 1985). Leucosomes are commonly medium- to very coarse-grained, are coarser-grained than adjacent nonleucosomes, and are typically hypidiomorphic-granular. Aplitic (allotriomorphic-granular) leucosomes also exist. Where leucosomes are deformed, polygonization of grain boundaries and other metamorphic textures overprint the igneous textures. Melanosomes and mesosomes are commonly fine- to medium-grained, lepidoblastic to nematoblastic foliated rocks, but granoblastic varieties occur in some terranes (Kenah and Hollister, 1983). Extraction of melt from some mesosomes apparently is accompanied by grain-size reduction in certain minerals, notably the micas (Dougan, 1983).

A number of processes have been suggested to account for the origin of various migmatites (table 27.4). These may be divided into single-stage and multistage processes. Both categories include processes that require open systems and those that do not (Ashworth, 1985a).[69] Magmatic injection involves the emplacement and crystallization of small layers, lenses, dikes, or sills of leucosomal magma between masses of darker material. Where one phase of magmatic activity is responsible for the migmatite, the migmatization is a single-stage, open-system process. Where injection has occurred more than once, at significantly different times, the origin of the migmatite is diachronous and multistage.

Recall that metasomatic processes involve chemical change promoted by chemically active fluids derived either from nearby intrusions or from the country rock.[70] By definition, metasomatism is also an open-system process of migmatization.

Processes that do not require an open system (at the scale of ~ 1 m^3) include anatexis and metamorphic differentiation, both of which may be single-stage or multistage, "semiclosed" system processes.[71] Additional multistage processes include combinations of the above processes, plus combined structural-metamorphic processes that involve early histories of fragmentation by ductile or brittle deformation, followed by later events involving either regional metamorphism or one or more of the four migmatite-forming processes mentioned above (Raymond, Yurkovich, and McKinney, 1989).

In summary, examination of the various processes invoked to explain the origin of migmatites reveals that igneous, metamorphic, and structural processes are involved. Magmatic injection and anatexis are magmatic by definition, whereas metasomatism and metamorphic differentiation are metamorphic by definition. Structural processes include ductile deformation, which yields folded structure and boudinage, and brittle deformation, which produces breccia structure and melange protoliths for the migmatites.[72] Although ductile deformation is usually associated with regional metamorphism and brittle deformation is not, both may predate the regional metamorphic event that gives the final textural, structural, and mineralogical character to the migmatite.

Magmatic injection may arise either through emplacement of magmas formed elsewhere and intruded into a terrane (Sawyer, 1987) or from crystallization of locally derived anatectic melts (S. N. Olsen, 1982). In the former case, there is not a genetic relationship between the mesosome or melanosome and the intruded leucosome. The leucosome may have any composition.

In the case of leucosomes of local anatectic-magmatic origin, an unambiguous genetic relationship will exist *if* (1) crystal fractionation did not occur during melt migration and (2) postcrystallization metasomatism is absent. An anatectic-magmatic origin for leucosomes must be demonstrated, not only by a genetic link, but by chemical and/or experimental data that indicate both melt compositions and a melt origin for the leucosome. The experimental work of Tuttle and Bowen (1958) clearly delineated the composition of ternary minimum melts in the haplogranite system $(SiO_2\text{-}NaAlSi_3O_8\text{-}KAlSi_3O_8\text{-}H_2O)$ (refer to figure 5.19, and Luth, Johns, and Tuttle, 1964).[73] The addition of calcic plagioclase to that system and increased degrees of partial melting may shift the compositions of anatectic melts away from the ternary minimum towards granodioritic or dioritic compositions (G. C. Brown and Fyfe, 1972). Similarly, melting experiments on shales yield melts of granitic to granodioritic composition (Wyllie and Tuttle, 1961). In any case, Johannes (1983b) found that the normative CIPW compositions of several leucosomes plot near either the eutectic or the quartz-orthoclase cotectic of the haplogranite system at 0.05 to 0.5 Gpa (0.5–5 kb) (figure 27.17a). This indicates near-minimum and cotectic melt compositions for the leucosomes and suggests that anatexis *may have been* a controlling process in the development of the migmatites studied.

Mass balance (the equivalence of *quantities* of elements) is also needed, in addition to genetic links between leucosome and mesosome or melanosome, to demonstrate a local anatectic origin. For example, a linear chemical relationship exists be-

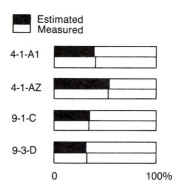

Figure 27.17 Evidence bearing on the origin of migmatites. Results of mass balance calculations on closed-system migmatites of the Front Range of Colorado (from S. N. Olsen, 1983). Rocks metamorphosed in closed systems will have estimated (calculated) volumes of leucosome plus marginal selvage material equal to the volume of the paleosome measured on slabs of the rock (as they do here). In open systems, the estimated and measured values will be significantly different, because some mass will have moved into or out of the system. See the works of S. N. Olsen (1977, 1982, 1983, 1985) for more detailed explanations of the procedures used to make such mass balance calculations.

tween leucosome, mesosome, and melanosome from migmatites of the Central Gneiss Complex of British Columbia (Kenah and Hollister, 1983), but simple anatectic formation of the leucosome is not indicated, because mass balance considerations preclude the derivation of the leucosome and melanosome solely and directly from the mesosome by simple separation of an anatectic melt. This is so, because the *amounts* of silica and other elements are not consistent with such a derivation. Instead, control of the composition of the migmatite layers either by the initial composition of the protolith layers or by postmelting fractionation processes is required to explain the differences in the bulk chemistry.

In some Colorado Front Range and Baltimore Gneiss Dome migmatites, mass balance *is* indicated (S. N. Olsen, 1977, 1982, 1985). The composition of the leucosome plus that of the melanosome selvage (bordering the leucosome) equals the composition of the mesosome, and their sum is equal to the predicted (estimated) value (figure 27.17b). That equality also suggests that the mesosome *was* the protolith for the migmatite and that the leucosome was derived directly from the mesosome. Yet it does not demonstrate an anatectic origin for the leucosome, because such chemical relationships are consistent with both anatectic and metamorphic differentiation origins for migmatites. Additional compositional and textural evidence is needed to demonstrate anatexis.

Demonstration of a metamorphic origin for a leucosome requires the demonstration that magmas were not involved. In addition, one must show either that metasomatism has occurred or that metamorphic differentiation is responsible for compositional equivalence of leucosome + melanosome and protolith (or mesosome). The absence of a magma is suggested by the lack of igneous textures in the rocks and a composition different than minimum melt compositions (Misch, 1968). Metasomatism is indicated where mass balance calculations show a lack of equivalence between leucosome + melanosome and mesosome (S. N. Olsen, 1983, 1985).

The preponderance of evidence, including that provided by recent major and trace element analyses, seems to indicate an anatectic origin or a combined anatectic-metasomatic origin for most migmatites.[74] Nevertheless, some migmatites of apparent metamorphic origin do exist (McLellan, 1983a).

SUMMARY

Buchan and Barrovian Facies series are moderate to high T/P facies series of regional metamorphism. Both types of facies series develop in orogenic belts formed at convergent plate margins. Here, regional heat flow in the orogen is increased by upward movement of magmas and migration of metamorphic fluids. Thrust faulting thickens the crust, increasing pressure and telescoping previously formed metamorphic zones. Buchan Facies Series may also develop in association with divergent (rift) margins or major fault zones where crust or mantle are thinned.

The high temperatures at relatively low pressures, characteristic of Buchan metamorphism, yield mineral assemblages typified by a critical aluminum silicate sequence of kaolinite → pyrophyllite → andalusite → sillimanite. The association of andalusite with cordierite in the middle grades of metamorphism is particularly definitive of Buchan metamorphism. In contrast, Barrovian Facies Series are characterized by the critical sequence of aluminum silicates: pyrophyllite → kyanite → sillimanite. Calcium silicates of the middle to upper grades of metamorphism include garnet and plagioclase feldspar, whereas the phyllosilicates biotite and muscovite are potassium-bearing phases. Both Buchan and Barrovian Facies series consist of various assemblages composed of these and other minerals representing Zeolite to Granulite Facies conditions. Together, the combined subassemblages for various bulk compositions of rock indicate geothermal gradients in the range 20° C to 40° C/km for the Barrovian Facies Series and 40° C to 80° C/km for the Buchan Facies Series.

Most of the important reactions that occurred during the formation of Barrovian Facies Series are continuous reactions involving such phases as white micas, chlorites, biotites, garnets, and Ca-amphiboles. Similar reactions occurred as Buchan Facies Series formed. Some discontinuous reactions, as well as polymorphic changes, such as kyanite <==> sillimanite, are important locally.

Buchan rocks develop at higher pressure and lower maximum temperature than typical Contact Facies Series rocks. Yet there is considerable overlap between the conditions of formation of the two facies series. Mineralogically, the rocks of the Buchan Facies Series are quite similar to those of the Contact Facies Series, but the Buchan rocks are typically foliated and distributed over terranes of regional extent.

Migmatites are common in high-grade zones of all of the facies series of moderate to high temperature. Various types of migmatites are recognized on the basis of the structural character of the leucosome-melanosome mixtures. Where hydrous fluids are available and temperatures exceed the minimum melting temperature of wet granitoid rocks, the low melting (granitoid) fraction of the rocks will melt, coalesce, and migrate as a magma, crystallizing to form a light-colored, granitoid leucosome interlayered or intermixed with a dark-colored, refractory melanosome. The latter is composed of various combinations of minerals including amphiboles, biotite, calcic plagioclase, and quartz. Some leucosomes, particularly some that have unusual compositions and/or exhibit metamorphic, rather than igneous textures, may have had an origin by metasomatism or metamorphic differentiation. Other migmatites form via intense ductile flow and boudinage, in some cases coupled with folding. Still other migmatites may have had a premetamorphic history of fragmentation and mixing of felsic and mafic materials. Whatever their early history, migmatites generally owe their final character to magmatism and/or metamorphism, with attendant recrystallization and neocrystallization, under Amphibolite or Granulite Facies conditions.

EXPLANATORY NOTES

1. This is a generalization, as there is a definite overlap in the conditions of regional and contact metamorphism (Pattison and Tracy, 1991). Nevertheless, it is *generally* true that most contact metamorphism occurs at $P < 4$ kb, whereas most regional metamorphic terranes develop at pressures in the range of 3–12 kb.
2. Refer to chapter 1. Also, Wyllie (1971a, pp. 30–32) summarizes older work on the geothermal gradient of the Earth.

3. See note 2 and Ernst (1972b).
4. Also see Harker (1932, p. 230ff.), Chinner (1966),
   J. A. Winchester (1974), Harte and Hudson (1979, 1980),
   A. J. Baker and Droop (1983), A. J. Baker (1985), and
   N. F. C. Hudson (1985).
5. For example, see Lux et al. (1986), Osberg et al. (1989), and
   the description of the Buchan Facies Series of New England
   in this chapter.
6. See Miyashiro (1973a) for references. See Bard (1967a,
   1967b, 1969, 1970) and Bard and Moine (1979) for
   descriptions of a Buchan Facies Series in the Aracena, Spain
   area. Also see Vernon (1978) and Morand (1990) for
   Australian examples.
7. The exact P-T history of the Buchan rocks in Scotland is not
   altogether clear, as there seems to be evidence, in the form of
   andalusite replacing and being replaced by other aluminum
   silicates, that episodes of Barrovian metamorphism both
   predated and postdated the Buchan event. For the details of
   this problem, the interested reader may refer to Harte and
   Hudson (1979), Chinner (1980), A. J. Baker (1985), and
   Beddoe-Stephens (1990).
8. Harker (1932), Chinner (1966), Porteous (1973),
   J. A. Winchester (1974), Harte and Hudson (1979), N. F. C.
   Hudson (1980, 1985), A. J. Baker (1985), Leslie (1988).
9. The assemblages listed here are those described from terranes
   that form under identical conditions. For example, see
   Dunoyer de Segonzac (1970), M. Frey (1970, 1986, 1987b),
   C. E. Weaver, Beck, and Pollard (1971), Zen (1974b), Zen
   and Thompson (1974), Bathurst (1975), Hower et al. (1976),
   Kubler et al. (1979), Seki and Liou (1981), C. E. Weaver et
   al. (1984), and Kisch (1987), for descriptions of assemblages
   in this grade of rocks and for related discussions.
10. Hower et al. (1976), J. Hoffman and Hower (1979), Eslinger
    and Sellars (1981), E. C. Thornton and Seyfried (1985). Also
    see Grim (1968).
11. Zen (1961), J. Hoffman and Hower (1979), Juster (1987).
12. M. Frey (1970), C. E. Weaver, Beck, and Pollard (1971),
    Ghent (1979), C. E. Weaver et al. (1984, p. 191).
13. This reaction is generally consistent with the observations of
    Weaver, Beck, and Pollard (1971) and Hower et al. (1976).
    Also see experimental work by B. M. Sass, Rosenberg, and
    Kittrick (1987) and Aja, Rosenberg, and Kittrick (1991).
14. Also see Carmignani et al. (1982). McNamara (1966) and
    E. H. Brown (1975) discuss various biotite-forming reactions.
    See Guidotti and Sassi (1976) for a discussion of the
    variations in white mica chemistry as related to metamorphic
    grade.
15. See Osberg (1968, 1971, 1979), Osberg, Moench, and
    Warner (1968), Guidotti (1966, 1968, 1970a,b, 1973, 1978),
    Guidotti, Robinson, and Conatore (1975), Guidotti and

Cheney (1989) Guidotti et al. (1991), and Ferry (1976,
1980a, 1981, 1982, 1983a, 1984), and Ferry and Osberg
(1989). Also see Billings (1937), Albee (1968), J. B.
Thompson and Norton (1968), Boone (1970), Cheney and
Guidotti (1973, 1979), Guidotti et al. (1973), Bickel (1974),
Osberg (1974), C. T. Foster (1977), Guidotti, Robinson, and
Guggenheim (1977), Dallmeyer (1979), Laird (1980),
Holdaway et al. (1982), Lux et al. (1986), Holdaway,
Dutrow, and Hinton (1988), DeYoreo et al. (1989),
A. A. Drake et al. (1989), Osberg et al. (1989), and Hatcher
et al. (1989).
16. W. J. Morgan (1972) depicts some areas with the
    subassemblage sillimanite + K-feldspar, as Granulite Facies
    terranes. However, the presence of coexisting white mica in
    the assemblage sillimanite–K-feldspar–muscovite
    (B. W. Evans and Guidotti, 1966; Cheney and Guidotti,
    1979; Ferry, 1980b) indicates that these rocks actually
    represent the Amphibolite Facies.
17. Pavlides (1962, 1973), T. H. Clark and Eakins (1968),
    D. S. Coombs, Horodyski, and Naylor (1970), Richter and
    Roy (1974), J. R. Walker (1989).
18. Dallmeyer (1979), Holdaway et al. (1982), Lux et al. (1986),
    Hubacher and Lux (1987), Holdaway, Dutrow, and Hinton
    (1988), Spear and Harrison (1989).
19. Guidotti (1970b), Cheney and Guidotti (1979), Ferry
    (1980b, 1984), Holdaway et al. (1982), Lux et al. (1986),
    Holdaway, Dutrow, and Hinton (1988).
20. Also see the references in note 19.
21. Cheney and Guidotti (1979), Lux et al. (1986). DeYoreo et
    al. (1989), recognizing the thermal conditions, refer to the
    metamorphism as "deep-level contact metamorphism."
22. Migmatites are complexly layered metamorphic rocks with
    light-colored layers that represent partial melts. They are
    discussed at the end of this chapter.
23. J. B. Thompson and Norton (1968), Albee (1968), Hietanen
    (1956, 1968), Grambling and Williams (1985), Dusel-Bacon
    and Foster (1983).
24. Barrow (1893, 1912), Tilley (1925), Harker (1932),
    Wiseman (1934), Billings (1937), Chinner (1960, 1966,
    1978, 1980), Zen (1960), Albee (1968), J. B. Thompson and
    Norton (1968), Porteous (1973), Vidale (1974), J. A.
    Winchester (1974b), Zen (1974b), Thompson et al. (1977),
    Mason (1978), Rumble (1978), Abbott (1979b), Harte and
    Hudson (1979), P. R. A. Wells (1979), Laird (1980), Laird
    and Albee (1981), A. J. Baker and Droop (1983), C. M.
    Graham et al. (1983), A. J. Baker (1985), McLellan
    (1985a,b), Moles (1985), Spear and Rumble (1986),
    Hollocher (1987), Juster (1987), K. P. Watkins (1987), A. A.
    Drake et al. (1989), Hatcher et al. (1989), and Osberg et al.
    (1989). For the central and southern Appalachian Orogen,

see Stose and Stose (1957), Dietrich (1959), Bryant and Reed (1970a), R. H. Carpenter (1970), Fisher (1970), Sundelius (1970), Hadley and Nelson (1971), B. A. Morgan (1972), J. R. Butler (1972, 1973b), Rankin, Espenshade, and Neuman (1972), Rankin, Espenshade, and Shaw (1973), V. J. Hurst (1973), Dallmeyer (1975b), Espenshade et al. (1975), Hatcher et al. (1979), Hatcher et al. (1980), W. A. Thomas et al. (1980), Bearce (1982), Nesbitt and Essene (1982), L. Glover et al. (1983), Abbott and Raymond (1984), C. E. Weaver et al. (1984), Absher and McSween (1985), Farrar (1985), Gully (1985), Secor et al. (1986b), Hatcher (1987), A. A. Drake et al. (1989), Hatcher et al. (1989), and Osberg et al. (1989).

25. Miyashiro (1973a) reviews some occurrences of Barrovian Facies Series rocks, including a possible occurrence in the Hida Complex of Japan (p. 361). Asian metamorphic belts are reviewed in Sobolev, Lepezin, and Dobretsov (1965). References to particular occurrences include the following reports and additional sources may be found in the reference lists of these works: *Rocky Mountains (U.S.)*—Hietanen (1968), J. Hoffman and Hower (1979), Hyndman (1980), Lang and Rice (1985a,b); *British Columbia*—Crosby (1968), Pigage (1976), Simony et al. (1980); *Southeast Alaska*—Himmelberg, Drew, and Ford (1991); *Venezuela*—B. A. Morgan (1970); *Sistema Central, Spain*—Ruiz, Aparicio, and Cacho (1978); *Calabria, Italy*—Paglionico and Piccarretta (1978), Maccarrone et al. (1983), Schenk (1984); *Alps*—Hoernes (1973), Fry et al. (1974), C. Miller (1977), Ernst (1979), Zingg (1980); *Menderes Massif, Turkey*—Evirgen and Ashworth (1984); *The Himalayan Range*—Misch (1964), Dobretsov (1965), Frank et al. (1973), Kumar (1978, 1981), Lal, Mukerji, and Ackermand (1981), various papers in Saklani (1981), and Pognante and Lombardo (1989); *The Ural Mountains, Taimyr Fold Belt, and other Asian occurrences*—Lepezin (1965b).

26. For additional information, see the following sources and the works listed in their reference lists. *Adirondack Mountains of New York*—Buddington (1939), A. E. G. Engel and Engel (1953, 1958, 1960a,b, 1962), Stoddard (1980), Bohlen, Valley, and Essene (1985), R. L. Edwards and Essene (1988). *Black Hills*—J. A. Noble and Harder (1948), Redden (1963), Redden et al. (1982). *Rocky Mountains*—Eslinger and Sellars (1981). *Labrador*—C. Klein (1978), Callahan (1980), Rivers (1983). *Quebec*—Indares and Martignole (1984). *Ontario*—E. W. Reinhardt (1968).

27. *The former Soviet Union* (e.g., in the Mama-Bodaibo Synclinorium north of Lake Baikal and along the margin of the Siberian Platform)—Lepezin (1965c). *India and Sri Lanka*—Lepezin (1965a), Ghosh (1978), Sharma and MacRae (1981), Hansen et al. (1987). *China*—Lepezin (1965a).

28. In addition to Barrow's work, see Tilley (1925), Harker (1932), Wiseman (1934), Chinner (1960, 1966, 1978, 1980), Mather (1970), J. A. Winchester (1974), Harte and Hudson (1979), A. J. Baker and Droop (1983), C. M. Graham et al. (1983), A. J. Baker (1985), McLellan (1985a,b), Moles (1985), K. P. Watkins (1987), and the summary in Miyashiro (1973a). Mather (1970) discusses Greenschist Facies assemblages and the reaction producing biotite. K. P. Watkins (1987) discusses biotite and garnet-forming reactions.

29. See the depictions of a complete set of topologies for just one type of phase diagram, for partial sets of conditions, in J. B. Thompson and Thompson (1976) and Abbott (1982).

30. Many workers who study low-grade rocks have adopted terms that are different from the traditional facies terms used by metamorphic petrologists. This is because many of the changes that occur in low-grade rocks are gradational changes in the crystallinity of illite, the rank of coal, the vitrinite reflectance, and the proportions of clay minerals. M. Frey (1986), Kisch (1987, 1990), and Blenkinsop (1988) review and compare the available data on these various measures of low-grade metamorphism. The major zones recognized include various zones of diagenesis, the "anchizone" or "anchimetamorphic" zone, and the "epizone" or "epimetamorphic" zone (Kubler, 1967; M. Frey, 1978, 1986, 1987b; Kubler et al., 1979; C. E. Weaver et al., 1984; Kisch, 1987, 1990). The epizone corresponds to the Greenschist Facies. The Prehnite-Pumpellyite Facies generally encompasses the anchizone and the uppermost zone of diagenesis, the middle to upper zones of diagenesis designated by these workers correspond to the Zeolite Facies, as it is defined here.

31. Work on rocks of this grade is summarized in Dunoyer de Segonzac (1970), Zen (1974b), Zen and Thompson (1974), Kubler et al. (1979), Seki and Liou (1981), M. Frey (1987b), and Kisch (1987). Also see, M. Frey (1970, 1986), C. E. Weaver, Beck, and Pollard (1971), Bathurst (1975), Hower et al. (1976), and C. E. Weaver et al. (1984).

32. Hower et al. (1976), J. Hoffman and Hower (1979), Eslinger and Sellars (1981), E. C. Thornton and Seyfried (1985). Also see Keller (1967), Grim (1968), and Kisch (1990).

33. M. Frey (1970, 1978), C. E. Weaver, Beck, and Pollard (1971), Hower et al. (1976), Ghent (1979), J. Hoffman and Hower (1979), C. E. Weaver et al. (1984, p. 191).

34. Also see McNamara (1966).

35. C. E. Weaver (1960), Kubler (1967, 1968), J. Hoffman and Hower (1979). Also see the review in Kisch (1987).

36. For example, see the descriptions in Harker (1932), Brown (1958), Zen (1960), M. L. Crawford (1966), Albee (1968), Frank et al. (1973), Hoernes (1973), C. Klein (1978), Ruiz, Aparicio, and Cacho (1978), Lal, Mukerji, and Ackermand (1981), Rivers (1983), C. E. Weaver et al. (1984), and Ghent, Stout, and Ferri (1989).

37. Also see Carmignani et al. (1982) and note 14.

38. Because the assemblage albite–epidote–hornblende occurs in the *mafic* rocks of some areas (i.e., hornblende appears with albite and epidote) before the appearance of garnet, some workers have proposed the names Epidote Amphibolite Facies and Albite Epidote Amphibolite Facies to designate the rocks with that assemblage (F. J. Turner, 1948, p. 88; Turner, 1958; also see Liou, Kuniyoshi, and Ito, 1974). Garnet-bearing *pelitic* rocks may occur in association with such hornblende–albite–epidote rocks. Here, I acknowledge that a transition zone exists between facies and that assemblages in mafic rocks are marked more by continuous changes in mineral composition and modal abundance of minerals than by discontinuous reactions in which new phases appear (Laird, 1980; J. B. Thompson, Laird, and Thompson, 1982; Spear, 1982). Consequently, a separate facies for the mafic rocks containing the hornblende–albite–epidote assemblage is unnecessary. Here, the disappearance of albite is considered to mark the maximum upper limit of the Greenschist Facies (cf. Fyfe et al., 1958; deWaard, 1959).

39. For example, see Harker (1932), Brown (1958), Chinner (1960), Albee (1968), Frank et al. (1973), Hoernes (1973), Ruiz, Aparicio, and Cacho (1978), Abbott (1979b), Tracy and Robinson (1979), Lal, Mukerji, and Ackermand (1981), Rivers (1983), Abbott and Raymond (1984), E. Wenk and Wenk (1984), Lang and Rice (1985b), and McLellan (1985a).

40. For a sample of alternative reactions, see J. B. Thompson and Norton (1968), Thompson et al. (1977), Karabinos (1985), Lang and Rice (1985), and McLellan (1985).

41. These rocks are discussed in more detail below, under "Migmatites."

42. The confusion surrounding the terms granulite and Granulite Facies is reviewed by Winkler (1979, ch. 16). Charnokite is a term applied in India and elsewhere to rocks of granitoid to ultramafic composition that contain orthopyroxene (see comments in F. J. Turner, 1981, p. 401, and Winkler, 1979, p. 263). Because igneous and metamorphic charnokites may be impossible to distinguish, because the term charnokite encompasses rocks of a wide compositional range, and because the term signifies little more than the fact that the rocks to which it is assigned contain orthopyroxene, the term charnokite is not used for metamorphic rocks in this text.

43. M. Frey (1987a) summarizes some assemblages in SAC rocks, as does F. J. Turner (1981, ch. 9). The P-T conditions at the lower grades of Greenschist Facies metamorphism of the Buchan, Barrovian, and Sanbagawa Facies series are so similar that the assemblages are essentially the same. Many reports on Prehnite-Pumpellyite and Greenschist Facies rocks are based on the works of Coombs (1954, 1960) and his successors (e.g., E. H. Brown, 1967) on the rocks of South Island, New Zealand. Similar work on rocks of the Great Valley Group (Sequence) of California (Dickinson et al., 1969) is also of interest here. The latter rocks actually represent a facies series more akin to the Franciscan Facies Series. Assemblages reported here are based, in part, on the author's observations in the southern Appalachian Orogen and the California Coast Ranges, on the works cited above,

and on additional relevant works including Schreyer and Chinner (1966), D. G. Bishop (1972), R. J. Stewart and Page (1974), Houghton (1982), Rivers (1983), Abbott and Raymond (1984), Absher and McSween (1985), Gulley (1985), Schenk (1984), and Vavra (1989).

44. Also see Liou (1971a, 1971b), Liou et al. (1974), S. E. Swanson and Schiffman (1979), Cho, Liou, and Maruyama (1986), and Cho and Liou (1987).

45. For additional phase assemblages and information, refer to Coombs, Horodyski, and Naylor (1970), Zen (1974b), W. Glassley (1975), Bevins (1978), Kirchner (1979), Offler et al. (1980), Bevins and Rowbotham (1983), Offler and Aguirre (1984), Liou, Maruyama, and Cho (1985, 1987), Cho, Liou, and Maruyama (1986), Cho and Liou (1987), Bettison and Schiffman (1988), Lucchetti, Cabella, and Cortesoguo (1990), Starkey and Frost (1990), and Ishizuka (1991).

46. Bevins and Rowbotham (1983), Liou, Maruyama, and Cho (1985, 1987).

47. Zen (1974b), Bevins (1978), Bevins and Rowbotham (1983).

48. E. H. Brown (1971), Coombs, Horodyski, and Naylor (1977), Offler et al. (1980), Laird (1980), Ernst, Liou, and Moore (1981), Liou (1981a), Houghton (1982), Offler and Aguirre (1984), Cho and Liou (1987), Bevins and Merriman (1988), Lucchetti, Cabella, and Cortesogno (1990). Additional assemblages are tabulated by F. J. Turner (1981, ch. 9).

49. Hornblende has also been reported from the Greenschist Facies (J. B. Lyons, 1955; E. H. Brown, 1971). Because the appearance of hornblende is generally considered to mark the transition to the Amphibolite Facies in mafic rocks (see note 17), the reported occurrences of hornblende in the Greenschist Facies raise some questions. For example: Is the amphibole actually hornblende? Is it a *stable* part of an equilibrium assemblage? J. B. Lyons (1955) reported such an occurrence, but amphibole compositions were not reported for the rocks he studied, the capacity to spot-analyze the compositions of amphiboles (thereby avoiding zones and impurities) did not exist at the time Lyons did his work, and the geologic history of the region in western New Hampshire in which Lyons worked is complex and polymetamorphic (F. S. Spear and Rumble, 1986). From an area near that in which Lyons worked, Brady (1974) reports coexisting hornblende and actinolite from rocks surrounding the staurolite isograd (the beginning of the Amphibolite Facies). Similarly, Hietanen (1968) reports coexisting actinolite and hornblende in calcareous rocks of the kyanite-staurolite zone from Idaho. These observations suggest that hornblende may appear, under certain conditions, either in the upper Greenschist Facies or in a transition zone between the Greenschist and Amphibolite Facies and that actinolite may occur in the Amphibolite Facies. (The problems associated with these gradational changes are discussed in detail by Laird, 1980, and J. B. Thompson, Laird, and Thompson, 1982.) All plagioclase reported by Brady is more calcic than albite, suggesting that the rocks are not of the Greenschist Facies, but rather those of a transition zone or the lower part of the Amphibolite Facies (cf. the lower-grade part of the Hornblende Hornfels Facies).

50. Refer again to notes 38 and 49. Note that some grossular-andradite garnets appear well within the Greenschist Facies, but the more common reactions yielding almandine-rich garnets occur near the top of the facies or at the Greenschist-Amphibolite Facies transition. Experimental analyses of this transition using real rock compositions have been carried out by Liou, Kuniyoshi, and Ito (1974), Spear (1981b), and Moody, Meyer, and Jenkins (1983). Also see Maruyama, Liou, and Suzuki (1982).

51. Also see A. F. Cooper (1972) and Maruyama, Liou, and Suzuki (1982).

52. For example, see Hietanen (1963, 1973a), J. P. Morgan (1970), Laird (1980), Spear (1982), Stephenson and Hensel (1982), Abbott and Raymond (1984), Sills and Tarney (1984), Gulley (1985), F. S. Spear and Rumble (1986), and Helms et al. (1987).

53. Granulite Facies assemblages are described in A. E. J. Engel and Engel (1958, 1960b), Paglionico and Piccarreta (1978), Schenk (1984), Absher and McSween (1985), Gulley (1985), and Rudnick and Taylor (1987). Hacker (1990) discusses problems of experimental analysis of this facies transition.

54. J. B. Lyons (1955), A. E. J. Engel and Engel (1958, 1960), Zen (1960), Misch (1964), Frank et al. (1973), Hoernes (1973), Ferry (1982; 1983), Absher and McSween (1985).

55. Chidester (1968), J. R. Carpenter and Phyfer (1969), Trommsdorff and Evans (1974), B. W. Evans (1977), Yurkovich (1977), Vance and Dungan (1977), Misra and Keller (1978), Raymond and Swanson (1981), Honeycutt and Heimlich (1980), S. E. Swanson (1980, 1981), Lan and Liou (1981), Sanford (1982), McElhaney and McSween (1983), Scotford and Williams (1983), Trommsdorff (1983), Abbott and Raymond (1984), Absher and McSween (1985), Raymond and Abbott (1985), Raymond (1987). See chapter 31 for additional information.

56. Greenwood (1971), Chernosky (1976), Hemley et al. (1977), Trommsdorff (1983).

57. Hadley (1964), Butler (1972, 1973b, 1991), Fullagar and Odom (1973), Dallmeyer (1975a,b, 1978), Fullagar and Dietrich (1976), Sinha and Glover (1978), R. D. Hatcher et al. (1979), R. D. Hatcher et al. (1980), W. A. Thomas et al. (1980), Tull (1980), Fullagar and Bartholemew (1983), L. Glover et al. (1983), Abbott and Raymond (1984), C. E. Weaver et al. (1984), Gulley (1985), A. S. Russell, Farrar, and Russell (1985), Secor et al. (1986a), Secor, Snoke, and Dallmeyer (1986b), Dallmeyer et al. (1986), R. D. Hatcher (1987), Dallmeyer (1988), A. A. Drake et al. (1989), Osberg et al. (1989), R. D. Hatcher et al. (1989), Kish (1990).

58. B. A. Morgan (1972), L. Glover et al. (1983), J. R. Butler (1984, 1991), C. E. Weaver et al. (1984), R. D. Hatcher (1987).

59. J. R. Butler (1973b), R. D. Hatcher et. al. (1979), Swanson et al. (1985), Butler (1991).

60. Hadley (1964), Abbott and Raymond (1984), and Absher and McSween (1985).

61. See Kulp and Poldervaart (1956), Stose and Stose (1957), P. B. King and Ferguson (1960), Brobst (1962), Hadley and Goldsmith (1963), Bryant and Reed (1970a,b), R. H. Carpenter (1970), Hadley and Nelson (1971), J. R. Butler (1972, 1973), Rankin, Espenshade, and Neuman (1972), Rankin, Espenshade, and Shaw (1973), V. J. Hurst (1973), Dallmeyer (1975b, 1978), Espenshade et al. (1975), Force (1976), Fullagar and Dietrich (1976), R. D. Hatcher (1978, 1987), R. D. Hatcher et al. (1979), R. D. Hatcher et al. (1980), Raymond and Abbott (1980), W. A. Thomas et al. (1980), Tull (1980), Raymond and Swanson (1981), Nesbitt and Essene (1982, 1983), Bartholomew et al. (1983), L. Glover et al. (1983), Abbott and Raymond (1984), Bartholomew and Lewis (1984), C. E. Weaver et al. (1984), Mohr and Newton (1983), Absher and McSween (1985), Gulley (1985), Raymond and Abbott (1985), Brewer (1986), Raymond (1987), Dallmeyer (1988), Eckert, Hatcher, and Mohr (1989), Gillon (1989), J. L. Hopson, Hatcher, and Stieve (1989), McSween, Abbott, and Raymond (1989), Raymond, Yurkovich, and McKinney (1989), Merschat and Wiener (1990), J. W. Miller (1990), J. R. Butler (1991), Raymond et al. (1991), and the references therein.

62. Hadley and Nelson (1971), Merschat (1977), Brewer and Woodward (1988), Raymond, Yurkovich, and McKinney (1989), Merschat and Wiener (1990). For a discussion of migmatites, see the last section of this chapter.

63. L. A. Raymond (unpublished data from the Spruce Pine District, North Carolina).

64. J. W. Miller (1990).

65. Yurkovich and Butkovich (1982), Z. A. Brown et al. (1985), J. L. Hopson, Hatcher, and Stieve (1989), Kish (1989), McSween, Spear, and Fullagar (1991).

66. Geothermometry and geobarometry on Blue Ridge rocks is reported by Nesbitt and Essene (1982), Mohr and Newton (1983), Absher and McSween (1985), Helms et al. (1987), Eckert, Hatcher, and Mohr (1989), and McSween, Abbott, and Raymond (1989).

67. Key sources that form the basis for this discussion and definitions include Sederholm (1967), Mehnert (1968), Atherton and Gribble (1983), and Ashworth (1985b). See the collected works in the latter two sources and the references therein for additional information. In particular, see Tuttle and Bowen (1958), Grant (1968, 1985b), Misch (1968), C. G. Brown and Fyfe (1972), S. N. Olsen (1977, 1983, 1985), Johannes and Gupta (1982), Dougan (1983), Johannes (1983a,b), Kenah and Hollister (1983), McLellan (1983a,b), Ashworth (1985a), and Ashworth and McLellan (1985). Also see Hedge (1972), Kays (1976), Sighinolfi and Gorgoni (1978), Touret and Dietvorst (1983), Abbott (1985), Barr (1985), Touret and Olsen (1985), Tracy (1985), C. Weber et al. (1985), P. Robinson et al. (1986), van Gaans et al. (1987), and Sawyer (1987). Note: The term paleosome is used by some workers in the same ways that protolith and mesosome are used here.

68. For example, see Raymond, Yurkovich, and McKinney (1989).

69. Here the terms open system and closed system are not used in a strict thermodynamic sense.

70. See the brief reviews in Ashworth (1985a) and S. N. Olsen (1985) describing the protracted debate on the origin of metasomatic fluids. Also see the works of Sederhom, in Sederhom (1967), and the works of H. Ramberg (1952, pp. 182–188) and Wickham (1987).

71. The term semiclosed system is used, following similar uses of closed system by S. N. Olsen (1983) and Ashworth (1985a), to refer to systems closed to all components other than the volatiles.

72. Raymond, Yurkovich, and McKinney (1989) propose that some migmatites are diachronous bodies of rock that have had an early history of fragmentation and mixing to yield a melange, followed by a later history involving high-grade metamorphism.

73. Also see Winkler, Boese, and Marcopoulos (1975) and Winkler and Breitbart (1978).

74. See the more recent references in note 67.

PROBLEMS

27.1. (a) Draw the necessary ACF topologies for a step-by-step series of reactions between the Zeolite and Greenschist Facies diagrams shown in figure 27.7. (b) Write and balance a series of reactions representing those topologic changes.

27.2. Using the CFM diagram of Abbott (1982, 1984), explain why epidote is not a typical constituent of Granulite Facies metabasites.

27.3. (a) What kinds of evidence would make a convincing case for the polymetamorphic model for Southern Appalachian Blue Ridge metamorphic history (figure 27.13a)? (b) What kinds of evidence would make a convincing case for the monotonic cooling model for Southern Appalachian Blue Ridge metamorphic history (figure 27.13b)?

28

High P/T Metamorphism: Franciscan and Sanbagawa Facies Series and the Origin of Blueschists

INTRODUCTION

Glaucophane in abundance imparts an attractive blue hue to rocks. This feature undoubtedly accounts for the considerable interest given to the relatively uncommon glaucophane schists (the "blueschists") of the California Coast Ranges, the Alpine-Himalayan orogenic belt, and some western Pacific Islands.[1] The blue color also serves as the basis for the name Blueschist Facies, even though this facies contains large volumes of rock that are neither blue nor schistose. It is also true that all rocks containing blue amphibole do not belong to the Blueschist Facies.[2]

The Blueschist Facies is one of six facies that develop in terranes in which the geothermal gradient is low (< 20° C/km) or the overall P/T is moderate to high (Ernst, 1971a, 1973, 1977, 1988).[3] Two types of facies series are recognized in such terranes—the Sanbagawa Facies Series and the Franciscan Facies Series. In the Sanbagawa Facies Series, in which the maximum temperatures are somewhat higher than in the Franciscan Facies Series, the facies sequence is Zeolite → Prehnite-Pumpellyite → Blueschist → Greenschist → Amphibolite.[4] In the Franciscan Facies Series, the facies sequence is Zeolite → Prehnite-Pumpellyite → Blueschist → Eclogite. As is the case with other facies series, not all metamorphic belts with a particular facies series exhibit all of the facies of that series.

The P-T conditions and histories represented by Sanbagawa and Franciscan facies series have been deduced from experimentally determined phase stabilities, geothermobarometry, vitrinite reflectance, apatite fission track analyses, and isotope studies.[5] These, together with microstructural and paleogeographic analyses, indicate that the rocks were metamorphosed where temperatures were maintained at low levels while pressures were elevated, and, in many cases, where deviatoric stresses were low to nonexistent. Rapid burial through sedimentation represents one process that will produce these static-type conditions and for that reason some petrologists have referred to such metamorphism as "burial metamorphism."[6] Yet Franciscan and Sanbagawa Facies series include rocks metamorphosed under both static and regional conditions; thus a name that implies that only static metamorphism is responsible for their petrogenesis is inappropriate. In this chapter, after assessing the occurrences, mineral assemblages, and textures of the rocks of these facies series, we examine a diversity of proposed origins, including metasomatic and tectonic burial hypotheses.

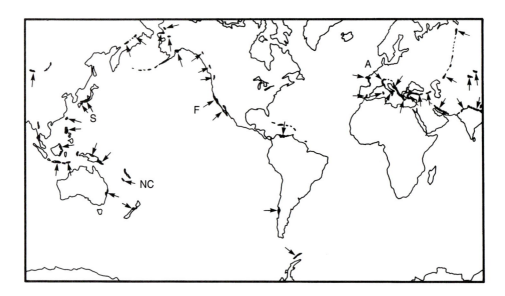

Figure 28.1   World map showing the location of high PT metamorphic belts, including those mentioned prominently in this chapter. A = Alps of southern Europe, F = Franciscan Complex of California, NC = New Caledonia, S = Sanbagawa Belt, Japan.
(Modified from R. G. Coleman (1972).)

OCCURRENCES

Franciscan and Sanbagawa facies series are widely distributed on the Earth (figure 28.1).[7] They occur in North, Central, and South America, in the Caribbean region, in Europe, especially in the Alps and along the northern margin of the Mediterranean Sea, in the Middle East, in Asia, and in the circum-Pacific region. Typically, these facies series form on the outer (trench) side of a paired metamorphic belt associated with a subduction zone (Miyashiro, 1961, 1973b). In some cases, high P/T (low-temperature) rocks form where subduction-induced collision between a continent and island arc, microcontinental block, or another continent is inferred.[8]

Young mountain belts contain the majority of Sanbagawa and Franciscan facies series rocks, but early Paleozoic and rare Precambrian Blueschist Facies rocks are known.[9] In many cases, Blueschist Facies mineral assemblages are overprinted by younger, higher-temperature or lower-pressure assemblages.[10] In part, such overprinting is made possible by the fact that, in general, low-temperature rocks contain hydrous phase assemblages that yield a fluid phase which facilitates recrystallization. In addition, during orogenesis, there is a tendency for cooler rocks to be heated, which also promotes recrystallization. In some cases, rocks formed under lower-pressure or higher-temperature conditions are remetamorphosed under Blueschist Facies conditions.[11]

The two facies series of high P/T metamorphism take their names from well-studied examples on opposite sides of the Pacific Ocean. The Franciscan Facies Series is named for the Franciscan Complex of western California and southern Oregon (E. H. Bailey, Irwin, and Jones, 1964; Berkland et al., 1972). Similar rocks are known from Baja California and Washington.[12] The Sanbagawa Facies Series takes its name from rocks exposed in southeastern Japan (Seki, 1958; Miyashiro, 1961; Ernst et al., 1970).[13]

MINERAL ASSEMBLAGES, FACIES, AND TEXTURES

Mineral assemblages, facies, and textures set the high P/T facies series apart from those of lower P/T. Minerals such as lawsonite occur only at high T and low P. In the lower grades of metamorphism, and locally in the Blueschist Facies, rock textures differ from the textures exhibited by rocks of other facies series.

Textures in Franciscan and Sanbagawa Facies Series Rocks

The textures of Franciscan and Sanbagawa Facies Series rocks are highly varied. In many localities, sedimentary or structural burial provides a P_{load} that is essentially hydrostatic. In addition, high pore-fluid pressures are common in rocks and sediments that descend from the ocean floor, where fluids are abundant, into a subduction zone. In the absence of high deviatoric stresses, these pressures promote recrystallization and neocrystallization without accompanying development of pronounced foliations. Under these circumstances, rocks commonly retain *relict textures* and structures, such as epiclastic texture, porphyritic texture, diabasic texture, amydaloidal structure, and pillow structure. The outlines of radiolaria may

Table 28.1 Textures in Sanbagawa and Franciscan Facies Series Rocks

Texture Type	Cataclastic and Related Textures (C types) (in SF-tectonites)	Foliated-Crystalline Textures (S types) (in S- and L-tectonites)
Texture I	*Relict textures (IC)*. Includes epiclastic, porphyritic, intergranular, ophitic, diabasic, and other textures, with or without slight breaking of grains.	*Relict texture (IS)*. Includes epiclastic, porphyritic, intergranular, ophitic, diabasic, and other textures, with some neocrystallized or recrystallized grains.
Texture II	*Breccia texture (IIC)*. Texture dominated by broken grains. Rocks lack foliation but may contain bands of microbreccia.	*Semischistose and semislaty textures (IIS)*. Weakly foliated, phaneritic, and aphanitic textures, respectively, produced by aligned, neocrystallied phyllosilicates, plus elongation and flattening of rock fragments.
Texture III	*Foliated cataclastic texture (IIIC)*. Foliation produced by microfolia composed of phyllosilicates that separate microlithons (m) and clasts (c)(m + c = 50–99% of rock).	*Schistose texture (IIIS)*. Well-developed foliation, with or without lineation of inosilicates and phyllosilicates. Incipient segregation of grains into foliated and granoblastic layers.
Texture IV	*Protomylonitic texture (IVC)*. Texture with well-developed microfolia with shear surfaces; syntectonically recrystallized bands of mica, quartz, or other minerals; with or without recovery (granoblastic) texture. (m + c > 50% of rock).	*Gneissose texture (IVS)*. Foliated texture in which segregation banding predominates. Dark bands typically dominated by ferromagnesian inosilicates and phyllosilicates. Light bands typically dominated by mosaic-textured quartz and feldspar.

even be preserved in cherts, in spite of the fact that the opal-A has thoroughly recrystallized to quartz. Where neocrystallization and recrystallization are pronounced and acicular to platy minerals crystallize, *diablastic textures* develop. In contrast, *foliated textures* form where deviatoric stresses are important.

Foliated textures belong to two main categories and range from weak to pronounced (table 28.1). Because of the tectonic setting in which high P/T rocks form, many of these rocks, are SF-tectonites that have *cataclastic-type textures*, including brecciated, foliated-cataclastic, or protomylonitic textures (figure 28.2a)(Raymond, 1975, 1984a; S. E. Lucas and Moore, 1986; J. C. Moore et al., 1986). The latter two textures are foliated types. Foliated-cataclastic textures are particularly characteristic of melanges and related rock bodies, which are common in subduction zones. In such rock bodies, the rocks may consist of sheared and broken clastic grains and microlithons separated by microfoliated zones that are characterized by aligned phyllosilicate grains (Cowan, 1982a; J. C. Moore et al., 1986). Shear-induced dislocation has occurred along many of these microfoliated zones. In contrast, in S- and L-tectonites, where recrystallization dominates, *crystalline-foliated textures* occur, ranging from aphanitic

semislaty and fine-grained semischistose types to coarse-grained gneissose types. A variety of textural classifications of foliated rocks have been proposed and used for mapping different textural zones in high P/T terranes (F. J. Turner, 1948, p. 38; 1968, pp. 31–32; Blake, Irwin, and Coleman, 1967, 1969).[14] Three types of foliated-crystalline textures are recognized here (table 28.1). These are semischistose/semislaty texture, schistose texture, and gneissose texture (figure 28.2b and d).

Characteristic Minerals, Mineral Assemblages, and Facies

Because the rocks in outer metamorphic belts are metamorphosed pieces of ocean crust and overlying sediments, metabasites, metawackes, and metapelites are the dominant rocks of the high P/T facies series. Of these, the metabasites and the metawackes are the most widely studied. The Sanbagawa Belt of Japan is exceptional, however, in that, rather than metawackes, there are abundant pelitic rocks that have been studied in detail. In the Sanbagawa Belt and elsewhere, the less abundant metacherts, metacarbonate rocks, and ultramafic rocks have been studied less thoroughly.

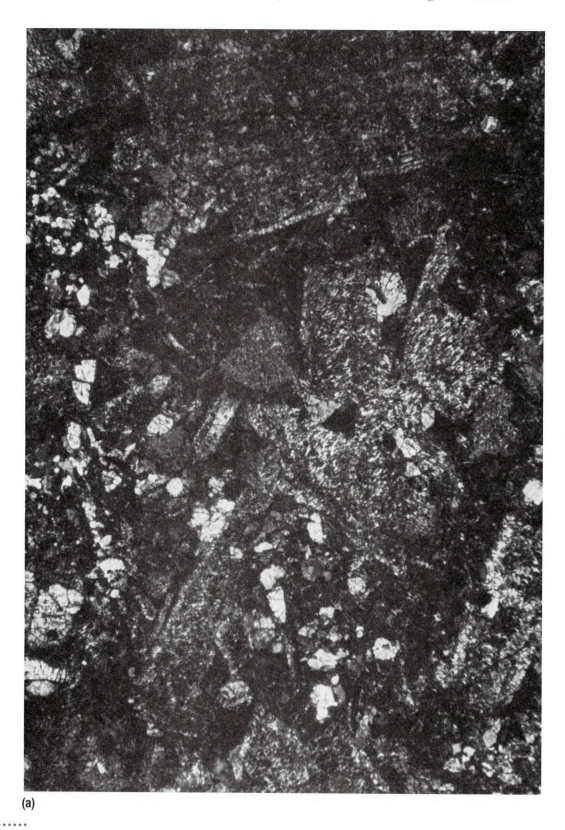

(a)

Figure 28.2   Photomicrographs of textures in high P/T rocks. a) Blastoporphyritic (relict porphyritic) texture (type IS) in metabasite, northeastern Diablo Range, California, (XN). b) Semischistose texture (type IIS) in metawacke, northeastern Diablo Range, California, (XN). c) Foliated cataclastic texture (type IIIC) in melange matrix, melange of Blue Rock, northeastern Diablo Range, California, (PL). d) Gneissose texture (type IVS), South Fork Mountain Schist, northern Coast Ranges, California, (XN). Long dimension of all photos is 3.25 mm.

(b)

(c)

(d)

Figure 28.2 Continued

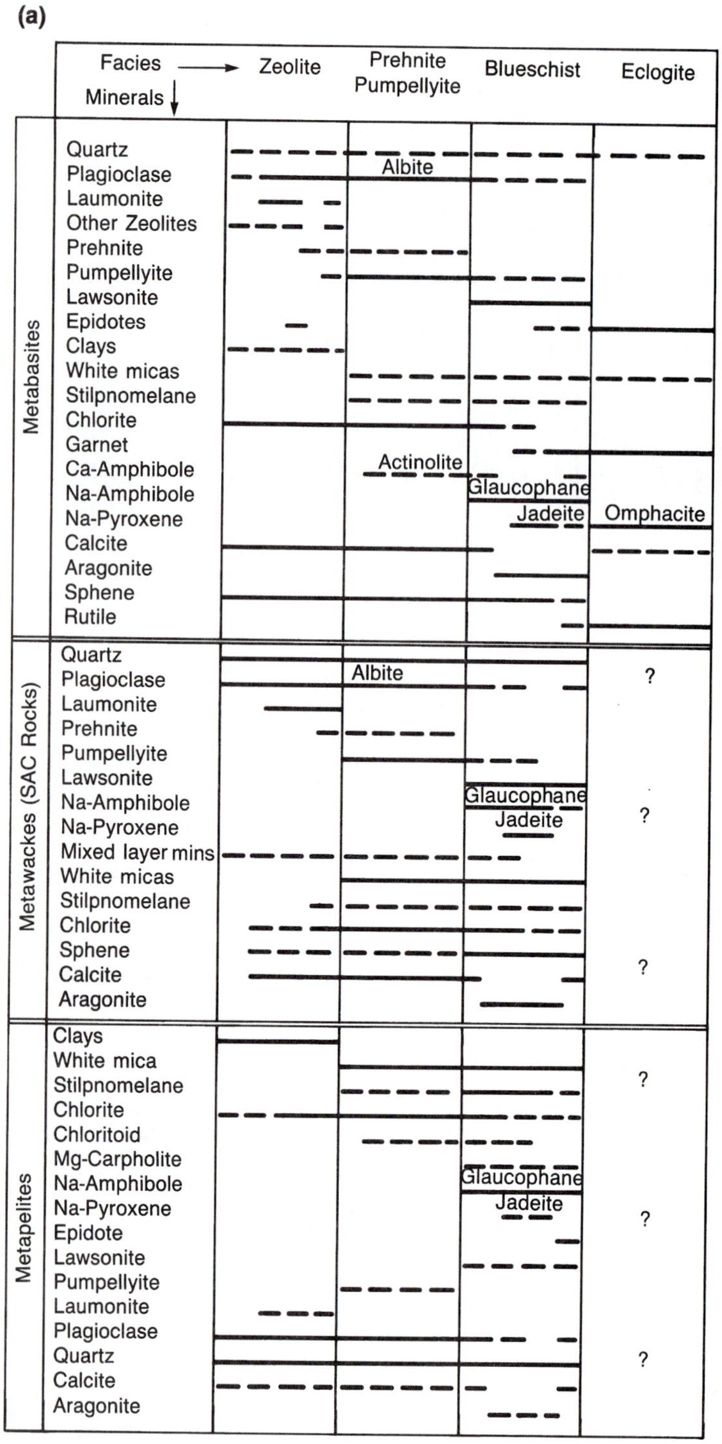

(a)

Figure 28.3 Idealized mineral-facies charts for Franciscan (a) and Sanbagawa (b) Facies series.

[Based in part on D. S. Coombs (1960), Seki et al. (1969), Ernst, Onuki, and Gilbert (1970), Ernst (1977a; 1984), Enami (1983), Blake et al. (1988), Higashino (1990), Otsuki and Banno (1990), Oh, Liou, and Maruyama (1991), observations of the author, and additional sources in notes 18 and 25–28]

Representative mineral-facies charts for the Sanbagawa and Franciscan Facies series are presented in figure 28.3. For the Franciscan example, note that all of the facies of the series—Zeolite, Prehnite-Pumpellyite, Blueschist, and Eclogite—are represented. Rocks of the Eclogite Facies, however, do not comprise large-scale units or regional terranes.

The most common of the critical minerals that appear include laumontite, pumpellyite, glaucophane/crossite, lawsonite, aragonite, jadeitic pyroxene, and omphacite (E. H. Bailey, Irwin, and Jones, 1964; Blake, Irwin, and Coleman, 1967; Ernst, 1971c, 1977a). As we shall see, jadeitic pyroxene occurs only in the highest-grade Blueschist Facies zones. In

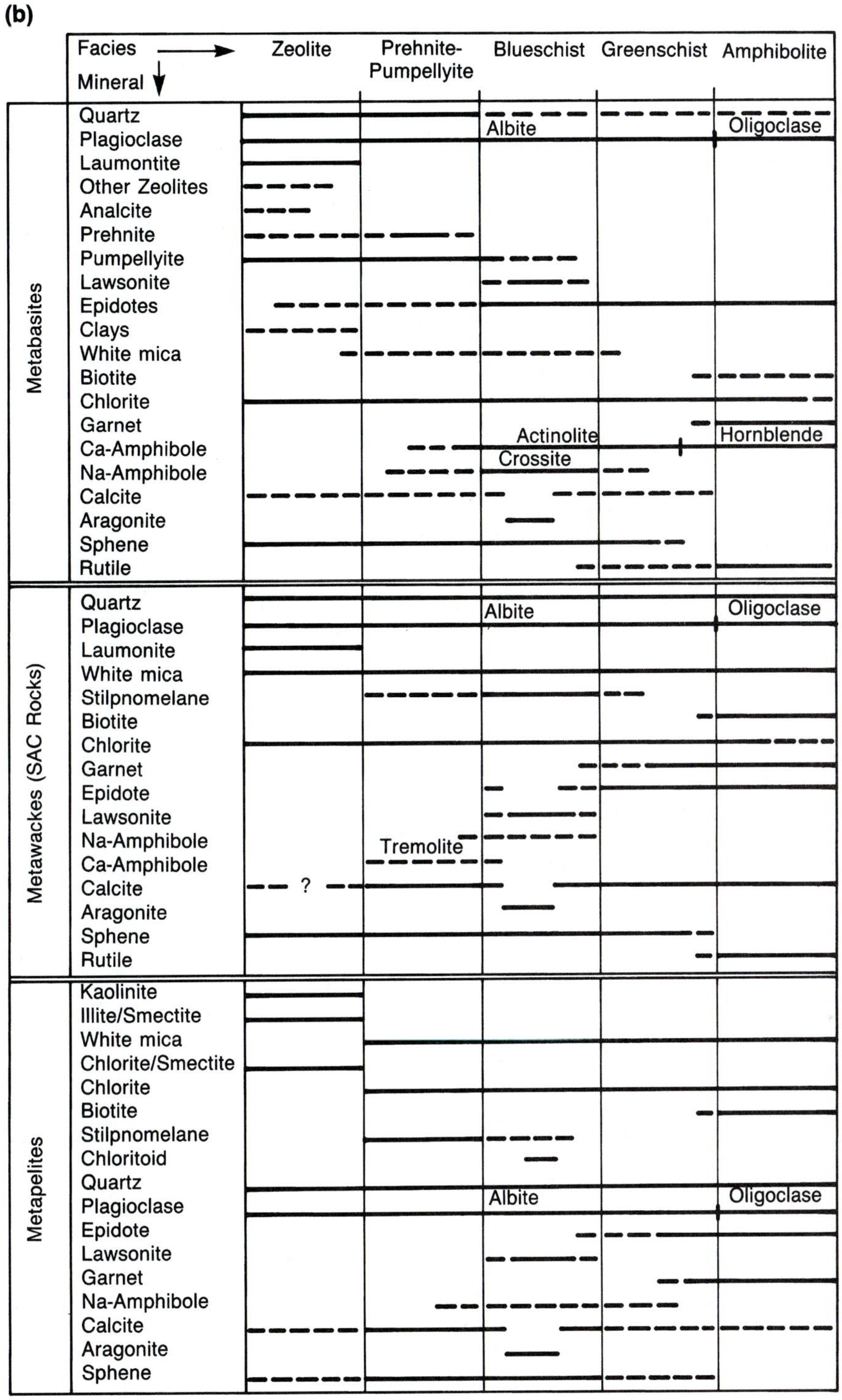

(b)

Facies → Mineral ↓	Zeolite	Prehnite-Pumpellyite	Blueschist	Greenschist	Amphibolite

Metabasites

Quartz
Plagioclase — Albite — Oligoclase
Laumontite
Other Zeolites
Analcite
Prehnite
Pumpellyite
Lawsonite
Epidotes
Clays
White mica
Biotite
Chlorite
Garnet
Ca-Amphibole — Actinolite — Hornblende
Na-Amphibole — Crossite
Calcite
Aragonite
Sphene
Rutile

Metawackes (SAC Rocks)

Quartz
Plagioclase — Albite — Oligoclase
Laumonite
White mica
Stilpnomelane
Biotite
Chlorite
Garnet
Epidote
Lawsonite
Na-Amphibole
Ca-Amphibole — Tremolite
Calcite — ?
Aragonite
Sphene
Rutile

Metapelites

Kaolinite
Illite/Smectite
White mica
Chlorite/Smectite
Chlorite
Biotite
Stilpnomelane
Chloritoid
Quartz
Plagioclase — Albite — Oligoclase
Epidote
Lawsonite
Garnet
Na-Amphibole
Calcite
Aragonite
Sphene

Figure 28.3 Continued

central California, the Blueschist Facies is extensive and is subdivided into several zones marked by the appearance (a) or disappearance (d) of several phases, including lawsonite (a), aragonite (a), albite (d), jadeitic pyroxene (a), and pumpellyite (d).[15]

Sanbagawa Facies Series rocks are assigned to the Zeolite, Prehnite-Pumpellyite, Blueschist, Greenschist, or Amphibolite Facies. In the Sanbagawa Belt of Japan, Zeolite Facies rocks apparently do not occur (Ernst, 1977a; Liou, Maruyama, and Cho, 1987; Otsuki and Banno, 1990), though they were reported earlier when the definition of the

Table 28.2   Representative Mineral Assemblages of Sanbagawa Facies Series Rocks

Zeolite Facies

quartz–white mica–chlorite–albite–hematite	(S)
quartz–albite–laumontite–white mica–chlorite–sphene	(SAC)
illite/smectite–kaolinite–chlorite/smectite–quartz–hematite	(P)
laumontite–analcite–pumpellyite–epidote–quartz–chlorite–white mica–sphene	(MB)
lizardite–magnetite–magnesite–dolomite	(U)
calcite–dolomite–quartz	(C)

Prehnite-Pumpellyite Facies

quartz–stilpnomelane–white mica–chlorite–hematite	(S)
quartz–albite–white mica–chlorite–calcite–graphite–sphene	(SAC)
white mica–chlorite–stilpnomelane–quartz–albite–Na-amphibole–sphene	(P)
Prehnite–pumpellyite–epidote–calcite–albite–quartz–chlorite–sphene	(MB)
antigorite–brucite–magnetite–dolomite	(U)
calcite–prehnite–chlorite–quartz	(C)

Blueschist Facies

quartz–crossite–aegirine-augite–stilpnomelane–garnet	(S)
quartz–albite–lawsonite–chlorite–white mica–calcite–glaucophane	(SAC)
white mica–chlorite–chloritoid–quartz–Na-amphibole–sphene–hematite	(P)
epidote–Na-amphibole–actinolite–chlorite–albite–quartz–sphene	(MB)
antigorite–magnetite–magnesite–dolomite	(U)
aragonite–chlorite–hematite	(C)

Greenschist Facies

quartz–albite–white mica–stilpnomelane–Na-amphibole–garnet	(S)
quartz–albite–epidote–white mica–chlorite–graphite	(SAC)
white mica–chlorite–quartz–albite–garnet–graphite	(P)
crossite–epidote–chlorite–albite–garnet–quartz–white mica–rutile	(MB)
antigorite–magnesite–brucite	(U)
calcite–tremolite–quartz	(C)

Amphibolite Facies

quartz–white mica–biotite–garnet	(S)
quartz–oligoclase–white mica–epidote–biotite–garnet–calcite–rutile	(SAC)
white mica–biotite–chlorite–quartz–garnet–oligoclase–epidote–graphite	(P)
hornblende–epidote–garnet–oligoclase–quartz–hematite	(MB)
olivine–diopside–tremolite–antigorite[1]	(U)
calcite–diopside–quartz–garnet[1]	(C)

Sources: See note 18.

[1]Assemblages are based on chemographic analyses. C = carbonate rock, MB = metabasite, P = pelitic rock, S = siliceous rock, SAC = siliceous-alkali-calcic (quartzo-feldspathic) rock, U = ultramafic rock.

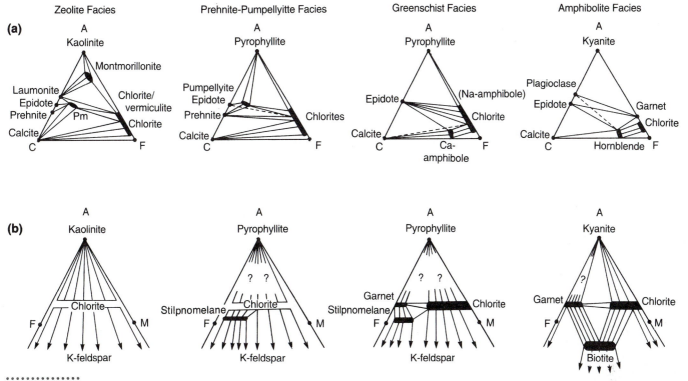

Figure 28.4 Selected schematic phase diagrams for the Sanbagawa Facies Series. (a) ACF diagrams for SAC and metabasic rocks. (b) AFM diagrams for pelitic rocks.

Sanbagawa Belt was broader.[16] In some Sanbagawa-like terranes (e.g., in the Alps), Eclogite Facies rocks occur in association with other rocks of the Sanbagawa Facies Series (Oberhansli, 1986). Jadeitic pyroxene is absent in the Sanbagawa Belt (but not in all rocks of the Sanbagawa Facies Series). Lawsonite is rare, occurring only locally in pelitic schists and gabbros.[17] Critical minerals in basic schists include pumpellyite with actinolite, epidote with actinolite, pumpellyite with actinolite and magnesioriebeckite (sodic amphibole), lawsonite and actinolite, winchite (sodic amphibole), crossite (sodic amphibole), barroisite (sodic amphibole), albite with hornblende, and oligoclase with hornblende (Banno, 1986).

Examination of common assemblages composed of the critical minerals and associated phases allows the construction of phase diagrams for each facies series. Typical phase assemblages for the Sanbagawa Facies Series are listed in table 28.2 and selected phase diagrams are shown in figure 28.4.[18] The rocks of the Zeolite, Prehnite-Pumpellyite, and Greenschist Facies are mineralogically like those of lower-pressure facies series. Prehnite-Pumpellyite and Greenschist Facies assemblages are virtually identical. Wairakite, however, is generally absent from Zeolite Facies rocks, whereas laumontite and heulandite are common.

In the Sanbagawa Belt of Japan (see figure 24.10a), four zones are recognized in pelitic rocks and seven or eight zones are defined in metabasites (Toriumi, 1975; Banno, 1986).[19] Characteristic assemblages of the pelitic zones are as follows:

Chlorite Zone

chlorite–white mica–quartz–albite–stilpnomelane–calcite–sphene–graphite

chlorite–white mica–quartz–albite–epidote–calcite–sphene–pyrite–graphite

Garnet Zone

chlorite–epidote–white mica–garnet–quartz–albite–sphene–graphite

Albite-Biotite Zone

white mica–chlorite–biotite–garnet–quartz–albite–epidote–graphite

Oligoclase-Biotite Zone

white mica–biotite–chlorite–garnet–oligoclase–quartz–rutile–graphite.[20]

The lowest part of the Chlorite Zone corresponds to the Prehnite-Pumpellyite Facies and the lower part of the

Blueschist Facies, whereas the middle and upper parts represent the lower Greenschist Facies.[21] The Garnet and Albite-Biotite zones correspond generally to the middle and upper Greenschist Facies, respectively, and the Oligoclase-Biotite Zone is equivalent to the lower Amphibolite Facies.

Critical reactions marking the isograds for the appearance of garnet, biotite, and oligoclase are probably continuous reactions[22] of the form

$$\text{chlorite}_1 + \text{epidote} \rightarrow \text{chlorite}_2 + \text{garnet} + \text{fluid} \quad (28.1)$$

in which the composition of chlorite becomes more magnesian,

$$\text{chlorite} + \text{white mica}_1 \rightarrow \text{biotite} + \text{white mica}_2 \quad (28.2)$$

in which the white mica composition changes (Brown, 1971), and

$$\text{albite} + \text{epidote} + \text{garnet}_1 + \text{biotite}_1 \rightarrow \text{oligoclase} + \text{garnet}_2 + \text{biotite}_2. \quad (28.3)$$

Alternative reactions for the appearance of garnet include

$$\text{chlorite} + \text{quartz} \rightarrow \text{garnet} + H_2O \quad (28.4)$$

(Banno, Sakai, and Higashino, 1986) and

$$\text{chlorite} + \text{quartz} + \text{magnetite} \rightarrow \text{garnet} + H_2O \quad (28.5)$$

(Hsu, 1968). The specific reactions for phase assemblage changes are not entirely resolved.

Two metabasite sequences are recognized in the Sanbagawa Belt, one with hematite-bearing rocks and one with rocks lacking hematite (table 28.3) (Banno, 1986; Otsuki and Banno, 1990). Assemblages in metabasites include the following:

Prehnite-Pumpellyite (Zone) Facies

prehnite–pumpellyite–chlorite–albite–calcite–quartz

Hematite-Pumpellyite-Actinolite Zone

chlorite–epidote–albite–quartz–Na-amphibole–sphene–hematite

pumpellyite–albite–chlorite–actinolite–epidote–stilpnomelane–quartz–sphene

chlorite–Na-amphibole–albite–quartz–calcite–white mica

chlorite–actinolite–lawsonite–epidote–albite–quartz

Hematite-Epidote-Actinolite Zone

chlorite–epidote–actinolite–albite–quartz–sphene

chlorite–epidote–crossite–albite–quartz–sphene–hematite

Winchite Zone

chlorite–epidote–actinolite–albite–quartz–sphene

chlorite–epidote–winchite–albite–quartz–sphene–hematite

Crossite Zone

chlorite–epidote–actinolite–albite–quartz–rutile

chlorite–epidote–crossite–albite–quartz–sphene–hematite–magnetite

Barroisite Zone

chlorite–barroisite–epidote–biotite–albite–quartz–sphene

chlorite–crossite–barroisite–epidote–albite–quartz–ilmenite–hematite

barroisite–chlorite–epidote–albite–quartz–rutile

Albite-Hornblende Zone

hornblende–chlorite–epidote–albite–quartz–hematite–ilmenite

hornblende–chlorite–epidote–albite–quartz–ilmenite

Oligoclase-Hornblende Zone

hornblende–epidote–oligoclase–quartz–magnetite–hematite

hornblende–epidote–chlorite–garnet–oligoclase–quartz–ilmenite[23]

Sodic amphibole and minor lawsonite show clearly that some Sanbagawa Belt rocks were metamorphosed under high-pressure, Blueschist Facies conditions (tables 28.2 and 28.3).

Reactions marking isograds between metabasite zones of the Sanbagawa Belt involve terminal and nonterminal, discontinuous and continuous reactions for the production of actinolite, epidote–actinolite–pumpellyite assemblages, winchite, barroisite, hornblende, and oligoclase. For example, the change from a Prehnite-Pumpellyite Facies assemblage to an actinolite-bearing, Hematite-Pumpellyite-Actinolite Zone assemblage may involve a reaction such as

$$\text{prehnite} + \text{chlorite} + \text{quartz} \rightarrow \text{pumpellyite} + \text{actinolite} + H_2O \quad (28.6)$$

(cf. Liou, Maruyama, and Cho, 1985). Lawsonite may be produced in this zone by a reaction such as

$$\text{prehnite} + \text{chlorite} + \text{quartz} + H_2O \rightarrow \text{lawsonite} + \text{actinolite} \quad (28.7)$$

(cf. Liou, Maruyama, and Cho, 1987). The transition to the Pumpellyite-Epidote-Actinolite Zone (in non-hematitic rocks) may be represented by the reaction

$$\text{pumpellyite} + \text{chlorite} + \text{quartz} \rightarrow \text{epidote} + \text{actinolite} + H_2O \quad (28.8)$$

(Toriumi, 1975; Brown, 1977b; Liou, Maruyama, and Cho, 1985). In successively higher-grade zones, the compositions of the amphiboles change via continuous reactions such as

$$\text{magnesioriebeckite} + \text{chlorite} + \text{quartz} + \text{albite} \rightarrow \text{glaucophane} + \text{hematite} + H_2O \quad (28.9)$$

(Hosotani and Banno, 1986),

$$\text{crossite} + \text{epidote} + \text{chlorite} + \text{albite} + \text{quartz} \rightarrow \text{barroisite} + \text{hematite} + H_2O \quad (28.10)$$

(Otsuki and Banno, 1990), or

$$\text{glaucophane} + \text{clinozoisite} + \text{quartz} + H_2O \rightarrow \text{chlorite} + \text{tremolite} + \text{albite} \quad (28.11)$$

Table 28.3 Mineral Zones in the Sanbagawa Belt of Japan

Pelitic Schists	Hematitic Metabasites	Metabasites Lacking Hematite	Facies
Oligoclase-Biotite Zone	Oligoclase-Hornblende Zone	Oligoclase-Hornblende Zone	Amphibolite Facies (lower)
Albite-Biotite Zone	Albite-Hornblende Zone	Albite-Hornblende Zone	Upper Greenschist Facies (high P, higher T)
	Barroisite Zone	Barroisite Zone	-----
Garnet Zone	Crossite Zone	Epidote-Actinolite Zone	Middle Greenschist Facies (high P, moderate T) -----
Chlorite Zone	Winchite Zone	Pumpellyite-Epidote-Actinolite Zone	Lower Greenschist Facies (high P, lower T)
	Hematite-Epidote-Actinolite Zone		
	Hematite-Pumpellyite-Actinolite Zone	Pumpellyite-Stilpnomelane Zone	Blueschist Facies
			Prehnite-Pumpellyite Facies

Source: From S. Banno, "The high-pressure metamorphic belts of Japan: A Review" in B. W. Evans and E. H. Brown, Eds., "Blueschists and Eclogites" in GSA Memoirs, 164:365–74, 1986. Copyright © 1986 Geological Society of America. Reprinted by permission of the author.

(Liou, Maruyama, and Cho, 1985). Hornblende is similarly produced by a complex reaction involving barroisite, epidote, chlorite, albite, and quartz (Otsuki and Banno, 1990). Albite and hornblende probably yield oligoclase through a continuous reaction like equation 26.18.

The Blueschist-Greenschist Facies transition exhibited in the Sanbagawa Facies Series has been of considerable interest to petrologists. E. H. Brown (1974, 1977b) and Laird (1980), in revealing mineralogical and topological changes that occur over this transition zone, laid the groundwork for many studies that followed.[24] Although at least 120 ternary topologies are theoretically possible, observed phase assemblages are more limited. A few topologies, based on observed assemblages, are shown in figures 28.4 and 28.5 and are implied by the petrogenetic grid for low- to moderate-temperature rocks that encompasses the transition (see appendix C). The upper thermal limit of the Blueschist Facies is marked by reactions in which epidote group minerals appear *at the expense of pumpellyite and lawsonite.*

The Franciscan Facies Series, which includes the Blueschist Facies but not the Greenschist Facies, exhibits more extensive development of the Blueschist Facies rocks than does the Sanbagawa Facies Series. In particular, assemblages containing jadeitic pyroxene, lawsonite, and aragonite are common. Franciscan Facies Series SAC rocks, typically metawackes, are more thoroughly studied than are other compositional types. The phase assemblages for these are represented on some ACF diagrams in figure 28.6, which presents selected ACF and other phase diagrams for the Zeolite, Prehnite-Pumpellyite, and Blueschist Facies of this facies series.

In the Zeolite Facies, common phase assemblages in SAC rocks are

heulandite–quartz–analcite–chlorite/vermiculite–white mica–sphene,
laumontite–quartz–albite–chlorite/vermiculite–white mica–calcite, and
laumontite–heulandite–quartz–albite–white mica–chlorite/vermiculite.[25]

These are replaced in the Prehnite-Pumpellyite Facies by assemblages such as

quartz–albite–pumpellyite–white mica–chlorite–stilpnomelane–calcite, and
quartz–albite–prehnite–chlorite–white mica–hematite.[26]

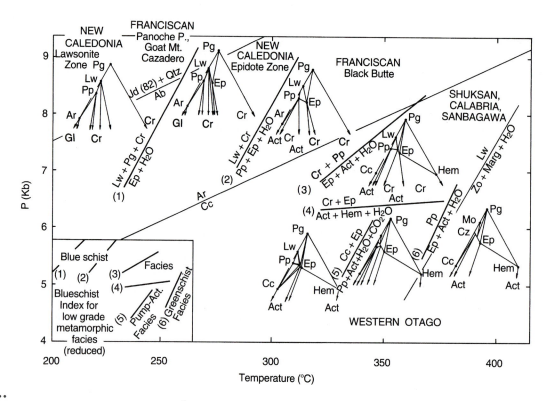

Figure 28.5   Petrogenetic grid with Al-Ca-Fe^3 phase diagrams for the Blueschist-Greenschist Facies transition. *(From E. H. Brown, 1977)*

Blueschist Facies SAC assemblages include

quartz–albite–lawsonite–pumpellyite–chlorite–white mica–calcite, and

quartz–jadeitic pyroxene–lawsonite–white mica–glaucophane–aragonite.[27]

Few Eclogite Facies rocks have SAC chemistries, but the assemblage

quartz–white mica–omphacite–glaucophane–garnet–epidote

is a representative phase assemblage under lower-grade Eclogite Facies conditions (Yokoyama, Brothers, and Black, 1986). More aluminous Eclogite Facies rocks may contain kyanite. Representative assemblages of the Franciscan Facies Series for other bulk compositions are presented in table 28.4.[28]

Some reactions that may yield critical phases in SAC rocks of the Franciscan Facies Series include the following:

$$\text{plagioclase (detrital)} + \text{quartz} + H_2O \rightarrow$$
$$\text{laumontite} + \text{albite}, \quad (28.12)$$

$$\text{laumontite} + \text{calcite} \rightarrow \text{prehnite} + \text{quartz} + H_2O + CO_2 \quad (28.13)$$

(A. B. Thompson, 1971),

$$\text{laumontite} \rightarrow \text{lawsonite} + \text{quartz} + H_2O \quad (28.14)$$

(Liou, 1971b),

$$\text{laumontite} + \text{prehnite} + \text{chlorite} \rightarrow \text{pumpellyite} + \text{quartz} + H_2O \quad (28.15)$$

(Cho, Liou, and Maruyama, 1986),

$$\text{plagioclase (detrital)} + \text{quartz} + \text{calcite} + \text{hematite} + H_2O$$
$$\rightarrow \text{pumpellyite} + \text{albite} + CO_2, \quad (28.16)$$

$$\text{plagioclase (detrital)} + H_2O \rightarrow \text{lawsonite} + \text{albite}, \quad (28.17)$$

$$\text{albite} \rightarrow \text{jadeite} + \text{quartz} \quad (28.18)$$

(Newton and Smith, 1967; Maruyama, Liou, and Sasakura, 1985), and

$$\text{albite} + \text{chlorite} + \text{pumpellyite} + \text{hematite} + H_2O \rightarrow$$
$$\text{jadeitic pyroxene} + \text{glaucophane} + \text{lawsonite} + \text{quartz} \quad (28.19)$$

(Patrick and Day, 1989). Note that some of these reactions involve the direct reaction of detrital plagioclase feldspar to form a phase present in a facies of higher grade than the Zeolite Facies. Petrographic evidence suggests that such reactions occur in Franciscan Facies Series rocks of the Franciscan Complex of California (Raymond, 1973c). These may occur because of very slow reaction rates in the rocks at low temperatures and rapid burial rates in convergent margin settings, a combination that allows sediments or rocks to reach higher-grade conditions before undergoing metamorphic changes.

Reactions in other bulk compositions, such as the carbonate rocks and metabasites, range from simple to complex.

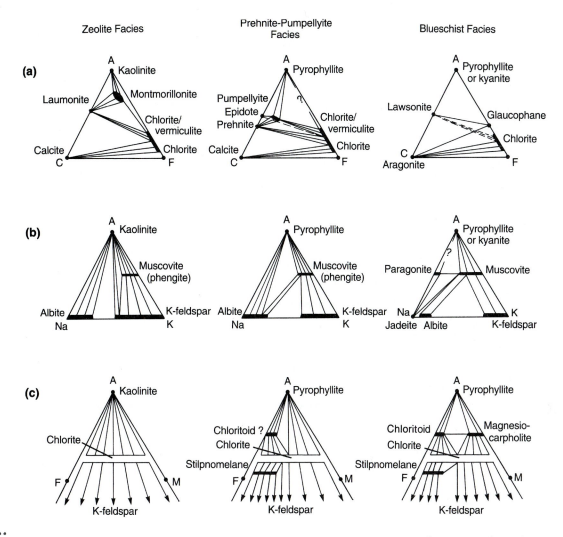

Figure 28.6 Selected phase diagrams for the Franciscan Facies Series. (a) ACF diagrams for SAC and metabasic rocks. (b) AKN diagrams for pelitic rocks. (c) AFM diagrams for pelitic rocks. Compare and contrast these with the diagrams in figure 28.4.

In the carbonate rocks, the most important reaction is the simple polymorphic phase transformation

$$\text{calcite} \rightarrow \text{aragonite.} \qquad (28.20)$$

Via this reaction, limestones are converted to aragonite marbles. In contrast, exchange of Fe and Mg among phases is indicated by complex reactions in metabasites, such as the discontinuous reactions

$$\text{laumontite + pumpellyite + quartz} \rightarrow \text{prehnite + epidote}$$
$$\text{+ chlorite + H}_2\text{O} \qquad (28.21)$$

(Cho, Liou, and Saskura, 1986), which marks the Zeolite/Prehnite-Pumpellyite Facies boundary;

$$\text{pumpellyite + epidote + chlorite + albite + quartz + H}_2\text{O}$$
$$\rightarrow \text{lawsonite + crossite} \qquad (28.22)$$

(Brown, 1977b), which marks the Prehnite-Pumpellyite Blueschist Facies boundary;

$$\text{lawsonite + Na-amphibole} \rightarrow \text{clinopyroxene +}$$
$$\text{pumpellyite + H}_2\text{O} \qquad (28.23)$$

(Maruyama and Liou, 1987);

$$\text{lawsonite + Na-amphibole} \rightarrow \text{epidote + chlorite + albite}$$
$$\text{+ quartz + H}_2\text{O} \qquad (28.24)$$

(E. H. Brown and Ghent, 1983); and the continuous reaction equation 28.10.

Table 28.4   Representative Phase Assemblages for the Franciscan Facies Series

Zeolite Facies

quartz–albite–calcite–hematite	(S)
illite/smectite–kaolinite–chlorite/smectite–quartz–hematite	(P)
laumontite–albite–pumpellyite–quartz–chlorite–sphene	(MB)

Prehnite-Pumpellyite Facies

quartz–albite–pumpellyite–chlorite	(S)
white mica–chlorite/smectite–stilpnomelane–quartz–albite - sphene	(P)
prehnite–pumpellyite–epidote–albite–quartz–chlorite–sphene	(MB)

Blueschist Facies

quartz–crossite–aegirine–stilpnomelane–garnet–hematite	(S)
white mica–quartz–lawsonite–Na-amphibole–garnet–hematite	(P)
glaucophane–lawsonite–jadeitic pyroxene–aragonite–magnetite–white mica–sphene	(MB)

Eclogite Facies

quartz–omphacite–white mica	(S)
quartz–sodic pyroxene–garnet–white mica–kyanite[1]	(SAC)
omphacite–garnet–kyanite–rutile	(MB)

Sources: See notes 25–28.

[1]From chemographic analysis.

MB = metabasites, P = pelitic rocks, S = siliceous rocks.

Experimental Investigations and Mineral Stabilities

Although the tectonic setting in which Blueschist Facies rocks occur suggests high pressures of metamorphism, alternative low-pressure models of formation have been proposed for these rocks. A high-pressure origin is supported by the available experimental data. The stability relations of several minerals important to both the Franciscan and Sanbagawa Facies series have been investigated in a number of experiments over the years. Representative curves are shown in figure 28.5 and in appendix C.

Petrogenetic grids such as figure 28.5, if based only on experimental data, are somewhat misleading, because (1) experiments have not been done on all of the important reactions, (2) the addition of components to these systems may alter the stability fields of the phases, and (3) the composition of the fluid phase can have a profound effect on the stabilities of phases. For example, the presence of various minerals including prehnite, pumpellyite, and other Ca-Al silicates is a function of X_{CO_2}.[29] In a rock of a certain bulk composition, the Prehnite-Pumpellyite Facies assemblage epidote–prehnite–pumpellyite, stable at lower values of X_{CO_2}, is replaced under conditions of high X_{CO_2} by the Greenschist Facies assemblage epidote–chlorite–calcite (W. Glassley, 1974; Ivanov and Gurevich, 1975; Liou, Maruyama, and Cho, 1987). The validity and scope of such grids can be enhanced via the addition of calculated reaction curves (Guiraud, Holland, and Powell, 1990; B. W. Evans, 1990).

The available data from the reactions shown in figure 28.5 and appendix C reveal that critical minerals of the Franciscan and Sanbagawa Facies series are stable under generally high P/T conditions.[30] The Zeolite Facies represents conditions of $P < 0.4$ Gpa (4 kb) and $T < 250°$ C. The Prehnite-Pumpellyite Facies occurs at slightly higher temperatures and similar pressures under conditions of about $P = 2$–3.5 kb and $T = 200$–350° C. At the low to moderate temperatures of $T = 100$–450° C and pressures of 0.35–1.0 Gpa (3.5–10 kb) the Blueschist Facies has an even higher P to T ratio. Greenschist Facies assemblages are stable between about 300° C and 550° C at pressures of about 0.2 to 0.8 Gpa (2 to 8 kb). The Eclogite Facies has the highest pressures, at $P > 0.8$ Gpa, but temperatures vary from about 200° C to more than 600° C.

Relating the experimental studies to natural occurrences is sometimes difficult. In addition to the limitations of the experimental studies noted above, natural phases are seldom pure (i.e., they do not have stoichiometric compositions), and many cases of metastable crystallization, incomplete recrystallization, and polymetamorphism are known.[31] For example, pyroxenes in some metabasites from the Franciscan Complex are zoned, reflecting incomplete neocrystallization under differing metamorphic conditions associated with progressive metamorphic events (Maruyama and Liou, 1987). Metastable persistence of lower-grade phases in higher-grade rocks is relatively common in the Sanbagawa and Franciscan Facies series. In part, this is the result of reactions being sluggish at low temperatures, but variations in the amount and character of a fluid phase may also control the neocrystallization process.[32] In addition, pseudomorphs and incomplete replacement of phases reflect polymetamorphic histories.[33] All of these factors make it difficult to relate mapped isograds in high P/T terranes to experimentally determined stability fields.

PETROGENETIC MODELS
• •

Increased knowledge about Blueschist Facies rocks obtained through field and laboratory studies during the 1950s and 1960s spurred interest in their petrogenesis. That interest led to a controversy that was outlined and discussed by van der Kaaden (1969) and Ernst (1971a). Four hypotheses for the origin of these rocks were proposed and advocated by various geologists. Two of them—the tectonic overpressure and the burial hypotheses—accepted the high pressures suggested by the experimental data. Two others—the metasomatic and the metastable recrystallization hypotheses—appealed to processes other than neocrystallization under high-pressure conditions.[34]

Metasomatic and Metastable Recrystallization Hypotheses

The metastable recrystallization and metasomatic hypotheses assign little importance to the experimental studies on the P-T stability ranges of Blueschist Facies minerals. Taliaferro (1943) recognized the common field association of glaucophane schists and serpentinites before the stabilities of minerals such as jadeitic pyroxene and glaucophane were known. He and other advocates of metasomatism[35] were followed by Gresens (1969, 1970), who argued that blueschists result from low-pressure metasomatism induced by highly concentrated, reducing, saline pore fluids created during serpentinization. Gresens cited several arguments and lines of evidence, including the following, in support of his thesis.

1. Glaucophane-rich rocks and serpentinites are commonly associated in field settings.

2. Minerals such as aragonite, glaucophane, pumpellyite, and lawsonite commonly occur in veins that presumably formed at low pressures.

3. The predominance of ferrous iron over ferric iron in Blueschist Facies minerals, especially in minerals lacking sodium, indicates the presence of a reducing ore fluid.

4. Sodium metasomatism is known, notably in the Green River Formation, where sodic pyroxene and blue amphibole formed in a concentrated, sodium-rich solution.[36]

5. Reactions that involve saline serpentinizing fluids, presumed to carry the appropriate chemical species, may be written to show the production of Blueschist Facies minerals.

Ernst (1971a) and others present convincing evidence and arguments against the metasomatic hypothesis.[37] First, although serpentinite and glaucophane schists are commonly associated in the field, large tracts of Blueschist Facies rocks in Washington, Italy, southwestern Japan, and California contain little or no ultramafic rock.[38] At some locales where serpentinite *is* present, serpentinization is known to have occurred *after* ultramafic rocks were emplaced; yet the contacts with surrounding rocks are not marked by an intervening layer of glaucophane schist.[39] Second, there is no reason to assume that veins can only form under low-pressure conditions. Third, ferrous iron does not predominate in all Blueschist Facies minerals. Rather, ferric iron dominates in some non- and low-sodium minerals, such as garnet, epidote, muscovite, and pumpellyite, that occur in the Blueschist Facies.[40] Fourth, although local metasomatism may have occurred in some blueschist-bearing terranes, the important questions, as noted by Ernst (1971a), are (1) is metasomatism required to produce Blueschist Facies mineralogies? and (2) did regional metasomatism take place? Available evidence, including the lack of appropriate metamorphic gradients away from serpentinite bodies, does not support an affirmative response to either question.[41] Finally, the fact that reactions may be written does not mean that those reactions took place in the rocks, nor is there reason to assume that the chemical species needed for the reactions will be present in serpentinizing or metasomatizing fluids. In short, because the metasomatic hypothesis is based on assumptions and conclusions not consistent with available data, it may be ruled out as a viable, general hypothesis for the petrogenesis of Blueschist Facies rocks.

Metastable recrystallization arguments have not been vigorously promoted. Ernst (1971a), however, summarized the two possibilities:

Metastable Recrystallization Model 1. Blueschist Facies mineral assemblages form from precursor mineral suites of higher energy state under conditions of low pressure.

Metastable Recrystallization Model 2. Blueschist Facies minerals, normally unstable under low-pressure conditions, are stabilized under these conditions by active pore fluids. The latter hypothesis is a version of the metasomatic hypothesis rejected above.

Model 1 was proposed to explain the occurrence of single mineral species, such as jadeitic pyroxene and aragonite, rather than an entire mineral assemblage. For example, simply stated, high albite has a higher energy state than low albite, and therefore the reaction

$$\text{high albite} = \text{jadeite} + \text{quartz} \qquad (28.25)$$

will occur (under low-temperature conditions) at lower pressures than will the reaction

$$\text{low albite} = \text{jadeite} + \text{quartz} \qquad (28.26)$$

Stated another way, at low pressures, the energy difference between the left and right sides of equation 28.25 is greater than that between the two sides of equation 28.26. This favors reaction equation 28.25. Therefore, high albite, which is metastable in relation to low albite, will react to form jadeite via the metastable reaction equation 28.25 *if* metastable high albite is present in the rocks.

In the Franciscan Complex of California, Ernst (1971a) found no evidence that (highly strained or disordered) high albite was an important premetamorphic component of jadeitic-pyroxene-bearing rocks. The absence of high albite and the fact that metastable recrystallization (according to model 1) only applies to single minerals, rather than to complete Blueschist Facies assemblages, argue against the general application of this hypothesis as an explanation for petrogenesis of Blueschist Facies rocks. The hypothesis also fails to explain regional distributions of facies and mineral zones.

The Tectonic Overpressure Model

Hypotheses that require relatively high pressures of formation for Blueschist Facies mineral assemblages seem more reasonable, considering the experimental data that indicate that relatively high pressures are required to stabilize jadeitic pyroxene, glaucophane, aragonite, and lawsonite in metamorphic rocks. These hypotheses become even more compelling in the absence of data supporting low-pressure hypotheses. Coleman, Blake, and others[42] argued that **tectonic overpressures,** i.e., pressures in excess of P_{load}, cause Franciscan Facies Series metamorphism (Blake, Irwin, and Coleman, 1967, 1969). They recognized relationships in Northern California and New

Caledonia in which metamorphic grade decreases away (and down) from major regional thrust faults (i.e., the metamorphism is "upside down," with higher grade rocks at shallower depths).[43] These relationships are consistent with the tectonic overpressure hypothesis. For the California case, Blake, Irwin, and Coleman (1967) argued that tectonic overpressures developed below a regional thrust fault that is capped by serpentinite. The impervious cap trapped water, creating the overpressures (figure 28.7).

The sum of the available evidence does not strongly support the tectonic overpressure hypothesis. Although the regional metamorphic patterns recognized in northern California *seem* to support the idea, lower-grade terranes, in many cases, are separated from adjoining higher-grade terranes by faults, and the isograds are cut by the faults (B. L. Wood, 1971; Jayko, Blake, and Brothers, 1986).[44] This suggests that the upside-down pattern may be a structural rather than a metamorphic phenomenon. In New Caledonia, in the South Pacific, the regional metamorphic patterns also apparently resulted from tectonic rather than strictly metamorphic processes.[45] Additional negative evidence is provided by regional and detailed studies in central California, which reveal that no relationship exists between the fault and the highest-grade (jadeitic) rocks or the associated isograds (Ernst, 1971c; Raymond, 1973a,c; Patrick and Day, 1989).[46] Furthermore, laboratory analyses of rock strengths suggest that the Franciscan rocks of California, under the appropriate metamorphic conditions, may not be able to maintain overpressures in excess of about 1 kb, particularly if the rocks contained a pore fluid (Brace, Ernst, and Kallberg, 1970; E. C. Robertson, 1972). Thus, field and laboratory evidence, taken together, suggest that tectonic overpressures are not generally responsible for regional Franciscan Facies Series metamorphism.

The Burial Metamorphism Hypothesis

Deep burial may result from either sedimentation or tectonic thickening of the crust via faulting. In either case, P_{load} will be high. Ernst (1965, 1971a, 1973a, 1975, 1984, 1988) and several other petrologists advocate some form of deep burial as a method of generating high pressures of metamorphism in Franciscan and Sanbagawa Facies series rocks (figure 28.8).[47] Experimental data, *which require high pressures* of metamorphism for Blueschist Facies mineral assemblages, are the primary evidence favoring the deep burial hypothesis. Thermal modelling also supports it (Oxburgh and Turcotte, 1971; C. Y. Wang and Shi, 1984; Dumitru, 1991). In addition, the paleogeographic and tectonic setting of high P/T metamorphic belts is consistent with the hypothesis. In particular, the presence of Sanbagawa and

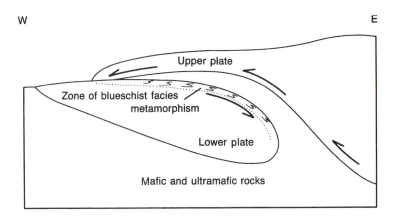

Figure 28.7 The tectonic overpressure model of Blueschist Facies petrogenesis.
(Redrawn from Blake, Irwin, and Coleman, 1967)

Franciscan Facies series rocks in paired metamorphic belts (which form in response to subduction) suggests that subduction and associated accretion of subducted sediments and rocks (i.e., tectonic burial), are generally responsible for the petrogenesis of Blueschist Facies and related rocks (figure 28. 8)(Ernst, 1984). Subduction-induced tectonic burial provides high pressures and low temperatures of metamorphism.[48] It also explains the common age progression in high P/T metamorphic belts, in which the age of the rocks decreases away from the associated high-temperature belt of the pair. Finally, the subduction burial hypothesis explains inverted sequences of Amphibolite, Greenschist, and Blueschist Facies rocks that apparently form on the hanging walls of subduction zones (Platt, 1975; Peacock, 1987a, 1988; Sorenson, Bebout, and Barton, 1991).

Although both sedimentation and tectonic burial can yield high pressures and low temperatures, the best model for any particular belt will be that which best explains regional patterns and other phenomena related to the metamorphic event. The regional patterns within both Franciscan and Sanbagawa Facies series belts typically are linear, coast-parallel patterns, with higher-grade rocks landward towards the core of the orogen (e.g., see figure 24.10). Tilting of a sedimentary sequence will produce such a pattern. Likewise, the imbricate faulting and accretion associated with subduction parallel to a coastline will yield a pattern of parallel metamorphic belts. In the former case, tilting and folding may yield a coast-parallel set of metamorphic zones without major, intervening faults—a condition that apparently exists in the Sanbagawa Belt of Japan.[49] Alternatively, individually subducted imbricate rock slabs and melange masses, separated by faults, may each develop with a distinctive metamorphic zone or facies, as is the case, in part, in northern California (Blake et al., 1988).[50]

One of the unresolved problems of Blueschist Facies metamorphism is the problem of preservation and uplift of Blueschist Facies rocks. Either prolonged burial or slow subduction may result in progressively higher temperatures, over time, in a buried mass of rock.[51] Progressive heating will result in conversion of Blueschist and Prehnite-Pumpellyite Facies rocks to Greenschist or Amphibolite Facies rocks. It follows that if blueschists are preserved at the surface, they must have been protected from heating, either by rapid uplift following metamorphism or by some form of "refrigeration" (i.e., continued cooling while at depth) (Peacock, 1987a; Ernst, 1988; Dumitru, 1991). If they were refrigerated, they must have remained cool *while they moved towards the surface.*

Two general types of subduction tectonics hypotheses that allow for the origin, preservation, and uplift of Blueschist Facies rocks have been proposed. One type of hypothesis advocates coeval subduction and uplift, that is synsubduction uplift, uplift that occurs while subduction continues. The second type assumes metmorphism during subduction, but relies on postsubduction isostatic rise for uplift of the blueschists. Both types of hypothesis accept the premise that, during subduction, masses of rock are successively underplated beneath the overriding plate. The fact that blueschists occur in two different structural settings, as regionally extensive slabs and as relatively small tectonic blocks and slabs in a matrix of serpentinite or metashale tectonite (i.e., in melanges), may mean that two kinds of uplift processes were operative.

Synsubduction models may be divided into five subtypes. The subduction zone is the avenue of uplift in the first, the S-type model. In S-type models, subduction zones are considered to be "two-way streets" along which materials are being subducted (carried down) *at the same time* that previously subducted materials are moved up (Suppe, 1972; Cloos, 1982; Ernst, 1984). Both Suppe (1972) and Ernst

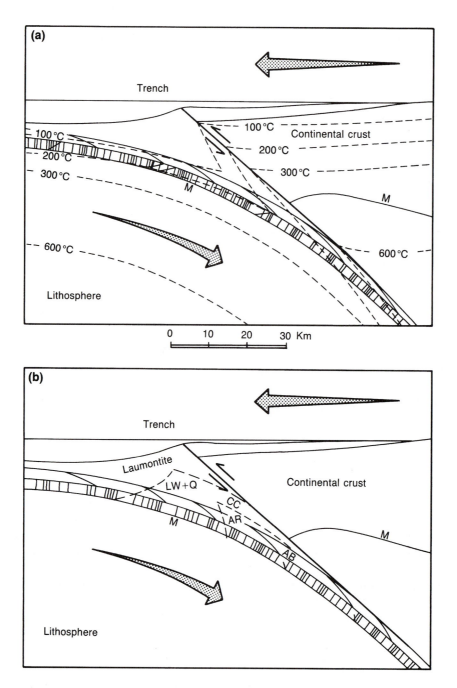

Figure 28.8   Tectonic burial (subduction zone) model of Blueschist Facies petrogenesis. (a) Thermal structure of an active subduction zone. M = Mohorovicic discontinuity. (b) Subduction zone showing location of selected isograds.

LW = lawsonite, Q = quartz, CC = calcite, AR = aragonite, AB = albite, JD = jadeite.
(Redrawn from Ernst, 1973a)

(1971a, 1977, 1988) adopt an imbricate thrust model of subduction accretion[52] with the proviso that large coherent *slabs* move up within the subduction zone to be emplaced at shallow levels of the crust, where they no longer would be subject to heating. Aragonite, which converts to calcite rather quickly upon heating (W. D. Carlson and Rosenfeld, 1981), is present in many blueschists, indicating that the

blueschists were never heated on their way to the surface. Because the subduction zone itself remains cool as long as the subduction rate does not fall below a certain limiting value or the heat content of the subducting plate does not exceed a specific level (C. Y. Wang and Shi, 1984; Peacock, 1987; Dumitru, 1991), materials moving *within* the subduction zone will remain cool. Ernst (1971a, 1988)

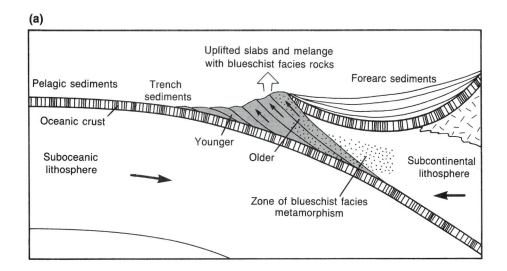

(a)

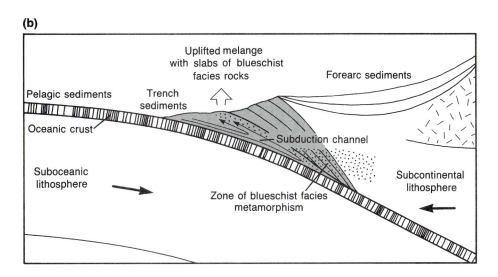

(b)

Figure 28.9 Various models for uplift of Blueschist Facies rocks (based on selected hypotheses). (a) Imbricate slab (S-type) model (cf. Ernst, 1977a, 1977b). (b) Subduction channel (S-type) model (cf. Cloos, 1982, 1984; Cloos and Shreve, 1988a,b). (c) Forearc extension (E-type) model (cf. Platt, 1986). (d) Postsubduction, diapiric uplift model.

suggests that, if the rate of subduction decreases, previously subducted slabs will move up due to buoyant forces (subducted rocks are considered to be of lower density than the rocks of the overlying plate). Clearly, such a decrease cannot fall below the limiting value, if refrigeration is to be maintained. Ernst (1971a, 1975) depicted the upward-moving slabs as fault-bounded, imbricate masses (figure 28.9a). These would appear at the surface as major (coherent) fault blocks.

The emplacement of relatively small tectonic blocks of eclogite and blueschist poses a different problem. In general, these are incorporated in shaly or serpentinous SF-tectonites. Adopting the view that the Central Belt of the Franciscan Complex (described below) is a single large melange and

drawing on the experimental work of D. S. Cowan and Silling (1978), Cloos (1982, 1984)[53] detailed an S-type synsubduction model, the subduction channel model, which requires that upward movement of blocks and slabs be a function of circulating flow in a shale-based tectonic melange that dominates the active part of the subduction zone (figure 28.9b). In this model, blueschists and eclogites metamorphosed deep within the subduction zone are fragmented and transported up the subduction zone, where they may mix with both descending Zeolite and Prehnite-Pumpellyite Facies rocks and ascending Amphibolite and Greenschist Facies rocks metamorphosed along the hotter hanging wall of the zone. Via subduction channel flow, some relatively small blocks, metamorphosed in the cool depths of the subduction zone, are

(c)

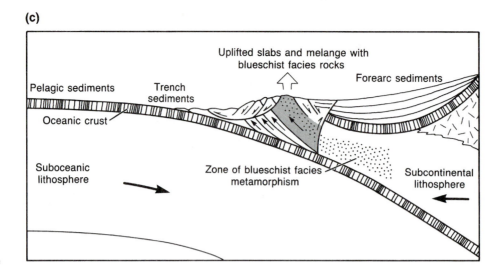

(d)

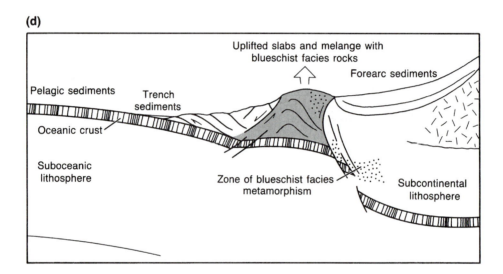

Figure 28.9   Continued

transported relatively quickly back to the surface. This rapid transport within the subduction zone prevents the reheating that yields retrograde or polymetamorphic effects in some blocks, while allowing those effects to develop in other blocks before they are transported up. The resulting melange contains a mix of low- to high-grade rocks, ranging from eclogites and high-grade blueschists metamorphosed at great depth to barely metamorphosed zeolitic metawackes that were altered at shallow depths.

In these S-type models, subduction must be fast and continuous enough to preserve the low geothermal gradients required for high P/T metmamorphism (C. Y. Wang and Shi, 1984; Dumitru, 1990, 1991). Rates of uplift, however, need

not be rapid. In some cases, uplift was relatively slow or prolonged, as indicated (1) by the absence of overprint assemblages, (2) by fission track data, and (3) by regional structural relationships coupled with a disparity between sedimentological and metamorphic uplift ages of the rocks (Raymond, 1981; S. L. Baldwin and Harrison, 1989; Dumitru, 1989).

A second type of synsubduction model, the E-type model, requires that accretion and underplating in the subduction zone generate an instability in the overlying accretionary wedge that causes *extension* and (listric) normal faulting (figure 28.9c) (Platt, 1986; Krueger and Jones, 1989; Harms, Jayko, and Blake, 1992).[54] The faulting removes the overburden from the buried Blueschist Facies rocks. Continued underplating (accretion) via imbricate thrusting of slabs below the extensionally faulted zone provides the driving force for uplift. Eventually, blueschist masses are driven to the surface and exposed by faulting and by erosion.

Three additional types of synsubduction, Blueschist Facies uplift models have been proposed, none of which is yet widely accepted. The T-type model, which involves *transpression* and strike-slip faulting, was proposed by Karig (1980).[55] According to this model, blueschists are exposed to erosion where an imbricate thrust margin is cut by strike-slip faults. Vertical movement, as well as lateral movement along the faults, is envisioned to transport deeply buried blueschists along strike to sites where less erosion is required to expose them at the surface. Such a model requires that blueschist terranes be bounded, at least in part, by high-angle (strike-slip) faults. Another model involving *oblique* plate convergence, the O-type model, was proposed by Avé Lallement and Guth (1990). In this model—which admittedly requires special circumstances—uplift of blueschists occurs when a decrease in subduction angle is caused by a change in the angle of plate convergence of an obliquely converging subducting plate. A final synsubduction model, the J-type model, like the T-type, involves lateral faulting, in this case associated with a *triple junction*. K. F. Fox (1976, 1983) proposed that, at a triple junction where obliquely converging plates are replacing a transform fault boundary along a continental margin, rocks of the overthrust continental edge would be deformed and uplifted.[56] Through these processes, tracts of Blueschist Facies rocks would be uplifted and exposed to gravity sliding that would yield blueschist-bearing melanges. Thus, both large terranes and isolated blocks of blueschist would be uplifted.

Post-subduction models of blueschist uplift rely on one of three processes. Uplift results from (1) bouyant rise and diapiric uplift of the subducted accretionary slabs and melanges caused by the disparity in density between subducted rocks and those of the overlying plate, (2) erosion induced isostatic uplift and diapirism, or (3) orogenic uplift (figure 28.9d)(Ernst, 1965, 1988; Dobretsov, 1991).[57] In the first and second processes, buoyancy would have to drive the uplifted material rapidly toward the surface to prevent a Greenschist Facies overprint from developing on the blueschists. Either entire deformed slabs or block-bearing melanges could be moved to the surface via this mechanism, but whether regional terranes of Blueschist Facies rocks could reach the surface *before* being overprinted by higher-grade assemblages is debatable (Draper and Bone, 1981). In the case of the Franciscan Complex of California, fission track data indicate that postsubduction uplift was, in fact, minor (Dumitru, 1989).

EXAMPLE: REGIONAL HIGH P/T METAMORPHISM OF THE FRANCISCAN COMPLEX, CALIFORNIA

The Franciscan Complex forms the structurally complicated, locally chaotic basement of much of the California and southwestern Oregon Coast Ranges (figure 28.10).[58] It is composed of a wide variety of rock types, not all of which are metamorphosed. As a group, however, metamorphic rocks dominate. Wacke and metawacke and associated shale and metashale are the most abundant rock types.[59] Red, green, and multicolored radiolarian chert; massive to pillowed basaltic rocks; gray and pink limestones; conglomerates; ultramafic rocks; and the metamorphic equivalents of all of these also occur at numerous localities. Well known among the metamorphic rocks are eclogites, glaucophane schists and gneisses, and actinolite and hornblende schist and gneiss that occur in isolated blocks and sheets.[60]

Both regional terranes and the isolated masses of metamorphic rock are present in the Franciscan Complex. The isolated masses—including glaucophane schists, eclogites, and related rocks, the first discovered and thoroughly studied of the high-pressure rocks (A. C. Lawson, 1895; Switzer, 1945, 1951; Borg, 1956)—most commonly occur in melanges. In addition, Eclogite, Blueschist, Amphibolite, and rare Greenschist Facies rocks form slabs and tectonic blocks along faults. The blue amphibole-bearing rocks, including eclogites, are typically divided into higher-temperature "high-grade" blocks and lower-temperature "low-grade" blocks. Both types may contain sodic pyroxene, but the high-grade blocks are generally coarser-grained, texture IIIS and IVS rocks, whereas the low-grade blocks are finer-grained, texture IIS to IIIS rocks. In contrast, texture I to texture III rocks are typical of the regional terranes. M. E. Maddock (1955) discovered and Bloxam (1956) and McKee (1962) reported that jadeitic pyroxene is widely distributed in rocks previously believed to be unmetamorphosed. Subsequent studies revealed that Blueschist, Prehnite-Pumpellyite, and Zeolite Facies assemblages characterize the rocks within large areas.[61]

In the northern Coast Ranges, rocks of the six metamorphic facies are distributed across three major, fault bounded belts—the Eastern Belt, the Central Belt, and the Coastal Belt—that are successively younger from east to west (figure 28.10) (Blake, Irwin, and Coleman, 1969; Berkland et al., 1972).[62] High-grade schists and gneisses, in tectonic blocks and slabs, form a fourth unit that locally caps the Franciscan Complex along its eastern edge. Each belt is

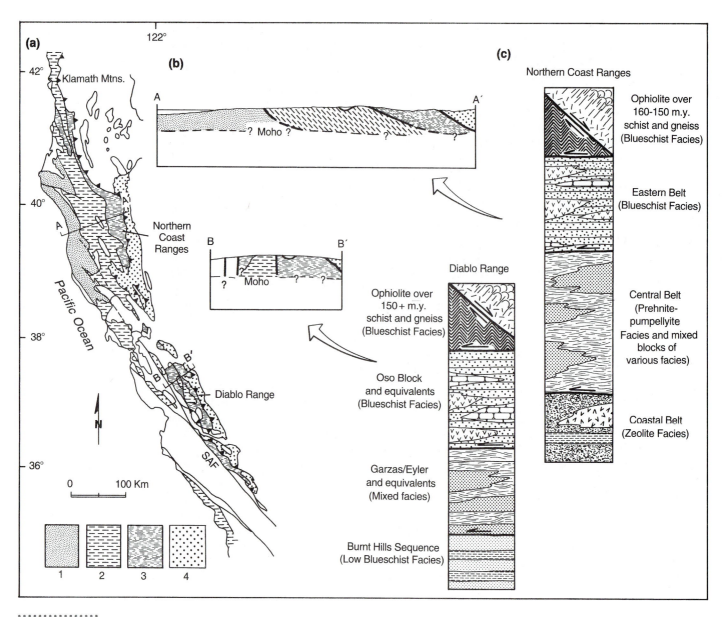

Figure 28.10 The Franciscan Complex of California.
a) Generalized map of western California showing the three structural-metamorphic belts of the Franciscan Complex. 1 = Coastal Belt, 2 = Central Belt, 3 = Eastern Belt, 4 = Great Valley Group (modified from Blake, Irwin, and Coleman, 1969; R. G. Coleman and Lanphere, 1971; Raymond and Swanson, 1980; Ernst, 1980). (b) Simplified, cross-sectional sketches across the northern coast ranges and the Diablo Range. (c) Diagrammatic, columnar structural sections showing the main structural units of the complex.

subdivided into several thrust sheets or fault blocks (commonly designated as terranes) that include various formations, broken formations, dismembered formations, and melanges.[63] The Central Belt is largely melange. In contrast, the adjoining Eastern and Coastal belts, though locally containing melange, consist predominantly of rock bodies with greater internal coherence. Similar structural units exist in the Diablo Range of central California, but relationships there are more complex and the Coastal Belt is missing.[64]

Some Franciscan rocks are also exposed along the south-central coast.[65] In the area at the southern end of the Northern Coast Ranges, in the San Francisco Bay area and to the north for several tens of kilometers, the structural and metamorphic patterns are highly disrupted by Cenozoic faulting.[66]

The metamorphic patterns of the northern Coast Ranges are more regular than the patterns in the south. In the north, the westernmost belt, the Coastal Belt, is a metawacke- and metashale-dominated, *Zeolite Facies* metamorphic belt

(a)

Figure 28.11 Photomicrographs of Franciscan Facies Series rocks from the Franciscan Complex of California. (a) Zeolite Facies metawacke from the Coastal Belt, near Willits, California, showing laumontite (Lm) replacing plagioclase (P), (XN). (b) Blueschist Facies metawacke from the northeastern Diablo Range containing lawsonite (Lw) and jadeitic pyroxene (J), (XN). (c) Blueschist Facies metachert containing crossite and aegirine needles (CR); from a

(figure 28.10).[67] The metawackes range from incipiently metamorphosed, texture IS rocks that retain their primary epiclastic textures and minerals to significantly neocrystallized texture IIS types. Locally, especially at the western edge of the area, broken formations and melanges with texture IIC to IVC rocks are present (R. J. McLaughlin et al., 1982). The metawackes in the north (and along the southern coast) contain laumontite, prehnite, or pumpellyite as incipient neocrystallized grains in feldspars, as complete replacements of feldspars, and as vein minerals. Thorough petrologic

analyses of these weakly and incompletely metamorphosed rocks have not been published, but representative phase assemblages (figure 28.11a) appear to include

quartz–albite–laumontite–chlorite–white mica–hematite,
quartz–albite–laumontite–chlorite/smectite–white mica–calcite–hematite,
quartz–albite–prehnite–chlorite–white mica–calcite, and
quartz–albite–alkali feldspar–chlorite–white mica–stilpnomelane–calcite.[68]

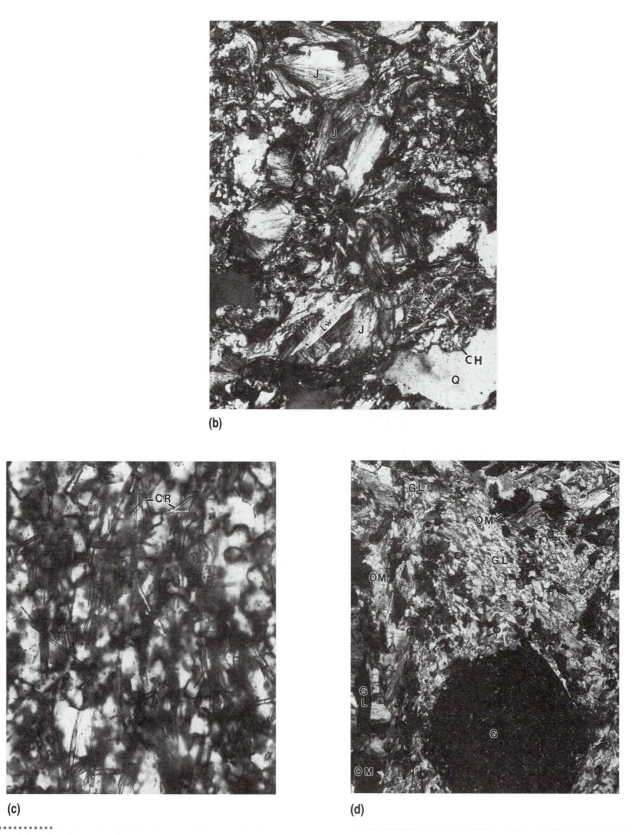

Figure 28.11 Continued
melange in the northeastern Diablo Range, California, (PL). (d) Glaucophane-bearing eclogite, Jenner, Sonoma County, California, (XN). CH = chlorite, G = garnet, GL = glaucophane, OM = omphacite, Q = quartz, W = white mica. Long dimension of photos (a) and (b) is 1.27 mm and of photos (c) and (d) is 1.06 mm.

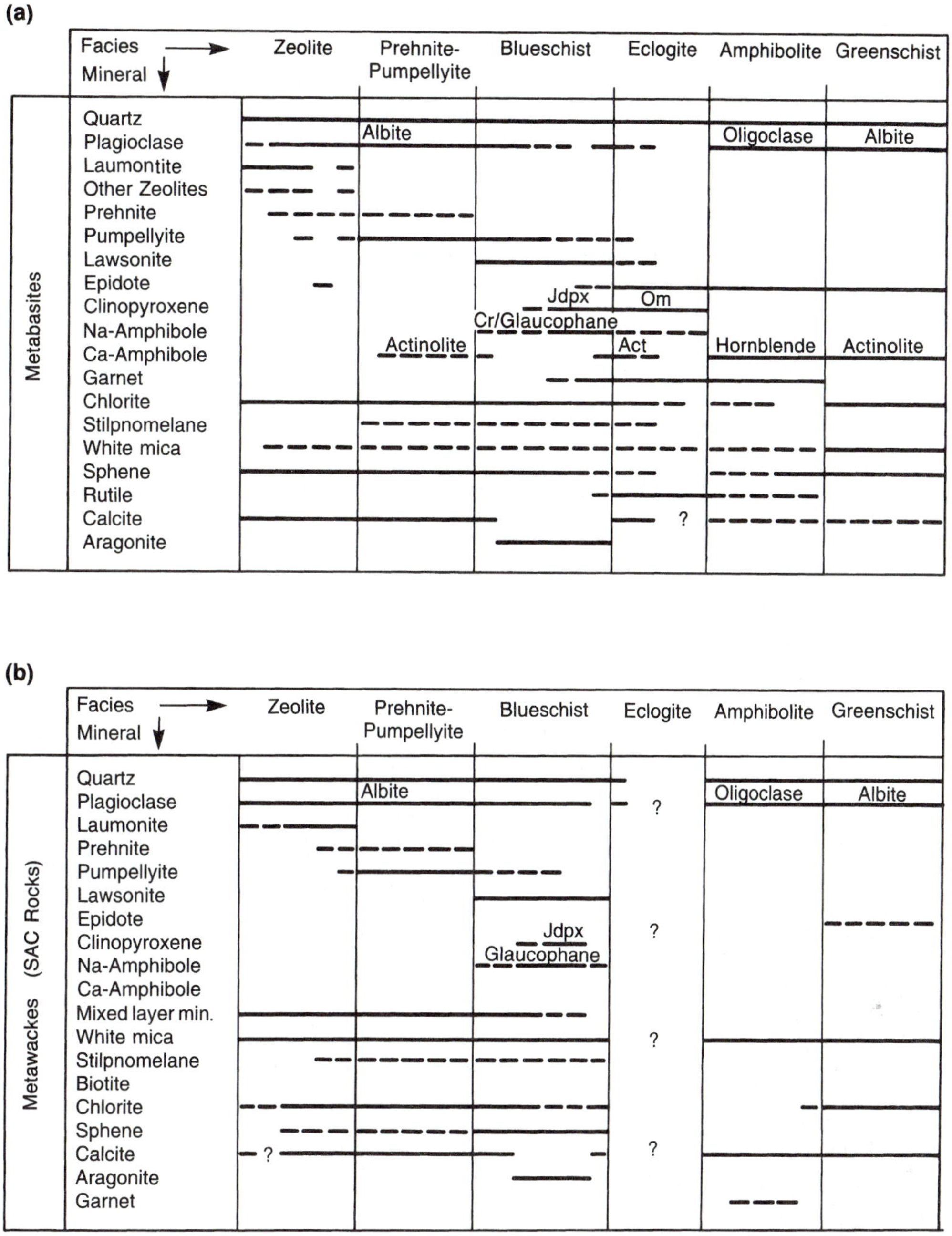

(a)

Facies → / Mineral ↓	Zeolite	Prehnite-Pumpellyite	Blueschist	Eclogite	Amphibolite	Greenschist
Quartz						
Plagioclase		Albite			Oligoclase	Albite
Laumontite						
Other Zeolites						
Prehnite						
Pumpellyite						
Lawsonite						
Epidote						
Clinopyroxene			Jdpx	Om		
Na-Amphibole		Cr/Glaucophane				
Ca-Amphibole	Actinolite			Act	Hornblende	Actinolite
Garnet						
Chlorite						
Stilpnomelane						
White mica						
Sphene						
Rutile						
Calcite				?		
Aragonite						

(Metabasites)

(b)

Facies → / Mineral ↓	Zeolite	Prehnite-Pumpellyite	Blueschist	Eclogite	Amphibolite	Greenschist
Quartz						
Plagioclase		Albite		?	Oligoclase	Albite
Laumonite						
Prehnite						
Pumpellyite						
Lawsonite						
Epidote				?		
Clinopyroxene			Jdpx			
Na-Amphibole			Glaucophane			
Ca-Amphibole						
Mixed layer min.						
White mica				?		
Stilpnomelane						
Biotite						
Chlorite						
Sphene						
Calcite	?			?		
Aragonite						
Garnet						

(Metawackes (SAC Rocks))

Figure 28.12 Mineral-facies charts for the Franciscan Complex: (a) metabasites, (b) metawackes, (c) metapelites, (d) metacherts.

(Sources include those in notes 25–28 and observations of the author)

Zeolite Facies metabasites include laumontite-, prehnite-, epidote-, and pumpellyite-bearing assemblages (figure 28.12). Reactions such as equation 28.12 probably produced the laumontite in metawackes.

The Central Belt melanges structurally overlie the Coastal Belt rocks. Most rocks of the Belt are considered to belong to the *Prehnite-Pumpellyite Facies*. As is the case for the Coastal Belt rocks, however, the petrogenesis of Central Belt rocks has not been adequately investigated. Furthermore, because the Central Belt consists primarily of an assemblage of melanges containing a variety of blocks and slabs of rock (J. C. Maxwell, 1974; Gucwa, 1975; Blake et al., 1988),[69] the metamorphic grade within the belt shows no regular pattern.

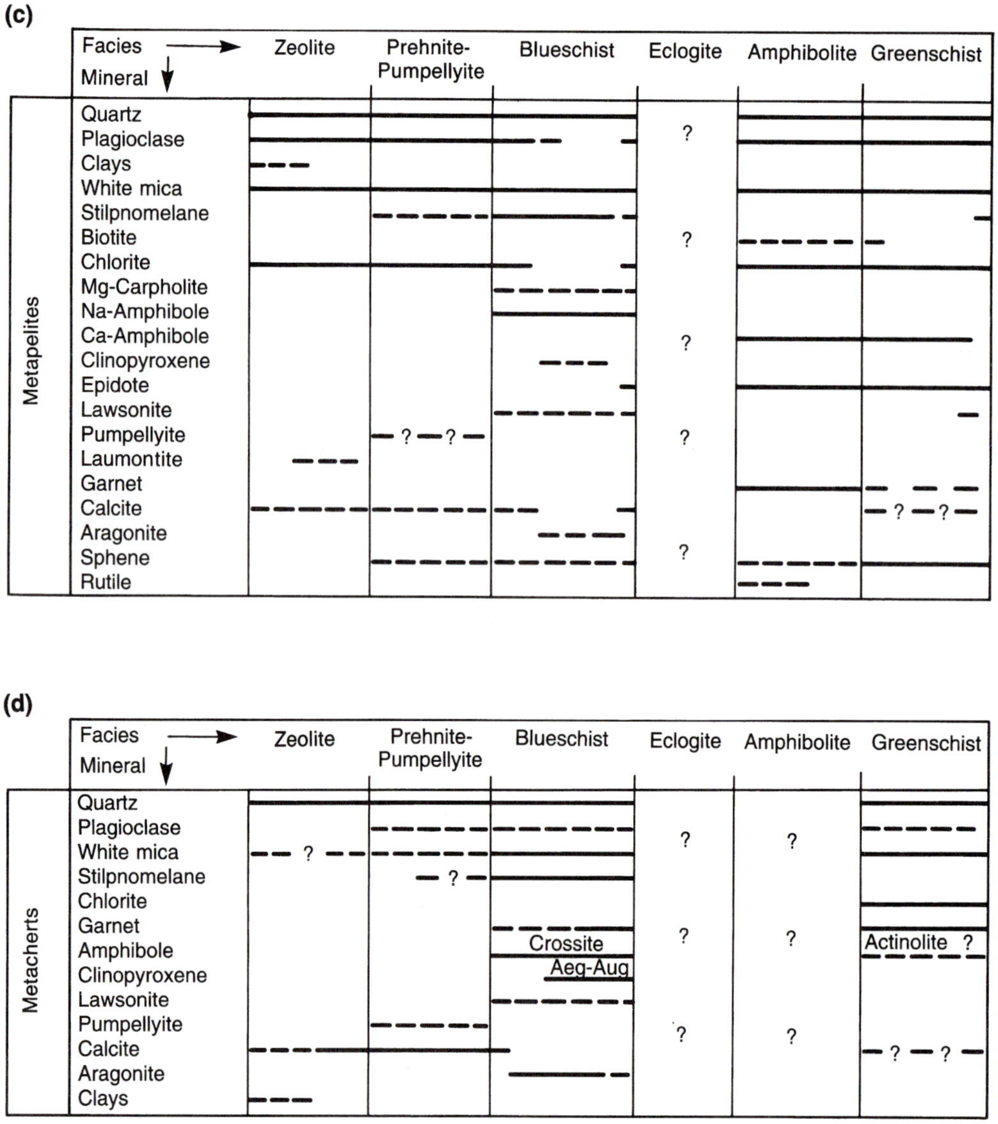

(c)

Facies → Mineral ↓	Zeolite	Prehnite- Pumpellyite	Blueschist	Eclogite	Amphibolite	Greenschist

Metapelites

Quartz, Plagioclase, Clays, White mica, Stilpnomelane, Biotite, Chlorite, Mg-Carpholite, Na-Amphibole, Ca-Amphibole, Clinopyroxene, Epidote, Lawsonite, Pumpellyite, Laumontite, Garnet, Calcite, Aragonite, Sphene, Rutile

(d)

Facies → Mineral ↓	Zeolite	Prehnite- Pumpellyite	Blueschist	Eclogite	Amphibolite	Greenschist

Metacherts

Quartz, Plagioclase, White mica, Stilpnomelane, Chlorite, Garnet, Amphibole (Crossite / Aeg-Aug), Clinopyroxene, Lawsonite, Pumpellyite, Calcite, Aragonite, Clays; Actinolite ?

Figure 28.12   Continued

Blocks in the melange range from very low-grade Zeolite Facies rocks to Eclogite and Amphibolite Facies metabasites. Most rocks retain some detrital or primary minerals, that is, they are only partly recrystallized and they exhibit a variety of phase assemblages. Texturally, metawackes range from IS and IC types to IIIS and IIIC types. Some rocks appear nearly unmetamorphosed, others contain assemblages such as

quartz–albite–alkali feldspar–chlorite–white mica (with veins of laumontite and prehnite),
quartz–albite–pumpellyite–chlorite–white mica–calcite, and
quartz–albite–prehnite–pumpellyite–chlorite–white mica–hematite,

whereas others, notably in the east, contain the Blueschist Facies assemblage

quartz–albite–white mica–chlorite–lawsonite–pumpellyite

(Jayko, Blake, and Brothers, 1986; Blake et al., 1988).[70] Prehnite is uncommon and jadeitic pyroxene is present in only a few metawacke blocks.[71] Metapelites that form the melange matrix contain the assemblages

white mica–chlorite–quartz–albite
white mica–chlorite–quartz–albite–pumpellyite, and
white mica–chlorite–quartz–albite–lawsonite

(Cloos, 1983). Like the metawackes, the metabasites show a range of mineral and metamorphic grades. The characteristic assemblage is

chlorite–pumpellyite–albite–quartz–calcite,

but actinolite-, lawsonite-, and glaucophane-bearing assemblages are present locally (see figure 28.12)(Blake et al., 1988).

To the east and structurally overlying the Central Belt is a faulted Blueschist Facies belt dominated by metasedimentary rocks and containing a variety of pumpellyite-, lawsonite-, and jadeitic-pyroxene-bearing assemblages (Blake, 1965; Suppe, 1973; Jayko, Blake, and Brothers, 1986).[72] Metawackes, metashales, and metacherts predominate. Typical metawacke assemblages include

quartz–albite–white mica–chlorite–lawsonite–aragonite,
quartz–albite–white mica–chlorite–lawsonite–jadeitic
pyroxene–aragonite, and
quartz–jadeitic pyroxene–white mica–chlorite–lawsonite–
aragonite–stilpnomelane–glaucophane.

In the Diablo Range of the central Coast Ranges (see figure 28.10), somewhat similar rocks are present. Blueschist Facies metawackes there include, in addition to the above assemblages, the assemblages

quartz–albite–white mica–chlorite–chlorite/vermiculite–
lawsonite–calcite,
quartz–albite–white mica–chlorite–chlorite/vermiculite–
lawsonite–glaucophane–pumpellyite, and
quartz–albite–white mica–chlorite–lawsonite–jadeitic
pyroxene–pumpellyite–aragonite–sphene–hematite

(see figure 28.11b) (Ernst, 1965; 1971c; Raymond, 1973a, 1973c; Patrick and Day, 1989).[73] In this region, jadeitic-pyroxene-bearing metawackes are widespread and well developed. Jadeitic pyroxenes have problably formed via reactions such as

$$\text{albite + chlorite/vermiculite + calcite + hematite} \rightarrow$$
$$\text{jadeitic pyroxene + chlorite + white mica + quartz + H}_2\text{O +}$$
$$\text{CO}_2, \qquad (28.27)$$

$$\text{albite + chlorite}_1 \text{ + lawsonite + hematite} \rightarrow \text{jadeitic}$$
$$\text{pyroxene + chlorite}_2 \text{ + quartz + H}_2\text{O} \qquad (28.28)$$

(cf. Kerrick and Cotton, 1971), equation 28.18 (Maruyama, Liou, and Sasakura, 1985), and equation 28.19 (Patrick and Day, 1989).

Other rocks include metapelites, marbles, and metacherts. Metapelites contain the assemblage

white mica–chlorite–quartz–albite–lawsonite

(Cloos, 1983). Metacherts contain such assemblages as

quartz–crossite–aegirine–hematite and
quartz–stilpnomelane–garnet

(see figures 28.11c and 28.12). Where a carbonate mineral is present in the cherts, the stable carbonate is typically aragonite. Aragonite marbles may also occur, but locally aragonite has reverted to calcite.

Blueschist Facies metabasites in the Eastern Belt and its equivalent in the Diablo Range contain such assemblages as

glaucophane–lawsonite–albite– sphene,
glaucophane–lawsonite–stilpnomelane–chlorite–albite–
quartz,
glaucophane–albite–quartz–garnet–white mica,
blue amphibole–lawsonite–pumpellyite–chlorite–albite–
garnet, and
chlorite–lawsonite–jadeitic
pyroxene–glaucophane–quartz–sphene.[74]

The assemblages reflect reactions in which plagioclase is converted to lawsonite (e.g., equation 28.17) and pumpellyite and pyroxenes incorporate Na to become sodic amphiboles. These rocks have textures that range from diablastic to schistose, but locally, relict diabasic and porphyritic textures are preserved.

In the Northern Coast Ranges, at the eastern edge of the Eastern Belt (figure 28.10) within a unit composed of metasedimentary and metavolcanic rocks, there is a small zone of *epidote-bearing,* textural zone IIIS–IVS rocks (Suppe, 1973; E. H. Brown and Ghent, 1983; Jayko, Blake, and Brothers, 1986). These rocks occur adjacent to the Coast Range Fault, which bounds the Franciscan Complex on the east. A lawsonite-out isograd separates the epidote-bearing rocks from other Blueschist Facies rocks to the west (Jayko, Blake, and Brothers, 1986). Typical assemblages in metabasites (figure 28.12) include

blue amphibole–albite–chlorite–pumpellyite–epidote,
blue amphibole–chlorite–pumpellyite–epidote–quartz–
sphene,
blue amphibole–albite–chlorite–epidote–quartz–white
mica–sphene,
blue amphibole–actinolite–albite–stilpnomelane, and
actinolite–albite–chlorite–pumpellyite–aragonite.

The reaction marking the isograd may be equation 28.24, in which lawsonite and Na-amphibole react to form epidote and other phases (E. H. Brown and Ghent, 1983). In this belt, typical metawacke phase assemblages include

quartz–albite–muscovite–paragonite–chlorite–lawsonite,
quartz–albite–white mica–chlorite–lawsonite–aragonite,
and
quartz–albite–white mica–chlorite–lawsonite–stilpnomelane.

Similar rocks occur in Oregon, to the north (B. L. Wood, 1971; R. G. Coleman, 1972; Blake and Jayko, 1986).

The high-grade blueschists and the Eclogite, Amphibolite, and Greenschist Facies metabasites that occur as blocks and slabs in Franciscan melanges have generally experienced polymetamorphism and a more complex history than the rocks in more coherent terranes (S. S. Sorenson, 1986;

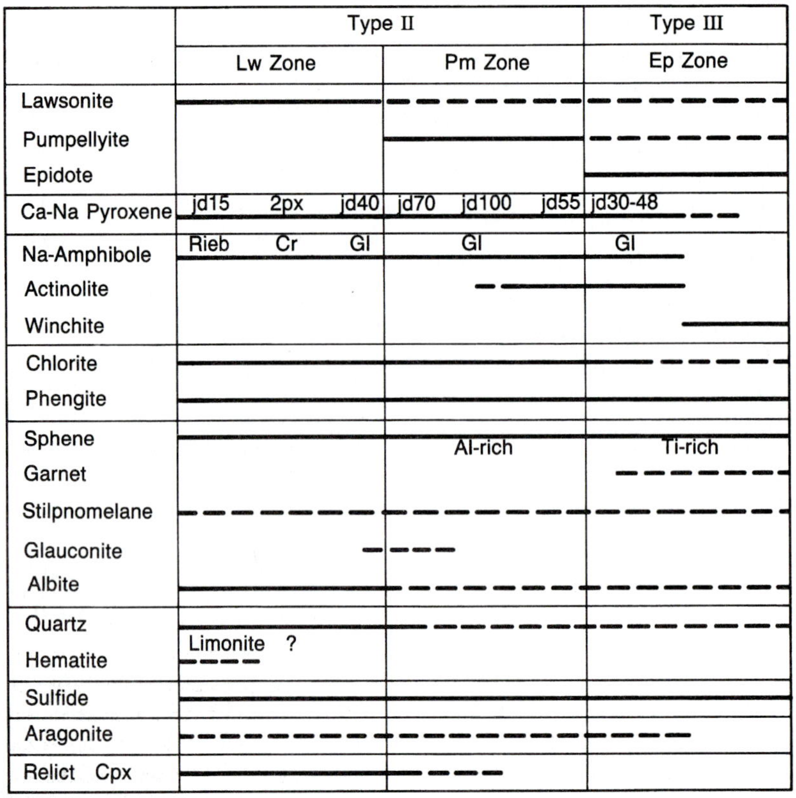

	Type II		Type III
	Lw Zone	Pm Zone	Ep Zone
Lawsonite			
Pumpellyite			
Epidote			
Ca-Na Pyroxene	jd15 2px jd40	jd70 jd100 jd55	jd30-48
Na-Amphibole	Rieb Cr Gl	Gl	Gl
Actinolite			
Winchite			
Chlorite			
Phengite			
Sphene			
Garnet		Al-rich	Ti-rich
Stilpnomelane			
Glauconite			
Albite			
Quartz			
Hematite	Limonite ?		
Sulfide			
Aragonite			
Relict Cpx			

Figure 28.13

Mineral-facies chart for the metabasites of Ward Creek, California.

(From S. Maruyama and J. G. Liou, "Petrology of Franciscan metabasites along the jadeite-glaucophane type facies series, Cazadero, California" in Journal of Petrology, 29:1-37, 1988. Copyright © 1988 Oxford University Press, Oxford England. Reprinted by permission.)

D. E. Moore and Blake, 1989; Wakabayashi, 1990). *In situ* terranes for all but the eclogites have been recognized in southern California (Platt, 1975; Bebout and Barton, 1989; S. S. Sorenson, Bebout, and Barton, 1991). Typical assemblages in the various facies are

glaucophane–lawsonite–chlorite–white
   mica–quartz–sphene (Blueschist Facies)
actinolite–epidote–albite–chlorite–quartz–white
   mica–sphene (Greenschist Facies)
hornblende–epidote–garnet–clinopyroxene–quartz–rutile
   (Amphibolite Facies)
omphacite–garnet–epidote–glaucophane–actinolite–
   phengite–sphene (Eclogite Facies)

(see figure 28.11d).[75] In many cases, early-crystallized assemblages formed in one facies are overprinted by assemblages formed later under conditions of another facies.

The most thoroughly studied of these high-grade blocks are those of the Ward Creek/Cazadero area of California (R. G. Coleman and Lee, 1963; Maruyama and Liou, 1988; Oh, Liou, and Maruyama, 1991).[76] At this locality, a large metabasite mass and some tectonic blocks are engulfed by Central Belt melange. R. G. Coleman and Lee (1963) divided metabasites here into four types, primarily on the basis of texture, but also on degree of metamorphism. These types include Type I (unmetamorphosed), Type II (metamorphosed, fine-grained, nonfoliated), Type III (metamorphosed, medium-grained, foliated), and Type IV (metamorphosed, coarse-grained, foliated; the high-grade tectonic blocks of blueschist and eclogite). The large metabasite mass at Ward Creek includes Type II and Type III blueschists and contains a series of three zones—a lawsonite zone, a pumpellyite zone, and an epidote zone—representing three metamorphic grades of increasing temperature (figure 28.13) (R. G. Coleman and Lee, 1963; Maruyama and Liou, 1988; Oh, Liou, and Maruyama, 1991). Reactions marking the zone boundaries are discontinuous and include a pumpellyite-in reaction, an actinolite-in reaction, and an epidote-in reaction.

Analyses of the metamorphic conditions that produced the metamorphic rocks in the Franciscan Complex are based on experimental phase equilibria, isotopic analyses, virtrinite

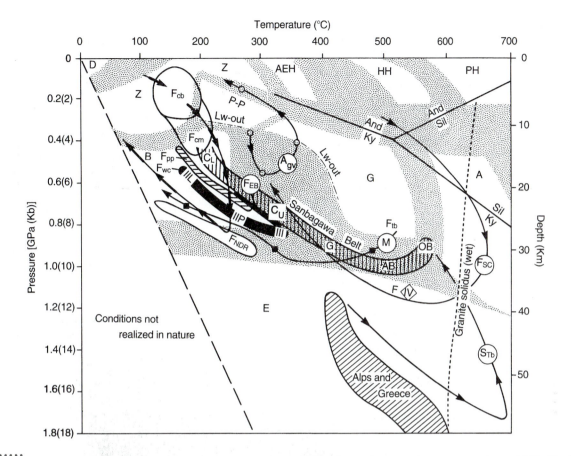

Figure 28.14 P-T-t paths and estimated P-T conditions under which Franciscan and Sanbagawa Facies series rocks were metamorphosed. Elliptical areas represent generalized conditions of metamorphism for various terranes or masses of rock.

(And = andalusite, Ky = kyanite, Sil = sillimanite. A_{gv} = P-T path for carpholite-bearing blueschists from the western Alps (Goffe and Velde, 1984). F_{Cb} = Franciscan Complex, Coastal Belt (Underwood, 1985; Underwood, Blake, and Howell, 1987; Blake et al., 1988; Dumitru, 1991). F_{CM} = Franciscan Complex, Central Belt melange (Cloos, 1983; Underwood, Blake, and Howell, 1987; Blake et al., 1988; Dumitru, 1991). F_{EB} (higher -T, dark band, left of center) = Franciscan Complex, Eastern Belt of the Northern Coast Ranges (Blake et al., 1988). F_{NDR} = Franciscan Complex rocks of the northern Diablo Range (Bostick, 1974; Patrick and Day, 1989). F_{pp} = Franciscan metawackes of Pacheco Pass, central Diablo Range, California. F_{WC} = Franciscan metabasites of Ward Creek. IIL = Lawsonite zone; IIP = Pumpellyite

Zone; III = Epidote Zone (Maruyama and Liou, 1988; Oh, Liou, and Maruyama, 1991). F-IV = Franciscan tectonic blocks of Type IV (Oh, Liou, and Maruyama, 1991). F_{SC} = Generalized path for Franciscan amphibolitic tectonic blocks (Moore and Blake, 1989; Wakabayashi, 1991) and Santa Catalina schists from subduction-zone hanging wall (Sorenson, Bebout, and Barton 1991). F_{tb} = Franciscan tectonic block retrograde metamorphic path (from D. E. Moore, 1984). In the Sanbagawa Belt, C_L = lower Chlorite Zone; C_U = upper Chlorite Zone; G = Garnet Zone; AB = Albite-biotite Zone; and OB = Oligoclase-biotite Zone for pelitic rocks (Takasu, 1989). S_{Tb} = generalized Sanbagawa tectonic block metamorphic path (Takasu, 1989). Field designated Alps and Greece represents P-T conditions indicated by some eclogitic blueschists from these regions (Schleistedt and Matthews, 1987; Brocker, 1990; Pognante, 1989; 1991). Facies fields are the same as those shown in figure 24.6. Facies designations as follows: A = Amphibolite, AEH = Albite-Epidote Hornfels, B = Blueschist, Ec = Eclogite, G = Greenschist, HH = Hornblende Hornfels, PH = Pyroxene Hornfels, P-P = Prehnite-pumpellyite, Z = Zeolite.)

reflectance, and fission tracks.[77] Estimated P-T conditions are shown in figure 28.14. In summary, Eclogite Facies rocks were metamorphosed under conditions of T = 290–540° C and P = 0.6–1.4 Gpa (6–14kb). Metamorphism of Eastern Belt rocks occurred at P = 0.4–1.0 Gpa and T = 125–350° C, whereas Central Belt melange metamorphism resulted from pressures of 0.2–0.6 Gpa and temperatures of 125–300° C. Zeolite Facies metamorphism of Coastal Belt rocks occurred at about

P = 0.1–0.3 Gpa and T = 100–200° C. Projected P-T-t paths are shown in figure 28.14. Note that the rocks of the eastern Diablo Range remained at low T conditions, whereas the metamorphic conditions for the Eastern Belt in the northern Coast Ranges reaches somewhat higher temperature conditions. Note also that there is not a single geothermal gradient for the entire complex, but rather, different path histories for different blocks or belts.[78]

The origin of the Franciscan Facies Series in California served as a focal point for the debates about the origin of blueschists that occurred during the 1960s and 1970s. It is now generally agreed that subduction carried the rocks of each of the belts (or terranes within the belts) to the depth appropriate for metamorphism of that belt. Among the petrologic problems that remain are the problems of preservation and uplift of these rocks and problems associated with the fact that in most of the rocks recrystallization is incomplete.

As noted above, evidence of disequilibrium, incomplete neocrystallization, and polymetamorphism abounds in Franciscan rocks.[79] Some of the evidence of disequilibrium gave rise to the suggestion that the jadeitic pyroxene and other minerals in the jadeitized metawackes of the Diablo Range are detrital rather than metamorphic (Brothers and Grapes, 1989). The disequilibrium and lack of homogeneity in jadeitic pyroxenes is well known,[80] as are detrital Blueschist Facies materials (O'Day and Kramer, 1972; D. S. Cowan and Page, 1975; D. E. Moore and Liou, 1980). In the case of the jadeitic pyroxenes of the Diablo Range, euhedral crystals that have grown across grain boundaries, sympathetic changes in entire phase assemblages coincident with inferred metamorphic grade, and jadeitic pyroxene aspect ratios clearly demonstrate that most (if not all) jadeitic pyroxene in these metawackes is metamorphic in origin (Raymond, 1991; Raymond and Nall, 1991).

How the Franciscan Blueschist Facies rocks reached the surface without being recrystallized under higher-T conditions is a matter of some debate. Models currently favored include both S- and E-type models of synsubduction uplift. Cloos (1982, 1984) has argued, using the S-type subduction channel model, that high-grade blocks and eclogites in the Central Belt have been uplifted via flow in a subduction channel, as well as by local diapiric intrusion of melange into layered sequences of sediment. Krueger and Jones (1989) and T. A. Harms, Jayko, and Blake (1992) favor shallowing of the subduction angle, resulting in E-type exhumation. Ernst (1988) accepts both of these models for given situations, but argues that buoyant rise of accreted wedges may also take place.

SUMMARY

Sanbagawa and Franciscan Facies series are high P/T facies series distinguished by the presence of blue, sodic amphiboles. The Sanbagawa Facies Series displays the facies sequence Zeolite → Prehnite-Pumpellyite → Blueschist → Greenschist → Amphibolite, corresponding to a geothermal gradient of about 10–20° C (figure 28.14). In the Franciscan Facies Series, the facies sequence Zeolite → Prehnite-Pumpellyite → Blueschist → Eclogite reflects lower temperatures at low to high pressures. These facies series occur worldwide in orogenic belts, but are almost entirely confined to orogens of Phanerozoic age.

The key facies for recognizing high P/T facies series is the Blueschist Facies. Critical minerals indicative of this facies are lawsonite, aragonite, and jadeitic pyroxene. In addition, blue, sodic amphiboles, abundant in blueschists and present in other rocks, reflect high-pressure conditions. A variety of continuous and discontinuous, terminal and nonterminal reactions involving phases such as plagioclase, laumontite, prehnite, pumpellyite, epidote, actinolite, glaucophane, crossite, jadeitic pyroxene, omphacite, chlorites, white micas, and garnet yield these and other index phases. Aragonite forms by a simple polymorphic transformation. Although Zeolite, Prehnite-Pumpellyite, Greenschist, and Amphibolite Facies rocks occur in the high P/T facies series, their phase assemblages are generally like those found in the same facies of other (lower P/T) facies series. Eclogites may occur in both Franciscan and Sanbagawa Facies series, but they are characteristic of Franciscan Facies Series. The metamorphic conditions under which the rocks of these facies series generally form are $P = 0.2$–1.4 Gpa and $T = 100$–550° C.

High P/T facies series form in subduction zone settings, where they characterize the rocks of the low-temperature outer belt of a paired metamorphic belt. How Blueschist Facies rocks are preserved over time and during uplift, when they would normally be heated and converted to Greenschist or other higher-temperature facies, is a problem. Postsubduction buoyant uplift, if quite rapid, can provide a mechanism for quick exposure of the rocks, but it is unlikely that this process is generally operative in the history of Franciscan Facies Series blueschists. The S-type imbricate thrust model of subduction, accretion, and uplift, in which subduction zones act as two-way streets, provides an explanation for ascent of slabs of rock up a cold subduction zone. As long as a relatively rapid subduction rate involving a cool slab is maintained, the rocks remain "refrigerated" during subduction. The S-type subduction channel model explains the similar upward movement of blueschist and eclogite blocks in tectonic melanges formed within the subduction zone. Alternatively, E-type exhumation of blueschists may occur where subduction and accretion create uplift and a resulting instability in the accretionary wedge, causing listric normal faulting at the top of the wedge and exposure of rocks formerly deeply buried in the subduction zone. In rare cases, other processes involving oblique subduction and strike-slip or transform faulting may also result in exposure of the blueschists along a plate margin.

EXPLANATORY NOTES

1. The term blueschist is used in this text in an informal way to refer to blue amphibole-bearing schists.
2. The use of the name Blueschist Facies in this text follows the original definition of E. H. Bailey (1962; E. H. Bailey, Irwin, and Jones, 1964). The critical mineral in SAC and metabasic rocks is lawsonite and assemblages such as lawsonite + albite + glaucophane and lawsonite + jadeitic pyroxene + glaucophane are characteristic. Brothers and Blake (1987)

noted and lamented the tendency of some petrologists to stray from this definition. In contrast, Taylor and Coleman (1968), B. W. Evans and Brown (1987), and B. W. Evans (1990) proposed broadening the definition of Blueschist Facies to include assemblages of glaucophane/crossite + epidote. Although B. W. Evans and Brown (1987) and B. W. Evans (1990) make a good case for distinguishing between high- and low-pressure Greenschist Facies rocks and perhaps for adding a new facies name for rocks formed at conditions of about T = 300–550° C and P = 0.6–1.1 Gpa (6–11 kb), the use of the name "epidote blueschist" and the extension of the name Blueschist Facies to encompass all glaucophane (or crossite)-bearing rocks is not a good idea and is bound to create confusion, because it redefines an established and widely used term. If a new facies term is needed, a name such as High-Pressure Greenschist Facies, Blue-Green Facies, or Epidote–Blue Amphibole Facies would distinguish such a facies without creating unnecessary confusion and controversy.

Other names that have been applied to facies defined similarly to the Blueschist Facies include Glaucophane Schist Facies (F. J. Turner and Verhoogen, 1960), Lawsonite-Glaucophane-Jadeite Facies, and Lawsonite-Albite Facies (Winkler, 1965, 1967). Winkler (1974, 1976, 1979) later abandoned the latter terms.

3. Also see Seki (1960), Essene, Fyfe, and Turner (1965), Ernst and Seki (1967), Landis and Coombs (1967), H. P. Taylor and Coleman (1968), Liou (1971b), Landis and Bishop (1972), Black (1974b), Bostick (1974), Platt (1975), Caron, Kienast, and Triboulet (1981), M. A. Carpenter (1981), Seki and Liou (1981), Brothers and Yokoyama (1982), Schleistedt and Matthews (1987), Schreyer (1988), Patrick and Evans (1989), Wakabayashi (1990), and Wallis and Banno (1990).

4. In some cases, Eclogite Facies rocks are found in Sanbagawa Facies Series, for example, in the Kargi Massif of Turkey (Okay, 1986) and the southern Alps (Barnicoat and Fry, 1989).

5. Refer to the references in note 3 and appendix C for phase stability studies and papers on isotopic analyses. Vitinite reflectance studies are studies on the light-reflecting quality of carbonaceous (coal) fragments. The reflectance increases with temperature (For applications in the Franciscan Facies Series, see Bostick, 1974; Underwood, Blake, and Howell, 1987; Underwood, O'Leary, and Strong, 1988). Fission track studies are studies of the microscopic trails made in minerals by radiation produced by radioactive decay of included elements (see Dumitru, 1988, for an example). P-T conditions are discussed by Seki (1960), Ernst (1965, 1971a), Ernst and Seki (1967), Landis and Coombs (1967), Newton and Smith (1967), H. P. Taylor and Coleman (1968), D. G. Bishop (1972), Landis and Bishop (1972), Black (1974b), Caron, Kienast, and Triboulet (1981), M. A. Carpenter (1981), Seki and Liou (1981), Brothers and Yokoyama (1982), Kienast and Ranguin (1982), Underwood (1985, 1989), Newton (1986), Yokoyama, Brothers, and Black (1986), Matthews and Schleistedt (1987), Pognante and Kienast (1987), Underwood and Howell (1984), Maruyama and Liou (1988), Sedlock (1988), Dumitru (1989; 1991), Patrick and Day (1989), Wakabayashi (1990), Oh, Liou, and Maruyama (1991), Pognante (1991), and others.

6. For example, see Winkler (1965, 1967), Dickinson et al. (1969), Miyashiro (1973a), and Yardley (1989). Offler et al. (1980) use the term "burial metamorphism," to refer to low-grade rocks metamorphosed under a high geothermal gradient.

7. The distribution of Franciscan and Sanbagawa Facies series rocks or the specifics of the petrology of these facies series at various localities are discussed in Seki (1958), Miyashiro (1961), E. H. Bailey, Irwin, and Jones (1964), Ernst (1965, 1973a, 1977a), van der Kaaden (1969), Brothers (1970, 1974), Guitard and Saliot (1971), Makanjuola and Howie (1972), Black (1973, 1974a,b, 1975, 1977), Hotz (1973), Miyashiro (1973a,b), Herve et al. (1974), Brothers (1974), Nagle (1974), Triboulet (1974), Roy (1977a), Wood (1979c), Okay (1980, 1986), Shams, Jones, and Kempe (1980), E. H. Brown et al. (1981), Caron, Kienast, and Triboulet (1981), Higashino et al. (1981), Liou (1981a,b), Brothers and Yokoyama (1982), Kienast and Rangin (1982), Lippard (1983), Forbes, Evans, and Thurston (1984), Goffe and Velde (1984), Banno, Sakai, and Higashino (1986), B. W. Evans and Brown (1986), Cotkin (1987), Pognante and Kienast (1987), Sedlock (1988), Maruyama and Liou (1988), Banno and Sakai (1989), Barnicoat and Fry (1989), D. E. Moore and Blake (1989), Patrick and Evans (1989), Pognante (1989a, 1991), Shibakusa (1989), Smelik and Veblen (1989), Takasu (1989), Brocker (1990), Higashino (1990), Lucchetti, Cabella, and Cortesogno (1990), Shenbao (1990), Wakabayashi (1990), Wallis and Banno (1990), El-Shazly and Liou (1991), and Oh, Liou, and Maruyama (1991). Also see the additional references in these sources.

8. For example, Zhang and Liou (1987), Patrick and Evans (1989), and Schermer (1990).

9. See Ernst (1972b, 1988) and Dobretsov et al. (1982) for reviews. Also see I. A. Paterson and Harakal (1974) and Armstrong et al. (1986).

10. For example, see Ernst (1973b, 1977a), Laird and Albee (1981), Goffe and Velde (1984), Dal Piaz and Lombardo (1986), Schliestedt and Matthews (1987), Patrick and Lieberman (1988), Takasu (1989), and Brocker (1990).

11. E. H. Brown et al. (1981), Ross and Sharp (1988), D. E. Moore and Blake (1989), Wakabayashi (1990).

12. W. Glassley (1974), E. H. Brown et al. (1981), Kienast and Ranguin (1982), E. H. Brown (1986), T. E. Moore (1986), Sedlock (1988).

13. Sanbagawa is sometimes spelled Sambagawa. The Sanbagawa Facies Series, as used here, differs from the Sanbagawa "jadeite-glaucophane type" originally defined by Miyashiro (1961), as it does not include the very low temperature rocks, which are assigned to the Franciscan Facies Series. Many lawsonite-glaucophane schists and the jadeite-bearing rocks, once assigned to the Sanbagawa belt in Japan, are now recognized to be blocks in a tectonic melange along the Kurosegawa zone (Maruyama et al., 1984; Banno, 1986). In fact, the Sanbagawa rocks more closely match Miyashiro's description of "high pressure intermediate type" facies series, in that the Sanbagawa belt, as now defined, lacks jadeite-quartz assemblages and contains sporadic occurrences of sodic amphibole. The blue amphibole, commonly crossite, occurs with epidote and actinolite (Ernst et al., 1970; Higashino, 1990; Otsuki and Banno, 1990). The Sanbagawa Facies

617

Series, as defined here, reflects the petrology of the Sanbagawa belt as it is now known. For reviews of the Sanbagawa rocks, see Banno (1986) and Banno and Sakai (1989), as well as Wallis and Banno (1990) and articles in the *Journal of Metamorphic Petrology*, v. 8, p. 393ff (1990).

14. Also see R. G. Coleman and Lee (1963), Raymond (1973a), and Jayko, Blake, and Brothers (1986).

15. McKee (1962), Ernst (1971c), Raymond (1973c), D. S. Cowan (1974), Maruyama, Liou, Sasakura (1985), Patrick and Day (1989).

16. Seki et al. (1971, in Liou, Maruyama, and Cho, 1987).

17. Ernst et al. (1970), Watanabe and Kobayashi (1984), Banno (1986), Liou, Maruyama, and Cho (1987), but see Higashino (1990) and Otsuki and Banno (1990). Toriumi (1975) reports the sodic pyroxene aegirine-augite in some basic rocks.

18. Phase assemblages in Sanbagawa-like Facies Series rocks are listed or discussed by Seki (1958), D. S. Coombs (1960), Ernst and Seki (1967), Seki, Ernst, and Onuki (1969), Brothers (1970, 1974), Ernst et al. (1970), Black (1973, 1974a,b, 1975, 1977), E. H. Brown (1974), Toriumi (1975), D. S. Coombs et al. (1977), Ernst (1977a, 1982, 1984), Roy (1977a), R. M. Briggs (1978), Liou (1979), Shams, Jones, and Kempe (1980), Higashino et al. (1981, 1984a,b), Brothers and Yokoyama (1982), Enami (1983), Lippard (1983), Goffé and Velde (1984), Watanabe and Kobayashi (1984), Sakai et al. (1985), Cho, Liou, and Maruyama (1986), Jayko, Blake, and Brothers (1986), Kunugiza, Takasu, and Banno (1986), Oberhansli (1986), Cotkin (1987), Cho and Liou (1987), Liou, Maruyama, and Cho (1987), Banno and Sakai (1989), Barnicoat and Fry (1989), Shibakusa (1989), Brocker (1990), Higashino (1990), Lucchetti, Cabella, and Cortesogno (1990), Otsuki and Banno (1990), Shenbao (1990), and El-Shazly and Liou (1991).

19. Prehnite-Pumpellyite Facies rocks, recognized in the Sanbagawa Belt of the Kanto Mountains of Central Japan are not specifically included in zones in the south (compare Toriumi, 1975 with Otsuki and Banno, 1990 and Wallis and Banno, 1990).

20. For example, see Seki (1958), Ernst et al. (1970), Toriumi (1975), Enami (1983), Watanabe and Kobayashi (1984), Sakai et al. (1985), and Higashino (1990). Zones are based on Banno (1986).

21. This lower Greenschist Facies is equivalent to the Pumpellyite-Actinolite Facies of Liou, Maruyama, and Cho (1987).

22. Ernst (1964) discusses exchange reactions involving some of these phases.

23. Ernst et al. (1970), Toriumi (1975), Banno (1964, in Brown, 1977b), Nakajima, Banno, and Suzuki (1977), Liou, Maruyama, and Cho (1987), Otsuki and Banno (1990).

24. Also see Nakajima, Banno, and Suzuki (1977), E. H. Brown et al. (1981), J. B. Thompson, Laird, and Thompson (1982), E. H. Brown (1986), Maruyama, Cho, and Liou (1986), Schliestedt and Matthews (1987), and B. W. Evans (1990). In addition see Owen (1989).

25. Zeolite Facies minerals and assemblages of Franciscan-type Facies series are described by D. S. Coombs (1960), R. J. Stewart and Page (1974), Ernst (1977a; 1984), R. J. McLaughlin et al. (1982), Liou et al. (1987), and Blake et al. (1988). Data also from L. A. Raymond (unpublished data, Franciscan Complex, figure 28.3 and table 28.4). Also see Cho, Liou, and Maruyama (1986).

26. Prehnite-Pumpellyite Facies minerals and assemblages from Franciscan-type Facies Series are discussed by Ernst (1971b, 1980), D. S. Cowan (1974), W. E. Glassley (1975), Raymond (unpublished data, Franciscan Complex, this text), Nakajima, Banno, and Suzuki (1977), E. H. Brown et al. (1981), J. B. Thompson, Laird, and Thompson (1982), and Blake et al. (1988). Also see Cho, Liou, and Maruyama (1986) and Liou et al. (1987).

27. Blueschist Facies minerals and assemblages of the Franciscan Facies Series, are discussed, for example, by M. E. Maddock (1955), E. H. Bailey, Irwin, and Jones (1964), Ernst (1965, 1971c), Seki, Ernst, and Onuki (1969), Guitard and Saliot (1971), Black (1973), Hotz (1973), Raymond (1973a,c), Suppe (1973), Caron, Kienast, and Triboulet (1981), Maruyama, Liou, and Sasakura (1985), Jayko, Blake, and Brothers (1986), S. S. Sorenson (1986), Cotkin (1987), Schliestedt and Matthews (1987), Liou and Maruyama (1987), Blake et al. (1988), Maruyama and Liou (1988), Sedlock (1988), Bebout and Barton (1989), Pognante (1991), and S. S. Sorenson, Bebout, and Barton (1991).

28. Additional discussion of Franciscan Facies Series phases or phase assemblages may be found in McKee (1962), Onuki and Ernst (1969), Black (1973, 1974a,b, 1975), C. Hoffman and Keller (1979), D. E. Moore and Liou (1979), Mevel and Kienast (1980), Wood (1980), Brothers and Yokoyama (1982), D. E. Moore (1984), Oh et al. (1991) Shau et al. (1991), and in other sources listed in notes 68, 70, 73, 74, and 75.

29. Ernst (1972a), W. Glassley (1974), Ivanov and Gurevich (1975), Liou (1981b), Liou, Maruyama, and Cho (1987). Also see Schliestedt and Matthews (1987), Bebout and Barton (1989), and S. S. Sorenson, Bebout, and Barton (1991) for discussions of the influence of a fluid phase on particular assemblages.

30. Refer to the references in appendix C and to sources such as Liou (1971a,b), Maresch (1977), Luckscheiter and Monteani (1980), M. A. Carpenter (1981), F. J. Turner (1981), Maruyama, Cho, and Liou (1986), Newton (1986), Heinrich and Althaus (1988), and Oh, Liou, and Maruyama (1991) for the data on which the P-T limits are based. Also see figure 28.13 and note 77 below.

31. For example, R. G. Coleman and Lee (1963), Ernst et al. (1970), R. G. Coleman and Lanphere (1971), Kerrick and Cotton (1971), Ernst (1973b), Raymond (1973a,b,c), S. E. Swanson and Schiffman (1979), Wood (1979b); D. E. Moore (1984), Maruyama, Liou, and Sasakura (1985), Maruyama and Liou (1988), Barnicoat and Fry (1989), D. E. Moore and Blake (1989), Pognante (1989a), Takasu (1989), Wakabayashi (1990).

32. For example, see Kerrick and Cotton (1971), Ernst (1972a), and Raymond (1973a).

33. Ernst (1971b), Raymond (1973a,c), D. E. Moore (1984), Maruyama and Liou (1988), D. E. Moore and Blake (1989).

34. For example, see Taliaferro (1943), Gresens (1969, 1970), Ernst (1971a), and also see Hlabse and Kleppa (1968).

35. Brothers (1954), Bloxam (1959, 1966), Essene, Fyfe, and Turner (1965).

36. See Milton and Eugster (1959).

37. See Suppe (1970, 1973), Raymond (1973a, 1974b).

38. Misch (1959, 1966), E. H. Bailey, Irwin, and Jones (1964), Ernst (1971a), E. H. Brown et al. (1981).

39. For example, see Raymond (1973a,c, 1974b), S. E. Swanson (1981), Abbott and Raymond (1984).

40. Seki (1958), Ernst (1964).

41. Experimental studies plus evidence that Blueschist Facies metamorphism was isochemical (R. G. Coleman and Lee, 1963; Ernst, 1963a; Ghent, 1965) indicate a no answer to question 1. There is simply no evidence available that supports a yes answer to question 2.

42. R. G. Coleman and Lee (1962), Blake (1965), Blake, Irwin, and Coleman (1967, 1969), Blake and Cotton (1969), Brothers (1970), R. G. Coleman (1972), van Bemmelen (1974). Also see de Roever (1967 in van der Kaaden, 1969).

43. Although thrust faulting apparently occurred in some places along the eastern boundary of the Franciscan Complex (Raymond, 1973a,c, Suppe and Foland, 1978; Roure, 1981; Roure and Blanchet, 1983; Jayko, Blake, and Harms, 1987), several studies suggest that the "regional thrust fault" has a complex history and may be a normal or strike-slip fault in many places (Raymond, 1969, 1970, 1973a,c, Worrall, 1981; C. M. Wentworth et al., 1984; Jayko, Blake, and Harms 1987; T. A. Harms and Dunlap, 1991; Unruh et al., 1991).

44. J. C. Maxwell (1974), Bachman (1978), Underwood (1983), Blake and Jayko (1986), Blake et al. (1988).

45. Brothers and Blake (1973).

46. Also see D. S. Cowan (1974), Crawford (1975), and Maruyama, Liou, and Sasakura (1985) whose work reveals no regular metamorphic patterns or upside-down metamorphism *related to the fault.*

47. For example, E. H. Bailey, Irwin, and Jones (1964), Miyashiro (1967), Suppe (1969), van der Kaaden (1969), Brothers and Blake (1973), and van Bemmelen (1974, 1976).

48. Ernst (1970, 1971a, 1984), R. G. Coleman (1972), C. Y. Wang and Shi (1984). Also see references in notes 30 and 77.

49. For example, see Ernst et al. (1970) and Wallis and Banno (1990).

50. Also see the description of the Franciscan Complex in the example at the end of this chapter.

51. Wood (1979a), Draper and Bone (1981), C. Y. Wang and Shi (1984), Cloos and Shreve (1988a), Ernst (1988).

52. Seely, Vail, and Walton (1974), Karig and Sharman (1975).

53. Also see Cloos and Shreve (1988a,b).

54. Also see Schermer (1990) and Schermer, Lux, and Burchfiel (1990).

55. Ernst (1971a) also suggested that strike-slip faulting may be important in exposing blueschists.

56. There are some structural problems with this model relating to sense of shear, thrust geometry, regional structure, and age of melange formation, but a discussion of these is beyond the scope of this text.

57. Van Bemmelen's mantle diapirism (undation) model (1974, 1976) also provides such an explanation.

58. This description of the Franciscan Complex is based on the works of Taliaferro (1943), M. E. Maddock (1955, 1964), R. G. Coleman and Lee (1963), E. H. Bailey, Irwin, and Jones (1964), Ernst (1965, 1970, 1971b,c, 1980), B. M. Page (1966, 1981), Blake, Irwin, and Coleman (1967, 1969), Ernst et al. (1970), Kerrick and Cotton (1971), Berkland et al. (1972), W. G. Gilbert (1973, 1974), Raymond (1973a,c, 1974a,b, 1977), Suppe (1973), D. S. Cowan (1974), J. C. Maxwell (1974), K. E. Crawford (1975, 1976), Suppe and Foland (1978), R. J. McLaughlin et al. (1982), Blake, Howell, and Jayko (1984), R. J. McLaughlin and Ohlin (1984), D. E. Moore (1984), D. E. Moore and Blake (1989), Underwood (1984, 1989), Maruyama, Liou, and Sasakura (1985), Blake and Jayko (1986), Cloos (1986), Jayko, Blake, and Brothers (1986), Liou and Maruyama (1987), Blake et al. (1988), Dumitru (1988, 1991), Maruyama and Liou (1988), Patrick and Day (1989), Wakabayashi (1990), Oh, Liou, and Maruyama (1991), and the many additional works referenced in the notes below.

59. Metawackes are usually referred to as metagraywackes in the literature.

60. These high-pressure rocks drew much attention during the 1950s and 1960s from petrologists and mineralogists at the University of California, Berkeley, and the U.S. Geological Survey (Brothers, 1954; Borg, 1956; Bloxam, 1959, 1960; G. A. Davis and Pabst, 1960; R. G. Coleman and Lee, 1963; Essene et al.,1965; R. G. Coleman and Papike, 1968). Their focus on glaucophane and related schists led many geologists who had not been to California to the mistaken view that the Coast Ranges were dominated by glaucophane schists. In fact, these "high grade" rocks comprise less than 1% of the rocks of the Coast Ranges.

61. E. H. Bailey, Irwin, and Jones (1964), Ernst (1965, 1971a,b,c, 1980), Blake, Irwin, and Coleman (1967, 1969), Blake et al. (1988).

62. E. H. Bailey, Irwin, and Jones (1964), Ernst (1971b), Suppe (1972), O'Day and Kramer (1972), Blake and Jones (1974), Evitt and Pierce (1975), Bishop (1977), R. J. McLaughlin et al. (1982), Underwood (1985), Blake et al. (1988). Whether or not all of the contacts are faults has been a matter of controversy. See Maxwell (1974) and J. C. Maxwell et al. (1981) for alternative interpretations.

63. For studies showing subdivision of the Franciscan Complex, see Hsu (1969), Raymond and Christensen (1971), Raymond (1973a, 1974a), Suppe (1973), D. S. Cowan (1974), Maxwell (1974) and the references therein, K. E. Crawford (1975, 1976), Gucwa (1975), M. A. Jordan (1978), B. M. Page (1981), R. J. McLaughlin et al. (1982), Blake, Howell, and Jayko (1984), R. J. McLaughlin and Ohlin (1984), Underwood (1984), Blake, Jayko, and McLaughlin (1985), Jayko, Blake, and Brothers (1986), and Wakabayashi (1990). Also see Wagner, Bortugno, and McJunkin (1990).

64. Ernst (1965, 1971c), Cotton (1972), Raymond (1973a,c), D. S. Cowan (1974), Crawford (1975). Also see summaries in B. M. Page (1981) and Wagner, Bortugno, and McJunkin (1990).

65. E. H. Bailey, Irwin, and Jones (1964), Hsu (1969), W. G. Gilbert (1973, 1974), Ernst (1980), Page (1981).

66. For example, see Blake, Howell, and Jayko (1984) and R. J. McLaughlin and Ohlin (1984).

67. E. H. Bailey, Irwin, and Jones (1964), R. J. McLaughlin et al. (1982), Underwood (1989).

68. Coastal Belt metawacke assemblages are based on observations by the author and the modes or generalized reports of E. H. Bailey, Irwin, and Jones (1964), O'Day and Kramer (1972), Ernst (1980), R. J. McLaughlin et al. (1982), Blake, Howell, and Jayko (1984), Cloos (1986), Underwood (1989), and Blake et al. (1988). Pumpellyite is reported from this belt by R. J. McLaughlin et al. (1982) and prehnite and pumpellyite are reported in metabasites by Blake et al. (1988), but complete phase assemblages are not reported, making it impossible to assess whether the assemblages are Zeolite Facies or Prehnite-Pumpellyite Facies assemblages (both can contain these phases).

69. Also see Raymond and Christensen (1971), F. W. McDowell et al. (1984), and Blake et al. (1988).

70. Central Belt assemblages are based on Raymond and Christensen (1971), Jayko, Blake, and Brothers (1986), Blake et al. (1988), Underwood (1989), L. A. Raymond (unpublished data).

71. For example, see Cloos (1986) and Blake et al. (1988).

72. Also see Ghent (1965), Blake, Irwin, and Coleman (1967, 1969), B. L. Wood (1971), J. C. Maxwell (1974), Suppe and Foland (1978), and Blake et al. (1988).

73. Also see Maddock (1964), Kerrick and Cotton (1971), D. S. Cowan (1974), Crawford (1975), and D. E. Moore and Liou (1979).

74. Eastern Belt metabasite phase assemblages are listed or discussed in Ernst (1965), Ghent (1965), Seki, Ernst, and Onuki (1969), B. L. Wood (1971), Suppe (1973), E. H. Brown and Bradshaw (1979), Cloos (1982), E. H. Brown and Ghent (1983), Jayko, Blake, and Brothers (1986), and Blake et al. (1988).

75. For descriptions of eclogites, hornblende schists, Greenschist Facies metabasites, and high-grade blueschists, see works such as Switzer (1945, 1951), Gealey (1951), Brothers (1954), Borg (1956), R. G. Coleman and Lee (1963), R. G. Coleman et al. (1965), E. H. Brown and Bradshaw (1979), D. E. Moore (1984), S. S. Sorenson (1986), Liou and Maruyama (1987), Maruyama and Liou (1987, 1988), Bebout and Barton (1989), D. E. Moore and Blake (1989), Wakabayashi (1990), and S. S. Sorenson, Bebout, and Barton (1991).

76. Also see R. G. Coleman and Lee (1962), R. G. Coleman and Clark (1968), R. G. Coleman and Papike (1968), H. P. Taylor and Coleman (1968), E. H. Brown and Bradshaw (1979), Liou and Maruyama (1987), and Maruyama and Liou (1987).

77. Newton and Smith (1967), Ernst (1963, 1971a, 1988), Liou (1971a,b), Nitsch (1972), Bostick (1974), Maresch (1977), Cloos (1983), D. E. Moore (1984), Liou, Maruyama, and Cho (1985), Maruyama, Liou, and Sasakura (1985), Underwood (1985), Liou et al. (1987), Underwood, Blake, and Howell (1987), Blake (1988), Underwood, O'Leary, and Strong (1988), Dumitru (1988, 1991), Maruyama and Liou (1988), D. E. Moore and Blake (1989), Patrick and Day (1989), Wakabayashi (1990), S. S. Sorenson, Bebout, and Barton (1991), Oh, Liou, and Maruyama (1991).

78. M. M. Earle (1980) discusses this problem. Geothermal gradients in orogenic belts are often composites rather than true path histories (P-T-t paths). The practice of referring to them as geothermal gradients should be abandoned, except in cases where the terrane involved is coherent.

79. R. G. Coleman and Lee (1963), Ernst et al. (1970), R. G. Coleman and Lanphere (1971), Kerrick and Cotton (1971), Ernst (1973b), Raymond (1973a,b,c), S. E. Swanson and Schiffman (1979), D. E. Moore (1984), Maruyama, Liou, and Sasakura (1985), Maruyama and Liou (1988), D. E. Moore and Blake (1989).

80. For example, see Kerrick and Cotton (1971), Raymond (1973a,b), D. E. Moore (1984), Maruyama, Liou, and Sasakura (1985), and D. E. Moore and Blake (1989).

PROBLEMS

28.1 In the low-pressure part of the Zeolite Facies, prehnite may appear with laumontite. Using the ACF diagram for the Zeolite Facies in figure 28.6 as a starting point, write a balanced reaction, draw the new topology, and determine the type of reaction (terminal or nonterminal) for the addition of this new phase.

28.2. Using a copy of figure 28.14 as a base, (a) plot the P-T-t path for the New Caledonian blueschist belt using the data in Yokoyama, Brothers, and Black (1986). (b) Compare and contrast this path and the phase assemblages of the rocks with the P-T conditions and phase assemblages of the Sanbagawa Belt (from this chapter). Suggest reasons for the differences.

29

Eclogites

INTRODUCTION

Eclogites are eye-catching rocks composed predominantly of green, omphacitic pyroxene [$(Na,Ca,Fe,Mg,Al)Si_2O_6$] and red or red-brown garnets (figure 29.1). Chemically, they are basic rocks (table 29.1). The distinctive character of these rocks—which in spite of their basic chemistry, lack plagioclase—probably led Eskola (1920, 1939) to assign them to a separate facies.

Eskola interpreted the eclogites to be high-pressure rocks. Yet, while experimental work and field relations generally support that interpretation, the question of whether or not the eclogites should be assigned to a separate facies of crustal metamorphism remains open. The opposing views hinge in part on assessment of the significance of the varying mineral assemblages of various eclogites and on the meanings of the relationships that exist between the eclogites and the rocks associated with them. This chapter reviews the nature of eclogites, the conditions of eclogite petrogenesis, and the question of the existence of an Eclogite Facies.

OCCURRENCES AND MINERALOGY

The term eclogite, as used here, refers to plagioclase-free rocks that are *dominated* by Na-bearing, Ca-Mg clinopyroxene (omphacite) and Mg-Fe (low-Ca) garnet. The two essential minerals together should comprise more than 67% of the primary phase assemblage and garnet alone typically makes up more than 30% of the rock . Varieties of eclogite may be distinguished on the basis of chemistry, textures, or the presence of accessory minerals (e.g., kyanite eclogite, quartz eclogite), but these accessory minerals do not dominate the mode (cf. R. G. Coleman et al., 1965; Banno, 1970; I. D. MacGregor and Carter, 1970). Note that *not all* rocks containing abundant clinopyroxene and garnet are eclogites, especially those rich in carbonate minerals or quartz, which are more appropriately called marble, skarn, or tactite.[1]

Eclogites are found in the crust in four different associations. These include occurrences as

1. Xenoliths in basalts and kimberlites,
2. Exotic blocks in serpentinite- and shale-matrix melanges and fault breccias,
3. Interlayers and lenses in high P/T blueschist-bearing terranes,
4. Interlayers and lenses in medium to high T/P terranes characterized by Granulite Facies, migmatitic, gneissic, and ultramafic rocks.

In the first two associations, the eclogites are exotic or xenolithic; that is, they are transported and petrogenetically distinct from the matrix in which they occur. In the latter two associations, the eclogites are

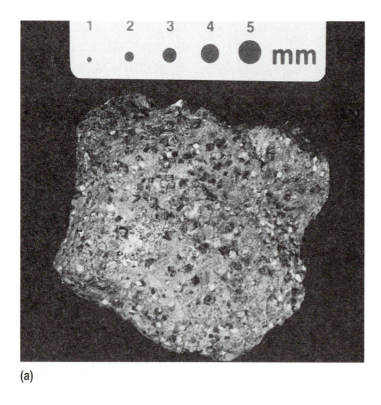

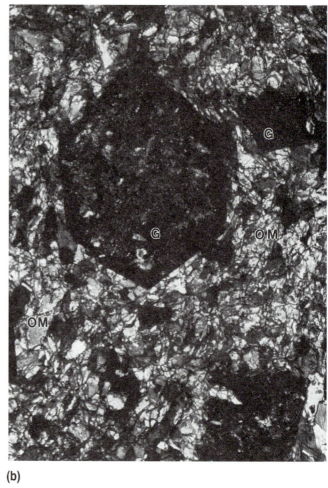

Figure 29.1   Photographs of eclogite. (a) Handspecimen, Franciscan Complex, Sonoma County, California. Note euhedral garnets in matrix of pyroxene. (b) Photomicrograph of specimen shown in (a).(XN). G = garnet, OM = omphacite. Long dimension of photo is 1.27 mm.

(a)

(b)

closely associated with, and in most cases petrogenetically related to, the enclosing rocks. Because some petrologists consider eclogite to be a major rock type in the upper mantle,[2] eclogites of occurrence 1, as well as some of those of occurrences 2, 3, and 4, may represent mantle rocks.

The four types of eclogite occurrence have been assigned by R. G. Coleman et al. (1965) to three major groups (Groups A, B, and C). Group A eclogites include the xenolithic types, whereas Group B eclogites occur in medium to high T/P terranes.[3] Associations 2 and 3 are combined as Group C eclogites. R. G. Coleman et al. (1965) use garnet chemistries in conjunction with *mode of occurrence* to divide eclogites into the three groups. These chemistries apparently form a continuum from low to high pyrope (Mg) content and do not exclusively distinguish rocks of the various groups. Nevertheless, these group names are used in some contemporary literature.

Xenolithic eclogites are rare rocks. Examples are reported from Hawaii,[5] the Colorado Plateau, the Colorado-Wyoming Front Ranges, Michigan, Japan, India, Australia, the former Soviet Union, South Africa, Botswana, and Angola.[6] These rocks typically occur in kimberlites, but are also reported from alkali olivine basalts. Mineralogically, xenolithic eclogites are composed of omphacitic to diopsidic pyroxene and Mg-rich garnet.[7] Additional minerals are listed in table 29.2. The garnets typically contain >55% pyrope component, 10–20% Ca-garnet component (grossularite-andradite), and 25–35% Fe-Mn-Al–garnet component (almandine-spessartite). In a few cases, the pyrope contents are lower (Hills and Haggerty, 1989). Omphacites contain from 1–37% jadeite component, a compositional range duplicated by pyroxenes from eclogites of high T/P terranes.

Table 29.1 Chemical Analyses and Modes of Selected Eclogites

	1	2	3	4	5
SiO_2	44.57[a]	44.6	45.21	47.2	51.51
TiO_2	1.76	1.8	1.43	2.1	1.20
Al_2O_3	13.61	16.3	14.90	15.9	13.92
Fe_2O_3	4.17	5.9	4.78	5.2	9.75[b]
FeO	8.49	8.8	10.15	4.5	—
MnO	0.21	0.26	0.19	0.15	0.15
MgO	13.34	5.9	10.91	6.1	8.36
CaO	11.42	11.7	7.98	7.3	11.23
Na_2O	1.69	3.1	2.83	3.5	3.26
K_2O	0.02	0.22	0.02	1.6	0.09
P_2O_5	0.02	0.12	0.24	0.43	0.23
Other	0.57	1.77	—	6.38	0.27
Total	99.87	100.35	98.63	100.36	99.98

Modes

	1	2	3	4	5
Cpx	61.2	42.6	57.0[d]	34.4[c]	30[c]
Gt	36.5	28.9	39.0	28.9	25
Ol	0.4	—	—	—	—
Qtz	—	—	—	8.2	trace
Ky	—	—	—	—	1
Sp	0.5	—	—	—	—
Rt	—	1.0	0.6	2.0	4
Sph	—	3.2	—	1.5	—
Ep	—	1.4	—	3.9	3
Lw	—	6.6	—	—	—
Il	0.3	—	?1.3	—	—
Mt	0.6	—	—	—	—
P-B	trace	—	0.5	—	1
Wm	—	2.2	—	0.7	—
Chl	—	14.1	—	0.5	—
CaA	—	—	?0.5	—	1
NaA	—	—	—	19.9	—
Pl	—	—	X	—	10
Sul	0.5	—	—	—	—
Other	—	—	—	—	25[e]

Sources:

1. Garnet pyroxenite (eclogite xenolith), sample 68SAL-11, Salt Lake Crater, Hawaii (Beeson and Jackson, 1970).
2. Eclogite 100-RGC-58, Tiburon, California (from melange) (R. G. Coleman et al., 1965).
3. Eclogite associated with gneisses, sample 82-42, Sunnmore, Norway (Jamtveit, 1987).
4. Eclogite from glaucophane-bearing schist terrane, New Caledonia, sample 36-NC-62 (R. G. Coleman et al., 1965).
5. Eclogite from garnet lherzolite, sample F-53, Alpe Arami, Switzerland (Ernst, 1977b).

[a] Values in weight percent

[b] Total iron as Fe_2O_3.

[c] Modal pyroxene and garnet do not always total 67+% because of secondary alteration.

[d] Pyroxene and amphibole include some plagioclase in symplectic intergrowths.

[e] Pyroxene-garnet symplectite.

X = plagioclase as symplectite with pyroxene.

Cpx = clinopyroxene, Gt = garnet, Ol = olivine, Qtz = quartz, Ky = kyanite, Sp = spinel, Rt = rutile, Sph = sphene, Ep = epidote-clinozoisite, Lw = lawsonite,
Il = ilmenite, Mt = magnetite, P-B = phlogopite-biotite, Wm = white mica, Chl = chlorite, CaA and NaA = Ca-and Na-amphiboles, Pl = plagioclase,
Su = sulfides.

Table 29.2   Minerals of Eclogites

Xenolithic Eclogites	Exotic Block Eclogites	Blueschist-related Eclogites	Granulite-related Eclogites
Primary			
omphacite	omphacite	omphacite	omphacite
garnet (Py$_{>30}$)	garnet (Py$_{<50}$)	garnet (Py$_{<30}$)	garnet (Py$_{5-60}$)
kyanite	glaucophane	quartz	quartz
corundum	barroisite	kyanite	coesite
rutile	lawsonite	talc	kyanite
diamond	epidote	white mica	talc
graphite	quartz	phlogopite	white mica
amphibole	white mica	epidote	phlogopite/biotite
phlogopite	rutile	lawsonite	Ca-amphibole
lawsonite	sphene	barroisite	orthopyroxene
	apatite	zoisite	zoisite
	pyrite	rutile	rutile
			sphene
Secondary			
serpentine	glaucophane	glaucophane	hornblende
phlogopite	actinolite	actinolite	actinolite
amphiboles	hornblende	hornblende	augite
ilmenite	lawsonite	epidote	plagioclase
plagioclase	epidote	lawsonite	biotite
spinel	jadeitic pyroxene	jadeitic pyroxene	sphene
analcite	plagioclase	plagioclase	white mica
calcite	pumpellyite	chlorite	epidote
chlorite	chlorite	quartz	diopside
goethite	quartz	calcite	ilmenite
pyrite	aragonite	aragonite	
pyrrhotite	calcite	rutile	
chalcopyrite	rutile	sphene	
	sphene		
	white mica		
	montmorillonite		

Sources: See notes 6, 8, 9, 12, and 14.

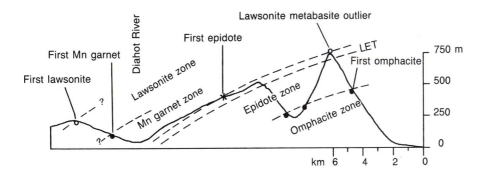

Figure 29.2 Cross section across part of northern New Caledonia showing metamorphic zones, including Omphacite Zone (Eclogite Facies) located structurally well below the Lawsonite Zone of the (high-pressure) Blueschist Facies. LET = lawsonite-bearing part of the Epidote Zone. *(From Yokoyama, Brothers, and Black, 1986)*

Serpentinite- and pelitic-matrix melanges and breccias exposed along fault zones in high P/T belts, locally contain exotic, tectonic blocks[8] of eclogite engulfed by matrix materials. R. G. Coleman et al. (1965) assigned eclogite blocks of this type to Group C. As required by the definition of eclogite, these rocks are composed of omphacitic pyroxene and garnet.[9] Garnets are high Fe-Al varieties (almandine) and are low in Mg (pyrope-component). The pyroxenes typically are relatively sodium-rich (Jd_{30-40}). Additional phases are listed in table 29.2.

Perhaps the best-known examples of exotic-tectonic eclogites occur in the Franciscan Complex of western California and southwestern Oregon.[10] Here, in association with blocks of many other rock types—including glaucophane schist, metabasalt ("greenstones"), metachert, metaconglomerate and metabreccia, limestone, ophiolite fragments, serpentinites, and voluminous metasandstones—the eclogites form mesoscopic-scale masses in pelitic- and serpentinite-matrix melanges. The blocks of eclogite are typically one to a few meters in diameter. Commonly they are veined and the primary mineralogy is partially replaced by more hydrous assemblages of the Greenschist and Blueschist facies. Similar eclogites have been reported from Japan and the southern Ural Mountains of Russia.[11]

Eclogites interlayered and closely associated with blueschists, ophiolites, and related rocks are reported from Greece, western France, the Alps, Asia, Venezuela, and New Caledonia (figure 29.2).[12] In size, the eclogite bodies range from thin lenses to macroscopic sheets. These rocks were also assigned to Group C by R. G. Coleman et al. (1965). Mineralogically, they contain primary omphacite and garnet ($Py_{<30}$), plus quartz, kyanite, rutile, zoisite, white mica, talc, and/or lawsonite.[13]

Eclogite interlayers and lenses also occur in medium to high T/P metamorphic terranes. Like the lenses and layers in glaucophane schists and related rocks, these occurrences range in scale from small mesoscopic masses to macroscopic sheets of a kilometer or more in length (W. L. Griffin, 1987; Jamtveit, 1987). Associated rocks include granitic gneisses, migmatites, marbles, metaquartz arenites, metamorphosed ultramafic rocks, anorthosites, and amphibole schists. Notable occurrences have been described in northern Europe from Sweden, Scotland, Poland, Germany, the former Soviet Union, and especially Norway, as well as in France and the Swiss and Austrian Alps and in Tasmania, Africa, Japan, and Newfoundland (Canada).[14]

The high T/P eclogites were assigned by R. G. Coleman et al. (1965) to Group B. Yet the garnets from some of these occurrences show a wide range of chemistry and plot in both Group A and Group B (Mysen and Heier, 1972). Pyrope contents range from about Py_5 to Py_{60}. The primary modes of the eclogites reveal, in addition to omphacite and garnet, such minerals as kyanite, talc, quartz, and orthopyroxene (see table 29.2).[15]

Because eclogites are composed primarily of *anhydrous* phases stable at medium to high temperatures and high pressures (see page 627), in cases where they are transported to higher levels of the crust and encounter hydrous, lower-grade environments, they tend to alter readily to *hydrous* phase assemblages. Hydrous phases commonly produced by such reactions include white micas (muscovite and paragonite), brown to black micas (phlogopite and biotite), sodic amphiboles (e.g., glaucophane and crossite), calcic amphiboles (hornblendes, including barroisite, and tremolite and actinolite), epidote-clinozoisite-zoisite, lawsonite, pumpellyite, and chlorite. Nonhydrous phases produced by these reactions may include aragonite, calcite, plagioclase, calcic pyroxenes, sphene, rutile, ilmenite, and magnetite. The particular phase assemblage produced by any of the retrograde or prograde reactions that yield these phases depends on the bulk chemistry of the eclogite, the nature of the fluid phase, and the P-T conditions present at the time of the reaction.

DO ECLOGITES REPRESENT A SEPARATE FACIES?

The Eclogite Facies is one of the facies originally defined by Eskola (1920). Yet, the existence of the Eclogite Facies as a *crustal* metamorphic facies (rather than a mantle facies) is debatable. Conditions of metamorphism in the mantle are generally beyond the scope of this text and the bulk chemistries of mantle rocks are largely restricted to basic and ultrabasic compositions.

Arguments Against the Existence of a Crustal Eclogite Facies

R. G. Coleman et al. (1965) suggested that the designation of a separate Eclogite Facies be abandoned. They argued that:

1. The interlayering of eclogites with glaucophane schists, Amphibolite Facies rocks, and Granulite Facies rocks indicates that eclogites form over a range of P-T conditions that overlap the conditions associated with other facies (eclogites, in part, may simply represent anhydrous equivalents of hydrous metabasic rocks, such as hornblende- or glaucophane-schist);[16]
2. Partitioning of Ca, Mg, and Fe between pyroxene and garnets varies significantly between eclogites of different associations, indicating various P-T conditions of formation; and
3. The eclogite facies cannot be mapped as a facies in regionally metamorphosed terranes, as can other facies.

An additional argument that may be raised is that:

4. The eclogite facies, as defined and recognized traditionally, involves *only one bulk composition*, basic rocks, and therefore does not represent the set of all mineral assemblages, representing all bulk compositions, indicative of a particular and restricted set of P-T conditions. Thus, its character is not consistent with the definition of a metamorphic facies.

The suggestion of R. G. Coleman et al. (1965) to abandon the use of a separate Eclogite Facies has not been universally accepted. Nevertheless, some petrologists have supported it. For example, Winkler (1979) implies that the use of an Eclogite Facies is unjustified and he simply describes eclogitic rocks and their origins without assigning them to a facies.

Arguments in Favor of the Existence of a Crustal Eclogite Facies

A number of petrologists have argued in favor of retaining the Eclogite Facies. Fyfe et al. (1958) argued strongly that the eclogites should be recognized as a facies because of the distinctive mineral assemblages of these rocks. Hyndman (1985, p. 537) and Yardley (1989) continue to use the Eclogite Facies, as does F. J. Turner (1981, p. 410), who explicitly rejects the suggestion of R. G. Coleman et al. (1965) to abandon it. Miyashiro (1973a, p. 310ff.) adopts the Eclogite Facies, but specifies that quartz- and kyanite-eclogites are petrographic indicators of the facies; that is, he restricts the definition of the facies. E. H. Brown and Bradshaw (1979) argue that eclogites represent distinct phase topologies and metamorphic conditions and that the concept of the Eclogite Facies is a valid one.

Arguments in favor of retaining an Eclogite Facies include the following.

1. The phase assemblages of eclogites are so distinctive and different from those of other metabasites that they must represent a unique facies (Fyfe et al., 1958, p. 235; F. J. Turner, 1981, p. 410).
2. Although rare, rocks representing *a range of bulk chemistries* and exhibiting some unique mineral assemblages that develop at conditions of $P \geq 1.0$ Gpa(10 kb) and $T > 200°$ C do exist and should be used to define an Eclogite Facies.
3. Rare regional terranes composed of or including eclogites do exist (e.g., Yokoyama, Brothers, and Black, 1986; X. Wang and Liou, 1991).

To evaluate the direct conflicts indicated by these pro and con arguments, it is necessary to address two questions. (1) Are there distinctive phase assemblages present in crustal rocks, which can be represented by phase topologies that *uniquely* define a high-pressure facies? (2) Are all eclogites representative of similar conditions of metamorphism?

Natural High-Pressure Phase Assemblages and Associated Phase Topologies

Natural occurrences, chemical data, and experimental data can be focussed on the solution to the first of these two questions. First, eclogites associated with both glaucophane schists and rocks of medium to high T/P terranes appear as concordant masses. As such, they are commonly assumed to form under similar P-T conditions, generally in excess of 0.8 Gpa (8 kb) and 200° C (refer to the petrogenetic grids in chapter 28 and appendix C). Although these P-T limits do not precisely constrain the petrogenetic conditions for eclogite formation, they do eliminate all low-pressure conditions. The presence of kyanite as the stable aluminum silicate phase in eclogites provides a similar constraint. Rare coesite in eclogites (X. Wang, Liou, and Mao, 1989) is an unequivocal indicator of high pressures of formation. From another perspective, an indication that pressures of formation are substantially higher than 0.3–0.4 Gpa (3–4 kb) is provided by occurrences such as that in New Caledonia, where eclogites and associated rocks form a metamorphic ("omphacite") zone that lies structurally *below* zones exhibiting typical, high-pressure phase assemblages of the Blueschist Facies (figure 29.2) (Yokoyama, Brothers, and Black, 1986; Ghent et al., 1987). Similarly, in eastern China, coesite-bearing eclogites are apparently part of a regional metamorphic terrane that is flanked by lower-grade rocks (X. Wang and Liou, 1991).

Specific experimental studies of systems relevant to the formation of eclogites provide better constraints on the P-T conditions under which eclogites form. The absence of plagioclase in rocks of basalt-gabbro composition provides an important constraint. The disappearance of plagioclase is governed by reactions[17] such as

$$\text{forsterite} + \text{anorthite} = \text{garnet (pyrope)}, \quad (29.1)$$

$$\text{enstatite} + \text{anorthite} = \text{garnet (pyrope)} + \text{diopside} + \text{quartz}, \quad (29.2)$$

$$\text{anorthite} = \text{garnet (grossularite)} + \text{kyanite} + \text{quartz, and} \quad (29.3)$$

$$\text{albite} = \text{jadeite} + \text{quartz}. \quad (29.4)$$

Experimental determinations of reactions 29.3 and 29.4 suggest limits for the conditions of formation of eclogite (at realistic temperatures) to pressures above about 0.5 or 0.6 Gpa (figure 29.3a).

A number of workers have evaluated the stability relations of eclogites and plagioclase-bearing basic rocks.[18] Experimental studies on basaltic compositions are synthesized in figure 29.3b. Under wet conditions in the crust and upper mantle, amphibole-free eclogite is stable only above about 1.8–2.7 Gpa, depending on the temperature and composition of the basalt. Amphibole-bearing eclogite is stable above 0.5–1.8 Gpa. Under dry conditions (above the wet basalt solidus), eclogite is only stable above about 650° C at pressures of 1.0–1.8 Gpa. The lower limit of amphibole-eclogite stability is well within the limits of crustal metamorphism, but the amphibole-free eclogite stability field occurs at pressures realized in the mantle, at mantle depths in subduction zones, or in the basal regions of thick crust like that beneath high mountain ranges.

Geothermometry and geobarometry document the high-pressure, low- to high-temperature conditions of formation of eclogites (Banno, 1970; Raheim and Green, 1974; Newton, 1986).[19] Pressure and temperature estimates for petrogenesis of Group B and C eclogites have been summarized by Newton (1986). Metamorphic temperatures ranged from 450° C to 750° C, whereas metamorphic pressures were between 1.2 and 2.0 Gpa. These values encompass those recently determined for specific localities (Schliestedt, 1986; Austrheim, 1987; Ghent et al., 1987), are a bit higher than older analyses suggested, and are quite compatible with the experimental results cited above.

ACF phase topologies for rocks metamorphosed under the wet and dry conditions necessary to yield eclogites are shown in figure 29.4. Recall that the bulk rock chemistry of basalts plots in the lower right third of the diagram. By comparing these topologies to ACF topologies for the Blueschist, Amphibolite, and Granulite facies, it is clear that a unique topology does exist for "dry" eclogites, one in which there is no tie line between Mg-garnet and plagioclase, epidote group minerals, or lawsonite. The diagram for "wet" eclogites does have such a tie line, making some of the phase assemblages of this topology like those of the Greenschist and Amphibolite facies. The latter, however, are characterized by plagioclase—albite in the Greenschist Facies and more calcic plagioclases in the Amphibolite Facies. The CFM diagram cannot be used to depict assemblages for eclogites, as the saturating phase plagioclase is absent from all eclogite assemblages, and the remaining saturating phases quartz, H_2O, and magnetite are commonly absent as well.

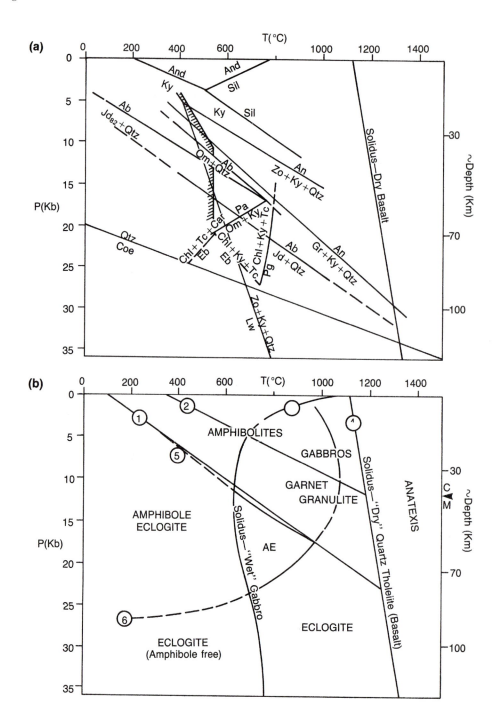

Figure 29.3 Eclogite stability. (a) Petrogenetic grid showing some reaction curves important to the stability of eclogites and related rocks.

And (andalusite), Ky (Kyanite), and Sil (sillimanite) curves from Holdaway (1971). An = Zo + Ky + Q+z (quartz) and Zo + Ky + Qtz = Lw (lawsonite) curves from Newton and Kennedy (1963). An = Gr + Ky + Qtz curve from Hariya and Kennedy (1968). Ab (albite) = Om (omphacite) + Qtz and Pa (paragonite) = Om + Ky curves (calculated) from Newton (1986). Ab = Jd₈₂ (jadeitic pyroxene, with 82% jadeite) + Qtz curve from Newton and Smith (1967). Ab = Jd + Qtz curve from Birch and LeCompte (1960) and Holland (1980).
Qtz = Coe (coesite) curve after Boyd and England (1960) and Ostrovsky (1966, 1967). Chl + Tc (talc) + Car (Mg-carpholite) = Eb (ellenbergite), Chl + Tc + Ky = Eb, and Chl + Ky + Tc = Py (pyrope)

curves are from Chopin (1986). The solidus for dry basalt is from Green and Ringwood (1967a).
(b) P-T grid showing experimentally determined melting and stability relations for rocks of basaltic composition.

Curves 1 and 2, bounding the upper and lower limits of garnet granulite stability are from Green and Ringwood (1972). Curve 3 is from I. B. Lambert and Wyllie (1972). Curve 4 is from D. H. Green and Ringwood (1967a) and Ito and Kennedy (1971). Curve 5 is from Essene, Henson, and Green (1970) and represents the lower limit of stability of amphibole-bearing eclogite (AE). Curve 6, which is a compound curve, represents the upper stability limit of amphibole in eclogite at pressures above 17 kb (from Essene, Henson, and Green 1970) and the upper limit of stability of hornblende in rocks of basaltic composition at pressures below 10 kb (from Holloway and Burnham, 1972).

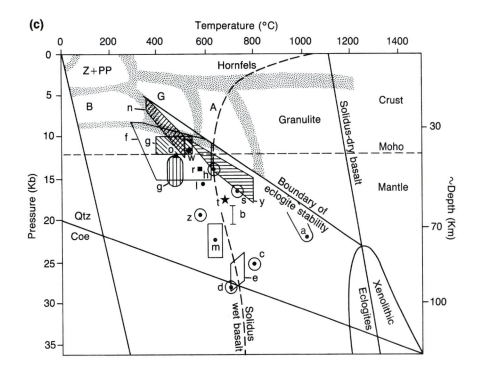

(c)

Figure 29.3 Continued

(c) P-T grid showing the crystallization fields of various eclogitic rocks superimposed on selected curves from figures 29.3a and b and various metamorphic facies fields.

Z + Pp = Zeolite + Prehnite-Pumpellyite Facies. G = Greenschist Facies. B = Blueschist Facies A = Amphibolite Facies. Stability fields are as follows: a = eastern Australian margin (Pearson et al., 1991); b = Bergen Arcs, Norway (Jamtveit et al., 1990); c = Cima di Gagnone, Ticino, Switzerland (Evans et al., 1979); d = Dora-Maira Nappe, Alps (Pognante, 1991); e = coesite and ellenbergite rocks of the western Alps (several sources, including Chopin, 1986); f = Franciscan Complex (see text and Oh et al., 1991); g = Sifnos Island, Greece (Matthews and Schleistedt, 1984; also see Schleistedt, 1986); h = Hareidland, Norway (Mysen and Heier, 1972); 1 = Alpine Sesia-Lanzo Zone (Castelli, 1991); m = Munchberg, Germany (Klemd, 1989); n = Newfoundland (Jamieson, 1990, and others); o = Alpine ophiolite eclogites (Pognante and Kienast, 1987); q = New Caledonia (Yokoyama et al., 1986); r = Rhodope Zone, northern Greece (Liati and Mposkos, 1990); s = Glenelg, Scotland (Sanders, 1989); t(star) = Troms, Norway (Krogh et al., 1990); w = Ward Creek Type IV (Oh et al., 1991); y = Norwegian eclogites (Griffin, 1985, 1987; Jamtveit et al., 1990); z = Zermatt-Saas Zone, Alps (Pognante, 1991). Refer to text for additional references.

Do other bulk compositions metamorphosed under conditions like those that yield eclogites occur in the crust and, if so, do they reflect unique phase topologies? The answer to these questions, which in the past seemed to receive a resounding "no," seems now to be a qualified "yes." A number of interesting occurrences of distinctive rocks are worth noting here. These include "whiteschists," coesite-bearing rocks, tactites, gneisses, garnet lherzolites, and sodic pyroxene + paragonite-bearing rocks. Some of these occurrences reveal unique phase assemblages. Others exhibit assemblages that are mineralogically indistinguishable from those of normal Blueschist, Amphibolite, or Granulite facies rocks, but geothermometry and geobarometry yield metamorphic P-T conditions equivalent to those of the mantle.

Whiteschists have been described by Schreyer (1973, 1977) and Munz (1990). These rocks, characterized by the assemblage kyanite + talc, are rare, but are found in Brazil, Tasmania, central Africa, Europe, and Afganistan. The bulk chemistry of the whiteschists is unusual and may represent that of evaporitic mudstones. The significance of these rocks to this discussion is that the assemblage kyanite + talc is stable between 600° C and 850° C at pressures above 1.0 Gpa. At lower and higher temperatures, the pressures of stability are higher. The presence of a hydrous fluid phase may lower the limit of stability somewhat to about 0.7–0.8 Gpa. Thus, whiteschists represent phase assemblages of nonbasic rocks that reflect P-T conditions like those under which many eclogites are formed.

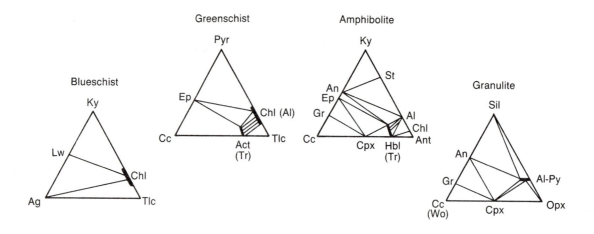

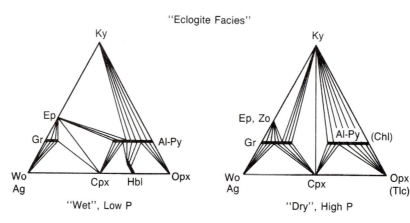

Figure 29.4 ACF topologies for "wet" and "dry" eclogites compared to typical topologies for Blueschist, Greenschist, Amphibolite, and Granulite facies rocks.
Act = actinolite, Ag = aragonite, Al = almandine, An = anorthite, ant = antigorite, Cc = calcite, Chl = chlorite, Cpx = clinopyroxene, Ep = epidote, Gr = grossularite, Hble = hornblende, Ky = kyanite, Lw = lawsonite, Opx = orthopyroxene, Py = pyrope, Pyr = pyrophyllite, Sil = sillmanite, St = staurolite, Tlc = talc, Tr = tremolite, Wo = wollastonite, Zo = zoisite

A pyrope-quartzite exposed in the western Alps near Parigi, Italy, contains both coesite (the high-pressure silica polymorph) and the recently discovered Mg-Al-Ti–silicate, ellenbergite (also stable only at high pressures)(Chopin, 1984, 1986). The assemblages pyrope (Py_{90-98}) + coesite + talc and kyanite + talc + chlorite + ellenbergite + rutile + zircon indicate conditions of metmorphism of about 2.5 Gpa and 700° C.

In the Lepontine Alps of Switzerland, lenses of eclogite occur with garnet-bearing lherzolite (Ernst, 1977b; Evans et al., 1979). Garnet lherzolite (an ultramafic rock) is stable under conditions like those of eclogite stability (>1.0 Gpa at $T \geq 500°$ C) (see chapter 31). The phase assemblage for the ultramafic rock is distinctive and reflects a unique topology. P-T conditions indicated for the eclogites are reported to be 800–1000° C and 2.5–5.0 Gpa (Ernst, 1977; Evans et al., 1979).

Carbonate rocks, SAC rocks, and pelitic schists were metamorphosed with eclogites in the Sesia-Lanzo Zone of the western Alps (Koons et al., 1987; Pognante, 1989, 1991; Castelli, 1991). Here, pelitic schists contain phases such as quartz, white mica, lawsonite, jadeitic pyroxene, and glaucophane, and SAC rocks are characterized by the high-pressure assemblage jadeitic pyroxene–zoisite–quartz–garnet–white mica. The carbonate rocks contain phases such as aragonite, dolomite, quartz, garnet, white mica, clinopyroxene, zoisite, and sphene. P-T conditions for the eclogite-forming event are estimated to have been 575° C and 1.5 Gpa (Castelli, 1991).

Calcite + garnet + clinopyroxene assemblages have been described by Swainbank and Forbes (1975) and E. H. Brown and Forbes (1986) from rocks of the Fairbanks, Alaska area. These rocks are calcite-bearing and/or quartz-rich and differ chemically from normal eclogites. They are more appropriately referred to as tactites or skarns. Associated with the tactites are garnet-glaucophane-barroisite-plagioclase schists and granoblastites (garnet-amphibolites), calcite granoblastites and schists, garnet-quartz-mica schists, and metaquartz arenites. None of the phase assemblages in these rocks seems to differ from those found in normal medium- to high-temperature crustal metamorphic rocks. Geothermometry and geobarometry on the various rocks, however, indicate P-T conditions of about 600° C and 1.5 Gpa (E. H. Brown and Forbes, 1986).

Eclogite of Hareidland, Norway, is associated with kyanite-bearing, dioritic gneiss and quartz-orthopyroxene-amphibole-plagioclase gneiss (Mysen and Heier, 1972). Mineralogically, these rocks are normal Amphibolite to Granulite Facies rocks, but thermobarometry reveals metamorphic P-T conditions of 625° C and 1.4 Gpa.

High-pressure, meta-SAC and metabasic rocks occur together in New Caledonia (Brothers and Blake, 1973; Yokoyama, Brothers, and Black, 1986; Ghent et al., 1987). Eclogites are associated with the more siliceous rocks, which exhibit phase assemblages such as quartz + omphacite + almandine + paragonite + muscovite + glaucophane + clinozoisite + rutile + sphene and quartz + omphacite + orthoclase + muscovite + ferroglaucophane + stilpnomelane. Except for the occurrence of pyroxenes more calcic than those of the Blueschist Facies and the presence of garnet in the assemblage, these rocks are mineralogically like those of the Blueschist Facies. These new phases, however, change the topologies. P-T conditions appear to have been about 1.0–1.2 Gpa and 400–550° C (Yokoyama, Brothers, and Black, 1986; Maruyama, Cho, and Liou, 1986; Ghent et al., 1987).

Like New Caledonia, Sifnos Island, Greece, contains eclogitic rocks associated with rocks having a range of bulk compositions (Matthews and Schliestedt, 1984; Schliestedt, 1986). These rocks include quartz-epidote-glaucophane schists, omphacite-actinolite-glaucophane schists, garnet-paragonite-glaucophane-quartz-jadeite gneisses, dolomite-calcite marble granoblastite, chloritoid metaquartz arenite, epidote-muscovite (phengite)-quartz-glaucophane-garnet schist, and a number of other rock types. The P-T conditions indicated by the associations are 470° C and 1.5 Gpa.

Coesite-bearing eclogites occur in eastern China (X. Wang, Liou, and Mao, 1989; X. Wang and Liou, 1991). The eclogites are present as interlayers in white mica schists, biotite gneiss, and marble, all of which also have quartz pseudomorphs after coesite. Marbles are composed of the high-pressure assemblage calcite–garnet–epidote–clinopyroxene–white mica–quartz (coesite)–rutile. Biotite gneiss contains assemblages such as biotite–muscovite–garnet–epidote–K-feldspar–plagioclase–quartz (coesite)–rutile–sphene. Mica and pyroxene compositions from these rocks yield P-T conditions compatible with the 600° C and >2.0 Gpa estimated for the eclogites.

These descriptions indicate that *locally* in the crust there are rocks of a range of bulk compositions that were metamorphosed under high-grade conditions different from those of the normal crust (see figure 29.3c). The phase assemblages of some of these rocks suggest unique topologies that represent these higher-grade conditions. Based on these observations and on theoretical projections, phase diagrams for a proposed Eclogite Facies may be constructed. Figure 29.5 presents schematic ACF, AKN, and AFM diagrams for such a facies. Reaction curves shown in the petrogenetic grid (see appendix C) constrain the probable P-T conditions for such a facies to conditions of $P \geq 1.0$ Gpa and $T = 300$–1200° C.

The formation of eclogites is not confined to the conditions of such an Eclogite Facies. Apparently many eclogites do form under these conditions in both anhydrous and H_2O-bearing environments (E. H. Brown and Bradshaw, 1979; Newton, 1986), but others reportedly occur in close association with Amphibolite, Greenschist, Blueschist, and Granulite Facies rocks. Eclogites found in the latter associations apparently owe their existence to different bulk compositions and/or to very low partial pressures of water (D. H. Green and Ringwood, 1967a; Fry and Fyfe, 1969; B. A. Morgan, 1970; De Wit and Strong, 1975). The latter eclogites represent the facies with which they are associated, rather than the Eclogite Facies.

In summary, there are some unique phase assemblages, representing *some* bulk compositions of crustal rocks, that formed under higher-grade conditions than did the phase assemblages of the typical crust. The existence of these rocks argues for the existence of an Eclogite Facies. Such a facies represents conditions indicative of mantle conditions. In crustal rocks, those "mantle conditions" are only developed in the deepest levels of thickened crust or in subduction zones, where sialic crust is transported to depths *in excess of* 35–40 km.

In spite of the above points, it may be argued that, given that the conditions necessary for the formation of most eclogites and similarly metamorphosed rocks *exceed* those typical of normal crust, a *separate* Eclogite Facies of *crustal* metamorphism does not exist. Clearly, in some rocks associated with eclogites, the phase assemblages do not differ from those of the Greenschist or Amphibolite facies (see figure 29.3c). In addition, rocks of bulk compositions other than metabasite compositions, metamorphosed under the conditions of the proposed Eclogite Facies, are extremely rare and may not form an adequate foundation for a separate facies.

The arguments for and against the designation of an Eclogite Facies remain unresolved and adoption of the Eclogite Facies as a facies of crustal metamorphism is still controversial. Here, the Eclogite Facies is recognized as a crustal metamorphic facies.

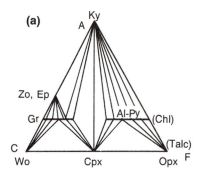

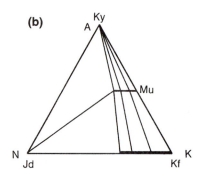

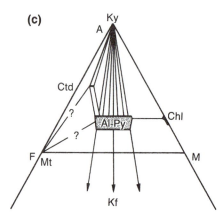

Figure 29.5 Schematic diagrams for the proposed Eclogite Facies: (a) ACF, (b) AKN, (c) AFM. Mu = muscovite, Kf = K-feldspar, Jd = jadeitic pyroxene, Ctd = chloritoid, Mt = magnetite. Other symbols as in figure 29.4.

EXAMPLES OF ECLOGITE OCCURRENCES

Eclogite of the East Pond Metamorphic Suite, Newfoundland

In northwest Newfoundland, a thick sequence of metamorphosed Eocambrian- to Ordovician-aged rocks, the Fleur de Lys Supergroup, overlies a Neoproterozoic to Cambrian metasedimentary sequence, that, in turn, overlies a Proterozoic basement composed of gneiss (figure 29.6).[20] Both the basement and the overlying metasediments contain metabasites and both are assigned to the East Pond Metamorphic suite. The metabasites apparently represent metamorphosed tholeiitic basalt dikes. Where they occur in the metasediments, they are now metamorphosed to amphibole schists, amphibole diablastites (?), and similar rocks, collectively referred to as amphibolites. Where they occur in the basement gneisses, they typically form boudins of amphibolite-eclogite rock that nowhere exceed 10 m in

diameter. Eclogite is not known to be in contact with the gneisses, being confined instead to the cores of the boudins, with a layer of amphibolite separating eclogite and gneiss.

The eclogites are of typical composition, with euhedral red to pink garnets in a granoblastic, fine-grained matrix of omphacite. Although this occurrence is of the medium to high T/P type, the garnet is 18–19% pyrope, and therefore plots in the Group C field of R. G. Coleman et al. (1965). The pyroxene has about 35% jadeite component. Additional phases include hornblende, quartz, white mica, zoisite, Mg-calcite, and rutile. Secondary minerals include diopside, hornblende, biotite, plagioclase, epidote, chlorite, calcite, apatite, and sphene. Hornblende, plagioclase, and epidote locally form wormy intergrowths (symplectite). Kelphytic rims of plagioclase, hornblende, and other minerals rim the garnets in altered eclogites.

The surrounding amphibolites are hornblende-plagioclase rocks with minor amounts of such minerals as epidote, quartz, biotite, sphene, and garnet. Some are schistose, whereas others are nonfoliated. DeWit and Strong (1975) report that phase

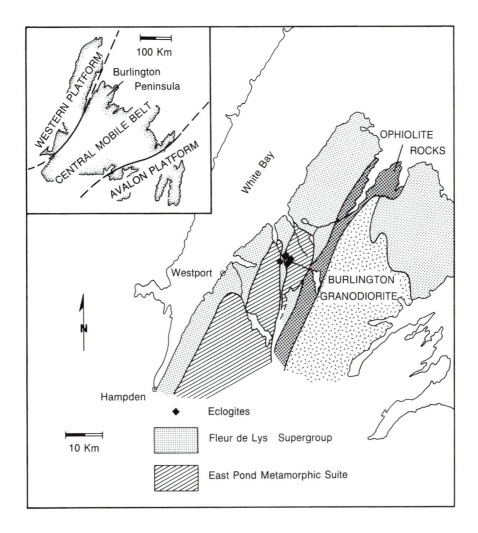

Figure 29.6 Regional geologic map showing terrane in which eclogite-bearing amphibolites occur in East Pond Metamorphic Suite of Newfoundland. Diamonds show locations of eclogites.
(Simplified from Jamieson, 1990)

relations indicate a temperature and pressure of formation of about 350° C and 0.5–0.7 Gpa, conditions that lie within the upper Greenschist Facies. Yet the phase assemblages of the amphibolites suggest amphibolite facies conditions. Reanalysis of the data using the plagioclase-hornblende geothermobarometer of Plyusnina (1982), in fact, does indicate lower Amphibolite Facies P-T conditions of about 450–500° C and 0.8–0.9 Gpa (figure 29.3c). For the eclogite, similar temperatures (T = 350–450° C), but higher pressures (P = 0.8–1.2 Gpa) were inferred. Jamieson (1990), using several geothermoneters, obtained a temperature in the range of T = 450–500° C for the eclogites, but amphibolite overprint temperatures of 600–750° C. Her geobarometric analysis produced a minimum pressure of 1.0–1.2 Gpa for the eclogite and 0.7–0.9 Gpa for the overprint in amphibole-rich rocks.

The origin of the Newfoundland eclogites is thought by DeWit and Strong (1975) to have occurred through metamorphism of basalts under "dry" conditions. The amphibolites are considered to be chemically equivalent, hydrous rocks, an assertion supported by major element chemistry and all of the minor element abundances, except Sr, for which analyses are available. Sr tends to be slightly lower in eclogites than in amphibolites. These chemical data and the petrographic data support the hypothesis that the eclogites and amphibolites were metamorphosed under similar conditions, except for the partial pressure of water, which was lower in those rocks that became eclogites. Jamieson (1990), however, points out that the eclogites occur in "wet" country rocks composed of metapelites and metasandstones, and further, that the eclogites contain hydrous phases such as zoisite and white mica. Consequently, she argues that while water was present and facilitated reaction progress, it did not serve as a thermodynamic agent in the metamorphic process.

Eclogites of the Franciscan Complex, California and Oregon

Franciscan eclogites occur as widely distributed blocks in melanges, fault zones, and serpentinites (figure 29.7).[21] They occur in association with blocks of glaucophane schist; glaucophane-metabasalt and -metagabbro; zeolitic-, pumpellyitic- lawsonitic-, and jadeitic-metabasalt ("greenstones"); chert and metachert; conglomerate, metaconglomerate, breccia and metabreccia; siliceous metavolcanic rocks; limestone and marble; ophiolite fragments; serpentinites; various schists; and voluminous metasandstones—all of which also form tectonic blocks. Commonly, the matrix materials originally surrounding the eclogites (and other rocks) have been eroded away and the eclogite masses rest on soil or other surficial deposits. The ages of the eclogite blocks and associated high-grade glaucophane and amphibole schists range from about 135 to 165 m.y.b.p. (late Jurassic), an age that is older than that of the enclosing Franciscan rocks.[22] No known source terrane exists for the eclogites, rendering their petrogenesis particularly enigmatic.

The eclogites appear to be of two types, those associated or interlayered with glaucophane schist and gneiss and those associated or interlayered with barroisitic amphibole schist and gneiss (Oh, Liou, and Maruyama, 1987, 1991). The former are more abundant and are characterized by primary assemblages such as

> omphacite + garnet + rutile,
> omphacite + garnet + epidote + rutile,
> omphacite + garnet + white mica + rutile, and
> omphacite + garnet + glaucophane + rutile

(R. G. Coleman et al., 1965).[23] The latter are characterized by assemblages such as

> omphacite + garnet + epidote, and
> omphacite + garnet + barroisitic hornblende

(Oh, Liou, and Maruyama, 1987, 1991). Geothermometry and geobarometry indicate that assemblages of the first type formed at temperatures between 290° C and 350° C and pressures between 0.8 Gpa and 0.9 Gpa (Oh, Liou, and Maruyama, 1991). P-T conditions under which the second type (Type IV tectonic eclogites)[24] formed were 400° C<T<715° C and P>0.7–1.3 Gpa (see figure 29.3c)

(D. E. Moore and Blake, 1989; Wakabayashi, 1990; Oh, Liou, and Maruyama, 1991). As is typical of eclogites, veins and rinds of retrograde minerals, as well as bands and patches of these minerals, have rendered many of the eclogites heterogeneous and their histories complex (figure 29.8) (cf. D. E. Moore, 1984; D. E. Moore and Blake, 1989; Wakabayashi, 1990).

Currently accepted tectonic models suggest that oceanic crust was subducted in western California during the late Jurassic Period.[25] Metamorphism of basaltic oceanic crust, under the conditions of high pressure and moderate temperature expected in a subduction zone environment, yields eclogite and amphibole eclogite (note that the stability field of Franciscan eclogites in figure 29.3c straddles the upper stability limit of glaucophane defined by Maresch (1977) and would thus include glaucophane-bearing assemblages at low T and hornblende-bearing assemblages at higher T). The Franciscan eclogites probably formed in such an environment and were then transported to the surface in thrust slices of serpentinite, in serpentinite diapirs, and in obducted, diapirically emplaced, subduction channel, or uplifted melanges. Either metamorphism during transport or resubduction with accompanying metamorphism may account for the various retrograde and prograde assemblages present in these rock masses.

PETROGENESIS OF ECLOGITES

Although the two examples described above provide some perspective on the origins of eclogites, a number of processes have been proposed to explain the origin of specific eclogite occurrences. These include (1) crustal or subcrustal metamorphism of basalt or gabbro under drier conditions than those that normally yield amphibole-rich rocks such as glaucophane or hornblende schist; (2) high-pressure metamorphism of a basaltic or gabbroic rock crystallized from the fusion-generated magma derived from a parental upper-mantle peridotite; (3) contact metasomatism of basalt at moderate P-T conditions; (4) direct (primary) crystallization of eclogite from alkaline magmas in the mantle, with or without subsequent recrystallization under lower grade conditions; (5) crystallization of pyroxenites from alkaline magmas at mantle depths, followed by exsolution of garnet at lower P-T conditions; and (6) fractional crystallization of alkaline magmas in the mantle, at the base of the crust or within the crust, to produce eclogitic crystal cumulates.[26] Some eclogites are polymetamorphic rocks with early histories that are at least partially obscure, making the nature of the protolith

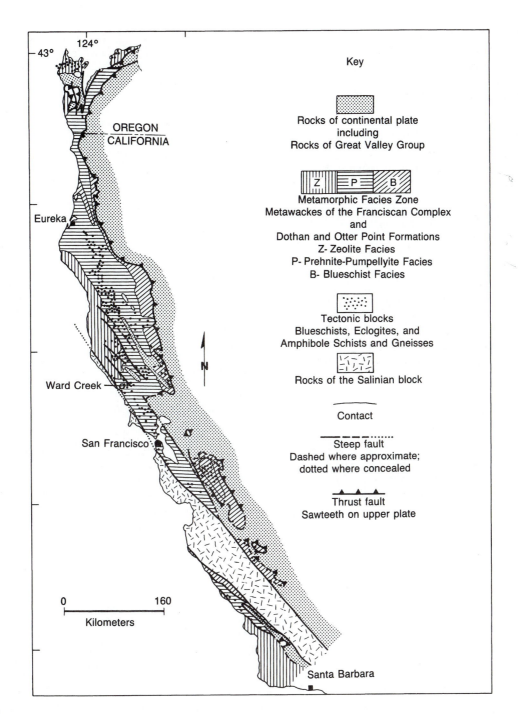

Figure 29.7 Generalized map of western California and southwestern Oregon showing the locations of the tectonic blocks of eclogite and associated glaucophane and amphibole schist and gneiss with respect to the regional belts of Blueschist Facies (B), Prehnite-pumpellyite Facies (P), and Zeolite Facies (Z) rocks.

[Modified from R. G. Coleman and Lanphere, 1971; based on data from E. H. Bailey, Irwin and Jones (1964), Blake, Irwin, and Coleman (1969), Raymond (1973c), Ernst (1980), and others]

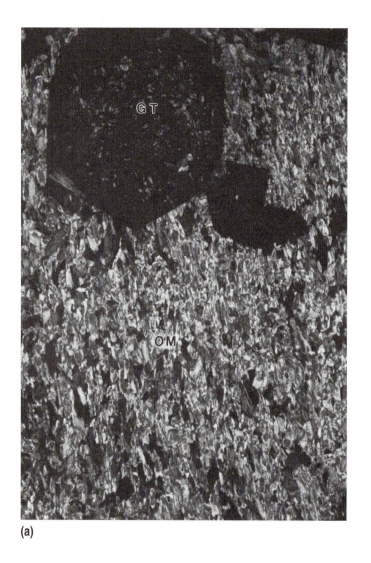

(a)

(b)

Figure 29.8 Photomicrographs of eclogites from the Franciscan Complex. (a) Relatively unaltered eclogite, Tiburon, California.(XN). (b) Glaucophane-bearing eclogite, Tiburon, California.(PL) (c) Veined and altered eclogite with pumpellyite (radiating fibrous mineral) in calcite-prehnite-pumpellyite veins, Narrows melange, near Hospital Creek, northeastern Diablo Range, California. (PL)

(C = calcite, CH=chlorite, E=epidote, G=garnet, GL=glaucophane, J=jadeitic pyroxene, L=lawsonite, OM=omphacite, PP=pumpellyite, PR=prehnite, W=white mica. Long dimension of all photos is 1.27 mm.)

(c)

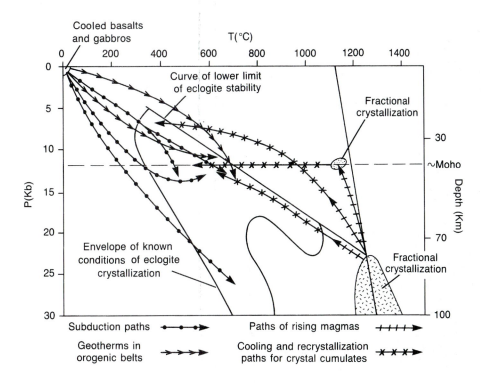

Figure 29.9
P-T grid showing the multiple paths by which rocks may reach eclogite compositions.

problematical (Jamtveit, Bucher-Nurminen, and Austrheim, 1990; Wakabyashi, 1990). Model 3 has been rejected, because mineral stability studies do not support the hypothesis. In models 1 and 2, which are very similar, the eclogites are derived from basaltic or gabbroic rocks that represent crystallized partial fusion products (the liquid separate) derived from mantle ultramafic rocks. In hypotheses 4, 5, and 6, the eclogites are derived from the products of mantle crystallization (the crystal separate) of all or fractions of alkaline magmas. Except for hypotheses 1, 3, and 6, as well as in some versions of those models, emplacement of the eclogites into the crust involves some kind of transport, either via magmatic transport (for xenoliths) or tectonic transport in thrust plates, diapirs, or tectonic melanges.

Considering the range of P-T conditions under which eclogites have formed, it is likely that different eclogites formed via different processes under different conditions. Which of the proposed petrogenetic hypotheses are valid is the question. Examination of figure 29.3c suggests that eclogites form along two general, curved geotherms. One extends down at 8 to 13° C/km through the New Caledonian, Sifnos, and Franciscan eclogite fields to the site of Alpine eclogite and coesite-pyrope generation (f, a, g to c, d, e of figure 29.3c). Such a geotherm is typical of subduction zones. The other geotherm has a slope of about 13–17° C/km and ex-

tends through the Newfoundland and Norwegian eclogite fields and on into the field of xenolithic eclogites. Such geotherms are representative of normal crust, some orogenic belts, and perhaps "hot" subduction zones. These observations, together with chemical and field data reported in the various sources cited in the notes, suggest the following.

Eclogite protoliths are of two kinds—those derived from fusion products and those derived from crystal cumulates. Eclogites are produced from these protoliths by *either* postcooling subduction, which yields high pressures, or postcrystallization cooling (± decompression), which yields lower temperatures (and pressures) than the crystallization temperatures (figure 29.9). Major and trace element chemistry and thermobarometric analyses (Menzies, 1983; F. A. Frey, 1980; Herzberg, 1978b) are consistent with a model of *xenolithic eclogite petrogenesis* in which the eclogites represent a crystal residuum derived from crystal fractionation of alkaline magmas (hypothesis 4 above). Fractionation occurs at mantle depths (*P*> 2.0 Gpa) and at temperatures in the range of 1200–1450° C. Although some re-equilibration and exsolution may have occurred at lower P-T conditions, those events likely occurred within the field of eclogite stability along the high-temperature geotherm (figure 29.9).

Some *layered to lenticular eclogites*, interlayered with ultramafic rocks, amphibole gneisses and schists, and anorthosites, may have originated as crystal cumulates in magma chambers (Griffin, 1987) in the mantle, the crust, or at or near the base of the crust. Ponding of basalt at or near the base of the crust is commonly suggested in models of rhyolite genesis and ocean crust formation. Fractional crystallization of the ponded mafic magmas (at $T = 1100$–$1200°$ C, $P = 0.1$–1.2 Gpa), followed by metamorphism in the eclogite field (at $T = 400$–$800°$ C, $P = 1.0$–1.5 Gpa, $P_{H_2O} < P_{total}$) will yield eclogites. Magmas crystallized at the base of the crust simply cool isobarically until they cross the eclogite stability boundary. Those crystallized at shallow depths (e.g., in the oceanic crust) are transported to depth via subduction.

Eclogites and related rocks formed from *crystallized fusion products*, that is, from typical basalts and from some gabbros, may occur in three associations. Some may form from protoliths crystallized from melts trapped in their parental, mantle ultramafic rocks (Beeson and Jackson, 1970; Dickey, Obata, and Suen, 1977), whereas others form from mafic dikes intruded into crustal country rock (DeWit and Strong, 1975; Griffin, 1987). Many, however, may form from magmas erupted and crystallized at the surface and later transported along the low-temperature geotherm by subduction, until they attain the pressures and temperatures necessary for the formation of eclogite.

Eclogite formation may have been isochemical in some cases, but the basaltic compositions revealed by analyses of eclogites commonly indicate losses of K, Sr, or other mobile elements, or other changes in chemistry.[27] In route along any of the postcrystallization paths, mafic crystal cumulates or crystalized fusion products may experience alteration; dynamothermal, contact, or other isochemical metamorphic changes; or metasomatism, leading to bulk compositions that differ from those of typical basalts. These changes in chemistry do not, however, appear to be necessary for the formation of eclogite.

eclogite as a result of increased pressure realized along subduction zones characterized by a low-temperature geotherm. Other eclogites form where mafic rocks, crystallized at depth, re-equilibrate to conditions between $300°$ C at 0.5 Gpa and $800°$ C at 1.8 Gpa, at $P_{H_2O} < P_{total}$. These conditions are, in part, typical of the crustal Greenschist and Amphibolite facies and, in part, represent mantle-like conditions that occur in both upper mantle rocks and in thickened or subducted crust.

The conditions under which eclogites form are clearly variable. This fact plus the observations that eclogites (1) represent a single kind of protolith (mafic igneous rock), (2) occur in some cases in close association with rocks of the Blueschist, Greenschist, and Amphibolite facies, (3) form over a wide range of conditions, and (4) cannot be mapped as regional terranes in most cases (as can other facies) have led some petrologists to suggest that the Eclogite Facies is not a legitimate facies of crustal metamorphism. Some eclogites seem to have formed under low fluid-pressure conditions at pressures and temperatures essentially identical to those that characterize the Greenschist and Amphibolite facies.

A contrasting view is founded on other data. In New Caledonia, Norway, and China, regional eclogite-bearing terranes are recognized. Geothermometry and geobarometry indicate that these and other eclogites crystallized under conditions typical of the mantle ($P > 1.2$ Gpa and $T > 400°$ C), but also characteristic of thickened crust. Rare, very high pressure metamorphic rocks of nonbasic character are known and are locally associated with eclogites. Together, the eclogites and these rocks may represent a range of protoliths formed under the conditions of an Eclogite Facies developed in thickened crust or in subduction zones. This evidence argues for the need to recognize the Eclogite Facies as a legitimate facies of high-pressure metamorphism. Because the legitimacy of the Eclogite Facies remains controversial, there is a clear need for further study of these interesting and unusual rocks.

SUMMARY

Eclogites (*sensu stricto*) are rocks composed of essential Mg-Fe-rich garnet and omphacitic pyroxene. Although they have basic protoliths, eclogites lack primary plagioclase feldspar. The absence of plagioclase, the presence of accessory minerals such as kyanite and coesite, and an association with rocks such as talc-kyanite schist, garnet lherzolite, and barroisitic hornblende-bearing gneiss indicate that eclogites are high-pressure rocks.

Eclogites apparently form from two types of mafic protolith—crystal cumulates derived from alkaline magmas and crystallized fusion products generated by partial melting of ultramafic rocks within the mantle. Some rocks develop into

EXPLANATORY NOTES

1. Fyfe et al. (1958, p. 236) note correctly that the presence of red garnet and green pyroxene in a rock does not make it an eclogite. They point out that skarns, ultramafic rocks, and granoblastites of the Granulite Facies may also contain that particular association of minerals. Nevertheless, some workers have called calcite-garnet-pyroxene rocks and other garnet-pyroxene-bearing rocks eclogites. For example, Swainbank and Forbes (1975) designate as eclogite some calcite-garnet-pyroxene and garnet-pyroxene-quartz rocks that are interlayered with marble and amphibole schist near Fairbanks, Alaska. None of the rocks for which modes are presented contains more than 62% garnet + pyroxene and all contain significant amounts of calcite or abundant quartz.

Chemically, as Swainbank and Forbes (1975) point out, the rocks differ from normal eclogites. Thus, regardless of the P-T conditions under which such rocks formed (E. H. Brown and Forbes, 1986), they are mineralogically and chemically different from eclogites and should not be called eclogites. A. J. R. White (1964) used pyroxene chemistry to distinguish between eclogites and basic "granulites."

2. For example, see Kuno (1969), Wyllie (1971a, ch. 6), and Kennedy and Ito (1972). I. D. MacGregor (1970) and Dawson (1981) provide brief reviews and Wyllie (1971, ch. 3, 5, 6) provides an extensive review that discusses eclogite as a mantle rock.

3. R. G. Coleman et al. (1965). See note 6 for references to xenolithic eclogites and note 14 for references to eclogite occurrences in high T/P terranes.

4. Mysen and Heier (1972) found that garnets from a high T/P Norwegian eclogite plot over a chemical range that spans the three groups. DeWit and Strong (1975) found eclogites associated with amphibolites and gneisses in a medium to high T/P terrane to contain garnets that plot in Group C, rather than Group B. J. Ferguson and Sheraton (1979) discovered that some xenolithic garnets plot in Group B rather than in Group A.

5. Hawaiian xenoliths composed primarily of clinopyroxene and garnet were called eclogite by Kuno (1969) and Shimizu (1975), but garnet pyroxenite by Green (1966), Beeson and Jackson (1970), and F. A. Frey (1980). D. H. Green (1966) suggested that the garnet in these rocks was exsolved from the pyroxene during cooling and that the exsolution occurred under conditions outside of the stability field of eclogite. Consequently, he suggested that the rocks are not eclogites *sensu stricto*. These rocks are here called eclogites for three reasons. First, the phase assemblage and chemistry are like those of other eclogites. Because the rocks have an eclogitic composition, in my view, they should be called eclogites and presumed genesis should not be used to disqualify them as eclogites. Second, at least some of these rocks may, in fact, contain primary garnet (see the data of Beeson and Jackson, 1970, and F. A. Frey, 1980). Third, the conditions of initial crystallization of clinopyroxenes that exsolved garnet may have been of higher P than supposed by D. H. Green (1966; see Herzberg, 1978) and subsequent exsolution may well have occurred *within* the stability field of eclogite. This case emphasizes the need to name rocks based on their textures and mineralogy or chemistry, leaving presumed genesis out of the naming process. See F. A. Frey (1980) for additional references to works on the much-studied Hawaiian xenoliths.

6. Xenolithic eclogites are discussed by Yoder and Tilley (1962), R. G. Coleman et al. (1965), M. J. O'Hara and Mercy (1966), Kuno (1969), Lovering and White (1969), I. D. MacGregor and Carter (1970), Sobolev (1970), Helmstaedt et al. (1972), Harte and Gurney (1975), Helmstaedt and Doig (1975), Lappin and Dawson (1975), M. E. McCallum, Eggler, and Burns (1975), M. J. O'Hara, Saunders, and Mercy (1975), Akella et al. (1979), J. Ferguson and Sheraton (1979), Boyd and Danchin (1980), Dawson (1980), McGee and Hearn (1984), D. N. Robinson, Gurney, and Shee (1984), Smyth,

McCormick, and Caporuscio (1984), Smyth and Caporuscio (1984), Schultze (1987), Hills and Haggerty (1989), Smyth, Caporuscio, and McCormick (1989), Pearson, O'Reilly, and Griffin (1991), and others. Also see I. D. MacGregor (1968, 1970) and Dawson (1981).

7. The minerals and chemistry cited in this paragraph are based on a compilation from the works cited in note 6. See particularly R. G. Coleman et al. (1965).

8. Berkland et al. (1972) define exotic and tectonic blocks. A *tectonic block* is a mass of rock that has been transported with respect to adjacent masses of rock through the operation of tectonic processes. An exotic block is a mass of rock occurring in a lithologic association foreign to that in which the mass formed. Also see Raymond (1984a). For descriptions of such blocks, see Brothers (1954), Borg (1956), Bloxam (1959), R. G. Coleman and Lee (1963), E. H. Bailey, Irwin, and Jones (1964), R. G. Coleman et al. (1965), Ernst et al. (1970), R. G. Coleman and Lanphere (1971), Raymond (1973a), D. E. Moore (1984), Cloos (1986), Dal Piaz and Lombardo (1986), N. V. Sobolev et al. (1986), D. E. Moore and Blake (1989), and Wakabyashi (1990).

9. The mineralogy cited here is based on descriptions in the references cited in note 8, on D. E. Moore (1984), and on the author's observations.

10. R. G. Coleman and Lanphere (1971) and Cloos (1986) provide reviews. See the examples described below for a more detailed discussion of the Franciscan-related occurrences.

11. R. G. Coleman et al. (1965), N. V. Sobolev et al. (1986), Dobretsov (1991).

12. Matthews and Schliestedt (1984), Schliestedt (1986), C. Miller (1977), N. V. Sobolev et al. (1986), Ohta, Hirajima, and Hiroi (1986), Ernst et al. (1970), B. A. Morgan (1970), Ridley (1984), Yokoyama, Brothers, and Black (1986), Ghent et al. (1987), Pognante and Kienast (1989), Bouchardon et al. (1989), Pognante, (1989, 1991), Pognante and Sandrone (1989), Liati and Mposkos (1990). Philippot and Selverstone (1991) discuss fluid compositions and veins in eclogites, and Castelli (1991) describes Eclogite Facies metamorphism in carbonate rocks.

13. Minerals are based on the references cited in note 12.

14. Eskola (1921, in R. G. Coleman et al., 1965), Yoder and Tilley (1962), R. G. Coleman et al. (1965), Mysen and Heier (1972), Mori and Banno (1973), DeWit and Strong (1975), Ernst (1977b), C. Miller (1977), N. V. Sobolev et al. (1986), Austrheim (1987, 1990), Griffin (1987), Jamtveit (1987), Bouchardon et al. (1989), Klemd (1989), I. S. Sander (1989), X. Wang, Liou, and Mao (1989), Krogh et al. (1990), Jamtveit, Bucher-Nurminen, and Austrheim (1990), X. Wang and Liou (1991). Also see Liati and Mposkos (1990) for a polymetamorphic example from northern Greece.

15. Minerals are based on the reports cited in note 14.

16. Contrast D. H. Green and Ringwood (1967a, 1972), Fry and Fyfe (1969), DeWit and Strong (1975), and Austrheim (1987) with B. A. Morgan (1970) and E. H. Brown and Bradshaw (1979).

17. See Dawson (1981) and Miyashiro (1973a, p. 313). Also see Newton and Smith (1967), Boettcher (1970), and Ghent et al. (1987).

18. For example, Yoder and Tilley (1962), D. H. Green and Ringwood (1967a, 1972), Essene, Hensen, and Green (1970), K. Ito and Kennedy (1970, 1971), Howells et al. (1975), and Saxena and Eriksson (1985).

19. Also see Ryburn, Raheim, and Green (1976) and Jamieson (1990).

20. This description is based essentially on the reports of DeWit and Strong (1975) and Jamieson (1990). DeWit and Strong assigned some eclogites to the Fleur de Lys Supergroup, but Jamieson assigns all eclogites to the underlying East Pond Metamorphic Suite. See also Church (1969) and Neale and Kennedy (1967).

21. E. H. Bailey, Irwin, and Jones (1964) and R. G. Coleman and Lanphere (1971) provide overviews. R. G. Coleman et al. (1965) give some detailed descriptions, as do Switzer (1945, 1951), Borg (1956), and Bloxam (1959). Also see Gealey (1951), Essene, Fyfe, and Turner (1965), Ernst et al. (1970), Raymond (1973a, 1977), D. E. Moore (1984), Cloos (1986), D. E. Moore and Blake (1989), Wakabyashi (1990), and Oh, Liou, and Maruyama (1987, 1991).

22. R. G. Coleman and Lanphere (1971), Suppe and Armstrong (1972), J. C. Maxwell (1974), Lanphere, Blake, and Irwin (1978), and Mattinson (1986, 1988).

23. Also see the references cited in note 21.

24. See the discussion of R. G. Coleman and Lee's (1963) subdivision of metabasites into Types I, II, III, and IV in chapter 28.

25. Ernst (1970, 1984), Schweickert and Cowan (1975), Evarts (1978), Dickinson (1981), Cloos (1982, 1984).

26. Most of these hypotheses are summarized by J. F. G. Wilkinson (1976), Ernst (1977b), or F. A. Frey (1980), all of whom supply additional references. See also B. A. Morgan (1970), DeWit and Strong (1975), and Griffin (1987) for discussions of hypothesis 1. See Kornprobst (1969), Beeson and Jackson (1970), Dickey, Obata, and Suen (1977), Cabanis and Godard (1987), Paquette, Menot, and Peucat (1989), and Pearson, O'Reilly, and Griffin (1991) for discussions of hypothesis 2. Switzer (1945) discusses hypothesis 3. Hypothesis 4 is discussed by Yoder and Tilley (1962) and Kuno (1969). See D. H. Green (1966) and Smyth, Caporuscio, and McCormick (1989) for expositions of hypothesis 5. Also see J. F. G. Wilkinson (1976). Hypothesis 6 is discussed or supported by Shimizu (1975), Dickey, Obata, and Suen (1977), and F. A. Frey (1980). Menzies (1983) provides a general discussion of xenolith-forming processes.

27. B. A. Morgan (1970), Mysen and Heier (1972), Evans, Trommsdorf, and Richter (1979), Evans, Trommsdorf, and Goles (1981), Jamtveit (1987), Jamtveit, Bucher-Nurminen and Austrheim (1990), Philippot and Selverstone (1991).

PROBLEMS

29.1. (a) Using ACF plots of eclogites, plot the position represented by the chemistry of sample 2, table 29.1, on a copy of the ACF diagram for "wet" eclogites in figure 29.4. (b) What minerals should occur in this rock? (c) Compare this predicted mineralogy to the mode reported in R. G. Coleman et al. (1965, p. 487).

29.2. (a) Write the discontinuous reactions necessary to explain the conversion of rocks of the Granulite Facies to rocks of the "dry" Eclogite Facies (see figure 29.4), assuming kyanite appears before zoisite. (b) Write reactions to explain the conversion of the Blueschist Facies assemblages shown in figure 29.4 to the "wet" Eclogite Facies assemblages shown in the same diagram.

29.3. Using a copy of figure 29.9 and the data of Wakabyashi (1990) for sample TEC2, (a) trace the P-T-t path for the sample. (b) Which of the six models of eclogite formation best explains the history of this rock?

30

Dynamic Metamorphism

INTRODUCTION

Dynamic metamorphism is metamorphism of rock masses caused primarily by deviatoric stresses that yield relatively high strain rates. Thus, it is metamorphism resulting from deformation. The deformation may be dominantly **brittle,** in which case rock and mineral grains are broken, or it may be dominantly **ductile,** in which case plastic behavior and flow occur via structural changes within and between grains.[1] Temperatures during dynamic metamorphism are typically elevated and may be created by the deformation process. Fluids commonly contribute to the metamorphic process, both by altering phase chemistry and by facilitating recrystallization.

Both local and regional dynamic metamorphism are recognized. At the local scale, in narrow zones from less than 1 mm to several meters wide, brittle or ductile deformation along localized zones of deformation (e.g., along faults and fold limbs) causes rock to break, to recrystallize, and even to melt locally. Similarly, both brittle and ductile deformation, as well as melting, occur during impacts of extraterrestrial bodies. Brittle and ductile deformation processes also operate at the regional scale, but at this scale, mappable rock masses of regional extent are affected.

The rocks produced at all scales by dynamic metamorphism are rocks composed of *clasts*, or fragments of preexisting material (porphyroclasts and microlithons), surrounded by a deformed matrix, the texture or mineral composition of which was produced by metamorphic processes. Such rocks, which fit into the broad category of clastic rocks, are here referred to as **dynamoblastic rocks.**[2]

OCCURRENCES OF DYNAMOBLASTIC ROCKS

Faults pervade the crust of the Earth. Inasmuch as faults are deformation zones, dynamoblastic rocks associated with faults are a ubiquitous feature of the crust. In addition, folds and related deformation zones are relatively common in the root zones of orogenic belts. Even in zones in which newly formed rocks are only partially lithified, for example, in soft sediments on the seafloor or in crustal rocks on lava lakes, deformation may yield dynamically metamorphosed rocks.[3] Particularly noteworthy among the local- to regional-scale zones of dynamoblastic rock are the mylonite zones associated with metamorphic core complexes (G. A. Davis, 1980, 1988; J. L. Anderson, 1988)[4] and the melanges of outer metamorphic belts.[5] Melanges are, in fact, mappable masses of dynamoblastic rock of local to regional dimensions. Impact structures with dynamoblastic rocks include Meteor Crater in Arizona; the Ries Basin of Germany, the Manicouagan impact structure of Quebec, and the Boltysh impact crater of Asia.[6]

Regional zones of dynamoblastic rocks occur at plate boundaries. Along spreading ridges, regional stress may be widespread enough, particularly at mantle depths beneath the spreading axis, to yield dynamically metamorphosed zones of rock. Perhaps more commonly, ductile deformation is concentrated in *narrow zones* within a regional terrane of semischistose ultramafic tectonite (Girardeau and Mercier, 1988). Most local and regional zones of this type are probably subducted and are not preserved. Nevertheless, evidence of their existence is preserved locally in mantle slabs and the basal tectonites of ophiolites.[7] More commonly, oceanic crustal rocks are deformed along transform faults, yielding regional zones (large scale ductile shear zones), localized zones, or bodies (melanges) of dynamoblastic rock.[8] Subaerial examples of rocks deformed in this way are exposed in the Sierra Nevada of California (Saleeby, 1982, 1984), in northern Italy (Gianelli, 1977), and on the island of Cyprus (Spray and Roddick, 1981; Murton, 1986). Exposures of transform faults that transect the continents also reveal brittly and ductily deformed rocks, such as those along faults of the San Andreas Fault System in California (J. L. Anderson, Osborne, and Palmer, 1983; Chester, Friedman, and Logan, 1985)[9] and faults in the British Isles (Flinn, 1977).

The most extensive development of regional, dynamically metamorphosed rocks occurs in the orogenic belts. Rocks of the transform fault zones may be accreted here, but most commonly, the regional zones of dynamoblastic rock are produced by deformation associated with the plate (and continent) collisions that yield the orogen. At the shallower and cooler levels of Phanerozoic orogens, *melanges*, formed by brittle deformation, ductile deformation, or both, are widespread (Raymond and Terranova, 1984). These rock bodies are less commonly recognized in Precambrian terranes and the internal zones of orogens, where they have been metamorphosed.[10] Well-known examples include the melanges of the Franciscan Complex of California, the Dunnage Melange of Newfoundland, the Gwna Melange in Wales, and the melanges of the Apennine Mountains of Italy.[11]

Ductile deformation zones of regional extent are common in the internal, high-temperature zones of the orogenic belts. Here, discrete fault lines are replaced by extensive zones of recrystallization and flow, attenuated limbs of folds may exhibit mylonitic fabrics, and mylonitic rocks may pervade entire terranes. Examples of such ductile deformation zones include some of the more regionally extensive mylonitic zones associated with metamorphic core complexes in the Rocky Mountain region, thrust faults of the Peninsular and Transverse ranges of California, the Brevard Zone of the Southern Appalachian Orogen, faults in the Grenville Front Tectonic Zone in Ontario, the Moine Thrust of the Scottish Highlands, thrust faults in southern Sweden and Norway, Hercynian deformation of the Pyrenees Mountains along the French-Spanish border, shear and nappe zones in the Armorican region of France, massifs of the Alps, Precambrian shear zones in the African craton, and the Woodroffe Thrust of central Australia.[12]

ROCK TYPES, TEXTURES, AND STRUCTURES

Dynamoblastic rocks exhibit either foliated or nonfoliated textures. Brittle deformation produces *cataclastic textures*, the generally nonfoliated metamorphic textures characterized by fragmented rock or mineral grains (figure 30.1a; also see figure 23.10d).[13] A matrix of fine-grained fragmental material may surround the clasts, frictionally produced glass may form a matrix, or there may be almost no matrix, the clasts being cemented by precipitated minerals. In some breccias and most gouges there are no cements and the rocks are incoherent and unlithified. *Foliated cataclastic textures*, a transitional type of texture between cataclastic textures *sensu stricto* and mylonitic textures, are characterized by sets of shear fractures, clasts and microlithons in preferred orientations, grains showing deformation bands, and compositional layering consisting of mineral and microlithon segregations (Raymond, 1975, 1985; Chester, Friedman, and Logan, 1985; S. M. Agar, Prior, and Behrmann, 1989).[14] Some gouges and the quasimylonites of many melanges and fault zones display these textures. *Mylonitic textures*, the foliated textures produced by ductile deformation, typically contain either numerous porphyroclasts of protolith minerals or microlithons of the protolith (figure 30.1c). Highly elongated "ribbon" quartz grains, grains with deformation bands, grains with mosaic-textured recrystallization zones and margins, and porphyroclasts occur in a fine mosaic of recrystallized minerals (figure 30.1c)(Higgins, 1971, Dell'Angelo and Tullis, 1989).[15] In both handspecimen and thin section, mylonites usually are characterized by intersecting planar fabric elements enhanced by phyllosilicates. These may include a schistocity (S), "cisaillement" or shear planes (C), and shear bands (SB) (figure 30.2)(Berthe, Choukroune, and Jegouzo, 1979; C. Simpson, 1986; Dell'Angelo and Tullis, 1989). These intersecting elements yield lenticular or phacoidal domains that differentially weather in many rocks to yield a "fish-scale" or "button-like" character.

Because dynamoblastic rocks are associated with fault zones and other zones of deformation, they exhibit structures characteristic of these zones. The mylonitic types commonly are banded and may contain crenulation cleavages, boudin, or flow folds (figure 30.3). Foliation and banding occur at all scales. The relationships between thin-section, handspecimen, outcrop, and regional textures and structures in mylonites are

(a)

(b)

Figure 30.1 Photomicrographs of textures in dynamoblastic rocks. (a) Manganese-oxide (black) cemented, cataclastic chert breccia, Rome Formation, Mountain City, Tennessee. Long dimension of photo is 6.5 mm. (XN). (b) Metafeldspathic arenite protomylonite, Stone Mountain Fault Zone, Highway 421, 4 miles north of Trade, Tennessee. (XN). (c) Orthomylonite, southwest flank of the Santa Catalina Mountains, Arizona. (XN). (d) Ultramylonite, Rosman Fault, Brevard Zone, North Carolina. Long dimension of photos in (b)–(d) is 1.27 mm.

((d) Stop 2 in Horton and Butler, 1986.)

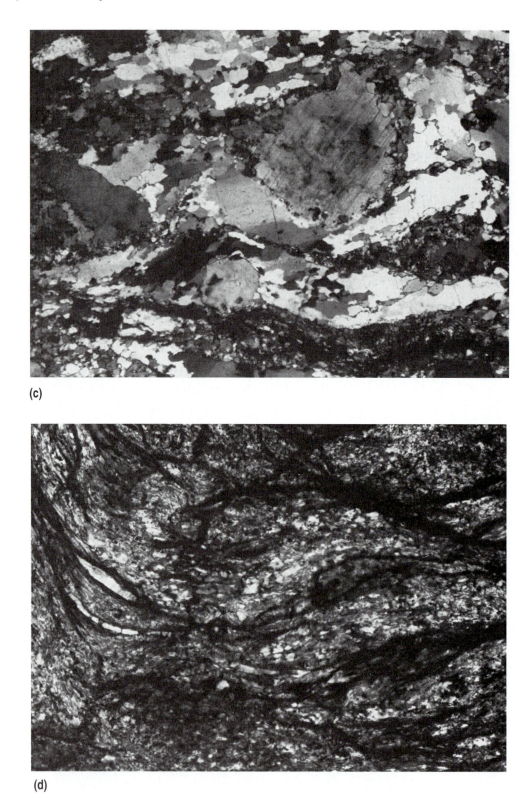

(c)

(d)

Figure 30.1 Continued

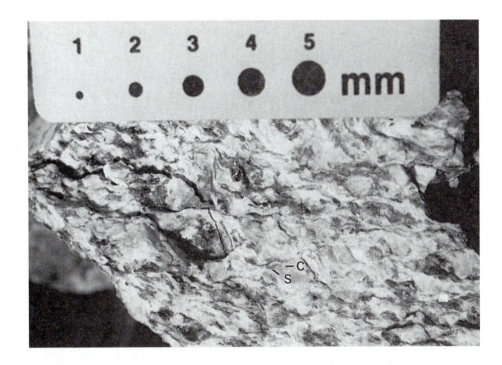

Figure 30.2 Mylonite showing S-C fabric. Southwest flank of Santa Catalina Mountains, Arizona. S = schistocity, C = spaced cleavage, shear planes.

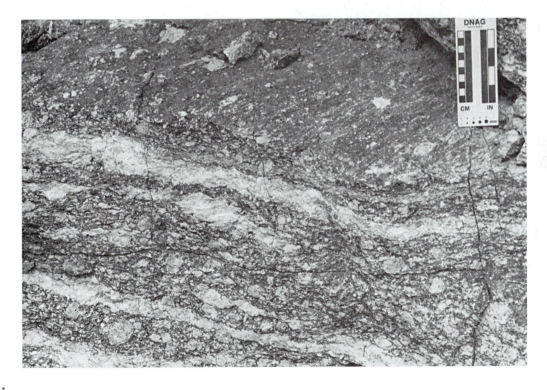

Figure 30.3 Structures in outcrop of mylonite on the southwest flank of the Santa Catalina Mountains, Arizona. Note lineation on the top surface and foliation, boudinage in veins, and augen on the front surface of the exposure.

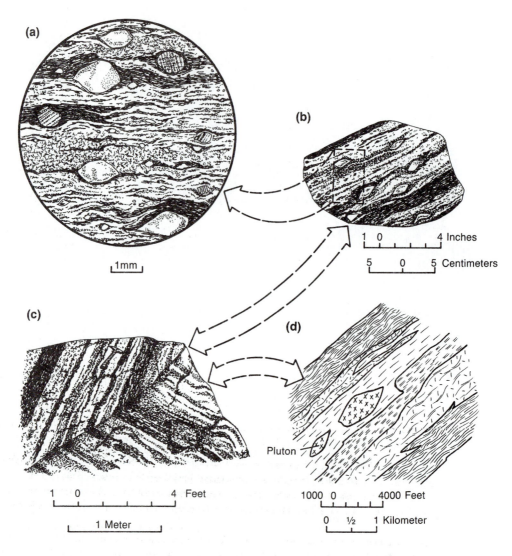

(a)

1mm

(b)

| 1 | 0 | | | 4 Inches |

| 5 | 0 | 5 Centimeters |

(c)

(d)

Pluton

| 1 | 0 | | | 4 Feet |

1 Meter

| 1000 | 0 | 4000 Feet |

| 0 | ½ | 1 Kilometer |

Figure 30.4   Relationship between textures and structures in mylonites at different scales: (a) thin section, (b) handspecimen, (c) outcrop, and (d) regional. Note that foliation and banding pervade all scales.
(Source: M. W. Higgins, "Cataclastic Rocks" in U S Geological Survey PP. 687, 1971)

shown in figure 30.4 and the site of mylonite formation in deep (hotter) fault zones is depicted in figure 30.5. Cataclastic rock types may appear more massive than mylonites, but commonly form tabular zones that contain associated slickensides, minor faults, and buckle folds. In fault zones, cataclasites form near the surface and exhibit increased milling of rock fragments and grains from the margins to the center of the zone (figure 30.5, top three circles). Transitional rocks with weak to moderate, foliated-cataclastic and protomylonitic foliations (quasimylonites and the protomylonites) form at moderate depths (J. L. Anderson, Osborne, and Palmer, 1983; Babaie et al., 1991).[16]

Names for rocks formed in zones of high strain were developed long before the processes involved in their origin were understood (Lapworth, 1885). Undoubtedly the most serious error made by early workers was that they thought that fine-grained rocks found in ductile shear zones owe their small crystal size to physical breakdown via brittle failure—cushing, grinding, and breaking—of the minerals of the protolith (Lapworth, 1885; Waters and Campbell, 1935). That such processes occur during near-surface faulting is evidenced by breccia and gouge that lack any coherence between grains. This is not the case at depth.

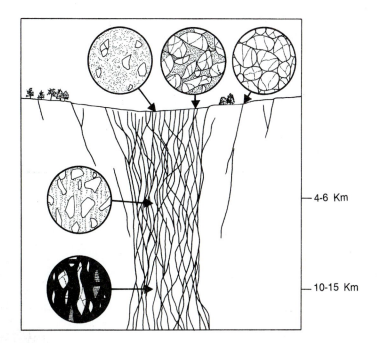

• • • • • • • • • • • • •

Figure 30.5 Relationship between depth of formation, position in shear zone, and dynamoblastic textures. Thin-section views are shown in circles. Upper right circle shows incipiently fractured sandstone. Upper center circle shows cataclastic breccia. Upper left circle shows gouge with a few larger clasts remaining. Weakly foliated quasimylonite occurs at intermediate depths. Well-developed mylonite (lowest circle) containing local ultramylonite bands (not shown) occurs at greatest depth.
(Based on the ideas of Sibson, 1977, and J. L. Anderson, Osborne, and Palmer, 1983)

In recent years, experimental deformation of rocks, plus the associated textural studies, have shown conclusively that the fine-grained textures of the mylonites result from syntectonic recrystallization associated with flow and intragranular deformation and from post-tectonic recovery and recrystallization, rather than from brittle failure of the constituent grains of the rock (N. L. Carter, Christie, and Griggs, 1964; T. H. Bell and Etheridge, 1973; Dell'Angelo and Tullis, 1989). Matrix textures are crystalline, not clastic. Thus, the term mylonite, based on the Greek root *mule* for mill (to grind down), is a misnomer, but it is retained because it is entrenched in the literature and is used widely for rocks of ductile deformation zones.

The main types of dynamoblastic rock have been defined and redefined by several workers, as knowledge of rock origins increased over the years.[17] Three classifications, one proposed by Higgins (1971), another proposed by Wise et al. (1985), and a third created by the author, are shown in figure 30.6. Higgins's (1971) classification, admirably, used observable criteria (cohesion, grain size, percent clasts, and foliation or fluxion structure) to define various types of dynamoblastic rocks. Because it was developed prior to our modern understanding of the roles of intragranular deformation and syntectonic recrystallization, however, it defines all dynamoblastic rocks as cataclastic (produced by fracture rather than ductile deformation processes) and therefore conveys meanings at odds with contemporary understanding of the rock-forming processes. Nevertheless, Higgins's (1971) subdivisions based on porphyroclast percentage are useful and have been adopted by D. U. Wise et al. (1984) and Raymond (1993). The classification of D. U. Wise et al. (1984) (1) focusses on fault-related rocks, (2) uses coherence, porphyroclast percentages, the presence or absence of foliation, and the presence or absence of evidence of "syntectonic crystal-plastic processes" as criteria for distinguishing various rock types, and (3) emphasizes relative rates of strain versus recovery in rock genesis. Yet the classification has drawbacks. It implies that dynamically metamorphosed rocks are not metamorphic, it does not clearly distinguish between recovery and recrystallization, it fails to recognize foliated quasimylonites, and it specifies that all mylonites are coherent.[18] The simplified classification proposed

(a)

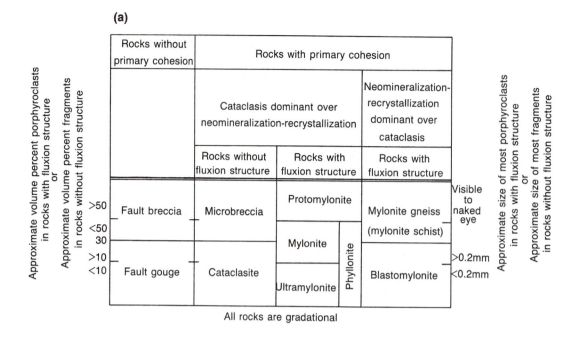

	Rocks without primary cohesion	Rocks with primary cohesion				
		Cataclasis dominant over neomineralization-recrystallization			Neomineralization-recrystallization dominant over cataclasis	
		Rocks without fluxion structure	Rocks with fluxion structure		Rocks with fluxion structure	
>50	Fault breccia	Microbreccia	Protomylonite		Mylonite gneiss (mylonite schist)	Visible to naked eye
<50 / 30			Mylonite	Phyllonite		
>10 / <10	Fault gouge	Cataclasite	Ultramylonite		Blastomylonite	>0.2mm / <0.2mm

All rocks are gradational

(left axis) Approximate volume percent porphyroclasts in rocks with fluxion structure or Approximate volume percent fragments in rocks without fluxion structure

(right axis) Approximate size of most porphyroclasts in rocks with fluxion structure or Approximate size of most fragments in rocks without fluxion structure

(b)

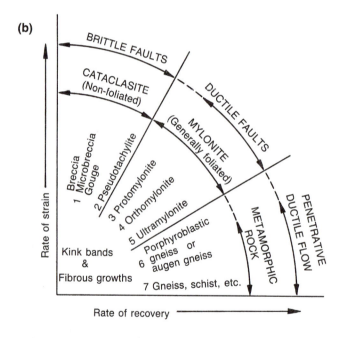

BRITTLE FAULTS

CATACLASITE (Non-foliated)

DUCTILE FAULTS

MYLONITE (Generally foliated)

METAMORPHIC ROCK

PENETRATIVE DUCTILE FLOW

Rate of strain →

1 Breccia / Microbreccia / Gouge
2 Pseudotachylite
3 Protomylonite
4 Orthomylonite
5 Ultramylonite
6 Porphyroblastic gneiss or augen gneiss
7 Gneiss, schist, etc.

Kink bands & Fibrous growths

Rate of recovery →

(c)

Foliated to Non-foliated Rocks with Glass in Matrix	CATACLASTIC (NON-FOLIATED)		Weakly Foliated with Broken Grains	MYLONITIC (FOLIATED)	
	Dominant grain size				
	<1/16	>1/16			
				(shaded)	100
Pseudotachylite	Gouge	Cataclastic breccia	Quasi-mylonite	Protomylonite	
					50
				Orthomylonite	
				Ultramylonite	10

PORPHYROCLASTS AND/OR LITHOCLASTS %

Figure 30.6 Three classifications of dynamoblastic rocks. (a) *Source: M. W. Higgins, "Cataclastic Rocks" in US Geological Survey pp. 687, 1971.* (b) Classification of Wise et al. (simplified from Wise et al., 1985). (c) Raymond's classification (1993).

by Raymond (1993) (figure 30.6c) is based on observable criteria alone. Nonfoliated dynamoblastic rocks, i.e., rocks with cataclastic textures, are subdivided into those with glass in the matrix, called **pseudotachylites**, those dominated by grains <1/16 mm in diameter called **gouges**, and those dominated by grains > 1/16 mm in diameter, **cataclastic breccias**(see figure 30.1a). Mylonitic (foliated) rocks are subdivided on the basis of the percentage of porphyroclasts plus microlithons they contain. **Protomylonites** contain > 50% porphyroclasts and microlithons (see figure 30.1b), **orthomylonites** contain between 10 and 50% (see figure 30.1c), and **ultramylonites** contain < 10% (see figure 30.1d). As is common in nature, there is a continuum between rocks of the two major textural and rock types. **Quasimylonites**, characterized by incipient foliations, some crystalline matrix materials, and microlithons or broken grains, mark the transition between cataclasites and mylonites.[19]

MINERALS AND FACIES OF DYNAMOBLASTIC ROCKS

Mineralogically, there is nothing unique about most dynamoblastic rocks. The rocks contain minerals typical of the various facies to which they belong. Although it is theoretically possible for dynamoblastic rocks to form under the P-T conditions of any facies, they most typically represent the Zeolite, Greenschist, or Amphibolite facies. Zeolite and Greenschist facies minerals are common in the cataclasites; Zeolite, Prehnite-Pumpellyite, Greenschist, and Blueschist facies assemblages occur in quasimylonites; and Greenschist and Amphibolite Facies phase assemblages usually characterize the mylonites (Cloos, 1983; Kamineni, Thivierge, and Stone, 1988; G. A. Davis, 1988; Hyndman and Myers, 1988).

The most unique phase assemblages in the dynamoblastic rocks are those containing glass and those containing high-pressure, shock-metamorphosed minerals, such as coesite and stishovite. Glass and the high-pressure phases occur in impact metamorphic rocks (Chao, Shoemaker, and Madsen, 1960; Grieve et al., 1987). Assemblages with glass also characterize the pseudotachylites (Sibson, 1975; R. H. Maddock, 1984). A representative assemblage in a pseudotachylite formed in an Amphibolite Facies quartz-feldspar gneiss is

quartz–plagioclase–hornblende–biotite–magnetite–
zircon–sphene–glass

(R. H. Maddock, 1984).

PROCESSES DURING DYNAMOBLASTIC ROCK FORMATION

Several process operate during the petrogenesis of dynamoblastic rocks. These include cataclasis, recrystallization, neocrystallization, pressure solution, and metasomatism. During the formation of various rock types, one or another of these process dominates.

Cataclasis

Cataclastic rocks form by brittle failure (breaking) of rock and mineral materials. Such failures occur under conditions in which the rocks (and mineral grains) are strained beyond the limits of their strength, typically under conditions of high deviatoric stress and high strain rate but relatively low temperature.[20] When the rocks or mineral grains can withstand no more strain, they break. Breaking begins as a series of fractures that extend across each grain from one point of contact with an adjoining grain to another point of contact (Blenkinsop and Rutter, 1986) (figure 30.5, upper right circle). Continued application of stress produces angular fragments. Additional grinding of the weakened and broken rock fragments between more rigid adjoining rock masses results in the milling of the broken clasts, eventually yielding a gouge (figure 30.5, upper left circle). Quasimylonites develop a foliation *either* where pervasive fracturing is accompanied by some ductile deformation or neocrystallization *or* where deformation induces alignment of inequant mineral grains or microlithons (J. C. Moore et al., 1986; Evans, 1988; Agar, Prior, and Behrmann, 1989; Babaie et al., 1991).

Different minerals fail under different conditions. Phyllosilicates tend to be weak and subject to flow at relatively low values of temperature and strain rate. In contrast, quartz and feldspars have greater strength. Feldspars may experience brittle failure even under temperatures of 800° C, strain rates of 2×10^{-5}, and confining pressures of 1.0 Gpa (10kb) (Marshall and McLaren, 1977). In dry rocks, quartz is quite strong, but water promotes weakening and plastic behavior (Blacic and Christie, 1984).

In general, neither fluids nor heat play a major role in cataclasis. Fluids may, and commonly do, transport materials that are precipitated to cement breccia fragments, but this is commonly a post-tectonic process. In some cases, elevated temperatures result in local melting and the formation of pseudotachylites. The processes are similar to those involved in the formation of breccias, but in addition to brittle failure, frictionally induced heating elevates the temperature of the deforming rocks enough to produce local melting (Sibson, 1975; 1977; R. H. Maddock, 1983; 1984). Sibson (1975) estimated the temperature of pseudotachylite formation in a Scottish fault zone to be about 1100° C. Pseudotachylites form under dry conditions.

Mylonitization

The formation of mylonites is more complex than the formation of cataclastic rocks and involves successive stages of deformation, recovery, and recrystallization. During deformation, pressure solution may contribute to fabric development (Vauchez, Maillet, and Songy, 1987), but deformation processes are basically mechanical in nature. Mylonitization also involves the chemical processes of metasomatism and neocrystallization. In these, as well as in the deformation processes, fluids are important. Other variables that control the nature of the mylonitization include the mineral composition of the protolith, the confining pressure, the temperature, and the homogeniety and continuity of the rock mass.

The deformation processes involved in mylonitization include microfracturing, twinning, dislocation glide, and grain-boundary sliding.[21] Microfracturing is a process in which microscopic fractures develop within and between grains, in response to applied stress; i.e. fractures are intragranular or intergranular, respectively (Nicolas and Poirier, 1976, pp. 43–44). In minerals with cleavage, the intragranular fractures may follow the cleavage. Feldspars, in particular, and zircon tend to fracture during mylonitization, even at high temperatures,[22] and in some cases, quartz, calcite, olivine, pyroxene, and biotite do so as well.[23]

Twinning is another mechanism by which crystals may reflect strain. The term twinning has its usual mineralogical meaning; it refers to the process in which one or more parts of a crystal lattice assume an orientation different from other parts. *Dislocation glide* refers to a shift in the position of a defect (i.e., a dislocation) within a crystal lattice.[24] The defect may change size or may simply change positions. Such dislocations are revealed in crystals by features such as deformation bands (see figures 30.1c and 23.15b).[25] *Grain-boundary sliding* is a process in which grains, in response to applied stress, shift positions relative to adjoining grains, with the shift occurring along the grain boundary.[26] All of these processes are granular adjustments made within rocks to accommodate an applied stress. The adjustments result in a foliated rock, shortened perpendicular to the foliation and lengthened parallel to it.

In addition to the mechanical processes of deformation involved in mylonite formation, three physical-chemical processes—recovery, recrystallization, and neocrystallization—and the chemical process of metasomatism are important in the development of the character of these rocks. **Recovery** is a process in which deformed grains containing a relatively high proportion of dislocations (and therefore a high-energy state) reduce the amount of intracrystalline deformation (and therefore the strain energy within the crystal) by internal reorganization or reduction of dislocations.[27] Recovery commonly results in the elimination of deformation bands, twins, and other deformation features.

Recrystallization is the process in which strain energy is reduced by the nucleation and growth of new crystals within and at the margins of host crystals.[28] The new crystals have orientations that differ from those of the deformed crystals and are typically small, irregular, and polygonal in shape. The process by which these new crystals form, called polygonization, is very common and is evidenced by mosaics of small polygonal crystals rimming and/or relacing porphyroclasts (figure 30.7). Recrystallization that occurs during deformation is referred to as *syntectonic recrystallization* or *dynamic recrystallization*. *Static recrystallization* occurs following deformation. Quartz, olivine, calcite, and micas commonly undergo syntectonic recrystallization during mylonite formation.

Neocrystallization, the process of new mineral formation, requires diffusion of new chemical species into the area of grain growth.[29] Where those chemical elements migrate in from regions outside of the local domain (or migrate out of the domain), metasomatism is considered to have occurred. The new minerals that form may simply be equilibrating with new P-T conditions or they may be responding to chemical potential gradients created by changes in the fluid composition.

Fluid flow in fault zones and ductile deformation zones is significant in promoting mechanical deformation, recrystallization, and neocrystallization.[30] Shearing associated with deformation may be accompanied by fluid flow that (1) facilitates deformation processes, such as dislocation glide, (2) transports heat, reducing the metamorphic temperature in the mylonite zone, and (3) changes the composition of the local fluid phase, promoting neocrystallization. As deformation opens channels for fluid flow, the fluids promote deformation, recrystallization, and neocrystallization, which, in turn, aid in the growth of the deformation zone. Thus, a kind of feedback loop exists. Another common consequence of fluid migration, the transportation of heat out of mylonite zones, results in the formation of retrograde effects along the

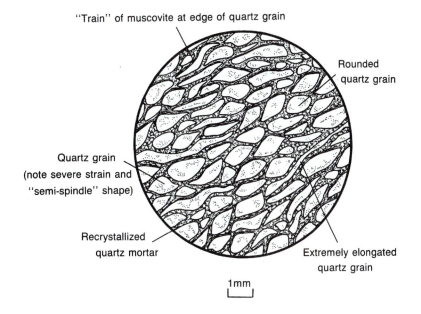

"Train" of muscovite at edge of quartz grain

Rounded quartz grain

Quartz grain (note severe strain and "semi-spindle" shape)

Recrystallized quartz mortar

Extremely elongated quartz grain

1mm

Figure 30.7 Sketch of micaceous metaquartz arenite protomylonite from the Weaverton Formation of the Blue Ridge Belt, Virginia.
(Source: M. W. Higgins, "Cataclastic Rocks" in US Geological Survey pp. 687, 1971.)

zones. For example, Amphibolite Facies mylonite zones may develop in Granulite Facies terranes, or Greenschist Facies mylonite zones may form in Amphibolite Facies terranes (Drury, 1974; Beach, 1980; Erslev and Sutter, 1990).[31] Major metasomatic effects are also produced by fluids.[32] For example, K. O'Hara (1988, 1990b) and Glazner and Bartley (1991) suggest that fluids have removed more than 60% of the volume of material in some mylonite zones, particularly by removing Si and alkali elements. Pressure solution promotes some of this volume loss. Reactions such as

5 plagioclase (andesine) + 2 microcline + 24 H$_2$O — clinozoisite + 2 muscovite + 10 H$_4$SiO$_4$ + 3 Na$^+$ + 3 (OH)$^-$

may explain the conversion of feldspar to white mica, the loss of soluble silica, and a loss of the alkali element sodium.[33] Tobisch et al. (1991), however, demonstrate that mylonitization of some granitoid rocks is accompanied by *gains* in Si, while alkalis are lost.

Together, combinations of the processes described above yield mylonitic rocks. The particular combination of processes that produces the specific fabric elements and mineral composition of any given mylonite is a function of the rock and fluid composition and the P-T and strain rate histories. Consequently, each mylonite requires individual study.

EXAMPLE: THE BREVARD ZONE, SOUTHERN APPALACHIAN OROGEN

The Brevard Zone is a major zone of deformation in the Southern Appalachian Orogen.[34] The zone extends from North Carolina to Alabama and generally separates the Blue Ridge Belt from the Inner Piedmont Belt (figure 30.8).

Definitive evidence indicates that the Brevard Zone is polymetamorphic and structurally complex. An early Amphibolite Facies metamorphic event was followed by a later Greenschist Facies event. These recrystallization events were accompanied by the formation of dynamoblastic rocks. More than thirty structural interpretations of the nature of the Brevard Zone have been proposed (Bobyarchick, Edelman, and Horton, 1988), including interpretations of the zone as a fold, a thrust fault, a normal fault, a strike-slip fault, and combinations of these various structures. Not only has there been disagreement about the nature of the zone, but there is no general agreement about the ages of the deformation events that produced the various features of the zone. What *is* clear is that the Brevard zone is characterized by mylonites.

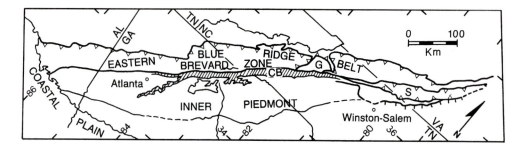

Figure 30.8 Map showing the location of the Brevard Zone in the southern Appalachian Orogen. CB = Chauga Belt, G = Grandfather Mountain Window, S = Smith River Allochthon.
(Modified from Bobyarchick, Edelman, and Horton, 1988)

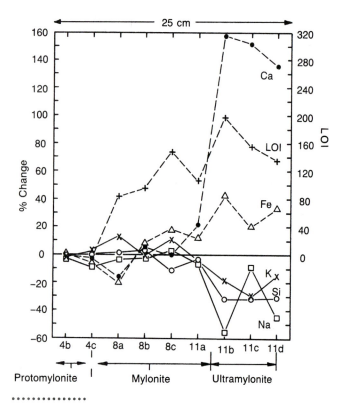

Figure 30.9
Chemical change as percentage change in oxides across a Brevard Zone mylonite sample. Values normalized to constant Al$_2$O$_3$, as measured in a subvolume of the mylonite.
(Modified from Sinha, Hewitt, and Rimstidt, 1986)

Traverses across the zone, as well as drilling, reveal that rock types and deformation styles vary within the zone (Hatcher, 1971; Bobyarchick, Edelman, and Horton, 1988; Christensen and Szymanski, 1988). Rock types vary from protomylonites to white mica ultramylonites (figure 30.1d). In the less deformed rocks, Greenschist Facies assemblages such as

quartz–white mica–biotite–plagioclase–epidote–chlorite–magnetite and

quartz–white mica–chlorite–pyrite–graphite

are typical (Bryant and Reed, 1970a; Bobyarchick, Edelman, and Horton, 1988). Such assemblages overprint older kyanite- and garnet-bearing assemblages.

Sinha, Hewitt, and Rimstidt (1986) obtained a 25 cm × 25 cm block of rock from the Brevard Zone near Rosman, North Carolina, that showed a complete gradation from protomylonite to ultramylonite, a range of lithologic change typical of the 15-km-wide Brevard Zone as a whole. Chemical analyses of small parts of this block (4b, 8c, etc., in figure 30.9) reveal that significant chemical changes occur across the block and are especially notable in ultramylonite (11b to 11d), where textural reconstitution was greatest. Fluid flow, concentrated in the ultramylonite, promoted losses of SiO$_2$, Na$_2$O, and K$_2$O, and gains of CaO, FeO, and H$_2$O. The amount of silica lost requires a fluid/rock weight ratio of 250. These data suggest that the ultramylonite zones serve as major channels for fluid migration during mylonitization and that there may be a feedback loop between fluid flow and recrystallization as the mylonite zone develops.

SUMMARY

Local to regional dynamic metamorphism is an important process in many orogenic belts and deformation zones throughout the world. Dynamic metamorphism, caused primarily by deviatoric stresses yielding moderate to high strain rates, results in the formation of dynamoblastic rocks of three major types—the nonfoliated cataclasites, the foliated quasimylonites, and the foliated mylonites. Cataclasites form at relatively shallow depths through physical breaking of grains. Coarser-grained cataclastic breccia and finer grained gouge result from this process. Where temperatures during milling are high enough, local melting occurs and pseudotachylite is formed. Fluids may facilitate postcataclasis cementation.

At increased temperatures or lower strain rates, quasimylonites form. Breaking of stronger grains (e.g., quartz and feldspar), recrystallization or neocrystallization of phyllosilicates, and shear dislocation to yield slickensides contribute to the formation of these rocks. Fluids may facilitate shearing and neocrystallization.

Temperature, confining pressure, fluids, and strain rate are important variables affecting mylonitization. Within developing mylonites, grains deform via various mechanisms, including microfracturing, twinning, dislocation glide, pressure solution, and grain-boundary sliding. The textures of the resulting rocks vary and depend on the impact of various recovery, recrystallization, and neocrystallization processes that also operate during mylonite formation. Most of these processes depend on diffusion and may be facilitated by fluid flow. In some cases, mylonite zones serve as fluid channel-ways. The fluids in these channel-ways function to transport chemical species into and out of the mylonite zone, yielding metasomatic effects. These fluids may also transport heat, reducing the temperature of the zone, with the result that retrograde metamorphism occurs. This retrograde metamorphism typically yields Greenschist and Amphibolite Facies mylonite zones within higher-grade terranes.

EXPLANATORY NOTES

1. Ductile behavior does not necessarily imply that crystal-plastic deformation has occurred (Rutter, 1986; Chester, 1988). "Cataclastic flow," or semibrittle flow, involves movement along mesoscopic to microscopic zones (rather than discrete planes) as a result of some combination of grain breaking, plastic deformation of grains, and recrystallization and neocrystallization.

2. This term is derived from the roots *dynamo*, meaning power (or in this case, stress), and *blasto*, meaning formative. Hence, *dynamoblastic* means formed by stress.

3. Soft sediments deform locally by cataclasis and recrystallization. Where this is the case, they fit the definition of metamorphic rocks. For more information on soft sediment deformation, see J. C. Moore (1986), for a collection of papers dealing with this and related subjects, and also see the references cited therein.

4. Refer to additional papers in Crittenden et al. (1980), and to Kerrich and Hyndman (1986), La Tour and Barnett (1987), J. L. Anderson (1988), Lister and Davis (1989), Guerin, Brun, and Vanden Driessche (1990), Hacker, Yin, and Christie (1990), Hodges and Walker (1990), and Palais and Peacock (1990).

5. See Raymond (1984c) for a series of papers dealing with various aspects of melanges and Raymond and Terranova (1984) for a review of melange distribution.

6. Metamorphic rocks associated with impact structures are unique and rare rocks, not treated in this text. For more information consult works such as Shoemaker (1960), Chao, Shoemaker, and Madsen (1960), Bunch and Cohen (1964), Englehardt and Stoffler (1966), N. M. Short (1966), Currie (1967), French and Short (1968), Floran et al. (1978), Simonds et al. (1978), and Grieve et al. (1987).

7. See various discussions in E. M. Moores and Vine (1971), Ave Lallemant (1976), Nicolas and Poirier (1976, ch.11), R. G. Coleman (1977), Juteau et al. (1977), Salisbury and Christensen (1978), Nicolas and Violette (1982), Boudier and Nicolas (1982a, 1982b, 1985), and Girardeau and Mercier (1988).

8. See Hebert, Bideau, and Hekinian (1983), and Honnorez, Mevel, and Montigny (1984) for examples from the ocean basins.

9. For additional mention and discussion of fault rocks of the San Andreas Fault System in California, see Waters and Campbell (1935), L. F. Noble (1954), K. J. Hsu (1955), C. R. Allen (1957), Proctor (1962), R. V. Sharp (1967), Higgins (1971), Sieh (1984), and Chester, Friedman, and Logan (1985).

10. Raymond (1974, 1977a,b), Horne (1979), E. H. Brown (1986), Horton et al. (1986), Lacazette (1986), Higgins et al. (1989), Lacazette and Rast (1989), and Raymond, Yurkovich, and McKinney (1989).

11. *Franciscan Melanges*—K. J. Hsu (1968, 1969); Berkland et al. (1972), Suppe (1973), Christensen (1973), Raymond (1973a), Cowan (1974, 1978, 1982a), J. C. Maxwell (1974), Crawford (1975), Cowan and Page (1975), Gucwa (1975), Page (1978), Aalto (1982), and Cloos (1982). *Dunnage Melange*—Horne (1969), Kay (1976), and Jacobi (1984). *Gwna Melange*—Greenly (1919), C. P. Wood (1974), and D. Wood and Schuster (1978). *Italian melanges*—Beneo (1956), Maxwell (1959), Boccaletti, Bortolotti, and Sagri (1966), Elter and Trevisan (1973), Naylor (1981, 1982), Agar, Prior, and Behrmann (1989).

12. Higgins (1971) provides an annotated bibliography of older literature and reviews of several occurrences of mylonites and other dynamoblastic rocks. In addition, see the following for particular references to regions mentioned in this text: *Rocky Mountain region* (G. A. Davis et al., 1980; Hyndman, 1980; Bykerk-Kauffman, 1986), *Peninsular Ranges region* (Southern California Batholith) of California (Alf, 1948; R. V. Sharp, 1967; C. Simpson, 1985; D. K. O'Brien et al., 1987), *San Gabriel Mountains of California* (K. J. Hsu, 1955); *Brevard Zone of the Southern Appalachian Orogen* (Higgins, 1971; Sinha, Hewitt, and Rimstidt, 1986), *Grenville Front Tectonic Zone in Ontario* (Themistocleous and Schwerdtner, 1977), *the Moine Thrust of the Scottish Highlands* (Higgins, 1971; Barr, Holdsworth, and Roberts, 1986; Blenkinsop and Rutter, 1986), *southern Sweden* (Zeck and Malling, 1974), *southern Norway* (Roy, 1977; Starmer, 1980), *the Pyrenees Mountains* (Carreras, Julivert, and Santanach, 1980; Lamouroux et al., 1980; McCaig, 1984), *the Armorican region of France* (Nicolas et al., 1977; Berthe and Brun, 1980; M. J. Watts and Williams, 1983; Vauchez, Maillet, and Songy, 1987), *massifs of the Alps* (Behr, 1980; Kerrich et al., 1980; C. Simpson, 1983), *the African craton* (Wakefield, 1977), *Woodroffe Thrust of central Australia* (T. H. Bell, 1979).

13. For example, see Lucas and Moore (1986), Kamineni, Thivierge, and Stone (1988), and Evans (1988).

653

14. Also see J. C. Moore et al. (1986), Evans (1988), Kano and Sato (1988), and Babaei et al. (1991).

15. Also see discussions or illustrations in Vauchez, Maillet, and Songy (1987), Hoogerduijn Strating (1988), T. H. Bell and Cuff (1989), K. A. Carlson, van der Pluijm, and Hanmer (1990), de Roo and Williams (1990), Ji and Mainprice (1990), K. O'Hara (1990a, 1990b), H. R. Wenk and Pannetier (1990), and van der Pluijm (1991).

16. The quasimylonites are referred to as *quasiplastic mylonites* by some workers.

17. K. J. Hsu (1955), Higgins (1971), Zeck (1974), D. U. Wise et al. (1984).

18. For discussions of the terminology of D. U. Wise et al. (1984), see Mawer (1985) and Raymond (1985).

19. See J. L. Anderson, Osborne, and Palmer (1983) and J. C. Moore et al. (1986) for descriptions of these kinds of rocks. One additional rock type, not shown on the classification, is the *blastomylonite*, a mylonitic rock that shows significant post-deformational crystal growth, yielding porphyroblasts or porphyroblastic overgrowths on porphyroclasts.

20. Consult standard texts on structural geology (G. H. Davis, 1984; Dennis, 1987; Suppe, 1985; Hatcher, 1990; Twiss and Moores, 1992), as well as Turcotte and Schubert (1982) and Sibson (1977), for more details and references on rock deformation studies relevant to the structural behavior of materials in fault zones. Also see Evans (1988) for an example.

21. For example, see T. H. Bell and Etheridge (1973), Boullier and Gueguen (1975), Nicolas and Poirier (1976), Roy (1977b), T. H. Bell (1979), Etheridge and Wilkie (1979), Beach (1980), Poirier (1980), C. J. L. Wilson (1980), M. J. Watts and Williams (1983), C. Simpson (1985), D. K. O'Brien et al. (1987), Vauchez (1987b), Dell'Angelo and Tullis (1989), de Roo and Williams (1990), and Ji and Mainprice (1990).

22. Alf (1948, pl. 3, fig. 2), Marshall and McLaren (1977), Boullier (1980), C. Simpson (1985), Vauchez (1987b), Vauchez, Maillet, and Songy (1987), Evans (1988).

23. Boullier and Gueguen (1975), Kerrich et al. (1980); Wojtal and Mitra (1986). Additional discussions of crack formation and annealing may be found in several papers in Chemical Effects of Water on the Deformation and Strengths of Rocks [special issue], *Journal of Geophysical Research*, v. 89, no. B6 (1984).

24. A dislocation is the *boundary* between areas of a crystal lattice in which a shift (or slip) of a set of atoms or atomic bonds has occurred and an area in which no such slip has occurred (Nicolas and Poirier, 1976, pp. 72–74). The slips may be due to the presence of point, linear, or planar defects that form initially because of the occurrence of extra atoms, the absence of atoms, or a stress-induced shift of part of the lattice. For more information, see the discussions of dislocations, slip, and twinning by Nicolas and Poirier (1976), Barber (1985), and Van Houtte and Wagner (1985).

25. See examples and discussions of experimentally and naturally deformed rocks in Carter, Christie, and Griggs (1964), Raleigh (1967), H. W. Green (1967), Avé Lallemant and Carter (1970), H. W. Green and Radcliffe (1972), J. Tullis, Christie, and Griggs (1973), C. J. L. Wilson (1973, 1980), Nicolas and Poirier (1976), Gueguen (1979a,b), Zeuch and Green (1984a,b), Mercier (1985), G. P. Price (1985), Toriumi and Karato (1985), Wenk (1985), Dell'Angelo and Tullis (1986, 1989), and Ji and Mainprice (1990).

26. For additional discussions, see Nicolas and Poirier (1976), Etheridge and Wilkie (1979), and van der Pluijm (1991).

27. Nicolas and Poirier (1976, p. 129ff.), Gottstein and Mecking (1985, p. 183).

28. Recrystallization is described in more detail by Nicolas and Poirier (1976), Bell (1978b), Gottstein and Mecking (1985), Dell'Angelo and Tullis (1989), R. W. Carlson, van der Pluijm, and Hanmer (1990), and van der Pluijm (1991).

29. For more detailed discussions of neocrystallization and metasomatic effects during mylonitization, see Drury (1974), Etheridge and Hobbs (1974), Wakefield (1977), Kerrich et al. (1977, 1980), Beach (1980), Spray and Roddick (1981), Watts and Williams (1983), McCaig (1984), Winchester and Max (1984), Sinha, Hewitt, and Rimstidt (1986), Hammond (1987), and K. O'Hara (1988).

30. The roles of fluids in promoting deformation and metamorphism are discussed in more detail by Drury (1974), Fyfe, Price, and Thompson (1978), Etheridge et al. (1984), Kerrich, La Tour, and Willmore (1984), and other authors whose work is published in the *Journal of Geophysical Research*, v. 89, no. B6 (1984), and by Ferry (1986), Wood and Walther (1986), Ridley and Thompson (1986), K. O'Hara (1988), and Glazner and Bartley (1991).

31. Also see Sinha and Glover (1978), Behr (1980), and Carreras, Julivert, and Santanach (1980).

32. Sinha, Hewitt, and Rimstidt (1986); K. O'Hara (1988, 1990a,b), T. H. Bell and Cuff (1989), K. O'Hara and Blackburn (1989), de Roo and Williams (1990), Erslev and Sutter (1990), Glazner and Bartley (1991), Tobish et al. (1991).

33. Compare this reaction to that of Bryant (1966) and see K. O'Hara (1988) for an additional reaction for the breakdown of alkali feldspar.

34. This review is based primarily on the works of Sinha, Hewitt, and Rimstidt (1986); and Bobyarchick, Edelman, and Horton (1988), and to a lesser extent on the works of J. C. Reed and Bryant (1964), Bryant and Reed (1970a), Hatcher (1971), Roper and Justus (1973), Sinha and Glover (1978), F. A. Cook et al. (1979), Horton and Butler (1986), Edelman, Lin, and Hatcher (1987), and Christensen and Syzmanski (1988). For additional references to the voluminous literature on the Brevard Zone, see Bobyarchick, Edelman, and Horton (1988).

PROBLEMS

30.1. Mylonitization and uplift of metamorphic core complexes in western North America along late detachment faults that separate a more brittle cover sequence from the underlying core may be rapid (G. A. Davis, 1988). (a) If mylonitization began at a pressure of about 0.515 Gpa (5.15 kb) at 25 m.y.b.p. and fission track, isotopic, and geobarometric analyses show that it was overprinted by a brittle fabric (mylonitization was completed) at about 23 m.y.b.p., at a pressure of about 3.30 kb, what was the rate of uplift in mm/year during this time? (b) If conglomerates reveal that the mylonites were exposed at the surface at 21 m.y.b.p., what was the rate of uplift in mm/year during the time interval 23 m.y. to 21 m.y.? (c) Considering that the mineral assemblage in the mylonite zone is

quartz–K-feldspar–oligoclase–hornblende–biotite–garnet

and the overprint assemblage is

quartz–K-feldspar–albite–white mica–biotite–chlorite,

what metamorphic facies does each assemblage represent? (d) Assuming the rock originally formed as an intrusive rock at 0.6 Gpa (6 kb) and 770° C at 26 m.y.b.p., the temperature of mylonitization was 575° C, and the overprint temperature was 350° C, sketch the P-T-t path for the core rocks of the complex on a copy of figure 24.6.

CHAPTER
31

Alpine Ultramafic Rocks
and the Mantle

INTRODUCTION

Alpine ultramafic rock bodies are irregular to elliptical bodies of ultramafic rock that occur in mountain belts (Benson, 1926; H. H. Hess, 1955; Moores and MacGregor, 1972). The rocks that comprise these bodies may form initially as (1) magmatic crystal cumulates (differentiates), (2) crystallized or recrystallized products of mantle diapirs, or (3) mantle tectonites. Thus, the bodies either form in the crust as crystallization products of mafic magmatic intrusions,[1] or they are emplaced into the crust by faulting, as mantle slabs, and by solid intrusion, as mantle diapirs. Small to large fragments of mantle or crustal rocks of any of these types may be incorporated into melanges (Moores, 1973). Evidence of these early formative and emplacement events are commonly obscured in alpine-type ultramafic rocks by subsequent metamorphism, making the histories of the rocks difficult to decipher.

Because alpine ultramafic rocks are derived from the mantle or form early in the developmental history of an orogen, they provide knowledge of generally obscure events. Mantle ultramafic rocks record histories of mantle deformation and recrystallization, mantle metasomatism, and mantle depletion resulting from partial melting, Alpine ultramafic rocks of crustal heritage may reveal the early histories of intrusion and crystallization, metamorphism, and deformation in the mountain belt as well as a synorogenic history, which is also portrayed by the textures, structures, and minerals of the more common rocks of the belt. These attributes provide ample incentive for the study of alpine ultramafic rocks.

OCCURRENCES OF ALPINE ULTRAMAFIC ROCKS

Alpine ultramafic rocks are found worldwide in Phanerozoic mountain belts.[2] In some orogens (e.g. in the Cordilleran Orogen of western North America), they define two or more parallel, linear zones. They also occur in Precambrian terranes.[3] Young alpine ultramafic rocks form and occur in Cenozoic orogenic belts, in volcanic arcs of both continental and oceanic type, and along spreading centers and transform faults, where they constitute layered segments of the oceanic crust and mantle (Dietz, 1963; Moores and Jackson, 1974).[4] In short, alpine-type ultramafic rocks appear at present and former plate boundaries of all types.

Occurrences of alpine ultramafic rocks are not evenly distributed among the three petrogenetic types. Mantle diapirs rarely enter the crust, and thus the associated types of alpine ultramafic rocks are rare. Mantle slabs, without an associated crustal carapace, are not abundant. Most commonly, mantle rocks, with their overlying crustal rocks, are emplaced by obduction into or onto the crust (Dewey and Bird, 1970, 1971; R. G. Coleman, 1971a, 1977a). Where that crust and mantle is oceanic, the alpine ultramafic rock bodies are recognized as ophiolites. Recall that the ophiolites contain both a basal mantle tectonite and an overlying, differentiated sequence of ultramafic to silicic magmatic rocks. Thus, inasmuch as they contain both mantle rocks and crustal differentiates, they exhibit a range of characteristics. Ultramafic crustal intrusions that occur in mountain belts, especially those that are subsequently deformed, may also be considered to be alpine ultramafic bodies.

Specific and notable occurrences of alpine ultramafic rocks are found at the Bay of Islands, Newfoundland; at Thetford Mines, Quebec; in the Roxbury District and at Grafton in central and southern Vermont; in the North Carolina Blue Ridge Belt at Daybook, Webster-Addie, and Buck Creek; in the California Coast Ranges, notably at Burro Mountain, Del Puerto Canyon, and Point Sal; in the eastern and western parts of the Klamath Mountains of northern California and Oregon; in the Cascade Range of Washington, in the Twin Sisters, Darrington, and Sultan areas; at the head of the Blue River in British Columbia; at Ronda, southern Spain; at Lizard, Cornwall, England; in the Swiss Alps; at Almklovdalen, Norway; and at Dun Mountain, New Zealand.[5] These specific occurrences include mantle slabs and mantle fragments in melanges, diapiric ultramafic bodies, ophiolites, and ultramafic bodies of unknown origin. In spite of the variations in their modes of occurrence, the alpine ultramafic rocks share certain characteristics that distinguish them from the various types of igneous ultramafic bodies.

DISTINGUISHING FEATURES OF ALPINE ULTRAMAFIC ROCKS AND ROCK BODIES

The primary characteristic of alpine ultramafic rock bodies is that they occur in orogenic belts. Typically, they are deformed. Because almost any kind of ultramafic rock may be emplaced into a mountain belt, the characteristics of alpine ultramafic bodies are diverse and may overlap those of igneous ultramafic-mafic complexes. In general, however, the rock bodies are characterized by (1) olivines and orthopyroxenes with moderate to high Mg numbers, (2) rocks with tectonite fabrics, (3) lenticular to lensoid shapes, and, with some exceptions, (4) a lack of chilled margins and contact metamorphism (Thayer, 1960; E. D. Jackson and Thayer, 1972; Moores, 1973). Perhaps more than the rocks of other ultramafic bodies, alpine ultramafic rocks have undergone serpentinization. Each of the individual petrogenetic types of alpine

ultramafic body—mantle slab, mantle diapir, and magmatic differentiate—is somewhat different chemically, mineralogically, texturally, and/or structurally (tables 31.1 and 31.2).

Minerals and Textures

Mineralogically, the alpine ultramafic rocks contain the same minerals as do the ultramafic igneous rocks,[6] but in addition, they may contain or consist entirely of distinctly metamorphic minerals. Typical minerals inherited from igneous protoliths or present in corresponding igneous rocks include olivine, orthopyroxenes, clinopyroxenes, Ca-rich plagioclase, and chromite and other spinels. These minerals are also stable in the mantle. Additional minerals, which occur in alpine ultramafic rocks and are stable in the mantle or are produced by crustal metamorphism, include Ca- and Mg-amphiboles (e.g., tremolite, anthophyllite, or hornblende), the phyllosilicates (talc, chlorite, the serpentines, and phlogopite), garnets, carbonate minerals (e.g., magnesite, dolomite, calcite, and aragonite), magnetite, and a host of less abundant minerals such as brucite, quartz, pyrrhotite, and garnierite.

The textures of alpine-type ultramafic rocks range from relict cumulate textures to mylonitic textures (Wicks, 1984a, 1984b; Nicolas and Poirier, 1976; Beslier, Girardeau, and Boillot, 1990).[7] Typically, alpine ultramafic tectonites dominated by olivines and pyroxenes have allotrioblastic-granular to semischistose textures, including the following:

1. *Protogranular texture*, a typically coarse, nonfoliated texture characterized by sinuous grain boundaries and intergrowths (figure 31.1a);
2. *Equigranular-mosaic texture*, a nonfoliated to very weakly foliated metamorphic texture characterized by equant polygonal grains with slightly curved to straight (equilibrium) grain boundaries, which meet at about 120° angles (figures 31.1b and 31.2a);
3. *Equigranular-tabular texture*, a semischistose texture characterized by elongate, deformed polygonal grains typified by deformation bands and related features (figure 31.2b); and
4. *Porphyroclastic texture*, a nonfoliated to semischistose texture characterized by a bimodal distribution of grain sizes, with larger, locally deformed porphyroclasts surrounded by a matrix of fine-grained minerals of the same type(s) as the porphyroclasts (figures 31.1c and 31.2c)(Mercier and Nicolas, 1975).[8]

Rocks dominated by phyllosilicates (e.g., talc and antigorite) range from diablastic to schistose in texture.[9]

Magmatic Differentiates

Magmatic differentiates that become alpine-type ultramafic rock bodies have the characteristics of the type of igneous protolith from which they were derived. These rocks are distinguished by (1) dominant or relict igneous textures, including

Table 31.1   Characteristics of Alpine-type Ultramafic Rocks

Characteristic	Magmatic Differentiates	Mantle Slabs	Mantle Diapirs
Typical Rock Types	Harzburgite, lherzolite, dunite, pyroxenites, talc-amphibole schist, serpentinite, chlorite schist	Lherzolite, harzburgite, dunite, serpentinite, talc-Amphibole schists, chlorite schist	Lherzolite, harzburgite, spinel and garnet pyroxenites, serpentinite, talc-amphibole schist, chlorite schist
Peridotite Chemistry	Al_2O_3 = 0.1–3.5 CaO = 0.1–17.4 Mg no. = 0.68–0.85	Al_2O_3 = 0.1–4.4 CaO = 0.0–7.7 Mg no. = 0.80–0.91	Al_2O_3 = 0.6–6.4 CaO = 0.4–5.7 Mg no. = 0.58–0.92
Typical Minerals	Orthopyroxene, clinopyroxene, olivine, talc, plagioclase, chlorite, hornblende, actinolite, tremolite, anthophyllite, serpentines, chromite, magnetite, phlogopite	Orthopyroxene, clinopyroxene, olivine, talc, hornblende, chlorite, actinolite, tremolite, anthophyllite, chromite, magnetite, serpentines	Orthopyroxene, clinopyroxene, olivine, plagioclase, garnet, chromite, spinel, magnetite, chlorite, serpentines, talc, tremolite, anthophyllite
Mineral Chemistry	Opx (Al_2O_3) = 0.0–1.4 Cpx (Al_2O_3) = 0.7–4.5 Olivine (Fo) = 53–95 Opx (MgO) = 43–92 Cr Spinel (Cr/Cr+Al) = 0.00–0.85	Opx (Al_2O_3) = 0.0–6.0 Cpx (Al_2O_3) = 0.3–4.5 Olivine (Fo) = 84–95 Opx (MgO) = 85–94 Cr Spinel (Cr/Cr+Al) = 0.05–0.90	Opx (Al_2O_3) = 0.5–6.7 Cpx (Al_2O_3) = 0.5–8.5 Olivine (Fo) = 75–92 Opx (MgO) = 88–91 Cr Spinel (Cr/Cr+Al) = 0.05–0.50
Typical Textures	Hypidiomorphic-granular, ophitic, diabasic, cumulate, diablastic, schistose	Protogranular, equigranular-mosaic, equigranular-tabular, porphyroclastic, mylonitic, diablastic, schistose	Porphyroclastic, equigranular-mosaic, protogranular, schistose, diablastic
Typical Structures	Primary layering, foliation, isoclinal folds, dikes (in some bodies), shear zones	Isoclinal folds, flow layering, foliation, lineation, podiform chromite bodies, dikes, ductile shear zones	Isoclinal folds, flow layering, foliation, lineation, dikes, ductile shear zones

Sources: See note 6.

cumulate, ophitic, diabasic, or hypidiomorphic-granular textures, (2) generally low amounts of alumina in the peridotites and their included orthopyroxenes, and, (3) primary layering in some bodies (table 31.1). Mafic to felsic differentiates are usually associated with these kinds of ultramafic rocks and many bodies may contain numerous dikes. In addition, some bodies consist of rocks with a wide range of olivine, orthopyroxene, and clinopyroxene compositions.[10]

The most common alpine-type, ultramafic, differentiated magmatic body is the basal cumulate section of ophiolites. Within such sections, dunite, lherzolite, harzburgite, wehrlite, and/or pyroxenites form interlayered sequences.[11]

Table 31.2 Chemical Analyses of Alpine Ultramafic Rocks

	1	2	3	4	5	6	7
SiO_2	27.2[a]	39.97	40.67	41.62	43.61	44.36	50.69
TiO_2	0.95	0.02	0.01	nd	0.06	0.12	0.06
Al_2O_3	19.6	0.37	0.75	1.08[b]	2.93	3.49	1.56
Fe_2O_3	35.3[c]	12.41[c]	1.15	9.28[c]	1.09	—	5.58[c]
FeO	—	—	6.56	—	6.74	8.29[c]	—
MnO	0.22	0.17	0.12	nd	0.11	0.14	0.11
MgO	10.4	46.75	48.77	46.40	39.86	39.38	23.99
CaO	0.60	0.26	0.00	0.20	2.66	2.96	16.21
Na_2O	0.05	0.09	0.00	nd	0.26	0.28	0.09
K_2O	0.04	0.00	0.03	nd	0.01	0.01	0.00
P_2O_5	0.03	0.00	0.02	nd	0.00	0.01	0.02
H_2O+	5.30	—	1.38	1.95	7.48	2.52	—
CO_2	0.10	—	0.12	nd	0.49	0.43	—
Other	tr	11.00[d]	0.71	nd	0.83	tr	4.17[d]
Total	99.8	100.70	98.91	100.53	100.05	101.99	99.01

Sources:

1. Magnetite-garnet-chlorite schist, Greer Hollow Ultramafic Body, Blue Ridge Belt, North Carolina. Analyst: XRAL (L. A. Raymond, previously unpublished data).
2. Dunite (cumulate), North Arm Mountain, Bay of Islands Ophiolite, Newfoundland (Komor et al., 1985).
3. Dunite, Day Book Dunite, North Carolina, Analyst: R. Stokes (Kulp and Brobst, 1954).
4. Harzburgite ("saxonite") sample A-1, Addie, North Carolina. (C. E. Hunter, 1941).
5. Plagioclase lherzolite, sample R224, Rhonda peridotite, Spain. Analyst: E. Jurosewich (Dickey, 1970).
6. Garnet lherzolite, sample R255, Rhonda peridotite, Spain, (F. A. Frey, Suen, and Stockman, 1985). $SiO_2 \rightarrow P_2O_5$ recalculated and listed here on a volatile free basis.
7. Clinopyroxene-rich wehrlite (cumulate), North Arm Mountain, Bay of Islands Ophiolite, Newfoundland (Komor et al., 1985).

[a]Values in weight percent

[b]Includes minor Cr_2O_3 and TiO_2.

[c]Total iron as this value.

[d]Loss on ignition.

nd - not determined.

tr - trace.

Mantle Slabs

Subcrustal mantle, faulted into the crust to form alpine-type ultramafic rock bodies, is characterized by high Mg numbers in both the peridotites (0.80–0.88) and in the constituent olivines and pyroxenes (table 31.1). Lherzolite and harzburgite are the most common rock types, along with their serpentinized equivalents. Dunites occur primarily as pods or dikes (Moores, 1969; Quick, 1981b; Nicolas, 1989). Similarly, minor associated mafic rocks occur only as dikes.

Many mantle slabs represent the subcrustal parts of ophiolites that were separated by faulting from the overlying oceanic crustal sections, when the rocks were obducted onto a continental margin. In some cases, fragments of the crustal rocks remain attached to the ultramafic rocks, revealing the ophiolitic character of the ultramafic body. Other slabs may represent mantle fragments thrust up between colliding continental masses (Nicolas and Poirier, 1976, ch. 11). Mantle slabs of either type are locally incorporated into melanges,

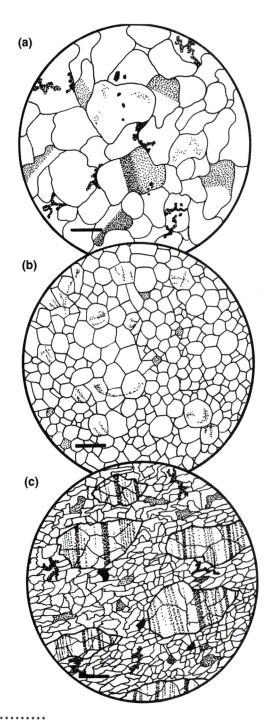

either during thrusting or during later submarine landsliding (Gansser, 1974; B. M. Page, 1981; Sarwar and DeJong, 1984).

The rocks of the mantle slabs may show a full range of tectonite fabrics, including those dominated by protogranular textures, equigranular-mosaic textures, equigranular-tabular textures, porphyroclastic textures, mylonitic textures, diablastic textures, and schistose textures (Den Tex, 1969; Nicolas and Violette, 1982). These textures are found in rocks with flow layering, lineation, lattice preferred orientations, foliations, isoclinal folds, faults, or combinations of these features (Nicolas and Boudier, 1975; Christensen and Lundquist, 1982; Nicolas, 1986b; 1989).

Mantle Diapirs

Garnet pyroxenites, garnet peridotites, and spinel peridotites are distinctive rock types in the mantle diapirs. Lherzolites of various types are also characteristic. The diapiric peridotites tend to be more aluminous than other alpine-type peridotites, and hence, so are the minerals. Pyroxenes are relatively aluminous and the chromium spinels are high in aluminum. In addition, the olivines and orthopyroxenes have a high Mg content. Typically, the diapiric rocks have porphyroclastic textures.

The mantle diapirs are distinct from most other alpine ultramafic rock bodies in that they are emplaced at high temperatures and produce a contact metamorphic aureole (D. H. Green, 1964; Loomis, 1972a). Their diapiric nature may be revealed by their structural relationships with surrounding rocks, by their internal structures, and by the mineral and rock chemistry. Internal structures may include flow layering, foliation, lineation, isoclinal folds, and dikes.

THE NATURE OF THE UPPER MANTLE: A BRIEF SURVEY

Several features indicate that many alpine ultramafic rock bodies represent fragments of the upper mantle.[12] Mantle samples are also provided by the ultramafic xenoliths that occur in some mafic magmas (Mercier and Nicolas, 1975; Basu, 1977b). Together with various geochemical and geophysical analyses, these rocks yield an image of the character of the upper parts of the mantle, which is structurally, texturally, and mineralogically complex.[13]

The uppermost mantle is a heterogeneous mass of rock dominated by peridotites.[14] Lherzolite and harzburgite are the main rock types, but dunite, "pyrolite," and eclogite may

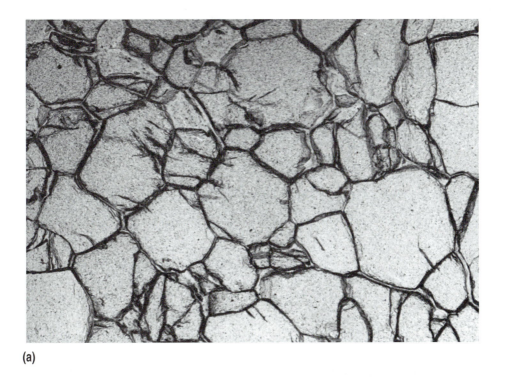

(a)

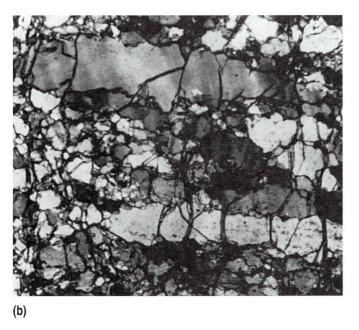

(b)

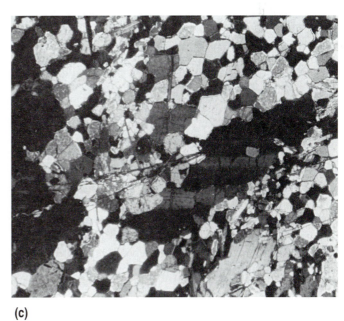

(c)

Figure 31.2 Photomicrographs of some textures found in alpine ultramafic rocks. (a) Equigranular-mosaic texture in dunite, Day Book Dunites, North Carolina. (PL). (b) Equigranular-tabular texture in dunite, Corundum Hill Ultramafic Body, North Carolina.(XN). (c) Porphyroclastic texture in dunite, Day Book Dunites, North Carolina, (XN). Long dimension of photo (a) is 3.25 mm, (b) is 2.75 mm, and (c) is 2.75 mm.

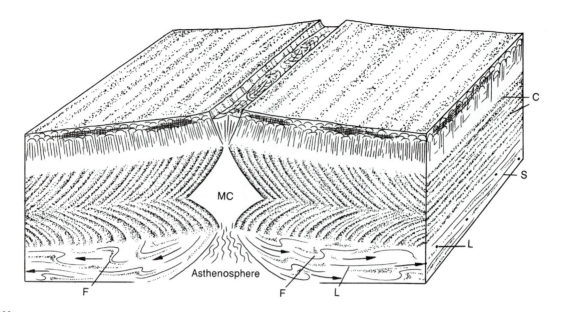

Figure 31.3
Schematic diagram showing orientations of structures in mantle rocks at and near a spreading center. C = cumulate layering, F = folds, L = lineation, MC = magma chamber, S = foliation.
(Based on Juteau et al., 1977; Girardeau and Nicolas, 1981; Christensen and Lundquist, 1982; Girardeau and Mercier, 1988; Casey et al., 1983; and Nicolas, 1986)

constitute parts of the upper mantle. Two main peridotite "subtypes" are recognized, the harzburgite subtype and the lherzolite subtype, (E. D. Jackson and Thayer, 1972; Boudier and Nicolas, 1985; Nicolas, 1986b; 1989). The harzburgite subtype is associated with ophiolites and represents depleted mantle rock that underlies fast-spreading ridges. The lherzolite subtype, less common in ophiolites, characterizes mantle diapirs and slow-spreading ridges, Dunite and chromite pods, dikes, and layers are largely confined to the upper (near-crustal) parts of the harzburgite subtype.

Textures of the mantle rocks range from protogranular to porphyroclastic and mylonitic. Foliations and lattice preferred orientations are commonly subhorizonatal to moderately inclined (Juteau et al., 1977; Christensen and Lundquist, 1982; Nicolas, 1986b; 1989), but in some areas are steeply inclined (Nicolas and Violette, 1982). Both layering and foliations may be folded (Nicolas and Boudier, 1975). Isoclinal folds are predominant. All of these features apparently form below spreading ridges. Lineations that form in mantle rocks away from the axial regions of spreading ridges may parallel and indicate the spreading direction (figure 31.3).

SERPENTINIZATION

Ultramafic rocks are prone to low-temperature metamorphism and alteration. This is the case, because their constituent high P-T phases, at near-surface and low-temperature conditions, are far from their fields of stability. In addition, mantle rocks are rather dry, and the upper levels of the crust are generally permeated with water, which readily combines with the anhydrous phases of the mantle rocks to produce new hydrous phases.

General discussions of regional and contact metamorphism of ultramafic rocks were presented in preceding chapters. In those kinds of metamorphism, hydrous phases such as tremolite, talc, anthophyllite, and serpentines are produced by metamorphism of ultramafic rock types. Although serpentine minerals were mentioned among these minerals previously—because they are the dominant phase, comprising more than 90% of the modes of most ultramafic rocks equilibrated at temperatures below about 500° C, and because serpentine is a nearly ubiquitous constituent of crustal ultramafic rocks—the processes of serpentinization warrant further comment.[15]

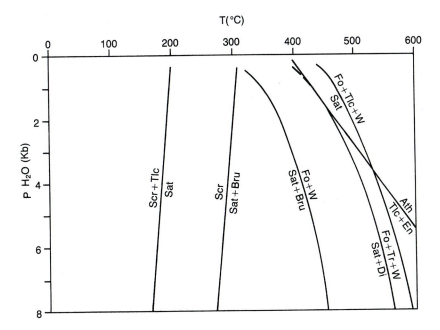

Figure 31.4   Petrogenetic grid for low- temperature, CO_2-free assemblages in ultramafic rocks.
Ath - anthophyllite, Bru - brucite, Di - diopside, En - enstatite, Sat - serpentine (antigorite), Scr - serpentine (chrysotile), Tlc - talc, Tr - tremolite, W - water.

(Modified from Mellini, Trommsdorf, and Compagnoni, 1987; based in part on Evans and Trommsdorf, 1970; Evans et al., 1976; and Day, Chernosky, and Kumin, 1985)

Serpentinization is the process or processes by which ultramafic or mafic rocks are transformed into serpentinites. These processes produce the three main varieties of serpentine that comprise the serpentinites—lizardite, chrysotile, and antigorite.[16] Lizardite, a planar-structured serpentine, and chrysotile, a cylindrically structured serpentine, are polymorphs with the composition $Mg_3Si_2O_5(OH)_4$. Antigorite is compositionally (as well as structurally) distinct, with Mg/Si values somewhat less than the 1.5 of lizardite and chrysotile.[17] Antigorite is generally stable at higher temperatures than the latter two phases, as discussed in preceding chapters and revealed by the petrogenetic grid in figure 31.4.

Inasmuch as alpine ultramafic rocks are composed primarily of one or more of the Mg-rich varieties of olivine, orthopyroxene, and clinopyroxene, serpentinization may result essentially from the addition of water. If that is the case, the serpentinization may be an isochemical, or *constant-chemical*, process. In serpentinization in which there is "constant chemical composition," the structural changes in the minerals require an increase in the volume of the ultramafic body. For example, in the equation

olivine + silica in solution → serpentine

$$3Mg_2SiO_4 + H_4SiO_4 + 2H_2O \rightarrow 2Mg_3Si_2O_5(OH)_4$$

(131 cc)   introduced   (220 cc)

there is a significant increase in volume.[18] For any given body, evidence suggestive of isochemical serpentinization may include:

1. Internal fracturing of the body (indicating deformation associated with a volume increase);
2. Deformation of the surrounding country rock (indicating deformation associated with a volume increase);
3. Identical chemistry of ultramafic protoliths and serpentinized equivalents (except for water content); and
4. The absence of evidence of metasomatism in the surrounding country rocks.[19]

Based on chemical evidence from the Burro Mountain, California, ultramafic body, from which only CaO was lost, R. G. Coleman and Keith (1971) argued that serpentinization there was nearly isochemical.

In contrast, serpentinization may occur as an allochemical, *constant-volume* process, in which Mg, Ca, Fe, Si, and other elements are transported by the serpentinizing fluids. Such a process is represented by the equation

olivine + water → serpentine + Mg ion + hydroxyl ion + silica in solution

$$5Mg_2SiO_4 + 10H_2O \rightarrow 2Mg_3Si_2O_5(OH_4) + 4Mg^{++} + 8(OH) + H_4SiO_4$$

(219 cc) introduced   (220 cc)   removed in solution

in which the aqueous fluid that facilitates the serpentinization removes excess ions (Turner and Verhoogen, 1960, pp. 318–319). Spring waters currently emanating from serpentinites do, in fact, carry a number of ionic species. I. Barnes and O'Neil 1969 I. Barnes, Rapp, and O'Neil 1972 . This observation is not only consistent with the hypothesis that some serpentinization is a constant-volume process, it suggests that serpentinization is an ongoing process, even at ambient temperatures at the Earth's surface. Evidence cited in support of constant-volume serpentinization includes:

1. Pseudomorphic replacement of olivine and pyroxene by serpentine;[20]
2. Density decrease with concomitant porosity increase in the serpentinized ultramafic body, relative to the unserpentinized rock;
3. Bulk rock chemical differences between serpentinites and their unserpentinized protoliths;
4. Precipitation of transported ions in metasomatically altered rock at a distance from the point of dissolution (i.e., the point of serpentinization); and
5. The presence of undisturbed primary layering.[21]

Condie and Madison (1969) demonstrated that serpentinization of the ultramafic body at Webster-Addie, North Carolina, involved loss of FeO and MgO and gains in SiO_2.

Serpentinizing fluids have a complex history that results in a variety of changes in the invaded rocks. Both major element and isotopic studies reveal that these solutions are of variable compositions and sources and may begin as ocean water, connate water, metamorphic water, or meteoric water.[22] During serpentinization the fluids may be highly reducing, as evidenced by the presence of native nickel-iron ($FeNi_3$), the mineral awaruite, which occurs in some serpentinites.[23] The pH values of fluids emanating from serpentinites are high, ranging from 8 to 12.

Metasomatism and vein formation in rocks adjacent to serpentinites is well known.[24] Notably, **rodingites,** calcium-metasomatized gabbros containing diopside and grossularite garnet, are distinctive metasomatically altered rocks formed in contact with serpentinite. Sandstones and granitoid rocks may experience similar metasomatism, and, in fact, serpentinites may even be remetamorphosed by fluids that have changed character during serpentinization.

Serpentinites commonly are texturally and structurally complex (O'Hanley, 1987, 1988).[25] They may have lepidoblastic, nematoblastic, and porphyroblastic schistose textures, diablastic textures, and, in some cases, blastoporphyritic textures. Mesh texture, in which olivine grains are replaced from grain margins and fractures inward, is common (figure 31.5). Cross-cutting veins of serpentine typify many serpentinites. Comb structure-like veins, with chrysotile grown perpendicular to the vein walls, are widespread. These complex textural and structural relations make interpretation of serpentinite metamorphic histories difficult.

EXAMPLES OF ALPINE-TYPE ULTRAMAFIC ROCKS

The Bay of Islands Ophiolite

The Cambrian-Ordovician Bay of Islands Ophiolite Complex[26] is an ophiolite that is considered to have formed at an oceanic spreading center.[27] The complex consists of four massifs—the Table Mountain, North Arm Mountain, Blow-Me-Down Mountain, and Lewis Hills massifs—exposed in the west-central coastal region of Newfoundland (figure 31.6). The North Arm Mountain and Blow-Me-Down Mountain massifs have a relatively complete ophiolite stratigraphy. Of the two incomplete sections, the Table Mountain section is the least complete, but it contains a thick section of ultramafic rock. The Bay of Islands Complex rocks are only moderately deformed, but were thrust into their present site in the orogen. To the west, along the coast itself, is the Coastal Complex, a highly deformed ophiolite complex thought to have been deformed in a transform fault environment adjacent to the Bay of Islands spreading center (Karson and Dewey, 1978; J. F. Casey et al., 1983).

Like all complete ophiolites, the Bay of Islands Ophiolite Complex contains both basal ultramafic tectonite and ultramafic cumulate (plutonic) sections (figure 31.7). The ultramafic tectonite is dominantly harzburgite, but includes lherzolite near the base, abundant dunite near the top, and local orthopyroxenite, websterite, and chromitite within the section (Christensen and Salisbury, 1979; Komor, Elthon, and Casey, 1985).[28] Serpentinization is widespread. The cumulate sections contain interlayered dunite, wehrlite, lherzolite, websterite, clinopyroxenite, minor harzburgite, and chromitite, with progressively more gabbro at stratigraphically higher levels. Some ultramafic rocks intrude the cumulates (Bedard, 1991). Both serpentinization and penetrative deformation have affected rocks at and near the base of the cumulate complex.

The basal harzburgite tectonite and associated dunite and lherzolite are interpreted to be residual mantle rocks. Mesoscopic tight to isoclinal folds, characteristic of other mantle peridotites, are present within this mantle section.[29] The rocks are lineated and foliated, with porphyroclastic, equigranular-mosaic, equigranular-tabular, and, especially near the base, mylonitic textures (Girardeau and Nicolas, 1981). Olivines show a limited range of composition (Fo_{89-92}) and orthopyroxenes are similarly Mg-rich (En_{88-92}) (J. Malpas, 1978; Suen, Frey, and Malpas, 1979).

Figure 31.5 Photomicrograph of mesh texture (above) and a vein of serpentine (v) in a serpentinite from the Webster-Addie Ultramafic Body, North Carolina. (XN). Long dimension of photo is 1.27 mm.

In the plutonic complex, cumulate textures are replaced near the base by equigranular-mosaic (equivalent to xenoblastic-granular) and porphyroclastic textures. Thus, the rocks have become tectonites. In local areas, mylonitic textures also occur. The olivines and pyroxenes that define these various textures have compositions that span a compositional range typical of differentiated igneous rocks, a range slightly greater than that of the residual tectonites.[30] Forsterite components of olivine range from Fo_{85} to Fo_{92}. Orthopyroxene Mg numbers range from 0.79 to 0.92 and

clinopyroxenes show a similar range (0.85–0.94). The peridotites have correlative magnesium numbers (0.77–0.91) (Elthon, Casey, and Komor, 1982). The Cr numbers of spinels range from 17 to 70.

The structures and textures of the Bay of Islands Complex rocks suggest that the complex was formed at an ocean spreading center. Major and trace element chemistry of the mafic and ultramafic rocks corroborate that interpretation.[31] Depleted mantle harzburgites and related rocks formed a foundation upon which the igneous rocks of the ophiolite formed.

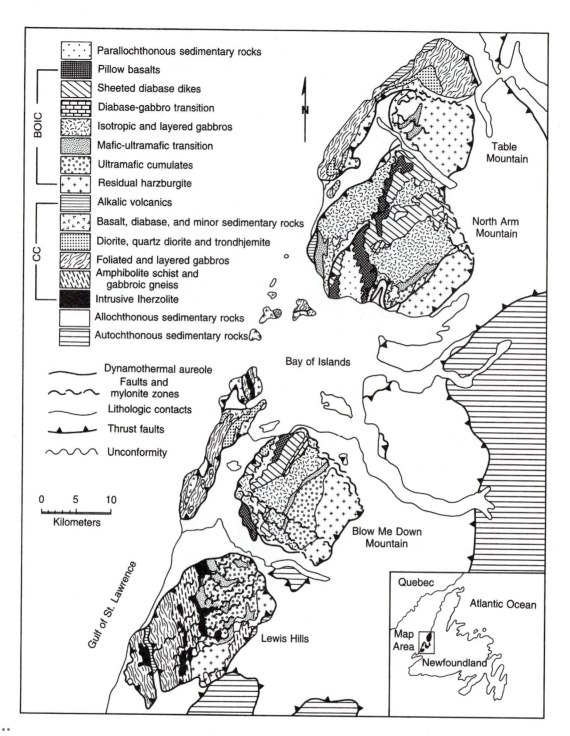

Figure 31.6
Map of the Bay of Islands Ophiolite Complex and surrounding areas, Newfoundland.
(From Casey et al., 1983)

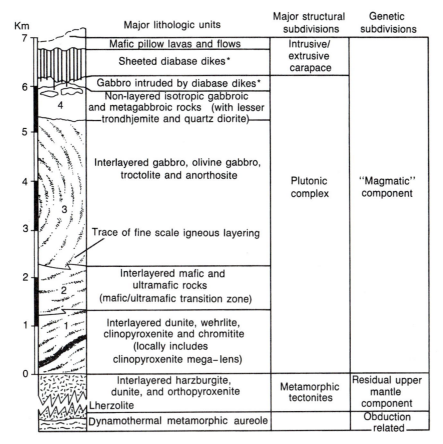

Figure 31.7 Generalized columnar section through the Bay of Islands Ophiolite Complex. *(Redrawn from Casey and Karson, 1981)*

After it was formed, the Bay of Islands Ophiolite Complex was thrust faulted (obducted) into the Newfoundland segment of the Appalachian Orogen (Dewey and Bird, 1970; W. R. Church and Stevens, 1971; Harold Williams, 1975), making it an alpine ultramafic-mafic complex. Clearly, some alpine ultramafic complexes represent oceanic crust and mantle. To the extent that ophiolites also represent both back-arc crust and mantle *and* island arc crust and mantle, rocks derived from these settings may also become alpine ultramafic rock bodies.

Alpine Ultramafic Mantle Slabs in the North Carolina Blue Ridge Belt

Alpine ultramafic rocks occur as scattered, small to large slabs and lenses (<100 m to >10 km) in two broad bands within the southern Appalachian Orogen (H. H. Hess, 1955; Larrabee, 1966; Misra and Keller, 1978). The western band, of interest here (figure 31.8), occurs primarily in the Eastern Blue Ridge Belt, but ultramafic rocks are also present in the Western Blue Ridge and Inner Piedmont Belts.[32] A number of studies have shown that the ultramafic rock bodies are polydeformed dunites, peridotites (primarily harzburgites), pyroxenites, serpentinites, and various talc, anthophyllite, tremolite, and chlorite schists.[33]

None of the North Carolina Blue Ridge bodies shows evidence of magmatic emplacement. There are no documented apophyses or dikes extending from the bodies into the country rocks, the bodies do not have chilled margins, and contact metamorphism has not been demonstrated. Instead, the rocks are generally concordant to slightly discordant masses with internal fabrics consistent with polydeformation and multiple periods of metamorphism, primarily of a retrograde nature (tables 31.3 and 31.4).[34] Isoclinal folds revealed locally by chromite and magnetite layers and an almost complete range of mantle slab textures have been recognized (Abbott and Raymond, 1984). Although some bodies are intimately associated with mafic rocks, including metatroctolites and amphibolites, few can be demonstrated to be ophiolites.[35] Virtually all olivine- and pyroxene-rich bodies have a reaction wallrock ("blackwall") of chlorite-talc-amphibole schist, like those described by R. F. Sanford (1982). Especially in the northern

667

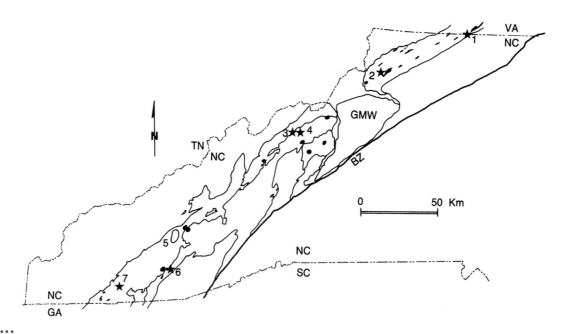

Figure 31.8 Map showing the locations of selected ultramafic bodies (stars, black dots, ellipses, and lines) scattered across the North Carolina segment of the eastern Blue Ridge Belt of the southern Appalachian Orogen. BZ = Brevard Fault Zone, GMW = Grandfather Mountain Window. Size of bodies is generally exaggerated. Many more ultramafic bodies exist than are shown. Numbered bodies (stars) correspond to those for which data are presented in table 31.4.

(Modified from Larrabee, 1966; Z. A. Brown et al., 1985)

part of North Carolina and in Virginia, olivines and pyroxenes in the bodies are extensively replaced by chlorite, serpentine, talc, tremolite, and anthophyllite. Representative CMS phase diagrams for the earlier and later phase assemblages are shown in figure 31.9a.

Chemically, the Blue Ridge peridotites are low in Al_2O_3 and CaO and relatively high in MgO. The latter results in high Mg numbers, typical of olivine-rich rocks and mantle rocks. The range of Mg numbers is small. Alkalis are nearly absent. These criteria are consistent with the rock masses being mantle slabs.

The high magnesium numbers of the peridotites, dunites, and pyroxenites are reflected in the compositions of the major minerals. Olivines have a limited range and high values of forsterite component (Fo = 84–95).[36] Orthopyroxenes are similarly magnesium-rich. Chromites, which are iron-rich and aluminum-poor, plot in a wide field that extends from the field of alpine-type chromites along a metamorphic chromite trend, indicating that metamorphic re-equilibration of the chromites and their compositions has occurred (figure 31.9b) (Lipin, 1984). Using the Evans and Frost (1975) olivine-spinel geothermometer, Lipin (1984) estimates that this metamorphic re-equilibration occurred at a temperature of about 700° C.

All of the features observed are consistent with the Blue Ridge ultramafic bodies being alpine ultramafic bodies, probably predominantly of the mantle slab type.[37] Although lherzolites and pyroxenites exist, where olivine and pyroxene mineralogies are preserved, it is evident that the dominant rock types are harzburgite and dunite. Metamorphism, however, has obscured and complicated the history of these rocks.

The bodies at four localities reflect the variations in petrology and structure within this suite of alpine ultramafic rocks (figure 31.10, table 31.4). In the north, the Greer Hollow body (locality 2 in figure 31.8) is a small, folded, concordant body composed predominantly of anthophyllite and chlorite schists. The Day Book bodies (locality 3) are dominantly dunite. The Webster-Addie body (locality 5), for which websterite is a namesake, is a folded, concordant, composite sheet containing abundant dunite and subordinate pyroxenites, serpentinites, and other rocks. In the south, the Buck Creek body (locality 7) is a part of the Chunky Gal Mountain Mafic-Ultramafic Complex.

The *Greer Hollow Ultramafic Body*, described by Raymond et al. (1988), consists of crudely teardrop-shaped mass of rock (figure 31.10a). Two mappable units exist within the body: (1) a chlorite schist unit consisting of interlayered magnetite + chlorite schist and anthophyllite ± tremolite + chlorite schist, and

Table 31.3 Mineral Associations in Southern Appalachian Blue Ridge Ultramafic Rocks

Association A-1 (Oli + Chr ± Opx ± Cpx ± Hbl ± Mtt)

1a	Olivine ± Chromite
1b	Olivine + Orthopyroxene ± Chromite
1c	Olivine ± Orthopyroxene ± Clinopyroxene ± Chromite
1d	Olivine + Clinopyroxene + Hornblende + Chromite + Magnetite (Mtt)

Association A-2 (Oli ± Opx ± Cpx ± Tre ± Chl ± Chr)

2a	Olivine + Chromite + Chlorite
2b	Olivine + Tremolite + Chlorite
2c	Olivine + Orthopyroxene + Tremolite
2d	Clinopyroxene + Tremolite ± Chlorite + Chromite

Association A-3 (± Ant ± Tre ± Tlc ± Chl ± Phl ± Mtt ± Mgt ± Grt)

3a	Anthophyllite ± Magnetite
3b	Talc ± Magnetite
3c	Anthophyllite + Talc ± Magnetite
3d	Anthophyllite ± Talc ± Tremolite ± Chlorite ± Magnetite ± Magnesite (Mgt)
3e	Talc + Chlorite ± Tremolite ± Magnetite ± Magnesite
3f	Tremolite + Chlorite ± Magnetite ± Magnesite
3g	Talc ± Tremolite + Magnesite + Phlogopite
3h	Chlorite ± Magnetite ± Garnet (Grt)

Association A-4 (± Sat ± Mtt ± Chl ± Bru ± Tlc ± Tr)

4a	Serpentine (antigorite = Sat) ± Magnetite ± Chlorite
4b	Serpentine (antigorite) + Brucite ± Magnetite[1]
4c	± Serpentine (antigorite) ± Talc ± Tremolite ± Chlorite ± Magnetite

Association A-5 (± Slz ± Scr ± Mtt ± Tlc ± Chl ± Tr ± Silica ± Mgt ± others)

5a	± Serpentine (lizardite, Slz) ± Serpentine (chrysotile, Scr) ± Magnetite
5b	± Serpentine (lizardite, Slz) ± Serpentine (chrysotile, Scr) ± Talc ± Tremolite ± Chlorite ± Magnetite
5c	± Serpentine (lizardite, Slz) ± Serpentine (chrysotile, Scr) ± Silica minerals ± Magnetite
5d	Silica minerals ± Magnesite ± Chlorite
5e	Miscellaneous vein minerals (e.g., aragonite, garnierite)

Sources: See note 33.

[1]This association was reported by Swanson et al. (1985) from the Day Book body, as an early association occurring between A-2 and A-3, but the author has been unable to confirm its existence.

Table 31.4 Observed Metamorphic Associations in Blue Ridge Ultramafic Rock Bodies

Localities[1]	Assemblages					Sources
	A-1[2]	A-2	A-3	A-4	A-5	
1. Edmonds	1a		3b, 3d, 3h	4a? 4c?		Scotford and Williams (1983), Raymond, unpublished data
2. Greer Hollow	1a, 1b, 1c,	2b	3a, 3d, 3h	4a? 4c?	5a	Raymond et al. (1988) Raymond (unpublished data)
3. Day Book	1a, 1b	2a, 2b	3b, 3e, 3g	4a, 4b	5a, 5d, 5e	S. E. Swanson (1981), Swanson et al. (1985), Raymond (unpublished data)
4. Woody	1a, 1c		3a, 3d, 3e	4a	5a? 5d	Kingsbury and Heimlich (1978), Raymond (unpublished data)
5. Webster-Addie	1a, 1c,	2b	3a, 3b, 3c, 3f	4a	5a, 5d	R. Miller (1951a), Raymond (unpublished data)
6. Corundum-Hill	1a, 1b		3d	4a?	5a	Yurkovich (1977), Raymond, (unpublished data)
7. Buck Creek	1a, 1d		3d	4a?	5a	McElhaney and McSween (1983), Raymond, unpublished data

[1] Localities correspond to those shown in figure 31.8.
[2] Assemblages are those listed in table 31.3.

(2) a porphyroblastic garnet ± magnetite + chlorite schist unit (analysis 1, table 31.2). Isolated pods of metamorphosed dunite and peridotite, boudins of hornblendite and epidote-hornblende schist, and veins of serpentinite also occur within the body. At least four, and possibly as many as six, stages of metamorphic recrystallization are reflected by various phase assemblages and textures. The olivines and pyroxenes in early-formed porphyroclastic dunites, harzburgites (?), and lherzolites (?) were replaced by tremolite, then by anthophyllite + chlorite and garnet + chlorite, and finally by magnetite + serpentine assemblages (tables 31.3 and 31.4). Available chemical analyses indicate Mg numbers for the rocks between 50 and 90.[38]

The Greer Hollow body is concordant and is infolded with hornblende schists and gneisses. At least three fold generations are evident in the body. An early isoclinal fold set has experienced postfolding extension and boudinage associated with the development of the dominant foliation in both the body and the country rock. That foliation has been refolded by two open to tight fold sets. Near its eastern margin, hornblende schist and gneiss, metagabbro tectonites, and a small mass of kyanite-bearing pelitic schist abut the body. Here and elsewhere, a thin talc + actinolite (?) selvage separates the ultramafic rocks from the country rock.

The P-T conditions of metamorphism can be estimated, assuming that the anthophyllite did not crystallize out of its field of stability and that the dominant foliation of the kyanite-bearing pelitic schist is coeval with the dominant foliation in the ultramafic rock. Overlap of the fields of stability of these phases suggests P-T conditions of about 0.7 Gpa (7 kb) and 700–800° C.

The *Day Book dunites* (there are two adjacent bodies) occur in a terrane composed of hornblende schists and gneisses, kyanite-bearing pelitic schists and gneisses, and quartz-feldspar semischists (figure 31.10b) (Brobst, 1962; S. E. Swanson, 1981; Raymond, unpublished data).[39] The two bodies are composed primarily of olivine, with minor amounts of orthopyroxene and chromite, plus a wide variety of other minerals formed during later metamorphic events. Thus, dunite is the dominant rock type, but minor harzburgite, chromitite, serpentinite, talc-anthophyllite schists, and phlogopite-talc-magnesite-olivine granoblastites also occur. Along the northwest margin of the larger of the two bodies, a granodiorite pegmatite has intruded along the contact between the country rocks and the dunite. Small dikes branch from the larger dike and cross-cut both dunite bodies. Along these dikes, as well as along the margins of the bodies, metasomatic reactions

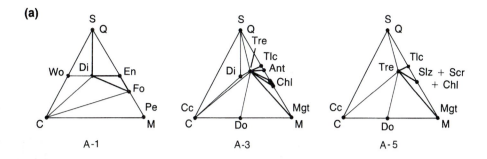

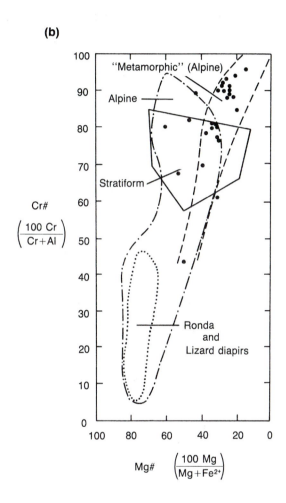

Figure 31.9 Selected phase diagrams and compositions of chromites from ultramafic rock bodies. (a) CMS phase diagrams for associations A-1, A-3, and A-5 of the Blue Ridge Belt, southern Appalachian Orogen. (b) Compositions of chromium spinels from diapiric, stratiform, alpine, and metamorphosed alpine bodies. Dots represent compositions of selected chrome spinels from Blue Ridge alpine-type ultramafic bodies.

[(b) Modified from Dick and Bullen, 1984; Lipin, 1984; and Agata, 1988]

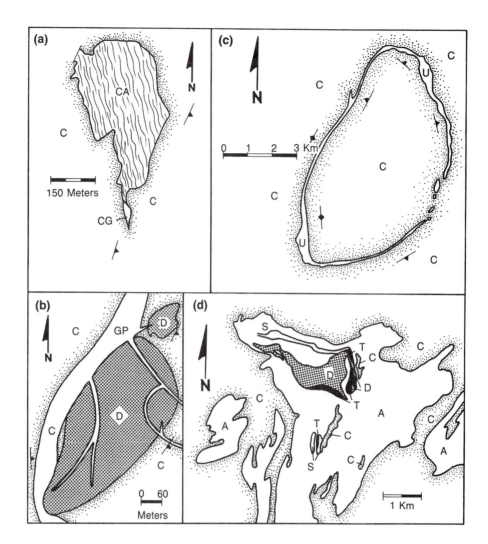

Figure 31.10   Sketch maps of Blue Ridge ultramafic bodies. (a) Map of Greer Hollow Ultramafic Body. CA = chlorite-anthophyllite unit, CG = chlorite-garnet unit, C = country rocks. (b) Map of the Day Book dunites (modified from S. E. Swanson, 1981). D = dunite, GP = granitoid pegmatite. C = country rocks. (c) Map of the Webster-Addie Ultramafic Body (modified from R. Miller, 1951a). U = ultramafic rocks, C = country rocks. (d) Map of the northern part of the Chunky Gal Mountain mafic-ultramafic complex, which includes the Buck Creek Dunite. (Modified from McElhaney and McSween, 1983). A = amphibolite, D = dunite, S = serpentinite, T = metatroctolite and related rocks, C = other country rocks.

have occurred, yielding a variety of rocks composed of such minerals as talc, anthophyllite, phlogopite, vermiculite, and chlorite (S. E. Swanson, 1981). Metasomatically altered parts of the interior of the larger dunite body contain patches of rock containing or composed of talc, tremolite, anthophyllite, chlorite, phlogopite, and magnesite. Shear zones and veins, formed later, are filled by serpentine with magnetite or less common minerals, including aragonite. Together, these minerals define a series of metamorphic assemblages formed during several successive metamorphic events (tables 31.3 and 31.4).

Structurally, the Day Book dunites are thick, concordant lenses. The earliest structures observed in the body are tabular to podiform chromite bodies. The tabular layers of chromite are locally folded into isoclinal folds. The generally massive-appearing interior of the body is characterized by a coarse porphyroclastic texture, consisting of large, cleavable, deformed, elongate olivine crystals up to 26 cm. in length (S. E. Swanson, pers. comm., 1979) in an equigranular-mosaic-textured matrix of olivine crystals that average about 0.2 mm in diameter. Disseminated subhedral to euhedral chromite and scattered orthopyroxene occur within the equigranular mosaic. Dikes, metasomatically altered zones, and shear zones are characterized by rocks of allotrioblastic-granular, diablastic, or schistose texture.

Chemical analyses of both rocks and minerals are available for the Day Book rocks. One harzburgite analysis gives an Mg number of about 0.86 (C. E. Hunter, 1941). As would be expected of rocks composed primarily of Mg-rich olivine, the dunites have high Mg numbers clustered around 0.88. The forsterite contents of olivines fall in the range 92 to 95 (J. R. Carpenter and Phyfer, 1975; S. E. Swanson, 1981). The Cr/(Cr + Al) values of the chromites fall in the range 0.7 to 0.9 at Mg numbers of 0.40 to 0.70, as is typical of alpine-type chromites (Irvine, 1967a; S. E. Swanson, 1981).[40]

The *Webster-Addie body* forms a distinctive, ring-shaped outcrop pattern (in map view) that is nearly 10 km in diameter (figure 31.10c) (R. Miller, 1951a,b).[41] The ring represents the erosional pattern of a thin sheet of ultramafic rock folded into a dome, from which the crest was removed by erosion. The body is essentially concordant with the surrounding biotite gneisses. Internally, however, the body contains compositional bands, some of which are isoclinally folded (Greenberg, 1976). As is the case at Day Book, the Webster-Addie body contains shear zones that cut across other features.

The Webster-Addie ultramafic body consists primarily of dunite, with lesser amounts of websterite and minor amounts of orthopyroxenite, harzburgite, serpentinite, and amphibole gabbro. The serpentinite and the gabbro form veins that cut the body. The dunite and serpentinized dunite exhibit a variety of textures, including equigranular-mosaic texture, equigranular-tabular texture, porphyroclastic texture, mesh texture, diablastic texture, and schistose texture. The serpentinite is dominantly schistose. Websterite has allotrioblastic-granular texture, in part similar to the equigranular mosaic of the dunite. As at Day Book and Greer Hollow, porphyroclastic and equigranular textures in rocks dominated by olivine and pyroxene are replaced successively by porphyroclastic and semischistose textures in which amphiboles are important, and schistose to diablastic textures in which amphiboles, chlorite, or serpentine dominate (tables 31.3 and 31.4).

The whole rock and mineral chemistries of the Webster-Addie ultramafic rocks are similar to those of the Day Book bodies. For example, relatively high Mg numbers (0.83–0.85) characterize the serpentinized dunites (Condie and Madison, 1969). The olivines are forsterite-rich (Fo = 84–95) and the orthopyroxenes are similarly Mg-rich (En = 87–95) (R. Miller, 1951b; Condie and Madison, 1969; Lipin, 1984). In the chromites, the Cr/(Cr + Al) values are about 0.8, but the Mg numbers are relatively low (0.3–0.6) (Lipin, 1984).

The *Buck Creek ultramafic bodies*, like the Greer Hollow ultramafite, are almost entirely enclosed within amphibole schist and gneiss (figure 31.10d) (McElhaney and McSween, 1983).[42] At the Buck Creek locality, one large dunite body is accompanied by at least four small serpentinite bodies. Minor lherzolite is present in the larger body (Kuntz and Hedge, 1981). Metatroctolites are associated with the ultramafic rocks, especially the serpentinites. The margins of the dunite and some veins within it consist of serpentinite and tremolite-talc-chlorite schist.

The Buck Creek dunite is a conformable mass within the amphibole schist and gneiss. At the contact, some shearing and minor folding have occurred, but there is no evidence of a major fault separating the dunite and the enclosing rocks.[43] The contacts and foliations in the contact zone generally parallel the foliation in the country rock. Lattice preferred orientations in the dunite (Sailor and Kuntz, 1973) apparently predate the dominant regional foliation and folding, which affected both the dunite and the country rocks during the Ordovician Taconic Orogeny (McElhaney and McSween, 1983).

Porphyroclastic, equigranular-mosaic, equigranular-tabular, mesh, and schistose textures characterize the rocks of the Buck Creek dunite. The oldest textures are porphyroclastic and equigranular textures. Mineralogically, the rocks consist of olivine (Fo = 88–89) and accessory hornblende, chromite, and/or Ca-pyroxene. Additional accessory minerals of probable retrograde origin include magnetite, pyrite, pyrrhotite, chlorite, tremolite, anthophyllite, talc, and serpentine minerals (tables 31.3 and 31.4).

Geothermometry performed on the amphibole schists and gneisses reveals that the early Taconic metamorphic event occurred at a temperature of about 725° C. Talc + chlorite + tremolite ± anthophyllite rocks may have developed at this time or during a subsequent Devonian (Acadian) metamorphic event, or both, but folding associated with the Acadian event reportedly affected the chlorite-talc schists (McElhaney and McSween, 1983). A Greenschist Facies event, which produced the serpentinites, followed the Acadian metamorphism.

In summary, all of the ultramafic bodies of the central Blue Ridge Belt apparently exhibit minerals and textures that reflect the same *sequence* of events (table 31.4). Pre- or early Taconic deformation and metamorphism in the mantle or lower crust produced porphyroclastic and equigranular textures in dunites, peridotites, and pyroxenites. Taconic and Acadian metamorphic events at Granulite and Amphibolite Facies grades produced tremolite, anthophyllite-,talc-,and chlorite-dominated assemblages that partially or completely replaced the earlier fabrics and minerals. The first of these events apparently occurred under high crustal pressures (6–9

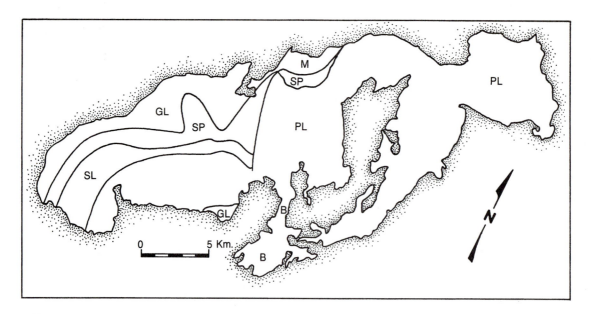

Figure 31.11   Sketch map of the Ronda Ultramafic Complex, Spain. PL = plagioclase lherzolite zone, SP = spinel pyroxenite zone, SL = spinel lherzolite zone, GL = garnet lherzolite zone, M = melange zone, B = breccia zone. *(Modified from Suen and Frey, 1987; after Lundeen, Obata, and Dickey, 1979; Obata, 1980; Frey, Suen, and Stockman, 1985)*

kb) and high temperatures (600–800° C). The second event occurred at somewhat lower P-T conditions. Later (late Paleozoic, Alleghenian?) Greenschist Facies metamorphism resulted in the replacement of earlier assemblages by diablastic to schistose serpentine + magnetite assemblages.

The Ronda Ultramafic Complex: Product of a Mantle Diapir

The Ronda ultramafic complex is an alpine-type peridotite exposed in the Betic Cordillera of southern Spain.[44] The mass is crudely elliptical in plan, is more than 35 km long, and is surrounded by metamorphic rocks, which, at least in part, are of the pyroxene hornfels facies (Loomis, 1972a; Lundeen, 1978). Internally, the Ronda Peridotite contains four major petrographic zones (figure 31.11) (Obata, 1980). These include a garnet lherzolite zone, a spinel pyroxenite zone (the "ariegite subfacies"), a spinel lherzolite zone (the "seiland subfacies"), and a plagioclase lherzolite zone. Reference to figure 6.2 will remind the reader that garnet peridotite, spinel peridotite, and plagioclase peridotite are stable at successively shallower depths (> 50 km, about 40km, and < 25 km, respectively) in the mantle. Consequently, the key petrographic zones, which are too thin to represent a cross section through the mantle (the total thickness of the Ronda body is < 2 km), are interpreted to represent partially equilibrated zones in an ascending mantle diapir. The zones are not homogeneous, as

variations in mineralogy result in the occurrence of harzburgites, amphibole-bearing lherzolites, and sepentinized variants of the dominant lherzolite rock types. In addition, there are melange and breccia zones within one or more of these units.

The mineral chemistry of the ultramafic and the mafic rocks reflect the complex history of the body. Although olivines have a restricted range of chemistry (Fo = 88–92), the pyroxenes and spinels are quite variable in composition (Obata, 1980; F. A. Frey, Suen, and Stockman, 1985). Orthopyroxene *porphyroclasts* have a restricted range of Mg numbers (0.89–0.91), but contain clinopyroxene blebs and exsolution lamellae and are strongly zoned in Al_2O_3. Orthopyroxene *porphyroblasts* also have a restricted range of Mg number (0.89–0.92) and show zoning in Al_2O_3. Similarly, clinopyroxenes show significant variations in Na_2O and Al_2O_3. The Cr/(Cr + Al) values of the spinels vary from about 0.05 to about 0.50 (see figure 31.9). The more aluminous varieties are present in the spinel and garnet peridotites.

Texturally and structurally, the body is complex. Textures range from protogranular and equigranular mosaic to porphyroclastic and cataclastic (Obata, 1980). Layering, foliation, and lineation characterize the metamorphic rocks. Layering in the body includes both metamorphic layers and magmatic layers of mafic to ultramafic composition, ranging from garnet pyroxenites to olivine gabbros (Dickey, 1970).

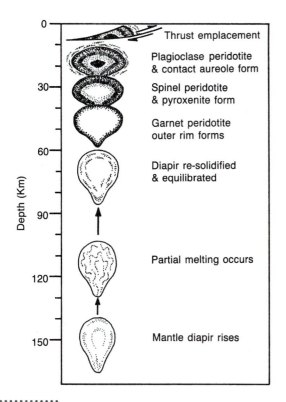

Depth (Km)

Thrust emplacement

Plagioclase peridotite & contact aureole form

Spinel peridotite & pyroxenite form

Garnet peridotite outer rim forms

Diapir re-solidified & equilibrated

Partial melting occurs

Mantle diapir rises

Figure 31.12 Schematic diagram depicting the history of the Ronda ultramafic complex.
(based on T. P. Loomis, 1972a; Lundeen, 1978; Obata, 1980; Frey, Suen, and Stockman, 1985; and Suen and Frey, 1987)

Overall, the Ronda ultramafic complex is now a tabular mass that constitutes part of a thrust sheet (Lundeen, 1978). Together, the data indicate a complex history for the body involving an early diapiric rise of a fertile mantle garnet lherzolite and later thrust fault emplacement into higher levels of the crust (figure 31.12). In transit from mantle depths of perhaps 200 km, the body experienced partial melting of both marginal and interior peridotites at depths of between 120 and 100 km and temperatures between 1700° C and 1500° C (F. A. Frey, Suen, and Stockman, 1985). At the margin, small degrees of partial melting left residues of garnet lherzolite, whereas in the interior, larger degrees of partial melting produced garnet-free peridotites. The melts crystallized as the diapir rose and the body equilibrated at 1100–1200° C and 2.0–2.5 Gpa at 60–75 km (Obata, 1980; Suen and Frey, 1987). Further rise of the body resulted in cooling of the body from the outside towards the interior, with consequent, incomplete textural and mineralogical re-equilibration. After emplacement in the crust and the creation of a high-temperature contact metamorphic aureole (about 22 m.y.b.p.), the

body was transported and emplaced at higher levels in the crust as part of a thrust sheet (Lundeen, 1978; Reisberg, Zindler, and Jagoutz, 1989).

SUMMARY

Alpine ultramafic rock bodies are irregular to elliptical, typically deformed masses of ultramafic rock that have been incorporated into mountain belts. They are characterized by tectonite fabrics and tight to isoclinal folds. The rocks usually have high Mg numbers, as do the olivines and pyroxenes that constitute them. Very commonly, alpine-type ultramafic rocks are serpentinized, some through nearly isochemical processes and others via constant-volume processes.

Any kind of igneous mafic-ultramafic complex may be incorporated into an orogenic belt and deformed with the rocks of that mountain system. Consequently, igneous ultramafic bodies of any kind may become alpine-type masses. Nevertheless, the most common type of igneous protoliths are ophiolitic. Ophiolitic mantle and other mantle slabs may also form alpine-type ultramafic bodies. In addition, albeit rarely, mantle diapirs emplaced into the upper mantle or lower crust become alpine-type ultramafic bodies. Because these three types of occurrences—magmatic bodies, mantle slabs, and mantle diapirs—form under widely varying conditions, the rocks that constitute them, and therefore alpine-type ultramafic bodies as a group, exhibit a wide range of characteristics. Metamorphism further diversifies the mineral composition and chemistry of these rocks.

EXPLANATORY NOTES

1. Refer back to chapter 10 for a review of the various forms of magmatic ultramafic rocks and rock bodies.
2. H. H. Hess (1955) schematically depicts these orogenic occurrences of ultramafic rocks. Also see Taliaferro (1943), R. Miller (1951b), Battey (1960), Larrabee (1966), Lappin (1967), Wyllie (1967b), Menzies (1973), Gansser (1974), Livingstone (1976), Pamic and Majer (1977), Maltman (1978), Varne and Brown (1978), C. A. Hall and Bennett (1979), Ozawa (1983), R. L. Christiansen (1984), Pognante, Rosli, and Toscani (1985), O'Hanley (1987), Bodinier (1988), Bodinier, Dupuy, and Dostal (1988), Girardeau and Mercier (1988), Girardeau, G. I. Ibarguchi, and Ben Jamaa (1989), Hebert Serri (1989), Mittwede and Stoddard (1989), Seyler and Mattson (1989), Harnois, Trottier, and Morency (1990), Spell and Norrel (1990), and additional references to specific localities in the following notes (e.g., notes 5, 26, and 33).

3. For example, see Friedman (1953) and Dymek, Brothers, and Schiffries (1988). Actually, some ultramafic rocks surrounded by Precambrian rocks, such as those in the Southern Appalachian Blue Ridge Belt, occur in mountain belts formed during the Phanerozoic eon.

4. Alpine ultramafic rocks formed at present and former plate boundaries are described by Bonatti, Honnorez, and Ferrara (1970), J. F. Dewey and Bird (1970), Moores and MacGregor (1972), C. G. Engel and Fisher (1975), Harold Williams (1975), DeWit et al. (1977), Sinton (1979), Hamlyn and Bonatti (1980), Nicolas and Le Pichon (1980), J. F. Casey and Karson (1981), Quick (1981a,b), Hebert, Bideau, and Hekinian (1983), Dick, Fisher, and Bryan (1984), Harper (1984), Kimball, Spear, and Dick (1985), Nicolas (1986), Shibata and Thompson (1986), Cabanes and Briqueu (1987), R. B. Miller and Mogk (1987), Girardeau and Mercier (1988), and Mevel et al. (1991). Also see note 26. See Beslier, Girardeau, and Boillot (1988, 1990) for descriptions of a deformed spreading-center ultramafic complex.

5. *Bay of Islands* (C. M. Smith, 1958; also see note 26); *Thetford Mines, Quebec* (Dresser, 1909; Cooke, 1937; Laurent, 1975; Laurent and Hebert, 1979; Cogulu and Laurent, 1984; O'Hanley, 1987; Hebert and Laurent (1989); Harnois, Trottier, and Morency, 1990); *Vermont* (Chidester, 1962; Jahns, 1967; Sanford, 1982); *North Carolina* (C. E. Hunter, 1941; R. Miller, 1951a, b; Kulp and Brobst, 1954; Condie and Madison, 1969; Misra and Keller, 1978; S. E. Swanson, 1981; McElhaney and McSween, 1983; Abbott and Raymond, 1984); *California Coast Ranges,* including Burro Mountain (Taliaferro, 1943; Burch, 1968; N. J. Page, 1967; R. G. Coleman and Keith, 1971; E. H. Bailey and Blake (1974); Evarts, 1977; 1978; Lienert and Wasilewski, 1979; C. A. Hopson, Mattinson, and Pessagno, 1981); *Klamath Mountains* (Irwin and Lipman, 1962; Ramp, 1975; Dick, 1977a; Lindsley-Griffin, 1977; Quick, 1981a; Harper, 1984; S. M. Peacock, 1987b); *Washington Cascade Range* (Ragan, 1967; M. R. W. Johnson, Dungan, and Vance 1977; Vance and Dungan, 1977); *Blue River, British Columbia* (Pinsent and Hirst, 1977); *Ronda, Spain* (Dickey, 1970; Obata, 1980; Suen and Frey, 1987; also see note 23); *Swiss Alps* (Trommsdorff and Evans, 1974; Pfeifer, 1978, 1981; Stille and Oberhansli, 1987); *Lizard, Cornwall, England* (D. H. Green, 1964); Norway (Lappin, 1967); *Dun Mountain, New Zealand* (F. J. Turner, 1942; Battey, 1960; Lauder, 1965; N. I. Christiansen, 1984).

6. Refer to chapter 10 for a review of this mineralogy. The chemistry of the minerals and other characteristics of alpine-type ultramafic bodies, listed in table 31.1, are based on the following sources: R. Miller (1951a,b); H. H. Hess (1955), E. D. Jackson (1961), D. H. Green (1964), Irvine (1967b), Wager and Brown (1968), Den Tex (1969), Moores (1969), Kornprobst (1969), Dickey (1970), O. B. James (1971), Moores and Vine (1971), Astwood, Carpenter, and Sharp (1972), Loomis (1972b), Medaris (1972), Moores and MacGregor (1972), E. D. Jackson and Thayer (1972),

Menzies (1973), J. Pike (1974), P. Henderson (1975), E. D. Jackson, Shaw, and Bargar (1975), Nicolas and Boudier (1975), Dribus et al. (1976), Loney and Himmelberg (1976), R. G. Coleman (1977b), Evarts (1977, 1978), Greenbaum (1977), Schubert (1977), Kingsbury and Heimlich (1978), Varne and Brown (1978), Dick and Sinton (1979), Ernst and Piccardo (1979), Nicolas and LePichon (1980), Obata (1980), Raymond and Swanson (1981), S. E. Swanson (1981), Garcia (1982), Hebert, Bideau, and Hekinian (1983), Ozawa (1983), Abbott and Raymond (1984), N. I. Christiansen (1984), Dick and Bullen (1984), Dick, Fisher, and Bryan (1984), Elthon, Casey, and Komor (1984), F. A. Frey, Suen, and Stockman (1985), Ishiwatari (1985), Kimball, Spear, and Dick (1985), Leroy (1985), Pognante, Rosli, and Toseani (1985), Nicolas (1986), R. B. Miller and Mogk (1987), Suen and Frey (1987), L. A. Raymond (unpublished data); Bodinier (1988), Bodinier, Dupuy, and Dostal, (1988), Hebert, Serri, and Hekinian (1989), O'Hanley (1989).

7. Also see the references in notes 8 and 9.

8. These textures are widely recognized in mantle xenoliths, as well as in ultramafic rocks derived from the mantle. For additional examples and descriptions, see Nicolas et al. (1971), Helmstaedt et al. (1972), Dribus et al. (1976), Nicolas and Poirier (1976), Loney and Himmelberg (1976), Nielson Pike and Schwarzman (1976), Basu (1977), Christensen and Salisbury (1979), Honeycutt and Heimlich (1980), Raymond and Swanson (1981), Boudier and Nicolas (1982a,b), Christensen and Lundquist (1982), Hebert, Bideau, and Hekinian (1983), Wicks (1984a), and R. B. Miller and Mogk (1987).

9. Maltman (1978), Wicks and Plant (1979), A. J. Williams (1979), Raymond and Swanson (1981), Wicks (1984a,b).

10. See chapter 10 for a discussion of these mineralogical variations.

11. The Bay of Islands Ophiolite Complex, described below, is a good example. See Ringwood (1966a, 1977), K. Ito (1974), Hervig, Smith, and Steele, (1980), Maaloe and Steel (1980), D. L. Anderson and Bass (1984), Menzies (1984), Arai (1987), Irifune and Ringwood (1987), Nicolas (1989), and recent issues of the *Journal of Geophysical Research* and *Earth and Planetary Science Letters* for more information on mantle compositions and structure.

12. An in-depth discussion of the nature of the mantle is beyond the scope of this text.

13. For example, see D. L. Anderson and Bass (1984) and Nicolas (1986, 1989). For discussions of mantle metasomatism, see Menzies and Hawkesworth (1987) and E. M. Morris and Pasteris (1987a).

14. For example, see Bonatti, Honnorez, and Ferrara (1970), Nicolas and Poirier (1976, ch. 11), Ernst and Piccardo (1979), Hamlyn and Bonatti (1980), Christensen and Lundquist (1982), Dick, Fisher, and Bryan (1984), Shibata and Thompson (1986).

15. In the early and middle parts of the twentieth century, a debate raged in the geologic community about whether serpentinite is a magmatic or a metamorphic rock. H. H. Hess (1938, 1955) championed the cause of a magmatic origin, largely on the basis of field relationships. Bowen and Tuttle (1949) demonstrated, via laboratory studies of the system MgO-SiO_2-H_2O, that serpentine was only stable at low temperatures and that magmas did not develop in that system, even at temperatures of about 900° C. Experimental evidence demonstrates that temperatures of 1800–2000° C are required to produce magmas from pure olivine compositions (Bowen and Anderson, 1914, Ohtani and Kumazawa, 1981). Additional references from the literature that provide a basis for our present understanding of serpentinization include Vaugnat (1963), Nauman and Dresher (1966), Thayer (1966), N. J. Page (1967, 1968), Whittaker and Wicks (1970), R. G. Coleman and Keith (1971), Wicks and Whittaker (1975), Moody (1976), DeWit et al. (1977), Vance and Dungan (1977), Caruso and Chernosky (1979), Coats and Buchan (1979), Dungan (1979a,b), Labotka and Albee (1979), Laurent and Hebert (1979), Ikin and Harmon (1983), Komor et al. (1985), Mellini, Trommsdorff, and Compagnoni (1987), and Mellini and Zanazzi (1987). Also see Wicks and O'Hanley (1988), O'Hanley, Chernosky, and Wicks (1989), and the review of Chernosky, Berman, and Bryudzig (1988).

16. Wicks and Whittaker (1975) and Moody (1976) list the various polytypes of serpentine minerals and Mellini, Trommsdorff, and Compagnoni (1987) discuss polytypism in antigorite. O'Hanley, Chernosky, and Wicks (1989), using a chemographic analysis, assess the stability of lizardite and chrysotile. Also see Middleton and Whittaker (1979), Yada (1979), Mellini and Zanazzi (1987), and Wicks and O'Hanley (1988) for additional information on the structures of serpentines. A. C. D. Newman and Brown (1987) discuss the chemistry of serpentines.

17. Whittaker and Wicks (1970), Wicks and Whittaker (1975), Mellini, Trommsdorff, and Compagnoni (1987).

18. Modified from F. J. Turner and Verhoogen (1960, pp. 318–319).

19. Hostetler, Coleman, and Evans (1966), R. G. Coleman (1971a), R. G. Coleman and Keith (1971), Moody (1976), Komor et al. (1985).

20. Serpentine pseudomorphs after pyroxenes, which typically mimic the pyroxene cleavage, are called "bastites" (see Dungan, 1979b; Wicks and Plant, 1979).

21. Thayer (1966, 1967), N. J. Page (1967), Condie and Madison (1969), Coleman (1971a), Moody (1976), Leach and Rodgers (1978).

22. I. Barnes and O'Neil (1969), I. Barnes, Rapp, and O'Neil (1972), Wenner and Taylor (1973), M. R. W. Johnson, Dungan, and Vance (1977), Janecky and Seyfried (1986), Peacock (1987b).

23. For example, E. H. Nickel (1959, 1961) and B. R. Frost (1985).

24. R. G. Coleman (1967, 1977a), Leach and Rodgers (1978), Schandl, O'Hanley, and Wicks (1989). Recall that Gresens (1969) argued that the highly reducing fluids involved in serpentinization produced Blueschist Facies rocks by metasomatic alteration of country rocks surrounding serpentinites (see chapter 28).

25. The textures of serpentinites are also discussed and/or depicted by Maltman (1978), Cressey (1979), Dungan (1979b), Morandi and Felice (1979), Wicks and Plant (1979), A. J. Williams (1979), Ikin and Harmon (1983), Wicks (1984a,b,c), Peacock (1987b), and A. E. Gates and Kambin (1990).

26. The Bay of Islands Complex has been described to varying degrees by a number of workers. This review, which focusses on the ultramafic rocks, is based on the works of C. M. Smith (1958), W. R. Church and Stevens (1971), Dewey and Bird (1970), Harold Williams (1973, 1975), J. Malpas (1978), Salisbury and Christensen (1978), Christensen and Salisbury (1979), S. B. Jacobson and Wasserburg (1979), Suen, Frey, and Malpas (1979), Casey and Karson (1981), Casey et al. (1981), Girardeau and Nicolas (1981). Christensen and Lundquist (1982), Elthon, Casey, and Komor (1982), Casey et al. (1983), Elthon, Casey, and Komor (1984), Casey et al. (1985), and Komor et al. (1985a,b).

27. For example, see W. R. Church and Stevens (1971), S. B. Jacobson and Wasserburg (1979), and Casey et al. (1985). Others have argued that the Bay of Islands Complex represents the crust of a marginal basin (Dewey and Bird, 1970).

28. Also see Girardeau and Nicolas (1981) and Casey et al. (1981).

29. Although not described in detail, such structures are mentioned by Casey et al. (1983).

30. Elthon, Casey, and Komor (1982); Elthon, Casey, and Komor (1984), Komor et al. (1985).

31. For example, see Jacobson and Wasserburg (1979) and Casey et al. (1985).

32. See Larrabee (1966), Rankin, Espenshade, and Neuman (1972), Hatcher (1978), Misra and Keller (1978), Scotford and Williams (1983), Abbott and Raymond (1984), Raymond and Abbott (1985), Butler (1989), Mittwede and Stoddard (1989), and Spell and Norrell (1990).

33. Reports on the petrology, petrography, structure, and petrofabrics of these bodies include the following: J. V. Lewis (1896); J. H. Pratt and Lewis (1905), C. E. Hunter (1941), Hadley (1949), R. Miller (1951a,b, 1953), Kulp and Brobst (1954), Stose and Stose (1957), Brobst (1962), Larrabee (1966), J. R. Carpenter and Phyfer (1969, 1975), Condie and Madison (1969), Stueber (1969), Bentzen (1970, 1975), Neuhauser and Carpenter (1971), Astwood, Carpenter, and Sharp (1972), Hartley (1973), Jones et al. (1973), Neuhauser

(1973), Sailor and Kuntz (1973), Dallmeyer (1974), Dribus et al. (1976), Greenberg (1976), Swanson and Raymond (1976), Swanson and Whittkop (1976), Alcorn and Carpenter (1976), Bluhm and Zimmerman (1977), Hearn et al. (1977), Palmer, Heimlich, and Kolb (1977), Tien (1977), Yurkovich (1977), Kingsbury and Heimlich (1978), Misra and Keller (1978), Sharpe and Whitney (1979), Heimlich et al. (1980), Honeycutt and Heimlich (1980), S. E. Swanson (1980, 1981), Honeycutt, Heimlich, and Palmer (1981), Kuntz and Hedge (1981), Penso, Heimlich, and Palmer (1981), Raymond and Swanson (1981), Dribus, Heimlich, and Palmer (1982), Neuhauser (1982), Schiering, Heimlich, and Palmer (1982), McElhaney and McSween (1983), Scotford and Williams (1983), Abbott and Raymond (1984), Hatcher, et al. (1984), Lipin (1984), Abbott and Raymond (1985), McSween and Hatcher (1985), Raymond and Abbott (1985), Swanson et al. (1985), Conley (1987), Meen (1988), and Raymond et al. (1988).

34. For example, see Astwood, Carpenter, and Sharp (1972), Dribus et al. (1976), Kingsbury and Heimlich (1978), Honeycutt and Heimlich (1980), Swanson (1981), McElhaney and McSween (1983), Abbott and Raymond (1984), and Raymond et al. (1988).

35. For discussions of ophiolites among these ultramafic bodies, see McSween and Hatcher (1985) and Conley (1987).

36. Miller (1951), Kulp and Brobst (1954), J. R. Carpenter and Phyfer (1975), Swanson (1981), Lipin (1984), Meen (1988).

37. The Lake Chatuge Complex on the North Carolina–Georgia border has a debated history, reportedly somewhat different from that of other Blue Ridge alpine ultramafic bodies (Hartley, 1973; Dallmeyer, 1974; Meen, 1988). Meen (1988) suggests that the body originally crystallized from a magma at a pressure of 1.0–1.3 Gpa (corresponding to a deep crustal or a mantle depth) at a temperature of $T > 900°$ C and was later emplaced via tectonic processes. The dunite-serpentinite masses in the Lake Chatuge Complex contain assemblages 1a and 4a, reflecting a history partly like those of other Blue Ridge ultramafic bodies.

38. Unpublished data. The low Mg numbers may reflect metasomatic effects. Compositions like analysis 1 in table 31.2 do not represent normal igneous bulk rock compositions, although some cumulates do have similarly odd bulk rock chemistries.

39. The Day Book dunites are widely known and visited bodies. Additional reports on their petrology and mineralogy may be found in C. E. Hunter (1941), Kulp and Brobst (1954), Phyfer and Carpenter (1969), Dribus et al. (1976), Swanson and Raymond (1976), Swanson and Whittkop (1976), Tien (1977), Abbott and Raymond (1984), Raymond and Abbott (1985), and Swanson et al. (1985). *Note:* Mining operations have dramatically changed the exposures reviewed here and described by Swanson (1981).

40. Also see Bentzen (1970), Lipin (1984), and Dick and Bullen (1984).

41. Additional references to the mineralogy and petrology of the Webster-Addie body include C. E. Hunter (1941), Condie and Madison (1969), Bentzen (1970), Dribus et al. (1976), and Greenberg (1976).

42. Petrologic and mineralogic information on the Buck Creek ultramafic rocks is also included in C. E. Hunter (1941), Sailor and Kuntz (1973), Kuntz and Hedge (1981), and McSween and Hatcher (1985).

43. Kuntz (1964, in McElhaney and McSween, 1983), McElhaney and McSween (1983).

44. This review is based on the works of Dickey (1970), Loomis (1972a b), Menzies, Blanchard, and Jacobs (1977), Schubert (1977), Lundeen (1978), Lundeen, Obata, and Dickey (1979), Obata (1980), M. A. Frey, Suen, and Stockman (1985), Suen and Frey (1987), and Reisberg, Zindler, and Jagoutz (1989). Somewhat similar spinel peridotites are described by Bodinier, Depuy, and Dostal (1988).

PROBLEMS

31.1. If an ultramafic body consists of equigranular-tabular-textured dunite and harzburgite and the harzburgite has an Al_2O_3 content of 3.1%, a CaO content of 6.01%, and a Mg number of 86, what type of alpine ultramafic body is it likely to be?

31.2. (a) Construct CMS phase diagrams for Blue Ridge alpine ultramafic rock associations A-2 and A-4, (see table 31.3). (b) If chlorite is assumed to be a saturating phase, only one reaction is needed to represent the change from A-1 to A-2. Write and balance that reaction. (c) Does this reaction represent a prograde or retrograde reaction? Explain.

Epilogue

Igneous, sedimentary, and metamorphic rocks of various types occur together at regional to local scales. In the photo at right, for example, an andesite dike and sill intrude Precambrian gneiss and both the andesite and gneiss are locally overlain by sediments deposited as a result of glacial and gravitational activity. On the regional scale, more diverse associations of rock types define the petrotectonic assemblages characteristic of plate settings, as outlined in this part of the text . These assemblages provide a foundation for understanding the history of the Earth.

PART V

Petrotectonic Assemblages

INTRODUCTION

Information on all of the major rock types within each of the classes—igneous, sedimentary, and metamorphic—has been provided in this text. Both descriptive details and petrogenetic theories were summarized. Together, the data give a view of the histories that have produced crustal rocks.

The igneous rocks are found to be crystallized from magmas, those liquids formed in the mantle and at the base of the crust by partial melting. Various primary basaltic magmas are the principal mantle melts, although andesitic, komatiitic, and other melts are generated under special conditions. The magmas rise towards the surface, losing heat along the way. The compositions of magmas may be modified as they migrate, by assimilation of country rock, by mixing with other magmas, or by various processes of differentiation. The loss of heat, if significant, results in crystallization of plutonic rocks at depth, but if minimal, allows magmas to rise to the surface where they form volcanic rocks.

The sedimentary rocks form at the surface through various combinations of sedimentation processes. The clastic rocks, both silicate and carbonate, are formed after weathering and erosion of a source terrane produces sediment, which is subsequently deposited via sediment rain, traction currents, or submarine or subaerial debris flows and landslides. The various

chemical sediments form through crystallization from aqueous solutions under varying conditions of pH and Eh. These sediments and their characteristics reflect the environments of deposition. On the continents, fluvial environments are preeminent. Deltas are the sites of the greatest accumulations of sediment at the continent-ocean interface, whereas shallow marine (shore, shelf, reef) and submarine fan environments contain great quantities of marine sediment. The sediment formed in each of these environments becomes rock through a number of diagenetic processes, including compaction, cementation, and recrystallization.

Where preexisting igneous, sedimentary, or metamorphic rocks are subjected to new conditions of pressure and temperature, especially in the presence of a fluid phase, they are metamorphosed. Such changes occur (1) in ocean crust, especially where hydrothermal activity is significant; (2) in incipient mountain ranges, where stratigraphic or structural burial adds pressure at low temperatures, yielding high P/T regional metamorphic terranes; (3) in arc and collisional mountain ranges, where intrusions of igneous rock and thermal conduction from the mantle result in heating, both local and regional, yielding contact and regional metamorphism, respectively; (4) beneath volcanoes, where magmas heat the surrounding rocks; (5) along fault zones, where dynamic metamorphism yields mylonitic and cataclastic rocks; (6) within the continental crust, both at depth and along fault zones; and (7) in the mantle. Metamorphic processes include various processes of recrystallization and neocrystallization, which generally result in both new textures and new mineral assemblages in the metamorphic rocks. Factors such as the bulk rock composition, the composition of the fluid phase, and the P-T conditions control the phase assemblages that result from these processes.

ROCKS AND PLATE BOUNDARIES

All of the processes that produce and transform crustal rocks occur at specific sites within or at the margins of tectonic plates. Each site is characterized by specific conditions that control the chemistry, minerals, textures, and structures of the rocks. Therefore, those features can be used to characterize the petrogenetic sites of ancient rocks and help us develop an understanding of their histories, one of the ultimate aims of petrology.

The major rock types that occur at petrogenetic sites within a plate or at a plate boundary are shown in figure 32.1 (a recasting of figure 1.4). Each petrogenetic site within a plate or at a plate boundary may yield the rocks shown for that site. At spreading centers, oceanic tholeitic basalts, the Mid-Ocean Ridge Basalts (MORBs), form from magmas developed by partial melting of mantle peridotites at shallow depths. The MORBs form a layered and pillowed cap on the rest of the rocks of the ocean crust, which together with the basalts constitute an ophiolite sequence. Below the basalts are the sheeted dike complex, nonlayered plutonic differentiates, cumulate gabbros and ultramafic rocks, and a basal, mantle ultramafic tectonite. Because the crust is faulted at the spreading center, a basin is formed into which mafic breccias, derived from adjoining upfaulted oceanic crust, and minor pelagic sediment typically accumulate. Hydrothermal alteration, resulting from the flux of seawater through the fractured crustal rocks, produces hydrothermal (contact) metamorphism near the areas above the spreading center magma chamber. The resulting rocks belong to the Zeolite, Prehnite-Pumpellyite, or Albite-Epidote Hornfels Facies. At depth, deformation and pervasive re-equilibration of magmatic rocks to crustal temperatures result in the development of metagabbro tectonites, amphibole schists, and related rocks of the Greenschist and Amphibolite facies. Mylonites and cataclasites form along faults.

Transform faults that cut the oceanic crust contain the same rocks as the spreading centers, but here the rocks are usually deformed by deviatoric stresses. Ophiolitic melanges and mylonite zones are characteristic of such fault zones (Karson, 1984; Saleeby, 1984). Where a prominent scarp develops along the transform fault, breccias eroded from the scarp may be deposited in the fault zone. Within the oceanic transform fault zones, pelagic sediments also are deposited. On the continents, a wide range of sedimentary rocks form in transform

fault zones, including slide breccias and debris flows, lacustrine shales, and fluvial conglomerate, sandstone, and shale.

Within the open ocean, pelagic sedimentation is the preeminent rock-forming process. Fine-grained sedimentary rocks such as shale, chert, and biogenic limestone are formed in environments controlled by water temperature, water depth, and proximity to land or volcanoes.

Several environments are associated with subduction zones. These include the trench, forearc basins, the subduction zone (sensu stricto), volcanic arcs and associated orogenic belts, back-arc basins, and back-arc passive margins. The trench and the forearc basins, although varied in nature, are characterized primarily by sedimentary rocks (Dickinson and Seely, 1979). These may include olistostromes, turbidite wackes and related rocks, and pelagic shales. The turbidites and olistostromes especially characterize submarine fans that form both in trenches and in trench-slope basins within the forearc. Sedimentary rocks that have undergone soft sediment deformation and dynamically recrystallized sedimentary rocks may also develop in the forearc region. Diapiric melanges intrude some forearc regions and may give rise to olistostromes (Cloos, 1984). Rarely, magmas may intrude into forearc and trench rocks, giving rise to plutons and contact metamorphism (Echeverria, 1980).

The subduction zone (sensu stricto) is a region of metamorphism and anatexis. Beneath the forearc region, rocks carried down on the subducting plate are subjected to progressively higher pressures, while the temperatures remain low. Depending on the temperature, the subducted sedimentary rocks and the underlying oceanic crust are metamorphosed under low to moderate temperature conditions of the Zeolite, Prehnite-Pumpellyite, Blueschist, Greenschist, Amphibolite, or Eclogite Facies. As the rocks descend beneath the overriding plate, they are heated, resulting in dehydration. The fluid phase generated migrates into surrounding rocks. In the depth range of about 100 to 200 km, minor melting of the subducted rocks and significant, but partial, melting of the rocks of the overlying mantle, result in the formation of magmas of tholeiitic, calc-alkaline, and alkaline character.

The magmas generated in the subduction zone provide the building material for the volcanic arc. These magmas rise upwards and intrude at depth to form plutons or erupt at the surface to form volcanic rocks. In transit, they may differentiate, assimilate country rocks, or melt the base of the crust to form siliceous magmas, with which they may mix to form andesites. The siliceous magmas erupt to form rhyolites and less siliceous volcanic rocks, or, if they fail to reach the surface, crystallize to form the rocks of the calc-alkaline plutons. The latter are typified by granodiorites and quartz monzonites. Crystallization of the most siliceous magmas yields granites. Differentiation of granitoid magmas, with the development of a fluid phase, may result in the crystallization of pegmatites. Surrounding the intrusions, at depth, entire regions are heated by the thermal flux, and together with the regional

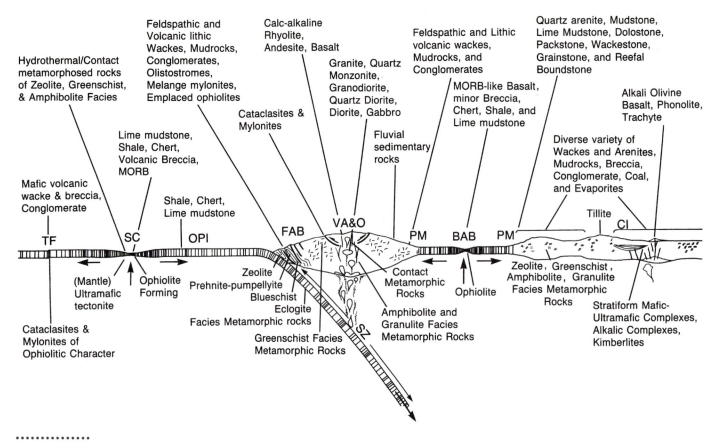

Figure 32.1 Petrotectonic assemblages of the principal plate-interior and plate-boundary sites. SC = spreading center, TF = transform fault, OPI = oceanic plate interior;

SZ = subduction zone, FAB = forearc basin, VA&O = volcanic arc and orogen, BAB = back-arc basin, PM = passive margin, CI = continental interior.

stresses caused by the plate collision, the temperatures (and pressures) yield regional terranes of dynamothermally metamorphosed rock of the Greenschist, Amphibolite, and Granulite Facies. At shallower depths, contact metamorphism associated with mesozonal and epizonal plutons produces rocks of the Zeolite and various hornfels facies. At the surface, fluvial, lacustrine, and glacial deposition will give rise to sediments, many of which are ephemeral, but some of which will be preserved and lithified to form sedimentary rocks.

The back-arc basin is formed by the rise of mantle, resulting partial melting, intrusion of magmas, and consequent spreading of the crust in the region on the side of the arc opposite the subduction zone (Karig, 1971). The principal rocks formed are similar to MORB, but differ slightly in their chemistry. Ophiolitic crust forms here, just as it does in oceanic spreading centers. Although the regional extent of back-arc spreading centers is smaller than that of oceanic spreading centers, the various rocks formed are very much like those of the mid-ocean ridge regions. Perhaps the most significant exception is that the sedimentary facies associated with the back-arc basin margins differ from the sedimentary

associations of the mid-ocean ridges. On the arc side of the basin, sedimentation (and diagenesis) produces volcanic wackes and related shales and conglomerates as the dominant lithologies (Karig and Moore, 1975). On the continent side of the basin, rocks typical of passive margins—arenites, limestones, dolostones, and shales—are typical rock types.

The passive margins of the continents, especially those with broad shelves and reefs, as well as inland seas, are the sites at which large quantities of the sedimentary rocks that occur within the present continents were formed. Here various limestones, ranging from boundstones to lime mudstones, and dolostones developed. The limestones result primarily from biochemical precipitation. The dolostones largely result from the flux of surface and near-surface fluids through limestones, altering their chemistry. Quartz arenites form from sediment deposited in vast inland seas and along strandlines. Ubiquitous shales are interlayered with these rocks. In environments transitional to the continental regions—estuaries, lagoons, and paludal deltas—precursor sediments for various mudrocks and associated sandstones, as well as coal, are deposited.

The continental interiors are diverse in rock types. Diatremes are created where carbonate-rich, ultramafic magmas blast through crustal rock to form kimberlite breccias, some of which are diamondiferous. Alkalic magmas of diverse types, differentiated from alkali olivine basalt parents, intrude the crust to form alkalic complexes that contain rocks ranging from alkali granites to jacupirangite and carbonatite. At depth within the continent, the ambient temperatures and pressures are such that Greenschist, Amphibolite, and Granulite Facies metamorphism are active. In contrast, at the surface under approximately STP conditions, sedimentary rocks develop. These include various mudrocks, wackes, arenites, evaporites, conglomerates, and breccias that are generated in the diverse environments of the continental surface.

CONCLUSION

The rock assemblages that form at various sites are distinctive *petrotectonic assemblages*. We use them, especially in conjunction with associated structures, to interpret past petrologic, tectonic, and Earth history.

Petrology, the study of rocks, is thus, a central subdiscipline within geology. Its utility within a variety of other subfields, such as structural geology, stratigraphy, paleontology, and geophysics, will be evident to the student who pursues knowledge in those areas. The rocks preserve the record of the past and are recording the history of the future.

APPENDIX

A

MINERALS IN COMMON ROCKS

Table A.1 Minerals in Igneous Rocks

	Grantd	Rhylt	Tra/Sy	Pho/NeS	An/D	Bas/Gab	UM
Quartz	X	X	m	–	o	o	–
Feldspars							
K-rich alkali	X	X	+	X	o	–	–
Sanidine	–	+	+	–	–	–	–
Anorthoclase	–	o	+	+	–	–	–
Na-plagioclase	X	+	+	+	X	o	–
Ca-plagioclase	X	–	–	–	+	X	m
Feldspathoids							
Nepheline	–	–	o	X	–	o	o
Leucite	–	–	–	o	–	o	–
Sodalite Group	–	–	–	+	–	–	–
Analcite	–	–	–	o	–	o	–
Zeolites	–	–	–	–	–	o	–
Olivines	o	–	o	o	o	+	X
Pyroxenes							
Orthopyroxene	o	–	o	–	o	+	X
Augite	o	o	o	–	+	X	+
Pigeonite	–	–	–	–	+	+	+
Diopside	–	–	–	–	+	+	+
Na-pyroxene	o	o	+	+	–	–	–

	Grantd	Rhylt	Tra/Sy	Pho/NeS	An/D	Bas/Gab	UM
Amphiboles							
Hornblende	+	o	+	–	+	+	o
Na-amphibole	o	o	o	+	–	–	o
Micas							
Muscovite	+	–	–	–	–	–	–
Biotite	+	+	+	+	+	+	–
Phlogopite	–	–	–	o	–	o	m
Zircon	m	m	m	m	m	m	–
Sphene	m	m	m	m	m	m	o
Garnets							
Pyralspites (Fe-Al)	o	o	–	–	–	–	o
Ugrandites (Ca)	–	–	–	o	–	o	–
Sil/Ky/Andalusite	o	–	–	–	–	–	–
Epidote Group	m	o	–	–	–	–	–
Cordierite	o	–	–	–	o	–	–
Tourmaline	o	–	–	–	–	–	–
Chlorites (sec.)	m	o	o	o	o	o	o
Clay minerals (sec.)	m	m	m	m	m	m	m
Carbonate Minerals							
Calcite (sec.)	o	o	o	o	o	o	o
Aragonite (sec.)	–	–	–	–	–	o	o
Magnetite	m	m	m	m	m	+	o
Ilmenite	m	–	m	–	o	m	o
Chromite	–	–	–	–	–	o	+
Spinel	–	–	–	–	–	o	+
Apatite	m	m	m	m	m	m	m
Pyrite (sec.)	o	o	o	o	o	o	o
Other sulfides	o	–	–	–	o	+	o

Grantd = granitoid rocks; Rhylt = rhyolite and other siliceous volcanic rocks; Tra/Sy = trachyte and syenite; Pho/NeS = phonolite, nepheline syenite and related rocks; An/D = andesite and diorite; Bas/Gab = basalt and gabbro; UM = ultramafic rocks; sec = secondary minerals.
X = abundant, key minerals; + = common minerals; m = minor minerals; o = occasionally present or uncommon minerals; – = generally absent.

Table A.2 Minerals in Sedimentary Rocks

	Mudrx	Wacke	Arenite	Ls	Dolost	Other
Quartz	m	X	X	m	m	m→X
Feldspars						
K-rich alkali	o	+	+	o	o	o
Na-plagioclase	o	X	+	o	o	o
Ca-plagioclase	–	o	o	–	–	–
Zeolites	o	–	–	–	–	o
Olivines	–	–	–	–	–	–
Pyroxenes	–	o	o	–	–	–
Amphiboles	–	o	o	–	–	o
Clay minerals	X	+	m	–	–	o
Micas						
White micas	m	m	m	–	–	o
Glauconite	o	o	–	o	–	o
Biotite	o	m	o	–	–	–
Chlorites	X	m	o	–	–	o
Zircon	–	m	m	–	–	–
Sphene	–	o	o	–	–	–
Garnets	–	o	o	–	–	–
Tourmaline	–	o	o	–	–	–
Staurolite	–	o	o	–	–	–
Aluminosilicates	–	o	o	–	–	–
Epidote Group	–	o	o	–	–	–
Carbonate Minerals						
Calcite	o	o	m	X	o	o
Aragonite	–	–	–	o	–	o
Dolomite	o	–	–	o	X	o
Magnetite	–	m	m	–	–	o→X
Hematite	o	o	o	o	o	o→X
Ilmenite	–	m	m	–	–	–
Apatite	–	m	m	–	–	–→X
Gypsum	o	–	–	o	–	–→X
Pyrite	o	–	–	–	–	–→o
Other sulfides	o	–	–	o	o	o

Mudrx = mudrocks; Wacke = wacke and polymict lithic conglomerates; Arenite = arenites, cherts, and quartz conglomerates; Ls = limestones; Dolost = dolostones; Other = other rocks, particularly precipitates.
X = abundant, key minerals; + = common minerals; m = minor minerals; o = occasionally present or uncommon minerals; – = generally absent.

Table A.3 Minerals in Metamorphic Rocks

	S	SAC	A	C	MB	UB
Quartz	X	X	X	m	m	–
Feldspars						
K-rich alkali	o	+	+	o	o (LTP)	–
Na-plagioclase	o	+	+	–	X	–
Ca-plagioclase	–	+	+	o	X	o
Analcite	–	+ (LTP)	–	–	+ (LTP)	–
Feldspathoids	–	o	–	–	–	–
Zeolites	–	X (LTP)	+ (LTP)	–	X (LTP)	–
Olivines	–	–	–	o	m (HT)	X (HT)
Pyroxenes						
Orthopyroxenes	–	m (HT)	m (HT)	m (HT)	+ (HT)	+ (HT)
Diopside	–	–	–	+	o	o
Augite	–	–	–	–	o (HT)	–
Na-pyroxenes	m (HP)	X (HP)	–	–	+ (HP)	o (HP)
Wollastonite	–	–	–	+	–	–
Amphiboles						
Hornblende	–	m (HT)	m (HT)	o (HT)	X	o
Actinolite	–	o	–	o	X	+
Tremolite	–	–	–	+	–	+
Orthoamphiboles	–	o	–	–	o	+
Na-amphiboles	m (HP)	m (HP)	X (HP)	–	X (HP)	–
Micas						
White micas	m	X	X	o	m	–
Biotite	o	+	X	–	+	–
Phlogopite	–	–	–	m	o	m
Stilpnomelane	m	m	m	–	m	–
Clay minerals	o (LTP)	+ (LTP)	X (LTP)	m (LTP)	o (LTP)	–
Prehnite	–	m (LTP)	o (LTP)	o (LTP)	X (LTP)	–
Chlorites	o	+	X	o	X	+
Serpentines	–	–	–	o	o	X
Talc	–	–	–	o	o	X

Table A.3 Continued

	S	SAC	A	C	MB	UB
ZZircon	m	m	m	–	–	–
Sphene	–	m	o	o	m	–
Garnets	+	+	X	o	o	o
Tourmaline	–	o	o	–	–	–
Staurolite	–	–	X	–	o	–
Chloritoid	–	–	m	–	–	–
Aluminosilicates						
Andalusite	o (HT)	–	+ (HT)	–	–	–
Kyanite	+	o	+	–	–	–
Sillimanite	+ (HT)	o (HT)	+ HT	–	–	–
Epidotes	m	m	m	o	+	–
Lawsonite	–	+ (HP)	m (HP)	–	+ (HP)	–
Pumpellyite	–	+	o	–	+	–
Cordierite	+ (HT)	o (HT)	+ (HT)	–	–	–
Carbonate Minerals						
Calcite	o	o	o	X	o	o
Aragonite	o (HP)	+ (HP)	o (HP)	X (HP)	o (HP)	o
Dolomite	–	–	–	X	–	o
Magnetite	m	m	m	o	m	m
Ilmenite	o	–	m	–	–	–
Chromite	–	–	–	–	–	m
Spinel	–	–	o	o	–	o
Apatite	o	m	m	o	m	–
Sulfides	o	o	o	m	m	o

S = siliceous rocks; SAC = siliceous-alkali-calcic rocks; A = aluminous (including pelitic) rocks; C = carbonate rocks; MB = metabasites; UB = ultrabasic rocks.
X = abundant, key minerals; + = common minerals; m = minor minerals; o = occasionally present or uncommon minerals; – = generally absent.
HT = in high-temperature rocks only; HP = in high-pressure rocks only; LPT = in low-temperature and low-pressure rocks only.

APPENDIX

B

CIPW NORMS

A rock norm is a list of standard minerals and their percentages calculated from the chemical analysis of a rock. The CIPW method of calculating a norm, developed by Cross, Iddings, Pirsson, and Washington (1903), has been described and modified by a number of petrologists (see Johannsen, 1931, p. 88ff.; Barth, 1962; Bickel, 1979). The procedure described here generally follows the original method, but is modified and slightly abbreviated in part, following Johannsen (1931) and Bickel (1979). The chemical analysis used in the norm calculation must be in weight percent oxides (e.g., weight percent SiO_2;

(a)

Step		SiO_2	TiO_2	Al_2O_3	Fe_2O_3	FeO	MnO	MgO	CaO	Na_2O	K_2O	P_2O_5	CO_2	Other Oxides			H_2O / %Min.
A	Wt.%																
	Mol. Wt.	60.09	79.90	101.96	159.69	71.85	70.94	40.31	56.08	61.98	94.20	141.95	44.01				
B	Moles																
C	Comb. M.						↵										
D	Mineral =																
	=																
	=																
	=																
	=																
	=																
	=																
E	=																
G	=																
	=																
	=																
L							Total Wt. % Oxides =				$\cong$Total Wt. % Minerals + Wt. % H_2O =						

Figure B.1

Charts for calculating CIPW norms. (a) Main chart for calculations (modified from and courtesy of Charles E. Bickel). (b) Chart for calculating diopside, hypersthene, and olivine. (c) Chart for calculating normative mineral percentages. See text for explanation of steps A–L.

(b)

Step	
F	$PRF = \dfrac{\text{Moles FeO}}{\text{Moles (FeO + MgO)}} =$ _____ $\qquad PRM = \dfrac{\text{Moles MgO}}{\text{Moles (FeO + MgO)}} =$ _____
I	Wt. % Wo in Diopside = (Mol. CaO in Diop.)(Mol. Wt. Wo) $\qquad$ = ($\quad$)(116.17) = Wo = _____ Wt. % En in Diopside = [Mol. (Fe, Mg)O in Diop.] (PRM) (Mol. Wt. En) $\qquad$ = [$\quad$] ($\quad$) (100.40) = En = _____ Wt. % Fs in Diopside = [Mol. (Fe, Mg)O in Diop.] (PRF) (Mol. Wt. Fs) $\qquad$ = [$\quad$] ($\quad$) (131.94) = Fs = _____ Total Diopside = Wo + En + Fs =
J	Wt. % En in Hypersthene = [(Fe, Mg)O in Hyp] (PRM) (Mol. Wt. En) $\qquad$ = [$\quad$] ($\quad$) (100.40) = En = _____ Wt. % Fs in Hypersthene = [(Fe, Mg)O in Hyp] (PRF) (Mol. Wt. Fs) $\qquad$ = [$\quad$] ($\quad$) (131.94) = Fs = _____ Total Hypersthene = En + Fs =
K	Wt. % Fo in Olivine = [(Fe, Mg)O in Ol] (PRM) (Mol. Wt. Fo/2) $\qquad$ = [$\quad$] ($\quad$) (70.355) = Fo = _____ Wt. % Fa in Olivine = [(Fe, Mg)O in Ol] (PRF) (Mol. Wt. Fa/2) $\qquad$ = [$\quad$] ($\quad$) (101.895) = Fa = _____ Total Olivine = Fo + Fa =

Figure B.1

Charts for calculating CIPW norms. (a) Main chart for calculations (modified from and courtesy of Charles E. Bickel). (b) Chart for calculating diopside, hypersthene, and olivine. (c) Chart for calculating normative mineral percentages. See text for explanation of steps A–L.

weight percent CaO, except for Cl, F, and S). Figure B.1 provides blank forms that may be copied for use in calculating a norm and figure B.2 shows a simple example. Standard minerals, their symbols, and their chemistries are shown in figure B.1c, a chart for calculating normative mineral percentages. The norm is calculated following the list of steps (A–L) listed here.

STEP A. List the values of the oxides from the chemical analysis in the column under the oxide heading. For minor element oxides, use the blanks at the end of the row.

STEP B. Calculate the number of moles of each oxide by dividing the weight percent of that oxide by the molecular weight of the oxide. Enter the value in the row marked B ("Moles").

STEP C. Combine certain molecular values of oxides that typically substitute for major oxides with those major oxide molecular values. Moles of MnO and NiO are added to the value of FeO, and are considered as FeO throughout the calculation. Moles of BaO and SrO are added to CaO. Moles of

Cr_2O_3 are added to moles of Fe_2O_3. In cases where these elements are abundant and may represent unusual minerals in a rock, they may be kept separate and used to calculate special mineral molecules (e.g., Cr_2O_3 may be used to calculate chromite in ultramafic rocks).

STEP D. Calculate the amounts of moles used for each standard mineral. Throughout this part of the calculation, the word "amount" refers to the number of moles. In general, two or three oxides are combined for the formation of a mineral. Each time a mineral is formed, one of the oxides will be completely used up. The amounts of the other oxides used are subtracted from the amounts available. Figure B.1a contains blanks for both the amount used (preceded by a negative sign) and the amount remaining (preceded by an equals sign). The amount remaining will determine, in many cases, which of two calculations will be completed for the formation of the next mineral. These alternative calculations are labelled a and b. *Never* do both calculations; always do one or the other. Complete the numbered steps in order. If an oxide is absent or has been used up, skip to the next successive instruction.

(c)

Step H - Calculate Wt. % normative mineral						
Mineral	Symbol	Formula	Reference oxide	$\left(\begin{array}{c}\text{Formula}\\\text{weight}\end{array}\right)$ X	$\left(\begin{array}{c}\text{Moles ref.}\\\text{oxide}\end{array}\right)$ =	Wt. % mineral
Quartz	Q	SiO_2	SiO_2	60.09		
Orthoclase	or	$K_2O \cdot Al_2O_3 \cdot 6SiO_2$	K_2O	556.70		
Albite	ab	$Na_2O \cdot Al_2O_3 \cdot 6\ SiO_2$	Na_2O	524.48		
Anorthite	an	$CaO \cdot Al_2O_3 \cdot 2\ SiO_2$	CaO	278.22		
Leucite	lc	$K_2O \cdot Al_2O_3 \cdot 4\ SiO_2$	K_2O	436.52		
Nepheline	ne	$Na_2O \cdot Al_2O_3 \cdot 2\ SiO_2$	Na_2O	284.12		
Kaliophilite	kp	$K_2O \cdot Al_2O_3 \cdot 2\ SiO_2$	K_2O	316.34		
Na-Metasilicate	ns	$Na_2O \cdot SiO_2$	Na_2O	122.07		
K-Metasilicate	ks	$K_2O \cdot SiO_2$	K_2O	154.29		
Acmite	ac	$Na_2O \cdot Fe_2O_3 \cdot 4\ SiO_2$	Na_2O	462.03		
Diopside	di	⟶ Obtain from Fig. B.1 (b) [Step I] =				
Hypersthene	hy	⟶ Obtain from Fig. B.1 (b) [Step J] =				
Olivine	ol	⟶ Obtain from Fig. B.1 (b) [Step K] =				
Wollastonite	wo	$CaO \cdot SiO_2$	CaO	116.17		
Larnite	la	$2\ CaO \cdot SiO_2$	SiO_2	172.25		
Sphene	sp	$CaO \cdot TiO_2 \cdot SiO_2$	CaO	196.07		
Perofskite	pf	$CaO \cdot TiO_2$	CaO	135.98		
Rutile	ru	TiO_2	TiO_2	79.90		
Ilmenite	il	$FeO \cdot TiO_2$	FeO	151.75		
Magnetite	mt	$FeO \cdot Fe_2O_3$	FeO	231.54		
Hematite	hm	Fe_2O_3	Fe_2O_3	159.69		
Chromite	cm	$FeO \cdot Cr_2O_3$	FeO	223.84		
Pyrite	pr	FeS_2	FeO	120.01		
Fluorite	fr	CaF_2	CaO	78.08		
Calcite	cc	$CaO \cdot CO_2$	CaO	100.09		
Halite	hl	$NaCl$	Na_2O	58.44		
Na-Carbonite	nc	$Na_2O \cdot CO_2$	Na_2O	105.99		
Thenardite	th	$Na_2O \cdot SO_3$	Na_2O	142.04		
Apatite	ap	$3\ (3\ CaO \cdot P_2O_5) \cdot CaF_2$	P_2O_5	336.22		
Corundum	c	Al_2O_3	Al_2O_3	101.96		
Zircon	z	$ZrO_2 \cdot SiO_2$	ZrO_2	183.31		

Figure B.1
Charts for calculating CIPW norms. (a) Main chart for calculations (modified from and courtesy of Charles E. Bickel). (b) Chart for calculating diopside, hypersthene, and olivine. (c) Chart for calculating normative mineral percentages. See text for explanation of steps A–L.

Do not skip steps unless the instructions expressly direct you to do so. Silica-poor rocks may result in negative values for SiO_2 early in the norm calculation. This is acceptable and will be corrected later in the calculation. *Do not* generate negative amounts for oxides other than SiO_2. Use four decimal places for your calculations, but use only the correct number of significant figures in reporting the final mineral percentages.

1. If Cr_2O_3 is present in more than trace amounts, allot all of the Cr_2O_3 and an equal amount of FeO to the formation of chromite. If Cr_2O_3 is absent or present only in trace amounts, skip to step 2 (next).
2. If TiO_2 is not present, skip to step 3. If TiO_2 exceeds the amount of FeO, do calculation 2a. If the amount of FeO exceeds that of TiO_2, do calculation 2b.
   2a. Allot all of the FeO and an equal amount of TiO_2 to the formation of ilmenite.
   2b. Allot all of the TiO_2 and an equal amount of FeO to the formation of ilmenite.

3. Allot all of the P_2O_5 and 3.333 times this amount of CaO to the formation of apatite. If F or Cl are present, use 3.000 times the amount of P_2O_5 and 0.333 times the amount of F plus Cl. If P_2O_5 is absent, skip to step 4.
4. If F is present, use the amount of F and one-half this amount of CaO to form fluorite. If F is absent or has been used up, skip to step 5.
5. If Cl is present or remains after making apatite, combine the amount remaining with an equal amount of Na_2O to form halite. If Cl is absent or used up, skip to step 6.
6. If SO_2 is present, allot all of it and an equal amount of Na_2O to thenardite. If SO_2 is absent, skip to step 7.
7. If S is present, allot all the S and one-half this amount of FeO to make pyrite. If S is absent, skip to step 8.
8. If CO_2 is present, allot all the CO_2 and an equal amount of CaO to calcite, or, if the rock is silica-undersaturated, allot the CO_2 and an equal amount of Na_2O to sodium carbonate. If CO_2 is absent, skip to step 9.

(a)

Step		SiO$_2$	TiO$_2$	Al$_2$O$_3$	Fe$_2$O$_3$	FeO	MnO	MgO	CaO	Na$_2$O	K$_2$O	P$_2$O$_5$	CO$_2$	Other Oxides		H$_2$O %Min.
A	Wt.%	72.06	0.28	12.38	1.00	1.44	0.04	0.35	0.96	2.88	5.74	0.04	0.01			2.68
	Mol. Wt.	60.09	79.90	101.96	159.69	71.85	70.94	40.31	56.08	61.98	94.20	141.95	44.01			—
B	Moles	1.1992	.0035	.1214	.0063	.0200	.0006	.0087	.0171	.0465	.0609	.0003	.0002			—
C	Comb. M.	—	—	—	—	.0206 ↵		—	—	—	—	—	—			—
D	Mineral — il =		.0035 / -0-			.0035 / .0171										0.53
	ap =								.0010 / .0161			.0003 / -0-				0.1
	cc =								.0002 / .0159				.0002 / -0-			0.02
	or =	.3654 / .8338		.0609 / .0605							.0609 / -0-					33.90
	ab =	.2790 / .5548		.0465 / .0140						.0465 / -0-						24.39
	an =	.0280 / .5268		.0140 / -0-					.0140 / .0019							3.90
	mt =				.0063 / -0-	.0063 / .0108										1.5
E	✕ =					(Fe,Mg)O .0195 ↵										——
G	di =	.0038 / .5230						.0019 / .0176	.0019 / -0-							0.44
	hy =	.0176 / .5054						.0176 / -0-								2.08
	Q =	.5054 / -0-														30.37
L								Total Wt. % Oxides = 99.86				≅Total Wt. % Minerals + Wt. % H$_2$O =				99.9

Figure B.2

Example of a simple norm calculation. (a), (b), and (c) as in figure B.1, but with illustrative numbers filled in appropriate blanks.

9. If ZrO$_2$ is present, allot all of it and an equal amount of SiO$_2$ to zircon. If ZrO$_2$ is absent, skip to step 10.

10. If the amount of K$_2$O exceeds the amount of Al$_2$O$_3$, do calculation 10a. If the amount of Al$_2$O$_3$ exceeds the amount of K$_2$O, do calculation 10b.

   10a. Allot all of the Al$_2$O$_3$, an equal amount of K$_2$O, and six times this amount of SiO$_2$ to provisional orthoclase. (The orthoclase is provisional until it is determined that the rock is not silica-deficient. In the latter case, leucite rather than orthoclase may need to be created.)

   10b. Allot all of the K$_2$O, an equal amount of Al$_2$O$_3$, and six times this amount of SiO$_2$ to the formation of provisional orthoclase. (The orthoclase is provisional until it is determined that the rock is not silica-deficient. In the latter case, leucite rather than orthoclase may need to be created.) If K$_2$O is absent, skip to step 12.

11. Use any remaining K$_2$O and an equal amount of SiO$_2$ to form potassium metasilicate. If potash is absent or if no potash remains, skip to step 12.

12. If the amount of Al$_2$O$_3$ remaining exceeds the amount of Na$_2$O, do calculation 12a. If the amount of Na$_2$O exceeds the amount of Al$_2$O$_3$, do calculation 12b.

   12a. Allot all the Na$_2$O, an equal amount of Al$_2$O$_3$, and six times this amount of SiO$_2$ to the formation of provisional albite.

   12b. Allot all of the Al$_2$O$_3$, an equal amount of Na$_2$O, and six times this amount of SiO$_2$ to the formation of provisional albite. Skip to step 15.

13. If the amount of Al$_2$O$_3$ remaining exceeds the amount of CaO remaining, do calculation 13a. If the amount of CaO remaining exceeds the amount of Al$_2$O$_3$ remaining, do calculation 13b.

   13a. Allot all of the CaO, an equal amount of Al$_2$O$_3$, and twice this amount of SiO$_2$ to the formation of anorthite.

(b)

Step	
F	$PRF = \dfrac{\overset{.0108}{\text{Moles FeO}}}{\underset{.0195}{\text{Moles (FeO + MgO)}}} = 0.55$ $\quad PRM = \dfrac{\overset{.0087}{\text{Moles MgO}}}{\underset{.0195}{\text{Moles (FeO + MgO)}}} = 0.45$
I	Wt. % Wo in Diopside = (Mol. CaO in Diop.)(Mol. Wt. Wo) $= (.0019)(116.17) = Wo = \underline{\ 0.22\ }$ Wt. % En in Diopside = [Mol. (Fe, Mg)O in Diop.] (PRM) (Mol. Wt. En) $= [.0019] (0.45) (100.40) = En = \underline{\ 0.08\ }$ Wt. % Fs in Diopside = [Mol. (Fe, Mg)O in Diop.] (PRF) (Mol. Wt. Fs) $= [.0019] (0.55) (131.94) = Fs = \underline{\ 0.14\ }$ Total Diopside = Wo + En + Fs = $\boxed{0.44}$
J	Wt. % En in Hypersthene = [(Fe, Mg)O in Hyp] (PRM) (Mol. Wt. En) $= [.0176] (0.45) (100.40) = En = \underline{\ 0.80\ }$ Wt. % Fs in Hypersthene = [(Fe, Mg)O in Hyp] (PRF) (Mol. Wt. Fs) $= [.0176] (0.55) (131.94) = Fs = \underline{\ 1.28\ }$ Total Hypersthene = En + Fs = $\boxed{2.08}$
K	Wt. % Fo in Olivine = [(Fe, Mg)O in Ol] (PRM) (Mol. Wt. Fo/2) $= [\quad] (\quad) (70.355) = Fo = \underline{\quad\quad}$ Wt. % Fa in Olivine = [(Fe, Mg)O in Ol] (PRF) (Mol. Wt. Fa/2) $= [\quad] (\quad) (101.895) = Fa = \underline{\quad\quad}$ Total Olivine = Fo + Fa = $\boxed{}$

Figure B.2

Example of a simple norm calculation. (a), (b), and (c) as in figure B.1, but with illustrative numbers filled in appropriate blanks.

13b. Allot all of the Al_2O_3, an equal amount of CaO, and twice this amount of SiO_2 to the formation of anorthite. Skip to step 15.

14. If Al_2O_3 remains, the Al_2O_3 is assigned to corundum. If none remains, skip to step 15.

15. If TiO_2 is absent or has been used up, skip to step 17. If TiO_2 remains and the amount exceeds that of CaO, do calculation 15a. If TiO_2 remains, but the amount of remaining CaO exceeds the amount of TiO_2, do calculation 15b.

   15a. Allot all of the CaO, an equal amount of TiO_2, and an equal amount of SiO_2 to the formation of provisional sphene.

   15b. Allot all of the TiO_2, an equal amount of CaO, and an equal amount of SiO_2 to the formation of sphene.

16. If TiO_2 remains, it is assigned to rutile. If no TiO_2 remains, skip to step 17.

17. If all of the Na_2O is used up, skip to step 18. If Na_2O remains and the amount exceeds the amount of Fe_2O_3, do calculation 17a. If the amount of Na_2O is less than the amount of Fe_2O_3, do calculation 17b.

17a. Allot all of the Fe_2O_3, an equal amount of Na_2O, and four times this amount of SiO_2 to the formation of acmite.

17b. Allot all of the Na_2O, an equal amount of Fe_2O_3, and four times this amount of SiO_2 to the formation of acmite. Skip to step 19.

18. If Na_2O remains, assign the remaining amount and an equal amount of SiO_2 to the formation of sodium metasilicate. If no soda remains, skip to step 19.

19. Compare the amounts of remaining FeO and Fe_2O_3. If the amount of FeO exceeds the amount of Fe_2O_3, do calculation 19a. If the amount of FeO is less than the amount of Fe_2O_3, do calculation 19b. If no Fe_2O_3 remains, skip to Step E.

   19a. Use all of the remaining Fe_2O_3 and an equal amount of FeO to make magnetite.

   19b. Use all of the remaining FeO and an equal amount of Fe_2O_3 to make magnetite.

20. Assign any remaining Fe_2O_3 to hematite. If no Fe_2O_3 remains, skip to Step E.

(c)

Mineral	Symbol	Formula	Reference oxide	Formula weight	X	Moles ref. oxide	=	Wt. % mineral
Quartz	Q	SiO_2	SiO_2	60.09		.5054		30.37
Orthoclase	or	$K_2O \cdot Al_2O_3 \cdot 6SiO_2$	K_2O	556.70		.0609		33.90
Albite	ab	$Na_2O \cdot Al_2O_3 \cdot 6\ SiO_2$	Na_2O	524.48		.0465		24.39
Anorthite	an	$CaO \cdot Al_2O_3 \cdot 2\ SiO_2$	CaO	278.22		.0140		3.90
Leucite	lc	$K_2O \cdot Al_2O_3 \cdot 4\ SiO_2$	K_2O	436.52				
Nepheline	ne	$Na_2O \cdot Al_2O_3 \cdot 2\ SiO_2$	Na_2O	284.12				
Kaliophilite	kp	$K_2O \cdot Al_2O_3 \cdot 2\ SiO_2$	K_2O	316.34				
Na-Metasilicate	ns	$Na_2O \cdot SiO_2$	Na_2O	122.07				
K-Metasilicate	ks	$K_2O \cdot SiO_2$	K_2O	154.29				
Acmite	ac	$Na_2O \cdot Fe_2O_3 \cdot 4\ SiO_2$	Na_2O	462.03				
Diopside	di	⟶ Obtain from Fig. B.1 (b) [Step I] =						0.44
Hypersthene	hy	⟶ Obtain from Fig. B.1 (b) [Step J] =						2.08
Olivine	ol	⟶ Obtain from Fig. B.1 (b) [Step K] =						
Wollastonite	wo	$CaO \cdot SiO_2$	CaO	116.17				
Larnite	la	$2\ CaO \cdot SiO_2$	SiO_2	172.25				
Sphene	sp	$CaO \cdot TiO_2 \cdot SiO_2$	CaO	196.07				
Perofskite	pf	$CaO \cdot TiO_2$	CaO	135.98				
Rutile	ru	TiO_2	TiO_2	79.90				
Ilmenite	il	$FeO \cdot TiO_2$	FeO	151.75		.0035		0.53
Magnetite	mt	$FeO \cdot Fe_2O_3$	FeO	231.54		.0063		1.5
Hematite	hm	Fe_2O_3	Fe_2O_3	159.69				
Chromite	cm	$FeO \cdot Cr_2O_3$	FeO	223.84				
Pyrite	pr	FeS_2	FeO	120.01				
Fluorite	fr	CaF_2	CaO	78.08				
Calcite	cc	$CaO \cdot CO_2$	CaO	100.09		.0002		0.02
Halite	hl	$NaCl$	Na_2O	58.44				
Na-Carbonite	nc	$Na_2O \cdot CO_2$	Na_2O	105.99				
Thenardite	th	$Na_2O \cdot SO_3$	Na_2O	142.04				
Apatite	ap	$3\ (3\ CaO \cdot P_2O_5) \cdot CaF_2$	P_2O_5	336.22		.0003		0.1
Corundum	c	Al_2O_3	Al_2O_3	101.96				
Zircon	z	$ZrO_2 \cdot SiO_2$	ZrO_2	183.31				

Step H - Calculate Wt. % normative mineral

Figure B.2
Example of a simple norm calculation. (a), (b), and (c) as in figure B.1, but with illustrative numbers filled in appropriate blanks.

STEP E. Combine FeO and MgO. At this point, all of the remaining amounts of FeO and MgO are added together and are called (Fe,Mg) O.

STEP F. Determine the porportions of each of the two oxides in (Fe,Mg)O. The proportion of each of the oxides in the combined oxide (Fe,Mg)O is calculated by dividing the amount of each oxide by the sum of the two oxides (figure B.1. The proportion of FeO is called PRF and that of MgO is called PRM. PRF + PRM = 1.0000.

STEP G. Continue calculating mineral percentages.
21. Compare the amounts of (Fe,Mg)O and CaO. If the amount of (Fe,Mg)O exceeds the amount of CaO, do calculation 21a. If the amount of (Fe,Mg)O is less than the amount of CaO, do calculation 21b. If no CaO remains, skip to step 22.
   21a. Allot all of the CaO, an equal amount of (Fe,Mg)O, and twice this amount of SiO_2 to the formation of diopside.

21b. Allot all of the (Fe,Mg)O, an equal amount of CaO, and twice this amount of SiO_2 to the formation of diopside. Skip to step 23.
22. Assign the remaining (Fe,Mg)O and an equal amount of SiO_2 to the formation of provisional hypersthene. If no (Fe,Mg)O remains, skip to step 23.
23. Allot any remaining CaO and an equal amount of SiO_2 to provisional wollastonite.
24. If silica remains at this point (i.e., there is a positive value for moles of SiO_2), it is assigned to quartz. If SiO_2 has a negative value, skip to step 26.
25. If quartz is created in step 24, skip to Step H.
26. If SiO_2 has a negative value, it will be necessary to break down one or more of the provisional minerals to make minerals that contain less silica (the negative value must be eliminated). Provisional hypersthene will be the first mineral to be broken down. If no hypersthene was created in step 22, skip to step 28. If hypersthene was created in step 22, move on to step 27.

27. Break down the hypersthene, adding the SiO_2 back into the SiO_2 column and the $(Fe,Mg)O$ back into the $(Fe,Mg)O$ column. If the amount of SiO_2 available now is a positive amount and that amount exceeds or equals one-half the amount of $(Fe,Mg)O$, do calculation 27a. If the amount of silica is still negative or is less than half of the value of $(Fe,Mg)O$, do calculation 27b.

27a. Olivine and hypersthene will be created using the available SiO_2 and $(Fe,Mg)O$. Use the following equations in which x is the amount of $(Fe,Mg)O$ allotted to hypersthene, y is the amount of $(Fe,Mg)O$ allotted to olivine, M is the amount of available $(Fe,Mg)O$, and S is the amount of available SiO_2.

$$x = 2S - M \qquad\qquad y = M - x$$

The amount of SiO_2 allotted to hypersthene is equal to x. The amount of SiO_2 allotted to olivine is $1/2(y)$. Go to Step H.

27b. Create olivine using all of the available $(Fe,Mg)O$ and one-half this amount of SiO_2.

28. If the amount of SiO_2 is still negative, break down the provisional sphene. Add the amount of SiO_2 back into the silica column. The remaining CaO and TiO_2 becomes (is assigned to) perofskite. If no sphene was created in step 15, skip to step 29.

29. Break down the provisional albite. Add the amounts of SiO_2, Na_2O and Al_2O_3 back into their respective columns. If the amount of available silica is positive and exceeds or equals twice the amount of Na_2O, do calculation 29a. If the amount of SiO_2 is still negative or is less than twice the amount of Na_2O, do calculation 29b.

29a. Create nepheline and albite using the equations below, in which x is the amount of Na_2O allotted to albite, y is the amount of Na_2O allotted to nepheline, N is the amount of available Na_2O, and S is the amount of available SiO_2.

$$x = (S - 2N)/4 \qquad\qquad y = N - x$$

Albite is composed of x, an equal amount of Al_2O_3, and $6x$ SiO_2. Nepheline is composed of y, an equal amount of Al_2O_3, and $2y$ SiO_2. Go to Step H.

29b. Create nepheline by combining all of the remaining Na_2O, an equal amount of Al_2O_3, and twice the amount of SiO_2.

30. If there is still a negative amount of SiO_2, break down the provisional orthoclase. Add the amounts of SiO_2 back into the SiO_2 column and the Al_2O_3 and K_2O into their respective columns. If the amount of available SiO_2 is now positive and exceeds or equals four times the amount of K_2O, do calculation 30a. If the amount of SiO_2 is still negative or is less than four times the amount of K_2O, do calculation 30b.

30a. Create leucite and orthoclase using the equation below, in which x is the amount of K_2O used to form orthoclase, y is the amount of K_2O used to form leucite, K is the amount of available K_2O, and S is the amount of available SiO_2.

$$x = (S - 4K)/2 \qquad\qquad y = K - x$$

Orthoclase consists of x, an equal amount of Al_2O_3, and $6x$ SiO_2. Leucite consists of y, an equal amount of Al_2O_3, and $4y$ SiO_2. Go to Step H.

30b. Form leucite from all of the K_2O, an equal amount of Al_2O_3, and four times that amount of SiO_2.

31. If there is still a negative amount of SiO_2, additional minerals must be broken down. If wollastonite was produced in step 23, break down the wollastonite, adding the silica thus obtained to the silica column and the lime to the CaO column. If silica is now positive and equal to one-half or more of the CaO, do calculation 31a. If the amount of SiO_2 is still negative or is less than one-half the amount of CaO, do calculation 31b. If wollastonite was not created in step 23, skip to step 32.

31a. Create larnite and wollastonite using the equations below, in which x is the amount of CaO allotted to wollastonite, y is the amount of CaO allotted to larnite, C is the amount of available CaO, and S is the amount of available SiO_2.

$$x = 2S - C \qquad\qquad y = C - x$$

Wollastonite consists of x and an equal amount of SiO_2. Larnite consists of y and one-half this amount of SiO_2. Go to Step H.

31b. Larnite is formed from all the CaO and one-half that amount of SiO_2.

32. If the amount of SiO_2 is still negative, break down all the provisional diopside. The silica is added to the silica column and the CaO and $(Fe,Mg)O$ are added to their respective columns. If the amount of available silica is positive and equal to more than the available amount of CaO, do calculation 32a. If the amount of silica is still negative or is less than the amount of CaO, do calculation 32b.

32a. Form olivine, diopside, and larnite using the equations below, in which x is the amount of CaO allotted to diopside, y is the amount of CaO allotted to larnite, C is the amount of available CaO, and S is the amount of available SiO_2.

$$x = S - C \qquad\qquad y = C - x$$

695

Diopside consists of x, an equal amount of $(Fe,Mg)O$, and $2x$ SiO_2. Larnite consists of y and one-half this amount of SiO_2. Olivine consists of an amount of $(Fe,Mg)O$ equal to y and $1/2$ that amount of SiO_2. Add the larnite to any larnite produced in step 21. Add the olivine to any olivine produced in step 27. Go to Step H.

32b. Form larnite and olivine. Allot to larnite all of the CaO and one-half that amount of SiO_2. Allot to olivine all the $(Fe,Mg)O$ and one-half that amount of SiO_2. Add the larnite and olivine to any larnite and olivine formed previously.

33. If the amount of SiO_2 is still negative, break down the leucite. Add the SiO_2, K_2O, and Al_2O_3 to their respective columns. Form kaliophilite and leucite using the equations below, in which x is the amount of K_2O allotted to leucite, y is the amount of K_2O allotted to kaliophilite, K is the amount of available K_2O, and S is the amount of available SiO_2.

$$x = (S - 2K)/2 \qquad y = (4K - S)/2$$

Leucite consists of x, an equal amount of Al_2O_3, and $4x$ SiO_2. Kaliophilite consists of y, an equal amount of Al_2O_3, and $2x$ SiO_2.

STEP H. Calculate the normative weight percentages of each of the minerals in the norm. Multiply the formula weight of each mineral times the number of moles of the reference oxide for that mineral (figure B.1c). The weight percentages of diopside, hypersthene, and olivine cannot be calculated in this way because these minerals contain both FeO and MgO. Compute the weight percentages of these minerals following Steps I, J, and K, in which the wollastonite (wo), enstatite (en), ferrosilite (fs), forsterite (fo), and fayalite (fa) components of the respective minerals are calculated using PRF and PRM determined in Step F.

STEP I. Calculate the weight percent of diopside using the formulae in figure B.1b and summing wo + en + fs.
Wt. % wo in di = [amount CaO in di](mol. wt. wo)
Wt. % en in di = [amount $(Fe,Mg)O$ in di](PRM)(mol. wt. en)
Wt. % fs in di = [amount $(Fe,Mg)O$ in di](PRF)(mol. wt. fs)

STEP J. Calculate the weight percent of hypersthene using the formulae in figure B.1b and summing the en + fs.
Wt. % en in hy = [amount $(Fe,Mg)O$ in hy](PRM)(mol. wt. en)
Wt. % fs in hy = [amount $(Fe,Mg)O$ in hy](PRF)(mol. wt. fs)

STEP K. Calculate the weight percent olivine using the formulae in figure B.1b and summing fa + fo.
Wt. % fo in ol = [amount of $(Fe,Mg)O$ in ol](PRM)(mol. wt. fo/2)
Wt. % fa in ol = [amount of $(Fe,Mg)O$ in ol](PRF)(mol. wt. fa/2)

STEP L. List the mineral weight percentages opposite the mineral symbol in the column marked L **and sum the minerals plus the water.** Use values with correct significant figures for the mineral weight percents. The total should be equal to the total of the original oxides within a few hundredths of a percent (rounding error). If the two totals do not match, you have made an error in the calculation.

PETROGENETIC GRIDS FOR METAMORPHIC ROCKS

Petrogenetic grids are pressure-temperature graphs that show reaction curves along which one mineral or mineral assemblage changes to another. Curves are determined experimentally or by calculation. Reactions for all curves are listed below (tables C.2, C.3, C.4, and C.5) and the mineral abbreviations are listed in table C.1. The equations are not balanced. These grids allow us to estimate approximately the metamorphic conditions for particular mineral assemblages found in metamorphic rocks.

Table C.1 Mineral Abbreviations Used in Petrogenetic Grids and Equations

A	=	Antigorite	Ez	=	Zoisite	P	=	Plagioclase
Ab	=	Albite	F	=	Fluid (H$_2$O)	Pe	=	Periclase
Ac	=	Acmite	Fo	=	Forsterite	Ph	=	Phlogopite
Af	=	Alkali feldspar	G	=	Garnet	Pr	=	Prehnite
Ag	=	Augite	Gl	=	Glaucophane	Pu	=	Pumpellyite
Ak	=	Akermanite	H	=	Hornblende	Py	=	Pyrophyllite
Am	=	Amphibole	Hm	=	Hematite	Q	=	Quartz
At	=	Actinolite	Hu	=	Heulandite	R	=	Rankinite
An	=	Anorthite	Hy	=	Hypersthene	S	=	Sphene
And	=	Andalusite	I	=	Ilmenite	Sa	=	Serpentine (antigorite)
Anl	=	Analcite	Id	=	Idocrase	Sb	=	Stilbite
Ant	=	Anthophyllite	Jd	=	Jadeite	Sc	=	Serpentine (chrysotile)
Ar	=	Aragonite	Jdpx	=	Jadeitic pyroxene	Sd	=	Sanidine
B	=	Brucite	K	=	Kaolinite	Sil	=	Sillimanite
Bio	=	Biotite	Ky	=	Kyanite	Sl	=	Serpentine (lizardite)
C	=	Chrysotile	L	=	Liquid = melt	Sp	=	Spinel
Ca	=	Carpholite	La	=	Larnite	Spp	=	Sapphirine
Cc	=	Calcite	Lm	=	Laumontite	St	=	Staurolite
Cd	=	Cordierite	Lw	=	Lawsonite	Stp	=	Stilpnomelane
Ch	=	Chlorite	M	=	"Mica"	Su	=	Spurrite
Cp	=	Clinopyroxene	Me	=	Merwinite	T	=	Talc
Ct	=	Chloritoid	Mg	=	Magnesite	Ti	=	Tillyite
CV	=	Chlorite-Vermiculite	Mm	=	Montmorillonite	Tr	=	Tremolite
Di	=	Diopside	Mo	=	Monticellite	W	=	Wairakite
Do	=	Dolomite	Mt	=	Magnetite	Wm	=	White mica
E	=	Epidote	Ol	=	Olivine	Wo	=	Wollastonite
Ec	=	Clinozoisite	Om	=	Omphacite	X	=	Extra components
En	=	Enstatite	Op	=	Orthopyroxene			

Table C.2 Equations and References for Petrogenetic Grid for SAC and Aluminous Rocks

1. Q + Anl = Ab + F — A. S. Campbell and Fyfe (1965)
2. Hu = Lm + Q + F — Cho, Maruyama, and Liou (1987)
3. Pr + Lm + Ch = Pu + Q + F — Liou et al. (1987)
4. Hu = Lw + Q + F — Liou et al. (1987)
5. Cc = Ar — Composite based on Jamieson (1953), Clark (1957), Crawford and Hoersch (1972), Irving and Wyllie (1975), Salje and Viswanathan (1976), and Brar and Schloessin (1979)
6. Ab + x = Jdpx + Q — Newton and Smith (1967)
7. Ab = Jd + Q — Newton and Smith (1967)
8. Lm + Pr = Ec + Q + F — Liou et al. (1987)
9. Lm = W + F — Liou (1971b)
10. Lm = Lw + Q + F — Nitsch (1968, in Liou, 1971)
11. Stability field of Gl — Maresch (1977)
    Gl = M + T
    Gl = M + T + Ab
    Gl = T + Ch
12. Pr = Ez + G + Q + F — Liou, Maruyama, and Cho (1985, 1986)
13. W = An + Q + F — Liou (1970, 1971b)
14. Pr + Pu + Q + Ch = At + Pu + Q + Ch — Nitsch (1971)
15. Pr + E + Ch = Pu + Q — Liou et al. (1987)
16. Py = And — Spear and Cheney (1989)
17. W = Lw + Q — Liou (1971b)
18. Py = Ky + Q + F — Haas and Holdaway (1973)
19. Lw = An + F — Newton and Kennedy (1963)
20. K + Q = Py + F — Chatterjee, Johannes, and Leistner (1984)
21. Lw + Ab = Ez + Wm + Q + F — Heinrich and Althaus (1988)
22. Py = Ky + F — Spear and Cheney (1989)
23. Lw = Ez + Ky + Q + F — Newton and Kennedy (1963)
    Between 0.5 and 1.0 Gpa, the same curve approximates the position of the reaction
    Ca = Ky + Ch — Guiraud, Holland, and Powell (1990)
24. Lw + Jd = Wm + Ez + Q + F — Heinrich and Althaus (1988)
25. Ab out (2P = P + x) — Maruyama, Liou, and Suzuki (1982)
26. And = Ky — Holdaway (1971)
27. Wm + G + Ch = St + Bi + Q + F — Spear and Cheney (1989)
28. Ct + Bi + Q + F = G + Ch + Wm — Spear and Cheney (1989)
29. Ez + Ky + Q = An + F — Newton and Kennedy (1963)
30. Ct + Ch + Wm = St + Bi + F — Modified from Spear and Cheney (1989)
31. And + Bi + Q = G + Cd + Af + F — Spear and Cheney (1989)
32. Wm + Q = Af + Sil + F — Chatterjee and Johannes (1974)
33. Ph + Q = En + Sd + F — Helgeson et al. (1978)
34. And = Sil — Hemingway et al. (1991)
35. St + Q = G + Sil + F — Dutrow and Holdaway (1989); 0.5–0.8 Gpa extension projected by the author
36. Cd + G = Hy + Q + Ky — Henson and Green (1973)
37. St + Q = Ky + G + F — Pigage and Greenwood (1982)
38. Gl + Wm + G = Ab + Ch — Gl stability from Guiraud, Holland, and Powell (1990)
    Gl + Ch = T + G + Wm
39. Cd = Sil + Q + Gt + F — Richardson (1968)
40. Cd + G = Bi + Sil — Spear and Cheney (1989)
41. Ez + Q = An + G + F — Newton (1966)
42. Ky = Sil — Composite curve based on Holdaway (1971), Bohlen, Montana, and Kerrick (1991), and Hemingway et al. (1991)

43. Cd = Sil + Q + Sp	Richardson (1968)
44. Cd + G = Ol + Q + Sp	Henson and Green (1973)
45. Af + Cd = Sil + Bi + Q	A. B. Thompson (1976a)
46. Cd + G = Hy + Q + Sil	Cd stability from Henson and Green (1973)
Cd + G = Hy + Q + Spp	
47. Cd + Ol = Hy + Q + Sp	Henson and Green (1973)
48. Cd + G = Hy + Q + Sp	Henson and Green (1973)
49. Minimum melting curve of muscovite granite	W.-L. Huang and Wyllie (1981)

Table C.3 Equations and References for Petrogenetic Grid for Metabasic Rocks

1. Hu = Lm + Q	Cho, Maruyama, and Liou (1987)
2. Pr + Lm + Ch = Pu + Q + F	Liou, Maruyama, and Cho (1985, 1987)
3. Sb = Lm + Q + F	Liou (1971c)
4. Cc = Ar	Composite based on Jamieson (1953), Clark (1957), Crawford and Hoersch (1972), Irving and Wyllie (1975), Salje and Viswanathan (1976), and Brar and Schloessin (1979)
5. Q + Anl = Ab + F	Liou (1971a)
6. Ab + x = Jd$_{50}$Di$_{50}$ + Q	Maruyama and Liou (1988)
7. Ab = Jd + Q	Newton and Smith (1967)
8. Lm + Pr = Ec + Q + F	Liou, Maruyama, and Cho (1987)
9. Pu + Hm + Ch + Q = Ec + Am + Ch	Moody, Meyer, and Jenkins (1983)
10. Lm = Lw + Q + F	Liou (1971b)
11. Lm + Pu = Ec + Ch + Q + F	Liou, Maruyama, and Cho (1985, 1987)
Pu + Q = Ec + Pr + Ch + F	Liou, Maruyama, and Cho (1985, 1987)
12. Pr + Ch + Q = Pu + Tr + F	Liou, Maruyama, and Cho (1985, 1987)
13. Pu + Ch + Q = Ec + Tr + F	Evans (1990)
14. Pu + Ch + Ab = Ec + Gl + F	Liou, Maruyama, and Cho (1987)
15. Lw + Jd/Di = Ec + Gl + Q + F	Evans (1990)
16. W = An + Q + F	Liou (1970)
17. W = Lw + Q + F	Liou (1971b)
18. Pu + Ch + Q = Ec + Tr + F	Liou, Maruyama, and Cho (1987)
19. Ab = Om + Q	Newton (1986b)
20. Pr + Ch + Q = Ec + Tr + F	Liou, Maruyama, and Cho (1987)
21. Ab + E + Ch + S + Hm = H + P + E + Ch + I	Moody, Meyer, and Jenkins (1983)
22. Lw = An + F	Newton and Kennedy (1963)
23. Lw = Ez + Ky + Q + F	Newton and Kennedy (1963)
24. Tr + Ch + Ab = Ec + Gl + Q + F	Evans (1990)
25. Tr + Ch + Ab = Ec + Gl + Q + F	Maruyama, Cho, and Liou (1986)
26. Ez + Q = An + G + F	Newton (1966), Boetcher (1970), Helgeson et al. (1978)
27. Ec + Ch + Q = G + Tr + F	Evans (1990)
28. Ab out (2P = P + x)	Maruyama, Liou, and Suzuki (1982)
29. Ec + G = An + Tr + F	Evans (1990)
30. Ch + Q = T + Ky + F	Massone (1989)
31. Gl + Ec + Q = G + Tr + Ab + F	Evans (1990)
32. G + Ky + Q = An	Hariya and Kennedy (1968)
33. Gl + Ec = G + Q + Jd/Di + F	Evans (1990)
34. Cp + An + Ch + F = Fo + An + F	Yoder (1967)
35. Tr = En + Di + Q + F	Jenkins, Holland, and Clare (1991)
36. Ez + Ky + Q = An + F	Newton and Kennedy (1963)
37. H + P + I = Cp + H + P + I [+ F]	Spear (1981b)
38. H + Cp + P + I = Op + H + Cp + P + I	Spear (1981b)

39. Am + Fo = Cp + Op +An + F Obata and Thompson (1981)
40. Ch + Cp + An + F = Ch + Cp + An + Am + F Yoder (1966)
41. Ch + Cp + An + Am + F = Sp + Cp + An + Am + F Yoder (1966)
42. Ch = Op + Fo + Sp + F Jenkins (1982), Jenkins and Chernosky (1986)
43. H + Cp + Op + Ol + P + I =
  Cp + Op + Ol + P + I [+ F] Spear (1981)
44. An + Fo = An + Sp + Cp + Op Yoder (1966)
45. An + Sp + Cp + Am + F = An + Sp + Cp + Op + F Yoder (1966)
46. Solidus (initial melting curve) for "wet" gabbro Lambert and Wyllie (1982)

Table C.4 Equations and References for Petrogenetic Grid for Metaultramafic Rocks

1. C + T = A B. W. Evans et al. (1976)
2. C = A + B O'Hanley, Chernosky, and Wicks (1989)
3. A + B = Fo + F O'Hanley, Chernosky, and Wicks (1989)
4. C = Fo + T + F Chernosky (1982)
5. B + L = Fo + Ch O'Hanley, Chernosky, and Wicks (1989)
6. T + En = Ant Greenwood (1971)
7. A = Fo + T + F O'Hanley, Chernosky, and Wicks (1989)
8. T + En = Ant Day, Moores, and Tuminas (1985)
9. A = Fo + T + F B. W. Evans et al. (1976)
10. A = Fo + T Johannes (1975)
11. L = Fo + T + Ch + F O'Hanley, Chernosky, and Wicks (1989)
12. T + En = Ant Hemley et al. (1977)
13. T + Fo = Ant + F Chernosky, Day, and Caruso (1985)
14. T + Mg = Fo + F Johannes (1969)
15. Ant + Fo = En + F Chernosky, Day, and Caruso (1985)
16. Ant + Fo = Op (En) + F Jenkins (1981)
17. Ant + Fo = En + F Hemley et al. (1977)
18. Ant = En + Q + F Chernosky, Day, and Caruso (1985)
19. Ch = Op + Fo + Sp + F Jenkins and Chernosky (1986)
20. Tr + Fo = Op + Cp + F Jenkins (1983)
21. T + En = Ant Chernosky, Day, and Caruso (1985)
22. Mg = Pe + F Trommsdorf and Connolly (1990)
23. Tr + Fo = Op + Cp + Ch + F Jenkins (1983)
24. T = En + Q + F Chernosky, Day, and Caruso (1985)
25. Ch = Op + Fo + Sp + F Jenkins and Chernosky (1986)
26. Tr = Op + Cp + Q + F Jenkins, Holland, and Clare (1991)
27. Tr + Fo = Op + Cp + Sp +F Jenkins (1983)
28. B = Pe + F Day, Chernosky, and Kumin (1985)

Table C.5 Equations and References for Petrogenetic Grid for Metamorphosed Carbonate Rocks

1. Cc = Ar	Composite based on Jamieson (1953), Clark (1957), Crawford and Hoersch (1972), Irving and Wyllie (1975), Salje and Viswanathan (1976), and Brar and Schloessin (1979)
2. Mg + Fo + F = T + F (X_{CO_2} = 0.5)	Johannes (1969)
3. Do + Q + F = T + Cc + F (X_{CO_2} = 0.5)	Eggert and Kerrick (1981)
4. Do + Q + F = T + F (X_{CO_2} = 0.5)	Metz and Puhan (1970)
5. T + Cc + Q = Di + F (X_{CO_2} = 0.5)	Skippen (1974)
6. Tr + Cc = Do + Di + F	Slaughter, Kerrick, and Wall (1975)
7. Tr + Cc + Q = Di + F (X_{CO_2} = 0.2)	Dachs and Metz (1988)
8. Do + Q = Di + F	Slaughter, Kerrick, and Wall (1975), Trommsdorf and Connolly (1990), Tracy and Frost (1991)
9. Tr + Do = Fo + Cc + F (X_{CO_2} = 0.3)	Metz (1976)
10. Di + Do = Fo + Cc + F	Kase and Metz (1980)
11. Di +Do = Fo + Cc + F	Trommsdorf and Connolly (1990), Tracy and Frost (1991)
12. Cc + Q = Wo + F	Trommsdorf and Connolly (1990)
13. An + Wo + Cc = G + F	Zharikov, Shmulovich, and Bulatov (1977)
14. Mg = Pe + F	Trommsdorf and Connolly (1990)
15. Do = Pe + Cc + F	R. I. Harker and Tuttle (1955), Tracy and Frost (1991)
16. Cc + Wo = Ti + F	Joesten (1974)
17. R + Su = La + F	Joesten (1974)
Ti = Su + F	Joesten (1974)
18. Cc + Di = Ak + F	Tracy and Frost (1991)
19. Ak + Cc = Me + F	Zharikov, Shmulovich, and Bulatow (1977)
20. Di + Fo + Cc = Mo + F	Zharikov, Shmulovich, and Bulatow (1977)
21. Di + Cc = Ak + F	Zharikov, Shmulovich, and Bulatow (1977)
22. Su = Cc + La	Joesten (1974)

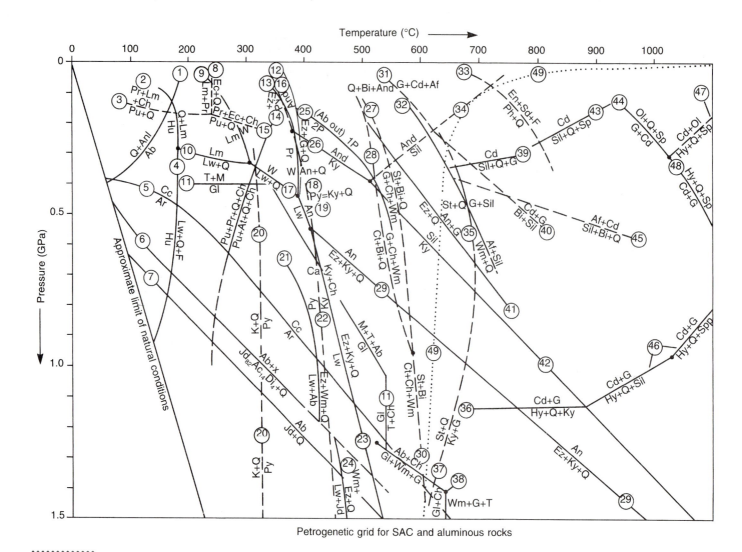

Petrogenetic grid for SAC and aluminous rocks

Figure C.1
Petrogenetic grid for SAC and aluminous rocks. See text for numbered reactions and symbols.

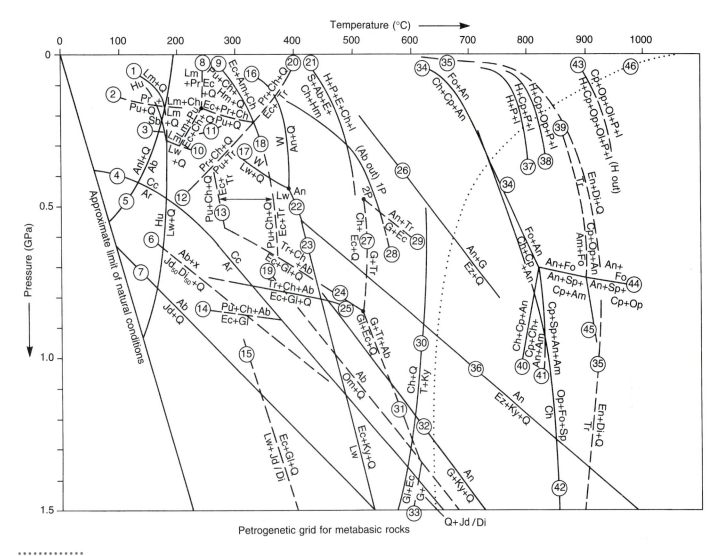

Figure C.2
Petrogenetic grid for metabasic rocks. See text for numbered reactions and symbols.

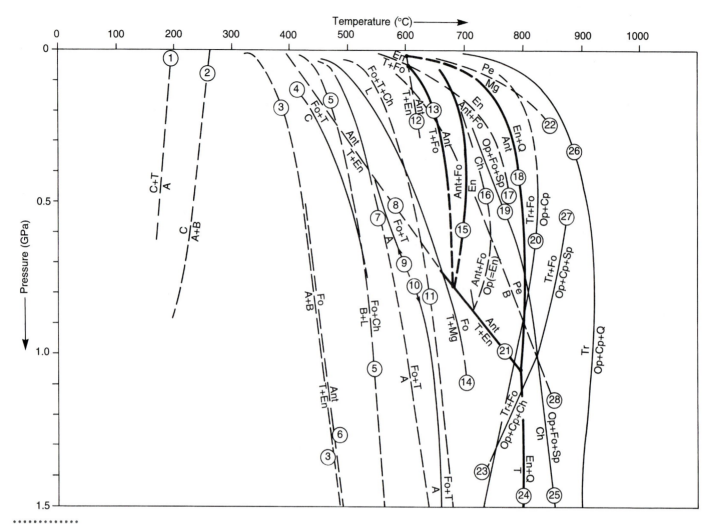

Figure C.3
Petrogenetic grid for metaultramafic rocks. See text for numbered reactions and symbols.

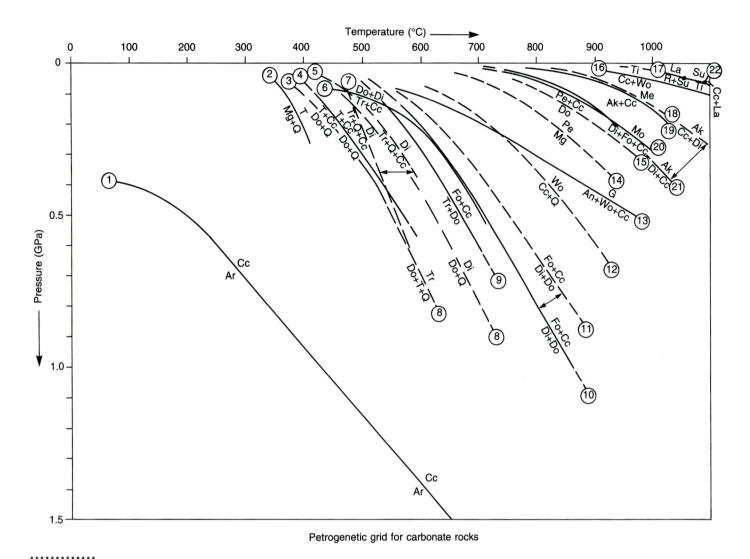

Petrogenetic grid for carbonate rocks

············

Figure C.4
Petrogenetic grid for metamorphosed carbonate rocks. See text for numbered reactions and symbols.

GLOSSARY

This glossary is based on definitions and references in the text. Refer there for references to original sources. For additional information, definitions of additional terms, and sources of terms, the reader is encouraged to consult Bates and Jackson (1987) and Mitchell (1985).

A

aa lava　A type of volcanic flow, typically basaltic in composition, characterized by sharp, angular fragments. Aa flows have flow tops that are rough and irregular. The rock of aa flows is referred to as aa lava.

abyssal　Of or referring to depth. Abyssal igneous rocks form at depth in the crust. In the ocean, abyssal depths are depths of greater than about 914 m (500 fathoms).

abyssal plain　The flatter, deep regions of the ocean floor extending between steeper slopes of the continental rises and other features such as mid-ocean ridges.

accessory mineral　A mineral in an igneous or metamorphic rock that is not used in deriving the rock name from a rock classification chart.

ACF diagram　A type of triangular pseudo-phase diagram used in metamorphic petrology, in which the A corner represents aluminous phases or compositions, the C corner represents CaO-bearing phases or compositions, and the F corner represents iron- and magnesium-rich phases or compositions.

adcumulate　An igneous rock with an adcumulate texture.

adcumulate texture　An igneous texture formed by fractional crystallization and consisting of a framework of touching crystals that continued to grow from the magma to the extent that less than 5% of the rock consists of *other* minerals crystallized from the intercrystalline liquid.

adcumulus　Of or relating to adcumulate texture.

adiabatic　A term used to describe a thermodynamic process in which rocks undergo neither a gain nor a loss of heat while undergoing changes in pressure.

aerobic layer　The upper, oxygenated layer of water in a water column.

AFM diagram　In igneous petrology, a triangular chemical plot of alkalis, iron oxide, and magnesia. In metamorphic petrology, a triangular phase diagram in which the respective corners represent aluminous, iron-rich, and magnesium-rich phases and compositions.

agglomerate　A fragmental volcanic rock dominated by rounded clasts larger than 64 mm in diameter or length.

agmatite　A type of migmatite that consists of angular fragments in a matrix; also called migmatite breccia.

A'KF diagram　A pseudo-phase diagram used in metamorphic petrology, in which aluminous, potash-rich, and iron- and magnesium-rich phases and compositions are plotted.

Alaska-type mafic-ultramafic rock body　A crudely ellipsoidal, zoned igneous complex containing a variety of silica-poor rocks including two-pyroxene gabbros, peridotites, hornblende pyroxenites, and dunites.

alaskite　A name applied by some petrologists to granitoid igneous rocks that contain less than 5% mafic phases.

Albite-epidote Hornfels Facies　The set of all of the metamorphic rocks containing phase assemblages indicative of conditions of crystallization in the general range of $P = 0–0.2$ Gpa (0–2 kb) and $T = 300–400°C$. The name is also applied to the P-T space represented by those conditions.

alkalic　An adjective used to describe both the rock suite and the rocks belonging to a rock suite with an alkali-lime index of less than 51. The term designates the rock suite consisting of basalt → hawaiite → mugearite → trachyte → phonolite.

alkali-lime index　For any suite of igneous rocks, the number determined by the intersection of a curve representing the sums of the alkalis ($K_2O + Na_2O$) for the rocks of the suite with a curve representing the CaO values for the rocks of the suite. The indices are used to divide rocks into one of four groups—alkalic, alkali-calcic, calc-alkali, and calcic. Also called the Peacock Index.

alkali olivine basalt　A basalt type characterized by abundant olivine and Ca-pyroxene (augite). Alkali olivine basalts are typically olivine- or nepheline-normative.

allo-　A combining form meaning other or extraneous (e.g., from an*other* place).

allochem　In a sediment or sedimentary rock, a fragment of a chemical or biochemical precipitate formed earlier in another place. These grains have been transported away from their sites of origin to the site of deposition.

allochemical metamorphism　The type of metamorphism in which, during the metamorphic process, there is a change in the bulk chemistry of the rock domain under consideration. (Literally, other-chemical metamorphism).

allolistostrome　A type of sedimentary melange, formed by submarine sliding, that is, a mappable body of rock lacking continuity of internal stata or contacts and containing both native and exotic fragments in a finer-grained matrix.

allomorphic　A process of diagenetic replacement in which an original sedimentary phase is replaced by a new phase of another crystal form. (Literally, other forms.) *Note:* There are other definitions and uses of this term (see Bates and Jackson, 1987).

allotrioblastic-granular texture　A metamorphic texture characterized by approximately equant, anhedral grains.

allotrioblastic texture　A metamorphic texture characterized by anhedral grains.

allotriomorphic-granular texture   An igneous texture characterized by approximately equant anhedral grains.

alluvial   Referring to sediment deposited by running water or features associated with such sediment.

alluvial fan   A typically semicircular (in plan), wedge-shaped (in longitudinal section) accumulation of sediment formed where a stream exits a valley or canyon and enters a broad valley or plain.

alluvium   Sediment deposited by running water.

alnoite   A type of mafic, porphyritic volcanic rock (a lamprophyre) with phenocrysts of euhedral biotite or melilite and commonly containing matrix phases such as nepheline, olivine, and clinopyroxene.

alpine-type ultramafic bodies   Irregular to elliptical bodies of rock that occur in orogenic belts and contain silica-poor rocks (with <45% SiO_2) composed of dark minerals. Typically they are deformed.

alpine-type ultramafic rock   Dark-shaded rock types containing less than 45% silica that occur in mountain belts.

alteration   A process of chemical change, typically associated with ore zones, in which the bulk chemistry and the mineralogy of rock masses are changed by passing, hydrothermal fluids.

amphibolite   A name applied by some petrologists to igneous or metamorphic rocks dominated mineralogically by hornblende.

Amphibolite Facies   The set of all metamorphic rocks containing phase assemblages indicative of conditions of crystallization in the general range of $P =$ 0.4–1.1 Gpa (4–11 kb) and $T =$ 550–750°C. The name is also applied to the P-T space defined by those conditions.

amygdaloidal   A textural term describing volcanic rocks containing mineral-filled, elliptically shaped vesicles.

amygdule   In a volcanic rock, a vesicle that has become filled with minerals such as calcite, quartz, or zeolites.

anaerobic   Oxygen-deficient.

anaerobic layer   An oxygen-deficient, bottom layer in a water column.

anastamosing cleavage   A type of metamorphic rock structure consisting of a curvi-planar fabric resulting from braided microzones of phyllosilicates or other minerals that give the rock a tendency to break into lens-shaped pieces. It is a type of spaced cleavage.

anatexis   The process of melting preexisting rock materials.

anchimetamorphism   A term used by some petrologists to designate very low-grade metamorphism under conditions ranging from those of diagenesis to those of the Zeolite Facies.

andesite   An intermediate-silica volcanic rock, characterized by a silica content between 52 and 65%. Plagioclase feldspar is the dominant feldspar, quartz is generally absent, and typical phenocrysts consist of plagioclase, hornblende, or augite. Plagioclase is typically light and has an An content of less than 50.

anhydrite evaporite   Aphanitic to phaneritic, soft, commonly layered, anhydrite-rich rock formed by evaporation of or crystallization from a brine.

anorthosite   A plutonic igneous rock composed almost entirely of plagioclase feldspar (a hyperleucocratic gabbro).

antidune   Elongate piles of sediment (a type of sand wave) of less than 1 m to tens of meters high, which may develop in currents of wind or water and may be preserved under appropriate conditions. In streams, antidunes form in phase with surface-water waves.

aphanitic   A descriptive term meaning that the grains in a rock are too small to see or identify with either the unaided eye or a low-power lens.

aphyric   A term used to describe volcanic rocks that lack phenocrysts.

aplite   A medium- to fine-grained, allotriomorphic-granular, hyperleucocratic, quartz-alkali feldspar (plutonic) rock.

apophysis   A short, irregular dike that extends from a pluton margin into the country rock.

appinite   An igneous rock of quartz diorite composition characterized by euhedral hornblende crystals.

Appinite-type mafic-ultramafic rock body   A type of rock complex consisting of small elliptical to lenticular plutons composed of rocks such as olivine hornblendite, norite, and hornblende quartz diorite. Appinite-type complexes are typically associated with granitoid rocks.

arc   A linear or curvilinear chain of volcanoes on land or in the sea.

arenite   A type of sandstone with matrix materials comprising 5% or less of the rock.

arkose   A root name used in some sandstone classifications to designate rocks relatively rich in feldspar.

ash   Fine-grained, unconsolidated, clastic volcanic material composed of grains less than 2 mm in diameter.

ash fall   A layer of volcanic ash formed by sediment rain from the atmosphere during or following a volcanic eruption. Also the process of sedimentation of volcanic ash.

ash flow   The rock mass, composed of ash and other pyroclastic materials, formed by a nuée ardente (a hot, gaseous, flowing volcanic cloud). Also called ignimbrite.

asthenosphere   A plastic zone in the mantle of the Earth separating the lithosphere and the mesosphere.

augen   An eye-shaped crystal or porphyroblast in a gneiss or schist.

augen texture   A metamorphic texture characterized by eye-shaped, larger grains or grain clusters in a finer-grained matrix.

aureole   A zone of contact metamorphic rock surrounding and associated with a pluton.

authigenesis   The process in which new mineral phases are crystallized in a sediment or rock during diagenesis. See also *neocrystallization*.

authigenic   Adjective describing minerals formed in place (i.e., precipitated rather than transported).

autolith   A small body or inclusion of rock found within and related to an igneous rock mass. (Literally, self-stone *auto* = self, *lith* = stone), that is, a piece of the enclosing body of rock.

automorphic-granular   An igneous texture that is dominated by euhedral crystals.

B

baked zone   A zone of thermal effects, typically a brick-red zone of oxidized soil, between older and younger lava flows.

ball-and-pillow structure   A primary sedimentary structure consisting of spherical and hemispherical to elliptical masses of deformed sediment.

bank benches A flat feature formed along the stream banks or margins over time as the stream channel migrates.

bar A ridgelike, linear accumulation of sediment in a stream or offshore marine environment.

barrier beach complex A linear beach (and dune) landform isolated from the land by a lagoon. Also called barrier island.

Barrovian Facies Series A medium-pressure, high-temperature sequence of metamorphic facies, with kyanite-sillimanite-bearing pelitic rocks. It is named for rocks of the Scottish Highlands described by Barrow (1893).

basalt A volcanic rock type that contains essential plagioclase feldspar and is characterized by a silica content between 45 and 52%. Plagioclase is characteristically dark and has an An content of more than 50. Phenocrysts are typically one or more of the phases plagioclase, augite, and olivine.

basaltic plain A major extrusive volcanic structure that is tabular in overall form and is composed primarily of lava flows of silica-poor (basaltic) volcanic rock erupted from overlapping centers of eruption that develop shield cones.

batch melting A process of fractional melting in which masses of magma repeatedly develop and then separate from a parent rock.

batholith A body of plutonic rock having aerial extent of 100 km^2 or more. Alternatively, a nonlayered body of intrusive rock, predominantly phaneritic in texture, with a minimum volume of 100 km^3.

beach An accumulation of sediment formed along the margin of a water body.

beach backshore A supratidal zone, behind a beach, composed of a berm—a sandy, somewhat flat area above the normal high tide level—and a dune.

beach foreshore The zone between high and low tide, including the swash zone where breaking waves run up the beach face.

beachrock A foreshore grainstone or conglomerate cemented on the beach by calcite, soon after deposition.

bed A layer of sediment or sedimentary rock, distinguished from adjoining layers on the basis of differences in color, texture, and composition. The most characteristic structure of sedimentary rocks.

bed form A sedimentary structure formed at the sediment-water interface, such as ripples, sand waves, dunes, and antidunes.

bed-load transport Movement of particles too large for extended suspension above the surface by rolling, bouncing (saltation), and temporary suspension in a current.

benmoreite A type of intermediate volcanic rock with a saturated to undersaturated silica content, a potash: soda value of less than 1:2, and modal pyroxene, alkali feldspar, and either andesine or oligoclase.

benthic Referring to the bottom of a water body (as in *benthic fauna*, the bottom-dwelling animals in the sea).

bentonite A soft, plastic mudrock composed primarily of smectite-group clays and derived from weathering of volcanic ash.

binary solid solution system A chemical system composed of two chemicals that may be mixed in any ratio to yield a single phase.

Bingham plastic material A material that has an initial strength, or yield strength, but in which, during deformation or flow, the viscosity remains constant after the yield strength is exceeded.

biochemical precipitate Material crystallized from a solution as a result of the activity of organisms.

biofacies A unit of sedimentary rock that is distinguished by its biota and is representative of specific environmental conditions.

biogenic element Any textural component of a rock formed by biological activity.

biolithic element A part of a biolithite or boundstone formed by organisms.

biolithite A fossiliferous reef rock formed *in situ* by biochemical precipitation and the intergrowth of organisms.

bioturbation The postdepositional process of mixing and disruption of lamination and bedding in sediment caused by the activities of organisms.

bioturbation structure A structure in sedimentary rock, formed where burrowing by organisms was extensive, characterized by highly swirled laminations or completely destroyed lamination.

bivariant system A system in which there are two degrees of freedom; that is, two variables may be changed independently without changing the number or nature of phases in the system.

blastomylonite A foliated metamorphic rock that has experienced ductile deformation and later recrystallization, so that it contains recrystallized microlithons, porphyroclasts that exhibit overgrowths, or new porphyroblasts.

block A large rock fragment that is typically angular. In volcanic rocks, blocks are more than 64 mm in diameter.

block-in-matrix structure A structure found in coarse, clastic sedimentary rocks and some igneous and metamorphic rocks consisting of large masses of rock of one or more types enclosed by generally finer-grained material of a different rock type.

Blueschist Facies The set of all metamorphic rocks containing phase assemblages indicative of conditions of crystallization in the general range of $P = 0.35–0.9$ Gpa (3.5–9 kb) and $T = 50–400°C$. The name is also applied to the P-T space defined by those conditions.

bomb A mass of volcanic rock larger than 64 mm in diameter that has a rounded form resulting from streamlining or deformation during flight, after ejection from a volcano.

boninite A type of andesite with a high magnesium content (Mg number > 0.7).

boudin A structure consisting of cylindrical masses of rock derived from a single bed or layer that has been stretched or pulled apart to resemble a row of sausages. (French for sausage.)

bouldery A term applied to sedimentary rocks in which clasts larger than 256 mm in diameter comprise less than 25% of the rock.

Bouma sequence A sedimentary unit consisting of a five-part series of beds and laminations that from bottom to top include a graded or structureless bed, a parallel-laminated layer, a convolute or cross-laminated layer, an upper parallel-laminated layer, and a mudrock layer. Bouma sequences occur in turbidites.

boundstone A reef rock consisting of intergrown and cemented organic structures.

breccia (sed.) A sedimentary rock composed of angular clasts larger than 2 mm in diameter in a sandy or gravelly matrix.

brittle deformation Deformation in which grains (and rocks) are broken before the material has attained 5% strain.

Buchan Facies Series   A low-pressure sequence of facies, with pelitic rocks containing andalusite to sillimanite, that was named after the Buchan area of northeastern Scotland.

burrow   An irregular to cylindrical, tubular structure found in sedimentary rocks, representing a filled hole that was dug by an organism.

bysmalith   An igneous intrusive structure consisting of a nearly vertical, cylindrical mass of rock bounded by faults.

C

caldera   An igneous structure consisting of a large circular depression produced by the eruption-induced collapse of a volcano.

caliche   A sedimentary rock that is chalky, light-colored, micritic or microsparitic and is formed by evaporation and precipitation of calcite in soil, sediment, or preexisting rock.

carbonate buildup   A sedimentary mass that is depositionally thickened, fossiliferous, and composed of carbonate rock. A general term that encompasses reefs, Walsortian mounds, and other similar structures.

carbonate rocks   Those rocks with 50% or more carbonate minerals (e.g., calcite, aragonite, and dolomite). The term is most commonly applied to sedimentary rocks.

carbonatite   An igneous rock composed primarily of one or more carbonate minerals, such as calcite, dolomite, and ankerite. Both volcanic and plutonic carbonatites are known.

cast   A sedimentary structure consisting of the filling in a depression or void.

cataclasis   The process of crushing and breaking the grains in a rock.

cataclasite   A nonfoliated metamorphic rock composed of angular fragments of rock in a matrix of finer-grained, but similarly composed materials.

cataclastic breccia   A nonfoliated dynamoblastic metamorphic rock dominated by broken, angular fragments greater than 1/16 mm in diameter in a matrix of more finely crushed material. Cataclastic breccias form along fault zones.

cataclastic texture   A generally nonfoliated, metamorphic texture characterized by broken rock and mineral grains. *Note:* Foliated-cataclastic texture is a transitional texture between (nonfoliated) cataclastic textures *sensu stricto* and (foliated) mylonitic textures.

catazonal   A descriptive term applied to plutons that form deep (>9 km) in the crust. Also spelled katazonal.

cement   Chemically precipitated material that fills the spaces between framework grains in sedimentary rocks.

cementation   The process in which chemical precipitates, in the form of new crystals, form in the pores of a sediment or rock, binding the grains together.

chalk   A soft, earthy, friable, porous, light-colored limestone.

channel   The concave-up, linear depression through which a stream flows.

channel bar   A cross-bedded mound of sediment within a stream channel.

characterizing accessory minerals   Minerals occurring in abundances of 5% or more that are not implied by the rock name but are used to modify the name of an igneous rock. Also called varietal minerals.

chelate   A compound in which a metal cation is bonded with and connects organic ring structures.

chelation   A chemical process, important in weathering, that results in the formation of a chelate, a chemical complex containing metal ions, via extraction of metal cations from a mineral.

chemical precipitates   In sedimentary petrology, a category of sedimentary rock that includes rocks that are crystalline-textured, generally fine-grained to aphanitic, and formed by inorganic crystallization of minerals from solution. In general, any materials formed by inorganic crystallization of minerals from solution.

chert   A hard, multi- or variously colored, waxy to grainy sedimentary rock composed predominantly of silica minerals.

chilled margin   A finer-grained phase of a pluton that forms within the pluton, along the pluton margin, where the magma was cooled by contact with cooler country rocks during emplacement of the pluton.

clast-supported   A descriptive term for sedimentary rocks composed of clasts abundant enough to be touching several neighboring clasts.

cleavage   A structure in rocks consisting of relatively closely spaced, parallel to subparallel surfaces of fracture or mineral alignment. Also, the tendency of rocks to break along parallel to subparallel surfaces. (Cleavage also refers to the tendency of minerals to break along parallel planes.)

closed system   A system that cannot exchange material but can exchange energy with the surroundings.

coarse-tail grading   A feature in medium- to coarse-grained clastic sedimentary rock beds in which there is a change in grain size from the bottom to the top of the bed (grading), but in which the change in grain size is defined only by a change in the sizes of the framework or coarsest clasts.

cobbly   A term applied to sedimentary rocks in which clasts between 64 mm and 256 mm in diameter comprise less than 25% of the rock.

colloid   Finely divided material in suspension.

color index   In an igneous rock, the total percentage of minerals in which iron, magnesium, or both are essential constituents of the mineral chemistry.

columnar joint   Pencil-like, polygonal mass of rock, several centimeters across and centimeters to meters in length, formed where fractures in volcanic rocks intersect.

comb texture   A texture in rocks characterized by a row of crystals that have grown perpendicular to a surface, like the teeth of a comb.

compaction   The process by which the volume of a rock mass is reduced, usually as a result of sediment or rock being compressed under a load of overlying materials.

complex composite pegmatite   A type of very coarse-grained, zoned or unzoned plutonic rock body with veined and/or replacement structures. Complex composite pegmatites show two distinct stages of development with the later stage consisting of replacements of early-formed minerals, fracture fillings that cross-cut early-formed minerals, or both.

complex pegmatite   A type of very coarse-grained plutonic rock body with a zoned, veined, or replacement structure.

component   A chemical species (e.g., SiO_2, OH^-). In the Phase Rule, the number of components is defined as the smallest number of chemical species needed to define the compositions of all phases in the system.

composite cone   Steep-sided volcanic structure consisting of layers of lava and pyroclastic material that radiate and thin away from a central vent area. Also called stratovolcanoes.

concordant   A term referring to intrusive structures (i.e., plutons) with contacts that parallel the layers in the intruded rock.

concretion A sedimentary structure consisting of a round to irregular mass of more resistant rock formed as a result of cement precipitating around a core material, commonly a fossil or grain of a different composition.

cone sheet An intrusive igneous structure (a dike) that is a discordant curvi-planar layer which diverges from a point at some depth in the Earth.

congelation crystallization A process of magmatic crystallization occurring on the walls and floor of the magma chamber which concentrates liquids enriched in residual elements in the top or center of the chamber.

conglomerate A sedimentary rock composed of rounded clasts greater than 2 mm in diameter and comprising more than 25% of the rock. The clasts are typically enclosed in a matrix of sand (sandstone) or gravel (conglomerate).

congruent melting A process of melting in which the melt has the same composition as the solid that is melted.

consanguineous An adjective used to describe igneous rocks derived from the same parent magma.

contact The boundary between two bodies of rock.

contact diablastite A fine- to medium-grained, nonfoliated rock composed of a combination of granular, acicular, and platy grains. The term is used as a substitute for hornfels where there is a need to imply genesis with the rock name.

Contact Facies Series A very low-pressure sequence of facies, represented by the rocks of contact metamorphic zones.

contact metamorphism Metamorphism caused primarily by heat in country rocks at and near their contact with an igneous rock mass.

continental environment A region of sedimentation on the Earth totally above sea level at high tide and beyond the direct influence of marine processes.

continental rise That gently sloping part of the seafloor that occurs between the abyssal plain and the more steeply sloping continental slope.

continental sabkha A salt flat along the margin of an ephemeral lake in a flat, vegetation-free desert basin.

continuous cleavage In rocks, a planar fabric element with a spacing of less than 0.01 mm or one that results from a penetrative preferred orientation of phyllosilicates or other minerals.

continuous melting A process in which melting and separation of magma from the parent rock are coeval and uninterrupted.

continuous reaction A reaction in which the compositions of one or more of the coexisting phases changes gradually over the course of the reaction.

continuous reaction series A group of reactions in which the compositions of phases formed earlier react with remaining liquids to change composition gradually, without abrupt changes in the phases present.

contourite A category of sandstone deposited in the sea by currents that flow parallel to the slope (contour currents).

convective fractionation A crystallization process in which convective flow, driven by heat within a magma chamber, feeds crystallization on the floor, walls, and roof of a magma chamber.

convolute lamination A sedimentary structure consisting of highly contorted, folded, and disrupted layers less than 1 cm thick. *Convolute beds* have layers greater than 1 cm thick.

corona A ring or shell of minerals, in some cases having a radial growth pattern, surrounding a core mineral grain. Some are reaction rims.

coronitic texture A texture in which larger grains have rims of generally finer-grained minerals.

cortlandite An ultrabasic rock consisting essentially of hornblende with poikilitically enclosed olivine grains.

cotectic line The line on a phase diagram that separates two separate phase fields and represents the line of liquid composition, or the temperature-composition curve, along which two solid phases crystallize simultaneously.

country rock The rock surrounding a rock body of interest, particularly, the rock surrounding an igneous intrusion.

crater A concave depression usually less than 1 km in diameter, and generally at the crest of a volcano, directly caused by eruptive activity.

critical mineral A mineral for which the reaction curves and stability field are known and which may be used to distinguish between one facies or zone and another.

critical mineral assemblage A group of minerals that occur together for which the reaction curves and stability field are known and which may be used to distinguish between one facies or zone and another.

cross-beds Layers of sedimentary rock inclined at an angle to the main plane of the stratification.

cryolacustrine environment A sedimentary setting consisting of a lake directly associated with a glacier and characterized by the unique physical, chemical, and biological features of such a setting.

cryptic layering Generally invisible tabular zones, within and typically parallel to the sides and bottom of plutons, that are marked by variations in the chemistries of the included minerals.

crystalline An adjective used to describe any material consisting of an internally ordered, long range array of atoms. The term is used to distinguish certain rocks composed entirely of crystalline phases (minerals) from those partly or entirely composed of glass or other noncrystalline materials.

crystalline texture A texture characterized by crystalline materials arranged in an interlocking array.

crystal-liquid fractionation A process of magmatic crystallization in which crystals form and separate from their parental liquid magma.

cumulate A rock with a cumulate texture. Also an adjective used to describe rocks and textures in which there is a framework of touching mineral grains that formed primarily through fractional crystallization.

cumulus crystals Crystals formed by fractional crystallization from a magma. They are concentrated in layers along the floor, walls, or top of a magma chamber.

cupola A large, convex-up bulge of plutonic rock on the roof of a magma chamber or pluton.

current deposition Bed-load or suspended load deposition by currents as they lose their ability to transport the sediment.

D

decomposition The general category of processes in which chemicals break down rock materials during weathering.

decussate texture A metamorphic texture consisting of platy or acicular crystals arranged in such a way as to produce no preferred orientation. Also called diablastic texture.

degree of freedom   A condition or variable that may be changed in a system (also called variance). In the Phase Rule, the *number* of degrees of freedom is of interest and is defined as the minimum number of variables needed to define a specific state of the system; in other words, the smallest number of conditions or variables that may be changed independently without altering the number or nature of phases in the system.

delta   An accumulation of sediment—locally shaped, in map view, like the Greek letter delta—that forms in a body of water where a stream enters the body and deposits its load.

derivative magma   A magma formed from another magma by a process of separation.

dessication crack   A crack formed in sediment as a result of drying.

deviatoric stress   A stress (force/unit area) that acts in a particular direction and exceeds the hydrostatic or mean stress.

diabasic   An igneous texture consisting of rectangular (lath-shaped) plagioclase feldspar crystals with smaller, intergranular crystals of pyroxene and other minerals. Diabasic texture is one member of the ophitic-subophitic-diabasic series and is a coarse-grained equivalent of the intergranular textures found in volcanic rocks.

diablastic texture   A metamorphic texture in which platy and acicular minerals are intergrown in a nonfoliated, interlocking, and locally radiating manner.

diablastite   A metamorphic rock with a diablastic texture.

diagenesis   A general term designating all of the surface-to-subsurface physical, chemical, and biological processes that collectively result in transformation of sediment into sedimentary rock and modification of the texture and mineralogy of a rock.

diamictite   A sedimentary rock composed of 25% or more clasts larger than 2 mm in diameter (or length) enclosed in a finer-grained matrix. Diamictites have rounded clasts, angular clasts, or both in a matrix dominated by mud.

diapir   A spherical, elliptical, or tail-down drop-shaped mass that rises towards the surface as a result of its low density compared to surrounding rocks. Diapirs may be solid or liquid.

diapiric   An adjective referring to a process in which a mobile core of material rises up (generally due to its low density), forcing the overlying materials into a domal or antiformal structure. The adjective is also used to describe features formed by diapirism.

diatomaceous chert   Diatomite that is well cemented.

diatomite   An aphanitic, light-colored, soft, friable, siliceous rock composed of diatoms (the fragments of certain plant cell walls). Where well cemented, diatomite is called diatomaceous chert.

diatreme   A cylindrical to dike-like volcanic structure generally composed of ultramafic to carbonate breccias.

dictytaxitic texture   An extrusive rock texture consisting of vesicular rock with vesicles that have crystals projecting into the cavity from the vesicle walls.

differentiation   A general term referring to all of those processes in which more than one rock type is formed from a magma.

diffuse   The action of diffusion.

diffusion   A process in which chemical species migrate between two sites, in a solvent phase, under the influence of a chemical potential gradient.

dike   A generally tabular, intrusive structure that is discordant to surrounding layers or that cross-cuts massive rocks.

dimensional preferred orientation   Also called shape preferred orientation (SPO). A texture in which inequant grains display some degree of parallel alignment.

discontinuous reaction   A reaction in which there is an abrupt change in the stable phase or phase assemblage present. In metamorphic petrology, both terminal and nonterminal reactions that involve univariant curves may be discontinuous.

discontinuous reaction series   A group of reactions in which reaction of earlier-formed crystals with a liquid phase results in abrupt changes in the phases present.

discordant   An adjective referring to igneous structures that cut across layering.

disintegration   The general category of processes of physical breakdown of rock materials during weathering.

dish structure   A sedimentary structure, typically found in sandstone, consisting of thin, oval- to circular-shaped, concave-up concentrations of finer-grained material.

dismicrite   A fine-grained limestone containing bird's-eye structures composed of sparry calcite and having less than 1% allochems.

disorganized bed   A type of layer composed of gravel or conglomerate that lacks internal grading or stratification.

distal   Located at a considerable distance from the source.

distal environment   A depositional setting on the surface of the Earth that is located at a considerable distance from the source of sediment.

distributary   A branch of a stream that occurs on a delta or alluvial fan where a major stream splits to form a series of smaller streams that do not reenter the main stream.

distribution grading   A change in grain size within a bed involving the gradual change of all of the grain sizes; contrasts with coarse-tail grading.

dolograinstone   A dolomite-rich rock composed of larger grains in a grain-supported array and a matrix of less than 5% fine-grained materials.

dolomitization   The process by which rocks are converted to dolostone. The process is primarily one of replacement.

dolomudstone   A rock composed of dolomite in grains of less than 1/16 mm.

dolopackstone   A dolomite-rich rock composed of larger grains in a grain-supported array with a fine-grained matrix.

dolosparite   A rock composed of dolomite in grains greater than 1/16 mm.

dolostone   A general term that refers to all aphanitic to phaneritic rocks composed dominantly of dolomite.

dolowackestone   A "mud"-supported, dolomite-rich rock consisting of scattered larger grains in a finer-grained matrix.

domain   A volume of rock being considered for a particular purpose or study.

double-diffusive convection   A process of material transport that occurs in magma chambers and involves the combined effects of physical movement (convection) and chemical transport (diffusion).

ductile   A descriptive term for deformation in which plastic behavior and flow occur via structural changes within and between grains.

ductile deformation   Deformation that is predominantly plastic in nature. Ductile materials experience strains of 5% or more before failing.

dune   A convex-up accumulation of sediment, deposited by wind or water currents, that may be domal, linear, curvilinear, stellate, elliptical, or irregular in shape.

dynamic metamorphism   A process of change in rocks caused primarily by deviatoric stress.

dynamoblastic rock   A metamorphic rock with micolithons or porphyroclasts surrounded by a deformed matrix in a texture resulting from the operation of deviatoric stress (i.e., from dynamic metamorphism).

dynamothermal metamorphism   A process of change in rocks caused primarily by elevated temperatures and pressures.

dysaerobic layer   An intermediate layer in a water column–between the surfical aerobic layer and the bottom anaerobic layer–that has variable oxygen content, salinity, and density.

E

eclogite   A plagioclase-free, silica-poor rock composed primarily of Mg,Fe garnet and the sodic clinopyroxene omphacite.

Eclogite Facies   The set of all metamorphic rocks containing phase assemblages indicative of conditions of crystallization in the general range of P = 0.9–2.0 Gpa (9–20 kb) and T = 150–900°C. The name is also applied to the P-T space defined by those conditions.

Eh   A measure of the ability of a solution to produce oxidation or reduction. Also called redox potential.

enclave   An inclusion in a plutonic rock, specifically, a xenolith, xenocryst, or autolith.

en echelon   A descriptive term referring to a group of features that form a diagonal array with each feature offset slightly from the next.

englacial environment   A depositional setting within a glacier (e.g., within tunnels and channels within the ice).

entrained   Carried by; as in, rock was *entrained* by a glacier.

eogenesis   The process of early diagenesis that occurs at or near the surface between deposition and burial.

epeiric sea   A large shallow sea that occurs within or along the margin of a continent.

epiclastic texture   A general term for all sedimentary textures formed at the surface (*epi-*) and consisting of accumulated rounded to angular grains (clasts) packed together. Grains are derived by normal processes of surficial weathering, erosion, and abrasion, and are bound together through recrystallization of grain boundaries, cementation, or matrix-grain amalgamation.

epilimnion   A layer of water in a lake consisting of less dense fresh water that overlies more dense saline water.

epimatrix   A type of intergrain material in sandstones and conglomerates consisting of a polymineralic mix of minerals produced by neocrystallization and alteration of framework grains.

epitaxial   An adjective describing a relationship between minerals in which one mineral overgrows another in such a way that some atomic (crystallographic) structural elements of the overgrowth parallel similar or identical elements in the core grain.

epizonal   Of or relating to the epizone.

epizone   A depth category, generally less than 6.5 km deep, in which generally composite, discordant plutons, with chilled margins and contact metamorphism, are characteristic.

equigranular   A textural term referring to textures in which the grains are all about the same size.

equigranular-mosaic texture   A metamorphic texture in which the grains are equant to subequant, polygonal, and fine-grained.

equigranular-tabular texture   A metamorphic texture characterized by aligned, elongate to tabular, subequant grains.

erg   A large, sand-covered desert terrain.

escape structure   A small, generally cylindrical structure, usually found in sandstones, consisting of a tube, now filled with sand or sandstone, that served as the path by which water or an organism escaped from a lower layer after being buried.

essential minerals   Minerals used in determining the root name of an igneous rock.

estuary   The wide mouth of a river that has been drowned by the sea, typically characterized by mixed fresh and marine waters.

eutectic   An invariant point representing the melt composition and the lowest temperature at which a mixture of two or more solid phases will melt. Initial melting does not change the composition of the solid phases. The eutectic point is the lowest point on the liquidus curve.

euxinic basin   A depression in the crust, filled at depth with anerobic (oxygen-deficient) water of limited circulation (i.e., stagnant water).

evaporite   A sedimentary rock formed by the crystallization of salts from concentrated solutions produced by evaporation of a watery solvent.

exotic block   A mass of rock that occurs in a lithologic association foreign to that in which the mass formed. Commonly, exotic blocks are also tectonic blocks.

extensive variable   A property of a system capable of change, the value of which is dependent on the mass of the system.

externally buffered   A descriptive term for reactions that have components that are provided by outside sources (e.g., a fluid phase) in enough abundance that the reaction is not controlled or limited by the availability of those components.

extrusive   A descriptive term for igneous bodies that erupt at the surface of the Earth.

extrusive structures   Structures of igneous rocks that are formed where magmas are forced out onto the surface of the Earth.

F

fabric   The sum of the structural and textural features of a rock that, taken together, define its geometrical character.

facies, metamorphic   See *metamorphic facies*.

facies, sedimentary   A body of sediments or sedimentary rock characterized by a set of physical, chemical, and biological features that together reflect a particular environment of formation.

fault   A fracture in rock or other material across which there is a total loss of cohesion and along which there has been significant movement parallel to the surface of failure.

feldspathic rocks   A collective term referring to igneous rocks characterized by the presence of feldspars in amounts greater than 10% and in which feldspars plus quartz or feldspathoids are the essential minerals.

felsic   Light-colored. Also refers to feldspar, feldspathoid, and silica minerals in rocks.

felty   A volcanic texture characterized by unoriented, microscopic crystals of plagioclase in a matrix of other minerals.

fenestrae   Holes, typically of bird's-eye shape, that occur in carbonate sediment. In ancient rocks, fenestrae are usually filled with calcite.

ferromagnesian rocks   Igneous rocks containing less than 10% feldspars and an abundance of minerals rich in iron and magnesium. Olivines, pyroxenes, amphiboles, and some micas are the essential minerals.

fibroblastic texture   A metamorphic texture consisting of nonaligned, radiating to randomly oriented acicular minerals of about the same size (i.e., a diablastic texture consisting of equally sized acicular minerals).

filter pressing   A process that consists of the separation of crystals from a magma via tectonic squeezing of the magma chamber, which drives off the magma, leaving the crystals behind.

fissility   The tendency of some rocks to split into thin pieces.

flame structure   A soft-sediment deformation feature consisting of a curved set of fine laminae that project into an overlying (sand) layer and have the general form of a small flame blowing in a breeze.

floatstone   A matrix-supported carbonate sedimentary rock in which large bioclasts comprise more than 10% of the rock.

flood basalt   Basalt formed by fissure eruptions in units that cover large regions. Also called plateau basalt.

floodplain   The flat geomorphic feature that occurs adjacent to the stream channel in a stream valley and on which floods occur.

flow-banding   Colored bands in volcanic rocks formed by flow of the magma and resulting from concentrations of crystals, vesicles, or inclusions.

flow differentiation   A process of separation of crystals and liquid in a magma in which crystals are concentrated in the center of a horizontal or inclined dike (or sill) by grain-dispersive pressure generated during flow of the magma.

fluid escape structure   A small, generally cylindrical structure, usually found in sandstones, consisting of a tube, now filled with sand or sandstone, that served as the path by which water escaped from a lower layer after being buried.

flute   A primary sedimentary structure consisting of an elongate or lobate depression or groove. Flutes occur on the top surface of beds and are probably produced by erosion during turbulent flow across the surface. Flutes are preserved in rocks where the flute is filled (usually by sand) to form a flute cast, a resistant mass on the sole of the overlying bed.

fluvial   A descriptive term referring to features or processes associated with streams and rivers.

flysch   A term used most commonly in Europe to refer to thin-bedded, calcareous sandstones and shales that form thick sequences. Turbidites are present in flysch, especially where the term has been applied to like sequences of noncalcareous rocks.

foid   A feldspathoid mineral.

foidite   An igneous rock in which feldspathoids make up 60–100% of the essential minerals of the rock. The term is used only for volcanic rocks by some petrologists.

foidolite   In the IUGS classification of igneous rocks, a plutonic rock in which feldspathoids constitute more than 60% of the essential minerals.

fold   A bend in a composition band or other planar structure in a rock.

foliated   A term used to describe rocks with an alignment of mineral grains that imparts to the rock a planar character or fabric.

foliated cataclastic texture   A texture consisting of broken grains of rock or mineral materials in elongate segregations (i.e., meso- to microlithons) arranged in preferred orientations and separated by anastamosing to subparallel shear surfaces or microfoliated laminae.

foliated texture   A texture characterized by an alignment of mineral grains in such a way as to give the rock the appearance of or the tendency for splitting into layers or flat pieces.

foliation   A planar feature of rocks that results from the parallel to subparallel alignment of inequant mineral grains or fractures.

forereef   The seaward, generally steep side of a reef.

foreset   A type of sedimentary bed formed in deltas on delta fronts and consisting of inclined layers between horizontally bedded bottomset and topset beds.

formation   A mappable body of rock of distinctive lithology or lithologies and unique stratigraphic position.

fossil   Any prehistoric evidence of past life.

fossiliferous   A descriptive term for rocks that contain fossils. The term is usually applied to rocks in which the fossils are abundant.

fractional crystallization   A process of magmatic differentiation in which a part of a magma is crystallized and the minerals thus formed are separated or isolated from further reaction with the remaining magma.

fractional fusion   A melting process in which a rock is melted in stages and the melt produced by heating is separated from the remaining rocks, preventing further reaction between the melt and the residual rock (the restite).

fractional melting   See *fractional fusion*.

framework (grains)   The larger grains in a clastic sediment or sedimentary rock (as distinguished from the matrix, the finer material).

Franciscan-type Facies Series   A high-pressure, very low temperature sequence of facies, represented by the rocks of the Franciscan Complex of California, in which jadeitic pyroxene and lawsonite are the key or critical minerals.

free energy   A thermodynamic characteristic of a system representing a measure of the internal energy of or the amount of work that can be done by the system.

friable   A descriptive term, usually applied to sedimentary rocks, meaning easily crumbled or broken.

funnel   A solid, cone-shaped, layered pluton in which the apex is down in the Earth.

fusion   Melting.

G

gabbro   A phaneritic igneous rock lacking essential quartz and containing plagioclase feldspar in excess of 90% of the feldspar. The An component of the plagioclase is greater than 50.

gabbroid   A descriptive term referring to those rocks of gabbroic or gabbro-like composition. Used as a field term in the IUGS classification.

gabbronorite   A rock name in the IUGS classification used to designate rocks containing 0–5% quartz, plagioclase in excess of 90% of the feldspar in the rock, plagioclase more calcic than An_{50}, and significant amounts of pyroxene, olivine, and/or hornblende (see the IUGS classification charts for numerical limits).

gas fluxing A magmatic process in which upward-streaming gases in a magma chamber promote melting of wall rocks.

gas streaming A magmatic differentiation process in which moving gases facilitate separation of magma and crystals.

geobarometry A quantitative study used to determine the pressures existing during metamorphic or igneous processes.

geothermal gradient The ratio of temperature to depth, typically plotted as a curve in P-T space.

geothermometry A quantitative study used to determine the temperatures existing during metamorphic or igneous processes.

glassy A textural term used to designate igneous rocks in which there are no crystals, only supercooled magma.

glomeroporphyritic A textural term applied to porphyritic textures in which the phenocrysts are grouped together in clusters that are scattered through the rock.

gneiss A name for phaneritic, foliated metamorphic rocks characterized by alternating bands of contrasting mineralogical composition, at least some of which are characterized by the preferred orientation of the included minerals. *Note:* Alternative definitions are in use (see note 34 in chapter 23).

gneissic structure A layered or banded structure in metamorphic rocks in which layers are between 1 mm and 1 m thick.

gneissose texture A metamorphic texture characterized by alternating bands of contrasting mineralogical composition, at least some of which are characterized by the preferred orientation of the included minerals.

Goldschmidt's Mineralogical Phase Rule The Phase Rule that indicates that under certain conditions, the number of phases in a rock will equal the number of components in the rock system ($P = C$).

gouge A soft, nonfoliated dynamoblastic rock dominated by grains less than 1/16 mm in diameter and characterized by cataclastic texture. Gouges are formed by extreme grinding and crushing of grains along a fault zone.

graded bedding A sedimentary structure in which individual beds show a gradual decrease in grain size from the bottom to the top of the bed.

graded stratified bed A sedimentary bed in which the lower part shows a gradual change in grain size from bottom to top and in which the top (finer-grained) part exhibits a series of layers.

grain flow A transport process (and the name given to the deposit formed by that process) in which a high-density, incohesive mix of sand and water moves downslope under the influence of gravity.

grain shape The dimensional descriptor for sedimentary grains in which the form may be classified as equant (subequal length, width, and height), tabular (two larger dimensions and one substantially smaller dimension), or rod-shaped (two short dimensions and one large one). Alternative descriptions of grain shape exist.

grainstone A type of limestone composed of clastic carbonate grains with little or no matrix.

granite A phaneritic igneous rock composed of essential quartz and feldspar. Alkali feldspar comprises two-thirds or more of the feldspar.

granitization A process of rock formation in which phaneritic rocks of hypidiomorphic-granular texture containing essential quartz and alkali feldspar are produced by metamorphic processes (without melting) via diffusion of ions.

granitoid A general term applied to all phaneritic plutonic rocks containing essential quartz and feldspar.

granoblastic texture A metamorphic texture consisting of equant to subequant grains in an aggregate.

granoblastic-polygonal texture A metamorphic texture characterized by polygonally shaped crystals of equal or nearly equal size with straight to slightly curved grain boundaries.

granoblastic-polysutured texture A metamorphic texture characterized by crystals of equal or nearly equal size with lobate or serrate grain boundaries.

granoblastite A metamorphic rock with granoblastic texture.

granophyre An igneous rock with granophyric texture.

granophyric An adjective used to describe rocks with irregular, microscopic intergrowths of quartz and alkali feldspar.

granular In sedimentary rocks a term applied, especially to sandstones, in which grains between 2 and 4 millimeters comprise less than 25% of the rock.

Granulite Facies A name used to designate the set of all metamorphic rocks containing phase assemblages indicative of conditions of crystallization in the general range of $P = 0.3–1.2$ Gpa (3–12 kb) and $T = 700–900°C$. The name is also applied to the P-T space defined by those conditions.

grapestone A sediment or rock composed of rounded clusters of carbonate grains cemented together in the form of clusters like bunches of grapes by micrite.

graphic A term describing phaneritic, granitoid igneous rocks with regular, poikilitic intergrowths of triangular or linear-angular quartz grains in larger alkali feldspar grains.

gravitational separation A process of magmatic differentiation in which crystals are separated from the liquid by sinking (or flotation) due to the relatively higher (or lower) density of the crystals.

graywacke A somewhat obsolete term used to designate sandstones in which matrix material composed of micas, chlorite, or clays typically constitutes more than 10 or 15% of the rock.

Greenschist Facies A name used to designate the set of all metamorphic rocks containing phase assemblages indicative of conditions of crystallization in the general range of $P = 0.3–1.0$ Gpa (3–10 kb) and $T = 300–550°C$. The name is also applied to the P-T space defined by those conditions.

greenstone A term used by some geologists to designate low-temperature metabasites, typically consisting of chlorite and other minerals.

groundmass The fine-grained material between larger grains.

gypsum evaporite Aphanitic to phaneritic, soft, commonly layered, gypsum-rich rock formed via evaporation of a saline solution.

H

halite evaporite Aphanitic to phaneritic, halite-rich rock formed via evaporation of a saline solution.

hardground A hard, lithified layer of sediment a few centimeters thick that forms on the seafloor during periods of nondeposition.

hawaiite A type of volcanic rock with a potash: soda value of less than 1:2, a moderate to high color index, and a modal composition that includes essential andesine and accessory olivine.

helicitic texture   A metamorphic texture consisting of bands of small inclusions arranged in spiral patterns or folds within porphyroblasts or other poikilitic grains.

hematitite   A rock in which hematite either constitutes more than 50% of the mineralogy of the rock or one in which hematite is the principle component and comprises more than 33% of the rock.

hemipelagic mud   Fine-grained to aphanitic marine sediment containing more than 25% material of greater than 5-micron size derived from silicate continental sources, volcanoes, or the shallow levels of the sea.

heterogeneous nucleation   The process in which a new crystal growth-center appears on a preexisting surface.

heterogranoblastic texture   A metamorphic texture characterized by equant or subequant minerals of various sizes.

heterogranular texture   A general metamorphic textural term for textures in which the grains are of notably different sizes.

HFSE   High field strength elements, a group of elements with a high charge to radius ratio, including elements such as Ti, Ni, Hf, Ta, and Y.

holocrystalline   A textural term applied to igneous rocks in which all of the physical components of the rock are crystalline. Contrast with *holohyaline* and *hypocrystalline*.

holohyaline   A textural term applied to igneous rocks composed entirely of glassy (noncrystalline) constituents.

homogeneous nucleation   The process in which new crystal growth-centers form within a melt spontaneously and independent of any preexisting crystal or surface.

homogranular texture   A general category of metamorphic texture in which all the grains in a rock are approximately the same size.

Hornblende Hornfels Facies   A name used to designate the set of all metamorphic rocks containing phase assemblages indicative of conditions of crystallization in the general range of $P = 0.0–0.3$ Gpa (0–3 kb) and $T = 400–575°C$. The name is also applied to the P-T space defined by those conditions.

hornfels   Fine-grained, usually dark-shaded rock with diablastic or granoblastic texture; produced by contact metamorphism.

HP-contact metamorphism   Metamorphism in which temperature is the dominant agent of metamorphism, but in which intrusions that cause the metamorphism are emplaced into high-pressure (generally low-temperature) zones of the crust.

HREE   Heavy rare earth elements, a group of metallic elements of high atomic number (65–71) and low abundance in the Earth's crust, including Dy, Yb, and Lu.

hybridization   An igneous process of magma modification and petrogenesis in which assimilation of country rocks alters magma compositions and consequent rock chemistries. Some petrologists use the term for processes involving magma mixing as well as assimilation.

hydration   A process of chemical weathering in which water combines with other components to yield a new phase.

hydrolysis   A process of chemical weathering in which excess H^+ or OH^- are produced in the associated solution.

hypabyssal   An archaic term referring to igneous rocks intruded at shallow depths. (The rocks are considered to be texturally intermediate between extrusive and intrusive rocks and characteristically have porphyritic textures.)

hyperleucocratic   A descriptive term for igneous rocks with a color index of 10 or less. (Literally, dominantly very light-colored; *hyper* = very, *leuco* = light-colored, *cratic* = ruled by.)

hypermelanocratic   A descriptive term for igneous rocks with a color index of 90 or more. (Literally, very dark-colored.)

hypidioblastic texture   A combining textural term used for metamorphic rocks to indicate that the grains are dominantly subhedral.

hypidiomorphic-granular texture   An igneous rock texture for phaneritic rocks in which equant to subequant, unaligned grains of subhedral character dominate the rock.

hypocrystalline   An igneous rock texture applied to volcanic rocks that consist of a mixture of crystals and glass.

hypolimnion   The oxygen-deficient, saline, bottom-water layer of a lake.

I

idioblastic texture   A combining textural term used for metamorphic rocks to indicate that the grains are dominantly euhedral.

igneous rock   A rock formed by crystallization or solidification of a magma.

ignimbrite   A volcanic rock formed by the lithification of ash flow deposits.

ijolite   A phaneritic igneous rock composed of abundant feldspathoids and 30–70% ferromagnesian minerals.

imbrication   A sedimentary structure consisting of disk-shaped, ovoid, or other inequant clasts stacked in inclined piles at an angle to the bedding.

incongruent melting   A process of melting in which heating of a solid phase produces a new solid phase plus a melt of a composition different from the composition of the original solid phase.

intensive variable   A property of a system that is capable of being changed and the value of which is independent of the mass (size) of the system (e.g., temperature).

inter-   A combining form meaning between.

intercumulus crystals   Crystals formed late in a fractional crystallization sequence from liquid trapped between early-formed cumulus crystals.

intercumulus liquid   The magmatic liquid trapped between crystals and with which postcumulus crystals interact when they crystallized or recrystallized to form cumulate textures.

intergranular   A holocrystalline, volcanic texture consisting of granular pyroxenes and other equant to subequant crystals filling the spaces between tabular plagioclase feldspar crystals. It is the volcanic equivalent of diabasic texture.

internally buffered   A description for systems or reactions in which the mineral phases control the chemistry of the fluid phase.

intersertal   A volcanic rock texture consisting of crystals, especially phenocrysts, in a matrix of glass.

intraclast   A fragment of sand to gravel size of a penecontemporaneous sediment or sedimentary rock included within (deposited to form a part of) a limestone.

intrusive   A descriptive term denoting magmatic bodies that invade other rocks below the surface of the Earth.

intrusive structures   Structures of igneous rocks that are formed below the surface of the Earth.

invariant A descriptive term denoting conditions in which there are no degrees of freedom.

invariant system A system in which there are no degrees of freedom; that is, no variable may be changed without altering the number or nature of phases in the system.

inversely graded bed A layer of sedimentary (or volcanic) rock in which particle size varies from fine at the bottom to coarse at the top.

inversely graded disorganized bed A layer of sedimentary rock in which the lower part shows a gradual change from fine to coarser material and the upper part lacks any grading or layering.

inversely-normally graded bed A layer of sedimentary rock that exhibits a gradual change from finer sediment at the base to coarser sediment in the middle and back to finer sediment at the top.

iron formation A cherty, aphanitic to phaneritic, thin-bedded, typically red to black, iron-rich rock containing 20% or more total iron oxides (FeO + Fe_2O_3).

ironstone An aphanitic to phaneritic, massive to bedded, commonly oolitic, yellow to maroon, silver, or black, iron-rich, noncherty sedimentary rock containing 20% or more total iron oxides (FeO + Fe_2O_3).

island arc A linear or curvilinear row of volcanic mountains that form islands in the sea.

isochemical metamorphism Metamorphism (a pressure-, temperature-, or fluid-induced change in mineralogy, chemistry, or structure of a rock) that occurs without change in the chemistry of the rock domain being considered. (Iso = equal; therefore, of equal or the same chemistry.)

isograd A line on a map representing the first appearance or the disappearance of any particular mineral. Originally, isograds were thought to represent all points connecting the exact same grade (iso- = same, *grad* = grade) of metamorphism, a supposition now generally known to be false.

isolated system A system that exchanges neither energy nor matter with its surroundings.

isotope An element that varies from others of the same atomic number in having more or less neutrons, and therefore a higher or lower mass number. (*Iso* = same, *tope* = place; hence, having the same place on the periodic chart.)

J

jacupirangite A phaneritic ultramafic rock composed of clinopyroxene and magnetite, with some nepheline.

joint A fracture in rock along which there has been a total loss of cohesion, across which there is some separation, and parallel to which there has been no significant movement.

K

kelphytic A descriptive term for a special corona texture involving garnet or olivine or the rock texture in which such mineral textures occur.

kelphytic rim A corona texture in which garnet or olivine are rimmed by amphibole, pyroxene, or both of these minerals.

keratophyre A sodic, metavolcanic rock, chemically equivalent to andesite and similar rocks, and containing phases such as albite, calcite, quartz, epidote, and chlorite.

kerogen A fine, brown to black, insoluble material composed of hydrogen, carbon, oxygen, and nitrogen, with or without sulfur, found in sedimentary rocks.

kimberlite A porphyritic ferromagnesian (ultramafic) rock composed of olivine and phlogopite phenocrysts in a groundmass of serpentine, chlorite, calcite, olivine, phlogopite, and other minerals.

kink band A small-scale, tight, mesoscopic fold developed in a rock such as a phyllite or schist that already has a fabric.

komatiite (1) A term used to designate a suite of MgO-rich igneous rocks containing ultramafic volcanic rocks. (2) An ultramafic volcanic rock characterized by spinifex texture; also called uv-komatiite.

L

laccolith A moderate-sized, concordant plutonic structure with a convex-up roof and a diameter/thickness ratio of less than 10.

lacustrine A descriptive term referring to lakes and associated environments.

lagoon A shallow water body along a coast that is separated from the sea by a small barrier, such as a bar.

lamination A layer in a sedimentary rock that is less than 1 cm thick.

lamprophyre A group of dark, aphanitic-porphyritic volcanic rocks containing an abundance of euhedral, ferromagnesian mineral phenocrysts (e.g., biotite, hornblende) in matix of similar minerals, plus feldspars and/or feldspathoids.

landslide A general category of subaerial mass wasting or slope failure that includes a wide variety of phenomena including earthflows, slumps, debris flows, rockslides, and other related processes and features. Also, the deposits produced by these processes.

lapilli Pyroclastic fragments 2–64 mm in diameter. (Singular *lapillus*.)

lapilli tuff A volcaniclastic (pyroclastic) rock composed of lapilli in a matrix of ash.

latite A volcanic rock with essential feldspar, little or no quartz or feldspathoid minerals, and an alkali feldspar to total feldspar ratio of 0.67–0.33.

lattice preferred orientation (LPO) A descriptive term for a metamorphic texture in which there is a special arrangement of mineral grains reflected by alignment of optical and crystallographic axes of those grains.

lava (1) Magma that has lost dissolved gases when it erupted at the surface. (2) The rock produced by solidification of magma that has lost its dissolved gases.

lava flow Tabular to lobate masses of erupted and solidified magma that has lost dissolved gases prior to or during eruption.

lava plateau A major, tabular, constructional volcanic structure consisting of a layered accumulation of extensive, tabular, generally basaltic lava flows dominantly erupted through fissures.

layering A group of tabular features, characterized by distinct mineralogical, textural, structural, or color attributes, that occur in igneous, sedimentary, or metamorphic rocks.

lepidoblastic texture A strongly foliated metamorphic texture characterized by aligned platy or sheet-structured minerals.

leucocratic (1) A descriptive term for light-shaded igneous rocks. (2) The color index designation for rocks with color indices between 11 and 50. (Literally, dominantly light-colored.)

leucosome A light-shaded, granitoid rock that occurs in the high-grade metamorphic rocks called migmatites. It is produced by partial melting and/or the local injection and crystallization of a lower-temperature (eutectic) melt between layers or foliation planes in a metamorphic rock.

liesegang rings Variously colored, highly regular rings that occur in weathered rocks and form via oxidation or reduction. Typically, they are dominated by yellow to brown iron oxides.

lime mudstone A calcite-dominated sedimentary rock (i.e., a limestone) in which the grains are less than 1/16 mm in diameter.

limestone A sedimentary rock dominated mineralogically by calcite.

lineation A structural feature of rocks, particularly metamorphic rocks, resulting from the alignment of acicular minerals, the intersection of planar features, the alignment of minor fold axes or kink-band axes, or parallel arrangement of other elongate structural elements.

lipids Long-chain carboxylic acids, particularly plant and animal fats, found in sediments.

liquidus (1) A line (or surface, in three dimensions) representing all points of equilibrium between particular solid phases and liquids of specific compositions and specific temperatures or pressures. (2) The curve or surface in a phase diagram separating a field containing only liquid (or melt) from one containing some solid phases.

liquidus curve See *liquidus*.

lithofacies A body of sediments or sedimentary rocks with distinct physical and chemical characteristics—rock types, textures, and structures—that represent a particular sedimentary environment.

lithosphere The outer, more brittle layer of the Earth, consisting of the crust and uppermost mantle.

littoral A descriptive term for the marine environment or zone between high and low tide, or features of that zone.

load cast A sole mark on a sedimentary bed consisting of a bulge or elongate knob formed during compaction.

local metamorphism Metamorphism that affects relatively small volumes of rock (less than 100 km^3).

loess A porous, friable, commonly calcareous siltstone that forms blanket deposits. Loess is generally formed by deposition of wind-blown material of glacial origin.

lopolith A layered pluton with a dish-shaped or concave-up shape.

LP-contact metamorphism Metamorphism in which temperature is the dominant agent of metamorphism and in which intrusions that cause the metamorphism are emplaced into low-pressure, low-temperature zones of the crust.

LPO See *lattice preferred orientation*.

LREE Light rare earth elements, notably those Lanthanides with atomic numbers between 57 and 64.

L-tectonite A rock dominated by linear features with a fabric that reflects the history of deformation during flow (i.e., ductile deformation).

M

mafic Dark-colored; a descriptive term used for igneous and metamorphic rocks and magnesium- and iron-rich minerals, which are typically dark in color and low in silica.

magma Melted rock material, with included gases, crystals, and rock fragments.

magnetitite A rock with more a than 50% magnetite or one with magnetite as the dominant phase and constituting more than 33% of the rock.

macrosparite A crystalline carbonate rock with grains larger than 0.06 mm in diameter.

major elements The most abundant elements that function as cations in the crust and mantle of the Earth, including Si, Al, Ti, Fe, Mn, Mg, Ca, Na, K, P, and H.

manganese nodule A globular mass of aphanitic, black to brown manganese oxide minerals typically formed on the seafloor by precipitation of manganese oxides and associated elements.

manganolite An aphanitic, black to brownish-black rock composed of manganese oxide minerals.

marine environment The environment in the sea below low tide level.

marl An old name used to refer to sediments or rocks composed of subequal amounts of clay minerals and calcite, that is, friable argillaceous limestones and calcareous mudrocks, which are intermediate between limestone and shale in composition.

marlstone A name used by some workers to refer to rocks composed of subequal amounts of clay minerals and calcite (also see marl), especially if the rock is well lithified.

mass deposition A process of deposition in which materials are deposited by various types of subaerial and submarine landslides (*sensu lato*) and grain flows.

matrix A term used to designate the finer-grained material that occurs between the coarser grains in a rock. In sedimentary rocks, matrix is *clastic* material. Matrix commonly consists of clay with silt-size particles of quartz and other minerals.

matrix-supported A descriptive term used to characterize sediments and sedimentary rocks containing clasts in a matrix, but in which the clasts are separated by matrix and do not generally contact one another.

meandering stream A stream in which water flows in one main channel that moves back and forth across the floodplain in a looping, snakelike fashion.

mechanical transportation Transportation by physical movement of solid materials via suspension, sliding, bouncing, rolling, dragging, or other related processes.

melange A body of rock mappable at a scale of 1:24,000 or smaller, characterized by a lack of internal continuity of contacts or strata and by the inclusion of fragments and blocks of all sizes, both exotic and native, embedded in a fragmental matrix of finer-grained material.

melanocratic (1) A descriptive term used to refer to igneous rocks that are dark in shade or color. (2) A term referring to a color index category with color indices of 50 to 89. (Literally, dominated by dark-shaded.)

melanosome The dark-shaded, ferromagnesian mineral-rich part of the high-grade metamorphic rock migmatite.

member A subdivision of a formation characterized by distinctive lithologic character and stratigraphic position.

mesocumulate A rock composed of crystals formed via fractional crystallization plus small amounts of minerals formed by crystallization of melt in the spaces between these crystals.

mesocumulate texture   An igneous texture consisting of cumulate crystals formed via fractional crystallization with small amounts of crystals formed by crystallization of melt between these cumulate crystals. Mesocumulate textures are intermediate between orthocumulate textures and adcumulate textures.

mesogenesis   The middle-stage process of conversion of sediment to rock (diagenesis) that occurs relatively soon after burial.

mesoscopic structures   Handspecimen- to outcrop-scale, nonpenetrative features of rocks.

mesosome   A metamorphic rock of intermediate shade associated with migmatites.

mesosphere   The more rigid part of the mantle that occurs below the plastic asthenosphere.

mesozonal   A descriptive term referring to a category of plutons and other features formed at intermediate depths (about 6 and 16 km).

metaluminous   A descriptive term applied to igneous rocks in which the mole percent alumina is less than the mole sum of alkalis and lime but more than the mole sum of alkalis.

metamorphic differentiation   The metamorphic process or processes that produce banded or lenticular segregations of minerals from an initially homogeneous rock.

metamorphic facies   A set of rocks representing the full range of possible rock chemistries, with each rock characterized by an equilibrium assemblage of minerals that reflects a specific, but limited, range of metamorphic conditions.

metamorphic facies series   The progression of facies across a metamorphic belt or across a petrogenetic grid.

metamorphic rock   A rock that formed originally as igneous or sedimentary rock but which has been changed mineralogically, texturally, or both—without undergoing melting—in response to heat, pressure, directed stress, or chemically active fluids or gases. Metamorphic rocks have textures or minerals, or both, that reflect cataclasis, recrystallization, or neocrystallization in response to conditions that differ from those under which the rock formed and that lie between those of diagenesis and anatexis.

metamorphism   A process or set of processes that affect rocks in such a way as to produce textural changes, mineralogical changes, or both under conditions in the Earth between those of diagenesis and weathering (at the lower limit) and melting (at the upper limit).

metasomatism   Metamorphism dominated by chemical changes induced primarily by chemically active fluids.

metastable system   A system that has no tendency to change if influenced by small pertubations, but which is not at the lowest possible energy state and is, therefore, subject to change if an energy barrier is overcome.

miarolitic cavity   A cavity in a plutonic rock into which crystals project.

micrite   (1) Sedimentary calcite in grains smaller than 0.004 mm in diameter; microcrystalline ooze. (2) Lime mudstone, a sedimentary rock composed of grains smaller than a specified threshhold value.

microdolostone   A rock composed predominantly of dolomite of crytocrystalline to aphanitic grain size.

micrographic   An igneous texture, visible only through a microscope, that consists of regular, angular crystals of quartz poikilitically enclosed in feldspar.

microlite   A crystal in a volcanic rock that is microscopic in size.

microlithon   A small mesoscopic to microscopic lens or piece of rock, typically granular, relatively undeformed, and quartz and feldspar-rich, that occurs between folia in metamorphic rocks that have cleavage, especially in quasimylonites and mylonites.

microlitic   A descriptive term for an igneous matrix texture consisting of microlites (very small fibrous crystals) with glassy or cryptocrystalline intercrystalline material.

microporphyritic   An igneous texture characterized by microscopic phenocrysts set in a finer-grained matrix.

microsparite   (1) Crystalline calcite grains between 0.004 and 0.06 mm in diameter. (2) A rock composed of calcite grains between 0.004 and 0.06 mm in diameter.

migmatite   A complex, mixed, high-grade metamorphic rock consisting of various proportions of dark, ferromagnesian mineral-rich rock and light, quartz- and feldspar-rich rock.

minette   A type of volcanic rock belonging to the lamprophyre group and composed of euhedral biotite phenocrysts in a groundmass of biotite and alkali feldspar.

minor accessory mineral   An igneous mineral that constitutes less than 5% of the rock and the presence of which is not implied by the name of the rock.

modal analysis   An analysis to determine the kinds and amounts of minerals in a rock by counting each type of mineral grain to determine the percentage of each.

mode   The list of minerals observed in a rock, with their volume percentages.

Moho   Short for Mohorovicic Discontinuity.

Mohorovicic Discontinuity   The seismic boundary between the mantle and the crust.

mold   A depression on a sedimentary surface, especially one that retains the detail of the surface of the structure or object that left the depression.

MORB   Abbreviation for Mid-Ocean Ridge Basalt, a type of tholeiitic basalt characteristic of mid-ocean ridges.

mortar texture   A nonfoliated metamorphic texture consisting of larger broken grains in a matrix of smaller broken grains.

mosaic texture   A metamorphic texture characterized by polygonally shaped crystals of equal or nearly equal size with straight to slightly curved grain boundaries.

mound   A small hill-like sedimentary structure built on the seafloor by organisms.

mudcrack   A crack formed in muddy sediment as a result of drying. Mudcracks typically form polygonal patterns on sedimentary surfaces. If filled by sediment different from the cracked sediment, the mudcracks may be preserved in sedimentary rocks and can indicate facing (the top of a bed).

mudrock   A sedimentary rock dominated by materials less than 0.06 mm in diameter.

mudstone   A sedimentary rock dominated by particles of less than 0.004 mm in diameter and lacking laminations.

mugaerite   A type of volcanic rock composed of essential alkali feldspar and oligioclase, plus the mafic accessory phases olivine and clinopyroxene.

mullion A columnar metamorphic structure in which the cylindrical columns are 2 cm to 2 m in diameter. The columns are composed of the same material as the surrounding rocks and may exhibit internal folding.

mylonite A metamorphic rock characterized by textures that typically show both plastic deformation features and evidence of syntectonic recrystallization and recovery.

mylonitic texture A foliated to porphyroclastic-foliated texture characterized by porphyroclasts in a fine-grained to very fine-grained matrix of materials that show syntectonic recrystallization and/or recovery textures.

myrmekitic texture An igneous texture characterized by an intergrowth of sodic plagioclase feldspar and wormlike grains of quartz.

N

nematoblastic texture A foliated metamorphic texture consisting of layers of needle-like crystals.

neocrystallization The process in which new minerals (that did not previously exist in a rock) are formed. Neocrystallization occurs during diagenesis and metamorphism (neo = new; thus new crystal formation).

neomorphic An adjective referring to a process of diagenesis in which minerals either recrystallize or change to a polymorph.

neosome A complex rock, including both a leucosome and a melanosome, formed through migmatization (a process involving partial melting and deformation). (Literally, new body.)

nepheloid layer In submarine environments, a turbid bottom-water layer in which mud is suspended.

Newtonian fluid A fluid that has no strength and does not change in viscosity as the shear rate increases.

nodularblastic texture A metamorphic texture consisting of nodular clusters of small grains of one or two minerals in a matrix of a different composition.

nodule, igneous A fragment of phaneritic igneous rock within an igneous matrix.

nodule, sedimentary A small rounded to irregular concretion characterized by a bumpy surface.

non-Newtonian fluid A fluid that lacks strength but will change in viscosity as the shear rate increases or decreases.

nonphyllosilicate cement Precipitated intergranular material that is not composed of sheet-structured minerals. Nonphyllosilicate cements consist of minerals such as calcite, quartz, dolomite, gypsum, hematite, phosphate minerals, manganese oxides, and zeolites.

nonterminal reaction A metamorphic reaction in which one mineral pair becomes stable as another pair becomes unstable, resulting in a tie-line flip on an appropriate phase diagram.

norm A list of hypothetical minerals and their percentages calculated from the chemical analysis of a rock.

normally graded bed A sedimentary bed in which there is a gradual change in grain size from coarse at the base to fine at the top of the bed.

normative analysis The procedure for calculating a list of hypothetical minerals and their percentages from a chemical analysis of a rock.

nucleation The formation of nucleii, a process in which a few atoms assume the same relationship to one another as they would have in a solid.

nuée ardente A hot cloud of volcanic ash and gas that erupts from and flows rapidly down the flanks of a volcano.

O

obsidian Volcanic glass; supercooled, solidified magma. High viscosity and rapid cooling combine during eruptions of high-silica magmas to produce its glassy texture.

olistostromal flow A high-density, high-viscosity, rapid, submarine, downslope movement of a cohesive mass of mud and rock fragments.

olistostrome (1) A submarine landslide or debris flow; an olistostromal flow. (2) The diamictite deposit of a submarine landslide or debris flow.

oncolite A small (generally <10 cm), concentrically laminated, spherical to irregular body formed during deposition by biochemical precipitation and trapping of carbonate mud by algae.

ooid A spherical to oval, concentrically layered particle between 0.25 and 2 mm in diameter that is composed of carbonate minerals or replacement phases.

oolite (1) An ooid. (2) A rock composed predominantly of ooids.

oolith Ooid.

oolitic A descriptive term applied to rocks containing ooids or ooidlike grains.

ooze A fine-grained sediment, like a clay in having less than 25% siliciclastic material of greater than 5-micron size, but different in containing more than 50% biogenic material.

opal-A Amorphous opaline silica that is a biochemical precipitate.

opal-CT A metastably crystallized, interlayered cristobalite-tridymite form of silica.

open system A system that exchanges both matter and energy with its surroundings.

ophiolite A type of igneous mafic-ultramafic body of specific structure consisting of an ultramafic tectonite at the base, a plutonic ultramafic-gabbro sequence that includes bodies of more siliceous differentiates, a sheeted dike complex composed of diabase, and an overlying basaltic lava flow sequence.

ophitic An igneous texture that occurs in gabbros and consists of large poikilitic pyroxenes enclosing smaller tabular plagioclase crystals.

orthocumulate A rock composed of cumulate crystals formed via fractional crystallization plus significant amounts of minerals formed by crystallization of melt between these cumulate crystals.

orthocumulate texture A texture in which crystals formed via fractional crystallization are interspersed with significant amounts of minerals formed by crystallization of melt that existed between these crystals.

orthomatrix Intergranular material in sandstones consisting of recrystallized, detrital, clay-rich mud.

orthomylonite A foliated metamorphic rock in which microlithons, porphyroclasts, or both, in a finer-grained, phyllosilicate-bearing matrix, comprise 10 to 50% of the rock.

orthomylonitic texture A metamorphic texture that is foliated to porphyroclastic-foliated and includes 10% to 50% porphyroclasts in a matrix of materials that are syntectonically crystallized or that show recovery textures.

oversaturated In igneous petrology, a term used to designate rocks that have free silica (silica in excess of that needed to compose feldspars) and hence contain quartz or another silica mineral.

oxidation A process of chemical weathering in which the valence of the cation increases.

P

packstone A limestone consisting of touching allochems (or other framework grains) in a matrix of lime mud.

pahoehoe lava Solidified volcanic flow rock with a relatively smooth but vesicular and ropy surface.

paired metamorphic belts In mountain systems, two roughly parallel, linear zones of metamorphic rock, one of higher P and lower T and the other of higher T and lower P. The inner belt (continent side) typically exhibits a Buchan- or Barrovian-type Facies Series, whereas the outer belt (ocean side) exhibits a Franciscan- or Sanbagawa-type Facies Series.

parental magma A magma that has given rise to other magmas via some process of differentiation.

partial fusion The process of partially melting a rock. See also *fractional fusion*.

partial melting See *partial fusion*.

pebbly A descriptive term applied to sedimentary rocks that contain less than 25% pebbles.

pebbly mudstone A rock name used by some geologists for rocks with rounded, pebble-size clasts in a muddy matrix.

pegmatite (1) A textural term used by some petrologists to designate igneous rocks dominated by large crystals (generally >3 cm in length). (2) A term used by some geologists to designate a granitic rock characterized by a very coarse-grained texture.

pegmatitic A descriptive term for igneous rocks dominated by crystals greater than 3 cm in length.

pelagic clay A type of muddy marine sediment with less than 25% silicate detritus of more than 5 microns and less than 50% biogenic material.

pelagic environment A tectonically passive environment that lies below the deep sea.

pelagic mud A type of muddy marine sediment with less than 50% biogenic material and more than 25% material of greater than 5-micron size derived from silicate continental sources, volcanoes, or the shallow levels of the seas.

pelitic A descriptive term used to designate mudrocks or rocks derived from mudrocks. Pelitic metamorphic rocks are typically aluminum-rich rocks characterized by abundant micas and containing aluminum silicates such as kyanite or andalusite.

pelitic schist A phaneritic, foliated, nonbanded metamorphic rock derived from a mudrock.

pellet A small (< 0.2 mm) rounded mass of very fine calcareous material (micrite). Pellets are peloids that were feces excreted by mud-eating worms, shrimp, and other organisms.

pelloid A small, rounded, carbonate allochem generally less than 1/4 mm in diameter.

penecontemporaneous fault A fracture, along which there has been movement, that occurs in and was formed during the depositional history of a rock sequence.

peralkaline A descriptive term used to designate igneous rocks in which the combined molecular percent of alkalis exceeds the molecular percent of alumina.

peraluminous A descriptive term used to designate igneous rocks in which the molecular percent of alumina exceeds the total combined molecular percentages of alkalis and lime.

peri-platform sediments Carbonate slope and basin margin sediments.

peritectic An inflection point on a liquidus curve representing the melt composition and temperature at which a discontinuous reaction (i.e., incongruent melting or crystallization) occurs.

peritectic system System exhibiting incongruent melting, or the melting of one solid phase to produce a liquid plus another solid phase.

permeability A measure of how well fluid will flow through a rock.

petrofacies Stratigraphic subdivisions based on similarity of sandstone petrology.

petrogenesis The study of the histories and origins of rocks.

petrogenetic grid A graph of subdivided P-T space with regions characterized by particular phases or phase assemblages.

petrography The study of the description and classification of rocks (also called lithology).

petrology The overall study of rocks, including petrography and petrogenesis.

petrotectonic assemblage A distinctive suite of rocks characterizing a type of plate boundary or a plate setting.

pH The negative logarithm of the hydrogen ion concentration of a solution.

phacolith A concordant, sill-like plutonic structure of generally modest size, that is lenticular and is emplaced into a site along a fold axis.

phaneritic A descriptive term applied to crystalline materials in which grains can be discerned without the aid of a microscope.

phase A homogeneous material that, because of its physical properties, can be separated by mechanical means from other phases with which it may occur.

Phase Rule An equation relating the number of components, the phases, and the variance of a system: $F = C - P + 2$, where F is the number of degrees of freedom, C is the number of components, and P is the number of phases.

phenocryst A larger grain surrounded by a population of grains of significantly smaller size; found in porphyritic (volcanic) textures.

phosphatic rock A rock that is a conglomerate, sandstone, or mudrock; is commonly dark and variously colored; and contains an apatite cement.

phosphorite (1) An aphanitic to phaneritic, typically brown to black rock that is oolitic, laminated, nodular, or fossiliferous, with more than 50% apatite. (2) A sedimentary rock containing more than 19.5% P_2O_5.

phyllitic texture An aphanitic to fine-grained metamorphic texture characterized by the presence of a crenulation cleavage, kink bands, or microfolds.

phyllosilicate cement A precipitated or recrystallized cement composed of one or more of the following minerals: smectites, chlorites, mixed-layer phyllosilicate minerals, chlorite-vermiculite, kaolinites, celadonite, illite, and muscovite.

phyric An adjective used to describe volcanic rocks with phenocrysts.

picritic basalt Olivine-rich basalt.

pillow lava Lava, extruded under water, that has tubular to elliptical structures.

pilotaxitic A volcanic texture consisting of randomly arranged, microscopic laths of plagioclase or other minerals.

pipe A crudely cylindrical channel, typically filled with breccia and lava, through which magmas have risen beneath the central vent of a volcano. *Note:* Other definitions are listed in Bates and Jackson, (1987).

pisolite A spherical, concentrically layered structure (similar to an oolite) that is larger than 2 mm in diameter. Pisolites occur in sedimentary and volcaniclastic rocks.

plate A major fragment of the lithosphere, thousands of kilometers in areal extent and about 100 km thick.

playa A flat, vegetation-free desert basin that occasionally contains an ephemeral lake.

playa lake An ephemeral, generally very shallow lake that forms in a playa.

pluton A general term denoting a rock body composed of plutonic rock.

plutonic rock Igneous rock that crystallized well below the surface of the Earth.

poikilitic A descriptive term designating an igneous texture in which a large crystal (an oikocryst) encloses irregularly scattered, smaller crystals of another mineral.

poikiloblastic texture A metamorphic texture in which a larger grain encloses several smaller grains.

poikilotopic A sedimentary texture in which framework grains are enclosed by larger crystals of cement materials (e.g., calcite).

porosity A measure of the amount of empty space existing between grains in a rock.

porphyritic An igneous texture in which there is a bimodal grain distribution; that is, a number of larger grains, called phenocrysts, are surrounded by a population of grains of significantly smaller size, which constitute the groundmass.

porphyroblastic-foliated texture A metamorphic texture in which there is a bimodal distribution of grains, consisting of some larger grains (the porphyroblasts) surrounded by a larger number of substantially smaller grains, including acicular or platey grains that are aligned to give a layered character to the rock.

porphyroblastic texture A metamorphic texture in which there is a bimodal distribution of grains, consisting of some larger grains (the porphyroblasts) surrounded by a larger number of substantially smaller grains.

porphyroclastic texture A weakly to strongly foliated metamorphic texture characterized by a bimodal distribution of grains, with larger, typically deformed protolith grains (the porphyroclasts) enclosed in a matrix of finer-grained, typically deformed grains, derived from protolith minerals. In some cases the matrix has recrystallized (recovered) and is strain free.

postcumulus crystal A crystal that crystallized from, or recrystallized through interaction with, an intercumulus liquid. Postcumulus crystals surround the cumulus crystals in cumulate-textured rocks.

preferred orientation The textural alignment of mineral grains. Two types of preferred orientations exist: (1) LPO, or lattice preferred orientation, a typically invisible preferred orientation resulting from the like orientation of lattices within the constituent crystals of a rock, and (2) SPO, or shape preferred orientation, the visible alignment of elongate or tabular mineral grains.

Prehnite-Pumpellyite Facies The set of all metamorphic rocks containing phase assemblages indicative of conditions of crystallization in the general range of $P = 0.1–0.3$ Gpa (1–3 kb) and $T = 200–350°C$. The name is also applied to the P-T space defined by those conditions.

pressure shadow A zone of lower stress and equant grain growth adjacent to a porphyroblast and generally parallel to the foliation in a metamorphic rock.

pressure solution A process in which pressure is concentrated at the point of contact between two grains, causing the contact area to dissolve and allowing for subsequent migration (diffusion) of ions or molecules away from the point of contact.

primary magma A chemically unchanged, anatectic melt derived from any kind of preexisting rock.

primary mineral A mineral that forms when a rock first forms.

primitive magma An unmodified magma, which forms through anatexis of mantle rocks that have not been melted or otherwise changed in composition since they formed.

proglacial aeolian environment A site of deposition downwind of a glacier, where winds deposit fine sediment.

proglacial fluvial environment A site of deposition where streams of meltwater flowing out from a glacial terminus deposit sediments originally derived from the glacier.

prograde metamorphism Metamorphism that follows a P-T path generally away from surface pressure and temperature conditions.

protolith The parent rock from which a metamorphic rock was derived. (Literally, first stone.)

protomatrix In a sedimentary rock, a type of matrix, or detrital, intergranular material, consisting of clay-rich mud.

protomylonite A foliated metamorphic rock composed of more than 50% porphyroclasts and microlithons enclosed in a matrix of finer-grained, typically phyllosilicate-bearing, deformed to syntectonically recrystallized material.

protomylonitic texture A porphyroclastic, foliated metamorphic texture consisting of more than 50% porphyroclasts and microlithons in a finer-grained matrix of deformed to syntectonically recrystallized minerals.

provenance The source area from which sediment is derived.

proximal A descriptive term referring to a site near to either the source of sediment or another reference point.

proximal environment A site of deposition near a source of sediment.

pseudomatrix Fine-grained, intergranular material in sandstones that looks like primary clastic material but is derived from deformed and recrystallized lithic fragments; false matrix.

pseudomorphic A descriptive term used to denote mineral replacements in which the new phase mimics the external crystal form of the replaced phase.

pseudo-phase diagram A triangular or other-shaped diagram that appears to be a phase diagram, but which does not meet the requirements of the Phase Rule, usually because of the way components are combined in defining the corners of the diagram.

pseudoplastic material A material in which viscosity varies during flow.

pseudotachylite A nonfoliated, cataclastic metamorphic rock that contains glass in the matrix. Such rocks form along faults.

P-T-t path Pressure-temperature-time path used in describing metamorphic rock histories.

pumiceous A term used to describe a texture or structure in volcanic rocks characterized by elongate, fine tubular vesicles.

pycnocline   A zone in a water body (or column) characterized by variable oxygen content, salinity, and density. Also called dysaerobic zone.

pyroclast   A clastic volcanic fragment produced by the eruption of material into the air (or water).

pyroclastic   A term used to describe rocks or textures that are characterized by pyroclasts.

pyroclastic breccia   A clastic volcanic rock dominated by angular clasts larger than 64 mm in diameter.

pyroclastic cone   A small, steep-sided structure composed largely of pyroclasts of various sizes with little, if any, lava. Also called cinder cone.

pyroclastic rock   Rock composed of fragments (clasts) of volcanic rock formed by explosive eruption.

pyroclastic sheet   A tabular accumulation, generally of silica-rich, fragmental volcanic rock, that forms where particles settle out of the atmosphere during and after explosive volcanic eruptions.

pyrolite   A hypothetical ultramafic mantle rock composed of one part basalt and three or four parts peridotite or dunite.

Pyroxene Hornfels Facies   The set of all metamorphic rocks containing phase assemblages indicative of conditions of crystallization in the general range of P = 0.0–0.4 Gpa (0–4 kb) and T = 550–750°C. The name is also applied to the P-T space defined by those conditions.

Q

quartzine   An optically length-slow form of chalcedony.

quartz keratophyre   A sodic, metavolcanic rock chemically equivalent to rhyolite or other siliceous volcanic rocks and characterized by phases such as quartz, albite, calcite, and epidote.

quasimylonite   A weakly to moderately foliated metamorphic rock characterized by incipient foliations, some crystalline matrix materials, and microlithons or broken grains. Aligned phyllosilicates or other mineral grains engulf the microlithons or porphyroclasts and provide the foliation. Evidence of cataclastic (grain-breaking) and crystal-plastic deformation is present. (*quasi* = to a degree; hence, mylonitic to a degree.)

R

radiolarian chert   An aphanitic to fine-grained sedimentary rock composed predominantly of silica minerals and containing significant numbers of radiolaria.

rain print   The preserved impact mark developed where a raindrop strikes a muddy surface.

rapikivi   A texture found in plutonic igneous and metaigneous rocks consisting of relatively large alkali feldspar grains, with cores of K-rich feldspar and rims of Na-rich (plagioclase) feldspar, surrounded by a matrix of finer-grained minerals.

reaction space   A hypothetical volume delimited by specific metamorphic reactions and containing one specific reaction path that was followed by a rock during metamorphism. The method of analysis used to determine the reaction space consists of an examination of the chemical data, minerals, and textures of a rock.

recovery   A process in which grains containing large amounts of strain (e.g., with deformation bands) reduce the amount of strain and intracrystalline deformation by internal reorganization via recrystallization.

recrystallization   A process of ion migration and lattice reorganization in which physical or chemical conditions induce a reorientation of intergrain relationships and the crystal lattices of mineral grains without accompanying breaking of grains.

redox potential   See *Eh*.

reduction   The weathering process in which the valence of cations in a mineral is decreased.

reefs   Domal to elongate, massive to bedded sedimentary structures built, during carbonate deposition, by organisms that biochemically precipitate carbonate minerals.

regional metamorphism   Metamorphism that affects thousands of cubic kilometers of rock.

relict structures   Primary structures, inherited from parent rocks, that remain in metamorphic rocks after metamorphism. Also called palimpsest features.

relict textures   Primary epiclastic, pyroclastic, or crystalline textures originally formed in a protolith and preserved in the derivative metamorphic rock.

replacement   An *in situ* diagenetic or alteration process in which a new mineral takes the place of an originally crystallized or sedimented phase.

replacement chert   A chert formed by a diagenetic process in which silica minerals take the place of preexisting minerals of various types in a rock.

restite   A residual (depleted) rock remaining after partial melting and melt extraction have affected a mass of rock. The term is applied to the rock remaining after leucosome has been removed from a protolith during the process of migmatization.

resurgent caldera   A large, generally circular, volcanic collapse structure within which later uplift of a central domal region occurred.

retrograde metamorphism   Metamorphism that follows a P-T path generally towards the conditions existing at the surface of the Earth.

reverse grading   A structure consisting of a layer exhibiting an increase in grain size from bottom to top.

rhyolite   Generally, siliceous volcanic rock that contains more than 69 weight percent silica and, where porphyritic, phenocrysts of quartz ± sanidine or another alkali feldspar. Various definitions exist.

rhythmic layering   A structure of plutonic igneous rocks that consists of repeated layer couplets, each layer of which is composed of a mineral assemblage with a different ratio of plagioclase to ferromagnesian minerals. Rhythmic layering is most distinct where the layers form dark and light bands.

ring dike   A special type of arcuate, sheetlike, discordant igneous intrusion, often large, that is vertical and cylindrical in orientation and form.

ripple marks   A series of regularly undulating shapes on bedding surfaces created either by oscillating water or by wind or water currents.

rip-up   A chip of rock that has been removed by erosion from an underlying bed and is preserved as a distinct fragment in the overlying bed, usually near its base.

rise, continental   See *continental rise*.

rock   (1) A solid aggregate of mineral grains. (2) A solid, naturally occurring mass of matter composed of mineral grains, glass, altered organic matter, or combinations of these components.

rock salt   An aphanitic to phaneritic, usually light-colored, salty-tasting, soft rock dominated by the mineral halite.

rod   A metamorphic structure that is long and cylindrical and is composed of segregated or introduced material (i.e., dike or vein material, such as quartz).

rodingite A calcium-rich metamorphic rock, typically containing diopside and grossular garnet, formed by metasomatism of gabbro.

roof pendant A mass of rock that formed part of the roof over a pluton and hung down into the magma at the time of intrusion but was later isolated from other connected roof rocks by erosion.

rubble Angular sediment dominated by clasts of coarse size (> 2mm).

S

sabkha An arid to semiarid, coastal plain sedimentary environment that lies above the normal high tide level but is subject to periodic tidal flooding and subsequent evaporite formation.

salinastone A rock composed of saline minerals of unspecified origin.

salinite An evaporite-like rock that forms via processes other than evaporation of saline solutions, e.g., by chemical precipitation from supersaturated solutions. Compare to *evaporite*.

salt marsh A flat, vegetated area, periodically flooded by salt water, that adjoins an estuary or lagoon.

Sanbagawa Facies Series A high-pressure, moderate-temperature sequence of metamorphic facies, characterized by rocks containing both blue sodic amphiboles and epidote at middle to high grades. The series is named for rocks of the Sanbagawa Metamorphic Belt of Japan.

sandstone dike A tabular, discordant structure composed of sandstone formed by the filling of a joint by fluidized sand and the lithification of the sand.

Sanidinite Facies The set of all metamorphic rocks containing phase assemblages indicative of conditions of crystallization in the general range of $P = 0.0–0.3$ Gpa (0–3 kb) and $T = 750–1100°C$. The name is also applied to the P-T space defined by those conditions.

saturated A descriptive term for igneous rocks with neither quartz nor an undersaturated mineral (e.g., a feldspathoid), but typically rich in feldspar.

schist A phaneritic, foliated metamorphic rock with abundant phyllosilicate or acicular minerals.

schistose texture A fine-grained to very coarse-grained metamorphic texture characterized by a subparallel arrangement of tabular, acicular, or flaky minerals.

schlieren Tabular, disklike concentrations of minerals with diffuse boundaries that occur within an igneous rock mass. Also called flow layers.

schlieren arch In a mesozonal pluton, an archlike pattern of aligned minerals and xenoliths.

schlieren dome A dome-shaped pattern of aligned minerals and xenoliths within a mesozonal pluton.

secondary mineral A mineral that forms after a rock has first formed, via alteration or weathering.

secondary porosity Porosity developed via postsedimentation processes.

sediment Material transported in water, air, or ice that accumulates and lithifies to form sedimentary rock.

sediment rain The falling of sediment, including siliciclastic, allochemical, and precipitated sediment, from the water column onto the floor of a depositional basin or other depositional site.

sedimentary environment A surface region of the lithosphere where sedimentation occurs, which is either above or below sea level and is distinguished by a particular set of chemical, physical, and biological characteristics.

sedimentary rock A rock that forms under surface conditions and consists of accumulations of (1) chemical precipitates, (2) biochemical precipitates, (3) fragments or grains of rocks, minerals, and fossils, or (4) combinations of these kinds of materials.

semischist A phaneritic metamorphic rock that is weakly foliated, generally because acicular to flaky minerals are poorly aligned or because such minerals, even though aligned, are a minor constituent of a quartz and/or feldspar dominated rock.

semislate An aphanitic, weakly foliated metamorphic rock with a poorly developed cleavage.

seriate A textural term used to describe igneous rocks consisting of grains of a wide range of sizes that grade from one size to another.

serpentinite A rock composed primarily of serpentine minerals produced by metamorphism of the magnesium-rich minerals of ultramafic or mafic rocks.

serpentinization The process or processes by which ultramafic or mafic rocks are transformed into serpentinites.

shale A laminated or fissile mudrock.

Shape preferred orientation (SPO) The textural alignment of elongate or tabular mineral grains.

sheaf texture A metamorphic texture resulting from an array of clusters of diverging acicular or platy grains.

sheeted dike complex An assemblage of dikes that are distinctly parallel in strike and may have only one chilled margin, indicating that successive dikes intruded earlier-formed dikes parallel to a regional fracture pattern.

shelf The part of the seafloor that occurs above the shelf-slope break at depths of about 124 m but below the normal wave base.

shield cone Flat cone-shaped accumulation of lava containing minor amounts of interlayered pyroclastic materials.

shoshonite An intermediate, porphyritic volcanic rock with phenocrysts of augite and olivine in a groundmass of leucite, labradorite with rims of alkali feldspar, olivine, augite, and glass.

sideritestone A sedimentary rock, the major component of which is siderite.

silexite A rock that is composed primarily of quartz and has a very high silica content.

silicate fragment A detrital or terrigenous grain or clast, including gravel-, sand-, silt-, and clay-size fragments of preexisting silicate minerals or rocks.

siliceous sinter An aphanitic to fine-grained, typically layered, variously colored, hard, typically porous, siliceous rock deposited by groundwater or surface water at or near a hot spring or geyser.

siliciclastic rock A general term for rock composed of silicate mineral and rock fragments and related clasts.

sill A concordant intrusive igneous structure of generally modest size and a tabular to elliptical shape, with a width to thickness ratio greater than 10.

siltstone A sedimentary rock composed of substantial, but variable, amounts of silt-size grains. The name is applied to rocks of varying structure, including laminated, fissile, nonlaminated, and nonfissile rocks.

simple pegmatite An igneous rock body which typically consists of (1) very coarse-grained areas within finer-grained granitoid plutons or (2) very coarse-grained lenses in high-grade (high T-P) metamorphic rocks.

skeletal   A descriptive term used to designate rounded to angular fragments of biochemically precipitated shell or hard parts of invertebrate organisms. Most skeletal fragments have a carbonate composition.

skeleton   The biochemically precipitated external shell (exoskeleton) or internal support (endoskeleton) of an organism.

skree   Angular sediment typically formed at the base of a cliff or steep slope and dominated by clasts of coarse size (>2mm).

slaty texture   An aphanitic, foliated metamorphic texture conditioned by the alignment of flaky (phyllosilicate) or acicular grains, which gives the rock a tendency to break into very flat, relatively smooth pieces.

soft-sediment fault   A break in sedimentary layers that forms where sediments undergo deformation before they are fully lithified.

soft-sediment fold   A bend in sedimentary layers formed before the sediments were fully lithified.

sole mark   A cast of a filled groove or flute, which is exposed on the base (sole) of an overlying bed when the weak underlying beds are eroded or removed.

solidus   (1) A line (or surface, in three dimensions) defined as the locus of all points for which a correspondence exists between specific temperatures (and/or pressures) and a particular liquid that is in equilibrium with one or more solid phases. (2) On a phase diagram, the line below which all materials are solid.

solute transportation   Stream transportation of dissolved materials that are moved in solution by the moving water.

solution   The process in which compounds are broken down and release ions.

solvus   A curve connecting the points representing compositions of pairs of coexisting minerals (notably the alkali feldspars) in a solid solution system.

sorting   A size parameter used to describe the textures of clastic rocks by relating the variation in grain sizes within the rock. In geology, well-sorted sediment has grains that are nearly all of identical size, whereas poorly sorted sediment consists of grains of varied sizes. In engineering, these definitions are reversed.

spaced cleavage   Rock cleavage (the tendency to break into tabular to lenticular pieces) with domain spacings of > 0.01 mm and which lacks a penetrative fabric.

spar   Crystalline calcite grains larger than 0.004 mm. In handspecimen study, spar may be defined as calcite grains larger than 0.16 mm.

sparite   A carbonate sedimentary rock composed of spar (crystalline calcite).

spherulitic   A descriptive term for a texture in which fibers that radiate from a point are clustered in spherical groups.

spheruloblastic texture   A metamorphic texture in which fibers that radiate from a point are clustered in spherical groups.

spilite   A metamorphosed basalt rich in sodium and containing such phases as albite, calcite, epidote, chlorite, sphene, and quartz. Spilites form at low pressures and temperatures of metamorphism.

spinifex   An igneous texture, characteristic of the komatiites, that consists of needle-like olivine grains forming a cross-hatched pattern enclosing intergranular pyroxenes.

spreading center   In plate tectonic theory, a kind of plate boundary where plates pull apart and new crust is formed by solidification of intruded and extruded magmas. Also called a divergent plate boundary.

stable system   A system that, under the specified conditions, is in its lowest energy state; i.e., the system has no tendency to change.

static metamorphism   Metamorphism that occurs primarily in response to pressure.

stock   A plutonic rock body with an aerial exposure of less than 100 km^2.

strand-plain   A wide beach.

stromatactis   A millimeter- to centimeter-scale, layered, plano-convex-up, lens-shaped structure consisting of lighter sparry calcite layers between darker carbonate layers. The structures are formed in carbonate rocks after deposition.

stromatolite   A sedimentary structure consisting of algal-generated, millimeter-scale micritic carbonate layers in flat, domal, columnar, conical, or nearly spherical forms larger than 10 cm in diameter.

structure   In rocks, a visible feature larger than a grain (but not penetrative at the handspecimen or larger scale) that results from the physical arrangement of grains, holes, fractures, or other entities in the rock mass.

stylolite   An irregular surface of dissolution that commonly appears as a dark, jagged line on exposed surfaces of carbonate rock (and rarely on other sedimentary rock types).

subduction zone   In plate tectonic theory, a zone of plate convergence and collision where one plate descends beneath another.

subglacial environment   A site of deposition beneath a glacier where sediments are formed as a glacier advances and retreats, depositing materials at its base.

submarine slope   The steep part of the seafloor that extends from the shelf break to the rise.

subophitic   An igneous texture in gabbros in which the augite and plagioclase assume similar sizes, with augite encompassing the ends of some plagioclase laths.

supraglacial environment   An environment of sedimentation along the sides and at the end of a glacier.

symplectic texture   A general category of texture in which wormy (vermicular) or irregular intergrowths of one mineral occur within another.

syntectonic recrystallization   A process of intracrystalline structural reorganization that occurs during (and as a result of) deformation of a rock and involves dissolution of detrital phyllosilicates and other minerals and recrystallization or neocrystallization of phyllosilicate and other grains parallel to the cleavage direction. Also called dynamic recrystallization.

system   Any part of the universe selected for study.

T

talus   Sediment dominated by angular clasts of coarse size (> 2mm) that accumulates at the base of cliffs and very steep, rocky slopes.

tectonic overpressure   Pressure in excess of P_{load} in metamorphic rocks. Tectonic overpressures typically develop beneath thrust faults that have shale or other impermeable rocks in the hanging wall.

tectonite   A rock with a fabric that reflects the history of its deformation.

tectonite fabric   A texture or structure, or both, that reflects a history of deformation.

telogenesis   Late-stage diagenesis that occurs after reexposure of formerly buried rocks.

tepee A small (centimeter- to meter-scale) angular anticline produced by compression of a layer towards a fracture in such a way as to produce two opposing, concave-up layers that come to a point like a Native American tepee.

terminal reaction (1) A metamorphic reaction in which a new phase is added to a phase diagram or a phase already present disappears from the phase diagram. (2) A metamorphic reaction in which one or more phases ceases to be possible on one side of the reaction.

texture The microscopic to small-scale mesoscopic character of a rock imparted by the size, shape, orientation, and distribution of grains and the intergrain relationships.

tholeiite A main type of basalt that is hypersthene-normative and contains little or no modal olivine but does contain modal Ca-poor pyroxene (hypersthene or pigeonite).

tidal flat A low-lying, subhorizontal geomorphic feature and sedimentary environment, which includes marshes and sabkhas, occurring along the edge of the continent.

tie-line In a phase diagram, a line drawn within a phase field connecting the phases coexisting (in equilibrium) at a fixed T (or P).

till Glacially deposited, very poorly sorted sediment (a diamict) composed of coarse, angular to rounded, locally faceted and striated clasts in a finer-grained matrix that typically consists of clay- to sand-size rock and mineral fragments.

tillite The rock that is equivalent to till; a glacial diamictite with a muddy or rock flour matrix enclosing larger clasts.

tool mark A groove on the top of a bed at the sediment-water interface made by currents carrying pebbles, sticks, shells, or other large objects. Tool marks may be filled with sediment and preserved as sole marks.

trace elements All elements other than the major elements.

trachyte A volcanic rock of intermediate silica content that is characterized modally by an abundance of alkali feldspar and a general lack of quartz and feldspathoid minerals.

trachytic A textural term for a volcanic rock texture in which the feldspars are aligned in subparallel arrangements.

trachytoidal A textural term for a plutonic rock texture in which the feldspars are aligned in subparallel arrangements.

tracks and trails Fossils that result from preservation of marks left by various organisms as they walk, crawl, or otherwise move across a sediment surface.

transform fault In plate tectonic theory, a shear boundary, where plates slide past one another; a major type of plate boundary.

transitional environment A sedimentary environment that occurs between the levels of the marine and continental environments and is influenced by both marine and continental agents (e.g., fresh and salt water, wind and wave action).

transposition A process in which primary layering is transected at a significant angle by a cleavage and through a history of movement and recrystallization is converted to compositional layering that parallels the cleavage direction.

travertine An aphanitic to phaneritic, layered carbonate rock, usually light-colored and concretionary, deposited by ground and surface waters emanating from springs.

triple junction In plate tectonic theory, a site where three plate boundaries intersect.

tufa Cellular travertine; a layered carbonate rock with holes, formed by springwater.

tuff Pyroclastic rock dominated by grains less than 2 mm in average diameter.

turbidite A rock unit deposited from moving, heavy sediment-water mixtures.

turbidity current A current of water (or air) that moves downslope by virtue of the gravitational influence on its mass, which is greater (denser) than that of the surrounding water (or air). In geology, significant turbidity currents develop in submarine environments, where sediment-rich waters, which have greater density than the surrounding waters, flow downslope.

U

ultrabasic rock A rock with an SiO_2 content of less than 45%.

ultramafic rock A dark rock with a color index of 90–100. Also called ferromagnesian rock.

ultramylonite A foliated metamorphic rock bearing significant evidence of crystal-plastic deformation and characterized by less than 10% microlithons and porphyroclasts in a fine-grained, typically phyllosilicate-bearing matrix.

ultramylonitic texture An aphanitic to phaneritic, foliated to porphyroclastic-foliated metamorphic texture with <10% microlithons and porphyroclasts, and evidence of crystal-plastic deformation or syntectonic recrystallization.

undercooling The difference between the temperature of the liquidus and the actual temperature of crystal growth for a given phase, expressed as $\Delta T = T_1 - T_c$.

undersaturated The term used to describe the chemistry of low-silica rocks containing minerals such as nepheline that are incompatible with quartz or other silica minerals.

uniform layering In igneous rocks, layering lacking conspicuous differences in mineralogy.

univariant A term designating a system with one degree of freedom or a reaction in which only one variable may be changed without changing the number or nature of phases.

univariant system A system with one degree of freedom.

unstable An adjective used to describe systems that are not in equilibrium; that is, they have a tendency to change spontaneously.

upwelling A process of vertical circulation in which deep waters, moving as currents, flow up to the surface in equatorial regions and along coastlines.

uv-komatiite High-Mg, low-Ti, olivine-rich, ultramafic volcanic rocks characterized by a distinctive herringbone-like texture called a spinifex texture.

V

varves Rhythmically repeated sediment couplets, each representing a year, consisting of silty, light-shaded summer layers and dark, clay-rich, organic winter layers.

varvite A laminated rock exhibiting varves, typically containing dropstones; diagnostic of a cryolacustrine environment of deposition.

vein A crudely tabular body that forms fracture fillings composed of one or more minerals.

vent   The site on a volcano from which magma erupts.

vesicle   A hole left when gas escapes from a lava flow.

vesicular   The adjective used to describe a volcanic rock in which the escape of gas has left holes.

vitriclastoblastic texture   A nonfoliated metamorphic texture consisting of broken fragments of rocks or minerals in a matrix of frictionally generated glass.

vitrophyric   A volcanic texture characterized by discrete phenocrysts in a glassy groundmass.

volcanic dome   A small, steep-sided structure shaped like an inverted cup or cone and composed of volcanic rocks.

volcanic neck   The eroded and exposed dike system that served as the feeder channel in a volcanic cone. Also called a pipe or vent.

volcanic rock   A rock produced by crystallization of magma, where magmas approach and break through to the surface in volcanic eruptions.

volcaniclastic rock
A sedimentary rock composed of fragments of volcanic material.

volcaniclastic texture   A texture consisting of angular grains of volcanic rock fragments, feldspar, quartz, and/or other minerals, generated by volcanic processes, that are cemented or otherwise stuck together.

vug   Cavities lined with crystals.

W

wacke   (1) A type of sandstone characterized by a matrix of 5% or more. (2) A type of sandstone characterized by a matrix of 10% or more.

wackestone   A type of limestone that contains more than 10% grains but is mud-supported (i.e., the larger grains do not generally touch one another).

weakly foliated   A texture in which some alignment of grains (usually phyllosilicate grains) yields a faint layered quality to the rock.

weathering   The general term assigned to the group of processes that result in the transformation of rock into soil.

welded tuff   A type of volcanic rock composed predominantly of fragments of volcanic material smaller than 2 mm in diameter, annealed soon after deposition by the heat still contained in the grains.

working   A term used to describe the condition of sediments that have experienced extensive transportation and abrasion.

X

xenocryst   A foreign crystal (*xeno* = foreign) formed in one rock but later incorporated into another by magmatic erosion or assimilation of the original rock.

xenolith   A piece of foreign rock (*xeno* = foreign, *lith* = stone) occurring as an inclusion within a plutonic or volcanic rock mass.

xenomorphic-granular   An igneous texture that is dominated by anhedral crystals.

xenotopic   A crystalline sedimentary texture dominated by anhedral crystals.

Y

yttrium   A trace element; the transition element of atomic number 39.

Z

Zeolite Facies   The set of all metamorphic rocks containing phase assemblages indicative of conditions of crystallization in the general range of P = 0.05–0.4, Gpa (0.5–4 kb) and T = 75° – 300°C. The name is also applied to the P-T space defined by those conditions.

zone, metamorphic   A specific mineral assemblage within a metamorphic facies and the region of geography and P-T space over which that assemblage is stable.

zoned   An adjective describing a texture in a mineral grain that exhibits a layered character resulting from changes in composition from the core to the rim of the grain.

zoned pegmatite   A very coarse-grained part of a plutonic body containing laycrs, lenses, shells, or irregular masses of rock of distinctive composition.

zoned texture   A texture characterized by zoned minerals.

CREDITS

Photo Credits

All photos not credited on-page are courtesy of the author, Loren A. Raymond.

Line Art Credits

Chapter 1
Figure 1.1 From: *Inside the Earth* by Bruce A. Bolt. Copyright © 1982 by W. H. Freeman and Company. Reprinted with permission; **Figure 1.3** From B. L. Isacks, et al., in *Journal of Geophysical Research*, 73:1968. Copyright © American Geophysical Union.

Chapter 2
Figure 2.1 From L. A. Raymond, *Petrology Laboratory Manual*, Vol. 1, Handspecimen Petrography. Copyright © 1984 GEOSI. Reprinted by permission; **Figure 2.3** From R. Greeley, "The Snake River Plain, Idaho: Representation of a New Category of Volcanism" in *Journal of Geophysical Research*, 87:2705–12, 1982. Copyright © 1982 American Geophysical Union; **Figure 2.11 A** From Arthur G. Sylvester, et al., "Papoose Flat Pluton: A granite blister in the Inyo Mountains, California" in *Geological Society of America Bulletin*, 89:1205–1219, 1978. Copyright © 1978 Geological Society of America. Reprinted by permission of the author; **Figure 2.11 B** From T. A. Richards and K. C. McTaggart, "Granatic rocks of Southern Coast Plutonic Complex and Northern Cascades of British Columbia" in *Geological Society of America Bulletin*, 87:935–53, 1976. Copyright © 1976 Geological Society of America. Reprinted by permission of the author; **Figure 2.11 C** From C. W. Williams and M. P. Billings, "Petrology and structure of the Franconia Quadrangle: New Hampshire" in *Geological Society of America Bulletin*, 49:1011–44, 1938. Copyright © 1938 Geological Society of America. Reprinted by permission of the author; **Figure 2.11 E** From W. A. Bothner, "Gravity study of Exeter Pluton: Southeastern New Hampshire" in *Geological Society of America Bulletin*, 85:51–6, 1974. Copyright © 1974 Geological Society of America. Reprinted by permission of the author; **Figure 2.11 F** From P. Kearey, "An interpretation of the gravity field of the Morin Anorthosite Complex, Southwest Quebec" in *Geological Society of America Bulletin*, 89:467–475, 1978. Copyright © 1978 Geological Society of America. Reprinted by permission of the author; **Figure 2.11 G** From J. F. Sweeney, "Subsurface distribution of Granitic rocks, South-central Maine" in *Geological Society of America Bulletin*, 87:241–249, 1976. Copyright © 1976 Geological Society of America. Reprinted by permission of the author; **Figure 2.12 A** From Arthur G. Sylvester, et al., "Papoose Flat Pluton: A granite blister in the Inyo Mountains, California" in *Geological Society of America Bulletin*, 89:1205–1219, 1978. Copyright © 1978 Geological Society of America. Reprinted by permission of the author; **Figure 2.12 B&C** From A. F. Buffington, "Granite Emplacement with Special Reference to North America" in *Geological Society of America Bulletin*, 70:641–47, 1959. Copyright © 1959 Geological Society of America; **Figure 2.19** From L. A. Raymond, *Petrology Laboratory Manual*, Vol. 1, Handspecimen Petrography. Copyright © 1984 GEOSI. Reprinted by permission; **Figure 2.25 B** Reprinted by permission from Samuel E. Swanson, "Relation of nucleation and crystal-growth rate to the development of granitic textures" in *American Mineralogist*, 62:(9/10):966–78, 1977. Copyright by the Mineralogical Society of America.

Chapter 3
Figure 3.2 B From T. H. Dixon and R. J. Stern, "Petrology, chemistry, and isotopic composition of submarine volcanoes in the Southern Mariana Arc" in *Geological Society of America Bulletin*, 94:1159–72, 1983. Copyright © 1983 Geological Society of America. Reprinted by permission of the author; **Figure 3.4 B** From A. Miyashiro, "Volcanic Rock Series and Tectonic Setting" in *Annual Review of Earth and Planetary Sciences*, 3:251–69, 1975. Copyright © 1975 Annual Reviews, Inc., Palo Alto CA. Reprinted by permission.

Chapter 4
Figure 4.1 A From L. A. Raymond, *Petrology Laboratory Manual*, Vol. 1, Handspecimen Petrography. Copyright © 1984 GEOSI. Reprinted by permission; **Figure 4.1 B** From D. W. Peterson, "Descriptive model classification of Igneous Rocks." Reprinted with permission from *Geotimes*, table 1, p. 32; March 1961; **Figure 4.1 C** From A. Streckeisen, "To Each Plutonic Rock its Proper Name" in *Earth Science Reviews*, 12:1–33, 1976. Copyright © 1976 Elsevier Science Publishers BV, Amsterdam, Netherlands. Reprinted by permission; **Figure 4.3** From A. Streckheisen, et al., "Plutonic rocks: Classification and Nomenclature Recommended by the IUGS Subcommission on the systematics of Igneous Rocks." Reprinted with permission from *Geotimes*, pp. 26–30, p. 28; October, 1973; **Figure 4.4** From A. Streckheisen, et al., "Plutonic rocks: Classification and nomenclature recommended by the IUGS Subcommission on the systematics of Igneous Rocks." Reprinted with permission from *Geotimes*, pp. 26–30, fig. 3, p. 27; October, 1973; **Figure 4.5** From L. A. Raymond, *Petrology Laboratory Manual*, Vol. 1, Handspecimen Petrography. Copyright © 1984 GEOSI. Reprinted by permission; **Figure 4.6** From A. Streckheisen, et al., "Plutonic rocks: Classification and nomenclature recommended by the IUGS Subcommission on the systematics of Igneous Rocks." Reprinted with permission from *Geotimes*, pp. 26–30, fig. 2, p. 26; October, 1973; **Figure 4.7 C** From L. A. Raymond, *Petrology Laboratory Manual*, Vol. 1, Handspecimen Petrography. Copyright © 1984 GEOSI. Reprinted by permission; **Figure 4.8** From M. A. Peacock, "Classification of Igneous Rock Series" in *Journal of Geology*, 39:54–67, 1931. Copyright © 1931 University of Chicago Press. Reprinted by permission; **Figure 4.10** From A. Miyashiro, "Volcanic Rock Series and Tectonic Setting" in *Annual Review of Earth and Planetary Sciences*, 3:251–69, 1975. Copyright © 1975 Annual Reviews, Inc., Palo Alto CA.

© 1982 American Geophysical Union; **Figure 8.3** From W. W. Hildreth, "Gradients in Silicic Magma Chambers: Implications for Lithospheric Magmatism" in *Journal of Geophysical Research*, 86:10153–92, 1981. Copyright © 1979 American Geophysical Union; **Figure 8.4 A** From W. W. Hildreth, "Gradients in Silicic Magma Chambers: Implications for Lithospheric Magmatism" in *Journal of Geophysical Research*, 86:10153–92, 1981. Copyright © 1979 American Geophysical Union; **Figure 8.4 B** From Robert L. Smith, "Ash flow magmatism" in E. E. Chapin and W. E. Elston, Eds., *Ash Flow Tuffs*, GSA *Special Paper 180*, 1979. Copyright © 1979 Geological Society of America. Reprinted by permission of the author; **Figure 8.5 A** From L. A. Morgan, et al., "Ignimbrites of the eastern Snake River Plain: Evidence of major caldera-forming eruptions" in *Journal of Geophysical Research*, 29:8665–78, 1984. Copyright © 1984 American Geophysical Union; **Figure 8.5 B** From R. L. Christiansen, "Cooling units and composite sheets in relation to caldera structure" in C. E. Chapin and W. E. Elston, "Ash Flow Tufts" in *Geological Society of America Special Paper* 180:29–42, 1979. Copyright © 1979 Geological Society of America. Reprinted by permission of the author; **Figure 8.8** From B. R. Doe, et al., "Lead and strontium isotopes and related trace elements as genetic tracers in the Upper Cenezoic rhyolite-basalt association of the Yellowstone Plateau Volcanic Field" in *Journal of Geophysical Research* 87:4785–4806, 1982. Copyright © 1982 American Geophysical Union.

Chapter 9

Figure 9.8 From J. Kienle and S. E. Swanson, "Volcanism in the Eastern Aleutian Arc: Late Quaternary and Holocene centers, Tectonic setting and petrology" in *Journal of Volcanology and Geothermal Research*, 17:393–44, 1983. Copyright © 1983 Elsevier Science Publishers, B V, Amsterdam, Netherlands, Reprinted by permission; **Figure 9.9** From A. R. McBirney, "Petrochemistry of the Cascade Andesite Volcanoes" in H. M. Dole, Ed., *Andesite Conference Handbook*. *Oregon Department of Geology and Mining Industries Bulletin*, 1968. Reprinted courtesy of Oregon Department of Geology and Mineral Industries; **Figure 9.11 B** From A. R. McBirney, "Some quantative aspects of orogenic volcanism in the Oregon Cascades" in R. B. Smith and G. P. Eaton, Eds., "Cenozoic tectonics and regional geophysics of the Western Cordilla" in *Geological Society of America Memoir* 152:369–88, 1978. Copyright © 1978 Geological Society of America. Reprinted by permission of the author; **Figure 9.11 C** From A. R. McBirney, ed., in *Procedures of the Andesite Conference*. *Oregon Department of Geological and Mining Industries Bulletin*, 65, 1969. Reprinted courtesy of Oregon Department of Geology and Mineral Industries.

Chapter 10

Figure 10.4 B From T. N. Irvine, "The ultramafic rocks of the Muskox Intrusion, Northwest Territories, Canada" in *Ultramafic and Related Rocks*. Copyright © 1967 John Wiley & Sons, Inc., New York. Reprinted by permission; **Figure 10.4 C** From G. D. Harper, "The Josephine ophiolite, northwestern California" in *Geological Society of America Bulletin*, 95:1009–26, 1984. Copyright © 1984 Geological Society of America. Reprinted by permission of the author; **Figure 10.6 A–C** From L. R. Wager, et al., "Types of igneous cumulates" in *Journal of Petrology*, 1:73–85, 1960. Copyright © 1960 Oxford University Press, Oxford England. Reprinted by permission; **Figure 10.8** From I. McDougall, "Differentiation of the Tasmanian Dolerites: Red Hill Dolerite-Granophyre association" in *Geological Society of America Bulletin*, 73:279–316, 1962. Copyright © 1962 Geological Society of America; **Figure 10.9** From I. McDougall, "Differentiation of the Tasmanian Dolerites: Red Hill Dolerite-Granophyre association" in *Geological Society of America Bulletin*, 73:279–316, 1962. Copyright © 1962 Geological Society of America; **Figure 10.10 A & B** From N. J. Page and M. L. Zientek, "Geological

and structural setting of the Stillwater Complex" in G. K. Czamanske, et al., Eds., "The Stillwater Complex, Montana: Geology and Guide." Montana Bureau of Mines Special Publication 92, 1985. Courtesy: Montana Bureau of Mines and Geology; **Figure 10.14** Reproduced with permission from Todd, et al., in *Economic Geology*, 1982, Vol. 77, p. 1459; **Figure 10.15 A** From E. M. Moores, "Petrology and structure of the Vourinos Ophiolite Complex of Northern Greece" in *Geological Society of America Special Paper 118*, 1969. Copyright © 1969 Geological Society of America. Reprinted by permission of the author; **Figure 10.15 C** From C. A. Hopson and C. G. Frano, "Igneous History of the Point Sal ophiolite, Southern California," in R. G. Coleman and W. P. Irwin, Eds., *North American Ophiolites*. *Oregon Department of Geology and Mineral Industries Bulletin*, 95, 1977. Reprinted courtesy of Oregon Department of Geology and Mineral Industries; **Figure 10.15 D** From R. C. Evarts, "The Geology and petrology of the Del Puerto ophiolite, Diablo Range, central California" in R. G. Coleman and W. P. Irwin, Eds., *North American Ophiolites*. *Oregon Department of Geology & Mineral Industries Bulletin 95*, 1977. Reprinted courtesy of Oregon Department of Geology and Mineral Industries; **Figure 10.15 E** From E. H. Brown, "Ophiolite on Faidalgo Island, Washington" in R. G. Coleman and W. P. Irwin, Eds., *North American Ophiolite*. *Oregon Department of Geology and Mineral Industries Bulletin*, 95, 1977. Reprinted courtesy of Oregon Department of Geology and mineral Industries; **Figure 10.16 A** From J. Auboin, *Geosynclines*. Copyright © 1965 Elsevier Science Publishers BV, Amsterdam, Netherlands. Reprinted by permission; **Figure 10.17** From R. G. Coleman, "Plate tectonic emplacement of Upper Mantle Periodotites along continental edges" in *Journal of Geophysical Research*, 76:1212–22, 1971. Copyright © 1971 American Geophysical Union; **Figure 10.18** From R. C. Evarts, "The geology and petrology of the Del Puerto ophiolite Diablo Range," in *North American Ophiolites*. *Oregon Department of Geology and Mineral Industries Bulletin*, 95, 1977. Reprinted courtesy of Oregon Department of Geology and Mineral Industries; **Figure 10.20** From R. C. Evarts, "The geology and petrology of the Del Puerto ophiolite Diablo Range," in *North American Ophiolites*. *Oregon Department of Geology and Mineral Industries Bulletin*, 95, 1977. Reprinted courtesy of Oregon Department of Geology and Mineral Industries; **Figure 10.21** From T. N. Irvine, "Petrology of the Duke Island ultramafic complex southeastern Alaska" in *Geological Society of America Memoir 138*; 1974. Copyright © 1974 Geological Society of America. Reprinted by permission of the author.

Chapter 11

Figure 11.4 From G. H. Anderson, "Grantization, albitization, and related phenomena in Northern Inyo Range of California-Nevada" in *Geological Society of America Bulletin*, 48:1–74, 1937. Copyright © 1937 Geological Society of America; **Figure 11.5 A&B** From O. F. Tuttle and N. L. Bowen, "Origin of granite in the light of experimental studies in the system $NaAlSi_3O_8$-$KAlSi_3O_8$-SiO_2-H_2O" in *Geological Society of America Memoir 74*, 1958. Copyright © 1958 Geological Society of America; **Figure 11.8** From R. A. Schweikert, "Shallow-level Plutonic Complexes and their Tectonic implications" in *Geological Society of America Special Paper*, 1976. Copyright © 1976 Geological Society of America. Reprinted by permission of the author; **Figure 11.9** From Paul C. Bateman and B. W. Chappell, "Crystallization, fractionation, and solidification of the Tuolumne Intrusive Series, Yosemite National Park, California" in *Geological Society of America Bulletin*, 90, Part I:465–82, 1979. Copyright © 1979 Geological Society of America. Reprinted by permission of the author; **Figure 11.10 B** From F. A. Frey, et al., "Fractionation of rare-earth elements in the Tuolumne Intrusive Series, Sierra Nevada batholity, California" in *Geology*, 6(4), 1978. Copyright © 1978 Geological Society of America. Reprinted by permission of the author; **Figure 11.10 C** From J. B. Reid, Jr., et al., "Magma mixing in granitic rocks of the Central

Sierra Nevada, California" in *Earth and Planetary Science Letters*, 66:243–61, 1983. Copyright © 1983 Elsevier Science Publishers, BV, Amsterdam, Netherlands. Reprinted by permission; **Figure 11.13** From R. A. Schweickert, "Triassic and Jurassic paleogeography of the Sierra Nevada and ancient regions, California and western Nevada" in D. G. Howell and K. A. McDougall, Eds., *Mesozoic Paleogeography of the Western United States*, 1978. Copyright © 1978 Society for Sedimentary Geology, Pacific Section, Stanford CA. Reprinted by permission.

Chapter 12

Figure 12.5 A–C From R. H. Jahns and C. W. Burnham, "Experimental studies of pegmatite genesis: I. A model for the derivation and crystallization of granite pegmatites" in *Economic Geology*, 64: 843–64, 1969. Copyright © 1969 Economic Geology Publishing Company. Reprinted by permission; **Figure 12.5 D** From E. N. Cameron, et al., "Internal Structure of Granitic Pegmatites" in *Economic Geology*, 1949. Copyright © 1949 Economic Geology Publishing Company. Reprinted by permission; **Figure 12.6** From P. C. Lyons and H. W. Krueger, "Petrology, chemistry, and age of the Rattlesnake pluton and implications for other alkali granite plutons of southern New England" in P. C. Lyons and A. H. Brownlow, Eds., *Studies in New England Geology. Geological Society of America Memoir* 146, 1976. Copyright © 1976 Geological Society of America. Reprinted by permission of the author; **Figure 12.8** From R. H. Jahns and C. W. Burnham, "Experimental studies of pegmatite genesis: I. A model for the derivation and crystallization of granite pegmatites" in *Economic Geology*, 64: 843–64, 1969. Copyright © 1969 Economic Geology Publishing Company. Reprinted by permission; **Figure 12.9** Reprinted by permission from R. J. Walker, et al., "Internal evolution of the Tin Mountain pegmatite, Black Hills, South Dakota" in *American Mineralogist*, 71(3/4):440–59, 1986. Copyright by the Mineralogical Society of America.

Chapter 13

Figure 13.3 From D. L. Hamilton and W. S. MacKenzie, "Phase-equilibrium studies in the system NaAlSiO$_4$ (hepheline) - KalSiO$_4$ (kalsilite) - SiO$_2$-H$_2$O" in *Mineralogical Magazine*, 34:214–31, 1965. Copyright © 1965 Mineralogical Society, London, Reprinted by permission; **Figure 13.5** From P. J. Wyllie, "Origin of Carbonatites: Evidence from phase equilibrium studies" in K. Bell, Ed., *Cartonatite: Genesis and Evolution*. Copyright © 1989 Chapman & Hall, London. Reprinted by permission; **Figure 13.6 B** From E. S. Larson, "Igneous Rocks of the Highwood Mountains, Montana, Part II: The Extrusive Rocks" in *Geological Society of America Bulletin*, 52:1733–52, 1941. Copyright © 1941 Geological Society of America; **Figure 13.8 A** From C. S. Hurlbut and D. T. Griggs, "Ingeous Rocks of the Highwood Mountains, Montana: Part I The Laccoliths" in *Geological Society of America Bulletin*, 50:1043–1112, 1939. Copyright © 1939 Geological Society of America. Reprinted by permission of the author; **Figure 13.8 B** From G. C. Kendrick and C. L. Edmond, "Magma immiscibility in the Shonkin Sag and Square Butte Laccoliths" in *Geology*, 9:615–619, 1981. Copyright © 1981 Geological Society of America. Reprinted by permission of the author; **Figure 13.11** Reprinted by permission from B. A. Olson, et al., "Petrogenesis of the Concord gabbro-syenite complex, North Carolina" in *American Mineralogist*, 68(2/3):315–33. Copyright by the Mineralogical Society of America; **Figure 13.12** From R. W. Chapman, "Criteria for the mode of emplacement of the alkaline stock at Mount Monadnock, Vermont" in *Geological Society of America Bulletin*, 65:97–114, 1954. Copyright © 1954 Geological Society of America.

Chapter 14

Figure 14.4 From L. A. Raymond, *Petrology Laboratory Manual*, Vol. 1, Handspecimen Petrography. Copyright © 1984 GEOSI. Reprinted by permission; **Figure 14.5 A** From A. H. Bouma, *Sedimentology of Some Flysch Deposits*. Copyright © 1962 Elsevier Science Publishing Co., Inc., New York, New York. Reprinted by permission; **Figure 14.5 B** From R. D. Kruse, "Storm-generated sedimentary structures in subtidal marine facies with examples from the Middle and Upper Ordovician of Southwestern Virginia" in *Journal of Sedimentary Petrology*, 51:823–48, 1981. Copyright © 1981 Society of Economic Paleontologists and Mineralogists, Tulsa, Oklahoma. Reprinted by permission; **Figure 14.7 A** From P. Hoffman, "Shallow and deepwater stromolites in Lower Proterozoic platform-to-basin facies change, Great Slave Lake, Canada" in *AAPG Bulletin*, 58:856–57, 1974. Copyright © 1974 American Association of Petroleum Geologists, Tulsa OK. Reprinted by permission; **Figure 14.17** From L. A. Raymond *Petrology Laboratory Manual*, Vol. 1, Handspecimen Petrography. Copyright © 1984 GEOSI. Reprinted by permission; **Figure 14.18** From L. A. Raymond, *Petrology Laboratory Manual*, Vol. 1, Handspecimen Petrography. Copyright © 1984 GEOSI. Reprinted by permission.

Chapter 15

Figure 15.1 From L. A. Raymond, *Petrology Laboratory Manual*, Vol. 1, Handspecimen Petrography. Copyright © 1984 GEOSI. Reprinted by permission; **Figure 15.2 A** From Robert R. Compton, *Manual of Field Geology*, 1962. Copyright © John Wiley & Sons, New York. Reprinted by permission of the author; **Figure 15.2 B** From R. L. Folk, *Petrology of Sedimentary Rocks*. Copyright © 1974 Hemphill Publishing Co., Austin Texas. Reprinted by permission of the author; **Figure 15.2 C** From A. C. M. Moncrieff, "Classification of poorly-sorted sedimentary rocks" in *Sedimentary Geology*, 65:191–4, 1989. Copyright © 1989 Elsevier Science Publishers BV, Amsterdam, Netherlands. Reprinted by permission; **Figure 15.3** From R. H. Dott, Jr., "Wacke, graywacke, and matrix: What approach to immature sandstone classification?" in *Journal of Sedimentary Petrology*, 34:625–32, 1964. Copyright © 1964 Society of Economic Paleontologists and Mineralogists, Tulsa, Oklahoma. Reprinted by permission; **Figure 15.4 A** From E. F. McBride, "Classification of common sandstones" in *Journal of Sedimentary Petrology*, 33:664–9, 1963. Copyright © 1963 Society of Economic Paleontologists and Mineralogists, Tulsa, Oklahoma. Reprinted by permission; **Figure 15.4 B** From R. L. Folk, *Petrology of Sedimentary Rocks*. Copyright © 1974 Hemphill Publishing Co., Austin Texas. Reprinted by permission of the author; **Figure 15.5 A** Figure 8.1, page 262 from *Sedimentary Rocks*, 2nd ed. by Francis J. Pettijohn. Copyright 1949, 1957 by Harper & Brothers, renewed 1985 by Francis J. Pettijohn. Reprinted by permission of HarperCollins Publishers, Inc; **Figure 15.5 B** From P. D. Lundegard and N. D. Samuels, "Field classification of fine-grained sedimentary rocks" in *Journal of Sedimentary Petrology*, 50:781–6, 1980. Copyright © 1980 Society for Sedimentary Geology, Tulsa OK. Reprinted by permission; **Figure 15.5 D** From L. A. Raymond, *Petrology Laboratory Manual*, Vol. 1, Handspecimen Petrography, 2d ed. Copyright © 1993 GEOSI. Reprinted by permission; **Figure 15.6 A & B** From R. J. Dunham, "Classification of carbonate rocks according to depositional texture" in W. E. Ham, ed., *Classification of Carbonate Rocks: A Symposium*, in *AAPG Memoir 1*, 108–21, 1962. Copyright © 1962 American Association of Petroleum Geologists, Tulsa OK. Reprinted by permission; **Figure 15.6 C** From L. A. Raymond, *Petrology Laboratory Manual*, Vol. 1, Handspecimen Petrography, 2d ed. Copyright © 1993 GEOSI. Reprinted by permission; **Figure 15.7** From R. L. Folk, "Spectral subdivision of limestone types" in W. E. Ham, ed., *Classification of Carbonate Rocks: AAPG Memoir 1*, 1962. Copyright © 1962 American Association of Petroleum Geologists, Tulsa OK. Reprinted by permission; **Figure 15.8** From R. J. Cuffey, "Expanded reef-rock textural classification and geologic history of bryozoan reefs" in *Geology*, 13:1985. Copyright © 1985 Geological Society of America. Reprinted by permission of the author.

Chapter 16

Figure 16.2 From T. R. Nardin, et al., "Review of mass movement processes, sediment and acoustic characteristics, and contrasts in slope and base-of-slope systems versus canyon-fan-basin floor systems" in L. H. Doyle and O. H. Pilkey, Eds., *Geology of Continental Slopes*. *Journal of Sedimentary Petrology Special Publication* 27:61–73:1979. Copyright © 1979 Society of Economic Paleontologists and Mineralogists, Tulsa, Oklahoma. Reprinted by permission; **Figure 16.3** From H. Blatt, et al., *Origin of Sedimentary Rocks*, 2e, © 1980, p. 187. Reprinted by permission of Prentice-Hall, Englewood Clifs, New Jersey; **Figure 16.5** From W. C. Krumbein and R. M. Garrels, "Origin and classification of chemical sediments in terms of pH and oxidation-reduction potentials" in *Journal of Geology*, 60:1–33, 1952. Copyright © 1952 University of Chicago Press. Reprinted by permission; **Figure 16.8** From B. B. Henshaw, et al., "A geochemical hypothesis for dolomitization by ground water" in *Economic Geology*, 66: 710–24, 1971. Copyright © 1971 Economic Geology Publishing Company. Reprinted by permission; **Figure 16.9** From K. Badiozamani, "The Doprag dolomitization model-application to the Middle Ordovician of Wisconsin" in *Journal of Sedimentary Petrology*, 43:965–84, 1973. Copyright © 1973 Society of Economic Paleontologists and Mineralogists, Tulsa, Oklahoma. Reprinted by permission; **Figure 16.13** From K. P. Helmold and P. C. van de Kamp, "Diagentic mineralogy and controls on albitization . . ." in *AAPG Memoir 37*, 1984. Copyright © 1984 American Association of Petroleum Geologists, Tulsa OK. Reprinted by permission.

Chapter 17

Figure 17.7 A From G. E. Reinson, "Barrier island and associated strand-plain systems" in R. G. Walkder, Ed., *Facies Models*, 2d ed. Copyright © 1984 Geological Association of Canada, Sudbury, Ontario, Canada. Reprinted by permission; **Figure 17.7 B–D** From S. D. Heron, et al., "Holocene sedimentation of a wave-dominated barrier-island shoreline: Cape Lookout, North Carolina" in *Marine Geology*, 60:413–34, 1984. Copyright © 1984 Elsevier Science Publishers B V, Amsterdam. Reprinted by permission; **Figure 17.9 A** From L. J. Doyle and T. N. Sparks, "Sediments of the Mississippi, Alabama, and Florida (MAFLA) Continental Shelf" in *Journal of Sedimentary Petrology*, 50:905–16, 1980. Copyright © 1980 Society of Economic Paleontologists and Mineralogists, Tulsa, Oklahoma. Reprinted by permission; **Figure 17.9 B** From B. D. Bornhold and C. J. Yorath, "Surficial geology of the continental shelf, Northwestern Vancouver Island" in *Marine Geology*, 57:89–112, 1984. Copyright © 1984 Elsevier Science Publishers, B V, Amsterdam, Netherlands. Reprinted by permission; **Figure 17.11 A** From "Paleoenvironment of late Mississippian fenestrate bryozoans, Eastern United States" by F K. McKinney and H. W. Gault from *Lethaia*, 13:127–46, 1980, by permission of Scandinavian University Press; **Figure 17.11 B** From W. E. Galloway and D. K. Hobday, *Terrigenous Clastic Depositional Systems: Applications to petroleum, coal, and uranium exploration*. Copyright © 1983 Springer-Verlag, New York. Reprinted by permission; **Figure 17.13 A** From R. V. Ingersoll, "Submarine fan facies of the Upper Cretaceous Great Valley Sequence, Northern and Central California" in *Sedimentary Geology*, 21:205–50, 1978. Copyright © 1978 Elsevier Science Publishers, BV, Amsterdam. Reprinted by permission; **Figure 17.13 B** From J. A. May, et al., "Role of submarine canyons on shelf break critical interference on continental margins" in *Journal of Sedimentary Petrology Special Publication* 33:315–82, 1983. Copyright © 1983 Society of Economic Paleontologists and Mineralogists, Tulsa, Oklahoma. Reprinted by permission; **Figure 17.13 D** From *AGI Reprint Series* 3, by E. Mutti and F. Ricci Lucchi, translated by Tor H. Milson; copyright © 1978, p. 141, figure 14. Courtesy of the American Geological Institute.

Chapter 18

Figure 18.3 From P. E. Potter, et al. *Sedimentology of Shale*. Copyright © 1980 Springer-Verlag, New York. Reprinted by permission; **Figure 18.5** From T. A. Davies and D. S. Gorsline, "Oceanic sediments and sedimentary procenes" in R. Riley, Ed., *Chemical Oceanography*, 1976. Copyright © 1976 Academic Press Ltd., London. Reprinted by permission; **Figure 18.7** From M. Ewing, et al., "Sediments and topography of the Gulf of Mexico" in L. G. Weeks, ed., *Habitat of Oil*. Copyright © 1958 American Association of Petroleum Geologists, Tulsa OK. Reprinted by permission; **Figure 18.8** From C. W. Holmes, "Geochemical indices of fine sediment transport, Northwest Gulf of Mexico" in *Journal of Sedimentary Petrology*, 52:307–321, 1982. Copyright © 1982 Society of Economic Paleontologists and Mineralogists, Tulsa, Oklahoma. Reprinted by permission; **Figure 18.9** From J. A. Lineback, et al. "Glacial and postglacial sediments in Lakes Superior and Michigan" in *Geological Society of America Bulletin*, 90:781–91, 1979. Copyright © 1979 Geological Society of America. Reprinted by permission of the author; **Figure 18.12** From Cook, T. D., et al., *Stratigraphic Atlas of North and Central America*. Copyright © 1975 Princeton University Press, Princeton NJ. Reprinted by permission; **Figure 18.14** From F. Ettonsohn and T. D. Elam, "Defining the nature and location of a Late Devonian-Early Mississippian pycnocline in Eastern Kentucky" in *Geological Society of America Bulletin*, 96:1313–21, 1985. Copyright © 1985 Geological Society of America. Reprinted by permission of the author.

Chapter 19

Figure 19.3 A From W. R. Dickinson and C. A. Suczek, "Plate Tectonics and Sandstone Compositions" in *AAPG Bulletin*, 63:2164, 1979. Copyright © 1979 American Association of Petroleum Geologists, Tulsa OK. Reprinted by permission; **Figure 19.3 B** From W. R. Dickinson, et al., "Provenance of North American phanerozoic sandstone in relation to tectonic setting" in *Geological Society of America Bulletin*, 94:222–35, 1983. Copyright © 1983 Geological Society of America. Reprinted by permission of the author; **Figure 19.6** Copyright © L. A. Raymond. Reprinted by permission; **Figure 19.13** Copyright © L. A. Raymond. Reprinted by permission; **Figure 19.14** From W. R. Dickenson, et al., "Provenance of Franciscan graywackes in Coastal California" in *Geological Society of America Bulletin*, 93:95–107, 1982. Copyright © 1982 Geological Society of America. Reprinted by permission of the author.

Chapter 20

Figure 20.2 From Loren A. Raymond, "Classification of melanges" in L. A. Raymond, Ed., "Melanges: Their nature, origin and significance" in *Geological Society of America Special Paper* 198, 1984. Copyright © 1984 Geological Society of America. Reprinted by permission of the author; **Figure 20.4 A–C** Unpublished data, courtesy of L. A. Raymond; **Figure 20.4 D–F** From John S. Schlee, "Upland gravels of Southern Maryland" in *Geological Society of America Bulletin*, 68:1371–1410, 1957. Copyright © 1957 Geological Society of America. Reprinted by permission of the author; **Figure 20.5 A** Figure 6.5, page 159 from *Sedimentary Rocks*, 2nd ed. by Francis J. Pettijohn. Copyright © 1949, 1957 by Harper & Brothers, renewed 1985 by Francis J. Pettijohn. Reprinted by permission of HarperCollins Publishers, Inc.; **Figure 20.5 B** From W. Nemec and R. J. Steel, "Alleuvial and coastal conglomerates: Their significant features and some comments on gravelly mass-flow deposits" in E. H. Koster and R. J. Steel, Eds., *Sedimentology of Gravels and Conglomerates, Memoir 10*. Copyright © 1984 Canadian Society of Petroleum Geologists, Calgary, Alberta, Canada. Reprinted by permission; **Figure 20.11** Copyright © L. A. Raymond. Reprinted by permission; **Figure 20.15** From F. J. Hein and R. G. Walker, "The Cambro-Ordovician Cap Enrage Formation,

Quebec, Canada: Conglomeratic deposits of a braided submarine channel with terraces" in *Sedimentology*, 29:309–29, 1982. Copyright © Blackwell Scientific Publications, Inc., Oxford, England. Reprinted by permission.

Chapter 21

Figure 21.2 B From N. P. James, "Reef Environment" in P. A. Scholle, et al., eds., *Carbonate Depositional Environment*. in AAPG Memoir 33:345–40, 1983. Copyright © 1983 American Association of Petroleum Geologists, Tulsa OK. Reprinted by permission; **Figure 21.7** From N. D. Newell, "Bahamian platforms" in A. Poldervaart, ed., "Crust of the Earth" in *Geological Society of America Special Paper*, 62:303–15, 1955. Copyright © 1955 Geological Society of America. Reprinted by permission of the author; **Figure 21.8** From E. G. Purdy, "Recent calcium carbonate facies of the Great Bahama Bank. 2. Sedimentary facies" in *Journal of Geology*, 71:472–97, 1963. Copyright © 1963 University of Chicago Press. Reprinted by permission; **Figure 21.10 A & B** From J. F. Read, "Carbonate ramp-to-basin transitions and foreland basin evolution, Middle Ordovician, Virginia Appalachians" in *AAPG Bulletin*, 64:1575–1612, 1980. Copyright © 1980 American Association of Petroleum Geologists, Tulsa OK. Reprinted by permission; **Figure 21.11** From J. F. Read, "Carbonate ramp-to-basin transitions and foreland basin evolution, Middle Ordovician, Virginia Appalachians" in *AAPG Bulletin*, 64:1575–1612, 1980. Copyright © 1980 American Association of Petroleum Geologists, Tulsa OK. Reprinted by permission.

Chapter 22

Figure 22.3 From A. Iijima, et al., "Mechanism of sedimentation of rhythmically bedded chart" in *Sedimentary Geology*, 41:221–233, 1985. Copyright © 1985 Elsevier Science Publishers B V, Amsterdam, Netherlands. Reprinted by permission; **Figure 22.4** From L. Paul Knauth, "A model for the origin of chert in Limestone" in *Geology*, 7:274–77, 1979. Copyright © 1979 Geological Society of America. Reprinted by permission of the author; **Figure 22.12** From J. M. Huh, "Depositional environments of Pinnacle Reefs, Niagara, and Salina Groups, northern shelf, Michigan basin" in J. H. Fisher, ed., *Reefs and Evaporites—Concepts and Depositional Models*. Copyright © 1977 American Association of Petroleum Geologists, Tulsa OK. Reprinted by permission; **Figure 22.13** From R. Budros and L. I. Briggs, "Depositional environment of Ruff Formation (Upper Silurian) in southeastern Michigan" in J. H. Fisher, ed., *Reefs and Evaporites—Concepts and Depositional Models*. Copyright © 1977 American Association of Petroleum Geologists, Tulsa OK. Reprinted by permission.

Chapter 23

Figure 23.3 From B. Bayly, *Introduction to Petrology*, © 1968, p. 225. Reprinted by permission of Prentice-Hall, Englewood Cliffs, New Jersey; **Figure 23.6 A & C** From C. McA. Powell, "A morphological classification of rock cleavage" in *Tectonophysics*, 58:21–34, 1979. Copyright © Elsevier Science Publishers, B.V., Amsterdam, Netherlands. Reprinted by permission; **Figure 23.9** From L. A. Raymond, *Petrology Laboratory Manual*, Vol. 1, Handspecimen Petrography. Copyright © 1984 GEOSI. Reprinted by permission; **Figure 23.11 G & H** From L. A. Raymond, *Petrology Laboratory Manual*, Vol. 1, Handspecimen Petrography. Copyright © 1984 GEOSI. Reprinted by permission; **Figure 23.12 E & F** From L. A. Raymond, *Petrology Laboratory Manual*, Vol. 1, Handspecimen Petrography. Copyright © 1984 GEOSI. Reprinted by permission.

Chapter 24

Figure 24.2 A Table 9.2 from *Guide to the Study of Rocks*, 2nd edition by Leslie E. Spock. Copyright © 1962 by Leslie E. Spock. Copyright renewed. Reprinted by permission of HarperCollins Publishers, Inc.;

Figure 24.2 B From: *Igneous and Metamorphic Petrology* by Myron G. Best. Copyright © 1982 by W. H. Freeman and Company. Reprinted with permission; **Figure 24.3** From J. P. Bard, *Microtextures of Igneous and Metamorphic Rocks*. Copyright © 1986 D. Reidel Publishing. Reprinted by permission of Kluwer Academic Publishers, Dordrecht, Netherlands; **Figure 24.4** From H. G. F. Winkler, *Petrogenesis of Metamorphic Rocks*, 5th ed. Copyright © 1979 Springer-Verlag, New York. Reprinted by permission.

Chapter 25

Figure 25.4 From N. L. Bowen, "Progressive metamorphism of siliceous limestone and dolomite" in *Journal of Geology*, 48:225–74, 1940. Copyright © 1940 University of Chicago Press. Reprinted by permission; **Figure 25.7** From H. G. F. Winkler, *Petrogenesis of Metamorphic Rocks*, 5th ed. Copyright © 1979 Springer-Verlag, New York. Reprinted by permission; **Figure 25.8 A & B** Reprinted by permission from James B. Thompson, Jr., "Graphical analysis of mineral assemblages in pelitic schists," in *American Mineralogist*, 42(11/12):842–58, 1957. Copyright by the Mineralogical Society of America; **Figure 25.9** Reprinted by permission from Richard N. Abbott, Jr., "A petrogenic grid for medium and high grade metabasites" in *American Mineralogist*, 67(9/10):865–76, 1982. Copyright by the Mineralogical Society of America; **Figure 25.10 A** From H. G. F. Winkler, *Petrogenesis of Metamorphic Rocks*. Copyright © 1965 Springer-Verlag, New York. Reprinted by permission; **Figure 25.10 B** From J. B. Thompson, Jr., and A. B. Thompson, "A model system for mineral facies in peletic schists." Reprinted by permission of the author; **Figure 25.11** From A. Miyashiro, *Metamorphism and Metamorphic Belts*. Copyright © 1973 Routledge Chapman & Hall, London. Reprinted by permission.

Chapter 26

Figure 26.4 Reprinted by permission from H. J. Greenwood, "Wollastonite: Stability in H_2O-CO_2 mixtures and occurrence in contact-metamorphic aureole near Salmo, British Columbia, Canada" in *American Mineralogist*, 52(11/12):1669–80, 1967. Copyright by the Mineralogical Society of America; **Figure 26.5** Reprinted by permission from H. J. Greenwood, "Wallastonite: Stability in H_2O-CO_2 mixtures and occurrence in contact-metamorphic aureole near Salmo, British Columbia, Canada" in *American Mineralogist*, 52(11/12):1669–80, 1967. Copyright by the Mineralogical Society of America. **Figure 26.6** From V. C. Hover Granath, et al., "The Notch Peak contact metamorphic aureole, Utah: Petrology of the Big Horse Limestone Member of the Orr Formation" in *Geological Society of America Bulletin*, 94:889–906, 1983. Copyright © 1983 Geological Society of America. Reprinted by permission of the author; **Figure 26.7** From E. W. Reinhardt, "Phase relations in corderite-bearing gneisses from the Gananoque area, Ontario: in *Canadian Journal of Earth Sciences*, 5:455–82, 1968. Copyright © National Research Council of Canada, Ottawa. Reprinted by permission; **Figure 26.10** Reprinted by permission from R. J. Tracy and B. R. Frost, "Phase equilibria and thermobarometry of calcareous, ultramafic and mafic rocks, and iron formations" in *American Mineralogist*, 26:207–89, 1991. Copyright by the Mineralogical Society of America; **Figure 26.11** Reprinted by permission from Kurt Buchner-Nurminen, "On the mechanism of contact aureole formation in dolomitic country rock by the Ademello intrusion (northern Italy)" in *American Mineralogist*, 67(11/12): 1101–17, 1982. Copyright by the Mineralogical Society of America; **Figure 26.12** From V. C. Hover Granath, et al., "The Notch Peak contact metamorphic aureole, Utah: Petrology of the Big Horse Limestone Member of the Orr Formation" in *Geological Society of America Bulletin*, 94:889–906, 1983. Copyright © 1983 Geological Society of America. Reprinted by permission of the author; **Figure 26.14** From C. W. Burnham, "Contact metamorphism of magnesian limestones at Crestmore, California" in *Geological Society of America*

Bulletin, 70:879–920, 1959. Copyright © 1959 Geological Society of America. Reprinted by permission of the author; **Figure 26.15** From C. W. Burnham, "Contact metamorphism of magnesian limestones at Crestmore, California" in *Geological Society of America Bulletin*, 70:879–920, 1959. Copyright © 1959 Geological Society of America. Reprinted by permission of the author.

Chapter 27

Figure 27.8 From J. Hoffman and J. Hower, "Clay mineral assemblages as low grade metamorphic geothermometers: Application to the thrust faulted disturbed belt of Montana, U.S.A." in A. Scholle and P. R. Schluger, Eds., *Aspects of Diagenesis. Journal of Sedimentary Petrology Special Publication*, 26:1979. Copyright © 1979 Society of Economic Paleontologists and Mineralogists, Tulsa, Oklahoma. Reprinted by permission; **Figure 27.13 A** From R. D. Hatcher, Jr., et al., "Guidebook for Southern Appalachian Field Trip in the Carolinas, Tennessee, and Northeastern Georgia," Project 27 Caledonide Oregon. International Geol Correl Program (IUGS), 1979. Copyright © 1979 North Carolina Geological Survey, Raleigh NC. Reprinted by permission; **Figure 27.13 B** From Richard N. Abbott, Jr., and Loren A. Raymond, "The Ashe Metamorphic Suite, Northwest North Carolina: metamorphism and observations on geologic history" in *American Journal of Science*, 284:1986. Copyright © 1986 American Journal of Science, New Haven, Connecticut. Reprinted by permission; **Figure 27.16** From K. R. Mehnert, *Migmatites and the Origin of Granitic Rocks*. Copyright © 1968 Elsevier Science Publishers, BV, Amsterdam, Netherlands. Reprinted by permission; **Figure 27.17** From S. N. Olsen, "A Quantative Approach to Load Mass Balance in Migmitites" in M. P. Atherton and C. D. Gribble, Eds., *Migmitites, Meeting and Metamorphism*. Copyright © 1983 Shiva Publishing Ltd. Reprinted by permission of the author.

Chapter 28

Figure 28.1 Sources: R. G. Coleman, "Blueschist Metamorphism and Plate Tectonics," 24th International Geological Congress, Rept Sec. 2, 1972 and R. W. Wood, "A Re-evaluation of the Blueschist Facies" in *Geological Magazine*, 116:24–25, 1979; **Figure 28.5** From E. H. Brown, "Phase equilibria among pumpellyite, lawsonite, epidote and associated minerals in low grade metamorphic rocks" in *Contributions to Mineralogy and Petrology*, 64:123–36, 1977. Copyright © 1977 Springer-Verlag GmbH & Co. KG, Berlin. Reprinted by permission; **Figure 28.8** From W. G. Ernst, "Blueschist metamorphism and P-T regimes in active subduction zones" in *Tectonophysics*, 17:255–72, 1973. Copyright © 1973 Elsevier Science Publishers B V, Amsterdam, Netherlands. Reprinted by permission.

Chapter 29

Figure 29.2 From K. Yokoyama, "Regional ecologite facies in the high-pressure metamorphic belt of New Caledonia" in B. W. Evans and E. H. Brown, Eds., *Blueschists and Eclogites*, in *Geological Society of America Memoir*, 164, 1986. Copyright © 1986 Geological Society of America. Reprinted by permission of the author; **Figure 29.6** From

R. A. Jamieson, "Metamorphism of an Early Paleozoic continental margin: Western Baie Veseta Peninsula, Newfoundland" in *Journal of Metamorphic Geology*, 8:269–288, 1990. Copyright © 1990 Blackwell Scientific Publications, Inc., Cambridge MA. Reprinted by permission.

Chapter 30

Figure 30.6 B From D. U. Wise, et al., "Fault-related rocks: Suggestions for terminology," in *Geology*, 12:391–94, 1984. Copyright © 1984 Geological Society of America. Reprinted by permission of the author; **Figure 30.6 C** From L. A. Raymond, *Petrology Laboratory Manual*, Vol. 1, Handspecimen Petrography, 2d ed. Copyright © 1993 GEOSI. Reprinted by permission; **Figure 30.8** From A. R. Bobyarchick, et al., "The role of dextral strike-slip in the displacement history of the Brevard Zone" in D. T. Secor, Jr., Ed., *Southeastern Geological Excursions*, 1988. Copyright © 1988 South Carolina Survey, Columbia, South Carolina. Reprinted by permission; **Figure 30.9** From A. K. Sinha, et al., "Fluid interaction and element mobility in the development of ultramylonites" in *Geology*, 14:883–86, 1986. Copyright © 1986 Geological Society of America. Reprinted by permission of the author.

Chapter 31

Figure 31.4 Based, in part, on Evans and Trommsdorf, 1970; Evans, et al., 1976; and Day, et al., 1985; **Figure 31.6** From John M. Casey, "Reconstruction of the geometry of accretion during formation of the Bay of Islands Ophiolite Complex" in *Tectonics*, 2:509–28, 1983. Copyright © 1983 American Geophysical Union; **Figure 31.7** From J. F. Casey and J. A. Karson, "Magma chamber profiles from the Bay of Islands ophiolite complex" in *Nature*, 292:295–301, 1981. Copyright © 1981 Macmillan Magazines Limited. Reprinted by permission; **Figure 31.10 B** From S. E. Swanson, "Mineralogy and petrology of the Day Book dunite and associated rocks, Western North Carolina" in *Southeastern Geology*, 22:53–77, 1981. Copyright © 1981 Southeastern Geology. Reprinted by permission; **Figure 31.10 C** Reprinted by permission from Rosewell Miller, III, "The Webster-Addie ultramafic ring, Jackson County, North Carolina, and secondary alteration of its chromite" in *American Mineralogist*, 38(1/12):1134–47, 1951. Copyright by the Mineralogical Society of America; **Figure 31.10 D** From M. S. McElhaney and H. Y. McSween, Jr., "Petrology of the Chunky Gal Mountain mafic-ultramafic complex, North Carolina" in *Geological Society of America Bulletin*, 94:855–74, 1983. Copyright © 1983 Geological Society of America. Reprinted by permission of the author; **Figure 31.11** From C. J. Suen and F. A. Frey, "Origins of the mafic and ultramafic rocks in the Rhonda peridotite" in *Earth and Planetary Science Letters*, 85:183–202, 1987. Copyright © 1987 Elsevier Science Publishers, Amsterdam. Reprinted by permission.

Appendix

Appendix B.1 A–C From Charles E. Bickel, unpublished manuscript. Reprinted by permission of the author.

INDEX